Flora of Alaska and Neighboring Territories

Flora of Alaska
and Neighboring Territories

A MANUAL OF THE VASCULAR PLANTS

Eric Hultén

STANFORD UNIVERSITY PRESS   STANFORD, CALIFORNIA

An extensive paper setting forth range extensions and other technical data supplementary to the information in this volume was published in 1973 (Hultén, "Supplement to Flora of Alaska and Neighboring Territories: A Study in the Flora of Alaska and the Transberingian Connection," *Botaniska Notiser*, vol. 126, pp. 459–512). The four errors in the initial printing of this volume that were noted in that paper (pp. 49 and 587, errors of scale on drawings; p. 440, "mm" for "cm"; and p. 749, drawings reversed) have been corrected for this printing; errors otherwise brought to light (drawings reversed, p. 469; running heads, pp. 33 and 52; four Index entries) have also been corrected. More recently noted, but not changed for this printing, are the following: p. xvi ("Meck" should be "Merck"), p. 2 ("Adianthaceae" should be "Adiantaceae"), p. 45 ("*phaegopteris*" should be "*phegopteris*"), p. 222 (for *Carex capitata*, "1,000 meters" should be "2,300 meters"), p. 257 ("*Morriseyi*" should be "*Morrisseyi*"), p. 307 (add Wrangel I. to range of *Allium schoenoprasum* var. *sibiricum*), p. 423 (add Wrangel I. to range of *Cerastium Regelii*), p. 829 (extend range of *Pinguicula vulgaris* subsp. *vulgaris* to Regina, Saskatchewan), p. 892 (add south Vancouver I. to range of *Chrysanthemum bipinnatum* subsp. *huronense*), p. 985 ("Palander of *Vega* (Adolf Arnold Louis)" should be "Palander af Vega, Adolf Arnold Louis"), and p. 5 of the color section ("*Carex Gmelinii*" should be "*Carex Mertensii*").

Stanford University Press
Stanford, California
© 1968 by the Board of Trustees of the
Leland Stanford Junior University
Printed in the United States of America
ISBN 0-8047-0643-3
Original edition 1968
Last figure below indicates year of this printing:
90 89 88 87 86 85 84 83 82 81

Preface

Because Alaska lies at the crossroads of past migrations between two continents and two oceans, its biogeography is of special interest. There can no longer be any doubt that several waves of land migration have passed over this region during successive periods when extreme glaciation produced low sea levels and consequent transgressions of the Bering Land Bridge. In arctic Alaska there are fossil remains of horses, camels, and mammoths; and the record of marine life in the region also appears to corroborate the postulated openings and closings of Bering Strait. The ancestors of the North and South American Indians appear to have spread from Asia during the penultimate transgression, those of the Eskimos and Aleuts during the final one.

The history of the Alaskan flora must have followed a similar course. Not a few Alaskan plant species occur over a broad but interrupted range, with large gaps between small isolated areas, thus demonstrating continuous distribution in earlier times from central Asia through much of North America. Analyses of fossil pollen samples have substantiated many of these presumed former ranges.

The central location of Alaska in the distributional pattern of many species is indeed critical, a point emphasized in my doctoral dissertation of 1937. In assembling this flora, I have accordingly prepared detailed, historically based, total-range maps for all known Alaskan plants; these maps, the result of more than 30 years' study, present global distribution patterns for 1,735 separate taxa, and are the first such maps to be used in a floral guide to a major region. A second map for each of these taxa indicates plant range in the region subsumed by this flora: the Chukchi (Chukotka) Peninsula in northeastern Siberia, all of Alaska and the Yukon, extreme northwestern British Columbia, and the Mackenzie District of the Northwest Territories. The maps will of course need to be corrected and the details filled in as further knowledge becomes available, but the basic features should remain stable.

My interest in northeast Asia began in 1920–22, when I took part as a botanist in an expedition to Kamchatka. The next three summers were taken up in a general study of the southern part of this peninsula, which led to publication of a flora. In pursuing these studies, I became intrigued by the relationships of plant distribution in the regions around the North Pacific—and conscious that my range accounts seemed invariably to run aground on the Aleutian Islands, a unique chain some 1,150 miles long whose flora was practically unknown.

In 1932 I decided to deal with this situation. A small fishing boat took me from Unalaska to Attu and back, with as many landings along the way as time and weather would permit. Later another trip was made, aboard the Coast Guard cutter *Haida*, to Attu, with stops at Bogoslof Island and Atka; and the revenue cutter *Montgomery* took me to St. Paul Island, in the Pribilofs. One result of these collections was a flora of the Aleutian Islands, which was published in 1937; a second was a synopsis of the distribution of the vascular plants then known to exist between the Lena and Mackenzie Rivers, published in 1937 as my doctoral dissertation. This synopsis, "Outline of the history of arctic and boreal biota during the Quaternary period," illustrated the present distribution of these plants in relation to their history during the Quaternary, Glacial, and Interglacial periods.

During the work on this dissertation it became evident that the existing knowledge of the flora of Alaska was far too deficient to constitute a basis for serious plant-geographical studies. Moreover, the available reports were scattered across a vast store of works of both past and modern authors writing in several languages, and the details of names and synonymy were frequently bewildering. The task of imposing order on this confusion, which I had undertaken in the process of assembling my Aleutian materials, was then extended to include the flora of all of Alaska and the Yukon. Research in the American and European herbaria, which contained extensive collections from these areas, was later supplemented by more recent collections of other botanists and by my own further collections, and in 1941–50 the *Flora of Alaska and Yukon* I-X was published. This work cited, below each accepted name, all plants examined, all places from whence collections had been reported in the literature, and all names given in these earlier reports; it did not provide plant descriptions or illustrations. The flora was thus intended primarily for the use of scientists.

My field experience at that time was limited to the Aleutians and a few points along the southern coast of Alaska. A series of grants then allowed me to pursue fieldwork in arctic Alaska, and later in the Mackenzie District and in Central Alaska and the Yukon, chiefly in areas not previously visited by botanists. Thus four summers were taken up in plant collecting. In the process, my understanding of habitat and distribution patterns became clearer, and the desirability of a flora of another character began to emerge—one that would serve a larger public. Short descriptions and drawings would be provided for each species, identifying keys would be provided, and distribution patterns would be given in the form of maps detailing the known or presumed range of every accepted taxon. The present work is the result.

Publication of this work would not have been possible without financial assistance. I am especially grateful to the National Science Foundation for a generous grant made to Stanford University Press. The grant covered, among other things, the preparation of plant drawings and range maps, clerical help with manuscript and artwork, and verification of English vernacular names and botanical usage.

Three grants from the Arctic Institute of North America, made through an arrangement with the U.S. Office of Naval Research, allowed me to carry out the fieldwork mentioned above in Alaska, the Yukon, and the Mackenzie areas. Trips were facilitated by the Arctic Research Laboratory at Barrow, Alaska, which arranged for pilots, aircraft, and assistants to be at my disposal. Without this generous help the range maps would have been far less complete. To these institutions I express my sincere gratitude. I especially wish to thank Dr. Max Britton of Washington, D.C., and Dr. Max Brewer of Barrow, for their encouragement and assistance in carrying out these collection trips.

The main work with the flora has been done at the Botanical Section of the State Museum of Natural History (Riksmuseum) at Stockholm, where a large herbarium from the Alaska–Yukon region has been assembled. The Alaska–Yukon collections in the National Herbarium at Washington, D.C., the Gray Herbarium at Harvard University, the National Herbarium of Canada at Ottawa, and the Herbarium of the University of Alaska at Fairbanks have also been studied. The Herbarium of the University of British Columbia at Vancouver has been used to improve the distribution particulars for that province on the circumpolar map. To the directors of these museums I convey my appreciation for permission to study the collections and for their courtesy and assistance in the course of this study.

Mr. Leslie A. Viereck has kindly given me access to two important collections, one from Upper Dry Creek south of Fairbanks, the other from Tonzona River, southwest of McKinley Park, for which I owe him sincere thanks.

To David F. and Barbara M. Murray of St. John's Memorial University of Newfoundland I am indebted for the opportunity afforded me to study their collections from the 2,000-meter level on Kaskawulsh Glacier. Many particulars of altitudinal limits are derived from these collections.

To my friend Dr. Robert Rausch of Anchorage, special thanks are due for his substantial and effective help and encouragement during my travels in Alaska and the Yukon.

I must also express gratitude to Dr. Th. Sørensen of Copenhagen for his treatment of the genus *Puccinellia*; to Dr. Aleksander Melderis of the British Museum for treatment of *Agropyron*; to Dr. Gunvor Knaben of Oslo for treatment of *Papaver*; to Dr. Stanley L. Welsh of Brigham Young University, Provo, for his assistance with *Oxytropis*; to Dr. E. C. Ogden of New York State Museum at Albany for valuable suggestions and for his editing of my material on *Potamogeton*; and to other botanists of many countries, too numerous to mention, who provided information on a great variety of matters.

Dr. John Hunter Thomas, Curator of the Dudley Herbarium, at Stanford, has read every word of this book, in proof; to him I owe special thanks for his improvements in botanical detail, his technical assistance to the publisher, and his kind offer

Preface

to publish my final comments on new combinations in the journal *Madroño*. Dr. Thomas's wife, Susan, prepared both Indexes and made various systematic checks.

My discussion of Alaska's geologic past, in the Introduction (and, briefly, above), I owe to a recent reading of *The Bering Land Bridge,* edited by David M. Hopkins (Stanford University Press, 1967). Dr. Hopkins, of the U.S. Geological Survey, Menlo Park, has kindly refined my synopsis of geologic events. The endpaper map, with certain adaptations, and the glacial-period maps in the Introduction have been taken from his volume.

Mrs. Dagny Tande-Lid of Oslo has prepared the plant drawings. Most of them are original, drawn from herbarium specimens collected in Alaska, the Yukon, or Siberia. Her drawings are at once faithful to nature, reflective of plant character, and lovely to look upon. The publishers join me in expressing appreciation for her splendid contribution.

Mrs. Maria Ferm, of the Riksmuseum, typed the manuscript. I am grateful to her for her care and patience.

The color photos are my own. (Circumstances required that I prepare the captions for these photos from memory. I might briefly emend four of them: the large photo on the first page was taken from calcareous mountains in the Yukon, toward Alaska; the upper righthand picture on the second page is of Mt. Chamberlain, as seen from Peters' Lake; in the lower lefthand picture on the second page, the polar timberline is in the foreground; and *Carex Gmelinii,* on the fifth page, is in fact *C. Mertensii.*)

Aktiebolaget Kartografiska Institutet in Stockholm prepared the range maps from my materials and photographed the maps and drawings for the publisher. To Mr. Olof Hedbom of that institute I convey my thanks for the considerable care he has exercised.

Finally, I must express my particular thanks to the many people at Stanford University Press who have worked so devotedly and for so long to render of my materials such an exemplary volume. They have understood my feelings about this work; that they have honored them is evident on every page. Mr. Emlen Littell, now living in British Columbia, was very helpful in our initial discussions. Elizabeth Spurr performed countless editorial tasks with skill and dispatch. Compositors, proofreaders, artists, cameramen, pressmen, and binders all worked with great care. Especially I wish to thank Mr. William W. Carver for his patient and critical editing of the manuscript, for valuable substantive suggestions and development of the book's structure, and for attentively guiding the book through proofs and production. My appreciation of his helpfulness cannot be adequately expressed.

E.H.

Stockholm, April 1968

Contents

Introduction	xi
Master Key	1
The Flora	**24**
Glossary of Botanical Terms	963
List of Authors	969
Bibliography	987
Index of Common Names	995
Index of Botanical Names	998

Eight pages of illustrations in color follow p. 448.

Introduction

This flora describes and illustrates all of the flowering plants and vascular cryptogams known to occur within Alaska and its neighboring territories—i.e., within the confines of the maps shown as Figs. 1–4, below. Keys based on physical characters are included for all genera, as well as for families and larger plant groups. The use of these keys in conjunction with the plant drawings and the plant and habitat descriptions should make it possible to identify any plant occurring in this region.

Accompanying each plant description are an Alaska range map and a circumpolar range map. The circumpolar (effectively worldwide) range maps, the first to be used in a major flora, are of special interest, suggesting as they do the affinities of plant groups and floral regions, and the spread of vegetation in the geologic past.

Some 1,974 distinct, taxonomically named plants, belonging to 1,559 species, 412 genera, and 89 families, occur in the region defined; all are described. For 1,735 of these (including all species and most subspecies), detailed descriptions, plant drawings, and range maps are provided. Some hundreds of hybrids, not included in these figures, are also distinguished. Finally, the text cites some 238 closely related taxa occurring in other (usually neighboring) floral regions, and provides range data for most of them.

LIMITS OF THE REGION COVERED

The configuration of Alaska, roughly a square with a long extension to the west, into the Aleutian Islands, and another to the southeast, down the Panhandle, makes it inconvenient to deal with as a biological unit. Plants and animals do not respect political boundaries. I have therefore settled on a region substantially greater in extent, one that includes the Yukon Territory, the narrow wedge of British Columbia penetrating between the Yukon and the Alaska Panhandle, parts of the Northwest Territories (chiefly the Mackenzie Delta region), and the Chukchi (Chukotka) Peninsula of Siberia. The inclusion of the Chukchi Peninsula, politically a part of the Soviet Union, helps to make clear the important overlapping of the floras of the two continents.

Alaska comprises approximately 571,050 square miles (1,478,960 km^2), the Yukon 205,350 square miles (531,727 km^2), and the remaining territories included in the flora roughly 246,000 square miles (637,140 km^2), or altogether about 1,022,400 square miles (2,648,000 km^2). The area considered is thus about four times as large as Texas, 16 times as large as New England, and somewhat less than one-third the size of Europe.

In degrees of latitude the area stretches from Point Barrow, at 71° 23′, to the southern end of the Panhandle, at 54° 40′, though the central Aleutian Islands reach farther south, to 51° 20′, and are actually the southernmost part of Alaska. In degrees of longitude the area reaches from Attu Island, at approximately 172° E., to the eastern end of the Yukon, at 124° W. The region is thus somewhat greater in extent north to south (roughly 2,035 statute miles, or 3,250 km) than the adjacent 48 states of the United States, and only slightly less in extent west to east (about 2,390 miles, or 3,825 km).

CLIMATIC VARIATION

In such a vast area the climate is necessarily varied. Figures 1 and 2 give mean annual temperature and mean July temperature. Mean annual air temperature ranges from +2° to +4° C. (35° to 29°F.) along the southern coast; from −2° to −6° (28° to 21°F.) in the interior; and from −10° to −12° (14° to 10°F.) on the Arctic slope. The extreme low temperature varies between −22° and −35° (−8° and −31°F.) along the southern

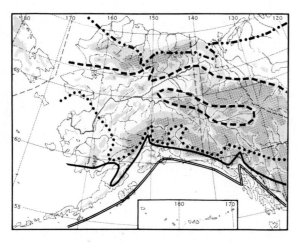 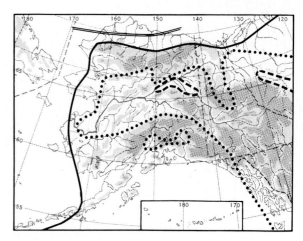

Fig. 1. Mean annual air temperature: dashed-double-dotted line, −10°C. (14°F.); dashed line, −8°C. (18°F.); dotted line, −4°C. (24°F.); solid line, −1°C. (30°F.); double line, +2°C. (36°F.).

Fig 2. Mean July air temperature: double line, 4°C. (40°F.); solid line, 10°C. (50°F.); dotted line, 13°C. (55°F.); dashed line, 16.°C. (60°F.).

coast, with −17 (+1.4°F.) in the Aleutian Islands, to between −50° and −55° (−58° and −67°F.) on the northern coast, and occasionally falls to −63° (−82°F.) in the interior. The extremes of high temperature vary between 23° and 35° (73° and 95°F.) in the interior. In July the mean daily temperature is about 13° (55°F.) on the southern coast, 10° (50°F.) in the Aleutians, 7° (45°F.) on the northern coast, and 16° (61°F.) in the interior. The length of the frost-free period varies from more than 200 days in the outer part of the Panhandle to 60 days in the interior and 40 in the Alaska Range and on the northern coast.

The period of growth is estimated at 180 to 200 days in the Panhandle, 120 to 140 days inland, and 60 days on the Arctic slope.

Figures 3 and 4 give total mean annual precipitation and total mean precipitation during the growing season (May 1 to August 31). Total precipitation is about 2,500 mm (100 inches) in the Panhandle, 250 to 300 mm (10 to 12 inches) inland, and 200 to 250 mm (8 to 10 inches) on the Arctic slope. Rainfall during the growing season amounts to 525 mm (21 inches) on the coast, between 150 and 375 mm (6 to 15 inches) in the interior, and only 150 to 175 mm (6 to 7 inches) on the Arctic slope. Annual days of rainfall are 240 along the southern coast, but only 120 in the interior. Days per year with one inch or more of snow-cover vary from 40 to 160 in the Panhandle, 200 along the southern coast, 200 to 240 in the interior, and 240 to 260 on the Arctic slope. The maximum depth of snow cover is least in the north, averaging between 250 and 500 mm (10 to 20 inches) on the Arctic slope, and greatest somewhat inland from the Pacific Coast, averaging between 1,000 and 1,250 mm (40 to 50 inches).

Above the northern reaches of the Yukon valley and the northern half of the Seward Peninsula, permafrost is the rule. In the Yukon's Tanana valley, weakly active to inactive ice wedges occur, with few or none south of the Alaska Range, in the Alaska Peninsula, in the Aleutian Islands, or along the Bering Sea north to Nunivak Island.

ECOLOGIC REGIONS

Alaska is also a region of diverse ecology, as the color illustrations suggest. Within the area, four major floral regions can be distinguished.

1. The narrow seaward strip bordering the coastal mountains in southeastern and central Alaska is blanketed by a coniferous forest composed of Sitka spruce, Alaska cypress, western hemlock, and mountain hemlock. Of the four, the Sitka spruce reaches westward to northern Kodiak Island and the opposite mainland, the other three to Kenai Peninsula but not to Kodiak. In the southwesternmost reaches of the Panhandle, western red cedar, western yew, and Pacific silver fir reach the area of the flora. A considerable number of species characteristic of American Pacific timberlands are found in the undergrowth of these forests. Until recently, the forests have remained largely primeval, but lumbering is now in full sway in many localities, and reforestation of the logged-over areas will be slow indeed. The subalpine belt in the coastal area is covered by dense thickets of the sinuate-leaved Sitka alder, which gives way gradually further inland to the lobeless-leaved mountain alder.

2. The lowlands of the interior northward to the mouth of the Mackenzie and the southern

Introduction

xiii

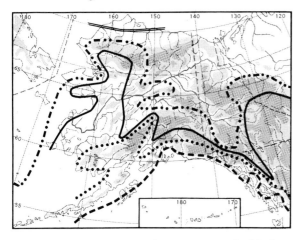

Fig. 3. Total mean annual precipitation: double line, 150 mm (6 in.); dashed-double-dotted line, 300 mm (12 in.); solid line, 400 mm (16 in.); dotted line, 600 mm (24 in.); dashed line, 2,500 mm (100 in.).

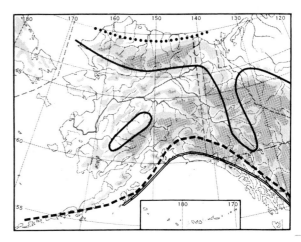

Fig. 4. Total mean precipitation during growing season (May 1–August 31): dotted line, 75 mm (3 in.); solid line, 150 mm (6 in.); dashed line, 300 mm (12 in.); double line, 450 mm (18 in.).

slope of the Brooks Range are covered by a boreal forest, chiefly of white spruce mixed with Alaskan paper birch (see the range map for *Picea glauca*, p. 61). In the muskeg, black spruce is common; and along the rivers, balsam poplar. Larch, best represented in the middle and lower Tanana valley, penetrates westward to the 160th meridian, but occurs scattered and seems nowhere to form pure stands.

Alpine fir and lodgepole pine extend into the Yukon, the coastal race of the latter also occurring in the Panhandle. The birches of the interior are taxonomically complicated, and hybridization and introgression between taxa are frequent. Their occurrence is best studied on the range maps.

Though none of the forest trees of the Alaskan region are identical with those of easternmost Asia, relationships and parallels are apparent. The American and Asiatic forests have been separated by tundra or muskeg since the beginning of Pliocene time.

In the subalpine belt of the interior, the alder thickets are less dense than in the coastal area. The many Rocky Mountain species that occur in the mountains diminish in number westward.

3. The Arctic slope of Alaska is a region of treeless tundra with an arctic flora. Many of the arctic plants have a circumpolar range. Alder and poplar, migrating over the passes in the Brooks Range during the Hypsithermal Interval (ca. 6,000–4,000 years ago), eventually reached the Arctic slope, where they occur today in small, isolated stands in only a few places.

4. The shores of Bering Sea, the Aleutian Islands, the Alaska Peninsula, and part of Kodiak Island sustain a treeless vegetation very much like that of the lowland arctic tundra. The flora here is rich in arctic species, especially in the north, though many of them reach the Pribilof Islands.

The Aleutian Islands should perhaps be regarded as a fifth floral region. Here, Pacific coastal plants predominate, only a few arctic plants reaching to the southernmost or middle part of the chain. Many arctic-montane circumpolar plants occur in both the eastern and the western parts of the Aleutian chain, but are lacking in the central islands.

GEOLOGIC FEATURES

The geology of the Alaskan region is extremely complicated, which is what one might expect, considering its vast extent and diverse topography. All major geologic periods are in fact represented.

From the Arctic Sea to the foothills of the Brooks Range lies the Arctic slope, a broad, flat, coastal plain covered by Quaternary deposits. It is dotted by innumerable lakes and characterized by tundra polygons with ice-wedges. The lakes often extend in the direction of the prevailing winds, and their shorelines are frequently redefined by wave erosion or by modification of drainage outlets.

The coastal plain is separated from the Yukon basin by the Brooks Range, a rugged mountain chain with peaks exceeding 6,000 feet and passes occasionally as high as 2,500 feet. Its northern slope is composed of Cretaceous rock; the southern is of Devonian age and is replaced farther south by still older rocks. The Cretaceous and Devonian stretches are separated by a more or less broad band of Mississippian age. At the eastern end of the Brooks Range lies the Mackenzie Delta, a vast, alluvial, forested plain.

The central part of Alaska and the Yukon is a plateau 2,000 to 3,000 feet high, cleft by the Yukon, Tanana, and Kuskokwim river valleys. This plateau consists mainly of Permian, Devonian, and other Paleozoic rocks, though large areas of Quaternary deposits occur around the northern bend of the Yukon River, along the middle and lower Yukon, in the Yukon–Kuskokwim estuary area, in the lower Nushagak valley, and in the northern half of the Alaska Peninsula. Seward Peninsula, at the western extremity of the plateau, consists of Paleozoic and glacial deposits and volcanic rock.

The central plateau reaches southward to the extension of the Pacific mountain system, the Alaska Range, a rugged mountain chain of complicated geologic structure rising parallel to the shoreline of the Gulf of Alaska. Though its summits vary generally from 10,000 to 12,000 feet, it is crowned by Mt. McKinley, at 20,320 feet (6,200 meters) the tallest mountain on the North American continent, and its sister peak, Mt. Foraker, 17,400 feet high (5,300 meters).

Along the southern coast rise the St. Elias Mountains, with peaks up to 16,500 feet. Their extensions to the west are the Chugach Mountains and the Kenai Mountains. The Alaska Range and this southern chain are separated by a lowland along the Susitna River and by a plateau flanking the upper Copper River, and united by the Talkeetna Mountains and the Wrangell Mountains. Quaternary and Tertiary deposits prevail in the Susitna lowland, Quaternary deposits in the Copper River plateau. Much of the Talkeetna Mountains are made up of Jurassic intrusive rocks; the Wrangells are volcanic in origin, and of Tertiary and Quaternary age. The Aleutian Islands and the southern part of the Alaska Peninsula are also of volcanic origin, with a row of high, often active, peaks, probably of late Eocene or early Miocene age. Kodiak Island is built up mostly of Cretaceous (in the southeast, Tertiary) rock, with intrusive, predominantly granitic rock in the central part.

The eastern extremity of the Alaska Panhandle and the southern part of the Yukon consist mostly of pre-Cambrian and intrusive rocks. The outer part of the Panhandle is of very complex structure.

Chukchi Peninsula is of fairly complex geologic structure: the northern part to the east of Kolyuchin Bay, and the southeastern corner, consist of Paleozoic, and to a large extent pre-Cambrian, sediments; in the center is Cretaceous rock; to the west, volcanic rock; and the northwestern and southwestern coasts are of Quaternary origin.

Glacial drift covers large parts of the bedrock in glaciated areas of Alaska. (Modern glaciers are indicated as white patches on the endpaper map.) In the unglaciated parts of the interior, the remains of prehistoric vegetation, or "muck," are sometimes found, frozen to a considerable depth.

THE GEOLOGIC PAST

In my thesis "Outline of the History of Arctic and Boreal Biota During the Quaternary Period" (1937), I emphasized the importance of unglaciated Alaska as a refugium for arctic and boreal plants during the glacial periods. The existence of a land bridge, Beringia, between Alaska and northeastern Asia during glacial periods, when the storage of water in the continental ice-sheets lowered the level of the sea, was postulated and discussed. Moreover, I observed that the present ranges of Alaskan plants seemed to demonstrate that Beringia had been a pathway for the interchange of biota and that its Asian and American remnants lie at the center of many present distributional patterns.

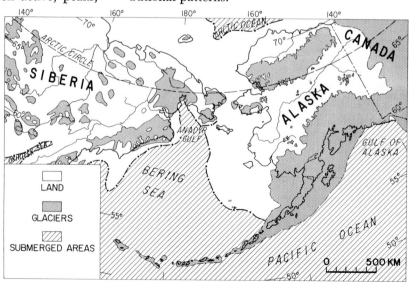

Fig. 5. Geography of Beringia during the height of the Illinoian or Riss Glaciation. Vegetation boundaries are omitted because of lack of adequate data. Reprinted from D. M. Hopkins, ed., *The Bering Land Bridge* (Stanford: Stanford University Press, 1967), p. 462.

Introduction

At that time all of this was speculation. In the thirty years that have elapsed since publication of that work, a large body of material has accumulated that sheds light on the Bering Land Bridge. Facts known up to 1967 bearing on the history of Beringia have recently been assembled from the fields of geology, oceanography, palynology, paleontology, plant geography, and anthropology in an excellent volume, *The Bering Land Bridge*; the history of the land bridge is summarized by the editor, D. M. Hopkins, in a final synthesis. The following review is based on his account.

In the early and middle Tertiary, a time with no pronounced climatic zonation, Beringia lay above sea level. During Oligocene through middle Miocene time, it belonged to a zone of mixed humid, mesophytic forests, stretching along the North Pacific from Japan to Oregon. The first opening of the Bering Strait, probably caused by tectonic movements, took place in the late Miocene, 12–15 million years ago, as testified by a comparison of fossil marine fauna from the shores of the North Atlantic and North Pacific Oceans.

At that time the forests of America and Asia became isolated from one another. A reduction in summer temperature as well as a change of climatic factors associated with the opening of the straits led to the formation of rich, dissimilar boreal forests in northeastern Asia and northwestern North America.

The land bridge seems to have been restored during much of the Pliocene epoch, but was submerged again in late Pliocene time, 3.5–4 million years ago, and the Bering and Chukchi Seas then assumed approximately their present form. In late Pliocene or early Pleistocene time, forests of *Chamaecyparis*, white and black spruce, and poplar, mixed with tundra, developed near Kotzebue Sound and in Seward Peninsula. In Siberia, hemlock, fir, linden, and walnut were present. And in the early Pleistocene, *Juglans cinerea* occurred at the mouth of the Lena River, in Siberia. On the land bridge, birch, aspen, and willow flourished.

About 2 million years ago, severe climatic conditions prevailed on the southern edge of the land bridge, as evidenced on St. George Island by frost-broken rubble intercalated by lava, and on St. Paul Island by the appearance of permafrost. Tundra vegetation was the rule on the Chukchi Peninsula.

During the Illinoian (Riss) Glaciation (Fig. 5), the climatic stress reached a climax in Beringia and the tree line must have retreated far beyond its present position. The region has in fact remained forestless to the present day.

The Illinoian Glaciation also produced large glaciers in Anadyr and easternmost Siberia, and the Cordilleran and Continental ice fields of America closed the earlier corridor between Alaska and central North America. Exchange of biota with central North America was thus prevented. Sea level fell considerably, probably to about −135 meters, creating a broad land bridge (Fig. 5). The climate became dry; and in Siberia, arctic tundra elements merged with southern steppe elements.

These conditions were changed during the Sangamon Interglaciation, about 100,000 years ago. The climate became warmer and more humid. The forests, which now differed little from those of the present species composition, expanded, and the continents parted again for a long interval.

The interglacial conditions were followed by the last, Wisconsin (Würm) Glaciation, which reached its maximum about 20,000 years ago (see Fig. 6). Sea level fell at least to −115 or −120

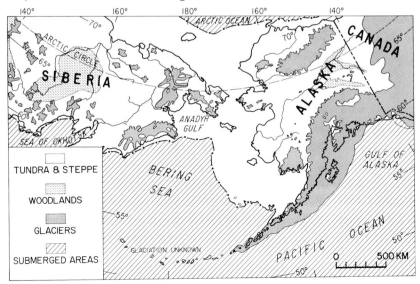

Fig. 6. Geography of Beringia during the height of the Wisconsin or Würm Glaciation. Reprinted from D. M. Hopkins, ed., *The Bering Land Bridge* (Stanford: Stanford University Press, 1967), p. 462.

meters, and the land bridge was restored. Timberline was lowered by at least 400 meters. A corridor probably emerged between the Cordilleran and Continental ice sheets in North America during a mid-Wisconsin interval of relatively warm climate, between about 38,000 and 25,000 years ago. The ancestors of the American Indians were probably in Alaska by the end of this mid-Wisconsin warming. The ice-free corridor closed again about 20,000 years ago and remained closed for 6,000 to 10,000 years. During this period Alaska had a broad connection with Siberia, but was isolated from central North America (Fig. 6). The climate again became drier, and tundra conditions prevailed on the land bridge. In Siberia, *Selaginella sibirica* is characteristic in the pollen spectra from this dry period. Dr. Hopkins notes that the skill of the hunters may have been responsible for the disappearance of many of the large land mammals during the period following the Wisconsin Glaciation, and in turn for marked changes in the character of Beringian landscapes.

About 18,000 to 20,000 years ago, long before there is any evidence of substantial shrinkage of the outer margins of the continental glaciers, and long before the paleobotanical records show any evidence of warming, sea level began an oscillating rise. About 14,000 years ago the Bering Strait opened, though St. Lawrence Island remained joined to Alaska.

The land bridge was probably restored about 13,000 years ago, very likely opening the way for the spread of the Eskimo and Aleuts into the New World, but the straits opened again about 12,000 years ago, this time separating St. Lawrence Island from the Alaska mainland. A brief land connection may have occurred 11,000 years ago, but about 10,000 years ago the Bering Strait opened and has remained open ever since.

Toward the end of the last land bridge period, spruce advanced to the flanks of the southern coast of the land bridge, but the forests of Alaska and Siberia did not merge.

Between 10,000 and 8,300 years ago, a pronounced climatic amelioration set in at Kotzebue Sound and Seward Peninsula, followed by a colder period. During the Hypsithermal Interval, some 6,000 to 4,000 years ago, the forests probably again reached positions north of their present limits.

Further paleobotanical and ecological studies of Beringia should find some relevance in the range information presented in this volume; certainly the extensive data on palynology and the like assembled by Dr. Hopkins and his colleagues will be of interest to students of the Alaskan flora.

PLANT COLLECTION AND DISTRIBUTION

The first botanical collections in Alaska were made by the German naturalist Georg Steller, who accompanied Bering on his voyage of discovery in 1741. For six hours he made collections on Kayak Island in the Gulf of Alaska. A few of his specimens later reached the hands of Linnaeus, but documentation of his Alaskan collections has fared badly: a manuscript list of the plants collected, written in an unknown hand, was found in a Russian archive and published by Stejneger in 1936, and another manuscript enumerates "twenty seeds collected by Steller in America."

Soon after Bering's discovery, many of the great voyages of exploration and circumnavigation touched Alaska, and their naturalist members made collections of Alaskan plants. Thus William Anderson, surgeon and naturalist of Captain Cook's expeditions, and David Nelson, a gardener from Kew, collected plants at several points on the southern and western coasts of Alaska in 1778. Carl Meck, who accompanied Billings aboard the *Slava Rossii*, collected in the Aleutian Islands, Prince William Sound, St. Lawrence Island, and the Seward Peninsula, in 1790–91. Thaddaeus Haenke, a naturalist with Malaspina's expedition, collected at Yakutat Bay in 1791. Archibald Menzies, surgeon and naturalist on Vancouver's voyage of 1793–94, collected in Prince William Sound and in the Panhandle. Georg Heinrich von Langsdorff and Wilhelm Tilesius, who accompanied Krusenstern on his voyage of circumnavigation on board the *Neva* and the *Nadeshda*, in 1803–06, collected on St. Paul Island, Unalaska, Kodiak, the Alaska Peninsula, and at Sitka.

The first more extensive collections in Alaska were made by Adalbert Ludwig von Chamisso, poet and naturalist, and Johann Friedrich von Eschscholtz, surgeon, who accompanied Kotzebue on his voyage of circumnavigation aboard the *Rurik*, in 1816–17. Several points along Kotzebue Sound and on St. Lawrence Island, the Pribilofs, and Unalaska were investigated, and the botanical findings were published in *Linnaea*, 1826–36, under the title "De Plantis in expeditione speculatoria Romanzoffiana observatis," the first extensive, scientific treatment of Alaskan plants.

During the ensuing years the Alaskan coast was visited by a number of botanical collectors, most of whom gathered but few specimens. Among those who made more serious contributions were George Lay, naturalist, and Alexander Collier, surgeon, who accompanied Capt. Beechey on board the sloop *Blossom* on its voyage in the

Introduction

Arctic in 1826–27; their collections, the first from the Arctic coast of Alaska, were published by Hooker and Arnott in 1841. The naturalist Carl Heinrich Mertens accompanied Lütke on the corvette *Senjavin* in 1827; his collections, taken especially at Sitka, were published by Bongard.

In 1867, Alaska was purchased from Russia by the United States. The transaction had no immediate noticeable effect on botanical investigations, however, the interior remaining as unknown as it had been before the purchase. Collections were made chiefly from the same coastal localities visited by earlier collectors. Lucien McShan Turner assembled a moderate collection from Nushagak, St. Paul Island, Unalaska, and Atka, later published under the title "Sketch of the Flora of Alaska." In 1879, Frans Reinhold Kjellman, professor of botany in Uppsala, and Ernst Bernhard Almquist, surgeon — both members of Nordenskjöld's expedition circumnavigating Asia aboard the *Vega*—gathered extensive materials from St. Lawrence Island and Port Clarence, which were published in the expedition's scientific reports. In 1881–82, the brothers Arthur and Aurel Krause of Bremen made important collections in the Haines and Chilcot areas, which were published in Engler's *Bot. Jahrbuch* in 1895. John Murdoch, in 1881, was the first to collect at Point Barrow, and in 1883 the first collection of any importance from the interior was made by Lt. Frederick Schwatka, along the Yukon River.

From then on, botanical collections were made almost yearly, usually by several collectors, and many were published. In 1898–1902, Frank Charles Schrader crossed the Brooks Range from the Yukon valley to the Arctic Sea, and made the first collection from that region. In 1899 the Harriman Alaska Expedition, a major undertaking with several scientific members and with Frederick Vernon Coville and Thomas H. Kearney as chief botanists, brought together the largest botanical collection from Alaska up to that time, though exclusively from the coastal areas. Part of the cryptogamic collection was published in 1904 in "Harriman Alaska Expedition," but though a manuscript on the flowering plants was prepared by P. C. Standley, it was never published.

During the first years of the new century, the flora of the interior came gradually to be known through the work of many collectors of diverse interests, though no new large collections were made or published. For about 20 years the more professional botanical activities were concentrated on the coasts or their vicinity. This circumstance persisted until the construction of highways began to open up the interior. Of collectors active in the interior before 1940, the following especially should be mentioned: Jacob Peter Anderson, from 1914 to his death in 1953, who collected extensively and indefatigably, amassing a great wealth of valuable material from many parts of Alaska; Alf Erling Porsild and his brother Robert Thorbjørn Porsild, who assembled a huge herbarium of some 6,000 specimens, while engaged in studies of reindeer; Ynez Mexia, who made a fine collection in 1928 in McKinley National Park; Edith Scamman, who made an excellent collection chiefly along Steese Highway in 1936–37; and Aven and Ruth Nelson, who investigated McKinley Park in 1939.

Elsewhere, Misao Tatewaki collected extensively in the Aleutians, in 1929; Otto Geist collected on St. Lawrence Island and in other localities in 1931; and in 1932 I collected some 2,500 specimens, chiefly in the Aleutian Islands, which were described in my "Flora of the Aleutian Islands," published in 1937.

A more detailed account of the botanical exploration of the Alaskan region prior to the year 1940 can be found in "History of botanical exploration in Alaska and Yukon territories from the time of their discovery to 1940," in *Bot. Notiser*, 1940, which includes a map of collection sites.

A first attempt to gather information on the flora of Alaska and the Yukon from the scattered botanical literature of many countries, and from examination of the herbarium collections on which the reports were founded, was published in my *Flora of Alaska and Yukon*, I–X, 1941–50. The tenth volume of this work (1950) includes a comprehensive bibliography listing the most important taxonomic works, of all periods, dealing with the flora of the area. The first part of that bibliography lists general works and works dealing with regions or major groups of plants; the second part lists articles and monographs dealing with smaller groups of taxa or individual species. The Bibliography in the present volume combines the general works from the 1950 listing with general works published since 1950. In my *Comments on the Flora of Alaska and Yukon* (Arkiv för Botanik, Serie 2, B and 7, nr. 1, 147 pp.), published in late 1967, changes in nomenclature and taxonomic treatment in the present work (as compared with the earlier *Flora*), as well as new name combinations, are discussed with reference to recent taxonomic literature. Supplements to these comments are given in an issue of *Madroño* scheduled to appear prior to publication of the present work.

Since publication of *Flora of Alaska and Yukon*, extensive new herbarium materials have been

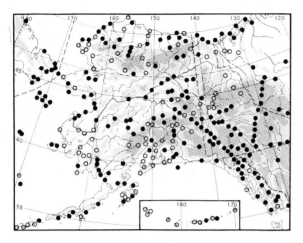

Fig. 7. Sites of botanical collections: open circles indicate localities where fieldwork (generally including collecting) has been done by the author; solid dots indicate localities where other notable botanical collections have been made.

amassed by a great many collectors. Arctic Alaska, for example, was known botanically only from the coast; the broad Arctic slope, from the Brooks Range to the coast, remained unknown. Establishment of the Arctic Research Laboratory at Barrow, by the U.S. Office of Naval Research, in 1946, led to an intense botanical exploration of that vast, roadless territory. Small aircraft landed investigators at many places, and the Arctic slope became one of the best-known regions of Alaska, second only to the southern coast. In 1962 all taxonomic knowledge about the area was collected by Wiggins and Thomas in their *A Flora of the Alaskan Arctic Slope*, which includes range maps of all species and a listing of all contributors to that flora.

More recently, generous assistance from the Arctic Institute of North America and the Arctic Research Laboratory at Barrow has permitted me to collect considerable new material and to study the vegetation in a great many places (some not easily reached), during four summers. These collections are principally from the Arctic slope of Alaska, the Mackenzie Delta, the Mackenzie Mountains, the central Yukon, Kotzebue Sound, Seward Peninsula, and the Bering Sea coast. The map shown as Fig. 7 indicates by open circles the localities where I have had the opportunity to pursue fieldwork; in all of these places I attempted, so far as possible, to collect or note down all the species I observed.

Places where other major botanical collections have been made—both recent and past—are marked on the collection map by solid dots. As the map indicates, major botanical collections are lacking from a large area north of the middle Yukon River, from below the mouth of the Koyukuk River to the mouth of the Porcupine River. Except for sporadic, isolated records, the flora of the fairly high mountains there is unknown. The upper Kuskokwim valley and the Mulchatna drainage are other little-investigated areas in this region. In the Yukon the area around the upper Porcupine River deserves more attention, especially as the northern and southern extremities presumed for many plant ranges might merge there. It is, though, an area decidedly poor in species: in brief visits at three points in this area, I found an extremely impoverished flora. Finally, from the mountains separating the Yukon from the Mackenzie only a few botanical collections have been made. This is an uninhabited and highly inaccessbile region, and future investigations will certainly reveal the existence there of a number of Rocky Mountain species not yet reported from our area of interest. Moreover, many of the gaps tentatively marked on the range maps in this book will probably be eliminated.

Source material at least three times the extent of that available to me in 1950 has been used. This has permitted me to formulate more precise impressions of the distribution of most species. Dots on the Alaska maps mark places where I have observed living specimens, collection points cited for herbarium material that I have examined, or points cited in reliable reports found in the literature. Alaska–Yukon materials in the following herbaria have been studied: the State Museum of Natural History (Riksmuseum), Stockholm; Gray Herbarium, Cambridge, Mass.; U.S. National Herbarium, Washington, D.C.; National Herbarium, Ottawa, Ont.; Herbarium of the University of British Columbia, Vancouver; and Herbarium of the University of Alaska, Fairbanks. Boundary lines on the Alaska maps, marking the probable distributional extent of the taxa, have been extrapolated from the available material in conjunction with a knowledge of the physiography and ecology of the region and of the localities where comprehensive botanical collections have been made. These range-limit lines, which will of course demand modification as further data accumulate, should be a challenge to collectors to extend the postulated plant ranges by new collections.

On the circumpolar maps, ranges are indicated only by boundary lines. In many cases, these ranges are composed of a fairly large number of small, more or less isolated areas. Only those areas where a species is actually known are indicated, often by small circles; gaps that frequently occur in worldwide ranges, as well as other important details, would otherwise disappear. Though the

Introduction

scale is quite small, the maps are not unduly generalized and should give a satisfactory representation of plant ranges as understood by the author. The occurrence of a plant outside the area of the circumpolar map is indicated on the map by such notes as "Also in southern South America" or "Introduced into New Zealand, Tasmania, and Australia." To this extent the circumpolar maps can be considered total-range maps.

The base map used for Figs. 1–4 and Fig. 7 is also the base map for the Alaska range maps accompanying the species descriptions in the body of the book. Both the Alaska maps and the circumpolar maps have been reproduced from large-scale originals drawn on base maps considerably more detailed than they appear to be in the printed reductions (details such as district lines and river systems were printed in blue, so as to filter out in the photoreduction). The larger map reproduced on the endpapers differs somewhat in extent, for reasons of appearance.

In some floras it is customary to treat introduced plants more superficially than native plants. In this flora they have been treated equally. For the introduced plants, especially the weeds, the total range indicated on the circumpolar map is often necessarily tentative; in many cases only the northern boundary can be indicated with any sort of precision. Often the occurrence of a plant in our area of concern is the northernmost in the entire range of the plant and therefore of special interest. With increasing worldwide communication, many of these introduced plants are spreading rapidly, and their distributional patterns can thus be expected to change accordingly. It has not been possible to distinguish between introduced and native occurrences on the maps; this is left to the judgment of the reader, though in some cases distinctions have been indicated in the text.

RACES, APOMICTS, HYBRIDS, AND ENDEMICS

A general condition of the flora of this region is that the morphological variation of a given taxon is greater in Alaska than in other parts of its range. The disparity is presumed to relate to the age of the flora: those parts of a plant's range that penetrate into areas glaciated during the last glacial period would seem to be of considerably more recent origin than the Alaskan part, where the flora of 22,000 years ago was similar to that of today, as demonstrated by the carbon-14 method. I have thought therefore to apply somewhat liberal use of the subspecies concept to the taxonomic treatment. Though this policy unfortunately involves a large number of new name combinations, I feel that this approach best conforms to conditions in nature and provides a sound nomenclatural expression of the close relationship between two slightly different populations. Within a given species, then, two subspecies are regarded herein as taxonomically separate but closely related populations in which all or some of the individuals in each population differ in minor morphological characteristics from those of the other, the two populations occupying large and partially or completely isolated geographic areas, and being potentially capable of interbreeding without substantial reduction of fertility. As thus defined, subspecies could be characterized simply as major geographical races, corresponding, for example, to the major human races. (In ornithology, to cite a further example, this concept of subspecies is in general use, the species often being differentiated into a number of geographically more or less isolated, more or less interbreeding, subspecies.) I anticipate much resistance to this view, especially where two populations have traditionally been regarded as separate species; on the other hand, it is apparent that the subspecies concept is gradually gaining favor in modern taxonomic treatments, and will, I believe, be generally accepted in the near future.

The recognition of major races is especially critical in Alaska, situated as it is on the path of the great migrations from Eurasia to America (and, in a few cases, from America to Eurasia), inasmuch as components in the Alaskan flora are closely akin to taxa from many very different floral regions: the Asiatic and American Arctic, the Asiatic and American boreal forest, the Asiatic and American Pacific shores, the Rocky Mountains, the Japanese mountains, and to a lesser extent, the Himalayas, the Siberian and western American steppes, and even eastern North America. The flora of Alaska and the Yukon thus proves to be very complex, as will be evident from a study of the circumpolar maps.

My treatment of varieties and forms is quite conventional, and no more need be said of them, save that I have cited very few taxa at the form level.

The apomictic taxa, which form viable seeds without fertilization, hold a special position and should not be treated as species. These taxa correspond essentially to the vegetatively multiplying apple varieties, which have never been regarded as species. Though the propagation of the apomicts does indeed take place through seeds—rather than through cuttings or graftings, as in the case of the apple—these are seeds that develop without normal sexual fertilization. In this flora the apomicts have been grouped to form taxa that correspond as nearly as possible to species in sexually reproducing plants. Apomixy is especially noticeable in *Taraxacum* and *Antennaria,* as well as in *Hieracium,* but probably also occurs in a number of other genera, including *Artemisia, Agoseris, Solidago, Aster, Erigeron,* and *Arnica.* The possibilities for different taxonomic treatment of these genera are endless.

Hybridization also occurs extensively in a number of genera, often resulting in very difficult taxonomic decisions. This is the case, for example, within the genera *Potamogeton, Sparganium, Agrostis, Calamagrostis, Agropyron, Elymus, Platanthera, Salix, Draba, Sanguisorba, Rubus, Potentilla, Dryas, Rosa, Lupinus, Phyllodoce,* and *Petasites.* Studies in the geographical distribution of the different types would seem to offer the best solution to these difficulties. Such hybrids as I have been able to distinguish are discussed in conjunction with their parent taxa, and are indicated in the customary manner; thus *Salix arctica* × *ovalifolia* is the presumed hybrid of *S. arctica* and *S. ovalifolia.*

In not a few cases, the morphological variation in a population supports the view that introgression between two species occurs—i.e., back-crossings of hybrids with parents and among the resulting hybrids can produce essentially uninterrupted intrograding from one species to the other. In such cases the plant descriptions usually make mention of the occurrence of *hybrid swarms.*

As can be seen from the distributional maps, a number of species occur only within the area of this flora and are thus endemic there. The most striking of the endemics is *Boykinia Richardsonii* (see color section), a magnificent plant belonging to the Saxifrage family, and doubtless a Tertiary relict. In spite of its large size, it is a very hardy plant, often growing in the meltwater between snow patches, right up to the Arctic coast.

Considering that large parts of interior Alaska–Yukon have never been glaciated, the endemic component must be said to be weak. Probably most species surviving the glacial period there have been able to spread in interglacial or postglacial times to both east and west, beyond the area of our flora—and thus to have relinquished their endemic status. A study of the circumpolar distributional maps suggests this conclusion.

ADDITIONAL SPECIES LIKELY TO BE OBSERVED

A number of species not described in this work have been reported close to the periphery of our Alaska–Yukon map, but are not yet known to occur within its limits. Since they can be expected to occur within these limits, they are listed below.

Liard Hot Springs, situated slightly outside the southeastern corner of the map, is apparently a northern outpost locality for several southern species. Its vegetation is described by A. E. Porsild and H. A. Crum (*Nat. Mus. Can. Bull.* 171 [1959], 131–97). Species known from this locality but not from the range of the Alaska–Yukon map are:

Oryzopsis asperifolia Michx.
Muhlenbergia cuspidata (Torr.) Rydb.
M. mexicana (L.) Trin.
M. racemosa (Michx.) BSP.
Carex Deweyana Schw.
Maianthemum canadense Desv. var. *interior* Fern.
Cypripedium calceolus L. var. *pubescens* (Willd.) Correll
Salix pyrifolia Anderss.
Parietaria pennsylvanica Muhlenb.
Drosera intermedia Hayne
Prunus pennsylvanica L.
P. virginiana L.
Viola nephrophylla Greene
V. rugulosa Greene
Epilobium leptophyllum Raf.
Aralia nudicaulis L.
Sanicula marilanica L.
Osmorhiza depauperata Phill.
Cynoglossum boreale Fern.
Lonicera glaucescens Rydb.
Lobelia Kalmii L.
Aster ciliolatus Lindl.

Close to the southern end of the Alaska Panhandle the following plants have been reported:

Dryopteris filix-mas (L.) Schott
Pseudotsuga Menziesii (Mirb.) Franco
Disporum Hookeri (Torr.) Britt. var. *oreganum* (S. Wats.) O. Jones
Maianthemum canadense Desv. var. *interior* Fern.
Lathyrus ochroleucus Hook.
Vaccinium ovatum Pursh
Convolvulus sepium L.
Gentiana sceptrum Griseb.
Dracocephalum Nuttallii Britt.
Pentstemon diffusus Douglas

A flora of the Queen Charlotte Islands is in preparation by Dr. J. A. Calder and his collaborators. The work will no doubt report a fairly large number of taxa that reach their northern limits in these so-far little-known islands. A few taxa now known to occur there but not yet reported within our range are worth mentioning:

Polypodium Scouleri Grev. & Hook.
Carex leptopoda Mack.
Saxifraga Taylorae Calder & Savile
S. Newcombei Small
Geum Schofieldii Calder & Taylor

Introduction

Ligusticum Calderi Mathias & Constance
Douglasia laevigata Gray
Lobelia dortmanna L.
Franseria Chamissonis Less.
Senecio Newcombei Greene

From the Commander Islands, the western continuation of the Aleutian chain, the following three taxa have been reported:

Solidago cuprea Juz.
Artemisia insulana Kraschen.
Saussurea pseudo-Tilesii Lipsch.

Just to the east of our range map, in the Mackenzie District, the following have been reported:

Salix Macalliana Rowlee
Cardamine parviflora L. var. *arenicola* (Britt.) Schultz
Rorippa chrystallina Rollins
Prunus pennsylvanica L.
Astragalus canadensis L.
Gentiana affinis Griseb.

Also to the immediate east, on the Arctic coast of the Northwest Territories, *Aster pygmaeus* Lindl. and *Salix fullertonensis* Schneid. occur.

A number of species reported from Chukchi Peninsula were felt to be too obscure taxonomically to be included in this manual; most of them are probably identical with Alaskan plants reported under other names. The questionable taxa occurring there are the following:

Bromus sibiricus Drobov
Roegneria villosa Vassiljev
Carex arakamensis C. B. Clarke
C. fuscidula Krecz. (*C. capillaris* group)
C. Ledebouriana C.A.M. (*C. capillaris* group)
Stellaria Fischeriana Ser. (*S. longipes* group)
S. peduncularis Bunge (*S. longipes* group)
S. Laxmannii Fisch. (*S. longipes* group)
Minuartia verna (L.) Hiern
Oxytropis Adamsiana (Trautv.) Vass.
Primula xanthobasis Fed.
Plantago asiatica L. (at a hot spring)
Erigeron Komarovii Botsch. (*E. peregrinus* group)
E. Thunbergii Gray (*E. peregrinus* group)
Saussurea tschuktschorum Lipsch.
Corydalis ambigua Cham. & Schlecht. (occurrence in Chukchi seems doubtful)

ORGANIZATION OF THE BOOK

A few additional comments are in order concerning the organization of the book and the treatment given each plant description. (Much of the following will be seen to be elementary, but may be found useful or interesting to nonbotanists.)

In studying the plant drawings, special attention should be given to the indications of plant size (the indication "× 1/3" means that the actual plant observed as the basis for the drawing was three times as tall as the printed image of the drawing). Relative sizes are better appreciated if these indications are borne in mind.

Artificial, analytical, dichotomous keys have been constructed for all genera comprising two or more taxa. A set of master keys, immediately preceding the main body of the work, has also been constructed: for the major groups of plants comprising the flora, for the 89 families making up these groups, and for the 412 genera within these families. Identification will probably proceed most quickly by combined consultation of these keys and the plant drawings.

The sequence of the genera of flowering plants is, with a single exception (*Empetrum*), that used in most herbaria, as based on Engler's natural system and enumerated in Dalla Torre & Harms, *Genera Siphonogamarum*. The vascular cryptogams are arranged according to Verdoorn, *Manual of Pteridology*.

It is of course desirable that each well-established taxon be referred to by the same name by all who deal with it—and the international rules of nomenclature have made it mandatory that the name thus used be the first legally published name. Varying concepts of species and the occasional lack of knowledge of, for example, exact dates of original publication, make this seemingly simple rule actually quite complex, and the results are often uncertain. For example, many Alaskan plants occur also in Siberia, and in many cases different names are used by Russian and American authors in discussing the same plant. The very narrow species concept adhered to in *Flora SSSR* (*Flora USSR*) is one cause of these difficulties: many taxa regarded in that work as species are in this flora treated as subspecies. Russian authors also acknowledge the names in Gilibert's *Flora Lithuanica* as having been legally published, though many other authors refuse to accept them, since the work does not in all cases employ binomial nomenclature; these names are not accepted in this flora.

Many nomenclatural differences arise from different concepts of what to include in certain genera. For example, most American authors do not distinguish the genus *Minuartia* from *Arenaria*, *Melandrium* from *Lychnis*, or *Platanthera* from *Habenaria*, as I have done here. On the genera centering around *Arenaria*, see a paper by McNeill in *Notes Bot. Gard. Edinb.* 24:2 (1962).

The current international rules recommend, but do not prescribe, that all species and subspecific names be spelled with lower-case initial letters.

In this flora, capital letters are used when the species name is derived from the name of a person (though not when it has been derived from a place name). No attempt has been made to introduce standardized spelling to such species names as *unalaskensis* and *kamtchatcensis*, which vary considerably in the original diagnoses.

The accents given for pronunciation of botanical names are roughly in accord with the principles elucidated by Rydberg in his *Flora of the Rocky Mountains and Adjacent Plains*, though in such a linguistic and phonetic swamp no set of principles can be made to suffice in all cases. Briefly, the grave accent (`) calls for the long form of the vowel and the acute accent (´) the short form. It is presumed that the English sounds of letters will be used in pronunciation. Where a botanical name has been derived from the name of a person, I have tended to retain some reflection of the person's name in the pronunciation. The use of pronunciation marks, felt by many to be simply an encumbrance, seems to me a useful addition, perhaps in the name of communication, and certainly harmless enough.

The usually abbreviated name following each botanical name identifies the author of the taxon in question (a comprehensive list of these authors, supplying basic biographical data, begins on p. 969). Where the name appears alone, as in *Salix arctica* Pall., what is implied is that the noted German botanist Peter Simon Pallas (1741–1811) first described the plant today known as *Salix arctica*, and assigned to it that name. Where the author's name is preceded by a second author's name in parentheses, the implication is that the second-named author has seen fit to partially re-name, but not to re-describe, the plant originally described by the first-named author—what he has done, in most cases, is simply to place the species in question in a new genus, or to raise or lower the subspecific rank of the taxon (from variety to subspecies, or whatever). Each of the synonyms given in small type is simply another name of the plant thus described, since abandoned for any of a variety of reasons—usually because it is not the oldest name for the taxon concerned, or because it has been preoccupied by some other plant or animal, or used in another rank (e.g., var., as opposed to subsp.), or because it does not conform with the international rules in some respect, such as not having been validly published. Where a subspecies (and/or variety) is given, there may be two, even three, lists of synonyms; in some of these cases, it is a moot point whether a given abandoned name is a synonym of the species or of the subspecies or variety. Finally, where a variety is mentioned within a plant description, it is assumed to be a variety of whatever taxon is being described; thus, on p. 369, var. **laciniata** is a variety of subsp. **sinuata**, *not* of subsp. **crispa** (p. 368).

Common, or colloquial, names, given at the right margin beyond the species name, and occasionally with the genus name, have been included only where such names are in common use. An index to these names begins on p. 995.

In addition to morphologic features (for which see the Glossary beginning on p. 963), the plant descriptions include typical habitat, economic importance (especially as food), toxicity (in not a few cases), and various other data of interest. In many floras, the flowering time of each species is indicated in months. Since flowering time is very short in Alaska (limited essentially to the months of July and August) and since the varying topography (generally broken ground, with varying exposure to the sun on different slopes) influences flowering to a considerable extent, it seemed to me meaningless to quote flowering times, except when a plant flowers exceptionally early.

The type locality, or place of first description, for each recognized taxon is also indicated, either in the text or, where possible, by an arrow on one of the two maps. In many cases, the text mentions and briefly describes certain closely related Eurasian or American taxa. Where convenient, the range of such a plant is indicated by a broken line on the circumpolar map. Doubtful occurrences reported in the literature are indicated on the range maps by question marks.

Recognition of these foreign relatives brings me, perhaps mercifully, to my final mention of a facet of botanical study I have personally found most absorbing. In other works it has been customary to account for the phytogeographic position of the area treated in a flora by stating statistically how many of its species it shares with other, usually neighboring, areas. Statistics of this sort are of course unsatisfactory for most purposes of botanical study. The circumpolar range maps given in this flora should offer a welcome opportunity to study the geographical affinities of the flora in detail, and to provide a basis for phytogeographic theory and speculation that has not earlier been available. For many purposes they may be found to be the most useful contribution I have made in preparing this flora.

Flora of Alaska and Neighboring Territories

Nomina si nescis,
perit & cognitio rerum.

If you do not know the names,
your knowledge of the things perishes.

Linnaeus, *Critica Botanica,* 1737

Master Key

The dichotomous keys that follow are: first, a key to the major groups of plants constituting the flora; second, a key to the families making up each of the major groups; and third, a key to the genera making up each of the families. The family numbers given in these keys are repeated in the running heads of the text (tops of pages), to permit rapid location of all families. Keys to the taxa making up each genus are included in the body of the book. All taxa named in the book, other than synonymy, are included in the Index of Botanical Names at the back of the book.

KEY TO THE MAJOR GROUPS

Plant without true flowers or seeds; reproduction by 1-celled spores; always herbaceous
.... A. PTERIDOPHYTA
Plant with true flowers bearing stamens or carpels or mostly both, and seeds containing a multicelled embryo (SPERMATOPHYTA):
   Trees or shrubs with needlelike or scalelike leaves; flowers mostly unisexual, lacking perianth; ovules naked, not enclosed in ovary .... B. GYMNOSPERMAE
   Trees, herbs, or shrubs with leaves of diverse form, needlelike, scalelike, or flat; flowers bisexual or unisexual, perianth usually present; ovules completely enclosed in ovary (ANGIOSPERMAE):
     Leaves usually parallel-veined, mostly alternate and entire; parts of flowers usually in threes, not in fives; embryo with single cotyledon ... C. MONOCOTYLEDONEAE
     Leaves net-veined, usually pinnately or palmately so; parts of flowers mostly in fours or fives; embryo with 2 cotyledons in most cases. Exceptions: NYMPHAEACEAE, some *Corydalis*, and some RANUNCULACEAE, with but 1. (DICOTYLEDONEAE):
        Petals distinct from each other or lacking. Exceptions (petals somewhat united near base): CRASSULACEAE and FUMARIACEAE, which appear in key to METACHLAMYDEAE as well. (CHORIPETALAE) ...... D. ARCHICHLAMYDEAE
        Petals united, at least at base. Exceptions: EMPETRACEAE, which occurs only in key to ARCHICHLAMYDEAE; PYROLACEAE and ERICACEAE, which occur in both keys. (SYMPETALAE) .... E. METACHLAMYDEAE

KEY TO THE FAMILIES

A. PTERIDOPHYTA

Stems of hollow joints; leaves whorled and forming sheath at nodes .... 4. EQUISETACEAE
Stems not jointed; leaves not fused into sheath:
   ■ Leaves not differentiated into lamina and petiole:
     Leaves forming basal rosette; sporangia at base of rosette leaves .... 3. ISOËTACEAE
     Leaves not forming basal rosette; sporangia in terminal cones:
       Sterile leaves 6–8 mm long, without ligule; spores of 1 kind, small ....
.... 1. LYCOPODIACEAE
       Sterile leaves less than 5 mm long, with minute, transverse ligule near base; spores of 2 kinds, minute (male) and large (female, fewer) .. 2. SELAGINELLACEAE

■Leaves with distinct lamina and petiole:
  Leaf single; sporangia in stalked spike or panicle from base of green blade, large, globular, without annulus .... 5. OPHIOGLOSSACEAE
  Leaves more than 1; sporangia borne on underside of leaf, small-stalked, with annulus of thick-walled cells on side:
    Fertile and sterile leaves dissimilar:
      Leaves 2–4-pinnate .... 9. CRYPTOGRAMMACEAE
      Leaves 1-pinnate:
        Pinnae pinnatifid .... 12. ATHYRIACEAE (*Matteuccia*)
        Pinnae entire .... 15. BLECHNACEAE
    Fertile and sterile leaves similar:
      Leaves mostly not more than one cell thick ....
        .... 8. HYMENOPHYLLACEAE
      Leaves more than one cell thick:
        Leaves pinnatifid, 1-pinnate or dichotomously forked 1–3 times:
          Leaves 1-pinnate or dichotomously forked .... 11. ASPLENIACEAE
          Leaves pinnatifid .... 14. POLYPODIACEAE
        Leaves pinnately divided:
          Sori covered by deflexed margin of leaf:
            Rhizome short; frond membranaceous, palmately forking ....
              .... 6. ADIANTHACEAE
            Rhizome coarse, extensively creeping; frond coarse, coriaceous, tripinnate .... 7. HYPOLEPIDACEAE
          Sori not covered by deflexed margin of leaf:
            Sori submarginal .... 10. THELYPTERIDACEAE
            Sori along veins:
              Indusium wanting or nearly obsolete .. 12. ATHYRIACEAE (*Athyrium*)
              Indusium present, at least in young fronds:
                Indusium borne beneath sorus, surrounding sorus in cuplike structure .... 12. ATHYRIACEAE
                Indusium spreading from above or from 1 side, over sorus ......
                  .... 13. ASPIDIACEAE

B. GYMNOSPERMAE

Female flowers solitary; seed single, bony, surrounded by fleshy cup (aril) . . 16. TAXACEAE
Female flowers in cones (in *Juniperus*, berrylike); seeds without fleshy cup:
  Leaves alternate or 2–5 together on short shoots or spurs .... 17. PINACEAE
  Leaves opposite or whorled .... 18. CUPRESSACEAE

C. MONOCOTYLEDONEAE

Plant minute, 1–10 mm long, free-floating, consisting of membranaceous stem or frond without obvious differentiation into stems and leaves .... 28. LEMNACEAE
Plant terrestrial or aquatic, rooted in mud; plant larger, with stem and (sometimes scale-like) leaves:
  Inflorescence a simple, fleshy spike (spadix) with small flowers, subtended by large bract (spathe); leaves broad, in dense rosettes .... 27. ARACEAE
  Inflorescence not a spadix; leaves grasslike or broad:
    ●Perianth lacking or reduced, not petal-like in color and texture, consisting of bristles or mere scales:
      Flowers in axils of regularly imbricated scales and concealed by scales:
        Leaf sheaths split lengthwise; stem mostly hollow and terete; leaves usually 2-ranked .... 25. GRAMINEAE
        Leaf sheaths forming closed tube around stem (sometimes rupturing in age); stem often triangular in cross section; leaves usually 3-ranked ....
          .... 26. CYPERACEAE
      Flowers in axils of bracts or, if subtended by bracts, exceeding or equaling them, not concealed:
        Flowers unisexual, in heads or dense spikes:
          Flowers in densely crowded terminal spikes, the lower half thick, dark brown, pistillate, the upper thinner, staminate .... 19. TYPHACEAE
          Flowers in globose heads, the lower pistillate, the upper staminate ....
            .... 20. SPARGANIACEAE
        Flowers perfect:
          Plant floating or submerged .... 21. POTAMOGETONACEAE
          Plant terrestrial:

Carpels free (except at base); leaves with conspicuous pore at apex .....
.... 23. Scheuchzeriaceae
Carpels more or less completely united, separating at maturity; leaves
without conspicuous pore at apex:
Flowers in unbranched racemes; styles short or lacking ....
.... 22. Juncaginaceae
Flowers in cymes or branched inflorescences; styles 3, distinct ....
.... 29. Juncaceae
● Perianth well developed, with at least the inner segments petaloid in color and texture:
Ovary wholly inferior:
Flowers regular; ovary with 3 rooms; stamens 3 .... 31. Iridaceae
Flowers irregular; 1 petal forming a lip with a spur; ovary 1-roomed; stamens 1,
in *Cypripedium* 2 .... 32. Orchidaceae
Ovary superior:
Outer 3 perianth segments sepal-like, the inner 3 petal-like ... 24. Alismaceae
Sepals of same color as petals .... 30. Liliaceae

D. Archichlamydeae (Choripetalae)

Plant parasitic on branches of western hemlock, lacking chlorophyll (witch's broom):
.... 37. Loranthaceae
Plant rooting in ground:
▲ Petals lacking or not evident (calyx sometimes petaloid, as in Ranunculaceae, *Polygonum*, and *Eriogonum*):
Plant definitely woody, with woody branches; tree or shrub:
Flowers in catkins; plant not covered with stellate hairs:
Fruit a many-seeded capsule; seeds with tuft of long hair; stamens with long
filaments .... 33. Salicaceae
Fruit a 1-seeded nutlet; seeds without tuft of hair:
Female flowers solitary, in axils of bracts; male flowers in small, lateral catkins; leaves dotted with resin glands .... 34. Myricaceae
Female flowers in erect, cylindrical or ovoid catkins, 2–3 in axils of bracts;
male flowers in drooping catkins .... 35. Betulaceae
Flowers not in catkins:
Stamens 4 or 8; plant covered with stellate hairs or scales; leaves broad, deciduous .... 61. Elaeagnaceae
Stamens 3; plant not covered with stellate hairs; leaves linear, needlelike, evergreen .... 68. Empetraceae
Plant herbaceous (sometimes slightly woody at base):
Plant aquatic, submerged or, as water recedes or evaporates, in wet mud:
Calyx petaloid, colored .... 46. Ranunculaceae
Calyx not petaloid:
Leaves opposite, entire, often crowded in terminal rosettes ....
.... 57. Callitrichaceae
Leaves whorled, entire or dissected:
Leaves dichotomously 2 or 3 times forked; ovary superior ....
.... 45. Ceratophyllaceae
Leaves entire or pinnatifid .... 63. Haloragaceae
Plant terrestrial (sometimes growing in wet places):
Ovary more or less inferior; calyx adnate to floral tube; leaves entire, oblong,
glabrous (root parasites) .... 38. Santalaceae
Ovary superior, free from calyx (although sometimes surrounded by it):
Pistils several; stamens more than 10; calyx petaloid, conspicuous ....
.... 46. Ranunculaceae
Pistil 1; calyx green or petaloid:
Style and stigma single; flowers in axillary clusters; stamens 4 ....
.... 36. Urticaceae
Styles and stigmas more than 1:
Flowers enclosed in campanulate involucrum; calyx bright yellow ....
.... 39. Polygonaceae (*Eriogonum*)
Flowers without involucrum; calyx sometimes petaloid, not bright yellow:
Stipules united into sheath above each node; calyx more or less petaloid .... 39. Polygonaceae
Stipules not united into sheath; calyx not petaloid:

Bracts not scarious; plant mostly mealy, scurfy, or fleshy ....
.... 40. CHENOPODIACEAE
Bracts subtending flowers, scarious; plant not mealy, scurfy, or
fleshy .... 41. AMARANTHACEAE
▲ Petals present, evident; calyx also present (caducous at anthesis in PAPAVERACEAE):
Stamens numerous, more than twice as many as petals:
Plants aquatic, with broad, cordate, floating leaves .... 44. NYMPHAEACEAE
Plants terrestrial, leaves various:
Sepals 2, caducous at anthesis:
Flowers regular .... 47. PAPAVERACEAE
Flowers irregular .... 48. FUMARIACEAE
Sepals more than 2:
Stamens inserted in receptacle, free from calyx .... 46. RANUNCULACEAE
Stamens borne on calyx or on rim of hypanthium .... 53. ROSACEAE
Stamens not more than twice as many as petals:
Stamens inserted into margin of woolly disk lining base of calyx ....
.... 53. ROSACEAE (*Sibbaldia*)
Stamens not thus inserted:
Pistils more than 1, nearly or quite separate:
Plant succulent; carpels with same number as calyx segments ....
.... 51. CRASSULACEAE
Plant not succulent; carpels fewer than calyx segments, mostly 2 ....
.... 52. SAXIFRAGACEAE
Pistil 1 (consisting of 1 or more, more or less united carpels):
Styles 2–5, separate near base:
Plant woody; shrub or small tree:
Ovary inferior; fruit a berry; leaves palmately lobed ....
.... 52. SAXIFRAGACEAE (*Ribes*)
Ovary superior; fruit winged .... 58. ACERACEAE
Plant herbaceous:
Plant submerged, aquatic; leaves finely dissected ....
.... 63. HALORAGACEAE (*Myriophyllum*)
Plant terrestrial:
Ovary more or less inferior:
Seeds many in each room of ovary; flowers not in umbels ....
.... 52. SAXIFRAGACEAE
Seeds solitary in each room of ovary; flowers in umbels:
Fruit a berry; densely prickly, decumbent shrub ....
.... 64. ARALIACEAE
Fruit a schizocarp, splitting at maturity into 2 mericarps; herb ..
.... 65. UMBELLIFERAE
Ovary superior:
Leaves reddish, with numerous stout glands on upper surface; insectivorous .... 50. DROSERACEAE
Leaves green, lacking stout glands; not insectivorous:
Sepals 2; plant more or less fleshy .... 42. PORTULACACEAE
Sepals more than 2, free or united into tube:
Ovary 1-celled; stamens twice as many as petals ....
.... 43. CARYOPHYLLACEAE
Ovary 5-celled; fertile stamens as many as petals ....
.... 56. LINACEAE
Style 1 (sometimes more or less divided toward apex):
Ovary inferior:
Shrub; flowers in lateral racemes or solitary; leaves alternate; stamens
mostly 5 .... 52. SAXIFRAGACEAE (*Ribes*)
Herb; stamens 2, 4, or 8:
Flowers axillary or in terminal racemes; stamens 2 or 8 ....
.... 62. ONAGRACEAE
Flowers in umbelliform cyme, surrounded by mostly 4 large, white,
petaloid bracts; stamens 4 .... 66. CORNACEAE
Ovary superior:
Plant a low shrub .... 69. ERICACEAE (*Ledum*)
Plant herbaceous:
▼Sepals 2:

Plant more or less fleshy .... 42. PORTULACACEAE
Plant not fleshy:
  Sepals early-caducous; leaves lobed .... 47. PAPAVERACEAE
  Sepals not early-caducous; leaves dissected .. 48. FUMARIACEAE
▼ Sepals more than 2:
  Flowers regular; all petals similar:
    Sepals and petals 4; stamens 6, 2 shorter than the others ......
      .... 49. CRUCIFERAE
    Sepals 5; stamens 5, 8, or 10, all the same length:
      Ovary 5-lobed with long beak bearing 5 stigmas; flowers solitary on long peduncles; leaves lobed or dissected ......
        .... 55. GERANIACEAE
      Fruit a many-seeded capsule; leaves entire .. 67. PYROLACEAE
  Flowers markedly irregular; petals not all similar:
    Flowers papilionaceous (as in the pea), with stamens forming tube around style; fruit a legume; leaves compound ......
      .... 54. LEGUMINOSAE
    Flowers not papilionaceous; fruit a 1-roomed, 3-valved capsule; leaves entire:
      Sepals 3, very unequal, 1 spurred; petals 3, not spurred; stipules lacking; flowers panicled ..... 59. BALSAMINACEAE
      Sepals 5, equal, not spurred; petals 5, 1 spurred; stipules present; flowers solitary .... 60. VIOLACEAE

E. METACHLAMYDEAE (SYMPETALAE)

✱ Ovary superior:
  Stamens more than 5:
    Petals united only near base:
      Corolla irregular; pistil 1; stamens 6 .... 48. FUMARIACEAE
      Corolla regular:
        Pistils 5; stamens 10; plant succulent .... 51. CRASSULACEAE
        Pistil 1; stamens 8–12, mostly 10 .... 67. PYROLACEAE
    Petals markedly united, often urn-shaped or tubular; style 1:
      Plant an evergreen herb, or saprophytic herb, lacking chlorophyll; anthers opening by longitudinal slits .... 67. PYROLACEAE
      Plant a shrub or dwarf shrub; anthers opening by an apical pore .. 69. ERICACEAE
  Stamens not more than 5:
    Plant lacking chlorophyll, brown, parasitic; corollas with 3-lobed lower lip; stamens 4 .... 81. OROBANCHACEAE
    Plant with chlorophyll, not parasitic, or partly so:
      Corolla regular:
        Pistils 2 (ovaries distinct, but styles united); plant with milky juice ....
          .... 74. APOCYNACEAE
        Pistil 1:
          Stamens as many as corolla lobes and opposite them:
            Style 1; fruit a several-seeded capsule .... 71. PRIMULACEAE
            Styles 5; fruit 1-seeded; leaves linear .... 72. PLUMBAGINACEAE
          Stamens as many as, or fewer than, corolla lobes and alternating with them:
            Corolla small, dry-scarious, veinless; capsule opening with a lid; stamens 2 or 4; leaves basal .... 83. PLANTAGINACEAE
            Corolla not dry-scarious, veiny; capsule not opening with a lid; stamens 5:
              Ovary with 4 rooms, usually 4-lobed (though some may abort); plant characteristically scabrid, hispid; inflorescence usually a scorpioid cyme .... 77. BORAGINACEAE
              Ovary with 1, 2, or 3 rooms:
                Anthers opening with apical pores; flowers large, red ....
                  .... 69. ERICACEAE (*Rhododendron*)
                Anthers opening with longitudinal slits:
                  Style 3-cleft; ovary 3-roomed; capsule 3-valved:
                    Plant an evergreen, dwarf shrub with entire, leathery leaves; filaments in notches of corolla, cohering with corolla; flowers solitary .... 70. DIAPENSIACEAE
                    Plant an herb with entire or often pinnately divided leaves; filaments not in notches of corolla ...... 75. POLEMONIACEAE

                Style not 3-cleft; ovary 1–2-roomed:
                    Calyx deeply 5-lobed; style often 2-cleft ....
                    .... 76. Hydrophyllaceae
                    Calyx 4–5-toothed or cleft; style 1, entire:
                        Stigmas 2; ovary 1-roomed .... 73. Gentianaceae
                        Stigma 1; ovary 2-roomed .... 79. Solanaceae
            Corolla irregular:
                Ovary 4-lobed; style arising between ovary lobes, cleft at apex .. 78. Labiatae
                Fruit a capsule:
                    Ovary 2-roomed .... 80. Scrophulariaceae
                    Ovary 1-roomed .... 82. Lentibulariaceae
✶ Ovary inferior, or partly so:
        Leaves ternately compound; flowers in cube-like head .... 86. Adoxaceae
        Leaves not ternately compound:
            Stamens more than 5 .... 69. Ericaceae
            Stamens 5 or fewer:
                Stamens distinct:
                    Leaves alternate; flowers regular; stamens 5 .... 88. Campanulaceae
                    Leaves opposite or whorled:
                        Stamens 1–3; flowers irregular; fruit 1-seeded .... 87. Valerianaceae
                        Stamens 4–5; flowers mostly regular, sometimes irregular:
                            Plant an herb; ovary 2-roomed; leaves whorled .... 84. Rubiaceae
                            Plant a shrub or rarely an herb (*Linnaea*); ovary mostly 3–5-roomed;
                                leaves opposite; flowers regular or irregular ... 85. Caprifoliaceae
                Stamens adnate to corolla, their anthers united to a tube; flower in involucrate
                    heads; fruit an achene, often with plumose pappus ...... 89. Compositae

KEY TO THE GENERA

1. **Lycopodiaceae** (Club Moss Family): *Lycopodium*

2. **Selaginellaceae** (Spikemoss Family): *Selaginella*

3. **Isoëtaceae** (Quillwort Family): *Isoëtes*

4. **Equisetaceae** (Horsetail Family): *Equisetum*

5. **Ophioglossaceae** (Adder's Tongue Family)

Lamina entire, with reticulate venation; sporangia in simple, 2-ranked spike ....
.... *Ophioglossum*
Lamina compound, with dichotomous free veins; sporangia in compound spike or panicle
.... *Botrychium*

6. **Adiantaceae** (Maidenhair Family): *Adiantum*

7. **Hypolepidaceae** (Bracken Family): *Pteridium*

8. **Hymenophyllaceae** (Filmy Fern Family): *Mecodium*

9. **Cryptogrammaceae** (Mountain Parsley Family): *Cryptogramma*

10. **Thelypteridaceae** (Marsh Fern Family): *Thelypteris*

11. **Aspleniaceae** (Spleenwort Family): *Asplenium*

12. **Athyriaceae** (Lady Fern Family)

Fertile leaves smaller, less dissected than sterile leaves, eventually dark brown ....
.... *Matteuccia*
Fertile and sterile leaves similar:
    Indusium rudimentary and caducous .... *Athyrium*
    Indusium present:
        Indusium a circumbasal ring of hairy scales .... *Woodsia*
        Indusium hoodlike, attached by 1 side .... *Cystopteris*

13. Aspidiaceae (Shield Fern Family)

Rhizome long, slender; indusium absent .... *Gymnocarpium*
Rhizome stout; indusium present:
    Indusium peltate without sinus .... *Polystichum*
    Indusium reniform or with deep sinus .... *Dryopteris*

14. Polypodiaceae (Licorice Fern Family): *Polypodium*

15. Blechnaceae (Deer Fern Family): *Blechnum*

16. Taxaceae (Yew Family): *Taxus*

17. Pinaceae (Pine Family)

Leaves solitary; short shoots absent:
    Leaves borne on persistent, peglike projections or cushions; cones not erect:
        Leaves tetragonal, pungent, petiolate .... *Picea*
        Leaves flattish, not pungent, sessile .... *Tsuga*
    Leaves not borne on peglike projections, but leaving circular, disklike scars; cones erect
    .... *Abies*
Leaves in fascicles of 2 or more at summit of short shoots:
    Leaves in fascicles of 2, evergreen .... *Pinus*
    Leaves many, at summit of truncate, short shoots, deciduous .... *Larix*

18. Cupressaceae (Cypress Family)

Fruit berrylike, indehiscent, with fleshy coalescent scales; leaves in 3's ...... *Juniperus*
Fruit a dehiscent cone with more or less woody scales; leaves opposite:
    Ripe cone scales flat, oblong, basally attached, imbricate .... *Thuja*
    Ripe cone scales peltate, valvate .... *Chamaecyparis*

19. Typhaceae (Cattail Family): *Typha*

20. Sparganiaceae (Bur Reed Family): *Sparganium*

21. Potamogetonaceae (Pondweed Family)

Flowers in spikes:
    Spikes pedunculate, free .... *Potamogeton*
    Spikes completely sheathed in leaflike spathe:
        Plants monoecious; ovary and fruit ovoid .... *Zostera*
        Plants dioecious; ovary and fruit heart-shaped .... *Phyllospadix*
Flowers in umbels:
    Fruits long-pedicellate .... *Ruppia*
    Fruits sessile .... *Zannichellia*

22. Juncaginaceae (Arrowgrass Family): *Triglochin*

23. Scheuchzeriaceae (Scheuchzeria Family): *Scheuchzeria*

24. Alismaceae (Water Plantain Family): *Sagittaria*

25. Gramineae (Grass Family)

Spikelets (all or most) distinctly pedicellate; inflorescence a panicle:
    Inflorescence an open panicle (partly) with long branches .... Group 1
    Inflorescence a panicle with most branches short: either short, dense, spikelike or composed of spikelike, agglomerate parts .... Group 2
Spikelets sessile; inflorescence a panicle or a spike .... Group 3

Group 1

■ Spikelets 1-flowered:
    Lemma indurate when mature, much harder than glumes .... *Oryzopsis*
    Lemma membranaceous or hyaline, not harder than glumes:
        Lemma with terminal or dorsal awn:
            Lemma with terminal twisted and bent awn, much longer than body ...... *Stipa*
            Lemma with dorsal awn less than twice as long as body:
                Lemma surrounded at base with copious, usually long hairs .... *Calamagrostis*

Lemma without long hairs at base:
  Rachilla elongated behind palea into minute bristle:
    Lemma bifid; floret stipitate; anther 1 .... *Cinna*
    Lemma not bifid; floret sessile; anthers 3 .... *Podagrostis*
  Rachilla not elongated behind palea .... *Agrostis*
Lemma awnless:
  Spikelet with 2 sterile, brushlike lemmas at base; terminal flower perfect ....
    .... *Phalaris arundinacea*
  Spikelet lacking sterile lemmas at base:
    Glumes nearly equal, longer than thin lemma .... *Agrostis*
    Glumes unequal, shorter than lemma:
      Glumes small, much shorter than small spikelet; lemma erose-truncate .....
      .... *Catabrosa*
      Glumes stout, nearly as long as spikelet; lemma not erose-truncate ....
      .... *Arctagrostis*

■ Spikelets 2-flowered to several-flowered:
  Lemma with terminal or dorsal awn:
    Spikelet 3-flowered, all at the same level, the central perfect, the lateral staminate ..
    .... *Hierochloë alpina*
    Spikelet 2-flowered or with 3 or more flowers on different levels:
      Inflorescence with 1 branch from each node, bearing a single long spike ....
      .... *Pleuropogon*
      Inflorescence with 2 to several branches from each node, often with more than 1 spike:
        Lemmas included in glumes:
          Plant a small annual with hyaline lemma, bifid at apex; leaves filiform .. *Aira*
          Plant perennial with coarser lemma, entire or bifid in apex:
            Lowest floret perfect, awnless, the upper staminate with hooked awn ....
            .... *Holcus*
            All florets perfect:
              Lemma keeled, entire in apex .... *Vahlodea*
              Lemma rounded at back, bifid in apex:
                Awn dorsal, terete; glumes with at least 7 nerves .... *Avena*
                Awn from notch between apical teeth of lemmas, flat .... *Danthonia*
         Lemmas reaching beyond glumes:
          Lowest floret staminate, awned, the upper perfect, awnless; lemmas acute ..
          .... *Arrhenatherum*
          Lowest floret, as well as others, perfect; lemmas awned on back or in apex:
            Lemmas awned on back:
              Lemmas keeled, deeply cleft in apex .... *Trisetum*
              Lemmas rounded on back, 3–4-toothed in apex .... *Deschampsia*
            Lemmas with apical or subapical awn or awn-pointed:
              Lemmas 2-toothed in apex, awned from notch or just below notch:
                Ovary hairy at summit; stigmas sessile, borne from below summit of ovary .... *Bromus*
                Ovary glabrous; styles approximately terminal .... *Schizachne*
              Lemmas not 2-toothed; awn terminal:
                Plant an annual; first glume very small .... *Vulpia*
                Plant a perennial; first glume larger:
                  Spikelets strongly compressed in dense fascicles at ends of stiff, naked, panicle branches, arranged in 1-sided panicle ......
                  .... *Dactylis*
                  Spikelets not strongly compressed, not crowded in 1-sided clusters
                  .... *Festuca*
  Lemma awnless:
    Spikelet 3-flowered, all at same level, the central perfect, the lateral staminate .....
    .... *Hierochloë*
    Spikelet 2-flowered or with 3 or more florets at different levels:
      Lemmas acute:
        Lemmas included in glumes; spikelet 2-flowered .... *Dupontia*
        Lemmas reaching beyond glumes; spikelets more than 2-flowered:
          Lemmas rounded on back; upper floret sterile .... *Melica*
          Lemmas usually keeled (sometimes somewhat rounded) on back ...... *Poa*
      Lemmas truncate, more or less toothed, rounded on back:
        ● Spikelets 1–2-flowered; glumes very small .... *Catabrosa*

- Spikelets more than 2-flowered:
    - Lemma with excurrent nerves; callus and ovary pubescent; very coarse grass .... *Scolochloa*
    - Lemma without excurrent nerves; callus and ovary glabrous:
        - Ligules lacerate, glumes spreading .... *Arctophila*
        - Ligules entire, glumes appressed:
            - Lemmas prominently nerved, glabrous or scabrous; styles definite .... *Glyceria*
            - Lemmas faintly nerved, often minutely hairy at base; stigmas sessile:
                - Glumes reaching beyond middle of lowest lemma; end of axis of spikelets platelike .... *Colpodium*
                - Glumes shorter; end of axis of spikelets not flattened .... *Puccinellia*

Group 2

Spikelets 1-flowered (in *Beckmannia*, often with a second sterile flower):
- Panicle cylindrical or capitate, not interrupted, very dense:
    - Inflorescence capitate, ovoid .... *Phalaris*
    - Inflorescence long, cylindrical:
        - Only lemmas awned .... *Alopecurus*
        - Only glumes awned:
            - Awn long, soft; plant annual .... *Polypogon*
            - Awn thick, stiff; plant perennial .... *Phleum*
- Panicle more or less interrupted, composed of more or less separate groups or spikelets:
    - Spikelets pyriform, laterally compressed, closely imbricate, forming short, unilateral spikes .... *Beckmannia*
    - Spikelets not pyriform or imbricate:
        - Glumes very unequal, second glume longer than lemma; stamens 2 .... *Anthoxanthum*
        - Glumes shorter than lemma; stamens 3:
            - Glumes very small or lacking; lemma not awn-tipped; small arctic grass .... *Phippsia*
            - Glumes normally developed; lemma awn-tipped .... *Muhlenbergia*

Spikelets 2-flowered to several-flowered:
- Lemma awn-pointed .... *Dactylis*
- Lemma not awn-pointed:
    - Glumes nearly alike .... *Koeleria*
    - Glumes dissimilar, the first narrow, the second broadly obovate ..... *Sphenopholis*

Group 3

Inflorescence a panicle; spikelets 1-flowered:
- Spikelets pyriform, laterally compressed .... *Beckmannia*
- Spikelets not pyriform, not laterally compressed .... *Arctagrostis*

Inflorescence a single spike on each culm; spikelets 2-flowered to several-flowered:
- Spikelets solitary at each node of rachis:
    - Spikelets with narrow side toward rachis and resting in concavity of rachis, the lateral with a single glume .... *Lolium*
    - Spikelets with broad side toward rachis; glumes normal:
        - Plant perennial; glumes lanceolate; spikelets compressed, native ..... *Agropyron*
        - Plant annual; spikelets not compressed; introduced cereal:
            - Glumes ovate, 3-nerved to several-nerved .... *Triticum*
            - Glumes subulate, 1-nerved .... *Secale*
- Spikelets normally more than 1 at each node of rachis:
    - Spikelets 3 at each node of rachis; 1-flowered, the lateral pair pedicellate, usually reduced to awns .... *Hordeum*
    - Spikelets 2 or more at each node of rachis, alike, 2–6-flowered .... *Elymus*

26. CYPERACEAE (Sedge Family)

Flowers unisexual, the staminate and pistillate in the same or in different spikes; achene enclosed in sac or spathe; male flowers in axis of scales (lacking sac or spathe):
- Achenes in bottlelike sac (perigynium) open only at tip .... *Carex*
- Achenes surrounded by scale or spathe, open on 1 side above middle ...... *Kobresia*

Flowers perfect (at least most of them), not in sac or spathe:
▲ Perianth with silky bristles, in age strongly elongating:
- Perianth bristles numerous, flat, smooth .... *Eriophorum*
- Perianth bristles 0–12, not flat .... *Trichophorum alpinum*

▲ Perianth bristles not flat or silky, not elongating in age:
    Spike single, apical:
        All leaves basal, reduced to sheaths, lacking blade .... *Eleocharis*
        Uppermost culm leaf with short blade; lowest scale of spike with elongated green apex .... *Trichophorum*
    Spikes 2 or more:
        Spikes in 2 rows; inflorescence flattened .... *Blysmus*
        Spikes not in 2 rows; inflorescence not distinctly flattened:
            Spikelets 2–3-flowered, with 3–4 short, empty basal scales ..... *Rhynchospora*
            Spikelets several-flowered, with 1–2 empty basal scales at least as long as upper scales of spikelet .... *Scirpus*

27. ARACEAE (Arum Family)

Leaves and spathe swordlike .... *Acorus*
Leaves not swordlike; spathe petaloid:
    Leaves cordate; spathe white or greenish white .... *Calla*
    Leaves large, oblong to elliptic; spathe yellow or greenish-yellow .... *Lysichiton*

28. LEMNACEAE (Duckweed Family): *Lemna*

29. JUNCACEAE (Rush Family)

Leaf sheaths open; plant glabrous; capsule with numerous seeds .... *Juncus*
Leaf sheaths closed; plant often hairy; capsule 3-seeded .... *Luzula*

30. LILIACEAE (Lily Family)

Flowers umbellate; inflorescence enclosed before flowering in spathe, splitting at flowering time; plant smelling of onion .... *Allium*
Flowers not umbellate; plant not smelling of onion:
    Fruit a capsule:
        Style 1, in *Fritillaria* with 3 stigmas:
            Flowers normally more than 1, purplish-black; leaves broad; stem leaves whorled .... *Fritillaria*
            Flowers normally solitary, white; leaves filiform .... *Lloydia*
        Styles 3, distinct:
            Leaves 2-ranked, swordlike; plant small .... *Tofieldia*
            Leaves not 2-ranked; plant stouter:
                Leaves broadly oval; plant with rootstock and many stem leaves .... *Veratrum*
                Leaves linear; plant with bulb and reduced, bractlike stem leaves .. *Zygadenus*
    Fruit a berry:
        Plant scapose; flower solitary .... *Clintonia*
        Plant with leafy stem; flowers in racemes or panicles:
            Perianth segments 4; leaves cordate at base .... *Maianthemum*
            Perianth segments 6; leaves narrowed at base:
                Inflorescence with several flowers, terminal .... *Smilacina*
                Inflorescence 1–2-flowered in axils or supra-axillary of alternate stem leaves .... *Streptopus*

31. IRIDACEAE (Iris Family)

Flowers large; outer perianth segments recurved, much larger than the inner ...... *Iris*
Flowers small; all perianth segments nearly equal, the outer not recurved .. *Sisyrinchium*

32. ORCHIDACEAE (Orchis Family)

Plant nearly lacking chlorophyll, brownish; stem leaves reduced, scalelike .. *Corallorrhiza*
Plant with normal green leaves:
    Flowers mostly single or 1–3, large, with saccate, inflated lip:
        Plant with leafy stem .... *Cypripedium*
        Plant with single basal leaf .... *Calypso*
    Flowers several; if occasionally 2–3, not with saccate lip:
        Plant small, with corm and small, yellowish-green perianth:
            Stem leaves 1–2; the blade densely several-nerved .... *Malaxis*
            Stem leaves 2–3; the blade with few nerves .... *Hammarbya*
        Plant lacking corm:

Plant with fleshy roots, tuberoids:
    Lip lacking spur; spike twisted; flowers small, white .... *Spiranthes*
    Lip spurred; spike not twisted:
        Spur short, saccate; lip 3-toothed at apex .... *Coeloglossum*
        Spur not saccate:
            Lip entire, narrow at apex; perianth white or greenish ...... *Platanthera*
            Lip broad, mostly toothed or 3-lobed at apex; perianth purplish ....
                .... *Dactylorhiza*
Plant with slender, creeping rhizome, lacking tuberoids:
    Plant with 2 (nearly) opposite, broad stem leaves .... *Listera*
    Plant with all leaves basal:
        Basal leaves several; raceme 1-sided; leaves white-reticulate .... *Goodyera*
        Basal leaf single; raceme not 1-sided .... *Amerorchis*

33. SALICACEAE (Willow Family)

Bracts lacerate; aments pendulous .... *Populus*
Bracts entire or merely toothed; aments ascending or divergent, rarely drooping .. *Salix*

34. MYRICACEAE (Wax Myrtle Family): *Myrica*

35. BETULACEAE (Birch Family)

Bracts of pistillate ament thin, 3-lobed, falling off from the fine, central spindle at maturity
.... *Betula*
Bracts of pistillate aments woody, the conelike inflorescence long-persisting ...... *Alnus*

36. URTICACEAE (Nettle Family): *Urtica*

37. LORANTHACEAE (Mistletoe Family): *Arceuthobium*

38. SANTALACEAE (Sandalwood Family)

Flowers in terminal corymbs or panicles .... *Comandra*
Flowers 2–4 in peduncled cymules from axils of middle leaves .... *Geocaulon*

39. POLYGONACEAE (Buckwheat Family)

Flowers subtended by an involucrum; calyx bright yellow; stamens 9 ...... *Eriogonum*
Flowers without involucre; stamens 4–8:
    Sepals 4; achene flat with orbicular wing; basal leaves reniform .... *Oxyria*
    Sepals 3, 5 or 6; achene not winged; leaves not reniform:
        Sepals 3, all of same size; small arctic or alpine plant with upper leaves more or less
            whorled around flowers .... *Koenigia*
        Sepals 5 or 6:
            Sepals 5, petaloid, joined at base, all of about same size; stamens 5 ... *Polygonum*
            Sepals 6, not petaloid, the outer 3 much smaller than the inner; achene included at
                maturity; stamens 6 .... *Rumex*

40. CHENOPODIACEAE (Goosefoot Family)

Leaves subterete, fleshy, and sublinear, or reduced to scales:
    Leaves scalelike; stems and branches jointed .... *Salicornia*
    Leaves fleshy and sublinear; stem not jointed .... *Suaeda*
Leaves foliaceous, flattened, not particularly fleshy:
    Fruit exserted beyond the 1–3, minute sepals .... *Corispermum*
    Fruit enclosed by calyx or by bracts:
        Plant pubescent with branched hairs .... *Eurotia*
        Plant glabrous, but often farinose:
            Calyx of 1 sepal, 1 stamen .... *Monolepis*
            Calyx of all flowers or of staminate flowers, 2–5 leaved:
                Fruit surrounded by 2–5 persistent perianth segments .... *Chenopodium*
                Fruit enclosed between 2 more or less vertical appressed bracteoles .. *Atriplex*

41. AMARANTHACEAE (Amaranth Family): *Amaranthus*

42. PORTULACACEAE (Purslane Family)

Petals 3–5, often connate at base; small herb with small, inconspicuous, white flowers ..
.... *Montia*

Petals 5–8; showy, taller plants:
    Capsule 2–3-valved, dehiscing from apex .... *Claytonia*
    Capsule circumscissile .... *Lewisia*

43. CARYOPHYLLACEAE (Pink Family)

Sepals and petals distinct:
    Petals 2-cleft, bifid or lacking:
        Petals bifid or lacking .... *Stellaria*
        Petals 2-cleft or notched .... *Cerastium*
    Petals entire:
        Leaves narrow, more or less linear:
            Scarious stipules present:
                Styles 5 .... *Spergula*
                Styles 3 .... *Spergularia*
            Scarious stipules lacking:
                Styles 4–5 .... *Sagina*
                Styles 3:
                      Capsule opening with 3 valves .... *Minuartia*
                      Capsule opening with 6 valves .... *Arenaria*
        Leaves broad; styles 3:
            Capsule inflated, 3–5-locular .... *Wilhelmsia*
            Capsule 1-locular:
                Capsule opening with 3 valves .... *Honckenya*
                Capsule opening with 6 valves:
                      Ovules and young seeds with pale, spongy appendage at hilum; seeds shiny
                          .... *Moerhingia*
                      Ovules and young seeds without spongy appendage at hilum; seeds dull ....
                          .... *Arenaria*
Sepals united, forming tube; petals distinct:
    Styles 2:
        Calyx with scarious or foliaceous bracts at base .... *Dianthus*
        Calyx without bracts at base .... *Vaccaria*
    Styles more than 2:
        Styles 3:
            Capsule 1-locular .... *Melandrium*
            Capsule 3-locular at base .... *Silene*
        Styles 5:
            Petals entire .... *Agrostemma*
            Limb of petals deeply cleft .... *Melandrium*

44. NYMPHAEACEAE (Water Lily Family)

Leaves centrally peltate .... *Brasenia*
Leaves with basal sinus:
    Flowers subglobose with 5 or more concave yellow sepals; petals stamenlike .. *Nuphar*
    Flowers widely expanding, with 4 green (or purplish) sepals and several white or
        reddish petals .... *Nymphaea*

45. CERATOPHYLLACEAE (Hornwort Family): *Ceratophyllum*

46. RANUNCULACEAE (Crowfoot Family)

Fruit dry capsules, pods, opening down 1 side; sepals petaloid:
    Flowers regular, yellow or white, without spurs:
        Leaves trifoliate, evergreen; scape with 1 white flower .... *Coptis*
        Leaves not trifoliate:
            Leaves palmately lobed; flowers globular .... *Trollius*
            Leaves not palmately lobed; flowers cup-shaped .... *Caltha*
    Flowers with helmet or with 1 or 5 spurs, blue or red:
        Flowers regular, with 5 long spurs .... *Aquilegia*
        Flowers irregular, with helmet or with 1 spur:
            Flowers hooded, with helmet, lacking spur .... *Aconitum*
            Flowers with 1 spur .... *Delphinium*
Fruit berrylike, or achenes borne in heads:
    Fruit black, whitish, or red, berrylike; flowers in thick racemes .... *Actaea*
    Fruit achenes borne in heads:

Flowers evidently with both sepals and petals:
    Plant small, single-flowered, scapose, glabrous, arctic, with entire ovate leaves and sepals persisting in fruit .... *Oxygraphis*
    Plant larger; sepals deciduous (except in R. *glacialis*) .... *Ranunculus*
Flowers with petaloid sepals only:
    Leaves all alternate; flowers in racemes, small and inconspicuous; leaves ternately compound .... *Thalictrum*
    Upper leaves forming an involucrum below flowers; flowers solitary or umbellate, larger:
        Styles long, plumose; stamens with glandlike staminodia .... *Pulsatilla*
        Styles nonplumose, shorter; staminodia none .... *Anemone*

47. PAPAVERACEAE (Poppy Family): *Papaver*

48. FUMARIACEAE (Earth Smoke Family)

Flowers regular; inflorescence subumbellate .... *Dicentra*
Flowers irregular; 1 petal spurred; flowers in racemes .... *Corydalis*

49. CRUCIFERAE (Mustard Family)

▼ Fruit at most 3(–4) times as long as broad (silicle):
  Plant with glabrous stem and leaves (fruit sometimes pubescent):
    Flowers yellow:
      Silicle valves flat .... *Draba*
      Silicle pear-shaped .... *Camelina*
    Flowers white or purplish:
      Flowers purplish .... *Cakile*
      Flowers white:
        Silicle broadly winged .... *Thlaspi*
        Silicle not winged:
          Plant small, submerged, with subulate leaves .... *Subularia*
          Plant with broad leaves:
            Plant coarse, with large, oblong, long-petiolated basal leaves .. *Armoracia*
            Plant low, with small leaves:
              Basal leaves rounded or reniform; seashore plant .... *Cochlearia*
              Basal leaves narrow .... *Draba*
  Plant pubescent with simple, forked, branched or stellate hairs:
    Plant with all hairs simple:
      Flowers purplish .... *Parrya*
      Flowers white or yellow:
        Flowers white:
          Lower leaves entire or toothed .... *Draba*
          Lower leaves pinnatifid or pinnate:
            Seeds solitary in each loculus; valves of silicle winged or strongly keeled
            .... *Lepidium*
            Seeds many; valves not winged .... *Rorippa*
        Flowers yellow:
          Lower leaves pinnatifid or pinnate .... *Rorippa*
          Lower leaves, as well as others, entire or toothed .... *Draba*
    Plant with forked, branched, or stellate hairs, sometimes mixed with simple hairs:
      Flowers purplish:
        Silicle orbicular, flat .... *Draba stenopetala* var. *purpurea*
        Silicle ovoid, oblong or linear, terete or quadrangular-cylindric .... *Braya*
      Flowers white or yellow:
        Flowers white:
          Silicle obcordate-triangular .... *Capsella*
          Silicle not obcordate-triangular:
            Plant densely pubescent with short, grayish hairs .... *Smelowskia*
            Plant not densely grayish-pubescent:
              Silicle torulose .... *Braya*
              Silicle not torulose .... *Draba*
        Flowers yellow:
          Plant densely pubescent with short, grayish hairs .... *Smelowskia*
          Plant not densely grayish-pubescent:
            ✶ Silicles flat:

Silicles orbicular with 1–2 seeds .... *Alyssum*
Silicles oblong, several-seeded .... *Draba*
★ Silicles globular or pear-shaped:
   Silicles pear-shaped; tall, introduced weed .... *Camelina*
   Silicles globular:
      Silicles single-seeded; tall, introduced plant .... *Neslia*
      Silicles several-seeded; alpine or arctic, low plant ...... *Lesquerella*
▼ Fruit more than 4 times (exceptionally only 3 times) as long as broad (silique):
  Plant with glabrous stem and leaves (fruit sometimes pubescent):
   Flowers purplish:
    Leaves pinnate; valves of silique veinless .... *Cardamine*
    Leaves not pinnate; valves veined:
     Silique broad, obtuse, blue-green; seashore plant .... *Cakile*
     Silique acute or long and narrow; inland plant:
      Silique broad, acute, with long style .... *Parrya*
      Silique linear, with short style .... *Arabis*
   Flowers white, pinkish white, or yellowish white:
    Leaves pinnate, digitate or trifoliate .... *Cardamine*
    Leaves simple:
     Basal leaves sessile or nearly so .... *Arabis*
     Basal leaves long-petiolate:
      Plant dwarfed, alpine, densely tufted .... *Cardamine bellidifolia*
      Plant taller, not densely tufted:
       Stem leaves clasping .... *Thellungiella*
       Stem leaves not clasping .... *Eutrema*
  Plant pubescent with simple, forked, branched or stellate hairs:
   All hairs simple:
    Flowers purplish:
     Basal leaves lyrate or pinnate:
      Stem leaves entire .... *Raphanus*
      Stem leaves pinnate:
       Valves of silique veined; leaves simple .... *Cardamine*
       Valves of silique veinless; leaves pinnate .... *Arabis*
     Basal leaves simple, entire, toothed or sinuate:
      Plant dwarfed, with spatulate-ovate leaves .... *Aphragmus*
      Plant taller, leaves not spatulate-ovate:
       Silique 4–6 mm broad .... *Parrya*
       Silique narrower:
        Stem leafless or with 1–3 leaves; silique torulose .... *Braya*
        Stem with several leaves; silique not torulose .... *Arabis*
    Flowers white or yellow:
     Flowers white:
      Plant dwarfed, with spatulate-ovate leaves .... *Aphragmus*
      Plant taller, leaves not spatulate-ovate .... *Arabis*
     Flowers yellow or pale yellow:
      Lower leaves bipinnate or tripinnate .... *Descurainia sophioides*
      Lower leaves, as well as others, neither bipinnate nor tripinnate:
       Siliques with stout beak; seeds globular:
        Beak of silique flat; sepals spreading horizontally .... *Sinapis*
        Beak of silique terete; sepals not spreading .... *Brassica*
       Siliques beakless, merely tipped with the style; seeds oblong:
        Stem angulate-corrugate; lower leaves lyrate .... *Barbarea*
        Stem terete:
         Siliques subulate, appressed to rachis .... *Sisymbrium officinale*
         Siliques not subulate, not appressed to rachis:
          Siliques 60–100 mm long .... *Sisymbrium altissimum*
          Siliques much shorter .... *Rorippa*
   Hairs forked, branched, or stellate, sometimes mixed with simple hairs:
    Flowers purplish:
     Stigma deeply 2-lobed; flowers large, 30–40 mm across .... *Hesperis*
     Stigma not deeply 2-lobed; flowers smaller:
      Stem leafless or with 1–3 leaves; silique torulose .... *Braya*
      Stem with several leaves; silique not torulose .... *Arabis*
    Flowers white or yellow:
     ■ Flowers white:
      Stem leaves clasping:

Seeds in 2 rows in each loculus .... *Turritis*
Seeds in 1 row in each loculus:
    Plant straight, erect .... *Arabis*
    Plant much branched; stems spreading .... *Halimolobus*
Stem leaves not clasping:
    Plant densely pubescent with short, grayish hairs:
        Siliques 18–30 mm long, flat .... *Christolia*
        Siliques shorter, inflated .... *Smelowskia*
    Plant not densely pubescent with short, grayish hairs:
        Siliques torulose .... *Braya*
        Siliques not torulose, flat:
            Siliques up to 15 mm long .... *Draba stenoloba*
            Siliques longer .... *Arabis*
■ Flowers yellow; lower leaves pinnate:
    Stem bearing closely appressed, straight, 2–3-pronged hairs attached near the middle .... *Erysimum*
    Stem lacking such hairs:
        Siliques linear, cylindric .... *Descurainia*
        Siliques flat or inflated:
            Siliques 18–30 mm long, flat .... *Christolia*
            Siliques shorter, inflated .... *Smelowskia*

50. DROSERACEAE (Sundew Family): *Drosera*

51. CRASSULACEAE (Stonecrop Family)

Stamens as many as sepals, petals, and carpels; dwarf plant, in wet places ...... *Crassula*
Stamens twice as many; much larger plant, in rocky places .... *Sedum*

52. SAXIFRAGACEAE (Saxifrage Family)

Plant a shrub; fruit a berry, wholly inferior .... *Ribes*
Plan an herb; fruit a capsule:
    Capsule ovate; stamens fertile, alternating with clusters of gland-tipped, sterile staminodia .... *Parnassia*
    Capsule with 2 horns; staminodia lacking:
        Stamens 4 or 8; petals absent; inflorescence flat-topped; flowers insubstantial .... *Chrysosplenium*
        Stamens 3, 5, or 10; flowers in elongate inflorescences:
            Stamens 10:
                Carpels soon markedly unequally 2-valved to extreme base; petals filiform, entire .... *Tiarella*
                Carpels not markedly unequal; petals not filiform:
                      Petals laciniate-pinnatifid .... *Tellima*
                    Petals entire:
                          Leaves leathery; carpels practically distinct .... *Leptarrhena*
                          Leaves not leathery; carpels united at least a fifth of their length. . *Saxifraga*
            Stamens 3 or 5:
                Stamens 3; petals filiform, entire .... *Tolmiea*
                Stamens 5:
                      Petals pinnately or ternately cleft .... *Mitella*
                    Petals entire:
                          Leaves reniform or cordate .... *Boykinia*
                          Leaves 3–5-lobed .... *Heuchera*

53. ROSACEAE (Rose Family)

Plant a tree or shrub:
● Leaves simple, entire or 3-lobed:
    Fruit a capsule:
        Leaves entire or serrate; plant not stellate-pubescent .... *Spiraea*
        Leaves 3-lobed; plant with stellate pubescence .... *Physocarpus*
    Fruit a pome:
        Flowers in few-flowered racemes; branches not thorny .... *Amelanchier*
        Flowers in terminal corymbs; branches thorny:
            Styles united at base; fruit with papery or leathery center .... *Malus*
            Styles entirely distinct at base; fruit with bony, central nutlets ...... *Crataegus*

- Leaves compound:
    - Flowers yellow .... *Potentilla fructicosa*
    - Flowers white or pink:
        - Plant a thorny shrub:
            - Receptacle convex; fruit a juicy group of drupelets .... *Rubus*
            - Receptacle strongly concave; fruit a hip .... *Rosa*
        - Plant not thorny; dwarf shrub, shrub, or small tree:
            - Plant a dwarf shrub; leaves dissected in linear divisions; fruit a capsule .... *Luetkea*
            - Plant a tall shrub or tree; leaves pinnate; fruit a pome .... *Sorbus*
- Plant herbaceous:
    - Leaves simple:
        - Leaves small, oblong to ovate, entire or crenate-dentate; plant matted; fruit with long, feathery awn .... *Dryas*
        - Leaves larger, orbicular or reniform, shallowly lobed; fruit a group of drupelets ... . *Rubus*
    - Leaves compound:
        - Inflorescence a dense spike or head; petals lacking; sepals 4, petaloid; pistil 1 .... *Sanguisorba*
        - Inflorescence not a dense spike; petals and sepals present, more than 4; pistils several:
            - Stamens 5; petals inconspicuous, yellow:
                - Leaves 3-foliate; leaflets 3-toothed .... *Sibbaldia*
                - Leaves 2–3-ternate; segments linear .... *Chamaerhodos*
            - Stamens more than 10; petals conspicuous:
                - Leaves 2–3-ternate, segments linear; flowers white .... *Luetkea*
                - Leaves 3-foliate, pinnate or digitate:
                    - Fruit with 3–4 dehiscent capsules; leaves 2–3-pinnate .... *Aruncus*
                    - Fruit not dehiscent; leaves 3-foliate or 3-pinnate:
                        - Fruit a fleshy strawberry; flowers white; plant with runners; leaves 3-foliate .... *Fragaria*
                        - Fruit not fleshy; flowers white, yellow, or purplish; plant lacking runners:
                            - Style persistent on fruit as a long, straight or jointed, simple or plumose awn; leaves large, pinnate or lyrate; flowers yellow or white. . *Geum*
                            - Style deciduous; leaves 3-foliate, digitate or pinnate:
                                - Flowers white; fruit a group of drupelets .... *Rubus*
                                - Flowers yellow or purplish; fruit a group of achenes ..... *Potentilla*

54. Leguminosae (Pea Family)

Leaves 3-foliate or palmately many-foliate:
- Leaves palmately many-foliate:
    - Flowers in terminal racemes or spikes; leaflets entire .... *Lupinus*
    - Flowers capitate; leaflets serrate .... *Trifolium lupinaster*
- Leaves 3-foliate:
    - Inflorescence in lax racemes .... *Melilotus*
    - Inflorescence dense, headlike:
        - Inflorescence capitate; leaves palmately 3-foliate; corolla persistent; pods not much exceeding calyx .... *Trifolium*
        - Inflorescence racemose; leaves pinnately 3-foliate; corolla deciduous; pods falcate or spirally coiled .... *Medicago*

Leaves pinnate:
- Leaves terminating in tendril:
    - Stem winged; keel free from wings; staminate tube transversely truncate .. *Lathyrus*
    - Stem not winged; keel adhering; tube obliquely truncate or nearly transverse .. *Vicia*
- Leaves odd-pinnate:
    - Legumen articulated with transverse joints .... *Hedysarum*
    - Legumen not articulated:
        - Keel tipped by sharp appendage or point .... *Oxytropis*
        - Keel not tipped by sharp appendage or point .... *Astragalus*

55. Geraniaceae (Geranium Family)

Leaves palmately lobed; beak of carpel rolling upward in dehiscence and releasing seeds .... *Geranium*
Leaves pinnate; dehiscent beak twisting spirally, seeds remaining attached .... *Erodium*

56. Linaceae (Flax Family): *Linum*

57. **Callitrichaceae** (Water Starwort Family): *Callitriche*

58. **Aceraceae** (Maple Family): *Acer*

59. **Balsaminaceae** (Touch-me-not Family): *Impatiens*

60. **Violaceae** (Violet Family): *Viola*

61. **Elaeagnaceae** (Oleaster Family)

Leaves alternate; flowers perfect; stamens 4 .... *Elaeagnus*
Leaves opposite; flowers dioecious; stamens 8 .... *Shepherdia*

62. **Onagraceae** (Evening Primrose Family)

Plant with 2 petals, sepals, and stamens; fruit indehiscent, bristly .... *Circaea*
Plant with 4 petals and sepals and 8 stamens; fruit a long, dehiscent capsule .. *Epilobium*

63. **Haloragaceae** (Water Milfoil Family)

Leaves simple, entire; stamen 1 .... *Hippuris*
Submerged leaves pinnatifid, emergent leaves entire or toothed; stamens 4 or 8 ....
.... *Myriophyllum*

64. **Araliaceae** (Ginseng Family): *Echinopanax*

65. **Umbelliferae** (Parsley Family)

Leaves entire, lanceolate to oblong-lanceolate .... *Bupleurum*
Leaves variously compound:
  Fruit with elongated beak, much longer than broad, hispid at base .... *Osmorhiza*
  Fruit not much longer than broad:
    Ultimate segments of leaves more than 2 cm long, linear or lanceolate, serrate; fruit orbicular or nearly so:
      Root tuberous-thickened, chambered; upper leaves twice pinnate; fruit with thick ribs .... *Cicuta*
      Root fibrous; upper leaves pinnate; fruit with winged ribs .... *Sium*
    Ultimate segments of leaves not long and linear-lanceolate:
      Ultimate segments of leaves large, ovate to ovate-lanceolate, entire or deeply lobed or toothed:
      Stem and leaves pubescent; fruit with conspicuous wings:
        Fleshy, prostrate, maritime plant; leaves surpassing the subacaulescent umbels .... *Ghlenia*
        Tall, straight, inland plant:
          Plant up to 2 meters tall; segments of leaves very broad; flowers white; dorsal ribs of fruit filiform .... *Heracleum*
          Plant much smaller; flowers yellow; dorsal ribs of fruit winged; introduced weed .... *Pastinaca*
      Stem and leaves glabrous (rays of umbel sometimes pubescent):
        Branches rooting at nodes; ultimate segments of leaves several, about 2 cm long .... *Oenanthe*
        Branches not rooting; ultimate segments longer:
          Fruit small, orbicular or ovate .... *Cicuta Mackenziei*
          Fruit larger:
            Upper sheaths of leaves inflated; fruit with broad, marginal wings .....
.... *Angelica*
            Upper sheaths not inflated; all ribs about equally broad; seashore plant
.... *Ligusticum scoticum* subsp. *Hultenii*
    Ultimate segments of leaves smaller, deeply cleft, divided or simple:
      Plant scapose:
        Ultimate segments of leaves simple .... *Podistera*
        Ultimate segments of leaves deeply cleft or divided:
          Fruit with broad marginal wings; basal leaves with several pairs of pinnae
.... *Phlojodicarpus*
          Fruit with all ribs of equal size; basal leaves with 2–3 pairs of pinnae ....
.... *Ligusticum mutellinoides*
      Plant with leafy stem:
        Fruit with all ribs of equal size; stem leaves small .... *Cnidium*
        Fruit with broad marginal wings; stem leaves longer .... *Conioselinum*

66. CORNACEAE (Dogwood Family): *Cornus*

67. PYROLACEAE (Wintergreen Family)

Plant a saprophytic herb with scaly stem and drooping flowers, without chlorophyll, yellowish or pink .... *Monotropa*
Plant with evergreen leaves in basal rosettes:
  Stem leafy; flowers in corymbs; style nearly obsolete .... *Chimaphila*
  Stem scapose; flowers solitary or in racemes; style definite:
    Flowers solitary; petals broadly spread .... *Moneses*
    Flowers in racemes; petals arching .... *Pyrola*

68. EMPETRACEAE (Crowberry Family): *Empetrum*

69. ERICACEAE (Heath Family)

Ovary inferior, forming a berry crowned by the calyx teeth:
  With petals nearly distinct, reflexed; plant trailing .... *Oxycoccus*
  With corolla campanulate or globular; stem erect; low shrub .... *Vaccinium*
Ovary superior:
  Fruit a berry or berrylike:
    Calyx becoming accrescent and fleshy, enveloping the fleshy capsule .... *Gaultheria*
    Calyx dry, not becoming fleshy; fruit a drupe or drupaceous berry, subtended by calyx .... *Arctostaphylos*
  Fruit a dry capsule:
    Petals distinct:
      Leaves glabrous beneath; flowers large, solitary, copper-colored .. *Cladothamnus*
      Leaves brown-woolly beneath; flowers small, white .... *Ledum*
    Petals more or less united:
      Flowers 4-parted; corolla urn-shaped, yellow .... *Menziesia*
      Flowers 5-parted:
        Anthers with 2 horns:
          Leaves imbricate; corolla campanulate; flowers white .... *Cassiope*
          Leaves not imbricate; corolla globose-urceolate; flowers pink ... *Andromeda*
        Anthers without horns:
          Stamens 5; corolla small, open .... *Loiseleuria*
          Stamens 10:
            Corolla open, funnel-shaped or saucer-shaped:
              Flowers from scaly buds; capsule longer than broad; corolla not saccate .... *Rhododendron*
              Flowers from axis of persistent bracts or from axis of leaves; capsule subglobose; corolla 10-saccate .... *Kalmia*
            Corolla cylindric, urceolate or globular:
              Leaves linear, rough-margined, crowded; flowers purple or yellow .... *Phyllodoce*
              Leaves about 3 times as long as wide, smooth in margin, not crowded; flowers white .... *Chamaedaphne*

70. DIAPENSIACEAE (Diapensia Family): *Diapensia*

71. PRIMULACEAE (Primrose Family)

Plant densely caespitose, forming dense, low cushions of persistent small, narrow leaves; flowering stem short, leafless; flowers showy, pink, solitary .... *Douglasia*
Plant not densely caespitose, not forming dense, low cushions:
  Plant scapose or acaulescent; flowers in umbels:
    Calyx deeply cleft; corolla leaves reflexed; anthers forming cone ..... *Dodecatheon*
    Calyx tubular; corolla lobes spreading; stamens distinct:
      Throat of corolla constricted; style very short .... *Androsace*
      Throat of corolla not constricted; style filiform-elongate .... *Primula*
  Plant with leafy stem; flowers not in umbels:
    Calyx campanulate, petaloid, pink; corolla lacking; leaves succulent, opposite; seashore plant .... *Glaux*
    Calyx and corolla present; corolla rotate or nearly so, deeply cleft:
      Stem with small, alternate leaves, upper leaves whorled; flowers white, mostly 7-parted .... *Trientalis*
      All leaves whorled; flowers yellow, 5–6-parted .... *Lysimachia*

72. Plumbaginaceae (Leadwort Family): *Armeria*

73. Gentianaceae (Gentian Family)

Leaves opposite, simple, entire:
 Corolla funnelform or campanulate; calyx with elongate tube .... *Gentiana*
 Corolla rotate; calyx deeply parted:
  Stigmas decurrent for half the length of ovary .... *Lomatogonium*
  Stigmas not decurrent on ovary .... *Swertia*
Leaves alternate, long-petioled, 3-foliate or reniform, crenate in margin:
 Leaves 3-foliate; capsules 2-parted .... *Menyanthes*
 Leaves reniform, crenate; capsules 4-parted .... *Fauria*

74. Apocynaceae (Dogbane Family): *Apocynum*

75. Polemoniaceae (Polemonium Family)

Leaves pinnate or pinnatifid:
 Flowers in dense, globular, long-pedunculate heads; leaves pinnatifid .... *Gilia*
 Flowers not in dense, globular, long-pedunculate heads; leaves pinnate .. *Polemonium*
Leaves entire:
 Calyx tube of uniform texture throughout; annual .... *Collomia*
 Calyx tube with green costae separated by hyaline intervals:
  Plant perennial; leaves opposite .... *Phlox*
  Plant annual; upper leaves alternate .... *Microsteris*

76. Hydrophyllaceae (Waterleaf Family)

Plant annual; with semirotate to bowl-shaped flowers, 1.5–4 cm broad, purple with light
 center .... *Nemophila*
Plant perennial; with smaller campanulate flowers:
 Leaves lanceolate, coarsely toothed or lobed; plant velutinous-pubescent .... *Phacelia*
 Leaves round to reniform, shallowly cleft or toothed, glabrous or viscid-pubescent beneath .... *Romanzoffia*

77. Boraginaceae (Borage Family)

Basal leaves large, broad, ovate, obovate or oblong, long-petiolated; flowers large, 10–15
 mm long (in *M. maritima*, only 5 mm long) .... *Mertensia*
Leaves smaller, not ovate or obovate; flowers much smaller:
 Flowers in slender racemes without leafy bracts; nutlets smooth; flowers blue.. *Myosotis*
 Flowers, at least the lower, leafy-bracted:
  Fruiting calyx greatly enlarged, strongly veined, with 5 broad, flat lobes; stem retrorsely prickly-hispid .... *Asperugo*
  Fruiting calyx not greatly enlarged; stem not retrorsely prickly-hispid:
   Nutlet armed with glochidiate prickles at least along rim of back, or with jagged teeth; throat of corolla closed with 5 scales:
    Nutlet with jagged teeth in margin; plant caespitose, matted with very short inflorescence at flowering time .... *Eritrichium*
    Nutlet with glochidiate prickles; plant not caespitose, with elongate inflorescence:
     Pedicels and fruiting calyx erect .... *Lappula*
     Pedicels and fruiting calyx reflexed .... *Hackelia*
   Nutlet smooth, granular, tuberculate or wrinkled:
    Corolla yellow or orange; plant pungently haired .... *Amsinckia*
    Corolla white:
     Nutlet with ventral groove scar; plant pungently haired .... *Cryptantha*
     Nutlet with ventral keel; plant not pungently haired .... *Plagiobothrys*

78. Labiatae (Mint Family)

Corolla regular or essentially so, almost equally 4-lobed:
 Flowers white; 2 fertile stamens .... *Lycopus*
 Flowers purplish; 4 fertile stamens .... *Mentha*
Corolla more or less 2-lipped:
 ▲Calyx 2-lipped with entire lips, the upper with a scale on back; flowers in 1-sided, interrupted racemes .... *Scutellaria*

▲ Calyx not with 2 entire lips:
   Calyx with 10 long, clawlike teeth .... *Marrubium*
   Calyx with 5 teeth:
      Upper (inner) stamens longer than the lower:
         Upper tooth of calyx much broader than others .... *Dracocephalum*
         Teeth of calyx subequal, narrow; leaves ovate-cordate to reniform:
            Stem erect; flowers crowded in dense, interrupted, spiciform, terminal clusters; flowers white .... *Nepeta*
            Stem trailing or creeping; flowers in axils of leaves, violet .... *Glechoma*
      Upper stamens shorter than the lower:
         Calyx with 3-toothed upper lip and 2-toothed lower lip:
            Inflorescence crowded and subtended by mere bracts, appearing terminal .... *Prunella*
            Flowers in axis of ordinary leaves, separated by internodes .... *Satureja*
         Calyx with 5 subequal teeth; inflorescence spikelike:
            Flowers in axis of reduced bracteal leaves .... *Stachys*
            Flowers in axis of well-developed leaves:
               Lateral lobes of lower lip of corolla short, obscure, with small teeth .... *Lamium*
               Lateral lobes of lower lip of corolla well-developed, rounded ... *Galeopsis*

79. SOLANACEAE (Nightshade Family): *Solanum*

80. SCROPHULARIACEAE (Figwort Family)

Plant diminutive, subaquatic, with only basal, oblong to subulate, entire leaves; flowers scapose .... *Limosella*
Plants taller, with stem leaves:
   Corolla tube with slender spur or saccate at base; leaves linear, entire:
      Corolla yellow with slender spur; throat closed by orange palate; calyx shorter than corolla .... *Linaria*
      Corolla pink-purple or white, saccate at base; calyx lobes surpassing corolla .... *Antirrhinum*
   Corolla tube without spur, not saccate at base:
      Stamens 2:
         Corolla rotate; leaves opposite, cauline .... *Veronica*
         Corolla not rotate; leaves chiefly basal, the cauline alternate, reduced in size:
            Plant villous-hirsute .... *Synthyris*
            Plant glabrous and glaucous .... *Lagotis*
      Stamens 4:
         Leaves opposite or whorled; corolla 2-lipped:
            Inflorescence umbelliform; upper leaves sometimes whorled .... *Collinsia*
            Inflorescence not umbelliform:
               Flowers small, up to 5 mm long, only slightly surpassing calyx ... *Euphrasia*
               Flowers much larger:
                  Upper lip forming hood or galea, including stamens:
                     Calyx flattened, much inflated in fruit; leaves opposite; corolla yellow .... *Rhinanthus*
                     Calyx not flattened, not conspicuously inflated in fruit; upper leaves verticillate, rarely opposite; corolla yellow or purplish .... *Pedicularis*
                  Upper lip not forming hood or galea:
                     Filaments 5; anthers 4; corolla blue or purplish .... *Pentstemon*
                     Filaments and anthers 4; corolla yellow .... *Mimulus*
         Leaves alternate:
            Corolla large, tubular, campanulate .... *Digitalis*
            Corolla forming hood or galea including stamens:
               Leaves dissected or toothed, often basal as well as cauline ...... *Pedicularis*
               Leaves entire or pinnatifid, not toothed, all cauline:
                  Galea much surpassing the flat lower lip .... *Castilleja*
                  Galea 1–2 mm longer than the inflated and 3-saccate lower lip .... *Orthocarpus*

81. OROBANCHACEAE (Broomrape Family)

Inflorescence a loose, flat-topped corymb; plant glandular, pubescent .... *Orobanche*
Inflorescence dense, spicate; plant essentially glabrous .... *Boschniakia*

82. Lentibulariaceae (Bladderwort Family)

Plant terrestrial; basal leaves entire; flowers solitary, scapose, blue; calyx 5-lobed ...... *Pinguicula*
Plant submerged; leaves strongly dissected; flowers solitary or racemose, yellow .... *Utricularia*

83. Plantaginaceae (Plantain Family): *Plantago*

84. Rubiaceae (Madder Family): *Galium*

85. Caprifoliaceae (Honeysuckle Family)

Plant an herb; low, trailing evergreen, with pink flowers, paired at ends of long, naked peduncles ...... *Linnaea*
Plant a shrub:
  Leaves pinnate ...... *Sambucus*
  Leaves simple or lobed:
    Flowers paired on axillary peduncles ...... *Lonicera*
    Flowers in short axillary or apical racemes or in umbelliform inflorescences:
      Flowers in axillary or apical racemes; corolla tubular ...... *Symphoricarpus*
      Flowers in umbels; corolla rotate ...... *Viburnum*

86. Adoxaceae (Moschatel Family): *Adoxa*

87. Valerianaceae (Valerian Family): *Valeriana*

88. Campanulaceae (Bluebell Family): *Campanula*

89. Compositae (Composite Family)

Heads with both tubular central disk flowers and ligulate marginal ray flowers .. Group 1
Heads with only tubular or ligulate flowers:
  Heads with only tubular flowers ...... Group 2
  Heads with only ligulate flowers ...... Group 3

Group 1

Plant acaulescent; head solitary, large ...... *Townsendia*
Plant with stem:
  Plant scapose; head solitary ...... *Bellis*
  Plant with leafy stem; heads mostly more than 1:
    Stem scaly; leaves large, cordate-sagittate ...... *Petasites*
    Stem with ordinary leaves:
      Achenes with 2–4 stiff bristles:
        Leaves large, truncate to cordate at base; achenes with 2 smooth bristles; involucral bracts all similar ...... *Helianthus*
        Leaves smaller, lanceolate, coarsely serrate; achenes with barbed bristles; involucral bracts in 2 dissimilar series ...... *Bidens*
      Achenes lacking stiff bristles; pappus lacking or with long, capillary hairs:
        Pappus lacking:
          Receptacle chaffy:
            Involucral bracts in 1 series, all on same level; plant subherbaceous.. *Madia*
            Involucral bracts of imbricate scales:
              Ray flowers 1–3 cm long; disk flowers yellow; heads large .... *Anthemis*
              Ray flowers 2–3 mm long; disk flowers not yellow; heads small . *Achillea*
          Receptacle naked:
            Involucral bracts in 1–2 rows, all equal; leaves finely divided into linear segments ...... *Matricaria*
            Outer involucral bracts shorter; leaves simple or pinnately dissected into broader lobes ...... *Chrysanthemum*
        Pappus with long, capillary hairs; involucral bracts more or less imbricate:
          ▼ Ray flowers yellow or orange:
            Heads small, numerous, in racemose panicles ...... *Solidago*
            Heads larger, fewer:
              Leaves, at least the lower, opposite ...... *Arnica*
              Leaves alternate ...... *Senecio*

▼ Ray flowers pink, purple, red, or blue:
    Heads on leafy branchlets; involucral bracts imbricated or with 1 or more foliaceous outer rows, nearly as long, or as long, as the inner; ligules mostly broad .... *Aster*
    Heads on naked peduncles or scapes, solitary or in corymbs or panicles; involucral bracts in a single equal row, or in a large inner row and a very short basal one; ligules mostly very narrow .... *Erigeron*

Group 2

Involucral bracts with prickly point, leaves prickly .... *Cirsium*
Involucral bracts or leaves not prickly:
  Pappus lacking:
    Heads small, in spicate-racemose panicles .... *Artemisia*
    Heads large, not in spicate-racemose panicles:
      Flowers 4-toothed; heads hemispherical, single or few on slender, terminal peduncles; achenes stipitate .... *Cotula*
      Flowers 4–6-toothed; achenes not stipitate:
        Involucral bracts in 1–2 rows, all equal:
          Fruit small, less than 1 mm long, with 3–4 thin ribs, no glands at tip .... *Matricaria*
          Fruit larger, with 3 thick cartilaginous ribs, 2 red glands at tip .... *Tripleurospermum*
        Involucral bracts of outer rows shorter .... *Chrysanthemum*
  Pappus with long, capillary bristles:
    Pappus plumose, at least the inner bristles .... *Saussurea*
    Pappus not plumose, sometimes barbellate:
      Leaves broad, cordate, sagittate, triangular or reniform:
        Stem scaly .... *Petasites*
        Stem with ordinary leaves:
          Leaves reniform; flowers white .... *Cacalia*
          Leaves triangular; flowers yellow .... *Senecio triangularis*
      Leaves narrower, simple or pinnatifid:
        Flowers yellow:
          Heads small, numerous, in racemose panicles .... *Solidago*
          Heads larger, fewer:
            Leaves, at least the lower, opposite .... *Arnica*
            Leaves alternate:
              Plant matted, woody at base; leaves needlelike .... *Haplopappus*
              Plant erect, not woody; leaves flat .... *Senecio*
        Flowers not yellow, but white, brownish, pink, purplish, or blue:
          Leaves woolly, white-pubescent, simple and entire:
            Heads in capitate, leafy-bracted clusters or in small spikes ... *Gnaphalium*
            Heads paniculate:
              Stem leaves small, 1–2 cm long .... *Antennaria*
              Stem leaves large, 5–10 cm long .... *Anaphalis*
          Leaves not woolly, white-pubescent:
            Involucral bracts imbricated .... *Aster*
            Involucral bracts in a single row .... *Erigeron*

Group 3

Plant scapose:
  Pappus plumose; scapes sometimes branched:
    Receptacle chaffy; at least the inner achenes extended in long, slender beak .... *Hypochoeris*
    Receptacle naked; achenes fusiform .... *Leontodon*
  Pappus of capillary, sometimes barbellate, bristles; heads solitary:
    Pappus tawny .... *Apargidium*
    Pappus white:
      Achenes completely lacking tubercles or spines; involucral bracts never gibbous in apex .... *Agoseris*
      Achenes tuberculate or spiny; involucral bracts often gibbous in apex.. *Taraxacum*
Plant with leafy stem:
✱Leaves prickly in margin; achenes beakless; outer pappus bristles stouter than others .... *Sonchus*

Master Key

✷ Leaves not prickly:
    Pappus lacking .... *Lapsana*
    Pappus of capillary or plumose bristles present:
        Pappus plumose; achenes terete, beakless .... *Picris*
        Pappus capillary, simple or rough:
            Achenes flat, strongly compressed, tapering to summit .... *Lactuca*
            Achenes not strongly compressed, columnar:
                Achenes beaked or narrowed above; involucral bracts in single row .. *Crepis*
                Achenes truncate above; involucral bracts in 2 or more rows:
                    Heads slenderly cylindrical, drooping; ligules white to purplish .... *Prenanthes*
                    Heads ascending; ligules yellow, in 1 species (*H. albiflorum*) white ..... *Hieracium*

The Flora

Lycopodium L. (Club Moss)

Stem ascending, dichotomously divided into more or less erect, stout branches; sporangia in axils of ordinary leaves, all leaves uniform:
    Leaves broad, appressed, often incurved .... 2. *L. selago* subsp. *appressum*
    Leaves longer, not appressed or incurved:
        Leaves yellowish-green, lanceolate, ascending ..... 1. *L. selago* subsp. *selago*
        Leaves dark green, linear, narrow, patulous, sometimes reflexed ....
        .... 3. *L. selago* subsp. *chinense*
Stem creeping; sporangia in terminal spikes:
    Leaves of sterile stem subulate, angular, curved upward; leaves of fertile stem always simple, green, leaflike, ovate to broadly lanceolate, scarious, toothed
    .... 4. *L. inundatum*
    Leaves of sterile stem not subulate:
        Stem flat, dorsiventral, leaves opposite or decussate, in 4 rows:
            Leaves of lower surface small, appressed, subulate; spikes normally pedunculate, often plural .... 10. *L. complanatum*
            Leaves of lower surface trowel-shaped; spikes subsessile, always single....
            .... 11. *L. alpinum*
        Stem not dorsiventral; leaves in all directions, all uniform:
            Stem subterranean, rhizome-like; scattered, treelike branches, with trunk and crown .... 7. *L. obscurum* var. *dendroideum*
            Stem creeping on ground; branches not treelike:
                Leaves thick, appressed, subulate, somewhat acute, 3–4 mm long; spikes normally solitary, pedunculated. .12. *L. sabinaefolium* var. *sitchense*
                Leaves thin, stiff, subulate or bristle-tipped, larger; 1–3 spikes:
                    Leaves bristle-tipped when young; 1–3 spikes on leafy, branched peduncle, at least lower fertile scales bristle-tipped:
                        Spikes 2 (to 3) .... 8. *L. clavatum* subsp. *clavatum*
                      Spike solitary .... 9. *L. clavatum* subsp. *monostachyon*
                Leaves all subulate; spikes solitary, sessile:
                    Leaves spreading, somewhat toothed ....
                    .... 5. *L. annotinum* subsp. *annotinum*
                  Leaves appressed, dorsally convex, smooth-margined..........
                    .... 6. *L. annotinum* subsp. *pungens*

1. LYCOPODIACEAE (Club Moss Family)

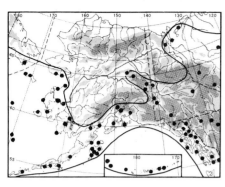
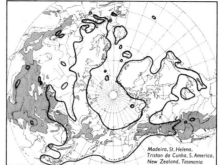
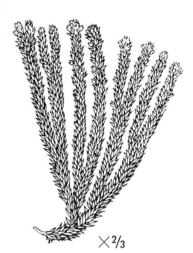

1. Lycopòdium selàgo L.   Fir Club Moss

Huperzia selago (L.) Bernh.

subsp. **selàgo**

Dichotomously branched, with thick branches reaching same height, often with young buds in their axes; spore-bearing leaves, similar to vegetative leaves, in zones alternating with vegetative leaves, each seasonal growth having a basal sterile zone and an apical fertile zone.

Woods, bogs, heaths, from lowlands up to alpine region. Described from North Europe. Circumpolar map indicates range of entire complicated species complex.

Contains a poisonous alkaloid causing pain in the mouth, vomiting, and diarrhea.

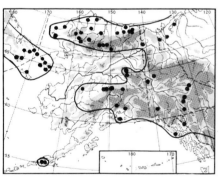
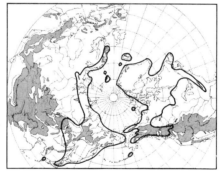

2. Lycopòdium selàgo L.

subsp. **appréssum** (Desv.) Hult.

Lycopodium selago var. *appressum* Desv.; *L. appressum* (Desv.) Petrov; *Huperzia selago* subsp. *appressum* (Desv.) D. Löve. Including *L. selago* subsp. *arcticum* Tolm.

Similar to subsp. *selago*, but leaves broad, appressed, often incurved. Passes freely into subsp. *selago*, but has own range.

On tundra and in mountains. Described from St. Paul Island and Newfoundland.

In extreme arctic or alpine localities, this plant has only a single, short stem, with strongly appressed short leaves.

×½

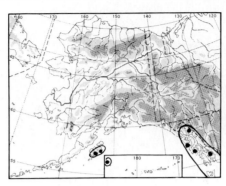 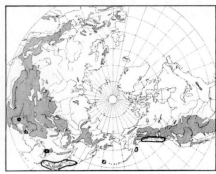

3. Lycopòdium selàgo L.
subsp. **chinénse** (Christens.) Hult.

Lycopodium chinense Christens.; *Huperzia selago* subsp. *chinense* (Christens.) Löve & Löve; *L. selago* var. *Miyoshianum* (Makino) Makino; *L. Miyoshianum* Makino; *L. selago* subsp. *Miyoshianum* (Makino) Calder & Taylor.

Differs from subsp. *selago* in having dark-green, linear, narrow, patulous leaves, somewhat reflexed at base.

Habitat probably woodlands, but largely indeterminate. Described from Shensi province, China.

×¾

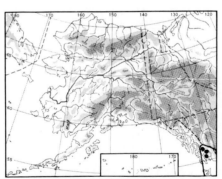 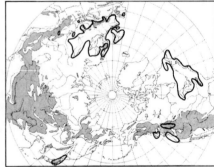

4. Lycopòdium inundàtum L.
Lepidotis inundata (L.) Börner.

Horizontal stem 5–10 cm. long, rooting at intervals and at tip, strongly anchored in ground, simple or branched, covered with upturned leaves. Fertile stem usually solitary in Alaskan specimens; vegetative and fertile leaves similar, soft, of yellowish-green color; spikes short, 1–2 cm long, erect.

Bogs and wet shores in lowlands. Described from Europe.

Barely reaches southern tip of Alaska, but is apparently not rare there.

1. Lycopodiaceae (Club Moss Family)

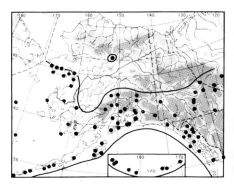

5. Lycopòdium annòtinum L. Stiff Club Moss
subsp. **annòtinum**

Stem long, creeping, and forking, with scattered, spreading leaves; branches on ascending part of stem annually increase in length and fork with age; leaves firm and stiff, distinctly toothed, subulate, and comparatively thin; spikes solitary, sessile at end of branches; spore-bearing leaves acute.

Woods and heaths from lowlands to lower alpine region. Described from Europe.

Var. **alpéstre** Hartm. [subsp. *alpestre* (Hartm.) Löve & Löve] differs in having broader, serrate leaves. Circumpolar map indicates range of entire species complex.

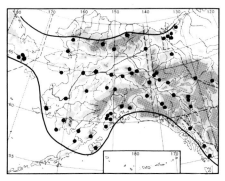
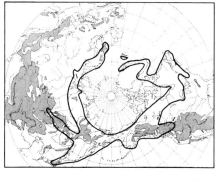
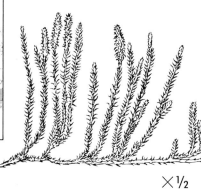

6. Lycopòdium annòtinum L.
subsp. **púngens** (La Pyl.) Hult.

Lycopodium pungens La Pyl.; *L. annotinum* var. *pungens* (La Pyl.) Desv.; *L. dubium* Zoega.

Similar to subsp. *annotinum*, but leaves more yellowish, small, narrow, thick, dorsally convex, with entire margin, ascending, and very acute.

Described from Newfoundland. Transitions to subsp. *annotinum* seem to occur.

×½

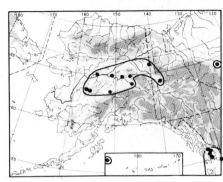
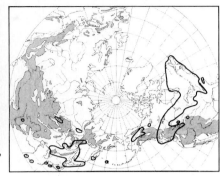

7. Lycopòdium obscùrum L. Ground Pine, Tree Club Moss
var. **dendroìdeum** (Michx.) D. C. Eat.
Lycopodium dendroideum Michx.

Subterranean, creeping, rhizome-like stem with scattered, upright, treelike branches, rounded in outline; leaves linear-attenuate, more or less spreading.

Woods, open places, margins of bogs in lowlands. *L. obscurum* described from Philadelphia, var. *dendroideum* from Canada, New England, and mountains of Carolina.

The typical *L. obscurum*, which does not occur in the area of interest here, has flattened branches with more appressed leaves. In much of the range shown on the circumpolar map, the two types occur together.

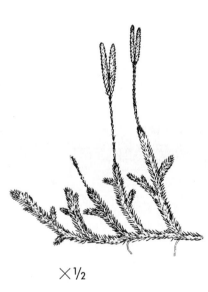
×½

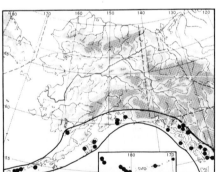

8. Lycopòdium clavàtum L. Common Club Moss
subsp. **clavàtum**

Stem long, densely covered with more or less appressed leaves, creeping on ground, with ascending branches repeatedly forking with age, branches terminated by a usually long peduncle covered with short bracts and bearing 2–3 spikes; at least the lower spore-bearing leaves tipped with a soft, hairlike bristle, best observed on young leaves at tops of branches.

Woods and rocky places in lowlands, ascending to lower alpine region; mostly on acid soil. Described from Europe.

Var. **integérrimum** Spring (common in British Columbia) differs in having leaves lacking bristles.

1. LYCOPODIACEAE (Club Moss Family)

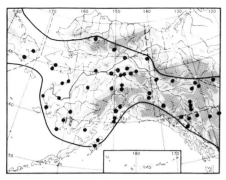 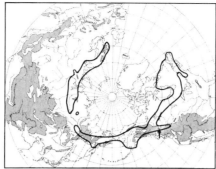 

9. Lycopòdium clavàtum L.
subsp. **monostáchyon** (Grev. & Hook.) Sel.

Lycopodium clavatum var. *monostachyon* Grev. & Hook.; *L. clavatum* var. *lagopus* Laest.

Differs from subsp. *clavatum* in having solitary, sometimes very short-peduncled spikes.

Described from Smoking River, lat 56°N, in Rocky Mountains.

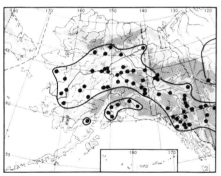 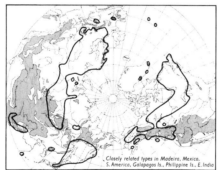 

10. Lycopòdium complanàtum L.   Christmas Green, Creeping Jenny
Diphasium complanatum (L.) Rothm.

Stem creeping in surface of soil, elongate, rooting, green; branches erect, forked, divergent, obviously flattened, with somewhat glaucous leaves in 4 rows; ventral leaves much smaller than others, subulate and appressed; spikes or groups of spikes usually pedunculate.

Dry woods, rarely above tree line. Described from Europe and America.

Specimens from interior differ from those from southern coast in having narrower branches and in having but one spike on each fertile branch.

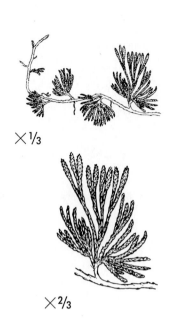
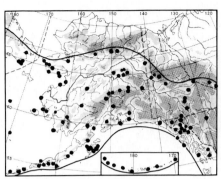
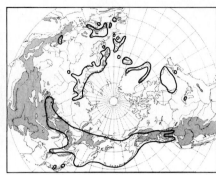

11. Lycopòdium alpìnum L. Alpine Club Moss

Diphasium alpinum (L.) Rothm.

Stem creeping in surface of soil, elongate, rooting, whitish-green; branches erect, blue-green, cylindrical or usually somewhat flattened, with leaves in 4 rows; spikes sessile, at tips of leafy branches.

Woods, meadows, and heaths, common from lowlands to mountains, absent from calcareous soil. Described from Lapland and Switzerland.

A good characteristic is the trowel-shaped ventral leaves, by which sterile specimens of this species can be distinguished from the sometimes similar *L. complanatum* or *L. sabinaefolium* var. *sitchense*.

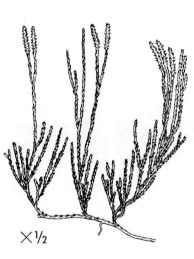
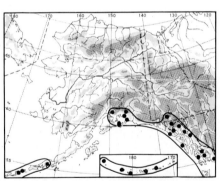
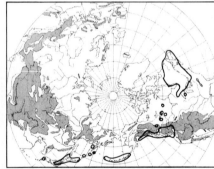

12. Lycopòdium sabinaefòlium Willd.

var. **sitchénse** (Rupr.) Fern.

Lycopodium sitchense Rupr.; *L. sabinaefolium* subsp. *sitchense* (Rupr.) Calder & Taylor; *Diphasium sitchense* (Rupr.) Löve & Löve.

Stem creeping in surface of soil, elongate, light-colored; branches erect or ascending, densely branching, cylindrical; leaves uniform, with the free tips usually longer than adnate portion; spikes sessile, at ends of elongate branches.

Mountain slopes, from lowlands to alpine zone. *L. sabinaefolium* described from Canada, var. *sitchense* from Sitka. Circumpolar map indicates range of entire species complex.

Similar to *L. alpinum*, but lacking the row of trowel-shaped leaves characteristic of that species.

2. SELAGINELLACEAE (Spikemoss Family)

Selaginella Beauv. (Spike Moss)

Herbaceous, with membranaceous, lanceolate-to-ovate, not bristle-tipped leaves .... 1. *S. selaginoides*
Evergreen, matted, with small, densely imbricated, bristle-tipped leaves:
   Setae of leaves white, branches intricate .... 2. *S. sibirica*
   Setae lutescent, branches discrete .... 3. *S. densa* var. *Standleyi*

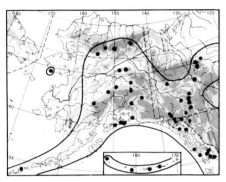
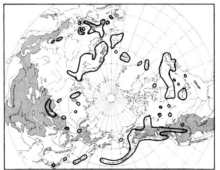
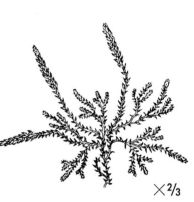

1. Selaginélla selaginoìdes (L.) Link
Lycopodium selaginoides L.

Yellowish-green perennial with creeping and forking stems forming small mats; leaves uniform, sparsely ciliate; fertile branches ascending, with macrosporangia on lower leaves and microsporangia on upper leaves.

Damp mossy ground and open damp woods, especially in calcareous soil. Described from Europe.

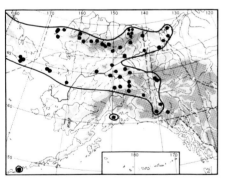
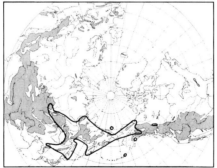
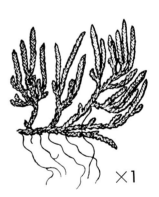

2. Selaginélla sibìrica (Milde) Hieron.
Selaginella rupestris f. *sibirica* Milde.

Stems forming open spreading mats; branches intricate; upper and under leaves equal or subequal in length, or the under slightly longer; leaves linear to ligulate, rounded at apex, with ciliate margin and white to tawny setae (of very different lengths in different specimens) at tip.

Dry exposed rocks, dry heaths, ridges, quite common in Brooks Range and interior Alaska. Described from "Dahuria, Ingodam, and Szita."

Very similar to *S. densa* var. *Standleyi*, which differs in having lutescent setae and more separated branches.

2. Selaginellaceae (Spikemoss Family) / 3. Isoëtaceae (Quillwort Family)

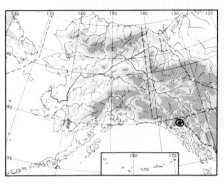
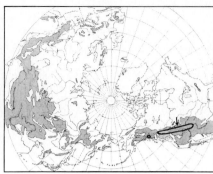

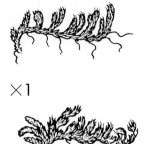

×1

3. Selaginélla dénsa Rydb.
var. Standléyi (Maxon) Tryon
Selaginella Standleyi Maxon.

Stems forming cushion-mats, branches discrete, lower leaves longer than upper or sometimes equal to upper on assurgent branch tips, leaves ligulate to ligulate-lanceolate, ciliate or eciliate with more or less lutescent setae at the tip.

No Alaskan specimens seen. Included on the authority of Tryon, who regards a specimen from north of Tlehini as belonging to this taxon.

Very closely related to *S. sibirica*.

Isoëtes L. (Quillwort)

Gynospore with broad spinules, forked or toothed; androspores smooth, stomata few .... 1. *I. muricata* var. *Braunii*
Gynospore covered with thick, blunt spines; androspores papillate, stomata numerous:
    Plant coarse, with leaves up to 13 cm long .... 3. *I. truncata*
    Plant more delicate .... 2. *I. muricata* subsp. *maritima*

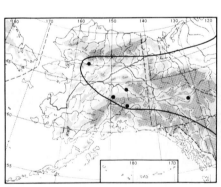
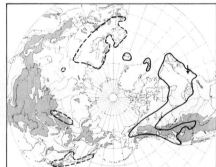

×3/4

1. Isoëtes muricàta Dur.
var. Braùnii (Dur.) Reed
Isoëtes Braunii Dur. not Unger.

Leaves 3–10 cm long, dark-green, half-erect in water, rigid, usually curved backward; stomata few, on tips of leaves; ligules deltoid; velum half-covering the sporangium; gynospores covered with broad, occasionally forked or toothed spinules; androspores smooth, very numerous.

Shallow water. Broken line on circumpolar map indicates range of the closely related **I. setàcea** Lam. (*I. echinospora* Dur.), of which *I. muricata* var. *Braunii* might well be regarded as a race.

3. Isoëtaceae (Quillwort Family)

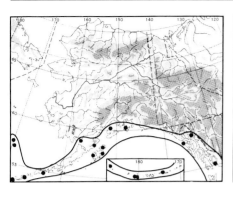

×¾

2. Isoëtes muricàta Dur.
subsp. **marítima** (Underw.) Hult.

Isoëtes maritima Underw.; *I. Braunii* var. *maritima* (Underw.) Pfeiffer; *I. Macounii* A. A. Eat.; *I. beringensis* Kom.

Leaves 2.5–5 cm long, green, slender, with fine-pointed tip and wide membranous border at base; stomata numerous; ligules triangular, a little longer than wide; velum covering half the sporangium or less; gynospores with thick, blunt spines, sometimes confluent into toothed ridges; androspores chiefly papillose.

Shallow water. In need of further detailed study.

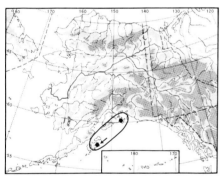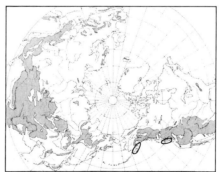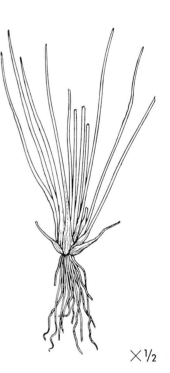

3. Isoëtes truncàta (A. A. Eat.) Clute

Isoëtes echinospora var. *truncata* A. A. Eat.; *I. Braunii* Hult. of Fl. Alaska & Yukon, in part; *I. asiatica* with respect to Alaskan specimens?

Leaves 6–13 cm long, stout, rigid, with almost setaceous apex and wide membranaceous margin at base; stomata numerous; ligules short, triangular; velum covering one-fourth to one-half of sporangium; gynospores thickly covered with truncate columns or blunt spines; androspores papillate.

Shallow water. In need of further detailed study.

×½

4. EQUISETACEAE (Horsetail Family)

Equisetum L. (Horsetail)

Spring phase with brown fertile stems lacking chlorophyll, lacking branches or tubercles formed by outgrowing branches .... 9. *E. arvense*
Spring phase with brown fertile stems lacking chlorophyll but with branches or tubercles; fertile or sterile green summer phase with or without branches:
  Central cavity more than half the diameter of stem; stem stout, round:
    Stem evergreen, persisting, scabrous, ridges flat-topped; unbranched; sheaths with dark-brown band at base .... 1. *E. hiemale* var. *californicum*
    Stem annual, smooth, ridges V-shaped; with or without branches; sheaths green at base .... 5. *E. fluviatile*
  Central cavity less than half the diameter of stem; stem thin, angular, less than 3 mm in diameter:
    Stem simple, sometimes with single branches, but not verticillate branches:
      Stem annual, green, soft (an occasional unbranched form of the usually branched *E. palustre*) .... 6. *E. palustre*
      Stem persistent, stiff, rough:
        Ridges V-shaped .... 4. *E. scirpoides*
        Ridges flat-topped, or with shallow furrow on top:
          Stem coarse, up to 45 cm tall and 4 mm broad .... 3. *E. variegatum* subsp. *alaskanum*
          Stem shorter, thinner .... 2. *E. variegatum* subsp. *variegatum*
    Stem with verticillate branches (or, in the spring phase, with tubercles):
      Secondary, verticillate branches present; teeth of sheaths reddish-brown, monochrome .... 7. *E. silvaticum*
      Secondary branches lacking or not verticillate, at least the upper branches simple; teeth of sheaths dark brown or blackish:
        Second internode of branches reaching to or beyond top of corresponding sheath of stem .... 9. *E. arvense*
        Second internode not reaching top of corresponding sheath of stem:
          Branches rigid, thick, upright, the first very short; internode of branches shiny, brownish-black; plant green .... 6. *E. palustre*
          Branches slender, pendulous, the first short; internode of branches dull, or light yellowish brown; plant grayish-green .... 8. *E. pratense*

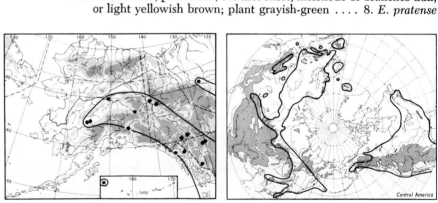
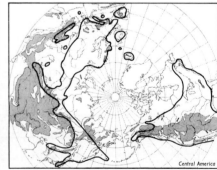

×⅓

1. Equisètum hiemàle L.
var. califórnicum Milde

Equisetum hiemale var. *affine* and var. *robustum* with respect to specimens in area of interest; *Hippochaete hiemale* var. *californica* (Milde) Farw.

Rhizome stout, long, branched, felted; stem branchless, thick, strongly striate, scabrous, evergreen, up to 50 cm tall, with a central cavity of two-thirds the diameter; sheaths cylindrical, whitish, with dark-brown band at base; teeth numerous, dark brown, with broad scarious margins falling off in old specimens; cones apiculate, short-stalked.

Sandy shores, woods, rare within our area. *E. hiemale* described from Europe, var. *californicum* from California.

An extremely variable species, composed of several not very distinct taxa. Alaskan specimens have two rows of silica tubercles on ridges of stem, and are referred to var. *californicum*, a western American type. Circumpolar map indicates range of entire, very complicated, species complex.

4. EQUISETACEAE (Horsetail Family)

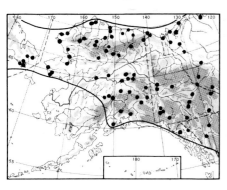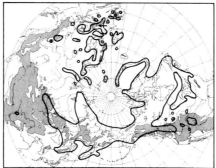

2. Equisètum variegàtum Schleich.
Hippochaete variegatum (Schleich.) Börner.
subsp. **variegàtum**

Creeping, black rhizome with ascending tufted evergreen stems about 2 mm thick, each with a narrow central cavity and 6–8 ridges; ridges flat-topped (in contrast to those of *E. scirpoides*), with 2 rows of silica tubercles at top and a shallow groove between them; ridges end in scarious-margined teeth of sheaths, which are equal in number to ridges; cones subsessile, apiculate.

Woods and tundra, scree slopes, in alpine zone.

Small specimens (var. **ánceps** Milde) are similar to *E. scirpoides*. In an area north of the southern coast, specimens occur that have thicker stems and more numerous ridges than typical *E. variegatum*. In southern Alaska, subsp. *variegatum* is replaced by a more robust race, subsp. *alaskanum* (see next description).

× ½

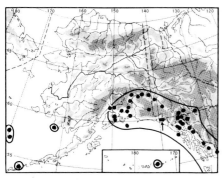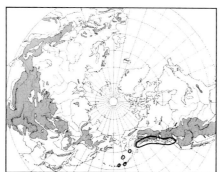

3. Equisètum variegàtum Schleich.
subsp. **alaskànum** (A. A. Eat.) Hult.
Equisetum variegatum var. *alaskanum* A. A. Eat.; *E. alaskanum* (A. A. Eat.) J. P. Anders.

Similar to subsp. *variegatum*, but coarser; stems up to 45 cm high and 4 mm broad, with 8–12 flat-topped ridges.

Woods and tundra, scree slopes, in alpine zone. Apparently a coastal race.

× ⅓

4. Equisetaceae (Horsetail Family)

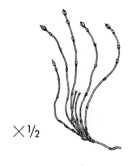

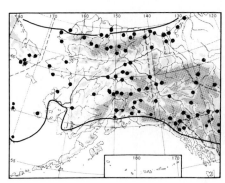
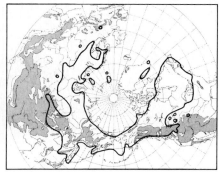

4. Equisètum scirpoìdes Michx.

Densely caespitose; stems solid, prostrate, ascending, 0.5–1 mm thick, evergreen, with usually 6 (occasionally 8) V-shaped, scabrous ridges; sheaths scarious-margined, usually with 3 teeth; 3 of the equally deep furrows of the stem end in the teeth and 3 between them; cones small, black, subsessile, apiculate.

Coniferous woods, tundra, *Dryas* heaths. Described from Canada.

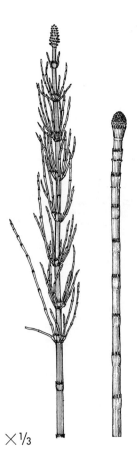
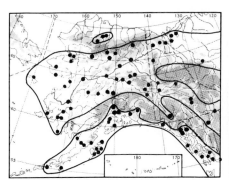
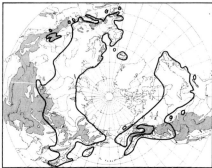

5. Equisètum fluviàtile L. ampl. Ehrh.
Equisetum limosum L.; *E. fluviatile* var. *limosum* (L.) Gilib.; *E. heleocharis* Ehrh.

Rhizone glabrous; stem simple or branched, with wide central cavity; stem finely striate, smooth to the touch, striae more conspicuous in dried specimens; branches single or in irregular whorls at internodes (var. **verticillàtum** Döll); sheaths green, with dark-brown teeth; cones obtuse, long-stalked.

Shallow water and marshy places in low altitudes. Described from Europe. Hybrids with *E. arvense* (*E. litorale* Kühlewein) occur.

4. EQUISETACEAE (Horsetail Family)

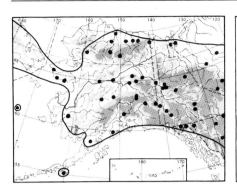
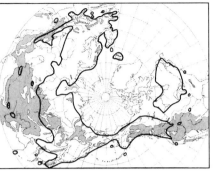
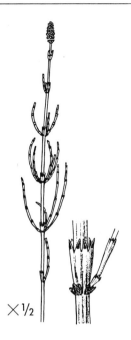

6. Equisètum palústre L.

Rhizome glabrous, thin, lustrous, dark reddish-brown; stem 2–3 mm thick, with usually 6 ridges, simple or (usually) with single to several thick branches of different lengths, top of stem usually lacking branches; first sheath of branches very short, dark brown, the next green, shorter than corresponding sheath of stem; cone long-stalked, obtuse, soon withering.

Wet or moist places, ponds, rare along shores, more common inland. Described from Europe.

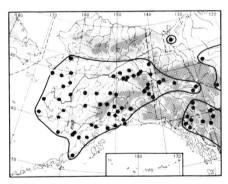
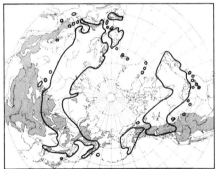

7. Equisètum silvàticum L.

Rhizome dark, creeping; stems of fertile spring phase brownish, 3–4 mm thick, with tubercles at base of sheaths soon developing into branches with inflated and upward-flaring sheaths, the sheaths with membranous, brownish, often coherent teeth; cone obtuse, long-peduncled, soon withering. Sterile stems green, with thinner, essentially simple or forking branches, at first recurving and later spreading; sheaths similar to those of fertile stems.

Lowland forests up to subalpine region. Described from North Europe.

In Alaska, typical specimens have scabrous branches, but specimens with smooth branches (var. **pauciramòsum** Milde) also occur.

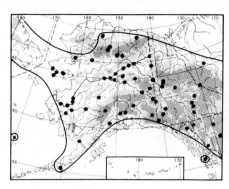
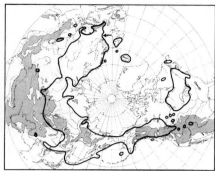

8. Equisètum praténse L.

Rhizome nearly black, creeping, with scattered stems; spring phase with light-brown stem and few short branches, later developing into whorls of branches; summer phase with grayish-green, thin, striated, very rough stems; sheaths green with scarious-margined teeth; branches numerous, slender, 3-angled; first sheath of branches shorter than corresponding sheath of stem; cone obtuse, long-peduncled, soon withering.

Common in woods of the interior.

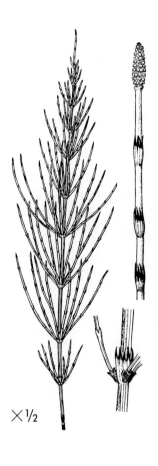
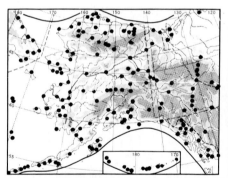
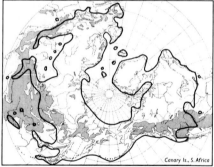

9. Equisètum arvénse L.

Rhizome creeping, with ovoid tubers and dark felt; spring phase brown, lacking chlorophyll, usually branchless with obtuse cones, soon withering; summer phase very different, with sterile, erect, depressed, or prostrate 3–5-angled stem, with simple or forking branches.

Extremely variable in habitat and soils, as well as in form; occurs sometimes as weed. Described from Europe.

Two variations deserve special attention. Var. **boreàle** (Bong.) Ledeb. [*E. boreale* Bong., *E. arvense* subsp. *boreale* (Bong.) Löve], with simple, long, spreading, 3-angled branches, is common. Especially in the north, the fertile spring phase sometimes develops green branches from the nodes; such plants have been called **E. Cálderi** Boiv. The taxonomic value of these variations is unclear.

5. Ophioglossaceae (Adder's Tongue Family)

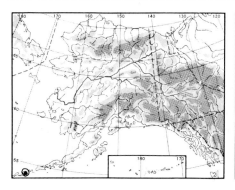
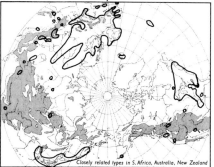

Ophioglossum L.

1. Ophioglóssum vulgàtum L. Adder's Tongue
   var. **alaskànum** (E. G. Britt.) Christens.
   Ophioglossum alaskanum E. G. Britt.

Rhizome short, erect, with long roots; sterile blade thin, pale, ovate, very distinctly veined. Typical plant has thicker, less distinctly veined blade.

On calcareous or saline soil, wet meadows, and seashores. Described from Europe.

O. vulgatum is highly variable in its worldwide range. The Alaskan plant therefore may well be regarded as merely a form, since it has been collected only twice and nothing is known of its variation. Circumpolar map indicates general range of entire species complex.

Botrychium L. (Moonwort)

Sterile blade basal, long-peduncled; bud hairy ... 4. *B. multifidum* subsp. *robustum*
Sterile blade borne on middle or upper part of stem, sessile or nearly sessile; bud glabrous:
  Height 20–60 cm; sterile blade tripinnate, horizontal....................
  ....5. *B. virginianum* subsp. *europaeum*
  Height 5–25 cm; sterile blade 1–2 times pinnate, erect:
    Sterile blade oblong in outline, borne on middle of stem; segments fan-shaped or more or less trapezoidal with cuneate base .... 1. *B. lunaria*
    Sterile blade triangular in outline, borne on upper part of stem, segments oblong to triangular:
      Segments broad, sparsely incised........................2. *B. boreale*
      Segments narrow, with more numerous incisions .... 3. *B. lanceolatum*

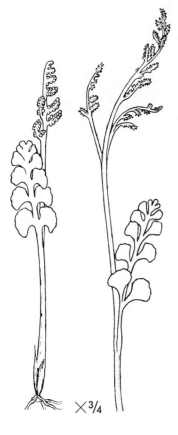
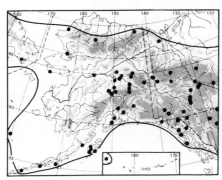
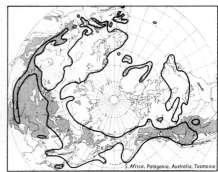

1. Botrýchium lunària (L.) Sw. — Moonwort
Osmunda lunaria L.

Light-green, oblong, sterile blade; pinnae flabellate or trapezoidal, with semilunar or cuneate base and semicircular or pointed apex, lacking midrib.

Grassy slopes. Described from Europe.

Variation in form of the segments is great: plants are seen with remote, roundish, cuneate-obovate, toothed, or notched pinnae, possibly caused by freezing of buds [f. **minganénse** (Vict.) Clute; *B. minganense* Vict.; *B. lunaria* var. *minganense* (Vict.) Döll; *B. lunaria* subsp. *minganense* (Vict.) Calder & Taylor]. Such specimens are here represented by the righthand drawing.

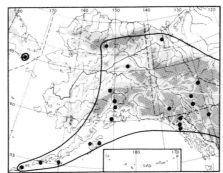
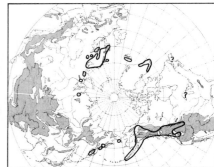

2. Botrýchium boreàle (E. Fries) Milde
Botrychium lunaria var. *boreale* E. Fries; *B. boreale* subsp. *typicum* and subsp. *obtusifolium* (Rupr.) Clausen.

Light-green, triangular sterile blade borne high on stem, with rhomboid, somewhat incised, acute or obtuse (var. **obtusilòbum** Rupr.) pinnae, having fairly distinct median vein; fertile blade much branched, broad.

Grassy slopes, alpine meadows. Described from northern Scandinavia.

Rare everywhere in its circumpolar range; possibly of hybrid origin, since it occurs in the area where ranges of *B. lanceolatum* and *B. lunaria* overlap.

5. Ophioglossaceae (Adder's Tongue Family)

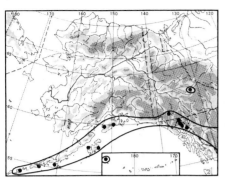
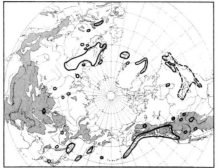
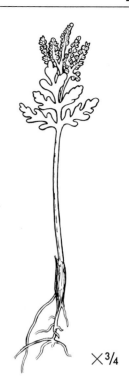

3. Botrýchium lanceolàtum (Gmel.) Ångstr.
Osmunda lanceolata Gmel.

Light-green, triangular sterile blade borne high on stem; pinnae longer than wide, basal pinnae with distinct midrib often bipinnate, or with several deep incisions; fertile blade short, much branched.

Usually very scattered. Grassy slopes, alpine meadows. Described from Europe.

Broken line on circumpolar map indicates range of var. **angustisegméntum**, which may be distinguished by its narrower, more widely spaced segments.

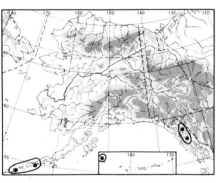
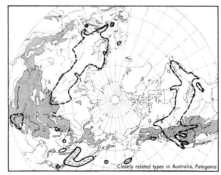

4. Botrýchium multífidum (S. G. Gmel.) Rupr.   Leathery Grape Fern
Osmunda multifida S. G. Gmel.
subsp. **robústum** (Rupr.) Clausen

Botrychium rutaceum var. *robustum* Rupr.; *B. multifidum* var. *robustum* (Rupr.) Christens.; *B. silaifolium* Presl.

Rhizome fairly long and thick, creeping, with several thick roots; up to 30 cm high, with 1 new green and 1 old overwintering basal, petiolated, thick, leathery, sterile blade, somewhat pubescent when young; fertile blade long-petioled, bipinnate.

Sandy meadows and woods. Described from Kamchatka.

Alaskan specimens, apparently very rare, belong to subsp. *robustum,* and are marked by somewhat more acute divisions and more hairy sterile blades than typical *B. multifidum.* Broken line on circumpolar map indicates range of other subspecies.

5. OPHIOGLOSSACEAE (Adder's Tongue Family) / 6. ADIANTACEAE (Maidenhair Family)

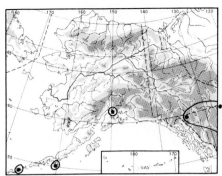
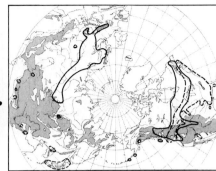
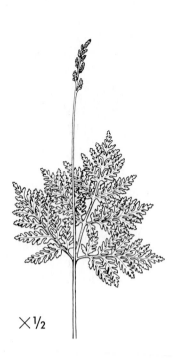

5. Botrýchium viginiànum (L.) Sw. Rattlesnake Fern
Osmunda virginiana L.
subsp. **europaèum** (Ångstr.) Clausen
Botrychium virginianum var. *europaeum* Ångstr.

From 20 to 60 cm tall, sterile blade light green, nearly sessile, borne above the middle of the stem; fertile blade long-stalked, often 2–3-pinnate.

Woods and meadows, with preference for calcareous soil. Described from "America," subsp. *europaeum* from Europe.

Apparently very rare within Alaskan area. Specimens observed belong to northern race, subsp. *europaeum*; sterile blades, more leathery than those of subsp. **viginià-num,** are pinnate, not so deeply divided, and ultimate divisions are often crowded or imbricate. Broken line on circumpolar map indicates range of subsp. *virginianum*.

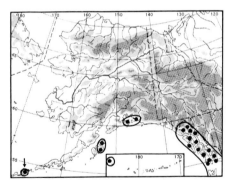
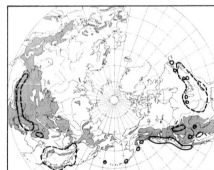
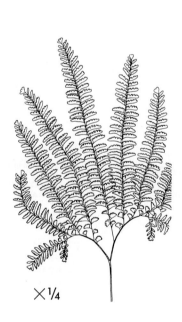

Adiantum L.

1. Adiántum pedàtum L. Maidenhair Fern
var. **aleùticum** Rupr.
Adiantum pedatum subsp. *aleuticum* (Rupr.) Calder & Taylor.

Up to 50 cm high; rhizome short, oblique, densely paleaceous; stipes dark red-brown, lustrous; fronds membranaceous, pinnules with marginal transverse sori. Often sterile in Alaskan area.

Cliffs and woods close to shore.

Alaskan plant, var. *aleuticum*, differs from type in its smaller size, more tufted growth, less spreading segments, and grayish-green color. This variety, which occurs in more exposed places, passes gradually into the type in less exposed localities. Broken line on circumpolar map indicates range of var. **pedátum.**

7. Hypolepidaceae (Bracken Family) / 8. Hymenophyllaceae (Filmy Fern Family)

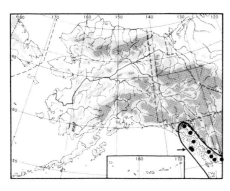

Pteridium Scop.

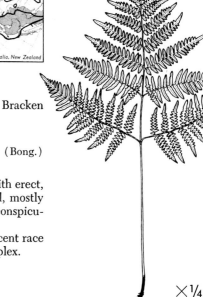

1. Pterídium aquilìnum (L.) Kuhn   Bracken
 Pteris aquilina L.
subsp. **lanuginòsum** (Bong.) Hult.
 Pteris aquilina var. *lanuginosa* Bong.; *Pteridium aquilinum* var. *lanuginosum* (Bong.) Fern.; *Pteris aquilina* var. *pubescens* Underw.

Rhizome branched, pubescent, buried deep in earth; fronds scattered, with erect, stout stipes, deciduous; blade light green, coarse, triangular; sori marginal, mostly continuous, covered by revolute margins of segments; indusium small, inconspicuous.

Dry, open places and woods. Alaskan specimens belong to a more pubescent race than European plants. Circumpolar map gives range of entire species complex.

The young, not-yet-developed frond can be eaten boiled.

×¼

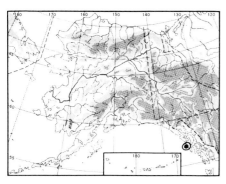

Mecodium Copeland

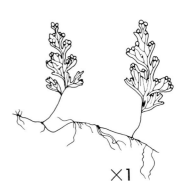

1. Mecòdium Wrìghtii (Bosch) Copeland
 Hymenophyllum Wrightii Bosch.

Small mosslike fern; rhizome filiform, thin, smooth; frond procumbent, forming mats, persistent after withering, pinnatisect; segments asymmetrical, with oblong, obtuse, dentate lobes; sori solitary near base of distal segments; indusium of 2 ovate valves.

Rocks, tree trunks, among moss, in the wettest maritime regions; certainly much overlooked. Described from Japan.

×1

9. CRYPTOGRAMMACEAE (Mountain Parsley Family)

Cryptogramma R. Br.

Fronds scattered along horizontal rhizome . . . . 3. *C. Stelleri*
Fronds tufted; blades thick, coriaceous:
    Ultimate segments ovate, obovate, or oblong, crenate . . . .
    . . . . 1. *C. crispa* var. *acrostichoides*
    Ultimate segments more dissected, deeply toothed . . . . 2. *C. crispa* var. *sitchensis*

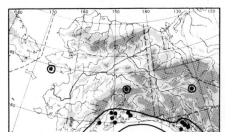
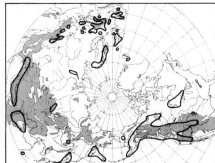

1. Cryptográmma críspa (L.) R. Br.   Parsley Fern
    Osmunda crispa L.
var. **acrostichoìdes** (R. Br.) Clarke
    Cryptogramma acrostichoides R. Br.; *Allosurus acrostichoides* (R. Br.) Spreng.; *A. foveolatus* Rupr.; *C. crispa* subsp. *acrostichoides* (R. Br.) Hult.

Up to 20 cm high; rhizome chaffy, short, ascending; fronds numerous, glabrous, yellowish-green, sterile ones shorter than fertile ones, with broader, shorter, ovate to obovate segments; sori under reflexed margin of fertile frond.

Rock crevices and rocky slopes. *C. crispa* described from England and Switzerland, var. *acrostichoides* from Mackenzie district, 56°N, and Nootka Sound.

American and eastern Asiatic plant is usually taken as a separate taxon, differing from European plant in having foveolate nerve ends and dark median scale nerves; the difference, however, is not clear, and the Alaskan plant is therefore here regarded as a variety.

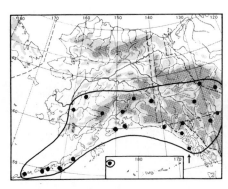
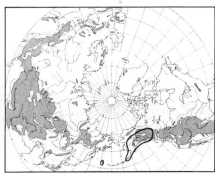

2. Cryptográmma críspa (L.) R. Br.
var. **sitchénsis** (Rupr.) Christens.
    Allosurus sitchensis Rupr.

Sterile leaves broadly deltoid, finely dissected, tripinnate, with small, obovate, toothed tertiary segments; nerve ends mostly not foveolate, basal scales with or without dark midvein.

Var. *sitchensis*, which is similar to var. **Raddeàna** (Fomin) Hult. (*C. Raddeana* Fomin) from Central Asia, seems markedly different from var. *acrostichoides*, but transitions do occur.

9. CRYPTOGRAMMACEAE (Mountain Parsley) / 10. THELYPTERIDACEAE (Marsh Ferns)

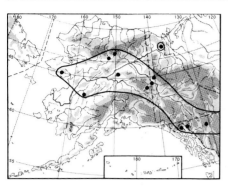
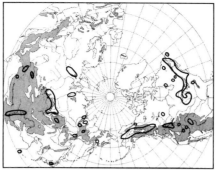
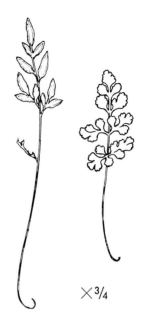

×¾

3. Cryptográmma Stélleri (S. G. Gmel.) Prantl
Pteris Stelleri S. G. Gmel.; *Allosurus Stelleri* (S. G. Gmel.) Rupr.; *Pellaea Stelleri* (S. G. Gmel.) Baker; *Pteris gracilis* Michx.; *A. gracilis* (Michx.) Presl; *Pellaea gracilis* (Michx.) Hook.

Up to 15 cm tall; rhizome very slender, cordlike, scaly, and pilose; fronds scattered, glabrous, fertile ones longer than sterile ones, pinnae few, thin; fertile frond with long stipe, and remote pinnae; ultimate segments linear-oblong, entirely or slightly crenulated; sori marginal, covered by revolute margins.

Crevices in calcareous rocks in shaded localities with dripping water. Usually very rare and scattered. Described from Siberia.

Thelypteris Schmidel

Fronds short-stalked, lanceolate to ovate-lanceolate; rhizome stout .... 1. *T. limbosperma*
Fronds long-stalked, triangular; rhizome slender, creeping.... 2. *T. phaegopteris*

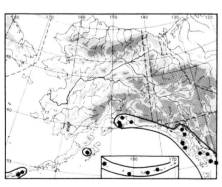
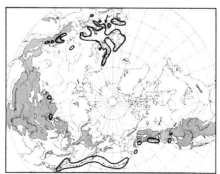
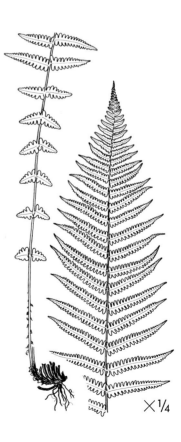

×¼

1. Thelýpteris limbospérma (All.) Fuchs
Polypodium limbospermum All.; *Dryopteris oreopteris* (Ehrh.) Maxon; *P. oreopteris* Ehrh.; *Aspidium oreopteris* (Ehrh.) Sw.; *Thelypteris oreopteris* (Ehrh.) Slosson; *D. montana* (Vogler) Ktze.; *P. montanum* Vogler.

Rhizome short, ascending, scaly, stipes grooved above, with dark base; fronds 30–80 cm tall, forming crown, yellowish-green, glandular and hoary white on nerves below, abruptly acuminate; pinnae pinnatisect, segments entire, blunt; sori small; indusium lobed, glandular.

Open, rocky slopes, reaching subalpine region; not common.
Often sterile in Alaska. Crushed leaves strongly scented.

10. THELYPTERIDACEAE (Marsh Fern Family) / 11. ASPLENIACEAE (Spleenwort Family)

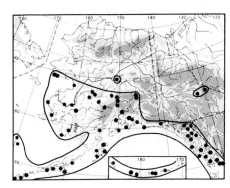

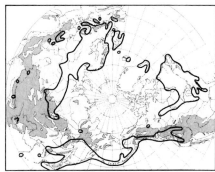

2. Thelýpteris phegópteris (L.) Slosson

Polypodium Phegopteris L.; *Dryopteris phegopteris* (L.) Christens.; *Phegopteris polypodioides* Fée; *P. dryopteris* (L.) Fée.

Rhizome slender, long, with light-brown, hairy, lanceolate scales; fronds solitary, widely separated, up to 40 cm tall; stipe twice as long as blade, hairy and scaly; pinnae hairy on both surfaces, scaly on veins, basal pair usually drooping in mature specimens; sori small, submarginal; indusium lacking.

Woods and stony slopes, common on southern coast from lowlands up to lower alpine region. Northernmost locality in Yukon valley is at a hot spring. Described from Europe and Virginia.

Asplenium L. (Spleenwort)

Stipe and rachis of frond black, stiff to top .... 1. *A. trichomanes*
Stipe brown on lower half, with broad green margins, not stiff at top ... 2. *A. viride*

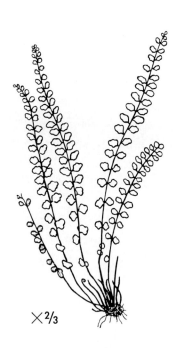

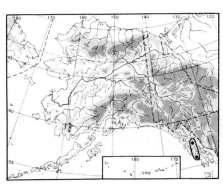

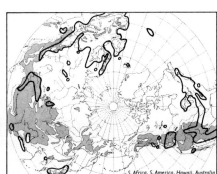

1. Asplènium trichòmanes L.

Up to 15 cm tall; rhizome short, scaly, scales with median nerve; fronds tufted, stipes dark chestnut to black, lustrous, glabrous; fronds 1–2 cm broad, evergreen, median nerve lacking sori.

Crevices and rock ledges, mostly in shadowy places. Described from Europe.

11. ASPLENIACEAE (Spleenwort Family) / 12. ATHYRIACEAE (Lady Fern Family)

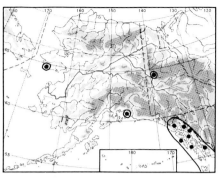
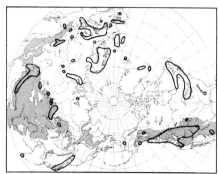
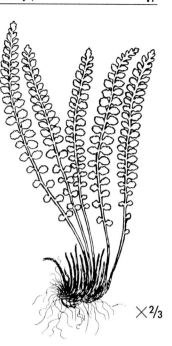

2. Asplènium víride Huds.

Up to 10 cm tall; rhizome short, scaly, but scales lack median nerve; fronds 1 cm broad, densely tufted; stipes with deep groove, brown in lower part only, with broad, green margins, glabrous; pinnae herbaceous, not evergreen; sori near indistinct midvein. Similar to *A. trichomanes*.

Crevices in calcareous rocks. Described from England.

The species name relates to the green rachis of the frond.

Athyrium Roth

Blades more or less tripinnate; sori small, round, indusium rudimentary or lacking
.... 1. *A. distentifolium* subsp. *americanum*
Blades bipinnate; sori large, about 1 mm in diameter, indusium present ....
.... 2. *A. filix-femina* subsp. *cyclosorum*

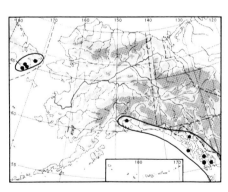
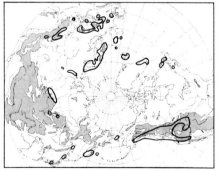

1. Athýrium distentifòlium Tausch
subsp. **americànum** (Maxon) Hult.

Athyrium alpestre (Hoppe) Rylands (*non* Clairv.) var. *americanum* Butt.; *A. americanum* Maxon.

Rhizome chaffy, forming large crown; fronds in vaselike clumps; stipe scaly, lowest primary segments shorter than middle segments.

Wet, rocky slopes, with preference for calcareous soil. *A. distentifolium* described from Bohemia, subsp. *americanum* from North America.

The Alaskan plant differs from the European plant in having generally more dissected fronds, with ultimate segments more widely separated; it is therefore regarded as a subspecies, subsp. *americanum*, and is apparently rare. The plant has long been known as *A. alpestre* (Hoppe) Rylands, a name antedated by *A. alpestre* Clairv. Circumpolar map indicates range of entire species complex.

12. ATHYRIACEAE (Lady Fern Family)

×¼

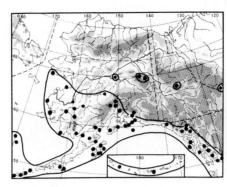

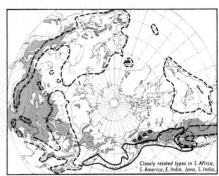

Closely related types in S. Africa, S. America, E. India, Java, S. India.

2. Athýrium fílix-fémina (L.) Roth   Lady Fern
Polypodium Filix-femina L.
subsp. **cyclosòrum** (Rupr.) Christens.
Athyrium cyclosorum Rupr.; *A. filix-femina* var. *sitchense* Rupr.

Up to 60 cm tall; rhizome chaffy, forming large crown terminating in tuft of fronds; stipe short, chaffy, blade dark green, bipinnate (rarely tripinnate), with highly variable dissection; lowest primary segments shorter than middle segments.

Common in woods and lowlands of southern shore, at least up to 1,200 meters, very rare northward, the northernmost localities being at hot springs. *A. filix-femina* described from Europe, subsp. *cyclosorum* from Unalaska and Kodiak.

Specimens from exposed places have broader, blunter, less dissected pinnules, and are usually sterile [f. **Híllii** (Gilb.) Butt.; *A. cyclosorum* f. *Hillii* Gilb.]; such specimens are sometimes mistaken for *Dryopteris filix-mas*. Alaskan plant has nearly round sori, and is therefore regarded as a race of the typical plant, which has oblong sori; variation in Alaska, however, corresponds to that in other parts of range. Broken line on circumpolar map indicates range of subsp. **filix-fémina**.

Cystopteris Bernh.

Fronds remote, arranged in rows from cordlike rhizome, deltoid in outline......
....3. *C. montana*
Fronds at apex of stout rhizome, lanceolate or lanceolate-oblong in outline:
   Spores spinose............................1. *C. fragilis* subsp. *fragilis*
   Spores rugose............................2. *C. fragilis* subsp. *Dickieana*

12. ATHYRIACEAE (Lady Fern Family)

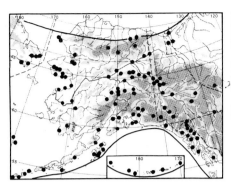
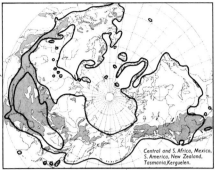

1. Cystópteris frágilis (L.) Bernh. Fragile Fern
subsp. **frágilis**
  Polypodium fragile L.; *Filix fragilis* Gilib.

Up to 30 cm high; fronds tufted, at tip of a stout rhizome, stipes shorter than blade, glabrous or with a few scales at base.

Rock crevices, talus slopes up to more than 2,000 meters—in Himalayas to over 5,400 meters. Described from Europe.

An extremely variable fern, with respect to both the dissection of the leaves and the sculpture of the spores. Alaskan population has been regarded as consisting of *C. fragilis*, with veins directed to the teeth, and transgressions between this species and **C. diàphana** (Bory) Blasdell (*Polypodium diaphanum* Bory), a South and Central American plant, with veins directed to the emarginations.

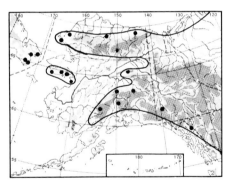
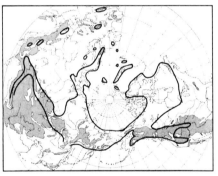

2. Cystópteris frágilis (L.) Bernh.
subsp. **Dickieàna** (Sim) Hyl.
  Cystopteris Dickieana Sim; *C. fragilis* var. *Dickieana* (Sim) Moore.

Similar to subsp. *fragilis*, but differing in having smoother, rugulose (not spiny) spores.

Arctic-montane circumpolar range. In the St. Elias Range, to at least 2,300 meters. Described from Scotland.

In Alaska, transgressions between this subspecies and subsp. *fragilis* occur.

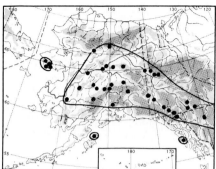
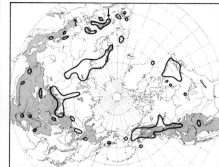

3. Cystópteris montàna (Lam.) Bernh.
Polypodium montanum Lam.

Up to 40 cm tall; fronds in rows from a branching, cordlike rhizome; stipes longer than blade, darker at base, with large, light-brown scales, glands in margin; sori and indusium round.

Damp woods and moist, rocky slopes, with preference for calcareous soil.

Woodsia R. Br.

Fronds glabrous, lacking scales and hairs (other than filaments of the indusia) . . . .
. . . . 4. *W. glabella*
Fronds with hairs, at least below:
   Fronds with hairs and scales . . . . 2. *W. ilvensis*
   Fronds lacking scales:
      Fronds lacking a basal joint, with long, flattish, white hairs on both sides, mixed with few, dark glands; segments much longer than broad . . 1. *W. scopulina*
      Fronds with joint at base, glabrous above; segments not much longer than broad . . . . 3. *W. alpina*

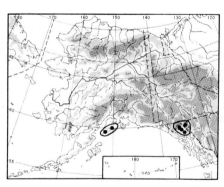
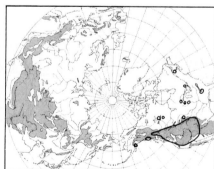

1. Woòdsia scopulìna D. C. Eat.

Up to 30 cm tall; rhizome stout, short, with numerous, persistent, brown, old stipe bases; new stipes stout, darker at base, pilose and slightly scaly at base; lower pinnae shorter than middle pinnae, with white, flattish, septate hairs and few dark glands; indusia with flat, linear-lanceolate or lanceolate filaments.

Dry rocks and rocky slopes, mostly on calcareous soil. Described from 40° to 49°N in western North America.

12. ATHYRIACEAE (Lady Fern Family)

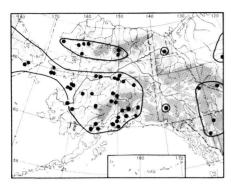 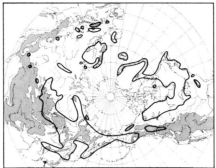 

2. Woòdsia ilvénsis (L.) R. Br.
Acrosticum ilvense L.

About 15 cm tall; rhizome ascending; stipe long, reddish-brown, midrib with narrow scales usually to tip; blade dull green, thick, with segments considerably longer than broad, and with scales and hairs below; indusia of hairlike segments.

Dry rocks, scree slopes. Described from North Europe. Hybrids between this species and *W. alpina* occur.

× ½

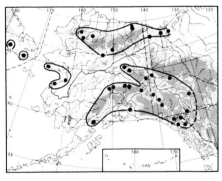 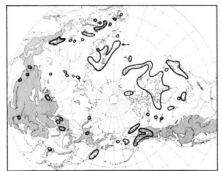 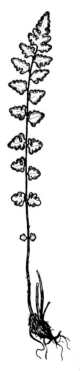

3. Woòdsia alpìna (Bolton) S. F. Gray
Acrosticum alpinum Bolton; *Woodsia hyperborea* R. Br.

About 10 cm tall; rhizome short, ascending; stipe with joint at base, and short, reddish-brown midrib, usually lacking scales; blades thin, last segments not much longer than broad, pubescent from septate hairs below, but lacking scales.

Prefers calcareous rocks in mountains. Hybrids with *W. ilvensis* occur.

× ¾

12. ATHYRIACEAE (Lady Fern Family)

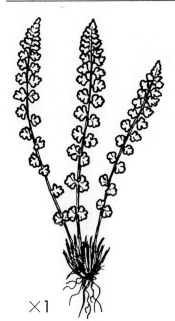
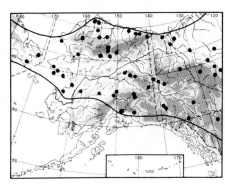
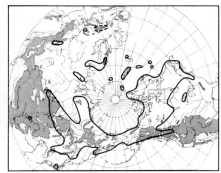

4. Woòdsia glabélla R. Br.

About 5 cm tall; rhizome short, ascending; stipe with joint at base, very short, light-colored, lacking scales or hairs above joint; blade yellowish-green, lacking scales or hairs other than filaments of indusia below.

Prefers calcareous rocks. Described from Mackenzie district.

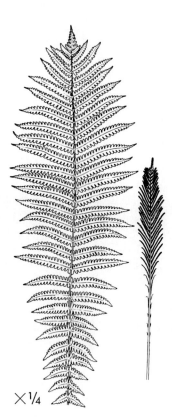
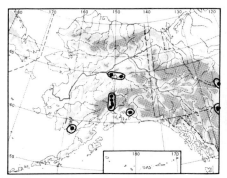
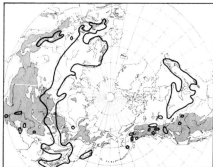

Matteuccia Todaro (Ostrich Fern)

1. Mattéuccia struthiópteris (L.) Todaro

Osmunda struthiopteris L.; *Struthiopteris filicastrum* All.; *Onoclea struthiopteris* (L.) Roth; *Pteretis nodulosa* (Michx.) Nieuwl.

Rhizome short, erect, stout; sterile fronds in distinctly vaselike arrangement more than 1 meter tall, bright green; stipe short, blade pinnate with pinnatifid pinnae; fertile fronds in center of sterile fronds, and much shorter, persistent, dark brown in age, with linear obtuse pinnae.

Woods and along creeks. In Alaska apparently restricted to small isolated areas; very common in Susitna valley. Described from Scandinavia, Switzerland, and Russia.

13. Aspidiaceae (Shield Fern Family)

Polystichum Roth

Blades simply pinnate:
  Mature fronds about 15 cm tall, thin; pinnae neither spinose nor aristate . . . .
  . . . . 1. *P. aleuticum*
  Mature fronds taller, coriaceous; pinnae with spinulose teeth:
    First pair of pinnae small, others gradually increasing in size; stipe short; pinnae oblong-lanceolate . . . . 2. *P. lonchitis*
    First pair of pinnae practically as large as the next; stipe long; pinnae linear-attenuate . . . . 3. *P. munitum*
Blades bipinnate:
  Upper basal pinnules not longer than the next . . . . . . 4. *P. Braunii* var. *alaskense*
  Upper basal pinnules conspicuously longer than the next . . . .
  . . . . 5. *P. Braunii* subsp. *Andersonii*

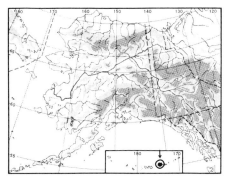 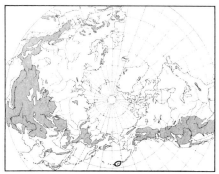 

×⅓

1. Polýstichum aleùticum Christens.

About 15 cm tall, with the general appearance of *Woodsia alpina*; rhizome oblique, fairly thick, dark brown, with thin brown scales and numerous chestnut-brown stipe bases; pinnae in about 15 pairs, sessile or short-petioled; sori at upper part of frond only; indusium orbicular or oblong, attached eccentrically, erose-laciniate, greenish, glabrous.

Known only from the type collection at Atka Island. Related to Chinese and Himalayan species—**P. lachenénse** (Hook.) Bedd. (*Aspidium lachenense* Hook.) and **P. sinénse** Christens.—but without close relatives in America.

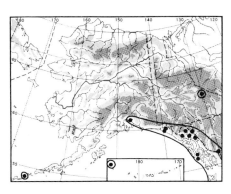 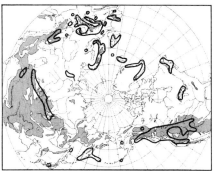 

×¼

2. Polýstichum lonchìtis (L.) Roth
Polypodium Lonchitis L.; *Aspidium lonchitis* Sw.; *Dryopteris lonchitis* (Retz.) Ktze.

Rhizome stout, ascending, covered with old stipe bases, terminating in crown of new fronds; stipes short, chaffy, with lustrous red-brown to castaneous scales; frond rigid, coriaceous, evergreen, dark green, with spreading teeth; sori in 2 single or double rows, later becoming confluent mostly on upper part; indusia large, more or less toothed.

Rocks and rocky slopes, mostly on calcareous soil. Described from Switzerland and "Virginicus."

13. Aspidiaceae (Shield Fern Family)

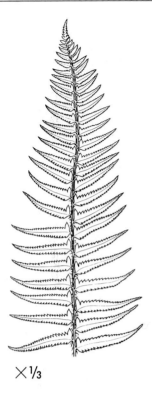
×⅓

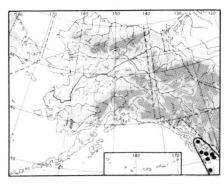
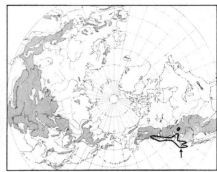

3. Polýstichum munìtum (Kaulf.) Presl
Aspidium munitum Kaulf.

Rhizome stout, ascending, chaffy, covered with old stipe bases, terminating in crown of new fronds; stipes long, densely chaffy, with lustrous, brown, dark-centered scales; frond rigid, coriaceous, evergreen, up to 1 meter long; teeth of pinnae incurved, long, aculeate; sori in 1 to several rows; indusia large, papillose to ciliate.

Moist woods. Variable in size and outline of frond; small forms with imbricate, overlapping pinnae occur in sunny places.

×⅓

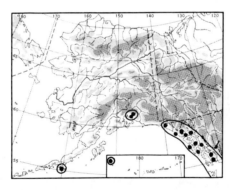
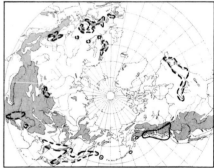

4. Polýstichum Braùnii (Spenn.) Fée
Aspidium Braunii Spenn.
var. alaskénse (Maxon) Hult.
Polystichum alaskense Maxon; *P. Braunii* subsp. *alaskense* Calder & Taylor.

Rhizome stout, terminating in vaselike crown of fronds; stipe furrowed above, densely chaffy, with thin, pale-brown, more or less attenuate scales; fronds elliptic-lanceolate, subcoriaceous, at least in some places evergreen, tapering at base and tip, with 20–40 pairs of alternating pinnae, the lowest rather short; pinnules with bristle-tipped teeth.

Woods and rocky slopes. Broken line on circumpolar map indicates range of other races or varieties.

Form of pinnules varies from ovate, with nearly peduncled base and distinct auricle, to ovate-ellipsoid, with base 2–3 mm broad attached to the rachises (var. *alaskense* in a narrow sense). These types often grow together, and a sharp differentiation is hardly possible. They are therefore here regarded as a variable population under the name var. *alaskense*.

13. ASPIDIACEAE (Shield Fern Family)

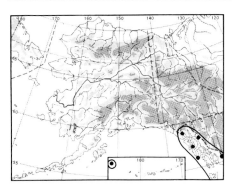
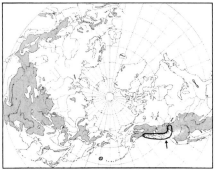

5. Polýstichum Braùnii (Spenn.) Fée
subsp. **Andersònii** (M. Hopkins) Calder & Taylor
   Polystichum Andersonii M. Hopkins; *P. Braunii* var. *Andersonii* (M. Hopkins) Hult.

Similar to var. *alaskense*, but differing in upper basal pinnae, which are conspicuously longer than succeeding pinnae.
Woods and rocky slopes.

Dryopteris Adans.

Stipe long; frond broad, more or less triangular in outline, tripinnate or nearly so, lowest pair of pinnae separated by 0.5–2 cm . . . 1. *D. dilatata* subsp. *americana*
Stipe very short; frond lanceolate, with pinnae very short at base and apex, long in middle . . . . 2. *D. fragrans*

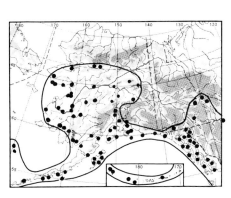

1. Dryópteris dilatàta (Hoffm.) Gray
   Polypodium dilatatum Hoffm.
subsp. **americàna** (Fisch.) Hult.
   Aspidium spinulosum var. *americanum* Fisch.; *Thelypteris dilatata* var. *americana* (Fisch.) House; *Dryopteris dilatata* var. *americana* (Fisch.) Benedict; *Polypodium dilatatum* Hoffm.; *D. austriaca* of authors.

Rhizome stout, ascending, covered with old stipe bases; stipe shorter than blade, chaffy, with pale-brown scales; frond not glandular, ovate to ovate-triangular, tripinnate; basal pinnae triangular, larger and broader than other pinnae; upper innermost pinnules of basal pair of pinnae much shorter than lower pinnules.

Woods, moist places, to about 800 meters in mountains in McKinley Park. Described from Germany.

13. ASPIDIACEAE (Shield Fern Family)

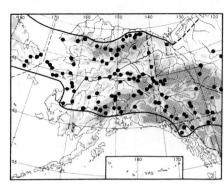 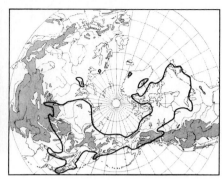

2. Dryópteris fràgrans (L.) Schott

Polypodium fragrans L.; *Polystichum fragrans* (L.) Ledeb.; *Lastrea fragrans* (L.) Presl; *Aspidium fragrans* (L.) Sw.; *Nephrodium fragrans* (L.) Richards.

Rhizome stout, covered with shriveled and curled old fronds; stipes short, glandular, chaffy, with brown or reddish-brown scales; frond coriaceous; pinnae chaffy and glandular, with inrolled margin.

Sunny, rocky slopes. Described from Siberia.

Typical plant with overlapping pinnae is common in most parts of Alaska, but some specimens differ in having more remote, non-overlapping pinnae: var. **remotiùscula** Kom.; var. *Hookeriana* Fern.; var. *aquilonaris* (Maxon) Gilb.; *D. aquilonaris* Maxon. The plant thus variously named is not well differentiated from var. **fràgrans** and occurs in less exposed habitats in scattered localities within species range.

Gymnocarpium Newm.

Fronds bright green, mostly glabrous, lateral primary divisions each nearly as large as terminal division .... 1. *G. dryopteris*
Fronds dull green, mostly glandular beneath, lateral primary divisions each considerably smaller than terminal division .... 2. *G. robertianum*

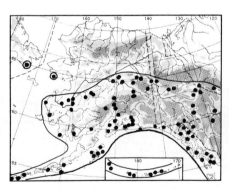 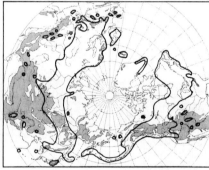

1. Gymnocárpium dryópteris (L.) Newm.

Polypodium dryopteris L.; *Phaegopteris dryopteris* (L.) Fée; *Lastrea dryopteris* (L.) Bory; *Nephrodium dryopteris* (L.) Michx.; *Dryopteris Linneana* Christens.; *D. disjuncta* (Rupr.) Morton; *Polypodium disjunctum* Rupr. Including *Gymnocarpium remotipinnatum* (Hayata) Ching.

Rhizome creeping; petioles glabrous; lamina light green, mostly glabrous; fronds deltoid, ternate, two lateral divisions each nearly as large as terminal division; lowest pinnule on lower side of first lateral divisions about equal to either pinna of third pair from base.

Woods, thickets. Described from Europe. Within range of interest, transitions to *G. robertianum* occur.

13. Aspidiaceae (Shield Fern Family) / 14. Polypodiaceae (Licorice Fern Family)

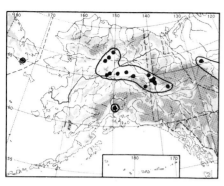
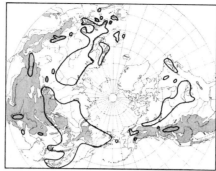

2. Gymnocárpium robertiànum (Hoffm.) Newm.

Polypodium robertianum Hoffm.; *Phaegopteris robertiana* (Hoffm.) A. Braun; *Lastrea robertiana* (Hoffm.) Newm.; *Nephrodium robertianum* (Hoffm.) Prantl. Including *Dryopteris continentalis* Petrov.

Similar to *G. dryopteris*; in typical specimens, fronds dull green, glandular below; two lateral divisions each smaller than terminal divisions; lowest pinnule on lower side of first lateral divisions considerably smaller than either pinna of third pair from base.

Woods. Described from Germany. Hybrids with, and transitions to, *G. dryopteris* are common.

× ⅓

Polypodium L.

Pinnae obtuse, often somewhat crenulated, midrib not pubescent . . . .
. . . . 1. *P. vulgare* subsp. *columbianum*
Pinnae more or less acute, sometimes serrulated, lower midrib slightly pubescent
from scattered, crisp, gray or brownish hairs . . 2. *P. vulgare* subsp. *occidentale*

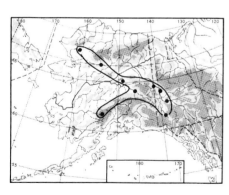
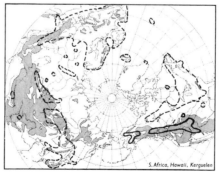

1. Polypòdium vulgàre L.
subsp. **columbiànum** (Gilb.) Hult.

Polypodium vulgare var. *columbianum* Gilb.; *P. hesperinum* Maxon.

Rhizome creeping, covered with dark-brown lanceolate scales; stipe shorter than blade; fronds yellowish-green, up to 20 cm tall, glabrous or with few scales; pinnae oblong to slightly obovate, rounded at apex, sometimes slightly crenate.

Rocks, tree trunks. *P. vulgare* described from Europe, subsp. *columbianum* from "Columbia."

Closely related to subsp. *occidentale,* but apparently restricted to interior of Alaska and very rare. Broken line on circumpolar map indicates range of *P. vulgare* races other than subsps. *occidentale* and *columbianum*.

× ½

14. POLYPODIACEAE (Licorice Fern Family) / 15. BLECHNACEAE (Deer Fern Family)

×⅓

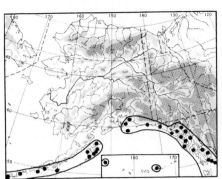
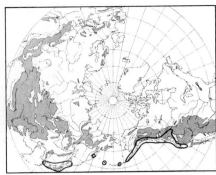

2. Polypòdium vulgàre L.
subsp. **occidentàle** (Hook.) Hult.

Polypodium vulgare var. *occidentale* Hook.; *P. occidentale* Maxon; *P. glycyrrhiza* D. C. Eat.

Similar to subsp. *columbianum*, but fronds generally taller (esp. in southern specimens), grayish-green; midrib of frond and segments sparsely pubescent, with crisp gray or brownish hairs, segments with more or less acute segments.

Rocks, tree trunks. Subsp. *occidentale* described from Sitka and Columbia River.

Most specimens have subacute leaves (var. **commùne** Milde), but specimens with lanceolate, acute segments (var. **occidentàle** Hook.) also occur, esp. southward.

×½

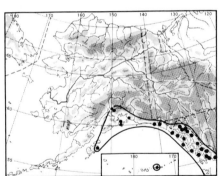

Blechnum L.

1. Bléchnum spìcant (L.) Roth

Osmunda spicant L.; *Lomaria spicant* (L.) Desv.; *Struthiopteris spicant* (L.) Weis; *Blechnum spicant* subsp. *nipponicum* (Kunze) Löve & Löve.

Up to 60 cm tall; rhizome short, creeping, woody; sterile fronds evergreen, coriaceous, forming circular crown, with pinnae 2–5 mm. broad; fertile fronds deciduous, central, erect, longer than sterile fronds and with narrower pinnae; at late stage of development, sori along margins of segments cover entire lower surface.

Forests along shore, usually calcifuge. Described from Europe. Occasionally common in southeast Alaska, rare in Aleutian Islands.

16. TAXACEAE (Yew Family) / 17. PINACEAE (Pine Family)

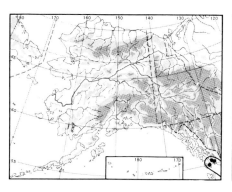

Taxus L.

1. Táxus brevifòlia Nutt.   Western Yew

Dioecious shrub or small tree, up to 10 meters tall (at least in British Columbia), with horizontal or drooping branches; needles in two rows, shiny dark green above, pale beneath; seeds surrounded by juicy, scarlet, cuplike disk (aril).

Undergrowth of dense woods and thickets, usually very scattered. Broken line on circumpolar map indicates range of *Taxus* sp. occurring in Eurasia and North America, all closely related to *T. brevifolia*.

Scarlet disks are not poisonous, but seeds contain poisonous toxin, causing vomiting, diarrhea, and inflammation of urinary ducts and the uterus.

Pinus L.

Needles 3–7 cm long; cones ovoid, prickly . . . . 1. *P. contorta*
Needles 1–3.5 cm long; cones usually curved, not prickly . . . . 2. *P. Banksiana*

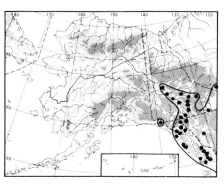
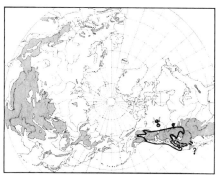

1. Pìnus contórta Dougl. *ex* Loud.   Lodgepole Pine

Low shrubby tree, 6–10 meters tall, with slender, often twisted dark-green needles in fascicles of 2; bark thin, furrowed; staminate flowers orange-red; cones 3–5 cm long, light yellow-brown, persisting for many years; scales thin, concave, prickly.

Muskegs, valley bottoms. Described from "NW America; Cape Dissapointment [Washington]; Cape Lookout [Oregon]."

The coastal plant belongs to subsp. **contórta**, the inland plant in Yukon and British Columbia to the taller subsp. **latifòlia** (Engelm.) Critchfield (*P. contorta* var. *latifolia* Engelm.; *P. Murrayana* var. *Sargentii* Mayr), which has very thin, scaly, unfurrowed bark. Westernmost lodgepole pines are found at mile 80 on highway from Dawson.

On the circumpolar map, continuous line indicates range of subsp. *contorta* and subsp. *latifolia*, broken line the range of subsp. **Murrayàna** (Balf.) Critchfield (*Pinus Murrayana* Balf.).

17. PINACEAE (Pine Family)

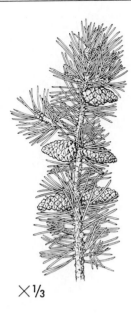

×⅓

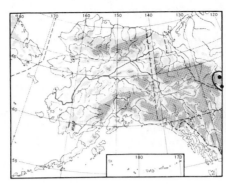
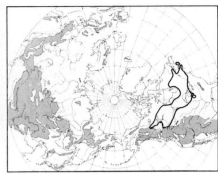

2. Pìnus Banksiàna Lamb.   Jack Pine

Small tree with spreading branches, dwarfed in alpine areas; needles 2–3 cm long in fascicles of 2; cones erect, often curved, asymmetrical, yellowish-brown, with pointless scales.

On rocky and sandy soil, barely reaching area of map in the Nahanni Mountains. Described from Arctic and Boreal America.

×½

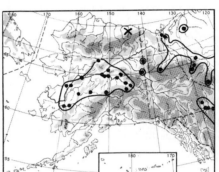
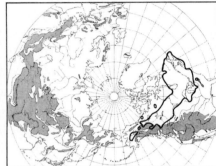

Larix Mill.

1. Làrix laricìna (Du Roi) K. Koch   Tamarack
Pinus laricina Du Roi.
var. alaskénsis (Wight) Raup
Larix alaskensis Wight.

Small tree, up to 10 meters tall, with brownish bark, horizontal branches, and thin crown extending far down on trunk; leaves deciduous, blue-green, in fascicles of 10–20 on short spur branches.

Muskegs, moist places, usually scattered; according to A. E. Porsild, a decided calciphile. Described from America.

Range within area of interest is still somewhat unclear, since known localities are rather scattered. Alaskan tamarack, not a well-marked variety, differs in having proportionately longer, more papery, cone scales, often enrolled at margins, and in having bracts lacking the projecting mucronate point. Circumpolar map indicates range of entire species complex; cross on Alaskan map marks a locality for fossil *L. laricina*.

Picea Dietr.

Needles with flat cross section (slightly keeled above, rounded beneath), white bands of stomata above, green beneath .... 2. *P. sitchensis*
Needles quadrangular in cross section, stomatiferous on all sides:
    Branchlets glabrous; cones 3–6 cm long, dropping at maturity ...... 1. *P. glauca*
    Branchlets pubescent; cones 1.5–3 cm long, persisting for several years ....
    .... 3. *P. mariana*

17. PINACEAE (Pine Family)

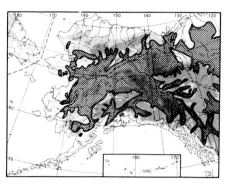
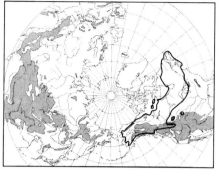

1. Pícea glaùca (Moench) Voss   White Spruce
Pinus glauca Moench; *Picea canadensis* of authors.

Tree 10–25 meters tall and up to 50 cm in diameter, with stiff, pointed, 4-angled, blue-green needles, standing out in all directions, stomatiferous on all sides; branchlets glabrous; deciduous cones subcylindrical, with thin, rounded, smooth-margined scales.

The most important tree in the interior spruce-birch forest, extending to at least 1,500 meters, becoming stunted and depressed in lower alpine zone. Described from cultivated specimens; var. *albertiana* (next paragraph) described from Bankhead, Alberta.

Two somewhat different types, with as yet unclear ranges and sometimes growing together, occur within range of interest: var. **albertiàna** (S. Brown) Sarg. (*P. albertiana* S. Brown), narrowly pyramidal or linear in outline, with rough bark and cone scales wedge-shaped at base; and var. **Porsíldii** Raup, conical or narrowly ovate in outline, with smooth bark beset with resin blisters, as on balsam fir, and more rounded cone scales. Typical *P. glauca* has broadly pyramidal form and occurs in central and eastern parts of range, with Alberta spruce (var. *albertiana*) replacing it from Keewatin and northern Saskatchewan west. On Alaskan map, approximate extent of the *P. glauca* forest is indicated by hatching.

(See color section.)

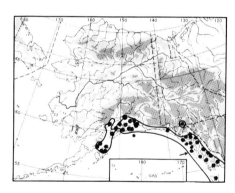
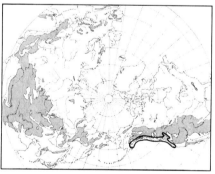

2. Pícea sitchénsis (Bong.) Carr.   Sitka Spruce
Pinus sitchensis Bong.; *Picea falcata* (Raf.) Valck.-Suringar; *Abies falcata* Raf.?

Up to 65 meters tall and 1.5 meters in diameter, occasionally larger, with dark, thin, somewhat scaly bark; needles pungent, flattened, dark green beneath, two white stomatiferous bands above; cones 5–10 cm long, cylindrical-to-oblong, with rhombic-oblong scales.

In pure stands or mixed with hemlock, chiefly below 500 meters in altitude, but extending occasionally to about 1,000 meters. Often propagates by layering.

A grove of Sitka spruce was planted in about 1805 at Unalaska, far west of natural range; it produces cones but does not multiply.

(See color section.)

17. PINACEAE (Pine Family)

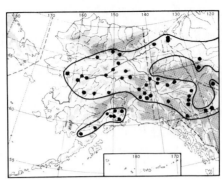
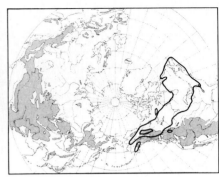

3. Pícea mariàna (Mill.) Britt., Sterns & Pogg.   Black Spruce
Abies mariana Mill.

Small, often shrubby tree, up to 10 meters tall, usually shorter, with short, sparse, often slightly drooping branches, and pubescent branchlets; needles 4-angled, pale bluish-green with whitish bloom; cones subglobose or short ovoid, 2–3 cm long, with rounded, slightly erose scales, persisting for several years.

Muskegs and lake margins, to about 700 meters in McKinley Park, to about 1,150 meters in mountains of eastern part of area of interest. Described from America.

Hybrids with *P. glauca* occur (*P. Lutzii* Little).

Tsuga Carr.

Needles flat, with 2 pale stomatiferous bands beneath; cones small, about 2 cm long
.... 1. *T. heterophylla*
Needles keeled, stomatiferous on both sides; cones 5–7 cm long .. 2. *T. Mertensiana*

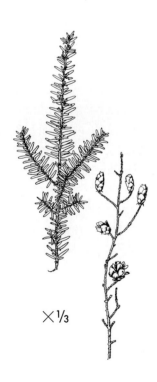
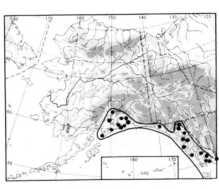

1. Tsùga heterophýlla (Raf.) Sarg.   Western Hemlock
Abies heterophylla Raf.

Up to 50 (occasionally 65) meters tall, with narrow crown, horizontal or drooping branches, and reddish-brown, furrowed bark; needles flattened, rounded at tips, shiny dark green above, with two white stomatiferous bands beneath; cones small, hanging, reddish-brown.

Forms about 70 per cent of the hemlock–Sitka spruce coastal forest in southern Alaska. Reaches an altitude of about 1,000 meters.

17. PINACEAE (Pine Family)

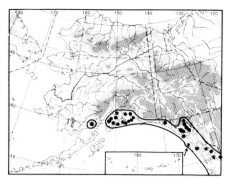

2. Tsùga Mertensiàna (Bong.) Sarg.   Mountain Hemlock
Pinus Mertensiana Bong.

× 1/3

Tree up to 20–30 meters tall, but often short and dwarfed, with horizontal or drooping branches and grayish to brownish furrowed bark; needles usually curved, stout, keeled above, blue-green, stomatiferous on both sides; cones cylindrical, longer than in *T. heterophylla*, hanging, purplish, later brown.

Muskegs and mountain slopes from sea level up to 1,200 meters, reaching beyond altitudinal limit of *T. heterophylla*, and most common in mountains.

Abies Mill.

Needles dark green above, silvery beneath from bands of stomata; cones glabrate or sparsely hairy .... 1. *A. amabilis*
Needles grayish-green, stomatiferous on both sides; cones pubescent .... 2. *A. lasiocarpa*

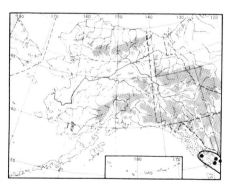

1. Àbies amábilis (Dougl.) Forbes   Pacific Silver Fir
Picea amabilis Dougl.

× 1/2

Tree up to 25 meters tall, with smooth, gray, white-spotted bark and pubescent twigs; needles curved upward, flat, shiny, grooved on upper side, silvery white and stomatiferous beneath; staminate flowers red; cones oblong, erect, purple-colored, with puberulent scales, rounded at apex.

In few localities in southernmost Alaska, up to about 300 meters.

17. PINACEAE (Pine Family) / 18. CUPRESSACEAE (Cypress Family)

× ½

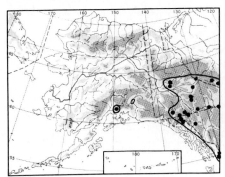 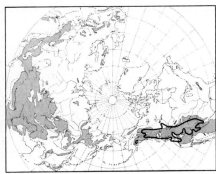

2. Àbies lasiocárpa (Hook.) Nutt.   Alpine Fir
Pinus lasiocarpa Hook.

Small or medium-sized tree with long narrow crown, but becoming shrubby and prostrate at its altitudinal limit. Young twigs reddish pubescent; needles curved upward, blue-green, obtuse, with stomata on all sides; staminate flowers bluish; cones oblong, erect, purple-colored, with fan-shaped scales.

Common, sometimes dominant, on mountain slopes, chiefly in acid soil in interior, extending to about 1,500 meters. Described from interior northwestern North America.

Reported from localities marked on map, and from Copper River valley.

× ½

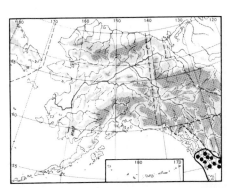 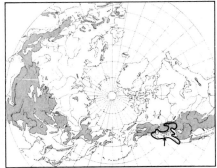

Thuja L.

1. Thùja plicàta D. Don   Western Red Cedar

A tall tree, up to 50 meters, with swollen base and conical crown; twigs flattened, fanlike; leaves decussate, yellowish-green above; cones small, brown when ripe, with up to 10 leathery scales, ripening in first year.

Mainly a lowland tree, but extending up to 1,000 meters, usually scattered in the hemlock–Sitka spruce forest, but sometimes in pure stands.

18. Cupressaceae (Cypress Family)

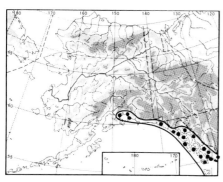

Chamaecyparis Spach

1. Chamaecýparis nootkaténsis (Lamb.) Spach — Alaska Cypress
Cypressus nootkatensis Lamb.

Medium-sized tree, 20–35 meters tall, with drooping branches; twigs 4-angled or somewhat flattened; leaves dull green, scalelike, appressed, pointed (occasionally acicular and patent on certain branches); cones small, globose, gray, with 4–6 scales, each tipped with conical projection, mostly ripening in second year.

Scattered in small groups in coastal forest from sea level up to about 1,000 meters, but best developed below 400 meters.

× ½

Juniperus L.

Leaves acicular in whorls of 3, divaricate . . . . 1. *J. communis* subsp. *nana*
Leaves scalelike, appressed, cuspidate . . . . 2. *J. horizontalis*

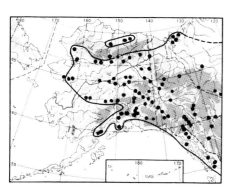
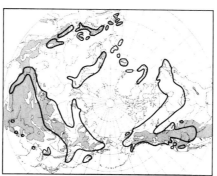

1. Juníperus commùnis L. — Common Mountain Juniper
subsp. **nàna** (Willd.) Syme

Juniperus nana Willd.; *J. communis* var. *montana* Ait.; *J. sibirica* Burgsd.; *J. communis* var. *saxatilis* Pall.

A usually prostrate shrub, forming large patches; leaves in whorls of 3, acicular, with white band above; cone ovoid to globose, green in first year, ripening to black in second or third year.

Dry slopes; extends to about 1,000 meters in McKinley Park, to 1,700 meters in eastern mountains. Described from Schlesien, Bohemia, and Altai.

Form of leaves varies considerably, from short, broad, curved, and abruptly short-pointed to straight and long-pointed. White band above is sometimes very broad, sometimes narrower (var. **depréssa** Pursh).

× ½

18. Cupressaceae (Cypress Family) / 19. Typhaceae (Cattail Family)

× ½

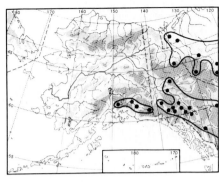

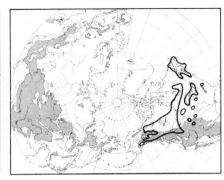

2. Juníperus horizontàlis Moench   Creeping Savin
Juniperus prostrata Pers.

A prostrate shrub; leaves blue-green, scalelike, appressed, cuspidate; fruits on short, recurved peduncles.

Rocky and sandy places. Described from cultivated specimens.

Young Alaskan specimens are said to have subulate, spreading leaves.

× ⅓

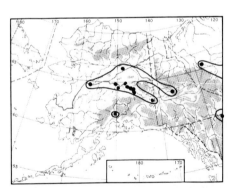

Typha L.

1. Tỳpha latifòlia L.   Cattail

Stout, more than 1 meter tall; leaves flat, 6–15 mm broad, light green, slightly longer than stem; staminate and pistillate spikes close together; pistillate spike thick, 2–3 cm in diameter, dark brown or reddish-brown; staminate spike narrower.

Shallow water, marshes, fairly common along Tanana River east to Big Delta. Described from Europe.

Young stems and spikes can be eaten.

20. Sparganiaceae (Bur Reed Family)

Sparganium L. (Bur Reed)

Stigmas 2 .... 1. *S. eurycarpum*
Stigma 1:
   Staminate head single; beak of fruit 0.5–1.5 mm long:
      Fruiting heads axillary; fruit with conical beak .... 4. *S. minimum*
      Fruiting heads, at least upper ones, supra-axillary; fruit nearly beakless ......
      .... 5. *S. hyperboreum*
   Staminate heads 2 or more; fruit with longer beak:
      Leaves rounded on back, 1.5–5 mm wide; upper leaves dilated at base ......
      .... 2. *S. angustifolium*
      Leaves flat, broader; upper leaves not dilated at base ....
      .... 3. *S. multipedunculatum*

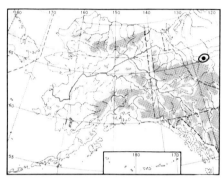

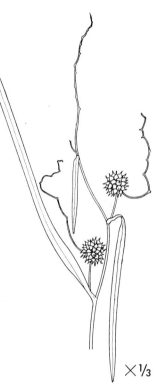

1. Spargànium eurycárpum Engelm.

Stout, ascending, with stiff, nearly flat leaves and forked inflorescence; branches with 1–3 pistillate and 2 to several staminate heads; stigmas 2; fruit cuneate, flat-topped, with stout beak in center.

Shallow water. Described from eastern United States.

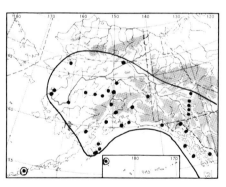
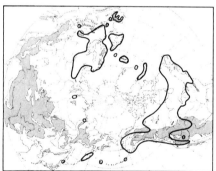

2. Spargànium angustifòlium Michx.

Sparganium affine Schnizl.; *S. angustifolium* Hult., of Fl. Alaska & Yukon, in part.

Slender, about 50 cm long, usually with submerged stems; leaves narrow, distinctly nerved, elongate, floating on surface of water, upper leaves dilated at base; staminate heads 2–3, congested; pistillate heads 2–3, lower heads peduncled; stigma long; fruit with beak about 2 mm long.

Deep or shallow water, from lowlands to alpine region. Described from Canada.

20. Sparganiaceae (Bur Reed Family)

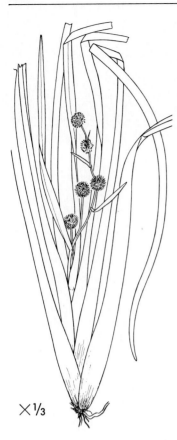

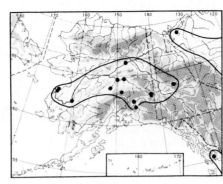 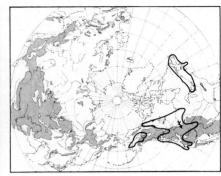

3. Spargànium multipedunculàtum (Morong) Rydb.

Sparganium simplex var. *multipedunculata* Morong; *S. simplex* and *S. angustifolium* Hult., of Fl. Alaska & Yukon, in part.

Similar to *S. angustifolium* but coarser, with large separated fruiting heads, larger stigmas, and broader leaves, scarcely dilated at base.

Deep or shallow water.

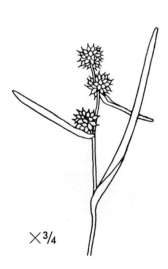

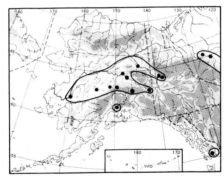 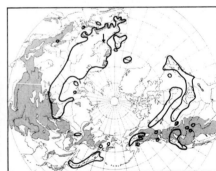

4. Spargànium mínimum (Hartm.) E. Fries

Sparganium natans var. *minimum* Hartm.

Up to 40 cm tall, leaves 3–5 mm broad, flat, lacking median nerve; staminate head short, peduncled, single; pistillate heads 2–3, remote, axillary, in Alaskan specimens the lower heads often long-peduncled; fruits elliptic to obovate, with short conic beak.

Shallow water, along creeks.

20. Sparganiaceae (Bur Reed Family) / 21. Potamogetonaceae (Pondweed Family)

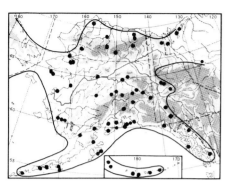 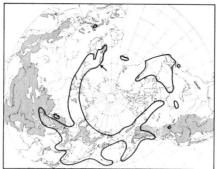 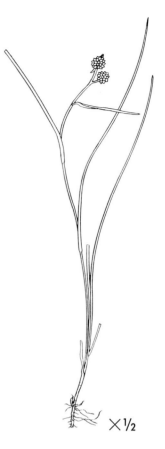

5. Spargànium hyperbòreum Laest.
Sparganium Williamsii Rydb.

About 15 cm tall, leaves 2–3 mm broad, thick, lacking median nerve; staminate head sessile, single; 1–3 pistillate heads close together, at least the upper heads supra-axillary, lower heads peduncled; fruits small, obovate, beakless, with short sessile stigma.

Shallow water, from lowlands to lower alpine region.

× ½

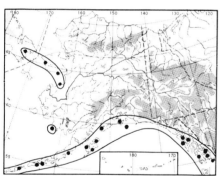 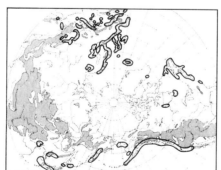 

Zostera L.

1. Zostèra marìna L.   Eelgrass

Plant monoecious; rootstock slender; stems branched, up to 2 meters long; leaves up to 10 dm long, 0.5–8 mm broad; in Alaskan specimens, 3–7-nerved, round at apex; seeds ovoid, ribbed.

Shallow shores in salt water. Described from "Baltic Sea, the Ocean."

Broad-leaved specimens [var. **latifòlia** Morong (*Z. pacifica* S. Wats.)] occur from Prince William Sound east. Critical study of larger sample, preferably with fruits, is desirable, in view of great variation.

× ⅓

21. POTAMOGETONACEAE (Pondweed Family)

×⅓

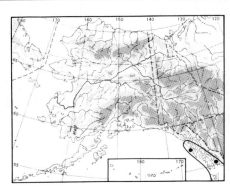
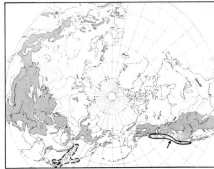

Phyllospadix Hook. (Surfgrass)

1. Phyllospàdix Scoulèri Hook.   Scouler's Surfgrass

Plant dioecious; rootstock thick; stem short, with inflorescence near summit, on short peduncle; leaves 1–2 meters long, 2–4 mm broad, with 3 primary nerves; spathe scarious-winged; fruit heart-shaped with winged margin.

Surf-beaten rocks in salt water.

Probably more common than could be expected from few collections taken. Broken line on circumpolar map indicates range of the closely related **P. iwaténsis** Makino.

Potamogeton L. (Pondweed)

Floating leaves present, submerged leaves always present on young plants (except in semiterrestrial forms), sometimes disappearing with age:
  Submerged leaves narrowly linear, more than 20 times as long as wide:
    Floating leaves cordate to rounded at base; petioles with brownish, curved, jointlike portion at junction with blade; submerged leaves nearly terete, linear, soon falling off .... 1. *P. natans*
    Floating leaves cuneate or tapering at base; petioles without jointlike portion at junction with blade; submerged leaves 2–10 mm wide, ribbonlike, flat, flaccid, and translucent .... 2. *P. epihydrus* var. *ramosus*
  Submerged leaves lanceolate to linear-lanceolate or oblong, less than 15 times as long as wide:
    Submerged leaves obtuse at apex, with margin entire; floating leaves gradually tapering into petioles; plant suffused with red .... 3. *P. alpinus* subsp. *tenuifolius*
    Submerged leaves acute to sharp-pointed at apex, with 1-celled, translucent denticles in margin, at least when young; floating leaves cuneate to rounded at base; plant greenish .... 4. *P. gramineus*
Floating leaves absent:
■ Submerged leaves narrowly linear, with parallel sides, more than 20 times as long as wide:
  Stipules adnate to leaf at base:
    Leaves 3–8 mm wide, stiffly 2-ranked, 15–35-nerved, finely and sharply serrate, blades auricled at juncture with stipule .... 7. *P. Robbinsii*
    Leaves 0.2–2 mm wide, 1–3-nerved, blades not auricled, margins entire:
      Primary stipular sheaths loose, conspicuous, 2–5 times as wide as stem, not connate; spike with 5–12 whorls of flowers .... 15. *P. vaginatus*
      Primary stipular sheaths tight, scarcely wider than stem; spike with 2–6 whorls of flowers:
        Leaves tapering to acute apex; fruits 2.5–4 mm long; margins of stipule overlapping, but not connate; fruit with short beak .... 13. *P. pectinatus*
        Leaves obtuse and blunt at apex; fruits 2.0–2.5 mm long; young stipules with connate margins; beak of fruit almost lacking, wartlike .... 14. *P. filiformis*
  Stipules free from leaf:
● Submerged leaves 2–10 mm wide, with prominent cellular median band .... 2. *P. epihydrus* var. *ramosus*

21. POTAMOGETONACEAE (Pondweed Family)

● Submerged leaves 0.5–5 mm wide:
  Leaves with 9–35 nerves:
    Stem strongly flattened; leaves 2–5 mm wide, 15–35-nerved; fruits 4–5 mm long with margined beak 0.6–1 mm long .... 8. *P. zosterifolius* subsp. *zosteriformis*
    Stem nearly terete; leaves 1.5–2 mm wide, 9–17-nerved; fruits 3–4 mm long with median beak 0.3–0.5 mm long ...... 9. *P. subsibiricus*
  Leaves with 1–7 nerves:
    Submerged leaves firm, semiterete, with obtuse apex; nerves obscure .... 1. *P. natans*
    Submerged leaves thin, flat, flaccid, with acute or mucronate apex; nerves evident:
      Stipules strongly fibrose, becoming whitish; spikes interrupted-cylindric, with 3–4 whorls of flowers; dorsal keel of fruit obscure .... 11. *P. Friesii*
      Stipules delicate; spikes capitate to short-cylindric, compact, with 1–3 whorls of flowers:
        Fruits with rounded or obscure keels; apex of leaves acute or obtuse, but mucronate; margins of stipules free; stems often with pair of glands at nodes .... 10. *P. Berchtoldii*
        Fruits with thin or acute, undulate-to-dentate dorsal keel; apex of leaves acute; margins of young stipules connate; stems usually without glands at nodes .... 12. *P. foliosus*
■ Submerged leaves without parallel sides, less than 15 times as long as wide:
  Submerged leaves sessile or petiolated, but not clasping the stem:
    Submerged leaves obtuse at apex, with entire margin; floating leaves gradually tapering into petiole; plant suffused with red .... 3. *P. alpinus* subsp. *tenuifolius*
    Submerged leaves acute to sharp-pointed at apex, with one-celled, translucent denticles in margin, at least when young; floating leaves cuneate to rounded at base; plant green.................... 4. *P. gramineus*
  Submerged leaves sessile and perfoliate, clasping the stem:
    Leaves ovate-oblong, mostly 10–20 cm long, apex often cucullate; stipules usually persistent, long, and conspicuous; fruits 4–5 mm long, with strongly developed dorsal keel .... 5. *P. praelongus*
    Leaves ovate or elongate-ovate, 3–10 cm long, apex rounded but not cucullate; stipules soon disintegrating to fibers; fruits 2.5–3.5 mm long, with weakly developed or obscure dorsal keel .... 6. *P. perfoliatus* subsp. *Richardsonii*

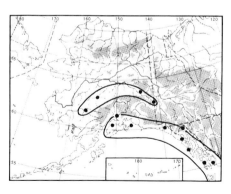
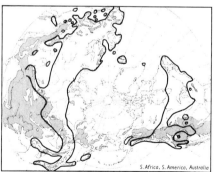

1. Potamogèton nàtans L.

Submerged leaves linear, semiterete, with no differentiation of blade and petiole, very different from ovate-to-oblong, long-petiolated, lustrous, grayish-green, coriaceous, many-nerved floating leaves; stipules long, longer in floating leaves than in submerged leaves; blade of submerged leaves soon perishes, leaving carinate, acute petioles, which persist; spike 4–5 cm long, with green fruits; beak short and broad.
  Lakes and streams with sluggish water. Described from Europe.
  Apparently rare in Alaska.

21. POTAMOGETONACEAE (Pondweed Family)

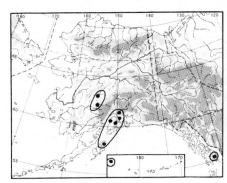 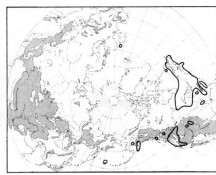

2. Potamogèton epihỳdrus Raf.
var. ramòsus (Peck) House

Potamogeton Nuttallii var. *ramosus* Peck; *P. Nuttallii* Cham. & Schlecht.; *P. epihydrus* var. *Nuttallii* (Cham. & Schlecht.) Calder & Taylor.

Stem compressed, simple or slightly branched; submerged leaves linear, ribbon-like, 2–8 mm broad, 3–7-nerved, strongly distichous on sterile shoots; floating leaves long-petiolated, narrowly oblong, or oblong-lanceolate, rounded at tip, transitional leaves often present; stipules hyaline, up to 4 cm long; fruits flattened laterally, 3-keeled.

Lakes and streams. *P. epihydrus* described from Ohio River, var. *ramosus* from Oswegatchie River (northwest New York State).

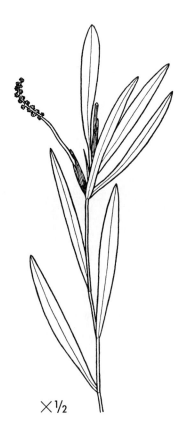

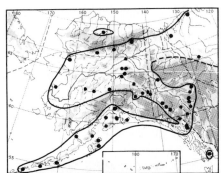 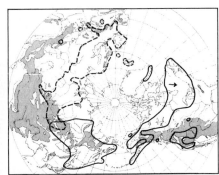

3. Potamogèton alpìnus Balb.
subsp. tenuifòlius (Raf.) Hult.

Potamogeton tenuifolius Raf.

Plant rust-colored, at least toward top; stem usually branching at apex, with linear-lanceolate, obtuse, submerged leaves, much longer than internodes; floating leaves, when developed (rarely present in Alaskan specimens) elliptic to obovate, blunt; peduncles not thicker toward apex; spike reddish-brown; fruit ovate, with distinct, bent beak.

Lakes and ponds. *P. alpinus* described from northern Italy, subsp. *tenuifolius* from Lake Mistassini.

Broken line on circumpolar map indicates range of subsp. **alpìnus.**

21. POTAMOGETONACEAE (Pondweed Family)

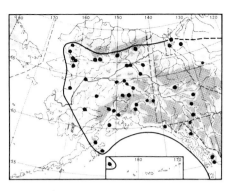  

4. Potamogèton gramíneus L.
Potamogeton heterophyllus of authors.

Stem usually branched; submerged leaves lanceolate to linear, acute, with long point; floating leaves (sometimes missing) long-petiolated, ovate-to-elliptic, acute or acutish, coriaceous, green; peduncles thicker toward apex; spike green, compact; fruits with short, recurved beak.

Lakes and ponds, usually in deeper water. Described from Europe.

Hybrids with *P. perfoliatus, P. praelongus,* and *P. alpinus* occur.

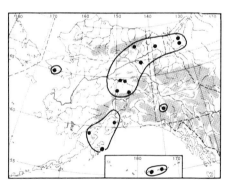 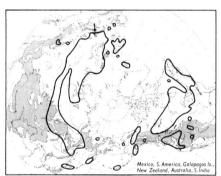 

5. Potamogèton praelóngus Wulf.

Coarse, light-colored, branched; stem bent zigzag when young; all leaves submerged, large, ligulate to lanceolate-oblong, translucent, often somewhat cordate at base; peduncles long, not thicker toward apex; fruit dark green. Stipules long, very prominent, persistent.

Deep water in lakes.

21. POTAMOGETONACEAE (Pondweed Family)

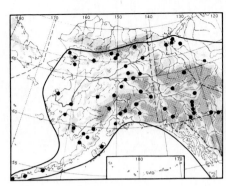 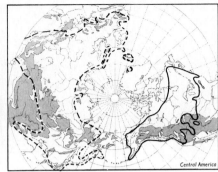

6. Potamogèton perfoliàtus L.
subsp. **Richardsònii** (Bennett) Hult.

Potamogeton perfoliatus var. *Richardsonii* Bennett; *P. Richardsonii* (Bennett) Rydb.

All leaves submerged; stem densely covered with ovate-lanceolate, dark-green, often undulate leaves clasping stem all around; stipules short, soon disintegrating into fibers; peduncles short, curved, not thicker toward apex; spike brown, 1–2 cm long.

Lakes. *P. perfoliatus* described from Europe, subsp. *Richardsonii* from the Great Lakes.

Alaskan fruiting specimens have cavity in endosperm loop, a characteristic of subsp. *Richardsonii*. Leaves shorter in some specimens, as in subsp. *perfoliatus* of Eurasia; longer in others, as in American plants. Hybrids with *P. gramineus* not uncommon. Broken line on circumpolar map indicates range of subsp. **perfoliàtus**.

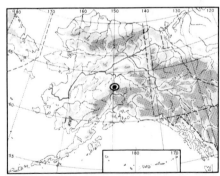 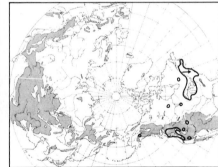

7. Potamogèton Robbínsii Oakes

Stem rooting at lower nodes, freely and widely branching, stiffly featherlike, closely invested by sheathing; stipules whitish; leaves auriculate at base, many-nerved, with thin, cartilaginous, minutely serrated margin; spike stiff, interrupted; fruit prominently keeled, with beak 1 mm long.

Muddy water. Only sterile specimens known from area of interest. Collected only once, at Summit.

21. POTAMOGETONACEAE (Pondweed Family)

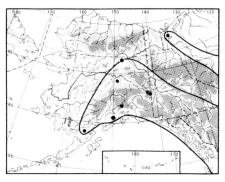 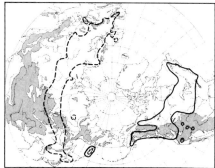 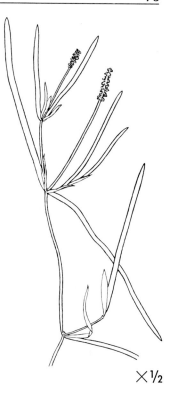

8. Potamogèton zosterifòlius Schum.
subsp. **zosterifórmis** (Fern.) Hult.
Potamogeton zosteriformis Fern.

Stem distinctly flattened, constricted at nodes; leaves all submerged, grasslike, obtuse or subacute, with 3–5 prominent and many fine nerves; stipules 1.5–3.5 cm long, fibrous; peduncles short; spike cylindrical; fruit with more or less dentate dorsal keel and short marginal beak.

Lakes and ponds. *P. zosterifolius* described from Sjaelland (Denmark), subsp. *zosteriformis* from America.

Broken line on circumpolar map indicates range of subsp. **zosterifòlius**.

× ½

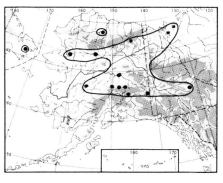  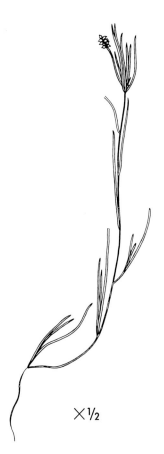

9. Potamogèton subsibìricus Hagstr.
Potamogeton Porsildorum Fern.

Stem filiform, compressed, simple or slightly branched in upper part; leaves linear, subacute or mucronate, with 9–17 nerves and prominent median nerve; stipules 1–2 cm long; peduncles short; spike cylindrical, up to 1.3 cm long, with 3–4 whorls; fruit 3–4 mm long, oblong-ovoid, carinate on back, with beak 0.3–0.5 mm long.

Shallow water. Range of this species still not well known. Described from lower Yenisei River.

× ½

21. POTAMOGETONACEAE (Pondweed Family)

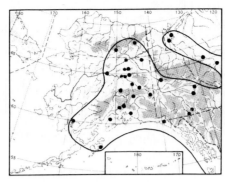
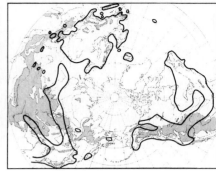

10. Potamogèton Berchtóldi Fieb.
Potamogeton pusillus auct.

An extremely variable species. Stem capillary, much branched, with short internodes; leaves 0.5–2 mm broad, with 2 glands at base, 3-nerved, obtuse or acute; stipules tubular, with margins united beyond middle, falling off in age; spikes compact or interrupted, on short peduncles.

Shallow water. Described from Bohemia.

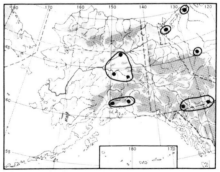
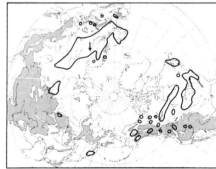

11. Potamogèton Frièsii Rupr.
P. mucronatus Schrad.

Similar to *P. Berchtoldii*, but leaves mucronate, narrower at base, with 5 nerves; spike usually with 3 somewhat remote whorls.

Shallow water.

21. POTAMOGETONACEAE (Pondweed Family)

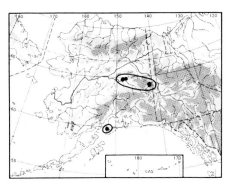
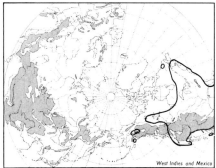

12. Potamogèton foliòsus Raf.

Stem filiform, compressed, simple or slightly branched; leaves narrowly linear, lacking basal glands, 3–5-nerved with prominent midrib; peduncles slightly thickened upward, 3–10 mm long; spike subcapitate with 2–3 closely spaced whorls of two flowers each; fruit suborbicular, compressed, 2–2.5 mm long, with undulate or dentate dorsal keel and very short beak.

Shallow water. Described from South Carolina.
Alaskan specimens belong to var. **macéllus** Fern.

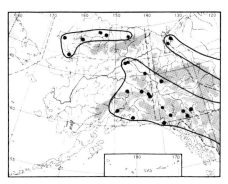
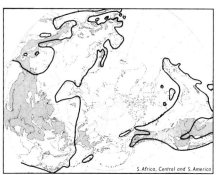

13. Potamogèton pectinàtus L.

Stem setaceous, long, mostly branching in upper part; leaves filiform, brownish-green, those of branches tapering to acute tip; peduncles filiform, long; stipules adnate to base of leaf; spike with about 5 remote whorls; fruit about 4 mm long, with slightly curved beak.

Saline, brackish, or calcareous waters. Described from Europe.

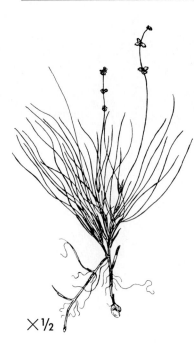
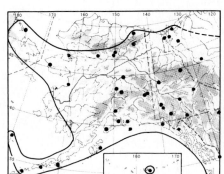

14. Potamogèton filifórmis Pers.

Stem filiform, branched especially at base; stipules adnate to base of leaf; sheaths tightly clasping, not much thicker than stem; leaves setaceous, brownish-green, blunt; peduncles filiform, long; spike with 3–4, usually widely separated whorls; fruit about 2 mm long, beakless.

Shallow water. Range in Asia preliminary; plant very scattered there.

Most specimens from area of interest are the northern var. **boreàlis** (Raf.) St. John (*P. borealis* Raf.), with shorter, more compact spike of greater length than leaves; low growth, which seems to be merely a growth form, is apparently an adaptation to northern conditions.

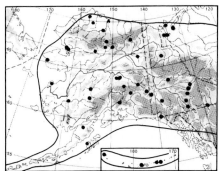
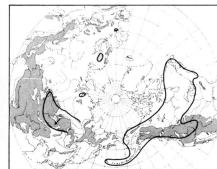

15. Potamogèton vaginàtus Turcz.

Potamogeton interior Hult., of Fl. Alaska & Yukon.

Stem coarser than in *P. filiformis* or *P. pectinatus*, long, mostly branching in upper part; leaves setaceous, 1–2 mm broad, with 3–5 nerves, blunt or acutish; stipules adnate to base of leaf; sheaths loose, 2–4 cm long, thicker than stem; spike with 5–12 whorls, fruit with short beak.

Deep water. Described from Dahuria and Selenga.

Hybrids with *P. filiformis* and *P. pectinatus* occur.

21. POTAMOGETONACEAE (Pondweed Family)

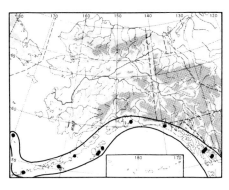

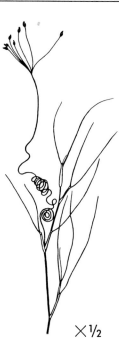

Ruppia L.

1. Rúppia spiràlis L. Ditch Grass
Ruppia occidentalis S. Wats.

Stem capillary, branched; leaves submerged, linear-capillary, with membranous sheaths; peduncles long, spiraling; fruit pear-shaped, about 2 mm long, tipped by straight style.

Salt and brackish water.

Since submerged plants of the seacoast are rarely collected, it seems probable that this species is common along southern coast of Alaska, though few collections have been taken. Broken line on circumpolar map indicates range of closely related **R. marítima** L. var. **lóngipes** Hagstr.

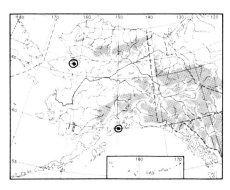
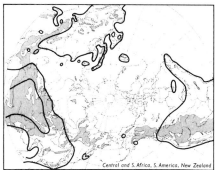

Zannichellia L.

1. Zannichéllia palústris L. Horned Pondweed

Stem branched, capillary, leaves submerged, linear, filiform, acute, with sheathed membranaceous stipules; male flowers with 1 stamen, female usually with 4 carpels; fruit short-stalked, straight or slightly curved, smooth or slightly dentate on back, about 2 mm long, with slender beak.

Mostly in brackish water. Described from Europe and Virginia.

Probably considerably more common than would be suggested by two collections known from area of interest.

22. Juncaginaceae (Arrow Grass Family)

Triglochin L. (Arrow Grass)

Fruit with 6 carpels, ovoid .... 1. *T. maritimum*
Fruit with 3 carpels, linear .... 2. *T. palustre*

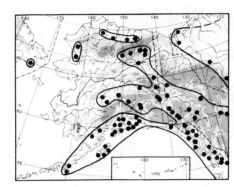
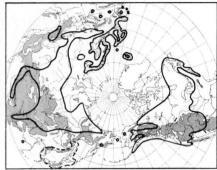

1. Triglòchin marítimum L.

Rootstock oblique, covered with whitish leaf bases, and with several fleshy leaves at top; scapes thick, in Alaska up to 90 cm tall, but usually much shorter; fruit ovoid, with 6 carpels.

Meadows and brackish marshes, common along coast, but also on marshes and shores inland. Described from Europe.

Broken line on circumpolar map indicates range of subsp. **asiáticum** Kitagawa.

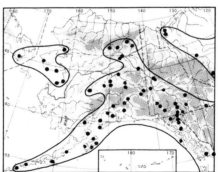
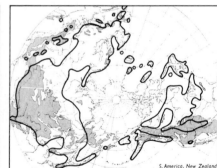

2. Triglòchin palústris L.

Rootstock short, bulblike, emitting filiform stolons with bulbs (rarely present on herbarium specimens), which grow out to new plants the following year; leaves filiform; scapes thin and short in flowering state, longer in fruit; fruit narrow, with 3 carpels (which separate from below upward when ripe), awl-pointed at base.

Wet places around fresh or brackish water. Described from Europe.

23. SCHEUCHZERIACEAE (Scheuchzeria Family) / 24. ALISMACEAE (Water Plaintain Family)

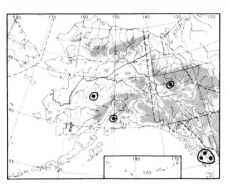

Scheuchzeria L.

1. Scheuchzèria palústris L.
subsp. **americàna** (Fern.) Hult.

Scheuchzeria palustris var. *americana* Fern.

Rootstock yellowish-gray, creeping, ending with single unbranched stem with terete, tubular, sheathed leaves; whole plant yellowish-green.

Bogs and quagmires, apparently rare in Alaska. S. *palustris* described from Lapland, Sweden, Switzerland, and Prussia.

Subsp. *americana* differs from European type in having somewhat longer follicles and a more distinct curving beak, and thus constitutes a slightly different race, but the difference is not marked in Alaskan specimens. Broken line on circumpolar map indicates range of subsp. **palústris**.

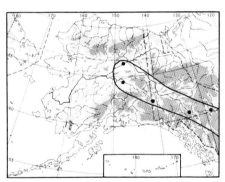

Sagittaria L.

1. Sagittària cuneàta Sheld.   Arrowhead

Terrestrial forms have numerous sagittate leaves on slender peduncles. Alaskan specimens seen are water plants with sagittate or lanceolate leaves, lacking basal lobes, with submerged leaves reduced to phyllopodia; lower whorl or whorls of flowers are pistillate, upper whorls staminate, with lanceolate or narrowly ovate bracts; filaments glabrous, achenes obovate, with wide, rounded, dorsal keel.

Muddy shores, in shallow water. Described from Otter Tail County, Minnesota.

Phalaris L. (Canary Grass)

Inflorescence lobed; plant perennial, with rhizome; glumes wingless ....
.... 1. *P. arundinacea*
Inflorescence spikelike, ovoid; plant annual; glumes winged:
  Wing with entire margin, 2 sterile lemmas .... 2. *P. canariensis*
  Wing of glumes toothed in margin, only 1 sterile lemma .... 3. *P. minor*

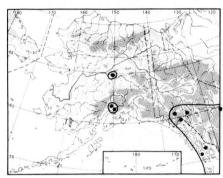 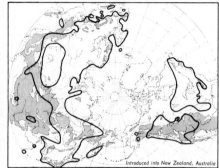

1. Phálaris arundinàcea L.   Reed Canary Grass
Digraphis arundinacea (L.) Trin.

Rhizome stout, creeping; leaves flat, broad; ligules 5–6 mm long; panicle with all spikelets turned in same direction; spikelets green, in age violet.

Wet meadows, along creeks, shores. Described from Europe.

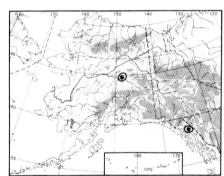 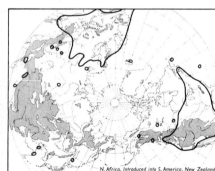

2. Phálaris canariénsis L.   Canary Grass; Birdseed Grass

Annual, 20–50 cm tall; leaves 5–6 mm broad, bluish-green, with inflated retrorsely scabrous upper sheath; glumes broadly wing-shaped, with broadly winged, smooth-margined keel; fertile lemma brownish, pubescent, lanceolate-acuminate; palea about 3 mm long.

Adventive or as weed, introduced so far only in few places. Described from Canary Islands.

25. GRAMINEAE (Grass Family)

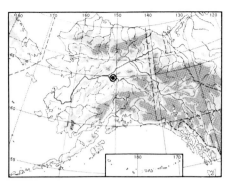
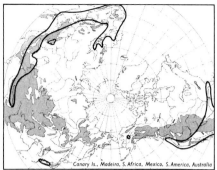

3. Phálaris mìnor Retz.

Resembling *P. canariensis*, but with narrower, grayish-green leaves, narrower panicle and spikelets; glumes with narrower wings, toothed in margin; palea very small.

Introduced at Manley Hot Springs, west of Fairbanks. Described from Europe.

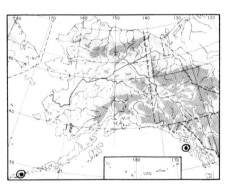

Anthoxanthum L.

1. Anthoxánthum odoràtum L.   Sweet Grass

Culms tufted, 20–40 cm tall, leaves light green, 2–4 mm broad, rough above; panicle spikelike, very acute; spikelets brownish-yellow with 1 perfect floret, 2 sterile lemmas, and only 2 stamens.

Introduced at Unalaska village and Sitka, probably also elsewhere along the southern coast. Described from Europe.

Dry plant is very sweet-scented (cumarin).

25. GRAMINEAE (Grass Family)

Hierochloë R. Br. (Holy Grass)

Second lemma bears an exserted geniculate and twisted awn . . . . 1. *H. alpina*
Lemmas with short point, but without exserted geniculate awn:
   Panicle broad, ovoid; leaves broad, flat; lemmas entire . . . . 2. *H. odorata*
   Panicle small, linear; leaves narrow, involute; lemmas erose . . . . . 3. *H. pauciflora*

× ⅓

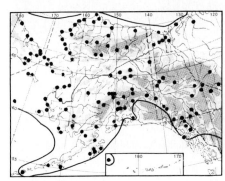
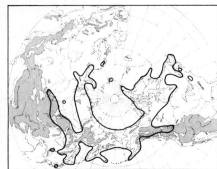

1. Hieróchloë alpìna (Sw.) Roem. & Schult.

Holcus alpinus Sw.

Culms about 20 cm tall, tufted, with reddish base; basal leaves involute; stem leaves with very short, finely pubescent blades; panicle contracted, 2–3 cm long; spikelets olivaceous to purple; second lemma bears exserted awn, bent below middle.

Alpine meadows and heaths and on rocky slopes, from arctic lowlands to mountains, to at least 1,800 meters. Described from Lapland.

Common all over area of interest, except in southeastern Alaska and middle Aleutians. Viviparous specimens occur rarely. Circumpolar map indicates range of *H. alpina*, in a broad sense, including **H. monticòla** (Bigel.) Löve & Löve (*Holcus monticola* Bigel.) and **H. orthántha** Sørens.

× ⅓

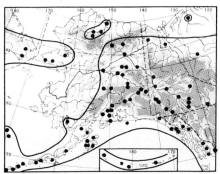
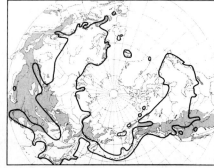

2. Hieróchloë odoràta (L.) Wahlenb.   Vanilla Grass

Holcus odoratus L. Including *Hierochloë fragrans* (Willd.) Roem. & Schult.; *Holcus fragrans* Roem. & Schult.

Culms about 20–40 cm tall, rising from slender, underground rhizome; basal leaves long, soft, cauline, short, usually 3; panicle pyramidal, longer and broader than in *H. alpina*; spikelets brownish to purple, 3–5 mm long, mostly awnless.

Spruce forests, riverbanks, to about 1,000 meters in McKinley Park. Described from Europe.

Common on southern coast, more rare to north. Circumpolar map indicates range of entire species complex, including subsp. **dahùrica** (Trin.) Printz (*H. dahurica* Trin.).

25. GRAMINEAE (Grass Family)

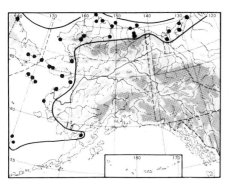

3. Hieróchloë pauciflòra R. Br.

Culms 10–20 cm tall, rising from slender, underground rhizome; basal leaves linear, involute, glabrous; panicle narrow, 1–3 cm long, few-flowered; spikelets rounded, more or less brownish; glumes broadly ovate, lemmas awnless.

Margins of freshwater pools on tundra.

Stipa L. (Feather Grass)

Lemma 8–12 mm long; awn 10–15 cm long .... 1. *S. comata*
Lemma less than 7 mm long; awn much shorter, twice geniculate:
    Panicle open, diffuse; branches divergent, lemma about 5 mm long; awn about 2 cm long .... 2. *S. Richardsonii*
    Panicle narrow; branches appressed; lemma 6–7 mm long; awn 2–2.5 cm long .. .. 3. *S. columbiana*

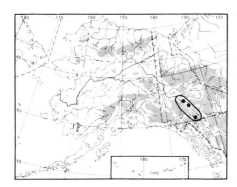
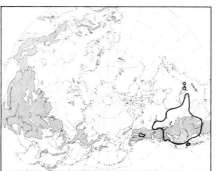

1. Stìpa comàta Trin. & Rupr.

Culms 2–6 dm tall, with crowded sheaths at base, upper sheaths including panicle, often loose; basal leaves involute, filiform, usually about half length of culm; culm leaves 0.5–1.5 dm long, 2–4 mm wide, flat or involute; panicle loose, branches erect or spreading, naked below; glumes tapering into slender awn; lemma sparsely pubescent, with long awn.

Roadsides; introduced into southern Yukon. Described from Carlton House and Missouri Portage.

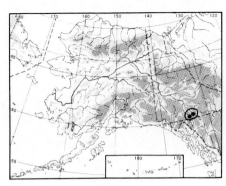 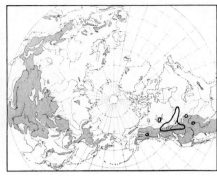

2. Stìpa Richardsònii Link

Culms 4–10 dm tall; basal leaves 13–25 cm long, involute, filiform, scabrous; panicle 10–20 cm long, open, with slender, widely spaced, spreading or sometimes drooping branches, naked below; glumes 8–9 mm long; lemma about 5 mm long, scabrous and pubescent, brown at maturity; awn 2.5–3 cm long.

Meadows, woods. Described from cultivated specimens from western North America.

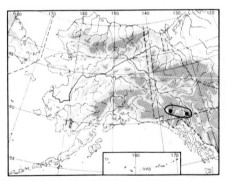 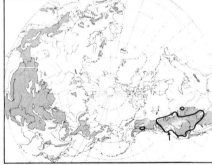

3. Stìpa columbiàna Macoun

Culms 3–5 dm tall or more; basal leaves filiform, involute; culm leaves involute or flat; panicle narrow, dense, purplish; glumes about 1 cm long, short-awned; lemma 6–7 mm long, pubescent, with awn 2–2.5 cm long.

Probably introduced.

25. GRAMINEAE (Grass Family)

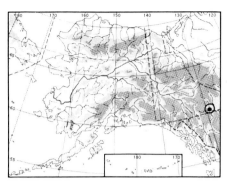
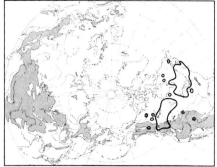

Oryzopsis Michx.

1. Oryzópsis púngens (Torr.) Hitchc.   Mountain Rice
Milium pungens Torr.

Culms 15–60 cm tall, forming dense tussocks; sheaths smooth or slightly scabrous, crowded at base; basal leaves filiform, involute, half as long as culm or longer, culm leaves short; branches of panicle ascending or erect; glumes obtuse, lemma as long as glumes, appressed-pubescent; awn 0.5–2 mm long, straight; palea as long as lemma.

Sandy or rocky soil. Described from New York ("Schenectady in Massachusetana").

Muhlenbergia Schreb.

Leaves filiform, 1–2 mm wide, involute; glumes not longer than lemma .... 1. *M. Richardsonis*
Leaves broader, flat; glumes much shorter than lemma .... 2. *M. glomerata* var. *cinnoides*

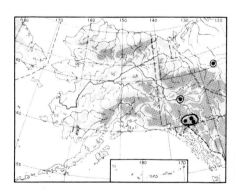
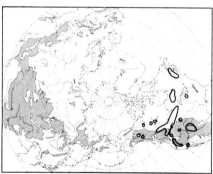

1. Muhlenbérgia Richardsònis (Trin.) Rydb.
Vilfa Richardsonis Trin.

Culms 2–4 dm tall, matted; leaves 1–2 mm wide, erect, ascending, involute in age; panicle linear, stiff, interrupted; glumes much shorter than attenuate, slender-pointed, tightly rolled, glabrous lemma.

Gravelly shores. Described from North America (Richardson).

×⅓

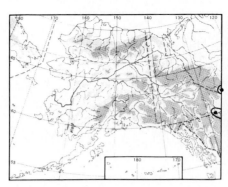
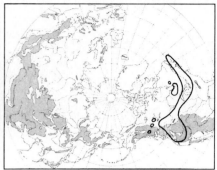

2. Muhlenbérgia glomeràta (Willd.) Trin.
Polypogon glomeratus Willd.
var. **cinnoìdes** (Link) Herm.
Dactylogramma cinnoides Link.

Culms 3–8 dm tall, from slender rhizome; leaves 2–5 mm wide, firm, erect, scabrous; panicle greenish or purplish, long-peduncled, interrupted, with remote lower branches; glumes with scabrous keel and awn; lemma villous below middle.

Hummocks in wet places. *M. glomerata* described from North America, var. *cinnoides* from cultivated specimens from western North America.

Total range little known, very approximate.

Phleum L.

Sheaths of upper culm leaf inflated; panicle short, cylindrical, or oblong ....
.... 1. *P. commutatum* var. *americanum*
Sheaths of upper culm leaf not inflated; panicle cylindrical .... 2. *P. pratense*

×⅓

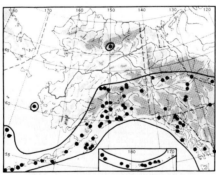
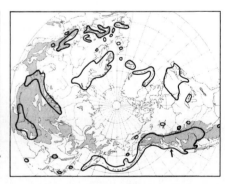

1. Phlèum commutàtum Gandoger   Mountain Timothy
var. **americànum** (Fourn.) Hult.
    Phleum alpinum L. in part; *P. Haenkeanum* Presl; *P. alpinum* var. *americanum* Fourn.

Culm up to 60 cm tall, usually lower; panicle short, cylindrical or oblong, rarely more than 4 times longer than broad, usually shorter, green to purplish; upper sheaths inflated; awns two-thirds to three-fourths as long as body of glumes; anthers light-colored.

Mountain slopes, alpine meadows, to about 1,700 meters; common along southern coast, more rare inland.

Var. *americanum* usually differs from circumpolar plant in having taller height, stronger inflated sheaths, and shorter awns on glumes; populations around Pacific belong to this type.

25. GRAMINEAE (Grass Family)

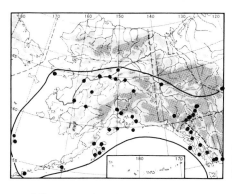
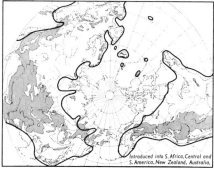

2. Phlèum praténse L. Timothy

Culm up to 1 meter tall; panicle slender-cylindrical, long; sheaths not inflated; awns about half as long as body of glumes; anthers dark-colored.

Fields, roadsides; introduced and partly naturalized in many inhabited places. Described from Europe.

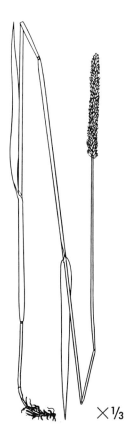

×½

Alopecurus L. (Foxtail)

Plant decumbent; spike narrow; anthers 1–1.5 mm long:
    Awn included in spikelet or nearly so, anthers orange-colored ..... 5 *A. aequalis*
    Awn long-exserted, geniculate, anthers yellow or violet .... 6. *A. geniculatus*
Plant erect, or decumbent at base; spike broader; anthers longer:
    Glumes with long hairs on keel, pubescent on nerves, long hairs occasionally alternating with short hairs; awn long-exserted .... 1. *A. pratensis*
    Glumes densely woolly:
        Plant 10–30 cm tall:
            Panicle oblong, narrow; spikelets grayish, villous; awn short, straight, fixed at middle of lemma or missing .... 2. *A. alpinus* subsp. *alpinus*
            Panicle thicker; spikelets yellowish, villous; glumes long, acute; awn long, geniculate, fixed close to base of lemma ....
            .... 4. *A. alpinus* subsp. *Stejnegeri*
        Plant taller, 30–60 cm .... 3. *A. alpinus* subsp. *glaucus*

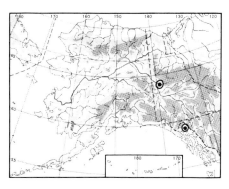
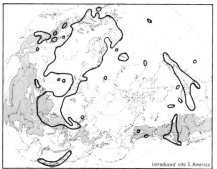

1. Alopecùrus praténsis L.

Culms 5–10 dm tall, from short subterranean runners; leaf blades 4–8 mm broad, attenuate, often drooping, scabrous in margin and on nerves; panicle 4–10 cm long, 6–10 mm broad; glumes united below middle, tapering to somewhat incurved apex, with green or often dark-violet nerves, glabrous or finely scabrous, and white-pubescent on nerves and keel; awn exserted 4–6 mm; anthers 3–3.5 mm long.

Cultivated fields, roadsides; introduced. Described from Europe.

×⅓

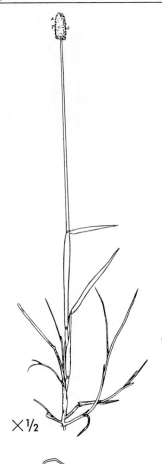
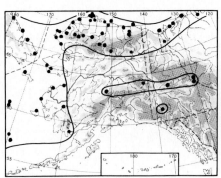
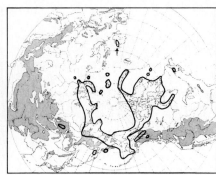

2. Alopecùrus alpìnus Sm.   Alpine Foxtail
subsp. **alpìnus**
Alopecurus behringianus Gandoger; *A. borealis* Trin.

Culms with 2–3 internodes, 10–30 cm tall, from slender rhizome; sheaths strongly inflated, glabrous; upper leaves short and broad; panicle oblong, narrow, grayish or purplish-gray; spikelets strongly woolly; glumes free nearly to base; anthers 1.5–2 mm.

Wet, sandy, and stony soil in tundra or mountains.

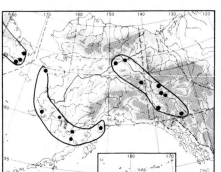
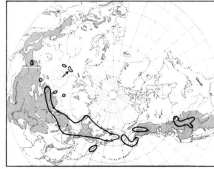

3. Alopecùrus alpìnus Sm.
subsp. **glaùcus** (Less.) Hult.
Alopecurus glaucus Less.; *A. occidentalis* Scribn. & Tweedy.

Similar to *A. alpinus*, but taller, 30–60 cm long; panicle longer; more internodes and stem leaves; prominent awn; scabrous leaves.

Wet, sandy, and stony soil.

Transgressions to subsp. *alpinus* occur in both Alaska and Asia.

25. GRAMINEAE (Grass Family)

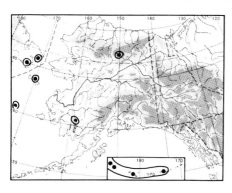 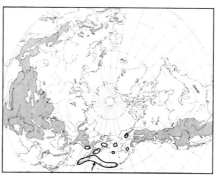 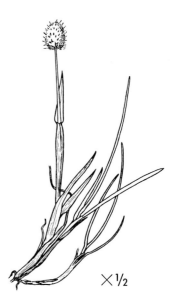

4. Alopecùrus alpìnus Sm.
subsp. **Stejnégeri** (Vasey) Hult.
  Alopecurus Stejnegeri Vasey; *A. alpinus* var. *Stejnegeri* (Vasey) Hult.

  Similar to *A. alpinus,* but panicle broader, ovoid, yellowish, villous; glumes long, acute; awn long, geniculate, fixed near base of lemma.
  Wet, sandy, and stony soil, chiefly in western Aleutian Islands.
  Sheaths very inflated in extreme specimens, upper leaf blades very short and broad. Transgressions to subsp. *alpinus* occur.

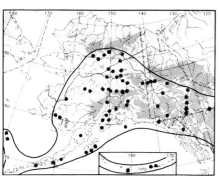 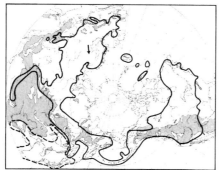 

5. Alopecùrus aèqualis Sobol.
  Alopecurus Macounii Merr.; *A. Howellii* var. *Merriamii* Beal.

  Lower part creeping, more or less glaucous, biennial or perennial; upper sheaths inflated; upper stem leaf about as long as lower; spike light-colored, with glumes 2 mm long; anthers orange-colored, in age yellowish-white.
  Wet soil or, with floating leaves, in ponds (f. **nàtans** Wahlenb.).
  Broken line on circumpolar map indicates range of subsp. **amurénsis** (Kom.) Hult. (*A. amurensis* Kom.).

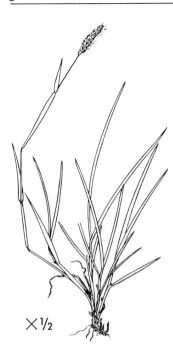
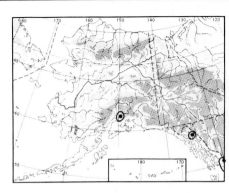
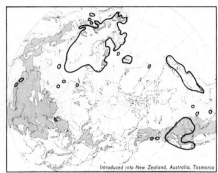

6. Alopecùrus geniculàtus L.

Lower part creeping, often rooting at nodes, grass-green, biennial or perennial; upper sheaths inflated; upper stem leaf short; spike dark green or violet with glumes 3 mm long; anthers yellow or violet, in age brownish.

Wet soil; introduced. Described from Europe.

Phippsia R. Br.

Lemma glabrous; fruit broadest above middle .... 1. *P. algida*
Lemma white-haired on nerves; fruit broadest at base .... 2. *P. concinna*

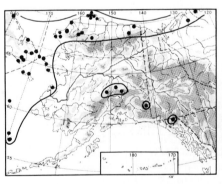
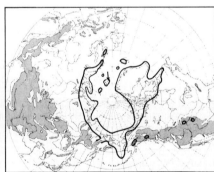

1. Phíppsia álgida (Soland.) R. Br.   Snow Grass

Agrostis algida Soland.; *Catabrosa algida* (Soland.) T. Fries; *Phippsia monandra* Hook.; *Vilfa monandra* Trin.

Culm 2–10 cm tall; plant densely tufted, more or less decumbent; leaves flat, shorter than culm; panicle narrow, up to 4 cm long, contracted; spikelets 1-flowered, about 1 mm long, greenish-yellow, sometimes purplish; lemma glabrous or very faintly pubescent; fruit broadest above middle; anthers 0.4 mm long.

Bogs and wet places, snow beds, mostly on tundra.

25. GRAMINEAE (Grass Family)

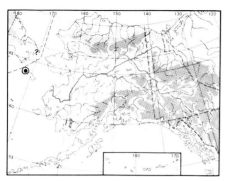 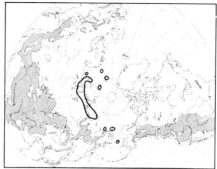 

2. Phíppsia concínna (T. Fries) Lindeb.

Catabrosa concinna T. Fries; *Phippsia algida* var. *concinna* (T. Fries) Richter.

Similar to *P. algida*, but often taller, panicle more branched, with divergent branches; lemma white-haired on nerves; palea densely ciliated; fruit broadest at base, usually longer than lemma; anthers 0.5 mm long.

Bogs and wet places, snow beds on tundra. Described from Spitzbergen.

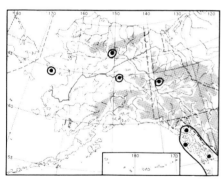 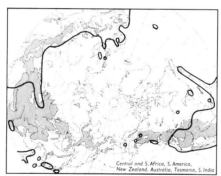 

Polypogon Desf. (Beard Grass)

1. Polypògon monspeliénsis (L.) Desf.

Alopecurus monspeliensis L.

Annual; plant branched at base; culm up to 6 dm tall; sheaths inflated; ligules long; panicle 3–5 cm long, dense, interrupted; spikelet 1-flowered; glumes about 2 mm long, 2-lobed, with awn 5 mm long.

Wasteland; introduced. Described from Montpellier, France.

25. GRAMINEAE (Grass Family)

Arctagrostis Griseb. (Polar Grass)

Anthers 2 mm long or longer, dark purple; spikelets 4 mm long or longer; panicle usually contracted .... 1. *A. latifolia* var. *latifolia*
Anthers about 1.5 mm long; spikelets usually smaller:
    Spikelets more or less purplish; glumes and lemmas narrow, not hyaline; panicle branches more or less scabrous............ 2. *A. latifolia* var. *arundinacea*
    Spikelets mostly pale, small; glumes and lemmas broad, hyaline; anthers 1.2–1.5 mm, yellow .... 3. *A. poaeoides*

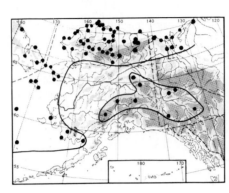

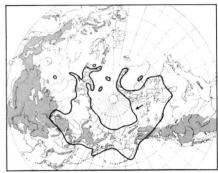

1. Arctagróstis latifòlia (R. Br.) Griseb.
Colpodium latifolium R. Br.
var. latifòlia

Culms often stout, from stout, creeping, branched rhizome; leaves mostly cauline, up to 1 cm broad, short, erect, flat, acute, scabrous; ligules long, up to 5 mm; panicle contracted, many-flowered, mostly purplish, with short, crowded branches; spikelets linear-lanceolate, short-pedicellated, 1-flowered; glumes lanceolate, slightly unequal, first shorter than lemma; lemma hispidulous, lacking or with indistinct lateral nerves.

Wet meadows, along rivers, on tundra.

Extremely variable. Circumpolar map indicates range of entire species complex.

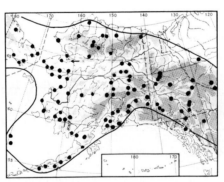

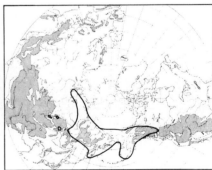

2. Arctagróstis latifòlia (R. Br.) Griseb.
var. arundinàcea (Trin.) Griseb.
Vilfa arundinacea Trin.; *Arctagrostis latifolia* subsp. *nahannensis* Pors.

Similar to var. *latifolia*, but taller, up to 1.5 meters tall; panicle open, with long branches, sometimes stout and stiff, sometimes capillary. Extremely variable.

Var. **angustifòlia** (Nash) Hult. (*A. angustifolia* Nash; *A. macrophylla* Nash), probably no more than an ecological variation, has long, narrow, flexible, green panicle, and long, lax, basal leaves and often short first glume. Specimens simulating *Cinna latifolia* occur, but glumes shorter, spikelets not geniculate, and awn lacking.

Meadows, sandbars along rivers.

25. GRAMINEAE (Grass Family)

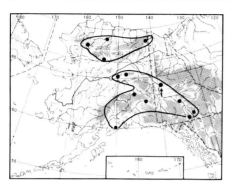
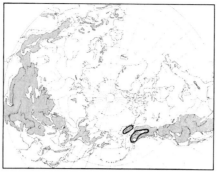

3. Arctagróstis poaeoìdes Nash

Similar to *A. latifolia* var. *arundinacea*, but spikelets smaller, pale; outer glume much shorter than inner; lemmas thin, translucent, broad and short; panicle branches densely scabrous.

Doubtfully distinct from *A. latifolia*. Habitat unknown; range given is tentative.

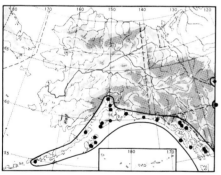

Cinna L. (Wood Reed Grass)

1. Cínna latifòlia (Trev.) Griseb.

Agrostis latifolia Trev.; *Cinna pendula* (Bong.) Trin.; *Muehlenbergia pendula* Trin.

Tall, sweet-scented grass, with short, broad leaf blades; ligules about 5 mm long; panicle loose, with slender spreading or drooping branches, naked at base; spikelets 1-flowered, falling off with joint; glumes acute, longer than lemmas, scabrous on keel; lemmas with short awn, as long as or longer than glumes.

Moist woods, clearings. Described from Scandinavia.

Podagrostis (Griseb.) Scribn. & Merr.

Spikelets 2 mm long; lemma considerably shorter than glumes; rachilla short; anthers about 0.5 mm .... 2. *P. Thurberiana*
Spikelets about 3 mm long; lemma only slightly shorter than glumes; rachilla longer; anthers about 1 mm .... 1. *P. aequivalvis*

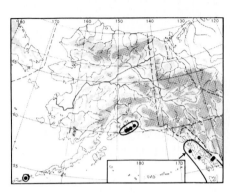 

1. Podagróstis aequivális (Trin.) Scribn. & Merr.
Agrostis aequivalvis Trin.

Culms 20–50 cm tall, tufted, with 1–2 culm leaves; leaves about 2 mm broad; panicle lax, branches glabrous, naked at base; spikelets about 3 mm long, violet; lemma and palea of about same length, or slightly shorter than glumes; rachilla minutely pubescent, long; anthers about 1 mm long.

Wet places.

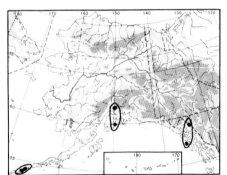 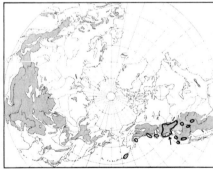

2. Podagróstis Thurberiàna (Hitchc.) Hult.
Agrostis Thurberiana Hitchc.; *A. atrata* Rydb.

Culm 15–30 cm tall, tufted; leaves about 2.5 mm broad; panicle narrow; spikelets about 2 mm long, violet; lemma considerably shorter than glumes; rachilla minutely pubescent, short; anthers about 0.5 mm long.

Wet places.

25. GRAMINEAE (Grass Family)

Agrostis L. (Bent Grass)

A very complicated genus within the area of interest, one that should be carefully studied when more material is available. It is apparent that hybridization between the different taxa is not uncommon, causing a high variability. It is also quite possible that still other taxa than those enumerated here, as for example the doubtful **A. anadyrénsis** Sotchava, occur within the area. The treatment here must therefore be regarded as tentative. Under such circumstances it is apparent that a working key for all cases cannot be constructed.

Lemma with prominent, geniculate awn, usually fixed below middle, exserted from spikelet; panicle branches smooth or somewhat scaberulous:
  Anthers 0.5–1 mm long; panicle usually contracted . . . . 1. *A. borealis*
  Anthers 1–1.5 mm long; panicle more open . . . . 2. *A. Trinii*
Lemma awnless or with short, mostly straight, awn, not or only slightly exserted from spikelet:
  Palea conspicuous, 2-nerved, about half the length of lemma, which usually lacks awn; anthers 1 mm long or longer; plant with rhizomes or stolons:
    Ligules of middle and lower leaves 0.5–1.5 mm, broader than long, blunt; panicle narrow, branches glabrous or scaberulous; leaves nearly glabrous above . . . . 4. *A. tenuis*
    Ligules 2–5 mm long, blunt or acute, leaves scabrous above:
      Plant with stolons (above ground); panicle contracted, with short branches (expanded in fruiting state); leaf blades 1–3 mm broad . . . . 5. *A. stolonifera*
      Plant with subterranean runners, tall, with large expanded panicle; leaf blades 1.5–6 mm broad, scabrous on both sides . . . . 6. *A. gigantea*
  Palea a minute, nerveless scale or absent; anthers shorter than 1 mm; plant tufted:
    Panicle dense, interrupted, with spikelets to base of at least the lower branches; lemma emarginate, with excurrent nerves . . . . 8. *A. exarata*
    Panicle loose, lower branches naked at base; lemma not emarginate; anthers less than 0.7 mm long:
      Anthers 0.5–0.6 mm long; glumes usually purplish; lemma hyaline, white; awn lacking or negligible . . . . 3. *A. alaskana*
      Anthers 0.2–0.4 mm long:
        Lemma nearly as long as glumes . . . . 7. *A. clavata*
        Lemma half to two-thirds as long as glumes, panicle branches very scabrous:
          Panicle very diffuse, one-third to two-thirds the entire height of plant, with widely divergent or reflexed branches . . . . 9. *A. scabra*
          Panicle broadly ovoid, one-sixth (to one-third) the entire height of plant, with merely arched branches; lemma usually with fine, long awn . . . . 10. *A. geminata*

×⅓

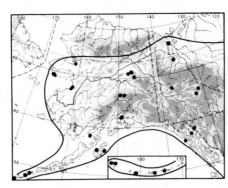
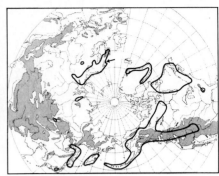

1. Agróstis boreàlis Hartm.

Culms 10–30 cm tall, mostly with only 1 joint; plant tufted, with erect leafy shoots at base; ligule 1.5–2 mm long; leaves flat, 1.5–3 mm broad, dark green; panicle contracted in age, ovoid or pyramidal, with usually glabrous branches; spikelets 2.5–4 mm long; lemma densely scabrous, with square apex and long, protruding, geniculate awn, fixed below middle; anthers 0.6–1 mm long and nearly as broad.

Moist, late-snow-free places, especially in mountains.

×⅓

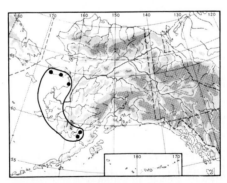
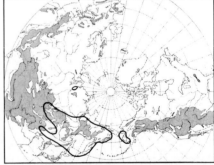

2. Agróstis Trínii Turcz.

Agrostis canina var. *rubra* Trautv.

Similar to **A. canìna** L., but densely tufted, with short subterranean runners; panicle branches less scabrous or glabrous; leaf blades 0.5–2 mm broad; ligules 0.5–2 mm long, shorter than in *A. canina*; panicle more or less contracted; lemma with conspicuous awn fixed at or below middle; anthers 1–1.5 mm.

Meadows and moist places. Described from Baikal.

25. GRAMINEAE (Grass Family)

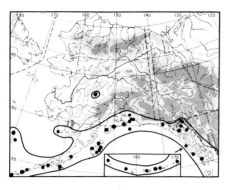
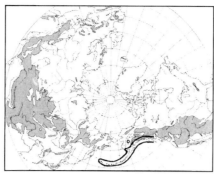

3. Agróstis alaskàna Hult.

Agrostis aenea Trin.; *A. melaleuca* Hitchc.

Up to 40 cm tall, tufted; basal leaves short, narrow; culm leaves broad at base, evenly tapering to point; panicle narrow, in age open, with numerous branches in lowest whorl; glumes usually dark purple, about 3 mm long; lemmas white, with short, straight awn, which fails to reach top of lemma; anthers 0.5–0.6 mm.

Wet places.

Plants infected by nematodes have elongated spikelets. Hybrids with *A. borealis* and *A. geminata* probably occur.

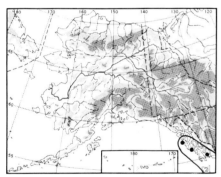
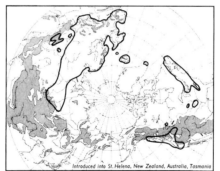
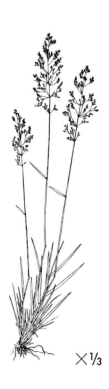

4. Agróstis ténuis Sibth.

Culm 20–40 cm tall, from long creeping rhizome with short subterranean runners and leaves in apex; culm leaves 1.5–4 mm broad, flat, scabrous; ligules of middle culm leaves broader than long, 0.5–1.5 mm long; panicle ovoid, with spreading capillary branches, floriferous chiefly from middle, not contracted in fruit state, usually purplish; lemma (usually) awnless, twice as long as palea; anthers 1–1.5 mm long.

Introduced; waste places, roadsides, yards, fields. Described from England.

×⅓

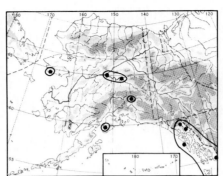 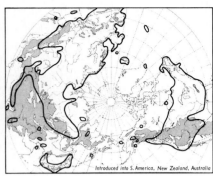

5. Agróstis stolonìfera L.
Agrostis alba L. in part.

Up to 50 cm tall, tufted, lacking subterranean runners but usually with creeping, leafy stolons; culm geniculate at base, leaf blades 2–5 mm broad; ligules of lower culm leaves 2–4 mm long; panicle ovoid-oblong, in fruit contracted, green or grayish-violet, with numerous capillary scabrous branches; spikelets 2–3 mm long; lemma 1.5–2 mm long, a quarter to a third longer than palea, lacking awn; anthers 1–1.4 mm long.

Introduced; waste places, roadsides, yards, fields. Described from Europe.

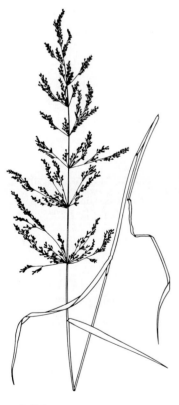
×⅓

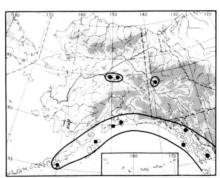 

6. Agróstis gigantèa Roth
Agrostis alba var. *gigantea* (Roth) Griseb.; *A. stolonifera* var. *major* (Gandoger) Farw.; *A. alba* subsp. *major* Gandoger.

Similar to *A. stolonifera*, but with taller, upright culms; lacking stolons, but with short, coarse, subterranean runners; panicle ovate; ligules 2–4 mm long; branches divaricate; leaf blades broad; anthers 1–2 mm long.

Introduced; waste places, roadsides, yards, fields. Described from Germany.
Range little known; circumpolar map highly tentative.

25. GRAMINEAE (Grass Family)

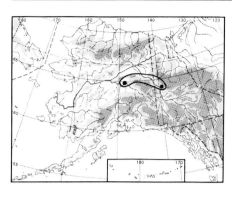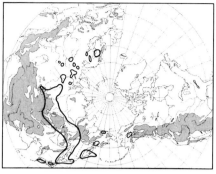

7. Agróstis clavàta Trin.

Agrostis idahoensis Nash, at least with respect to Alaskan plant.

Tufted culm about 30–50 cm tall, with 2–3 internodes; ligules 2–3 mm, lacerate at apex; culm leaves 3–6 mm broad; panicle lax, with long capillary branches, branched only in upper part; spikelets about 2 mm long on clavate peduncles; glumes ovate-lanceolate; lemma about as long as spikelet, lacking awn; anthers 0.3–0.4 mm.

Bare soil, wet meadows. Described from Kamchatka.
American range little known; circumpolar map tentative.

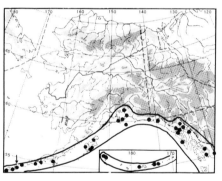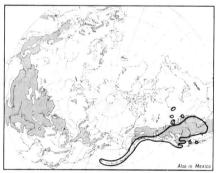

8. Agróstis exaràta Trin.

Culm up to 70 cm tall, tufted; sheaths smooth or slightly scabrous; panicle mostly spikelike, dense, with numerous spikelets crowded to base of branches; glumes acuminate or awn-pointed, long and narrow, scabrous; lemma emarginate, with excurrent nerves, with or without awn; palea very small.

Open ground at low altitude, common along southern coast of mainland.

Var. **purpuráscens** Hult., with spikelets less crowded at base of panicle branches, violet spikelets, and less scabrous glumes, probably represents hybrid form with *A. alaskana*.

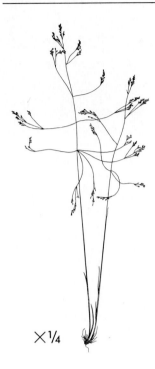
×¼

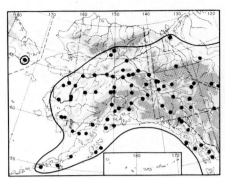 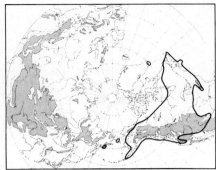

9. Agróstis scàbra Willd.

Agrostis Michauxii Trin.; *A. hiemalis* of authors, not Britt.

Up to 50 cm tall, tufted, with numerous narrow leaves at base; panicle at first narrow, in mature state very diffuse, one-third to two-thirds as long as entire plant, lacking spikelets at base of harshly hispid, capillary branches; glumes 2–3 mm long, lanceolate-attenuate, scabrous on keel; lemma 1.2–1.7 mm long, awnless; anthers 0.4–0.7 mm long.

Dry, open slopes, alluvial flats. Described from North America.

Common in southern half of range of interest (except in westernmost part). Forms with awned lemma occur.

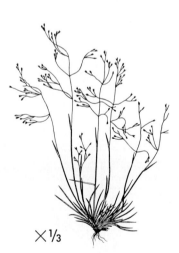
×⅓

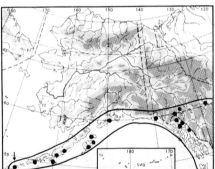 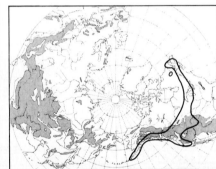

10. Agróstis geminàta Trin.

Agrostis hiemalis var. *geminata* (Trin.) Hitchc.; *A. scabra* var. *geminata* (Trin.) Hult.

Similar to *A. scabra*, but mature panicle broadly ovoid, one-third to one-sixth the height of entire plant, with shorter, divaricate branches; lemma awned, two-thirds as long as glumes.

Habitat unknown, range incompletely known, circumpolar map tentative.

Calamagrostis Adans. (Reed Bent Grass)

A critical genus within area of interest, as it is in entire circumpolar area. Variation within the recognized taxa is considerable, partly because of hybridization, and taxa not yet recognized must be expected to occur within our area; for example, specimens suggestive of **C. crassiglùmis** Thurb. and **C. hyperbòrea** Lange occur there. Authors differ markedly on question of taxa to be regarded as specifically distinct species. To construct a universally adequate key is hardly possible.

Awn much longer than glumes, distinctly geniculate:
 Culms tall; panicle large; glumes hyaline, abruptly pointed, grayish-violet .... 8. *C. purpurascens* subsp. *purpurascens*
 Culms shorter; panicle 1–4 cm; glumes not hyaline, long-pointed, dark purplish .... 9. *C. purpurascens* subsp. *arctica*
Awn included or only slightly exserted, straight or geniculate:
 Callus hairs (except few short, basal hairs) all of about same length, longer than lemma; panicle loose and open in anthesis:
  Panicle large, nodding; glumes long, narrow, very acute, herbaceous, strongly scabrous on keel; leaves flaccid ..... 2. *C. canadensis* subsp. *Langsdorffii*
  Panicle shorter; glumes shorter and broader, more abruptly pointed, more scarious, less scabrous on keel; leaves stiffer .... 1. *C. canadensis* subsp. *canadensis*
 Callus hairs some short, others longer, longest much shorter than lemma or barely as long:
  Culms tall, coarse, 3–4 mm thick at base (including sheath); leaves up to 1 cm broad; glumes long (5.5–8 mm), narrow, lanceolate, evenly scabrous along entire keel; longest callus hairs about one-third of lemma .... 3. *C. nutkaënsis*
  Culms thinner; leaves narrower:
   Longest callus hairs equaling or only slightly shorter than lemma; awn fixed below middle of lemma; panicle contracted; leaves flat:
    Culm leaves and panicle stiff; glumes green on back, dark purple on tip, thick and firm; longest callus hairs slightly shorter than lemma .... 4. *C. inexpansa*
    Culm leaves and panicle soft; glumes purplish on back, brown on tip, more scarious; longest callus hairs as long as lemma .... 5. *C. lapponica*
   Longest callus hairs distinctly shorter than lemma; panicle contracted or open:
    Awn thin, straight, not twisted:
     Culms tall; glumes dull, somewhat scabrous on sides; panicle grayish-brown .... 6. *C. neglecta*
     Culms shorter; glumes shiny, completely glabrous on sides; panicle purplish-black when young .... 7. *C. Holmii*
    Awn larger, twisted, bent or geniculate; panicle open .... 10. *C. deschampsioides*

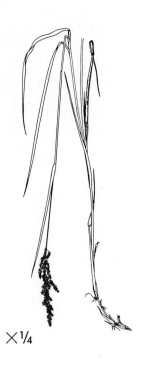

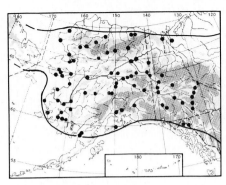 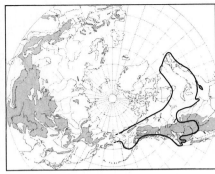

1. Calamagróstis canadénsis (Michx.) Beauv. Bluejoint

Arundo canadensis Michx.

subsp. **canadénsis**

Calamagrostis angustifolia Kom.?

Up to 1 meter tall, culms from creeping rhizomes forming tussocks; leaf blades flat, scabrous, 3–8 mm broad; panicle during anthesis usually loose, open, erect, with spreading branches; glumes lanceolate or narrowly ovate, scarious, scabrous on keel; lemma nearly as long as glumes; callus hairs as long as or longer than lemma; awn delicate, straight.

Meadows, wet places. Common in interior, passing into subsp. *Langsdorffii* (probably forming hybrid swarms). Described from Canada.

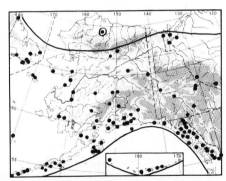 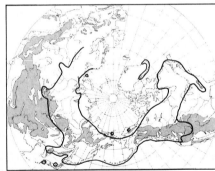

2. Calamagróstis canadénsis (Michx.) Beauv.

subsp. **Langsdórffii** (Link) Hult.

Arundo Langsdorffii Link; *Calamagrostis Langsdorffii* (Link) Trin.; *C. canadensis* var. *Langsdorffii* (Link) Inman; *C. purpurea* subsp. *Langsdorffii* (Link) Tsvel.

Similar to subsp. *canadensis*, but taller; glumes longer and narrower, herbaceous, very acute, very scabrous on keel; panicle larger, nodding; leaves softer, flaccid.

Alpine and subalpine meadows and thickets. Described from cultivated specimens grown from seeds, possibly from Kamchatka.

Common along coast, where subsp. *canadensis* is lacking. Passes inland into that taxon, probably through introgression; distinction between two taxa somewhat arbitrary there.

(See color section.)

25. GRAMINEAE (Grass Family)

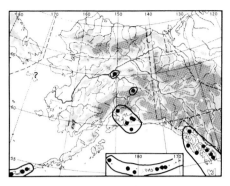
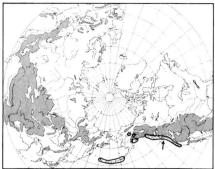
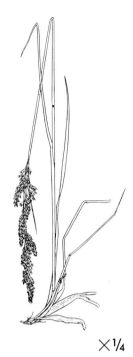

3. Calamagróstis nutkaénsis (Presl) Steud.

Deyeuxia nutkaënsis Presl; *Calamagrostis aleutica* Trin.; *D. aleutica* (Trin.) Munro.

Culms from short rhizomes, stout, up to 1 meter tall, coarse, 3–4 mm thick at base; ligules about 6 mm long; leaves up to 1 cm broad, scabrous, flat, becoming involute; panicle long, narrow, with stiff ascending branches; glumes 5–8 mm long, acuminate, scabrous along entire keel; lemma with rather stout, geniculate awn from near base, shorter or about as long as lemma; callus hairs not exceeding half the length of lemma.

Wet places along coast.

Hybrids with *C. canadensis* subsp. *Langsdorffii* occur. Inland specimens marked on map not seen by the writer.

×¼

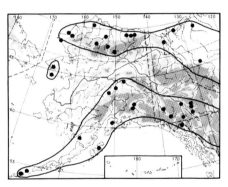
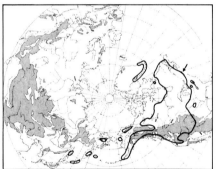

4. Calamagróstis inexpánsa Gray

Culms up to 80 cm tall from subterranean runners, usually with only two nodes; ligules 3–6 mm long; leaf blades flat, harshly scabrous, 2–8 mm broad; panicle 5–12 cm long, usually spikelike, with harsh, ascending branches; glumes opaque, thick, short-acuminate, green on back, with dark-purple tips; callus hairs shorter than lemma.

Wet places. An extremely variable plant.

×¼

× ⅓

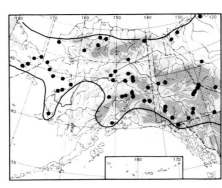 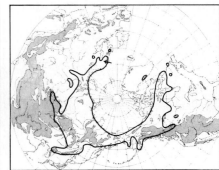

5. Calamagróstis lappónica (Wahlenb.) Hartm.
Arundo lapponica Wahlenb.; *Calamagrostis alaskana* Kearney.

Culms up to 1 meter tall, usually with only 2 nodes, from subterranean runners; ligules 3–5 cm long; leaves narrow, scabrous; panicle 5–15 cm long, brownish-violet in mature state, cylindrical or narrowly pyramidal (var. **alpìna** Hartm.); glumes broadly lanceolate, only slightly longer than lemma, scarious at apex; awn usually fixed below middle of lemma; callus hairs about as long as lemma.

Dry places, chiefly in mountains.

× ½

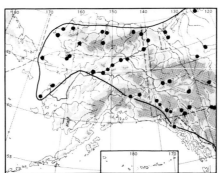 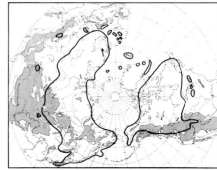

6. Calamagróstis neglécta (Ehrh.) Gaertn., Mey. & Schreb.
Arundo neglecta Ehrh.

Culms slender, up to 60 cm tall, from subterranean runners, with 1–2 nodes; ligules 1–2 mm long; leaf blades mostly involute, filiform, nearly glabrous; panicle cylindrical to narrow pyramidal, grayish-brown; glumes short-acuminate or blunt, hyaline at least at tip; lemma with short awn, fixed at or usually below middle; callus hairs much shorter than lemma.

Shores, wet places. An extremely variable plant, probably with local races.

25. GRAMINEAE (Grass Family)

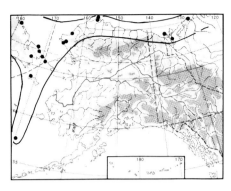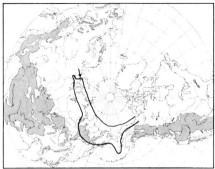

7. Calamagróstis Hólmii Lange
Calamagrostis neglecta var. *borealis* of authors; *C. kolymaensis* Kom.; *C. chordorrhiza* Pors.?.

Similar to *C. neglecta*, but culm usually shorter; panicle purplish-black when young; glumes with gradually narrowed tip, lustrous, glabrous or with few short spines on keel only; lemma more membranous at tip; ligules very short; panicle branches often more sparsely scabrous to nearly glabrous.

Mossy tundra.

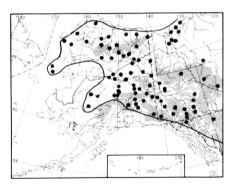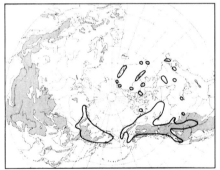

8. Calamagróstis purpuráscens R. Br.
subsp. purpuráscens
Calamagrostis yukonensis Nash; *C. purpurascens* var. *sylvatica* Thurb.; *Deyeuxia purpurascens* (R. Br.) Kunth.

Caespitose, with dry, whitish marcescent leaves at base; culms stiff, 2–5 dm tall; leaves 2–4 mm broad, flat, in age involute, harsh; panicle spikelike, stiff, thick, grayish-purple, in age whitish-brown; glumes scarious, scabrous, ovate-lanceolate, longer than lemma; lemma emarginate, dentate at apex, with coarse, geniculate, exserted awn from near base; callus hairs shorter than lemma.

Rocks and cliffs, preferably on calcareous soil. Common in interior. In St. Elias Mountains, to at least 2,000 meters. Described from Mackenzie district and Arctic Sea.

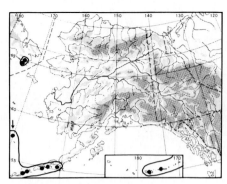
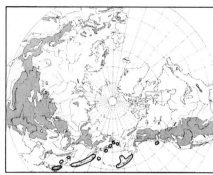

9. Calamagróstis purpuráscens R. Br.
subsp. **árctica** (Vasey) Hult.

Calamagrostis arctica Vasey; *C. purpurascens* var. *arctica* (Vasey) Kearney; *C. purpurascens* subsp. *tasuensis* Calder & Taylor?; *Trisetum sesquiflorum* Trin.; *Avena sesquiflora* (Trin.) Griseb.; *Calamagrostis sesquiflora* (Trin.) Tsvel.

Similar to subsp. *purpurascens*, but more low-growing; panicle shorter; glumes thicker, not hyaline; awns narrow, purplish, long-pointed.

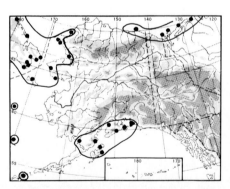
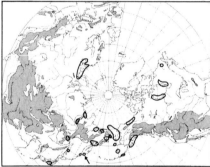

10. Calamagróstis deschampsioìdes Trin.

Up to 45 cm tall on southern coast, in arctic localities much smaller; culms geniculate at base, single, from thin, creeping rhizome; ligules 1–2 mm long, toothed; culm leaves short; panicle pyramidal, with most spikelets toward ends of glabrous panicle branches; glumes 3–4.5 mm long; lemmas of about same length as glumes, with fine, somewhat bent awn, fixed at or above middle, reaching to top of lemma; callus hairs about half the length of lemma, or somewhat longer.

Brackish marshes, wet meadows along seashore, cliffs along sea. Can be expected to occur all along western and northern shore.

25. GRAMINEAE (Grass Family)

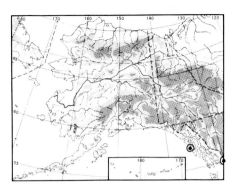
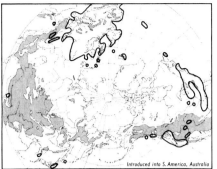

Holcus L.

1. Hólcus lanàtus L. Velvet Grass

Densely tufted, culms coarse, soft, up to 80 cm tall, pubescent at and below nodes from soft, downward-pointing hairs; ligules 2–4 mm long; culm leaves short, 2–4 mm broad, grayish, puberulent; panicle oblong to broadly ovate, whitish or grayish-green to violet, with pubescent branches; glumes scaberulous, upper floret with hooked awn usually close to apex, lower floret with short hairs at base.

Not native within area considered; waste places, roadsides, yards, fields. Described from Europe.

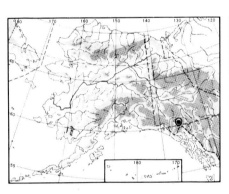
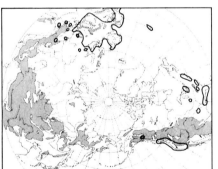

Aira L.

1. Aíra caryophýllea L. Hair Grass
Aspris caryophyllea (L.) Nash.

Culms mostly single, erect, up to 15 cm tall; panicle violet, open, with fine branches at oblique angle, bearing spikelets at tops only; glumes acute, exceeding florets; lemmas with geniculate awn.

Sandy and waste places. Occasionally introduced at Alaska Highway mile 988. Described from Europe.

Deschampsia Beauv.

Awn twice as long as floret or nearly so, long-exserted; florets included in glumes:
    Spikelets 3–5 mm; awn nearly straight; plant perennial, densely tufted ....
    .... 1. *D. elongata*
    Spikelets 5–8 mm; awn distinctly geniculate; plant annual, with few basal leaves
    .... 2. *D. danthonioides*
Awn considerably shorter than double the length of floret, slightly or not at all exserted; florets as long as glumes, usually longer:
    Awn geniculate; 2 flowers in spikelet close together, reaching about same height
    .... 9. *D. flexuosa*
    Awn not geniculate; upper flower in spikelet situated higher on axis:
        Glumes narrow, lanceolate, 5–7 mm long; anthers 2 mm long or longer ......
        .... 8. *D. beringensis*
        Glumes broader in comparison to length, not longer than 5 mm; anthers less than 2 mm long:
            Plant tall; glumes and lemmas thin, translucent; panicle open at maturity:
                Leaves broad, folded, scabrous, green or yellowish-green ....
                .... 3. *D. caespitosa* subsp. *caespitosa* var. *caespitosa*
                Leaves involute, filiform, more or less glabrous, blue-green ....
                .... 4. *D. caespitosa* subsp. *caespitosa* var. *glauca*
            Plant low, rarely exceeding 30 cm; glumes and lemmas more herbaceous and firm; panicle contracted or open:
                Panicle contracted, with flowers also close to base of branches; blade of upper stem leaf very short .... 6. *D. brevifolia*
                Panicle open, with spikelets toward ends of branches only:
                    Plant matted with capillary leaves; panicle with delicate culm, and capillary branches with few very small spikelets .... 7. *D. pumila*
                    Plant caespitose, with flat, yellowish-green leaves; awn usually fixed at middle of lemma, spikelets larger ....
                    .... 5. *D. caespitosa* subsp. *orientalis*

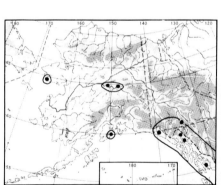

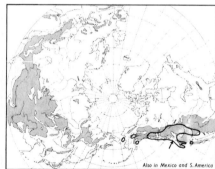

1. Deschámpsia elongàta (Hook.) Munro
Aira elongata Hook.

Perennial, densely tufted culm up to 60 cm tall, basal leaves filiform, culm leaves flat, 1–1.5 mm broad; panicle long, narrow, with appressed, capillary branches; glumes equaling or slightly exceeding the florets; lemma smooth, shining, with a geniculate awn up to 4 mm long, and pilose base.

Wet places. Found at inhabited places only, thus doubtless introduced into our area.

25. GRAMINEAE (Grass Family)

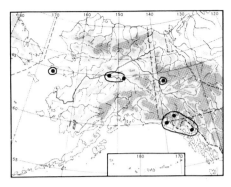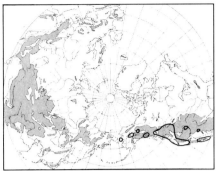

2. Deschámpsia danthonioìdes (Trin.) Munro
Aira danthonioides Trin.

Related to *D. elongata*, but annual and more low-growing, with few basal leaves; panicle more open; spikelets only at ends of branches; glumes exceeding florets; lemma with long (up to 6 mm) awn and pilose base.

Wet places. Found at inhabited places only, thus doubtless introduced into our area. Described from northwestern North America.

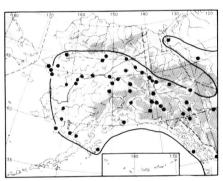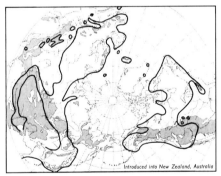

3. Deschámpsia caespitòsa (L.) Beauv.
Aira caespitosa L.
subsp. **caespitòsa**
var. **caespitòsa**

Culm up to 70 cm tall, tufted; leaves flat, folded, in typical specimens very scabrous; panicle large, open at maturity, often somewhat drooping; panicle branches slightly scabrous; florets 2–3, extending beyond glumes; awn short, sometimes fairly long, fixed at middle or lower part of lemma.

Moist places, mostly near habitations. Described from Europe.

A highly complex and variable species, with genetic affinity to several closely related species. Apparently forms hybrids with *D. beringensis* and *D. brevifolia* in area of interest. Specimens with glabrous leaves occur (hybrids with subsp. *orientalis*?).

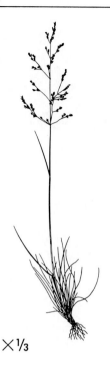
× ⅓

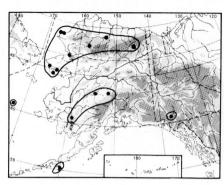 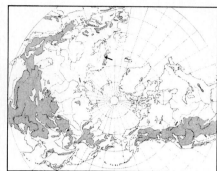

4. Deschámpsia caespitòsa (L.) Beauv.
subsp. **caespitòsa**
var. **glaùca** (Hartm.) Sam.
  Deschampsia glauca Hartm.

  Similar to var. *caespitosa*, but differs in having capillary, involute, bluish-green leaves, and in having awn often fixed near base of lemma.
  Along occasionally inundated shores. In need of further study.
  Circumpolar plant, but range details unknown, and map cannot be constructed.

× ½

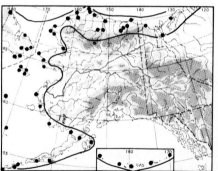 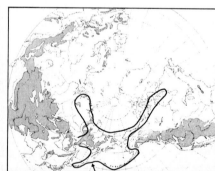

5. Deschámpsia caespitòsa (L.) Beauv.
subsp. **orientàlis** Hult.
  Aira Sukatschewii Popl.; *Deschampsia Sukatschewii* (Popl.) Roshev.; *D. Komarovii* Vassiljev; *D. hudsonica* Abbe.

  Similar to subsp. *caespitosa*, but of lower growth, yellowish-green; panicle branches stouter, scales firmer.
  Riverbanks, meadows, shores. Range incompletely known.
  Transitions to subsp. *caespitosa* occur along southern border of range; subsp. *orientalis* can be regarded as an arctic race of that plant.

25. GRAMINEAE (Grass Family)

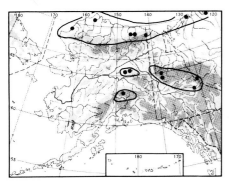 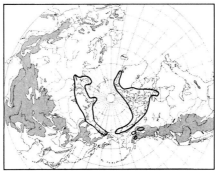 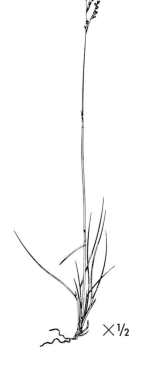

6. Deschámpsia brevifòlia R. Br.

Deschampsia caespitosa var. *brevifolia* (R. Br.) Trautv.; *Aira brevifolia* (R. Br.) Lange; *A. arctica* Spreng.; *Deschampsia arctica* (Spreng.) Ostenf.

Similar to *D. caespitosa*, but low-growing; panicle contracted, oblong or ovate, with short, upright branches bearing spikelets close to base; leaves of basal tuft short, narrow; blade of stem leaf very short.

Wet places in tundra, riverbanks, solifluction soil. Described from Melville Island.

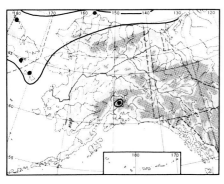 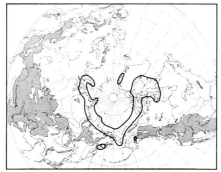 

7. Deschámpsia pùmila (Trin.) Ostenf.

Deschampsia brevifolia var. *pumila* Trin.; *D. borealis* (Trautv.) Roshev.; *D. caespitosa* var. *borealis* Trautv.

Similar to *D. brevifolia*, but matted, with few short, delicate culms; panicle with capillary branches and few small spikelets.

Wet places. Described from Greenland and Ellesmere Land.

×¼

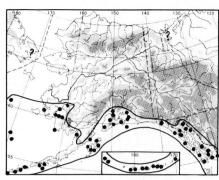
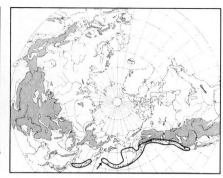

8. Deschámpsia beringénsis Hult.
Deschampsia Mackenzieana Raup?; *D. holciformis* with respect to Alaskan plant.

Tufted, up to and exceeding 1 meter in height; leaves glabrous, mostly flat, 1.5–4 mm broad, with long, acute ligules; panicle ample, open, up to 40 cm long, with capillary, very scabrous branches, and numerous spikelets; glumes narrow, 4–5 mm long, florets usually 3; lemma scarious with 4 teeth at apex; awn mostly fixed close to base, somewhat longer than floret; rachis pubescent. In typical specimens, panicle is yellowish-green, but many specimens have darker panicle, caused at least in part by hybrid influence from *D. caespitosa*.

Muddy shores. The very closely related **D. obénsis** Roshev. is reported from Chukchi Peninsula; the report needs confirmation. A plant probably belonging to **D. Mackenzieàna** Raup and very closely related to *D. beringensis* occurs at mouth of Mackenzie River. Both reports are indicated on Alaskan map by question marks.

×⅓

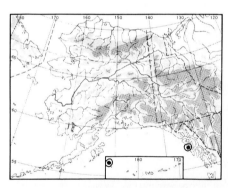

9. Deschámpsia flexuòsa (L.) Trin.
Aira flexuosa L.; *Avenella flexuosa* (L.) Drej.

Loosely tufted, culm geniculate at base, smooth, shiny, usually with 2 joints; basal leaves filiform, scabrous; culm leaves short; ligules about 2 mm long; panicle ovate in mature state, with undulating, somewhat scabrous branches; spikelets yellowish-brown to violet, florets close together; glumes shorter than spikelet; lemma with somewhat exserted awn.

Dry slopes and woods. Probably not native within area of interest. Described from Europe.

25. GRAMINEAE (Grass Family) 115

Vahlodea E. Fries

Leaves pubescent, with long, soft hairs .. 1. *V. atropurpurea* subsp. *paramushirensis*
Leaves glabrous or very slightly scabrous on nerves ....
.... 2. *V. atropurpurea* subsp. *latifolia*

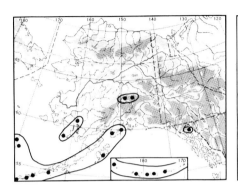
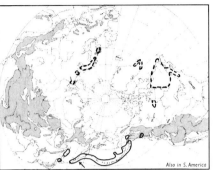
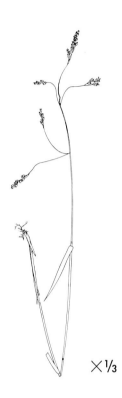

1. Vahlòdea atropurpùrea (Wahlenb.) E. Fries
  Aira atropurpurea Wahlenb.
subsp. **paramushirénsis** (Kudo) Hult.
  Deschampsia atropurpurea var. *paramushirensis* Kudo.

Loosely tufted, culms erect, slender; leaves flat, soft, more or less pubescent with long, soft hairs; panicle few-flowered, with capillary, flexuous, scabrous branches, naked below; glumes equally long, exceeding floret; lemmas scabrous and erose at tip, with tuft of hairs reaching less than halfway from base; awn inserted about middle of lemma.

Moist places, meadows, snow beds. *V. atropurpurea* described from Lapland, subsp. *paramushirensis* from Paramushiro Island, N. Kurile Islands.

Broken line on circumpolar map indicates the range of subspecies that do not occur within area of interest.

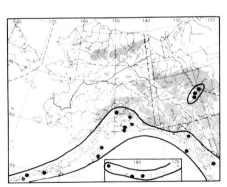
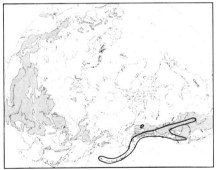

2. Vahlòdea atropurpùrea (Wahlenb.) E. Fries
subsp. **latifòlia** (Hook.) Pors.
  Aira latifolia Hook.; *Deschampsia latifolia* (Hook.) Vasey; *D. atropurpurea* var. *latifolia* (Hook.) Scribn.

Similar to subsp. *paramushirensis*, the two sometimes growing together; differs in having glabrous leaves (or very slightly scabrous on nerves).

Moist places, meadows, snow beds. Described from Rocky Mountains.

Trisetum L.

Culms glabrous below panicle:
  Panicle open, lax, drooping .... 1. *T. cernuum*
  Panicle dense, spikelike or ovate:
    Panicle spikelike, yellowish-green .... 5. *T. spicatum* subsp. *majus*
    Panicle ample, ovoid or broadly spikelike, yellowish-brown, more or less mottled .... 6. *T. sibiricum* subsp. *litoralis*
Culms densely pilose to nearly tomentose below panicle:
  Florets extend considerably beyond glumes; glumes and lemmas with violet color-zone and brown margins; plant up to 30 cm tall .... 2. *T. spicatum* subsp. *spicatum*
  Florets as long as glumes or slightly longer; plant usually taller:
    Culms finely pubescent above, glabrous and strongly furrowed in lower part, slender; panicle narrow; spikelets 5–6 mm long .... 4. *T. spicatum* subsp. *molle*
    Culms strongly pubescent, coarser; panicle broader, often somewhat lobed; spikelets larger .... 3. *T. spicatum* subsp. *alaskanum*

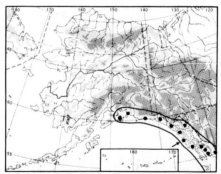 

1. Trisètum cérnuum Trin.

Culms lax, up to 120 cm tall; leaf blades flat, lax, up to 12 mm wide; panicle open, lax, often drooping, with slender branches bearing spikelets mostly toward ends; spikelets usually with 3 distal florets; first glume narrow, acuminate, second longer, broader, and abruptly pointed; lemmas with setaceous teeth and geniculate, long-exserted awn.

Moist woods.

25. GRAMINEAE (Grass Family)

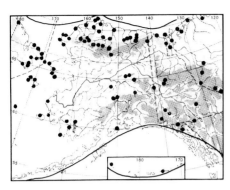
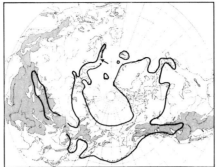

2. Trisètum spicàtum (L.) Richter
  Aira spicata L.
subsp. **spicàtum**

Tufted, culm up to 30 cm tall, densely pubescent, with long, soft hairs pointing down except at top; leaves 2–3 mm broad, with soft, gray pubescence; panicle spikelike, dense, with pubescent branches; glumes and lemmas usually with violet color-zone and brown margins; awn bent and twisted, affixed at upper third or fourth of lemma; floret considerably longer than glumes.

Tundra, dry places in the mountains, snow beds, to at least 2,000 meters.
Viviparous plants occur very rarely.

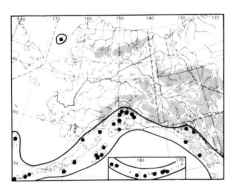
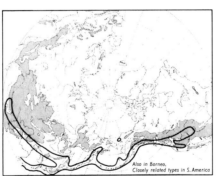

3. Trisètum spicàtum (L.) Richter
subsp. **alaskànum** (Nash) Hult.
  Trisetum alaskanum Nash; *T. molle* subsp. *alaskanum* (Nash) Rebr.

Similar to subsp. *spicatum*, but taller, with long, dense, contracted, somewhat lobed panicle; narrower, longer, more aristate glumes and lemmas, with longer, straighter awns; inner glume mostly extends beyond florets; culm and sheaths strongly pubescent.

Meadows and woods in coastal belt. Described from Alaska.

A specimen similar to the South American subsp. **phleoìdes** (D'Urv.) Hult. (*T. phleoides* D'Urv.) has been collected at Mendenhall (close to Juneau).

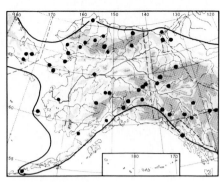
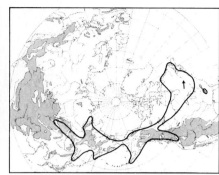

4. Trisètum spicàtum (L.) Richter subsp. mólle (Michx.) Hult.

Avena mollis Michx.; *Trisetum molle* (Michx.) Kunth; *T. spicatum* var. *molle* (Michx.) Beal; *T. triflorum* (Bigel.) Löve & Löve subsp. *molle* (Hult.) Löve & Löve.

Similar to subsp. *spicatum*, but usually taller, with slender culms, finely short-pubescent above, glabrous and strongly furrowed in lower part; panicle usually yellowish-green; florets nearly included within glumes; scarious margin of lemma not reaching apex of glumes.

Woods, meadows.

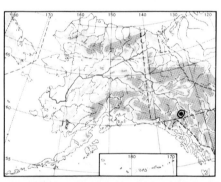
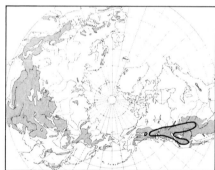

5. Trisètum spicàtum (L.) Richter subsp. május (Vasey) Hult.

Trisetum majus (Vasey) Rydb.; *T. subspicatum* var. *"major"* Vasey; *T. majus* (Vasey) Rydb.

Similar to subsp. *molle*, but taller, with glabrous culm and panicle glabrous or nearly so; leaves scabrous in margin; lower sheaths finely pubescent or nearly glabrous; scales longer and narrower, very acute.

Described from Colorado.

25. GRAMINEAE (Grass Family)

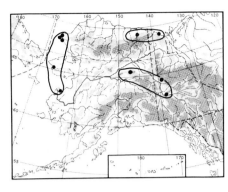
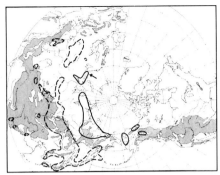

6. Trisètum sibìricum Rupr.
subsp. **litoràlis** (Rupr.) Roshev.
  Trisetum sibiricum var. *litorale* Rupr.; *Avena Ruprechtii* Griseb.

Culm 15–30 cm tall, glabrous, from short, creeping rhizome; culm leaves short and broad, nearly glabrous; panicle ovate or spikelike, dense, 3–5 cm long; spikelets about 8 mm long, 2–3-flowered, yellowish-brown; lemma two-pointed, with long, bent and twisted awn attached to middle or upper part.
  Meadows, along creeks.
  Subsp. *litoralis* differs from the type in its low culms and dense, compact panicle. Broken line on circumpolar map indicates range of subsp. **sibìricum.**

Avena L.

Plant perennial; panicle spikelike	3. *A. Hookeri*
Plant annual; panicle loose and open, with long branches:	
Spikelets 3-flowered, with 2–3 awns	1. *A. fatua*
Spikelets 2-flowered, with 1 awn	2. *A. sativa*

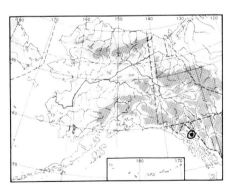
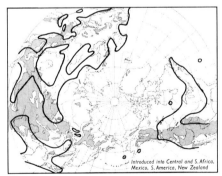

1. Avèna fátua L.   Wild Oat

Up to 1 meter tall, stout, tufted; panicle loose, open mostly on one side; spikelets pendulous; florets mostly 3, falling off separately when brittle axis of spikelets disintegrates; spikelets with 2 or more awns; lemma deeply 2-cleft in apex.
  Waste places; introduced at Juneau. Described from Europe.
  Questionable whether persistent within area of interest. Map highly tentative.

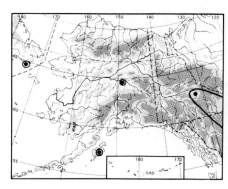
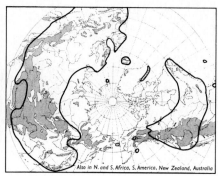

2. Avèna satìva L. Oat

Similar to *A. fatua*, but panicle usually more symmetrical; 2 florets; axis of spikelet not disintegrating; spikelets with 1 awn or lacking awn; lemma shallowly cleft in apex.

Certainly occurring occasionally in area of interest, but uncertainly persistent there. Waste places. Described from Europe.

Can be regarded as a cultivated race of *A. fatua* [*A. fatua* L. subsp. *sativa* (L.) Thell.]. Map highly tentative.

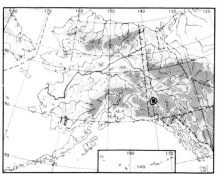
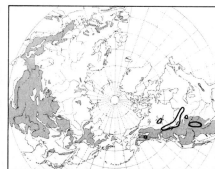

3. Avèna Hòokeri Scribn.

Helictotrichon Hookeri (Scribn.) Henrard.

Densely tufted, up to 40 cm tall; leaf blades flat, up to 4 mm broad, glabrous, scabrous in margin; panicle narrow; branches short, with 1–2 spikelets; spikelets 1.5 mm long, 3–6-flowered; lemmas firm, scaberulous, with geniculate awn about 1 cm long.

Dry places, prairies. Described from Rocky Mountains.
Collected at Duke River, Alaska Highway mile 1098.

25. GRAMINEAE (Grass Family)

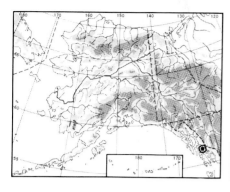
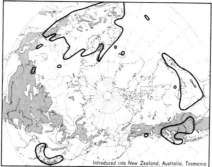

Arrhenatherum Beauv.

1. Arrhenátherum elàtius (L.) J. & C. Presl
Avena elatior L.

Tufted, with short runners; culms up to more than 1 meter tall; leaf blades long-acuminated, flat at maturity, scabrous, and on upper side finely pubescent 4–7 mm broad; panicle purplish or yellowish, open, with branches 2–6 cm long, capillary, scabrous, congested at base; florets 2, lowest staminate, with twisted and geniculate awn, upper perfect with short straight awn.

Collected at Petersburg, where it has been introduced. Roadsides, fields, yards, waste places. Described from Europe.

Danthonia Lam. & DC.

Lemmas 7–10 mm long, glabrous on back, bearded on base and margins .... 1. *D. intermedia*
Lemmas shorter, densely to sparsely pilose on back .... 2. *D. spicata*

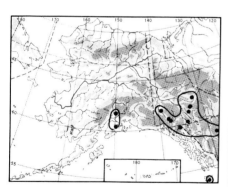
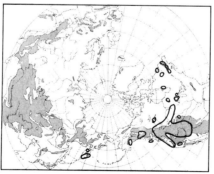

1. Danthònia intermèdia Vasey
Trisetum Williamsii Louis-Marie.

Densely tufted, culms up to 60 cm tall, mostly considerably shorter; leaves usually more or less involute, basal ones much shorter than culms; spikelets purplish; glumes broad, oblong-lanceolate, with weak lateral ribs; lemmas glabrous except for bearded base and margin, teeth at tip with broad base and short awn.

Meadows and bogs. Described from Canada.

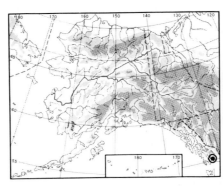
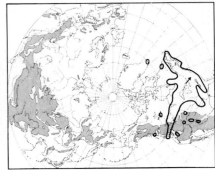

2. Danthònia spicàta (L.) Beauv.

Avena spicata L.

Similar to *D. intermedia*, but with shorter lemmas, more or less pilose on back. Dry places. Described from Pennsylvania.

Var. **pinètorum** Piper (*D. thermalis* Scribn.), to which western American plant is often referred, differs in having glumes with weak lateral nerves, broadest at middle and covering or nearly covering florets, and with basal leaves less curved and twisted. Seems to be a very weakly differentiated taxon.

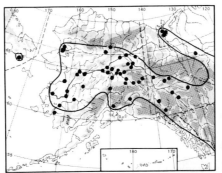

Beckmannia Host

1. Beckmánnia erucaefórmis (L.) Host

Phalaris erucaeformis L.

subsp. **baicalénsis** (Kuzn.) Hult.

Beckmannia erucaeformis var. *baicalensis* Kuzn.; *B. syzigachne* (Steud.) Fern. subsp. *baicalensis* (Kuzn.) Koyama & Kawano.

Plant light green, with solitary or tufted culms up to 90 cm tall and 6 mm broad; flat, scabrous leaf blades; panicle 1–2 dm long, narrow, with upright branches, or, often, ovate-lanceolate with branched lower branches; spikelets with usually 1 perfect and often 1 imperfect floret; glumes rounded, triangular, broadest toward summit; mucronate apex of lemma projecting beyond glumes.

Wet ground. Introduced and partly naturalized. *B. erucaeformis* described from Siberia, Russia, and southern Europe, subsp. *baicalensis* from Lake Baikal.

Broken line on circumpolar map indicates range of subsp. **erucaefórmis**, dotted line that of subsp. **syzigáchne** (Steud.) Breitung (*Panicum syzigachne* Steud.).

25. GRAMINEAE (Grass Family)

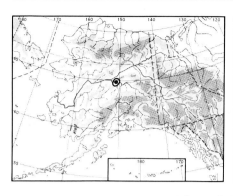 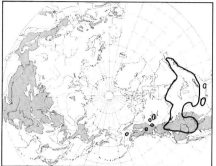

Sphenopholis Scribn.

1. Sphenópholis intermèdia (Rydb.) Rydb.
Eatonia intermedia Rydb.

Tufted; culms about 40 cm tall; leaf blades 1–2 mm broad, scabrous on nerves; panicle contracted, often nodding; first glume linear-attenuate; second glume oblanceolate to narrowly obovate; lemmas glabrous except on keel, awnless or with short awn near apex.

Meadows, shores. The single locality is at Manley Hot Springs. Occurs also at Liard Hotsprings, just east of our area. Described from eastern Gallatin County, Montana.

Koeleria Pers.

Plant loosely tufted, with long subterranean runners; culms pubescent .... 1. *K. asiatica*
Plant densely tufted; culms glabrous .... 2. *K. gracilis*

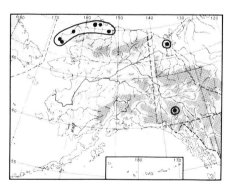 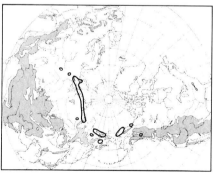

1. Koelèria asiática Domin
Koeleria Cairnsiana Hult.

Tufted, with subterranean runners; culms pubescent, slender, 10–35 cm tall, surrounded by old leaf bases; ligules 0.4–0.6 mm long, erose; basal leaves flat, 1.5–2.5 mm wide, some involute, appearing nearly filiform; culm leaves short; panicle cylindrical to obovoid, contracted; glumes glabrous or nearly so; lemmas finely pilose, without awn.

Dry, windswept places on tundra, sandy soil. Described from Taimyr.

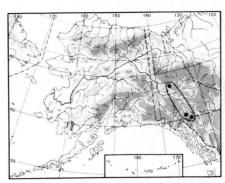
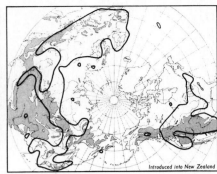

2. Koelèria gràcilis Pers. sens. lat.

Koeleria cristata (L.) Pers. (nomen ambig. rejiciend.); *Aira cristata* L.; *Koeleria yukonensis* Hult.

Tufted; culms glabrous, 2–5 dm high; blades involute, with glabrous sheaths; panicle cylindrical, pale, and shining; glumes glabrous, with scabrous keel; lemmas long, scabrous.

Described from Europe.

For the present it might be best, pending a detailed study, to include the plant from area of interest in this collective species.

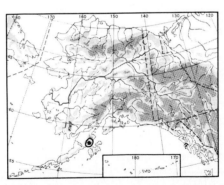
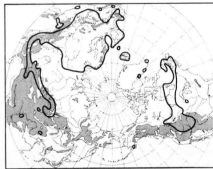

Catabrosa Beauv.

1. Catabròsa aquática (L.) Beauv.

Aira aquatica L.

Culms decumbent, rooting at basal nodes, or floating, often branched at base; leaf blades short, abruptly pointed or rounded at apex, 3–8 mm broad; panicle ovoid, light green tinged with violet, with capillary, scabrous, often somewhat down-pointing branches, floriferous from close to base; first glume about 1 mm, second 2 mm long, much shorter than spikelet; lemma glabrous, 3-nerved.

Wet places. Described from Europe.

Circumpolar map gives range of entire species complex.

25. GRAMINEAE (Grass Family)

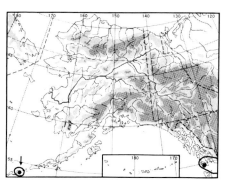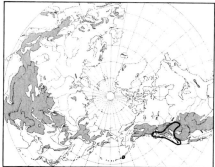

Melica L.

1. Mèlica subulàta (Griseb.) Scribn.

Bromus subulatus Griseb.

Culms glabrous, up to 1 meter tall, bulbous at base; sheaths retrorsely scabrous; leaves 3–4 mm broad; panicle narrow, branches appressed, single or in pairs; spikelets 1.5–2 cm long, 3–4-flowered; glumes of unequal length, much shorter than spikelet; lemma 7-nerved, awnless, evenly tapering to acuminate point.

Meadows.

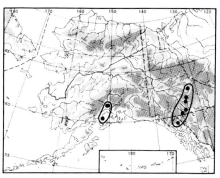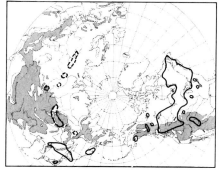

Schizachne Hack.

1. Schizáchne purpuráscens (Torr.) Swallen

Trisetum purpurascens Torr.; *Avena striata* Michx. not Lam.; *Melica striata* (Michx.) Hitchc.; *Schizachne striata* (Michx.) Hult.

Stoloniferous, culms slender, up to 80 cm tall, much longer than basal leaves, decumbent at base; leaves flat, about 2 mm broad; panicle open, branches usually in pairs, capillary drooping; spikelets 3–5-flowered; first glume 3-nerved, second 5-nerved, much shorter than adjacent lemma; lemmas purplish, bifid at apex, 7-nerved, with long, divergent, bent awn fixed close to tip.

Rocky slopes, alder thickets. Described from Massachusetts and Montreal.

Specimens belong to subsp. **purpuráscens**. Broken line on circumpolar map indicates range of subsp. **callòsa** (Turcz.) Koyama & Kawano (*S. callòsa* Turcz.).

25. GRAMINEAE (Grass Family)

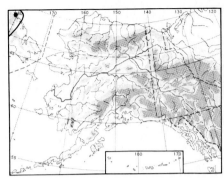 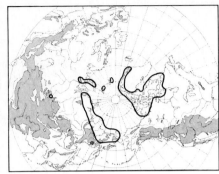

Pleuropogon R. Br.

1. Pleuropògon Sabìnei R. Br. — Sabine Grass

Stoloniferous; culms 20–30 cm tall, smooth; leaves flat, narrow; culm leaves short; panicle 1-sided, racemose, with few, long, dark-purplish, 5–8-flowered, drooping spikelets; glumes small, hyaline, toothed at tip; lemmas scarious-margined, 7-nerved.

Muddy shores and shallow water in tundra. Described from Melville Island.
Not yet found in American region of map, and barely reaching Siberian part.

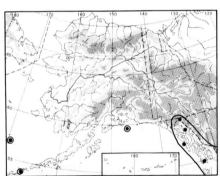 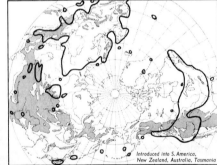

Dactylis L.

1. Dáctylis glomeràta L. — Orchard Grass

Tufted, with short, subterranean stem, culms up to 1 meter tall; leaves grayish-green, 4–8 mm broad, their sheaths with sharp keel; panicle compound, of dense, 1-sided, congested clusters, of extremely variable appearance, at ends of long, stiff, scabrous branches; spikelets 6–7 mm long, with 3–4 florets and highly variable pubescence; lemma 5-nerved, with short awn.

Waste places, fields, yards, roadsides. Introduced. Described from Europe.

25. GRAMINEAE (Grass Family)

Poa L.

This large genus presents numerous doubtful forms not yet properly understood—neither within the area of this flora nor within other parts of the arctic and boreal belts. In several cases, key, nomenclature, and ranges are to be regarded as tentative.

■ Anthers 1 mm long or longer:
● Tuft of long (often plicated) cobweb hairs at base of (lowest) lemma:
Lemmas lacking long or lanate hairs between keel and marginal nerve, glabrous or scabrous:
Grass very coarse, with leaves 5–10 mm broad; numerous panicle branches bearing spikelets to base; anthers more than 2 mm long .. 6. *P. eminens*
Grass ordinary, with narrower leaves; anthers shorter than 2 mm:
Glumes long, narrow, highly acute, reaching almost to top of often 2-flowered spikelet .... 7. *P. macrocalyx*
Glumes shorter relative to spikelet:
First glume narrow, curved, highly acute; nerve between keel and marginal nerve of lemma very prominent; upper ligules 4–5 mm long, tufted .... 8. *P. trivialis*
First glume narrow or broader; nerve between keel and marginal nerve of lemma less prominent; upper ligules shorter:
Plant with creeping rhizome; lemma with distinct nerve between keel and marginal nerve:
Panicle large, open; culms retrorsely scabrous ..... 9. *P. laxiflora*
Panicle smaller; culms glabrous:
Panicle with 2 (–4) branches in lowest whorl:
Low-growing; sheaths glabrous .... 12. *P. subcoerulea*
Taller; sheaths minutely scabrous .... 10. *P. Eyerdamii*
Panicle with (3–) 5 branches in lowest whorl:
Both glumes lanceolate or ovate-lanceolate; leaves flat, as broad as culm or broader; culm thick ... 11. *P. pratensis*
First glume narrow, lanceolate, curved; leaves involute, narrow; culm tall, slender; basal leaves narrower than culm:
Spikelet more or less violet; culms not forming mats .....
.... 13. *P. alpigena*
Spikelets green or grayish-brown-green; culms with numerous basal shoots, with long narrow leaves, forming mats .... 14. *P. angustifolia*
Plant tufted; lemma lacking nerve or with very thin nerve between keel and marginal nerve:
Culm with 2 (rarely 3) nodes; uppermost culm leaf short, fixed below middle of culm; spikelets usually glaucous ....
.... 15. *P. glauca*
Culm with 4–5 nodes, some usually close to base; spikelets usually not glaucous:
Panicle large, open, 15–30 cm long; leaves 3–4 mm broad ....
.... 16. *P. occidentalis*
Panicle smaller; leaves narrower:
Upper ligules broader than long, about 1 mm long, blunt; panicle often flexuose .... 18. *P. nemoralis*
Upper ligules longer than broad, mostly 2–3 mm long, somewhat acute; panicle usually more rigid and open ......
.... 17. *P. palustris*
Lemmas with long or lanate hairs between keel and basal part of marginal nerve:
Spikelets violet or grayish-violet, with brown-tipped scales; lemma with often profuse lanate hairs below:
Glumes lanceolate, not pruinose .... 19. *P. lanata*
Glumes ovate-lanceolate, pruinose .... 20. *P. Norbergii*
Spikelets green, or green with brown-tipped scales; lemma with scattered, less lanate, often straight hairs below:
▲ Panicle nodding; glumes and lemmas about 7 mm long, narrow; marginal nerve prominent, reaching almost to top of lemma ... 21. *P. Turneri*

▲ Panicle erect or nearly so; glumes and lemmas shorter; marginal nerve shorter, less prominent:
  Base of culm surrounded by cylinder of hyaline sheaths; leaves broad; subterranean runners short, curved .... 22. *P. malacantha*
  Base of culm not surrounded by cylinder of hyaline sheaths:
    Plant stout, with broad leaves, lacking subterranean runners .... 23. *P. hispidula*
    Plant slender, with narrow leaves:
      Plant tufted, with short runners .... 5. *P. arctica* subsp. *caespitans*
      Plant with slender subterranean runners:
        Culms tall, slender; basal leaves long, filiform; lemmas not very pubescent at base .... 4. *P. arctica* subsp. *longiculmis*
        Culms shorter; basal leaves flat; lemmas more pubescent at base:
          Lemmas nearly lanate at base; tuft of cobweb hairs prominent .... 3. *P. arctica* subsp. *Williamsii*
          Lemmas less pubescent at base; tuft of cobweb hairs not prominent .... 2. *P. arctica* subsp. *arctica*
● Tuft of cobweb hairs at base of lemma lacking:
  Lemmas lacking long, lanate hairs between keel and marginal nerve:
    Culm compressed, 2-edged above; plant tufted .... 24. *P. compressa*
    Culms not compressed:
      Lemma glabrous; keel and nerve minutely scabrous:
        Leaves filiform, scabrous .... 25. *P. Cusickii*
        Leaves about 1 mm broad, glabrous .... 26. *P. vaseyochloa*
      Lemma scabrous or puberulent:
        Lemmas with obscure keel, convex on back:
          Lemmas crisp, puberulent in lower part:
            Sheaths scabrous .... 27. *P. scabrella*
            Sheaths glabrous .... 28. *P. Canbyi*
          Lemmas scabrous below:
            Leaves involute; sheaths scaberulous; ligules long .... 29. *P. nevadensis*
            Leaves flat, 1–3 mm broad; sheaths glabrous; ligules short .... 31. *P. ampla*
        Lemmas compressed, distinctly keeled; plant usually low-growing, with 2, rarely 3, joints .... 15. *P. glauca*
      Lemmas with long, lanate hairs between keel and marginal nerve:
        Plant with slender subterranean runners; leaves narrow ..... 2–5. *P. arctica*
        Plant tufted, with broader leaves:
          Plant 20–30 cm tall; panicle short and broad with glabrous branches; glumes ovate-lanceolate; lemmas distinctly keeled; spikelets broad and short .... 1. *P. alpina*
          Plant taller; panicle narrow, elongate; branches scabrous; lemmas rounded on back; spikelets long, narrow, acute .... 32. *P. stenantha*
■ Anthers less than 1 mm long:
  Tuft of long cobweb hairs at base of lemma distinct:
    Plant annual; first glume clawlike, only slightly more than half as long as second; nerve between keel and marginal nerve very distinct; plant yellowish-green .... 33. *P. annua*
    Plant perennial; 2 glumes not very different in size; nerve between keel and marginal nerve indistinct:
      Plant about 25 cm tall loosely tufted; panicle open, with capillary branches; spikelets green or grayish-green .... 34. *P. leptocoma*
      Plant 10–15 cm tall; panicle compact; spikelets purple .. 35. *P. paucispicula*
  Tuft of cobweb hairs at base of lemma not distinct:
    Panicle branches thick; leaves about 2 mm wide .... 30. *P. Merrilliana*
    Panicle branches capillary; leaves about 0.5 mm wide:
      Panicle branches long, slender; panicle open; spikelets green .... 36. *P. brachyanthera*
      Panicle branches shorter; panicle contracted; spikelets purplish:
        Lemma pubescent at base between keel and marginal nerve .... 37. *P. abbreviata*
        Lemma glabrous between keel and marginal nerve .... 38. *P. pseudoabbreviata*

25. GRAMINEAE (Grass Family)

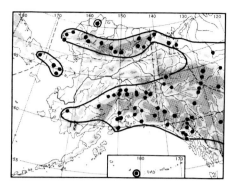
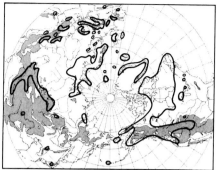

1. Pòa alpìna L.

Tufted; rootstock clothed with light-gray, papery sheaths forming cylinder; leaves short, flat, abruptly pointed, 2–5 mm broad; panicle broad, ovoid to oblong, nearly smooth, lowest branches 2 together; spikelets ovate, often reddish-violet; glumes ovate, broadly hyaline; lemma ovate, obscurely 5-nerved, pubescent on keel, with 2 lateral nerves; palea about as long as lemma.

Dry slopes, meadows, rocks. Described from the Alps of Lapland and Switzerland.

Plants with viviparous spikelets occur [var. **vivípara** (L.) Willd. (*P. alpina* f. *vivipara* L.)] at Hope and Juneau.

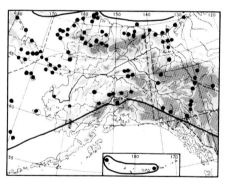
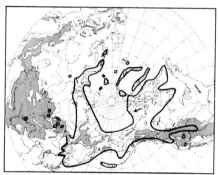

2. Pòa árctica R. Br.
subsp. **árctica**

Culms 10–20 cm tall, thin, glabrous, straight, mostly with 2 nodes, from long, light-colored, bowlike subterranean runners; leaves flat; panicle pyramidal, open, with 2 branches together; spikelets often violet; glumes lanceolate; lemma with indistinct nerves, pubescent in lower half, and with more or less distinct tuft of cobweb hairs at base.

Dry places on tundra, in meadows and thickets, and in the mountains to at least 2,150 meters.

Plants with viviparous spikelets occur.

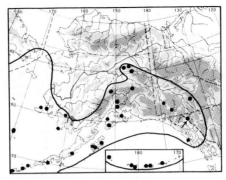 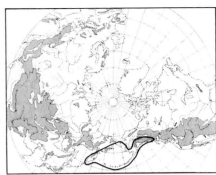

3. Pòa árctica R. Br.
subsp. **Williámsii** (Nash) Hult.
Poa Williamsii Nash.

Culm up to 20 cm tall; scales broad, grayish-violet, variegated with yellowish-brown; lemma nearly lanate, with long, soft hairs; tuft of cobweb hairs at base prominent.

Meadows, thickets.

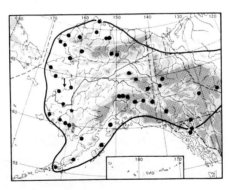 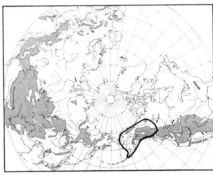

4. Pòa árctica R. Br.
subsp. **longicúlmis** Hult.
Poa brintnellii Raup.

Culms tall, slender; lower panicle branches often more than 2 together; basal leaves long, nearly filiform; lemmas less pubescent between nerves, more or less scabrous toward apex.

25. GRAMINEAE (Grass Family)

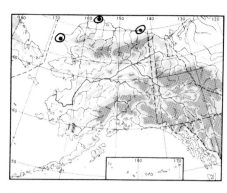
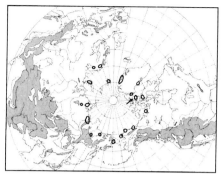

5. Pòa árctica R. Br.
subsp. **caèspitans** (Simmons) Nannf.
  Poa cenisea f. *caespitans* Simmons; *P. Tolmatchewii* Roshev.

  Coarse, low-growing, caespitose; forming dense tufts with arcuate, sterile shoots; leaves broad.
  Described from Ellesmere Land.

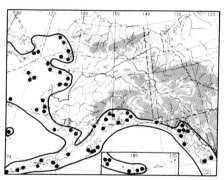
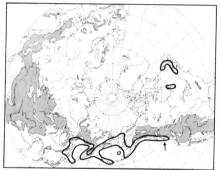

6. Pòa èminens Presl
  Poa glumaris Trin.; *Glyceria glumaris* (Trin.) Griseb.; *P. Trinii* Scribn. & Merr.

  Culm stout; up to 1 meter tall; glabrous, from thick rhizome with runners; leaves up to 10 mm broad, flat, scabrous in margin; ligules ciliate, 1–2 mm long; panicle elliptic, with numerous spikelets and glabrous branches; spikelets light green when young; glumes broadly lanceolate; lemma pubescent on keel and lateral nerves; basal cobweb hairs lacking.
  Seashores.

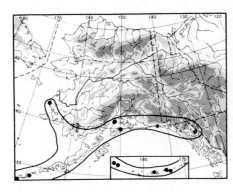
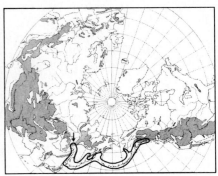

7. Pòa macrocàlyx Trautv. & Mey.

Culms glabrous, from short rhizome; leaves 2–4 mm broad, scabrous above, with ligules up to 4 mm long; panicle oblong with scabrous branches; spikelets broadly lanceolate, 2–3-flowered; glumes lanceolate, acute, as long as or nearly as long as spikelet; lemma lacking lanate hairs between keel and marginal nerve, with tuft of cobweb hairs at base.

Seashores in the *Elymus* belt.

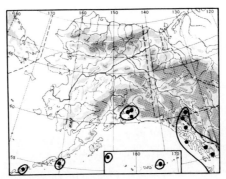
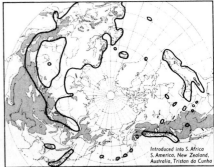

8. Pòa triviàlis L.

Tufted, with culms geniculate or decumbent at base; ligules of upper leaves acute, 4–5 mm long (in large specimens); panicle lax, broadly ovate to oblong, with scabrous branches, 3–5 together at base; glumes narrow, acute, the first short and 1-nerved, the second longer and 3-nerved, scabrous on keel; lemma prominently 5-nerved, glabrous except for silky keel, with distinct tuft of cobweb hairs at base.

Waste places, roadsides, yards; an introduced weed. Described from Europe.

25. GRAMINEAE (Grass Family)

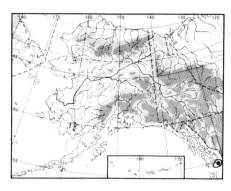 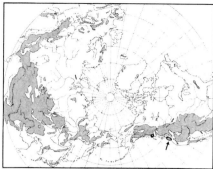 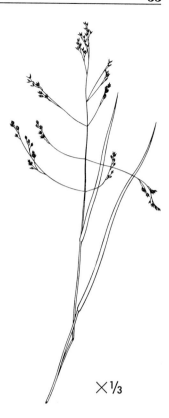

9. Pòa laxiflòra Buckl.

Culms from creeping rhizome, retrosely scabrous, tall; sheaths slightly retrosely scabrous; ligules 3–5 mm long; leaves 2–4 mm wide; panicle loose, open, nodding or drooping; lower branches 3–4 together; spikelets 3–4-flowered; lemmas sparsely pubescent on nerves below, with tuft of cobweb hairs at base.

Known only from Cape Fox Springs, in region of interest. Described from Oregon.

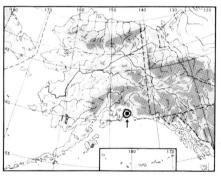 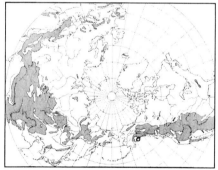 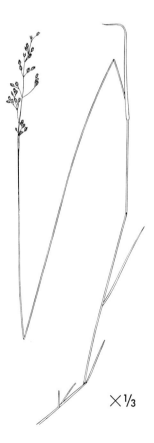

10. Pòa Eyerdàmii Hult.

Culms glabrous, from long creeping rhizome; leaves about 2 mm wide, somewhat scabrous above, glabrous beneath; ligules 1–2 mm long; panicle open, with scattered spikes, naked below; glumes scabrous on keel; lemma 3-nerved, finely white-pilose on keel and marginal nerve, glabrous between them, but with tuft of cobweb hairs at base; anthers 1.4–1.9 mm long.

× ½

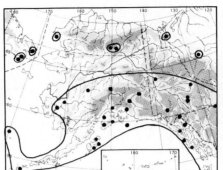
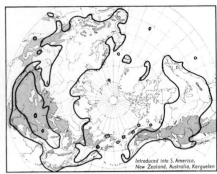

11. Pòa praténsis L.

Culms from stoloniferous rhizome; leaves flat or folded, broad; panicle pyramidal to ellipsoid, with scabrous branches, 3–5 at lower node; spikelets coarse, large; glumes broadly lanceolate to ovate; lemmas copiously webbed at base.

Waste places, roadsides, yards; an introduced weed. Described from Europe.

× ½

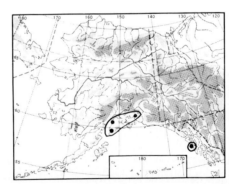
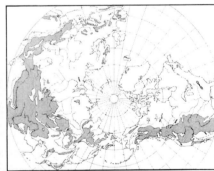

12. Pòa subcoerùlea Sm.

Including *Poa irrigata* Lindm.; *P. pratensis* subsp. *subcoerulea* (Sm.) Tutin.

Similar to *P. pratensis*, but less tufted; culms short; panicle pyramidal, with more or less deflexed branches.

Waste places, roadsides, yards; introduced. Described from England.
Total range unknown.

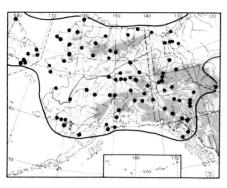 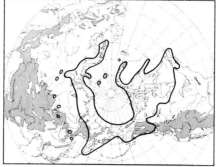 

13. Pòa alpìgena (E. Fries) Lindm.
Poa pratensis var. *alpigena* E. Fries.

Culm up to 40 cm tall, from slender rhizome with runners; leaves narrow, up to 2 mm wide; panicle dense, oblong to pyramidal, with 3–5 nearly glabrous, more or less erect branches at lower nodes; first glume narrow, lanceolate, bent; lemma with cobweb hairs at base.

Grassy slopes, gravel bars. Described from mountains of Norway.
Viviparous plants occur in the north and, rarely, in the mountains.

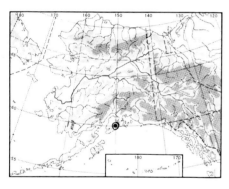 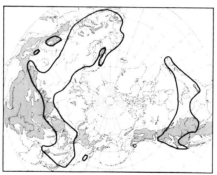 

14. Pòa angustifòlia L.

Similar to *P. pratensis* and *P. alpigena*, but forming small dense tufts with short runners; leaves filiform, narrower than culm.

Waste places, roadsides, yards; introduced. Described from Europe.
Circumpolar map tentative.

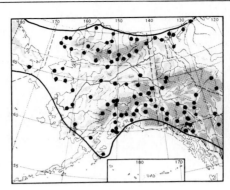
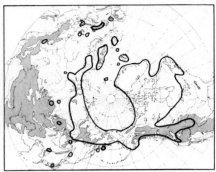

15. Pòa glaùca M. Vahl

Including *Poa rupicola* Nash, *P. anadyrica* Roshev., *P. bryophila* Trin., and *P. ammophila* Pors.

Tufted, mostly glaucous; culms mostly stiff, often curved, with only 2–3 joints, uppermost internode long; leaves spreading, long-acuminate, the uppermost short, at or below the middle of culm; 1–2 branches of panicle together, scabrous, straight, spreading, or dense and erect [var. **confértа** (Blytt) Nannf. (*P. conferta* Blytt)]; glumes broadly lanceolate; lemma scabrous; keel and lateral nerves pubescent, mostly without cobweb hairs at base.

Extremely variable. Dry slopes, sandy places, in the St. Elias Range to at least 2,300 meters. Described from Norway.

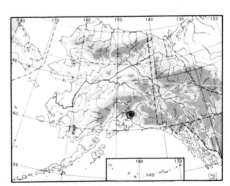
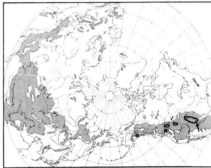

16. Pòa occidentàlis Vasey

Tufted; culms stout, tall; sheaths more or less retrorsely scabrous; leaves 3–4 mm broad, scabrous; panicle open, with branches 15–20 cm long, the lower 3–5 together, naked below; spikelets 3–6-flowered; glumes acute, scabrous on keel; lemmas glabrous, somewhat pubescent on keel and marginal nerve, with tuft of cobweb hairs at base.

25. Gramineae (Grass Family)

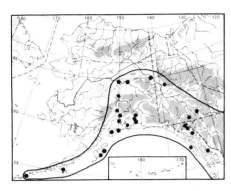

17. Pòa palústris L.

Poa rotundata Trin.; *P. triflora* Scribn.

Tufted; culm smooth, tall, erect, creeping at base; ligules acute, about 2 mm long; leaves scabrous, narrow; panicle oblong to ovate, with scabrous branches, the lowest 4–6 together; spikelets ovate; glumes lanceolate, subequal, obscurely nerved; lemma lanceolate, with purplish band, obscurely nerved; keel and marginal nerve silky, tip scarious, tuft of cobweb hairs at base.

Moist places. Described from Europe.

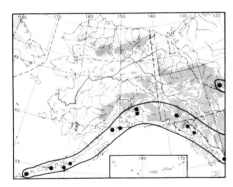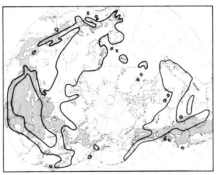

18. Pòa nemoràlis L.

Similar to *P. palustris*, but more slender; panicle narrow, mostly green, with smaller, narrow, and sometimes only 2-flowered spikelets; ligules shorter than broad, blunt.

Shady places. Described from Europe.

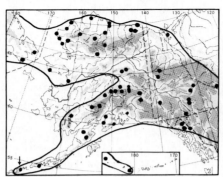
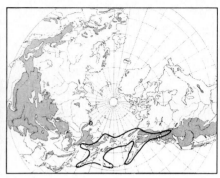

19. Pòa lanàta Scribn. & Merr.

Including *Poa platyantha* Kom. and *P. petraea* Trin.(?).

Culms from creeping rhizome; leaves short, rigid, 2–4 mm wide, the upper much shorter than its sheath; ligules about 4 mm long; panicle pyramidal, few-flowered, lower branches 2 together; spikelets ovate, purplish, 3–6-flowered; glumes acute, 3-nerved, scabrous on keel; lemmas with broad hyaline margins, obtuse, 5-nerved, densely lanate in lower half, strigose above.

Meadows.

A highly variable plant. Viviparous specimens also occur (var. **vivípara** Hult.).

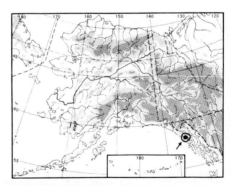
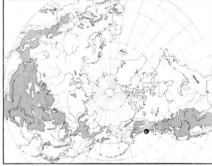

20. Pòa Norbérgii Hult.

Culms glabrous, from creeping rhizome; stem leaves 3.5–4 mm broad, scabrous; spikelets short, peduncled; glumes 1-nerved, glabrous, glaucous, hyaline-margined, very scabrous on keel toward purplish, acute apex, the lower lanceolate, the upper ovate to lance-ovate; lemma broad, finely scaberulous, lanate toward base, with long straight hairs at keel and lateral nerves.

25. Gramineae (Grass Family)

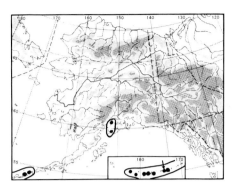 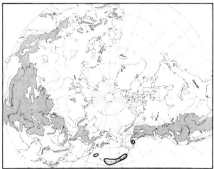 

21. Pòa Tùrneri Scribn.

Culm stout, glabrous, from stoloniferous rhizome; leaves soft, flat; panicle large, oblong to pyramidal, more or less nodding, with essentially glabrous, capillary branches; glumes 3-nerved, acute, as long as lowest lemma; lemma densely long-pilose on keel and marginal nerves and between them at base, with tuft of cobweb hairs at base; anthers 2.5 mm long.

Meadows.

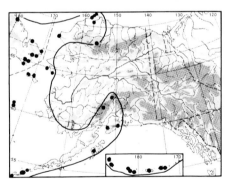 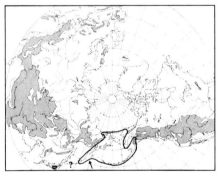 

22. Pòa malacántha Kom.

Poa Komarovii Roshev.

Densely tufted, with short runners, glabrous; culm surrounded by old sheaths at base; leaves flat, 2–4 mm broad, with ligules 2–4 mm long; panicle pyramidal or ovate; spikelets lanceolate; glumes lanceolate; lemma distinctly nerved, pubescent below, with longer pubescence on nerves.

Meadows and stony slopes on tundra and in the mountains.

25. GRAMINEAE (Grass Family)

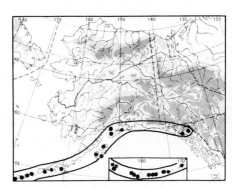 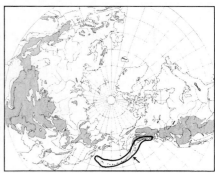

23. Pòa hispídula Vasey

Culms decumbent at base, from creeping rhizome; leaves mostly cauline, short, 3–4 mm wide; ligules about 4 mm long; panicle ovate with scabrous branches, 2 to several together; spikelets ovate-lanceolate, green or slightly purplish, 3–5-flowered; glumes with prominent nerves and fine spines along keel; lemma acute, narrow, 5-nerved, scabrous, long-hairy on nerves in lower half and with distinct tuft of cobweb hairs at base.

Meadows.

In Aleutian Islands, small specimens with small spikes, narrow leaves, and more or less glabrous panicle branches occur (var. **aleùtica** Hult.), as well as viviparous specimens (var. **vivípara** Hult.).

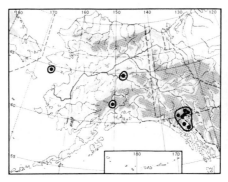 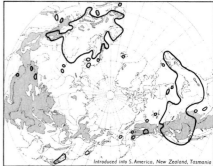

24. Pòa compréssa L.

Bluish-green; culm strongly flattened, geniculate, from creeping rhizome; ligules about 1 mm long; panicle stiff, branches spikelet-bearing to below middle; spikelets green; glumes rounded at apex; lemmas with obscure intermediate nerve.

Waste places, roadsides, yards; an introduced weed. Described from Europe and North America.

25. Gramineae (Grass Family)

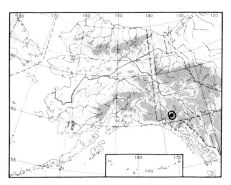
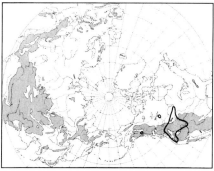

25. Pòa Cusíckii Vasey

Densely tufted; culms tall, erect, naked above; basal leaves numerous, filiform, erect, more or less scabrous; ligules long, acute; panicle narrow, pale green, shining; spikelets 7–9 mm long; glumes broad, unequal; lemmas thin, smooth, or somewhat scabrous.

Grassy slopes, probably introduced. Described from Oregon.

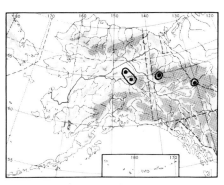

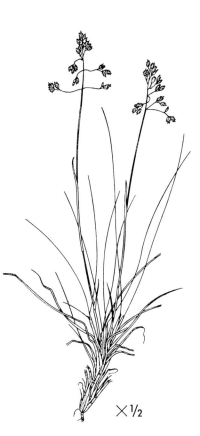

26. Pòa vaseyòchloa Scribn.

Poa pulchella Vasey not Salisb.; *P. Porsildii* Gjaerevoll; *Colpodium Wrightii* with respect to Yukon plant.

Tufted; culms glabrous, surrounded by old sheaths at base; basal leaves about 1 mm broad; ligules 1.5–2 mm long, laciniate; panicle open, branches smooth, naked below; lowest branches 1–2 together; spikelets dark purplish, 3–5-flowered; glumes acute, hyaline-margined; lemma 5-nerved, with minutely scabrous keel and nerves, lacking hairs; anthers 2.5–3.2 mm long.

Snow beds.

×⅓

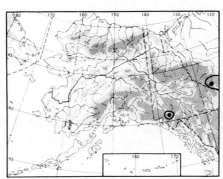 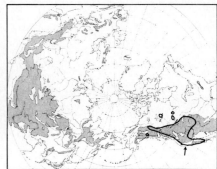

27. Pòa scabrélla (Thurb.) Benth.
Atropis scabrella Thurb.; *Poa Buckleyana* Nash.

Tufted; culms slender, scabrous; leaves mostly basal, soft, slender; ligules 3–7 mm long, acuminate; panicle narrow, contracted; spikelets 3–7-flowered; lemmas mostly puberulent toward base; anthers about 2 mm long.

Meadows, open woods. Highly variable.

×⅓

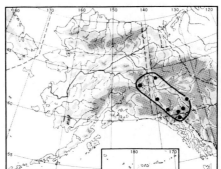 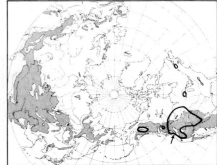

28. Pòa Canbỳi (Scribn.) Piper
Glyceria Canbyi Scribn.

Densely tufted; culms tall; basal leaves numerous, bright green, flat or folded, up to 2 mm wide; ligules 2–5 mm long; panicle narrow, compact, with appressed branches; spikelets 3–6-flowered, whitish-green; lemmas with rounded back, narrowly lanceolate, with obscure nerves, crisply puberulent on lower half; anthers 1.5–2.5 mm long.

Sandy soil. Described from Cascade Mountains of Washington.

25. GRAMINEAE (Grass Family) 143

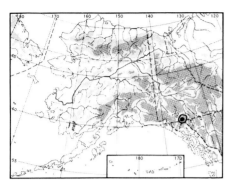
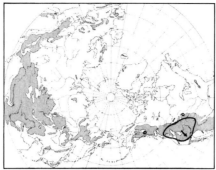

29. Pòa nevadénsis Vasey

Tufted; culms tall, erect, several, leafy throughout; leaves narrow, involute, stiff, mostly with scabrous sheaths; ligules 3–6 mm long, acute; panicle narrow, pale, with appressed branches; spikelets 4–8 mm long; glumes narrow; lemmas obtuse, scarious at tip, more or less scabrous, with obscure keel, lacking tuft of cobweb hairs at base.

Meadows, wet places.

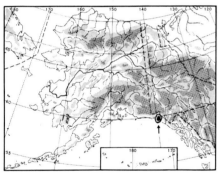
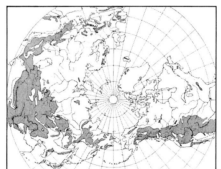

30. Pòa Merrilliàna Hitchc.

Poa glacialis Scribn. & Merr., not Stapf.

Densely caespitose, glabrous; leaves 3–4 mm broad, flat, glabrous; pale green, abruptly acute; panicle open, with few large, purple spikelets at end of flexuous branches; ligules about 2 mm long; glumes unequal, broadly lanceolate, 3-nerved; lemmas 5-nerved, not webbed at base, pubescent on keel and marginal nerve; anthers 0.6–0.7 mm long.

A doubtful species known only from Hubbard Glacier, Yakutat Bay.

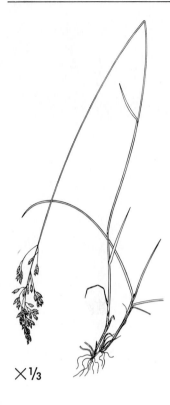
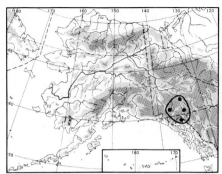
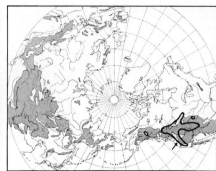

31. Pòa ámpla Merrill

Tufted, more or less glaucous; culms tall, erect, slender; sheaths glabrous; basal leaves numerous, flat, 1–3 mm broad; panicle narrow, elongated, dense; spikelets narrow, 4–7-flowered; glumes acuminate, scabrous on keel; lemmas scabrous in upper part, often more or less purplish, lacking tuft of cobweb hairs at base; anthers 2 mm long.

Dry grassland.

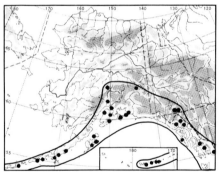
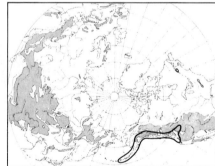

32. Pòa stenántha Trin.

Poa flavicans Ledeb.; *P. acutiglumis* Scribn.

Tufted, glabrous or essentially so; basal leaves short, narrow; ligules long; panicle oblong, open in age; lowest panicle branches 2–3 together, slender, scabrous, naked below; spikelets lanceolate, 3–6-flowered; glumes ovate-lanceolate, much shorter than spikelet; lemmas rounded on back, oblong-lanceolate, acute, broadly scarious-margined at top, with white hairs on keel and marginal nerve.

Meadows. Described from Kamchatka (where it does not occur), Unalaska, and Sitka. Much misunderstood; circumpolar map tentative.

Viviparous specimens often occur.

25. Gramineae (Grass Family)

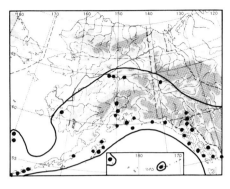 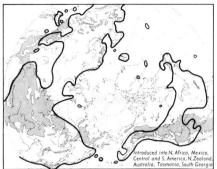 

33. Pòa ánnua L.

Tufted, light green; leaves soft, smooth, flat, abruptly contracted at tip; panicle pyramidal, with spikelets along branches; spikelets yellowish-green, with narrow, acute, unequal, boat-shaped glumes, the lower 1-nerved, the upper 3-nerved; lemmas 5-nerved, glabrous or pubescent on nerves toward base; anthers 0.7 mm long.

Roadsides, yards; an introduced, mostly annual weed. Described from Europe.

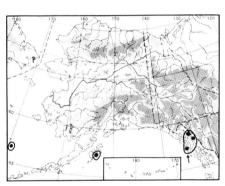 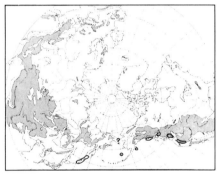 

34. Pòa leptocòma Trin.

Loosely tufted, culm up to 25 cm tall; leaves 2–3 mm broad, somewhat scabrous in margin, panicle nodding, few-flowered, with capillary branches, the lowest mostly 2 together; spikelets green or grayish-violet-green; glumes narrow; lemma pubescent on keel and marginal nerves, with tuft of cobweb hairs at base.

Wet places, snow beds. Circumpolar map tentative.

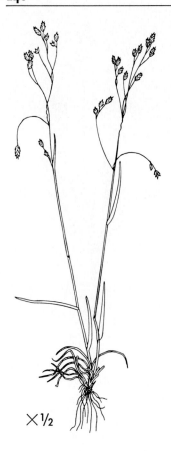
×½

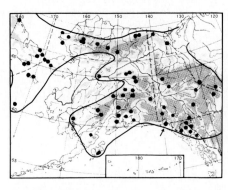
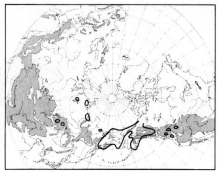

35. Pòa paucispìcula Scribn. & Merr.

Densely tufted, culm mostly 10–15 cm tall; leaves nearly glabrous; spikelets and glumes as in *P. leptocoma,* but dark violet-colored.

Rocky slopes, snow beds.

Doubtfully distinct from *P. leptocoma.* Circumpolar map tentative.

×½

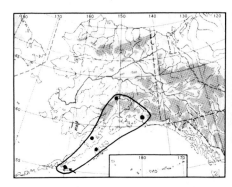
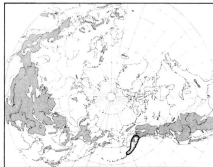

36. Pòa brachyánthera Hult.

Loosely tufted; culms glabrous; basal leaves short, narrow; panicle ovate, open, with long, slender, glabrous branches, naked below; spikelets long-pedicellated, mostly green; glumes glabrous, the lower 1-nerved, the upper 1–2-nerved; lemmas glabrous between keel and puberulent marginal nerve; anthers 0.5 mm long.

Rocky slopes, snow beds.

25. Gramineae (Grass Family)

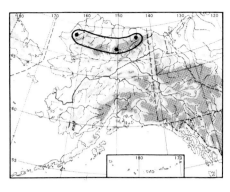 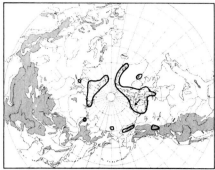 

×⅔

37. Pòa abbreviàta R. Br.

Densely tufted; culms glabrous, short; leaves linear, glabrous; panicle dense, with short branches; spikelets ovate, purple; glumes lanceolate; lemmas obscurely nerved, densely pubescent in lower part, lacking tuft of cobweb hairs at base.
Arctic tundra.

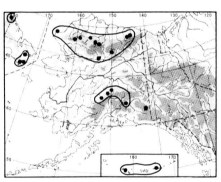 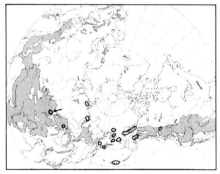 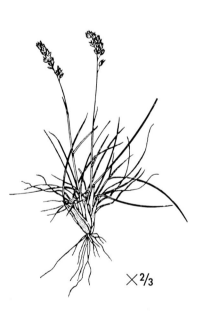

×⅔

38. Pòa pseudoabbreviàta Roshev.
Poa Jordalii Pors.?

Densely tufted; culms glabrous; leaves linear, short, with ligules 1.5–2 mm long; panicle oblong to pyramidal, with scabrous branches; spikelets long-pedicellated, purple; lemmas obscurely nerved, glabrous between keel and puberulent marginal nerve; tuft of cobweb hairs lacking at base; anthers 0.3–0.5 mm long.
Dry ridges.

Colpodium Trin.

Plant up to 15 cm tall; panicle branches erect, appressed, short ... 2. *C. Vahlianum*
Plant usually taller; panicle branches spreading or reflexed, with spikelets only at apex .... 1. *C. Wrightii*

×½

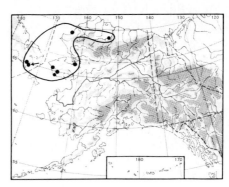
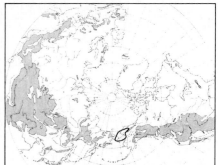

1. Colpòdium Wrìghtii Scribn. & Merr.

Poa Wrightii (Scribn. & Merr.) Hitchc.; *Puccinellia Wrightii* (Scribn. & Merr.) Tsvel.

Caespitose, glabrous, up to 50 cm tall; basal leaves numerous, short, linear, involute with marcescent sheaths; ligules about 3 mm long, acute; panicle with spreading or ascending, glabrous branches, bearing few spikelets at apex; spikelets 3–4-flowered, purplish; first glume shorter than the obtuse, 3-nerved second; lemmas lanceolate, 4.5–5 mm long, prominently 5-nerved, appressed, silky-pubescent at base; keels of palea scabrous.

Wet meadows.

×½

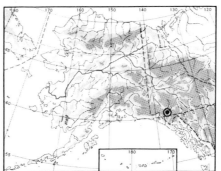
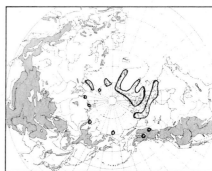

2. Colpòdium Vahliànum (Liebm.) Nevski

Poa Vahliana Liebm.

Caespitose, yellowish-green, up to 15 cm tall; leaves glabrous; stem leaves 1–3; ligules acute, up to 4 mm long; panicle narrow; spikelets lanceolate, violet, with 2–4 flowers; glumes thin, the lower 1-nerved, the upper longer, 3-nerved; lemma ovate-lanceolate, 3-nerved, pubescent in the lower half.

Stony tundra. Described from Niokornak (western Greenland).

Hybridizes with *Phippsia algida* (see p. 92) [*Puccinellia vacillans* (T. Fries) Schol.].

25. GRAMINEAE (Grass Family)

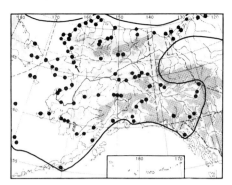 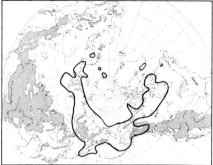

Arctophila R. Br.

1. Arctòphila fúlva (Trin.) Anderss.   Pendent Grass

Poa fulva Trin.; *Colpodium fulvum* (Trin.) Griseb.; *Graphephorum fulvum* (Trin.) Gray; *Poa pendulina* Laest.; *Arctophila pendulina* (Laest.) Anderss.; *Colpodium pendulinum* (Laest.) Griseb.; *Graphephorum pendulinum* (Laest.) Gray; *Arctophila effusa* Lange; *Colpodium mucronata* (Hack.) Beal; *Arctophila mucronata* Hack.; *A. brizoides* Holm; *A. chrysantha* Holm; *A. trichopoda* Holm.

Up to 9 dm high, yellowish-green; culms from stout but very brittle subterranean runners; leaves of sterile shoots arranged in 2 rows; leaves up to 20 cm long, 3–8 mm broad; panicle ample, open, with capillary branches; spikelets with 1–7 flowers; glumes broadly scarious-margined, shorter than adjacent lemma.

Arctic polygon tundra; lakeshores and banks of streams in shallow water.

Highly variable. Specimens from interior have larger panicle and many-flowered spikelets.

Dupontia R. Br.

Glumes blunt; lemmas subsericeous . . . . 1. *D. Fischeri* subsp. *Fischeri*
Glumes long-acuminate; lemmas glabrous . . . . 2. *D. Fischeri* subsp. *psilosantha*

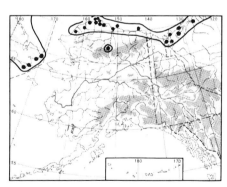 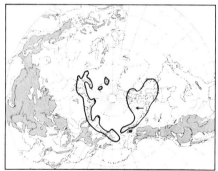

1. Dupóntia Físcheri R. Br.   Tundra Grass
subsp. **Físcheri**

Culms up to 30 cm high, from creeping rhizome; panicle narrow, with ascending, smooth, appressed, branches; glumes smooth, blunt, yellowish-green or purple, with gold tips; lemmas much shorter than glumes, obtuse, subsericeous or slightly hirsute, with longer hairs at callus; anthers 1.7–2.5 mm long.

Sandy shores, wet meadows.

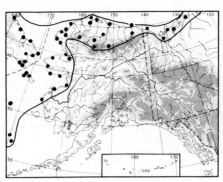 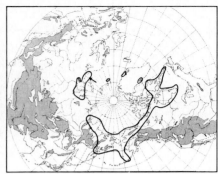

2. Dupóntia Físcheri R. Br.
subsp. **psilosántha** (Rupr.) Hult.
   Dupontia psilosantha Rupr.

Similar to *D. Fischeri*, but taller, usually with longer panicle and, at maturity, spreading lower branches; glumes longer, long-acuminate; lemma longer, glabrous; anthers about 3 mm long.

Wet tundra.

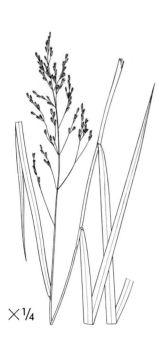

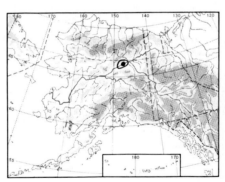 

Scolochloa Link

1. Scolòchloa festucàcea (Willd.) Link   Sprangletop
   Arundo festucacea Willd.

Culm coarse, glabrous, often branched, about 1 meter tall, from stout, creeping subterranean stems; leaves flat, 30–50 cm long, 6–12 mm broad, long-acuminate, drooping; panicle open, with fine scabrous branches and few spikelets toward ends; glumes 7–10 mm long, somewhat shorter than spikelet; lemmas with 3–5 short awns at tip, lacking awn on back; short tufts of hair at base of the lemma.

Open water along lakes and creeks. Described from Europe.

25. GRAMINEAE (Grass Family)

Glyceria R. Br. (Manna Grass)

Spikelets linear, nearly terete when young, 10 mm long or longer when fully developed; panicle narrow, erect:
  Lemmas glabrous except on keel .... 1. *G. borealis*
  Lemmas scabrous on nerves, somewhat so between nerves ... 2. *G. leptostachya*
Spikelets ovate or oblong, shorter, more or less compressed; panicle nodding:
  Lemmas apparently 5-nerved (not including marginal nerves); nerves markedly scabrous .... 3. *G. pauciflora*
  Lemmas 7-nerved, with distinct margin along lemma outside marginal nerve:
    First glume 0.6–0.8 mm long; lemma small, about 2 (1–3) mm. long, very prominently nerved .... 5. *G. striata* subsp. *stricta*
    First glume longer than 2 mm, not so prominently nerved:
      Plant stout; panicle 20–40 cm long, with numerous spikelets; lemmas nearly entire at apex, without or with inconspicuous scarious margin .... 6. *G. maxima* subsp. *grandis*
      Plant slender; panicle usually less than 25 cm long, not so rich in spikelets; lemmas with broad, scarious margin, distinctly erose at top .... 4. *G. pulchella*

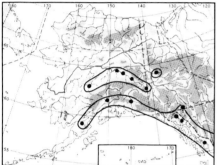 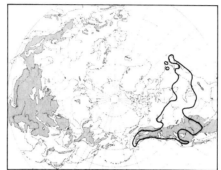

×⅓

1. Glycèria boreàlis (Nash) Batchelder
Panicularia borealis Nash.

Culm up to 1 meter tall, from decumbent base; leaves glabrous or somewhat scabrous above; panicle long and narrow, with linear-cylindrical spikelets, 1–1.5 cm long, 5–12-flowered; glumes scarious-margined; lemma glabrous, scabrous on keel, strongly 7-nerved, obtuse.

Wet places, in shallow water, sometimes with floating leaves. Described from North America.

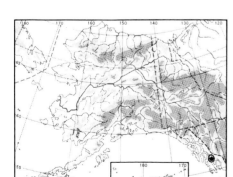 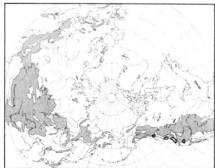

×⅓

2. Glycèria leptostáchya Buckl.

Similar to *G. borealis*, but stouter; leaf blades scaberulous on upper surface, broader (4–10 mm broad); lemmas scaberulous between nerves, especially between keel and next nerve.

Shallow water. Described from Oregon.

25. GRAMINEAE (Grass Family)

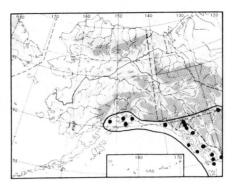 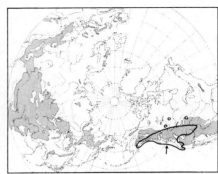

3. Glycèria pauciflòra Presl
Puccinellia pauciflora (Presl) Munz; *Panicularia pauciflora* (Presl) Ktze.

Culms up to 1 meter tall, from subterranean runners; leaf blades up to 15 cm long and 1 cm wide, flat, scaberulous; panicle open, nodding, with ascending or spreading, flexuous branches, with spikelets toward ends; spikelets 5–6-flowered; lemmas with 2 prominent nerves on each side of keel, exclusive of the prominent marginal nerve.

Wet meadows, shallow water.

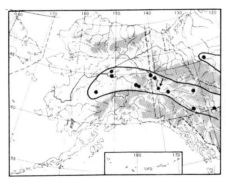 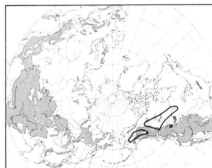

4. Glycèria pulchélla (Nash) Schum.
Panicularia pulchella Nash.

Stoloniferous, up to 1 meter tall; leaves 4–5 mm broad, scabrous; panicle lax, 15–25 cm long, with capillary, ascending branches and spikelets toward ends; spikelets 4–6-flowered; glumes short, obtuse; lemmas purplish, with rounded apex and broad, erose, scarious margin.

Wet places.

25. GRAMINEAE (Grass Family)

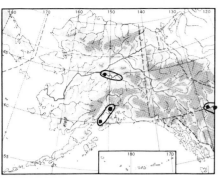 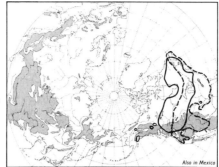 

5. Glycèria striàta (Lam.) Hitchc.
  Poa striáta Lam.
subsp. **strícta** (Scribn.) Hult.
  Panicularia nervata var. *stricta* Scribn.; *Glyceria striata* var. *stricta* (Scribn.) Fern.

Loosely tufted, stiff, up to 50 cm tall; leaves flat, up to 5 mm broad; panicle lax, open, with ascending branches; spikelet purplish, about 3 mm long; lemmas with broad scarious tips, very prominently nerved.

Moist ground. *G. striata* described from Virginia and the Carolinas, subsp. *stricta* from Colorado–Wyoming boundary.

Broken line on circumpolar map indicates range of subsp. **striàta.**

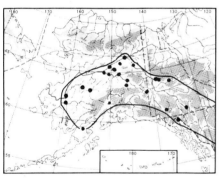 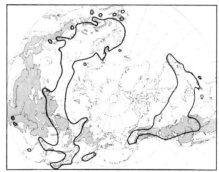 

6. Glycèria máxima (Hartm.) Holmb.
  Molinia maxima Hartm.
subsp. **grándis** (S. Wats.) Hult.
  Glyceria grandis S. Wats.; *G. grandis* var. *Komarovii* Kelso; *G. Hulteniana* Löve.

Culms stout, up to 1.5 meters tall; panicle 2–4 dm long, highly compound, with numerous, usually purplish, spikelets on smooth, capillary branches; leaf blades more than 2 dm long, up to 12 mm broad; spikelets 4–8-flowered, 5–6 mm long; glumes acute; lemma 7-nerved, with blunt apex, lacking scarious margin.

Wet ground, banks of streams. *G. maxima* described from Stockholm, subsp. *grandis* from Quebec.

A link in a circumpolar chain of very closely related taxa. Alaskan plants have somewhat smaller spikelets and smaller second glumes than subsp. **máxima.** Circumpolar map gives range of entire species complex.

Puccinellia Parl. (Alkali Grass)
(Key and identifications by Dr. Th. Sørensen, Copenhagen)

Glumes and lemmas entire, not evidently erose-ciliolate:
  Anthers 1.2–2.0 mm long:
    Glumes and lemmas firm, not translucent; panicle meager, with few spikelets; stoloniferous:
      Anthers commonly without pollen, never developing fruit .... 1. *P. phryganodes*
      Anthers polliniferous, developing fruit .... 2. *P. geniculata*
    Glumes and lemmas translucent, glossy; panicle compound:
      Lemmas distinctly nerved, often with a few spines at back; keels of palea densely ciliate .... 3. *P. arctica*
      Lemmas obscurely nerved, not spiny at back; keels of palea sparsely ciliate to almost glabrous .... 4. *P. agrostoidea*
  Anthers 0.5–1.0 mm long:
    Lemmas herbaceous throughout; veins prominent; anthers about 0.5 mm long:
      Lemmas about 3 mm long, acute to acuminate, conspicuously hairy in lower part .... 5. *P. Langeana* subsp. *alaskana*
      Lemmas 1.8–2.5 mm long, bluntly acute to obtuse, sparsely hairy at base or glabrous:
        Lemmas incurved, veins conspicuously converging toward apex, sparsely hairy at extreme base .... 5. *P. Langeana* subsp. *asiatica*
        Lemmas not incurved or slightly incurved, veins slightly converging, commonly without hairs at base ...... 5. *P. Langeana* subsp. *Langeana*
    Lemmas firm throughout, or scarious-translucent in upper part; veins not prominent; anthers 0.6–1.0 mm long:
      Pedicels with pearly luster from tumid epidermis cells; spikelets commonly greenish:
        Pedicels commonly thickened, their glumes inserted at different levels; panicle branches stiffly ascending .... 6. *P. nutkaënsis*
        Pedicels not thickened, glumes apparently opposite; panicle branches spreading or reflexed:
          Culms short, 10–30 cm; lemmas acute, entirely firm ..... 7. *P. pumila*
          Culms tall, 30–40 cm; lemmas subobtuse, translucent and glossy toward apex .... 8. *P. Hultenii*
      Pedicels not distinctly pearly-lustrous; spikelets purple:
        Lemmas conspicuously hairy in lower part; keels of palea commonly woolly-haired .... 9. *P. angustata*
        Lemmas glabrous or only sparsely hairy on nerves at base; keels of palea glabrous or sparsely spinulose near summit; glume and lemmas commonly coarse-toothed .... 10. *P. Andersonii*
Glumes and lemmas erose-ciliolate:
  Palea distinctly exceeding lemma:
    Panicle branches and pedicels scabrous .... 11. *P. kamtschatica*
    Panicle branches and pedicels glabrous . See 11. *P. kamtschatica* var. *sublaevis*
  Palea not exceeding lemma:
    Pedicels pearly-lustrous, commonly thickened .... 6. *P. nutkaënsis*
    Pedicels not lustrous, not or scarcely thickened:
      Keels of palea and lower part of lemma woolly-haired ...... 9. *P. angustata*
      Keels of palea smooth or finely spinulose; lemmas glabrous or faintly hairy (sericeous, not woolly) at base:
        Plant low-growing; culms mostly geniculate; panicle about half the length of entire plant; glumes extremely unequal, the first acute, the second of double length, dilated, more or less truncate ..... 12. *P. vaginata*
        Plant taller; culms more or less erect; panicle commonly not exceeding one-third the length of entire plant; glumes less different in shape and size:
          ■ Lemmas 3–4 mm long; anthers 1.0–1.5 mm long:
            Panicle branches obviously scabrous .... 13. *P. grandis*

25. GRAMINEAE (Grass Family)

  Panicle branches glabrous or sparsely spinulose:
   Panicle branches loosely ascending; spikelets 5–7-flowered .... 14. *P. glabra*
   Panicle branches stiffly spreading or reflexed; spikelets 2–3-flowered .... 15. *P. triflora*
■ Lemmas 1.8–2.5 mm long (rarely up to 3 mm, anthers then less than 1 mm long):
  Anthers 1.2–1.5 mm long:
   Lemmas distinctly nerved .... 3. *P. arctica*
   Lemmas obscurely nerved .... 4. *P. agrostoidea*
  Anthers commonly 0.5–0.8 mm long:
   Glumes more or less keeled; lemmas acute or subacute:
    Glumes and lemmas glossy, translucent; keels of glumes often sparsely spinulose .... 16. *P. Nuttalliana*
    Glumes and lemmas firm, pruinose; keels of glumes without spinules .... 17. *P. deschampsioides*
   Glumes not keeled; lemmas obtuse or truncate:
    Spikelets loose; rachilla joints about 0.8 mm long; lemmas commonly 2.2–2.4 mm long, distinctly hairy on veins in lower fourth .... 18. *P. borealis*
    Spikelets rather tight; rachilla joints about 0.5 mm long; lemmas commonly 1.8–2.2 mm long, sparsely hairy at base, or almost glabrous:
     Anthers 0.8–1 mm long; leaves 2–5 mm broad . 21. *P. distans*
     Anthers 0.5–0.6 mm long; leaves 1–2 mm broad:
      Glumes obtuse; lemmas thin, commonly opaque, truncate, hairy on nerves at base; panicle cylindrical .... 19. *P. Hauptiana*
      Glumes somewhat acute; lemmas rather firm, shining, obtuse, not truncate, almost glabrous; callus hairs present; panicle pyramidal .... 20. *P. interior*

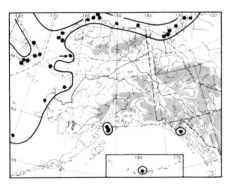 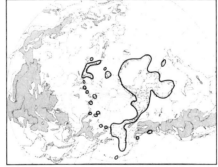 

×¾

1. Puccinéllia phryganòdes (Trin.) Scribn. & Merr.
Poa phryganodes Trin.

 Forming loose tufts, with long stolons particularly appressed to ground; stolons with short, sprouting, extra-axillary shoots, opposite bases of flaccid leaves; rarely flowering; panicle purplish, glabrous, with 2–3 branches from lower node; spikelets oblong, 3–6-flowered; glumes obtuse, the first 1-nerved, the second 3-nerved; lemmas oblong, obtuse, obscurely 5-nerved, entire, glabrous; callus hairs absent; keels of palea glabrous; anthers 1.5–2 mm long.
 Salt and brackish marshes along shores.

×¾

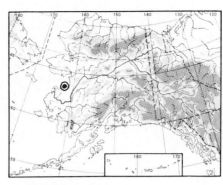
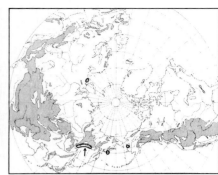

2. Puccinéllia geniculàta (Turcz.) Krecz.
Poa geniculata Turcz.

Loosely tufted; culm up to 30 cm tall; leaves 1.5–2 mm wide, glabrous; panicle with glabrous branches, at first erect, later spreading; spikelets 2–5-flowered, oblong; glumes obtuse, with the second longer than the first; lemma purplish, ovate, 3–3.5 mm long, indistinctly nerved, blunt, glabrous at base; palea with glabrous keels; anthers 1.3–2.0 mm.

Salt and brackish marshes along shores.

×⅓

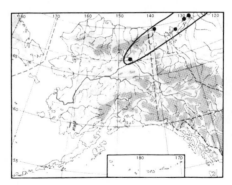
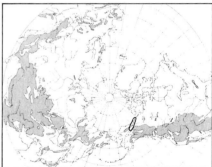

3. Puccinéllia árctica (Hook.) Fern. & Weath.
Glyceria arctica Hook.

Caespitose; culms stout, erect or geniculate at base, leaves furrowed, 1.5–2.0 mm wide at base, tapering toward apex; ligules 1.5–2 mm long; panicle 6–9 cm long, lanceolate-oblong, with short, slender, slightly scabrous, ascending branches, 3–5 from each node; spikelets 5–9-flowered; glumes and lemmas very thin, translucent, distinctly nerved, sometimes dentate (not erose-ciliate); first glume ovate, obtuse, 1-nerved, the second 3-nerved; lemmas broadly oblong, obtuse, 5-nerved, slightly pilose on nerves below; callus hairs short; anthers 1.8–2 mm long.

Seashores. Described from "Arctic Sea-coast" (Richards.).

25. GRAMINEAE (Grass Family)

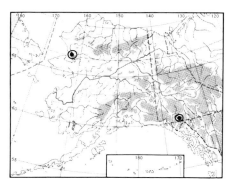
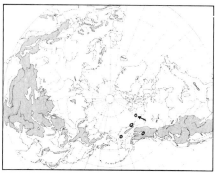

4. Puccinéllia agrostoìdea Sørens.

Caespitose, purplish-tinged; culms erect, up to 20 cm tall; stem leaves about 1.3 mm wide at base, with strongly scabrous margins, involute; upper sheaths widened; ligules about 2 mm long, truncate, often lacerate; panicle linear-lanceolate, with fasciculate, appressed, capillary, nearly glabrous branches; spikelets 4–5 mm long, 2–4-flowered; bracts purple, translucent, obscurely nerved, not ciliolate; first glume ovate, acute, 1-nerved, the second oblong-ovate, acutish, 3-nerved; lemmas oblong, 5-nerved, rounded or abruptly pointed, slightly hairy on nerves near base; anthers 1.2–1.5 mm long.

Salt or alkaline soil.

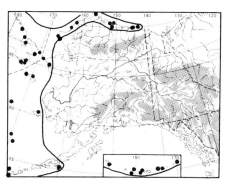
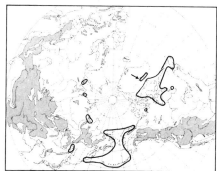

5. Puccinéllia Langeàna (Berl.) Sørens.

Glyceria Langeana Berl.; *G. paupercula* Holm; *Puccinellia paupercula* (Holm) Fern. & Weath.; *P. alaskana* Scribn. & Merr.; *Atropis laeviuscula* Krecz.

Caespitose, glabrous, glaucous; culms tall, erect or recurved; ligules 1.5–2 mm long; cauline leaves 1–2 mm wide; panicle branches in pairs; spikelets appressed, short-pedicellate; pedicels not or very slightly thickened; spikelets more or less purplish, 4–6 mm long, 3-flowered; glumes distinctly nerved, narrowly white-margined, the second longer; lemmas 5-nerved, glabrous or more or less hairy at base; callus hairs absent; anthers 0.4–0.6 mm long.

Shores, in salt and brackish water.

Sørensen distinguishes 3 very closely related subspecies (**Langeàna, alaskàna,** and **asiática**); for differences, see the key.

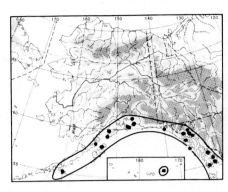 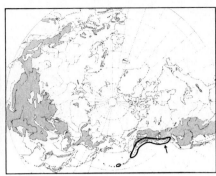

6. Puccinéllia nutkaénsis (Presl) Fern. & Weath.
Poa nutkaensis Presl.

Densely tufted; culms erect, slender, up to 30 cm (rarely 45 cm) tall; ligules 1–2 mm long, obtuse; leaves 1–2 mm wide, soft, flat, folded, glabrous or sparsely scabrous above; panicle with few, slender, glabrous, appressed branches, naked at base; spikelets 5–6-flowered, 7–8 mm long; first glume subobtuse, 1-nerved, the second ovate, obscurely ciliate, 3-nerved; lemmas elliptic, glabrous or with a few hairs at base; anthers 0.8–1.2 mm long.

Seashores.

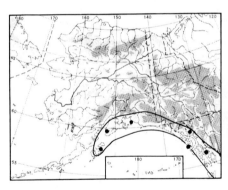 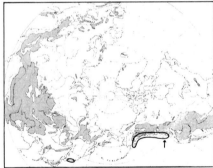

7. Puccinéllia pùmila (Vasey) Hitchc.
Glyceria pumila Vasey.

Tufted; culms erect or decumbent at base, up to 30 cm tall; ligules 1.5–2.3 mm long; leaves flat, 1–1.25 mm wide, scaberulous; panicles with glabrous branches; branches stiffly ascending to reflexed, naked in lower half; spikelets 4–6-flowered, appressed; first glume 1-nerved, subacute, the second 3-nerved; lemma with conspicuous nerves, abruptly narrowed to subacute apex; palea with glabrous keels; callus hairs present; anthers 0.8–1.2 mm long.

Seashores.

25. GRAMINEAE (Grass Family)

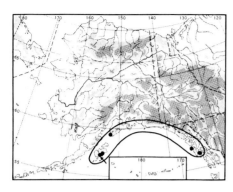  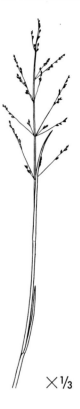

8. Puccinéllia Hulténii Swallen

Tufted; culms up to 40 cm tall, stiffly erect from a decumbent base; ligules 2.5–3 mm long; leaves 2–2.5 mm wide, stiff, erect, glabrous, strongly nerved; panicle with mixed short and long branches; branches stiffly ascending or spreading, glabrous or obscurely scabrous, 2 to several at each node; spikelets 3–4-flowered; first glume acute or subacute, 1-nerved, the second broader, 3-nerved; lemma somewhat acute, obscurely pubescent on lateral nerves at base; palea strongly scabrous on keels; anthers 0.8 mm long.

Seashores.

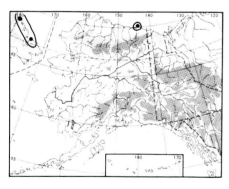 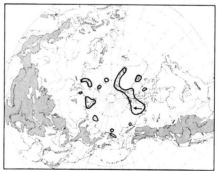 

9. Puccinéllia angustàta (R. Br.) Rand & Redf.
Poa angustata R. Br.

Caespitose; old basal sheaths whitish; culms stout, rigid, decumbent at base; cauline leaves 1.5–2.5 mm wide, flat, nearly glabrous, upper sheaths longer than leaves; ligules 2.5–4.0 mm long; panicle dense, erect, with stout, ascending branches, scabrous above, longer and shorter ones from the same node; spikelets purple, pedicelled to subsessile, 3–4-flowered; first glume 1-nerved, oblong-lanceolate, acute, the second 3-nerved, oblong, indistinctly toothed; lemmas obscurely 5-nerved, copiously pilose on back below, translucent at apex; keels of palea long-hairy below, spinulose at apex; callus hairs profuse; anthers 0.5–0.8 mm long.

Wet places along shore.

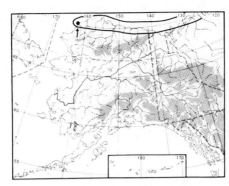
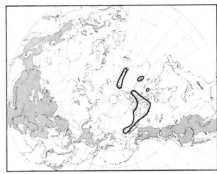

10. Puccinéllia Andersònii Swallen

Caespitose, glaucescent, reddish-tinged; culms robust, geniculate, 2-leaved; leaves 2 mm wide, folded, abruptly contracted at apex; ligules 1.5–2 mm long; panicle contracted, with 2–3 glabrous, ascending branches at lower nodes; spikelets subsessile or short-pedicellate; pedicels thickened; spikelets pink to purplish, 4–7-flowered; glumes acute, the first 1-nerved, the second 3-nerved, irregularly dentate; lemmas entire at apex, somewhat acute, thin, translucent at apex, slightly pilose at base; callus hairs present; anthers 0.8–1 mm long.

Wet places along shore.

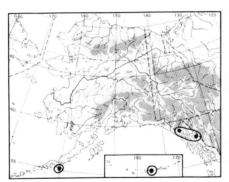

11. Puccinéllia kamtschática Holmb.

Densely tufted; culms erect, decumbent at base, up to 25 cm tall; ligules about 2 mm long; leaves smooth, soft, flat, drying involute, not more than 2 mm wide; panicle with ascending to spreading branches; branches sparsely scabrous, bearing spikelets on upper half; spikelets 3–4-flowered; first glume acute, the second obtuse and much broader, with hyaline tip; lemmas obtuse, glabrous; anthers 0.6–0.8 mm long.

Wet places.

Var. **sublaèvis** Holmb. differs from the type plant in having glabrous (rather than scabrous) panicle branches and pedicels.

25. GRAMINEAE (Grass Family)

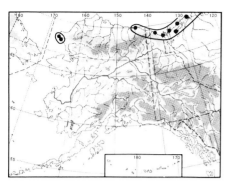
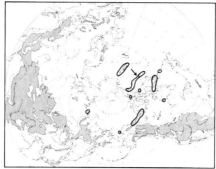

12. Puccinéllia vagináta (Lange) Fern. & Weath.
Glyceria vaginata Lange.

Caespitose; culms geniculate, 15–20 cm tall; cauline leaves 2–3 mm wide, nearly smooth, slightly purplish-tinged; ligules 0.8–1 mm long; panicle many-flowered, somewhat drooping, with 2–3 scabrous, slender branches, floriferous nearly to base from each node; spikelets linear-oblong, glossy; bracts strongly erose-ciliate, thin, translucent; first glume broadly ovate, obtuse, 1-nerved, the second suborbicular, obscurely 3-nerved; lemmas broadly ovate-truncate, obscurely 5-nerved, sparsely pilose on nerves below; keels of palea sparsely spinulose at apex; anthers 0.6–0.8 mm long.

Wet places along coast.

Some of the specimens from our area of interest belong to var. **paradóxa** Sørens., with nerves of glumes and lemmas evident, subcarinate glumes, and greenish panicle, whereas others belong to a closely related "western American" variety (Sørens.) with nerves of glumes and lemmas obscure, no carinate glumes, and purplish panicle.

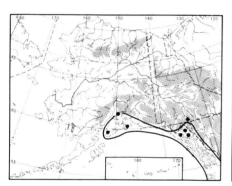

13. Puccinéllia grándis Swallen

Densely caespitose; culms up to 90 cm tall, erect or geniculate; ligules 2–3 mm long, obtuse; leaves flat, drying involute, 2–3 mm broad, those of the innovations soft; panicle pyramidal, with scabrous branches; branches at first appressed, later spreading; spikelets purple-tinged; first glume 1-nerved, obtuse or subacute, the second 3-nerved, obtuse, broader than the first, often minutely toothed; lemmas with obscure nerves, abruptly narrowed to obtuse or subacute apex, sparsely pilose at base; anthers mostly 1.3–1.5 mm long.

Salt marshes.

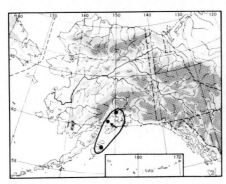
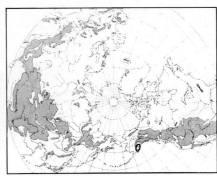

14. Puccinéllia glàbra Swallen

Densely tufted; culms glabrous, up to 40 cm tall, erect or decumbent at base; ligules 3–5 mm long; leaves 1.5–3 mm wide, glabrous; panicle long, with glabrous, ascending branches, naked at base; spikelets 5–7-flowered, appressed; first glume 1-nerved, acute or subobtuse, the second 3-nerved, obtuse, minutely ciliate; lemma 3.5–4 mm long, thin and shiny, with obscure nerves, obtuse, glabrous or with few hairs at base; palea with glabrous keels; anthers 1.3–1.5 mm long.

Wet places.

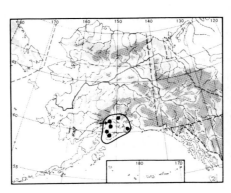

15. Puccinéllia triflòra Swallen

Densely tufted; culms erect, up to 60 cm tall; ligules 4–5 mm long; obtuse; leaves 1–1.5 mm wide, soft, glabrous, flat; panicle with glabrous, stiff, spreading or reflexed branches, 2–4 at each node, naked at base; spikelets 2–3-flowered, appressed, purple-tinged; glumes acute to subobtuse, the first 1-nerved, the second 3-nerved; lemmas 3.5–4 mm long, broad, obtuse with evident nerves, nearly glabrous at base; palea with prominent glabrous keels.

Wet places.

25. GRAMINEAE (Grass Family)

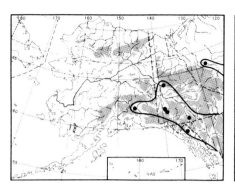
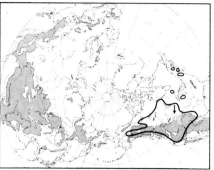
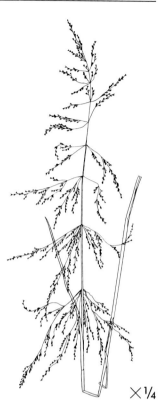

16. Puccinéllia Nuttalliàna (Schult.) Hitchc.
Poa Nuttalliana Schult.; *Puccinellia airoides* (Nutt.) Wats. & Coult.

Tufted; culms slender, stiff, erect, up to 60 cm tall; leaves 1–2 mm wide; panicle with ascending or spreading, strongly scabrous branches; spikelets 3–6-flowered, with rather scattered florets; glumes thick, erose-ciliate, 1-nerved, the second longer than the first, 3-nerved; lemma ovate, abruptly contracted to blunt or subacute apex, erose-serrulate, with nerves pubescent in lower part; anthers 0.6–0.7 mm long.
Alkaline soil.

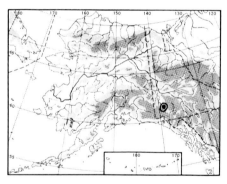
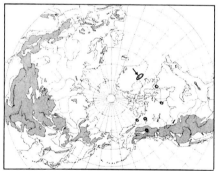

17. Puccinéllia deschampsioìdes Sørens.

Densely caespitose, with persisting old, withered sheaths; culms stout, erect, 2-leaved, naked above; leaves rigid, involute, scabrous at margin; ligules 2 mm long, truncate; branches in pairs, densely scabrous, ascending, in age spreading; spikelets 3–5-flowered, reddish-purple, variegated with yellow, shining; glumes and lemmas indistinctly nerved, faintly ciliate; first glume subacute, the second obtuse or subacute; lemmas obtuse, delicately pilose below, aureate-translucent at apex; keels of palea spinulose above, glabrous below; anthers 0.7–0.9 mm long.
Dry places.

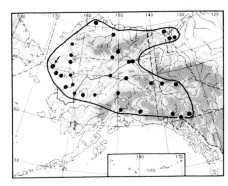
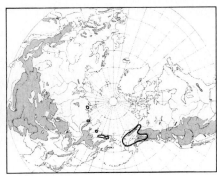

18. Puccinéllia boreàlis Swallen

Densely tufted, up to 35 cm tall; culm decumbent; ligules about 2 mm long, obtuse; leaves 1–2 mm wide, flat, scabrous above, glabrous below; panicle with slender scabrous branches; branches ascending to reflexed, 2–4 at nodes; spikelets appressed, 4–6-flowered, purple-tinged, short-pedicelled; first glume 1–1.5 mm long, acute, the second 1.5–2 mm long, obovate, obtuse; lemmas obtuse or subtruncate, minutely erose-ciliate; keels of palea hispid-ciliate; anthers 0.6–0.7 mm long.

Moist places.

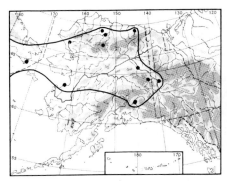
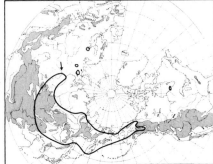

19. Puccinéllia Hauptiàna (Krecz.) Kitagawa

Atropis Hauptiana Krecz.

Caespitose; culms up to 60 cm tall; leaves 1–2 mm wide; panicle contracted, open in age, with long, capillary, scabrous branches; spikelets elongate, 6–8-flowered; glumes ovate; lemma obovate to broadly rounded, obtuse with obscure nerves, ciliolate, more or less pilose at base; anthers 0.5–0.6 mm long.

Wet places.

25. GRAMINEAE (Grass Family)

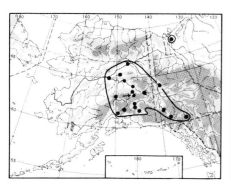 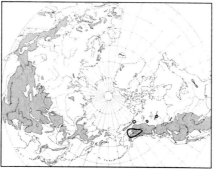 

20. Puccinéllia intèrior Sørens.

Densely caespitose, glaucous; culms erect, up to 55 cm tall; upper sheaths much longer than lower ones; leaves 1.5–2 mm broad, involute with scarious margin; ligules 1.8–2.3 mm long, obtuse; panicle pyramidal, with about 4 capillary, flexuous, scabrous branches to each node; branches in age reflexed; pedicels somewhat thickened; spikelets purple-variegated, shiny, mostly 3–4-flowered; glumes erose-ciliate, somewhat acute, oblong-lanceolate, the first 1-nerved, the second 3-nerved; lemmas obscurely 5-nerved, obtuse, broadly hyaline-margined, erose, minutely hairy at base, callus hairs present, anthers 0.5–0.6 mm long.

Wet places.

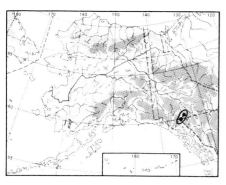 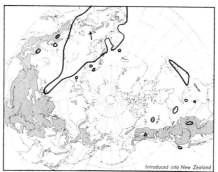 

21. Puccinéllia dístans (L.) Parl.
Poa distans L.

Tufted; culms slender, up to 60 cm tall; leaves 2–6 mm wide, flat; panicle pyramidal, with branches at first ascending, later deflexed, scabrous, fasciculated; spikelets 3–6-flowered; glumes ovate, blunt erose-ciliate, the first 1-nerved, the second 3-nerved; lemmas about 2 mm long, broadly ovate, obscurely nerved, obtuse or subtruncate, erose-ciliate, hyaline at apex, with a few hairs at base; palea erose-ciliate at apex; anthers 0.7–0.9 mm long.

Roadsides, yards, waste places; an introduced weed.

Festuca L. (Fescue Grass)

The taxa belonging to this genus form very intricate complexes that are still not known in detail. It is therefore probable that taxa not yet recognized occur within our area. **F. duriùscula** L., reported in Fl. Alaska & Yukon, seems doubtful and is therefore excluded here.

Basal leaves 3–10 mm broad, flat, lax:
    Lemmas awnless or with an awn not exceeding 2 mm .... 1. *F. arundinacea*
    Lemmas long-awned, 5–20 mm .... 2. *F. subulata*
Basal leaves narrow, folded or involute, more or less stiff:
    Plant tall; panicle open; spikelets 10–15 mm long; basal leaves with cylindrical, yellowish-brown sheaths .... 3. *F. altaica*
    Plant usually shorter, panicle spikelike; spikelets shorter; sheaths not conspicuous:
        Spikelets viviparous .... 5. "*F. vivipara*"
        Spikelets normal:
            Anthers small, 0.5–1 mm long:
                Culms glabrous .... 4 *F. brachyphylla*
                Culms pubescent below panicle .... 7. *F. baffinensis*
            Anthers longer, usually 1.5 mm long or longer:
                Culms decumbent at base; plant loosely caespitose, with short, creeping rootstocks; leaves more or less flat:
                    Culm leaves 2–3 mm broad; spikelets glaucous; upper sheaths loose.. .... 10. *F. rubra* subsp. *aucta*
                    Culm leaves narrower; spikelets usually not glaucous; upper sheaths not loose .... 9. *F. rubra* coll.
                Culms erect at base; plant densely caespitose; leaves filiform:
                    Culms 30–40 cm tall .... 6. *F. saximontana*
                    Culms usually less than 8 cm tall ..... 8. *F. ovina* subsp. *alaskensis*

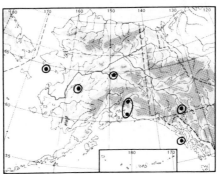

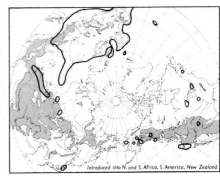

1. Festùca arundinàcea Schreb.
Festuca elatior L., in part.

Culms up to 1 meter tall, densely caespitose; leaves 5–10 mm wide, grayish-green, about 7 mm broad; panicle 15–30 cm long, open, drooping, with more than 3 spikelets on the shorter branch at lowest node; glumes narrow, acute; lemma ovate-lanceolate, short-awned, with scarious margin; anthers about 3 mm long.

Roadsides, yards, waste places; introduced. Described from Germany.

25. GRAMINEAE (Grass Family)

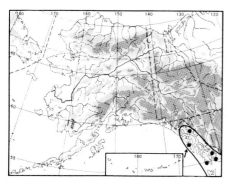
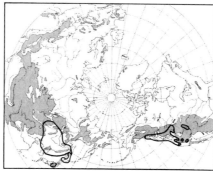

2. Festùca subulàta Trin.
Festuca Jonesii Scribn.

Up to 130 cm tall, leaf blades flat, 10 mm wide; panicle about 25 cm long or longer, loose, open, drooping, with long slender branches; spikelets 3–5-flowered; glumes acuminate; lemmas attenuate into awn, 5–20 mm long.
Woods.

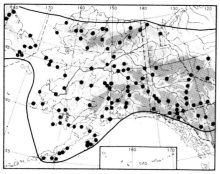
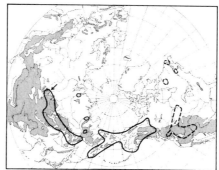

3. Festùca altaìca Trin.

Forming compact tufts; culms smooth, up to 80 cm high; old basal sheaths long-persistent, blades early disarticulating; basal leaves numerous, stiff, filiform, highly scabrous; 1–2 cauline leaves with short, broader blade; panicle open, with thin, scabrous, branched branches, bearing spikelet toward ends; spikelets 2–5 flowered; glumes broadly lanceolate to ovate, acuminate; lemmas 5-nerved, purplish-brown or sometimes yellowish-green (f. **pállida** Jordal).
Very common, with wide ecological tolerance, from sea level to at least 2,200 meters in mountains.
Broken line on circumpolar map indicates range of closely related **F. scabrélla** Torr.

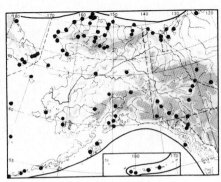
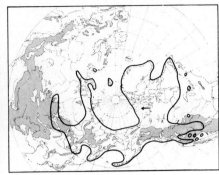

4. Festùca brachyphýlla Schult.

Festuca ovina of authors; *F. brevifolia* R. Br. not Muhl.

Densely tufted; culms up to 30 cm high, capillary; basal leaves capillary, soft; panicle cylindric to lanceolate-ovoid, compact; spikelets usually purplish, sometimes yellowish-green; lemmas not strongly involute, short-awned; anthers 0.5–1 mm long.

Sandy and rocky places on tundra, and in mountains to at least 2,300 meters.

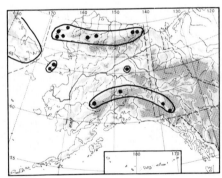
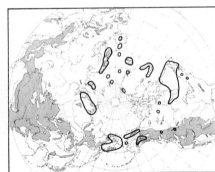

5. "Festùca vivípara"

Similar to *F. brachyphylla* but with viviparous spikelets.

According to some authors, this represents a distinct species. It seems more probable that it is a viviparous form of *F. brachyphylla* or *F. baffinensis*.

Circumpolar map gives range of viviparous *Festuca* of this type, in a broad sense, including **F. prolífera** (Piper) Fern. var. **lasiolèpis** Fern. (*F. rubra* var. *prolifera* Piper).

25. GRAMINEAE (Grass Family)

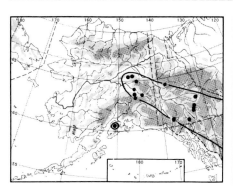
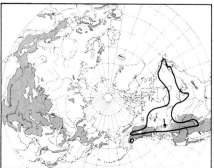

6. Festùca saximontàna Rydb.
Festuca brachyphylla subsp. *saximontana* (Rydb.) Hult.

Similar to *F. brachyphylla*, but taller, and with scabrous leaves; anthers 1.2–1.7 mm long.
Dry mountain slopes.

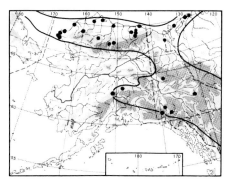
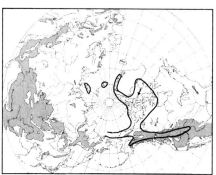

7. Festùca baffinénsis Polunin

Similar to *F. brachyphylla*, but culms more or less pubescent, at least below panicle; very densely tufted; panicle often broader and with very dark color.
Sandy and rocky places.

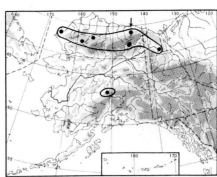
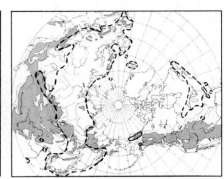

8. Festùca ovìna L.
subsp. **alaskénsis** Holmen

Similar to a small *F. brachyphylla*, but panicle more open during flowering and anthers longer than 2 mm; panicle usually purplish, sometimes yellowish-green (f. **pállida** Holmen).

Alpine slopes. Broken line on circumpolar map indicates range of *F. ovina* in a broad sense.

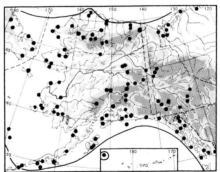
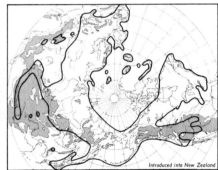

9. Festùca rùbra L. coll.

Including var. *barbata* and var. *arenaria* of authors, subsp. *Richardsonii* (R. Br.) Hult., and subsp. *cryophila* (Krecz. & Bobr.) Löve & Löve.

Loosely caespitose, with short, creeping rootstocks, sometimes forming dense tufts; culms glabrous, decumbent at base, 30–70 cm tall; basal leaves setaceous; culm leaves often flat; panicle with divergent lower branches; first glume 1-nerved, second 3-nerved; lemmas usually awn-tipped, narrowly scarious-margined; anthers 2–3 mm long.

Moist and sandy places. Described from Europe.

Extremely variable. Circumpolar map indicates range of *F. rubra* in a broad sense. On the Alaska map, all varieties occurring within area of interest are included except subsp. *aucta*.

25. GRAMINEAE (Grass Family)

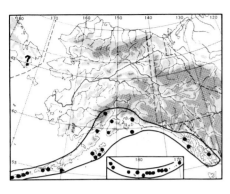 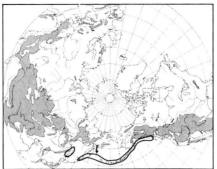 

10. Festùca rùbra L.
subsp. **aùcta** (Krecz. & Bobr.) Hult.
  Festuca aucta Krecz. & Bobr.

Similar to *F. rubra*, but with culm leaves 2–3 mm broad; glaucous; spikes large and dense; upper sheaths loose.
  Wet places.
  A coastal race. The only type occurring in the middle Aleutian Islands, where it is quite characteristic.

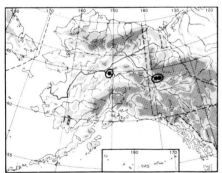 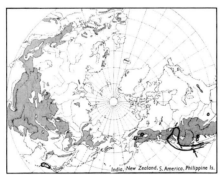 

Vulpia C. C. Gmel.

1. Vúlpia megalùra (Nutt.) Rydb.
  Festuca megalura Nutt.

Culms 20–50 cm tall; leaf blades narrow; panicle with appressed branches; spikelets 4–5-flowered; lemma glabrous or with short hairs on back, sometimes ciliated at margin, shorter than its awn; anthers 0.3–0.4 mm long.
  A garden weed at Dawson; also found at Manley Hot Springs.
  Range incompletely known; introduced in eastern North America and in Europe.

Bromus L. (Brome Grass)

First glume with keel nerve only, lacking lateral nerves:
    Awn as long as lemma or longer .... 1. *B. tectorum*
    Awn shorter than lemma:
        Lemma with row of long cilia on submarginal nerve; plant lacking stolons:
            Lemma glabrous on back; sheaths and nodes pubescent .... 2. *B. ciliatus*
            Lemma pubescent on back; sheaths and nodes glabrous ....
                .... 3. *B. Richardsonii*
        Lemma not long-ciliated, pubescent all over, or at least puberulent on lower part:
            Awn lacking or very short, less than 2 mm; lemma puberulent at base....
                .... 4. *B. inermis*
            Awn longer than 2 mm:
                Panicle large, open (about 20 cm long), with long, drooping branches; plant caespitose .... 8. *B. pacificus*
                Panicle about 15 cm long, narrow, with short branches, plant stoloniferous:
                    Glumes glabrous .... 5. *B. Pumpellianus* var. *Pumpellianus*
                    Glumes pilose or villous:
                        Glumes and lemmas pilose ...... 6. *B. Pumpellianus* var. *arcticus*
                      Glumes and lemmas densely villous ....
                        .... 7. *B. Pumpellianus* var. *villosissimus*
First glume with keel nerve and 1 or 2 pairs of lateral nerves:
    Lemmas awnless or nearly so, very broad, obtuse, inflated, glabrous........
        .... 9. *B. brizaeformis*
    Lemmas with long awn, not inflated, glabrous or hairy:
        Lemmas compressed, keeled from apex to base:
            Panicle branches elongate, drooping .... 10. *B. sitchensis*
            Panicle branches not greatly elongated:
                Spikelets glabrous or slightly pilose .. See 10. *B. sitchensis* var. *aleutensis*
                Spikelets coarsely pubescent .... 11. *B. marginatus*
        Lemmas not compressed and keeled, at least not at base:
            Sheaths glabrous; awn shorter than half the lemma ...... 12. *B. secalinus*
            Sheaths pilose; awn longer than half the lemma:
                Lemma pubescent .... 13. *B. hordeaceus*
                Lemma glabrous:
                    Margin of lemma forming even bow from base to apex..........
                    .... 14. *B. racemosus*
                  Margin of lemma forming angle above middle .... 15. *B. commutatus*

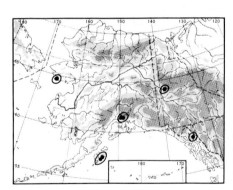

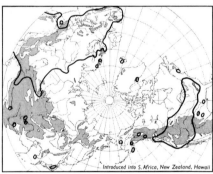

1. Bròmus tectòrum L.

Annual, tufted; culm and leaves usually pubescent; panicle nodding, with long flexuous branches; spikelets pubescent; glumes pilose, with long hairs; lemmas slenderly 5–7 nerved, hispid, with awn 3 cm long; anthers brown, about 0.5 mm long.

Waste places, roadsides, yards. Described from Europe.

25. GRAMINEAE (Grass Family)

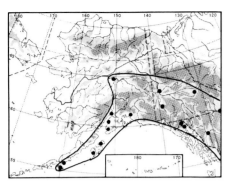
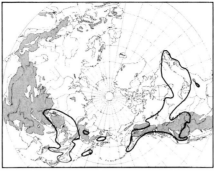

2. Bròmus ciliàtus L.
Bromus Dudleyi Fern.

Tufted; culms up to 1 meter tall; leaf blades up to 1 cm broad; panicle open, with slender, spreading, and drooping branches; first glume 1-nerved, the second 3-nerved; lemmas involute, glabrous, except ciliation along marginal nerve; palea linear; awn 3–5 mm long; anthers 1–2.5 mm.

Rocky slopes, thickets. Described from eastern Canada.

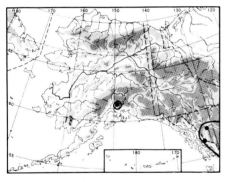
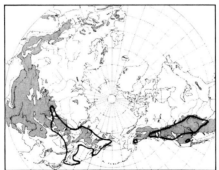

3. Bròmus Richardsònii Link

Similar to *B. ciliatus*, but lemma with short, appressed hairs across back; nodes and sheaths glabrous.

Described from cultivated specimens grown from seed collected in northwest North America. Closely related taxon in Asia.

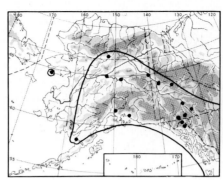
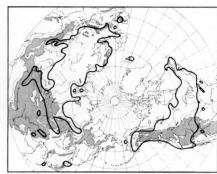

4. Bròmus inérmis Leyss.

Stoloniferous, with long runners and numerous sterile shoots and straight, stiff glabrous culms; leaves 5–10 mm broad; spikelets flat, narrow, about 2 cm long, often purplish-brown, with imbricate lemmas lacking awn; first glume 1-nerved; anthers about 4 mm long.

Roadsides, yards, waste places. Described from Europe.

Farther south, where *B. inermis* is more common, hybrid swarms with *B. Pumpellianus* occur.

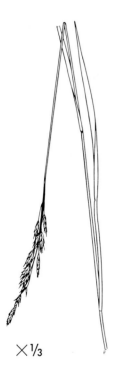
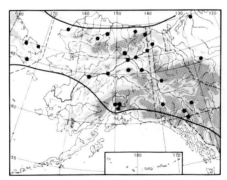
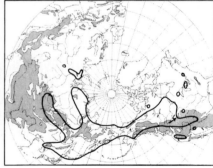

5. Bròmus Pumpelliànus Scribn.

Zerna Pumpelliana (Scribn.) Tsvel.; *Bromus inermis* subsp. *Pumpellianus* (Scribn.) Wagnon; *B. inermis* subsp. *Pumpellianus* var. *purpurascens* (Hook.) Wagnon; *B. purgans* var. *purpurascens* Hook.

var. Pumpelliànus

Stoloniferous, with long runners; culms stiff, up to 80 cm tall; panicle erect; leaves 4–8 mm broad; spikelets 1.5–2.5 cm long, usually purplish; first glume 1-nerved, the second 3-nerved, both glabrous; lemma more or less pubescent, especially at margin, with short awn 2–3 mm long; anthers about 4 mm long.

Meadows, dry grassy slopes, from lowlands and tundra up into mountains. In southern part of its range, hybrids with *B. inermis* occur.

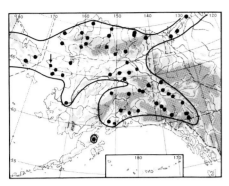
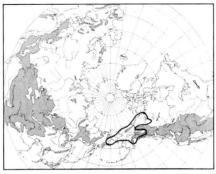

6. Bròmus Pumpelliànus Scribn.
var. **árcticus** (Shear) Pors.

Bromus arcticus Shear; *Zerna arctica* (Shear) Tsvel.; *B. inermis* subsp. *Pumpellianus* var. *arcticus* (Shear) Wagnon.

Similar to var. *Pumpellianus,* but with pilose glumes; the pubescence varies from quite prominent to just noticeable.

Meadows, dry grassy slopes, from lowlands and tundra up into mountains.

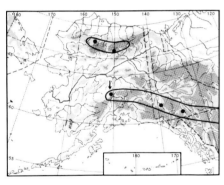
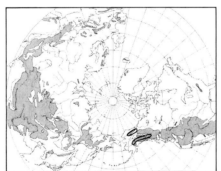

7. Bròmus Pumpelliànus Scribn.
var. **villosíssimus** Hult.

Similar to var. *Pumpellianus* and var. *arcticus,* but glumes and lemmas densely covered with long, gray, sericeous hairs.

This taxon may represent a separate species, mixing through introgression with var. *Pumpellianus.* If so, var. *arcticus* is the resulting hybrid population.

Habitat probably same as that of type variety.

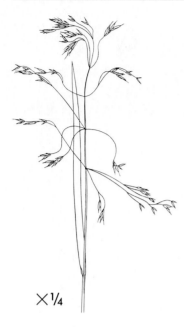
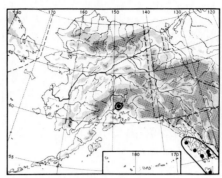
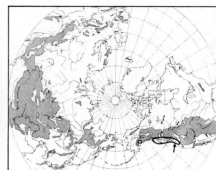

8. Bròmus pacíficus Shear
Bromus magnificus Elmer.

Culms tufted; up to more than 1 meter tall, stout, pubescent at nodes; sheaths pilose; leaf blades lax, 8–10 mm wide, pilose above; ligules 3–4 mm long; panicle markedly open, large, with slender, drooping branches; spikelets pubescent, 6–12-flowered; lemmas with awn 4–6 mm long, evenly pubescent on back, more densely so at margin.

Moist places, mostly near coast.

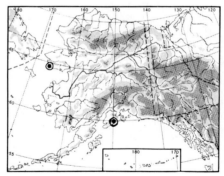
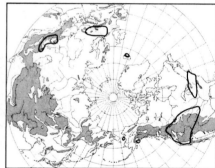

9. Bròmus brizaefórmis Fisch. & Mey.

Annual; culms up to 60 cm tall; leaf blades pubescent; panicle lax, drooping; spikelets flat, single or few, at ends of branches; glumes broad, obtuse, the first 2-nerved, the second 5-nerved; lemmas about 10 mm long, very broad, inflated, obtuse, smooth, with broad scarious margin, awnless or nearly so.

Waste places, roadsides, yards; introduced in Europe. Described from Europe. Circumpolar map tentative.

25. GRAMINEAE (Grass Family)

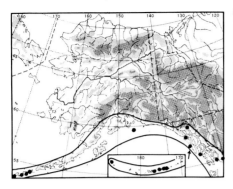
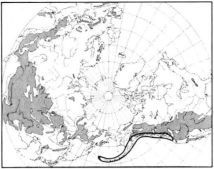

×¼

10. Bròmus sitchénsis Trin.

Culms stout; up to 1.5 meters tall, glabrous; leaves with sparse white hairs above, 7–12 mm broad; panicle large, lax, drooping, with long branches bearing spikelets toward ends; spikelets large, up to 3.5 cm long, 6–12-flowered, flattened; first glume 3-nerved, the second 5–7-nerved; lemmas compressed-keeled, glabrous to evenly pilose, with awn 5–10 mm long; anthers 5–6 mm long.

Woods and meadows.

Var. **aleuténsis** (Trin.) Hult. (*B. aleutensis* Trin.) is similar in all details except in having shorter, erect or appressed panicle-branches. Possibly only an early state of development occurring in the barren Aleutian Islands and at the altitudinal limit of the species.

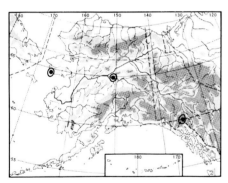
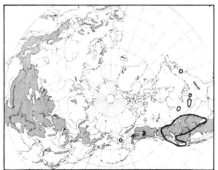

×⅓

11. Bròmus marginàtus Nees

Tufted; culms stout, pubescent; panicle erect, narrow, with stiff, spreading or ascending branches; spikelets compressed, 2.5–4 cm long, 5-9-flowered; first glume 3–5-nerved, scabrous, the second 5–7-nerved; lemmas subcoriaceous, pubescent, with awn 4–7 mm long; palea about equalling the lemma.

Waste places, roadsides, yards. Described from Columbia River.

×⅓

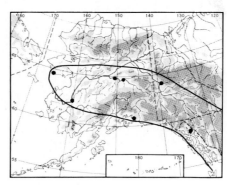
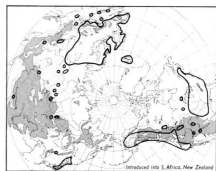

12. Bròmus secalìnus L.

Annual; with coarse, upright culm, and, at maturity, nodding, yellowish-green panicle, with long, more or less divergent, branches; leaf blades light-green, about 6 mm broad; spikelets large; glumes scabrous, scarious-margined, the first with 3–5 nerves, the second with 7; lemma firm, strongly inrolling, without awn or with short awn; anthers 2 mm long.

An introduced weed. Described from Europe.

×½

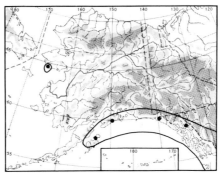
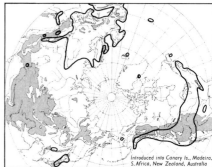

13. Bròmus hordeàceus L.
Including *Bromus mollis* L.

Culm soft, pubescent at least at top; leaves grayish-green, soft-pubescent, abruptly pointed, 3–4 mm wide; sheaths soft-pubescent; panicle short, with few, erect branches; spikelets thick, grayish-green, about 2 cm long; lower glume about 5 mm long; lemma, with rounded back and awn 7 mm long, fixed 1 mm below apex; anthers about 1 mm long.

Waste places, roadsides, yards. Described from Europe.

25. GRAMINEAE (Grass Family)

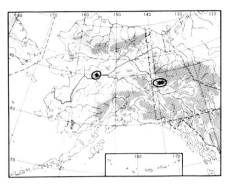 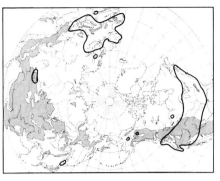 

14. Bròmus racemòsus L.

Panicle with erect branches, solitary or in pairs, usually with only 1 spikelet. Similar to *B. commutatus,* but lemma with evenly bowlike margin in upper part; awns of all florets about 7 mm long; anthers about 2–2.5 mm long.

Waste places, roadsides, yards. Described from Europe.

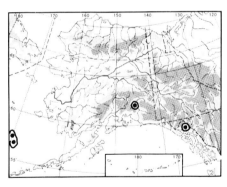 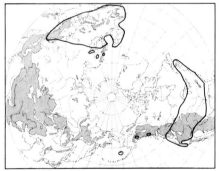 

15. Bròmus commutàtus Schrad.

Similar to *B. racemosus,* but spikelets longer and less dense; panicle with short, upright branches and spikelets 2 cm long; lemma stiff; margin of lemma, above middle, angular and not evenly bowlike; awns of lower florets 3 mm long and, of the upper, 7 mm long; anthers about 1.5 mm long.

Roadsides, waste places, yards. Described from Germany.

Lolium L.

Second glume as long as spikelet; lemma awned . . . . 3. *L. temulentum*
Second glume much shorter than spikelet:
   Lemma awnless . . . . 1. *L. perenne*
   Lemma awned . . . . 2. *L. multiflorum*

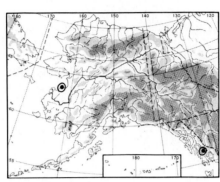 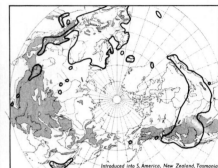

1. Lòlium perénne L.   Ryegrass

   Culm decumbent at base, from creeping rhizomes; leaves thick, dark green; spikelets 6–8-flowered, about 10 mm long; glumes 7-nerved, as long as adjacent lemma.
   Waste places, roadsides. Described from Europe.

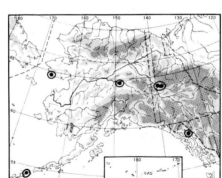 

2. Lòlium multiflòrum Lam.   Italian Ryegrass

   Culm decumbent from creeping rhizomes, retrorsely scabrous above; leaves thin, light green; spikelets 10–20 mm long, usually more than 10-flowered; glumes 7-nerved, not reaching top of adjacent lemma.
   Waste places, roadsides. Described from France.

25. Gramineae (Grass Family)

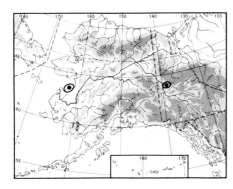
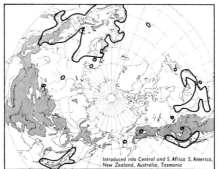

× ½

3. Lòlium temuléntum L.   Poison Darnel

Culms solitary, often branched at base; annual; culms stiff, scabrous above; leaves grayish green; spikelets (awns not counted) 10–15 mm long, 4–9-flowered; the stout awn longer than the lemma.

Waste places, roadsides. Described from Europe.

The Latin and vernacular names allude to the presence of temuline in the grain, a narcotic that inhibits the power of locomotion.

Agropyron Gaertn. (Wheatgrass)
(Key and identifications by Dr. A. Melderis, British Museum)

Spikes pectinate; spikelets very crowded, divergent from rachis, much compressed;
    joints of rachis usually 1–2 mm long . . . . 1. *A. pectiniforme*
Spikes not pectinate; spikelets widely spaced or slightly crowded, ascending, not
    much compressed; joints of rachis more than 4 mm long:
    Plants forming loose tufts or patches, spreading extensively by long-creeping, wiry
        rhizomes; anthers usually 4–5 mm long:
    Lemmas densely hairy . . . . 4. *A. yukonense*
    Lemmas glabrous:
        Glumes rigid, gradually tapering into short awn; leaves glaucous, firm, in-
            rolled at maturity . . . . 2. *A. Smithii*
        Glumes not rigid, acute or more or less abruptly narrowed into an awn;
            leaves green, usually thin, flat . . . . 3. *A. repens*
    Plants forming dense tufts, without rhizomes, or sometimes with short rhizomes:
    ■ Glumes narrowly lanceolate or elliptic, gradually tapering from middle toward
        tip, with narrow hyaline-scarious margin; margin not wider near tip:
    Glumes about half length of spikelet, usually 3-ribbed:
        Lemmas glabrous, usually with awn up to 20 mm long, arcuate at ma-
            turity; anthers 3–6 mm long; leaves usually 1–2 mm wide, densely
            short-hairy on ribs above, also sometimes with long, scattered hairs
            . . . . 5. *A. spicatum*
        Lemmas hairy in lower part, becoming more or less glabrous toward tip,
            awnless or with awn not longer than body of lemma; anthers up to
            2 mm long; leaves wider, not densely short-hairy on ribs above . . . .
            . . . . 6. *A. macrourum*
    Glumes about three-quarters length of spikelet:
        Spikes usually tinged with purple; glumes usually 3–5-ribbed, not keeled;
            lemmas hairy in lower part, becoming more or less glabrous toward
            tip . . . . 8. *A. violaceum* subsp. *andinum*
        Spikes usually green; glumes usually 5–7-ribbed, keeled; lemmas scabrous
            at tip, glabrous toward base:
        ● Lemma with awn longer than body of lemma . . . . . 9. *A. subsecundum*

● Lemma awnless or with awn shorter than body of lemma:
    Spikes slender; spikelets rarely reaching bases of those above on same side; rachilla joints usually scabrous or strigose, tightly embraced:
        Glumes acute; nodes of culm glabrous ....
        .... 13. *A. pauciflorum* subsp. *pauciflorum*
        Glumes blunt; nodes of culm short-hairy ....
        .... 14. *A. pauciflorum* subsp. *teslinense*
    Spikes thick; spikelets overlapping; rachilla joints nearly always hairy, free:
        Glumes (excluding awns when present) 7–10 mm long; spike 3–6 mm thick when mature ....
        .... 15. *A. pauciflorum* subsp. *novae-angliae*
        Glumes (excluding awns when present) 10–16 mm long; spike 5–12 mm thick when mature ....
        .... 16. *A. pauciflorum* subsp. *majus*
■ Glumes usually narrowly obovate or broadly oblanceolate, with broad hyaline-scarious margin; margin conspicuously wider near tip:
    Glumes large, only slightly shorter than spikelet, roundish acuminate, not abruptly narrowed toward tip, with distinctly asymmetrical hyaline-scarious margin; lemmas hairy in lower part, becoming less hairy or glabrous toward tip; nodes of culm always glabrous ....
    .... 7. *A. violaceum* subsp. *violaceum*
    Glumes one-half to two-thirds length of spikelet, abruptly narrowed toward tip, with symmetrical hyaline-scarious margin; lemmas glabrous or hairy in upper part, becoming less hairy or glabrous toward base; nodes of culm nearly always hairy:
        Glumes and lemmas densely hairy .. 11. *A. boreale* subsp. *hyperarcticum*
        Glumes glabrous:
            Lemmas glabrous .... 10. *A. boreale* subsp. *boreale*
            Lemmas more or less hairy in upper part ....
            .... 12. *A. boreale* subsp. *alaskanum*

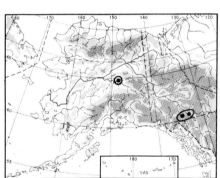 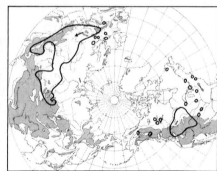

1. Agropyron pectiniforme Roem. & Schult.

Agropyron cristatum of American authors, in part.

Tufted; culms usually scabrous below spike; leaves flat or inrolled, scabrous or hairy above; spikes pectinate, ovate-oblong, dense; spikelets much compressed, 3–10-flowered, divergent from rachis, horizontal or ascending; joints of rachis usually 1–2 mm long; glumes ovate-lanceolate, 3–5 mm long, scabrous on keel, with short awn 2–3 mm long; lemmas glabrous or slightly hairy, 5–7 mm long, with short awn 3–4 mm long; joints of rachilla scaberulous; anthers 4–5 mm long.

Introduced, planted.

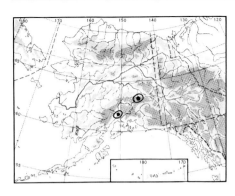
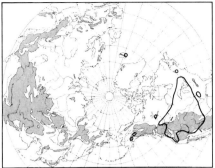

2. Agropỳron Smíthii Rydb.

Agropyron glaucum var. *occidentale* Scribn.; *A. occidentale* (Scribn.) Scribn.

Glaucous, loosely tufted, with wiry rhizomes; leaves flat, inrolled when dry, prominently ribbed, scabrous or sparsely villous above; spikes stiff; spikelets closely overlapping, 15–20 mm long, 6–10-flowered; glumes 10–12 mm long, 3–6-ribbed, gradually tapering into short awn; lemmas about 10 mm long, firm, glabrous, awnless or with very short awn; joints of rachilla scaberulous; anthers 3.5–4 mm long.

Introduced. Described from "Upper Missouri."

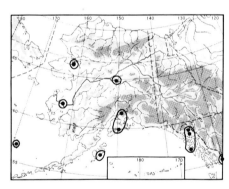
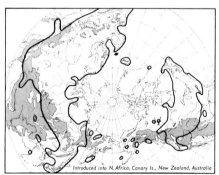

3. Agropỳron rèpens (L.) Beauv.

Triticum repens L.

Green or glaucous, loosely tufted, with long, creeping rhizomes; sheaths of leaves often hairy; leaves flat, with thin ribs, glabrous or sparsely hairy above; spikes 7–15 cm long, slender; spikelets 10–20 mm long, 3–7-flowered, falling entire at maturity; glumes 7–12 mm long, not rigid, 3–7-ribbed, acute or abruptly awn-pointed; lemmas 8–11 mm long, glabrous, awnless or with awn up to as long as body; joints of rachilla scaberulous; anthers 3.5–5.5 mm long.

An introduced weed; waste places, roadsides. Described from Europe.

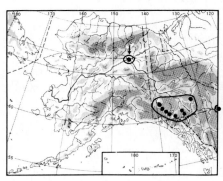
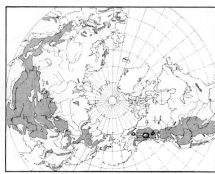

4. Agropỳron yukonénse Scribn. & Merr.

Glaucous, with creeping rhizomes; leaves 6–7 mm wide, soft, prominently ribbed, scabrous above; spikes 4–10 cm long, loose to rather dense; spikelets 10–15 mm long, 3–6-flowered, often overlapping; glumes broadly lanceolate, 5–7 mm long, prominently 3-ribbed, densely hairy, acute or short-acuminate; lemmas 8–11 mm long, densely villous, awnless or with awn up to 10 mm long; joints of rachilla hairy; anthers 3–5 mm long.

Stable dunes, sand flats, sandy hillsides.

Hybrids occurring with *A. spicatum*, *A. subsecundum*, and *A. violaceum* are recognized by narrow indehiscent anthers and by empty (sterile) pollen grains.

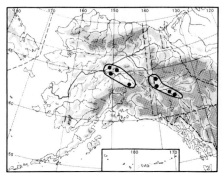
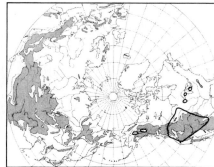

5. Agropỳron spicàtum (Pursh) Scribn. & Sm.

Festuca spicata Pursh; *Agropyron divergens* (Nees.) Vasey; *A. Vaseyi* Scribn. & Sm.

Green or glaucous, forming compact tufts; leaves 1–2(–4) mm wide, flat or loosely inrolled, usually minutely and densely hairy above; spike slender; spikelets 12–15 mm long, rather scattered, 6–8-flowered; glumes 5–8 mm long, glabrous, usually 4-ribbed, obtuse to acute; lemmas about 10 mm long, glabrous, with strongly divergent awn, up to 2 cm long; joints of rachilla scaberulous; anthers 4–5 mm long.

Dry soil, riverbanks, dry slopes. Described from the Columbia River.

25. GRAMINEAE (Grass Family)

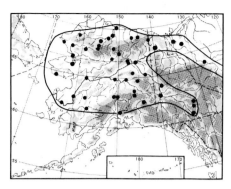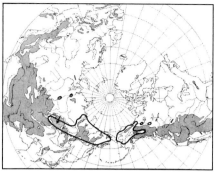

6. Agropỳron macroùrum (Turcz.) Drobov

Triticum macrourum Turcz.; *Agropyron sericeum* Hitchc.; *A. dasystachyum* with respect to Alaska-Yukon plant.

Tufted, sometimes with short rhizomes; leaves 3–7 mm wide, glabrous, scabrous, or sparsely long-hairy above; spikes loose; spikelets 12–20 mm long, 5–7-flowered, pale green or purplish; glumes 5–9 mm long, 3–5-ribbed, scabrous on ribs, acuminate or short-pointed, with narrow hyaline-scarious margins; lemmas 8–11 mm long, hairy especially in lower part, becoming more or less glabrous toward tip, awnless or with awn up to 7 mm long, joints of rachilla with spreading hairs; anthers 1–2 mm long.

Sandy and gravelly bars, riverbanks. Described from the upper Angara River in Siberia.

Hybrids with *Elymus sibiricus* (*Agroelymus palmerensis* Lepage) have been found at Palmer; and with *Hordeum brachyantherum* (*Agrohordeum Jordalii* Melderis), at Bettles.

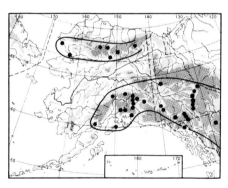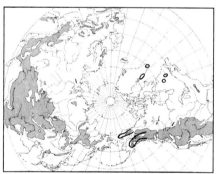

7. Agropỳron violàceum (Hornem.) Lange
subsp. violàceum

Triticum violaceum Hornem.; *A. violaceum latiglume* Scribn. & Sm.; *A. latiglume* (Scribn. & Sm.) Rydb.; *A. caninum* var. *latiglume* (Scribn. & Sm.) Pease & More.

Tufted; nodes of culm glabrous; lower sheaths of leaves usually with minute hairs; leaves 2–5 mm wide, scabrous or sparsely short-haired on ribs above; spikes dense; spikelets overlapping, 9–15 (–17) mm long, 3–5-flowered, purple; glumes 7–11 mm long, broadly oblanceolate or narrowly obovate, roundish-acuminate toward tip, 3–5-ribbed, often slightly scabrous on ribs, with broad hyaline-scarious margin 0.5–1 mm wide; margin conspicuously wider near tip, more so on one side than other; lemmas 9–11 mm long, hairy in lower part, more or less glabrous toward tip; awnless or with short awn up to 5 mm long; joints of rachilla hairy; anthers 1–1.5 mm long.

Sandy and gravelly river bars, subalpine meadows. Described from southern Greenland. Circumpolar map very incomplete, range little known.

Var. **alboviride** (Hult.) Melderis (*A. latiglume* var. *alboviride* Hult.) is characterized by green or yellowish-green spikelets, and by glumes having a broad hyaline-scarious margin extending to lower part.

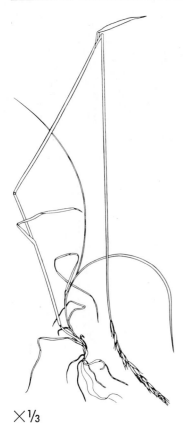
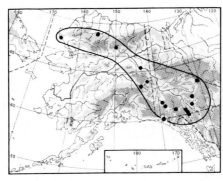
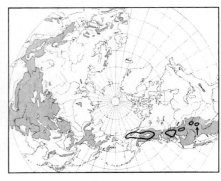

8. Agropỳ**ron viol**à**ceum** (Hornem.) Lange
subsp. **and**ì**num** (Scribn. & Sm.) Melderis

Agropyron violaceum andinum Scribn. & Sm.; *A. andinum* (Scribn. & Sm.) Rydb.

Similar to subsp. *violaceum*, but glumes narrowly lanceolate or elliptic, gradually tapering from middle toward tip, the hyaline-scarious margin narrow, up to 5 mm wide, not wider near tip.

In mountains to at least 1,200 meters.

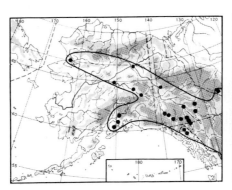
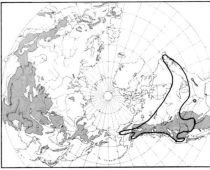

9. Agropỳ**ron subsec**ú**ndum** (Link) Hitchc.

Triticum subsecundum Link; *T. Richardsonii* Schrad.; *Agropyron unilaterale* Cass.; *A. trachycaulum* var. *unilaterale* (Cass.) Malte; *A. trachycaulum* var. *glaucum* Malte.

Green or glaucous, tufted; sheaths of leaves usually glabrous; leaves flat, scabrous or sparsely long-hairy above; spikes dense, erect or slightly nodding, sometimes unilateral; spikelets 10–20 mm long, usually green, 5–7-flowered; glumes 10–17 mm long, broadly lanceolate, 5–7-ribbed, keeled, scabrous on ribs; lemmas about 10 mm long, glabrous or more or less scabrous toward tip, with erect awn 1–3 cm long; joints of rachilla strigose or villous; anthers 1.5–2 mm long.

Woods, hillsides, riverbanks. Described from cultivated specimens raised from seeds collected in "Western North America."

25. GRAMINEAE (Grass Family)

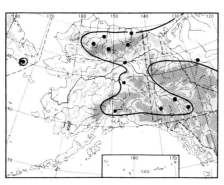
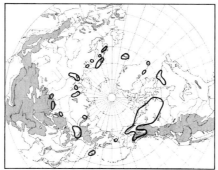
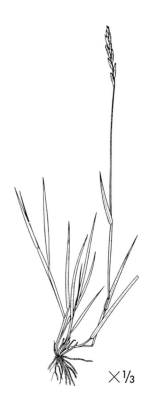

10. Agropỳron boreàle (Turcz.) Drobov
Triticum boreale Turcz.
subsp. **boreàle**
Agropyron latiglume subsp. *eurasiaticum* Hult.; *Roegneria borealis* (Turcz.) Nevski; *R. scandica* Nevski.

Tufted; nodes of culms nearly always short-hairy; lower sheaths of leaves usually glabrous; leaves 2.5–6 mm wide, flat or inrolled, usually scaberulous; spikes rather dense; spikelets 12–20 mm long, 4–6-flowered, purple or greenish-purple; glumes 5–8 mm long, oblong or broadly oblanceolate, usually abruptly narrowed into short awn up to 3 mm long, 3–5-ribbed, glabrous, with hyaline-scarious margin wider near tip; lemmas 8–10 mm long, glabrous, with short awn 1.5–5 mm long; joints of rachilla strigose; anthers 1.5–2 mm long.

River bars and flats; dry, sandy banks of rivers, hillsides. In mountains to at least 1,000 meters. Described from the Aldan valley in eastern Siberia.

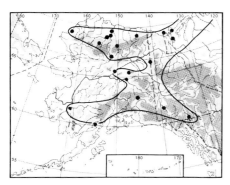
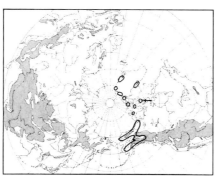

11. Agropỳron boreàle (Turcz.) Drobov
subsp. **hyperárcticum** (Polunin) Melderis
Agropyron violaceum var. *hyperarcticum* Polunin; *A. latiglume* var. *pilosiglume* Hult.; *Triticum repens* var. *nanum* Hook.

Similar to subsp. *boreale*, but glumes and lemmas densely hairy; joints of rachilla hairy.

Riverbanks, hillsides.

×⅓

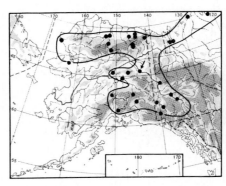 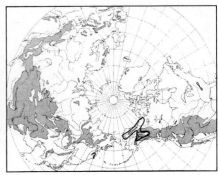

12. Agropỳron boreàle (Turcz.) Drobov
subsp. **alaskànum** (Scribn. & Merr.) Melderis

Agropyron alaskanum Scribn. & Merr.

Similar to subsp. *boreale,* but with glumes glabrous or scabrous on ribs; lemmas more or less hairy in upper part; joints of rachilla strigose or hairy.

Gravel bars.

Hybrids with *Elymus arenarius* subsp. *mollis* (*Agroelymus Hultenii* Melderis) have been found at Deering.

×⅓

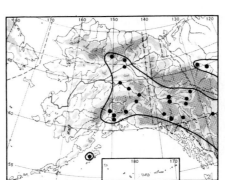 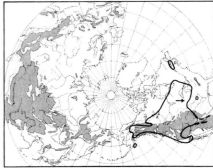

13. Agropỳron pauciflòrum (Schwein.) Hitchc.

Triticum pauciflorum Schwein.; *T. trachycaulum* Link; *Agropyron trachycaulum* (Link) Malte; *A. tenerum* Vasey.

subsp. **pauciflòrum**

Tufted; sheaths of leaves usually glabrous; nodes of culm glabrous; leaves 2–4 mm wide, flat, usually scabrous above; spikes slender, 6–25 cm long; spikelets 10–16 mm long, usually green, rarely reaching bases of those above on the same side; glumes nearly as long as spikelet, keeled, acute or acuminate or tapering into very short awn, 4–7-ribbed, scabrous on ribs, usually coriaceous with hyaline-scarious margin 0.4–0.6 mm wide; lemmas 9–13 mm long, glabrous or more or less scabrous toward tip, awnless or with very short awn; joints of rachilla usually scabrous or strigose, tightly embraced; anthers 1.3–2 mm long.

Dry, open soil, subalpine meadows, riverbanks, hillsides. Described from "prairies of St. Peter" (Minnesota).

25. GRAMINEAE (Grass Family)

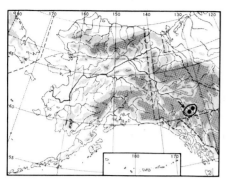 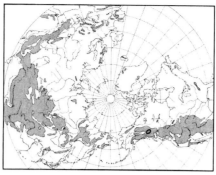 

14. Agropỳron pauciflòrum (Schwein.) Hitchc.
subsp. **teslinénse** (Pors. & Senn) Melderis
   Agropyron teslinense Pors. & Senn.

   Similar to subsp. *pauciflorum*, but glumes blunt, 3–5-ribbed; nodes of culm hairy.

× ⅓

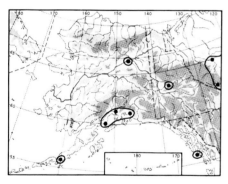 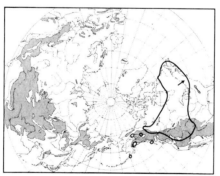 

15. Agropỳron pauciflòrum (Schwein.) Hitchc.
subsp. **nòvae-ángliae** (Scribn.) Melderis
   Agropyron tenerum var. *novae-angliae* Scribn.; *A. trachycaulum* var. *novae-angliae* (Scribn.) Fern.

   Similar to subsp. *pauciflorum*, but spikes dense, 3–6 mm thick when mature; spikelets overlapping; glumes usually herbaceous, 7–10 mm long, with hyaline-scarious margin 0.1–0.4 mm wide; joints of rachilla usually villous.

× ⅓

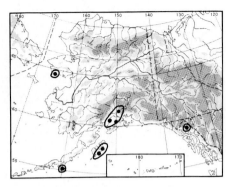
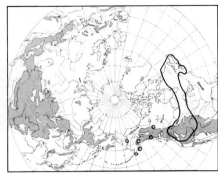

16. Agropỳron paucifloŕum (Schwein.) Hitchc.
subsp. május (Vasey) Melderis

Agropyron violaceum var. *majus* "*major*" Vasey; *A. trachycaulum* var. *majus* (Vasey) Fern.; *A. angustiglume* of American authors.

Resembling subsp. *novae-angliae,* but glumes (excluding awns, when present) 10–16 mm long; spikes 5–12 mm thick when mature.

Described from Oregon. Circumpolar map tentative.

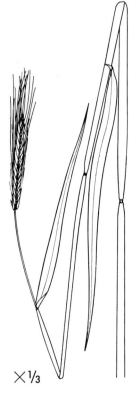
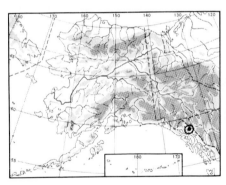

Secale L.

1. Secàle sereàle L. Rye

Annual; culm pubescent below 2-sided spike; leaves bluish-gray, scabrous above; spikelets sessile, 2-flowered; glumes subulate, 1-nerved, about 10 mm long.

Cultivated, sometimes escaping into waste places and fields; not self-maintaining. Native of southwest Asia. Described from Europe.

No circumpolar map can be constructed.

25. GRAMINEAE (Grass Family)

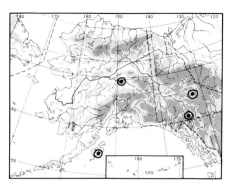
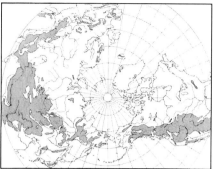

Triticum L. (Wheat)

1. Tríticum aèstivum L.

Triticum sativum Lam.; *T. vulgare* Vill.

Annual; culms tufted, hollow, leaf blades auriculate, pubescent when young; section of spike quadrangular; rachis of spike disintegrating; lemma with short, or sometimes prolonged, awn.

Waste places; not self-maintaining. Described from Europe. No circumpolar map can be constructed.

Hordeum L.

Glumes broad, blunt; awn lacking .... 3. *H. vulgare*
Glumes awnlike:
  Awns about 1 cm long .... 1. *H. brachyantherum*
  Awns 2–8 cm long .... 2. *H. jubatum*

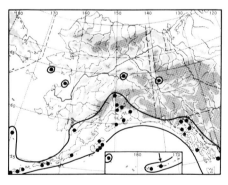
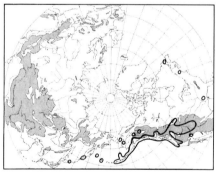

1. Hórdeum brachyántherum Nevski

Hordeum jubatum subsp. *breviaristatum* Bowden; *H. boreale* Scribn. & Sm., not Gandoger; *H. nodosum* of authors.

Culm glabrous, geniculate at base; leaf blades 3–8 mm wide; spikes up to 5 cm long, green or purplish, with extremely brittle rachis and awns about 1 cm long; glumes setaceous, scabrous; lemma of fertile spikelet glabrous, lanceolate, lemma of the lateral spikelets, rudimentary.

Meadows, grassy slopes, shores. Sometimes found as weed.

The hybrid *H. brachyantherum* × *jubatum* (*H. jubatum* var. *caespitosum* (Scribn.) Hitchc.; *H. jubatum* subsp. *intermedium* Bowden; *H. caespitosum* Scribn.), with awn only 2–3 cm long, is not rare. Hybrids with *Agropyron* species, such as *A. macrourum*, also occur.

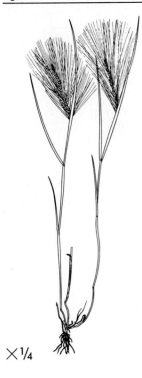
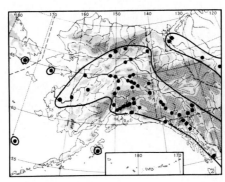
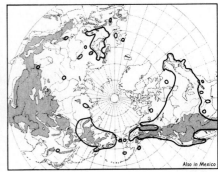

2. Hórdeum jubàtum L. — Squirreltail Grass

Sitanion jubatum (L.) J. G. Sm.; *Critesion jubatum* (L.) Nevski.

Annual or biennial; culms geniculate at base; leaf blades 3–5 mm wide, scabrous; spike very brittle, nodding, green to purple; lateral pairs of spikelets each reduced to 1–3 spreading awns; glumes awnlike, spreading, 4–8 cm long; lemma 4–5 mm long, with awn 4–8 cm long; anthers short, less than 1 mm long.

Sandy soil, river banks; also occurs as weed along roadsides. Described from Canada.

Hybrids occur with *H. brachyantherum* and with *Agropyron pauciflorum* [*Elymus Macounii* Vasey; *Agrohordeum Macounii* (Vasey) Lepage].

(See color section.)

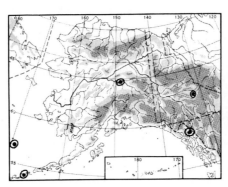
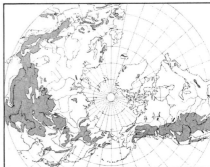

3. Hórdeum vulgàre L. — Barley

Culm 60–100 cm tall, bluish-green; leaf blades auriculated at base; spike not disintegrating, with all spikelets of same kind, long-awned; all glumes subulate.

Sometimes found in waste places as an escape from cultivation. Type locality not given. Circumpolar map cannot be constructed.

25. GRAMINEAE (Grass Family)

Elymus L. (Lyme Grass)

Glumes setaceous or subulate (sometimes reduced or lacking); plant with long, slender rhizomes .... 4. *E. innovatus*
Glumes linear-lanceolate to linear; plant tufted or with rhizomes:
  Lemma with awn much longer than lemma; plant tufted:
    Back of lemma short- to long-hirsute in the margin; body of glumes much shorter than body of adjacent lemma .... 7. *E. hirsutus*
    Back of lemma glabrous or scabrous:
      Back of lemma glabrous at base, scabrous toward apex; body of glumes approximately equaling body of adjacent lemma ....
      .... 5. *E. glaucus* var. *glaucus*
      Back of lemma densely scabrous; body of glumes much shorter than body of adjacent lemma .... 8. *E. sibiricus*
  Lemma awnless, or with awn much shorter than lemma:
    Back of lemma glabrous at base, scabrous toward apex; plant tufted ....
    .... 6. *E. glaucus* var. *virescens*
    Back of lemma fine-pubescent, sericeous or densely coarse-pubescent; plant with rhizomes:
      Glumes densely hirsute, about 5–6 mm long; lemmas densely coarse-pubescent .... 3. *E. interior*
      Glumes 10–35 mm long:
        Glumes hirsute or sometimes nearly glabrous; lemmas softly hirsute ....
        .... 1. *E. arenarius* subsp. *mollis* var. *mollis*
        Glumes lanate; lemmas densely pubescent ....
        .... 2. *E. arenarius* subsp. *mollis* var. *villosissimus*

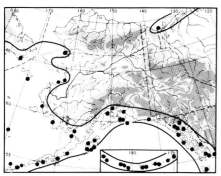  

×⅓

1. Èlymus arenàrius L.
subsp. **móllis** (Trin.) Hult.
var. **móllis**

Elymus mollis Trin.; *Leymus mollis* (Trin.) Pilger; *Elymus capitatus* Scribn.; *E. arenarius* var. *villosus* H. E. Meyer.

Culms from long, stout, creeping rootstocks, with marcescent old leaves at base; leaves firm, flat or involute, 3–15 mm wide; spike stiff, dense, 1–3 dm long; spikelets 3–7-flowered; glumes lanceolate, hirsute on back; lemma awnless, softly hirsute.

Sandy beaches, forming belt along shore; rare on dunes inland. *E. arenarius* described from Europe, subsp. *mollis* from Kamchatka and the Aleutian Islands.
Broken line on circumpolar map indicates range of subsp. **arenàrius**.
E. aleuticus Hult. is the hybrid *E. arenarius* subsp. *mollis* × *E. hirsutus*.

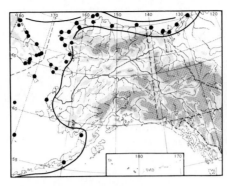
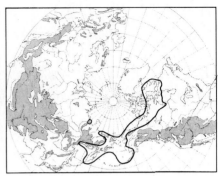

2. Èlymus arenàrius L.
subsp. **mòllis** (Trin.) Hult.
var. **villosíssimus** (Scribn.) Hult.
> *Elymus villosissimus* Scribn.; *E. arenarius* subsp. *villosissimus* (Scribn.) Löve.

Similar to var. *mollis,* with which it intergrades, but glumes densely long-villous or lanate on back.
Sandy seashores. Described from St. Paul Island.

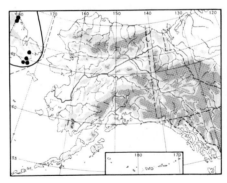
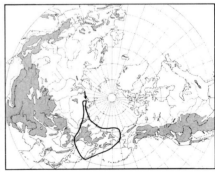

3. Èlymus intèrior Hult.
> *Leymus interior* (Hult.) Tsvel.; *Elymus mollis* subsp. *interior* (Hult.) Bowden.

Culm 30–50 cm tall; leaves scabrous above; spike grayish-violet, erect, stiff, 6–10 cm long, 1–1.7 cm wide; glumes lanceolate, with narrow, scarious margin, 1–3-nerved, lanate; lemma with excurrent median nerve, densely pubescent.
Similar to *E. innovatus,* but with broader glumes.
Sandy soil.

25. Gramineae (Grass Family)

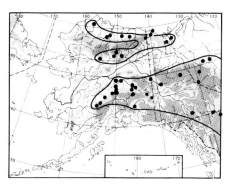 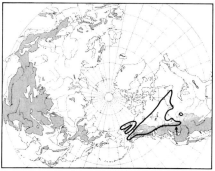 

4. Èlymus innovàtus Beal

Culms from creeping rhizomes; culms with short, cauline leaves; spikes erect, cylindrical, up to 16 cm long, usually shorter; spikelets 1–1.5 cm long, 3–4-flowered; glumes narrow, setaceous or subulate, scabrous or villous, at least 1 often reduced, rudimentary, or absent; lemma longer than body of glumes, with short awn.

Sandy soil, in McKinley Park up to at least 1,400 meters.

Plant from Big Delta, with back of lemma velutinous, is var. **velùtinus** (Bowden) Hult. (subsp. *velutinus* Bowden).

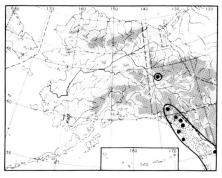 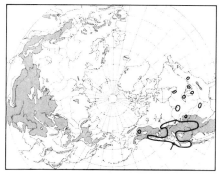 

5. Èlymus glaùcus Buckl.
var. glaùcus

Tufted or with short runners; culm 50–100 cm tall; leaf blades thin, flat, glabrous or somewhat scabrous; spikes slender, erect, up to 20 cm long; spikelets greenish-purplish, appressed, subsessile; glumes linear to linear-lanceolate; lemma with membranaceous margin indistinctly 5-nerved, with straight or bent awn longer than lemma, glabrous toward base, scabrous toward apex.

Meadows, rocky places, woods.

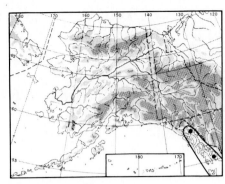
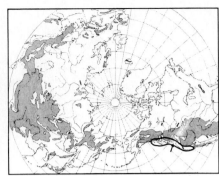

6. Élymus glaúcus Buckl.
var. viréscens (Piper) Bowden
   Elymus virescens Piper; *E. glaucus* subsp. *virescens* (Piper) Gould.

Similar to var. *glaucus*, but awn of lemma shorter than its body; awn of glumes shorter than in typical plant.

Meadows, rocky places.

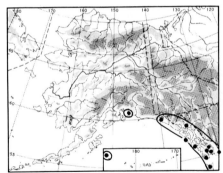

7. Élymus hirsùtus Presl
   Elymus borealis Scribn.

Tufted or with short runners; culms up to 1 meter tall; leaf blades thin, flat, villous or pilose on veins above; spike slender, straight or arched; spikelets appressed, subsessile; glumes linear to linear-lanceolate, 3–5-nerved, with awn shorter than bodies or slightly longer; lemma 5-nerved, glabrous or sparsely scabrous in central area, scabrous to long-hirsute near margins, with awn much longer than body.

Woods, rocky slopes, shores. Described from Yakutat Bay or Nootka Sound, Vancouver Island.

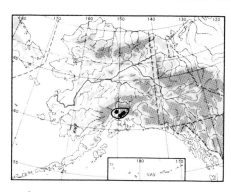
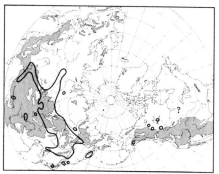

×¼

8. Élymus sibìricus L.

Clinelymus sibiricus (L.) Nevski; *Elymus canadensis* with respect to Alaskan plant; *Elymus pendulosus* Hodgson.

Tufted or with short runners; leaf blades thin, pilose on veins above; spikes pendulous, up to 30 cm long; spikelets 3–6-flowered, subsessile; glumes 3(–5)-nerved, scabrous on nerves, the body much shorter than adjacent lemma; lemma 5-nerved, scabrous, glabrous between nerves at apex, with awn 1–1.5 cm long.

Roadsides, clearings; introduced around experimental station at Palmer. Described from Siberia.

Eriophorum L. (Cotton Grass)

A taxonomically very intricate genus; hybrids between taxa are apparently not uncommon.

Spikelets more than 1, subtended by one or more foliaceous bracts; cauline leaves well developed:
  Leaves linear, triangular in cross section, channeled; involucre a single 1-colored bract, shorter than the inflorescence .... 4. *E. gracile*
  Leaves flat; involucre of 2 or more leaflike bracts, longer than the inflorescence:
    Midrib of scales prominent to very tip .... 5. *E. viridi-carinatum*
    Midrib of scales not prominent at tip:
      Inflorescence congested; scales nearly black ....
      .... 3. *E. angustifolium* subsp. *triste*
      Inflorescence with long, pedunculated, drooping spikes:
        Peduncles glabrous .... 1. *E. angustifolium* subsp. *subarcticum*
        Peduncles scabrous .... 2. *E. angustifolium* subsp. *scabriusculum*
Spikelets solitary, without foliaceous bracts; cauline sheaths lacking blade, or with very short blade:
  ■ Plant stoloniferous; bristles white or rust-colored; not more than 7 empty scales at base of spikelet:

Anthers 0.5–1 mm long; middle scales lance-attenuate, acute, grayish-black; bristles white:
Culm thick, stiff; fruiting head subglobose .... 6. *E. Scheuchzeri* var. *Scheuchzeri*
Culm slender, often bent bowlike; spike smaller, more lax, with less dense bristles .... 7. *E. Scheuchzeri* var. *tenuifolium*
Anthers 1.5 mm long or longer; scales broader, with broader pale margin; fruiting heads globose or oblong; bristles white or rust-colored:
Plant stout; the first empty scale very prominent, thick; middle scales lanceolate, more or less acuminate, with a narrower, pale margin; bristles pale rust-colored .... 11. *E. russeolum* subsp. *rufescens*
Plant slender; middle scales broad, ovate-lanceolate, blunt, with broad, scarious margin:
Bristles rust-colored .... 9. *E. russeolum* var. *majus*
Bristles white (or very pale rust-colored) .. 10. *E. russeolum* var. *albidum*
■ Plant densely tufted; bristles white or sordid, sometimes very slightly rusty:
Scales grayish- to greenish-black, not clearly translucent; empty scales not reflexed:
Plant 6–20 cm high; uppermost sheath situated below middle of culm, inflated, usually bearing a short blade; leaves rigid, spreading; bristles shiny white .... 8. *E. callitrix*
Plant taller, slender; uppermost sheath situated above middle of culm, less inflated, usually leafless; leaves softer, longer, not spreading .... 12. *E. brachyantherum*
Scales gray, with or without dark center and base; empty scales reflexed:
Spike oblong; scales light-colored; sheaths distinctly disintegrating into shreds; culm tall .... 13. *E. vaginatum* subsp. *vaginatum*
Spike depressed, globose; scales with dark center and base; sheaths not, or only slightly, disintegrating into shreds; culm shorter, mostly 15–20 cm long; tufts very compact .... 14. *E. vaginatum* subsp. *spissum*

× 1/3

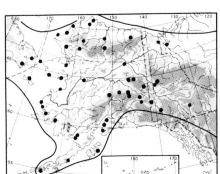
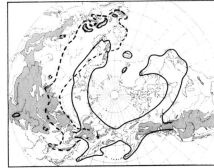

1. Eriòphorum angustifòlium Honck.
subsp. **subárcticum** (Vassiljev) Hult.

Eriophorum subarcticum Vassiljev; ? *E. angustifolium* var. *majus* K. F. Schultz.

Culms from creeping rhizome; leaves dark green, triangular, channeled toward tip, broad; inflorescence with 2–3 long, involucral leaves; spikelets on flattened peduncles; anthers 2.5–3 mm long; bristles white, exceptionally pinkish (var. **coloràtum** Hult.).

Wet bogs and shores. The lower part of the stem is sometimes eaten by natives, raw or mixed with seal oil.

Var. **gigánteus** Hult. is an extremely coarse type, up to 1 meter tall, with leaves up to 8 mm broad, occurring especially on the west coast of Kenai Peninsula. The hybrid *E. angustifolium* × *Scheuchzeri* has been called *E. Rousseauianum* Raymond; the hybrid *E. angustifolium* var. *majus* × *Chamissonis* has been called *E. beringianum* Raymond.

Broken line on circumpolar map indicates range of subsp. *angustifolium*.

26. CYPERACEAE (Sedge Family)

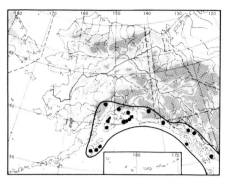
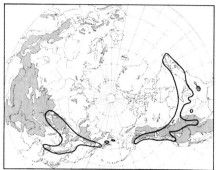

2. Eriòphorum angustifòlium Honck.
subsp. **scabriùsculum** Hult.
Eriophorum Komarovii Vassiljev.

Similar to subsp. *subarcticum*, but more slender; panicle branches more or less scabrous; leaves narrow.
Wet bogs and shores. Described from Alaska.
The hybrid *E. angustifolium* subsp. *scabriusculum* × *vaginatum* subsp. *spissum* has been called *E. Churchillianum* Lepage.

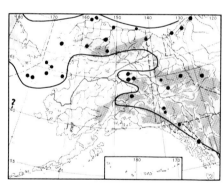
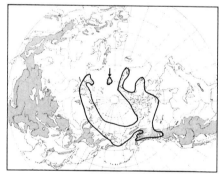

3. Eriòphorum angustifòlium Honck.
subsp. **tríste** (T. Fries) Hult.
Eriophorum angustifolium var. *triste* T. Fries; *E. triste* (T. Fries) Löve & Hadač.

Similar to subsp. *subarcticum*, but more low-growing; inflorescence subcapitate; panicle branches scabrous; leaves narrow; scales blackish.
Wet tundra. Described from Spitzbergen.
Numerous transitions connecting this type with the other types of the complex (not represented on the map) occur over large parts of the area.
The hybrid *E. angustifolium* subsp. *triste* × *Scheuchzeri* has been called *E. Sorenseni* Raymond.

×⅓

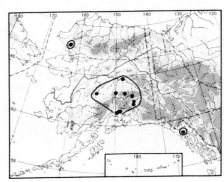
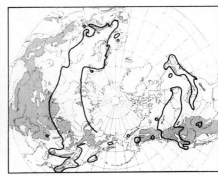

4. Eriòphorum gràcile W. D. J. Koch

Culm glabrous, from creeping rhizome; leaves narrow, triangular in cross section, channeled, early-withering; involucral bract short; spikes 2–4, on crisp-puberulent peduncles; scales short-pointed or rounded in the dark-colored apex, green at base; anthers about 1.5 mm long; bristles white.

Peaty soil. Described from Germany.

×¼

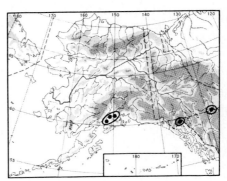
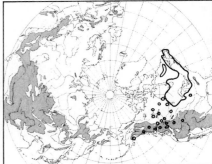

5. Eriòphorum víridi-carinàtum (Engelm.) Fern.
Eriophorum latifolium var. *viridicarinatum* Engelm.

Tufted, with numerous basal leaves; culm glabrous, trigonous; leaves 2–5 mm broad, flat, with narrow, involute tips; 2–3 involucral bracts, longer than the inflorescence; scales greenish-drab to lead-colored, with midrib prominent to extreme tip; anthers 1–1.5 mm; bristles creamy white.

Peat bogs, wet meadows. Described from Massachusetts and Ohio.

26. Cyperaceae (Sedge Family)

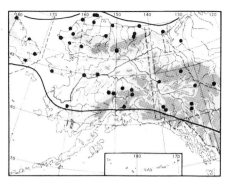
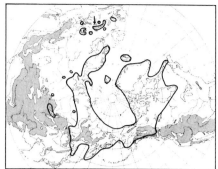

6. Eriòphorum Scheúchzeri Hoppe
   Eriophorum capitatum Torr.
var. **Scheúchzeri**

Loosely stoloniferous; culm solitary, thick, stiff, with 1 bladeless, black-tipped sheath; leaves filiform, somewhat channeled; spike subglobose; outer scales ovate-lanceolate, inner lance-attenuate, evenly tapering from base to tip, grayish-black, not translucent; anthers 0.5–1 mm long; bristles bright white.

Wet places, peaty soil.

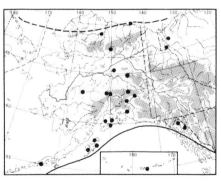
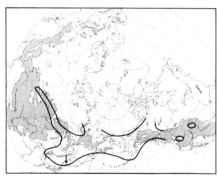

7. Eriòphorum Scheúchzeri Hoppe
var. **tenuifòlium** Ohwi
   Eriophorum altaicum Meinsh.?

Similar to var. *Scheuchzeri*, but culm slender, less stiff, often bowlike; spike smaller with less dense bristles.
Wet places. A variety not well marked.
Circumpolar map tentative.

26. Cyperaceae (Sedge Family)

× ½

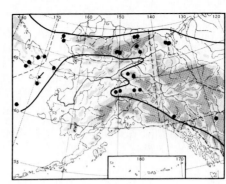
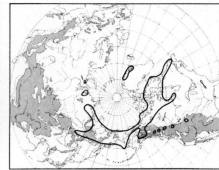

8. Eriòphorum cállitrix Cham.

Densely tufted; culms 10–15(–20) cm tall, thick, stiff, curved; cauline sheath at lower half of culm or close to its base, somewhat inflated; basal leaves numerous, coarse, stiff; involucral scale ovate, acute; scales lanceolate, dark, translucent on margin and tip; anthers 0.7–1 mm long; bristles shiny white.

Peaty soil. In var. **moràvium** Raymond (var. *pallidus* Hult.), the scales are yellow.

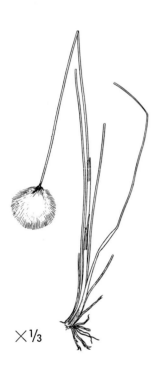
× ⅓

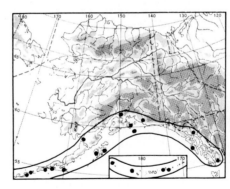
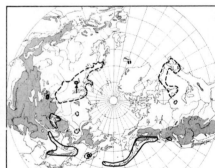

9. Eriòphorum russèolum E. Fries
var. **május** Sommier

Eriophorum mandschuricum Meinsh.

Culms long, slender; leaves filiform, canaliculate; spikes oblong or obovate; first scale dark brown, clasping, the inner scale grayish-black in middle, hyaline in margin; bristles rust-colored.

Wet places.

The hybrid *E. russeolum* × *E. vaginatum* subsp. *spissum* has been called *E. Pylaieanum* Raymond.

Broken line on circumpolar map indicates range of var. **russèolum**.

26. Cyperaceae (Sedge Family)

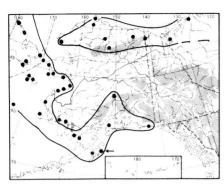

10. Eriòphorum russèolum E. Fries
var. **albìdum** Nyl.

Eriophorum russeolum var. *leucothrix* (Blomgr.) Hult.; *E. russeolum* f. *leucothrix* Blomgr.

Similar to var. *majus*, but culms slender and shorter; spike smaller; bristles white or very pale rust-colored.

Wet places.

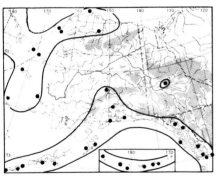
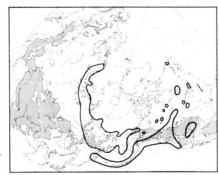
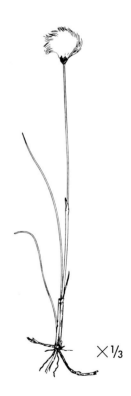

11. Eriòphorum russèolum E. Fries
subsp. **ruféscens** (Anders.) Hyl.

Eriophorum rufescens Anders.; *E. medium* Anders.; *E. Chamissonis* of authors, not C. A. Mey.

Culms stout, straight, 20–30 cm tall, in far arctic specimens 8–10 cm; leaves filiform, canaliculate; spikes obovate or globose; outer scales prominent, triangular, broad, up to half as long as the head, dark brown; upper scales with dark center, hyaline in margin; bristles pale, rust-colored; anthers 0.5–1 mm long.

Wet places. Described from northern Scandinavia.

The hybrid *E. Chamissonis* × *vaginatum* subsp. *spissum* has been called *E. Porsildii* Raymond.

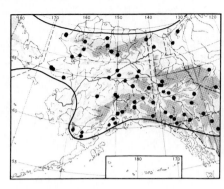
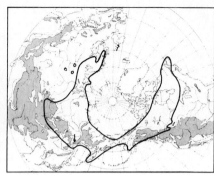

12. Eriòphorum brachyántherum Trautv. & Mey.

Eriophorum opacum (Björnstr.) Fern.; *E. vaginatum* var. *opacum* Björnstr.

Tufted; slender culm about 30–40 cm tall, with 2 sheaths; basal leaves filiform, surrounded by grayish-brown sheaths; spike with more or less cuneate base, and with lance-attenuate, dark, not translucent scales; bristles slightly yellowish-white; anthers 1.5 mm long.

Wet places.

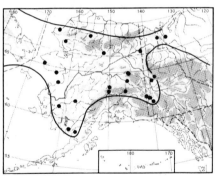
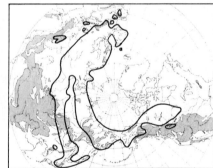

13. Eriòphorum vaginàtum L.   Hare's-Tail Grass
subsp. **vaginàtum**

Tufted, forming tussocks; basal sheaths pale rust-brown; culms scabrous at tip, with 2–3 sheaths; leaves filiform, trigonous; spike oblong, becoming more or less globular; scales lead-colored, translucent, the basal ones reflexed at maturity; bristles white; anthers long.

Wet places, peaty soil. Described from Europe.

Distinction from subsp. *spissum* not clear. Circumpolar map tentative.

26. Cyperaceae (Sedge Family)

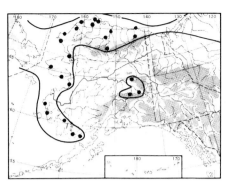
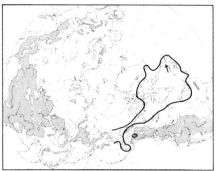

14. Eriòphorum vaginàtum L.
subsp. **spíssum** (Fern.) Hult.
  E. spissum Fern.

Similar to subsp. *vaginatum*, but culms 15–20 cm tall; tussocks very compact, tall; spike globose to depressed-globose; scales with darker center.

Arctic and mountain tundra, predominant in large areas in somewhat higher and dryer parts of the northern coastal plain. Range little known, in Siberia probably at least to the Lena River.

Trichophorum Svens.

Culms densely tufted . . . . 3. *T. caespitosum*
Culms with creeping rhizome:
  Culms straight, from short, ascending rhizome; spike in fruit with long setae . . . .
  . . . . 1. *T. alpinum*
  Culms bowlike, from long rhizome; no long setae . . . .
  . . . . 2. *T. pumilum* var. *Rollandii*

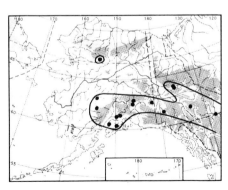
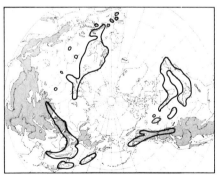

1. Trichòphorum alpìnum (L.) Pers.
  Eriophorum alpinum L.; *E. hudsonianum* Michx.; *Scirpus hudsonianus* (Michx.) Fern.; *S. trichophorum* Aschers. & Graebn.

Culms yellowish-green, stiff, triangular in cross section, scabrous, in dense row from ascending, short, creeping rhizome; upper sheath with square mouth and blade 1 cm long; involucral scale with excurrent median nerve, not exceeding spike; bristles long, white.

Peat bogs, wet meadows. Described from Europe.

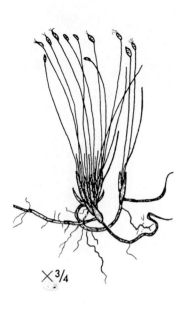
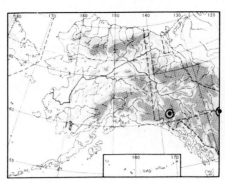
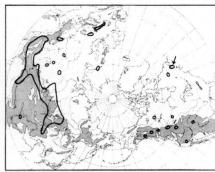

2. Trichòphorum pùmilum (M. Vahl) Schinz & Thell.

Scirpus pumilus M. Vahl; *Baeothryon pumilum* (M. Vahl) Löve & Löve.

var. **Rollándii** (Fern.) Hult.

Scirpus Rollandii Fern.

Culms stiff, several together, grayish-green, about 5 cm tall and 0.5 mm thick, from creeping rhizome; basal sheaths blackish-brown, the uppermost green, with blade 5–10 mm long; spike ovate, 2–3 mm long, with 2–3 flowers; scales chestnut-brown, the first with green, not excurrent, median nerve.

Wet places. *T. pumilum* described from Switzerland.

Circumpolar map indicates range of entire species complex.

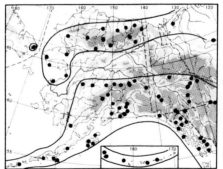
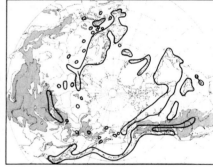

3. Trichòphorum caespitòsum (L.) Hartm.

Baeothryon caespitosum (L.) Dietr.; *Scirpus caespitosus* L.; *Trichophorum caespitosum* subsp. *austriacum* (Pall.) Hegi; *T. austriacum* Pall.; *Scirpus caespitosus* L. subsp. *austriacus* (Pall.) Aschers. & Graebn.; *S. caespitosus* L. var. *callosus* Bigel.

Culms grayish-green, densely tufted, with numerous, stramineous-to-brown sheaths at base, lacking a blade and usually with 2 cauline sheaths, the upper with square mouth and blade about 5 mm long; involucral scale about as long as spike; bristles about twice as long as achene.

Peaty soil, wet places. Described from Europe, America, and Himalayas.

26. Cyperaceae (Sedge Family)

Scirpus L. (Bulrush)

Involucral leaf solitary, appearing as continuation of culm:
  Spikelet solitary; rhizome soft and weak .... 1. *S. subterminalis*
  Spikelets more than 1; rhizome firm and hard:
    Culms leafy; sharply triangular spikelets few .... 2. *S. americanus*
    Culm leaves reduced to sheaths, with short blade; culm terete; spikelets numerous .... 4. *S. validus*
Involucral leaves 2–5, divergent:
  Spikelets large, few, capitate .... 3. *S. paludosus*
  Spikelets small, numerous, in umbelliform panicles .... 5. *S. microcarpus*

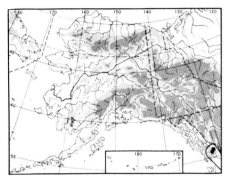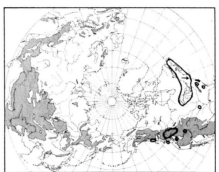

1. Scírpus submerinàlis Torr.

Submerged or terrestrial; culm from slender rhizome with tuber-bearing stolons; leaves of submerged plant long, capillary; leaves of terrestrial plant short or missing; involucre a single filiform leaf with excurrent median nerve; achenes lustrous, trigonous-obovoid, abruptly short-beaked; bristles shorter than achenes.

Ponds, quagmires.

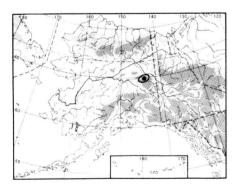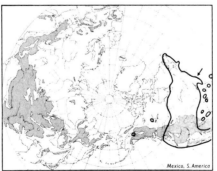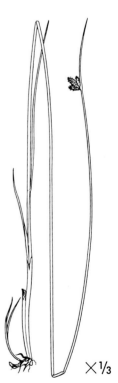

2. Scírpus americànus Pers.

Culms triangular, with concave sides, from stout, elongate, horizontal rhizome; basal leaves linear, the upper keeled, about 2 mm broad; involucral bract erect, acute; spikelets oblong or ovoid, fulvous to brown, cleft at apex, short-awned.

Wet places at Circle Hot Spring.

×⅓

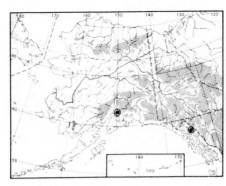 

3. Scírpus paludòsus Nels.

Scirpus pacificus with respect to Alaskan plant.

In Alaska less than 1 meter tall; culms from horizontal rhizome; stem leaves 3–4, shorter than culm, 3–4 mm wide; inflorescence densely compressed, with 1–2 long, involucral bracts; spikelets ovoid; anthers about 4 mm long, long-exserted.

Seashores, probably not native within our area. Broken line on circumpolar map indicates range of var. **atlánticus** Fern.

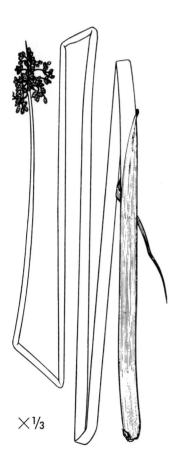
×⅓

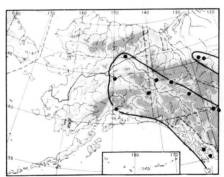 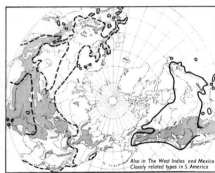

4. Scírpus valìdus M. Vahl

Culms stout, up to 2 cm thick at base, erect, in rows from thick, stout, scaly, reddish-brown rhizome; basal sheaths membranaceous, bladeless; involucral leaf shorter than the compressed inflorescence; spikelets dark brown; scales ovate, with excurrent keel, equaling or slightly exceeding achenes.

Shallow water, marshes. Described from islands of the Caribbean. Broken line on circumpolar map indicates range of the closely related **S. Tabernaemontàni** S. G. Gmel.

26. CYPERACEAE (Sedge Family)

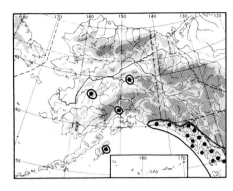
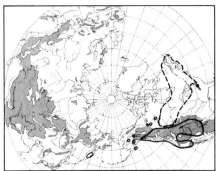
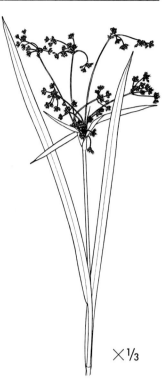

5. Scírpus microcárpus Presl
Scirpus silvaticus subsp. *digynus* (Boeck.) Koyama; *S. silvaticus* var. *digynus* Boeck.

Stout, about 1 meter tall, erect; culms leafy, from stout rhizome; leaves flat, 1–1.5 cm broad; involucral leaves 2–5, exceeding panicle; scales green to brown, rounded-ovate, with prominent midrib and scarious margin, longer than the small, mucronate achenes; achenes about 1 mm long and obovate.

Wet places. Described from Nootka Sound or Yakutat Bay.

Broken line on circumpolar map indicates range of var. **rubrotínctus** (Fern.) M. E. Jones (*S. rubrotinctus* Fern.).

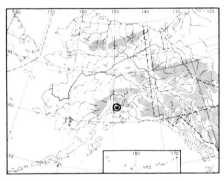
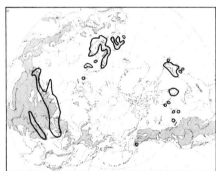
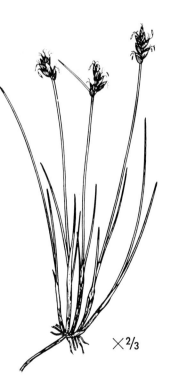

Blysmus Endl.

1. Blýsmus rùfus (Huds.) Link
Schoenus rufus Huds.; *Scirpus rufus* (Huds.) Schrad.

Culms terete, smooth, 10–30 cm tall, in a row from creeping, forking rhizome; filiform, canaliculate leaves from culm bases; spike flat, with 2 rows of shiny, castaneous to fulvous, 2–5-flowered spikelets; scales acuminate; style with 2 stigmas.

Seashores, saline soil. Described from England.

26. CYPERACEAE (Sedge Family)

Eleocharis R. Br. (Spike Rush)

Culms 6–7 cm tall; capillary, 0.5 mm or less in diameter:
  Tubercle flat, with short, central projection . . . . 5. *E. nitida*
  Tubercle conical-triangular . . . . 6. *E. acicularis*
Culms taller, stouter, 10 cm tall or more:
  Basal scale of spike solitary, completely encircling top of culm:
    Tubercle depressed-deltoid or lanceolate, much shorter than achene . . . .
    . . . . 2. *E. uniglumis*
    Tubercle about as large as achene, or slightly smaller . . . . . 3. *E. kamtschatica*
  Basal scales of spike 2(–3), each encircling half the culm only:
    Basal scale shorter than half the length of the spike; base of style tuberculate
    . . . . 1. *E. palustris*
    Basal scale half as long as spike or longer; base of style not distinctly tuberculate . . . . 4. *E. quinqueflora*

×½

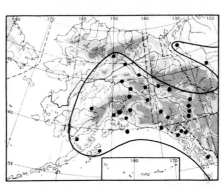 

1. Eleócharis palústris (L.) Roem. & Schult.
Scirpus palustris L. Including *E. macrostachya* Britt.

Culms stiff, with reddish basal sheaths from stout, brownish-black rhizome; uppermost sheath with green, oblique mouth; spike lanceolate to ovate, brown, up to 13 mm long, with two basal scales, each clasping half the culm; achenes with obovoid, conical, or pyriform tubercle of considerably varying form and size; style with 2 stigmas; bristles 4, reaching the tubercle.

Wet places, shallow water. Described from Europe.
Circumpolar map indicates range of entire species complex.

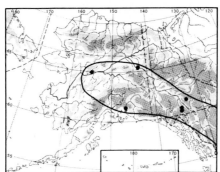 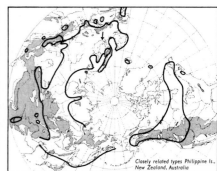

2. Eleócharis uniglùmis (Link) Schult.
Scirpus uniglumis Link.

Culms 10–15 cm tall, in groups from creeping rhizome; basal sheaths reddish-brown; basal scale solitary, ovate, entirely clasping top of culm; spike dark brown; tubercle depressed-deltoid or lanceolate, much shorter than achene.

Wet places. Circumpolar map tentative.

×⅓

26. CYPERACEAE (Sedge Family)

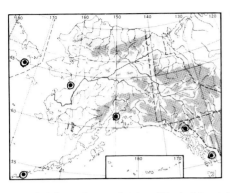 

3. Eleócharis kamtschática (C. A. Mey.) Kom.
Scirpus kamtschaticus C. A. Mey.

Similar to *E. uniglumis*, but tubercle ovoid, about as large as the achene; basal scale blackish-brown, ovate, entirely clasping top of culm.

Wet places.

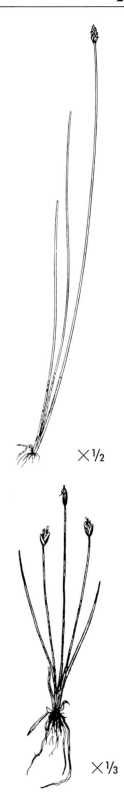

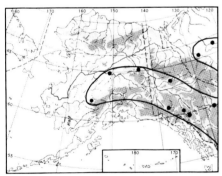 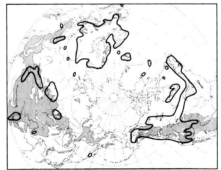

4. Eleócharis quinqueflòra (F. X. Hartm.) O. Schwarz
Scirpus quinqueflorus F. X. Hartm.; *S. pauciflorus* Lightf.; *Eleocharis pauciflora* (Lightf.) Link.

Loosely tufted, with short runners; culm leafless, with 2 basal sheaths, the upper with oblique mouth; spike ovate, 5–7-flowered; scales ovate, obtuse to somewhat acute, brown with hyaline margin, the lower about half as long as the spike and encircling half the culm; bristles 6; achenes trigonous-obovoid, with barely distinguishable tubercle, conical above and black at tip.

Swamps. Described from Berne, Switzerland.

Circumpolar map includes range of the eastern North American subsp. **Fernáldii** (Svens.) Hult. (*E. pauciflora* var. *Fernaldii* Svens.).

×3/4

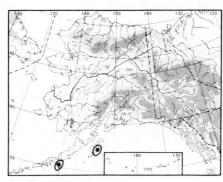 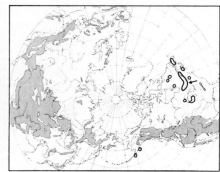

5. Eleócharis nìtida Fern.
Scirpus nitidus (Fern.) Hult.

Culms capillary, up to 8 cm tall, from thin, creeping, dark-purplish rhizomes; spikelets ovate, about 4 mm long; scales about 1 mm long, with rounded tips, dark purplish with narrow scarious margin; achenes triangular; tubercle flat, with short central projection.

Peaty or sandy places. Quite easily overlooked.

×3/4

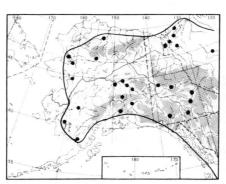 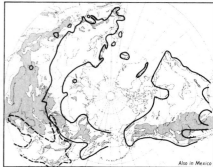

6. Eleócharis aciculàris (L.) Roem. & Schult.
Scirpus acicularis L.

Acicular culms with reddish-brown sheaths in small tufts; spikes 3–4 mm long; scales dark brown with greenish midrib and scarious margin; achenes triangular, with minute conical tubercle.

Muddy places close to water; shallow water. Submerged forms, often sterile, with elongated culms occur. Described from Europe.

Probably much overlooked. Broken line on circumpolar map indicates range of subsp. **longisèta** (Svens.) Hult. (*E. acicularis* var. *longiseta* Svens.).

26. Cyperaceae (Sedge Family)

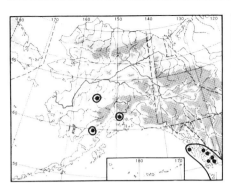
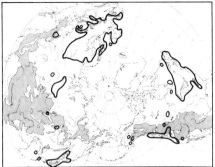

Rhynchospora M. Vahl

1. Rhynchóspora álba (L.) M. Vahl — Beak Rush
Schoenus albus L.

Loosely tufted, with short runners; culm leafy, triangular; leaves filiform, grayish-green, about equaling the inflorescence, with long, cylindrical sheaths; involucral bract barely exceeding inflorescence; spikelets 2-flowered, with white scales, in age brownish; bristles 10–12; achenes tapering to triangular-lanceolate tubercle; stamens commonly 2.

Bogs, peaty or sandy soil. Described from Europe.

Kobresia Willd.

Spike narrow, linear, 2–2.5 mm broad . . . . 1. *K. myosuroides*
Spike ovate or oblong:
    Culm slender; basal sheaths with withered blades at flowering time; scales small
    . . . . 2. *K. simpliciuscula*
    Culm stout; basal sheaths stout, lacking blades at flowering time; scales larger,
    4–5 mm long . . . . 3. *K. sibirica*

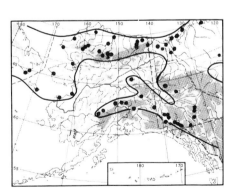
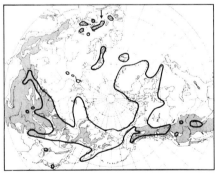
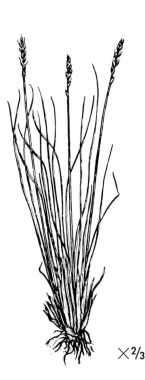

1. Kobrèsia myosuroìdes (Vill.) Fiori & Paol.
Carex myosuroides Vill.; *Elyna spicata* Schrad.; *Kobresia Bellardii* (All.) K. Koch; *K. scirpina* Willd.; *Carex Bellardii* All.

Densely tufted; culm stiff, 0.5–0.7 mm thick, up to 30 cm tall, with numerous, bladeless, brown sheaths at base; leaves filiform, stiff, in small specimens as long as culm, in tall specimens shorter than culm; spike 1–2 cm long, narrow, with brown scales; spikelets with 1 staminate and 1 pistillate flower; stigmas 3.

Dry, mostly calcareous slopes, gravel bars, lichen tundra, up to at least 1,400 meters.

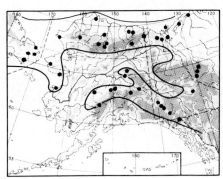
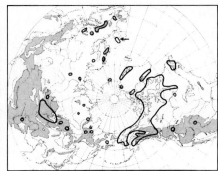

2. Kobrèsia simpliciùscula (Wahlenb.) Mack.

Carex simpliciuscula Wahlenb.; *Kobresia caricina* Willd.; *K. simpliciuscula* var. *americana* Duman.

Tufted, with short rhizomes; culm about 1 mm thick, firm, surrounded by blade-bearing sheaths at base; leaves filiform, shorter than culm; spike lanceolate to oblong, lobed, with several spikelets with 1 pistillate and 1 staminate flower and with staminate terminal spikelet; stigmas 3.

Calcareous rocky slopes, heaths; in the mountains to at least 2,000 meters.

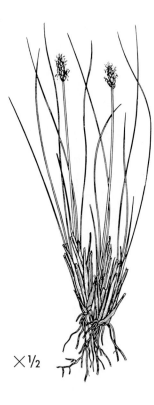
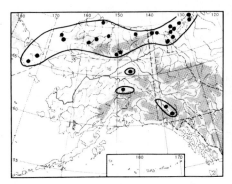
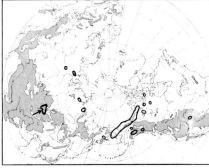

3. Kobrèsia sibìrica Turcz.

Kobresia arctica Pors.; *K. hyperborea* Pors. (var. *alaskana* Duman; var. *Lepagei* Duman); *K. macrocarpa* Clokey; *K. schoenoides* with respect to eastern Siberian plant.

Densely tufted; culms up to 35 cm tall, stiff, smooth, surrounded by dark-brown, leaf-bearing sheaths, bladeless in age; leaves canaliculate, slightly scabrous in margin; spikes ovate or ellipsoid, 1–1.8 cm long, with up to 20 spikelets, bearing 2–3 pistillate flowers and 1 staminate flower; scales dark chestnut-brown with paler center; stigmas 3, varying with lateral spikelets each bearing 2–3 sterile scales (*K. hyperborea* var. *alaskana* Duman), or with narrower lateral spikelets each bearing only 1 sterile scale (*K. hyperborea* var. *Lepagei* Duman).

Wet places.

26. Cyperaceae (Sedge Family)

Carex L. (Sedge)

Key to the Groups

Spike single, terminal .... 1–16. *Group 1*
Spikes more than 1 (in Group 2, often densely congested into headlike inflorescence):
  Spikes sessile, bisexual .... 17–55. *Group 2*
  Spikes peduncled, usually unisexual, sometimes bisexual:
    Stigmas 2, perigynia plano-convex or biconvex .... 56–69. *Group 3*
    Stigmas 3, perigynia mostly trigonous:
      Beak of perigynia with truncate mouth .... 70–104. *Group 4*
      Beak of perigynia with more or less bidentate mouth ... 105–20. *Group 5*

Group 1

Staminate and pistillate flowers on different plants:
  Stigmas 2; perigynia smooth .... 5. *C. dioica* subsp. *gynocrates*
  Stigmas 3; perigynia pubescent .... 6. *C. scirpoidea*
Staminate and pistillate flowers on same spike:
  Stigmas 2:
    Staminate flowers at base of spike; small beach plant .... 3. *C. ursina*
    Staminate flowers at top of spike:
      Culm straight, slender; spike globular, apiculate .... 4. *C. capitata*
      Culm shorter, mostly curved; spike oblong:
        Perigynia slender, spreading or reflexed at maturity; pistillate scales falling early .... 13. *C. pyrenaica* subsp. *micropoda*
        Perigynia short and broad:
          Leaves filiform; perigynia with scabrous beak .... 1. *C. nardina*
          Leaves more or less flat; perigynia with smooth beak .... 2. *C. Jacobi-Peteri*
  Stigmas 3:
    Perigynia with short beak or beakless:
      Perigynia beakless, flat .... 10. *C. leptalea*
    Perigynia with short beak, trigonous:
      Perigynia ovate, shiny, dark brown .... 9. *C. obtusata*
      Perigynia obovate:
        Leaves filiform, convolute; plant tufted .... 7. *C. filifolia*
        Leaves keeled or flat; plant with rhizome; perigynia yellowish-brown.. .... 8. *C. rupestris*
    Perigynia lanceolate, with long beak:
      Perigynia reflexed; pistillate scales falling early:
        Perigynia 6–7 mm long; culm triangular in cross section, curved at base.. .... 16. *C. pauciflora*
        Perigynia mostly shorter; culm circular in cross section, straight at base.. .... 15. *C. microglochin*
      Perigynia not reflexed:
        Plant tufted or with short rhizome; leaves canaliculate, thin, curved, withering early at apex .... 12. *C. circinnata*
        Plant with long, creeping rhizome; leaves flat:
          Spike slender, scales light brown with pale midvein .... 11. *C. anthoxanthea*
          Spike ovate, dark, with dark-brown scales .... 14. *C. nigricans*

Group 2

Spikes with pistillate flowers at base, staminate at apex:
  Stigmas 3; spikes forming very large head, 2–3 cm long and about 2½ cm in diameter .... 23. *C. macrocephala*
  Stigmas 2; spikes much smaller:
    ■ Rhizome long, creeping:
      Perigynia wing-margined, beak acutely bidentate .... 19. *C. chordorrhiza*
      Perigynia not winged, beak obliquely cut or sometimes bidentate:
        Spikes few, densely aggregated, forming one solitary, broadly ovoid inflorescence; perigynia inflated .... 17. *C. maritima*
        Spikes 5 or more; perigynia not inflated:
          Rootstock slender; culms terete, smooth; leaves 1–1.5 mm broad, canaliculate to flattened .... 18. *C. stenophylla* subsp. *eleocharis*

                Rootstock stout; culms sharply triangular; leaves 1.5–3 mm broad, flat
                .... 20. *C. praegracilis*
    ■ Rhizome short; plant tufted:
        Leaves 4–8 mm broad .... 21. *C. stipata*
        Leaves 1–2.5 mm broad .... 22. *C. diandra*
Spikes with staminate flowers at base, pistillate at apex:
    Lower bracts very long, leaflike .... 24. *C. sychnocephala*
    Lower bracts not prominent:
        Margin of perigynia more or less winged:
            Spikes aggregate, forming dense head:
                Lower bracts leaflike, exceeding head in length ..... 25. *C. atrostachya*
                Lower bracts shorter than head:
                    Plant about 20 cm tall; perigynia not conspicuous in spikes ....
                        .... 28. *C. phaeocephala*
                    Plant taller; perigynia conspicuous in spikes:
                        Scales chestnut-brown; perigynia dark-brown ....
                            .... 26. *C. macloviana* subsp. *pachystachya*
                        Scales light-brown; perigynia green .... 27. *C. Preslii*
            Spikes not distinctly aggregated into 1 head:
                Perigynia lanceolate or ovate-lanceolate:
                    Perigynia lanceolate-subulate .... 31. *C. Crawfordii*
                    Perigynia ovate-lanceolate:
                        Perigynia 6–8 mm long .... 32. *C. petasata*
                        Perigynia 3.5–4.5 mm long .... 33. *C. microptera*
                Perigynia ovate:
                    Beak of perigynia terete, not margined toward tip ... 35. *C. praticola*
                    Beak of perigynia flattened, serrulate and margined to tip:
                        Scales shorter than perigynia .... 29. *C. Bebbii*
                        Scales concealing perigynia, or nearly so:
                            Perigynia widest near middle, several-nerved ventrally; beak hya-
                                line-tipped; scales silvery green .... 30. *C. foena*
                            Perigynia widest near base, nerveless ventrally; beak reddish-
                                brown-tipped; scales brownish .... 34. *C. aenea*
        Margin of perigynia not winged:
        ● Perigynia white-puncticulate, beakless or with short, not decidedly biden-
            tate beak:
            Plant loosely tufted, with stolons; spikes few, few-flowered; perigynia
                beakless or very short-beaked:
                Spikes with staminate flowers at top, pistillate at base; perigynia slightly
                    inflated, biconvex, abruptly short-beaked, with prominent nerves
                    .... 47. *C. disperma*
                Spikes with pistillate flowers at top, staminate at base; perigynia com-
                    pressed, beakless:
                    Spikes aggregated at top of culm; perigynia obovate-oval, with slight-
                        ly prominent nerves .... 48. *C. tenuiflora*
                    Spikes mutually remote; perigynia slightly conical at tip, with promi-
                        nent nerves .... 49. *C. loliacea*
            Plant tufted, lacking stolons; spikes usually several, densely flowered;
                perigynia (sometimes very shortly) beaked:
                Spikes 2–4, aggregate:
                    Leaves 2 mm broad, flat; culm glabrous .... 36. *C. Lachenalii*
                    Leaves narrower, more or less canaliculate:
                        Culms scabrous; leaves grayish-green; scales yellowish-brown:
                            Perigynia ovate to ovate-elliptic, with short, more or less sca-
                                brous, slightly emarginate beak ...... 37. *C. heleonastes*
                            Perigynia oblong-elliptic, with broadly conical, glabrous beak,
                                or nearly beakless .... 38. *C. amblyorhyncha*
                        Culms glabrous or less scabrous; scales darker:
                            Leaves 0.75–1.5 mm broad .... 39. *C. glareosa* subsp. *glareosa*
                            Leaves 1.5–2.5 mm broad .. 40. *C. glareosa* subsp. *pribilovensis*
                Spikes 4–8, the lower separated from others:
                    Beak of perigynia scabrous on margins:
                        Beak and upper part of perigynium with distinct hyaline-margined
                            suture dorsally:

Densely tufted; culm 20–40 cm tall, stiff, erect ....
.... 45. *C. brunnescens* subsp. *alaskana*
Loosely tufted; culm 30–60 cm tall, weak, curved ....
.... 46. *C. brunnescens* subsp. *pacifica*
Beak sometimes slightly cleft at apex, but lacking distinct suture down body of perigynium:
Perigynia 1.5–1.8 mm long; heads small, few-flowered, subglobose; scales dark-colored .... 44. *C. bonanzensis*
Perigynia 2–3 mm long; heads (usually) larger, elongated, many-flowered; scales yellowish or yellowish-brown .....
.... 42. *C. canescens*
Beak of perigynia smooth:
Scales yellowish-brown, concealing perigynia, which is about 3 mm long .... 41. *C. Mackenziei*
Scales yellowish-green, shorter than the much smaller perigynia
.... 43. *C. lapponica*
● Perigynia not white-puncticulate, with long, bidentate beak:
Perigynia tapering toward base, more or less appressed:
Spikes small, subglobose, few-flowered; perigynia several-nerved ....
.... 53. *C. laeviculmis*
Spikes ovate; perigynia indistinctly nerved ...... 54. *C. Deweyana*
Perigynia broadest near base, spreading:
Leaves flat, soft, 2–4 mm broad .... 50. *C. arcta*
Leaves canaliculate, rigid, 1–3 mm broad:
Perigynium 3.75–4.5 mm long, many-nerved .. 55. *C. phyllomanica*
Perigynium 2.5–4 mm long, few or hardly any nerves:
Culm straight, about 40 cm tall, longer than leaves ....
.... 51. *C. interior*
Culm shorter, curved, often shorter than leaves ... 52. *C. echinata*

Group 3

Perigynia long-beaked .... 115. *C. saxatilis* subsp. *laxa*
Perigynia beakless or very short-beaked:
Inflorescence nearly flat-topped, formed by 3–5 dark spikes at top of culm; lowest bract with sheath .... 59. *C. eleusinoides*
Inflorescence otherwise:
Lowest bract with sheath; perigynia almost beakless:
Sheath of lowest bract 2–4 mm long, with black auricles at mouth ......
.... 67. *C. bicolor*
Sheath of lowest bract lacking black auricles:
Perigynia fleshy, translucent when ripe; terminal spike mostly staminate
.... 68. *C. aurea*
Perigynia dry; terminal spike pistillate above ....
.... 69. *C. Garberi* subsp. *bifaria*
Lowest bract sheathless or very short-sheathed:
Lowest bract shorter than inflorescence:
Aphyllopodic runners lacking; leaves narrow, about 1.5 mm broad; pistillate spikes linear .... 58. *C. lugens*
Aphyllopodic runners present; leaves broader; pistillate spikes oblong:
Perigynia spreading .... 56. *C. scopulorum*
Perigynia appressed .... 57. *C. Bigelowii*
Lowest bract as long as inflorescence or longer:
Spikes ovate, aggregate at top of culm, the upper sessile, the lower stipitate, gynaecandrous .... 82. *C. Enanderi*
Spikes cylindrical or prolonged, stipitate, the upper staminate:
▲ Culm phyllopodic (sterile shoots often aphyllopodic); spikes erect:
Perigynia nerved, ovate; sheaths brown or yellowish-brown ....
.... 60. *C. Kelloggii*
Perigynia nerveless, rounded; sheaths purplish:
Staminate apical spike only 1; stem short ....
.... 62. *C. aquatilis* subsp. *stans*
Staminate apical spike more than 1; stem taller ....
.... 61. *C. aquatilis* subsp. *aquatilis*

▲ Culm aphyllopodic; spikes erect or drooping:
　Plant normally tall; pistillate spikes long-peduncled:
　　Pistillate spikes normally erect (the lower sometimes drooping), linear-cylindrical, large and narrow, on long, more or less stiff peduncles .... 63. *C. sitchensis*
　　Pistillate spikes drooping, oblong, ovoid, short and thick, on capillary peduncles .... 66. *C. Lyngbyaei*
　Plant low or medium-sized; pistillate spikes sessile or short-peduncled:
　　Plant more or less decumbent; leaves and culm curved; spikes few-flowered; lower bracts spathaceous ... 64. *C. subspathacea*
　　Plant medium-sized; leaves and culm straight; spikes many-flowered .... 65. *C. Ramenskii*

Group 4

Bracts without sheaths, or with only very short sheaths:
　Spikes small, short, few-flowered, aggregated, sessile; plant low-growing, with narrow leaves:
　　Perigynia glabrous; beak cylindrical ...... 93. *C. supina* subsp. *spaniocarpa*
　　Perigynia pubescent:
　　　Scales ciliate .... 86. *C. melanocarpa*
　　　Scales not ciliate:
　　　　Subradical pistillate spikes not present; perigynia copiously hirsute .... 92. *C. Peckii*
　　　　Lower pistillate spikes on elongated, subradical peduncles; perigynia finely pubescent:
　　　　　Loosely caespitose; rhizome thin; perigynia 2.5 mm long, with short, thick beak .... 85. *C. deflexa*
　　　　　Densely caespitose; rhizome stout; perigynia 2.5–3 mm long, with lanceolate beak .... 91. *C. Rossii*
　Spikes larger, with more numerous flowers; plant mostly taller:
　　Terminal spike staminate (rarely with 1–2 undeveloped perigynias at base):
　　　Pistillate scales broad, obovate, small, 1.5–2.5 mm long; perigynia small; spikes approximate; leaves narrow:
　　　　Terminal staminate spike very small, hidden between pistillate spikes; scales and perigynia black .... 71. *C. holostoma*
　　　　Terminal spike overtopping pistillate spike; scales purplish-black; style terete, thick, long-exserted .... 76. *C. stylosa*
　　　Pistillate scales larger, acute; spikes more or less remote; leaves broader:
　　　　Scales long-aristate .... 84. *C. macrochaeta*
　　　　Scales blunt, acute, or short-acuminate:
　　　　　Perigynia ciliate-serrulate in margin .... 104. *C. atrofusca*
　　　　　Perigynia smooth in margin:
　　　　　　Culms aphyllopodic, with short, scalelike leaves below, and well-developed stem leaves:
　　　　　　　Scales obtuse, with midvein obscure toward apex .... 87. *C. podocarpa*
　　　　　　　Scales light colored, with excurrent midvein ... 88. *C. spectabilis*
　　　　　　Culms phyllopodic, lacking stem leaves:
　　　　　　　Spikes ovate, on slender, capillary peduncles, drooping in age .. .. 89. *C. microchaeta*
　　　　　　　Spikes narrow, linear, on stiff, erect peduncles; midvein of scales conspicuous .... 90. *C. nesophila*
　Terminal spike with pistillate flowers at top:
　▼ Pistillate scales awned, or at least cuspidate:
　　　Spikes distinctly peduncled; scales abruptly slender-awned .... 77. *C. Gmelini*
　　　Spikes sessile; scales lanceolate, acuminate or awned:
　　　　Perigynia beakless; lower scales shorter than perigynia .... 74. *C. adelostoma*
　　　　Perigynia short-beaked; lower scales longer than perigynia .... 75. *C. Buxbaumii*

▼ Pistillate scales not awned or cuspidate:
   Perigynia 5 mm long, much longer than scales, membranous, broad, light
      green-yellowish, abruptly contracted into very narrow beak; spikes
      numerous, cylindrical; culms nearly winged ...... 83. *C. Mertensii*
   Perigynia shorter, dull green to brown when ripe; culms not winged:
      Perigynia nerved, sparsely spinulose in margin ...... 82. *C. Enanderi*
      Perigynia nerveless (except for marginal nerves), smooth in margin:
         Rhizome long; leaves smooth .... 73. *C. sabulosa*
         Rhizome short; leaves scabrous at apex:
            Scales with conspicuous white-hyaline margins:
               Culm slender; scales small, shorter than perigynia; perigynia
                  subinflated, with slightly serrulate beak .... 70. *C. media*
               Culm stiff; scales about as long as perigynia:
                  Scales obtuse, rounded, reddish-brown, with hyaline margin
                     and green midvein .... 72. *C. Parryana*
                  Scales somewhat acute, purplish-black, with hyaline tip and
                     margin .... 78. *C. albo-nigra*
            Scales lacking conspicuous white-hyaline tip or margins:
               Culms scabrous above; spikes cylindrical ....
                 .... 81. *C. atratiformis* subsp. *Raymondii*
               Culms glabrous; spikes oblong-ovoid:
                  Scales as long as perigynia, black; perigynia with emarginate,
                     purplish beak .... 79. *C. atrata* subsp. *atrata*
                  Scales shorter than perigynia, brownish; perigynia with
                     shorter and broader, less emarginate beak ....
                     .... 80. *C. atrata* subsp. *atrosquamea*
Bracts with distinct sheaths:
  Perigynia pubescent .... 94. *C. concinna*
  Perigynia glabrous:
     Leaves narrow, 0.2–1 mm broad, canaliculate or involute:
        Lowest bract with setaceous blade; scales purplish-black .. 95. *C. glacialis*
        Lowest bract without blade; scales yellowish-white ...... 96. *C. eburnea*
     Leaves broader, flat (or, in *C. limosa*, canaliculate):
        Pistillate spikes drooping, more or less densely flowered:
           Sheath of lowest bract long, tubiform .... 101. *C. laxa*
           Sheath of lowest bract short, spathiform:
               Lowest bract leaflike, overtopping spikes; scales narrower than peri-
                 gynia, long-acuminate ...... 100. *C. magellanica* subsp. *irrigua*
               Lowest bract subulate, not overtopping spikes:
                  Pistillate spike mostly solitary; leaves canaliculate .. 99. *C. limosa*
                  Pistillate spikes 2–3; leaves flat:
                     Pistillate spikes 2–10-flowered; perigynia 3–3.5 mm long; culm
                       obtusely triangular .... 97. *C. rariflora*
                     Pistillate spikes 10–25-flowered; perigynia 4–4.5 mm long; culm
                       sharply triangular .... 98. *C. pluriflora*
        Pistillate spikes erect, loosely flowered:
           Leaves glaucous; perigynia elliptic, nearly beakless ..... 102. *C. livida*
           Leaves green; perigynia ovate, long-beaked .... 103. *C. vaginata*

Group 5

Leaves not septate-nodulose:
✶ Perigynia flat, ciliate-serrulate in margin:
   Terminal spike with staminate flowers at top; plant with slender rhizomes:
      Perigynia oblong-ovate, abruptly beaked; spikes more numerous, the ter-
         minal strongly pistillate at base; scales reddish-brown ....
      .... 106. *C. Franklinii*
      Perigynia lanceolate; spikes 2–5, the terminal with few perigynia at base;
         scales purplish-black .... 107. *C. petricosa*
   Terminal spike with pistillate flowers at top; plant caespitose or with ascending
      stolons:
     ■ Perigynia with rounded base, about as long as scales, beak with indistinctly
         bidentate mouth; scales purplish-black .... 104. *C. atrofusca*

■ Perigynia narrowly lanceolate, long-beaked, longer than scales; beak distinctly bidentate; scales dark-brown .... 105. *C. misandra*
★ Perigynia trigonous, not ciliate-serrulate in margin; scales light brown or hyaline:
  Spikes on capillary peduncles, drooping when ripe; beak indistinctly bidentate, hyaline-margined:
   Terminal spike with pistillate flowers at top .... 109. *C. Krausei*
   Terminal spike staminate:
    Leaves flat; perigynia nerveless except for marginal nerve; beak ciliate on margin .... 108. *C. capillaris*
    Leaves setiform, involute; perigynia with few nerves; beak smooth .... 110. *C. Williamsii*
  Spikes short, erect, on stout peduncles; beak distinctly bidentate:
   Lower perigynia deflexed; beak curved, about as long as body .... 111. *C. flava*
   Lower perigynia not deflexed; beak nearly straight, distinctly shorter than body .... 112. *C. Oederi* subsp. *viridula*
Leaves septate-nodulose:
 Perigynia firm, subcoriaceous; teeth of beak subulate, 1 mm long or longer .... 120. *C. atherodes*
 Perigynia membranaceous; teeth of beak shorter:
  Perigynia pubescent:
   Leaves filiform; culm smooth; teeth of beak 0.2–0.5 mm long .... 118. *C. lasiocarpa* subsp. *americana*
   Leaves flat, 2–5 mm broad; culm scabrous; teeth of beak 0.3–0.8 mm long .... 119. *C. lanuginosa*
  Perigynia glabrous:
   Leaves involute, narrow; plant with long stolons ...... 116. *C. rotundata*
   Leaves flat, broader; plant lacking long stolons:
    Plant tall; pistillate spikes 5–12 cm long; perigynia and scales yellowish-green:
     Perigynia divaricate, inflated, abruptly beaked; leaves green, flat; culm scabrous .... 114. *C. rhynchophysa*
     Perigynia ascending, tapering to shorter beak; leaves grayish-green, more or less channeled; culm often smooth .... 113. *C. rostrata*
    Plant shorter; pistillate spikes 1–2 cm long; scales purplish-black:
     Spikes long-peduncled, drooping .... 115. *C. saxatilis* subsp. *laxa*
     Spikes sessile, erect .... 117. *C. membranacea*

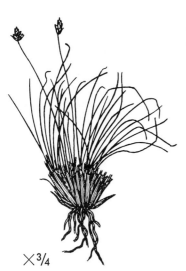
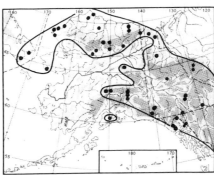
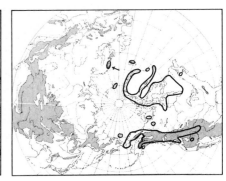

1. Càrex nardìna E. Fries

Including *Carex Hepburnii* Boott.

Densely tufted; culms often arcuate, shorter or somewhat longer than leaves; basal sheaths light brown; leaves setaceous, usually curved; spike ovate, dark brown; perigynia stipitate, light brown, with short, ciliate-to-scabrous, dark-brown, bidentate beak; 2 stigmas.

Dry ground in the mountains, in the St. Elias Range to at least 2,600 meters.

26. Cyperaceae (Sedge Family)

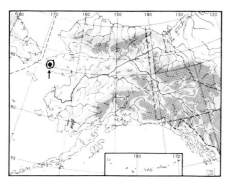 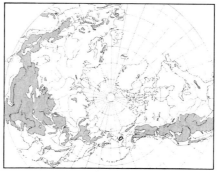 

2. Càrex Jacòbi-Petéri Hult.

Tufted, with recurving leaves longer than culms and brown sheaths; leaves up to 1.5 mm broad, scabrous in margin; spike with staminate flowers at top, pistillate at base, ovate or spherical; scales acute, with green midvein; utricles ovate, prolonged into smooth beak; 2 stigmas.

Similar to *C. ursina*, but differing in the long beak of the perigynia. Known only from the type specimens from Tin City.

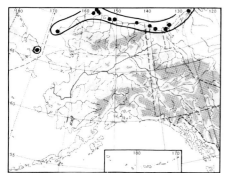 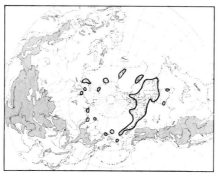 

3. Càrex ursìna Dew.

Densely tufted, with glabrous culms, usually somewhat shorter than leaves; leaves canaliculate or involute, scabrous in margin; spikes obovate or spherical, with few pistillate flowers at base; scales ovate, obtuse, dark brown with lighter midvein, shorter than ovate, nearly beakless perigynia.

Grassy places near salt or brackish water. Described from arctic America.

×½

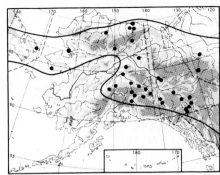 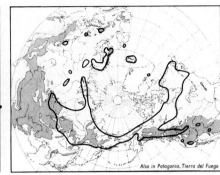

4. Càrex capitàta Soland. in L.

Grayish-green, densely tufted; basal sheaths purplish-red; culm as long as involute leaves, or considerably longer; spike with a few staminate flowers at top; scales light brown, sometimes darker, shorter than smooth, sessile, ovate, short-beaked perigynia.

Bogs and marshes, poplar forests from lowlands to at least 1,000 meters. Described from northern Scandinavia.

×½

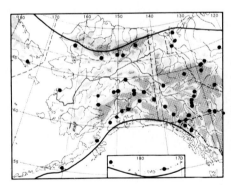 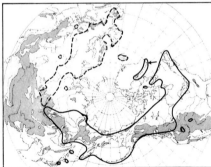

5. Càrex diòica L.
subsp. **gynocràtes** (Wormsk.) Hult.

Carex gynocrates Wormsk.; *C. alaskana* Boeck.

Staminate and pistillate spikes usually on separate plants; culms smooth, often curved, from slender, yellowish-brown rhizome with capillary runners; leaves filiform, canaliculate; staminate spike linear to lanceolate-linear; pistillate spike elliptic or broadly cylindrical, bractless; perigynia oblong-ovate, with short, somewhat serrulated beak, at first erect, in age spreading or reflexed.

Wet meadows, marshes. Broken line on circumpolar map indicates range of subsp. **diòica**.

26. Cyperaceae (Sedge Family)

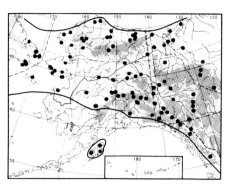
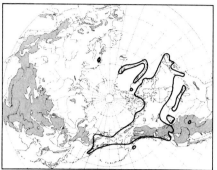

6. Càrex scirpoìdea Michx.

Including *Carex stenochlaena* Holm.

Culms in dense rows from stout, dark-brown, scaly rhizome; leaves flat, very acute, scabrous, about 3 mm broad; staminate spikes broad, with light-brown scales in middle, about 2.5–3 mm long; anthers conspicuous during flowering time; pistillate spikes cylindrical, with oblong-ovate, purplish-brown or darker, scarious-margined, ciliate scales, and very hairy, fusiform to obovate perigynia, with short beak; 3 stigmas.

Meadows, heaths, wet places; common in mountains, to at least 2,000 meters. Described from Hudson Bay.

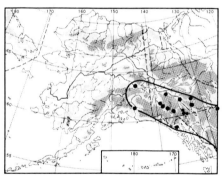
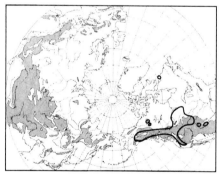

7. Càrex filifòlia Nutt.

Densely tufted, with numerous, conspicuous, leaf-bearing, striate, yellowish-brown-to-brown sheaths at base; culms often curved; leaves filiform, acute; grayish-green; spike solitary, with staminate flowers at top, pistillate at base, bractless; scales broadly ovate, or obovate, reddish-brown, with broad, hyaline margin or sometimes entirely hyaline; perigynia somewhat puberulent at top, with short beak, about 0.4 mm long; 3 stigmas.

Dry ridges up to about 1,500 meters. Described from "Hills of Missouri River."

×½

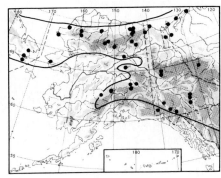 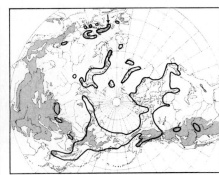

8. Càrex rupéstris All.

Culms from short, creeping rhizome; basal sheaths of leafy shoots brown; leaves recurving; spike with several staminate flowers at tip and few pistillate flowers at base; scales persistent, broad, obtuse, dark, with hyaline margin; perigynia erect, yellowish-brown, with beak 0.3 mm long; 3 stigmas.

Dry ridges in the mountains, to at least 2,000 meters.

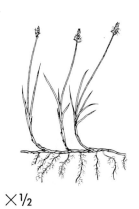
×½

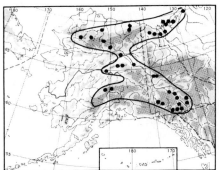 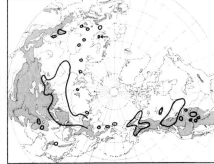

9. Càrex obtusàta Lilj.

Culms in rows from long, creeping, purplish-black rhizome; leafy shoots with purplish-black sheaths; spike with pistillate flowers at base, staminate flowers at tip; scales persistent, light brown with pale midvein, acute; perigynia ovate, shiny, brown, darker at tip, with glabrous, slightly bidentate, hyaline-tipped beak; 3 stigmas.

Sandy soil, dry ridges.

26. Cyperaceae (Sedge Family)

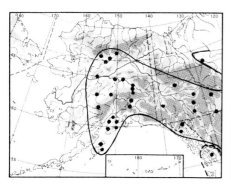
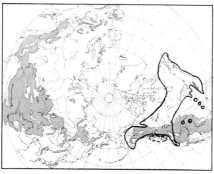

10. Càrex leptàlea Wahlenb.
Carex polytrichoides Holm.

Tufted; culms scabrous at top, very slender; leaves capillary; spikes very small, with staminate flowers at top, pistillate flowers at base; scales yellowish-green with hyaline margin and green midrib; perigynia less than 4 mm long, light green, fusiform, beakless; 3 stigmas, reddish-brown.

Bogs, meadows, shores, up to at least 700 meters. Described from North America.
Plant from southernmost Alaska belongs to subsp. **pacífica** Calder & Taylor, with leaves over 1.2 mm broad and perigynia more than 4 mm long; its range is indicated by broken line. Range of southeastern subsp. **Harpéri** (Fern.) Calder & Taylor (var. *Harperi* Fern.) is included on circumpolar map.

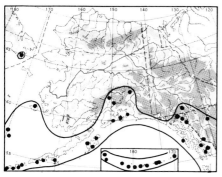
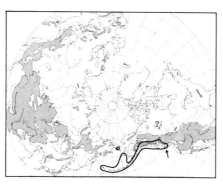

11. Càrex anthoxánthea Presl

Culms with bladeless sheaths at base, in rows from creeping, yellowish-brown rhizome; solitary spike pistillate or staminate, more rarely androgynous; perigynia fusiform, smooth, yellowish-green; stigmas usually 3.

Grassy slopes.

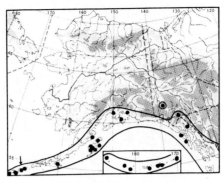
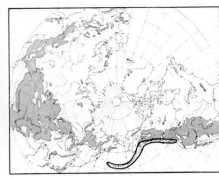

12. Càrex circinnàta C. A. Mey.

Forming dense tufts; culms not longer than curved, stiff, light-green, involute leaves, with light-brown sheaths at base; spikes narrow, with staminate flowers at top, pistillate flowers at base; pistillate scales brown, the lower acute, the upper obtuse, with prominent green midvein; perigynia long and narrow, passing into serrulate beak; 3 stigmas.

Stony places in mountains, to at least 1,000 meters.

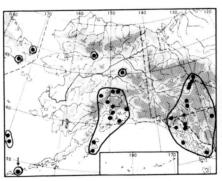
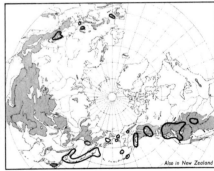

13. Càrex pyrenáica Wahlenb.
subsp. **micròpoda** (C. A. Mey.) Hult.
Carex micropoda C. A. Mey.

Densely caespitose; culms short, slender; leaves strongly convolute, 0.5–1 mm broad; spike oblong-ovate, solitary, bractless, with staminate flowers at top; scales obtuse or somewhat acute, reddish-brown with paler midvein; perigynia ovate-lanceolate, longer than scales, shining, brownish, stipitate, spreading at maturity, with short, glabrous beak; stigmas mostly 2.

Meadows, snow beds. *C. pyrenaica* described from the Pyrenees, subsp. *micropoda* from Unalaska.

Circumpolar map indicates range of entire complex.

26. Cyperaceae (Sedge Family)

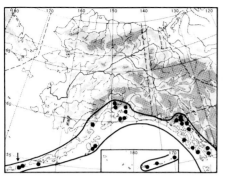 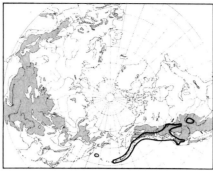 

×½

14. Càrex nìgricans C. A. Mey.

Culms and leaves densely packed, from stout, creeping rhizome; leaves 1.25–2 mm broad, flat, with light-brown sheaths; spike with staminate flowers at tip, pistillate flowers at base; pistillate scales dark brown with lighter midvein, falling off in age; perigynia long-stipitated, in age reflexed; 3 stigmas.

Stony places in the mountains, to at least 1,000 meters in southeast Alaska.

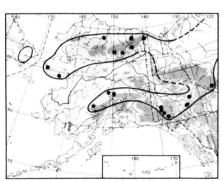 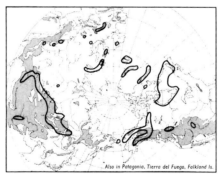 

×⅔

15. Càrex microglóchin Wahlenb.

Loosely tufted, with short, brown, scaly runners; culms about 10 cm tall (exceptionally up to 30 cm), straight at base; lowest sheath blade-bearing; spike with staminate flowers at tip, pistillate flowers at base; pistillate scales blunt, drooping; perigynia greenish, lance-subulate, slenderly tapering into beak, with stiff rachilla projecting beyond perigynia and stigmas, which can be seen just at mouth of grayish-green perigynia; 3 stigmas.

Moist, preferably calcareous soil. Described from northern Lapland.

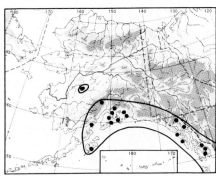
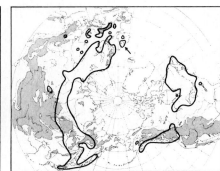

16. Càrex pauciflòra Lightf.

Similar to *C. microglochin,* but usually taller; culms curved at base; lowest sheath bladeless; style (recognized from stigmas at tip) projecting beyond mouth of brownish-green perigynia; rachilla lacking.

Peat bogs, preferably acid soil.

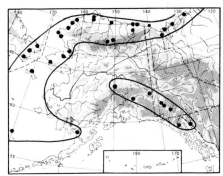
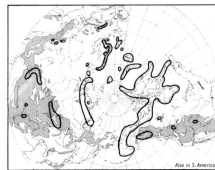

17. Càrex marítima Gunn.

Carex incurva Lightf. Including *C. incurvaeformis* Mack.

Culms grayish-green, often curved, in row from creeping rhizome; leaves more or less curved, longer than culm; staminate flowers at tip of spike, pistillate flowers at base; scales light brown to brown; perigynia ovoid, dark, nearly nerveless, tapering into scabrous beak.

Shores, sandy soil; in the St. Elias Range, sterile specimens reach probably to 2,800 meters. Described from the coast of Norway.

26. CYPERACEAE (Sedge Family)

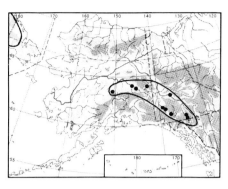 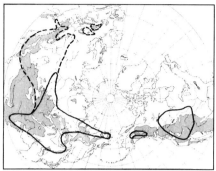 

18. Càrex stenophýlla Wahlenb.
subsp. **eleócharis** (Bailey) Hult.

Carex eleocharis Bailey; *C. stenophylla* var. *enervis,* at least with respect to American plant; *C. duriuscula* C. A. Mey.

Culms smooth, slender, extending beyond leaves, from slender, dark, fibrous rhizomes; leaves narrow; spikes with staminate flowers at tip, pistillate flowers at base, forming head; perigynia ovate, distinctly nerved, with short beak, scabrous in margin.

Grassy meadows, rocks, sandy soil; in mountains to at least 1,200 meters. *C. stenophylla* described from Pannonia (Austria), subsp. *eleocharis* from Saskatchewan.

Broken line on circumpolar map indicates range of other subspecies of this complicated complex.

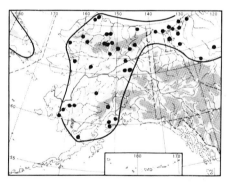 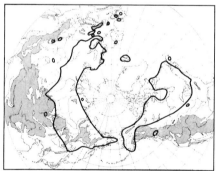 

19. Càrex chordorrhíza Ehrh.

Hollow culms, curved at base and straight at top, from slender, branching rhizome; leaves short, straight, thick; spikes densely compressed, forming head; staminate flowers at tip of spike, pistillate flowers at base; scales light brown; perigynia ovoid to subglobose, strongly nerved, with short, glabrous beak.

Bogs, lake margins; quagmires in lowlands, ascending to at least 1,200 meters in valleys. Described from Sweden.

×⅓

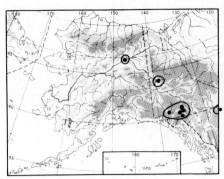
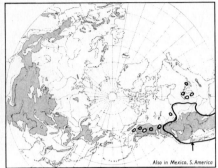

20. Càrex praegràcilis W. Boott

Culms scabrous, acutely angled, from elongate, forking, dark-colored rhizomes; leaves 1.5–3 mm broad in scattered tufts; head cylindrical, interrupted; lowest bract with short projection; spikes with staminate flowers at tip, pistillate flowers at base; perigynia ovate, slightly nerved, with serrulate beak half as long as body.

Open, grassy ground.

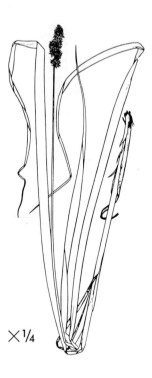
×¼

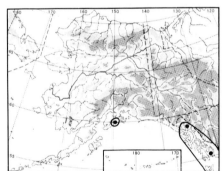

21. Càrex stipàta Muhl.

Caespitose, very coarse and stout, yellowish-green; culms triangular, with concave sides, light brown at base; leaves flaccid, 4–10 mm broad; head paniculate-spiciform, dense, yellowish-green, in age brown, spikes with inconspicuous staminate flowers at top and pistillate flowers at base; perigynia 4–5 mm long, strongly nerved, evenly tapering into sharp-edged, serrulated, tridentate beak; teeth of beak reddish-brown.

Swamps and meadows.

26. Cyperaceae (Sedge Family)

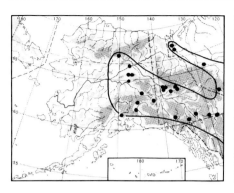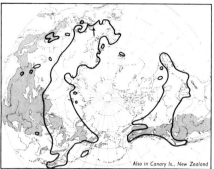

22. Càrex diándra Schrank

Tufted, with short rhizome; culms obtuse-angled at top, terete at base; leaves grayish-green, 1–2 mm broad; head 3–4 cm long; spikes sessile with staminate flowers at tip, pistillate flowers at base; scales light brown, the lowest long, acute; perigynia dark brown, broadly ovate, highly convex on back, long-beaked, lustrous, nerveless, about 3 mm long.

Swamps, mires, bogs, borders of ponds.

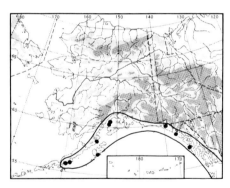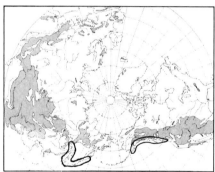

23. Càrex macrocéphala Willd.

Carex anthericoides Presl; *C. macrocephala* subsp. *anthericoides* (Presl) Hult.

Long, horizontal, scaly rootstock, buried deep in sand; culms stout and stiff, with old leaves at base; leaves yellowish-green, thick, serrulate, pungent; plant dioecious; pistillate spikes crowded into large ovate head, 3–5 cm long and about 3 cm broad; staminate heads 3–4 cm long, 1 cm broad, with numerous spikes; perigynia lance-ovate, serrulate in margin, with long, bidentate beak.

Sandy seashores. Described from Siberia.

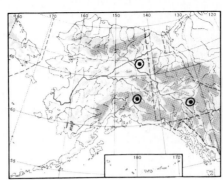
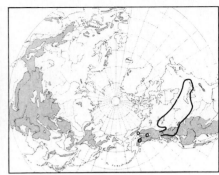

24. Càrex sychnocéphala Carey

Caespitose; culm with old leaves at base; leaves 1.5–3 mm broad, scabrous in margin; spikes with inconspicuous staminate flowers at base, pistillate flowers at tip; lower 2–4 bracts up to 20 cm long, leaflike; scales acuminate; perigynia (including bidentate beak) lanceolate, serrulate in margin, longer than scales, greenish with reddish-brown tip.

Meadows, grassy slopes. Described from "Nov. Ebor. Comitat." (New York).

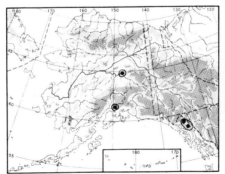
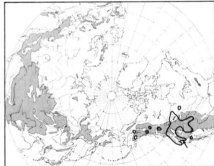

25. Càrex atrostáchya Olney

Caespitose; culms with old leaves at base; leaves 2–3 mm broad, flat; inflorescence capitate, spikes with staminate flowers at base, pistillate flowers at tip, the lowest subtended by a foliaceous bract; pistillate scales somewhat shorter and narrower than perigynia; perigynia lanceolate-ovate, with serrulate margin, tapering into somewhat flattened beak.

Dry, grassy meadows.

26. Cyperaceae (Sedge Family)

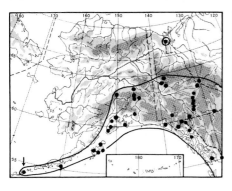
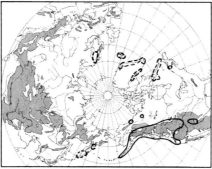

26. Càrex macloviàna d'Urv.
subsp. **pachystáchya** (Cham.) Hult.

Carex pachystachya Cham.; *C. pyrophila* Gandoger.

Densely caespitose; culm coarse, sharply angled, and scabrous at summit; leaves firm, shorter than culms, about 4 mm broad; inflorescence capitate, dark brown; spikelets tightly compressed, with staminate flowers at base, pistillate flowers at tip; scales dark brown or chestnut-brown; perigynia ovate, dark brown, serrulated in margin, with long, slender beak.

Meadows, gravelly shores. Described from Falkland Islands, subsp. *pachystachya* from Unalaska.

× ⅓

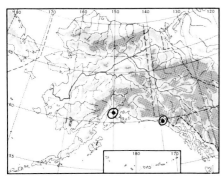
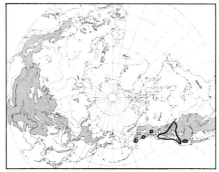

27. Càrex Préslii Steud.

Similar to *C. macloviana* subsp. *pachystachya*, but scales light brown and perigynia green when young.

Meadows. Described from Yakutat Bay or Nootka Sound, Vancouver Island.

× ⅓

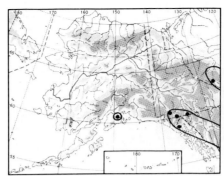
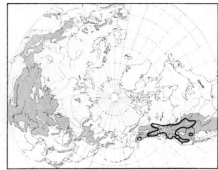

28. Càrex phaeocéphala Piper

Densely caespitose, with brownish, old leaves at base; culms longer than leaves, smooth; leaves canaliculate, 1.5–2 mm wide; inflorescence about 2 cm long, compressed; spikes with conspicuous staminate flowers at base, pistillate flowers at tip; scales reddish-brown, the lower ones often cuspidate; perigynia oblong-ovate, with beak 1 mm long, concealed by scales; 2 stigmas.

Wet meadows. Described from Oregon.

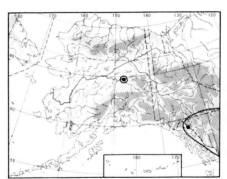
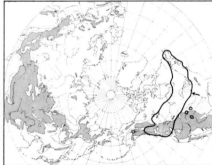

29. Càrex Bébbii Olney

Densely caespitose; culm sharply angled, with old leaves at base; leaves 2–4.5 mm broad, flat, yellowish-green; spikes densely compressed, with staminate flowers at base, pistillate flowers at tip; inflorescence subtended by short, setaceous bract; perigynia crowded, ascending, ovate, with bidentulate beak 1 mm long; 2 stigmas.

Wet meadows. Described from Illinois.

26. Cyperaceae (Sedge Family)

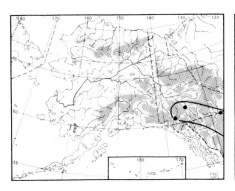

30. Càrex foèna Willd.
C. siccata Dew.

Culms slender, from creeping, thin, scaly, branching, dark-brown rhizome; leaves 1–2 mm broad; 2–8 spikes in inflorescence 1.5–3.5 cm long, gynaecandrous, androgynous, or staminate; scales ovate-lanceolate, with green center and whitish margin; perigynia ovate to ovate-lanceolate with narrow, greenish, serrulate margin, tapering to long beak.

Woods, riverbanks, sandy soil. Described from North America.

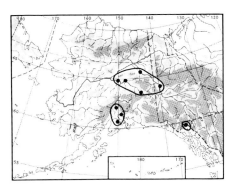
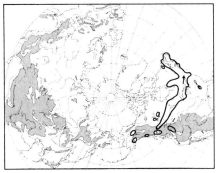

31. Càrex Crawfórdii Fern.

Caespitose; culms sharply angled; leaves 1–4 mm broad, yellowish-green; spikes in ellipsoid head, 1–3 cm long; scales pale brown, nearly as broad as perigynium, but much exceeded by the prominent, slender beak; perigynia narrowly lanceolate, gradually tapering to long, narrow beak.

Dry grassland, roadsides. Described from New Hampshire.

× 1/3

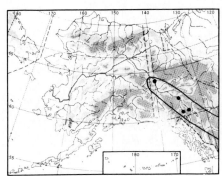
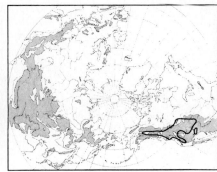

32. Càrex petasàta Dew.

Caespitose; culms with old leaves at base; leaves light green, 2–3 mm broad; spikes in head 1–1.5 cm long, with few staminate flowers at base and numerous pistillate flowers at tip; bracts scalelike, the lowest slightly prolonged; scales acute, light reddish-brown, with hyaline margins nearly covering perigynia; perigynia ovate-lanceolate, wing-margined, serrulate, striate, with bidentate beak 2 mm long.

Meadows, woods. Described from the Rocky Mountains.

× 1/3

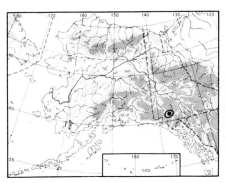
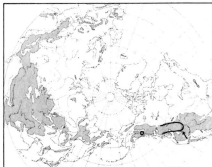

33. Càrex micróptera Mack.

Caespitose; culms with old leaves at base; leaves 2–4.5 mm broad; spikes densely compressed, with staminate flowers at base, pistillate flowers at tip; lowest bract awned; scales acute; perigynia membranaceous, light-colored, ovate-lanceolate, narrowly margined.

Meadows.

26. Cyperaceae (Sedge Family)

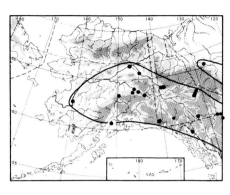
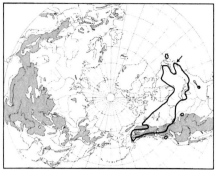

34. Càrex aenèa Fern.

Caespitose; culms smooth, longer than leaves; leaves 2–4 mm broad, flat; inflorescence nodding, with at least lower spikes remote; lower bracts cuspidate; scales yellowish-brown with hyaline margins; perigynia (nearly concealed by scales) ovate, serrulate in margin, nerveless, tapering to flat, bidentate, serrulated beak.

Woods, dry slopes in the St. Elias Mountains to 2,000 meters.

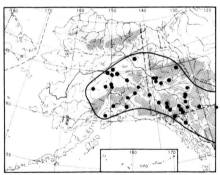
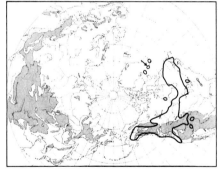

35. Càrex praticola Rydb.
Carex pratensis Drej. *non* Host.

Similar to *C. aenea*, but culms more slender; leaves narrower; heads silvery green to pale ferruginous; perigynia with longer, terete beak not serrulate at tip.

Meadows, grassland.

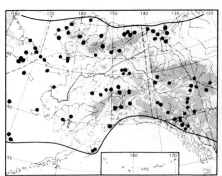
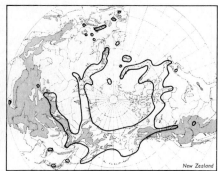

36. Càrex Lachenàlii Schkuhr
Carex tripartita All.; *C. bipartita* All.?; *C. lagopina* Wahlenb.

Loosely caespitose; culms stiff, 10–30 cm tall, usually curved; leaves yellowish-green, 2 mm broad; spikes 2–4, close together, dark-colored, with staminate flowers at base, pistillate flowers at tip, terminal spikes longer than others; scales blunt, hyaline-margined; perigynia brown, ellipsoid or fusiform, with short, smooth, nearly nerveless beak.

Alpine tundra, snow beds, above tree line. Type locality not given.

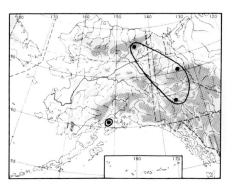
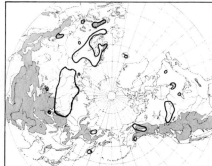

37. Càrex heleonástes Ehrh.
Carex cryptantha Holm.

Grayish-green, tufted; culm scabrous; leaves 1–2 mm wide; spikes 3–4, ovate, crowded, with pistillate flowers at base; scales ovate, acutish, light-brown with green midrib, scarious-margined, shorter than perigynia; perigynia ovate to ovate-elliptic, with 8–10 indistinct nerves, and short, conical, more or less scabrous, slightly emarginate beak.

Peat bogs, swamps. Described from Sweden.

Plants with shorter beak have been called *C. neurochlaena* Holm [*C. heleonastes* subsp. *neurochlaena* (Holm) Böcher].

26. Cyperaceae (Sedge Family)

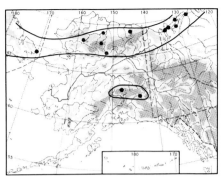 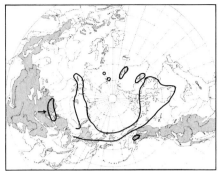 

38. Càrex amblyorhýncha Krecz.

Similar to *C. heleonastes*, but perigynia oblong-elliptic, short-stipitate, with broadly conical, glabrous beak, or nearly beakless.

Bogs, swamps.

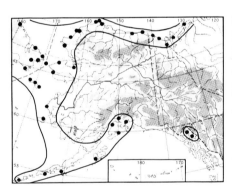 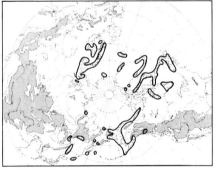 

39. Càrex glareòsa Wahlenb.

subsp. **glareòsa**

Including *C. marina* Dew.; *C. glareosa* var. *amphigena* Fern.

Grayish-green, loosely tufted; leaves 0.75–1.5 mm broad, shorter than culm; spikes 2–3, terminal spikes with staminate flowers at base, lower spikes with pistillate flowers; bracts scalelike; scales shorter than perigynia, obtuse, brown, with hyaline margin; perigynia oblong, elliptic to lanceolate, distinctly nerved, stipitate, with minute beak; 2 stigmas.

Brackish marshes. Described from northern Scandinavia.

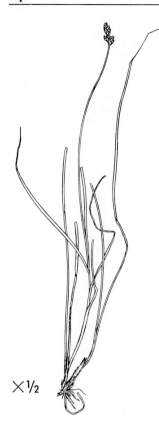
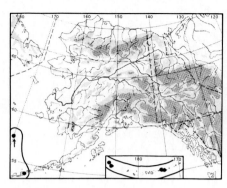
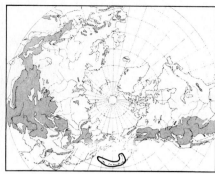

40. Càrex glareòsa Wahlenb.
subsp. **pribilovénsis** (Macoun) Chater & Halliday ined.
Carex pribilovensis Macoun; *C. lagopina* var. *pribilovensis* (Macoun) Kükenth.

Similar to subsp. *glareosa*, but leaves 1.5–2.5 mm broad; scales longer than perigynia.

Brackish marshes.

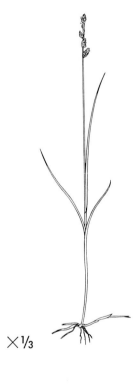
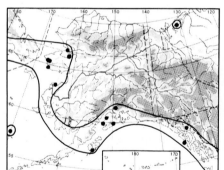

41. Càrex Mackenzièi Krecz.
Carex norvegica Willd. not Retz.

Loosely caespitose, culm triangular; leaves 2–3 mm broad, flat, about as long as culm; spikes 3–6, ovoid, scattered, fulvous, apical spike with staminate flowers at base, pistillate flowers at tip; lower spikes pistillate; perigynia fusiform to ovoid, densely nerved.

Wet places along seashore. Described from northern Norway.

26. Cyperaceae (Sedge Family)

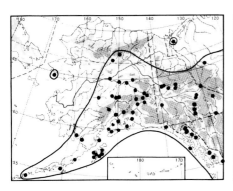

42. Càrex canéscens L.

Caespitose, grayish-green; culms scabrous above; leaves flat, about 2.5 mm broad; spikes more or less distant, subtended by stiff, short bracts; scales light-colored, with green central nerve shorter than perigynia; perigynia yellowish-green, ovate, slightly scabrous toward tip, with short beak.

Bogs, swamps. Described from northern Europe.

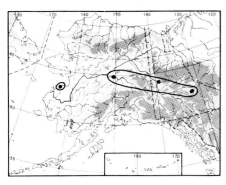
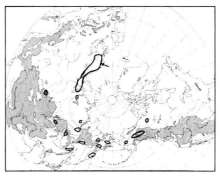

43. Càrex lappónica Lang

C. canescens var. *subloliacea* Laest.

Similar to *C. canescens*, but less distinctly caespitose; culms gracile, leaves narrow, about 1 mm broad; spikes globose, few-flowered, widely separated, lowest spikes with stiff, setaceous bract; perigynia with short but distinct, smooth beak.

Wet, grassy places. Total range very incompletely known, probably throughout northern parts of North America.

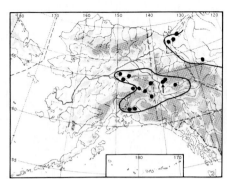
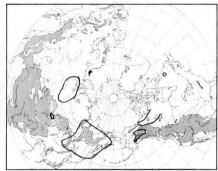

44. Càrex bonanzénsis Britt.

Similar to *C. canescens*; upper heads clustered, the lower separated, about 8 mm long and 3 mm thick, the lowest subtended by a flattened, serrulate bract about 2 cm long; scales brown with hyaline margin, the lower acute, the upper obtuse; perigynia 1.5 mm long, 1 mm wide, strongly nerved; beak 0.25 mm long, smooth.

Wet, grassy places. Total range very uncertain.

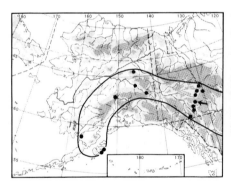

45. Càrex brunnéscens (Pers.) Poir.
Carex curta var. *brunnescens* Pers.
subsp. **alaskàna** Kalela

Densely tufted, grayish-green when young; culm 20–40 cm tall, erect, stiff; leaves short and stiff, 1.5–2 mm wide; spikes rounded to oval, the uppermost with numerous staminate flowers at base, the lower somewhat removed; scales ovate, obtuse to somewhat acute, broadly scarious-margined, olivaceous, later reddish-brown with green center; perigynia 2–2.5 mm long, ovate to oblong, thin-walled, often bursting in age, tapering into conical beak, which is somewhat scabrous in margin.

Meadows, dry, stony slopes, in mountains to at least 1,500 meters. *C. brunnescens* described from Switzerland.

Broken line on circumpolar map gives range of *C. brunnescens* in a broad sense.

26. Cyperaceae (Sedge Family)

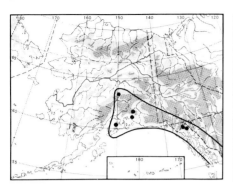 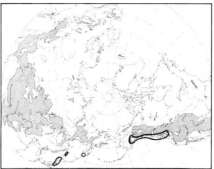 

46. Càrex brunnéscens (Pers.) Poir.
subsp. **pacífica** Kalela

Similar to subsp. *alaskana*, but less densely tufted; culm 30–60 cm tall, weaker and often curved; perigynia oval to broadly oval, more abruptly beaked; beak and upper part of perigynia somewhat scabrous.

Meadows, dry slopes. Described from Azouzetta Lake, British Columbia.

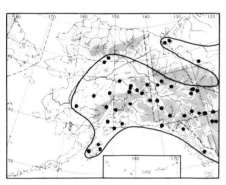 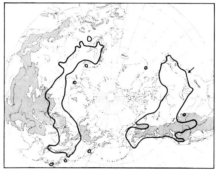 

47. Càrex dispérma Dew.
Carex tenella Schkuhr not Thuill.

Culms single, very slender; leaves light green, about 1 mm broad; spikes green, widely separated, with 1–3 perigynia; minute staminate flowers at summit of spike; perigynia ellipsoid-ovoid, faintly nerved, with very short beak.

Moist places.

×½

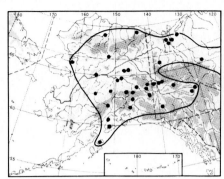
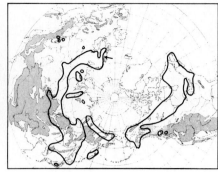

48. Càrex tenuiflòra Wahlenb.

Loosely caespitose; culms scabrous, firm; leaves grayish-green, 0.5–1 mm broad, shorter than culm; spikes usually 3, subglobose, bractless, densely compressed, with staminate flowers above; scales blunt, yellowish-white; perigynia green, ellipsoid, beakless, finely puncticulate.

Wet, grassy places, bogs.

×⅔

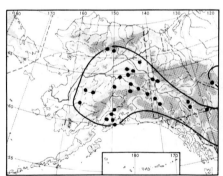
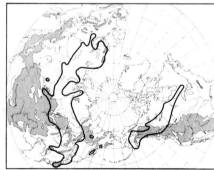

49. Càrex loliàcea L.

Culm slender, slightly scabrous above; leaves yellowish-green, 1–1.5 mm broad, flat; spikes 3–5, semiglobular, few-flowered, scattered; scales blunt; perigynia light-colored, scarious-margined, oblong-ovate, strongly nerved, beakless.

Moist places, peaty soil, muskeg. Described from Sweden.

26. Cyperaceae (Sedge Family)

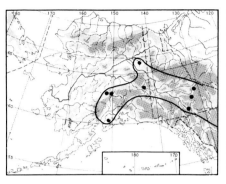
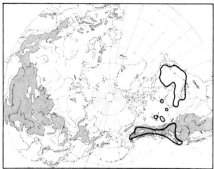

50. Càrex árcta Boott

Caespitose; culms as long as the soft, flat, 2–3-mm-broad leaves, or shorter; spikes in dense, prolonged head; perigynia broadest near base, cordate-ovate, with long, serrulate beak; scales acute; staminate flowers inconspicuous at base of spikes.

Wet places. Described from Canada.

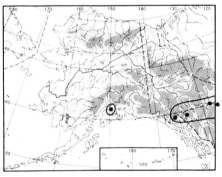

51. Càrex intèrior Bailey

Densely caespitose; culms tall, straight, slender; leaves flat to canaliculate, rough; spikes 2–4, the terminal staminate or with pistillate flowers at top, the lower pistillate; bracts scalelike; scales obtuse, yellowish-brown, with hyaline margin and green midvein; perigynia spreading, olive-green to brownish, with bidentate, broad, serrulate beak; 2 stigmas.

Wet meadows. Described from Penn Yan, New York.

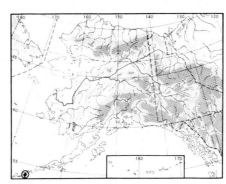 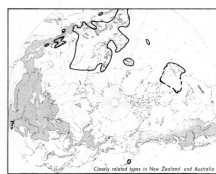

52. Càrex echinàta Murr.

Carex stellulata Good.

Similar to *C. interior*, but with shorter culms, not extending beyond leaves; perigynia smaller, indistinctly nerved.

Wet places. Described from England.

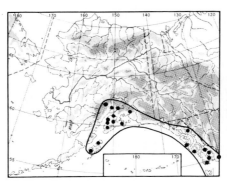 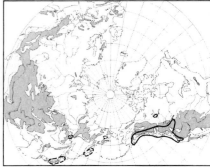

53. Càrex laevicúlmis Meinsh.

Caespitose; culms slender, weak; leaves flat, weak, 1–2 mm broad, rough on margins and at apex; spikes 3–8, the terminal with pistillate flowers at top, the lower pistillate; lowest bract prolonged; scales yellowish-brown with hyaline margins and green midvein, somewhat obtuse; perigynia appressed, green to brownish, several-nerved, tapering to beak 0.5–1 mm long; 2 stigmas.

Wet meadows. Described from Kamchatka (where it does not occur) and Sitka. Broken line on circumpolar map indicates range of the closely related **C. Kreczetòviczii** Egor.

26. Cyperaceae (Sedge Family)

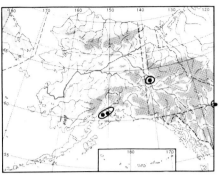 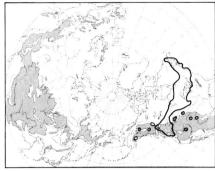 

54. Càrex Deweyàna Schwein.

Caespitose; culms weak, slender; leaves flat, thin, about 2.5 mm wide, yellowish-green; spikes 3–4, ovate, the terminal with pistillate flowers at top, the lateral pistillate; lower bract prolonged; scales thin, whitish, hyaline, with 3-nerved green center; perigynia light green, obscurely nerved, serrulate at apex, tapering into serrulate, bidentate beak 2 mm long.

Probably introduced. Described from New England.

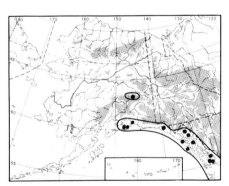  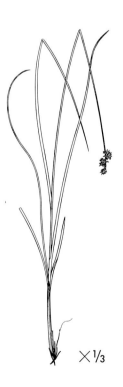

55. Càrex phyllomànica W. Boott

Caespitose, with thick, short rootstock; leaves canaliculate, 2–3 mm broad, stiff; spikes 3–4, the terminal with pistillate flowers at top, the lateral mostly pistillate; lowest bract setaceous; scales brown with hyaline margin and green midvein, obtuse; perigynia 3.75–4.5 mm long, green, many-nerved, tapering into bidentate, hyaline-tipped beak 1.25–1.5 mm long; 2 stigmas.

Sphagnum bogs.

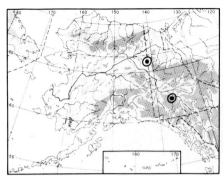
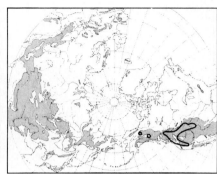

56. Càrex scopulòrum Holm

Stoloniferous, from scaly runners; culm stiff, sharply triangular, with dried-up leaves at base; leaves 3–7 mm wide; staminate spike solitary, linear-clavate, often partly pistillate; pistillate spikes 2–6, closely aggregated; lower bract shorter than culm; scales obovate, obtuse, black, with pale midvein; perigynia turgid, orbicular or broadly obovoid, nerveless, purplish-black-spotted with short, often bent beak; 2 stigmas.

Moist woods. Described from Colorado.

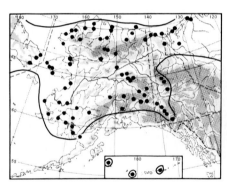
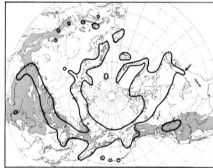

57. Càrex Bigelòwii Torr.

Carex consimilis Holm; *C. rigida* Good. not Schrank, including *C. hyperborea* Drej.; *C. Bigelowii* subsp. *hyperborea* (Drej.) Böcher.; *C. ensifolia* Turcz.; *C. rigidioides* Gorodk.

Stout runners, with shiny, dark-reddish-brown sheaths; culms stiff, sharply triangular, with dried-up leaves at base; leaves stiff, flat, light green, 3–4 mm wide; terminal spike staminate, the lateral thick, pistillate, the upper approximate; lower bract shorter than inflorescence, with black auricles at base; scales black, with lighter midvein; perigynia flat or slightly biconvex, nerveless, green or purplish-black-spotted, with short beak; 2 stigmas.

Dry places, solifluction soil, mostly in mountains. Circumpolar map gives range of entire, very complicated, and still unclear complex.

26. Cyperaceae (Sedge Family)

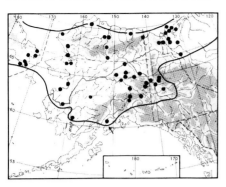
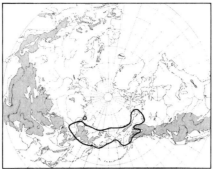

58. Càrex lùgens Holm
Carex yukonensis Holm; *C. cyclocarpa* Holm.

Densely caespitose, lacking long, scaly runners; culms slender, sharply triangular, with numerous dried-up leaves at base; leaves thin, 1–2 mm wide; terminal spike staminate, the lateral narrowly cylindrical; lowest bract short; scales purplish-black, obtuse; perigynia slightly biconvex, nerveless, mostly purplish-black, with short beak.

Tundra bogs, spruce muskeg.

Similar to *C. Bigelowii*, with which it hybridizes, but differs in lacking long, scaly runners, and in being more slender, with narrower leaves and narrow spikes.

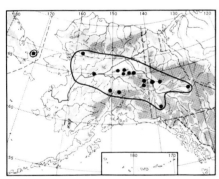
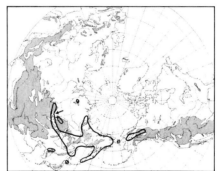

59. Càrex eleusinoìdes Turcz.
Carex kokrinensis Pors.

Densely caespitose; culm with purplish, bladeless sheaths at base; leaves flat, 2–3 mm broad; spikes 3–5, cylindrical, 1–2.5 cm long, the lowest longer-stipitated than the upper, the apical with pistillate flowers above, staminate below, the lateral pistillate, sometimes with staminate flowers at extreme tip; lowest bract as long as inflorescence, or longer; scales elliptical, somewhat acute, brownish-black, narrower than perigynia; perigynia broadly ovate, grayish-green, indistinctly nerved, with short beak; 2 stigmas.

Wet places, gravel bars.

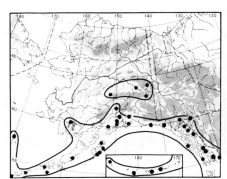
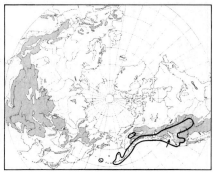

60. Càrex Kellóggii W. Boott

Including *C. Hindsii* Clarke; *C. aleutica* Akiyama.

Caespitose; culms slender, sharply triangular, with brown, bladeless sheaths and withered leaves at base; leaves 2–3 mm broad; apical spike staminate, dark brown, lateral pistillate, sessile, cylindrical; lowest bract leaflike, longer than inflorescence; scales obtuse or somewhat acute, black with green midrib, shorter and narrower than perigynia; perigynia green, more or less nerved, stipitate, tapering to short beak or more abruptly contracted to beak; 2 stigmas.

Wet places.

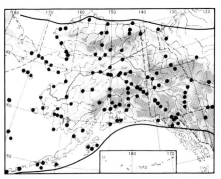

61. Càrex aquàtilis Wahlenb.
subsp. **aquàtilis**

Rhizome coarse, scaly, brown or reddish-brown, cordlike; culms thick, smooth, rounded; leaves 4–5 mm wide, as long as culm, or somewhat shorter; spikes stiff, erect, the upper sessile, the lower short-peduncled, the apical clublike, long, staminate, the lateral pistillate, often staminate at tip; bracts leaflike, extending beyond spikes; scales brown or blackish with pale midvein; perigynia light-colored, nerveless, with minute, entire beak; 2 stigmas.

Shallow water, marshes, along rivers. Described from Lapland.

26. Cyperaceae (Sedge Family)

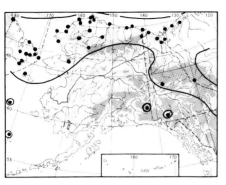
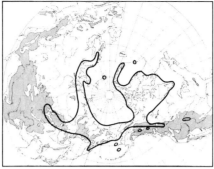

62. Càrex aquàtilis Wahlenb.
subsp. stáns (Drej.) Hult.

Carex stans Drej.

Shorter, more light green, and with more glabrous leaves than subsp. *aquatilis*; only 1 staminate apical spike.

Habitat same as that of subsp. *aquatilis*; common on tundra. Described from Greenland.

Intermediates with subsp. *aquatilis*, by introgression, are common.

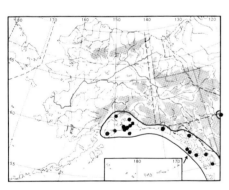

63. Càrex sitchénsis Prescott

Rhizome very coarse, short, scaly, brown or purplish; culms acutely triangular, smooth; leaves 4–8 mm wide; sheaths with colored mouth; staminate spikes 2–5, slender, erect, pistillate strongly separate, more or less drooping, on long, slender peduncles; bracts leaflike, extending beyond spikes; scales brownish, with pale center, acute in age, distinctly hyaline-tipped; perigynia oval, indistinctly nerved, with short beak; 2 stigmas.

Swamps.

×2/3

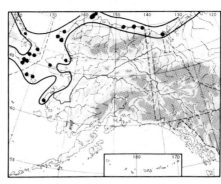
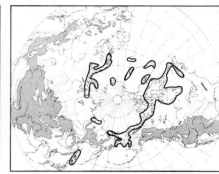

64. Càrex subspathàcea Wormsk.

Loosely caespitose, with thin, yellowish-green runners; culm curved, glabrous; leaves 1–2 mm wide, curved, glabrous; spikes 2–4, the upper staminate, up to 10 mm long, the lateral pistillate, sessile or short-stipitate, few-flowered, 5–10 mm long; lower bract broad at base, subspathulate, as long as inflorescence; scales dark brown, with lighter center, 3-nerved, shorter than perygnia; perygnia elliptical with indistinct beak.

Coastal salt marshes. Described from Greenland.

×1/3

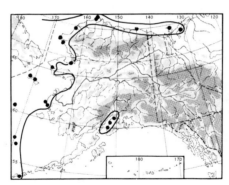
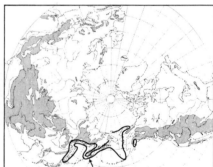

65. Càrex Raménskii Kom.
Carex salina var. *cuspidata* Holm.

Loosely caespitose with brown runners; culm triangular, glabrous, up to 30 cm tall, with reddish-brown sheaths at base; leaves flat, often extending beyond culms, 2–3 mm broad; spikes 3–5, the apical 1–2-staminate, the pistillate ovate or prolonged, short-peduncled, often with staminate flowers at tip; lower bract extending beyond inflorescence; scales dark brown with lighter midvein, somewhat shorter than perygnia; perygnia elliptical or ovate, slightly nerved, with short beak.

Coastal salt marshes.

26. Cyperaceae (Sedge Family)

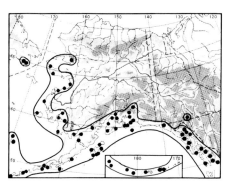 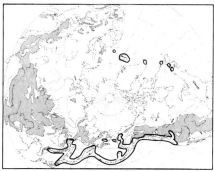 

66. Càrex Lyngbyaèi Hornem.
Carex cryptocarpa C. A. Mey.; *C. Lyngbyaei* subsp. *eryptocarpa* (C. A. Mey.) Hult.

Stoloniferous, tufted; culms triangular, glabrous, longer than leaves; leaves flat, light green, abruptly pointed; staminate spikes 2–3, pistillate 2–4 drooping, long-peduncled, usually staminate at tip; lower bracts extending slightly beyond spikes; scales lanceolate, long-accuminate, dark reddish-brown with lighter center, much longer than perigynia; perigynia ovate, 6–8-nerved with short beak.

Coastal salt marshes, rarely inland. Described from Faeroe Islands.

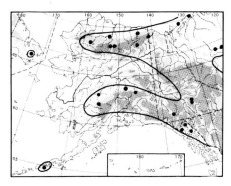 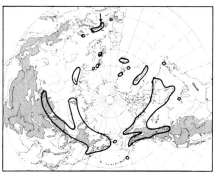 

67. Càrex bìcolor All.

Rhizome with short runners; culms flexuous, curving; spikes lying on ground when ripe; leaves flat, blue-green, shorter than culm; spikes crowded on lower part, the terminal pistillate above, the others pistillate; bract of lowest spike shorter than inflorescence, with sheath 2–4 mm long, with black or dark-brown auricles at mouth; scales purplish-black with green midrib; perigynia bluish-white, beakless; 2 stigmas.

Wet, gravelly places in mountains.

× ½

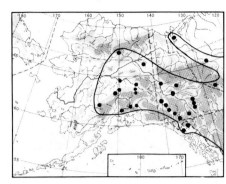
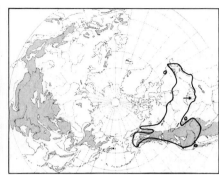

68. Càrex aùrea Nutt.

Rhizome with short runners; culms flexuous, triangular; leaves shorter than culm, flat, pale green. Similar to *C. bicolor*, but terminal spike usually staminate throughout, clavate; lowest bract longer than inflorescence, its sheath lacking dark auricles; perigynia fleshy, translucent.

Meadows, moist places.

× ½

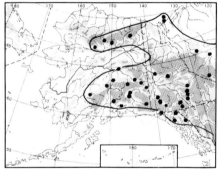
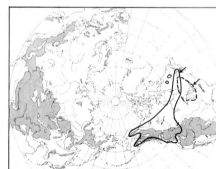

69. Càrex Gárberi Fern.
subsp. **bifària** (Fern.) Hult.
Carex Garberi var. *bifaria* Fern.

Similar to *C. aurea*, but terminal spike pistillate above; scales darker, with rounded summit.

Wet places.

Broken line on circumpolar map indicates range of subsp. **Gárberi.**

26. Cyperaceae (Sedge Family)

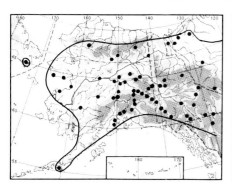 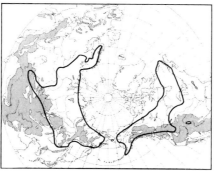 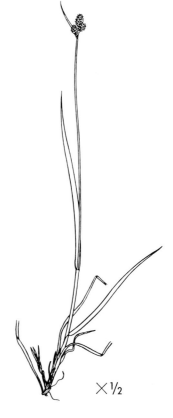

70. Càrex mèdia R. Br.

Carex alpina var. *inferalpina* Wahlenb.; *C. norvegica* subsp. *inferalpina* (Wahlenb.) Hult.; *C. angarae* Steud.

Loosely caespitose; culms with dark, purplish-red sheaths at base, leaves in lower third; leaves light green, soft; spikes 3–4 of equal size, short, compressed; scales narrowly ovate, shorter than perigynia; perigynia bluish-green when young, later olive-brown, lighter-colored than scales, ellipsoid, with short beak.

Moist places in lowland and subalpine zones. Described from northwest Canada, 54–64° N.

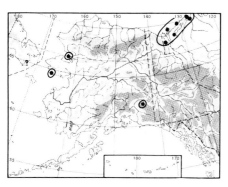 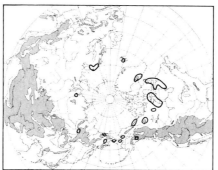 

71. Càrex holóstoma Drej.

Carex arakamensis Clarke?

Similar to *C. media* but stoloniferous; terminal spike staminate, inconspicuous between the two pistillate spikes; perigynia barely beaked, with orifice entire.

Carex marshes, mountain slopes. Described from western Greenland.

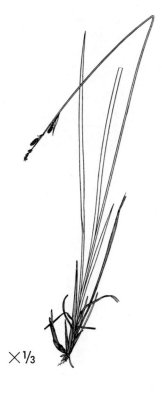

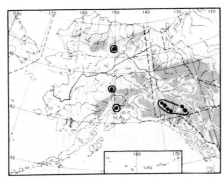 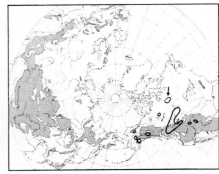

72. Càrex Parryàna Dew.

Stoloniferous, with long, horizontal, scaly runners; culm stiff, with numerous basal leaves; leaves stiff, light green, 2–3 mm wide, much shorter than culm; spikes 3–5, erect, the terminal with pistillate flowers above, the lateral pistillate, the upper sessile, the lower short-peduncled; lowest bract shorter than head, with very short sheath; scales suborbicular, dark brown with green midvein, hyaline-margined, nearly concealing perigynia; perigynia obovate, nerveless, very short-beaked; 3 stigmas.

Wet places.

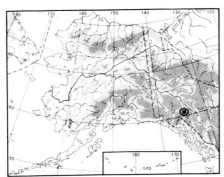 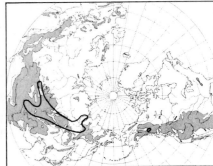

73. Càrex sabulòsa Turcz.
Carex leiophylla Mack.

Plant grayish-green, with long, slender rootstock; culm glabrous, curved, with purplish, hyaline-margined sheaths at base, surrounded by numerous weak leaves, 1.5–2.5 mm broad, shorter than culm; spikes 3–5, the terminal club-shaped, with pistillate flowers above, the lateral pistillate; scales ovate-lanceolate, acute, dark brown, as long as perigynia, or longer; perigynia yellowish-green, purplish-blotched, ovate to suborbicular, stipitate, with bidentate, dark beak 1 mm long.

Sandy places. Described from Baikal. Circumpolar map tentative.

26. Cyperaceae (Sedge Family)

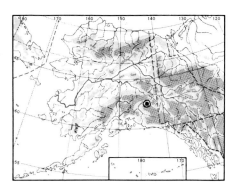 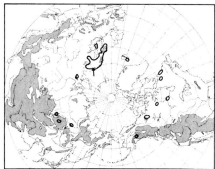 

74. Càrex adelóstoma Krecz.

Carex Morriseyi Pors.; *C. Buxbaumii* Wahlenb. subsp. *alpina* (Hartm.) Liro; *C. Buxbaumii* var. *alpina* Hartm.

Plant grayish-green, with thin, subterranean runners; culm scabrous, with purplish, bladeless sheaths at base; leaves 2–3 mm wide, shorter than culm; spikes 3–4, sessile, the lower remote, the apical staminate, with pistillate flowers above, or entirely staminate, the lateral pistillate; scales ovate, acute, dark brown with green midvein, shorter than perigynia, or as long; perigynia grayish-green, elliptical, glabrous in margin, indistinctly nerved.

Wet places.

×⅓

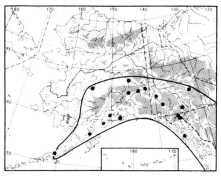 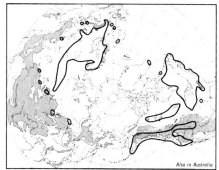 

75. Càrex Buxbaùmii Wahlenb.

Loosely tufted, stoloniferous; sheaths shiny reddish-brown to wine-red; culm sharply angled, rough at summit; leaves 1–2 mm broad, bluish-green, long; spikes 3–4, short, sessile or short-peduncled, the terminal pistillate above, staminate at base, clublike; pistillate scales purplish-black, lanceolate, acuminate or usually awned, equaling or longer than perigynia; perigynia indistinctly nerved, nearly beakless, with wide opening.

Swamps and bogs. Described from Sweden.

×⅓

×⅓

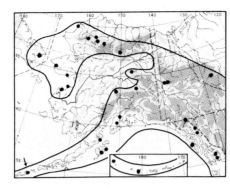
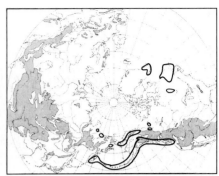

76. Càrex stylòsa C. A. Mey.

Densely caespitose; culms scabrous; leaves flat, 1.5–2 mm wide, subrigid, shorter than culm; lowest bract shorter than inflorescence, with short sheath; terminal spike staminate, lateral 1–3 thick-cylindrical, erect, compressed, more or less peduncled; scales dark purple with paler midrib, blunt or somewhat acute; perigynia elliptical or ovate, nerveless, dark with short beak, in age with thick, very characteristic, long, protruding style.

Moist places, muskeg.

In eastern North America, var. **nigritélla** (Drej.) Fern. (*C. nigritella* Drej.) occurs.

×⅓

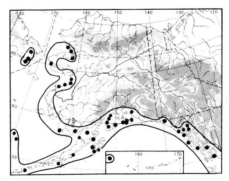
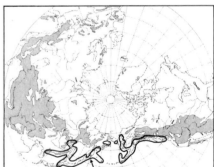

77. Càrex Gmelìni Hook. & Arn.

Densely caespitose, with short runners; culm coarse, curved at top, with purplish, bladeless sheaths at base; leaves 3–5 mm broad; upper spikes sessile, with pistillate flowers above, the lateral stipitate, pistillate; scales ovate, dark brown with pale, long, excurrent midvein, as long as perigynia or longer; perigynia dark, ovate, stipitate, nerved, abruptly contracted into short, emarginate beak.

Sandy, saline shores.

26. Cyperaceae (Sedge Family)

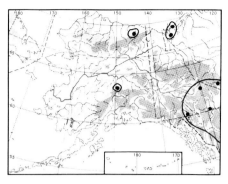
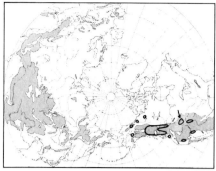

78. Càrex albo-nìgra Mack.

Caespitose; culms stiff, with conspicuous old leaves at base; blades light green, 2.5–5 mm broad; spikes usually 3, the apical with pistillate flowers above, the lateral pistillate; lowest bract shorter than inflorescence; scales obtuse or somewhat acute, with white-hyaline margins at tip; perigynia obovate, purplish-black, granular, with short, somewhat bidentate beak, 0.5 mm long; 3 stigmas.

Mountainsides.

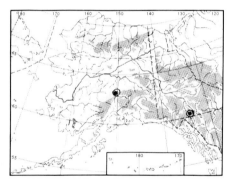
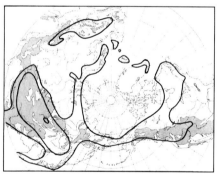

79. Càrex atràta L.
subsp. **atràta**

Caespitose, with short runners covered with purplish-brown sheaths; culms longer than leaves; leaves 3–5 mm broad; spikes 3–5, the apical with pistillate flowers above, the lateral pistillate, the lower long-peduncled; lowest bract foliaceous, extending beyond inflorescence; scales somewhat acute, black, as long as perigynia or slightly shorter; perigynia light-colored, papillose, with short, slightly emarginate, purplish beak; stigmas 3.

Mountain meadows. Described from European Alps. Circumpolar map gives range of *C. atrata* in a broad sense.

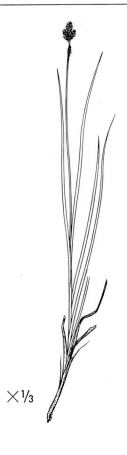
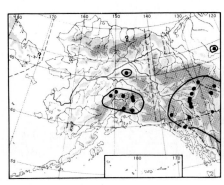
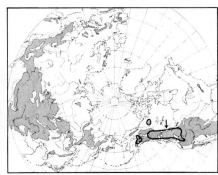

80. Càrex atràta L.
subsp. **atrosquàmea** (Mack.) Hult.

Carex atrosquamea Mack.

Similar to *C. atrata*, but scales shorter than perigynia, more brownish; perigynia with shorter and broader, less emarginate beak.

Mountain meadows.

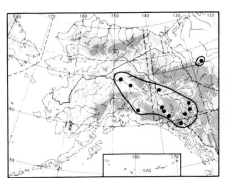
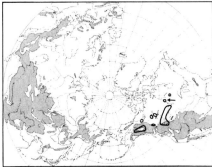

81. Càrex atràtiformis Britt.
subsp. **Raymóndii** (Calder) Pors.

Carex Raymondii Calder.

Caespitose; culm scabrous above, slender, triangular, much longer than leaves, phyllopodic; leaves about 3–5 mm wide; spikes 3–4, approximate, ellipsoid-cylindrical, the apical with pistillate flowers above, the lateral pistillate or with few basal, staminate flowers; lowest bract foliaceous, shorter than inflorescence; scales narrowly ovate, acute, usually shorter than perigynia, purplish-red, more or less hyaline-margined; perigynia broadly ovate, nerveless, pale, sometimes purplish below the 0.4-mm-long, bidentate, purplish beak; 3 stigmas.

Moist places.

26. Cyperaceae (Sedge Family)

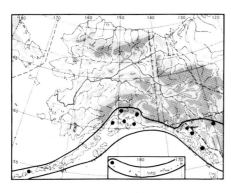
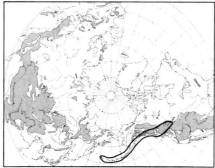

82. Càrex Enánderi Hult.

Carex eurystachya Herm.

Loosely caespitose; culms longer than leaves, glabrous, with bladeless, brownish sheaths at base; leaves 1.5–2 mm wide; spikes 3–5, oblong, the lower long-peduncled, the apical with staminate flowers at base, the lateral pistillate; lowest bract foliaceous, lacking sheath, extending beyond inflorescence; scales ovate, obtuse or somewhat acute, purplish-black, with conspicuous green midvein; perigynia ovate, longer and broader than scales, nerved, sparsely ciliate in margin, with very short, slightly bidentate beak; stigmas 2(–3).

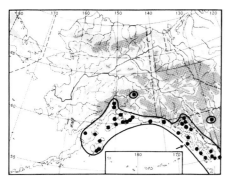
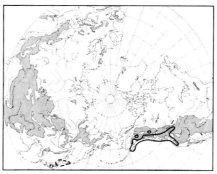

83. Càrex Merténsii Prescott

Densely caespitose; culms tall, longer than leaves; leaves 4–7 mm wide, flaccid, the basal with short blade; spikes 5–10, approximate, drooping, on slender stalks, the terminal with (sometimes few) staminate flowers above, the lateral pistillate or with few staminate flowers at base; lower bract leaflike, extending beyond inflorescence; scales in mature specimens much narrower and shorter than perigynia; perigynia ovate to roundish, light-colored, nerved, with beak 0.25–0.5 mm long.

Wet, rocky slopes. A beautiful and very characteristic sedge.

Broken line on circumpolar map indicates range of subsp. **urostáchys** (Franch.) Calder & Taylor (*C. urostachys* Franch.).

(See color section, under *C. Gmelinii*.)

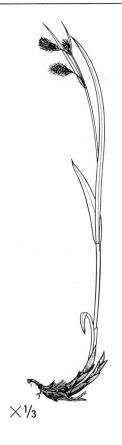
× ⅓

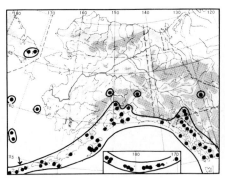 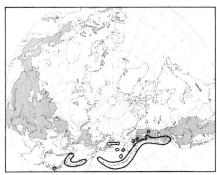

84. Càrex macrochaèta C. A. Mey.

Loosely caespitose, with short runners; culm longer than leaves, nearly glabrous, with purplish-brown, bladeless sheaths disintegrating into shreds at base; leaf blades light green, 1–3 mm wide; spikes 3–4, oblong, the apical staminate, the lateral pistillate (rarely staminate above) on thin, glabrous peduncles; scales lanceolate, emarginate, with light-colored midvein, excurrent into yellowish awn up to 1 cm long; perigynia elliptic-lanceolate, 5–7-nerved, tapering into glabrous beak; stigmas 3 (rarely 2).

Wet places, very common along coast, rare inland. Described from Unalaska.

× ½

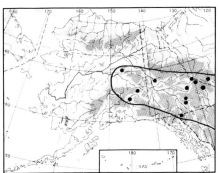 

85. Càrex defléxa Hornem.

Stoloniferous, with creeping, dark-brown, horizontal rhizomes; culms curved, filiform; leaves shorter than culms, 1.5–2 mm wide, with purplish sheaths at base; lowest bract leaflike, sheathless; spikes 3–4, approximate, the apical small, staminate, linear, the lateral pistillate, small, globose, few-flowered; scales ovate, acute, brown with pale center and hyaline margin; perigynia slightly longer than scales, ovate to orbiculate, pale green, pubescent, long-stipitated, with conical beak; style incrassated at base.

Gravel bars. Described from Greenland.

26. Cyperaceae (Sedge Family)

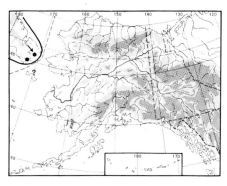 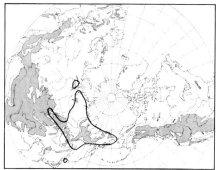 

86. Càrex melanocárpa Cham.
Carex ericetorum subsp. *melanocarpa* (Cham.) Kükenth.

Plant grayish-green, tufted, with short runners; culms slender, often curved, glabrous, with reddish-brown sheaths at base; leaf blades 1–2 mm wide, flat, much shorter than culm; spikes 2–3, linear, the apical staminate, the lateral pistillate, sessile; lower bract scalelike, apiculate; scales broadly ovate, narrowly hyaline-margined, ciliate, brown, yellowish-pubescent; perigynia longer than scales, ovate, nerveless, short-pubescent in margin and at short, cylindrical beak.

Stony slopes. Old, doubtful report from St. Lawrence Island.

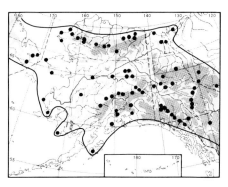 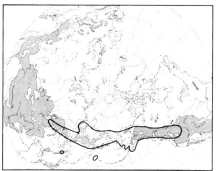 

87. Càrex podocárpa C. B. Clarke
Carex montanensis Bailey; *C. behringensis* C. B. Clarke.

Plant yellowish-green with long, slender, brown, scaly, fibrillose rootstock; culms stout, with scalelike leaves at base, longer above; leaves 3–4 mm broad, abruptly pointed; spikes 3–4, the apical 1–2 staminate, the lateral pistillate, on long, slender, capillary peduncles, in mature specimens drooping; lower bracts foliaceous, shorter than inflorescence; scales acute, purplish-black, as long as or slightly shorter than perigynia; perigynia elliptic, blackish-tinged, nerveless, abruptly contracted into short, mostly bidentulate beak; stigmas 3.

Meadows, moist places. Described from northwest Canada 64–69°N.

×⅓

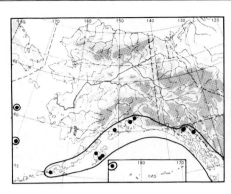
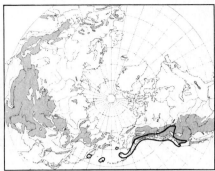

88. Càrex spectàbilis Dew.

Similar to *C. podocarpa*, but scales light-colored, with yellowish median nerve excurrent into short tip or awn.

Moist places within range of *C. macrochaeta*. Described from "Arctic Regions."
More or less intermediate between *C. macrochaeta* and *C. podocarpa*.

×½

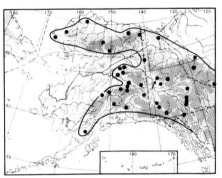
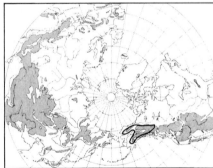

89. Càrex microchaèta Holm
Carex podocarpa Hult. of Fl. Alaska & Yukon.

Plant with long, slender, brown, scaly, fibrillose rootstock; culms with numerous basal leaves, the outer withered and brown; leaves 3–5 mm wide, much shorter than culm; spikes 2–4, the apical 1–2-staminate, the lateral pistillate, ovate, on slender, capillary peduncles, in mature specimens drooping; lower bract foliaceous, not extending beyond inflorescence; scales acute, purplish-black with lighter midrib; perigynia elliptic, blackish-tinged, with short, bidentulate beak; stigmas 3.

Meadows, wet places, heaths to over 2,100 meters. Very similar types occur in northeast Siberia under the name "**C. melanóstoma.**"

26. Cyperaceae (Sedge Family)

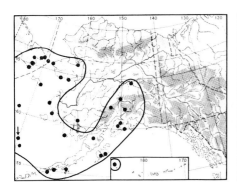
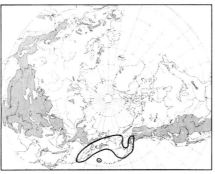

90. Càrex nesòphila Holm

Very similar to *C. microchaeta* and doubtfully distinct from that species, but spikes are narrower, linear, on stiff, erect peduncles; midrib of scales more conspicuous.

Wet places. Asiatic range unclear.

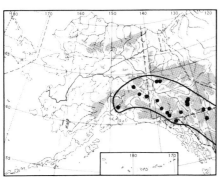

91. Càrex Róssii Boott

Densely caespitose; culms triangular, scabrous, as long as leaves or shorter, with purplish sheaths at base; bracts leaflike, longer than inflorescence; spikes 4–5, the apical staminate, linear, prominent, the lateral ovate, few-flowered, some long-peduncled, subradical; scales lanceolate-ovate, acuminate, brown, with green, more or less excurrent median vein; perigynia longer than scales, pale green, nerveless, pubescent, long-stipitated, abruptly contracted into conical, bidentate, hyaline-margined beak.

Rocky slopes. Described from the northwest coast of North America.

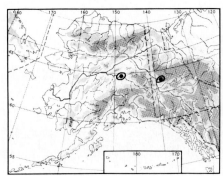
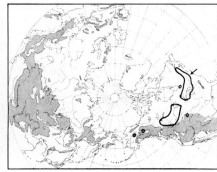

92. Càrex Péckii Howe

Pale green, loosely caespitose; culms greatly exceeding leaves; leaves soft, 1.5–2 mm wide; bract narrow or lacking; spikes 2–4, the apical small, sessile, staminate, the lateral pistillate, globose; scales brownish, with broad hyaline margins; perigynia much longer than scales, ellipsoid to fusiform, short-hirsute, with slender, bidentate, hyaline-margined beak.

Dry slopes, woods. Described from New York State.

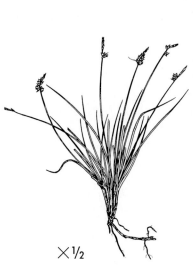

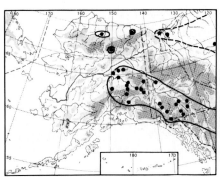
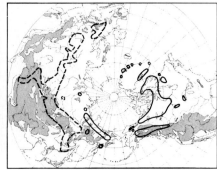

93. Càrex supìna Willd.
subsp. **spaniocárpa** (Steud.) Hult.
Carex spaniocarpa Steud.

Loosely caespitose, with long, reddish-brown, subterranean runners; culms thin, longer than leaves, with reddish sheaths at base; leaves 1–1.5 mm wide; spikes 2–3, the apical staminate, linear, the lateral pistillate, sessile, broad; lower bract scale-like with apical awn; scales about as long as perigynia, ovate, acute, rust-brown, with prominent lighter center, scarious-margined; perigynia ovate to oval-elongate, nerveless, rust-colored, evenly tapering into short, cylindrical beak.

Dry, sandy places. *C. supina* described from Europe, subsp. *spaniocarpa* from Greenland. Broken line on circumpolar map indicates range of other subspecies.

26. Cyperaceae (Sedge Family) 267

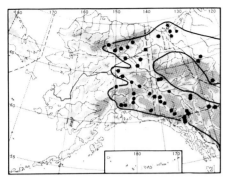
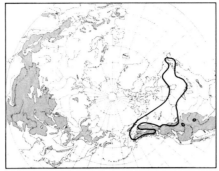

94. Càrex concínna R. Br.

Loosely caespitose, with short runners; culms thin, curved, with purplish basal sheaths; leaves dark green, 1–3 mm wide; lowest bract short, with long sheath; spikes approximate, 2–3, sessile or short-peduncled, the apical staminate, the lateral pistillate, subglobose; scales ovate, obtuse, dark brown with green center, ciliate, hyaline-margined; perigynia twice as long as scales, elliptic or obovate, nerveless, pubescent, long-stipitated, with very short, brown beak.

Stony, dry places in forest, preferably in calcareous soil. Described from northwest Canada from 54°N to the Arctic Ocean.

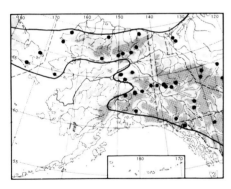
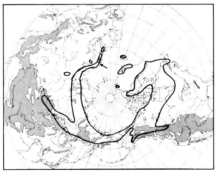

95. Càrex glaciàlis Mack.

Carex pedata Wahlenb. not L.

Forming small tufts with capillary, light-green leaves; culms longer than leaves, stiff, with reddish-brown sheaths at base (hidden in tuft); bracts short, subulate; terminal spike staminate, sessile, the lateral 1–3, very small, with 1–4 perigynia; scales inconspicuous, exceeded by perigynia, broadly ovate, purplish-black, with hyaline margins; perigynia obovate to nearly spherical, glabrous, nerveless, darker above, abruptly contracted into cylindrical, hyaline-tipped beak 0.5 mm long.

Dry, calcareous soil. Easily presumed to have but one spike, because of its aggregate spikes; superficially similar to *C. rupestris* (sp. no. 8, above). On circumpolar map, range of **C. terrae-nòvae** Fern. is included.

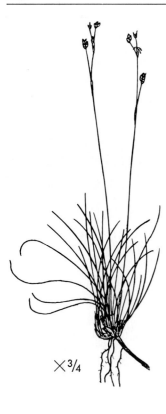
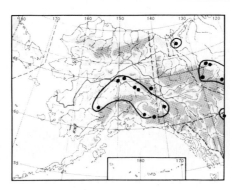
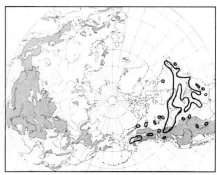

96. Càrex ebúrnea Boott

Caespitose, with short stolons; culms capillary, smooth; leaves involute-filiform, shorter than culm; lowest bract with tubular sheath; inflorescence corymbose; apical spike staminate, lateral 2–4 pistillate, on slender, erect peduncles; scales whitish to pale brown, thin, shorter than perigynia, falling off in age; perigynia elliptical, olivaceous to blackish, with short beak longer than scales; base of style bulbous, thickened.

Dry sand or rocky places, preferably on calcareous soil. Described from western Canada.

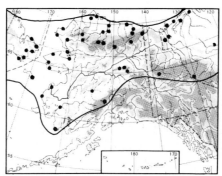

97. Càrex rariflòra (Wahlenb.) J. E. Sm.
Carex limosa var. *rariflora* Wahlenb.

Plant with long, dark brown, scaly rhizomes and rust-colored tomentum on roots; culms obtusely triangular, smooth; leaves about 2 mm wide; bracts with short, dark, sheaths and slender blades; terminal spike staminate, the lateral 2–3 pistillate, few-flowered, the lowest long-peduncled, drooping; scales purplish-black, broadly ovate, obtuse or mucronate, shorter but at base broader than the light-colored perigynia; perigynia elliptic, obscurely nerved, with inconspicuous beak.

Bogs, margins of ponds.

26. Cyperaceae (Sedge Family)

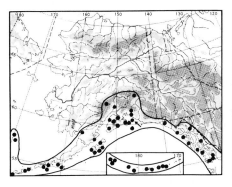
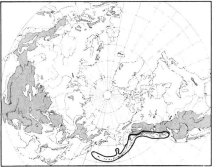

98. Càrex pluriflòra Hult.

Carex stygia of authors; *C. rariflora* subsp. *stygia* Hult.; *C. limosa* var. *stygia* Vasey.

Plant with long, scaly, purplish-black rhizome and yellowish, tomentose roots; culm triangular; leaves about as long as culm, about 2 mm broad, flat, grayish-green; bracts short, setaceous, with short sheath; spikes 2–3, the terminal staminate, long-peduncled, the lateral pistillate, drooping, on capillary peduncles, 10–20-flowered; scales broadly ovate, acute or cuspidate, black, with somewhat paler median nerve; perigynia ovate to ovate-lanceolate, slightly shorter than scales, beakless, in age blackish-brown.

Bogs, margins of ponds along coast.

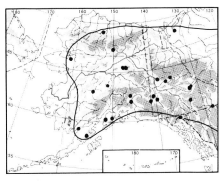
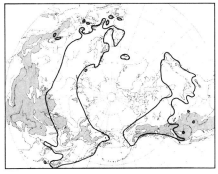

99. Càrex limòsa L.

Plant with long rhizomes and rust-colored tomentum on roots; culm sharply angled, rough above; leaves grayish- or bluish-green, 1–2 mm wide; bracts with short, dark, auriculated sheaths; terminal spike staminate, long-peduncled, and usually single- (rarely 2–3-) pistillate, drooping spikes; scales stramineous to brown, acute, covering the perigynia; perigynia ellipsoid, bluish-green, beakless.

Bogs, quagmires, often in shallow water. Described from Europe.

×⅓

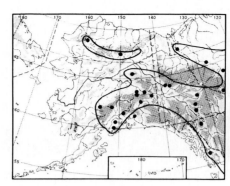
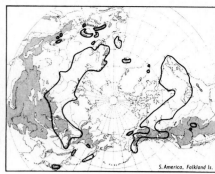

100. Càrex magellànica Lam.
subsp. **irrígua** (Wahlenb.) Hult.

Carex limosa var. *irrigua* Wahlenb.; *C. magellanica* with respect to Alaskan plant; *C. paupercula* Michx.

Caespitose; roots with yellowish tomentum; culms acutely angled above; leaves about 3 mm broad, pale green, shorter than culm; lowest bract foliaceous, usually exceeding inflorescence; terminal spike mostly staminate, the lateral 2–4 pistillate, short, drooping, on capillary peduncles; scales castaneous to stramineous, long-acuminate, much longer but narrower than the perigynia, falling off in age; perigynia 4 mm long, broadly ovate, pale, persistent, longer than scales.

Peaty soil. *C. magellanica* described from Magellan Sound, subsp. *irrigua* from northern Sweden.

×¾

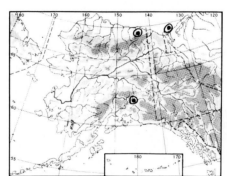
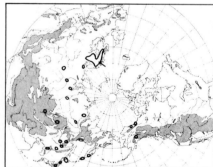

101. Càrex láxa Wahlenb.

Bluish-green, with thin subterranean runners; roots lacking rust-colored tomentum; culms thin, slack; leaves 1–2 mm broad, shorter than culm; lower bract with sheath 1–2 cm long and blade 1–3 cm long; apical spike staminate, long-peduncled, the lateral 1–2 pistillate, on long, thin, glabrous peduncles; scales broad, light brown with greenish midvein; perigynia grayish-green, elliptical to lanceolate, with short beak, slightly longer than scales.

Wet places, mostly in woods.

26. Cyperaceae (Sedge Family)

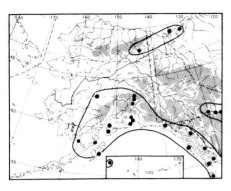 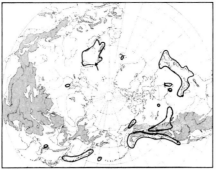 

102. Càrex lìvida (Wahlenb.) Willd.

Carex limosa var. *livida* Wahlenb.

Stoloniferous; culms smooth; leaves light-bluish-gray, usually longer than culm; lowest bract nearly as long as inflorescence, with long sheath; terminal spike usually staminate, the lateral 1–3, nearly sessile, few-flowered; scales obtuse, with green center and margin; perigynia light-colored, blunt, beakless.

Wet places.

Plants in southern part of American range are var. **Grayàna** (Dew.) Fern. (*C. Grayana* Dew.)

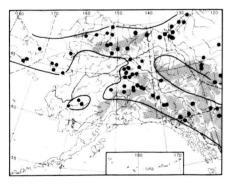 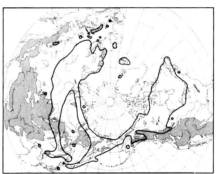 

103. Càrex vaginàta Tausch

Carex sparsiflora Steud.; *C. saltuensis* Bailey; *C. altocaulis* (Dew.) Britt.; *C. vaginata* var. *altocaulis* Dew.; *C. quasivaginata* Clarke?; *C. algida* Turcz.?

Plant with long, slender, grayish-brown, scaly stolons; culms smooth; leaves much shorter than culms, flat, yellowish-green; lowest bract short, with long, somewhat inflated, sheath; apical spike staminate, club-shaped, light brown, long-peduncled, the lateral 1–3 pistillate, peduncled, erect; scales grayish-brown, acute, shorter than perigynia; perigynia ovate, light-colored, nerveless, with slender, cylindrical, bidentate beak, hyaline at tip.

Mossy woods, bogs, wet places, in mountains to at least 700 meters.

26. CYPERACEAE (Sedge Family)

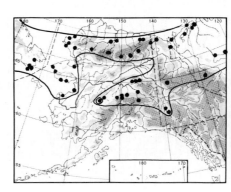
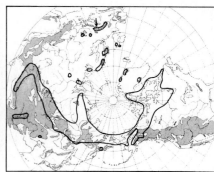

104. Càrex atrofúsca Schkuhr
Carex ustulata Wahlenb.

Caespitose; culms stiff, smooth; leaves 3–4 mm wide, grayish-green, with light-brown sheaths; lowest bract short, setaceous, with brown sheath; apical spike staminate, obovate, the lateral 2–3 pistillate, ovate, on slender pedicels, drooping in age; scales ovate, acuminate, dark; perigynia broader than scales, thin, nerveless, scabrous in margin, tapering into long, bidentate, distinctly hyaline-margined beak.

Wet places, in mountains to at least 1,000 meters, preferably on calcareous soil.

A very tall plant (up to 60 cm) has been called var. **màjor** (Boeck.) Raymond (*C. ustulata* var. *major* Boeck.).

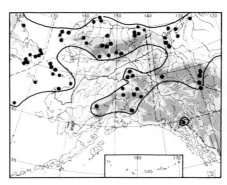
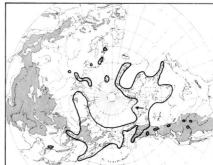

105. Càrex misándra R. Br.
Carex fuliginosa var. *misandra* Kükenth.

Caespitose; culm glabrous, nodding above, with conspicuous dried-up leaves at base; leaves usually recurved, much shorter than culm, canaliculate at base, with recurved, light-brown sheaths; lowest bract with long sheath and short blade; terminal spike drooping, with pistillate flowers above, the lateral 2–3 pistillate; scales ovate, somewhat acute, dark reddish-brown with lighter midvein, scarious-margined; perigynia longer than scales, narrowly lanceolate, purplish-black above, strongly ciliate on margins of prolonged, bidentate apex.

Sandy and stony places in mountains, marshes.

26. Cyperaceae (Sedge Family)

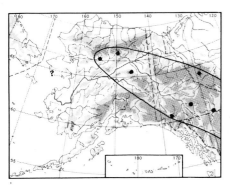
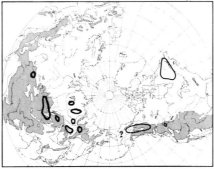

×⅓

106. Càrex Fránklinii Boott

Carex Lepageana Raymond?; *C. macrogyna* Turcz.; *C. misandroides* Fern.

Plant with long, slender stolons; culm smooth or rough above; leaves light green, somewhat stiff, 1–3 mm wide, shorter than culm, deeply channeled; sheaths cinnamon-brown; lowest bract foliaceous, shorter than inflorescence, with sheath 5–15 mm long; terminal 3–4 spikes staminate or with staminate flowers above, at least the uppermost with several perigynia at base; pistillate 2–3 spikes drooping, on slender, somewhat scabrous peduncles; scales oblong-ovate, short-awned, reddish-brown, with yellowish midrib to tip, scarious-margined; perigynia oblong-ovate, ciliate on margins, minutely beaked, hyaline-tipped, longer than scales.

Alpine slopes, preferably on calcareous soil. Similar to *C. petricosa* but taller; scales with scabrous point, and perigynia yellowish-brown. Described from the Rocky Mountains (Drummond).

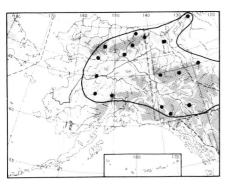
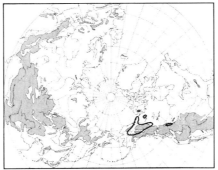

×⅓

107. Càrex petricòsa Dew.

Carex Franklinii var. *nicholsonis* Boiv.

Plant with horizontal stolons; culm smooth, with dried-up leaves at base; leaf blades light green, stiff, flat, 2–2.5 mm wide; lowest bract foliaceous, with sheath 1 cm long; terminal spike short-peduncled, with staminate flowers at tip, the lateral 2–3 pistillate, erect, the lower long-peduncled; scales oblong-obovate, not awned, purplish-black with lighter center and hyaline apex, slightly ciliate; perigynia lanceolate, ciliate in margin, darker above, finely nerved, beakless; white-hyaline and slightly bidentulate at apex.

Dry slopes in the mountains. Described from the Rocky Mountains (Drummond).

×⅓

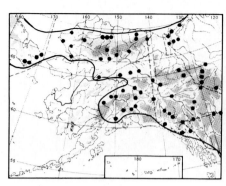
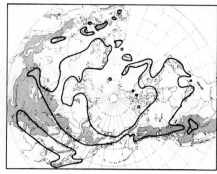

108. Càrex capillàris L.
Carex chlorostachys Stev. Including *C. fuscidula* Krecz.

Densely caespitose; culms filiform, smooth; leaves shorter than culms, flat, 1–2 mm wide; apical spike small, staminate, pale, the lateral 2–4, drooping, on capillary peduncles; scales pale, blunt or short-tipped, falling off in age; perigynia light brown, lance-ovoid to fusiform, tapering to conical beak.

Moist or dry places. Described from Sweden.

The typical plant is low-growing, below 20 cm. For the area concerned, the taller var. **màjor** Drej. [var. *elongata* Olney; subsp. *chlorostachys* (Stev.) Löve & Löve & Raymond] is common and occupies the same range.

×½

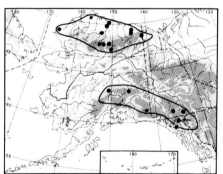
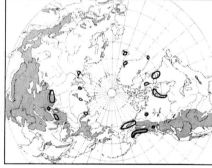

109. Càrex Kraùsei Boeck.
Carex capillaris var. *Krausei* Kurtz; *C. capillaris* subsp. *Krausei* (Boeck.) Böcher; *C. Krausei* subsp. *Porsildiana* Löve & Löve & Raymond.

Similar to *C. capillaris*; differs chiefly in the apical spike, which has pistillate flowers at top, staminate flowers at base.

Moist places.

26. Cyperaceae (Sedge Family)

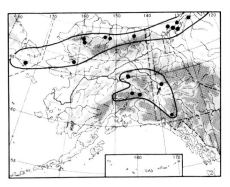 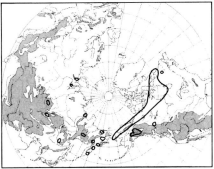 

110. Càrex Williámsii Britt.
Carex Novograblenovii Kom.

Similar to *C. capillaris*, but differs in having capillary, canaliculate leaves. Wet, grassy places.

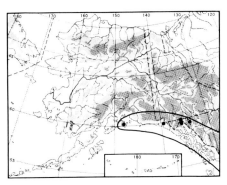 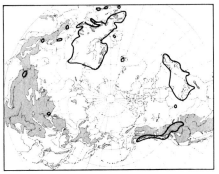 

111. Càrex flàva L.

Caespitose, light green; culms firm, tall, acutely triangular, glabrous; leaves flat, shorter than culm; lowest bract foliaceous, much longer than inflorescence; apical spike staminate, short-peduncled, the lateral pistillate, aggregated, subglobose; scales yellowish-brown with green midvein, lanceolate, acute; perigynia yellowish-green, shiny, with coarse nerves, 5–6 mm long, with curved, bidentate beak of the same length as the body or slightly shorter, the lowest reflexed.

Moist places, preferably on calcareous soil. Described from Europe.

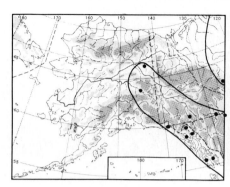
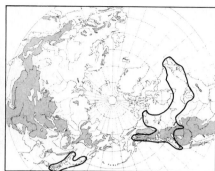

112. Càrex Oèderi Retz.
subsp. virídula (Michx.) Hult.

Carex viridula Michx.; *C. Urbani* Boeck.; *C. Oederi* var. *oedocarpa* with respect to Alaskan plant.

Caespitose; culm obtusely triangular; leaves yellowish-green, about 2 mm broad; lowest bract much longer than inflorescence; apical spike staminate or partly pistillate, the lateral subglobose, compressed; scales yellowish-brown, obtuse; perigynia yellowish, in age brown, 2–3.5 mm long, not reflexed, with beak one-third as long as body.

Wet, grassy slopes. *C. Oederi* described from Sylt and Eiderstedt (western Schleswig), subsp. *viridula* from Canada.

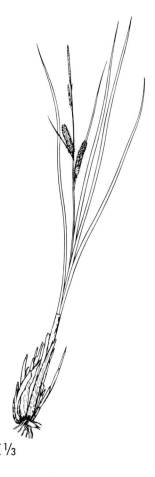
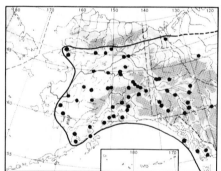
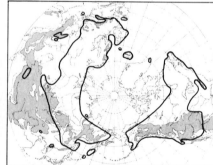

113. Càrex rostràta Stokes

Carex utriculata Boott; *C. vesicaria* and *C. exsiccata* with respect to Alaskan plant.

Light green, with thick, horizontal rhizomes penetrating deep into mud (rarely present in herbarium specimens); culms glabrous, thick, spongy, with broad, grayish-to-reddish sheaths at base; leaves 4–8 mm wide; lowest bract longer than inflorescence; 2–4 apical spikes staminate, the lateral 2–3 cylindrical, pistillate, in age somewhat drooping; scales ovate, acute; perigynia longer than scales, ovate or flask-shaped, nerved, with slenderly conical, bidentate beak, ascending, or in age spreading.

Wet shores, swamps. Described from Great Britain.

The lower part of the plant is eaten by the Eskimos.

26. CYPERACEAE (Sedge Family)

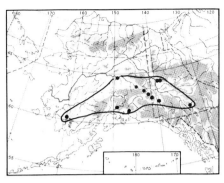
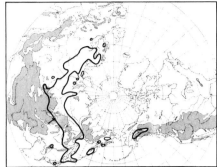

114. Càrex rynchophỳsa C. A. Mey.
Carex laevirostris (E. Fries) Blytt & Fr.

Very similar to *C. rostrata*, but staminate spikes thicker, denser, with divaricate perigynia, often staminate at tip; perigynia inflated, spherical to ovate, abruptly contracted into slender, more cylindrical, strongly bidentate beak; pistillate scales narrow, inconspicuous in spike.

Wet shores, swamps.

The lower part of the plant is eaten by the Eskimos. Hybrids with *C. rostrata* occur.

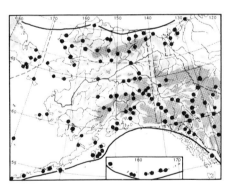
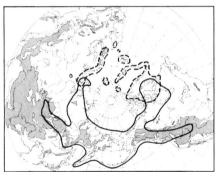

115. Càrex saxàtilis L.
subsp. láxa (Trautv.) Kalela
Carex pulla var. *laxa* Trautv.; *C. physocarpa* Presl; *C. ambusta* Boott; *C. procerula* Krecz.

Yellowish-green, with short, thick, subterranean runners; culm slender, often curved, with purplish sheaths at base; leaf blades 2–3 mm wide, slightly shorter than culm; lowest bract as long as inflorescence or slightly shorter; apical 1–2 spikes staminate, the lateral 1–2 pistillate, oblong on long capillary peduncles, drooping; scales acute, dark brown; perigynia broader and longer than scales, elliptical, reddish-brown, shiny, abruptly contracted into short, emarginate beak.

Wet places, common to at least 1,300 meters. *C. saxatilis* described from Europe, subsp. *laxa* from Lapland.

Hybrids with *C. rostrata* occur. Broken line on circumpolar map indicates range of subsp. **saxatìlis**.

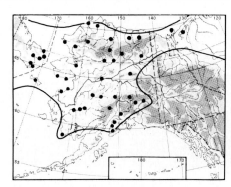
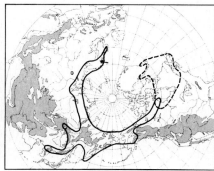

116. Càrex rotundàta Wahlenb.

Carex rostrata subsp. *rotundata* Kükenth.; *C. melozitensis* Pors.

Grayish-green, with long runners deep into mud (often lacking in herbarium specimens); culms stiff, smooth, with grayish or reddish sheaths at base; leaves narrow, often curved; lowest bract often divaricate or reflexed, shorter than inflorescence; 1 staminate spike, rarely 2; pistillate spikes usually 2, short, round or ovate, sessile; scales blunt, dark; perigynia in age blackish-brown, shiny, obovate, abruptly contracted into cylindrical, emarginate beak.

Wet places, swamps, muskeg. Broken line on circumpolar map indicates range of the closely related **C. miliàris** Michx.

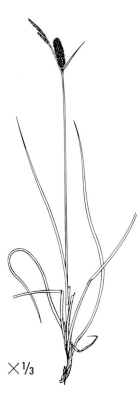
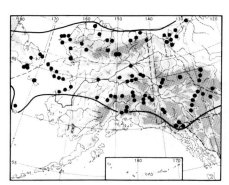
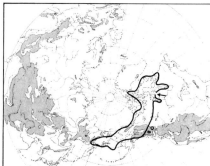

117. Càrex membranàcea Hook.

Carex campacta R. Br.; *C. physochlaena* Holm.

Yellowish-green, with long, thick, scaly, dark, horizontal stolons; culms stiff, smooth, with dried-up leaves and brown or purplish sheaths at base; leaf blades firm, 3–4 mm broad, slightly shorter than culm; terminal spike staminate, often with smaller staminate spikes at base; lateral spikes pistillate, sessile or short-peduncled, short-cylindrical, erect, dense, dark; lowest bract leaflike, somewhat shorter than inflorescence; scales ovate, acute, purplish-black; perigynia ovate to globose, shiny, inflated, purplish-black, abruptly contracted into short, emarginate beak.

Wet places.

26. Cyperaceae (Sedge Family)

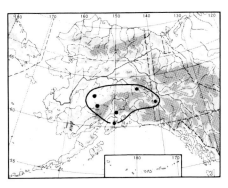
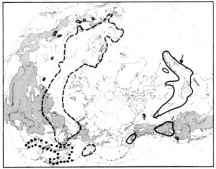
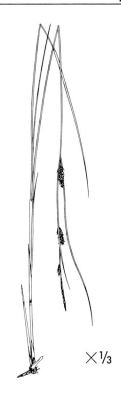

118. Càrex lasiocárpa Ehrh.
  Carex filiformis Good.
subsp. **americàna** (Fern.) Hult.
  Carex lasiocarpa var. *americana* Fern.

Grayish-green, with thick, horizontal rhizomes; culms slender, obtusely angled, smooth, with reddish-brown sheaths at base; leaf blades long, filiform, convolute, 1–2 mm broad, about as long as culms; lowest bract longer than inflorescence, short-sheathed or sheathless; upper 1–3 spikes staminate, linear, the lower 2–3 pistillate, ovate, or ovate-lanceolate, sessile or short-peduncled, ascending; scales ovate to lanceolate, 3-nerved, cuspidate or awn-tipped; perigynia ovate, covered with thick, grayish-brown pubescence, tapering into short, sharply bidentate beak with teeth about 0.2–0.5 mm long.

Shallow water, bogs. *C. lasiocarpa* described from Germany, subsp. *americana* from Argyl, Nova Scotia.

Broken line on circumpolar map indicates range of subsp. **lasiocárpa**, dotted line that of subsp. **occúltans** (Franch.) Hult. (*C. filiformis* var. *occultans* Franch.).

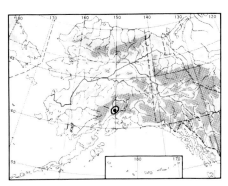
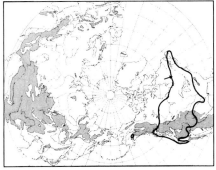

119. Càrex lanuginòsa Michx.

Similar to *C. lasiocarpa* subsp. *americana*, but leaves flat, 2–5 mm broad; culm acutely angled, scabrous; teeth of beak 0.3–0.8 mm long.

Wet places.

26. CYPERACEAE (Sedge Family) / 27. ARACEAE (Arum Family)

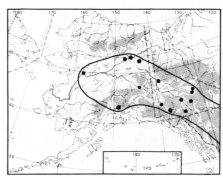 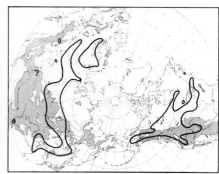

120. Càrex atheròdes Spreng.
Carex orthostachys C. A. Mey.

Yellowish-green, with long, thick, dark, subterranean runners; culm coarse, sharply angled, with long, purplish-black sheaths at base; leaves 4–10 mm wide; upper 1–3 spikes staminate, clublike to lanceolate, the lower 2–5 pistillate, ascending, more or less peduncled; scales ovate-lancolate, aristate; perigynia ovate to lanceolate, glabrous, strongly ribbed, green, tapering to slender beak; beak with slender, often outward-curved teeth.

Wet meadows. Described from arctic America.

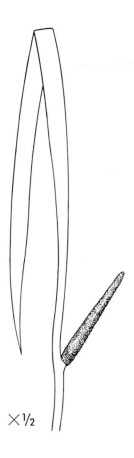

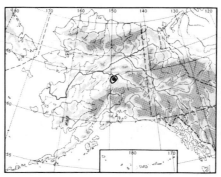 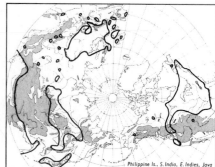

Acorus L.

1. Ácorus cálamus L. Calamus

Plant up to 1 meter tall, with thick, creeping rhizome; leaves ensiform, up to 2 cm wide; stem triangular, bearing cylindrical, green spadix 4–8 cm long and leaflike, ensiform, green spathe, appearing as continuation of stem.

Wet places, borders of quiet water. Described from Europe.

Entire plant has aromatic taste and smell; rarely flowers in Alaska. Probably not native within area of interest.

27. ARACEAE (Arum Family)

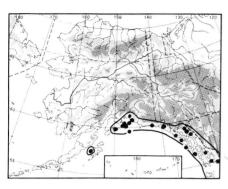
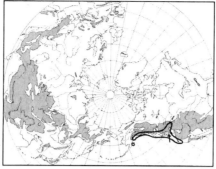
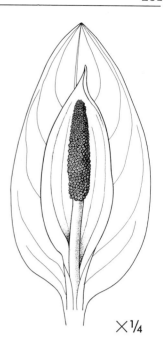

Lysichiton Schott

1. Lysichìton americànum Hult. & St. John — Yellow Skunk Cabbage
Lysichiton (Lysichitum) kamtschatkense with respect to American plant.

Plant acaulescent, with stout, erect rhizome; leaves large, up to 1 meter long, oblong to elliptic, glabrous, in basal cluster; spathe nearly sessile, yellow, appearing before leaves; spadix cylindrical, green in fruit.
Swampy woods. Entire plant has strong skunklike odor.

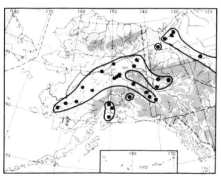
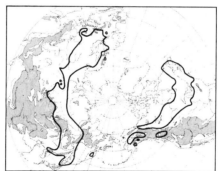

Calla L.

1. Cálla palústris L. — Wild Calla

Plant with long, thick, creeping rhizome, rooting at nodes; spathe white on upper surface, green on lower; leaves cordate, shiny; fruit light red, few-seeded berries.
Bogs, margins of ponds; a very distinctive plant. Described from northern Europe. Entire plant, especially the berries, contains poisonous acids and burning saponin-like substances—all of which, however, are neutralized by drying or boiling.

28. Lemnaceae (Duckweed Family)

Lemna L. (Duckweed)

Fronds thick, round or ovate, floating on surface .... 1. *L. minor*
Fronds thin, acute, long-stalked, submerged .... 2. *L. trisulca*

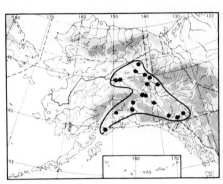
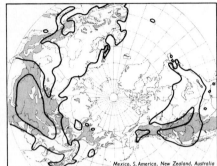

1. Lémna mìnor L.

Frond a green, round or elliptical, thick disk, 2–3 mm broad, with a single root from lower surface, floating on surface of water; flowers inconspicuous in margin of disk, usually 2 staminate, consisting of single stamen, and 1 pistillate, consisting of pistil, all surrounded by spathe.

Ponds, standing water. Described from Europe.

Flowers very rarely in northern regions.

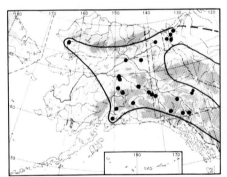

2. Lémna trisúlca L.

Fronds grayish-green, oblong-lanceolate, denticulate, attenuate at base in a slender stalk, often lacking root; submerged in water; several fronds usually hang together, new fronds emerging crosswise; flowers inconspicuous, similar to those of *L. minor*.

Ponds, standing water. Described from Europe.

29. JUNCACEAE (Rush Family)

Juncus L. (Rush)

Inflorescence appearing as if lateral, subtended by terete bract appearing as continuation of culm:
  Bract of inflorescence about as long as stem; stem thin, about 1–1.2 mm in diameter .... 1. *J. filiformis*
  Bract shorter; stem usually thicker:
    Involucral bract short, 0.5–2 (–3) cm, extending not or only slightly beyond inflorescence; seeds conspicuously tailed at both ends; inflorescence 1–3 (–5)-flowered .... 6. *J. Drummondii*
    Involucral bract longer; seeds not conspicuously tailed; inflorescence many-flowered:
      Culms tufted; stamens 3, opposite outer perianth leaves; anthers shorter than filaments .... 2. *J. effusus*
      Culms in rows from horizontal rhizome; stamens 6:
        Anthers much longer than filaments .... 5. *J. arcticus* subsp. *ater*
        Anthers about as long as filaments:
          Involucral bract 12–30 cm long; inner perianth leaves acute, narrowly scarious-margined .... 3. *J. arcticus* subsp. *sitchensis*
          Involucral bract 2–5 cm long; inner perianth leaves short-acuminate, broadly scarious-margined ...... 4. *J. arcticus* subsp. *alaskanus*
Inflorescence appearing as if terminal, with lower bract flat or channeled; stem leafy:
  Leaves gladiate, compressed from sides, with one edge toward stem........ .... 7. *J. ensifolius*
  Leaves terete, flat or channeled:
    Inflorescence a loose, open cyme with many separate flowers; perianth longer than capsule:
      Plant annual; outer perianth leaves longer than inner; leaves only on stem.. .... 22. *J bufonius*
      Plant perennial; all perianth leaves of same length; leaves also from base.. .... 15. *J. tenuis*
    Inflorescence with 1 or several compact heads:
      ■ Heads few, 1–3 (normally 1, except in *J. oreganus* and *J. castaneus*):
        Leaves all basal; heads small, 1–3 (–5)-flowered:
          Capsule retuse; inflorescence of usually 2 flowers, one above other; involucral bract exceeding flowers .... 20. *J. biglumis*
          Capsule conical, obtuse or mucronate; inflorescence a single head with 3–5 flowers:
            Perianth segments dark, shorter than capsule .... 18. *J. triglumis* subsp. *triglumis*
            Perianth segments often paler, of same length as capsule .... 19. *J. triglumis* subsp. *albescens*
        Leaves in middle of stem:
          ● Leaves terete, septate, septa noticeable as somewhat elevated darker spots at regular intervals:
            Capsule as long as, or shorter than, perianth; stem normally simple, with one dark-brown head .... 8. *J. Mertensianus*
            Capsule much longer than perianth; stem usually branched; heads few-flowered; perianth greenish-brown ...... 13. *J. oreganus*

● Leaves not septate:
  Stem thin (usually less than 1 mm in diameter); heads narrow, pale, 1–5-flowered; leaves less than 1 mm broad .... 21. *J. stygius* subsp. *americanus*
  Stem thicker, heads dark brown, large, many-flowered; leaves broader:
    Leaves flat, 2 (rarely 3) mm wide, linear; head single .... 9. *J. falcatus* subsp. *sitchensis*
    Leaves convolute, broader at base; heads 1–3; plant taller and coarser .... 10. *J. castaneus* subsp. *castaneus*
■ Heads several, forming branched inflorescence:
  Leaves distinctly septate, terete:
    Stems from slender, tuber-bearing rhizome; heads spherical, with echinate-spreading flowers, 7–10 mm in diameter; capsule subulate .. .. 12. *J. nodosus*
    Stems tufted; heads conical, with erect or ascending flowers:
      Perianth 3.5–5 mm long or longer, considerably shorter than acute capsule .... 13. *J. oreganus*
      Perianth 2–3 mm long, equaling or slightly shorter than obtuse or broadly acute, mucronate capsule:
        Outer perianth leaves acute, inner obtuse .... 16. *J. alpinus*
        Outer and inner perianth leaves acute .... 17. *J. articulatus*
  Leaves flat or convolute, not septate:
    Leaves 0.7–1.5 mm broad .... 14. *J. Leschenaultii*
    Leaves broader:
      Plant with several long-peduncled, pale heads; lowest bract long .. .. 11. *J. castaneus* subsp. *leucochlamys*
      Plant with 1–3 dark heads; lowest bract shorter .... 10. *J. castaneus* subsp. *castaneus*

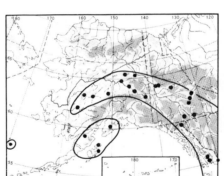 

1. Júncus filifórmis L.

Culms filiform, about 1 mm thick, in dense rows, from creeping, slender, dark-brown rhizome; involucral bract about as long as stem, erect; cymes single, with mostly simple branches and few, light-brown flowers; perianth leaves narrow, lanceolate, acute, as long as or somewhat longer than the greenish-brown, nearly spherical capsule; anthers much shorter than filaments.

Bogs, marshes, sandy soil. Described from Europe.

29. Juncaceae (Rush Family)

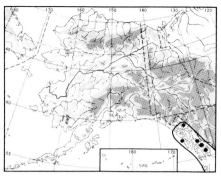
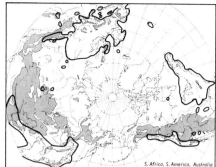

2. Júncus effùsus L.

Culms with dark-brown sheaths at base, up to over 1 meter tall, finely striate, green, tufted, from stout, subterranean rhizome; cymes mostly open, with forking branches, many-flowered; involucral bract erect; flowers greenish; stamens usually only 3.

Margins of ponds, swamps. Described from Europe.

Circumpolar map gives range of *J. effusus* in a broad sense.

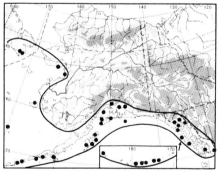
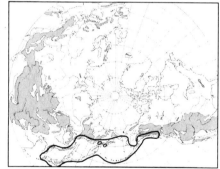
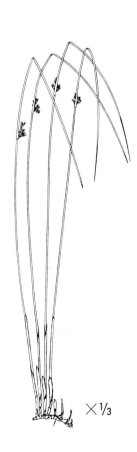

3. Júncus árcticus Willd.
subsp. sitchénsis Engelm.

Juncus balticus var. *Haenkei* (E. Mey.) Buchenau; *J. Haenkei* E. Mey.; *J. balticus* subsp. *sitchensis* (Engelm.) Hult.

Culms stout, thick, with yellowish-brown, lustrous basal sheaths in rows, from thick, horizontal, dark rootstocks; bract 12–30 cm long; inflorescence contracted, up to 2 cm long, many-flowered; flowers sessile or short-pedicellate; outer perianth leaves linear-lanceolate, brown, with greenish center, narrowly scarious-margined, longer than the inner; filaments about as long as anthers.

Sandy shores, tidal marshes.

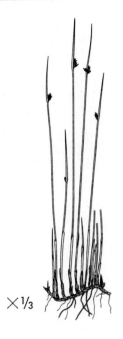
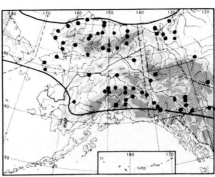
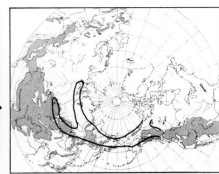

4. Júncus árcticus Willd.
subsp. **alaskànus** Hult.

Juncus balticus var. *alaskanus* (Hult.) Pors.

Similar to subsp. *sitchensis*, but involucral bract shorter, 2–5 cm long; inner perianth leaves short-acuminate, broadly scarious-margined.

Wet places, river flats. *J. arcticus* described from Lapland, subsp. *alaskanus* from Wiseman, Alaska.

Circumpolar map tentative.

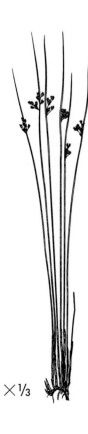
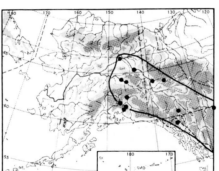
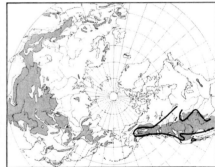

5. Júncus árcticus Willd.
subsp. **àter** (Rydb.) Hult.

Juncus ater Rydb.; *J. balticus* var. *littoralis* with respect to Alaskan plant; *J. balticus*, of Fl. Alaska & Yukon.

Similar to subsp. *sitchensis*, but more slender and with longer bract and open cyme; anthers much longer than filament.

Wet places. Described from "Western plains and mountains."

Circumpolar map tentative.

29. JUNCACEAE (Rush Family)

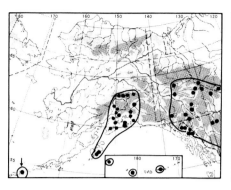
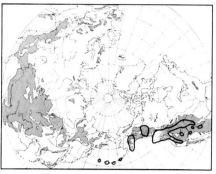
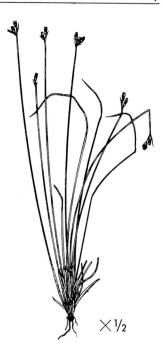

6. Júncus Drummóndii E. Mey.

Caespitose from very short rootstocks; stem terete, with straw-colored sheaths at base; lowest bract as long as, or slightly longer than, inflorescence; perianth segments subequal, lanceolate, acuminate; capsule dark brown, about as long as perianth or longer.

Moist, gravelly slopes, alpine meadows, snow beds, to about 2,000 meters.

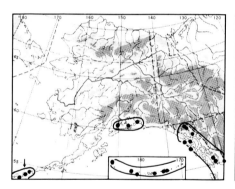
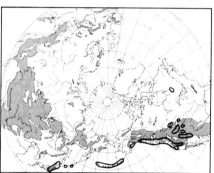

7. Júncus ensifòlius Wikstr.

Juncus xiphioides var. *triandrus* Engelm.

Culms in rows from thick, creeping, light-brown or purplish-brown rootstock; stem compressed, 2-edged, with 3–4 laterally flattened leaves, about 4 mm broad; inflorescence of 1–3 large heads, in age dark-brown; perianth segment lanceolate, acuminate, subequal; stamens 3; anthers shorter than filaments; capsule oblong, slightly longer than perianth.

Wet meadows.

× ½

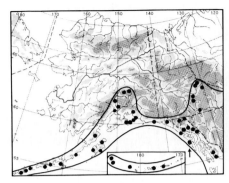
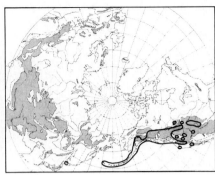

8. Júncus Mertensiànus Bong.

Culms in rows from short, brown rootstocks; stems 5–15 cm tall, with 2–3 leaves and brown or purplish basal sheaths; heads solitary, blackish-brown, hemispherical; bract spathelike, equaling or often extending well beyond head; perianth segments subequal, lanceolate, subulate-tipped; anthers 6, much shorter than filaments; capsule oblong-obovate, emarginate or obtuse, about as long as perianth.

Wet subalpine meadows, snow beds.

× ½

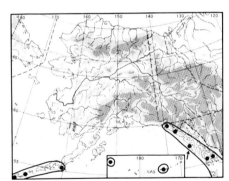

9. Júncus falcàtus E. Mey.
subsp. **sitchénsis** (Buchenau) Hult.

Juncus falcatus var. *sitchensis* Buchenau.

Culms up to 25 cm tall, from long, light-brown, scaly runners; leaf blades flat, stiff, thickish, about as long as stem; heads mostly single, compact, hemispherical; lowest bract lanceolate, extending slightly beyond head; outer perianth segments acute, the inner obtuse, all of about same length, dark brown with lighter center; anthers 6, about as long as filaments; capsule oblong to obovate, emarginate, about as long as perianth.

Moist, sandy places, near shores.

J. falcatus is a complicated complex occurring also in Japan (var. **próminens** Buchenau) and in Australia and Tasmania.

29. JUNCACEAE (Rush Family)

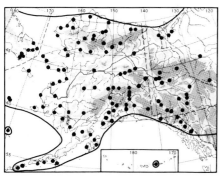 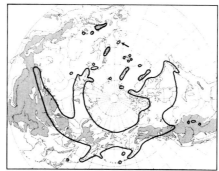 

10. Júncus castàneus Sm.
subsp. **castàneus**

Culms from short, light-brown, scaly rhizome with stolons; stem stiff, coarse, leafy below; leaves erect, canaliculate; heads 1–3; lower involucral leaf generally extending beyond inflorescence; sepals brownish-black, linear-lanceolate, acute, as long as or longer than the narrow, obtuse petals; capsule shiny, castaneous to purplish-black, mucronate, much longer than perianth.

Common in wet places in tundra and mountains. Var. **pállidus** Hook., with pale perianth bracts and capsule and often with few-flowered heads, occurs in scattered localities.

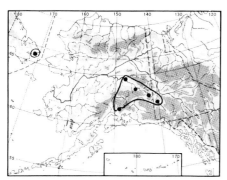 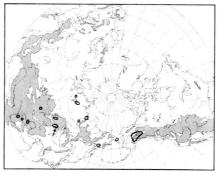 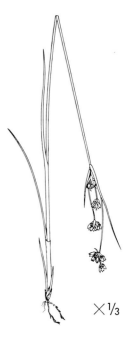

11. Júncus castàneus Sm.
subsp. **leucochlàmys** (Zinz.) Hult.
Juncus leucochlamys Zinz. ex Krecz.; *J. leucochlamys* var. *borealis* Tolm.

Similar to subsp. *castaneus* but taller, with several heads; lowest bract longer; heads long-peduncled; scales and capsule paler; capsule large, somewhat acute.
Wet places on tundra and in the mountains.

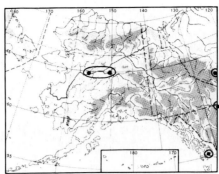 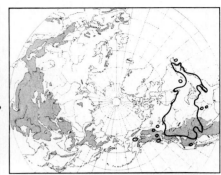

12. Júncus nodòsus L.

Stems 1–4 dm tall, from slender, creeping, tuber-bearing rhizome, mostly with 2–3 slender leaves, the uppermost often exceeding inflorescence; heads spherical, reddish-brown, exceeded by involucral leaf, in open inflorescence; capsule about as long as the lance-acuminate perianth leaves; anthers 6, shorter than filaments; style extremely short.

Swamps, hot springs. Described from North America.

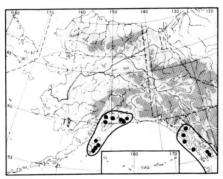 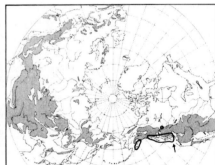

13. Júncus oregànus S. Wats.
Juncus paucicapitatus Buchenau.

Caespitose, with slender rootstock, up to 30 cm tall; stem with 2–4 leaves and basal sheaths; leaves terete, septate; inflorescence of 2–5 few-flowered heads; perianth segments narrowly lanceolate, acute, nearly equal; anthers much shorter than filaments; capsule ovate-lanceolate, mucronate, much longer than perianth.

Bogs, ponds.

29. JUNCACEAE (Rush Family)

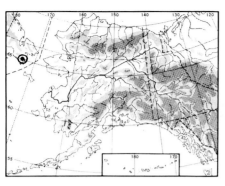
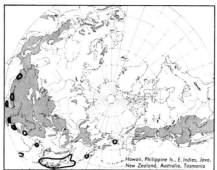

14. Júncus Leschenáultii Gay
Juncus prismatocarpus var. *Leschenaultii* (Gay) Buchenau.

Caespitose; stem with 1–3 leaves; leaves flat, thin, indistinctly septate; inflorescence with few, 3–8-flowered stellate heads; perianth segments linear-lanceolate, subulate, nearly equal; stamens 3; anthers about half as long as filaments; capsule triangular-conical, as long as perianth, or slightly longer.

Exclusively at hot springs. Because the hot springs of Alaska are very little known botanically, this plant might also occur there. Described from Nilgiri Hills, India.

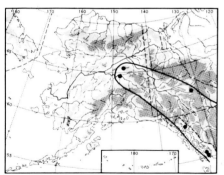
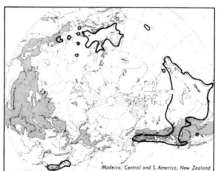

15. Júncus ténuis Willd.
Juncus macer S. F. Gray; *J. Dudleyi* with respect to Yukon plant.

Densely caespitose; leaves flat, in dry specimens involute, as long as or shorter than culms; sheaths with broad, scarious margin, auricles much longer than broad; lower bracts longer than cyme; cyme compact or loose; sepals light brown, lanceolate, highly acute, much longer than the retuse capsule; anthers much shorter than filaments.

Roadsides, open ground. Described from North America.

29. JUNCACEAE (Rush Family)

× ½

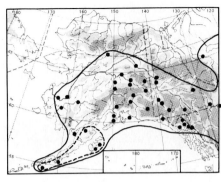
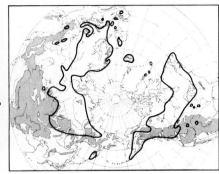

16. Júncus alpìnus Vill. (in broad sense)

Culms from short, creeping rhizome, with 1–2 slender, stiff, erect leaves and yellowish sheaths at base; cyme with few, elongate branches; heads elongate to hemispherical, few-flowered; flowers sessile or pedicellate; sepals longer than petals.

Wet places.

An extremely variable aggregate species. Most of the specimens belong to subsp. **nodulòsus** (Wahlenb.) Lindm. (*J. nodulosus* Wahlenb., *J. Richardsonianus* Schult.), with straw-colored, scarious-margined perianth leaves, rounded in apex, and with slightly exserted, acute capsules. West of Kodiak Island and elsewhere, a type closer to subsp. **alpìnus** [var. *alpestris* Hartm.; var. *mucroniflorus* (Clairv.) Aschers. & Graebn.] occurs, with sessile, dark flowers in compact heads.

J. alpinus described from Dauphiné, subsp. *nodulosus* from Sweden. Broken line on Alaskan map indicates range of subsp. *alpinus*. Circumpolar map indicates range of entire species complex.

× ⅓

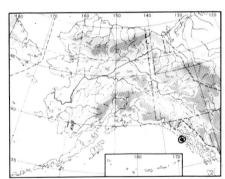

17. Júncus articulàtus L.
Juncus lamprocarpus Ehrh.

Similar to *J. alpinus*, but heads larger, with more flowers; branches of inflorescence more spreading, and both sepals and petals acuminate, brown; leaves with distinct knots.

Wet places. Described from Europe.

Var. **Turczannìnòvii** Buchenau, with smaller flowers and with leaves having less distinct knots, occurs in the eastern Asiatic part of the range.

29. JUNCACEAE (Rush Family)

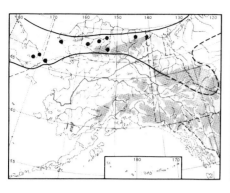 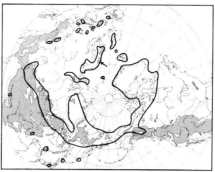 

18. Júncus triglùmis L.
subsp. **triglùmis**

Caespitose, with short rootstock; culm straight, stiff, up to 15 cm tall, with basal filiform leaves and reddish-brown sheaths; leaves shorter than culm; head single, with 3–5 flowers on same level, subtended by 2–3 dark, spathiform bracts, mostly shorter than head; perianth segment ovate-lanceolate, dark reddish-brown in age, shorter than the mature, apiculate, blackish-brown capsule.

Moist places in open ground, heaths. Described from the "Alps of Lapland."

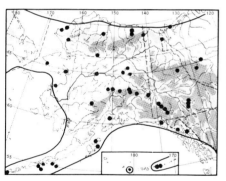 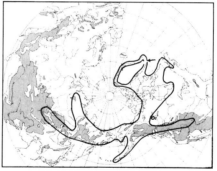 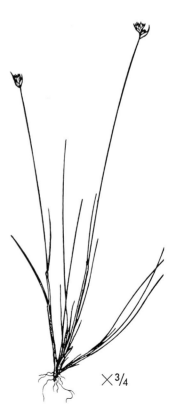

19. Júncus triglùmis L.
subsp. **albéscens** (Lange) Hult.

Juncus triglumis var. *albescens* Lange; *J. albescens* (Lange) Fern.; *J. Schischkini* Kryl. & Serg.

Similar to subsp. *triglumis*, but somewhat more slender; heads smaller; lowest bract often longer than inflorescence, narrower and lighter-colored; capsule of same length as perianth.

Moist places, heaths. Described from western Greenland.

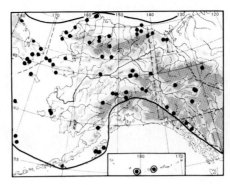
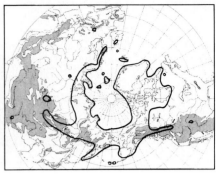

20. Júncus biglùmis L.

Loosely caespitose; culms 5–15 cm long, with single leaf and bladeless sheaths at base; leaves about 1 mm wide; inflorescence usually with 2 flowers, one above other; bract single, dark brown, ovate-lanceolate, reaching above flowers; perianth soon becoming nearly black; capsule emarginate.

Moist gravel, margins of tundra ponds, snow beds, in mountains, to at least 1,200 meters. Described from Lapland.

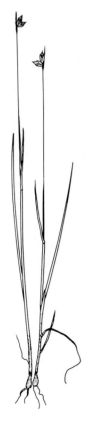
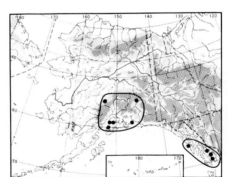
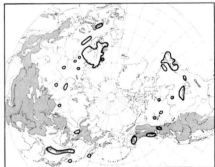

21. Júncus stýgius L.
subsp. americànus (Buchenau) Hult.

Juncus stygius var. *americanus* Buchenau.

Culms filiform, 1–3-leaved, in small tufts; leaves filiform, the basal shorter than culm; heads 1 (–3), each 1–4-flowered; lowest bract shorter than inflorescence; perianth pale, light brown, striped; sepals acute; petals somewhat obtuse; capsule ovate, light yellowish-brown, longer than perianth; anthers much shorter than filaments.

Wet bogs. *J. stygius* described from Europe, subsp. *americanus* from America. American plant differs from Eurasiatic plant in having larger, more distinctly mucronate capsules. Circumpolar map indicates range of entire species complex.

29. JUNCACEAE (Rush Family)

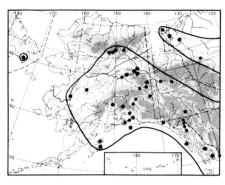
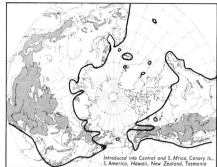

22. Júncus bufònius L.
Including *J. ranarius* Perr. & Song. and *J. nastanthus* Krecz. & Gontsch.

Annual; light green, in age brownish; stems simple or branching from the base, erect or decumbent; cymes open, dichotomously branched; flowers remote, scattered among branches; perianth leaves green, linear-lanceolate to lanceolate, scarious-margined, subulate-tipped, the inner somewhat shorter and broader; capsule chestnut-brown, shorter than perianth leaves.

Wet places. Described from Europe.

Var. **ranàrius** (Perr. & Song.) Hayek (*J. ranarius* Perr. & Song.) has inner perianth leaves obtuse, about as long as capsule; in several localities within area of interest, it occurs with the type variety.

Luzula DC. (Wood Rush)

- Flowers solitary (or 2, rarely 3) at tips of ultimate branches of inflorescence:
    - Inflorescence umbelliform, peduncles 1 (–2)-flowered; leaves with long, white hairs in margin of lower half .... 1. *L. rufescens*
    - Inflorescence a loose, compound cyme; leaves glabrous (or almost so), or with long, white hairs at mouth of sheath only:
        - Stem leaves 1–3, short, 2–5 mm wide; bracts of flowers lacerate and abundantly ciliate; plant tufted; perianth dark brown:
            - Stem leaves 2–3, broad; lowest bract more than 1 cm long; inflorescence many-flowered .... 3. *L. Wahlenbergii* subsp. *Piperi*
            - Stem leaves 1–2, narrower; lowest bract less than 1 cm long; inflorescence few-flowered .... 2. *L. Wahlenbergii* subsp. *Wahlenbergii*
        - Stem leaves usually 3–4, long and broad, 5–7 mm wide; bracts less lacerate, the ultimate glabrous or with few ciliae; plant often with short runners:
            - Cyme many-flowered; branches not extended; perianth pale brown .... 4. *L. parviflora*
            - Cyme few-flowered; branches strongly extended; perianth greenish .... 5. *L. parviflora* subsp. *divaricata*

29. JUNCACEAE (Rush Family)

■ Flowers crowded in spikes or glomerules:
   Flowers in dense, nodding, often interrupted, spikelike panicle; bracts of flowers conspicuous, silvery-hyaline, longer than flowers .... 15. *L. spicata*
   Flowers in glomerules (or sometimes in erect spike); bracts not longer than flowers:
      Leaves involute, channeled; base of basal leaves purplish; plant with short runners:
         Top of culm has 1 sessile or short-peduncled head; mostly also 2–3 long-peduncled lateral heads .... 10. *L. confusa*
         Top of culm lacking head; inflorescence of 2–8 capillary, arched branches, with 3–5 flowers at each tip .... 6. *L. arcuata* subsp. *arcuata*
      Leaves flat; base of basal leaves brown or light brown, sometimes slightly purplish:
         Leaves with long, soft, white hairs at margin; inflorescence aggregate, or with 2–6 heads .... 11–14. *L. multiflora* complex
         Leaves glabrous in margin, or nearly so:
            Inflorescence one head, at top of culm (rarely 2 aggregate heads), no capillary branches; bract of inflorescence inconspicuous; leaves 2–3 mm wide .... 9. *L. arctica*
            Inflorescence with capillary branches:
               Inflorescence similar to that of *L. arcuata*, lacking central head, with 3–5 flowers at tip of capillary, arched branches; sheaths dark brown or reddish-brown .... 7. *L. arcuata* subsp. *unalaschcensis*
               Inflorescence with central head and capillary, arched branches; leaves 4–8 mm broad; sheaths light brown .... 8. *L. tundricola*

×½

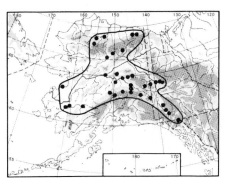 

1. Lúzula ruféscens Fisch.

Luzula japonica with respect to Alaskan plant.

Loosely caespitose; culm thin, usually with 2 short, reddish-sheathed leaves at base; leaves with long, white hairs in the lower part; inflorescence umbellate, rays with few branches; lowest bract much shorter than inflorescence; flowers single; inner perianth leaves somewhat longer than the outer, chestnut-brown, scarious-margined; capsule pale, acuminate, shorter than perianth.

Wet places, woods, gravel bars.

29. JUNCACEAE (Rush Family)

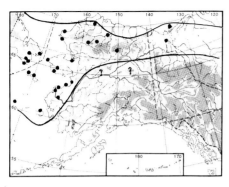 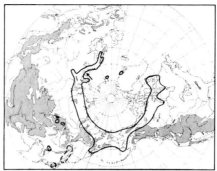 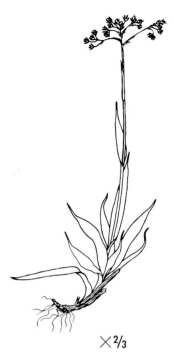

2. Lúzula Wahlenbérgii Rupr.
subsp. **Wahlenbérgii**

Tufted, dark green or sometimes purplish-green; culm with generally 2 short, narrow, not very conspicuous stem leaves; lowest bract mostly not over 1 cm long; cyme with few-flowered, slender, arching branches; perianth leaves broadly lanceolate, acute, rust-colored, with paler margin; mature capsule as long as perianth, brown.

Wet tundra, snow beds. Broken line on circumpolar map indicates range of subsp. **yezoénsis** (Satake) Hult. (*L. parviflora* var. *yezoensis* Satake).

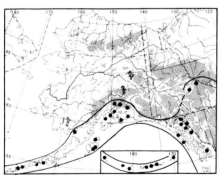 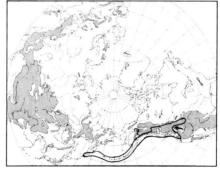 

3. Lúzula Wahlenbérgii Rupr.
subsp. **Pìperi** (Cov.) Hult.
Juncoides Piperi Cov.; *Luzula Piperi* (Cov.) M. E. Jones.

Similar to subsp. *Wahlenbergii*, but cauline leaves mostly 3, broader, the lowest 4–6 mm wide; lowest bract longer, mostly 1.5–1.8 cm long; inflorescence more ample, with reddish-brown perianth leaves.

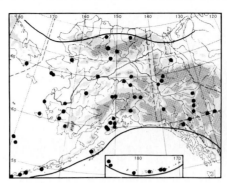
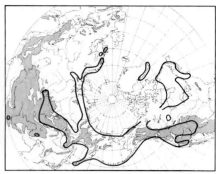

4. Lúzula parviflòra (Ehrh.) Desv.

Juncus parviflorus Ehrh.; *Juncoides parviflorum* Cov.; *Luzula melanocarpa* (Michx.) Desv.; *L. parviflora* var. *melanocarpa* (Michx.) Buchenau; *Juncus melanocarpus* Michx.; *L. labradorica* Steud.

subsp. **parviflòra**

Yellowish-green with short runners; culm with 3–5 large stem leaves; basal leaves flat, 6–12 mm broad, nearly glabrous, abruptly pointed; cyme with long, slender, arching rays, mixed with shorter rays; perianth leaves broadly lanceolate, acute, castaneous or brownish, scarious-margined; mature capsule as long as perianth or slightly longer, shiny, chestnut-brown.

Moist places in forest and tundra. Described from Europe.

Some specimens belong to var. **melanocárpa** (Michx.) Buchenau (*Juncus melanocarpus* Michx.), with lax cyme, pale perianth, and very dark capsule. Circumpolar map indicates range of *L. parviflora*, in broad sense.

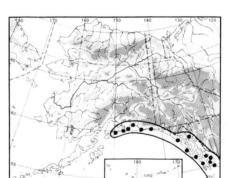
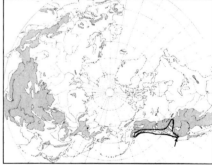

5. Lúzula parviflòra (Ehrh.) Desv.
subsp. **divaricàta** (S. Wats.) Hult.

Luzula divaricata S. Wats.

Similar to subsp. *parviflora*, but cyme very large, few-flowered, with much-extended branches; perianth greenish, light-colored.

Moist places in forests.

29. JUNCACEAE (Rush Family)

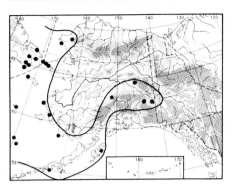
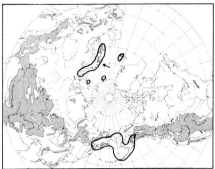

6. Lúzula arcuàta (Wahlenb.) Sw.
subsp. **arcuàta**
  Juncus arcuatus Wahlenb.; *Luzula beringensis* Tolm.

Caespitose, with short runners; culm thin, more or less arched, acute, with 1 stem leaf and brown sheaths, purplish at base; leaves narrow, involute, much shorter than culm; inflorescence with 2–8 capillary, arched branches, sometimes branched in apex, with 3–5 flowers in glomerules; floral bracts acute, ciliate; perianth leaves lanceolate, acute, dark, whitish in margin and at tip; capsule awned, shorter than perianth.

Dry places in mountains, snow beds. Very similar to the typical European *L. arcuata*, despite broad separation between ranges.

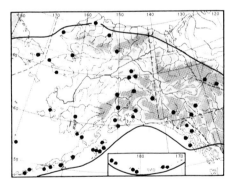
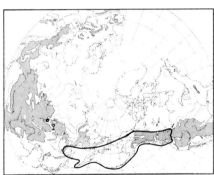

7. Lúzula arcuàta (Wahlenb.) Sw.
subsp. **unalaschcénsis** (Buchenau) Hult.
  Luzula arcuata var. *unalaschcensis* Buchenau; *L. unalaschkensis* (Buchenau) Satake.

Similar to subsp. *arcuata*, but leaves broader, flat; sheaths brown or slightly purplish at base.
  Described from Unalaska and St. Paul Island.
  Dry places in the mountains.

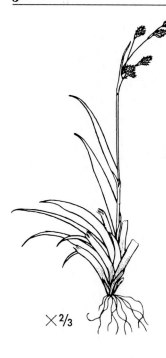 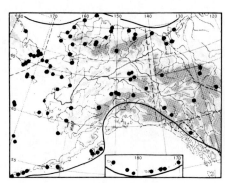 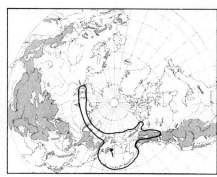

8. Lúzula tundrícola Gorodk.

Luzula arcuata f. *latifolia* Kjellm.; *L. nivalis* var. *latifolia* (Kjellm.) Sam.; *L. confusa* var. *latifolia* Buchenau; *L. beeringiana* Gjaerevoll.

Loosely caespitose, light-colored; leaves flat, 4–8 mm wide, abruptly pointed, glabrous; sheaths broad, light brown; inflorescence with few, capillary, arched branches; lowest bract shorter than inflorescence; capsule as long as perianth.

Tundra and mountain tundra. Hybrids with *L. arctica* probably occur.

 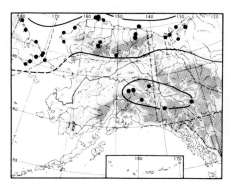 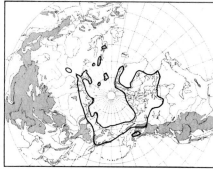

9. Lúzula árctica Blytt

Luzula nivalis (Laest.) Beurl.; *L. campestris* var. *nivalis* Laest.

Densely caespitose, often dark-colored, grayish-violet; leaves flat, acute, short, glabrous or nearly so, 2–3 mm wide; culm stiff, straight, with 1–2 aggregate, small heads; lowest bract inconspicuous; floral bracts not ciliated; capsule longer than perianth.

Tundra and mountain tundra, moist slopes. Described from Scandinavia.

29. JUNCACEAE (Rush Family)

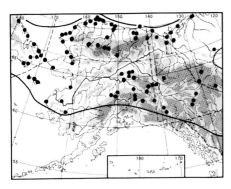 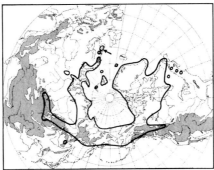 

10. Lúzula confùsa Lindeb.

Caespitose, with short runners; culm stiff, with purplish, lustrous sheaths at base; leaves narrow, channeled, acute; spikes crowded in subglobose glomerule (var. **eradiàta** Hult.) or clustered at tips of 1–4 filiform peduncles; bractlets and perianth fimbriate-ciliate; capsule about as long as perianth.

Dry heaths in mountains and on tundra, in McKinley Park to at least 2,100 meters.

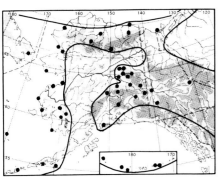 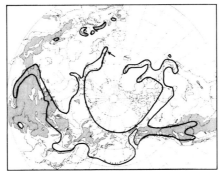 

11. Lúzula multiflòra (Retz.) Lej.
subsp. **multiflòra**
  Juncus multiflorus Retz.
var. **frígida** (Buchenau) Sam.
  Luzula frigida Buchenau.

Densely caespitose, lacking runners; leaves 2–4 mm broad, with long, soft, white cilia in margin; cyme mostly subumbellate, with straight, stiff branches; heads cylindrical, oblong or ovate; perianth leaves lanceolate or broadly lanceolate, acuminate, more or less scarious-margined and scarious-tipped; capsule rounded, longer or shorter than perianth.

Woods, grassy places, from lowlands to mountains. An extremely complicated complex.

Var. *frigida* has aggregate inflorescence consisting of 3–4 small, blackish heads, one of which is often long-peduncled; and blackish-brown perianth, longer or shorter than black capsule.

×⅓

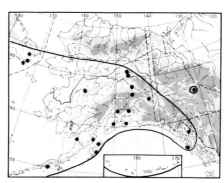
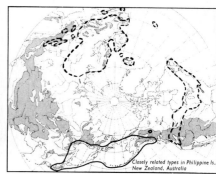

12. Lúzula multiflòra (Retz.) Lej.
subsp. **multiflòra**
var. **Kjellmaniàna** (Miyabe & Kudo) Sam.
  L. Kjellmaniana Miyabe & Kudo.

Tall; leaves narrower than in subsp. *Kobayasii*; cyme more or less contracted; perianth dark-colored; flowers small, about 3 mm long.
  Grassy slopes. Described from Shumshu Island (northern Kurile Islands).
  Broken line on circumpolar map indicates range of other races not present within area of interest.

×⅓

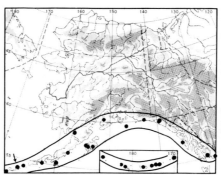
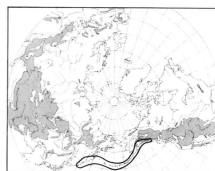

13. Lúzula multiflòra (Retz.) Lej.
subsp. **Kobayásii** (Satake) Hult.
  Luzula Kobayasii Satake; *L. multiflora* var. *Kobayasii* (Satake) Sam.

Tall, with very broad leaves; heads many-flowered, oblong; bracts of inflorescence long; perianth exceeding capsule, chestnut-brown, conspicuously hyaline-margined.
  Grassy places. Transitions to subsp. *multiflora* var. *Kjellmaniana* not uncommon.

29. Juncaceae (Rush Family)

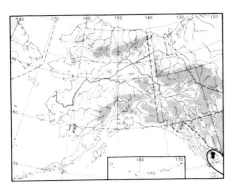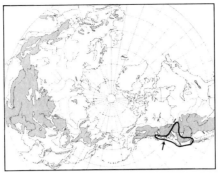

14. Lúzula multiflòra (Retz.) Lej.
subsp. **comòsa** (E. Mey.) Hult.

Luzula comosa E. Mey.; *L. campestris* var. *comosa* Fern. & Wieg.

Tall; leaves broad; bracts of inflorescence large, leaflike; inflorescence more or less compressed; bracts strongly ciliated; perianth yellowish, slightly exceeding capsule.

Grassy places in the lowlands. Described from Nootka Sound or Yakutat Bay.

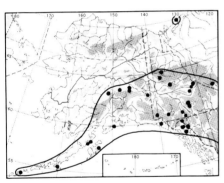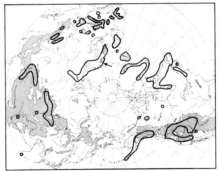

15. Lúzula spicáta (L.) DC.

Juncus spicatus L.

Densely caespitose; leaves acute, canaliculate, pubescent at base, with coarse, pale sheaths having hairs at mouth; stem slender, arching; spike nodding in age, on capillary peduncles; floral bracts long-ciliate; perianth with broad scarious margin; capsule brownish-black, shorter than perianth.

Gravel, stony places, in mountains; to about 2,300 meters in the St. Elias Mountains. Circumpolar map indicates range of *L. spicata*, in broad sense.

Tofieldia Huds. (False Asphodel)

Plant stout, about 25 cm tall; stem viscid, glandular ....
.... 3. *T. glutinosa* subsp. *brevistyla*
Plant tender, smaller; stem glabrous, not viscid:
  Stem with 1 or more leaves; involucre at top of pedicel .... 1. *T. coccinea*
  Stem leafless or with only 1 small bract; involucre at base of pedicel ....
.... 2. *T. pusilla*

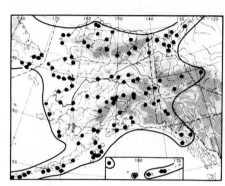
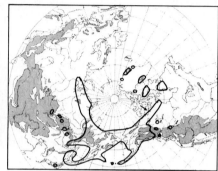

×½

1. Tofiéldia coccínea Richards.

Tofieldia nutans Willd.

Scape with 1–2 (or more) leaves; basal leaves numerous, flabelliform, mostly 5-nerved; raceme at first short and compact, later elongated and loose, in fruit with pedicels 1–3 mm long; flowers yellowish; anthers darker; involucre at top of pedicel.
Stony, dry places, heaths, polygon tundra.

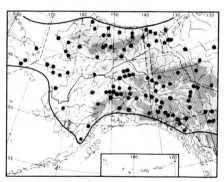
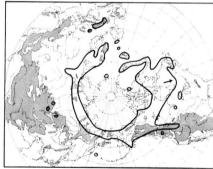
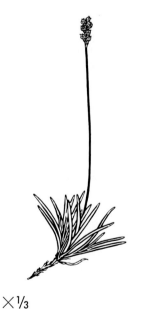

×⅓

2. Tofiéldia pusílla (Michx.) Pers.

Narthecium pusillum Michx.; *Tofieldia minima* (Hill) Druce; *Phalangium minimum* Hill; *Tofieldia palustris* of authors; not Huds.

Glabrous; scape leafless or with 1 small bract; basal leaves numerous; leaves flabelliform, mostly 3-nerved; raceme subglobose to cylindrical; flowers whitish to greenish; involucre at base of pedicel.
Wet places, heaths, in mountains to at least 1,700 meters.

30. LILIACEAE (Lily Family)

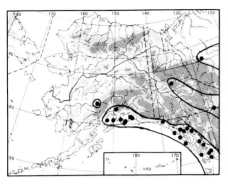

3. Tofiéldia glutinòsa (Michx.) Pers.
Narthecium glutinosum Michx.; *Triantha glutinosa* (Michx.) Baker.
subsp. **brevistỳla** Hitchc.
Tofieldia occidentalis Hult., of Fl. Alaska & Yukon.

Scape with short, dark glands, leafless or with bract; leaves one-half to two-thirds the length of the scape; spike long, often interrupted; pedicels long, with dark glands; sepals and petals yellowish; capsule stramineous or red; style short, about 1 mm long.

Marshes, shores. On circumpolar map, the range indicated in eastern North America is that of subsp. **glutinòsa,** which has longer styles and somewhat larger flowers and capsules. Alaskan specimens of subsp. *brevistyla* sometimes approach subsp. *glutinosa*.

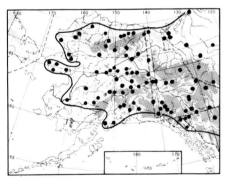
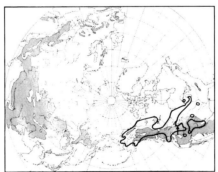

Zygadenus Michx.
(Original spelling *Zigadenus*)

1. Zygadènus élegans Pursh   White Camass
Anticlea glauca Kunth; *A. elegans* (Pursh) Rydb.; *Zygadenus glaucus* Nutt.

Bulbous, the outer bulb-coat fibrous; glaucous; stem with 1–2 leaves; inflorescence a loose, cylindrical raceme; middle and upper bracts scarious-margined; perianth greenish or yellowish-green; capsule twice as long as perianth.

Open woods, heaths, grassy slopes, *Populus* forests. To at least 2,000 meters in the St. Elias Mountains.

Contains the poisonous alkaloid zygadenine, which causes vomiting, lowered body temperature, difficult breathing, and finally, coma.

30. LILIACEAE (Lily Family)

Veratrum L. (False Hellebore)

Inflorescence open, with few long, drooping branches; leaves evenly pubescent beneath, on nerves and between nerves .... 1. *V. viride* subsp. *Eschscholtzii*
Inflorescence spikelike or with short upright branches; leaves pubescent beneath, on nerves only .... 2. *V. album* subsp. *oxysepalum*

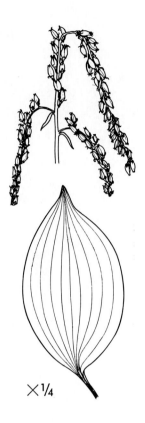

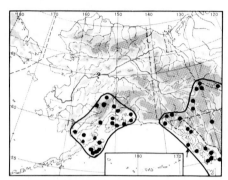 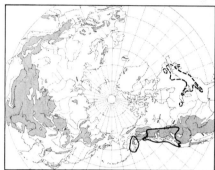

1. Veràtrum víride Ait.

subsp. **Eschschòltzii** (Gray) Löve & Löve

Veratrum Eschscholtzii Gray; *V. viride* var. *penduliflorum* Harshb.

Inflorescence open, at maturity with few long, drooping branches; leaves woolly-pubescent below; perianth green; segments oblong, green, pubescent in margin.

Meadows, moist places. *V. viride* described from North America.

Contains a poisonous alkaloid, which causes vomiting, purging, general paralysis, and death from asphyxia.

Broken line on circumpolar map indicates range of subsp. **víride.**
(See color section.)

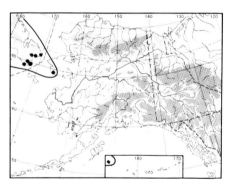 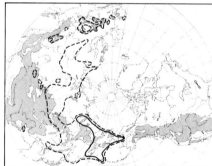

2. Veràtrum álbum L.

subsp. **oxysèpalum** (Turcz.) Hult.

Veratrum oxysepalum Turcz.

Inflorescence spikelike or with short upright branches; leaves pubescent on nerves below, or sometimes nearly glabrous; perianth green, campanulate, the segments lanceolate, denticulate; bracts about as long as perianth.

Meadows, moist places. *V. album* described from Russia, Siberia, Austria, Switzerland, Italy, and Greece; subsp. *oxysepalum* from Kamchatka.

Broken line on circumpolar map indicates range of subsp. **álbum.**

30. LILIACEAE (Lily Family)

Allium L.

Leaves linear; bulb without reticulate cover; flowers rose-colored . . . .
. . . . 1. *A. schoenoprasum* var. *sibiricum*
Leaves ovate, broad, flat; bulb with reticulate cover; flowers white . . . .
. . . . 2. *A. victorialis* subsp. *platyphyllum*

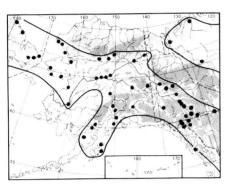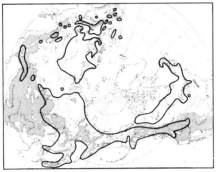

1. Állium schoenóprasum L.
var. sibìricum (L.) Hartm.
Allium sibiricum L.

Wild Chive

Bulbs 1 to several, oblong-ovate, within papery coating; leaves coarse, semiterete, hollow at base, about as long as the coarse scape or shorter; umbel dense, nearly spherical; perianth leaves pink or rose-violet with darker veins, lanceolate or linear-lanceolate, attenuate; capsule 2–3 times shorter than perianth.

Meadows, grassy slopes. *A. schoenoprasum* described from Siberia and the Island of Oeland (Sweden), var. *sibiricum* from Siberia.

The leaves are eaten in early spring, the bulbs in summer and fall. The Siberian Eskimo keep the plant for long periods in airtight sealskin bags.

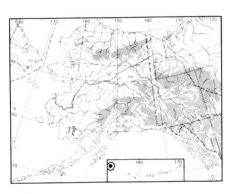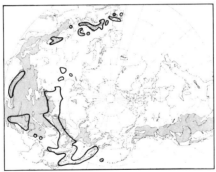

2. Állium victoriàlis L.
subsp. platyphýllum Hult.
Allium ochotense Prokh.

Victory Onion

Bulbs 1 to several, surrounded by a grayish-brown, netlike coating; leaves 1–3, glabrous, broadly elliptic, 5–8 cm broad; umbel lax, nearly spherical; perianth leaves whitish-green, elliptical, obtuse; capsule rounded.

Meadows; in Siberia, where the young shoots are eaten, also in the woods. *A. victorialis* described from Switzerland and Italy, subsp. *platyphyllum* from Kamchatka.

Circumpolar map indicates range of *A. victorialis* in a broad sense.

30. Liliaceae (Lily Family)

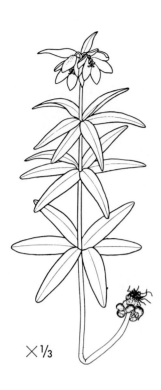

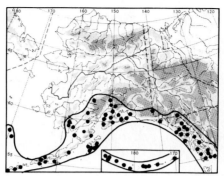 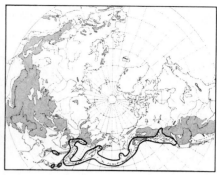

Fritillaria L.

1. Fritillària camschatcénsis (L.) Ker-Gawl.   Kamchatka Fritillary, Saraná

Lilium camschatcense L. (different spellings of the species name are used by different authors; above spelling is that of Linnaeus).

Bulbs of several fleshy scales and their bladelike petioles, which disintegrate into numerous ricelike bulblets; leaves in 2–3 whorls; flowers purplish-black, 1 to several close together, on short pedicels; capsule obtusely angled. The perianth segments are often striped with green on the outside in American specimens, less so in Asiatic. In meager soil the plant is often sterile, with only 1, ovate, basal leaf.

Meadows. Described from "Canada, Kamtschatka." The smaller subsp. **alpìna** Mats. & Toyok. occurs in the alpine zone of Japan.

The bulbs, which contain starch and sugar, were a staple food of the prehistoric natives; the taste is bitter. The bulblets are dug in the fall, dried, and used in stews or powdered into flour.

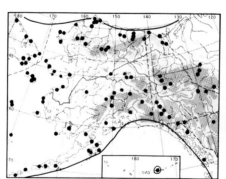 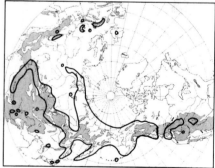

Lloydia Salisb.

1. Llóydia seròtina (L.) Rchb.   Alp Lily

Bulbocodium serotinum L.

Bulb oblong, covered with a membranaceous gray coat from cordlike rootstock; flowers solitary or more rarely 2 together, about 1 cm long, the segments creamy-white, with prominent purple midvein and numerous shorter, lateral purplish nerves; capsule rounded, bluntly triangular in cross section, 3-celled.

Rocky places, polygon tundra, alpine meadows and heaths, to at least 1,800 meters. Described from England and Switzerland.

Subsp. **flàva** Calder & Taylor, which is taller and has petals with yellow base and green or yellow venation, occurs on the Queen Charlotte Islands.

30. LILIACEAE (Lily Family)

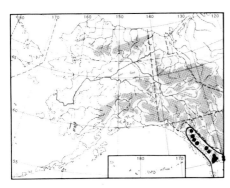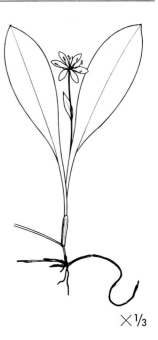

Clintonia Raf.

1. Clintònia uniflòra (Schult.) Kunth Single-Flowered Clintonia
Smilacina borealis var. *uniflora* Schult.; *Smilacina uniflora* (Schult.) Menzies.

Leaves 2–3, obovate to oblanceolate, more or less pubescent; peduncle bractless or with 1–2 bracts; flowers pubescent, white or yellowish-white; the berry blue and about 1 cm in diameter.
Shaded woods. Described from northwest North America.

× ⅓

Smilacina Desf. (False Solomon's Seal)

Flowers in terminal racemose panicle .... 1. *S. racemosa*
Flowers in simple raceme:
  Raceme peduncled; leaves glabrous beneath .... 2. *S. trifolia*
  Raceme nearly sessile; leaves pubescent beneath .... 3. *S. stellata*

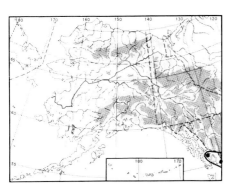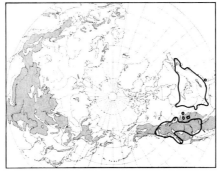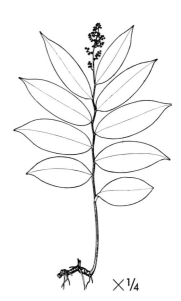

1. Smilacìna racemòsa (L.) Desf.
Convallaria racemosa L.

Rootstock stout, fleshy; stem pubescent above, slightly zigzag; leaves pubescent beneath; inflorescence a many-flowered panicle, with pilose rachis; flowers short-pedicelled; perianth segments pubescent, shorter than stamens.
Shaded woods. Described from Virginia and Canada.
Alaskan specimens, which belong to var. **amplexicaùlis** (Nutt.) S. Wats. (*S. amplexicaulis* Nutt.), have leaves with dilated, clasping petioles.
Circumpolar map indicates range of entire species complex.

× ¼

×½

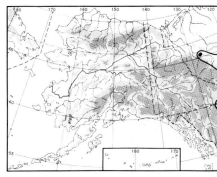
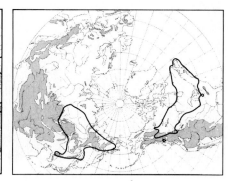

2. Smilacìna trifòlia (L.) Desf.

Convallaria trifolia L.; *Vagnera trifolia* (L). Morong.

Rhizome forking, filiform, sending up both sterile and fertile shoots; leaves 2–4, oblanceolate to elliptic, glabrous; inflorescence a simple, peduncled raceme; flowers long-pedicelled; perianth segments longer than stamens.

Bogs, peaty soil. Described from Siberia.

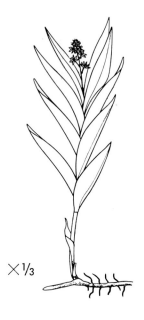
×⅓

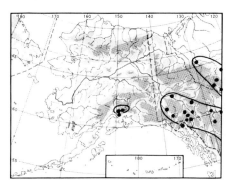
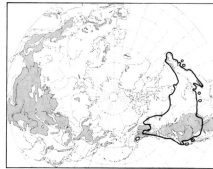

3. Smilacìna stellàta (L.) Desf.

Convallaria stellata L.; *Vagnera stellata* (L.) Morong.

Stoloniferous; leaves sessile, minutely pubescent beneath; inflorescence a simple raceme; flowers long-pedicelled; perianth segments longer than stamens.

Meadows, wet places. Described from Canada. Introduced and partly naturalized in Europe.

Range indicated on circumpolar map includes that of var. **sessifòlia** (Baker) Henders. (*Tovaria sessilifolia* Baker).

30. LILIACEAE (Lily Family)

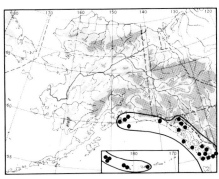

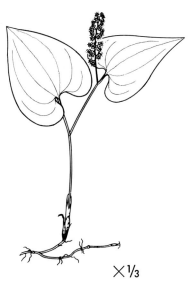

Maianthemum Web. (False Lily-of-the-Valley)

1. Maiánthemum dilatàtum (How.) Nels. & Macbr.

Unifolium dilatatum How.; *Majanthemum bifolium* var. *dilatatum* (How.) Wood; *Unifolium bifolium kamtschaticum* Piper.

Stems from forking, filiform rhizome; stem leaves 2–3, cordate, scabrous; perianth segments obtuse, longer than stamens; the berry red, globose.

Moist woods, meadows, alder thickets. Described from Oregon.

Contains glycosides (derivatives of sugars), active on the heart.

Streptopus Michx. (Twisted-Stalk)

Perianth rotate; stigma on filiform style; stem simple . . . . 2. *S. streptopoides*
Perianth campanulate; style filiform:
   Stem forking; leaves clasping; nodes glabrous . . . . 1. *S. amplexifolius*
   Stem simple; leaves sessile but not clasping; nodes fringed . . . .
   . . . . 3. *S. roseus* subsp. *curvipes*

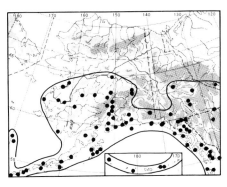
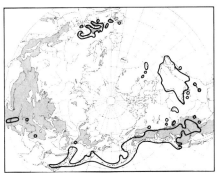
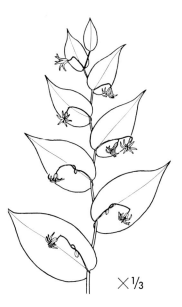

1. Stréptopus amplexifòlius (L.) DC.

Uvularia amplexifolia L.

Stem mostly hispid at base, forking, from short, stout rhizome with thick pubescent roots; peduncles simple or forked; perianth white or yellowish-white (never pink); fruit red.

Moist woods, alder thickets. Described from Europe. Young shoots can be eaten raw; the berries are also edible.

Within area of concern, 4 weakly differentiated varieties occur: var. **denticulàtus** Fassett, with minutely denticulated leaf margin (the only variety common in the Aleutians); var. **americànus** Schult. [subsp. *americanus* (Schult.) Löve & Löve], with leaf margin entire; var. **papillàtus** Ohwi [subsp. *papillatus* (Ohwi) Löve & Löve], with denticulate leaf margin and smaller flowers (Hoonah); and var. **chalazàtus** Fassett, with leaf margin entire and with numerous small papillae on lower side of leaf (Curry). Circumpolar map indicates range of entire species complex; var. **amplexifòlius** occurs only in Europe.

×½

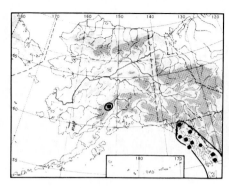
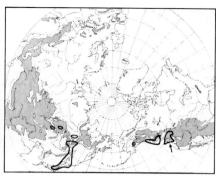

2. Stréptopus streptopoìdes (Ledeb.) Frye & Rigg
Smilacina streptopoides Ledeb.

Stem simple, glabrous, from slender, creeping rootstock; leaves glabrous; flowers 1–5 on recurved pedicels; perianth rotate, the segments twice as long as stamens; style lacking; the berry red.

Moist places, mostly in forests.

Alaskan plant belongs to var. **brévipes** (Baker) Fassett (*S. brevipes* Baker, *S. streptopoides* subsp. *brevipes* Calder & Taylor), doubtfully distinct from the typical plant. *S. streptopoides* described from Ajan, Sitka; var. *brevipes* from the Cascade Mountains of Oregon.

×⅓

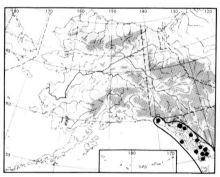
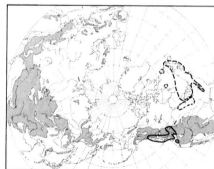

3. Stréptopus ròseus Michx.
subsp. **cúrvipes** (Vail) Hult.

Streptopus curvipes Vail; *S. roseus* var. *curvipes* (Vail) Fassett; *S. roseus* var. *longipes* Fassett.

Stems simple, glabrous, from slender rootstock; margin of leaves finely glandular-pubescent; flowers campanulate, rose-colored; perianth segments minutely glandular-pubescent on inner surface; style 3-cleft; the berry globose, red.

Moist woods. *S. roseus* described from North Carolina and Canada; subsp. *curvipes* from Asulkan Pass (B.C.). Broken line on circumpolar map indicates range of other races.

31. IRIDACEAE (Iris Family)

Iris L.

Bracts herbaceous, not distinctly violet-colored; leaves usually more than 8 mm broad .... 1. *I. setosa* subsp. *setosa*
Bracts scarious, distinctly violet-colored; leaves about 5–8 mm broad ....
.... 2. *I. setosa* subsp. *interior*

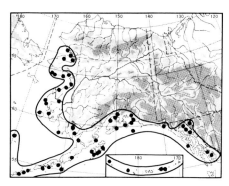
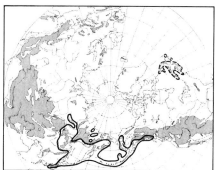
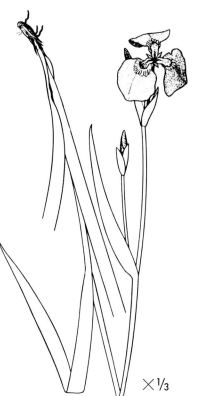

1. Íris setòsa Pall.   Wild Flag
subsp. **setosa**
Iris sibirica with respect to Alaskan plant; *I. arctica* Eastw.

Rootstock short, thick, covered with fibrous remains of disintegrated leaves; stem mostly with 1–2 branches; bracts herbaceous, green or somewhat purplish, as long as pedicels or longer; sepals blue with dark veins (rarely purplish or white), broad, abruptly contracted into short claw; petals small, up to 2 cm long, with narrow blade, abruptly contracted into lanceolate, acute or subulate tip; capsule ovate, obtusely angled. Poisonous, causes vomiting.

Meadows, shores. Described from eastern Siberia.

The form of the petals varies considerably; specimens with dilated, somewhat acute or even obtuse petals occur (var. **platyrhýncha** Hult.). The eastern North American counterpart to this plant is **I. Hoòkeri** Penny, often regarded as a race of *I. setosa*; its range is indicated by the broken line on the circumpolar map.

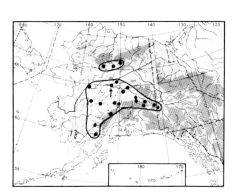
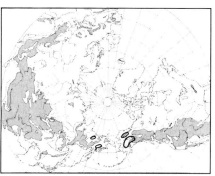

2. Íris setòsa Pall.
subsp. **intèrior** (Anders.) Hult.
Iris setosa var. *interior* E. Anders.

Similar to subsp. *setosa*, but differs in having shorter, more scarious and violet-colored bracts and narrower, less arched leaves.

Meadows, shores; a slightly differentiated inland race.

31. Iridaceae (Iris Family)

Sisyrinchium L. (Blue-Eyed Grass)

Plant darkened in drying; leaves few, short and broad . . . . 2. *S. litorale*
Plant not darkened in drying; leaves more numerous, longer and narrower . . . .
. . . . 1. *S. montanum*

×½

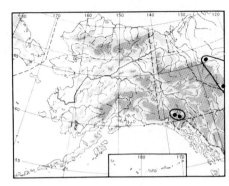 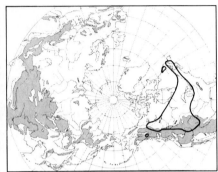

1. Sisyrínchium montànum Greene

Sisyrinchium angustifolium with respect to Yukon and Mackenzie plant.

Tufted, not darkened in drying; leaves 1–3 mm broad; stem simple, wing-margined; spathe single; outer bract narrow and long, with margins united 2–6 mm above base; perianth with mucronulate-aristulate segments, blue-violet; pedicels only slightly exceeding the inner bract.

Moist places. Described from southern Colorado.

×½

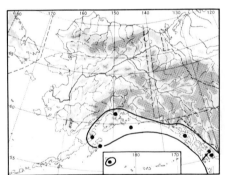 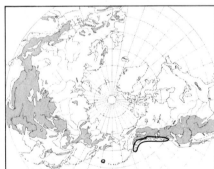

2. Sisyrínchium litoràle Greene

Tufted, darkened in drying; leaves few, short, 3–4 mm broad; stem simple, wing-margined; spathe single; the outer short and broad bract with margins united, the inner as long as pedicels or longer; perianth 12–15 mm long, with subulate-cuspidate segments.

Moist places along shores.

32. ORCHIDACEAE (Orchis Family)

Cypripedium L. (Lady's Slipper)

Plant 15–20 cm tall, with only 2 broad, elliptical basal leaves; flower solitary, scapose:
    Flowers purple-blotched, lateral petals tapering to blunt apex; lip as long as dorsal sepal or shorter .... 1. *C. guttatum* subsp. *guttatum*
    Flowers of different coloring; lateral petals ending in a more or less rounded disk; lip longer than dorsal sepal .... 2. *C. guttatum* subsp. *Yatabeanum*
Plant taller, with several ovate to elliptical stem leaves:
    Sepals shorter than lip; lip obovoid, white to pale purple, 1–1.5 cm long; flowers small .... 5. *C. passerinum*
    Sepals longer than lip; flower larger:
        Lip yellow, purple-veined, slipper-shaped .. 3. *C. calceolus* subsp. *parviflorum*
        Lip white, purple-veined, globose .... 4. *C. montanum*

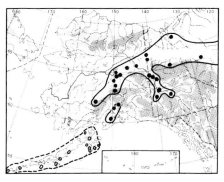
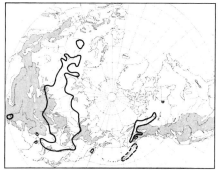
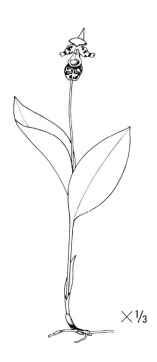

× ⅓

1. Cypripèdium guttàtum Sw.
subsp. guttàtum

Stems from long, slender rootstock, with 2 leaves at about the middle and 2–3 clasping sheaths at base; flowers subtended by foliaceous bract, solitary, blotched with purple; lateral petals ovate, tapering to blunt apex; lip as long as the elliptical dorsal sepal, or shorter.

Meadows, woods. Described from eastern Siberia.

The color of the flower is highly variable, on the coast, where the population is considered to be an introgression between subsp. *guttatum* and subsp. *Yatabeanum*. Flowers can be nearly white, pinkish, blotched with greenish-purple, or purple; collection localities of specimens thus flowered are marked with open circles on the Alaskan map. Broken line on maps indicates range of forms more or less intermediate between subsp. *guttatum* and subsp. *Yatabeanum*.

(See color section.)

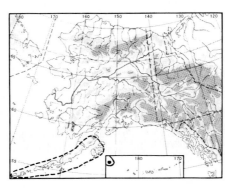
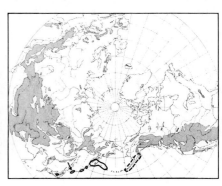

2. Cypripèdium guttàtum Sw.
subsp. **Yatabeànum** (Makino) Hult.
Cypripedium Yatabeanum Makino.

Similar to subsp. *guttatum*, but differs in having flowers blotched with yellowish-green; lateral petals terminate with a more or less rounded disk, slightly broader leaves, and larger lip, usually longer than dorsal sepal.

Meadows, woods.

Thought to form hybrid swarms with subsp. *guttatum* on the southwestern coast. Broken line on maps indicates range of forms more or less intermediate between subsp. *Yatabeanum* and subsp. *guttatum*.

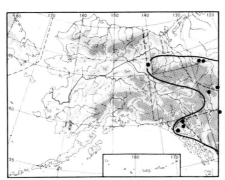
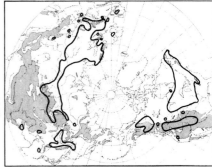

3. Cypripèdium calcèolus L.
subsp. **parviflòrum** (Salisb.) Hult.
Cypripedium parviflorum Salisb.; *C. montanum,* of Fl. Alaska & Yukon, in part.

Stems glandular-pubescent from stout rootstock; cauline leaves 3–5, ciliate on veins and margin; flowers 1–3, in axis of foliaceous bracts, fragrant; sepals usually purplish, the upper ovate to ovate-lanceolate, the lower united; petals 3.5–4 cm long, twisted spirally; lip 2–4 cm long, yellow, purple-veined.

Woods, swamps. *C. calceolus* described from Europe, Asia, and America, subsp. *parviflorum* from Virginia. Circumpolar map indicates range of entire species complex.

32. ORCHIDACEAE (Orchis Family)

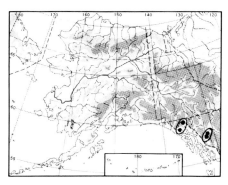 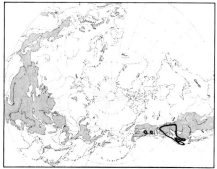 

4. Cypripèdium montànum Dougl.

Stems glandular-pubescent, from stout rootstock; cauline leaves 3–5, glandular-pubescent; flowers 1–3 in axis of foliaceous bracts, sweet-scented; sepals green to purplish, lanceolate, the dorsal long-acuminate, the lateral united almost to apex; petals similar but linear-lanceolate, twisted; lip 2–3 cm long, white with purple veins; capsule ascending, oblong, 2–3 cm long.

Moist woods. Described from northwest North America.

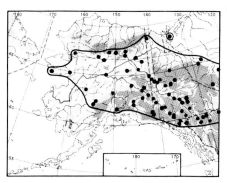 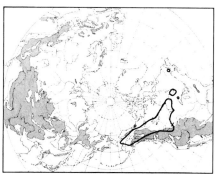 

5. Cypripèdium passerìnum Richards.

Stems from stout, dark rootstock; cauline leaves 3–4, viscid-pubescent on both sides; flowers 1–3, mostly solitary, fragrant; bract foliaceous; sepals about 1.5 cm long, pubescent on outer surface, the lateral free almost to base or nearly united to apex; petals yellowish-green, the upper green on outside, obtuse; lip about 2 cm long, yellowish-green to pale magenta, more or less purple-spotted inside; capsule elliptical, erect, 2–3 cm long.

Woods, bogs. Described from Alberta and Saskatchewan.

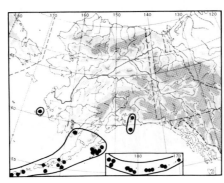
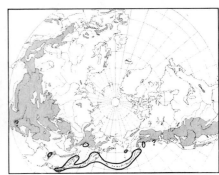

Dactylorhiza Neck. (Key Flower)

1. Dactylorhìza aristàta (Fisch.) Soó
Orchis aristata Fisch.; *O. latifolia* var. *Beeringiana* Cham.

Tuberoids 2–5-lobed; leaves 2–5, with or without dark spots, the lower rounded in apex, the upper somewhat acute; bracts broadly lanceolate, often more or less violet-colored, the lower longer than flowers, the upper shorter; flowers violet to purplish, with darker lip, sometimes with darker spots; sepals lanceolate or narrowly lanceolate with nearly aristate tip; lip obovate with entire, 3-lobed or dentate tip; spur thick, shorter than lip.

Meadows.

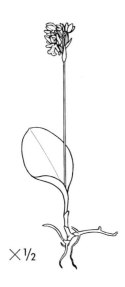
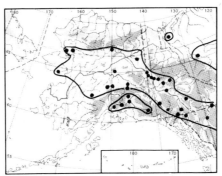
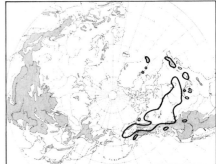

Amerorchis Hult. (Rhizome Orchis)

1. Ameròrchis rotundifòlia (Banks) Hult.
Orchis rotundifolia Banks.

Stoloniferous, with thick roots, glabrous; scape naked, with few, rarely up to 10 flowers in a lax raceme; single orbicular to elliptical subbasal leaf; floral bracts shorter than capsule; sepals and upper petals pale rose-colored; lip white, with few purple spots, 3-lobed, with larger, dilated terminal lobe, more or less 2-cleft in apex; spur slender, much shorter than lip; capsule elliptical.

Moist woods, in McKinley Park to at least 600 meters. Described from Hudson Bay.

32. ORCHIDACEAE (Orchis Family)

Coeloglossum Hartm.

Lower bracts as long as flowers or slightly longer; lower stem leaves close to base of stem .... 1. *C. viride* subsp. *viride*
Lower bracts 2–6 times longer than flowers; lower stem leaves not close to base of stem .... 2. *C. viride* subsp. *bracteatum*

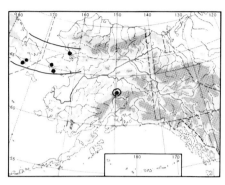
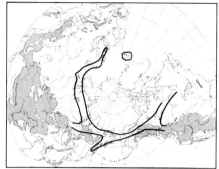

1. Coeglóssum víride (L.) Hartm.   Frog Orchis
subsp. **víride**
   Satyrium viride L.
var. **islándicum** (Lindl.) Schulze
   C. viride subsp. *islandicum* (Lindl.) Sel.; *Peristylus islandicus* Lindl.

Stem from thick, fleshy, palmately lobed tuberoids; lower stem leaves close to base of stem; spike lax, few-flowered; lower bracts as long as flowers or slightly longer, the upper shorter than the flowers; flowers green or brownish; lip 3-toothed or sometimes nearly entire; remaining 5 perianth leaves together form hemispherical hood over lip.

Meadows, mostly in mountains.

The Alaskan plant belongs to the arctic-montane var. *islandicum*, with few-flowered spike; the outer perianth leaves much broader than the inner, and lip entire, or nearly so. *C. viride* described from northern Europe, var. *islandicum* from Iceland.

×⅓

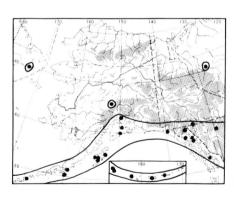
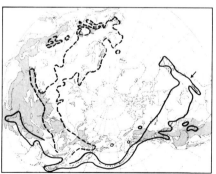

2. Coeglóssum víride (L.) Hartm.
subsp. **bracteàtum** (Muhl.) Hult.
   Orchis bracteata Muhl.; *Habenaria viridis* var. *bracteata* (Muhl.) Gray; *Peristylus bracteatus* Ledeb.

Similar to subsp. *viride*, but usually taller; lower stem leaves not so close to base of stem; bracts very prominent, the lower 2–6 times longer than the flowers.

Woods, meadows. Subsp. *bracteatum* described from Pennsylvania. Broken line on circumpolar map indicates range of subsp. *viride* (see above).

×⅓

32. ORCHIDACEAE (Orchis Family)

Platanthera L. C. Rich. (Bog Orchis)

Spur much longer than ovary:
    Plant 30 cm tall or taller; 2 basal orbicular leaves; flowers 2 cm in diameter or larger .... 1. *P. orbiculata*
    Plant smaller; leaves not distinctly basal, oblong to lanceolate-oblong; flowers much smaller .... 7. *P. tipuloides* var. *behringiana*
Spur as long as ovary, or usually shorter:
    Stem leaves lacking, or reduced to scales:
        Plant 2–5 dm high, with rounded tuberoids; basal leaves 2–3, oblanceolate, withering before flowers open; stem leaves scalelike, 1-nerved; sepals single-nerved .... 9. *P. unalaschcensis*
        Plant smaller, with fleshy, rootlike tuberoids; basal leaves not withering early; stem leaves lacking or single; sepals 3-nerved:
            Basal leaves 2; stem leaf single; lip orbicular; flowers very small .... 8. *P. Chorisiana*
            Basal leaf solitary; stem leaves lacking; lip linear-lanceolate.. 10. *P. obtusata*
    Stem with several normal leaves:
    Lip linear:
        Spike dense; lip parallel to spur, not decidedly dark-colored in herbarium specimens; spur filiform or very slightly clavate:
            Flowers green .... 2. *P. convallariaefolia*
            Flowers greenish-white to white .... See 2. *P. convallariaefolia* var. *dilatatoides*
        Spike lax; lip forming right angle to axis formed by spur and upper sepal, dark-colored in herbarium specimens; spur clavate to saccate:
            Spur saccate or broadly clavate, shorter than lip .... 5. *P. saccata*
            Spur narrow, slightly clavate, longer than lip .... 6. *P. gracilis*
    Lip obtusely triangular, broader at base, or linear but distinctly dilate at base:
        Lip obtusely triangular, not papillate in margin; flowers small, green .... 3. *P. hyperborea*
        Lip linear at tip, but distinctly dilate at base, papillate in margin; flowers larger:
            Flowers greenish; lip somewhat dilate at base .... See 4. *P. dilatata* var. *chlorantha*
            Flowers white; lip strongly dilate at base:
                Leaves 0.5–2.5 cm broad; spur usually shorter than lip:.. 4. *P. dilatata*
                Leaves narrower; spur longer than lip .... See 4. *P. dilatata* var. *angustifolia*

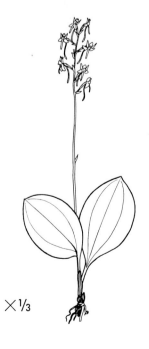

×⅓

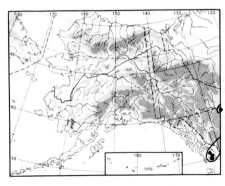

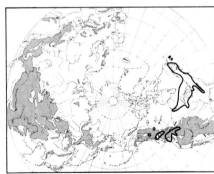

1. Platanthèra orbiculàta (Pursh) Lindl.

Orchis orbiculata Pursh; *Habenaria orbiculata* (Pursh) Torr.; *Lysias orbiculata* (Pursh) Rydb.

Plant with fleshy, elongate tuberoids; scape with 1–5 bracts, leaves 2; raceme lax, open; flowers greenish-white; lip narrowly lanceolate or ligulate; spur cylindrical to slenderly clavate, much larger than ovary.

Dry to moist woods. Described from Pennsylvania, Virginia.

32. ORCHIDACEAE (Orchis Family)

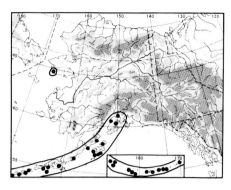
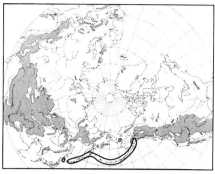

2. Platanthèra convallariaefòlia (Fisch.) Lindl.

Orchis convallariaefolia Fisch.; *Habenaria borealis* var. *viridiflora* Cham.; *Platanthera dilatata* var. *viridiflora* Ledeb.; *Limnorchis convallariaefolia* (Fisch.) Rydb.

Stems thick, coarse, and fleshy, from fleshy, rootlike tuberoids; stem leaves 5–7, crowded at middle and upper part of stem; spike dense and many-flowered; bracts lanceolate, the lower longer than the flowers; flowers greenish; lip linear; spur parallel to lip, slender, filiform, blunt, about as long as lip.

Wet meadows. Described from Kamchatka.

Forms hybrid swarms with *P. dilatata*. Specimens of this introgression series similar to *P. convallariaefolia* but with white or greenish-white flowers represent var. **dilatatoìdes** Hult. and occur where the ranges of the two species overlap.

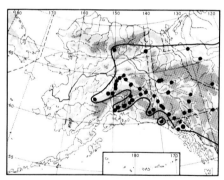
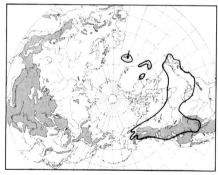

3. Platanthèra hyperbòrea (L.) Lindl.

Orchis hyperborea L.; *Habenaria hyperborea* (L.) R. Br.; *Limnorchis hyperborea* (L.) Rydb.

Stems from fleshy, rootlike tuberoids; leaves oblong-lanceolate to lanceolate; spike few-flowered, cylindrical, open or dense; lower bracts as long as flowers or longer; flowers small, greenish or greenish-yellow; lip obtusely triangular or lanceolate, with dilate base; spur slender, strongly curved; ovary large.

Wet places, peat bogs in lowlands.

Forms hybrid swarms with *P. dilatata* where ranges overlap.

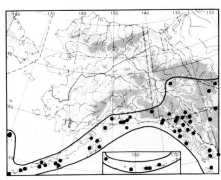
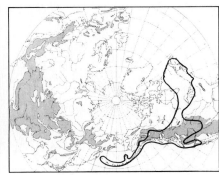

4. Platanthèra dilatàta (Pursh) Lindl.

Orchis dilatata Pursh; *Habenaria borealis* var. *dilatata* Cham.; *H. dilatata* (Pursh) Hook.; *Limnorchis dilatata* (Pursh) Rydb.

Stems from fleshy, rootlike tuberoids; leaves ovate-lanceolate to lanceolate, the lower obtuse; spike many-flowered; flowers white, sweet-scented; lip papillate-margined, rhomboid at base, with linear apex; spur filiform, usually shorter than lip.

Wet meadows, bogs. Described from Labrador.

A narrow-leaved form, with spur longer than lip, is common (var. **angustifòlia** Hook.). *P. dilatata* forms hybrid swarms with *P. convallariaefolia* on the western coast (var. **chloràntha** Hult., with greenish flowers) and with *P. hyperborea* on the North American continent.

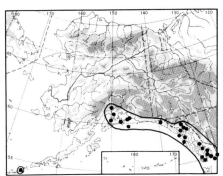
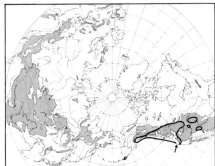

5. Platanthèra saccàta (Greene) Hult.

Habenaria saccata Greene; *Limnorchis stricta* Rydb.; *Platanthera stricta,* of Fl. Alaska & Yukon (not Lindl.).

Stem tall, slender, leafy, from fleshy, rootlike tuberoids; the lower leaves oblanceolate, obtuse, gradually diminishing upward, the upper lanceolate; spike slender, lax, long, many-flowered; flowers green, tinged with purplish-brown; lip linear or oblong-linear, often dark-colored, longer than spur; spur obtuse, saccate, purplish.

Habitat uncertain; probably wet meadows.

32. ORCHIDACEAE (Orchis Family)

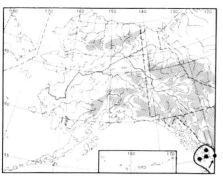

6. Platanthèra gràcilis Lindl.

Platanthera stricta var. *gracilis* (Lindl.) Hult.; *Limnorchis gracilis* (Lindl.) Rydb.; *Habenaria gracilis* (Lindl.) S. Wats.

Similar to *P. saccata*, but spur filiform, longer than lip.

Wet meadows. Range unknown; known only from a few places in the southernmost part of the "Panhandle."

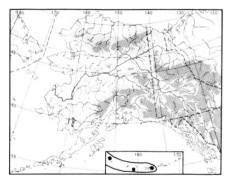

7. Platanthèra tipuloìdes (L. f.) Lindl.
Orchis tipuloides L. f.
var. behringiàna (Rydb.) Hult.

Limnorchis behringiana Rydb.; *Habenaria behringiana* (Rydb.) Ames.

Stem about 10 cm tall, from fleshy, rootlike tuberoids; stem leaves 2, not distinctly basal; spike few-flowered, loose; bracts as long as, or longer than, flowers; flowers green or yellowish-green with narrow, whitish margin, the petals in age often brownish; lip ovate-lanceolate, obtuse; spur slender, filiform, longer than ovary.

Wet places. The Aleutian plant, supposed to be dwarfed by the exposed habitat in the barren islands, represents a variation of the taller Asiatic plant.

×½

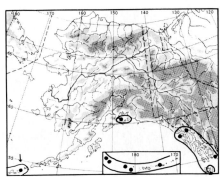
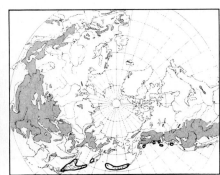

8. Platanthèra Chorisiàna (Cham.) Rchb.

Habenaria Chorisiana Cham.; *Peristylus Chorisianus* (Cham.) Hook.; *Pseudodiphryllum Chorisianus* (Cham.) Nevski; *Limnorchis Chorisiana* (Cham.) J. P. Anders.

Stem from fleshy, rootlike tuberoids, 6–12 cm tall, with bladeless sheath at base and 2 broad leaves at middle, the lower with round apex, the upper somewhat more acute, sometimes also with small bractlike leaf in upper part; spike few-flowered; bracts lanceolate, the lower longer than the flowers, acute; flowers very small, greenish; perianth leaves oblong, ovate, or obovate; lip orbicular, fleshy; spur short, sacklike.

Wet places, *Sphagnum* bogs.

The corresponding Asiatic plant is taller and many-flowered, and grows in meadows and birch forests (var. **elàta** Finet).

×⅓

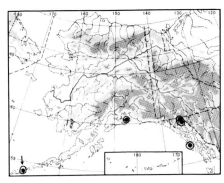
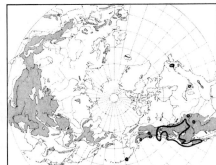

9. Platanthèra unalaschcénsis (Spreng.) Kurtz

Spiranthes unalaschcensis Spreng.; *Habenaria unalaschcensis* (Spreng.) S. Wats.; *H. schischmareffiana* Cham.; *Platanthera schischmareffiana* (Cham.) Hook.; *Piperia unalaschcensis* (Spreng.) Rydb.

Stems scapose, with few, narrow bracts, from 2 ovoid tuberoids; leaves 2–3, oblanceolate, obtuse or somewhat acute, soon withering; spike remotely flowered; flowers green, with disagreeable odor; petals and upper sepals forming hood; lip oblong; spur slender, about as long as lip.

Meadows, woods. Circumpolar map indicates range of *P. unalaschcensis* in a broad sense.

32. ORCHIDACEAE (Orchis Family)

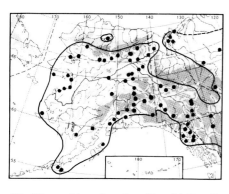
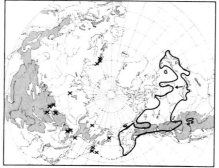

10. Platanthèra obtusàta (Pursh) Lindl.

Orchis obtusata Pursh; *Habenaria obtusata* (Pursh) Lindl.; *Lysiella obtusata* (Pursh) Britt. & Rydb.

Stem extending beyond leaf, scapose, up to 20 cm tall, usually shorter, from somewhat thickened roots; leaf single (rarely 2), basal, blunt; spike slender, remotely flowered; flowers greenish-white, lip linear-lanceolate, deflexed; spur slender, curved, about as long as lip; ovary stipitate.

Mossy forests, wet places, to at least 1,000 meters in the Yukon.

Subsp. **oligántha** (Turcz.) Hult. [*Lysiella obtusata* subsp. *oligantha* (Turcz.) Tolm.; *L. oligantha* (Turcz.) Nevski] occurs in scattered localities in Eurasia. On the circumpolar map the known localities for subsp. *oligantha* are marked with crosses.

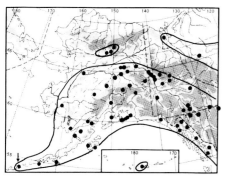
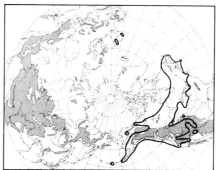
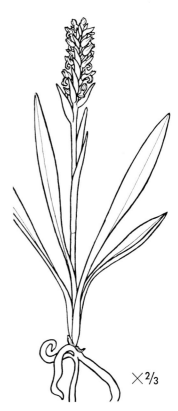

Spiranthes L. C. Rich. (Ladies' Tresses)

1. Spiránthes Romanzoffiàna Cham.

Gyrostachys Romanzoffiana (Cham.) Britt. & Rydb.

Stem leafy in lower part, with bracts above, from fleshy, tuberoid-thickened roots; flowers in 3 spiral ranks; bracts longer than flowers; flowers fragrant, white to creamy; sepals and 2 petals forming hood; lip constricted above middle with terminal, round-to-ovate lobe.

Bogs, marshes, in mountains to at least 1,000 meters.

The European plant is considered by some authors to be specifically different from the American.

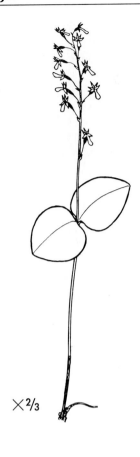

×2/3

Listera R. Br. (Twyblade)

Rachis of inflorescence glabrous; lip deeply cleft into linear, acute lobes ....
.... 3. *L. cordata*
Rachis of inflorescence glandular; lip entire, or shallowly cleft into obtuse lobes:
    Lip with auricles at base, rectangular or obtuse .... 4. *L. borealis*
    Lip not auricled at base, wedge-shaped, with pair of lateral teeth:
        Ovary glabrous; lip 2–5 mm long, not ciliate in margin .... 1. *L. caurina*
        Ovary glandular; lip longer, ciliate in margin .... 2. *L. convallarioides*

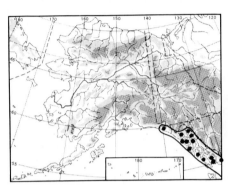
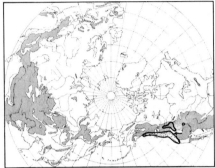

1. Lístera caurìna Piper
Ophrys caurina (Piper) Rydb.

Stem up to 20 cm tall, glandular-pubescent above, from creeping rootstock; leaves 2, ovate, somewhat acute; raceme loose, bracts small, ovate-lanceolate, nearly glabrous; flowers small, yellowish-green; lip glabrous in margin, rounded or retuse at apex, with pair of small teeth near base; ovary glabrous.

Coniferous forests.

×2/3

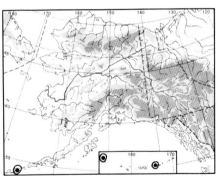
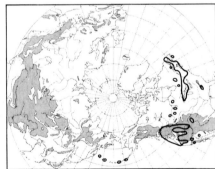

2. Lístera convallarioìdes (Sw.) Nutt.
Epipactis convallarioides Sw.; *Listera Eschscholtziana* Cham.

Stem up to 15 cm tall, glandular-pubescent above, from creeping rootstock; leaves 2, ovate; raceme loose; bracts small, ovate-lanceolate, glandular; flowers yellowish-green; lip ciliated in margin, with pair of triangular teeth, the apex dilate and emarginate; capsule glandular.

Moist places. Described from North America.

The closely related **L. Yatábei** Makino and **L. brévidens** Nevski occur in eastern Asia.

32. ORCHIDACEAE (Orchis Family)

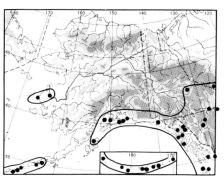

3. Lístera cordàta (L.) R. Br.
Ophrys cordata L.

Usually about 10 (occasionally up to 30) cm tall, very slender; the 2 leaves in the middle of the stem, cordate to cordate-ovate; raceme glabrous, open, few-flowered; lip deeply divided into 2 linear, spreading lobes, with pair of hornlike teeth at base. Some specimens have greenish flowers, others dark purple to purplish-black flowers; intermediates have not been observed.

Moist, mossy places in woods or meadows. Described from northern Europe.

The form of the leaves of the Alaskan plant differs from that of the typical plant in being broader in comparison to length and in having longer pedicels [var. **nephrophýlla** (Rydb.) Hult.; subsp. *nephrophylla* (Rydb.) Löve & Löve; *Listera nephrophylla* Rydb.]. This is especially noticeable in specimens from the mainland coast. (Var. *nephrophylla* is an obscure taxon, perhaps a weak coastal race; plants occurring, for example, in eastern Siberia and in the Yukon are the typical *L. cordata*.)

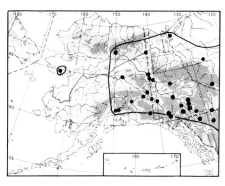
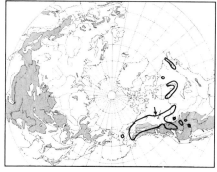

4. Lístera boreàlis Morong
Ophrys borealis (Morong) Rydb.

Stem up to 20 cm tall; leaves 2, ovate or ovate-lanceolate, obtuse; raceme loose, few-flowered, with glandular rachis; flowers green or yellowish-green; lip rectangular or oblong, shallowly 2-cleft, broad and auriculated at base.

Moist woods.

32. ORCHIDACEAE (Orchis Family)

Goodyera R. Br. (Rattlesnake Plantain)

Plant tall; leaves 5–10 cm long; flowers in spiral; lip not distinctly saccate, with incurved margins .... 1. *G. oblongifolia*
Plant shorter; leaves smaller; flowers 1-sided; lip saccate, with recurved margins.. .... 2. *G. repens* var. *ophioides*

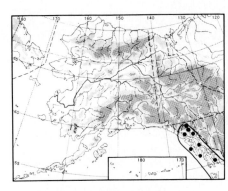 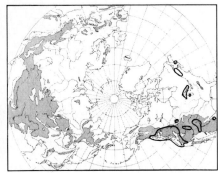

1. Goodyèra oblongifòlia Raf.

Spiranthes descipiens Hook.; *Goodyera descipiens* (Hook.) Hubbard; *G. Menziesii* Lindl.; *Epipactis descipiens* (Hook.) Ames; *Peramium descipiens* (Hook.) Piper.

Stem stout, glandular above, from short rootstock with fibrous roots; leaves in basal rosette; raceme 1-sided, lax; flowers white, tinged with green; perianth segments glandular; petals and dorsal sepals forming hood; lip saccate with long, sulcate beak and involute margins.

Mossy forests. Described from mountains of Oregon.

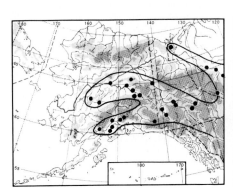 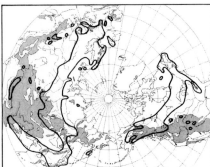

2. Goodyèra rèpens (L.) R. Br.

Satyrium repens L.; *Peramium repens* (L.) Salisb.
var. **ophioìdes** Fern.

Creeping, with slender, forking rootstock; scape pale green, finely pubescent above; leaves in basal rosette, their veins bordered with pale tissue, lacking chlorophyll (var. *ophioides*); raceme 1-sided, lax; flowers yellowish-white; lip inflated, entire, with recurved tip.

Mossy woods. Described from Europe, Siberia. Circumpolar map gives range of entire species complex.

32. Orchidaceae (Orchis Family)

Corallorrhiza Châtelain (Coral Root)

Flowers large, about 20 mm in diameter, purple; lip white with purplish spots; sepals and petals 3-nerved .... 1. *C. maculata* subsp. *Mertensiana*
Flowers small, about 8 mm in diameter, yellowish-green, spotted with purple; sepals and petals 1-nerved .... 2. *C. trifida*

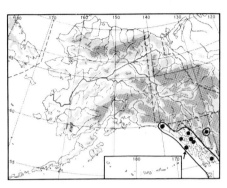
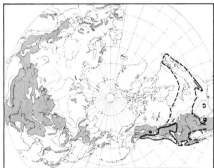

1. Corallorrhìza maculàta Raf.
subsp. **Mertensiàna** (Bong.) Calder & Taylor
Corallorrhiza Mertensiana Bong.

Plant reddish or brownish-purple, erect, glabrous, lacking green leaves, from coral-like underground rhizome; flowers purplish; sepals 3-nerved, linear, 9–12 mm long when fully expanded, widely spreading; lip oblong to broadly obovate, 3-nerved, with a pair of teeth near base.

Moist, coniferous woods. Broken line on circumpolar map indicates range of subsp. **maculàta**.

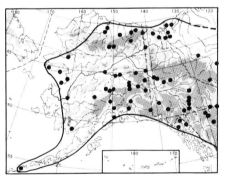
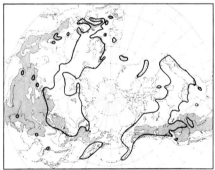

2. Corallorrhìza trífida Châtelain
Corallorrhiza innata R. Br.; *Ophrys corallorrhiza* L.

Rhizome coral-like, brittle, pale, lacking roots; scape yellowish-green, with long, brown sheaths; raceme few-flowered, lax; flowers erect; the fruit reflexed; sepal brownish-green; lip white, with red or purplish spots, notched at tip, and with pair of conspicuous basal lobes.

Wet places, woods, bogs, in mountains to about 900 meters. Described from Europe.

32. ORCHIDACEAE (Orchis Family)

×½

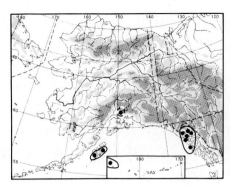 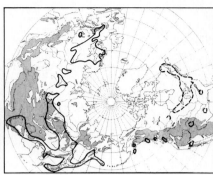

Malaxis Sw. (Adder's Mouth)

1. Maláxis monophýlla (L.) Sw.

Ophrys monophyllos L., *Malaxis diphylla* Cham.; *Microstylis monophyllos* (L.) Lindl.

Pale green, stem from an ovoid corm, surrounded by grayish-white sheaths; leaf solitary or 2, subbasal, long-sheathing; raceme slender; flower yellowish-green, with divergent segments; lip ovate or ovate-triangular, with indistinct basal lobes and slender, pointed tip.

Marshes, bogs. Described from Prussia, Sweden, province of Medelpad.

Typical plant has lip uppermost in the flower, and most Alaskan specimens are of this form. Specimens from Cook Inlet, however (see map), have the lip lowermost in the flower and belong to var. **brachypòda** (Gray) Morris & Ames (*Microstylis brachypoda* Gray). Broken line on circumpolar map indicates range of var. *brachypoda*.

×⅔

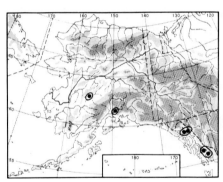 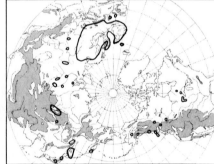

Hammarbya Ktze.

1. Hammarbýa paludòsa (L.) Ktze.

Ophrys paludosa L.; *Malaxis paludosa* (L.) Sw.

Small, inconspicuous plant; scape filiform from a globose corm, surrounded by whitish scales; leaves 2–3 at base of stem; raceme very slender; flowers small, yellowish-green; petals half as long as sepals; lip ovate, about 2 mm long, much shorter than the lateral sepals, curved upward.

Wet *Sphagnum* bogs, quagmires. Described from Sweden.

32. ORCHIDACEAE (Orchis Family) / 33. SALICACEAE (Willow Family)

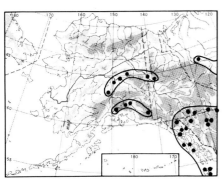
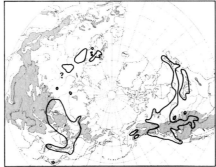

Calypso Salisb. (Calypso)

1. Calýpso bulbòsa (L.) Rchb. f.

Cypripedium bulbosum L.; *Cythera bulbosa* (L.) House; *Cymbidium boreale* Sw.; *Calypso borealis* (Sw.) Salisb.; *C. bulbosa* f. *occidentalis* Holz.; *C. bulbosa* subsp. *occidentalis* (Holz.) Calder & Taylor; *C. occidentalis* (Holz.) Heller.

Stem from a fleshy, oblong or globose corm, with 1 basal, soon-wilting leaf, and 2–3 brown sheaths; flower solitary, large, with mostly pink, acute, perianth leaves and broad, pale, purple-spotted lip. Several color forms occur.

Shady, moist woods. Described from Lapland, Russia, Siberia.

Populus L.

Leaves small, 3–5 cm long, suborbicular; petioles laterally flattened .... 3. *P. tremuloides*
Leaves larger, ovate to rhombic-oblong; petioles not laterally flattened:
  Ovary and capsule glabrous .... 1. *P. balsamifera* subsp. *balsamifera*
  Ovary and young capsule pubescent ...... 2. *P. balsamifera* subsp. *trichocarpa*

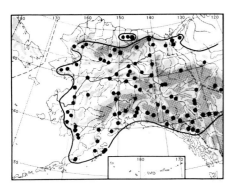
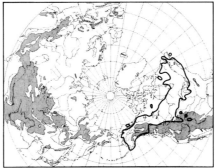

1. Pópulus balsamífera L.   Balsam Poplar, Cottonwood
Populus tacamahacca Mill.
subsp. **balsamífera**

Tree, usually about 15 meters tall, exceptionally up to 30 meters tall and 60 cm in diameter; bark grayish-brown, deeply furrowed; leaves broadly lanceolate to ovate, acuminate or subcordate, paler beneath; overwintering buds large, pointed, covered with yellow, fragrant resin; aments with fringed bracts; capsule glabrous, thick-walled, ovoid.

Common in river valleys and alluvial flats over large parts of interior Alaska and the Yukon; reaches to 1,200 meters in McKinley Park. Described from North America.

Hybridizes rarely with *P. tremuloides*.

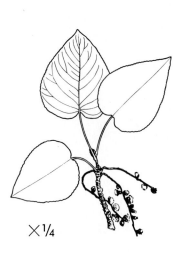 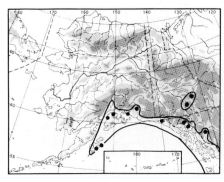 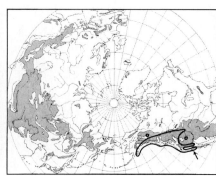

2. Pópulus balsamífera L. Black Cottonwood
subsp. **trichocárpa** (Torr. & Gray) Hult.
Populus trichocarpa Torr. & Gray.

Similar to subsp. *balsamifera*, but young capsules pubescent. Constitutes a coastal race, and forms hybrid swarms with subsp. *balsamifera*.

If young capsules are not available, it seems hardly possible that subsp. *trichocarpa* could be recognized.

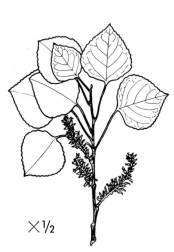 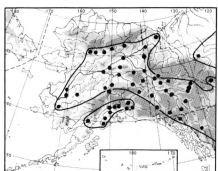 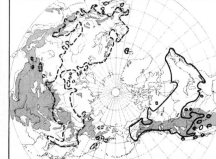

3. Pópulus tremuloìdes Michx. Quaking Aspen, American Aspen

A tree 6–12 meters tall, with smooth, pale, yellowish-green bark; leaves ovate, rhombic, suborbicular, crenate-serrulate, short acuminate; capsules smooth, seeds maturing in early spring.

In open woods and along creeks, to about 1,300 meters in McKinley Park; pioneer on burned slopes. Described from Canada and "Novoboraca" (New Brunswick).

The Eurasiatic counterpart, **P. trémula** L., has irregularly sinuate-dentate, obtuse or somewhat acute leaves, not noticeably overlapping; its range is marked with a broken line on the circumpolar map. Hybrids of *P. tremuloides* with *P. balsamifera* occur, but are rare (1 clump occurs at Lignite; also at Mission Bluff, Eagle).

Salix L. (Willow)

■Young capsules glabrous (in S. *interior* sometimes pubescent at top):
 Adult leaves small, usually not exceeding 2.5 cm in length; decumbent dwarf shrubs or low erect shrubs:
  Margin of leaves entire, or with single glands or teeth:
   Leaves very small, 4–10 mm long; catkins containing only few flowers; entire plant forming dense clumps:
    Leaves oblong, acute .... 6. S. *Dodgeana*
    Leaves round, ciliated .... 7. S. *rotundifolia*
   Leaves larger; plant with elongated branches:
    Styles long, 1.5–2 mm; leaves obovate .... 19. S. *stolonifera*
    Styles short, about 0.5 mm long:
     Catkins long, cylindrical; stipules serrulate .... 14. S. *saxatilis*
     Catkins short; stipules lacking or entire:
      Leaves rounded, with square or subcordate base:
       Leaves large, thick, orbicular; catkins many-flowered; capsules large and plump .... 17. S. *cyclophylla*
       Leaves smaller; catkins few-flowered, attenuated .... 18. S. *nummularia*
      Leaves obovate-lanceolate to elliptical, with cuneate base:
       Leaves dark green, elliptical, mostly blunt, small .. 16. S. *ovalifolia*
       Leaves yellowish-green, obovate-lanceolate, acute, long ...... 21. S. *arctolitoralis*
  Margin of leaves distinctly serrate, at least in lower half of leaf; style short:
   Leaves serrulate only at base, rhombic-obovate, blackened in drying ..... 15. S. *fuscescens*
   Leaves serrulated or crenulated along entire margin:
    Leaves thick, yellowish-green, broader toward apex, serrulate from coarse teeth; branches with blue bloom .... 3. S. *Setchelliana*
    Leaves thin, usually dark green, shiny, usually broadest at middle, crenate; branches without blue bloom .... 34. S. *myrtillifolia*
 Adult leaves longer (in S. *hastata* sometimes as small as 2.5 cm); erect or decumbent shrubs:
  Leaves linear, 3–6 cm long, about 0.5 cm broad, remotely denticulate ...... 54. S. *interior*
  Leaves not linear:
   Catkins sessile, lacking leafy peduncles (or occasionally with very small leaves on short peduncles):
    Young twigs not densely pubescent; stipules broad; young, undeveloped leaves yellowish-red .... 38. S. *padophylla*
    Young twigs densely pubescent; stipules persistent, long, narrow, glandular-margined .... 39. S. *lanata* subsp. *Richardsonii*
   Catkins on long, leafy peduncles:
    ●Margins of leaves distinctly serrate:
     Style 1–2 mm long .... 35. S. *Barclayi*
     Style shorter:
      Capsules on pedicels 3–4 mm long:
       Twigs reddish to brown; bracts persistent ... 32. S. *Mackenzieana*
       Twigs yellowish; bracts soon falling off .... 33. S. *lutea*
      Capsules on shorter pedicels:
       ▲Peduncles and catkins together 4–10 cm, with several large leaves:
        Leaves elliptic to broadly oblanceolate, pubescent on both sides .... 36. S. *commutata*
        Leaves lanceolate, glabrous, shiny above:
         Leaves acuminate at apex, about 4 times as long as broad, or longer .... 55. S. *lasiandra*

Leaves acute, scarcely acuminate, about 3 times as long as broad .... 56. *S. serissima*
▲ Peduncles with catkins shorter, rarely exceeding 3.5 cm; decumbent dwarf shrubs:
Leaves thick, yellowish-green, broader toward apex, serrate from coarse teeth; branches with blue bloom .... 3. *S. Setchelliana*
Leaves thin, usually dark green, usually broadest at middle; branches without blue bloom ...... 34. *S. myrtillifolia*
● Margins of leaves entire, or denticulate only at base:
Style 1–2 mm long .... 35. *S. Barclayi*
Style shorter, about 0.5 mm, or shorter:
Style nearly obsolete; leaves elliptical to oblanceolate, obtuse .... 30. *S. pedicellaris* var. *hypoleuca*
Style about 0.5 mm long:
Leaves pubescent on both sides .... 36. *S. commutata*
Leaves glabrous on both sides:
Leaves lanceolate or broadly lanceolate to ovate-lanceolate; plant with serrate stipules; erect low shrub .. 31. *S. hastata*
Leaves elliptic or obovate-cuneate; low dwarf shrub:
Branches thick, woody; leaves obovate-cuneate .... 12. *S. sphenophylla*
Branches slender, elongate; leaves elliptical .... 21. *S. arctolitoralis*
■ Young capsules pubescent (mature capsules sometimes glabrescent):
Catkins on long, leafless peduncles; leaves orbicular or broadly obovate, dark green above and strongly reticulate below .... 1–2. *S. reticulata*
Catkins on leafy peduncles or sessile:
Leaves serrate along entire margin (in S. *boganidensis* sometimes nearly entire):
Leaves small, 1–2 cm long, sharply serrate; numerous skeletonized leaves at base of plant .... 8. *S. tschuktschorum*
Leaves larger:
Leaves obovate or elliptical, glabrous beneath; peduncles with large leaves .... 37. *S. Chamissonis*
Leaves lanceolate or oblanceolate:
Leaves slightly pubescent when young, glabrescent in age .... 46. *S. boganidensis*
Leaves minutely sericeous beneath ...... 53. *S. arbusculoides*
Leaves with entire or indistinctly toothed margin:
▼Adult leaves glabrous on both sides, or essentially so:
✶ Plant a decumbent dwarf shrub:
Leaves small, rarely exceeding 1.5 cm in length:
Plant with pale, thin, subterranean branches, lacking marcescent leaves at base; leaves pale below .... 4. *S. polaris* subsp. *pseudopolaris*
Plant base covered by marcescent leaves; leaves green on both sides .... 5. *S. phlebophylla*
Leaves larger:
Bracts light brown, oblong or pointed; capsules light brown, with styles 1–1.5 mm long:
Leaves large, light green, 3–4 cm long, oblanceolate, cuneate at base; catkins 4–5 cm long, on long pedicels with several large leaves .... 11. *S. arctica* subsp. *torulosa*
Leaves smaller, about 2.5 cm long, elliptical to obovate-elliptical; catkins about 2 cm long on small-leaved peduncles:
Basal gland broad, short .... 13. *S. arctophila*
Basal gland slender, longer .... 20. *S. hebecarpa*
Bracts orbicular or rounded, in age becoming black:
Leaves orbicular or obovate, on long petioles; catkins large, thick, on long, leafy peduncles; capsules short peduncled, with styles up to 2 mm long .... 10. *S. arctica* subsp. *crassijulis*
Leaves narrower, on shorter petioles; catkins smaller, short-peduncled; capsules sessile, with shorter styles .... 9. *S. arctica* subsp. *arctica*

33. SALICACEAE (Willow Family)

✱ Plant an erect shrub or small tree:
    Capsules on pedicels 2–5 mm long ..... 45. *S. depressa* subsp. *rostrata*
    Capsules on shorter pedicels, or sessile:
        Catkins on leafy peduncles...... See 27. *S. niphoclada* var. *Muriei*
        Catkins sessile:
            Stipules not persistent after growing season ....
            .... 47. *S. phylicifolia* subsp. *planifolia*
            Stipules persistent, prominent .... 48. *S. pulchra*
▼ Adult leaves distinctly pubescent below:
    Adult leaves glabrous above:
        Leaves densely tomentose below, from curled, dull, white hairs:
            Young twigs glabrous, covered with blue bloom ....
            .... 42. *S. alaxensis* subsp. *longistylis*
            Young twigs densely tomentose:
                Petioles thinly pubescent; midvein of leaves not white-woolly above .... 41. *S. alaxensis* subsp. *alaxensis*
                Petioles densely white-woolly; midvein of leaves white-woolly above .... 43. *S. Krylovii*
        Leaves pubescent below, from straight, white hairs:
            Leaves glaucous and thinly pubescent below, glabrate in age ....
            .... 50. *S. athabascensis*
            Leaves not glaucous, more pubescent below (glabrescent forms of the *S. glauca* group):
                Shrub upright; leaves small, 1–3 cm long, pubescent below, with straight, white hairs; catkins short, narrow; bracts large, round, light brown .... 24. *S. glauca* subsp. *desertorum*
                Shrub decumbent; leaves usually larger, pubescent only on nerves below .... 23. *S. glauca* subsp. *callicarpaea*
    Adult leaves pubescent on both sides:
        Shrub decumbent, with creeping branches; bracts ovate, nearly black at tip .... 29. *S. reptans*
        Shrub upright; bracts mostly narrower, not black at tip:
            Leaves silvery-tomentose or satiny below, thinly pubescent above:
                Twigs pruinose .... 49. *S. subcoerulea*
                Twigs not pruinose .... 52. *S. sitchensis*
            Leaves not silvery or satiny below:
                Catkins sessile or with very short, leafy peduncles:
                    Leaves, when young, white-tomentose on both sides, oblong to lanceolate, acute .... 44. *S. candida*
                    Leaves short-pubescent, often with reddish hairs, oblanceolate, obtuse or acute .... 51. *S. Scouleriana*
                Catkins on long, leafy peduncles:
                    Capsules on slender pedicels, 2–5 mm long ....
                    .... 45. *S. depressa* subsp. *rostrata*
                  Capsules short pedicellated or sessile:
                      Leaves sessile or with very short petioles:
                        Catkins spherical or short-oblong .... 26. *S. brachycarpa*
                      Catkins cylindrical, more than twice as long as thick; leaves narrow:
                          Leaves sessile, grayish-green, usually acute ....
                          .... 27. *S. niphoclada* var. *niphoclada*
                        Leaves with short petioles, yellowish-green, blunt ....
                          .... 28. *S. niphoclada* var. *Mexiae*
                    Leaves with petioles more than 2 mm long:
                      Leaves white-tomentose below; stipules glabrous, persistent, glandular-margined; catkins stout ....
                      .... 40. *S. Barrattiana*
                    Leaves not white-tomentose below; stipules not glandular-margined; catkins smaller:
                        Leaves grayish-pubescent, mostly acute; young twigs densely grayish-villous-pubescent; capsules densely gray-hairy .... 22. *S. glauca* subsp. *acutifolia*
                      Leaves yellowish-green, often glabrescent above; young twigs less densely pubescent; capsules glabrescent
                      .... 25. *S. glauca* subsp. *glabrescens*

33. Salicaceae (Willow Family)

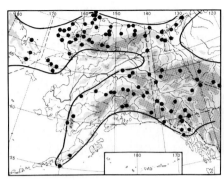
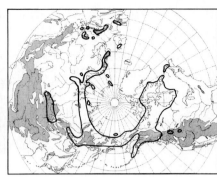

1. Sàlix reticulàta L.
subsp. **reticulàta**

Netted Willow

Prostrate trailing shrub; young twigs greenish-brown, glabrous; leaves dark green, leathery, stiff, long-peduncled, with raised reticulate nerves below and revolute margin; leaves extremely variable in form and pubescence, oblong-obovate or rounded, varying in size from about 5 mm in diameter to 60 mm (var. **gigantifòlia** Ball), and in being pubescent on both sides, with long, silky, white hairs (f. **villòsa** Kitamura, rare), or glabrous on both sides (var. **semicálva** Fern.); scales dark red; capsules pubescent, with short style and purplish stigmas; filaments pubescent.

Common on both dry and moist areas of the tundra, and in mountains to over 2,000 meters. Described from Lapland and Switzerland.

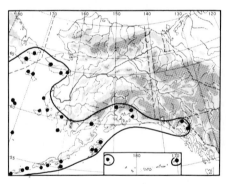

2. Sàlix reticulàta L.
subsp. **orbiculàris** (Anderss.) Flod.
Salix orbicularis Anderss.

Similar to subsp. *reticulata*, but leaves more rounded or even cordate, glabrous or nearly so, with long, glabrous petioles and glabrous branches.

A coastal race, poorly delimited, but obvious if many specimens from the coast are compared with inland specimens. Described from Kamchatka.

Subsp. **glabellicárpa** Argus, possibly a hybrid, with oblong pistillate bracts and glabrous or partly pubescent capsules, occurs in the Queen Charlotte Islands.

33. SALICACEAE (Willow Family)

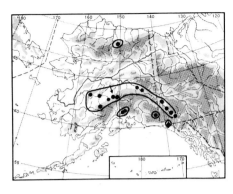 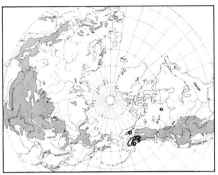 

× ⅓

3. Sàlix Setchellàna Ball
Salix aliena Flod.

Prostrate shrub, pruinose, glabrate in age; leaves bluish-green, thick, leathery, obovate to oblanceolate, blunt, serrated with coarse teeth; catkins on leafy peduncles; bracts yellowish, ciliate; capsules glabrous, with style 0.2–0.3 mm long; stamens hairy.

Gravel bars, shores, sandy slopes. A very characteristic species, endemic to our area.

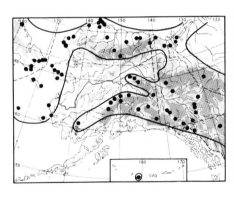 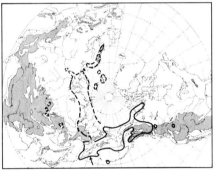 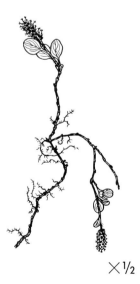

4. Sàlix polàris Wahlenb.
subsp. pseudopolàris (Flod.) Hult.

Salix venusta Anderss. not Host; *S. pseudopolaris* Flod.; *S. polaris* var. *selwynensis* Raup.

Dwarf shrub, with greenish or greenish-brown, more or less subterranean branches; leaves rounded to obovate or oblong, entire in the margin, often emarginate, ciliated with long, white, wavy hairs, shiny green on both sides; catkins obovate or prolonged; bracts ovate or obovate, dark brown with long, white hairs; capsules with thin, grayish pubescence, sometimes nearly glabrous (var. **glabràta** Hult.), greenish or purplish, with a slender style, about 1.5 mm long.

Below snow beds in mountains, to at least 2,000 meters. *S. polaris* described from Lapland.

Large plants with distinctly obovate leaves and long, many-flowered catkins are considered to be hybrids with the *S. arctica* series (nos. 9–11). Broken line on circumpolar map indicates range of subsp. **polàris**.

× ½

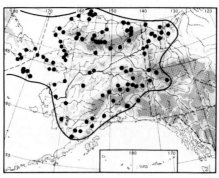 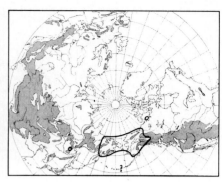

×2/3

5. Sàlix phlebophýlla Anderss.

Salix palaeoneura Rydb.

Dwarf shrub forming mats; leaves somewhat pubescent beneath when young, glabrous in age, obovate-lanceolate, acute, shiny, with entire margin and veins prominent on both sides, persistent for at least 3 years, the old, skeletonized leaves covering the basal parts; catkins short, upright; bracts rounded to obovate, black, with very long, white hairs; capsules pubescent, sometimes glabrescent in age, with short styles; stamens 2, filaments glabrous.

Arctic and alpine lichen and moss tundra, windswept ridges to 1,750 meters in McKinley Park. Described from northwest North America, Cape Mulgrave, Arctic coast.

Hybridizes with S. *rotundifolia* (no. 7).

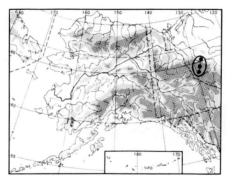 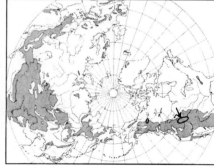

×2/3

6. Sàlix Dodgeàna Rydb.

Similar to S. *phlebophylla,* but with few-flowered, often only 2-flowered, catkins and glabrous capsules.

Alpine tundra, on limestone.

33. SALICACEAE (Willow Family)

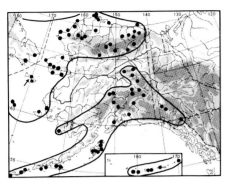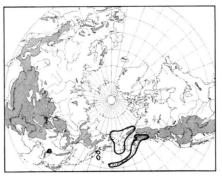

7. Sàlix rotundifòlia Trautv.

Salix polaris var. *leiocarpa* Cham.; *S. leiocarpa* (Cham.) Cov.; *S. fuscescens* var. *reducta* Ball.

Dwarf shrub forming mats, with very thin, somewhat pubescent annual shoots; leaves small, leathery, round or ovate, with obtuse or somewhat cordate base, entire in the margin or nearly so, with lateral veins raised on both sides and prominent veinlets; catkins very short, few-flowered; bracts obovate, yellowish-brown, glabrous or with few, long hairs; capsules glabrous, with style about one-third its length.

Arctic and alpine lichen tundra, rocky places, to at least 2,000 meters. Hybridizes (or perhaps introgrades) with S. *phlebophylla* (no. 5) and S. *ovalifolia* (no. 16).

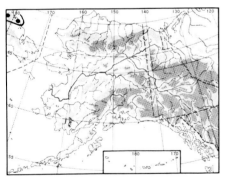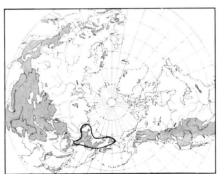

8. Sàlix tschuktschòrum Skvortz.

Salix berberifolia of authors.

Glabrous dwarf shrub, densely covered with marcescent leaves at base; leaves elliptical, shiny, with prominent veins and serrated or spinulose margin; catkins on leafy peduncles, up to 4 cm long; bracts brown, with long white hairs; capsules glabrous or pubescent. Similar to S. *phlebophylla* (no. 5), but easily recognized by the serrated leaves.

Arctic and alpine lichen and moss tundra. Described from Chukchi Peninsula.

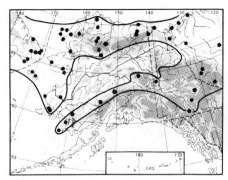 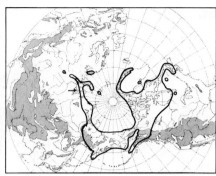

9. Sàlix árctica Pall.   Arctic Willow
subsp. **árctica**

Dwarf shrub, with thick, trailing, glabrous branches; leaves ovate to obovate, entire in the margin, dark green above and paler beneath, somewhat pubescent when young, later glabrescent; catkins 2–3 cm long, borne on long, leafy, peduncles at maturity; bracts broad, blunt, dark, with long, white hairs; capsules pubescent, in age often glabrescent, with styles about 1 mm long; stamens 2, filaments glabrous.

Dry tundra, mostly in mountains.

The *Salix arctica* group is extremely variable, owing to a large extent to introgression between the 3 races subsp. *arctica*, subsp. *crassijulis*, and subsp. *torulosa*, but also to hybridization or introgression with several other, not very closely related species, including *S. pulchra*, *S. Barclayi*, *S. Richardsonii*, *S. ovalifolia*, *S. stolonifera*, *S. rotundifolia*, *S. phlebophylla*, *S. polaris* subsp. *pseudopolaris*, and *S. glauca*. *S. glacialis* Anderss. is considered to be the hybrid *S. arctica* × *ovalifolia*; it differs from *S. ovalifolia* in having more or less pubescent capsules, and occurs where the 2 species overlap.

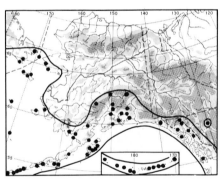 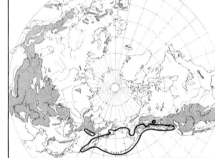

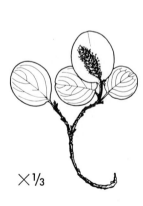

10. Sàlix árctica Pall.
subsp. **crassijùlis** (Trautv.) Skvortz.
Salix crassijulis Trautv.; *S. Pallasii* Anderss.

Similar to subsp. *arctica*, but more sturdy; branches and leaves more pubescent, especially when young; leaves more long-petiolated; catkins larger; styles longer and capsules practically sessile.

Described from Kamchatka, Unalaska. Forms introgression with subsp. *arctica* where ranges overlap.

(See color section.)

33. Salicaceae (Willow Family)

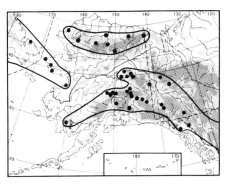
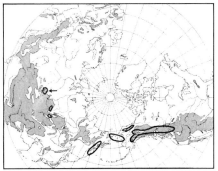

11. Sàlix árctica Pall.
subsp. **torulòsa** (Trautv.) Hult.
  Salix torulosa Trautv.; *S. sphenophylla* subsp. *pseudotorulosa* A. Skvortz. in herb.

  Similar to subsp. *arctica* and subsp. *crassijulis*, but young twigs light brown; leaves light green, narrowly oblanceolate, glabrous on both sides; capsules light brown, narrow and acute, somewhat short-pubescent when young, but glabrescent in age; style 1–1.5 mm long.
  Similar also to *S. sphenophylla*, which differs in having young capsules completely glabrous. Represents perhaps an introgression between that plant and *S. arctica*.

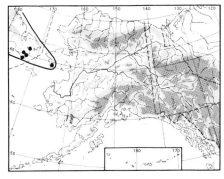
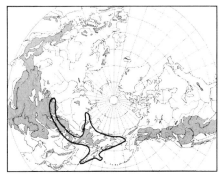

12. Sàlix sphenophýlla A. Skvortz.
  Salix cuneata Turcz. not Nutt.; *S. arctica* var. *nervosa* subvar. *cuneata* (Turcz.) Anderss.; *S. cuneatifolia* Flod. in herb.

  Prostrate shrub; twigs glabrous; leaves long-petiolated, obovate, blunt or short-pointed, entire or somewhat serrated in the margin, dull green above, glaucous beneath, slightly pubescent when young, glabrous in age, with prominent nerves, especially beneath; catkins erect, 3–4 cm long, on pubescent, leafy peduncles; bracts elliptical, brown, with long hairs; capsules ovate-conical, brown or purplish, glabrous, with style 0.6–1.5 mm long.
  Stony places in mountains. Described from eastern shore of Lake Baikal.
  Closely related to *S. arctica* subsp. *torulosa*, but differing in having completely glabrous capsules.

×⅓

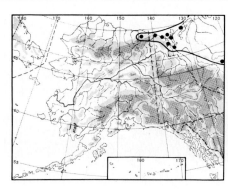
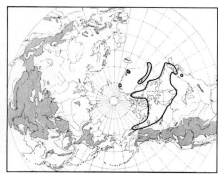

13. Sàlix arctòphila Cockerell

Prostrate shrub, with dark, trailing branches; leaves dark green and shiny above, pale bluish-green below, elliptic to obovate, rounded or somewhat acute in apex, glabrous, entire in the margin or essentially so; catkins erect on leafy peduncles; bracts obovate, purplish to blackish toward tip, long-haired; capsules often reddish, pubescent, or [f. **leiocárpa** (Anderss.) Fern. (*S. arctica* subsp. *groenlandica* var. *leiocarpa* Anderss.)] glabrous, with styles 1–1.5 mm long; gland short and broad, about half as long as pedicels.

Wet places on tundra and in mountains. Described from eastern and western Greenland.

×½

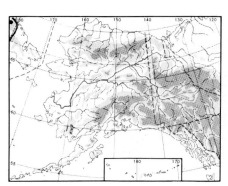
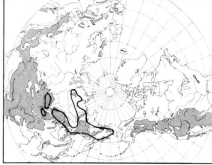

14. Sàlix saxàtilis Turcz.

Salix fumosa Turcz.

Shrub up to 50 cm tall; young twigs pubescent, glabrescent in age; stipules ovate to ovate-lanceolate, serrated; leaves obovate, blunt, somewhat serrulated or entire, green above, distinctly glaucous below, glabrous in age, blackened in drying; catkins up to 7 cm long; bracts small, obovate, blunt, dark brown, with long, white hairs; capsule about 4 mm long, glaucous, with style about 0.5 mm long.

Tundra, riverbanks. Described from Tunkinsk district, Baikal.

Similar to *S. ovalifolia* (no. 16), but differing in having serrated stipules.

33. Salicaceae (Willow Family)

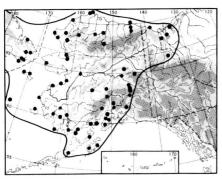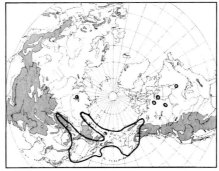

15. Sàlix fuscéscens Anderss.

Salix arbutifolia and *S. rhamnifolia*, with respect to Alaskan plant.

Trailing dwarf shrub with glabrous twigs; leaves obovate to elliptical, usually obtuse, or very short-pointed, dark green above, with a few glandular-tipped teeth at base; catkins broad and short on leafy peduncles, with sparsely distributed capsules; scales obovate or obovate-lanceolate, brown; capsules long-attenuated, abruptly truncate in apex, with very short and thick styles and stigmas.

Wet meadows, muskeg, tundra bogs, chiefly inland. Described from Lake Baikal, eastern Siberia, Kamchatka.

Hybridizes with *S. ovalifolia*.

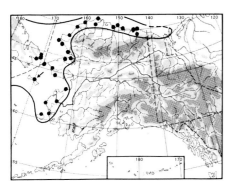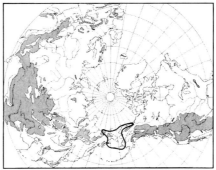

16. Sàlix ovalifòlia Trautv.

Salix flagellaris Hult., with respect to type.

Dwarf shrub, with slender, creeping or subterranean branches; stipules lacking; leaves obovate to elliptical, blunt or somewhat acute, pale beneath, entire in the margin, slightly white-haired in lower margin; catkins short and thick; bracts obovate or rounded, brown, with long, white hairs; capsules glabrous, often glaucous, with broad, round base, abruptly contracted to narrow, blunt apex; style about 0.5 mm long, slender.

Salt marshes, shores.

Similar to *S. cyclophylla*, but differs in the form of the leaves and the capsule; also similar to *S. fuscescens*, from which it differs in having slender branches and entire-margined leaves, and in having a coastal habitat. Hybridizes with these species, with *S. rotundifolia*, and with *S. arctica* (*S. glacialis* Anderss.). Total range very uncertain.

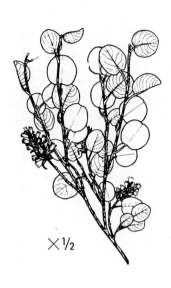
×½

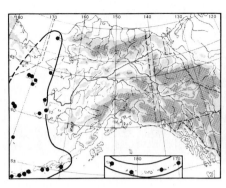 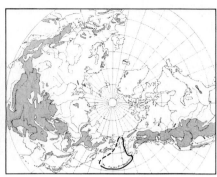

17. Sàlix cyclophýlla Rydb.

Salix ovalifolia cyclophylla Ball; *S. ovalifolia* of Fl. Alaska & Yukon, for the most part.

Dwarf shrub, with stout branches, somewhat pubescent when young, glabrescent in age; leaves orbicular, often subcordate to obovate-oblong, thick, glaucous beneath, entire in the margin, glabrous or somewhat pubescent in lower margin; catkins short and thick; bracts obovate, obtuse, brown, with long, white hairs; capsules glabrous, large, plump, distinctly truncate and glaucous, with style up to 1 mm long.

Mountain tundra. Described from St. Paul Island, Hall Island, and Cape Vancouver.

Hybridizes with *S. ovalifolia*.

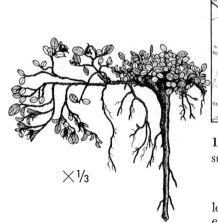
×⅓

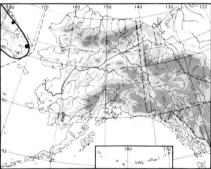 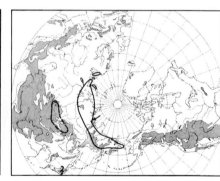

18. Sàlix nummulària Anderss.
subsp. **tundrícola** (Schljakov) Löve & Löve
Salix tundricola Schljakov.

Dwarf shrub, with thick central root and long subterranean, not rooting, branches; leaves ovate, with 4–6 pairs of nerves; rounded or somewhat cordate base and entire or somewhat obtusely denticulated margin, glabrous or sparsely long-pilose below; catkins few-flowered on leafy peduncles; capsules attenuated, sessile, with style 0.5 mm long.

Stony tundra, polygon tundra.

The similar subsp. **nummulària**, described from Transbaikalia, occurs in central Asia; its range is indicated by the broken line on the circumpolar map.

33. Salicaceae (Willow Family)

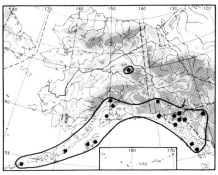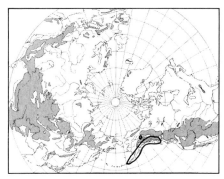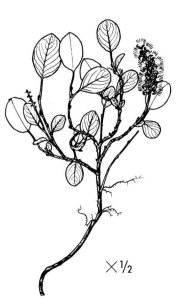

19. Sàlix stolonífera Cov.

Dwarf shrub, with thick branches; leaves obovate or elliptical, blunt or somewhat acute, entire in the margin, slightly white-haired, pale below; catkins short and broad, bracts obovate or rounded, brown, white-haired; capsules glabrous, with style about 2 mm long.

Alpine tundra.

Similar to *S. ovalifolia* (no. 16, above), but differs in having long styles and thick branches and in having an alpine habitat; also similar to *S. fuscescens* (no. 15, above), but differs in having long slender styles and entire leaf margins.

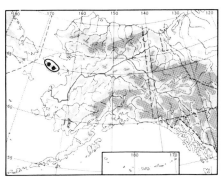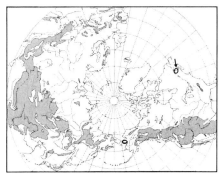

20. Sàlix hebecárpa Fern.

Prostrate dwarf shrub; twigs glabrous; leaves ovate, glabrous, entire in the margin, paler beneath; catkins on pubescent, leafy peduncles; bracts ovate, obtuse, light brown, with long, white hairs; capsules short-pedicellated, somewhat pubescent when young, later glabrescent, light brown with short style.

Alpine regions. Known only from 2 fragmentary specimens, one from Teller and one from Salmon Lake, north of Nome. Further collections desirable. Identification necessarily tentative.

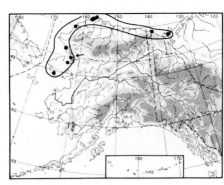
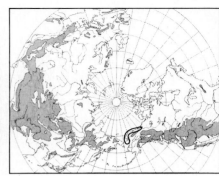

21. Sàlix arctolitoràlis Hult.

Salix arctophila Cockerell × *ovalifolia* Trautv.?

Prostrate shrub, with long, trailing, light-brown branches; leaves yellowish-green above, paler below, ovate to obovate-lanceolate, glabrous, entire in the margin; catkins on leafy peduncles; bracts small, light-colored to purplish, ciliated, shorter than, or as long as, the pedicels; capsules glabrous, greenish to purplish, with style about 0.5 mm long.

Chiefly along the shore.

Similar to S. *arctophila* (no. 13, above), but yellowish-green; capsules always glabrous, style shorter.

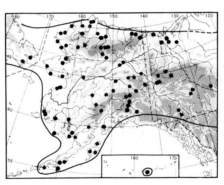
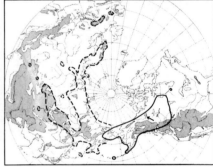

22. Sàlix glaùca L.
subsp. **acutifòlia** (Hook.) Hult.

Salix villosa var. *acutifolia* Hook.; S. *glauca* var. *acutifolia* Schneid.; S. *Seemannii* Rydb.; S. *glauca* var. *Seemannii* (Rydb.) Ostenf.; S. *glauca* "Bering Sea phase" and "Western phase" of Argus (for the most part).

Shrub up to 1.5 meters tall; young twigs densely grayish-villous, pubescent; leaves grayish-pubescent on both sides, elliptical, obovate to oblanceolate, mostly acute, entire in the margin or somewhat glandular-denticulated at base; catkins cylindrical, densely flowered, with villous rachis; bracts oblong to obovate-oblong, obtuse, darker in apex, gray-pubescent on both sides; capsules subsessile, ovate-oblong to conical, densely grayish-tomentose, with distinct style.

Forms thickets along rivers, both on tundra and in mountains.

Broken line on circumpolar map indicates range of subsp. **glaùca**.

33. SALICACEAE (Willow Family)

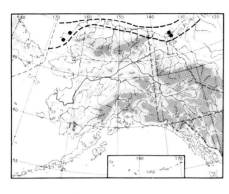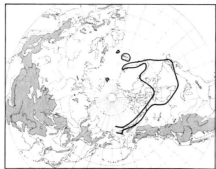

23. Sàlix glaùca L.
subsp. **callicarpáea** (Trautv.) Böcher

Salix callicarpaea Trautv.; *S. cordifolia* Pursh var. *callicarpaea* (Trautv.) Fern.; *S. cordifolia* subsp. *callicarpaea* (Trautv.) Löve.

Similar to subsp. *acutifolia*, but prostrate shrub with leaves glabrate or somewhat silky on nerves below, oblong to obovate.

The arctic representative of the *S. glauca* group. Total range incompletely known. Described from Labrador.

×½

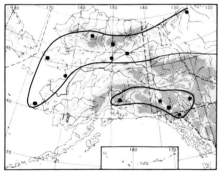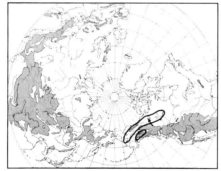

24. Sàlix glaùca L.
subsp. **desertòrum** (Richards.) Anderss.

Salix desertorum Richards.

Similar to subsp. *acutifolia*, but leaves nearly glabrous on upper side, or with hairs on midvein in younger leaves; catkins small, short, sessile, very tomentose; capsules small and light brown; bracts large; round in apex.

A somewhat obscure taxon; habitat unknown. Range indicated on map is tentative.

×½

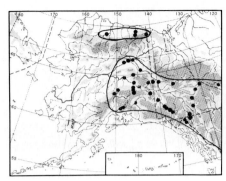
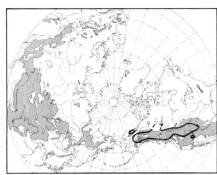

25. Sàlix glaùca L.
subsp. **glabréscens** (Anderss.) Hult.

Salix glaucops var. *glabrescens* Anderss.; *S. glauca* var. *glabrescens* Schneid.; *S. glauca* × *pseudomonticola* of Fl. Alaska & Yukon, in part; *S. glauca* var. *Alicae* Ball; *S. glauca* "Rocky Mountain phase" of Argus.

Similar to subsp. *acutifolia*, but differs in having young twigs less densely villous or subglabrous in age; leaves glabrescent, olivaceous, more or less shiny, indistinctly denticulated, elliptic-lanceolate, often glabrous above, yellowish-green; capsules glabrescent.
Described from the Rocky Mountains.
Forms introgression with subsp. *acutifolia* where ranges overlap.

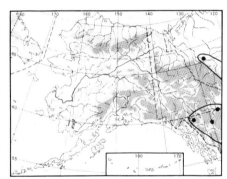
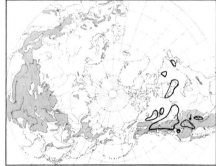

26. Sàlix brachycárpa Nutt.

Shrub up to 1 meter tall; twigs grayish-villous; leaves sessile or nearly so, ovate, obovate to oblanceolate, entire in the margin, densely gray-hairy, sometimes glabrescent; catkins on leafy peduncles, nearly spherical to short-oblong, usually not longer than 2 cm; bracts yellowish to straw-colored or brownish, short-pubescent; capsules sessile, grayish-woolly, with very short style; stamens 2, filaments hairy toward base.
Wet places, riverbanks.
Similar to *S. glauca* and *S. niphoclada*, but differs in having short, thick catkins.

33. Salicaceae (Willow Family)

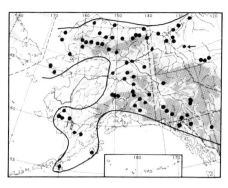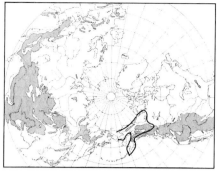

27. Sàlix niphoclàda Rydb.

?*Salix lingulata* Anderss.; *S. brachycarpa* subsp. *niphoclada* (Rydb.) Argus.

var. **niphoclàda**

Usually upright shrub, up to 1 meter tall; twigs grayish-villous; leaves grayish-green, sessile, narrowly oblong or obovate to lanceolate, usually acute or somewhat acute, entire in the margin, paler on lower side, grayish-pubescent on both sides, in age sometimes glabrescent above; catkins on leafy peduncles, cylindrical, up to 4 cm long; bracts brownish, short-pubescent; capsules sessile, grayish-woolly, with short style.

Along rivers, in wet meadows; in the St. Elias Mountains to at least 2,000 meters.

Leaves of var. **Mùriei** (Hult.) Hult. (*S. Muriei* Hult.) are glabrous, with bluish bloom below.

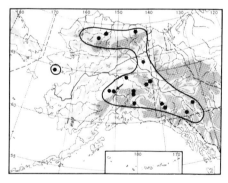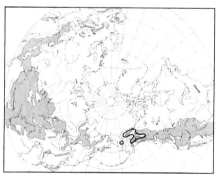

28. Sàlix niphoclàda Rydb.
var. **Méxiae** (Ball) Hult.

Salix brachycarpa var. *Mexiae* Ball; *S. fullertonensis* of Fl. Alaska & Yukon.

Similar to var. *niphoclada*, but leaves short-pedicellated, yellowish-green, blunt; catkins short.

33. SALICACEAE (Willow Family)

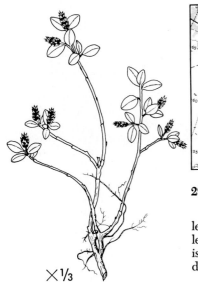 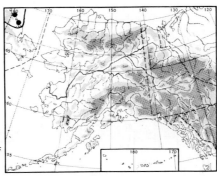 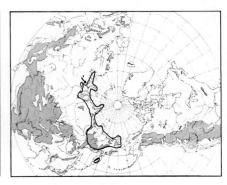

29. Sàlix réptans Rupr.

Decumbent, stem with rooting branches, somewhat pubescent when young; leaves nearly sessile, spatulate-ovate with cuneate base, entire in margin, more or less soft-pubescent; catkins long, pedunculate, catkin scales broad, rounded, brownish black, filaments glabrous; capsule short, blunt, sessile, densely pubescent, style deeply divided, with long, cleft, spreading stigmas.

Wet meadows.

 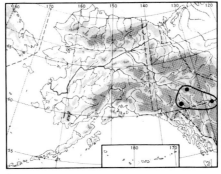 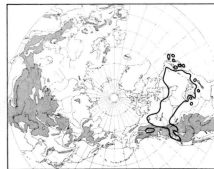

30. Sàlix pedicelláris Pursh
var. **hypoleùca** Fern.

Creeping shrub, up to 1 meter tall, with glabrous twigs; leaves entire, elliptical, oblong-ovate or oblanceolate, obtuse, glaucous beneath with revolute margin; catkins more than 2 cm long, on leafy peduncles; bracts yellow, glabrous or nearly so; capsules glabrous with obsolete style and 4 stigmas.

Wet woods, muskeg. S. *pedicellaris* described from the Catskill Mountains, New York, var. *hypoleuca* from West Roxbury, Massachusetts.

33. SALICACEAE (Willow Family)

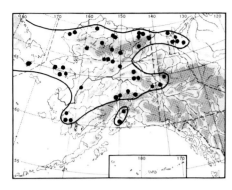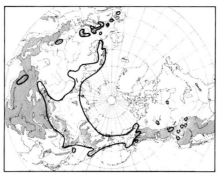

31. Sàlix hastàta L.
Salix Farrae Ball var. *Walpolei* Cov. & Ball; *S. Walpolei* (Cov. & Ball) Ball.

Shrub up to 2 meters tall, usually lower; young twigs reddish-brown, pubescent, glabrescent in age; stipules reniform, serrated; leaves thin, glabrous, ovate, oblong, or elliptical, sharply serrated, paler below, in alpine forms small, lanceolate, and nearly entire, lacking stipules; catkins reddish-brown on short, leafy peduncles; bracts oblong, blunt or acute, light brown with white wavy hairs; capsules green or brownish-green, glabrous, acute, short-pedicellated, with short, green style; stamens 2; filaments glabrous.

Woods, river bars, reaching up to alpine region and onto tundra. Described from Lapland, Switzerland.

Specimens with nearly entire leaf margin have been called var. **subintegrifòlia** Flod. [subsp. *subintegrifolia* (Flod.) Flod.]; subalpine specimens with narrow leaves and small stipules, var. **subalpìna** Anderss.; and those lacking stipules, var. **alpéstris** Anderss. *S. hastata* hybridizes with several other species. Broken line on circumpolar map indicates range of var. **Fárrae** (Ball) Hult. (*S. Farrae* Ball).

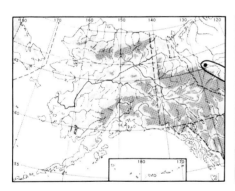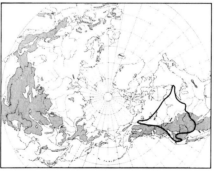

32. Sàlix Mackenzieàna Barratt

Shrub up to 3 meters tall; young twigs pubescent, glabrescent in age; stipules reniform to semilunar; leaves glabrous, lanceolate or obovate-lanceolate, acuminate, usually subcordate at base, finely glandular-serrate, green above, glaucous beneath; catkins on leafy peduncles, appearing with leaves; bracts dark brown; capsules glabrous, on pedicels 3–4 mm long; styles about 0.5 mm long.

Sand bars, along streams. Described from Great Slave Lake and the Mackenzie River.

×⅓

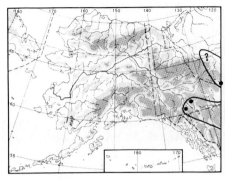 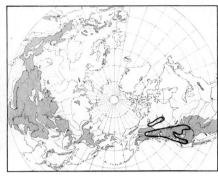

33. Sàlix lùtea Nutt.

Similar to S. *Mackenzieana* and probably more properly considered a subdivision of that species. Differs in having early-deciduous, yellow catkin scales and yellowish twigs.

Described from the Rocky Mountains. Circumpolar map tentative.

×½

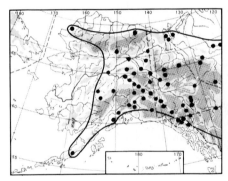 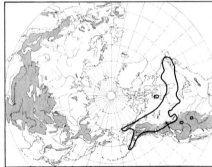

34. Sàlix myrtillifòlia Anders.

Low shrub, usually less than 30 cm tall; young twigs smooth or finely pubescent; leaves yellowish-green on both sides, comparatively small, oblong, oblong-ovate or oblanceolate, glabrous, obtuse or somewhat acute, regularly serrated to base; catkins on leafy peduncles, small, narrower than in S. *Barclayi*, with short, blunt capsules and shorter styles; bracts dark, short, gray-hairy.

Wet places in river valleys. Intergrades with S. *Barclayi*.

Var. **pseudomyrsinìtis** (Anderss.) Ball [S. *pseudocordata* (Anderss.) Rydb.], with acute, long leaves and longer styles, seems to represent hybrid forms.

33. SALICACEAE (Willow Family)

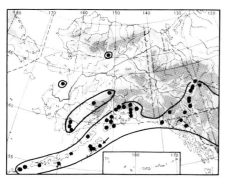

35. Sàlix Barclàyi Anderss.

Shrub up to 2–3 meters tall; young shoots tomentose; leaves blackened in drying, broad, ovate to obovate or elliptical, somewhat acute, crenate-serrate in the margin; catkins on leafy peduncles; capsules green when young, 4–4.5 mm long, pedicellated, long-attenuate in apex, with style about 1–2 mm long; scales small, dark, ovate, acute, white-woolly.

Forms thickets along the central and western parts of the southern Alaskan coast and in the interior Yukon.

An extremely variable plant, doubtfully distinct from several other species, with which it apparently forms hybrids, its common identity resting on its having long, glabrous capsules, pubescent twigs, darkening leaves, and long styles. Hybridizes (probably introgrades) with *S. pulchra* (no. 48, below) and *S. arctica* subsp. *crassijulis* (no. 10, above); hybridizes also with *S. glauca* (no. 22 *et seq.*), *S. myrtillifolia* (no. 34), and probably other species.

The leaves are highly variable: as described above in the typical plant (var. *rotundifolia* Anderss., var. *grandifolia* Anderss.), but oblanceolate or ovate-lanceolate in the commonly occurring var. **angustifólia** Anderss.

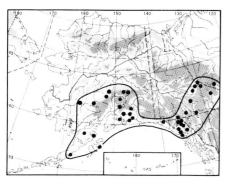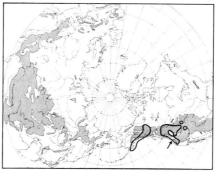

36. Sàlix commutàta Bebb

Shrub up to 3 meters tall; young twigs densely gray-hairy; leaves elliptical or obovate to broadly oblanceolate, abruptly pointed, densely grayish-tomentose on both sides when young, in age glabrescent; catkins on leafy peduncles; bracts densely woolly; capsules reddish, glabrous; stamens 2; filaments glabrous.

Alpine meadows, to at least 1,800 meters in the Yukon.

Forms with more serrated, less pubescent leaves, which have been referred to var. **denudàta** Bebb, are probably of hybrid origin.

× ⅓

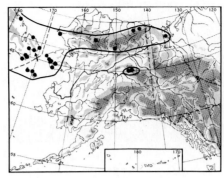
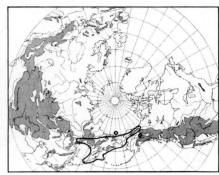

37. Sàlix Chamissònis Anderss.

Dwarf shrub; branches prostrate, light-brown, thick, pubescent when young, later glabrous; leaves thin, obovate or elliptical, obtuse or somewhat acute, slightly paler below, regularly glandular-serrate along entire margin; peduncles with large leaves; catkins cylindrical, 3–4 cm long, in fruit longer; bracts brownish-black, acute, gray-haired; capsules conical, green when young, gray-haired, short-pedicellated, with undivided style about 1 mm long.

Dry, alpine tundra, snow beds.

Preferred habitats are those of the *S. arctica* complex, with which hybrids are formed.

× ⅓

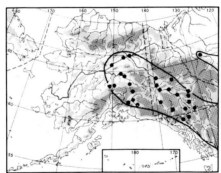
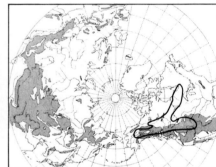

38. Sàlix padophýlla Rydb.
Salix pseudomonticola Ball.

Shrub at least 3 meters tall; twigs reddish-brown, shiny, pubescent when young, glabrescent in age; leaves ovate to obovate, acute, often cordate at base, glandular-serrate along entire margin, glabrous, paler beneath; young, undeveloped leaves reddish; catkins sessile; bracts dark brown, somewhat acute, long-hairy; capsules glabrous, lanceolate when young, long-pedicellated, with style about 1 mm long.

Along streams, in forests, at low altitudes. Described from Turkey Creek, Colorado.

33. SALICACEAE (Willow Family)

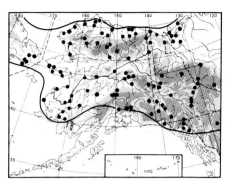 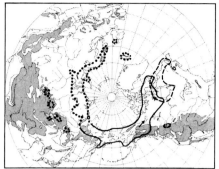 

39. Sàlix lanata L.
subsp. **Richardsonii** (Hook.) A. Skvortz.
Salix Richardsonii Hook.

Shrub 1–2 meters tall; young twigs densely pubescent; leaves glabrous, glandular, serrate in lower part, small in arctic specimens, obovate-lanceolate to ovate and nearly as broad as long, acute, yellowish-green beneath; stipules persistent, long and narrow, glandular along entire margin; catkins sessile, thick, stiff, appearing before leaves; capsules glabrous, with long styles; stamens 2; filaments glabrous.

Wet places, heaths, riverbanks, both on tundra and in mountains, to at least 1,700 meters in the southern Yukon.

Dotted line on circumpolar map indicates range of subsp. **lanàta**, broken line that of subsp. **calcicòla** (Fern. & Wieg.) Hult. (*S. calcicola* Fern. & Wieg.).

(See color section.)

×⅓

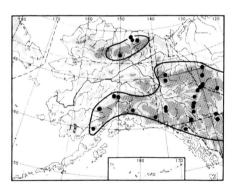 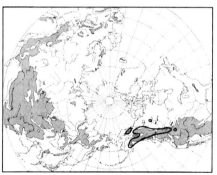 

40. Sàlix Barrattiàna Hook.
Salix albertana Rowlee.

Shrub up to 1 meter tall; twigs pubescent; leaves densely silky-pubescent on both sides, elliptical to obovate or oblanceolate (var. **angustifòlia** Anderss.), acute, nearly entire or usually serrate in margin; stipules short and broad, nearly glabrous, glandular-margined, sometimes persistent (var. **marcéscens** Raup); catkins sessile, upright, appearing before leaves; bracts acute, black, silky-hairy; capsules white-silky; stamens 2; filaments glabrous.

Forms dense thickets on river flats and in alpine meadows, to 1,700 meters in the Yukon, and to about 1,400 meters in McKinley Park. Described from the Rocky Mountains.

×½

×¼

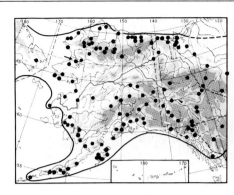

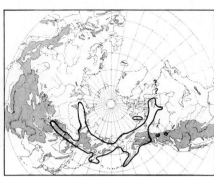

41. Sàlix alaxénsis (Anderss.) Cov.   Alaska Willow
subsp. **alaxénsis**

Salix speciosa var. *alaxensis* Anderss.; *S. speciosa* Hook. not Host.

Shrub or small tree, up to 6–8 meters tall; young twigs woolly; leaves oblanceolate to obovate, acute (rarely lanceolate), with dense, white felt beneath; catkins stout, sessile, erect, appearing at same time as young leaves; stipules glandular on margin, tomentose; capsules densely white-haired; stamens 2; filaments glabrous.

Common along creeks and rivers, in the Yukon to at least 1,800 meters.

The inner bark, or "keeleeyuk," is scraped together and eaten by the natives. (See color section.)

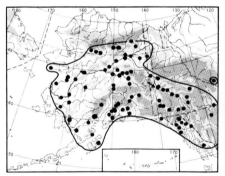

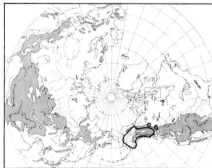

42. Sàlix alaxénsis (Anderss.) Cov.
subsp. **longistỳlis** (Rydb.) Hult.

Salix longistylis Rydb.; *S. alaxensis* var. *longistylis* (Rydb.) Schneid.

Similar to subsp. *alaxensis,* but young twigs show a bluish bloom, and lack the felt of the typical plant.

In McKinley Park to 600 meters.

Replaces subsp. *alaxensis* at low altitudes, but often occurs in company with it, and in Alaska is regarded simply as an altitudinal race, with similar habitats.

×⅓

33. SALICACEAE (Willow Family)

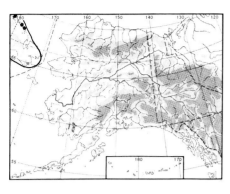 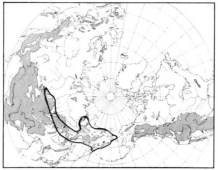 

43. Sàlix Krylòvii E. Wolf
Salix helvetica Vill. subsp. *Krylovii* E. Wolf.

Shrub up to 2.5 meters tall, with dark, glabrous branches; young twigs white-woolly; leaves oblong-obovate, dark green above, white-woolly beneath; catkins appearing at same time as leaves, on white-woolly pedicels; bracts ovate, acute, black, running to brown at base, with long, white pubescence; capsules prolonged, conical, pedicellated, densely white-woolly, with style 0.5 mm long.

Along creeks in the subalpine region and on tundra. Described from Altai.

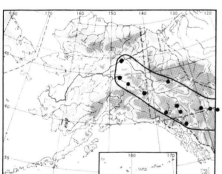 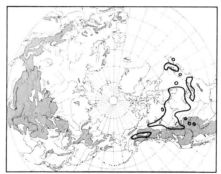 

44. Sàlix cándida Flügge

Upright shrub, up to 1 meter tall; twigs white-woolly when young, glabrescent in age; leaves oblong or oblong-lanceolate, acute, with entire or undulate, inrolled margin, white-tomentose on both sides when young, later sometimes glabrescent and dark green above; catkins appearing before leaves, sessile; bracts pale, brown toward apex; capsules densely white-woolly, with styles 1 mm long; stamens 2, glabrous.

Wet meadows, muskeg. Type locality not given.

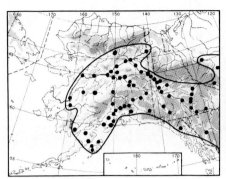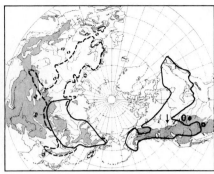

×⅓

45. Sàlix depréssa L. Long-Beaked Willow
subsp. **rostràta** (Anderss.) Hiitonen

Salix vagans var. *rostrata* Anderss.; *S. rostrata* Richards. not Thuill.; *S. perrostrata* Rydb.; *S. Bebbiana* Sarg.

Shrub or (rarely) small tree up to 10 meters tall; twigs grayish-pubescent when young, more or less glabrescent in age; leaves elliptical, ovate, oblong or oblanceolate, acute, entire, or sparsely toothed, pubescent when young, glabrescent in age, dull green above, paler beneath; catkins on short, leafy peduncles, often lax and nodding; bracts yellowish-brown, narrow, somewhat pubescent; capsules slender, on pedicels 2–5 mm long, finely pubescent; stamens 2, filaments hairy at base.

Dry places in woods, up into subalpine region. *S. depressa* described from the mountains of Lapland.

Some specimens from the Yukon are nearly glabrous (var. **depílis** Raup). Asiatic specimens from the Anadyr and Udsk districts to the Yenisei River are considered to belong to subsp. *rostrata*; the rest of the Eurasiatic range belongs to other very closely related races.

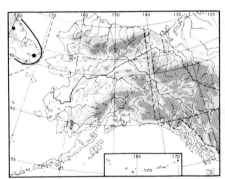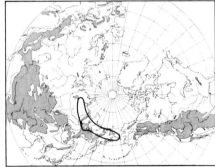

×⅓

46. Sàlix boganidénsis Trautv.

Salix boganidensis var. *angustifolia* Herder; *S. kolymensis* K. O. von Seem.

Tall shrub with thin branches; young twigs pubescent, glabrescent in age, dark reddish-brown, shiny; leaves oblong-lanceolate, flat, finely and sharply serrated in margin to nearly entire, slightly pubescent when young, later glabrous, dark green above, glaucous beneath; catkins densely flowered, short-peduncled, developing before leaves, about 4 cm long and 4 mm broad; bracts ovate, brown, paler at base, with long, white or grayish pubescence; capsules ovate-conical, grayish-pubescent.

Wet places on tundra and forest tundra. Described from the Boganida River.

33. SALICACEAE (Willow Family)

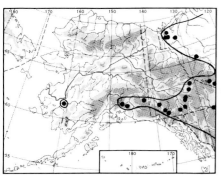
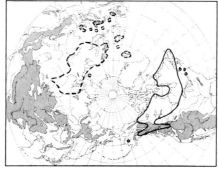

47. Sàlix phylicifòlia L.
subsp. **planifòlia** (Pursh) Hiitonen
   Salix planifolia Pursh.

Shrub up to 3 meters tall, with glabrous, purplish twigs; leaves elliptical-lanceolate to oblong or oblong-lanceolate, acute, glabrous, green above, glaucous beneath, with entire or irregularly toothed margin; catkins sessile, or with a few small leaves; bracts blackish, hairy; capsules silky-pubescent; stamens 2; filaments glabrous.

Stream banks, wet forests. *S. phylicifolia* described from northern Sweden, subsp. *planifolia* from Labrador.

Broken line on circumpolar map indicates range of subsp. **phylicifòlia**.

Introgression with *S. pulchra* occurs in the upper Yukon valley.

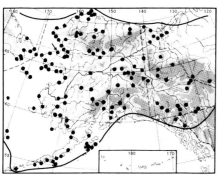
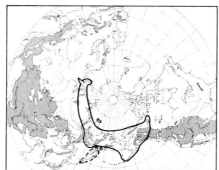

48. Sàlix pùlchra Cham.

Shrub up to 3 meters tall, slightly pubescent, with long, soft hairs when young, but soon glabrous; leaves (highly variable in form) are broad, diamond-shaped or elliptical in the typical form, obovate-oval (var. **Loóffiae** Ball), or lanceolate, green above, more or less glaucous beneath, entire in the margin or slightly serrulate at base; stipules narrowly linear, acute, persisting (the best characteristic of the species); previous year's leaves often persisting as brown, withered remains; catkins sessile, erect; scales blackish; capsules silky-pubescent; stamens 2.

The most common willow in the area of interest, often forming extensive thickets above timberline in the mountains of the Yukon, to at least 1,750 meters.

Possibly better regarded as a subspecies of the very closely related *S. phylicifolia*. Forms hybrid swarms with *S. phylicifolia* subsp. *planifolia* in the upper Yukon valley, and hybrids with *S. arbusculoides* (no. 53, below) and *S. arctica* (no. 9, above). Var. **Palméri** Ball, with narrowly oblong, elliptical-oblong, or elliptical-lanceolate leaves, occurs in extreme arctic or alpine situations; var. **yukonénsis** Schneid., with pubescent twigs, occurs at scattered localities (probably indicating a hybrid influence). Broken line on circumpolar map indicates range of subsp. **parallelinérvis** (Flod.) A. Skvortz. (*S. parallelinervis* Flod.).

The young leaves are eaten raw or mixed with seal oil by the natives of Siberia.

× ⅓

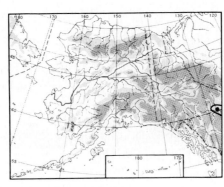
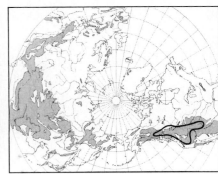

49. Sàlix subcoerùlea Piper
Salix Drummondiana Barr. var. *subcoerulea* (Piper) Ball.

Shrub 2–3 meters tall, with dark, glabrous, pruinose twigs; leaves oblanceolate to oblong-lanceolate, acute or obtuse, with margin entire or nearly so, green and thinly pubescent above, densely silvery-pubescent below; catkins sessile, appearing before leaves; bracts brown to blackish, somewhat hairy; capsules on short pedicels, silver-silky, with style about 1.5 mm long.

Wet meadows, margins of streams. Described from Wallowa County, Oregon.

× ½

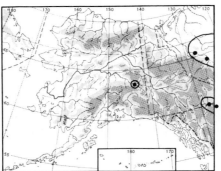
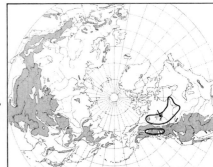

50. Sàlix athabascénsis Raup
Including *Salix fallax* Raup.

Erect shrub, up to 6 dm tall, with gray, hairy twigs; leaves elliptical or obovate, acute or acuminate, entire or minutely glandular-dentate on margin and silky-pubescent above when young, glaucous and with appressed hairs below; catkins on leafy peduncles; capsules with appressed pubescence; bracts obtuse, brown, silky-haired; styles about 0.3 mm long.

Margins of pools, muskeg.

33. SALICACEAE (Willow Family)

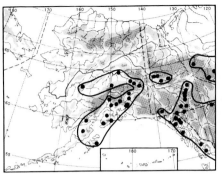 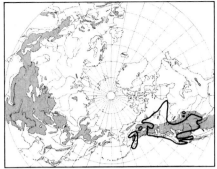 

51. Sàlix Scouleriàna Barratt

Shrub or small tree up to 5 meters tall; twigs densely pubescent, in age glabrescent; leaves crowded at ends of twigs, obovate to oblanceolate, obtuse to acute, entire or somewhat crenate-serrate, dark green and glabrate above, paler and more or less reddish, short-pubescent, below; catkins sessile or on very short, leafy peduncles; bracts obovate, black; capsules tomentose, short-pedicellated; stamens 2, glabrous.

Woods, muskeg, burned areas. Described from the Columbia River, Fort Vancouver.

Our form is generally var. **coetànea** Ball, with catkins appearing at the same time as the leaves and with short, leafy peduncles. Plants with dense, persistent pubescence have been called f. **poìkila** Schneid.

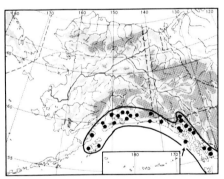 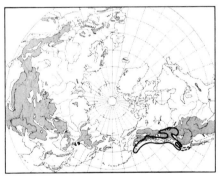 

52. Sàlix sitchénsis Sanson   Sitka Willow

Shrub or (rarely) small tree, up to 10 meters tall, with trunk 30 cm in diameter, sometimes prostrate shrub; twigs pubescent when young, more or less glabrescent in age; leaves obovate or oblong-obovate, obtuse or somewhat acute, more or less silky-pubescent, with shiny, short, appressed hairs; catkins long, slender, densely flowered, on short, leafy peduncles; bracts brown, pubescent; capsules silky-pubescent, nearly sessile; stamen single, with glabrous filament.

Along streams and shores.

Broken line on circumpolar map indicates range of the very closely related var. **ajanénsis** Anderss., in eastern Asia (Ajan), and of the closely related **S. Coultéri** Anderss., from California to Washington.

×⅓

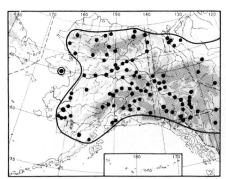
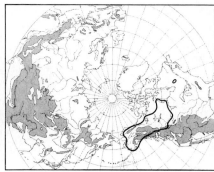

53. Sàlix arbusculoìdes Anderss.

Shrub or tree, up to 3–4 meters tall; twigs reddish, shiny, glabrous; leaves elliptical or elliptic-lanceolate, acute, usually finely and regularly serrate in the entire margin, minutely sericeous beneath, with short, white, appressed hairs; catkins appearing with leaves, with 2–3 small leaves on the short peduncle; capsules pubescent; bracts blackish, pubescent; styles about 0.8 mm long; stamens 2; filaments glabrous.

Along streams, in woods, in the Yukon to at least 1,200 meters. Described from Prince Albert Sound, Rae River, Labrador.

Plants with leaves nearly glabrous below (var. glàbra Anderss.) and with less serrulated leaf margin probably represent hybrids with S. *pulchra* (no. 48, above).

(See color section.)

×⅓

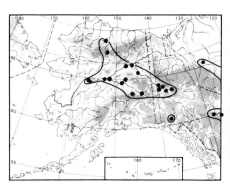
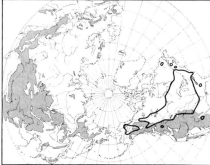

54. Sàlix intèrior Rowlee
Salix longifolia Muhl. not Lam.

Shrub up to 4 meters tall; twigs smooth; leaves sessile or nearly so, narrowly lanceolate to linear, glabrous or sometimes sericeous [f. Wheèleri (Rowlee) Rouleau (S. *interior* var. Wheeleri Rowlee)], entire in the margin when young, later with widely separated teeth; catkins on leafy peduncles; bracts pale yellow, deciduous; capsules silky, glabrate in age, on pedicels 1.5 mm long; stigmas sessile.

Sandbars in rivers, along shores. Described from the upper Mississippi Valley, the Great Lakes, and the eastern slope of the Allegheny Mountains.

Most specimens belong to var. pedicellàta (Anderss.) Ball (S. *longifolia pedicellata* Anderss.), with very narrow leaves, usually not over 6 mm wide.

33. Salicaceae (Willow Family)

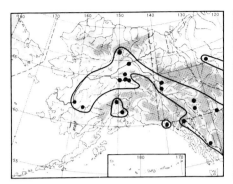  

55. Sàlix lasiándra Benth.

Tall shrub or small tree with grayish-brown bark, up to 6 meters tall, with reddish-brown, shiny twigs; leaves thick, lanceolate or ovate-lanceolate, long-acuminate, finely and regularly crenate-serrate, dark green above, paler beneath, glabrous or glabrate; petioles with a pair of glands at base; catkins appearing with leaves, on leafy peduncles; bracts light-colored, ovate-lanceolate, small; capsules light brown, glabrous, with style about 0.5 mm long; stamens about 5, pubescent.

Sandbars along streams. Described from the Sacramento River.

Most specimens show densely pubescent young branches and belong to var. **lancifòlia** (Anderss.) Bebb (*S. lancifolia* Anderss.), described from Vancouver Island, but some belong to the typical plant (Yukon Flats, Nevada, Dawson).

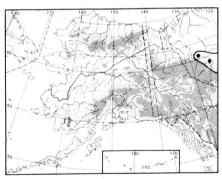 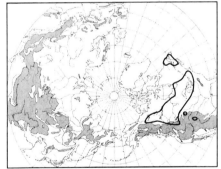 

56. Sàlix seríssima (Bailey) Fern.
Salix lucida var. *serissima* Bailey.

Shrub up to 4 meters tall, with glabrous, shiny twigs; leaves elliptical to oblong-lanceolate, short-acuminate, glabrous, finely and regularly glandular-serrate, dark green above, somewhat glaucous beneath; petioles with glands on both sides near base of blade; pistillate catkins 2–3.5 cm long, on leafy peduncles; bracts yellowish, short-haired; capsule glabrous, conic-subulate, with style 0.5 mm long, the pedicel twice as long as the upper gland.

Marshes and bogs. Described from Minnesota.

34. MYRICACEAE (Wax Myrtle Family) / 35. BETULACEAE (Birch Family)

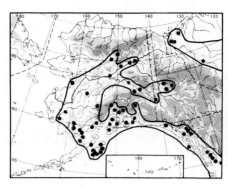
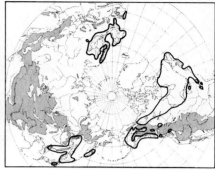

Myrica L.

1. Myrìca gàle L. Sweet Gale
var. **tomentòsa** C. DC.

Shrub up to 1 meter tall, with odorous resin-dots; branches reddish-brown; leaves deciduous, cuneate-oblanceolate, somewhat serrate toward apex, grayish-green with more or less dense grayish-green pubescence on both sides; aments developed before leaves; male and female flowers mostly on different shrubs; male flower with 4 anthers, female with 1 style.

Swamps, shallow water. *M. gale* described from Europe, var. *tomentosa* from Kamchatka, Sitka.

The Pacific plant belongs to var. **tomentòsa**. Circumpolar map indicates range of *M. gale* in a broad sense, i.e., of the entire, not yet clearly subdivided, species complex.

Betula L. (Birch)

Leaves rounded in apex:
    Leaves small, up to 2 cm long, crenate, serrate, petioles short, densely fine-pubescent; wing of nutlets narrow:
        Leaves orbicular, with truncate or cordate base, often broader than long, crenulated all around; decumbent dwarf shrub .. 1. *B. nana* subsp. *exilis*
        Leaves obovate, with cuneate toothless base; upright shrub ....
        .... 2. *B. glandulosa*
    Leaves large .... See 1–2. Hybrid *B. nana* subsp. *exilis* × *glandulosa*
Leaves acute or somewhat acute in apex:
    Leaves small, 2–3 cm long, mostly ovate; twigs densely glandular ....
    .... 3. "*B. occidentalis*"
    Leaves larger; twigs not densely glandular:
        Leaves yellowish-green, especially below, with distinctly prolonged caudate apex .... 5. *B. papyrifera* var. *humilis*
        Leaves dark green, paler below, lacking caudate apex:
            Leaves triangular or diamond-shaped, with white hairs in margin; wings of nutlet as broad as body or narrower .... 4. *B. kenaica*
            Leaves ovate, double-serrate; wings of nutlet broader than body ....
            .... 6. *B. papyrifera* var. *commutata*

35. BETULACEAE (Birch Family)

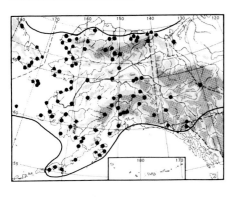 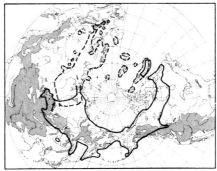 

1. Bétula nàna L. Dwarf Birch
subsp. **exìlis** (Sukatsch.) Hult.

Betula exilis Sukatsch.; *B. nana* var. *sibirica* Ledeb. in part; *B. glandulosa* var. *sibirica* (Ledeb.) Blake. Including *B. tundrarum* Perfiljev.

Strongly branching, decumbent dwarf shrub; young twigs covered with resin spots; leaves dark green above, paler beneath, short-petiolated, 0.5–1 cm long, orbicular or somewhat broader than long, sometimes with subcordate base, crenate all around, glabrous or with a few glands on nerves; bracts lacking resiniferous hump on back, with 3 nearly parallel lobes, the median longer than the lateral; nutlets with wings about half as broad as body.

Tundra, bogs. *B. nana* described from the mountains of Lapland, the bogs of Sweden and Russia. Broken line on circumpolar map indicates range of subsp. **nàna**.

Subsp. *exilis* forms complete introgression with *B. glandulosa* where ranges overlap. Hybrids with the large-leaved tree birches, recognized by large, more or less rounded, more or less crenate leaves, are not rare.

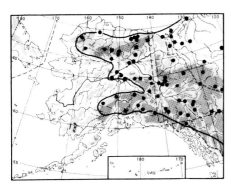 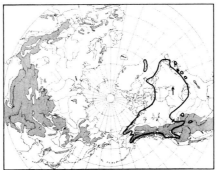 

2. Bétula glandulòsa Michx. Shrub Birch

Upright shrub up to 2.5 meters tall, forming brush-wood; young twigs densely dotted with resin spots; petioles puberulent; leaves 1–2 cm long, somewhat glutinose, glabrous, dark green above, yellowish-green beneath, crenate except near the cuneate base; bracts with resiniferous hump on back at maturity, 3-lobed, the median lobe longer than the lateral; nutlets with narrow wings often broader toward apex.

Wet places, swamps, bogs.

Forms complete introgression with *B. nana* subsp. *exilis* where ranges overlap, as well as hybrids with the tree birches—usually tall shrubs, but sometimes small trees with larger, more or less rounded, more or less crenate-serrate leaves.

35. Betulaceae (Birch Family)

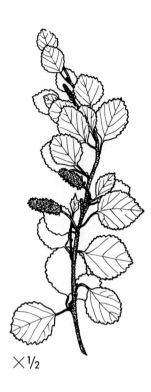
×½

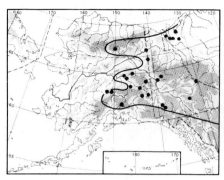 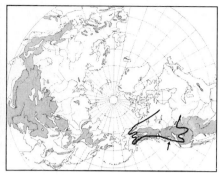

3. "Bétula occidentàlis Hook."

Betula microphylla of American authors; *B. glandulosa* × *resinifera* of Fl. Alaska & Yukon; *B. commixta* Sarg.; *B. Eastwoodae* Sarg.; *B. Beeniana* Nels.

Under this name is included a number of shrubs (or, rarely, small trees), smaller than the tree birches but larger than *B. glandulosa*, which are characterized by highly variable form and serrulation of the leaves; variable pubescence and short catkins; leaves that are mostly ovate and acute, often with finely pubescent petioles; twigs that are very densely glandular; and not uncommon occurrences of specimens lacking fructification.

In Alaska, "*B. occidentalis*" occupies the area common to *B. glandulosa* and *B. papyrifera* subsp. *humilis*; it seems reasonable, therefore, to regard the group as a hybrid population between these taxa. It is also very similar to **B. mìnor** Fern. of eastern North America, which might in fact be the corresponding *B. glandulosa* hybrid series with the eastern American birches.

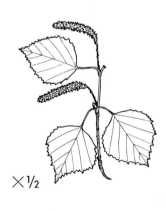
×½

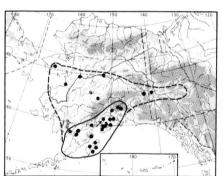 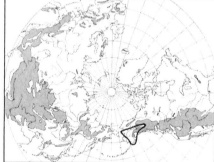

4. Bétula kenáica Evans   Kenai Birch

Betula papyrifera var. *kenaica* (Evans) Henry; *B. Perfiljevii* Vassil. with respect to Alaskan plant.

Tree up to 10(–12) meters tall, with trunk up to 30 cm in diameter and dark bark; young twigs resiniferous; leaves ovate, with cuneate or truncate base, double-serrate, more or less pubescent above and in margin toward base, slightly glandular-dotted, and with pilose nerves below; catkins brown, narrow; bracts with lobes of about equal length, rounded in apex, the lateral slightly diamond-shaped; nutlets with wings about as broad as body.

Subalpine zone, often in the alder belt or at the boundary toward the treeless tundra zone.

Forms hybrids with *B. papyrifera* and with *B. nana* subsp. *exilis*; such hybrids are common in the Kenai Peninsula (*B. Hornei* Butler) and on Kodiak Island. Within the area indicated by the broken line, the hybrid influence of *B. kenaica* can be traced.

35. BETULACEAE (Birch Family)

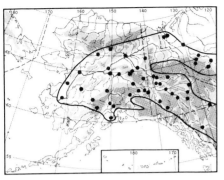
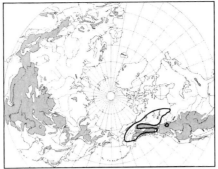

5. Bétula papyrífera Marsh.
subsp. **hùmilis** (Regel) Hult.   Paper Birch

Betula alba subsp. *papyrifera* var. *humilis* Regel; *B. papyrifera* var. *humilis* (Regel) Fern.; *B. papyrifera* var. *neoalaskana* (Sarg.) Raup; *B. alaskana* Sarg.; *B. resinifera* Britt.; *B. neoalaskana* Sarg.

Tree, usually 10–15, rarely up to 20, meters tall; with trunk up to 60 cm in diameter and white bark; young twigs strongly resiniferous; leaves yellowish-green, ovate, with elongated apex and cuneate or truncate base, serrate, glabrous above and in margin, resin-dotted below, with tufts of hairs in angles of nerves below; catkins short, thick, greenish-brown; bracts with median lobe usually longer than the blunt, diamond-shaped, lateral lobes; nutlets with wings broader than body.

Common in the lowlands; to 800 meters in McKinley Park; to 1,200 meters in the Yukon. No type locality given for *B. papyrifera*; subsp. *humilis* described from Saskatchewan.

Forms hybrid swarms with *B. kenaica* as well as hybrids with *B. nana* subsp. *exilis* and *B. glandulosa*.

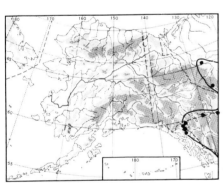

6. Bétula papyrífera Marsh.
var. **commutàta** (Regel) Fern.   Paper Birch

Betula alba subsp. *occidentalis* (Hook.) Regel, var. *commutata* Regel; *B. papyrifera* var. *occidentalis* Sarg. (not *B. occidentalis* Hook.); *B. papyrifera* subsp. *occidentalis* Hult.

Tree up to 15 meters tall; leaves dark green, ovate, rounded at base, coarsely double-serrate, with apex not markedly elongated; leaves glabrous above, somewhat pilose below, especially in angles of nerves; bracts with long median lobe, strongly ciliate; wings of nutlet broader than body.

Replaces *B. papyrifera* subsp. *humilis* in the Lynn Canal area and toward the southeast.

Alnus Mill. (Alder)

Leaves lustrous; peduncles longer than cones; nuts with membranaceous, broad wing-margin:
    Leaves ovate, often with slightly cuneate base, not (or only occasionally) lobed .... 1. *A. crispa* subsp. *crispa*
    Leaves broadly ovate, with truncate or sometimes slightly cordate base, more or less irregularly lobed .... 2. *A. crispa* subsp. *sinuata*

Leaves dull; peduncles shorter than cones; nuts wingless or narrowly wing-margined:
    Leaves revolute in margin, rusty-pubescent beneath; nutlets narrowly winged ... . 3. *A. oregona*
    Leaves flat, not rusty-pubescent; nutlets wingless .. 4. *A. incana* subsp. *tenuifolia*

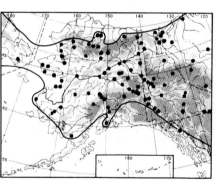 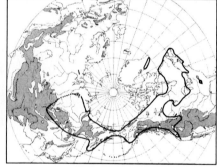

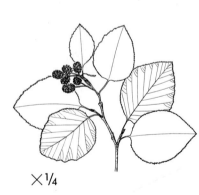

1. Álnus críspa (Ait.) Pursh
subsp. **críspa**

Mountain Alder

Betula crispa Ait.; *Alnus viridis* Vill. subsp. *crispa* (Ait.) Löve & Löve.

Ascending shrub, up to 3 meters tall; leaves glutinous, fragrant, ovate to elliptic, glabrous above, pubescent on nerves below, finely and sharply serrulate or biserrulate; cones on long, slender peduncles.

Occupies a somewhat indistinct subalpine region, diminishing in height with increasing altitude, becoming, at its altitudinal limit, a prostrate dwarf shrub (f. **strángula** Fern.). Introgrades with subsp. *sinuata* in the southern part of its range.

Described from North America.

(See color section.)

35. BETULACEAE (Birch Family)

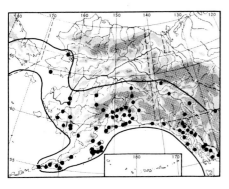 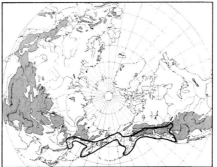 

2. Álnus críspa (Ait.) Pursh
subsp. **sinuàta** (Regel) Hult. Sitka Alder
Alnus viridis var. *sinuata* Regel; *A. sinuata* (Regel) Rydb.

Similar to subsp. *crispa*, but leaves larger and broader, more or less lobed.

Occurs in a more distinct subalpine region along the coast. The branches are bowlike and pressed along the slopes by the snow, forming almost impenetrable thickets on the mountainsides, especially in the coastal zone. Described from Kamchatka, and the coast of Alaska and Canada.

An extreme variation with laciniated, sharply serrulated or double-serrulated leaves is var. **laciniàta** Hult. Subsp. *sinuata* introgrades with subsp. *crispa*; the approximate northern boundary for its influence is marked by a line on the Alaskan map.

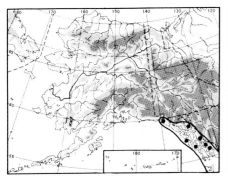 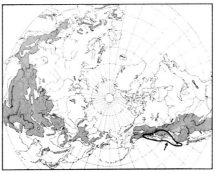 

3. Álnus oregòna Nutt. Red Alder

Tree up to 10 (rarely 15) meters tall, with stem up to 40 cm in diameter, sometimes a shrub forming into dense thickets; bark smooth, grayish; leaves dark green and nearly glabrous above, paler and rusty-pubescent below, elliptic to ovate, with revolute margin, shallowly lobed, the lobes coarsely toothed and glandular-denticulated; cones short-peduncled; nutlets narrowly winged.

Wet places, river bottoms, along creeks.

35. Betulaceae (Birch Family) / 36. Urticaceae (Nettle Family)

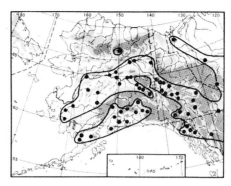

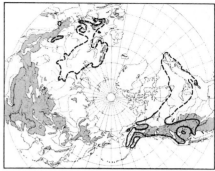

4. Álnus incàna (L.) Moench
Betula Alnus var. *incana* L.
subsp. **tenuifòlia** (Nutt.) Breitung
Alnus incana var. *virescens* S. Wats.; *A. tenuifolia* Nutt.

Large shrub or small tree with first grayish, later reddish, bark; leaves roundish or oblong, ovate, dull grayish-green above, paler beneath, pubescent when young, glabrescent in age, closely toothed, the teeth finely serrated, rounded or subcordate at base; cones short-peduncled; nutlets with thin, narrow margin.

Forms thickets along streams. *A. incana* described from Europe, subsp. *tenuifolia* from the Rocky Mountains and the Blue Mountains.

Very closely related to **A. rugòsa** (Du Roi) Spreng. (*Betula rugosa* Du Roi) of eastern North America and to **A. incàna** subsp. **incàna** of Europe. Broken line on circumpolar map indicates range of these relatives.

Urtica L. (Nettle)

Inflorescences shorter than petioles . . . . 3. *U. urens*
Inflorescences much longer than petioles:
    Leaves lanceolate to ovate; stipules greenish to brown, often acute or acuminate
    . . . . 1. *U. gracilis*
    Leaves cordate, the uppermost with more or less caudate apex; stipules broad,
      blunt, in age brown . . . . 2. *U. Lyallii*

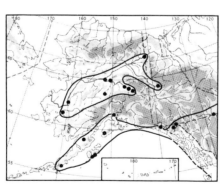

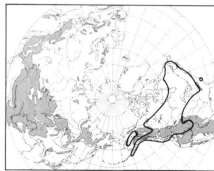

1. Urtìca gràcilis Ait.

Stem glabrous or sparingly pilose above; leaves lanceolate to ovate-cuneate or slightly cordate at base, glabrous or sparingly pilose beneath, coarsely toothed; stipules mostly acute; inflorescence monoecious.

Thickets, stream banks. Described from Hudson Bay.

The young buds, boiled, make an excellent soup.

36. Urticaceae (Nettle Family)

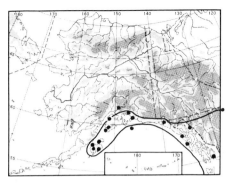
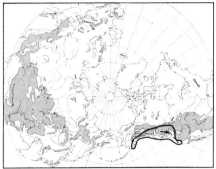
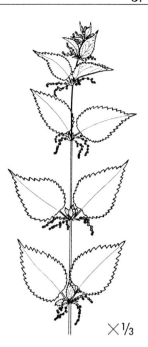

2. Urtìca Lyállii S. Wats.

Stem glabrous or sparingly pilose; leaves ovate to cordate, very coarsely toothed, the uppermost with caudate apex, glabrous or sparingly pilose beneath; stipules oblong, rounded.

Thickets, moist places.

In some specimens, the leaves are more pubescent, and lack, or nearly lack, the caudate apex; these correspond to var. **califórnica** Jepson.

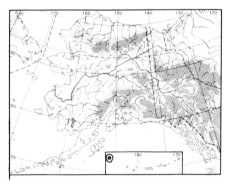
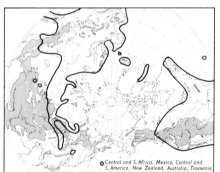

3. Urtìca ùrens L.   Burning Nettle

Annual; stem usually branched; leaves ovate, long-petiolated; pistillate and staminate flowers mixed in the inflorescence; inflorescence shorter than petioles.

Waste ground; once found introduced on Attu Island. Described from Europe.

37. Loranthaceae (Mistletoe Family) / 38. Santalaceae (Sandalwood Family)

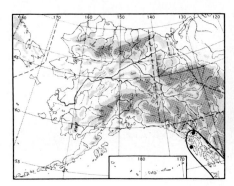
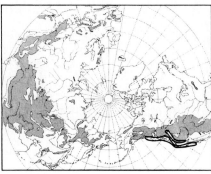

Arceuthobium Bieb.

1. **Arceuthòbium campylòpodum** Engelm.   Dwarf Mistletoe

Razoumofskya campylopoda (Engelm.) Ktze.; *Arceuthobium Douglasii* var. *tsugensis* of Fl. Alaska & Yukon.

Stem segments yellowish to olive-green or brown, 5–10 times longer than thick; shoots often much branched, mostly tufted; sepals mostly 4, oblong-ovate; male flowers light green to yellow, in pairs at nodes, female flowers axillary, in lateral pairs.

Parasite on *Tsuga heterophylla*. Much overlooked. Described from Oregon.

Alaskan plant belongs to f. **tsugénsis** (Rosend.) Gill (*R. tsugensis* Rosend.).

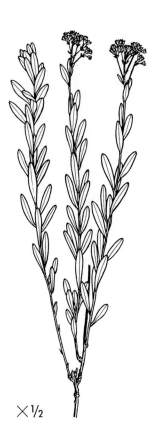
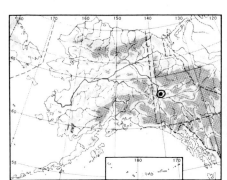

Comandra Nutt.

1. **Comándra umbellàta** (L.) Nutt.
subsp. **pállida** (DC.) Piehl

Comandra pallida DC.; *C. umbellata* (L.) Nutt. var. *pallida* (DC.) M. E. Jones; *Thesium umbellatum* L.

Stems striate, glaucous, branched from subterranean rootstock; leaves subsessile, alternate, acute or cuspidate; cymes small, few-flowered, axillary; flowers purplish-green; fruit green, oblong to ovoid.

Dry places; a root parasite.

Broken line on circumpolar map indicates range of other subspecies.

38. Santalaceae (Sandalwood Family) / 39. Polygonaceae (Buckwheat Family)

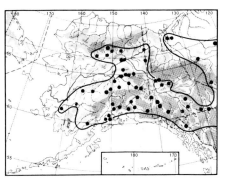
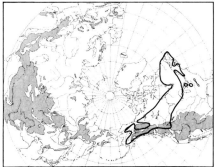

Geocaulon Fern.

1. **Geocaùlon lívidum** (Richards.) Fern.
   Comandra livida Richards.

Rootstock filiform, creeping, reddish; stem simple; leaves alternate, membranaceous, elliptic to narrowly obovate; flowers sessile; drupe solitary, scarlet when ripe, oblong to globose.

Dry places, poplar flats, in McKinley Park to about 600 meters. Described from 54°N to Great Slave Lake.

The fruit is edible.

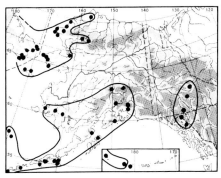
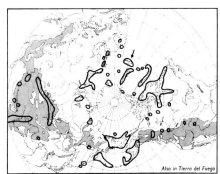

Koenigia L.

1. **Koenígia islándica** L.

Annual; plant more or less reddish (in arctic dwarf specimens, entirely red), glabrous; stem from a few mm up to (in southern localities) 15 cm long, often branched; leaves elliptic to ovate, blunt; flowers single or in small heads; calyx greenish-white with 3, occasionally 4, lobes; stamens as many as lobes.

Wet places with bare earth, snow beds, in McKinley Park to at least 1,300 meters.

39. POLYGONACEAE (Buckwheat Family)

×3/4

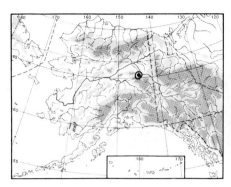
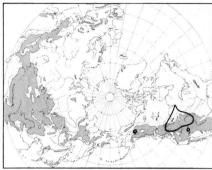

Eriogonum Michx. (Umbrella Plant)

1. Eriógonum flàvum Nutt.
var. **aquilìnum** Reveal

Stems tomentose from branching caudex, short and stout; leaves elliptic or oblanceolate to spatulate, densely whitish-tomentose below, less so above; inflorescence umbelliform; involucrum turbinate, deeply divided, 4-lobed; perianth short-peduncled, yellow, pubescent on outside, with elliptical segments; achenes villous above.

Dry hills. A single collection from Mission Bluff, Eagle area. *E. flavum* described from the Rocky Mountains to the headwaters of the Missouri River.

Alaskan specimens belong to var. *aquilinum* Reveal, with deeply divided 4-lobed involucrum.

Rumex L. (Sorrel, Dock)

■ Basal leaves hastate or linear; flowers mostly dioecious:
  Leaves broad, with broad, downward-pointing basal lobes:
    Leaves ovate-triangular, with broad sinus; ocrea long, acute, integrate or slightly lacerate .... 5. *R. acetosa* subsp. *alpestris*
    Leaves with narrower sinus; ocrea short, broad, strongly lacerate ....4. *R. acetosa* subsp. *acetosa*
  Leaves narrow, with narrow, outward- or upward-pointing basal lobes; or leaves linear, lacking basal lobes:
    Leaves linear, lacking basal lobes (in large specimens sometimes somewhat hastate); valves broad, up to twice as long as nutlet ... 3. *R. graminifolius*
    Leaves hastate, with large basal lobes; valves small, not longer than nutlet:
      Perianth leaves easily separable from nutlet ....1. *R. acetosella* subsp. *acetosella*
      Perianth leaves not easily separable from nutlet ....2. *R. acetosella* subsp. *angiocarpus*

39. POLYGONACEAE (Buckwheat Family)

■ Basal leaves not hastate or linear:
  Valves distinctly denticulate or subulate-dentate:
    Leaves cordate, not more than 2.5 times as long as broad; teeth of valves not longer than body .... 6. *R. obtusifolius*
    Leaves with cuneate, truncate or slightly cordate base, long and narrow; teeth of valves longer than body:
      Median cauline leaves with cuneate base, flat; fruit yellowish .... 15. *R. maritimus* subsp. *maritimus*
      Median cauline leaves with truncate base, undulate; fruit brownish .... 16. *R. maritimus* subsp. *fueginus*
  Valves entire or slightly erose-crenulate:
    Stem tall, erect, lacking axillary shoots:
      Leaves with cuneate base; at least 1 valve with a distinct grain .... 7. *R. crispus*
      Leaves with truncate or cordate base; valves grainless:
        Leaves broadest at middle, undulate; valves broad, often broader than long .... 8. *R. longifolius*
        Leaves broadest near base, not undulate; valves ovate or cordate, usually longer than broad:
          Panicle not branched or with few short, erect branches; stem leaves narrow, lanceolate; plant tinged with purple ...... 9. *R. arcticus*
          Panicle with several, more patent branches; stem leaves broad; plant not tinged with purple .... 10. *R. fenestratus*
    Stem decumbent or ascending, with axillary shoots:
      Valves lacking grain, regularly reticulated .... 14. *R. utahensis*
      Valves with grain:
        Margin of valves broader than grain .... 12. *R. mexicanus*
        Margin of valves narrower than grain:
          Valves 2.5–3 mm long; leaves grayish-green, narrow, thick, flat .... 13. *R. sibiricus*
          Valves 3–4 mm long; leaves green, broader, shorter, thinner, undulate .... 11. *R. transitorius*

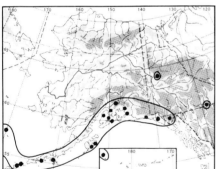  

1. Rùmex acetosélla L.   Sheep Sorrel
subsp. acetosélla

Stem thin, erect, from slender rootstock; leaves long and narrow, hastate with divergent basal lobes; inflorescence loose, reddish or yellowish; pedicels jointed; flowers dioecious; valves separable from nut and scarcely exceeding it; ripe fruit about 1.5 mm long.

Introduced weed. Described from Europe.

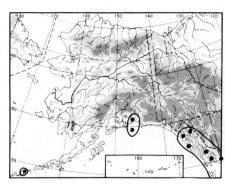 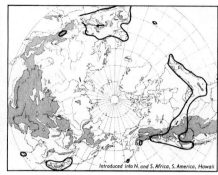

2. Rùmex acetosélla L.
subsp. **angiocárpus** (Murb.) Murb.
Rumex angiocarpus Murb.

Similar to subsp. *acetosella*, but valves not separable from nut; ripe fruit about 1 mm long.
An introduced weed.

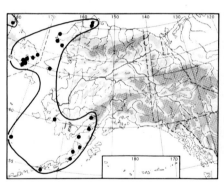 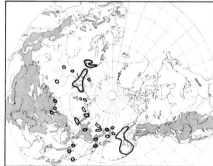

3. Rùmex graminifòlius Lamb.

Similar to R. acetosella, but all leaves narrowly linear; valves from about as long as nut to twice as long.
Sandy places on tundra. Described from "Kuriles, Kamchatka, Mare Glacialis."
The leaves are eaten raw by the Siberian Eskimo.

39. POLYGONACEAE (Buckwheat Family)

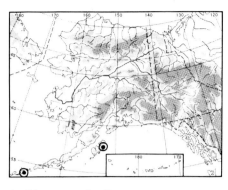 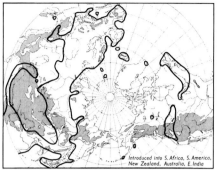 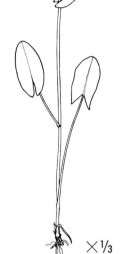

4. Rùmex acetòsa L.   Garden Sorrel
subsp. **acetòsa**

Stem from taproot; ocrea short, broad, fimbriate; basal leaves 2–3 times as long as broad, with divergent basal lobes and lanceolate or oblanceolate terminal lobe; branches of inflorescence reddish, erect, flowers unisexual.

An introduced weed. Described from Europe.

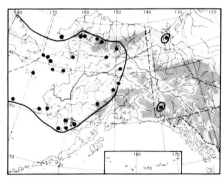 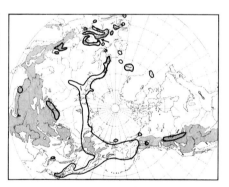 

5. Rùmex acetòsa L.
subsp. **alpéstris** (Scop.) Löve
   Lapathum alpestre Scop.

Similar to subsp. *acetosa*, but leaves ovate-triangular, with broad sinus; ocrea acute, only occasionally somewhat lacerated.

Wet places in the mountains. Type locality given as "In Alpibus Vochinensibus" (northern Croatia).

39. POLYGONACEAE (Buckwheat Family)

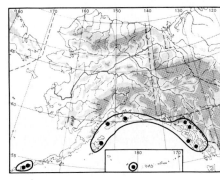 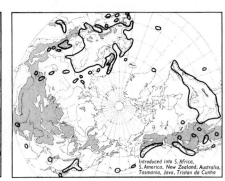

6. Rùmex obtusifòlius L. Blunt-Leaved Dock

Rumex obtusifolius subsp. *agrestis* (E. Fries) Danser.

Stem erect, branched; leaves ovate-oblong, cordate at base, obtuse in the apex, undulate in the margin; inflorescence open, with spreading branches; whorls of flowers distinct; pedicels jointed; valves with 3–5 long teeth, one of them with a grain.

Introduced weed. Described from Germany, Switzerland, France, England.

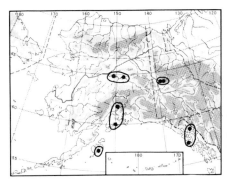 

7. Rùmex críspus L.

Stem tall, erect; leaves with undulate margin, lanceolate or oblong-lanceolate, cuneate or rounded at base, tapering from middle to an obtuse point; inflorescence narrow, simple or little-branched; pedicels jointed; valves round-cordate, entire or slightly denticulated, all three with ovate grain.

Introduced weed. Described from Europe.

39. POLYGONACEAE (Buckwheat Family)

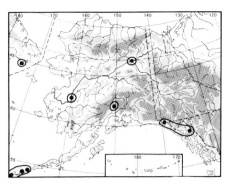
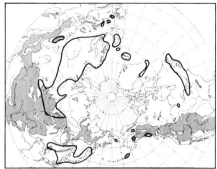

8. Rùmex longifòlius DC.
Rumex domesticus Hartm.

Stem tall, erect; lower leaves ligulate, 3–5 times as long as broad, with square or rounded base, the upper narrower; inflorescence dense, branched, few-leaved; pedicels jointed; fruiting valves reniform-ovate, as broad as long, without grain.

An introduced weed. Described from France.

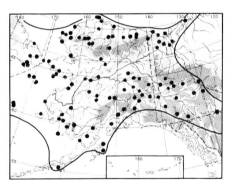
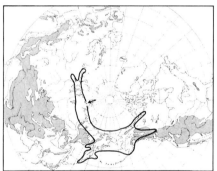

9. Rùmex árcticus Trautv.

Stem up to 50 cm (in the Arctic about 10 cm) tall, reddish-violet at base; basal leaves long-petiolated, lanceolate or ovate-lanceolate, blunt, with cuneate or square base, often reddish below; stem leaves acute; inflorescence leafless or nearly so; unbranched or with few upright branches; flowers bisexual; valves entire, without grain.

Common in wet places, snow beds. The arctic counterpart of **R. aquáticus** L.

Extensively collected, boiled, and preserved for use in the winter by the natives. The leaves are eaten chopped and cooked with fat and sugar by the Siberian Eskimo.

Var. **perlàtus** Hult., with ovate, broad leaves, occurs in the Bering Sea area.

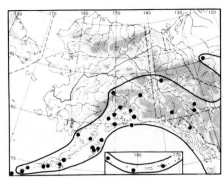
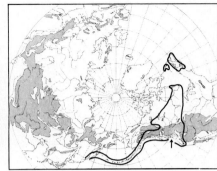

10. Rùmex fenestràtus Greene

Including *Rumex occidentalis* of Fl. Alaska & Yukon.

Stem erect, stout, from taproot; leaves lance-ovate, the lower long-petiolated with cordate base, up to 30 cm long, the upper smaller; panicle dense; valves 7–10 mm long, round-cordate, reticulate, lacking grain.

Wet places. Closely related to **R. aquáticus** L.

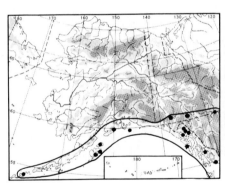
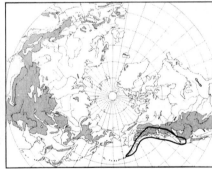

11. Rùmex transitòrius Rech. f.

Stem ascending; leaves green, thin, broadly lanceolate, somewhat undulate; inflorescence open; valves 3–4 mm long, ovate to ovate-lanceolate, brownish, nearly entire in the margin, each with an ovate, prominent grain.

Salt marshes. Described from British Columbia to California.

39. POLYGONACEAE (Buckwheat Family)

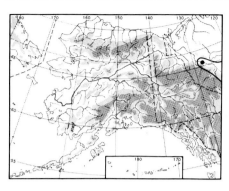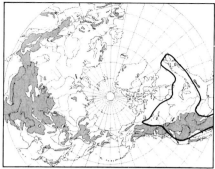

12. Rùmex mexicànus Meisn.
Including *Rumex triangulivalvis* Rech. f.

Stem erect or ascending, suffused with brown or purple; leaves pale green, glabrous, the lower linear-lanceolate; inflorescence with elongated, rigid, ascending branches; valves 3–5 mm long, broadly triangular to subcordate, olivaceous to brown, entire in the margin or crenulate-erose at the base, all with a brown grain, narrower than margin.

Rich, often brackish soil. Described from "Mexico circa Leon."

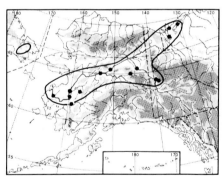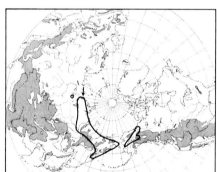

13. Rùmex sibìricus Hult.

Stem ascending; leaves thick, grayish-green, linear-lanceolate, flat; inflorescence open; valves 2.5–3 mm long, ovate-lanceolate, with entire margin, each with a prominent grain.

Sandy or clayey soil along brooks and rivers.

× ¼

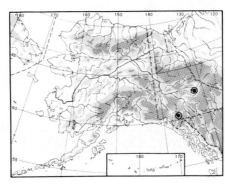
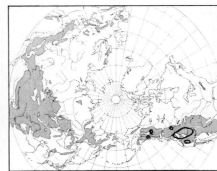

14. Rùmex utahénsis Rech. f.

Stem erect or ascending; leaves pale yellowish-green, glabrous; lower stem leaves lanceolate, tapering to both ends; inflorescence with short branches, compact; valves 2.5–3 mm long, rounded-deltoid, somewhat acute, pale brown, nearly entire in the margin, regularly reticulated, lacking grain or with trace of a grain.

Shores. Described from Oregon, Utah, California.

× ⅓

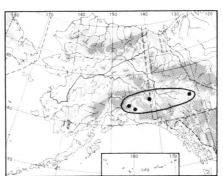
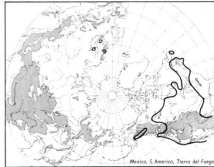

15. Rùmex marítimus L.
subsp. **marítimus**

Often annual; upright or decumbent; leaves long and narrow, tapering to both ends; inflorescence many-leaved, whorls confluent toward ends of branches; valves ovate-triangular usually with 2 setiform teeth on each side, longer than body of valve. Entire plant yellowish when in mature fruit.

Wet places; introduced. Described from Europe.

39. POLYGONACEAE (Buckwheat Family)

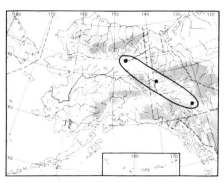 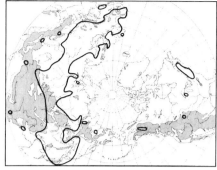 

16. Rùmex marítimus L.
subsp. **fuèginus** (Phill.) Hult.
  Rumex fueginus Phill.

Similar to subsp. *maritimus*, but median cauline leaves curled, with truncate base; fruit brownish.
  Waste places. Described from "Fuegia orientali" (eastern Tierra del Fuego).

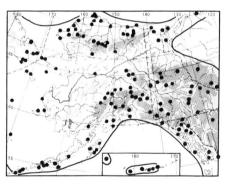 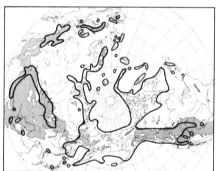 

Oxyria Hill

1. Oxýria dígyna (L.) Hill   Mountain Sorrel
  Rumex digynus L.

Stems up to 35 cm tall, usually shorter, from stout rootstock; leaves reniform, long-petiolated, basal; 2 outer sepals broad and flat, the 2 inner appressed, boat-shaped, about half as broad; stamens 6; fruit with red wing about as broad as body.
  Wet places, snow beds, in mountains and on tundra, in the Yukon to over 2,000 meters, in the Brooks Range to over 1,000 meters. Described from Lapland, Wallis (Switzerland).
  The leaves, but not the root, can be eaten raw.

Polygonum L.

Flowers in ample, loose panicles:
  Leaves ovate; fruit wingless .... 4. *P. alaskanum*
  Leaves lanceolate; fruit with 3 wings .... 5. *P. tripterocarpum*
Flowers in spikes or axillary clusters:
  Stem twining, leaves cordate .... 1. *P. convolvulus*
  Stem not twining:
    Flowers in spikes:
      Stem not branched; inflorescence dense; plant with short, stout rhizome and radical leaves:
        Inflorescence slender, with bulblets in lower part; calyx white or pinkish
        .... 2. *P. viviparum*
        Inflorescence about 1 cm thick, lacking bulblets; calyx pink ....
        .... 3. *P. bistorta* subsp. *plumosum*
      Stem branched; inflorescence dense or interrupted; plant lacking stout rhizome and radical leaves:
        Plant with creeping rhizome, usually floating; base of leaves ovate or cordate .... 6. *P. amphibium* subsp. *laevimarginatum*
        Plant not floating; base of leaves cuneate:
          Perianth with brownish glands .... 11. *P. hydropiper*
          Perianth lacking glands:
            Ocreae ciliate; peduncles and stem lacking glands:
              Raceme slender, loosely flowered, about 5 mm thick in fruit ....
              .... 8. *P. hydropiperoides*
              Raceme ovate, stout and compact, broader ...... 9. *P. persicaria*
            Ocreae not ciliate:
              Peduncles with stipitate glands ....
              .... 7. *P. pennsylvanicum* subsp. *Oneillii*
              Peduncles glabrous or with sessile glands .... 10. *P. lapathifolium*
    Flowers in axillary clusters:
      Achenes lustrous:
        Achenes included; leaves linear-oblong .... 13. *P. prolificum*
        Achenes exserted; leaves elliptic to obovate, succulent .... 14. *P. Fowleri*
      Achenes dull:
        Leaves elliptic-obovate; fruiting calyx constricted below subrostrate tip
        .... 15. *P. achoreum*
        Leaves linear to oblong or elliptical; fruiting calyx not constricted below tip:
          Stem filiform; leaves small; calyx with distinct purplish margins ......
          .... 16. *P. caurianum*
          Stem thicker; leaves larger; calyx green or purplish ... 12. *P. aviculare*

×½

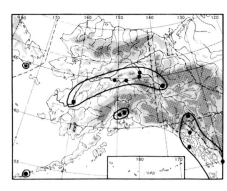

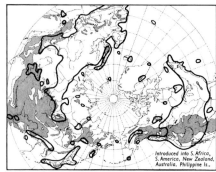

Introduced into S. Africa, S. America, New Zealand, Australia, Philippine Is.

1. Polýgonum convólvulus L.   Black Bindweed
Bilderdykia convolvulus (L.) Dumort.

    Stem twining, leaves ovate, dull green; flowers greenish-white, short-peduncled, pedicels 1–2 mm, jointed above middle; fruiting calyx rhomboid-ovoid, not winged. An introduced weed. Described from Europe.

39. POLYGONACEAE (Buckwheat Family)

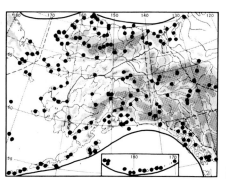
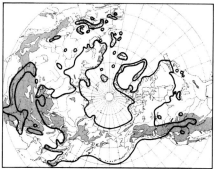

2. Polýgonum vivíparum L.
Bistorta vivipara (L.) S. F. Gray.

Rhizome thick, hard, usually contorted; radical leaves lanceolate, lance-oblong or linear, lustrous above, grayish below, glabrous; spike with lower flowers replaced by bulblets, sometimes growing out to small plants; calyx of upper flowers 5-parted, white or pink.

Dry meadows, heaths, in the Yukon to at least 2,200 meters. Described from Europe.

Highly variable within area of interest. Inland specimens are tall, with long, petiolated basal leaves, linear stem leaves, and long, narrow spike (*P. fugax* Small; *Bistorta ophioglossa* Greene). A gigantic plant, with large leaves, compact spike, and long-persistent bulblets, is var. **Macoùnii** (Small) Hult. (*P. Macounii* Small; *Bistorta Macounii* Greene; *B. littoralis* Greene).

The rhizome is collected in early spring and eaten raw by the natives of Siberia. It is said to taste like almonds.

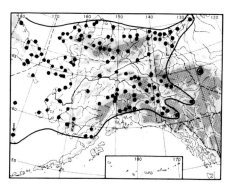
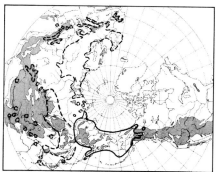
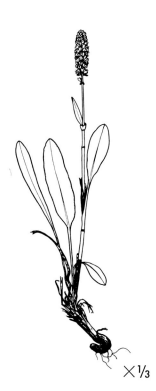

3. Polýgonum bistórta L.   Bistort
subsp. plumòsum (Small) Hult.

Polygonum plumosum Small; *P. ellipticum* Willd.; *P. bistorta* subsp. *ellipticum* (Willd.) Petrovsky; *Bistorta plumosa* Greene; *B. major* subsp. *plumosa* (Small) Hara.

Rhizome thick, hard, usually contorted; radical leaves elliptic or elongated, with rounded or mostly cuneate base and winged petioles, dark green above, grayish beneath; spike thick, with pink flowers and dark anthers.

Meadows, heaths, tundra bogs; in the Yukon to at least 2,000 meters. *P. bistorta* described from Switzerland, Austria, France. Broken line on circumpolar map indicates range of other subspecies.

The rhizome and the leaves are eaten boiled by the natives.
(See color section.)

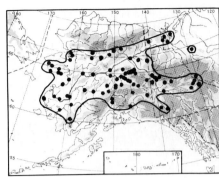
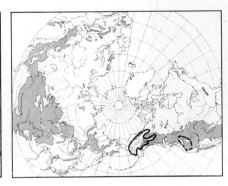

4. Polýgonum alaskànum (Small) Wight

Polygonum alpinum All. *alaskanum* Small; *P. alpinum* All. var. *lapathifolium* Cham. & Schlecht.

Root thick, woody; stem hollow, up to 2 meters tall, branched; leaves ovate, acute or ovate-lanceolate, with attenuate, acute apex, dark green above, paler beneath, crisp-pubescent in margin; inflorescence with divaricate, strigose-pubescent branches, many-flowered; calyx yellowish-white; fruiting calyx membranous, elliptic; fruit ovate, triangular in cross section, light brown.

Spruce woods, along rivers. Described from Alaska to Washington.

Closely related to **P. phytolaccaefòlium** Meisn., which occurs south of the continental glaciation, and to **P. alpìnum** All., of Eurasia. Yukon specimens are almost always glabrous (var. **glabréscens** Hult.).

Young stems and leaves are eaten boiled by the natives.

Broken line on circumpolar map indicates range of *P. phytolaccaefolium*.

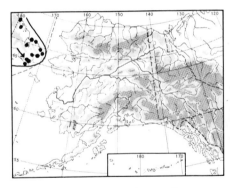
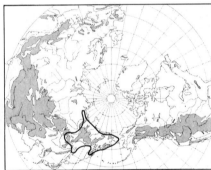

5. Polýgonum tripterocárpum Gray

Polygonum Pawlowskianum Glehn.

Plant with stout rootstock; stem up to 40 (50) cm tall, often reddish; stem leaves lanceolate, acute, with cuneate or rounded base; inflorescence leafy below; flowers on pedicels 1.5–2 mm long; calyx 2.5–3 mm long; achenes with 3 wings, up to 1.5 mm wide.

Tundra, alpine meadows. Described from "Arakamtchatch Island" (Arakamchechen Island, southern Bering Strait).

39. POLYGONACEAE (Buckwheat Family)

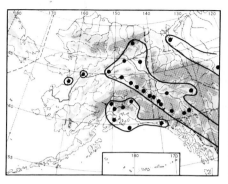 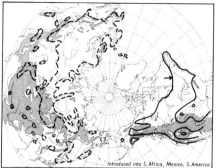 

6. Polýgonum amphíbium L.
subsp. **laevimarginàtum** Hult.

Water Smartweed

Polygonum amphibium var. *stipulaceum* Colem.; *Persicaria amphibia* var. *stipulacea* (Colem.) Hara.

Creeping, from branched rhizome; leaves with smooth margin, the lower long-petiolated; spikes thick; calyx pink.

Wet places. Occurs in aquatic forms with very long-petiolated leaves and in terrestrial forms with shorter-petiolated leaves. *P. amphibium* described from Europe, subsp. *laevimarginatum* from lake "S. Joannis."

Broken line on circumpolar map indicates range of subsp. **amphíbium**.

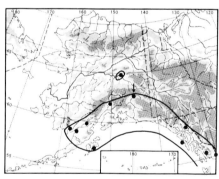 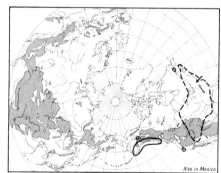 

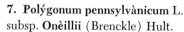

7. Polýgonum pennsylvànicum L.
subsp. **Onèillii** (Brenckle) Hult.

Persicaria Oneillii Brenckle; *P. pennsylvanicum* of Fl. Alaska & Yukon.

Ascending or decumbent; peduncles with scattered, short-stipitate glands; leaves lanceolate to elliptic, pubescent below; spikes dense, erect, pink, subglobose to ovoid or short cylindric; achenes lenticular, biconcave when dry, brown, somewhat shiny.

Wet places. Broken line on circumpolar map indicates range of subsp. **pennsylvànicum**.

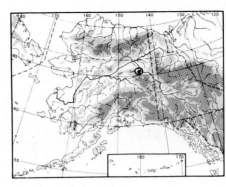 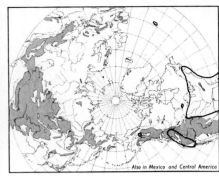

8. Polýgonum hydropiperoìdes Michx.

Plant with slender, creeping rootstock, and ascending flowering branches; ocreae short, strigose, ciliate with bristles; leaves linear-lanceolate to lance-oblong, scabrous to ciliate; spikes slender, cylindric, interrupted at base, about 5 mm thick in fruit; calyx rose-colored.

Reported from Circle Hot Springs. Range very unclear. Described from Pennsylvania, Virginia, Carolina.

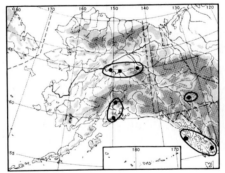 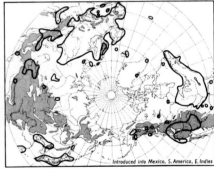

9. Polýgonum persicària L.

Stem ascending or decumbent, sometimes submerged; ocreae pubescent, with red ring below, ciliate; leaves narrowly to broadly lanceolate, glabrous, often with dark spot above; spike thick, dense, rachis and pedicels not glandular; calyx white or reddish, not glandular.

An introduced weed. Described from Europe.

39. POLYGONACEAE (Buckwheat Family)

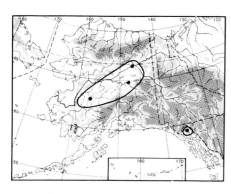
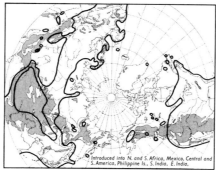

10. Polýgonum lapathifòlium L.
Including *Polygonum nodosum* Pers., *P. tomentosum* Schrank, and *P. scabrum* Moench.

Stem procumbent or erect, sometimes red-spotted; ocreae not ciliate, or with very short ciliae; leaves ovate to linear-lanceolate, acute or obtuse, glabrous or tomentose beneath, often with yellow glands beneath; spikes usually dense; peduncles (sometimes also pedicels) with yellow, subsessile glands; perianth greenish-white or pink.

An introduced weed. Described from "Gallia."

Extremely variable; includes several variations sometimes regarded as species or subspecies: extremes are **P. nodòsum** Pers. (*P. lapathifolium* subsp. *nodosum* E. Fries at least in part), which has glabrous leaves with yellow glands, pink flowers, and lax spike; and **P. tomentòsum** Schrank [*P. lapathifolium* subsp. *pallidum* (With.) E. Fries; *P. pallidum* With.], which has densely tomentose leaves and greenish-white flowers.

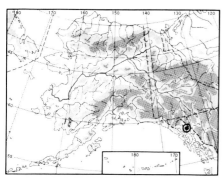
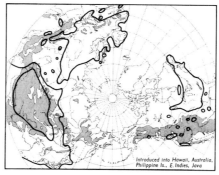

11. Polýgonum hydròpiper L.

Stem erect or depressed; leaves lanceolate, glabrous, undulate in margin; lower ocreae glabrous, truncate, with cilia 1–2 mm long; spike slender, lax, few-flowered; calyx greenish, covered with dark, sessile glands; pedicels and rachis glandular.

An introduced weed. Described from Europe.

39. POLYGONACEAE (Buckwheat Family)

× ⅓

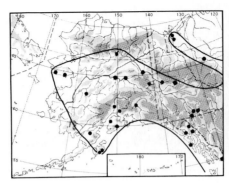

12. Polýgonum aviculàre L. Knotweed

Prostrate or ascending, much branched; leaves blue-green, linear to oblong or elliptic; flowers sessile; calyx green, with whitish or reddish margin; achenes dull, mostly included within calyx.

Introduced. Described from Europe.

Includes several variations sometimes regarded as species (*P. heterophyllum* Lindm.; *P. neglectum* Bess., *P. buxiforme* Small).

× ⅓

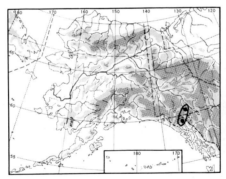
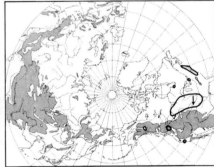

13. Polýgonum prolíficum (Small) Robins.

Polygonum ramosissimum prolificum Small; *P. ramosissimum* of Fl. Alaska & Yukon.

Similar to *P. aviculare*, but achenes black, smooth and shiny, included; leaves lanceolate to linear, firm; sepals green, with white margin.

Waste places, saline soil, prairies; probably introduced. Total range incompletely known.

39. POLYGONACEAE (Buckwheat Family)

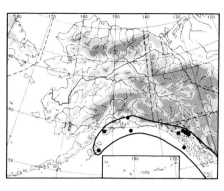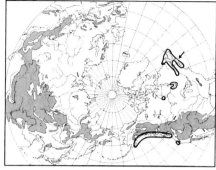

14. Polýgonum Fówleri Robins.
Polygonum boreale with respect to Alaskan plant.

Similar to *P. aviculare*, but stout, suffrutescent; leaves succulent, oblong to elliptic or obovate; achenes lustrous, olive-brown, strongly exserted at maturity.

Sandy seashores.

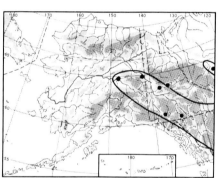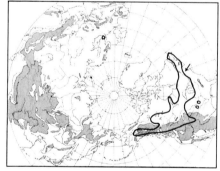

15. Polýgonum achòreum Blake

Similar to *P. aviculare*, but leaves crowded, elliptic to obovate, broadly rounded above; fruiting calyx large, strongly constricted below the subrostrate tip; inner sepals white-margined; achenes olivaceous, dull.

Waste places. Introduced in Denmark (Svendborg).

39. POLYGONACEAE (Buckwheat Family) / 40. CHENOPODIACEAE (Goosefoot Family)

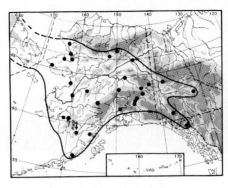

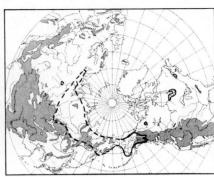

16. Polýgonum cauriànum Robins.

Similar to *P. aviculare*, but stem filiform, leaves smaller, blunt, distinctly stipitate; pedicels often as long as calyx; calyx with distinct purplish margin. A fairly characteristic relative of *P. aviculare*.

Gravel bars along rivers, waste places; probably native. Known only from localities marked on map.

Broken line on circumpolar map indicates range of the closely related or possibly identical **P. humifùsum** Pall.

Chenopodium L.

Flowers in globose, sessile heads forming an interrupted spike, and becoming red and berrylike in fruit . . . . 1. *C. capitatum*
Flowers not in globose heads, not berrylike in fruit:
    Larger stem leaves cordate or truncate at base; flowers in loose, panicled racemes
    . . . . 2. *C. hybridum* subsp. *giganthospermum*
    All leaves more or less cuneate at base:
        Leaves sinuate-toothed, pale green above, white-farinose beneath . . . .
        . . . . 3. *C. glaucum* subsp. *salinum*
        Leaves green on both sides:
            Stem with leaves to top; plant often reddish . . . . 4. *C. rubrum*
            Stem leafless at top:
                Leaves linear, 1-nerved, mostly entire . . . . 7. *C. leptophyllum*
              Leaves broader, mostly toothed:
                    Seeds nearly smooth, shiny, 1.2–1.7 mm in diameter . . . . . 5. *C. album*
                    Seeds pitted, up to 1 mm in diameter . . . .
                    . . . . 6. *C. Berlandieri* subsp. *Zschackei*

40. Chenopodiaceae (Goosefoot Family)

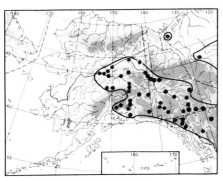 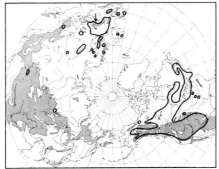 

1. Chenopòdium capitàtum (L.) Aschers. Strawberry Blite
Blitum capitatum L.

Glabrous, yellowish-green, erect annual; branches leafless at tip; lower leaves triangular, sinuate, sometimes hastate; flowers of two kinds, the one with 3–4 stamens, the other with 1 or no stamen, in spherical glomerules; calyx lobes oblong, fleshy in fruit, bright red.

Waste places, cultivated soil, river bars.
The young leaves can be eaten boiled.

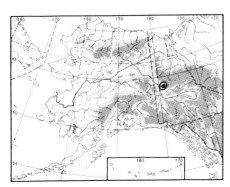 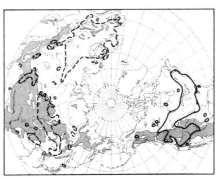 

2. Chenopòdium hỳbridum L.
subsp. **giganthospérmum** (Aellen) Hult.

Chenopodium giganthospermum Aellen; *C. hybridum* var. *giganthospermum* (Aellen) Rouleau.

Glabrous, much branched annual; leaves thin, bright green, ovate, with rounded or subcordate base and triangular lobes; panicle loose, with leafless branches; calyx thin, with lanceolate to ovate segments; seeds black, lenticular, lustrous, 1.2–2 mm in diameter.

Waste places; weed at Dawson. *C. hybridum* described from Europe, subsp. *giganthospermum* from Canada, United States.
Broken line on circumpolar map indicates range of subsp. **hỳbridum**.

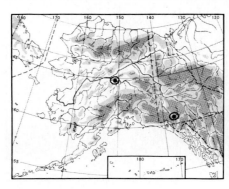
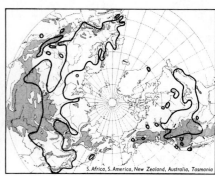
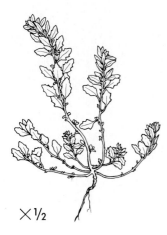

3. Chenopòdium glaùcum L.
subsp. **salìnum** (Standl.) Aellen
Chenopodium salinum Standl.

Prostrate or ascending, branching from base, glabrous or nearly so; leaves ovate to broadly lanceolate, sinuately toothed, pale green above, densely farinose beneath; flowers in small, axillary spikes, shorter than subtending leaves; calyx green; seeds finely tuberculate.

Waste places, saline or alkaline soil. *C. glaucum* described from Europe, subsp. *salinum* from northern New Mexico.

Plant from hot springs on Tanana River, which has entire upper leaves and small (1 mm in diameter), foveolate-punctate seeds, has been described as var. **pùlchrum** Aellen. Circumpolar map indicates range of *C. glaucum* in a broad sense; subsp. *salinum* occurs in northern and western North America.

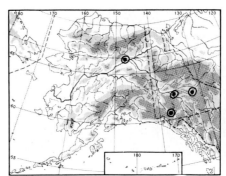
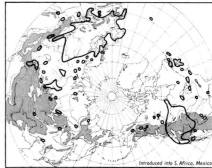

4. Chenopòdium rùbrum L.

Stem often reddish, erect, simple or branched, glabrous, leafy to top; leaves not farinose, with cuneate base, rhombic to deltoid, very coarsely toothed, the upper narrow; flowers in axillary spikes; calyx with obtuse lobes; seeds lustrous, about 0.9 mm in diameter.

An introduced weed. Described from Europe.

40. CHENOPODIACEAE (Goosefoot Family)

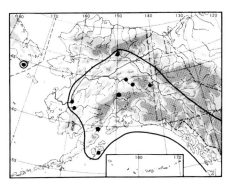
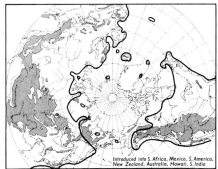
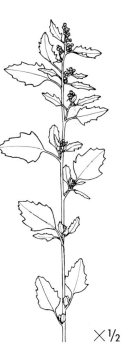

5. Chenopòdium álbum L. Pigweed

Including *Chenopodium paganum* Rchb.

Plant more or less mealy, with short, strict branches; leaves longer than broad, rhomboid to lanceolate, toothed, highly variable in form; inflorescence lacking leaves at top; flowers in densely crowded glomerules; perianth segments keeled; seeds 1.2–1.7 mm in diameter, lustrous, not pitted.

An introduced weed. Described from Europe.

The leaves can be eaten raw or boiled.

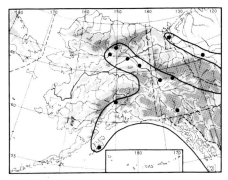
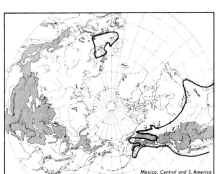

6. Chenopòdium Berlandièri Moq.
subsp. **Zscháckei** (Murr.) Zobel

Chenopodium Zschackei Murr.

Similar to *C. album*, but highly variable in leaf form; seeds smaller, up to 1 mm in diameter, pitted on upper surface.

An introduced weed. *C. Berlandieri* described from Mexico, subsp. *Zschackei* from Anhalt.

Circumpolar map gives range of entire species complex.

40. CHENOPODIACEAE (Goosefoot Family)

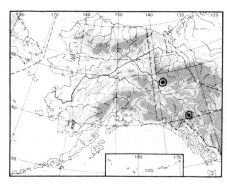 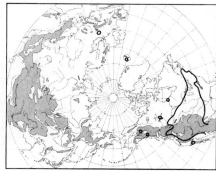

7. Chenopòdium leptophýllum Nutt.

Erect annual, simple or branched from base, whitish-farinose above; leaves thick, linear to linear-oblong, 1-nerved, entire, farinose; ends of branches leafless; flowers in dense glomerules; calyx with keeled lobes, completely enclosing fruit; seed 1 mm in diameter, smooth, lustrous.

Waste places, sandy soil; introduced. Described from Colorado, Nevada, New Mexico.

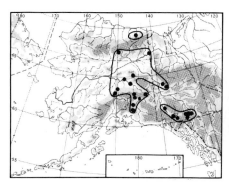 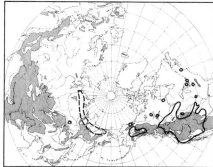

Monolepis Schrad.

1. Monólepis Nuttalliàna (Schult.) Greene
Blitum Nuttallianum Schult.

Ascending or erect, pale green, glabrous, somewhat fleshy annual, branched from base; leaves narrowly rhombic, usually hastate, sometimes linear; only 1 bract-like, obtuse, fleshy sepal and 1 stamen; seeds dark, papillate.

Saline and alkaline soil; in McKinley Park to about 600 meters. Described from the banks of the Missouri River.

Not native in eastern North America; introduced in a few places in Europe. Broken line on circumpolar map indicates range of the closely related **M. asiática** Fisch. & Mey.

40. CHENOPODIACEAE (Goosefoot Family)

Atriplex L. (Orach)

Leaves sessile .... 1. *A. drymarioides*
Leaves (at least the lower) petiolated:
    Fruiting bracts with 2 lateral teeth; basal leaves denticulate, with forward-turned hastate lobes .... 3. *A. patula*
    Fruiting bracts smooth on sides, entire or slightly dentate:
        Thin, ovate, entire fruiting bracts, not jointed below; leaves cordate or triangular .... 5. *A. hortensis*
        Thick, triangular or ovate-rhombic fruiting calyx, jointed below:
            Leaves narrow; fruiting bracts small, obtuse, ovate-rhombic, sometimes with somewhat attenuate apex .... 2. *A. Gmelini*
            Leaves broad; fruiting bracts large, acute, ovate or rhombic, with broad, attenuate apex .... 4. *A. alaskensis*

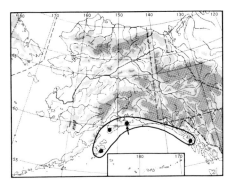
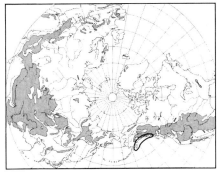
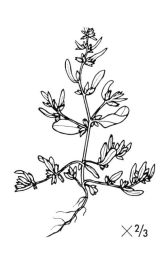

1. Átriplex drymarioìdes Standl.

Branched from base, branches slender; lower leaves opposite, upper alternate, all sessile, cuneate-obovate to oblong, obtuse, entire, somewhat farinose; flowers in few-flowered axillary glomerules; calyx 4-cleft; fruiting bracts elliptic-oblong or oblanceolate, entire, on long pedicels.

Seashores.

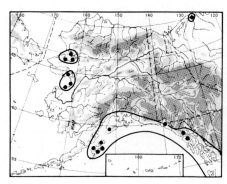 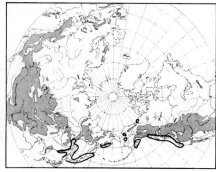

2. Átriplex Gmélini C. A. Mey.

Decumbent or erect, branched from base; lower leaves opposite, oblong to lance-oblong, obtuse, sometimes with a few teeth near base; upper leaves alternate, linear, somewhat acute; flowers in simple spikes and axillary glomerules; calyx 5-cleft; fruiting bracts ovate-rhombic to oblong, sessile, united at base, not appendaged.

Seashores. Described from Kamchatka and Kotzebue Sound.

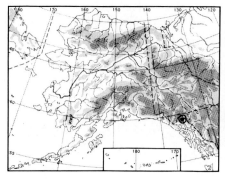 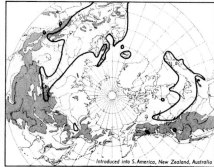

3. Átriplex pátula L.

Erect or decumbent; stem usually much branched, leafless above; leaves short-peduncled, triangular-hastate, cuneate at base, 3–4 times as long as wide, usually not mealy, the upper entire; flowers in interrupted spikes; fruiting bracts rhombic, acute or somewhat acute, often tuberculate on back, with 2 lateral teeth.

An introduced weed. Described from Europe.

40. Chenopodiaceae (Goosefoot Family)

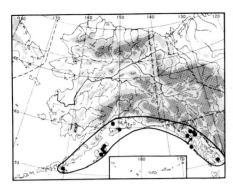 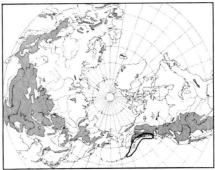 

4. Átriplex alaskénsis S. Wats.

Erect or ascending, with stout branches; lower leaves opposite, petiolated, ovate, ovate-oblong or deltoid-oblong, entire or repand-denticulate; upper leaves alternate, lanceolate, entire or subhastate; flowers in leafy spikes and axillary glomerules; fruiting bracts large, sessile, oblong, sometimes broader than long, united and thickened at base, with attenuate apex, not appendaged.

Seashores.

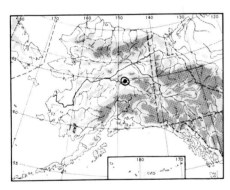 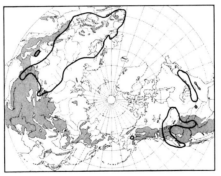 

5. Átriplex horténsis L.

Erect annual; leaves cordate or hastate-triangular, slightly farinose when young, glabrous when mature, often purple-brown; inflorescence a terminal spike; fruiting bracts thin, ovate, entire in margin, completely free from each other.

Escaped from cultivation at Fairbanks; probably not persistent. Described from Tataria.

40. Chenopodiaceae (Goosefoot Family)

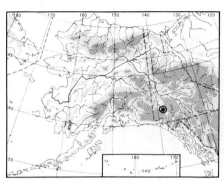
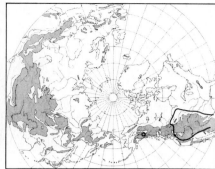

Eurotia Adans.

1. Euròtia lanàta (Pursh) Moq.
Diotis lanata Pursh.

Plant woody at base, with stout, erect branches; plant pubescent, with coarse, white-to-reddish, stellate hairs mixed with straight hairs; leaves linear, the upper sessile, obtuse with revolute margin; fruiting bracts lanceolate, densely villous.

Rocky places. One collection from Lake Kluane. Described from the banks of the Missouri River.

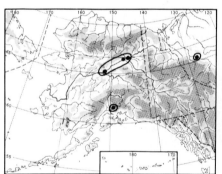
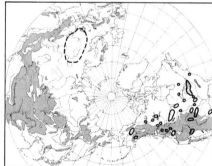

Corispermum L.

1. Corispérmum hyssopifòlium L. Bugseed

Erect, slender, much branched, stellate-villous when young, glabrescent in age, often reddish; leaves linear, cuspidate; spikes 4–8 mm thick, with imbricate, lanceolate to ovate bracts, the median and upper bract mostly broader than the fruit; fruit with thick, narrow margin.

Sandy beaches. Described from the Volga, Prussia, Montpellier.

Very unclear taxonomically. Circumpolar map tentative; broken line indicates range of closely related taxa in Europe.

40. CHENOPODIACEAE (Goosefoot Family)

Salicornia L.

Plant annual, central flower far exceeding lateral flowers .... 1. *S. europaea*
Plant perennial, from creeping rootstock; all flowers in each group of 3 of nearly
    same size .... 2. *S. virginica*

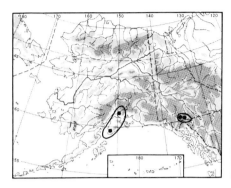
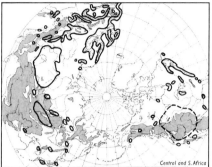

1. Salicórnia europaèa L.   Glasswort
Salicornia herbacea L.

Simple or branched annual; segments articulate, longer than broad, with convex sides, often reddish when in fruit; leaves succulent, translucent, glabrous; flowers in terminal, more or less branched, spikes, with axillary, 3-flowered cymes; lateral flowers smaller than central flowers.

Saline and alkaline soil. Described from Europe.

Circumpolar map tentative; broken line indicates range of subsp. **rùbra** (Nels.) Breitung (*S. rubra* Nels.).

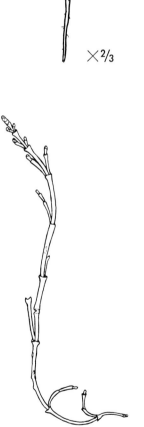

×²⁄₃

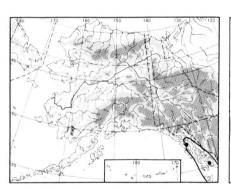
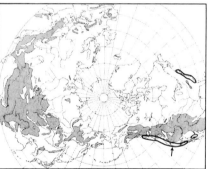

2. Salicórnia virgínica L.
Salicornia pacifica Standl.

Plant from creeping rootstock, perennial, suffrutescent; stem with stout branches; segments longer than broad; 3 flowers in the cymes, about equally large.

Seashores, alkaline soil.

×²⁄₃

40. Chenopodiaceae (Goosefoot Family)

Suaeda Forsk. (Sea Blite)

Upper leaves distinctly broader at base; calyx lobes unequal, one or more corniculated .... 1. *S. depressa*
Upper leaves less distinctly broader at base; calyx lobes with lobed transversal wing .... 2. *S. occidentalis*

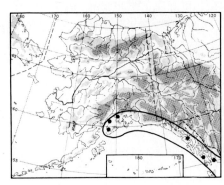

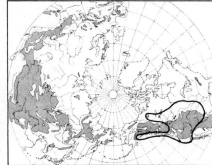

1. Suaèda depréssa (Pursh) S. Wats.
Salsola depressa Pursh; *Dondia depressa* (Pursh) Britt.

Decumbent or erect, glabrous, branching from base; leaves linear, semiterete, acute, at least the upper broadest at base; calyx lobes acute or obtuse, unequal, 1 or more longer than the other and corniculately appendaged; seeds black, shiny.

Seashore. Described from the Missouri plains.
Circumpolar map highly tentative.

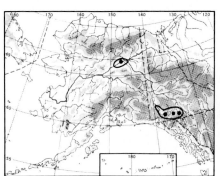

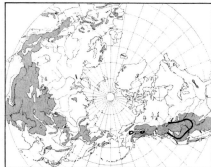

2. Suaèda occidentàlis S. Wats.
Dondia occidentalis (S. Wats.) Heller.

Similar to *S. depressa*, but leaves less broadened at base; calyx lobes obtuse, with more or less lobed, transversal wings in mature state.

Alkaline soil. Described from Ruby Valley, Nevada.
Circumpolar map highly tentative.

41. AMARANTHACEAE (Amaranth Family)

Amaranthus L.

Perianth segments 5; stem erect .... 1. *A. retroflexus*
Perianth segments 3; stem prostrate or decumbent .... 2. *A. graecizans*

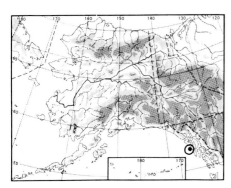
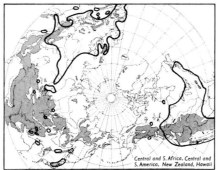

1. Amaránthus retrofléxus L.   Green Amaranth

Stem erect, lanate above; leaves rhombic-ovate; inflorescence spicate, leafless at tip, short and dense; bracteoles stout and spinescent; perianth segments 5, enlarged in upper part.

Introduced weed at Sitka; certainly not persistent. Described from Pennsylvania.

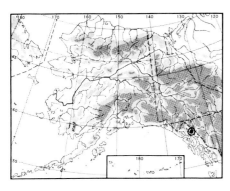
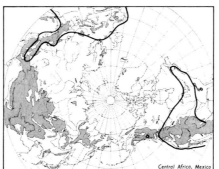

2. Amaránthus graecìzans L.   Tumbleweed

Stem ascending or prostrate, glabrous; leaves ovate to rhombic; inflorescence of axillary, cymose clusters; bracteoles mucronulate; perianth segments 3.

Introduced weed at Juneau; certainly not persistent. Described from Virginia.

Claytonia L. (Spring Beauty)

Inflorescence shorter than basal leaves or about equaling them .... 1. *C. megarhiza*
Inflorescence longer than basal leaves:
   Stem leaves alternate:
      Basal leaves ovate or obovate .... 4. *C. parvifolia* subsp. *flagellaris*
      Basal leaves linear .... 5. *C. Bostockii*
   Stem leaves opposite:
      Cauline leaves in several pairs .... 10. *C. Chamissoi*
      Cauline leaves in 1 pair:
         Cauline leaves united, forming ring around stem; flowers small ....
         .... 11. *C. perfoliata*
         Cauline leaves separated; flowers larger:
            Cauline leaves narrow, lanceolate, linear, or linear-lanceolate:
               Stem 1(–3) flowered; plant with thin, filiform rootstock; flowers purple
               .... 8. *C. Scammaniana*
               Stem 3–many-flowered; plant with thick, fleshy root, or bulblike corm; flowers white, pink, or yellowish:
               Leaves fleshy; plant with thick, fleshy rootstock ....
                   .... 2. *C. acutifolia* subsp. *graminifolia*
               Leaves thin; plant with ovoid corm .... 3. *C. tuberosa*
            Cauline leaves broad, ovate, obovate, or rhombic:
               Inflorescence many-flowered; pedicels with bracts; plant usually annual .... 6. *C. sibirica*
               Inflorescence 1–5-flowered, pedicels bractless; plant perennial:
                 Plant with fleshy taproot, lacking stolons .... 7. *C. arctica*
                 Plant with filiform rootstock and bud-bearing stolons ....
                 .... 9. *C. sarmentosa*

×½

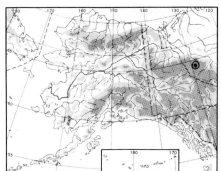 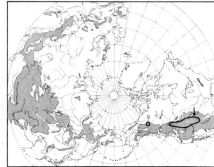

1. Claytònia megarhìza (Gray) Parry
Claytonia arctica megarhiza Gray.

Plant with large, thick, fleshy taproot; stems several, fleshy; basal leaves fleshy, with winged petioles, numerous, spatulate; stem leaves linear to spatulate, usually opposite; inflorescence shorter than leaves; sepals ovate, somewhat acute; petals clawed, white to pink.

Crevices in rocks.

42. Portulacaceae (Purslane Family)

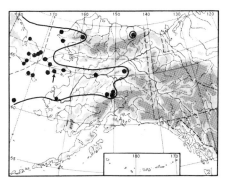
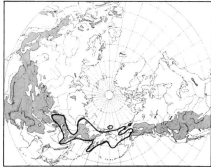

2. Claytònia acutifòlia Pall.
subsp. graminifòlia Hult.
Claytonia Eschscholtzii Cham.

Plant with thick, long taproot; stems several, ascending or more or less prostrate; basal leaves several, surrounded by hyaline, leafless sheaths, linear to lanceolate-linear, rarely broader; stem leaves 2, opposite, lanceolate; inflorescence few-flowered, subtended by bracts; sepals ovate; petals obovate, white or pinkish, with yellowish base.

Stony slopes in the mountains. Circumpolar map indicates range of *C. acutifolia* in a broad sense.

The root is eaten raw or boiled by the natives of Siberia; it keeps, uncooked, over the winter.

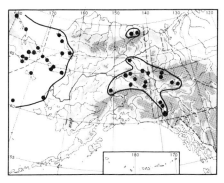
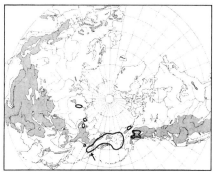

3. Claytònia tuberòsa Pall.

Plant with ovoid corm from which extends long, filiform, very brittle runners, very rarely present in herbarium specimens; basal leaves lacking or few, long-petiolated, lanceolate; stem mostly single, buried deep in ground, very thin and brittle close to corm; stem leaves 2, lanceolate to ovate-lanceolate; raceme few-flowered, usually with 1 rounded bract; sepals ovate; petals obovate, more or less retuse, white with yellowish base.

Wet places, stony slopes in the mountains.

The corm is eaten boiled or roasted by the natives; the leaves are also eaten as salad. Var. **czukczòrum** (Volk.) Hult. (*C. czukczorum* Volk.), with 2–4 stems from each corm and several basal leaves, the leaves usually short and broad, occurs in exposed places.

42. Portulacaceae (Purslane Family)

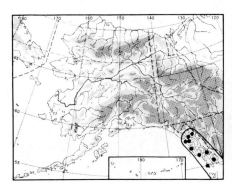
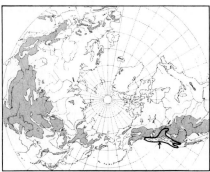

4. Claytònia parvifòlia Moq.

Montia parvifolia Greene; *Naiocrene parvifolia* (Moq.) Rydb.

subsp. **flagellàris** (Bong.) Hult.

Claytonia flagellaris Bong.; *C. parvifolia* and *C. flagellaris* of Fl. Alaska & Yukon; *C. parvifolia* var. *flagellaris* (Bong.) R. J. Davis; *Montia flagellaris* Robins.; *M. parvifolia* subsp. *flagellaris* (Bong.) Ferris.

Plant with branched, fleshy rootstock; basal leaves obovate, with broad, scarious base, fleshy; stems decumbent or ascending, often with bulblike buds in the axils; stem leaves several, alternate, ovate to rhombic, petiolated, reduced in size upward; flowers in racemes at ends of branches; sepals rounded, ovate, up to 4 mm long; petals truncate or emarginate, white or pink, up to 15 mm long.

Wet places.

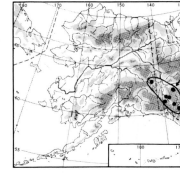
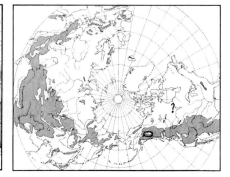

5. Claytònia Bostóckii Pors.

Claytonia sp. of Fl. Alaska & Yukon.

Plant with thin root and long leafy stolons; basal leaves somewhat fleshy, linear, 1–2 cm long, those of the stolons linear-oblanceolate, petiolated; inflorescence few-flowered, with single lanceolate bract at base; calyx leaves broadly ovate, somewhat rose-colored; petals about 1 cm long, truncate or slightly emarginate, white or pale rose-colored.

Wet places in the mountains.

42. Portulacaceae (Purslane Family)

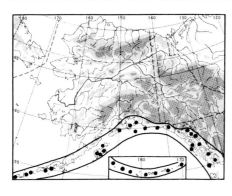
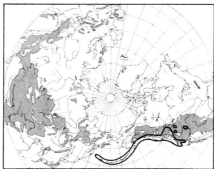
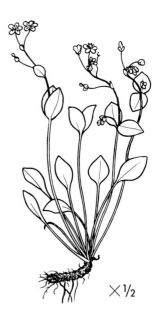

6. Claytònia sibìrica L.

Claytonia unalaschkensis Fisch.; *C. alsinoides* Sims; *C. asarifolia* Bong.; *Montia sibirica* (L.) How.; *Limnia sibirica* (L.) Harv.; *L. asarifolia* (Bong.) Rydb.; *L. alsinoides* (Cham.) Rydb.

Plant with a slender rootstock; basal leaves several, long-petiolated, broadly to narrowly ovate to rhombic or subreniform, mostly acute or somewhat so; stems several; stem leaves 2, opposite, broadly ovate; raceme many-flowered, open, with 1 to few bracts; sepals ovate; petals obcordate, pink, white, or white with pink lines.

Moist, shady places; common along coast. Described either from Bering Island or from Kayak Island, Alaska. Occurs rarely as an introduced weed in Europe.

The leaves are eaten raw or boiled by natives.

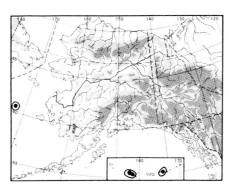
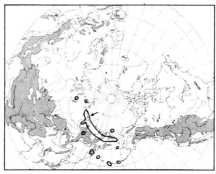
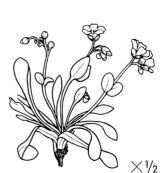

7. Claytònia árctica Adams

Plant with thick, fleshy taproot; basal leaves several, thick, nearly nerveless; stems several; stem leaves 2, sessile, ovate, at middle of stem; bracts lacking; calyx leaves broadly elliptic; petals white with yellowish base, emarginate.

Alpine tundra.

×½

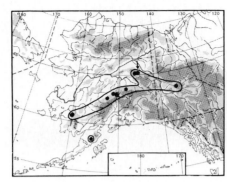 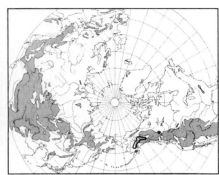

8. Claytònia Scammaniàna Hult.

Claytonia Koliana Gombócz.

Plant with filiform rootstock forming loose clumps; basal leaves linear to oblanceolate, obtuse, fleshy, dilated at base into broad hyaline sheath; stems 1–2-flowered (rarely 3-flowered); stem leaves 2, ovate, opposite; sepals orbiculate; petals obovate-emarginate, deeply purple-colored.

Stony slopes and scree slopes in the mountains, to at least 2,000 meters.

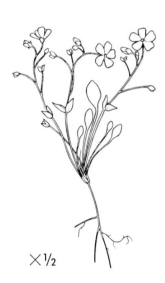

×½

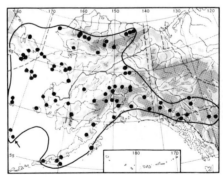 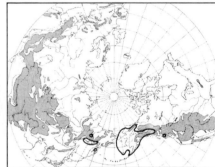

9. Claytònia sarmentòsa C. A. Mey.

Limnia sarmentosa (C. A. Mey.) Rydb.; *Montia sarmentosa* (C. A. Mey.) Robins.

Plant with thin, filiform rootstock and filiform runners; basal leaves several, obovate or spatulate, petiolated; stem leaves 2, opposite, sessile, broadly ovate, somewhat acute; inflorescence few-flowered, lacking bracts; sepals broadly ovate; petals obovate, retuse, white or rose-colored.

Rocky slopes, snow beds, in the mountains to at least 2,000 meters.

42. PORTULACACEAE (Purslane Family)

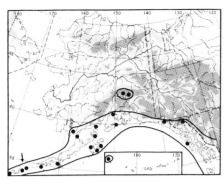
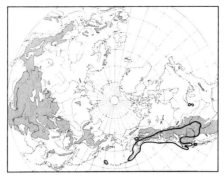

10. Claytònia Chamíssoi Esch.

Claytonia Chamissonis Esch.; *Montia Chamissonis* (Esch.) Gray; *Montia Chamissoi* (Esch.) Durand & Jackson; *Crunocallis Chamissonis* (Esch.) Rydb.

Plant with thin root and long filiform stolons, ending with buds; stolons procumbent or decumbent, rooting at nodes; leaves entire, opposite, in several pairs, oblanceolate to obovate-lanceolate; inflorescence bractless; sepals obovate to orbicular; petals obovate, white or pinkish; capsule 3-valved.

Wet places, gravel bars.

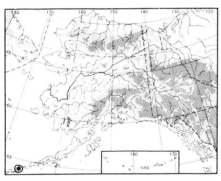
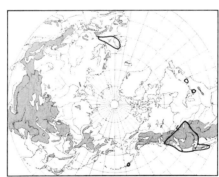
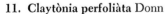
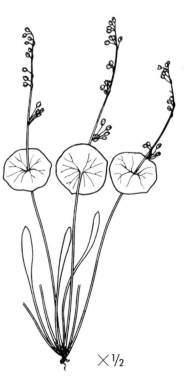

11. Claytònia perfoliàta Donn

Limnia perfoliata (Donn) How.; *Montia perifoliata* (Donn) How.

Annual, glabrous; basal leaves petiolated, rhombic-ovate to elliptic-obovate; stem leaves connate, forming oblique, suborbicular disk; racemes elongate, sometimes with small bract; flowers more or less verticillate; sepals rounded-ovate or suborbicular; petals obovate, setose, clawed, white or pink.

Shady places. Found introduced once at Unalaska; introduced also in Europe, eastern North America, and in the Southern Hemisphere. Described from the west coast of North America.

42. Portulacaceae (Purslane Family)

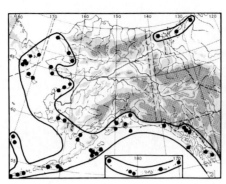
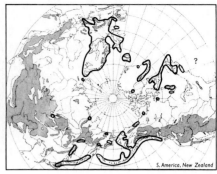

Montia L.

1. Móntia fontàna L.
subsp. **fontàna**

Montia lamprosperma Cham.

Water Blinks

Stem weak, glabrous, simple or forking, prostrate or floating; leaves opposite, narrow; flowers inconspicuous, sepals persistent, petals 5, white; seeds 1–1.5 mm long, reniform, more or less lustrous, slightly tuberculate on keel.

Springs, pools, wet places. Described from Europe.

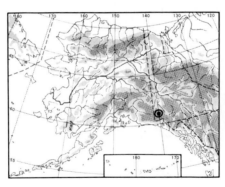
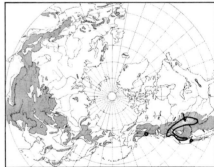

Lewisia Pursh

1. Lewísia pygmaèa (Gray) Robins.

Talinum pygmaeum Gray.

Plant with fleshy taproot; leaves linear to oblanceolate, few to several, with underground petioles; bracts hyaline, lanceolate, connate at base; flowers mostly single; sepals glandular-dentate, strongly veined; petals white to pink; capsule ovoid, circumscissile near base.

Rocky soil in high mountains to at least 2,000 meters.

43. Caryophyllaceae (Pink Family)

Stellaria L. (Chickweed)

Flowers in axis of green leaves, similar to stem leaves:
  Lower leaves long-petiolated .... 1. *S. media*
  Lower leaves, as well as others, sessile:
    Leaves lustrous, more or less firm, green or glaucous, carinate; plant usually with single terminal flower:
      Sepals completely glabrous .... 14. *S. monantha*
      Sepals ciliated in margin, more or less pubescent on back .... 16. *S. laeta*
    Leaves not lustrous, flat; plant 1- to several-flowered:
      Margin of leaves smooth, glabrous:
        Leaves ovate, very acute, thin, with somewhat undulating, translucent margin; flowers few, axillary .... 2. *S. crispa*
        Leaves ovate or narrow, lacking undulating translucent margin:
          Leaves thick, coriaceous, rigid, cordate-ovate or broadly lanceolate, glaucous .... 12. *S. ruscifolia* subsp. *aleutica*
          Leaves not coriaceous or rigid, ovate or lanceolate or narrowly lanceolate, not glaucous:
            Sepals blunt, about as long as mature capsule; seeds smooth .... 3. *S. humifusa*
            Sepals acute, small, distinctly shorter than mature capsule; seeds rugose .... 4. *S. crassifolia*
      Margin of leaves tuberculate-serrulate, ciliate at least at base:
        Plant coarse; sepals 3–4 mm long; leaves 2.5–8 cm long:
          Upper leaves small, strongly reduced; flowers numerous, in terminal cymes .... 9. *S. sitchana* var. *sitchana*
          Upper leaves not considerably reduced; flowers few, terminal or axillary; leaves long, narrow ...... 10. *S. sitchana* var. *Bongardiana*
        Plant more delicate; sepals 2–3 mm long; leaves 1–2 cm long:
          Leaves ovate, flowers solitary or in few-flowered cymes .... 6. *S. calycantha* subsp. *calycantha*
          Leaves elliptic-lanceolate:
            Upper leaves not much reduced; flowers few, axillary or terminal .... 7. *S. calycantha* var. *isophylla*
            Upper leaves much reduced; cymes strongly branched, many-flowered; flowers small .... 8. *S. calycantha* subsp. *interior*
Flowers in axis of scarious or scarious-margined bracts (in single-flowered specimens often somewhat obscure):
  Inflorescence subumbellate, with several hyaline bracts at base.. 11. *S. umbellata*
  Inflorescence not subumbellate:
    Leaves linear-lanceolate, broadest above middle, scabrous in margin from nearly microscopic tubercles; stem scabrous .... 5. *S. longifolia*
    Leaves lanceolate or broader, not scabrous in margin; stem not scabrous:
      Leaves dull, flat; sepals 5 mm long or longer; petals equaling sepals or shorter; plant 1–3-flowered .... 13. *S. alaskana*
      Leaves rigid, lustrous, carinate; sepals shorter; petals usually longer than sepals:
        Sepals glabrous in margin .... 15. *S. longipes*
        Sepals ciliated in margin .... 17. *S. Edwardsii*

43. Caryophyllaceae (Pink Family)

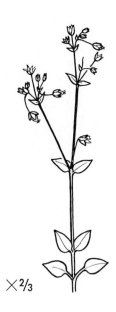

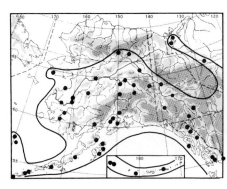

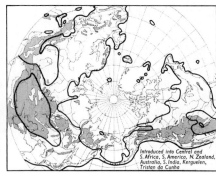

1. Stellària mèdia (L.) Vill.
Alsine media L.

Yellowish-green plant, trailing or loosely ascending; stem pubescent on one side; lower leaves ovate, acute, glabrous, with petioles pubescent on one side, the upper sessile; calyx leaves ovate-lanceolate, hyaline-margined, acute, longer than the 2-cleft petals, which are sometimes lacking; capsule as long as calyx or longer; seeds tuberculate.

An introduced weed. Described from Europe.

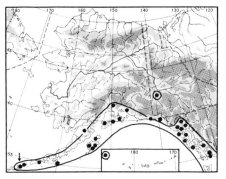

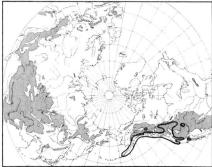

2. Stellària críspa Cham. & Schlecht.
Alsine crispa (Cham. & Schlecht.) Holz.

Plant glabrous, with slender rootstock; stems weak, mostly simple, with numerous pairs of ovate, sessile, sharply acuminate leaves, slightly undulate in margin; pedicels axillary; sepals lanceolate, acute, scarious-margined; petals shorter than sepals or lacking; capsules pale, exceeding calyx.

Moist places.

43. CARYOPHYLLACEAE (Pink Family)

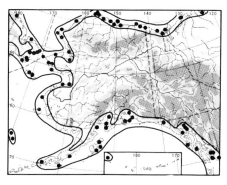
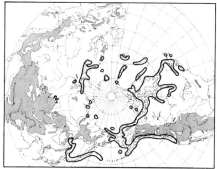

3. Stellària humifùsa Rottb.
Alsine humifusa (Rottb.) Britt.

Plant glabrous, yellowish-green; arctic specimens densely matted with short internodes and ovate, small, fleshy leaves; southern specimens loosely caespitose, with larger lanceolate-oblong, somewhat acute, sessile leaves and long internodes; branches rooting at lower internodes; flowers lateral or apical, short-peduncled, single or in few-flowered, leafy cymes; sepals ovate-lanceolate, blunt, shorter than petals but longer than, or as long as, capsule; seeds smooth.

Wet seashores. Described from Scandinavia.

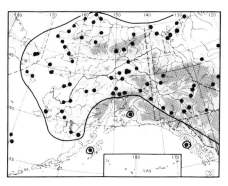
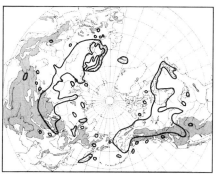

4. Stellària crassifòlia Ehrh.
Alsine crassifolia (Ehrh.) Britt.

Stems matted, depressed or diffuse, sometimes ending in buds (f. **gemmíficans** Norman); leaves sessile, slightly fleshy, highly variable in size and form, in arctic specimens small, broadly elliptic-lanceolate, in southern specimens larger, narrowly oblong-lanceolate, somewhat acute (var. **lineàris** Fenzl); flowers solitary or few, in apical or lateral leafy cymes; sepals broadly lanceolate, acute, glabrous or somewhat pubescent in margin (var. **eriocalýcina** Schischk.), slightly shorter than petals; petals cleft nearly to base; capsule conical, pale, longer than calyx; seeds wrinkled or rugose.

Wet places, snow beds, from lowlands to alpine region. Described from Germany.

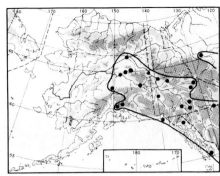
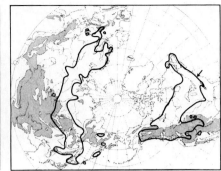

5. Stellària longifòlia Muhl.

Stellaria mosquensis Bieb.; *S. diffusa* Willd.; *S. atrata* (J. W. Moore) Boiv.; *S. longifolia* var. *atrata* J. W. Moore.

Plant yellowish-green; stem slender, quadrangular, ascending, usually glabrous or slightly scabrous; leaves sessile, linear, glabrous, minutely scabrous in margin, broadest at middle; cymes with scarious bracts, terminal or lateral, open; sepals lanceolate, acute, glabrous; petals about as long as sepals or slightly longer; capsules pale; seeds smooth.

Moist places.

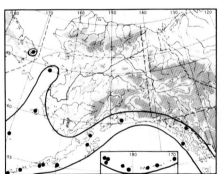
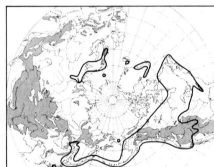

6. Stellària calycántha (Ledeb.) Bong.

Arenaria calycantha Ledeb.; *Stellaria borealis* Bigel.; *Alsina borealis* (Bigel.) Britt. subsp. **calycántha**

Stem ascending, weak, branching, glabrous or slightly scabrous; leaves sessile, ovate-lanceolate, acute, minutely ciliate near base; flowers solitary in forks of stem or in few-flowered cymes, with glabrous bracts; sepals ovate-lanceolate, acute; petals shorter than sepals or usually lacking; capsule much longer than sepals.

Wet places. Described from eastern Siberia.

A polymorphous plant. Circumpolar map indicates range of the entire species complex.

43. Caryophyllaceae (Pink Family)

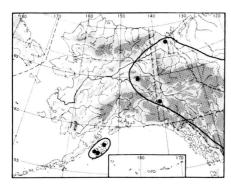 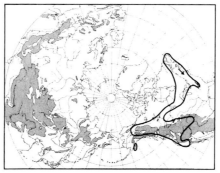 

7. Stellària calycántha (Ledeb.) Bong.
var. **isophýlla** (Fern.) Fern.
  Stellaria borealis var. *isophylla* Fern.

Upper leaves only slightly reduced; flowers few, axillary or terminal. Wet places. Circumpolar map tentative.

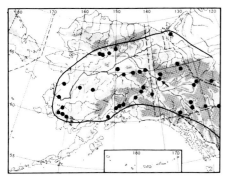 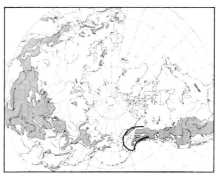 

8. Stellària calycántha (Ledeb.) Bong.
subsp. **intèrior** Hult.

Differs from subsp. *calycantha* in having longer and narrower leaves, strongly branched inflorescence, and small flowers.
  Wet places.

43. Caryophyllaceae (Pink Family)

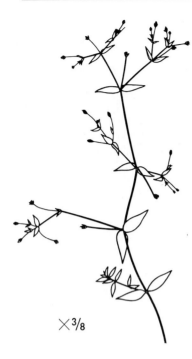
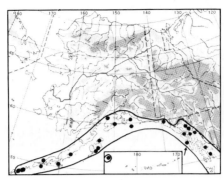
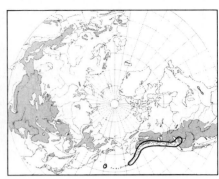

9. Stellària sitchàna Steud.
var. **sitchàna**
Stellaria borealis var. *sitchana* (Steud.) Fern.

Stems fairly stout, ascending, branching, scabrous; leaves sessile, ovate-lanceolate, acute, somewhat ciliated in margin; cymes terminal, several-flowered, with leafy bracts; sepals acute, scarious-margined, longer than the petals, which are, in fact, often missing; capsules dark, longer than calyx.

Moist places.

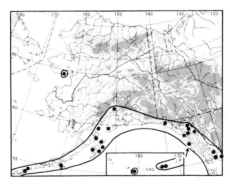
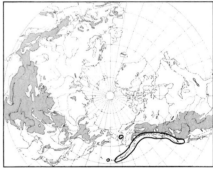

10. Stellària sitchàna Steud.
var. **Bongardiàna** (Fern.) Hult.
Stellaria borealis var. *Bongardiana* Fern.; *S. brachypetala* Bong.

Similar to var. *sitchana*, but stem simple, few-flowered; leaves serrulated, but scarcely ciliated in the margin; upper leaves nearly as large as median leaves.
Moist places.

43. Caryophyllaceae (Pink Family)

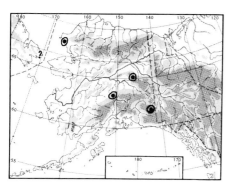
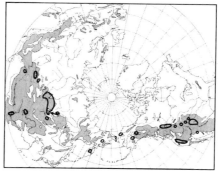

11. Stellària umbellàta Turcz.

Stellaria gonomischa Boiv.; *S. Weberi* Boiv.; *Alsine baicalensis* Cov.

Roots thin, stems several, thin, glabrous; leaves elliptic to oblong, somewhat acute; inflorescence apical, simple or compound, subumbellate, with several hyaline bracts at base; pedicels long, slender, more or less divaricate, in age reflexed; sepals ovate, acute, hyaline-margined, glabrous; petals lacking; capsule ovate, twice as long as sepals at maturity; seeds smooth.

Wet places and stony slopes, snow beds, in the mountains.

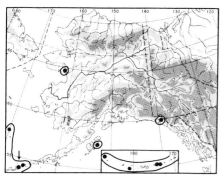
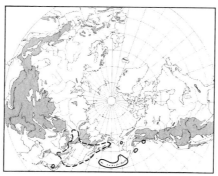

12. Stellària ruscifòlia Pall.
subsp. **aleùtica** Hult.

Glabrous; loosely tufted; leaves thick, coriaceous, densely crowded, ovate-lanceolate to ovate or subcordate, acute; flowers solitary, terminal, long-pedicellated; sepals lanceolate-ovate, acute, scarious-margined, half as long as petals.

Along creeks in the mountains. Broken line on circumpolar map indicates range of subsp. **ruscifòlia.**

43. Caryophyllaceae (Pink Family)

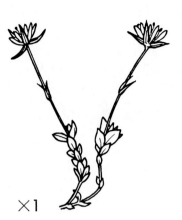
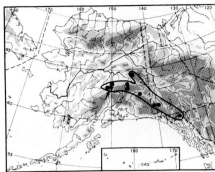
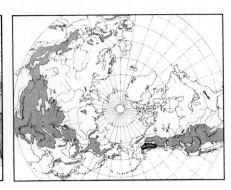

13. Stellària alaskàna Hult.

Glabrous; leaves compressed at base of the flowering stem or on separate sterile shoots, sessile, ovate-elliptic, somewhat acute, 1-nerved, thick; stems solitary or in loose tufts, 1–2-flowered; bracts scarious or broadly scarious-margined, acute; sepals triangular-lanceolate, scarious-margined, markedly acute, longer than, or as long as, the petals, which are cleft nearly to base; styles 3.

Stony slopes in the mountains, above 1,000 meters.

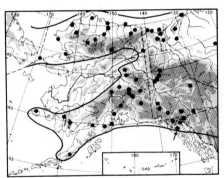
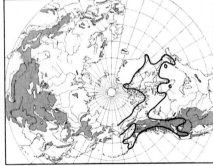

14. Stellària monántha Hult.

Stem ascending, loosely tufted, glabrous or somewhat pubescent in lower part; leaves sessile, ovate-lanceolate to narrowly lanceolate, acute, glabrous; flowers solitary, in axis of green, leaflike bracts; sepals scarious-margined, ovate to lance-ovate, glabrous, somewhat shorter than petals; capsule pale brown to black, longer than sepals.

Stony places in the mountains and on tundra, to at least 2,600 meters in the St. Elias Range.

43. CARYOPHYLLACEAE (Pink Family)

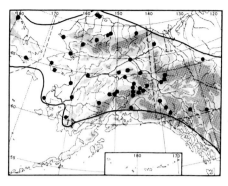 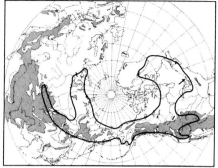 

15. Stellària lóngipes Goldie
Including *Stellaria subvestita* Greene and *S. stricta* Richards.

Erect or decumbent, loosely tufted; stem glabrous or pubescent; leaves sessile, ovate-lanceolate to linear, acute, mostly glabrous; flowers in few-flowered, terminal cymes; bracts small, scarious or broadly scarious-margined; sepals scarious-margined, ovate to lance-ovate, glabrous, shorter than petals; capsule longer than sepals, brown to black.

Stony slopes in the mountains and on tundra.

A highly variable species. Specimens with pubescent stem have been called *S. subvestita;* and specimens with stiff ascending branches and small flowers, *S. stricta*. *S. longipes* forms, along with *S. Edwardsii, S. laeta, S. monantha,* and *S. "Laxmannii"* (see *S. Edwardsii*), a very critical group, not yet well understood. Siberian plants reported as **S. peduncularis** Bunge, **S. dahùrica** Turcz., **S. árctica** Schischk., and **S. Fischeriàna** Ser., are very closely related and belong to the same complex.

×²⁄₃

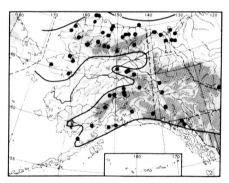 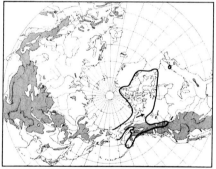 

16. Stellària laèta Richards.

Stem ascending, loosely tufted, glabrous or somewhat pubescent; leaves sessile, ovate-lanceolate to narrowly lanceolate, acute, often somewhat pubescent; flowers solitary, in axis of green, leaflike bracts; short branches or buds often in same axis; sepals scarious-margined, ovate to lance-ovate, ciliated in margin and often somewhat pubescent on back; capsules longer than sepals, pale brown to black.

Stony places in the mountains and on tundra.

×³⁄₄

43. Caryophyllaceae (Pink Family)

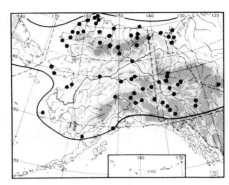
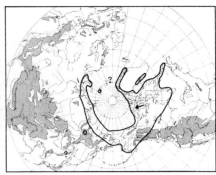

17. Stellària Edwárdsii R. Br.

Stellaria ciliatosepala Trautv.; *S. arctica* Schischk.

Stems ascending, loosely tufted, glabrous or somewhat pubescent in lower part; leaves sessile, ovate-lanceolate to lanceolate, acute, glabrous or sometimes somewhat pubescent; flowers in few-flowered cymes, in axis of scarious or scarious-margined bracts, sometimes solitary; sepals scarious-margined, ovate, mostly glabrous except in the ciliated margin, somewhat shorter than petals; capsule dark-colored, longer than sepals.

Stony places in the mountains and on tundra.

A similar plant occurring in the interior, with the back of the sepals profusely pubescent, has been called **S. "Láxmannii"**; its status is obscure.

Cerastium L. (Mouse-Ear Chickweed)

Plant annual, with thin root, lacking basal shoots; petals shorter or only slightly longer than sepals .... 8. *C. glomeratum*
Plant perennial, with thicker root and sterile basal shoots; petals short or long:
    Plant tall (25–30 cm), coarse; stem single; capsule strongly recurved .... 1. *C. maximum*
    Plant smaller, more or less caespitose; petals much shorter; teeth of capsule erect:
        Plant glabrous or nearly so, pulvinate, often sterile; at least some of branches ending with buds .... 6. *C. Regelii*
        Plant pubescent, flowers usually present; no buds at end of branches:
            Petals about as long as sepals .... 9. *C. fontanum* subsp. *triviale*
            Petals longer than sepals:
                Plant with sterile shoots in axis of uppermost leaves; leaves narrow .... 10. *C. arvense*
                Plant lacking sterile shoots in axis of upper leaves; leaves usually broader:
                      Plant coarse, with thick, yellowish, hirsute stem; leaves large; flowers in dense cymes; pedicels of fruit reflexed ..... 7. *C. Fischerianum*
                      Plant more delicate, with thin stem; flowers single or in open dichotomous cymes:
                          Plant lax, delicate, weak, with long basal shoots ending in budlike tufts of leaves .... 5. *C. jenisejense*
                          Plant with rigid, erect stem slightly matted or densely tufted, with short, flowering stem:
                              Plant small, matted, often single-flowered; leaves oblong, glabrescent, long-ciliated; scarious bracts lacking; pedicels with simple, soft hairs .... 4. *C. aleuticum*
                              Plant loosely tufted or, in arctic specimens, densely tufted; flowers in dichotomous cymes; leaves oblong to ovate, hirsute; pedicels with simple and glandular hairs mixed:
                                Petals about as long as sepals .... 2. *C. Beeringianum* var. *Beeringianum*
                                Petals about twice as long as sepals .... 3. *C. Beeringianum* var. *grandiflorum*

43. Caryophyllaceae (Pink Family)

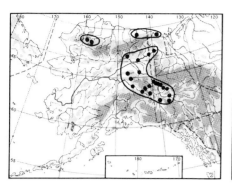
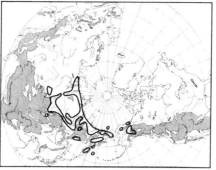

1. Cerástium máximum L.

Stem erect, tall, viscid, pubescent; leaves lanceolate or ovate-lanceolate, in sterile shoots linear-lanceolate, markedly acute, pubescent on both sides and ciliated in margin; inflorescence aglomerate, terminal; sepals ovate-lanceolate, pubescent, scarious-margined; petals white, about 2 cm long, cleft to about one-fourth the length; capsule slender, much longer than sepals, its teeth rolled backward.

Woods, thickets, meadows. Described from Siberia.

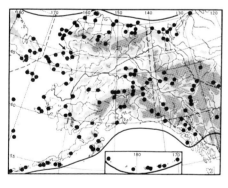
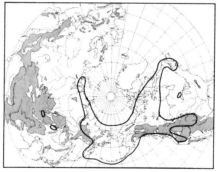

2. Cerástium Beeringiànum Cham. & Schlecht.
Cerastium Bialynickii Tolm.
var. Beeringiànum

In arctic specimens matted; stems several, ascending, pubescent from mixed simple and glandular hairs; lowest internodes glabrescent or retrorsely hirsute; leaves ovate to oblong-lanceolate, with short strigose pubescence; inflorescence a dichotomously branched cyme; pedicels long, with short, patent, glandular or viscid hairs; bracts slightly scarious-margined; inner sepals more broadly scarious-margined than the outer; petals not much longer than sepals.

Gravel and cliffs, from the lowlands at least to 2,000 meters in the St. Elias Range and in the mountains of the central Yukon.

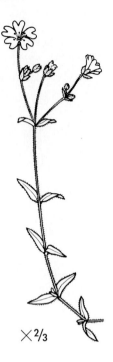
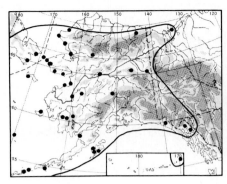
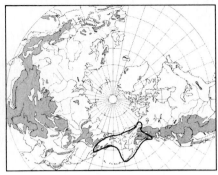

3. Cerástium Beeringiànum Cham. & Schlecht.
var. **grandiflòrum** (Fenzl) Hult.

Cerastium vulgatum var. *grandiflorum* Fenzl; *C. Scammaniae* Polunin; *C. arcticum* of Fl. Alaska & Yukon, in part.

Similar to var. *Beeringianum*, but usually coarser; flowers large, about twice as long as sepals.

Gravel and cliffs. Described from Kamchatka and Alaska.

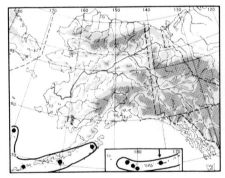
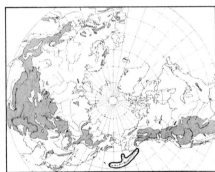

4. Cerástium aleùticum Hult.

Similar to *C. Beeringianum*, but plant small, matted; leaves glabrescent, oblong, long-ciliated; stems single-flowered, covered with withered leaves at base; scarious bracts lacking; pedicels with soft, simple hairs.

Mountain slopes.

43. Caryophyllaceae (Pink Family)

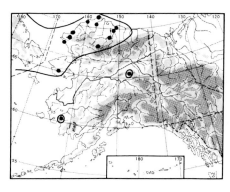 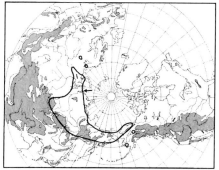 

5. Cerástium jenisejénse Hult.

Similar to *C. Beeringianum*, but prostrate, lax, delicate, weak, with long basal shoots ending in budlike tufts of leaves; petals about twice as long as sepals.

On mountain slopes and tundra.

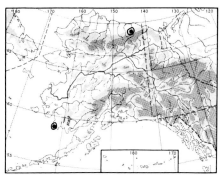 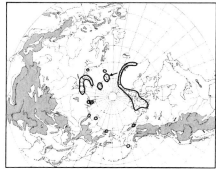 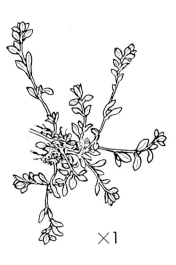

6. Cerástium Regélii Ostenf.

Densely pulvinate; leaves glabrous or slightly ciliated; branches ending with buds, which fall off and root; sepals obtuse, glandular, one-third as long as petals.
Solifluction soil.

This high-arctic plant is often sterile, propagating by means of the above-mentioned buds; one sterile specimen has been collected at Peters' Lake (eastern Brooks Range), another at Cape Newenham.

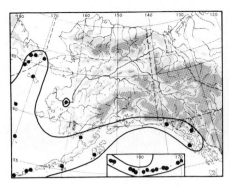
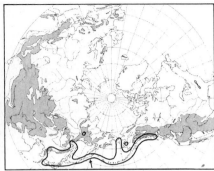

7. Cerástium Fischeriànum Ser.

Cerastium rigidum Cham. & Schlecht.; *S. unalaschcense* Takeda.

Coarse, densely pubescent, with stiff, divaricate or retrorse hairs; stem thick, erect, with leafy runners at base; leaves lanceolate or ovate-lanceolate to ovate; flowers in terminal, fairly dense, subumbellate inflorescence; bracts herbaceous; sepals lanceolate, acute, narrowly scarious-margined, hirsute; petals deeply notched, longer than sepals; capsule much longer than sepals, somewhat curved.

Wet shores. A variable species forming hybrids with *C. Beeringianum*.

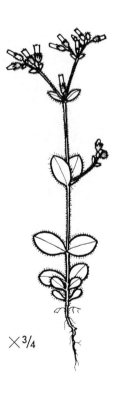
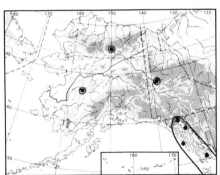
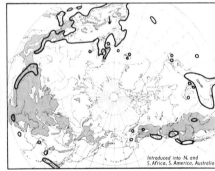

8. Cerástium glomeràtum Thuill.

Cerastium viscosum L. in part.

Viscid annual; stem erect, usually branched; leaves ovate; inflorescence with subumbellate terminal clusters; bracts small, herbaceous; sepals ovate-lanceolate, acute, scarious-margined; petals white, deeply notched, as long as, or longer than, sepals, sometimes missing; capsule slender, about twice as long as sepals.

An introduced weed. Described from Paris.

43. CARYOPHYLLACEAE (Pink Family)

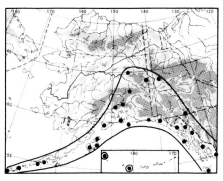 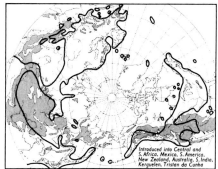 

9. Cerástium fontànum Baumg.
subsp. **triviàle** (Link) Jalas

Cerastium triviale Link; *C. vulgatum* L. in part; *C. caespitosum* Gilib.; *C. holosteoides* E. Fries.

Grayish-green, short-pubescent, sometimes viscid, loosely caespitose, with depressed, leafy basal shoots; stem with long internodes; leaves oblong to narrowly ovate; flowers in terminal cymes; bracts broadly scarious-margined; sepals ovate-lanceolate, acute, hirsute, scarious-margined, of about same length as petals; styles 5; capsule curved, twice as long as sepals.

An introduced weed. *C. fontanum* described from Transylvania, subsp. *triviale* from Europe, Asia, and North Africa.

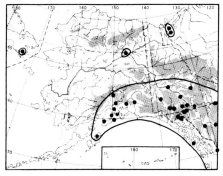 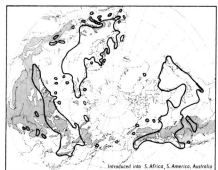 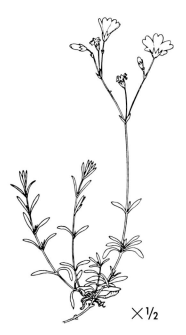

10. Cerástium arvénse L.

Cerastium campestre Greene.

Grayish-green, finely pubescent, tufted; stem branched from base, with axillary fascicles or leafy tufts; leaves narrowly ovate or linear to linear-subulate, the lower leaves marcescent; bracts broadly scarious-margined; petals two or three times as long as sepals; capsule cylindric, usually longer than calyx.

Dry slopes. Described from southern Sweden, southern Europe.

Highly variable in pubescence and in form of leaves.

Sagina L. (Pearlwort)

Petals much longer than sepals; bulblike axillary fascicles of leaves in upper axils
.... 5. *S. nodosa*
Petals about as long as sepals or shorter; no bulblike fascicles of leaves:
  Plant lacking basal rosettes of leaves; stem straight, not succulent ....
.... 4. *S. occidentalis*
  Plant with basal rosettes of leaves:
    Sepals margined with purple; pedicels not recurved at apex ... 2. *S. intermedia*
    Sepals with white margin:
      Plant not fleshy, forming mats; stems often rooting at lower internodes;
      pedicels recurved during some period of development ....
.... 1. *S. saginoides*
      Plant caespitose, more or less fleshy; pedicels straight ....
.... 3. *S. crassicaulis*

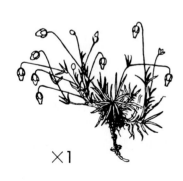

×1

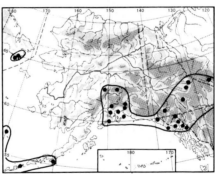 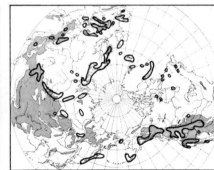

1. Sagina saginoìdes (L.) Karst.

Spergula saginoides L.; *Sagina Linnaei* Presl.

Matted, yellowish-green; stems often rooting at basal nodes; leaves markedly acute; pedicels long; fruiting pedicels at first recurved at tip, later erect; sepals with white margin, mostly 5; petals shorter than sepals.

  Grassy slopes, snow beds. Described from Germany, Siberia.

  Forms hybrids with *S. intermedia*.

43. CARYOPHYLLACEAE (Pink Family) 427

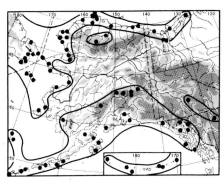 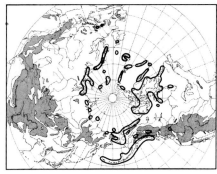 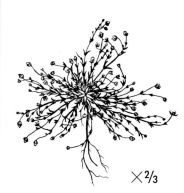

×⅔

2. Sagìna intermèdia Fenzl

Sagina nivalis (Lindbl.) E. Fries; *Spergula nivalis* Lindbl.

Forms dense, dark-green tufts, with broader leaves in center; stems never rooting at nodes; pedicels straight; sepals with purple margin or purplish spots in margin, usually 4; petals shorter than sepals.

Sandy places, snow beds; to about 2,800 meters in the St. Elias Mountains.

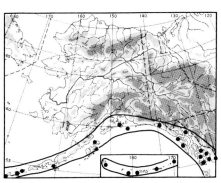 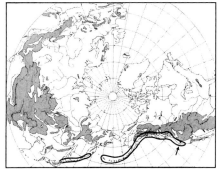 

×⅔

3. Sagìna crassicaùlis S. Wats.

Loosely caespitose, glabrous; stem and leaves more or less fleshy; pedicels straight; sepals 5, broadly ovate, somewhat shorter than petals; capsule broadly ovate, somewhat longer than sepals.

Seashores.

Plants with more or less glandular pedicels and calyx represent var. **litoràlis** (Hult.) Hult. (*S. litoralis* Hult.).

×2/3

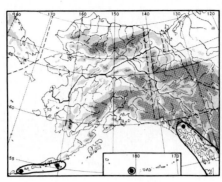
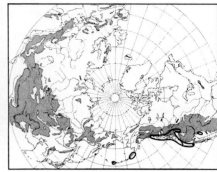

4. Sagìna occidentàlis S. Wats.

Plant loosely branched, lacking persistent basal rosette; stem erect, not fleshy, glabrous; leaves filiform; pedicels erect or recurved at apex; sepals 5, oblong-ovate, longer than petals; capsules longer than sepals.

Moist places. Described "from Oregon to San Francisco."

×2/3

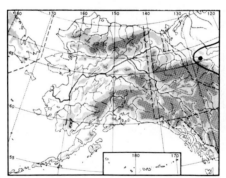
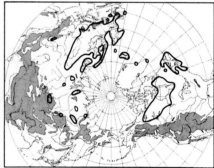

5. Sagìna nodòsa (L.) Fenzl
Spergula nodosa L.

Dark green, caespitose, with short, sterile stems and procumbent or erect, glabrous or glandular, flowering stems; leaves linear, short-mucronate; cauline leaves diminishing upward, the upper shorter than the internodes, with dense axillary leaf fascicles; sepals 5, ovate-oblong, obtuse; petals white, about twice as long as sepals; capsule longer than sepals.

Moist places. Described from Europe.

43. CARYOPHYLLACEAE (Pink Family)

Minuartia Loefl.

Stem and leaves glabrous:
    Leaves short, about 3 mm long; sepals patent in anthesis; stem normally single-flowered .... 9. *M. Rossii*
    Leaves longer; sepals not patent; stem single-flowered or branched:
        Plant tufted, with distinct taproot; stem erect, 1–3-flowered ..... 5. *M. stricta*
        Plant with weak taproot, stem several-flowered, spreading .. 6. *M. dawsonensis*
Stem pubescent:
    Sepals markedly acute .... 8. *M. rubella*
    Sepals obtuse:
        Leaves markedly acute .... 7. *M. yukonensis*
        Leaves obtuse or somewhat acute:
            Leaves 3-nerved, flat, with long cilia in margin .... 1. *M. macrocarpa*
            Leaves nerveless or 1-nerved, glabrous or ciliated:
                Ripe seeds smooth or nearly so; long basal shoots lacking; calyx glabrous or nearly so; leaves flat; petals oblong, only slightly longer than sepals .... 4. *M. biflora*
                Ripe seeds tuberculate, at least in margin; basal shoots comparatively long; calyx mostly glandular; leaves linear; petals short or long:
                      Flowers large, often 6–7 mm broad and 10 mm long; petals about twice as long as calyx .... 2. *M. arctica*
                      Flowers small, often about 5 mm broad and 5 mm long; petals as long as calyx or slightly longer .... 3. *M. obtusiloba*

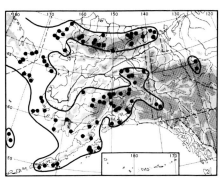 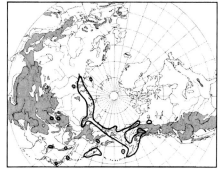 

1. Minuártia macrocárpa (Pursh) Ostenf.
Arenaria macrocarpa Pursh; *A. crilloniana* W. B. Drew; *Alsine macrocarpa* (Pursh) Fenzl.

Matted, in high-alpine specimens densely so, with densely leaved runners; leaves flat, ciliate, distinctly 3-nerved, variable in form from long and linear to short and broad; pedicels glandular; flowers solitary; sepals oblong, blunt, glandular; petals about twice as long as sepals, white, very rarely rose-colored (var. **ròsea** Hult.); capsule oblong, two to two and a half times as long as sepals.

Sandy and rocky slopes and ridges, moist places, mostly in the mountains, to at least 2,150 meters. Described from the west coast of Alaska.

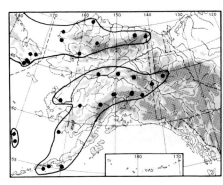
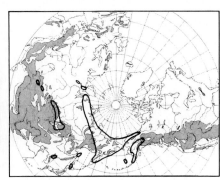

×⅔

2. Minuártia árctica (Stev.) Aschers. & Graebn.

Arenaria arctica Stev.; *Alsine arctica* (Stev.) Fenzl.

Matted; stem glandular, at least above, mostly single-flowered; basal leaves narrowly linear, in lowland specimens long, in exposed places short and crowded, blunt, glabrous or sometimes slightly glandular, glabrous or sparsely and shortly ciliated in margin, nerveless; stem leaves shorter and often broader; calyx leaves ovate, oblong, 3-nerved, obtuse, more or less glandular, often purplish in margin; flowers 6–7 mm broad, about 10 mm long; petals white, about twice as long as sepals; capsule one and a half to two times as long as calyx; ripe seeds tuberculate in margin.

Dry ridges, rocky slopes in the mountains and on tundra. Described from "littore Sibirico maris glacialis."

Highly variable; transitions to *M. obtusiloba* common.

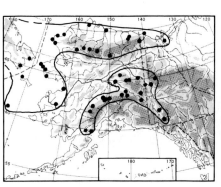
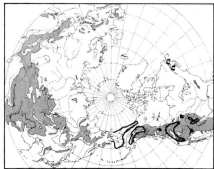
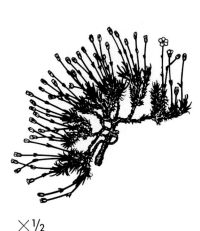

×½

3. Minuártia obtusilòba (Rydb.) House

Alsinopsis obtusiloba Rydb.; *Arenaria obtusiloba* (Rydb.) Fern.; *Arenaria obtusa* Torr. not All.; *Arenaria biflora* S. Wats. not L.; *Arenaria biflora* var. *obtusa* (Torr.) S. Wats.

Similar to, and forming hybrid swarms with, *M. arctica*, but less clearly matted; thick, woody root often present; flowers much smaller, about 5 mm high and 5 mm broad; leaves, at least in alpine specimens, very short and densely crowded.

Dry mountain slopes, snow beds. Described from the higher Rocky Mountains.

Broken line on circumpolar map indicates range of the very closely related **M. marcéscens** (Fern.) House (*Arenaria marcescens* Fern.), which differs chiefly in having a glabrous calyx.

43. Caryophyllaceae (Pink Family)

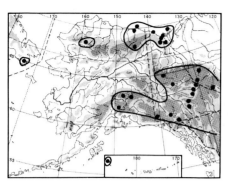
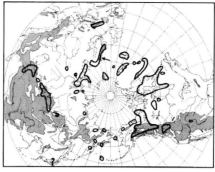

4. Minuártia biflòra (L.) Schinz & Thell.

Stellaria biflora L.; *Alsine biflora* (L.) Wahlenb.; *Alsinanthe biflora* (L.) Rchb.; *Arenaria sajanensis* Willd.

Tufted, light green, lacking long basal shoots; leaves obtuse or somewhat acute, nerveless, glandular, ciliated; stem glandular, single- or 2-flowered; sepals obovate-oblong, blunt, 3-nerved, somewhat glandular-pubescent, shorter than petals; capsule longer than petals; seeds glabrous or slightly tuberculate.

Dry places, snow beds, in the mountains.

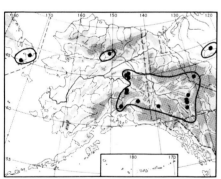
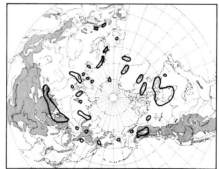

5. Minuártia strícta (Sw.) Hiern

Spergula stricta Sw.; *Arenaria uliginosa* Schlecht. (not *Arenaria stricta* Michx.).

Glabrous, tufted, with taproot; stem erect, simple, rarely 3-flowered; leaves narrowly linear, blunt, indistinctly nerved, often proliferous in axils; sepals ovate, somewhat acute, 3-nerved; petals as long as or slightly longer than sepals; capsule somewhat longer than sepals; seeds 0.6–0.7 mm long, granulate.

Moist places. Described from northern Sweden.

×2/3

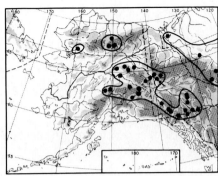
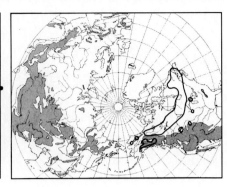

6. Minuártia dawsonénsis (Britt.) Mattf.
Arenaria dawsonensis Britt.; *A. stricta* Michx. subsp. *dawsonensis* (Britt.) Maguire.

Similar to (and sometimes somewhat obscurely distinct from) *M. stricta*, but with less prominent taproot; tall, branched stem; and more strongly granulate seeds.
Moist places in the lowlands.

×1/2

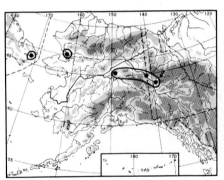
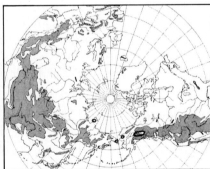

7. Minuártia yukonénsis Hult.
Minuartia laricifolia of Fl. Alaska & Yukon; not Schinz & Thell.

Loosely caespitose; stems several, erect, branched or unbranched, glandular; leaves linear-subulate, glabrous, nerveless, sparsely and shortly ciliated in margin; pedicels glandular; sepals ovate-lanceolate, blunt, 3-nerved, scarious at tip; petals one and a half times as long as sepals; capsule somewhat longer than calyx; seeds reniform, tuberculate on back.

Dry places, scree slopes. Total range unknown; probably occurs also in the Rocky Mountains.

43. Caryophyllaceae (Pink Family)

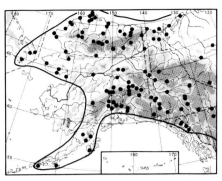
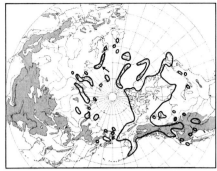
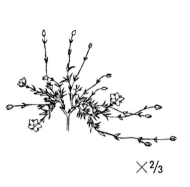

8. Minuártia rubélla (Wahlenb.) Graebn.

Alsine rubella Wahlenb.; *Alsine hirta* (Wormsk.) Hartm.; *Arenaria rubella* (Wahlenb.) Sm.; *Arenaria propinqua* Richards.; *Alsinopsis propinqua* (Richards.) Rydb.; *Arenaria verna* L. var. *rubella* (Wahlenb.) S. Wats.; *Arenaria verna* var. *propinqua* Fern.; *Arenaria verna* var. *pubescens* Fern.

Densely tufted, bluish-green; basal shoots with densely crowded, subulate, 3-ribbed leaves; stem filiform, branched, densely glandular; sepals lanceolate, acuminate, 3-nerved, longer than the white, obtuse, sometimes reddish petals; capsule slightly longer than sepals.

Rocky and sandy places in arctic lowlands and in the mountains. Regarded as calciphile.

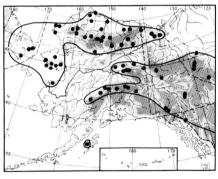
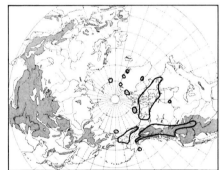

9. Minuártia Róssii (R. Br.) Graebn.

Arenaria Rossii R. Br.; *Alsine Rossii* (R. Br.) Fenzl; *Minuartia Rolfii* Nannf.

Glabrous or nearly so; pulvinate, with short pedicels, in exposed arctic or high-alpine specimens; loosely tufted, with longer pedicels, in less exposed habitats [var. **élegans** (Cham. & Schlecht.) Hult. [*Arenaria elegans* Cham. & Schlecht.; *M. elegans* (Cham. & Schlecht.) Schischk.)]; leaves linear, blunt, triangular in cross section; fascicles of leaves in axils, which sometimes form globular buds [var. **orthotrichoídes** (Schischk.) Hult. (*M. orthotrichoides* Schischk.)]; sepals ovate, acute, 3-nerved, purplish; petals from slightly shorter to slightly longer than calyx; capsule spherical, about as long as sepals.

Rock slides, dry ridges in the mountains, and moist places on tundra. Described from Melville Island.

43. Caryophyllaceae (Pink Family)

Honckenya Ehrh.

Leaves ovate to lanceolate; flowers scattered in upper axils .... 1. *H. peploides* subsp. *peploides*
Leaves longer and narrower; flowers mostly in several-flowered cymes; plant coarser .... 2. *H. peploides* subsp. *major*

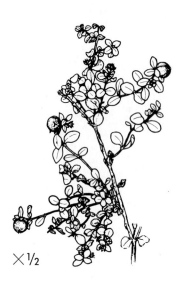

×½

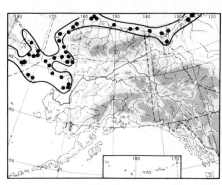

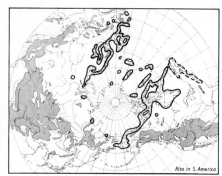

1. Honckénya peploìdes (L.) Ehrh.   Seabeach Sandwort

Arenaria peploides L.; *Ammodenia peploides* (L.) Rupr.; *Halianthus peploides* (L.) E. Fries; *Minuartia peploides* (L.) Hiern. Including *Honckenya frigida* Pobed.

subsp. **peploìdes**

Glabrous, dioecious; stem branching, buried in sand; plant fleshy; leaves yellowish-green, ovate, oblong or lanceolate, somewhat acute; flowers scattered in upper axils; sepals ovate to lanceolate, acute, about as long as the greenish-white petals; styles 3; capsule globose, longer than sepals, poorly developed in staminate flowers; stamens undeveloped in pistillate flowers.

Sandy seashores. Described from northern Europe. Broken line on circumpolar map indicates range of subsp. **robústa** (Fern.) Mattf. (*Arenaria peploides* var. *robusta* Fern.

×½

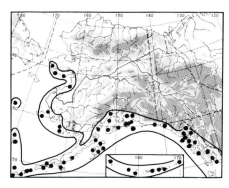

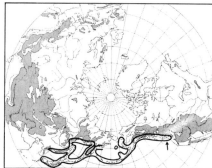

2. Honckénya peploìdes (L.) Ehrh.
subsp. **màjor** (Hook.) Hult.

Arenaria peploides var. *major* Hook.; *A. peploides* subsp. *major* (Hook.) Calder & Taylor; *Honckenya oblongifolia* Torr. & Gray.

Similar to subsp. *peploides,* but coarser; stem long; flowers often in several-flowered cymes; leaves long and comparatively narrow.

Sandy seashores.

43. CARYOPHYLLACEAE (Pink Family)

Arenaria L.

Plant 5–20 cm high; leaves long, capillary, acuminate . . . . 1. *A. capillaris*
Plant matted or pulvinate; leaves lanceolate to ovate:
    Plant pulvinate, with stout central taproot . . . . 4. *A. Chamissonis*
    Plant matted:
        Flowers long-peduncled; capsule ovate . . . . 2. *A. longipedunculata*
        Flowers short-peduncled; capsule cylindrical . . . . 3. *A. humifusa*

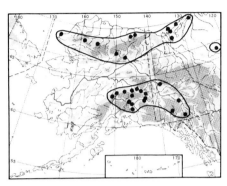
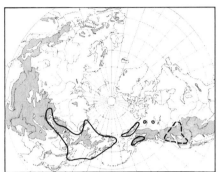
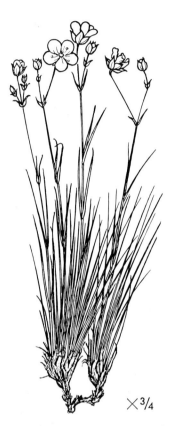

1. Arenària capillàris Poir.

Arenaria nardifolia with respect to Alaskan plant.

Densely tufted, with basal runners, covered with withered leaves at base; stems several, straight, glabrous; leaves light green, capillary, glabrous or more or less scabrous in margin, acuminate; inflorescence open, 1–3-flowered; bracts narrowly lanceolate, scarious-margined; sepals ovate to ovate-lanceolate, acute, violet-margined; petals longer than sepals; capsule somewhat longer than sepals.

Dry, sandy hillsides. Described from Siberia.

Alaskan specimens belong to subsp. **capillàris**; subsp. **americàna** Maguire occurs south of the glaciation in western North America (broken line on circumpolar map). The closely related **A. tschuktschòrum** Regel, with shorter leaves, occurs in easternmost Asia.

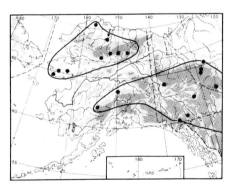
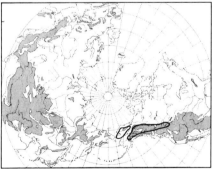
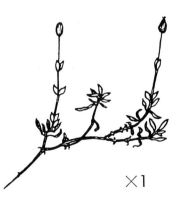

2. Arenària longipedunculàta Hult.

Arenaria humifusa of western American authors.

Matted, with filiform, subterranean, light-colored runners; leaves lanceolate to ovate-lanceolate, blunt or somewhat acute, slightly ciliated at base; flowering stems thick, erect, glandular-puberulent, always 1-flowered, with pair of ovate, acute bracts below middle; sepals ovate, obtuse, or somewhat acute, glabrous or glandular-puberulent, indistinctly 3-nerved about as long as petals; capsule ovate, narrower at tip, slightly longer than calyx, opening with 3 or 6 teeth; styles 3; seeds brown, suborbicular, shiny, rugulose, 0.7–0.8 mm long.

Gravel, moist places in the mountains.

43. Caryophyllaceae (Pink Family)

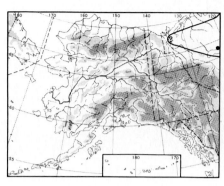
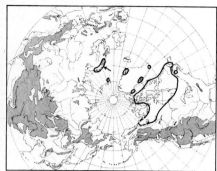

3. Arenària humifùsa Wahlenb.

Similar to *A. longipedunculata,* but smaller; flowers very short-peduncled; capsule cylindric, considerably longer than calyx; seeds somewhat smaller.

Gravel, alpine ledges.

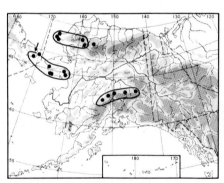
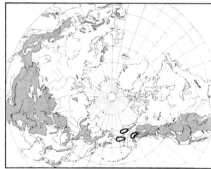

4. Arenària Chamissònis Maguire

Cherleria dicranoides Cham. & Schlecht.; *Stellaria dicranoides* Fenzl; *Arenaria dicranoides* (Cham. & Schlecht.) Hult. not HBK.

Pulvinate, with central taproot; stems branched, densely covered with imbricate leaves; leaves lanceolate to oblong, glabrous, 1-nerved, the lower marcescent; flowers single, short-pedicellated, barely reaching above leaves; sepals oblong-lanceolate acute, 1-nerved, scarious-margined, gibbous; petals lacking or rudimentary; disk scales large, nearly petaloid; capsule ovate-conical, about 2 mm long, as long as sepals, opening with 6 teeth.

Rocks, scree slopes.

43. Caryophyllaceae (Pink Family)

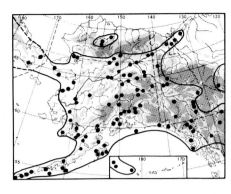
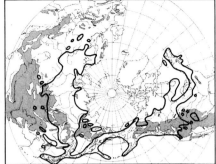

Moehringia L.

1. Moehríngia lateriflòra (L.) Fenzl — Grove Sandwort
Arenaria lateriflora L.

Rhizome filiform, branched, more or less horizontal; flowering stems erect, thin, usually branched, retrorsely puberulent; leaves ovate to oblong or lanceolate, obtuse or somewhat acute, glabrous or puberulent, ciliated, pellucid-punctate; bractlets small, blunt or somewhat acute; sepals glabrous, ovate or oblong, obtuse, scarious-margined; petals 2–3 times as long as sepals; filaments pubescent at base; capsule ovate, twice as long as sepals.

Woods, thickets, dry meadows; common in many places; to at least 600 meters in McKinley Park. Described from Siberia.

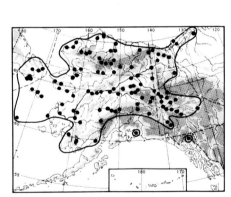
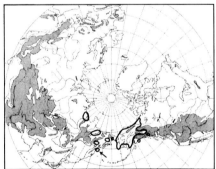
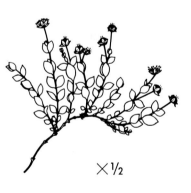

Wilhelmsia Rchb.

1. Wilhélmsia physòdes (Fisch.) McNeill
Arenaria physodes Fisch. ex Ser. in DC.; *Merckia physodes* (Fisch.) Fisch.

Stem creeping, strongly branched; flowering stems ascending, glandular-puberulent; leaves ovate or oblong, acute, glabrous, ciliated in margin; flowers solitary; pedicels glandular; sepals ovate, blunt, often reddish; petals somewhat longer than calyx; stamens as long as, or longer than, petals; capsule spherical, with several deep grooves, inflated, larger than sepals, falling in 3 parts at maturity.

Moist, sandy places, gravel bars; common in many areas; in the central Yukon to at least 1,100 meters.

43. Caryophyllaceae (Pink Family)

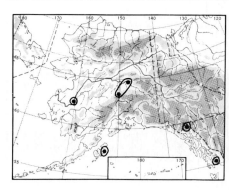

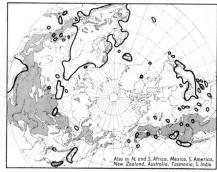

Spergula L.

1. Spérgula arvénsis L.   Spurry

Glandular-pubescent, yellowish-green; stem weak, with long upper internodes; leaves in whorls of 6, filiform, channeled at base below; petals white.

An introduced weed. Described from Europe.

Spergularia J. & C. Presl

Leaves not densely fascicled; stipules connate, deltoid; seeds smooth on sides . . . . .
. . . . 1. *S. canadensis*
Leaves densely fascicled; stipules free, lance-acuminate; seeds minutely papillate
. . . . 2. *S. rubra*

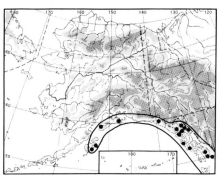

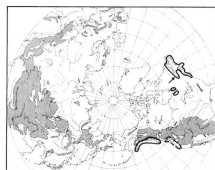

1. Spergulària canadénsis (Pers.) G. Don
Arenaria canadensis Pers.

Glabrous nearly throughout; stems erect or prostrate; leaves fleshy, blunt; stipules truncate or apiculate, broader than long; sepals ovate, mostly longer than the white or pink petals; capsules longer than sepals; seeds brown, 0.8–1.4 mm long, with erose wing or wingless.

Saline or brackish soil. Described from the mouth of the St. Lawrence River.
Broken line on circumpolar map indicates range of var. **occidentàlis** Rossb.

43. Caryophyllaceae (Pink Family)

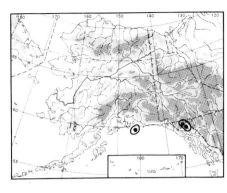 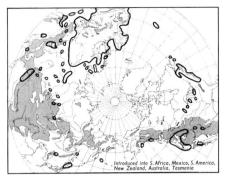 

2. Spergulària rùbra (L.) J. & C. Presl   Sand Spurry
Arenaria rubra L.; *Tissa rubra* (L.) Britt.

Densely glandular-pubescent, dark green; stem prostrate; leaves opposite, fleshy, linear-filiform, acute; stipules lustrous, silvery; bracts long; flowers pale red; capsule as long as sepals.

An introduced weed. Described from Europe.

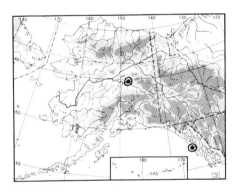 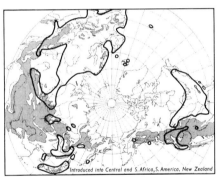 

Agrostemma L.

1. Agrostémma githàgo L.   Corn Cockle

Annual, with tall, erect stem and narrow, opposite leaves; flowers solitary or few; sepals formed into 10-ribbed tube with 5 narrow, spreading teeth, much longer than petals; petals 5, purplish-red, with long, pale claw; capsule ovoid.

Waste places, roadsides; an occasionally introduced weed, probably not persistent. Described from Europe.

Silene L.

Plant densely caespitose-pulvinate; flowers solitary, pink:
  Leaves flat, short, strongly ciliate . . . . 1. *S. acaulis* subsp. *acaulis*
  Leaves linear, long, sparsely ciliate with short cilia . . . .
  . . . . 2. *S. acaulis* subsp. *subacaulescens*
Plant not pulvinate; flowers in cymes, white or pink:
  Plant with numerous linear, long-petiolated, basal leaves, 1–2 cm long; flowers pink . . . . 3. *S. stenophylla*
  Plant with broader basal leaves; flowers white or pink:
    Calyx 0.5–0.7 cm long; leaves mostly ovate-lanceolate . . . .
    . . . . .6. *S. Menziesii* subsp. *Menziesii*
    Calyx longer; leaves lanceolate or oblanceolate:
      Calyx purple-colored, cylindrical or even clavate; leaves lanceolate; plant glabrous or puberulent . . . . 4. *S. repens*
      Calyx green, urceolate-ovoid:
        Plant puberulent; leaves narrow, oblanceolate; nerves of calyx prominent
        . . . . 5. *S. Douglasii*
        Plant viscid; leaves lanceolate; nerves of calyx not prominent . . . .
        . . . . 7. *S. Menziesii* subsp. *Williamsii*

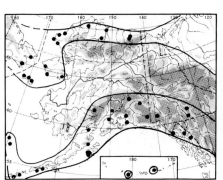 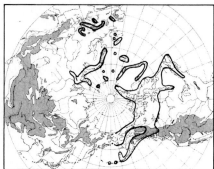

1. Silène acaùlis L.   Moss Campion
subsp. **acaùlis**

Silene acaulis subsp. *arctica* Löve & Löve; *S. acaulis* var. *excapa* (All.) DC.

Densely tufted to pulvinate; stem densely covered with short, flat leaves, ciliated with stiff hairs; calyx glabrous, mostly reddish; peduncles short; petals purple (rarely white); capsule as long as calyx or somewhat longer.

Sandy soil, ridges, to at least 2,200 meters. Described from Lapland, Austria, Switzerland, and the Pyrenees.

Circumpolar map indicates range of the entire species complex, subsp. *subacaulescens* excluded.

43. Caryophyllaceae (Pink Family)

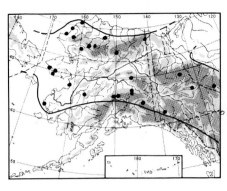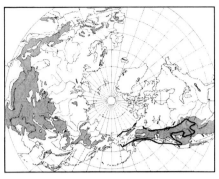

2. Silène acaùlis L.
subsp. **subacauléscens** (F. N. Williams) Hult.
  Silene acaulis f. *subacaulescens* F. N. Williams.

Similar to subsp. *acaulis*, but less densely tufted; leaves long, linear, more or less terete, sparsely ciliated with short cilia; capsule longer.

Sandy soil, ridges. Transitions to subsp. *acaulis* frequent where the two ranges overlap.

Described from Colorado and Arizona.

(See color section.)

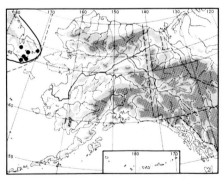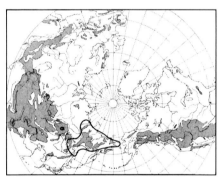

3. Silène stenophýlla Ledeb.

Densely tufted, with stout taproot; stems several, straight, glabrous; basal leaves numerous, slightly ciliated, linear, with long petioles; stem leaves in pairs, connate at base; flowers single or in few-flowered cymes; calyx inflated, glabrous, with triangular, blunt or somewhat acute teeth; petals pale lilac, one and a half times as long as sepals; capsule ovate.

Rocky and sandy slopes. Reported from "Arctic America." Described from Siberia.

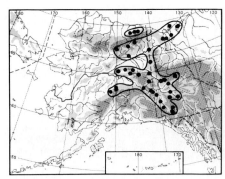
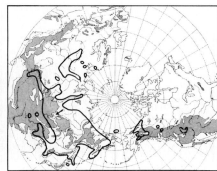

4. Silène rèpens Patrin

Silene purpurata Greene; *S. repens* subsp. *purpurata* (Greene) Maguire & Hitchc.

Short-pubescent; stems several, erect, up to 30 cm tall, sometimes reddish; leaves linear-lanceolate, acute, ciliated, in several pairs; inflorescence branched, sometimes congested; calyx purplish, pubescent, with blunt or somewhat acute teeth, hyaline-tipped; petals about twice as long as calyx, cleft to one-third the length; capsule with pubescent carpophore.

Rocky slopes, meadows, subalpine forests. Described from Siberia.

Broken line on circumpolar map indicates range of subsp. **austràle** Hitchc. & Maguire.

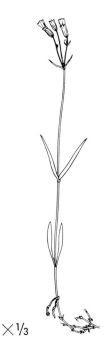
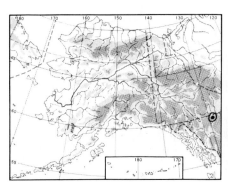
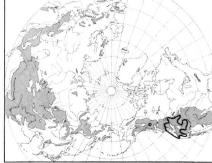

5. Silène Douglásii Hook.

Several pubescent or nearly glabrous stems from taproot, stems decumbent at base; leaves oblanceolate, acute, puberulent or glabrous, with prominent midrib; cauline leaves in pairs; flowers in few-flowered cymes; calyx tubular-urceolate, strongly nerved, with ovate-triangular teeth, the teeth with broadly scarious, ciliate margin; corolla yellowish-white or tinged with pink; carpophore 3–4 mm long, pubescent; seeds rugose-tesselate.

Dry places. Described from Grand Rapids of Columbia River.

43. Caryophyllaceae (Pink Family)

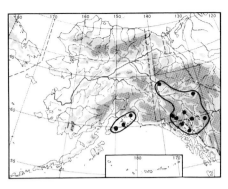 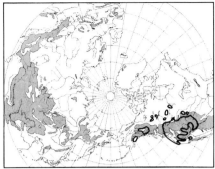 

6. Silène Menzièsii Hook.
subsp. **Menzièsii**

Stems several, from long, slender, branched rootstock, glandular-pubescent; leaves elliptic-lanceolate to ovate- or obovate-lanceolate, acute, glandular-pubescent, ciliolate; pedicels slender, 1–4 cm long; calyx tubular-campanulate, 5–7 mm long, glandular, with acute teeth; petals small, somewhat longer than calyx, with claws narrower than blades, the blades 2–3 mm long; carpophores about 1.5 mm long; seeds black, shining; some flowers staminate, others pistillate.

Wet places. Described from Straits of Juan de Fuca, Slave River.

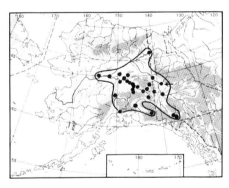 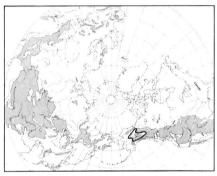 

7. Silène Menzièsii Hook.
subsp. **Williámsii** (Britt.) Hult. comb. nov.
Silene Williamsii Britt. in Bull. N.Y. Bot. Gard. 2:6, 1901, p. 168.

Similar to subsp. *Menziesii*, but leaves lanceolate; calyx longer; corolla larger; claws of petals considerably broadened above; carpophores shorter and seeds tuberculate; all flowers bisexual in most plants.

Roadsides.

Melandrium Roehl.

Styles 3, capsule opening with 6 teeth .... 1. *M. noctiflorum*
Styles 5, capsule opening with 5 or 10 teeth:
  Seeds large, 1.5–2.4 mm in diameter, conspicuously winged:
    Seeds 1.8 mm in diameter or more; calyx inflated in young flowers; petal pink to purple, not much longer than calyx:
      Flowers solitary (rarely 2), nodding when young, erect in fruit, seeds light brown, with thin wing .... 2. *M. apetalum* subsp. *arcticum*
      Flowers 1–3, erect; seeds grayish-brown, with thick wing .... 3. *M. macrospermum*
    Seeds less than 1.8 mm in diameter; calyx not inflated in young flowers; petals long-exserted:
      Seeds small, cuneate, with narrow wing; plant tall, slender; cauline leaves as long as basal leaves; petals pink .... 6. *M. Taylorae*
      Seeds rounded, larger, with broader wing; plant rarely exceeding 20 cm, coarser; cauline leaves shorter than basal leaves; petals white ...... 4. *M. affine*
  Seeds small, less than 1 mm in diameter, lacking wings:
    Seeds cordate or reniform, tuberculate; stem coarse, purplish above; flowers sessile .... 5. *M. triflorum*
    Seeds angular, granulate; stem slender, green; flowers more or less pedunculate .... 7. *M. taimyrense*

×⅓

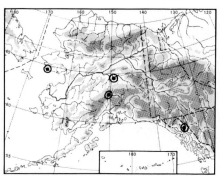

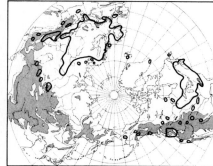

1. Melándrium noctiflòrum (L.) E. Fries
Silene noctiflora L.

Viscid-villous; lower leaves spatulate; stem leaves lanceolate to ovate-lanceolate, the upper sessile, acute; flowers few, in open cyme, opening at night, fragrant; calyx cylindric to conical, in age inflated; petals creamy white, dirty white, or rose-colored, one and a half times longer than calyx, deeply cleft into obcordate lobes; capsule ovate.

Introduced weed. Described from Europe.

43. CARYOPHYLLACEAE (Pink Family)

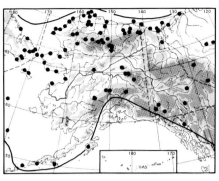
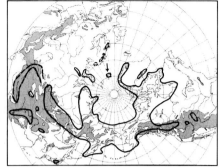

2. Melándrium apètalum (L.) Fenzl
   Lychnis apetala L.
subsp. **árcticum** (E. Fries) Hult.
   Wahlbergella apetala var. *arctica* E. Fries; *Gastrolychnis uralensis* Rupr.; *Melandrium Soczovianum* Schischk. at least with respect to Alaskan plant; *M. apetalum* subsp. *attenuatum* with respect to Alaska–Yukon plant; *Lychnis Soczovianum* (Schischk.) J. P. Anders., with respect to Alaskan plant.

Stem viscid-pubescent above, usually up to 20 (occasionally 30) cm tall; basal leaves linear-lanceolate to oblanceolate, more or less pubescent, especially in margin; stem leaves in 1–3 pairs, narrow, smaller than the basal leaves; flowers single (rarely 2), drooping when young, erect in fruit; calyx ovate, viscid-pubescent; calyx inflated also when young; petals lilac, slightly longer than calyx; seeds large, light brown, with broad wing.

Dry, grassy slopes in the mountains. *M. apetalum* described from Lapland and Siberia, subsp. *arcticum* from Spitzbergen.

Var. **glàbrum** Regel, a local race with robust growth, petals about as long as the calyx, and nearly glabrous leaves, occurs on St. Paul Island. Broken line on circumpolar map indicates range of subsp. **apètalum** in Scandinavia, and of other closely related taxa in the southern mountains.

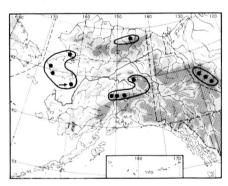
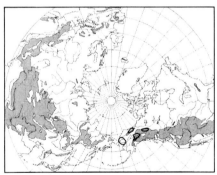

3. Melándrium macrospérmum Pors.
   Lychnis macrosperma (Pors.) J. P. Anders.

Similar to *M. apetalum* and *M. affine*, but flowers erect; stem often forked; seeds large, grayish-brown, with thick wing.

Dry, grassy slopes in the mountains.

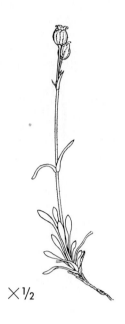
×½

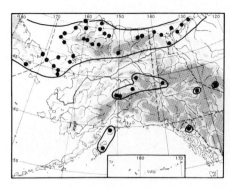
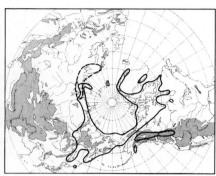

4. Meléndrium affíne J. Vahl

Lychnis furcata Fern. (probably not *Silene furcata* Raf.); *Melandrium furcatum* of Fl. Alaska & Yukon. Including *Lychnis apetala* var. *elatior* Regel.

Stem 5–20 cm tall or sometimes taller, short-pubescent; basal leaves oblanceolate or oblong spatulate, pubescent, with small, light-colored hairs, especially in margin; stem leaves in 1 to 2 pairs, oblong, shorter than basal leaves; flowers erect, single or 2–3 together; calyx longer than broad, campanulate, densely viscid-pubescent, inflated in age, in var. **brachycàlyx** (Raup) Hult. (*M. brachycalyx* Raup) about as long as broad or broader; petals white, one-third longer than calyx; seeds with thin wing.

Dry places. Described from Greenland.

Broken line on circumpolar map indicates range of the closely related **M. angustiflòrum** (Rupr.) Walp. (*Gastrolychnis angustiflora* Rupr.), which has longer stem leaves.

×⅔

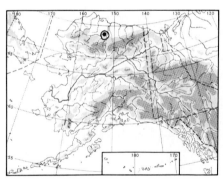
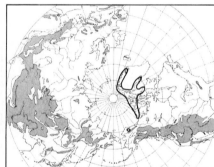

5. Meléndrium triflòrum (R. Br.) J. Vahl

Lychnis triflora R. Br.; *L. Sorensenis* Boiv.

Low-growing; leaves ovate-lanceolate to broadly lanceolate, pubescent, especially in margin; stem with 1 to 2 pairs of leaves, covered with glandular hairs; flowers erect, mostly 3 together; calyx campanulate, with broad, reddish-violet stripes, not inflated in age; petals white to rose-colored.

Dry places. Described from Greenland.

43. Caryophyllaceae (Pink Family)

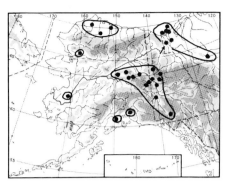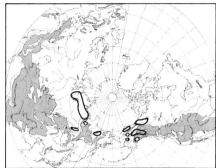

6. Melándrium Taylòrae (Robins.) Tolm.

Lychnis Taylorae Robins.; *Melandrium affine* var. *tenellum* Tolm.; *M. tenellum* Tolm.; *Lychnis furcata* subsp. *elatior* Maguire in part.

Plant light green; stem tall, straight, thin, pubescent; basal leaves linear-oblanceolate; stem leaves sparsely pubescent, in 2–3 pairs, as long as or sometimes longer than basal leaves; flowers mostly 2–3 together; calyx viscid-pubescent, somewhat inflated in age; petals longer than calyx, white to dirty white; seeds small, with narrow wing.

Meadows, dry places. Described from Peel River, Mackenzie Delta.

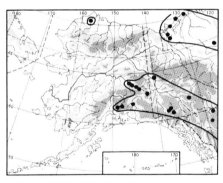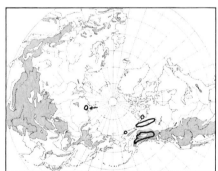

7. Melándrium taimyrénse Tolm.

Lychnis triflora var. *Dawsonii* Robins.; *L. triflora* subsp. *Dawsonii* (Robins.) Maguire; *L. Dawsonii* (Robins.) J. P. Anders.; *Melandrium Dawsonii* (Robins.) Hult.; *M. Ostenfeldii* Pors.

Stem tall; leaves linear, somewhat blunt; flowers 2–4 together; calyx viscid-pubescent; petals somewhat longer than calyx, white or pinkish; seeds small, wingless.

Sandy soil.

43. Caryophyllaceae (Pink Family)

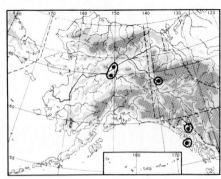
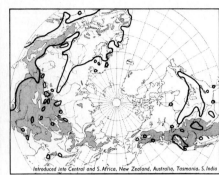

Vaccaria Medic.

1. Vaccària pyramidàta Medic.   Cow Cockle

Saponaria vaccaria L.; *S. segetalis* Neck.; *Vaccaria segetalis* (Neck.) Garcke.

Glabrous annual up to 60 cm tall; basal leaves oblong-lanceolate, upper ovate-lanceolate, cordate, acute; calyx tube inflated, with 5 sharp angles or wings and 5 triangular teeth; petals pale rose, somewhat emarginate, toothed; capsule globular, opening with 4 teeth.

Waste places; occasionally introduced. Described from Germany.

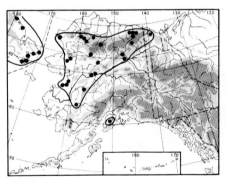
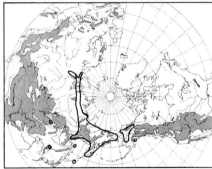

Dianthus L.

1. Diánthus rèpens Willd.   Carnation

Glabrous, stems several from stout taproot, 1- (to few-) flowered; leaves glaucous, lanceolate, acute, 1-nerved; calyx cylindrical with obtuse to somewhat acute teeth, blotched with purple; petals rose-colored, 10–20 mm long, finely toothed in apex.

Sandy and rocky places. Described from Siberia.

Tundra, lower Kuskokwim River. *Calamagrostis canadensis* subsp. *Langsdorffii*, in nearly pure stands, covers large areas. Willow thickets, of *Salix alaxensis*, *S. arbusculoides*, and *S. lanata* subsp. *Richardsonii*, edge the rivers.

Alaska Range, west of Mt. Spurr. The mountains are snow-covered to late August, and large areas are open soil totally lacking vegetation. The region has been visited rarely, if ever, and no botanical collections are known.

Firth River Valley, seen from Alaskan side toward calcareous mountains in the Yukon. Ice covers much of the river until August. A small *Braya* is the only plant able to flower and fruit on the newly ice-free areas.

Landscape northeast of Fairbanks, characterized by open spruce forest.

Peters' Lake, Franklin Mountains, in arctic eastern Alaska, seen from the north. The mountains are covered with alpine heaths and meadows, alternating with open solifluction soil.

White spruce forest, upper Copper River north of the Wrangell Mountains.

Skilak Lake, Kenai Peninsula, seen from the west. Sitka spruce forests predominate at low altitudes. Glaciated mountains rise in the background.

Firth River Valley at the mouth of Mancha Creek, on the Alaska–Yukon boundary. On the floodplain are alluvial meadows; in the background, the polar timberline.

Arctic polygon tundra, Seward Peninsula. Large parts of the arctic slope are of this character. The pattern is caused by frost action.

Vegetation in the high mountains of Alaska often suggests planted rock-gardens. On this slope in McKinley Park, *Dryas octopetala, Geum Rossii, Potentilla uniflora, Synthyris borealis,* and other species adorn the open rocky soil.

Picea sitchensis woods at their altitudinal limit north of Homer, Kenai Peninsula. Above the spruce are subalpine thickets of *Alnus crispa* and rich subalpine meadow studded with *Calamagrostis* and *Veratrum viride* subsp. *Eschscholtzii.*

The long, entangled, prickly branches of the Devil's Club (*Echinopanax horridum*) form nearly impenetrable thickets in moist lowland woods.

Geum Rossii, a beautiful plant not uncommon in the mountains of Alaska and the central Yukon, and especially common in the Aleutian Islands.

Primula tschuktschorum, a primrose common on the shores of Bering Sea. The form of the leaves is highly variable.

Pale Corydalis or Rock Harlequin (*Corydalis sempervirens*), an annual, sometimes overwintering, evergreen plant occurring in the interior.

Chrysanthemum bipinnatum, an elegant arctic-Siberian plant especially common in Western Alaska.

Geranium erianthum, a wild geranium common in woods and mountains far above timberline in the southern part of Alaska.

Bearberry (*Arctostaphylos rubra*), with juicy scarlet drupes, otherwise similar to its black-fruited relative, *A. alpina*. Both turn scarlet in autumn, sometimes painting entire slopes red.

Bluebell (*Mertensia paniculata*), common in woods and mountains in interior and western Alaska, sometimes completely covering burned areas.

Arnica frigida, a common plant throughout interior Alaska, from the lowlands up into the mountains.

Lupinus nootkatensis, the common lupine of the Aleutian Islands and the southern coast. One of the parents of the hybrid garden lupines.

Spiderplant (*Saxifraga flagellaris* subsp. *platysepala*), a high-arctic plant that propagates from bulblets at ends of long runners, thus spreading in areas where seeds cannot ripen.

Salix arctica subsp. *crassijulis*, a Pacific large-leaved race of the circumpolar Arctic Willow, common in the coastal mountains.

Lady's Slipper (*Cypripedium guttatum*). Rare in Alaska, this little plant occurs chiefly in Asia. Flower color is quite variable, owing to hybridization between two races.

Carex Gmelinii, perhaps the most showy sedge of Alaska. A coastal species, up to one meter tall.

Roseroot (*Sedum rosea* subsp. *integrifolia*), common in the high mountains. The fleshy rhizome is not edible, but the leaves and young shoots are often eaten.

Squirreltail Grass (*Hordeum jubatum*), a beautiful plant, particularly so when the wind disturbs its lustrous spikes.

Castilleja elegans, one of several closely related species of Indian Paint Brush occurring in Alaska. This species grows on rocky slopes in the high mountains.

Yellow Dryas (*Dryas Drummondii*), often forming large masses on gravelbars in rivers. The flowers do not expand, and the seeds have long feathery plumes.

Crepis nana, a peculiar little species of Hawk's-Beard, intermediate between the genus *Crepis* and the exclusively Asiatic genus *Youngia*.

Bistort (*Polygonum bistorta* subsp. *plumosa*), common over most of Alaska. The name derives from the contorted rhizome, which can be eaten raw or boiled.

Soapberry (*Shepherdia canadensis*), common in interior western Alaska. The fruit is eaten mixed with sugar and water and beaten frothy. The leaves, covered with silvery scales, appear silver-plated.

Alaska Boykinia (*Boykinia Richardsonii*), a splendid, hardy plant of alpine meadows, certainly a relic from the late Tertiary. Occurs only within our area of interest.

Elephant's Head (*Pedicularis groenlandicus*), which occurs in southeastern Yukon only, within our area of interest. The flower resembles an elephant's head with lifted trunk.

Bunchberry or Canadian Dwarf Cornel (*Cornus canadensis*), common in woods. The small dark flowers are surrounded by large, white, petal-like bracts.

Oxytropis Maydelliana, a characteristic legume occurring in most Alaskan mountains, easily recognized by its yellow flowers and brown stipules.

Draba stenopetala, a rare, remarkable plant of the high mountains of Alaska, the Yukon, and easternmost Siberia.

Moss Campion (*Silene acaulis*), one of the most striking plants of the Alaskan mountains and tundra.

Siberian Aster (*Aster sibiricus*), common in mountains and at low altitudes throughout Alaska, except for the southeastern coast. Variable, sometimes single-flowered but often branched.

Senecio pseudo-Arnica (which looks like a large *Arnica*—thus the name) is characteristic of all sandy or gravelly shores of the southern coast and Bering Sea.

Eritrichium aretioides, an alpine plant closely related to the Forget-Me-Not, growing in patches on stony slopes.

Pedicularis Kanei subsp. *Kanei*, a lousewort completely covered with white wool when young. Common in most mountains of interior, arctic, and western regions.

Pedicularis capitata, a peculiar lousewort, markedly different from most other species of the genus. Grows on stony slopes in the mountains and on tundra.

Potentilla villosa, a beautiful species with densely silky-pubescent leaves. Characteristic of the cliffs along the southern and western coast.

Papaver alaskanum, a large-flowered, low-growing species occurring only in Alaska. Poppies are among the finest adornments of the arctic tundra and mountains.

Dryas octopetala subsp. *alaskensis*, distinguished by large, strongly ribbed leaves, with midvein glands below. This genus presents a number of puzzling forms in Alaska.

Phyllodoce aleutica subsp. *glanduliflora*, a common heath plant on the mountains of the southern coast.

Phyllodoce empetriformis, a rare plant occurring in the interior western part of Alaska.

Cassiope Mertensiana, first collected by Carl Mertens at Sitka. Cassiope heaths are characteristic western-American plant societies.

44. Nymphaeaceae (Water Lily Family)

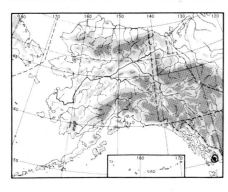
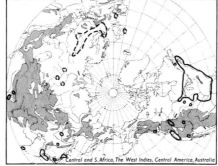

Brasenia Schreb.

1. Brasènia Schrèberi J. F. Gmel.   Water Shield

Rootstock creeping; leaves long-petiolated, centrally peltate, floating, entire or somewhat crenate; sepals 3–4; petals 3–4, linear, sessile, dull purple; stamens 12–18; pistils 4–10; fruit clavate, indehiscent; stems, petioles, and lower surface of leaves covered with viscid jelly.

Ponds and streams. Type locality not given. Broken line on circumpolar map indicates fossil range in Europe.

× ¼

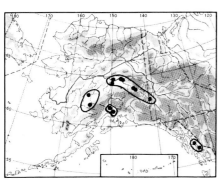
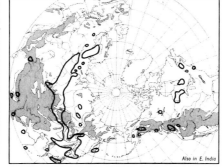

Nymphaea L.

1. Nymphaèa tetragòna Georgi   Dwarf Water Lily

Castalia Leibergii Morong; *Nymphaea tetragona* subsp. *Leibergii* (Morong) Pors.; *N. tetragona* var. *Leibergii* (Morong) Schuster.

Flowers and leaves from submerged, horizontal or erect rhizome; leaves ovate to obovate; flowers 4–5 (–8) cm in diameter, with quadrangular base; petals 10–12, white, sometimes pinkish or with crimson lines; stamens 12–40, the inner with broad filaments.

Ponds and swamps. Described from the Lena River, Siberia.

× ½

44. NYMPHAEACEAE (Water Lily Family)

Nuphar Sm.

Sepals 7–9, yellow; anthers reddish; petioles terete .... 1. *N. polysepalum*
Sepals 6 or fewer, the inner red at base, inside; petioles flattened .. 2. *N. variegatum*

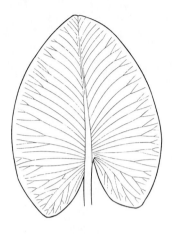

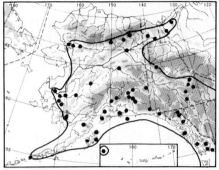

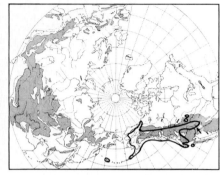

1. Nùphar polysèpalum Engelm. Yellow Pond Lily

Nymphaea polysepala (Engelm.) Greene; *Nymphozanthus polysepalus* (Engelm.) Fern.; *Nuphar luteum* subsp. *polysepalum* (Engelm.) Beal.

Stout submerged rhizome; submerged, basal leaves thin; floating leaves long-petiolated; petioles terete; sepals 7–9, yellow; petals narrowly cuneate, hidden by stamens; anthers reddish; fruit ovoid to subglobose.

Ponds and slow streams. More common than indicated by map; probably common along the Yukon River and its tributaries.

The rhizome can be eaten boiled or roasted.

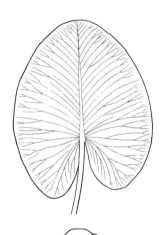

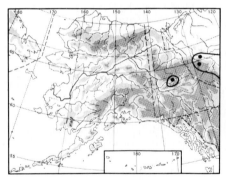

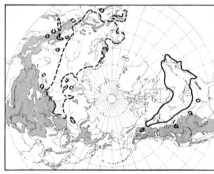

2. Nùphar variegàtum Engelm.

Nymphaea americana Mill. & Standl., not *Nuphar americanum* Prov.; *Nuphar luteum* subsp. *variegatum* (Engelm.) Beal.

Similar and closely related to *N. polysepalum*, but sepals not more than 6, the inner sepals red at the base inside; petioles flattened; anthers yellow.

Ponds and slow streams. Described from "St. John to Glorie Lakes, Quebec." Broken line on circumpolar map indicates range of the related **N. lùteum** L.

45. Ceratophyllaceae (Hornwort Family) / 46. Ranunculaceae (Crowfoot Family)

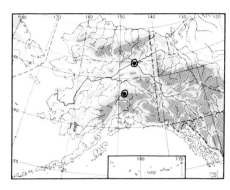
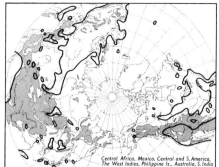

×¾

Ceratophyllum L.

1. Ceratophýllum demérsum L. Hornwort

Root lacking; dark-green, submerged stems with leaves in whorls, dichotomously dissected into stiff, linear, serrate divisions; flowers in axils, monoecious, lacking corolla, the staminate with numerous anthers, the pistillate with 1 style; fruit with 2 long spines and the remaining style.

Quiet water. Described from Europe.

Caltha L.

Stem erect, scapose or single-leaved; sepals white or bluish at base:
    Leaves oblong-cordate, with coarsely dentate basal lobes . . . . . . 1. *C. leptosepala*
    Leaves reniform-orbicular, crenate . . . . 2. *C. biflora*
Stem decumbent, with several leaves; flowers white or yellow:
    Stem mostly floating; flowers small, white, or reddish in margin; follicles straight
    . . . . 3. *C. natans*
    Stem decumbent; flowers mostly larger, yellow; follicles curved:
        Leaves reniform, with open sinus, crenate-dentate at lobes . . . .
        . . . . 4. *C. palustris* subsp. *arctica*
        Leaves ovate-reniform, with narrow, evenly crenate sinus . . . .
        . . . . 5. *C. palustris* subsp. *asarifolia*

46. Ranunculaceae (Crowfoot Family)

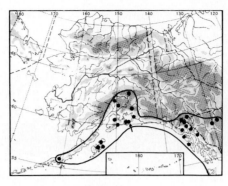
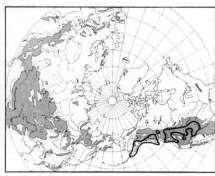

1. Cáltha leptosèpala DC.

Glabrous; stems single, from fibrous root; stem simple, scapose or 1-leaved; stem single-flowered in small specimens, 2-flowered in large specimens; basal leaves cordate, with shallow sinus, longer than broad, repand-dentate; sepals 6–12, white, bluish at base; follicles with straight beak, markedly short-stipitated, divergent.

Running water, bogs.

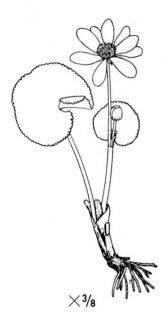
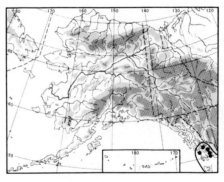
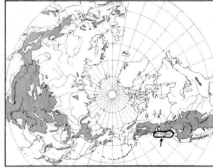

2. Cáltha biflòra DC.

Glabrous; basal leaves several, broader than long, crenate; stems with 1 leaf; sepals 6–9, oblong, white; follicles upright, stipitated at maturity, with thin, straight beak.

Bogs. In California, var. **Howéllii** Huth.

46. Ranunculaceae (Crowfoot Family)

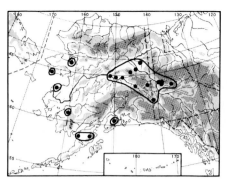 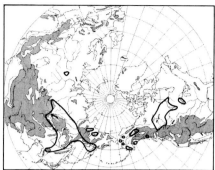 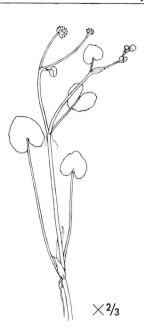

3. Cáltha nàtans Pall.

Glabrous; stem floating or creeping in mud, rooting at nodes; leaves cordate-reniform, bluntly dentate; sepals white, sometimes reddish in margin, ovate, 3–5 mm long; follicles with short, straight beaks, about 4 mm long, in globular clusters.
Ponds, moist mud.

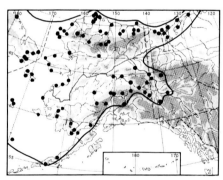 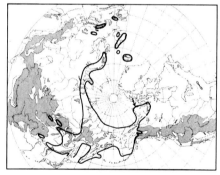 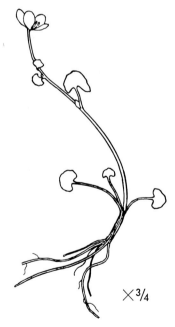

4. Cáltha palùstris L.   Marsh Marigold
subsp. árctica (R. Br.) Hult.

Caltha arctica R. Br.; *C. palustris* var. *arctica* (R. Br.) Hult.; *C. confinis* Greene.

Glabrous; highly variable in size, from delicate arctic specimens to coarse southern specimens; stem decumbent, rooting at nodes; leaves rounded or reniform, with open sinus and more or less crenate-dentate, basal lobes; leaves nearly entire in the most arctic specimens, distinctly crenate-dentate in the more southern ones; flower very small (from 10 mm in diameter) in arctic specimens, to large (35 mm in diameter) in the more southern specimens; sepals 5, yellow; follicles with hooked beak.

Moist places. *C. palustris* described from Europe, subsp. *arctica* from Melville Island.

Subsp. *arctica* differs from typical *C. palustris* in its stem, which roots at the nodes. Contains a poison, protoanemonin, which is broken down by boiling.

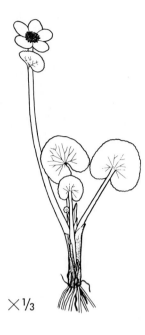
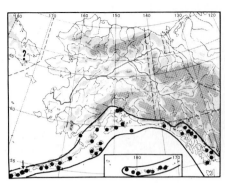
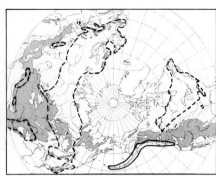

5. Cáltha palùstris L.
subsp. **asarifòlia** (DC.) Hult.

Caltha asarifolia DC.; *C. palustris* var. *asarifolia* Huth and var. *aleutensis* Huth; *C. Funstonii* K. C. Davis; *C. fistulosa* Schipczinskij with respect to plant from Chukchi Peninsula?

Similar to large specimens of subsp. *arctica*, but basal leaves more reniform, evenly crenate; sinus narrow, sometimes even with overlapping basal lobes; beaks of mature follicles shorter, less curved.

Moist places. Broken line on circumpolar map indicates range of other subspecies not present in area of interest.

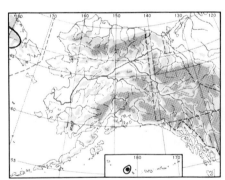
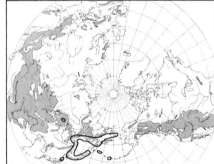

Trollius L.

1. Tròllius Riederiànus Fisch. & Mey.   Globeflower

Stem simple, erect, 1-flowered, with withered leaves at base, from thick rootstock with numerous dark-brown roots; leaves 3–5-parted, with lobed and dentated parts; flowers up to 3 cm in diameter; sepals orange-yellow; petals 5–7 mm long, about as long as stamens; follicles about 10 mm long (the 3-mm-long beak included), congested into a head.

Moist meadows. Described from Kamchatka.

46. Ranunculaceae (Crowfoot Family)

Coptis Salisb. (Goldthread)

Rhizome thick; leaves ternate, with pinnate, toothed divisions . . . 1. *C. aspleniifolia*
Rhizome filiform; leaves with 3 toothed leaflets . . . . 2. *C. trifolia*

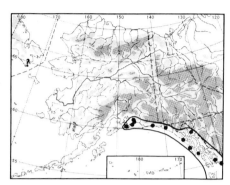
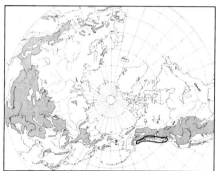
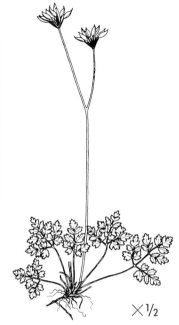

1. Cóptis aspleniifòlia Salisb.

Leaves basal, from thick rootstock; stem up to 35 cm high, often branched above; leaves terete, the divisions pinnate, with sharply toothed ultimate segments; scape with hyaline scales at base, mostly 2-flowered; sepals linear-lanceolate, gradually narrowed to acute tip; follicles long-stipitated, up to 12 in a head, membranaceous, with very short beak.

Moist places, woods. Described from the northwest coast of North America.

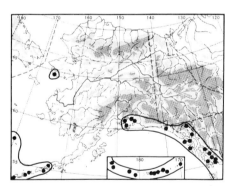
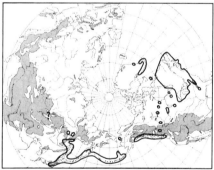

2. Cóptis trifòlia (L.) Salisb.
Helleborus trifolius L.

Rhizome filiform, bright yellow; leaves persistent through winter, lustrous, dark green above, paler below, with 3 sharply toothed leaflets; scape single-flowered, with short bract; sepals white, pinkish on outside, elliptic-lanceolate; petals rounded-obovate; follicles 3–6, long-stipitated at maturity, with long straight beak.

Woods and thickets; mossy places. Described from Canada, Siberia.

C. groenlándica (Oeder) Fern. (*Anemone groenlandica* Oeder) is also included in range shown on circumpolar map, since it seems impossible to distinguish it from the Alaskan plant.

46. Ranunculaceae (Crowfoot Family)

Actaea L.

Ultimate leaf segments not caudate; plant sparsely pubescent ....
....1. *A. rubra* subsp. *rubra*
Ultimate leaf segments caudate, very acute; plant more pubescent ....
....2. *A. rubra* subsp. *arguta*

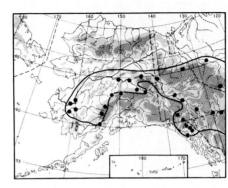

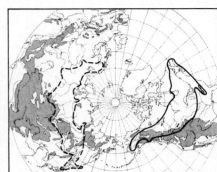

1. Actaèa rùbra (Ait.) Willd.   Baneberry
   Actaea spicata L. var. *rubra* Ait.
subsp. **rùbra**

Stems from thick root, branching above, sparsely pubescent; leaves 2–3, ample, all cauline, ternately compound, the ultimate segments cleft and toothed but not caudate; raceme subcylindric; pedicels filiform; petals whitish; fruit red or white, berrylike, oblong-ovoid.

Woods, dry slopes. Described from North America.

Very closely related to, or doubtfully distinct from, **A. erythrocárpa** Fisch. of Eurasia, the range of which is indicated by the broken line on the circumpolar map.

Both the berries and the root contain a poison, protanemonin, which causes vomiting, bloody diarrhea, and finally paralysis of the respiration.

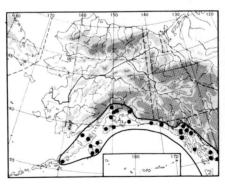

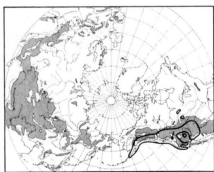

2. Actaèa rùbra (Ait.) Willd.
subsp. **argùta** (Nutt.) Hult.
   Actaea arguta Nutt.; *A. spicata* var. *arguta* Torr.

Similar to subsp. *rubra*, but leaves somewhat more dissected; ultimate segments of leaves caudate and more acute; plant somewhat more pubescent; and fruit more rounded.

Woods.

46. Ranunculaceae (Crowfoot Family)

Aquilegia L. (Columbine)

Flowers bluish-purple; style of mature follicles shorter than one-third of body; lamina of petals longer than spur .... 1. *A. brevistyla*
Flowers orange-red or yellow; style of mature follicles longer than one-third of body; lamina of petals much shorter than spur .... 2. *A. formosa*

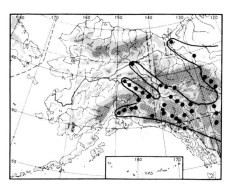
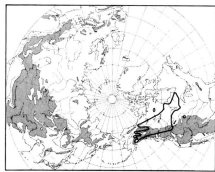
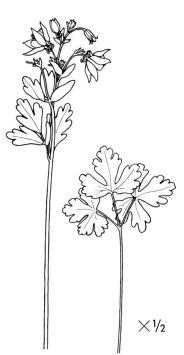

1. Aquilègia brevistỳla Hook.

Stems up to 1 meter tall from stout taproot, glabrous below, glandular above, simple or branched; basal leaves bi-ternate, green above, glaucous beneath, glabrous or pubescent; flowers spreading or pendant, pubescent; sepals and spurs blue or purple; laminae yellowish-white, shorter than petals; follicles upright, with straight styles 2–5 (–10) mm long.

Moist woods, meadows. Described from the western part of Canada, Bear Lake.

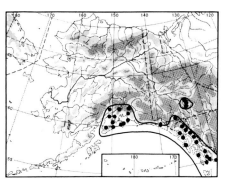
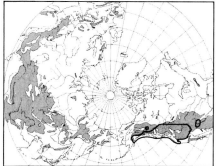
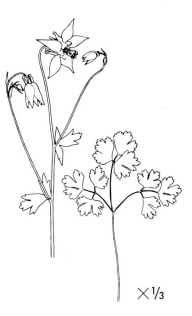

2. Aquilègia formòsa Fisch.   Western Columbine

Aquilegia canadensis L. var. *formosa* (Fisch.) S. Wats.

Stems up to 1 meter tall from stout taproot, glabrous below, glandular above, branched; basal leaves bi-ternate, green above, glaucous beneath, glabrous or somewhat pubescent; stem leaves reduced; flowers pendant, somewhat pubescent; sepals and spurs red, spreading; laminae yellow; stamens longer than lamina, glandular-pubescent; follicles with divergent styles 10–15 mm long.

Moist woods, mountain slopes. Described from Kamchatka, where, however, it does not occur; probably from collections in Sitka.

Large-flowered forms have been called var. **megalántha** Boiv.

Delphinium L. (Larkspur)

Delphinium species contain a poison similar to that of *Aconitum,* but less active.

Pedicels villous; follicles densely pubescent; calyx leaves large, about 2 cm long, clear blue, with darker spot near apex; plant 10–40 cm tall .... 1. *D. brachycentrum*
Pedicels pilose; follicles glabrous; calyx leaves smaller, 1–1.5 cm long, light to dark purple; plant taller .... 2. *D. glaucum*

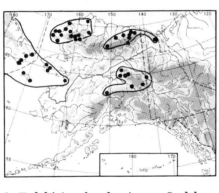 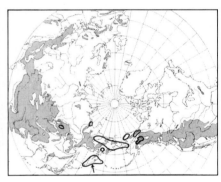

1. Delphínium brachycéntrum Ledeb.

Delphinium Chamissonis Pritz.; *D. pauciflorum* Rchb. (name only); *D. Maydelianum* Trautv.; *D. Blaisdellii* Eastw.; *D. Ruthae* Nels.; *D. alatum* Nels.; *D. Menziesii* of authors, with respect to Alaskan plant.

Pubescent from soft, white, curved or curled hairs; root thick, dark brown, woody; stem up to 60 cm tall, usually much shorter; lower leaves with rounded outline, cleft to base into 3 parts, which are again cleft or lobed into acute or somewhat acute segments, narrow in small specimens from exposed places, broader in taller specimens from more sheltered localities (though both types sometimes occur in the same specimen); bracts linear; pedicels villous; flowers several, blue or creamy white, suffused with blue (f. **pállidum** Lepage), about 2 cm long; calyx pubescent on outside, glabrous on inside, the lobes acute, somewhat acute, or blunt; spur straight, horizontal, with somewhat curved apex; petals white, tinged with blue; capsules pubescent, with beak about 2 mm long.

Rocky slopes, meadows along tundra, rivers, solifluction soil, scree slopes; in McKinley Park to at least 1,200 meters.

A plant from McKinley Park, suggestive of the hybrid *D. brachycentrum* × *glaucum,* has been described as **D. nùtans** Nels.

46. RANUNCULACEAE (Crowfoot Family)

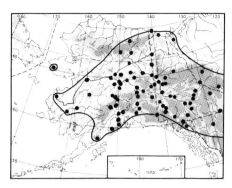
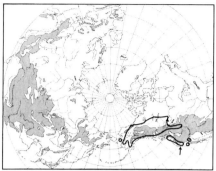
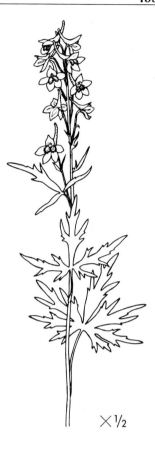

2. Delphínium glaùcum S. Wats.

Delphinium Brownii Rydb.; *D. scopulorum* var. *glaucum* (S. Wats.) Gray.

Stems up to 2 meters tall, coarse and glabrous, from woody, fibrous caudex; leaves palmatifid; segments incised and toothed, acute; racemes many-flowered, long, glabrous or somewhat pubescent; lower bracts leaflike, reduced upward; flowers dark violet-purple; spur 8–10 mm long; follicles glabrous or sometimes puberulent, especially when young.

Wet meadows, thickets.

Aconitum L. (Monkshood)

The *Aconitum* species are highly poisonous. The tubers contain aconitin, an ester alkaloid, which paralyzes the nerves and lowers the body temperature and blood pressure.

Leaves not cleft to base; nectaries (petals) about 8 mm long; helmet about as high as broad; plant coarse .... 4. *A. maximum*
Leaves cleft to base; nectaries small, about 3–4 mm long; helmet broader than high:
    Plant low-growing, single-flowered; nectaries not hooked at apex ....
    .... 3. *A. delphinifolium* subsp. *paradoxum*
    Plant tall, with several flowers; nectaries hooked at apex:
      Plant slender; flowers small; leaves very narrow-lobed ....
      .... 1. *A. delphinifolium* subsp. *delphinifolium*
      Plant coarser; flowers larger; leaves more broad-lobed ....
      .... 2. *A. delphinifolium* subsp. *Chamissonianum*

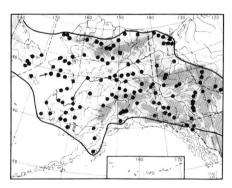
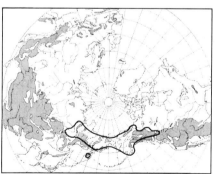
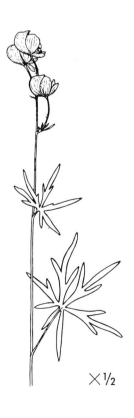

1. Aconìtum delphinifòlium DC.
subsp. delphinifòlium

Stem glabrous, straight, thin, up to 70 cm tall; leaves few, digitately cleft into 5-lobed parts, round in outline, glabrous; inflorescence few-flowered; flowers dark blue; hood rounded, about 2 cm long and a little broader, with beak; petals hooked in apex; stamens lacking teeth, glabrous.

Meadows, thickets, along creeks; common over large parts of the area, to at least 1,700 meters in the central mountains.

The form of the leaves is highly variable (the ultimate lobes are usually linear, sometimes broader); and the flowers are sometimes white (var. **albiflòrum** Pors.).

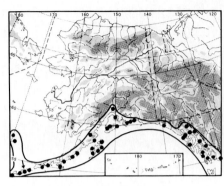
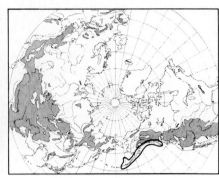

2. Aconìtum delphinifòlium DC.
subsp. **Chamissoniànum** (Rchb.) Hult.

Aconitum Chamissonianum Rchb.

Similar to subsp. *delphinifolium*, but taller, many-flowered, and with relatively broad-lobed leaves.

Meadows, thickets, woods. Intergrades with subsp. *delphinifolium* where the ranges overlap.

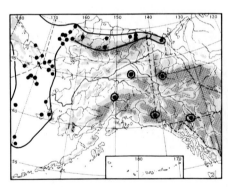
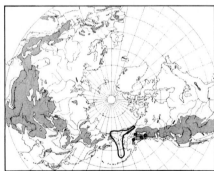

3. Aconìtum delphinifòlium DC.
subsp. **paradóxum** (Rchb.) Hult.

Aconitum paradoxum Rchb.; *A. nivatum* Nels.

Similar to subsp. *Chamissonianum*, but usually 1-flowered, rarely few-flowered; petals not hooked at apex.

Rocky slopes, alpine tundra. A reduced arctic-montane race.

46. Ranunculaceae (Crowfoot Family)

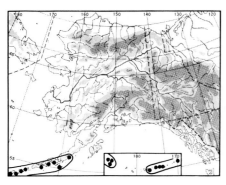
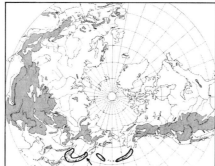

4. Aconìtum máximum Pall.
Aconitum kamtschaticum Rchb.

Upper part of root bulblike; stem glabrous below, pubescent above; up to 70 cm tall; basal leaves divided to base into many-cleft, acute, linear or lanceolate segments; upper leaves reduced, merely lobed; inflorescence at first terminal, capitate, later prolonged, with longer pedicels; flowers pubescent, grayish-blue; hood nearly beakless.

Subalpine thickets and meadows.
Root very poisonous.

Anemone L.

Many *Anemone* species contain a poison, anemonin.

Achenes, young as well as old, glabrous:
  Plant with slender, horizontal rootstock; leaves cleft to middle in 5 divisions; flowers yellow; styles not reflexed; achenes not winged .. 2. *A. Richardsonii*
  Plant with thick, oblique rootstock; leaves cleft to base in 3–5, twice-cleft sections; achenes wing-margined:
    Leaves round or reniform in outline, quinate; segments of leaves sessile; plant markedly villous; stem many-flowered .... 6. *A. narcissiflora* subsp. *villosissima*
    Leaves more or less pentagonal in outline, usually ternate; segments of leaves petiolulated; plant less villous; stem 1- to few-flowered:
      Plant tall, few-flowered; leaves twice ternate; ultimate segments toothed .. .. 7. *A. narcissiflora* subsp. *alaskana*
      Plant more low-growing; stem normally 1-flowered; leaves ternate; segments toothed or 3-cleft at middle:
        Segments of leaves narrow; sepals rhomboid, much broader at middle; flowers large, white on outside .... 5. *A. narcissiflora* subsp. *interior*
        Segments of leaves broader; sepals oval, sometimes slightly rhomboid; flowers smaller, often bluish on outside .... 4. *A. narcissiflora* subsp. *sibirica*
Achenes more or less densely villous:
  Leaves 3-foliate; segments not dissected .... 1. *A. deltoidea*
  Leaves ternate; segments more or less dissected:
    Leaves ternate; segments flabelliform, broad, crenate or sometimes cleft to middle .... 3. *A. parviflora*
    Leaves 3 times ternate; segments linear to oblong:
      Plant generally tall (20–40 cm long), silky-villous; flowers white, reddish, or red; styles short, 1–1.3 mm .... 8. *A. multifida*
      Plant shorter, sparsely silky-hirsute to glabrate; flowers white, bluish, or blue; styles longer, 1.5–2 mm .... 9. *A. Drummondii*

46. RANUNCULACEAE (Crowfoot Family)

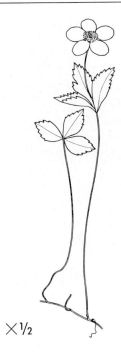
× ½

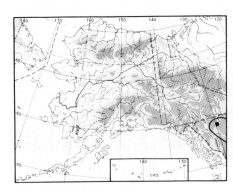 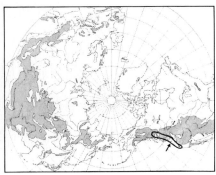

1. Anemòne deltoìdea Hook.

Rootstock slender, creeping; stem glabrous or with scattered hairs; basal leaves mostly solitary, 3-foliate; leaflets crenate-dentate; involucral leaves 3, subsessile, dentate; flowers solitary; sepals white; achenes glabrous above, more or less short-hirsute below.

Dry forests. Described from the mouth of the Columbia River.

× ½

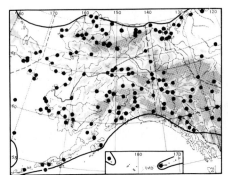 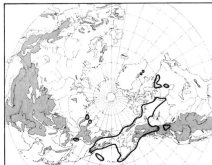

2. Anemòne Richardsònii Hook.

Stems from thin, filiform, horizontal rhizome, from which leaves emerge each some distance from the next; stem pubescent below; leaves 3-lobed, rounded-reniform in outline, the lobes shallowly divided and acutely toothed; involucre of 2 or 3 sessile, sharply toothed leaves; flowers single; sepals bright yellow; fruiting head globose; achenes with very long beak, recurved at tip.

Thickets, meadows, snowbeds in the mountains, to at least 1,000 meters. Described from Hudson Bay, Rocky Mountains, Unalaska, Siberia.

46. RANUNCULACEAE (Crowfoot Family)

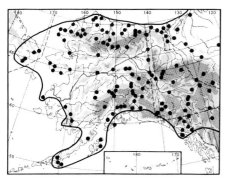 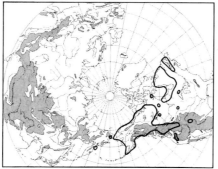 

3. Anemòne parviflòra Michx.

Stems from long, slender, scaly rootstock, often branched at tip; basal leaves dark green, lustrous, glabrous, divided into 3 lobed and bluntly toothed parts; involucre sessile; stem white-pubescent, 1-flowered; sepals mostly 6, white, appressed-pubescent and bluish at base beneath, much longer than stamens; fruits acute, densely woolly, in globose or ovoid head.

Meadows, heaths, stony slopes, snow beds, up to at least 2,000 meters in the St. Elias range, and down to the upper reaches of the forest region. Described from Hudson Bay.

Most specimens belong to var. **grandiflòra** Ulbr., with flowers 30–40 mm in diameter, but specimens with flowers 15 mm in diameter also occur.

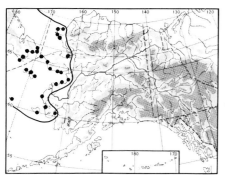 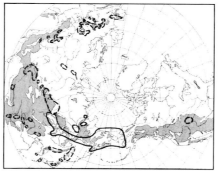 

4. Anemòne narcissiflòra L.
subsp. **sibìrica** (L.) Hult.

Anemone sibirica L.; *A. narcissiflora* var. *monantha* of some authors.

More or less villous; smaller than subsp. *villosissima*; basal leaves 3-parted, the parts cleft into fairly broad, more or less petiolulated ultimate segments; flowers mostly small, sepals oval, bluish on the outside.

Snow beds, heaths. *A. narcissiflora* described from the Alps of Austria and Switzerland and from Siberia; subsp. *sibirica* from Siberia.

Broken line on circumpolar map indicates range of other races of *A. narcissiflora* not occurring within area of interest.

The rootstock is eaten raw or boiled by the Siberian Eskimo.

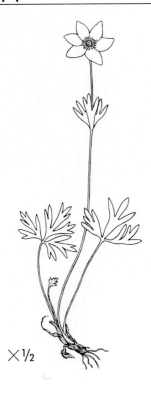
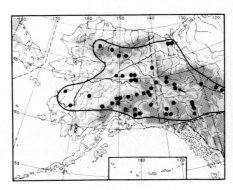
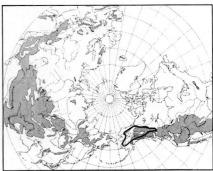

5. Anemòne narcissiflòra L.
subsp. **intèrior** Hult.

Similar to subsp. *sibirica*, but plant taller and more slender, mostly single-flowered; sepals rhomboid, white on the outside.

Dry heaths, stony slopes.

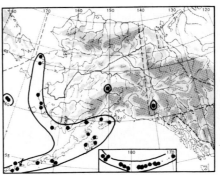
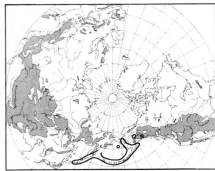

6. Anemòne narcissiflòra L.
subsp. **villosíssima** (DC.) Hult.

Anemone narcissiflora var. *villosissima* DC.; *A. villosissima* (DC.) Juz.

Coarse, up to 35 cm tall; basal leaves several, spreading, ciliated in margin, with orbicular or reniform outline, divided into 5 sessile parts, which in some measure cover each other and which are cleft into linear-oblong, somewhat acute segments; petioles densely pubescent from long, soft, patent hairs; stem densely long-haired; flowers several, large, white or creamy-white, in subumbellate inflorescence.

Meadows; common in the Aleutian Islands. Transitions to subsp. *sibirica* occur.

46. RANUNCULACEAE (Crowfoot Family)

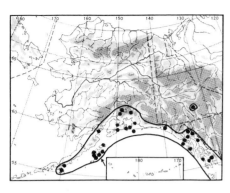
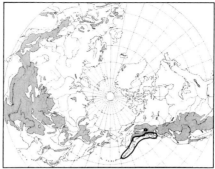

7. Anemòne narcissiflòra L.
subsp. **alaskàna** Hult.

Plant usually tall, slender, few-flowered; basal leaves pentagonal in outline, twice ternate, the ultimate segments toothed; plant less villous than subsp. *villosissima*.

Grassy mountain slopes.

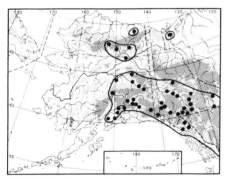
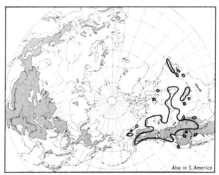

8. Anemòne multífida Poir.
Anemone hudsoniana DC.; *A. multifida* var. *hudsoniana* DC.

Loosely caespitose, with thick, many-headed caudex; stems silky-villous, up to 3 dm high; basal leaves 2–3 times ternately divided into narrowly lanceolate, acute lobes; involucral leaves 3; sepals 5–10 mm long, silky on the outside, highly variable in size and color, from yellowish-white to grayish or purplish; fruiting head subglobose to short-cylindric; achenes woolly, with short beak.

Dry slopes. Described from Magellan Sound.

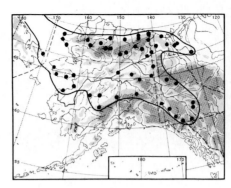
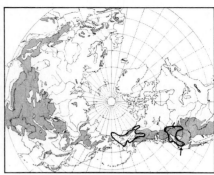

9. Anemòne Drummóndii S. Wats.
Pulsatilla multiceps Greene; *Anemone multiceps* (Greene) Standl.

Plant with stout, branched, many-headed dark-brown rootstock; basal leaves several, sparsely villous to nearly glabrous, twice ternate, ultimate segments linear, in arctic specimens oblong-linear, acute or somewhat acute; petioles long, villous, reddish at base; involucral leaves sessile; peduncles usually solitary, villous; flowers 10–35 mm in diameter; sepals ovate, blunt or somewhat acute, white, tinged with blue, or blue, villous outside; fruiting head rounded, densely woolly; styles slender, often wine-red.

Dry, rocky ledges, scree slopes.

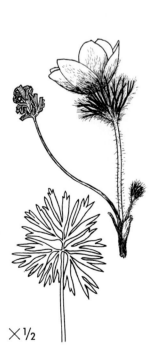

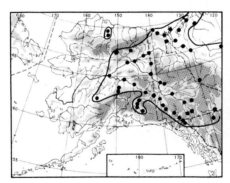
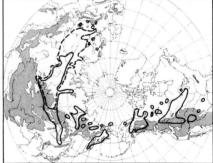

Pulsatilla Mill.

1. Pulsatílla pàtens (L.) Mill. Pasqueflower
Anemone patens L.
subsp. **multifida** (Pritz.) Zamels

Anemone patens var. *multifida* Pritz.; *A. patens* subsp. *multifida* of Fl. Alaska & Yukon; *A. patens* var. *Wolfgangiana* (Bess. ex W. D. J. Koch) Trautv. & Mey.; *A. patens* var. *Nuttalliana* Gray; *A. multifida* (Pritz.) Juz.; *A. ludoviciana* Heller; *A. Wolfgangiana* Bess.; *Pulsatilla hirsutissima* Britt.

Silky-villous; stems from vertical, dark-brown, many-headed caudex; basal leaves developing after floration, ternately divided, the divisions 2- or 3-parted into numerous narrow segments; the very villous involucrum deeply cleft into narrow, lanceolate, acute lobes; sepals blue or purple; staminodia glandlike; fruit plumose when ripe.

Dry, sandy soil. *P. patens* described from "Tobolsko Sibiriae, Lusatia inferiore" (Lausitz); subsp. *multifida* from Siberia.

Broken line on circumpolar map indicates range of other subspecies.

46. RANUNCULACEAE (Crowfoot Family)

Ranunculus L. (Buttercup, Crowfoot)
Many *Ranunculus* species contain a poison, protoanemonin.

Plant with submerged, capillary leaves, with or without broader-lobed floating or aerial leaves:
  Plant aquatic; flowers white with yellow spot at base; achenes transversely wrinkled:
    Floating leaves present .... 2. *R. trichophyllus* var. *hispidulus*
    Floating leaves lacking:
      Plant dark green; leaves large, divided several times; achenes with short but distinct beak; flowers about 2 cm in diameter .... 1. *R. trichophyllus* var. *trichophyllus*
      Plant light green; leaves small, 2–3 cm in diameter, 3–4 times divided; achenes with almost indistinct beak; flowers about 1 cm in diameter .... 3. *R. confervoides*
  Plant amphibious; flowers white or yellow; achenes not transversely wrinkled:
    Leaves 3-lobed, with entire lobes or lateral lobes sometimes again slightly cleft; calyx with 3 leaves:
      Flowers white, about 2 cm in diameter; plant large .... 8. *R. Pallasii*
      Flowers yellow, 0.6 cm in diameter or smaller; plant minute:
        Plant very small; leaves deeply cleft into linear lobes, directed forward .... 7. *R. hyperboreus* subsp. *Arnelli*
        Plant larger; leaves less deeply cleft; lobes broad .... 6. *R. hyperboreus* subsp. *hyperboreus*
    Leaves 3–5-lobed, the lobes again cleft into narrow, often more or less linear or capillary segments; calyx with 5 leaves; flowers yellow:
      Leaves 0.5–1.5 cm in diameter; end lobes long and narrow; petals 1–2(4) mm long .... 4. *R. Gmelini* subsp. *Gmelini*
      Leaves larger; end lobes broader and shorter; petals 4–8 mm long........ .... 5. *R. Gmelini* subsp. *Purshii*
Plant erect or decumbent, not aquatic or amphibious:
  Flowers large, white, or in age red; leaves 3-parted, the lobes again trifid; sepals persistent, hirsute from dark-brown hairs .... 9. *R. glacialis* subsp. *Chamissonis*
  Flowers yellow; sepals deciduous:
    Plant scapose:
      Sepals 3; plant with long, filiform rhizome .... 10. *R. lapponicus*
      Sepals 5; plant with thick, short rhizome .... 11. *R. Cooleyae*
    Plant not scapose:
    ■ Plant decumbent, rooting in nodes, or with radicant runners:
      Basal leaves 3-foliate:
        Head of achenes ovoid; petals less than 6 mm long .. 25. *R. Macounii*
        Head of achenes spherical; petals 6–10 mm long ...... 26. *R. repens*
      Basal leaves entire or 3-lobed; flowers much smaller:
        Basal leaves entire; calyx leaves 5:
          Basal leaves linear or oblong .... 12. *R. reptans*
          Basal leaves ovate, cordate, or cuneate at base, coarsely toothed... .... 13. *R. cymbalaria*
        Basal leaves 3-lobed; calyx leaves 3:
          Plant very small; leaves deeply cleft into linear lobes directed forward .... 7. *R. hyperboreus* subsp. *Arnelli*
          Plant larger; leaves less deeply cleft, lobes broad .... 6. *R. hyperboreus* subsp. *hyperboreus*

■ Plant erect (in *R. pygmaeus* and *R. gelidus* sometimes more or less decumbent); stem not rooting in nodes; runners lacking:
  Radical leaves round, reniform, crenate (later developed leaves often 3-cleft); tall plants with very small flowers (about 5 mm in diameter); stigma nearly sessile .... 21. *R. abortivus*
  Radical leaves all lobed; flowers small or large:
    Leaves glabrous or ciliated in margin; smaller or medium-sized plants (*R. sceleratus* often tall), with glabrous stem, or in some cases slightly hairy:
      Achenes beakless, basal leaves 3-lobed, again lobed .... 22. *R. sceleratus* subsp. *multifidus*
      Achenes distinctly beaked; basal leaves 3–5-lobed or 3-foliate:
        Sepals densely black- or dark-brown-hairy:
          Receptacle naked .... 15. *R. nivalis*
          Receptacle hispid, with brown hairs .... 16–17. *R. sulphureus*
        Sepals glabrous or with yellowish hairs:
          Peduncles glabrate .... 14. *R. Eschscholtzii*
          Peduncles pubescent, especially above:
            Plant 15–35 cm tall, several-flowered; bracts sessile, with linear lobes, forming 2 or 3 whorls .... 23. *R. pedatifidus* subsp. *affinis*
            Plant shorter, single-flowered:
              Basal leaves simply or pedately 3–5-cleft, with rounded lobes; flowers small .... 19–20. *R. pygmaeus*
              Basal leaves biternately or pedately divided and parted, with oblong to spatulate lobes; flowers about 1 cm in diameter; plant stouter .. 18. *R. gelidus* subsp. *Grayi*
    Leaves pubescent, especially on lower surface; mostly tall plants:
      Flowers less than 1.5 cm in diameter:
        Receptacle glabrous; beak slender, hooked at tip; head of achenes globose .... 27. *R. Bongardi*
        Receptacle hispid; beak broad, not distinctly hooked at tip; head of achenes ovoid or cylindrical:
          Head of achenes cylindrical; sepals longer than petals .... 24. *R. pennsylvanicus*
          Head of achenes ovoid; sepals as long as petals, or slightly shorter .... 25. *R. Macounii*
      Flowers 1.5–3 cm in diameter:
        Basal leaves 3-foliate, with stipitate divisions, or pinnate, with 3–7 leaflets; receptacle hispid; beaks of achenes straight:
          Basal leaves 3-foliate, with stipitate divisions; beaks of achenes about 1.5 mm long .... 28. *R. pacificus*
          Basal leaves pinnate, with 3–7 leaflets; beaks of achenes about 2 mm long ...... 29. *R. orthorhynchus* subsp. *alaschensis*
        Basal leaves palmately 3–5-lobed:
          Beaks of achenes slender, long, falcate, hooked at apex:
            Sepals spreading; basal leaves 3-lobed; lobes narrow, with 2–3 long, acute teeth; flowers often large, 2.5–3.7 cm in diameter .... 35. *R. Turneri*
            Sepals reflexed in full anthesis; basal leaves 3–5-lobed; flowers about 2.5 cm in diameter:
              Plant glabrous or soft-pubescent, slender .... 30. *R. occidentalis* subsp. *occidentalis* var. *brevistylis*
              Plant densely coarse-pubescent:
                Pubescent, with brownish, patent hairs .... 31. *R. occidentalis* subsp. *Nelsoni*
                Pubescent, with appressed, gray, silky hairs .... 32. *R. occidentalis* subsp. *insularis*
          Beaks of achenes very short (0.3–0.6 mm), broad, nearly straight:
            Plant lacking stolons; leaves pedately 5–7-parted .... 33. *R. acris*
            Plant with subterranean stolons; leaves 3-parted, the parts divided or lobed .... 34. *R. grandis* var. *austrokurilensis*

46. Ranunculaceae (Crowfoot Family)

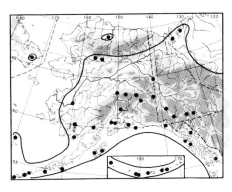
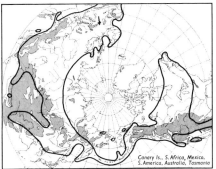

1. Ranúnculus trichophýllus Chaix.
var. **trichophýllus**

White Water Crowfoot

Ranunculus aquatilis var. *capillaceus* (Thuill.) DC.; *R. subrigidus* with respect to arctic plant.

Stem long, floating; leaves dark green, finely dissected, falling together like a soft brush when removed from water; floating leaves lacking; pedicels arching; flowers white, 1–2 cm in diameter; petals not overlapping; stamens 10–15; achenes glabrous, rugose, with short beak; receptacle glabrous or evenly pubescent.

Shallow water.

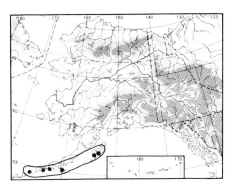
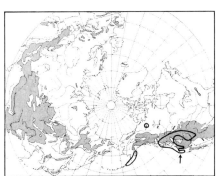

2. Ranúnculus trichophýllus Chaix.
var. **hispídulus** (E. R. Drew) W. B. Drew

Ranunculus aquatilis var. *hispidulus* E. R. Drew; *Batrachium Grayanum* (Freyn) Rydb.; *R. Grayanum* Freyn.

Similar to var. *trichophyllus*, but with floating leaves; receptacle pubescent with tufted hairs.
Shallow water. Very closely related taxa occur in Eurasia.

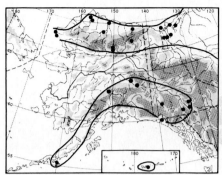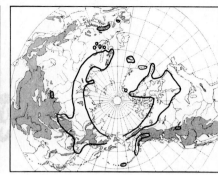

×3/4

3. Ranúnculus confervoìdes (E. Fries) E. Fries

Batrachium confervoides E. Fries; *Ranunculus trichophyllus* var. *eradicatus* (Laest.) W. B. Drew; *R. aquatilis* var. *eradicatus* Laest.; *R. trichophyllus* subsp. *lutulentus* (Perrier & Songeon) Janchen.

Similar to *R. trichophyllus,* but more delicate; leaves with very narrow segments; flowers smaller, 5–6 mm in diameter; petals broad; stamens 4–5; achenes glabrous; receptacle with tufted hairs.

Shallow water. Described from Lapland and from Uleåborg, Finland.

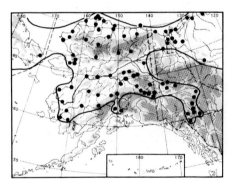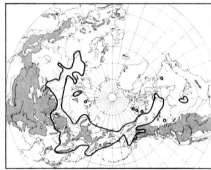

×3/4

4. Ranúnculus Gmélini DC.
subsp. **Gmélini**

Ranunculus Gmelini var. *typicus* Benson; *R. Gmelini* var. *yukonensis* Benson; *R. Purshii* var. *Gmelini* D. Don; *R. Purshii* subsp. *yukonensis* (Britt.) Pors.; *R. yukonensis* Britt.

Creeping in mud or floating in water; stem sparsely hirsute, rooting at nodes; leaves divided into 3 parts, which are again divided or lobed and finely dissected into linear, elongated divisions, acute in submerged forms; flowers 6–10 mm in diameter; petals longer than sepals, yellow; achenes glabrous, with broad, recurved beak.

Wet mud, ponds. Described from Siberia.

46. RANUNCULACEAE (Crowfoot Family)

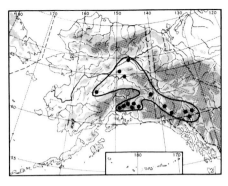
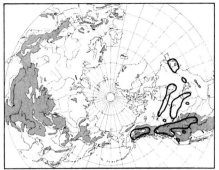

5. Ranúnculus Gmélini DC.
subsp. **Pùrshii** (Richards.) Hult.

Ranunculus Purshii Richards.; *R. Gmelini* var. *Hookeri* (D. Don) Benson; *R. Purshii* var. *Hookeri* D. Don; *R. Gmelini* var. *Purshii* Hara.

Similar to subsp. *Gmelini*, but coarser, mostly glabrous; leaves larger; ultimate segments broader and shorter; flowers 8–16 mm in diameter.

Pools; wet, muddy shores. Described from the southern Mackenzie district.

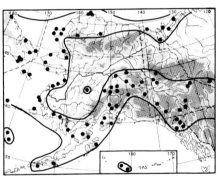
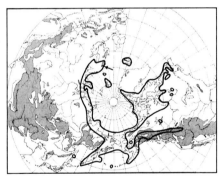
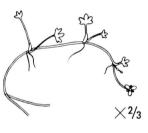

6. Ranúnculus hyperbòreus Rottb.
subsp. **hyperbòreus**

Glabrous; stem creeping in mud or floating; leaves small, 3–5-lobed, in submerged specimens sometimes divided and again parted; peduncles arched; sepals and petals 3 (–4); sepals glabrous; petals about 4 mm long, with short claw; fruiting head ovate; fruits with short, straight beak, hooked in apex.

Wet meadows, ponds, wet tundra, snow beds.

Plants occurring in nitrogen-rich soil (in inhibited areas) are much larger and more luxuriant.

46. RANUNCULACEAE (Crowfoot Family)

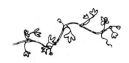
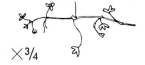
× ¾

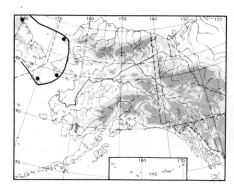
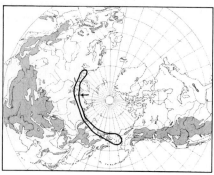

7. Ranúnculus hyperbòreus Rottb.
subsp. **Arnélli** Scheutz.

Ranunculus hyperboreus subsp. *samojedorum* (Rupr.) Hult.; *R. samojedorum* Rupr.

Similar to subsp. *hyperboreus*, but leaves only about 2 mm long, cleft to base, with linear, forward-turned lobes.

Wet tundra. A high-arctic race.

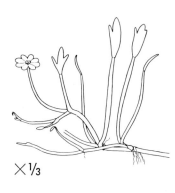
× ⅓

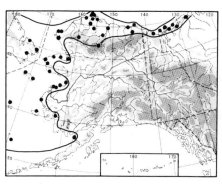
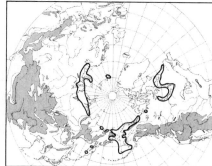

8. Ranúnculus Pallàsii Schlecht.

Glabrous; flowers and leaves scattered, from long floating rhizome, rooting at nodes; leaves highly variable in form, from simple oblong-linear to cuneate, broadly 2–3-lobed, with blunt lobes; flowers single, up to 2.8 mm in diameter; petals 5 to numerous, usually white or sometimes reddish; achenes with short, recurved beak, in semispherical head.

Ponds on tundra. Described from St. George Island and Kotzebue Sound.

The young shoots are eaten boiled by the natives.

46. Ranunculaceae (Crowfoot Family)

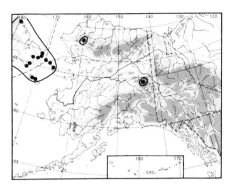
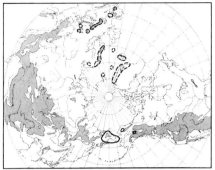
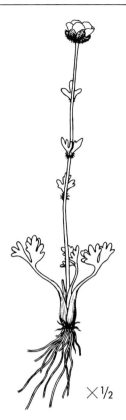

9. Ranúnculus glaciàlis L.
subsp. **Chamissònis** (Schlecht.) Hult.

Ranunculus Chamissonis Schlecht.; *R. glacialis* var. *Chamissonis* (Schlecht.) Benson; *Oxygraphis Chamissonis* (Schlecht.) Freyn.

Stems single, mostly 1-flowered, glabrous below, hirsute above, with brown hairs; basal leaves divided into 3 segments, which are again 2–5-lobed, the ultimate lobes varying from linear-acute to ovate, somewhat acute, or blunt; cauline leaves hirsute; sepals and petals long-persisting; sepals densely covered with dark-brown, coarse, stiff hairs; petals longer than sepals, at first white, soon dark red.

Grassy slopes. *R. glacialis* described from the Alps of Lapland and Switzerland; subsp. *Chamissonis* from Anadyr and the Chukchi Peninsula.

Broken line on circumpolar map indicates range of subsp. **glaciàlis**.

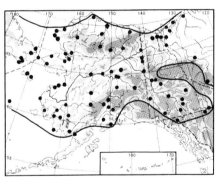
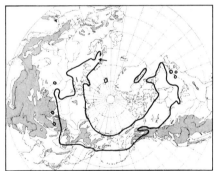
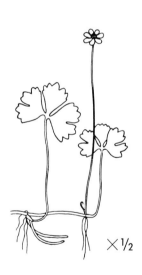

10. Ranúnculus lappònicus L.

Glabrous; stems and leaves spread from white, creeping rhizome; leaves long-petiolated, deeply 3-parted, the middle part 3–5-toothed, the lateral parts 2-lobed and toothed; stem single-flowered, with sessile stem leaf; sepals 3; petals 6–8, somewhat longer than sepals, yellow; achenes constricted at middle, easily falling off; beak of achene shorter than body, hooked.

Wet forests, bogs, black-spruce muskeg, peaty soil.

46. RANUNCULACEAE (Crowfoot Family)

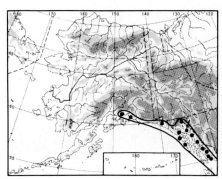

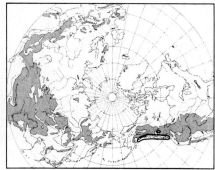

11. Ranúnculus Coolèyae Vasey & Rose

Kumlienia Cooleyae (Vasey & Rose) Greene; *Arcteranthis Cooleyae* (Vasey & Rose) Greene.

Glabrous, with short caudex; basal leaves circular in outline, divided into 3 crenately lobed parts; cauline leaf lacking or solitary, small, scalelike; flowers solitary; sepals glabrous, slightly longer than the numerous, narrow, yellow petals; fruiting head conical-ovate; fruits glabrous, with evenly tapering, long, hooked beak.

Snow banks.

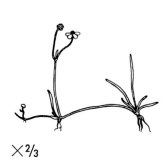

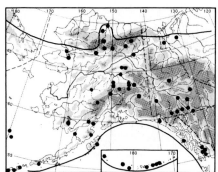

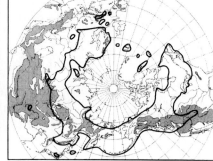

12. Ranúnculus réptans L.   Creeping Spearwort

Ranunculus filiformis Michx.; *R. flammula* L. var. *filiformis* (Michx.) Hook.; *Flammula unalaschcensis* Piper; *R. flammula* var. *reptans* (L.) Rchb.

Glabrous or nearly so; stem filiform, arched, creeping in mud, rooting at nodes; basal leaves filiform, linear or linear-lanceolate; flowers solitary, about 5 mm in diameter; sepals and petals 5; achenes plump, with short beak.

Wet or muddy shores and riverbanks. Described from Sweden and Russia.

Most of the specimens belong to var. **intermèdius** (Hook.) Torr. & Gray (*R. flammulus* var. *intermedius* Hook.), which has more or less lanceolate leaves.

46. Ranunculaceae (Crowfoot Family)

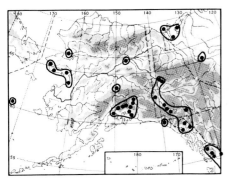 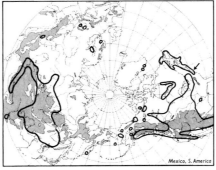 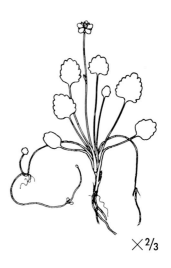

13. Ranúnculus cymbalària Pursh
Halerpestes cymbalaria (Pursh) Greene.

Tufted, with filiform stolons, rooting at nodes; leaves rounded, reniform, or cordate, crenate; in var. **alpìna** Hook., small, rectangular, 3-toothed in apex; flowers small; sepals and petals usually 5, 3–5 mm long; sepals glabrous; petals yellow, about as long as sepals; fruiting head ovate to cylindrical; fruits with short beak; receptacle hairy.

Moist places, brackish water; sometimes apparently spread by human activity.

Circumpolar map indicates range of the entire species complex; part of this range represents introduced populations, as for example, in Europe.

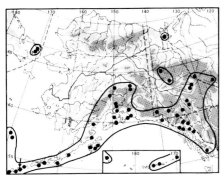 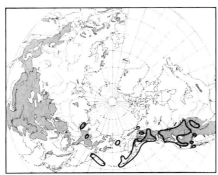 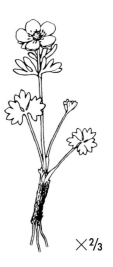

14. Ranúnculus Eschschòltzii Schlecht.
Ranunculus nivalis var. *Eschscholtzii* (Schlecht.) S. Wats.

Stems glabrous, from short caudex; basal leaves glabrous, highly variable, circular or semicircular in outline, often cordate at base, 3-cleft to divided, the middle lobe again 3-lobed, ultimate segments usually blunt; sepals 5, yellow, glabrous or with yellowish hairs; petals of highly variable size, yellow; receptacle cylindric, glabrous.

Meadows, along creeks.

Hybrids with *R. nivalis* probably occur. Specimens approaching var. **Suksdórffii** (Gray) Benson (*R. Suksdorffii* Gray), with acute end lobes on the basal leaves, occur rarely.

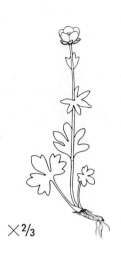
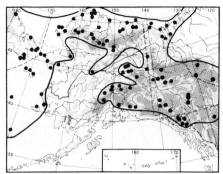
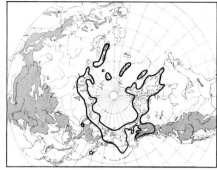

15. Ranúnculus nivàlis L. — Snow Buttercup

Stems single from short caudex, glabrous below, more or less brown-hispid above; basal leaves several, glabrous, with reniform outline, cleft into 3–5 lobes, which are usually again lobed or toothed; stem leaves 1–3, sessile; flower solitary, highly variable in size; sepals 5, pubescent, from brown, stiff hairs; petals longer than sepals, yellow, glabrous; fruiting head elliptic-oblong; fruit with long, somewhat curved beak; receptacle glabrous.

Wet meadows, along brooks in the mountains and on tundra, snow beds. Described from Lapland and Switzerland.

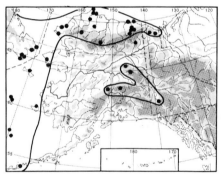

16. Ranúnculus sulphùreus Soland.
var. **sulphùreus**

Similar to *R. nivalis,* but leaves rounded in outline, shallowly cleft or toothed; receptacle with stiff, brown hairs; flowers large.

Stony places, solifluction soil, below snowbeds, to at least 1,000 meters in the Brooks Range.

46. Ranunculaceae (Crowfoot Family)

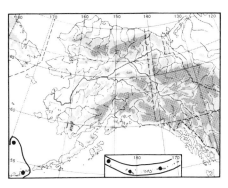

17. Ranúnculus sulphùreus Soland.
var. **intercèdens** Hult.

   Similar to var. *sulphureus,* but leaves deeply cleft into 3 distinct lobes. Described from Kamchatka.

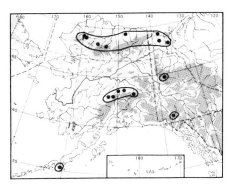
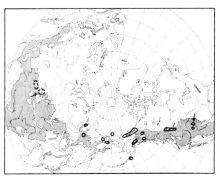
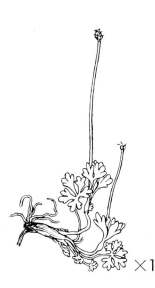

18. Ranúnculus gelìdus Karel. & Kiril.
subsp. **Gràyi** (Britt.) Hult.
   Ranunculus Grayi Britt.

   Flowering stem from short caudex, 1–3-flowered, often bent at right angle just above surface of soil, pubescent above; basal leaves 3-parted and again cleft and lobed, ultimate segments blunt; sepals grayish-pubescent on back; petals yellow, longer than sepals; fruit with stout beak.
   Wet places, scree slopes.
   The typical plant has a glabrous receptacle; plants with pubescent receptacle occur at Shumagin Islands (var. **shumaginénsis** Hult.).
   Broken line on circumpolar map indicates range of subsp. **gelìdus.**

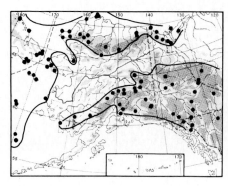

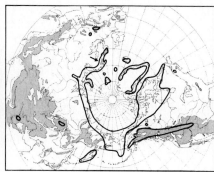

19. Ranúnculus pygmaèus Wahlenb.   Dwarf Buttercup
subsp. **pygmaèus**

Small plant with vertical caudex; stems ascending, 1-flowered, pubescent with white hairs; radical leaves glabrous, somewhat ciliated in margin, flabelliform to reniform, deeply cleft into 3–5 lobes; upper leaves reduced; pedicels pubescent; sepals and petals 5; sepals yellowish, short-haired; petals 2–4 mm long; achenes about 1 mm long, glabrous, with beak hooked in apex; receptacle cylindric, glabrous.

Moist places, snow beds in the mountains, to at least 2,000 meters.

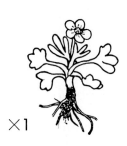

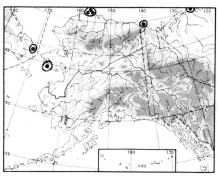

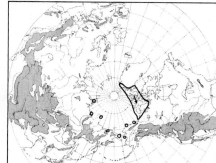

20. Ranúnculus pygmaèus Wahlenb.
subsp. **Sabìnei** (R. Br.) Hult.
Ranunculus Sabinei R. Br.

Similar to subsp. *pygmaeus*, but larger and stouter, with more or less erect, stout, stiff stem; petals 5–8 mm long.

Moist places on tundra.

46. Ranunculaceae (Crowfoot Family)

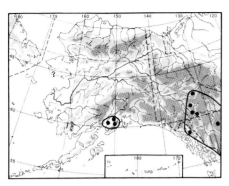
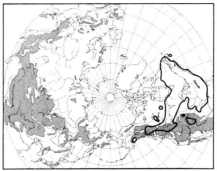
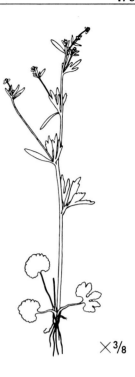

21. Ranúnculus abortìvus L.

Stem glabrous or minutely pilose, branching; radical leaves reniform to round-ovate or orbicular, or lobed and toothed; lower cauline leaf petioled; flowers small; sepals reflexed; petals narrow, about as long as sepals, yellow; achenes suborbicular, lustrous, with minute style; receptacle pubescent.

Woods, thickets; sometimes adventive. Described from Virginia, Canada.

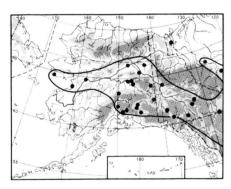
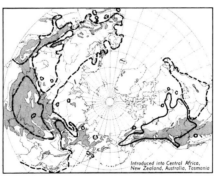
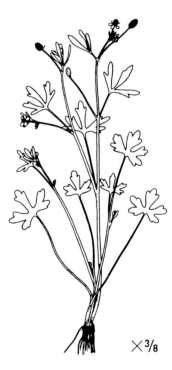

22. Ranúnculus sceleràtus L. Cursed Crowfoot
subsp. **multífidus** (Nutt.) Hult.

Ranunculus sceleratus var. *multifidus* Nutt.

Yellowish-green, annual; stem branched, pubescent above; basal leaves deeply 3-parted and again deeply parted or divided; sepals yellowish-green; petals pale yellow; fruiting head ovoid; fruit with minute depressions on sides.

Wet places. *R. sceleratus* described from Europe, subsp. *multifidus* from the Platte River.

Broken line on circumpolar map indicates range of other subspecies.

Contains much protoanemonin; if the leaves are rubbed against the skin, wounds may occur.

 ×3/8

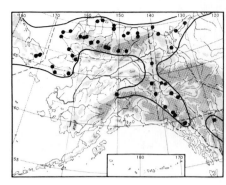 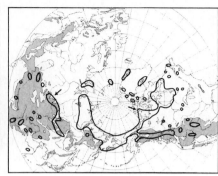

23. Ranúnculus pedatífidus Sm.
subsp. **affinis** (R. Br.) Hult.

Ranunculus affinis R. Br.; *R. affinis* var. *leiocarpa* Trautv.; *R. pedatifidus* var. *leiocarpa* (Trautv.) Fern.; *R. pedatifidus* var. *affinis* (R. Br.) Benson; *R. verticillatus* Eastw.; *R. Eastwoodianus* Benson.

Stem tall, slender, glabrous or somewhat pubescent; basal leaves glabrous or somewhat ciliate, deeply cleft; lateral leaf lobes again cleft into 3 to several lobes or teeth; bracts sessile, divided into linear lobes, forming 2 or 3 verticillate whorls; sepals pubescent, rarely glabrous; flowers highly variable in size, 10–25 cm in diameter; petals longer than sepals; fruit glabrous or (rarely) pubescent.

Dry, grassy slopes. The range in central Asia belongs to subsp. **pedatífidus.**

 ×1/3

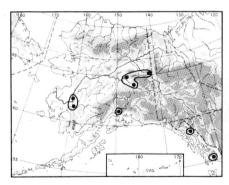 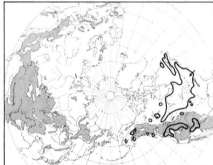

24. Ranúnculus pennsylvànicus L. f.

Stem erect, tall, hispid-pubescent, branching; leaves hirsute, ternate, the leaflets petiolulated, deeply cleft and acutely toothed; sepals bristly, reflexed; petals pale yellow, as long as, or shorter than, sepals; achenes 2–2.7 mm long, with deltoid beak; receptacle hirsute.

Wet meadows.

46. Ranunculaceae (Crowfoot Family)

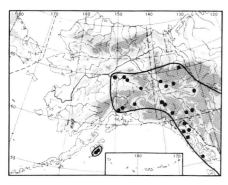
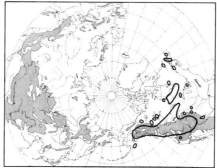

25. Ranúnculus Macoùnii Britt.

Stem hispid, ascending or trailing, often with stolons, rooting at nodes; radical leaves ternate, the leaflets petiolulated, cleft and acutely toothed; sepals hispid, reflexed; petals deep yellow, as long as sepals; achenes larger than in *R. pennsylvanicus,* with long, subulate beak.

Woods, thickets. Described "from Canada to Mackenzie River, lat. 67°; shores of Hudson Bay to the Pacific."

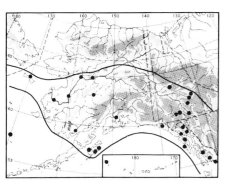
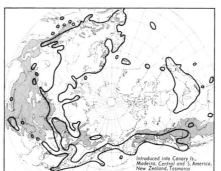
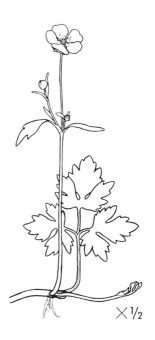

26. Ranúnculus rèpens L. Creeping Buttercup

Dark green; stem glabrous or pubescent, trailing, with creeping, elongate branches, rooting at nodes; leaves ternate; leaflets rhombic, petiolulated, deeply cleft or lobed; sepals hirsute, spreading; petals bright yellow, large, up to 17 mm long, much longer than sepals; achenes glabrous, with short, deltoid-subulate beak; receptacle villous.

An introduced weed. Described from Europe.

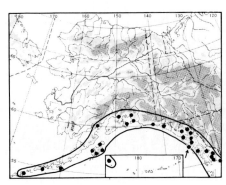
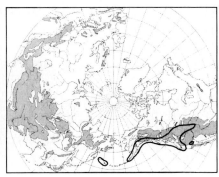

27. Ranúnculus Bongárdi Greene

Ranunculus uncinatus var. *parviflorus* (Torr.) Benson; *R. occidentalis* var. *parviflorus* Torr.

Stout; stem branched, hispid, from often reddish-brown hairs; basal leaves 3-parted, the parts lobed and acutely toothed, more or less appressed-hispidulous; petioles hispid; sepals 5, pubescent, reflexed, soon falling off; petals 5, yellow, longer than sepals; achenes glabrous or hispid, with long, slender, hooked beak; receptacle subglobose, glabrous.

Moist places.

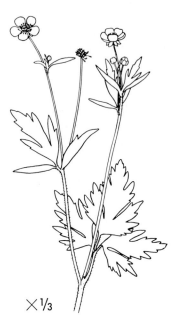
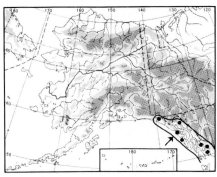
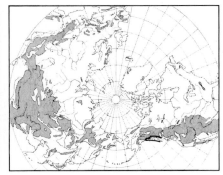

28. Ranúnculus pacíficus (Hult.) Benson

Ranunculus septentrionalis subsp. *pacificus* Hult.

Stem suberect, hirsute; basal leaves appressed, hirsute beneath, with 3 stipitate leaflets, again 3-parted and laciniately toothed; sepals 5, purplish-tinged, reflexed, sparsely appressed-hispidulous; petals 5, longer than sepals, yellow, dorsally tinged with purple; achenes in ovoid-globose head, smooth, glabrous, with stout, subulate beak about 1.5 mm long, hooked at tip; receptacle hispidulous.

Moist places.

46. Ranunculaceae (Crowfoot Family)

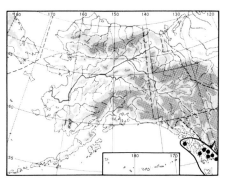
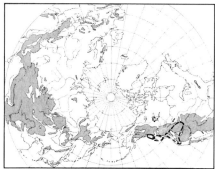

29. Ranúnculus orthorhýnchus Hook.
subsp. **alaschénsis** (Benson) Hult.
Ranunculus orthorhynchus var. *alaschensis* Benson.

Tall, glabrous or hirsute, suberect; basal leaves with 3–7 alternate or opposite leaflets, again lobed, ultimate lobes acute or obtuse, appressed-pubescent to glabrous; petioles sparsely appressed-hispidulous; sepals 5, reflexed, glabrous or pilose; petals longer than sepals, yellow; achenes glabrous, with straight, slender beak about 3 mm long; receptacle cylindroid, hispidulous.

Meadows, forests. *R. orthorhynchus* described from northwest North America. Broken line on circumpolar map indicates range of subsp. **orthorhýnchus.**

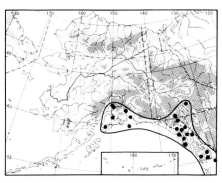
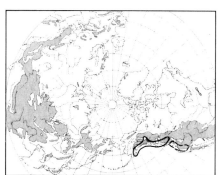

30. Ranúnculus occidentàlis Nutt.
subsp. **occidentàlis**
var. **brevistỳlis** Greene

Glabrous or soft-pubescent; stem erect; basal leaves deeply 3-parted, the parts again 3–4-lobed or toothed; pedicels appressed-pubescent; sepals pubescent, reflexed; petals yellow, longer than sepals; achenes glabrous or hispid, with falcate beak; receptacle globose, glabrous.

Moist places. Described from Sitka and Oregon; var. *brevistylis* from Yes Bay, Alaska.

Broken line on circumpolar map indicates range of subspecies other than those occurring in Alaska.

46. RANUNCULACEAE (Crowfoot Family)

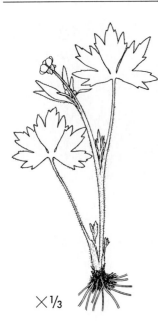

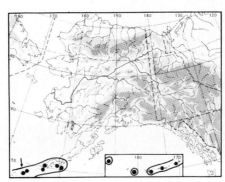 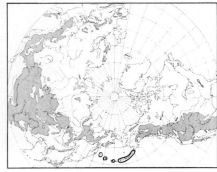

31. Ranúnculus occidentàlis Nutt.
subsp. **Nelsòni** (DC.) Hult.

Ranunculus recurvatus var. *Nelsoni* DC.; *R. Nelsoni* (DC.) Gray; *R. occidentalis* var. *Nelsoni* (DC.) Benson, in part.

Similar to var. *brevistylis,* but much coarser and taller; stem and petioles densely pubescent from brownish, patent hairs.

Moist places.

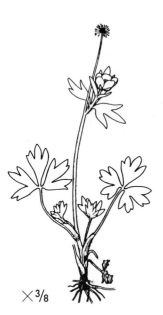

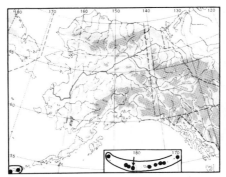 

32. Ranúnculus occidentàlis Nutt.
subsp. **insulàris** Hult.

Very similar to subsp. *Nelsoni,* but less robust; leaves with appressed, gray, silky pubescence; beak of achenes shorter and broader.

Moist places.

46. Ranunculaceae (Crowfoot Family)

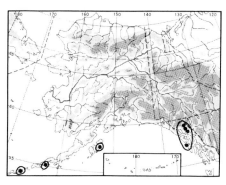
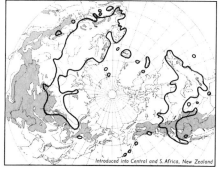
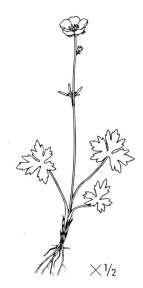

33. Ranúnculus ácris L. Common Buttercup

Hirsute, with spreading hairs, rhizome erect; stem erect, branching above; basal leaves pentagonal in outline, pedately 5–7-parted, the divisions cleft or parted into lanceolate or linear segments; sepals villous; flowers 1.5–2.5 cm in diameter; petals yellow, shiny, much longer than sepals; achenes glabrous, with short, straight beak, broader at base; receptacle glabrous.

Meadows. Not native. Described from Europe.

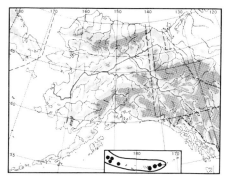
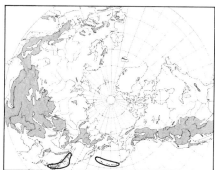

34. Ranúnculus grándis Honda
var. **austrokurilénsis** (Tatew.) Hara

Ranunculus acris var. *austrokurilensis* Tatew.; *Ranunculus transochotensis* Hara; *R. acris* var. *frigidus* with respect to Aleutian plant.

Tall, sparsely hirsute; similar to *R. acris*, but with subterranean rhizomes; leaves 3-parted, the parts divided and lobed, the ultimate lobes acute; pedicels appressed-pubescent; flowers large; sepals pubescent; petals yellow, twice as long as sepals; achenes glabrous, with very short, falcate beak.

Meadows. *R. grandis* described from Rikushu Province, Honshu, var. *austrokurilensis* from the southern Kuriles.

46. RANUNCULACEAE (Crowfoot Family)

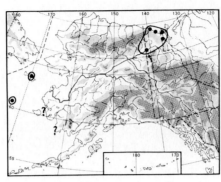
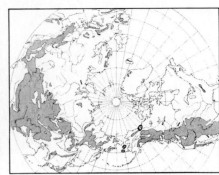

35. Ranúnculus Túrneri Greene

Ranunculus occidentalis var. *Turneri* (Greene) Benson; *A. acris* var. *frigidus* with respect to arctic Mackenzie plant.

Sparsely hirsute; basal leaves appressed-pubescent, 3-parted and again cleft or lobed; flowers large; sepals pubescent; petals 10–15 mm long; achenes glabrous, with falcate beak, 1–2 mm long; receptacle glabrous.

Meadows.

Closely related to **R. boreàlis** Trautv.

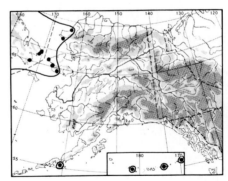
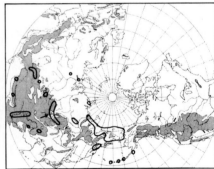

Oxygraphis Bunge

1. Oxygráphis glaciàlis (Fisch.) Bunge

Ficaria glacialis Fisch.; *Caltha glacialis* (Fisch.) Spreng.; *Ranunculus kamtschaticus* DC.

Small, glabrous plant, lacking stem; pedicels and leaves from thick, short, vertical caudex; basal leaves thick, orbicular-elliptic to oblong, crenate at apex or subentire; sepals 5, thick, greenish to purplish-black, persistent in fruit; petals oblanceolate, yellow; fruiting head hemispherical; achenes with short, somewhat hooked beak.

Polygon tundra in the mountains; stony slopes.

46. Ranunculaceae (Crowfoot Family)

Thalictrum L. (Meadow Rue)

Plant low-growing, scapose or with single stem leaf; inflorescence a simple raceme, rarely branched; filaments not dilated .... 1. *T. alpinum*
Plant tall; stem branched, with several stem leaves; inflorescence paniculate:
    Carpels markedly oblique, long-stipitated, with long style; flowers perfect; filaments dilated above, broader than anthers .... 2. *T. sparsiflorum*
    Carpels not particularly oblique, sessile, or short-stipitated:
        Carpels small, in Alaska specimens about 1.5 mm long, sessile, with style less than 1 mm long; flowers perfect; filaments slightly dilated ....
        .... 3. *T. minus* subsp. *kemense*
        Carpels larger, short-stipitated, with long style; flowers unisexual; filaments not dilated .... 4. *T. occidentale*

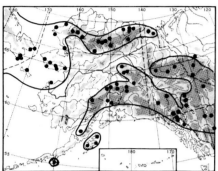
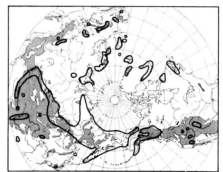

×1

1. Thalíctrum alpìnum L.

Glabrous; rhizome slender; stem capillary, leafless or with single leaf; basal leaves dark green, glossy, bluish-green below, 2–4 times ternate, with flabellate leaflets; inflorescence without leafy bracts, often nodding; flowers reddish-brown; filaments violet; anthers brown.

Alpine meadows, stony slopes in the mountains, to at least 2,000 meters. Described from "in alpibus Lapponiae, Arvoniae."

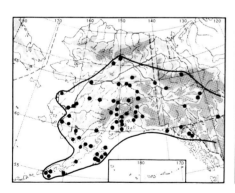
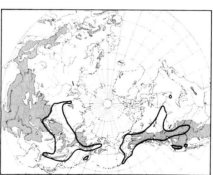

2. Thalíctrum sparsiflòrum Turcz.

Thalictrum sparsiflorum var. *Richardsonii* (Gray) Boiv.; *T. Richardsonii* Gray.

Glabrous; leaves 2–3 times ternate, the leaflets with rounded or cordate base; sepals 4, pinkish-white; filaments dilated above; carpels stipitate, strongly compressed, half-moon-shaped, with long, thin, straight style.

Meadows, woods. Described from Dahuria.

×½

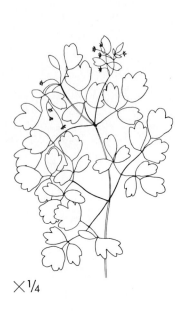
×¼

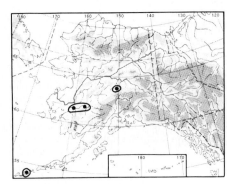
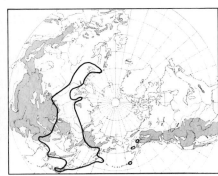

3. Thalíctrum mìnus L.
subsp. **keménse** (E. Fries) Hult.

Thalictrum kemense E. Fries; *T. Hultenii* Boiv.

Glabrous, up to 80 cm tall; leaves thick, petiolated, 2–3 times ternate, the leaflets rounded-obovate, coarsely 3-toothed; sepals reddish-green; filaments slightly dilated; carpels in hemispherical heads, small, sessile, elliptic-ovate to ovate, strongly nerved, with straight style.

Meadows. *T. minus* described from Europe, subsp. *kemense* from "Lapponia Kemensi."

×⅓

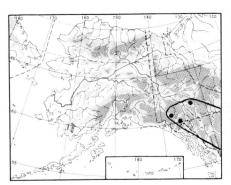
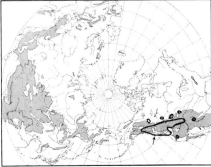

4. Thalíctrum occidentàle Gray
Thalictrum Breitungii Boiv.

Glabrous, up to 70 cm tall; leaves 3–4 times ternate; leaflets round to obovate-cuneate, 3-lobed, the lobes coarsely dentate; inflorescence branched; peduncles more or less divaricate; filaments not dilated; carpels short-stipitated, ovate, with slender style about 2 mm long.

The Alaska-Yukon plant, which has shorter filaments and stigma than the southern plant, has been described as *T. Breitungii* Boiv.

Meadows, thickets.

47. Papaveraceae (Poppy Family)

Papaver L. (Poppy)

Stem with leaves; flowers scarlet .... 1. *P. rhoeas*
Stem scapose:
  Plant low-growing; scapes up to 15 cm tall:
    Leaves bipinnate; capsule with marked central projection ... 3. *P. McConnellii*
    Leaves less dissected; capsule flat, vaulted, or acute:
      Caudex thick, long, densely covered with light-brown bases of petioles; flowers large, yellow; capsule short and broad .... 6. *P. alaskanum*
      Caudex less prominent, with darker remnants of petioles:
        Leaves glabrous, with revolute margin, small; flowers small .... 2. *P. Walpolei*
        Leaves pilose or setose, with flat margin; flowers larger:
          Flowers white or rose, with yellow spot at base; capsule broad, ovate to globose .... 4. *P. alboroseum*
          Flowers yellow; capsule broadest at top or middle:
            Flowers large; capsule longer than broad, broadest at middle .... 5. *P. Macounii*
            Flowers smaller; capsule broadest at top .... 8. *P. lapponicum* subsp. *occidentale*
  Plant taller:
    Plants mostly more than 25 cm tall; petioles nearly glabrous or with long, light-colored, patent hairs; capsule club-shaped, with many rays; flowers large .... 10. *P. nudicaule*
    Plants less than 25 cm tall; flowers large or smaller:
      Capsule pear-shaped to elliptical, or nearly spherical:
        Leaves with broad, ovate lobes, decurrent on petioles .... 9. *P. lapponicum* subsp. *Porsildii*
        Leaves with ovate-lanceolate lobes .. 8. *P. lapponicum* subsp. *occidentale*
      Capsule much longer than broad:
        Capsule broader in middle .... 5. *P. Macounii*
        Capsule cylindric, tapering at base .... 7. *P. Hulténii*

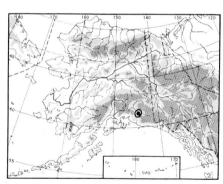
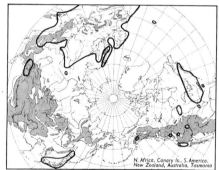

1. Papàver rhoèas L. — Corn Poppy

Stem erect or ascending, usually branched, from slender taproot, with stiff, spreading hairs; basal leaves petiolated, once to twice pinnately cut or divided into more or less toothed segments; upper leaves sessile, 3-lobed, central lobe elongate, lanceolate; flowers large, petals scarlet; capsule glabrous, subglobose to obovate; stigma mostly with 10 lobes.

An introduced weed. Described from Europe.

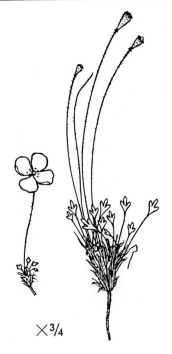
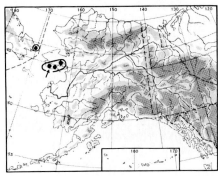
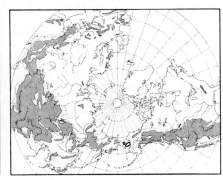

2. Papàver Walpòlei Pors.

Densely caespitose; leaves with revolute margins, glabrous, short-petiolated, up to 4 cm long, usually smaller, entire or mostly 3-lobed, the lobes obtuse, the median sometimes shallowly 2–3-lobed; persistent petioles dark brown; scape sparsely pubescent above; flowers 1.5–4 cm in diameter, yellow or white with yellow basal spot (var. **sulphùreo-maculàta** Hult.); capsule conical when young, in age obovate-conical, sparsely yellowish-hirsute; stigma pyramidal, mostly with 5 rays.

Solifluction soil.

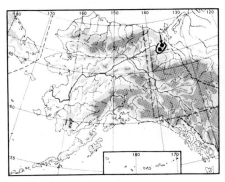
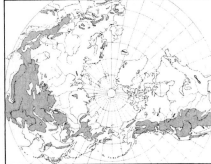

3. Papàver McConnéllii Hult.

Caespitose; leaves bipinnate, with 2–3 pairs of pinnae, the segments obtuse, narrow, glaucous, sparsely pilose on both sides; scape pilose; small, yellow flowers about 2 cm in diameter; stamens shorter than capsule; capsule obovate with pale hairs; stigmatic disk convex, with central projection about 1 mm long; long subclavate rays and prominent membrane between bases of rays.

Sandy, gravelly soil.

47. Papaveraceae (Poppy Family)

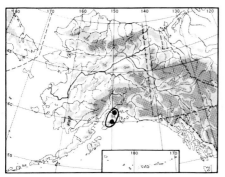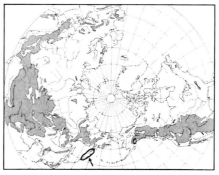

4. Papàver alboròseum Hult.

Caespitose, with short caudex; leaves white-setose on both sides, pinnate, the segments 2–5-lobed, mucronate; scapes bowlike, ascending, up to 15 cm tall, white- to brown-setose; flower buds with light-colored hairs; petals white or rose, with yellow basal spot; capsule ovate to globose, with white to brown, stiff setae, tuberculate at base; stigma with 5–6 rays.

Sandy, gravelly soil.

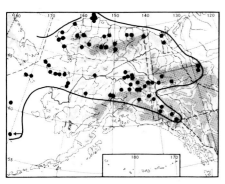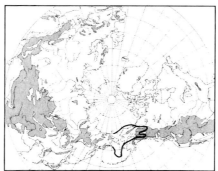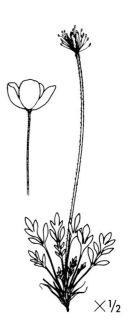

5. Papàver Macoùnii Greene

Papaver alaskanum var. *macranthum* Hult.; *P. Keelei* Pors.; *P. Macounii* var. *discolor* Hult.; *P. microcarpum* with respect to Alaskan plant, at least in part; *P. Scammanianum* D. Löve.

Densely caespitose; leaves long-petiolated, ovate in outline, pinnate secondary lobes rare, pubescent to nearly glabrous; scapes up to 30 cm tall; flower buds short, ovate to almost globose; flowers large, yellow; capsule broadest at middle; stigma rays few, 4–5, forming an acute top.

Sandy, gravelly soil; in the mountains to over 2,100 meters. Total range unknown.

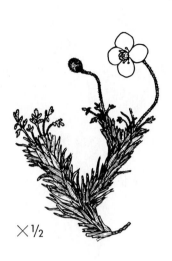
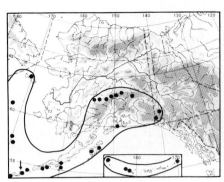
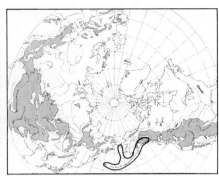

6. Papàver alaskànum Hult.

Papaver nigroflavum D. Löve; *P. denalii* Gjaerevoll.

Similar to *P. Macounii*, but low-growing, densely caespitose; caudex densely covered with the light-brown, dilated bases of previous year's petioles; leaf blade short, with blunt to somewhat acute lobes; flower buds short, ovate or almost globose; capsule shorter and broader, often as broad as long.

Sandy, gravelly soil.

(See color section.)

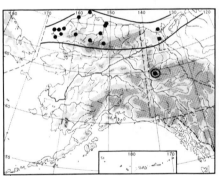
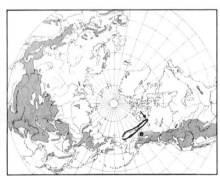

7. Papàver Hulténii Knaben

Leaves long-petiolated, with long, widely separated, linear-lanceolate, often bipinnate, acute lobes; scapes up to 40 cm long; flowers yellow or salmon pink (var. **salmonìcolor** Hult.); buds long, oval, drooping; capsules cylindric, tapering at base, up to 3 times as long as broad; stigmatic disk as broad as capsule; stigma with 6–7 rays.

Sandy, gravelly soil.

47. Papaveraceae (Poppy Family)

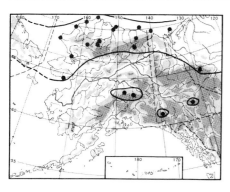
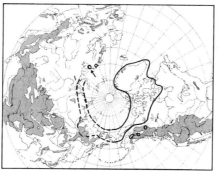

8. Papàver lappónicum (Tolm.) Nordh.
subsp. **occidentale** (Lundstr.) Knaben

Papaver radicatum subsp. *lapponicum* Tolm.; *P. radicatum* subsp. *occidentale* Lundstr.; *P. kluanensis* D. Löve; *P. Freedmanianum* D. Löve.

Leaves ovate or ovate-oblong in outline, pinnate, the lowest pair of pinnae somewhat widely separated; pinnae ovate-lanceolate; flowers cup-shaped or campanulate; petals deciduous, mostly yellow, rarely white; capsule pear-shaped to elliptical, seldom shorter or nearly spherical; stigmatic disk vaulted or flat, the rays sometimes decurrent.

Sandy, gravelly soil; to at least 2,600 meters in the St. Elias Range.
Circumpolar map gives tentative range of entire species complex.

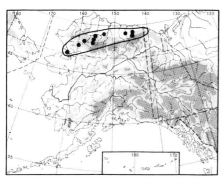
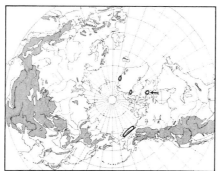
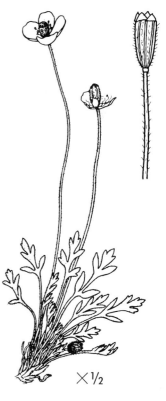

9. Papàver lappónicum (Tolm.) Nordh.
subsp. **Porsíldii** Knaben

Similar to subsp. *occidentale*, but with taller scapes; leaves ovate in outline with few broad, ovate lobes; margins of lower lobes decurrent on petioles.

Sandy, gravelly soil. Total range imperfectly known.

47. Papaveraceae (Poppy Family) / 48. Fumariaceae (Earth Smoke Family)

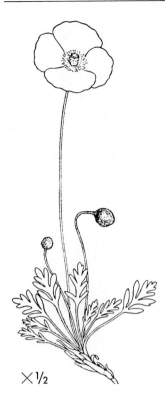

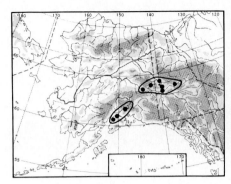

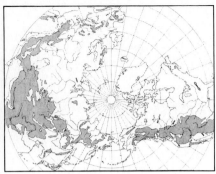

10. Papàver nudicaùle L.

Loosely caespitose; leaves setose or nearly glabrous, with 2–3 pairs of broadly rounded pinnae, obtuse or acute, entire or toothed; scapes up to 50 cm tall, coarse; flowers yellow, white, or red; stamens numerous, long; capsule large, club-shaped, setose, with flat disk, almost lacking membrane between rays.

Roadsides, waste places; escaped from cultivation in several localities. Described from Siberia.

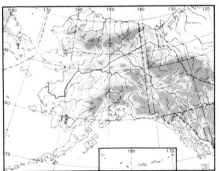

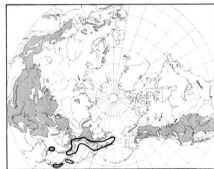

Dicentra Bernh.

1. Dicéntra peregrína (Rudolph) Makino
Fumaria peregrina Rudolph.

Stem from short rootstock; leaves all basal, bluish-green, long-petiolated, ternately dissected into linear lobes; inflorescence 2–5-flowered; flowers about 20 mm long; sepals broadly ovate, small; petals purplish to rose, the outer pair strongly saccate at base, with reflexed tips; capsule oblong.

Stony places. Described from "Selenga" (where it probably does not occur).

48. Fumariaceae (Earth Smoke Family)

Corydalis Vent.

Plant perennial with tuberous root; leaves ternately divided, divisions again lobed;
    flowers bluish-violet (rarely white) .... 1. *C. pauciflora*
Plant annual or biennial, branched, with fibrous root; leaves pinnately dissected:
    Corolla yellow; plant low-growing; stem ascending, diffusely branching ......
    .... 2. *C. aurea*
    Corolla pink with yellow tip; plant erect, taller .... 3. *C. sempervirens*

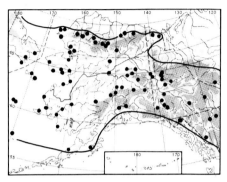
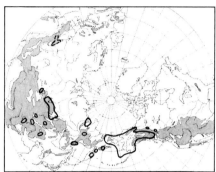
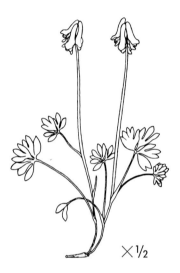

1. Corýdalis pauciflòra (Steph.) Pers.

Fumaria pauciflora Steph.; *Capnoides pauciflorum* (Steph.) Ktze.; *Corydalis arctica* Popov.

Stem very thin and brittle at base, from dark-brown tuber buried deep in soil, with 1–3 scales below; leaves ternate, the segments again 3–4-cleft, ultimate segments blunt; inflorescence dense, few-flowered; flowers horizontal or sometimes vertical, bluish-violet; spur slightly bent.

Heaths, meadows, snow beds, moist spruce forests, in the mountains to at least 1,400 meters. Described from the Altai Mountains.

White-flowered specimens (var. **albiflòra** Pors.) occur, as well as specimens with bluish-white flowers (in the Ogilvie Mountains).

Broken line on circumpolar map indicates range of the closely related *C. Emanuéli* C.A.M.

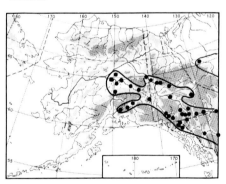
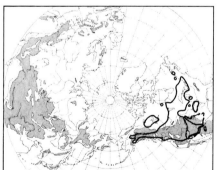

2. Corýdalis aùrea Willd.   Golden Corydalis

Glaucous, annual or biennial; stem diffusely branching; leaves bipinnate, the leaflets pinnatifid; flowers golden yellow, recurving or ascending; capsules linear or lanceolate-linear, with long style; seeds smooth.

Roadsides; sandy places. Described from Canada.

Broken line on circumpolar map indicates range of subsp. **occidentàlis** (Engelm.) Owneby (*C. aurea* var. *occidentalis* Engelm.). Several closely related species occur in China.

48. Fumariaceae (Earth Smoke Family) / 49. Cruciferae (Mustard Family)

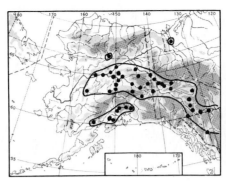
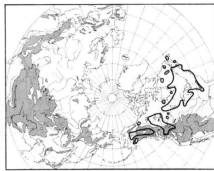

3. Corýdalis sempervìrens (L.) Pers.   Pale Corydalis, Rock Harlequin
Fumaria sempervirens L.

Glaucous, annual or biennial; stem erect, branched; leaves petiolate, 1–3-pinnate, ultimate segments blunt; flowers pink with yellow tips; spur saccate; capsules erect, or ascending, linear-cylindric, with long style.

Rocky places, roadsides; occurs sometimes as a weed, as in Inuvik in the Mackenzie Delta. Described from Canada, Virginia.

(See color section.)

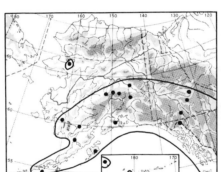
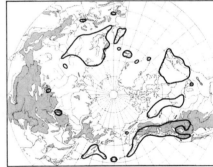

Subularia L.

1. Subulària aquática L.   Awlwort

Small aquatic or littoral plant lacking stem; leaves subulate; scape bractless; flowers very small, petals white; silicles globular to ovate or obovate; flowers in submerged plants cleistogamous.

Along shores and streams, often submerged. Described from Europe.

The American plant, which has been distinguished as subsp. **americàna** Mulligan & Calder, has persistent sepals and oblong silicles; the distinction is obscure within the area of interest, and some specimens from western Alaska are identical to the Eurasiatic plant.

49. CRUCIFERAE (Mustard Family)

Lepidium L.

Silicles broadly wing-margined .... 1. *L. sativum*
Silicles not or narrowly wing-margined:
    Petals lacking or rudimentary; seeds not wing-margined ...... 2. *L. densiflorum*
    Petals longer than sepals; seeds narrowly wing-margined ...... 3. *L. virginicum*

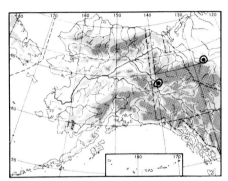
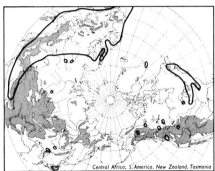

1. Lepídium satìvum L.

Glabrous; stem branched; leaves light green, the lower bipinnatifid, the upper with few divisions; petals white or pale reddish; silicles oblong-ovate, broadly winged.

Waste places; escaped from cultivation. Described from Europe. Circumpolar map tentative.

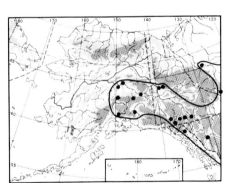
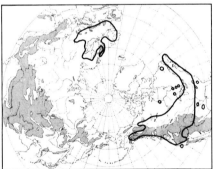

2. Lepídium densiflòrum Schrad.

Lepidium apetalum and *L. ruderale* of authors.

Annual; stem puberulent; basal leaves oblanceolate, pinnatifid, the divisions toothed; cauline leaves entire or slightly toothed; racemes numerous, ascending; pedicels slightly flattened; petals lacking or shorter than calyx; silicles about 2–5 mm long, round-obovate to elliptic-obovate, glabrous; seeds lacking wing margin.

An introduced weed. Described from North America (?).

Var. **Bourgeauànum** (Thell.) C. L. Hitchc. (*L. Bourgeauanum* Thell.) differs in having rather flat pedicels and silicles averaging 3–3.5 mm long. Both types occur within area of interest.

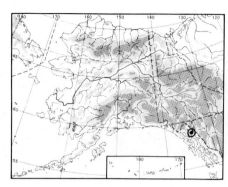 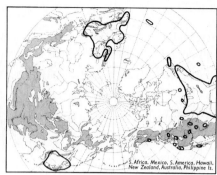

3. Lepídium virgínicum L.

Similar to *L. densiflorum*, but pedicels terete; petals longer than sepals; silicles orbicular; seeds narrowly wing-margined.

Waste places. Described from Virginia, Jamaica. Introduced in Europe.

Thlaspi L.

Stem leaves not sagittate; silicles clavate, small .... 1. *T. arcticum*
Stem leaves sagittate; silicles suborbicular or oblong, large .... 2. *T. arvense*

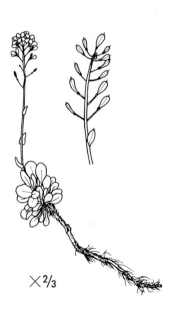

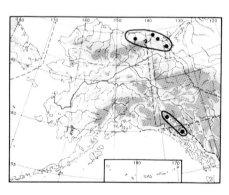 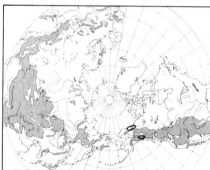

1. Thláspi árcticum Pors.

Glabrous, with short, many-headed caudex; basal leaves spatulate, with prominent midvein, reddish below; stems short in flowering state, elongated in fruit; stem leaves 3–5, linear, sessile; pedicels divaricate, about as long as silicles; petals twice as long as sepals, white or faintly lilac; silicles clavate, with filiform style about 1 mm long.

Very rare; only single specimens collected. Described from Arctic coast of Yukon Territory.

49. CRUCIFERAE (Mustard Family)

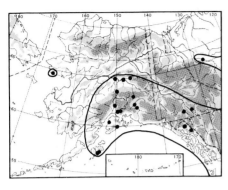 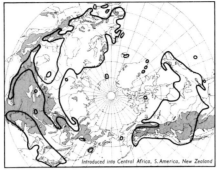 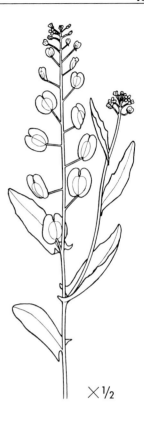

2. Thláspi arvénse L.   Pennycress

Glabrous, yellowish-green; stem mostly simple; lowest leaves petiolated, soon withering; stem leaves with sagittate base; petals white; silicles suborbicular or oblong.

An introduced weed. Described from Europe.

Cochlearia L. (Scurvy Grass)

Silicles broadly elliptic . . . . 1. *C. officinalis* subsp. *arctica*
Silicles globose to ovate-globose . . . . 2. *C. officinalis* subsp. *oblongifolia*

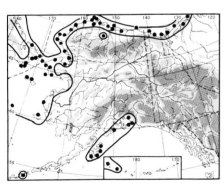 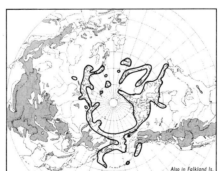 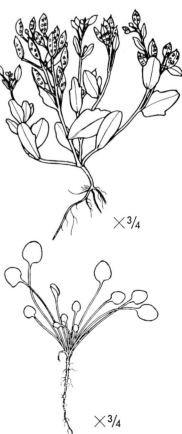

1. Cochleària officinàlis L.

subsp. **árctica** (Schlecht.) Hult.

Cochlearia arctica Schlecht.; *C. fenestrata* R. Br. Including *C. groenlandica* L.

Somewhat fleshy; basal leaves soon withering, long-petiolated, ovate, with cuneate, rounded or reniform base; stem leaves short-petiolated or sessile, entire or toothed; petals white; silicles broadly elliptic, with short style.

Along the coast, also rarely on tundra inland. *C. officinalis* described from northern Europe, subsp. *arctica* from northern Siberia.

A plant from Kodiak Island with sessile leaves has been described as **C. sessilifòlia** Rollins [var. *sessilifolia* (Rollins) Hult.].

The leaves are eaten raw or boiled and are a valuable antiscorbutic.

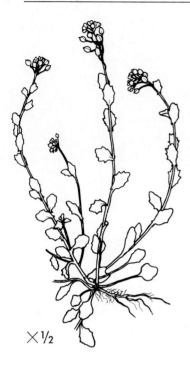

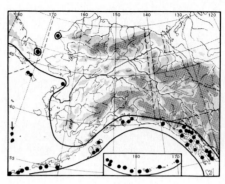
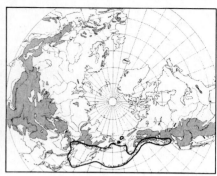

2. Cochleària officinàlis L.
subsp. **oblongifòlia** (DC.) Hult.
Cochlearia oblongifolia DC.; *C. arctica* var. *oblongifolia* (DC.) N. Busch.

Similar to subsp. *arctica*, but coarser; silicles globose or ovate-globose.

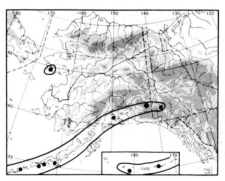
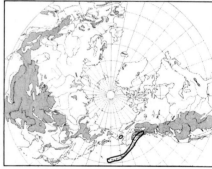

Aphragmus Andrz.

1. Aphrágmus Eschscholtziànus Andrz.
Oreas involucrata Cham. & Schlecht.; *Braya Eschscholtziana* (Andrz.) Benth. & Hook.; *Eutrema Eschscholtzianum* (Andrz.) Rollins.

Plant with long root and subterranean stolons, with persistent, dilated petioles at base; stems short, underground, glabrous or very nearly so; basal leaves spatulate-ovate, with long, underground petioles strongly dilated at base; stem leaves crowded at top of stem; inflorescence subumbellate; flowers short-pedicellated; petals longer than sepals, obovate, with long claw, white or purplish with purple veins; silicles oblong-ellipsoid with short style, septum lacking.

Very rare, in solifluction soil. Described from the Aleutian Islands.

49. CRUCIFERAE (Mustard Family) 501

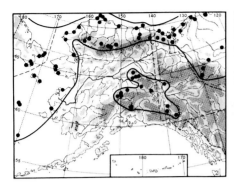
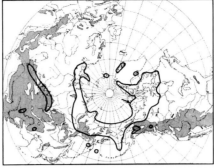

Eutrema R. Br.

1. Eutrèma Edwárdsii R. Br.

Root thick, the high-arctic plant short, the southern plant up to 35 cm tall; stems 1 to several; basal leaves long-petiolated, ovate; stem leaves lanceolate, the upper sessile; petals white, obovate, longer than sepals; siliques elliptic, with more or less complete septum.

Solifluction areas; moist places in the mountains and on tundra.

Sisymbrium L.

Leaves pinnate, with large apical lobe; siliques short, markedly acute, appressed to stem .... 1. *S. officinale*
Leaves pinnatifid, with long, linear segments; siliques long, slender .... 2. *S. altissimum*

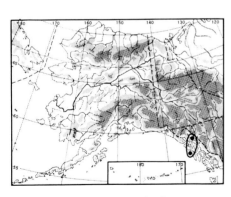
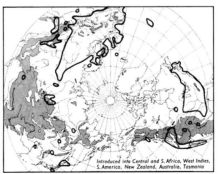

1. Sisýmbrium officinàle (L.) Scop.
Erysimum officinale L.

Stiffly hirsute; lower leaves pinnate, with large, apical lobes; branches spreading; flowers pale yellow; siliques appressed to branches, tapering from base to tip, mostly pubescent.

Roadsides, waste places; an introduced weed. Described from Europe.

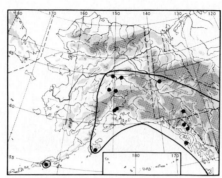 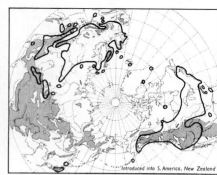

2. Sisýmbrium altíssimum L.   Tumble Mustard

Up to 1.5 mm tall; stems sparsely hirsute at base; leaves pinnatifid, with long, linear segments; flowers pale yellow; siliques long, glabrous, hardly thicker than pedicels.

Roadsides; an introduced weed. Described from Italy, Germany, Siberia.

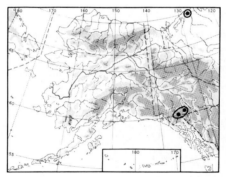 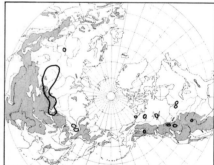

Thellungiella O. E. Schulz

1. Thellungiélla salsugìnea (Pall.) O. E. Schulz

Sisymbrium salsugineum Pall.; *Turritis salsuginosa* DC.; *Arabidopsis glauca* (Nutt.) Rydb.; *Sisymbrium glaucum* Nutt.

Glabrous and glaucous; stem simple or branched; no rosette or poorly developed basal rosette; basal leaves small, long-petiolated, ovate; cauline leaves lanceolate to oblong-lanceolate, clasping; petals white, 2–3 mm long; siliques linear, with very short, thick style.

Alkaline soil. Reported also from North China. Described from the Irtysh River and Baikal.

49. Cruciferae (Mustard Family)

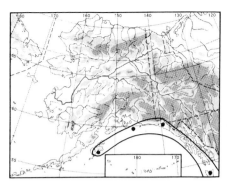
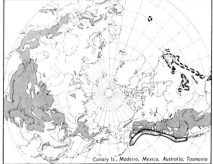

Cakile Hill

1. Càkile edéntula (Bigel.) Hook.   Sea Rocket
  Bunias edentula Bigel.
subsp. **califórnica** (Heller) Hult.
  Cakile californica Heller.

Glabrous, fleshy, branched from base; branches decumbent; leaves sinuate-dentate; pedicels stout; petals pink; silicles ribbed, flattened at apex, the lower joint obovoid, the upper ovoid.

Sandy beaches. *C. edentula* described from coast of Canada and Great Lakes, subsp. *californica* from Monterey, California.

Broken line on circumpolar map indicates range of subsp. **edéntula**.

Sinapis L.

Leaves lyrate-pinnatifid or entire; pedicels shorter than calyx . . . . 1. *S. arvensis*
Leaves pinnatifid; pedicels longer than calyx . . . . 2. *S. alba*

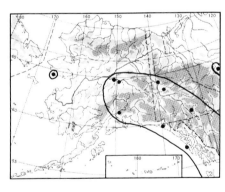
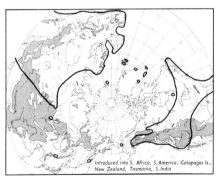

1. Sinàpis arvénsis L.   Charlock
  Brassica kaber (DC.) Wheeler; *Sinapis kaber* DC.

Leaves obovate, lyrate-pinnatifid or entire; pedicels shorter than calyx; sepals spreading; petals yellow; silicles glabrous or sometimes pubescent, with scarcely compressed beak about half as long as body; seeds 8–15.

An introduced weed. Described from Europe.

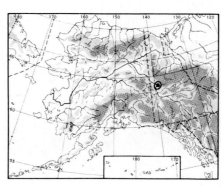
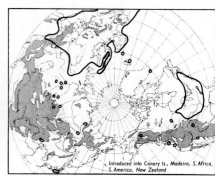

2. Sinàpis álba L.
Brassica hirta Moench.

Hirsute; leaves pinnatifid; sepals spreading; petals yellow; pedicels longer than sepals; siliques with stiff, white hairs, with flat, somewhat curved beak of about same length as body; seeds 4–8.

An introduced weed. Described from Belgium, England, and Germany.

Brassica L.

Upper stem leaves not clasping	1. *B. juncea*
Upper stem leaves clasping:	
Lower leaves pubescent; petals about 8 mm long	2. *B. rapa*
Glabrous; petals about 12 mm long	3. *B. napus*

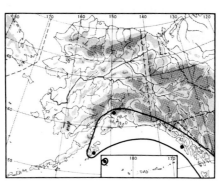
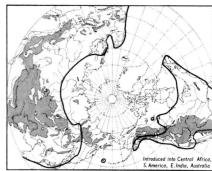

1. Brássica júncea (L.) Czern.
Sinàpis juncea L.

Lower leaves runcinate-pinnatifid, petiolated, the upper ligulate or lanceolate, entire or dentate, nearly sessile; petals yellow, 7–9 mm long; siliques sessile, constricted at internodes, 3–4 cm long, with seedless beak 5–8 mm long; buds and petals on same level.

An introduced weed. Described from Asia.

49. CRUCIFERAE (Mustard Family)

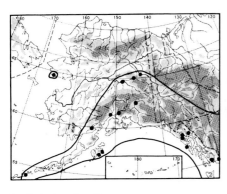 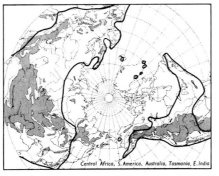

2. Brássica ràpa L. Bird's Rape
Brassica campestris L.

Annual, with thin root; stem with few branches; lower leaves bright green, with setose hairs, toothed to pinnatifid, with large ultimate lobe, the upper glaucous, clasping; inflorescence short, the flowers extending beyond buds; sepals spreading; petals pale yellow; siliques gradually narrowed into slender beak.

Waste places; escaped from cultivation. Described from England and Belgium.

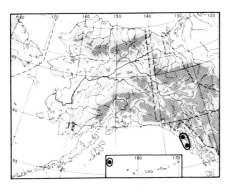 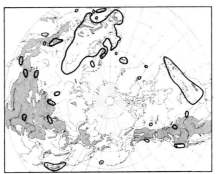

3. Brássica nàpus L. Turnip

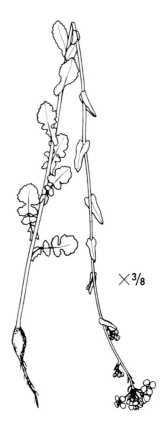

Similar to *B. rapa*, but glaucous and in most cases nearly glabrous; petals longer; buds usually extending beyond flowers.

Waste places; an introduced weed. Described from Gotland, Belgium, England.

49. CRUCIFERAE (Mustard Family)

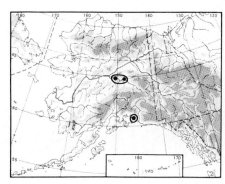
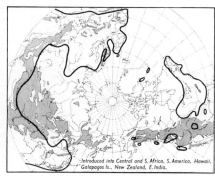

Raphanus L.

1. Ráphanus satìvus L.   Radish

Root thick, napiform or cylindric; plant hirsute; lower leaves lyrate; stem leaves entire; petals large, reddish-violet; siliques thick, not opening.

Escaped from cultivation. Type locality not given.

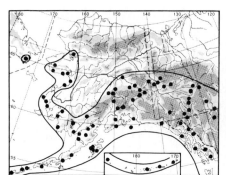
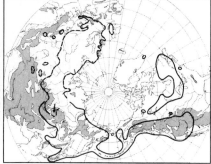

Barbarea R. Br.

1. Barbarèa orthocèras Ledeb.   Winter Cress

Barbarea americana Rydb.

Glabrous; stem tall, strict, not much branched; lower leaves lyrate, with large, apical lobe, upper lyrate-pinnatifid or entire, ovate, coarsely toothed; petals yellow or pale yellow; siliques erect, appressed, with thick beak about 0.5–1 mm long.

Banks of streams, moist places, to at least 1,000 meters in McKinley Park and the Yukon.

The young leaves are eaten raw or boiled by the natives.

Broken line on circumpolar map indicates range of **B. strícta** Andrz.

49. Cruciferae (Mustard Family)

Rorippa Scop. (Yellow Cress)

Petals white, 3–6 mm long, twice as long as sepals . . . . 1. *R. nasturtium-aquaticum*
Petals yellow:
    Plant with slender rhizome petals, about 4 mm long . . . . 8. *R. calycina*
    Plant lacking rhizome petals, shorter:
        Pedicels short; siliques 2–4 times as long as pedicels:
            Siliques not curved; leaf segments rounded; style 1–2 mm . . . . 6. *R. obtusa*
            Siliques strongly curved, leaf segments usually acute; style 0.5 mm . . . .
            . . . . 7. *R. curvisiliqua*
        Pedicels longer; siliques shorter with respect to pedicels:
            Stem glabrous or essentially so; siliques long and thick, plump . . . .
            . . . . 2–3. *R. islandica*
            Stem hispid, especially below:
                Siliques with 2 valves . . . . 4. *R. hispida* var. *hispida*
                Siliques with 4 valves . . . . 5. *R. hispida* var. *barbareaefolia*

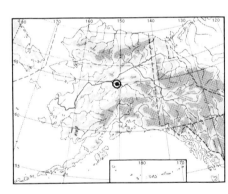 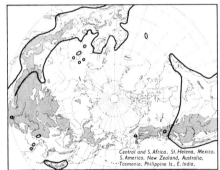 

×3/8

1. Roríppa nastúrtium-aquáticum (L.) Hayek

Sisymbrium nasturtium-aquaticum L.; *Nasturtium officinale* R. Br.

Glabrous, aquatic; leaves evergreen, pinnate; terminal leaflet oval, lateral leaflet rounded at base; petals white, twice as long as sepals; siliques about 15 mm long, with thick, short style.

Growing at Manley Hot Springs. Described from Europe and North America.

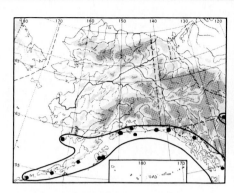
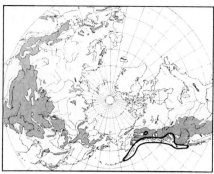

2. Roríppa islándica (Oeder) Borb.

Sisymbrium islandicum Oeder; *Nasturtium palustre* (L.) DC. not Crantz; *N. terrestre* R. Br.; *Rorippa palustris* (L.) Bess.; *Sisymbrium amphibium* var. *palustre* L.

subsp. **islándica**

Stem erect or decumbent, simple or branched, glabrous or nearly so; lower leaves petiolate, pinnatifid, with irregularly toothed segments and large, dentate or lobed terminal segment; upper leaves pinnatifid to entire; petals yellow, 1–2 mm long, as long as or shorter than sepal.

Wet places, gravel, roadsides.

The Alaskan coastal plant differs slightly from the European *R. islandica* in having thick, obtuse, sausagelike silicles, slightly longer than the pedicels [var. **occidentàlis** (S. Wats.) Butt. & Abbe; *Nasturtium terrestre* var. *occidentale* S. Wats.].

R. islandica described from Iceland, var. *occidentalis* from Shumagin Islands, etc.

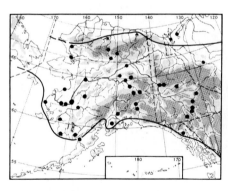
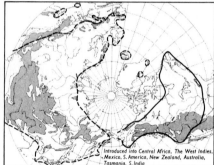

3. Roríppa islándica (Oeder) Borb.
subsp. **Fernaldiàna** (Butt. & Abbe) Hult.

Rorippa islandica var. *Fernaldiana* Butt. & Abbe; *R. palustris* var. *Williamsii* (Britt.) Hult.; *R. Williamsii* Britt.; *R. islandica* var. *microcarpa* (Regel) Fern.; *Nasturtium palustre* var. *microcarpa* Regel.

Similar to subsp. *islandica*, but often branched at top only; stem glabrous or somewhat hispid at base; upper leaves less deeply cleft; silicles vary in form from nearly orbicular to subcylindric, on pedicels of varying length.

Wet places, roadsides, waste places. Described from Fort Fairfield, Maine.

Forms hybrid swarms with *R. hispida*.

Broken line on circumpolar map indicates range of other subspecies.

49. CRUCIFERAE (Mustard Family)

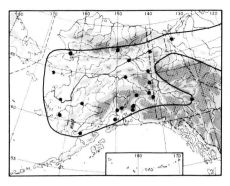
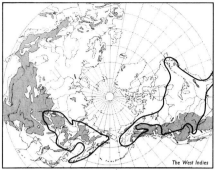
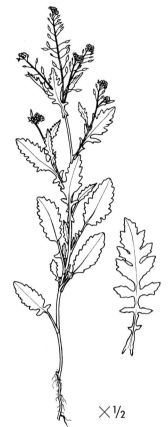

4. Roríppa híspida (Desv.) Britt.
Brachilobus hispidus Desv.; *Rorippa islandica* var. *hispida* (Desv.) Butt. & Abbe.
var. híspida

Similar to *R. islandica*. In its most extreme form, the stem is straight, upright, tall, branched above only, and densely hispid, with whitish hairs; the silicles are nearly spherical, much shorter than the pedicels.

Waste places, roadsides; wet places. Described from Pennsylvania.

Forms hybrid swarms with *R. islandica*. All transitions between the two taxa occur; those with densely hispid stem are here transferred to *R. hispida*.

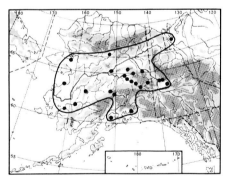
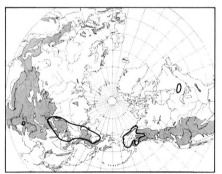
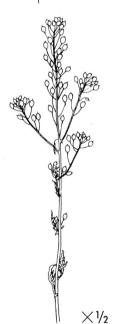

5. Roríppa híspida (Desv.) Britt.
Brachilobus hispidus Desv.
var. barbareaefòlia (DC.) Hult.

Camelina barbareaefolia DC.; *Tetrapoma pyriforme* Seem.; *T. barbareaefolia* (DC.) Turcz.; *Rorippa barbareaefolia* (DC.) Pors.

Differs from var. *hispida* in having 4-valved silicles.
Waste places; open soil. Described from Irkutsk, Doroninsk.
Since it is an interesting taxon, subject to different interpretation by different authors, it is mapped separately.

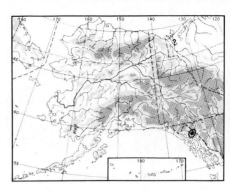
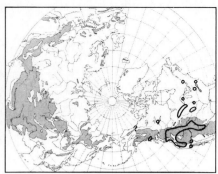

6. Roríppa obtùsa (Nutt.) Britt.
Nasturtium obtusum Nutt.

Leaves pinnately parted or divided, obtusely toothed; flowers minute, short-pedicelled; petals about 1 mm long; siliques 2 to 4 times as long as pedicels, with short style.

Shores, wet places. Described from the banks of the Mississippi River.

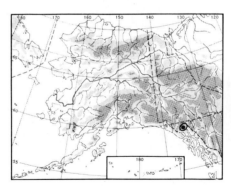
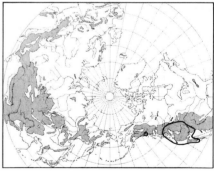

7. Roríppa curvisíliqua (Hook.) Bess.
Sisymbrium curvisiliquum Hook.; *Nasturtium lyratum* Nutt.; *Radicula Nuttallii* (S. Wats.) Greene.

Glabrous; lower leaves short-petiolated, pinnatifid, lobes oblong or lanceolate, acute, the upper sessile; pedicels 1–2 mm long; petals yellow, shorter than sepals; siliques terete, curved.

Wet places. Found but once, at mile 37, Haines Highway. Described from the northwest coast of North America, lat. 47–48°.

49. Cruciferae (Mustard Family)

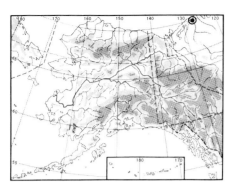
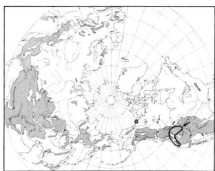

8. Roríppa calýcina (Engelm.) Rydb.
Nasturtium calycinum Engelm.

Stem decumbent to erect, hirsute, from thin rhizome; leaves sinuate to pinnatifid, the lower petiolate, with ovate to oblong, toothed segments; sepals not saccate, often persisting; petals yellow; siliques oblong to ovate, somewhat arcuate, soft-pubescent; style 1–2 mm.

Single specimen known from the mouth of the Anderson River. Described from Yellowstone River. Total range tentative.

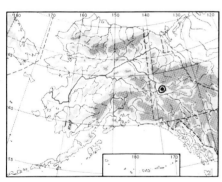
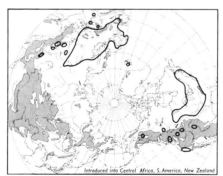

Armoràcia Gaertn., Mey. & Scherb.

1. Armoràcia rusticàna Gaertn., Mey. & Scherb. Horseradish
Armoracia lapathifolia Gilib.

Stems from thick taproot; cauline leaves lanceolate, the lower large, oblong, ovate, long-petiolated; petals white; silicles subglobose.

Escaped from cultivation. Described from Europe.

49. CRUCIFERAE (Mustard Family)

Cardamine L. (Bitter Cress)

Leaves entire, ovate or obovate .... 1. *C. bellidifolia*
Leaves pinnate, digitate, trifoliate, or linear:
  Flowers small, petals 4 mm long or less:
    Upper leaves larger than basal leaves .... 6. *C. Regeliana*
    Upper leaves smaller than basal leaves:
      Inflorescence elongate; basal leaves mostly with more than 2 pairs of leaflets; stem mostly somewhat hirsute .... 3. *C. pennsylvanica*
      Inflorescence subumbellate; basal leaves with 2 pairs of leaflets; stem glabrous .... 5. *C. umbellata*
  Flowers larger:
    Plant 30–60 cm tall; leaves 3-foliate, rarely 5-foliate, leaflets 3-toothed .... 2. *C. angulata*
    Plant shorter; leaves pinnate or digitate:
      Stem (upper part) and pedicels pubescent .... 10. *C. purpurea*
      Stem glabrous:
        Stem leaves with several pairs of linear leaflets .... 4. *C. pratensis* subsp. *angustifolia*
        Stem leaves with 1 or 2 pairs of leaflets or digitately 3–5-parted or linear:
          Leaves entire, narrowly linear .... 7. *C. Victoris*
          Leaves with 1–3 pairs of leaflets:
            All leaflets narrow, linear to oblong-lanceolate .... 8. *C. hyperborea*
            At least the basal leaflets with broad, ovate or slightly lobed leaflets .... 9. *C. microphylla*

× ½

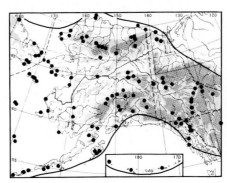

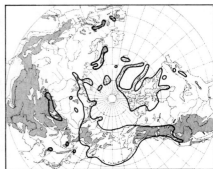

1. Cardàmine bellidifòlia L.

Cardamine bellidifolia var. *beringensis* Pors.

Tufted, dwarf plant; leaves long-petiolated, ovate, entire or rarely with lateral tooth, rarely with some leaves pinnatifid (Cape Beaufort, var. **pinnatifida** Hult.); stem leafless; flowers small, white; siliques linear, upright, with short, thick style.

Gravelly places below melting snow, along creeks in the mountains, to at least 2,000 meters. Described from the Alps of Lapland, Switzerland (and England, where it does not occur).

49. CRUCIFERAE (Mustard Family)

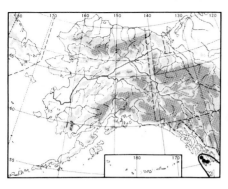 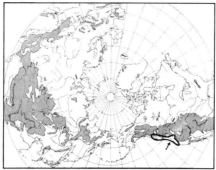 

2. Cardàmine angulàta Hook.

Glabrous; stems from slender rhizome, erect, up to 60 cm tall; leaves pinnate, with 3–5 leaflets, the terminal 3–5-toothed; petals white, 8–12 mm long; siliques with stout style 2 mm long, spreading.

Wet places.

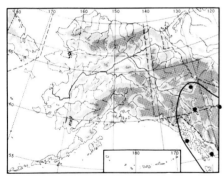 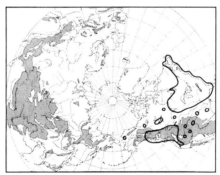 

3. Cardàmine pennsylvànica Muhl.

Cardamine flexuosa subsp. *Regeliana* f. *sitchensis* O. E. Schulz.

Stems glabrous or usually somewhat pubescent below; basal leaves 5–9-foliate; lateral leaflets ovate or obovate, terminal somewhat larger, often 3-lobed; upper leaves with narrower leaflets; inflorescence elongated in fruit; fruiting pedicels ascending; petals white; siliques erect, less than 1 mm wide, with very short style; seeds not winged.

Moist places. Described from Pennsylvania.

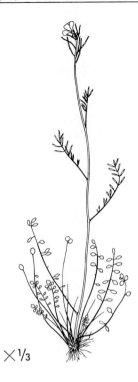
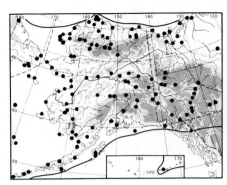
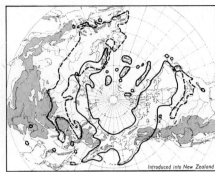

4. Cardàmine praténsis L.   Cuckoo Flower
subsp. **angustifòlia** (Hook.) O. E. Schulz
Cardamine pratensis L. var. *angustifolia* Hook.; *C. Nymani* Gandoger.

Glabrous; stems terete, from short rhizome; leaves thick with embedded veins; basal leaves with several stalked, rounded leaflets, soon withering, the upper pinnatisect, with linear divisions; petals white or rose, with darker veins, 2 to 3 times longer than sepals; mature siliques 3–4 cm long, with short style.

Wet places, along creeks, thickets, to at least 500 meters. *C. pratensis* described from Europe, subsp. *angustifolia* from Igloolik and other islands in the Arctic Sea.

Broken line on circumpolar map indicates range of subsp. **praténsis**.

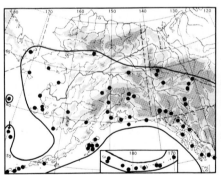
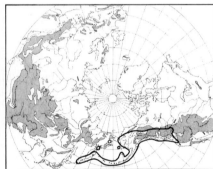

5. Cardàmine umbellàta Greene

Stems glabrous, from slender rootstock; basal leaves with 1–3 pairs of rounded or ovate, stipitate leaflets, and large 3-lobed ultimate lobe; upper stem leaves with lanceolate or oblong leaflets; flowers crowded at top of stem, subumbellate; pedicels strongly ascending; petals white, 3–4 mm long; siliques with short style.

Wet places. Very variable.

49. CRUCIFERAE (Mustard Family)

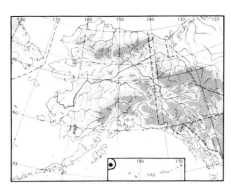

6. Cardàmine Regeliàna Miq.

Cardamine hirsuta var. *Regeliana* (Miq.) Maxim.; *C. flexuosa* var. *Regeliana* (Miq.) Kom.

Basal leaves glabrous or somewhat pubescent, with 1–3 pairs of leaflets, the apical leaflet much larger than the lateral, irregularly lobed; upper leaves with small basal lobes or none; sometimes all leaves entire, irregularly lobed; flowers small; sepals often purplish; petals twice as long as sepals; siliques linear with short style.

Wet places. Described from Japan.

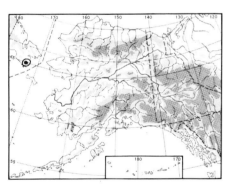
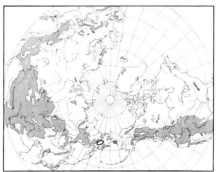
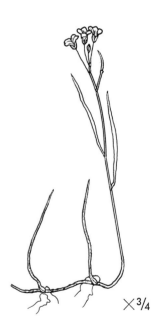

7. Cardàmine Victòris N. Busch

Glabrous; stem simple from thin white rhizome; leaves narrowly linear; inflorescence 3–6-flowered; sepals oblong-obovate with narrow white margin; petals obovate, white-clawed, 5–6.5 mm long, 2–2.5 mm broad; siliques flat, long, with beak up to 3 mm long.

Wet places. Described from Penzhina and Anadyr.

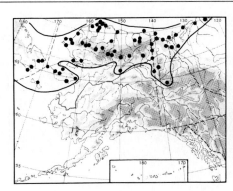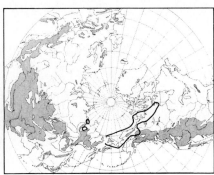

8. Cardàmine hyperbòrea O. E. Schulz
Cardamine digitata Richards.; *C. Richardsonii* Hult.

Glabrous; basal leaves in row from slender, horizontal rhizome, which breaks off easily and is often missing in herbarium specimens; basal leaves with 2–3 pairs of linear-lanceolate to oblong-lanceolate, blunt or somewhat acute leaflets, often with an extra unpaired leaflet at base; apical leaflet similar to lateral; stem leaves irregularly lobed, with linear, acute leaflets; petals 5–8 mm long, white or pinkish, twice as long as sepals; siliques ascending, with style about 2 mm long.

Wet slopes in the mountains, to at least 1,100 meters, and on tundra. Described from Mackenzie district, Barren Grounds.

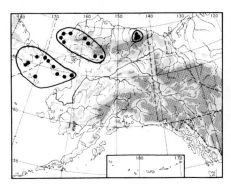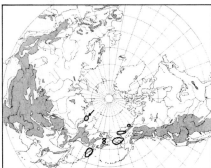

9. Cardàmine microphýlla Adams
Cardamine minuta Willd.; *C. Blaisdellii* Eastw.

Glabrous; leaves and stems from slender rhizome; stem ascending; basal leaves with 1–2 pairs of rounded, stipitate leaflets; apical leaf similar to lateral; stem leaves with narrower, acute lobes, sometimes with 3 leaflets, or entire, 3-lobed; flowers large, white or pinkish; pedicels glabrous; style short in young silique, elongated at maturity.

Wet places.

49. CRUCIFERAE (Mustard Family)

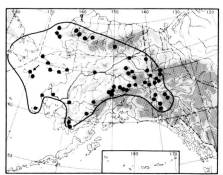
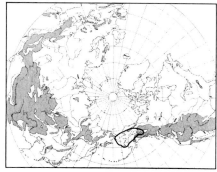

10. Cardàmine purpùrea Cham. & Schlecht.

Stems 1 to several, from tip of long rhizome; upper stem and pedicels pubescent; leaves sparsely white-hirsute; basal leaves lyrate-pinnate, with 1 to 3 pairs of round leaflets and broad, short apical lobe; stem leaves similar, but smaller, the uppermost entire; petals purple, or white (var. **albiflòra** Hult.), rarely pinkish; siliques 10–15 mm long, glabrous, tapering to short, thick style, with broad stigma.

Wet hillsides, heaths, scree slopes, in the mountains to at least 1,500 meters.

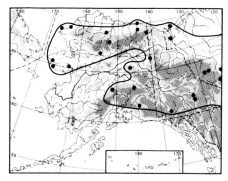
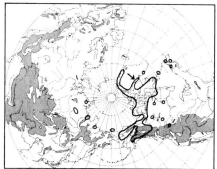
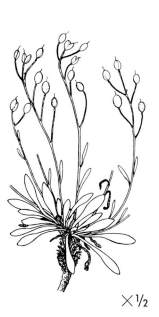

Lesquerella S. Wats.

1. Lesquerélla árctica (Wormsk.) S. Wats.

Alyssum arcticum Wormsk.; *Vesicaria arctica* (Wormsk.) Richards.; *V. arctica* subsp. *leiocarpa* Trautv.; *V. leiocarpa* (Trautv.) N. Busch; *Lesquerella arctica* var. *Purshii* S. Wats.; *L. arctica* subsp. *Purshii* (S. Wats.) Pors.; *L. arctica* var. *Scammanae* Rollins.

Several decumbent or ascending stems from short root; leaves silvery, stellate-pannose, the basal more or less petiolated, oblanceolate to spatulate, the upper small, oblanceolate; petals yellow; silicles globose to obovate, glabrous or lepidote-pubescent.

Dry slopes and ridges in the mountains.

The southern plant, with larger, long-petiolated leaves and often slightly lepidote-pubescent silicles, may be referred to var. **Pùrshii** S. Wats.

Capsella Medic.

Silicles shallowly emarginate, with straight or somewhat convex sides .... 1. *C. bursa-pastoris*
Silicles deeply emarginate, the sides convex toward the apex .... 2. *C. rubella*

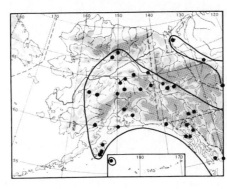
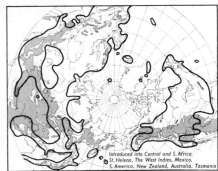

1. Capsélla bùrsa-pastòris (L.) Medic. Shepherd's Purse
Thlaspi bursa pastoris L.

Basal leaves variable in form from entire, toothed, to pinnatifid; stem leaves sessile, sagittate; petals white, about 2 mm long, longer than sepals; silicles triangular, with straight or somewhat convex sides, shallowly emarginate.

An introduced weed. Described from Europe.

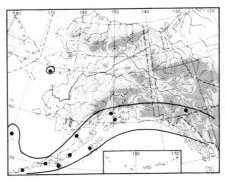
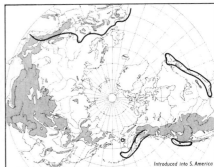

2. Capsélla rubélla Reut.

Similar to *C. bursa-pastoris*, but petals somewhat shorter, sides of silicles convex toward summit, deeply emarginate.

An introduced weed. Described from Europe.

49. CRUCIFERAE (Mustard Family)

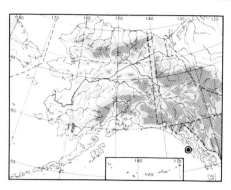
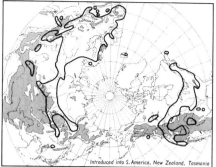

Camelina Crantz

1. Camelìna satìva (L.) Crantz — False Flax
Myagrum sativum L.

Stem glabrous or somewhat stellate-pubescent; leaves lanceolate, lowest leaves petiolated, upper leaves sessile, sagittate; petals yellowish-white, about 4 mm long; silicles pyriform, with long style and several seeds, opening when ripe.

An introduced weed. Described from Europe. Circumpolar map includes range of **C. microcárpa** Andrz.

Many doubtful, closely related taxa, sometimes regarded as separate species, also occur.

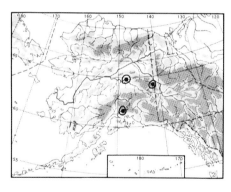
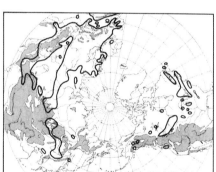

Neslia Desv.

1. Néslia paniculàta (L.) Desv. — Ball Mustard
Myagrum paniculatum L.; *Vogelia paniculata* (L.) Hornem.

Stellate-pubescent; stem simple; leaves oblong, sagittate at base; racemes short in flower, elongate in fruit; petals yellow, about 2 mm long; silicles spherical, 1-seeded, not opening, reticulated.

An introduced weed. Described from Europe.

Circumpolar map indicates range of *N. paniculata* in a broad sense, including the southern **N. apiculàta** Fisch. & Mey.

Draba L.

This genus presents many puzzling forms, permitting a variety of interpretations, doubtless owing at least partly to gene exchange. A number of species are obscurely limited, and it has proved impossible to divide all of the available material among the taxa accepted here. To construct a key workable for all cases is not possible under these conditions.

Plant annual, with basal leaves soon withering, lacking sterile rosules; racemes bractless; flowering stems with several leaves; flowers yellow .... 32. *D. nemorosa*
Plant perennial, with permanent basal leaves, often with sterile rosules:
  Leaves ligulate, narrow, completely glabrous or ciliated, fleshy, the midrib not persisting after leaf withers; flowers yellow .... 1. *D. crassifolia*
  Leaves more or less pubescent, often also ciliate, midrib usually persisting after leaf withers:
    Flowering stem very short; flowers not exceeding leaves; plant pulvinate .... 20. *D. aleutica*
    Flowering stem much exceeding leaves:
      ■ Plant scapose, with leafless, comparatively short flowering stem, or stem sometimes with 1 leaf:
        Leaves and scapes with minute stellate hairs, lacking simple or forked elongate hairs, or ciliated with simple hairs near base:
          Pubescence of leaves formed by sessile, stellate, entangled hairs, concentrated toward leaf tip; flowers yellowish-white; leaves narrow, carinate .... 2. *D. oligosperma*
          Pubescence not formed by sessile hairs concentrated toward leaf tip:
            Plant densely caespitose:
              Flowers yellow; silicles pubescent; style very short .. 5. *D. exalata*
              Flowers white; silicles glabrous:
                Silicles oblong to elliptic; stem pubescent .... 3. *D. nivalis*
                Silicles linear, much longer; stem glabrous .. 4. *D. lonchocarpa*
            Plant loosely caespitose; flowers yellow:
              Leaves with very minute, canescent, stellate hairs, not ciliated at base .... 6. *D. caesia*
              Leaves with coarser stellate hairs, mostly ciliated at base .... 7. *D. incerta*
        Leaves with long, simple, forking or branched hairs, or mixed with short stellate hairs:
          Style about 1.2–2 mm long; leaves nearly glabrous above; flowers yellow:
            Silicles glabrous; plant not pulvinate .... 8. *D. Eschscholtzii*
            Silicles pubescent from simple or stellate hairs; plant pulvinate .... 18. *D. densifolia*
          Style shorter:
            Style about 0.5 mm or shorter:
              Flowers yellow; leaves glabrous above, with coarse, forked hairs below; silicles as long as pedicels; withered leaves remain several years .... 9. *D. pilosa*
              Flowers white; leaves with simple or stellate hairs:
                ● Stem glabrous or nearly so; silicles pubescent, longer than the pedicels; lower side of leaves with stellate hairs .... 10. *D. pseudopilosa*

- Stem densely pubescent, with branched hairs; leaves narrow, ciliated with simple hairs; flowers small; style 0.2–0.3 mm long; silicles glabrous .... 13. *D. subcapitata*
  - Style 0.5–1 mm:
    - Scape and pedicels glabrous:
      - Leaves ciliate, with mixed simple and forked hairs, pubescent on both sides, with mixed simple, forked, branching, or even stellate hairs .... 11. *D. lactea*
      - Leaves ciliate, with simple hairs, the surface glabrous or with simple hairs .... 12. *D. fladnizensis*
    - Scape pubescent; flowers yellow:
      - Petals very narrow, linear; leaves with simple and branched hairs; silicles glabrescent or pubescent, orbicular; plant densely tufted .... 19. *D. stenopetala*
      - Petals broader:
        - Flower small; petals 2–3 mm long, hardly longer than sepals .... 14. *D. micropetala*
        - Flowers larger; petals 4–5 mm long:
          - Leaves about 1 mm broad, carinate ...... 15. *D. barbata*
          - Leaves flat, broader:
            - Silicles glabrous .... 16. *D. alpina*
            - Silicles pubescent .... 17. *D. macrocarpa*
- Plant with flowering stem with 1 to many leaves (occasionally without leaves in *L. stenoloba* and *D. longipes*); flowering stem often taller:
  - Plant with long, leafy runners; flowers large, yellow .... 33. *D. ogilviensis*
  - Plant caespitose, lacking runners:
    - Basal leaves large, 8–16 cm long, yellowish-green, with cuneate base and few coarse teeth; stem usually extending barely beyond basal leaves; flowers yellow .... 34. *D. hyperborea*
    - Basal leaves smaller, not yellowish; stem several times longer than basal leaves; flowers white or yellow:
      - Silicles glabrous:
        - Silicles linear, long and narrow; stigma sessile or style very short; petals yellow .... 21. *D. stenoloba*
        - Silicles oblong to elliptic; style longer; petals white:
          - Pedicels much longer than silicles .... 23. *D. longipes*
          - Pedicels usually not longer than silicles:
            - Pedicels glabrous; leaves with coarse, branched or stellate hairs .... 22. *D. hirta*
            - Pedicels pubescent; mostly 1 or 2 broad stem leaves:
              - Leaves canescent from small, dense, stellate hairs, similar to those of *D. nivalis* .... 24. *D. kamtschatica*
              - Leaves with coarse, forked, branched, or stellate hairs, sometimes ciliated with simple hairs .... 25. *D. Chamissonis*
      - Silicles pubescent:
        - Pedicels hirsute, with divaricate, simple hairs .... 26. *D. borealis*
        - Pedicels with forked, branched, or stellate hairs:
          - Basal part of stem hirsute, with long, coarse, simple hairs, mixed with forked or branched hairs:
            - Flowers white; plant tall, with several large, coarsely toothed stem leaves .... 27. *D. maxima*
            - Flowers yellow; numerous small stem leaves ... 29. *D. aurea*
          - Basal part of stem with chiefly forked, branched, or stellate hairs:
            - Pedicels of ripe silicles much longer than silicles .... 22. *D. longipes*
            - Pedicels of ripe silicles not much longer than silicles:
              - Plant tall, with compact rosette and large stem leaves ... . 28. *D. praealta*
              - Plant smaller, with smaller stem leaves:
                - Stem with several leaves; silicles short-pedicellated, acute, appressed to stem ...... 30. *D. lanceolata*
                - Stem with 2–5 leaves; silicles more long-pedicellated, elliptic to ovate, more divaricate ... 31. *D. cinerea*

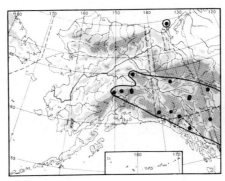 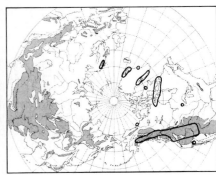

×1

1. Dràba crassifòlia Graham

Yellowish-green; leaves thick, glabrous, or ciliated with long simple hairs; rosette leaves linear, spatulate, obtuse, the upper oblong, sessile; stem glabrous; pedicels with spreading hairs; petals pale yellow; silicles oblong to elliptic, with short style.

Snow beds, in the mountains to at least 2,000 meters. Described from the Rocky Mountains.

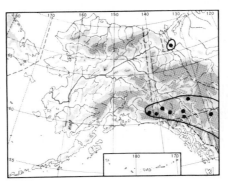 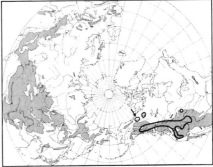

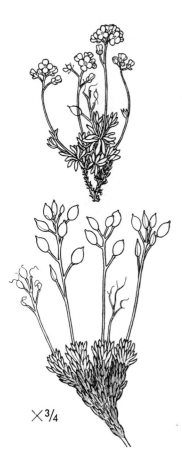

×¾

2. Dràba oligospérma Hook.

Matted; leaves linear-lanceolate, keeled, pubescent, with short, sessile, branched hairs, usually concentrated toward tip; flowering stems leafless, glabrous; petals yellow, fading in drying; siliques elliptic, with short style, pubescent, with simple or branched hairs.

Dry hillsides and rocks.

49. CRUCIFERAE (Mustard Family)

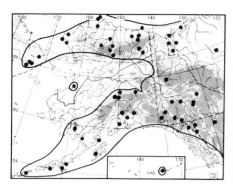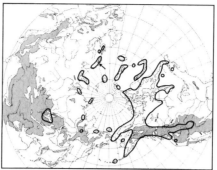

3. Dràba nivàlis Liljebl.

Caespitose; leaves oblong-obovate, densely grayish-pubescent, with short, stellate hairs, sometimes with few simple hairs in lower margin; flowering stem leafless (rarely with single leaflet); inflorescence dense, elongated in fruit; sepals densely pubescent with simple or forking hairs; petals white; silicles elliptic to lanceolate, glabrous, with very short style.

Dry mountains; to at least 2,600 meters in the St. Elias Range.

Forms with glabrous scapes and pedicels and also a few simple hairs might represent the hybrid *D. lactea* × *nivalis*.

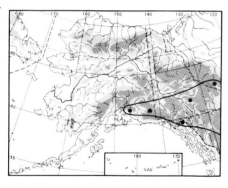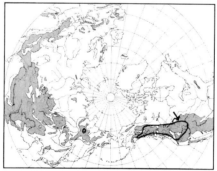

4. Dràba lonchocárpa Rydb.
Draba nivalis var. *elongata* S. Wats.

Similar to *D. nivalis*, but stellate hairs more distinctly stalked, more simple hairs in margin of basal leaves; glabrous scapes and pedicels, and longer pedicels and silicles.

Rocky places.

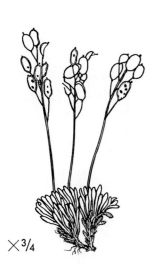
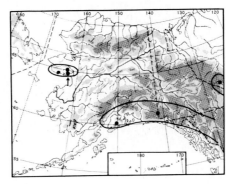
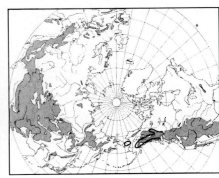

5. Dràba exalàta Ekman

Draba ventusa var. *ruaxes* (Payson & St. John) Hitchc., at least with respect to Alaskan plant.

Densely caespitose, branches with marcescent leaves at base; leaves obovate, densely covered with soft, stellate hairs on both sides, and with few simple hairs at base; stem with short, branched and simple hairs; petals yellow; silicles ovate to oblong-ovate, pubescent when young, glabrescent in age, with style 0.5 mm long.

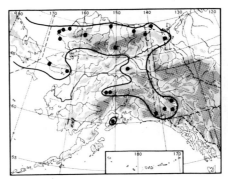
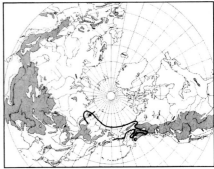

6. Dràba caèsia Adams

Draba Palanderiana Kjellm.

Loosely caespitose; leaves oblanceolate, small, densely packed, cinereous, with short, stellate hairs; flowering stems leafless, pubescent with stellate hairs; sepals pubescent; petals yellow; silicles glabrous, elliptic, more or less acute, with style about 1 mm long.

Polygon soil on tundra and in the mountains; dry rocks.

49. CRUCIFERAE (Mustard Family)

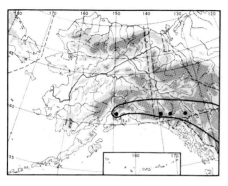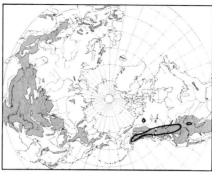

7. Dràba incérta Payson

Loosely tufted; leaves linear-oblanceolate, with stellate hairs; often ciliate at base; flowering stems with stellate or branched hairs, leafless or with single leaf; petals yellow; silicles as long as pedicels, glabrous or somewhat pubescent with simple or branched hairs; style 0.5–1 mm long.

Rocky slopes. Described from Yellowstone National Park.

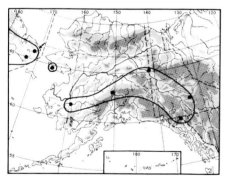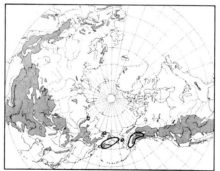

8. Dràba Eschschòltzii Pohle

Caespitose; leaves flat, oblong-obovate, ciliate in margin, with few simple or forked hairs above, more dense, forked or branched hairs below; stem with branched or forked hairs, nearly glabrous above; petals yellowish-white; silicles shorter than pedicels, glabrous, with long style.

In the mountains to over 1,700 meters.

×3/4

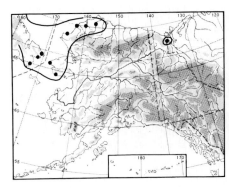
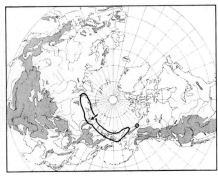

9. Dràba pilòsa DC.

Densely caespitose branches covered with old marcescent leaves, leaves lanceolate to linear-lanceolate, with elevated midrib below, ciliate in margin, nearly glabrous above and with coarse branched hairs below; stem glabrous, or somewhat pubescent with simple or forked hairs; pedicels glabrous; petals broad, yellow; silicles with style 0.5–0.8 mm long, about as long as pedicels, glabrous or with some simple hairs in margin.

Dry places.

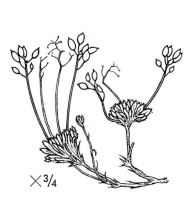
×3/4

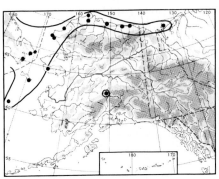
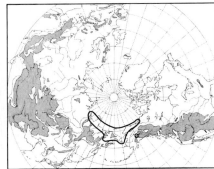

10. Dràba pseudopilòsa Pohle

Densely caespitose; branches with remains of withered leaves; leaves linear-lanceolate, ciliate, glabrous above, with small, stellate hairs below; pedicels glabrous; petals white, 3.5–4 mm long, silicles lanceolate, longer than pedicels, with style 0.5 mm long.

In the mountains to over 1,700 meters. Described from northern Yakutsk province.

49. CRUCIFERAE (Mustard Family)

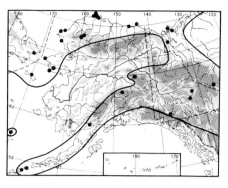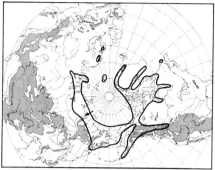

11. Dràba láctea Adams

Loosely caespitose; leaves lanceolate, with simple or forked hairs in margin of lower part, and fine stellate hairs on apical half; flowering stem glabrous, leafless (rarely with single leaflet); inflorescence dense, elongated in fruit; silicles ovate-lanceolate, glabrous, with style about 0.5 mm long.

Dry and moist places in the mountains and on tundra.

Forms hybrid swarms with *D. fladnizensis*.

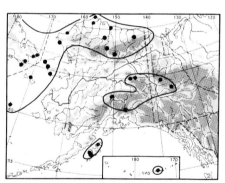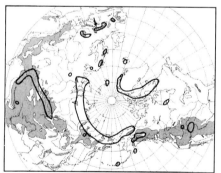

12. Dràba fladnizénsis Wulf.

Similar to *D. lactea*, but leaves with long, simple hairs, lacking stellate hairs. Forms hybrid swarms with that species.

Rocks, scree slopes.

×1

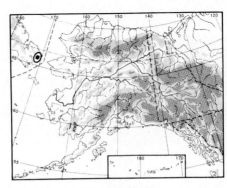
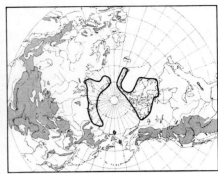

13. Dràba subcapitàta Simm.

Plant densely tufted; branches with dense, marcescent leaves; leaves densely ciliate, linear-lanceolate, consisting mostly of the median nerve, glabrous above or nearly so, with small, dense, branched hairs and, toward apex, simple hairs below; pedicels densely pubescent; petals about 2 mm long, yellowish-white, about as long as sepals; silicles elliptic to ovate, glabrous, with barely noticeable style 0.2–0.3 mm long.

Stony tundra, rocks. Described from Ellesmereland.

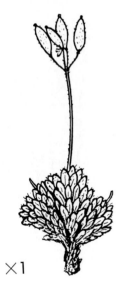
×1

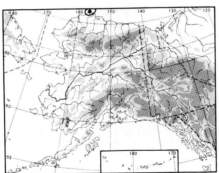
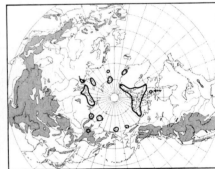

14. Dràba micropétala Hook.
?*D. oblongata* R. Br.

Densely caespitose; leaves oblong, with small, branched or stellate hairs; flowering stems leafless, with short, branched hairs or nearly glabrous; petals narrow, oblong-spatulate, not overlapping, 2.5–3.5 mm long, yellow; silicles ovate or broadly ovate, pubescent with simple hairs; style very short.

Dry tundra.

49. CRUCIFERAE (Mustard Family)

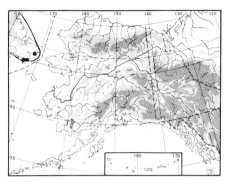
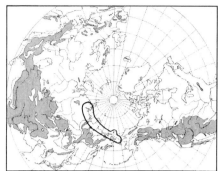

15. Dràba barbàta Pohle

Densely caespitose; branches covered with imbricate, marcescent leaves; leaves linear-oblong, about 1 mm broad, carinate, ciliate in margin, densely pubescent on both sides with simple, forked and branched hairs; stem and pedicels densely covered with similar hairs; petals yellow; silicles ovate, with style 0.5–1 mm long, glabrous or pubescent.

Dry tundra. Described from Chukchi Peninsula. A doubtful species reported from arctic America.

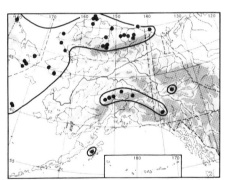
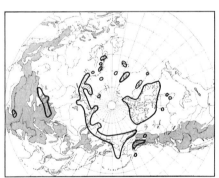

16. Dràba alpìna L.

Caespitose; leaves oblong-lanceolate, with forked or branched hairs, ciliate with long, simple hairs; flowering stem leafless, pubescent with long, soft, simple or forked hairs; petals yellow or pale yellow; silicles ovate to oblong-ovate, glabrous.

Stony places in the mountains and on tundra to at least 1,800 meters. Described from the European Alps.

Highly variable; hybridizes with several other species, such as *D. macrocarpa, D. microcarpa,* and *D. pilosa.*

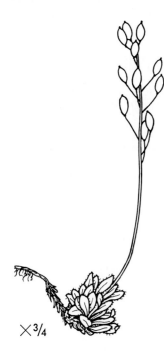
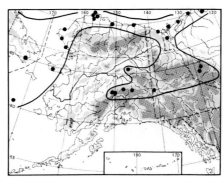
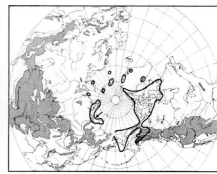

17. Dràba macrocárpa Adams
Draba Bellii Holm.

Densely caespitose; leaves oblong-lanceolate, with simple or forked hairs, somewhat ciliate with simple hairs; flowering stems leafless, pubescent with soft, simple or forked hairs; petals yellow; silicles large, 7–12 mm long, with short style, densely pubescent with simple hairs.

Rocky slopes, to about 2,800 meters in the St. Elias Mountains.

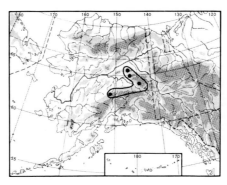
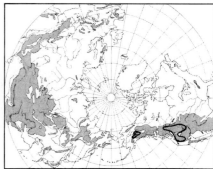

18. Dràba densifòlia Nutt.

Caespitose; leaves linear-oblanceolate, glabrous or with few forked or stellate hairs, ciliate with simple hairs; flowering stems leafless, glabrous or with simple or branched hairs; petals yellow; silicles ovate to ovate-elliptic, with style 0.5–1 mm long, coarsely pubescent with simple hairs.

Scree slopes. Described from the Rocky Mountains toward the Lewis River.

49. Cruciferae (Mustard Family)

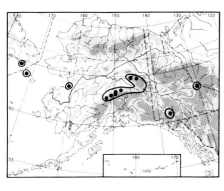 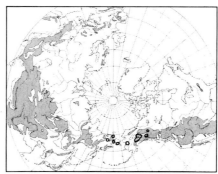 

19. Dràba stenopétala Trautv.
Draba densifolia with respect to Alaskan plant, in part.

Densely tufted; leaves spatulate, with simple and branched hairs; petals narrow, 2–2.5 mm long, yellow; silicles orbicular, glabrous (f. **leiocárpa** Pohle) or pubescent (f. **hebecárpa** Pohle), with fairly long style. Var. **purpùrea** Hult. has purplish petals. On rocks in high mountains. Described from Anadyr.
(See color section.)

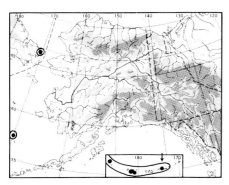 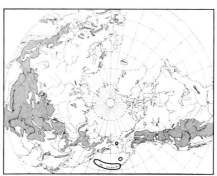 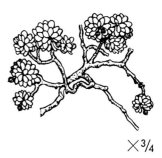

20. Dràba aleùtica Ekman
Draba Behringii Tolm.

Very densely tufted; leaves oblong-ovate, short-petiolated, obtuse, nearly glabrous or with some branched hairs toward apex, ciliate with simple hairs; flowering stem very short; flowers pale yellow; silicles obovate to broadly pyriform, inflated, glabrous, with short style.
Solifluction areas in high mountains.

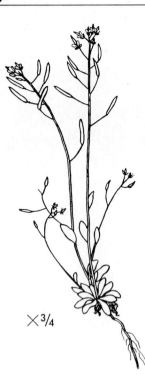
× ¾

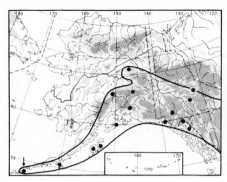 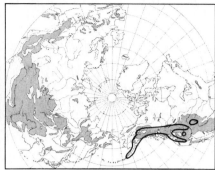

21. Dràba stenolòba Ledeb.
Including *Draba nitida* Greene.

Basal leaves ovate to obovate-lanceolate, entire or denticulate, with more or less dense, forked or branched hairs, ciliate with simple hairs; stem sparsely pubescent with forked or branched hairs; stem leaves 1 to 3, ovate, dentate; pedicels as long as, or longer than, silicles; petals yellow, turning reddish when dried; silicles linear, glabrous, lacking style or with very short style.

Dry mountain slopes.

× ½

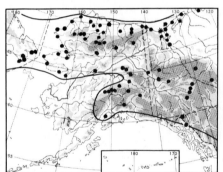 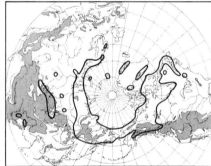

22. Dràba hírta L.
Draba daurica DC.; *D. glabella* Pursh. Including *D. juvenilis* Kom.

Loosely caespitose; basal leaves oblong, blunt or somewhat acute, entire or with coarse teeth, pubescent with branched or stellate hairs, sometimes with simple hairs at base; stem leaves dentate; stem with stellate hairs; pedicels glabrous or with few simple hairs; petals white; silicles about as long as pedicels, glabrous or nearly so; style 0.25–0.5 mm long.

Dry, stony places in the mountains. Type locality not given.

49. CRUCIFERAE (Mustard Family)

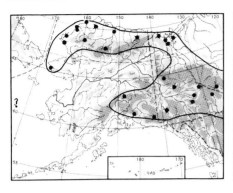 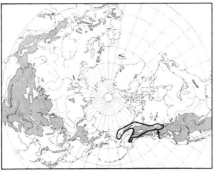 

23. Dràba lóngipes Raup

Similar to *D. hirta*, but pedicels often pubescent and silicles much shorter than pedicels.

Below snowbeds in the mountains and on tundra.

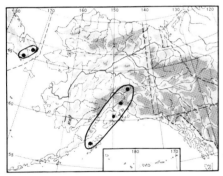 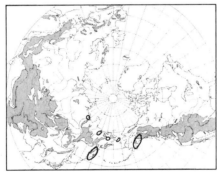 

24. Dràba kamtschática (Ledeb.) N. Busch

Draba frigida var. *kamtschatica* Ledeb.; *D. lonchocarpa* subsp. *kamtschatica* (Ledeb.) Calder & Taylor.

Caespitose; basal leaves oblong to oblong-obovate, entire or with small teeth, densely canescent, pubescent with small, branched and stellate hairs; stem with small, stellate hairs; stem leaves broad, 2–5, more or less denticulated; pedicels densely pubescent with simple and branched hairs; petals white; silicles mostly longer than pedicels, glabrous.

Habitat unknown. Type locality presumably Kamchatka.

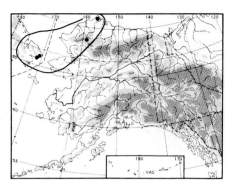
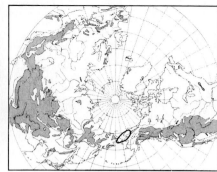

25. Dràba Chamissònis G. Don

Caespitose; basal leaves oblong, entire, pubescent with coarse, forked, branched and stellate hairs, sometimes ciliate with simple hairs; stem leaves 1–2, ovate, more or less denticulated, with forked or branched hairs; pedicels with forked, branched, long, simple hairs; petals white; silicles erect, appressed to stem, narrowly elliptic, blunt, glabrous, about 3 times longer than pedicels, with short style.

Dry tundra. Described from Chukchi Peninsula.

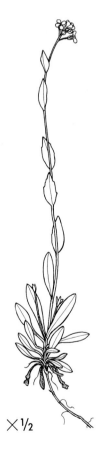
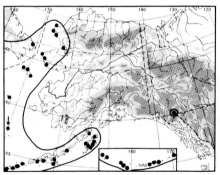
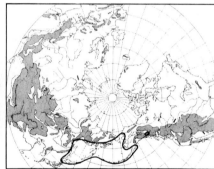

26. Dràba boreàlis DC.
Draba unalaschcensis DC.

Loosely tufted; basal leaves ovate to obovate-oblong, entire or with few coarse teeth, pubescent with simple, forking or branched hairs; stem with simple hairs at least below, mixed with forked or branched hairs; stem leaves 2 to several, broad, coarsely toothed; pedicels densely pubescent with divaricate, long, simple hairs; petals white; silicles ovate-oblong to elliptic, often twisted; more or less densely pubescent with simple or forking hairs; style about 0.5 mm.

Grassy slopes in the mountains, rocks.

49. CRUCIFERAE (Mustard Family)

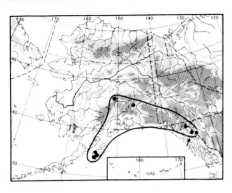 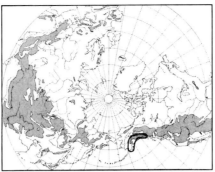 

27. Dràba máxima Hult.

Similar to *D. borealis*, but usually taller and coarser; pedicles and silicles with branched to stellate hairs; stem leaves ovate, grossly dentate; flowers white; mature silicles often twisted.

Scree slopes.

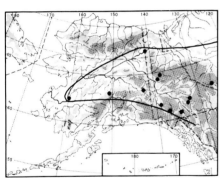 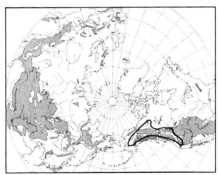 

28. Dràba praeálta Greene

Plant with compact rosettes; basal leaves oblanceolate, pubescent with simple, forking or branched, 4–7-rayed hairs; stem with simple and branched hairs at base and branched hairs at top, with several lanceolate leaves; inflorescence many-flowered; pedicels with branched hairs; petals white; silicles linear-lanceolate to lanceolate, longer than pedicels, with soft, simple or forked hairs.

Grassy slopes, rocks. Described from Banff.

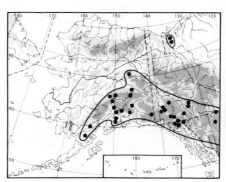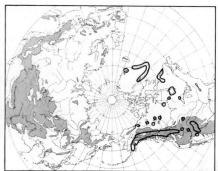

29. Dràba aùrea Vahl
Including *Draba luteola* Greene.

Stem tall, straight, several-leaved, pilose-hirsute, somewhat branched or unbranched; basal leaves oblanceolate to spatulate, entire or dentate, somewhat acute, often red below, densely covered on both sides with simple, forked and stellate hairs; petals pale yellow to bright yellow; silicles lanceolate to oblong-lanceolate, often twisted, pubescent with simple, branched and stellate hairs; style 0.5–1.5 mm long.

Dry, gravelly slopes. Described from Greenland.

A highly variable species.

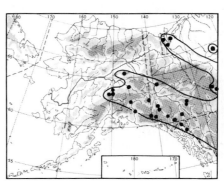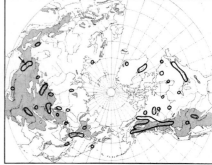

30. Dràba lanceolàta Royle

Tufted; basal leaves lanceolate, canescent with small, branched or stellate hairs, sometimes also ciliate with simple or forked hairs; stem pubescent, with small, branched hairs, with several oblong, ovate to ovate-lanceolate, toothed or entire leaves; inflorescence many-flowered; pedicels densely pubescent with branched or stellate hairs; petals white; silicles lanceolate, 2–4 times as long as pedicels, with dense, branched and stellate hairs.

Stony slopes, dry places.

49. CRUCIFERAE (Mustard Family)

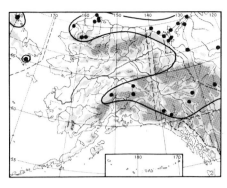 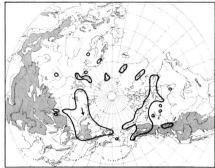 

31. Dràba cinèrea Adams
Draba arctica J. Vahl.

Loosely tufted; basal leaves oblong, entire or somewhat dentate, canescent with small, dense, stellate or forking hairs; stem pubescent from small, stellate or forked hairs, with 1 to several densely pubescent leaves; inflorescence much elongated in fruit; petals white; silicles elliptic or ovate, the upper usually longer than pedicels, densely pubescent with small, stellate hairs.

Dry places in the mountains and on tundra, scree slopes.

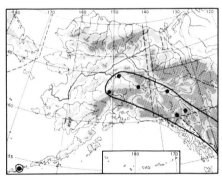 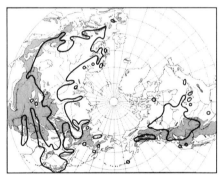 

32. Dràba nemoròsa L.

Pubescent with hispid, bifurcate or stellate hairs, sometimes also simple hairs; basal leaves soon withering, elliptic to obovate; stem leaves ovate, sessile, entire or dentate; inflorescence many-flowered, leafless, elongated in fruit; sepals with simple hairs; petals pale yellow; silicles linear-elliptic to elliptic, shorter than pedicels, glabrous or pubescent with simple hairs, with short style.

Dry slopes. Described from Sweden.

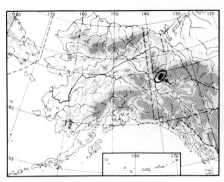
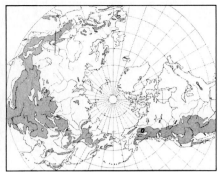

33. Dràba ogilviénsis Hult.

Draba sibirica with respect to Yukon plant, not Thell.

Loosely matted; stem branched, with long, slender, leafy branches; leaves glabrous or with very few, simple or branched hairs in margin, elliptic-lanceolate; flowering stem ascending, glabrous or with few, simple hairs, with 1 to 2 pairs of opposite or alternating leaves similar to those of sterile shoots; pedicels glabrous, divaricate in fruit; sepals ovate, 3-nerved, blunt, glabrous, scarious-margined; petals 4–6 mm long, golden yellow; silicles oblong, glabrous, reticulated, with slender beak 1 mm long.

Known only from the Ogilvie Range, from about 1,200 to 2,100 meters, in alpine meadows below snow melts.

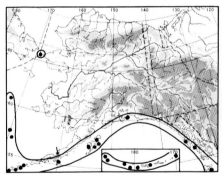
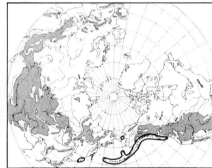

34. Dràba hyperbòrea (L). Desv.

Alyssum hyperboreum L.; *Cochlearia spathulata* Cham.; *Draba grandis* DC.; *Nesodraba grandis* (DC.) Greene.

Loosely tufted; leaves with short, branched hairs, the basal large, ligulate to obovate, grossly serrate, long-petiolated, the upper entire; stems ascending; inflorescence many-flowered; petals pale yellow, 4–5 mm long; siliques large, dark brown, elliptic to ovate, glabrous, with style about 0.5–1 mm long.

Cliffs along beaches.

49. CRUCIFERAE (Mustard Family)

Smelowskia C. A. Mey.

Siliques pyriform .... 3. *S. pyriformis*
Siliques oblong, ovate or obovate:
   Caudex mostly branched; mature siliques oblong, tapering to both sides ....
   .... 1–2. *S. calycina*
   Caudex mostly simple; mature siliques long, papery, irregularly twisted ....
   .... 4. *S. borealis*

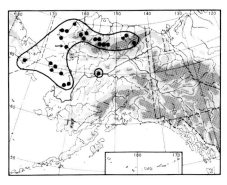 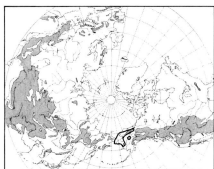 

1. Smelówskia calýcina (Steph.) C. A. Mey.
  Lepidium calycinum Steph.
subsp. **integrifòlia** (Seem.) Hult.
  Hutchinsia calycina var. *integrifolia* Seem.; *Smelowskia calycina* var. *integrifolia* (Seem.) Rollins.

Root thick, woody, with branched, many-headed caudex; basal part of stems covered with bases of old leaves; basal leaves oblong to linear, entire or with blunt, shallow teeth at apex, whitish-puberulent with short, branched hairs; inflorescence at first capitate, elongated in age; sepals oblong; petals yellowish; siliques oblong, with style 0.5 mm long.
  Rocky hillsides, gravel.
  The plant with entire, linear leaves is subsp. **integrifòlia** var. **Porsíldii** (Drury & Rollins) Hult. (*S. calycina* var. *Porsildii* Drury & Rollins).

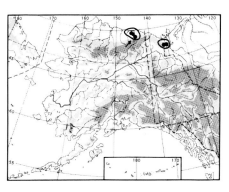 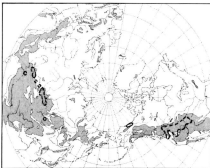 

2. Smelówskia calýcina (Steph.) C. A. Mey.
subsp. **calýcina**
var. **mèdia** (Drury & Rollins) Hult.
  Smelowskia calycina var. *media* Drury & Rollins.

Differs from subsp. *integrifolia* in having pinnately lobed basal and cauline leaves. Rocky hillsides, gravel. *S. calycina* described from Altai.
Very closely related to the Asiatic and American var. **calýcina**; broken line on circumpolar map indicates range of var. *calycina*.

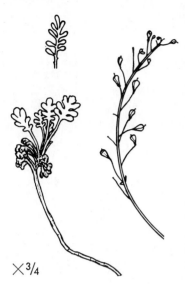
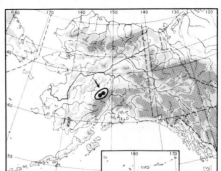
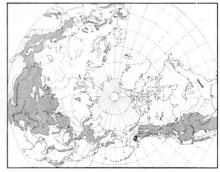

3. Smelówskia pyrifórmis Drury & Rollins

Root long, fibrous; caudex covered with withered, old leaves; basal leaves broadly oblong to ovate, deeply pinnately divided, whitish, with branched hairs; cauline leaves ovate-spatulate to linear; stem much branched; sepals oval, sparsely hirsute; petals white or cream-colored; siliques pyriform, reticulate, with style 0.5 mm long.

Rock slides in high mountains.

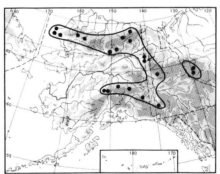
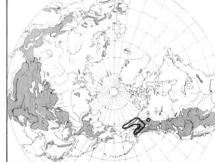

4. Smelówskia boreàlis (Greene) Drury & Rollins

Melanidion boreale Greene; *Acroschizocarpus Kolianus* Gombócz; *Ermania borealis* (Greene) Hult.

Plant with long, branched, fibrous root; caudex mostly simple, covered by withered leaves; basal leaves oblong, ovate to cuneate, entire or mostly 3-toothed in apex, hirsute and densely canescent with short, branched hairs; stems decumbent, with small, oblong or linear leaves; sepals oblong to ovate, usually purplish, scarious-margined, hirsute; petals spatulate, nearly twice as long as sepals, lavender or cream-colored; siliques at first ovate or obovate, in age papery, long, nearly falcate, often broader in apex and irregularly twisted, with slender style.

Rock slides in high mountains, to at least 2,000 meters; prefers calcareous rocks. Four varieties have been described by Drury and Rollins: var. **Jordalii**, with caducous calyx and style shorter than 0.5 mm.; var. **villòsa**, with persistent calyx petals 4–6 mm long, style longer than 0.5 mm, and white tomentose leaves; var. **boreàlis**, described above, differing from var. *villosa* in having fewer, predominantly white-villous leaves, petals 3–4.5 mm, and uninflated siliques; and var. **Koliàna**, differing from var. *borealis* in having membranaceous inflated siliques. To some extent, at least, these varieties represent different stages of development.

49. CRUCIFERAE (Mustard Family)

Descurainia Webb & Berthelot (Tansy Mustard)

Leaves and stem with glandular hairs . . . . 2. *D. sophioides*
Leaves and stem with stellate hairs:
   Pedicels short, strongly ascending; siliques with 4–8 seeds in each loculus . . . . . .
   . . . . 3. *D. Richardsonii*
   Pedicels longer, spreading:
      Siliques long, curved, with 10–20 seeds in each loculus . . . . 1. *D. sophia*
      Siliques short; style obsolete . . . . 4. *D. pinnata* subsp. *filipes*

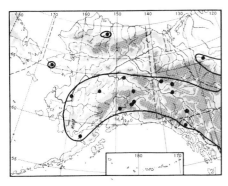
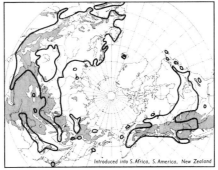

1. Descurainia sòphia (L.) Prantl
Sisymbrium sophia L.

Stem branched above, stellate-pubescent, especially below; leaves 2–3-pinnate, with 5 linear or oblanceolate segments; pedicels divaricate, capillary; petals greenish-yellow; siliques about 1 mm thick, mostly arcuate, with 10–20 seeds in each loculus.

Waste places, roadsides; an introduced weed. Described from Europe.

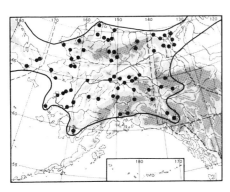
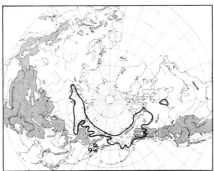

2. Descurainia sophioìdes (Fisch.) O. E. Schulz
Sisymbrium sophioides Fisch.

Very similar to *D. sophia*, but more or less densely glandular-pubescent with capitate glands; upper leaves sometimes simply pinnate; siliques sometimes slightly thicker.

Gravel bars, disturbed soil; native. Described from Hudson Bay to Alaska.

Specimens with both stellate and glandular hairs, that occur especially around villages, are considered to be the hybrid *D. sophia* × *sophioides*.

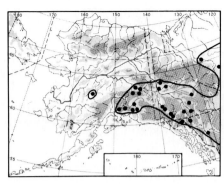
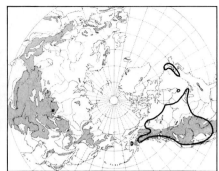

3. Descuraìnia Richardsònii (Sweet) O. E. Schulz
Sisymbrium Richardsonii Sweet.

Canescent from stellate hairs; basal leaves bipinnate or tripinnate or pinnatifid, upper leaves less dissected; fruiting pedicels strongly ascending; pedicels short; siliques short, with 4–8 seeds in each loculus.

Mountain slopes, roadsides. Described from North America.

Circumpolar map indicates range of the entire species complex.

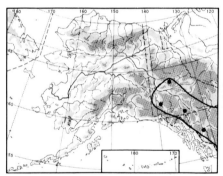
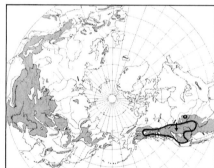

4. Descuraìnia pinnàta (Walt.) Britt.
Erysimum pinnatum Walt.
subsp. fílipes (Gray) Detling
Sisymbrium incisum Engelm. var. *filipes* Gray.

Sparsely stellate-pubescent; similar to *D. Richardsonii*, but fruiting pedicels horizontally divergent; siliques sometimes clavate, biseriate, but in subsp. *filipes* often uniseriate; style obsolete.

Disturbed soil.

49. CRUCIFERAE (Mustard Family)

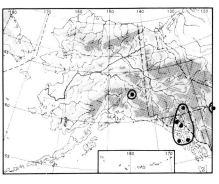
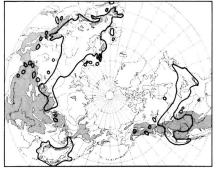

Turritis L.

1. Turrìtis glàbra L.   Tower Mustard
Arabis glabra (L.) Bernh.

Bluish-green stem usually unbranched, erect, stout, hirsute at base; basal leaves spatulate to oblanceolate, entire or repand-dentate; cauline leaves imbricate below, sagittate-amplexicaul; petals cream-colored; siliques slender, terete, stiffly erect, with short, thick style.

Waste places; an introduced weed. Described from Europe.

×⅓

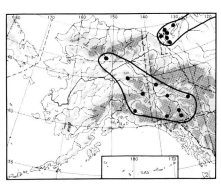
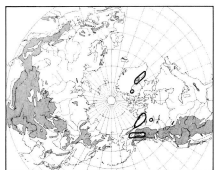
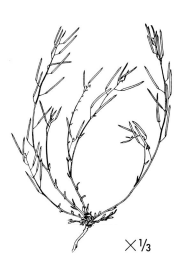

Halimolobus Tausch

1. Halimolòbus móllis (Hook.) Rollins
Turritis mollis Hook.; *Arabis Hookeri* Lange; *Arabidopsis mollis* (Hook.) O. E. Schulz.

Stem more or less branched, somewhat pubescent; rosette leaves wither early; lanceolate, more or less toothed, pubescent with branched hairs; stem leaves clasping, sagittate; pedicels spreading; sepals scarious-margined, pubescent, reddish; petals white, twice as long as sepals; siliques glabrous, linear, with fairly long style.

Very dry slopes. Described from the Northwest Territories.

×⅓

Arabis L. (Rock Cress)

Stem leaves attenuate at base, not square, or auriculate, or clasping:
  Flowers about 5 mm long; young rosulate leaves lyrate .... 2. *A. lyrata* subsp. *kamchatica*
  Flowers smaller, about 4 mm long; young rosulate leaves spatulate, dentate or entire .... 1. *A. arenicola*
Stem leaves with square base, or auriculate, or clasping:
  Basal leaves densely covered with minute, stellate hairs:
    Plant low-growing, with several stems; pedicels divaricate .... 4. *A. Lemmoni*
    Plant tall, with single stem; pedicels reflexed, appressed to stem .... 10. *A. Holboellii*
  Basal leaves hirsute or strigose, with simple, branched or sometimes also stellate hairs, or glabrous:
    Outer sepals saccate; basal leaves broad and obtuse; seeds wingless:
      Stem leaves not auriculate; plant glabrous or sparsely hirsute .... 3. *A. Nuttallii*
      Stem leaves auriculate; plant strongly hirsute at base:
        Petals 5–9 mm long, white or pinkish; stem leaves remote .... 6. *A. hirsuta* subsp. *Eschscholtziana*
        Petals 3–5 mm long, white or yellowish-white; stem leaves more approximate .... 7. *A. hirsuta* subsp. *pycnocarpa*
    Outer sepals not saccate; basal leaves more or less acute; seeds winged:
      Plant small, with branched caudex; basal leaves glabrous or sparsely pubescent, narrow, lanceolate .... 5. *A. Lyallii*
      Plant tall, caudex not branched; basal leaves strigose or pubescent, sometimes glabrous, oblanceolate:
        Siliques erect, broad, obtuse, biseriate; flowers white; basal leaves strigose, with coarse stellate hairs, or glabrous .... 8. *A. Drummondii*
        Siliques ascending or divaricate, somewhat acute, narrow, uniseriate; flowers pinkish; basal leaves sparsely pubescent, with coarse, appressed, branched hairs .... 9. *A. divaricarpa*

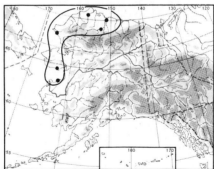 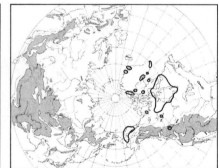

× ½

1. Árabis arenícola (Richards.) Gelert

Eutrema arenicola Richards.; *Arabis humifusa* S. Wats.

var. **pubéscens** (S. Wats.) Gelert

Arabis humifusa var. *pubescens* S. Wats.

Strongly branched; branches decumbent or ascending, pubescent at base; rosette leaves spatulate, entire or grossly dentate, somewhat pubescent with few simple hairs at base; flowers small; sepals reddish, scarious-margined; petals twice as long as sepals, white or reddish; siliques linear, with distinct style, erect to divaricate.

Sandy places. Described from shores of arctic America.

49. CRUCIFERAE (Mustard Family)

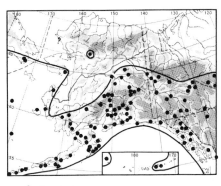
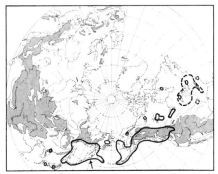

2. Árabis lyràta L.
subsp. **kamchática** (Fisch.) Hult.

Arabis lyrata var. *kamchatica* Fisch.; *A. kamchatica* Fisch.; *Cardamine lyrata* (L.) Hiitonen.

Stem more or less branched from base, glabrous or somewhat pubescent with simple or forked hairs; rosette leaves lyrate, pinnatifid; stem leaves pinnatifid, spatulate or linear, glabrous or slightly pubescent; petals white or pinkish; siliques linear, glabrous, with short style.

Moist, stony places, scree slopes; common along the southern coast. *A. lyrata* described from Canada.

The young leaves have a radish flavor and are eaten raw or boiled by the natives. Broken line on circumpolar map indicates range of subsp. **lyràta**.

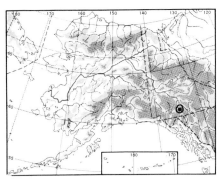
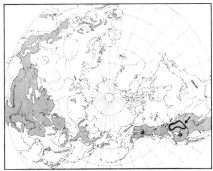

3. Árabis Nuttállii Robins.

Stems several, from branching caudex, glabrous above, hirsute with long, simple or forked hairs below; basal leaves spatulate to obovate, ciliate, hirsute on both sides to glabrous; cauline leaves sessile, oblong, not auriculate; pedicels glabrous, divaricately ascending; outer pair of sepals saccate; petals white; siliques glabrous, with slender style 1 mm long; seeds wingless.

Found introduced at Whitehorse.

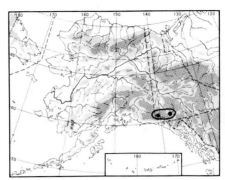
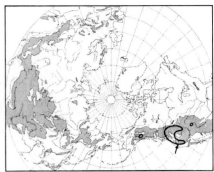

4. Árabis Lemmòni S. Wats.

Caespitose, with branched caudex; stems several, more or less pubescent at base with simple and branched hairs; basal leaves rosulate, oblanceolate, entire or remotely toothed, with short petioles, densely pubescent with stellate or branched, short hairs; cauline leaves oblong, sessile, more or less auriculate; sepals elliptic, with violet, scarious margin; petals violet to purple; siliques glabrous, spreading, almost without style; seeds with wing 0.2 mm broad.

Alpine meadows at about 2,000 meters.

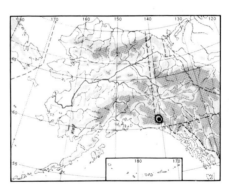
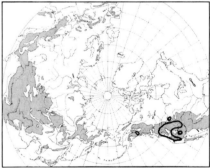

5. Árabis Lyállii S. Wats.

Arabis Drummondii var. *Lyallii* Jepson; *A. Drummondii* var. *alpina* S. Wats.

Caespitose, with branched caudex; stems simple or several, glabrous; basal leaves numerous, fleshy, lanceolate, glabrous or sparsely pubescent with short, branched hairs, ciliate with simple hairs at base; cauline leaves several, linear-lanceolate; sepals obovate, glabrous; petals purplish, twice as long as sepals; siliques glabrous, linear, with short, thick style; seeds with narrow wing.

Alpine meadows and ridges. Described from "Fort Colville to the Rocky Mountains" (Ashnola River, Cascade Mountains, Washington).

49. Cruciferae (Mustard Family)

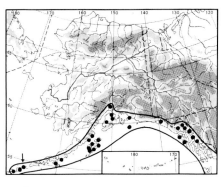
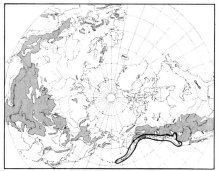

6. Árabis hirsùta (L.) Scop.
Turritis hirsuta L.
subsp. **Eschscholtziàna** (Andrz.) Hult.
Arabis Eschscholtziana Andrz.; *A. hirsuta* var. *Eschscholtziana* (Andrz.) Rollins.

Stem simple, hirsute with simple or forked hairs; basal leaves obovate to oblanceolate, coarsely dentate, more or less pubescent with simple or forked hairs; cauline leaves comparatively remote, several, ovate to oblong, somewhat acute, auriculate, dentate; outer sepals saccate; petals white or pinkish, 5–9 mm long; mature siliques erect to spreading, often with bifid stigma.

Rocky slopes.

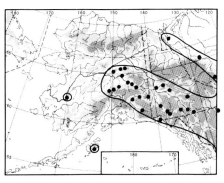
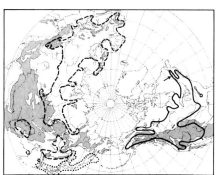
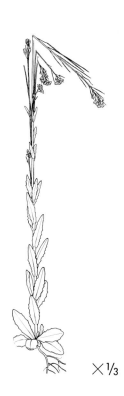

7. Árabis hirsùta (L.) Scop.
Turritis hirsuta L.
subsp. **pycnocárpa** (M. Hopkins) Hult.
Arabis pycnocarpa M. Hopkins; *A. hirsuta* var. *pycnocarpa* (M. Hopkins) Rollins.

Similar to subsp. *Eschscholtziana*, but flowers smaller; petals 2–5 mm long; flowers white or creamy-white, and siliques strictly erect.

Dry, rocky places. *A. hirsuta* described from Sweden, Germany, and England.

Broken line on circumpolar map indicates range of subsp. **hirsùta**; dotted line, range of subsp. **Stélleri** (DC.) Hult. (*A. Stelleri* DC.).

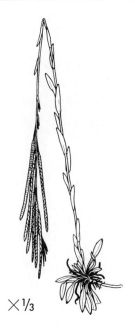
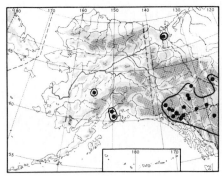
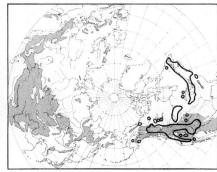

8. Árabis Drummóndii Gray

Stems single or several, from short caudex, glabrous or sparingly appressed-pubescent at base; basal leaves oblanceolate, entire or dentate, somewhat or decidedly acute, glabrous or pubescent, with short, appressed, simple or 2–3 forked hairs; cauline leaves oblong to lanceolate, acute, auriculate; pedicels glabrous, erect; sepals not saccate; petals twice as long as sepals, white or pinkish; siliques broad, biseriate, glabrous; style obsolete.

Dry, rocky slopes. Described from the Rocky Mountains.

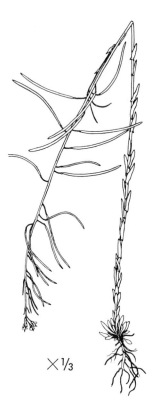
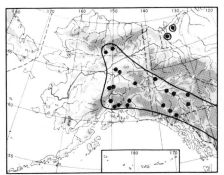
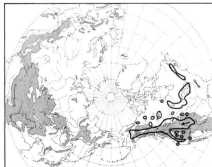

9. Árabis divaricárpa Nels.

Stem simple or branched, appressed-pubescent with forking hairs at base; basal leaves oblanceolate to spatulate, acute, dentate, stellate-pubescent; cauline leaves oblong to linear-lanceolate, entire, strongly ascending, with sagittate base, glabrous; petals pink or purplish; siliques glabrous, divergent; style obsolete.

Dry slopes.

49. CRUCIFERAE (Mustard Family)

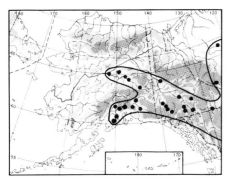
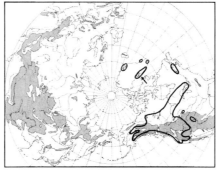

×⅓

10. Árabis Holboéllii Hornem.

Stem simple, in lower part densely pubescent with fine, appressed, stellate hairs in the upper, glabrous; basal leaves entire, somewhat acute, canescent with fine, appressed, stellate hairs; pedicels strongly reflexed; petals pinkish; siliques linear, glabrous or sometimes finely pubescent, reflexed; style lacking.

Dry places, open soil.

Most specimens belong to var. **retrofrácta** (Graham) Rydb. (*A. retrofracta* Graham), with densely pubescent to pannose basal leaves, pubescent upper cauline leaves, and narrow siliques (1–2 mm wide). Var. **pendulocárpa** (Nels.) Rollins (*A. pendulocarpa* Nels.), with basal leaves less than 3 mm broad and cauline leaves lacking auricles, occurs at Whitehorse.

Erysimum L.

Pubescence of leaves chiefly with hairs having 3 branches; petals 4–5 mm long:
    Plant annual, lacking thick root; stem simple or branched in middle part, with about 5 basal internodes and 5–10 nodes .... 2. *E. cheiranthoides* subsp. *cheiranthoides*
    Plant biennial, with thick root; stem not branched in middle part, with 8–20 basal internodes and 20–40 nodes .... 3. *E. cheiranthoides* subsp. *altum*
Pubescence of leaves chiefly with hairs having only 2 branches; petals larger:
    Petals 6–8 mm long .... 4. *E. inconspicuum*
    Petals 12–20 mm long, yellow or purple:
        Petals yellow; beak of siliques 3–5 mm long .... 5. *E. angustatum*
        Petals purple; beak of siliques shorter .... 1. *E. Pallasii*

×¾

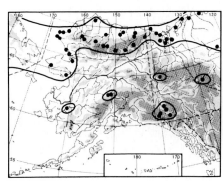
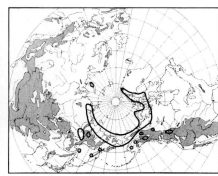

1. Erýsimum Pallàsii (Pursh) Fern.

Cheiranthus Pallasii Pursh; *C. pygmaeus* Adams; *Hesperis pygmaea* (Adams) Hook.; *Erysimum pygmaeus* (Adams) Gray; *Hesperis Hookeri* Ledeb.; *H. Pallasii* (Pursh) Seem.

Stems 1 or several, sometimes very short at flowering time, from thick root; leaves oblong-lanceolate to linear, entire or somewhat dentate, densely pubescent from 2-parted hairs; sepals ovate, narrowly scarious-margined; petals with claw about as long as sepals, and obovate blade, purplish or, rarely, yellowish (var. **ochroleùcum** Tolm., not seen in Alaska); siliques linear, extending above late flowers; style 1–3 mm long.

Stony places on tundra; in the mountains to at least 2,000 meters. Described from "Northwest coast" (of America).

×⅓

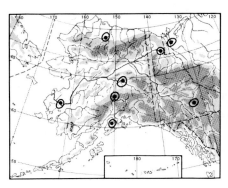
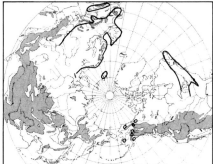

2. Erýsimum cheianthoìdes L.

Cheirinia cheiranthoides (L.) Link.

subsp. **cheianthoìdes**

Annual; stem simple or with ascending branches, with about 5 basal internodes and 5–10 nodes; leaves lanceolate to broadly lanceolate, entire in margin or nearly so, pubescent with chiefly trifid hairs; petals small, 3.5–5 mm long; pedicels yellow, filiform, pubescent with scattered trifid hairs; siliques subterete, glabrous or sparsely pubescent.

Roadsides; an introduced weed. Described from Europe.

Total range incompletely known, since subsp. *altum* (see next description) has been distinguished only recently.

49. CRUCIFERAE (Mustard Family)

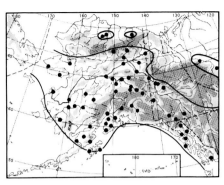
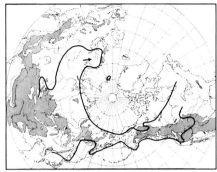

3. Erýsimum cheiranthoìdes L.

subsp. áltum Ahti

Similar to subsp. *cheiranthoides*, but biennial, taller, with 8–20 basal internodes, which in flowering specimens form a "bulb" and 20–40 nodes of the stem; stem not branched at its middle.

Grassy slopes; native. Total range incompletely known.

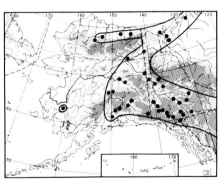
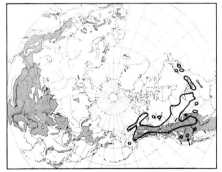

4. Erýsimum inconspícuum (S. Wats.) MacM.
Erysimum asperum var. *inconspicuum* S. Wats.

Similar to *E. cheiranthoides*, but pedicels thick and short, densely covered with 2-parted hairs; petals 3–5 mm long; pubescence from mostly 2-parted hairs.

Dry places.

× ⅓

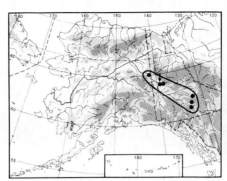
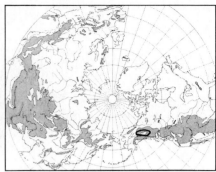

5. Erýsimum angustàtum Rydb.

Root thick, long; stems several, from many-headed caudex; basal leaves linear, acute, entire, canescent from dense, 2-parted hairs; stem leaves linear or linear-lanceolate; sepals oblong, scarious-margined; petals yellow, with claw about as long as sepals, and oblong-obovate blade; siliques linear, curved, densely pubescent, with style about 3 mm long.

Dry slopes.

× ¾

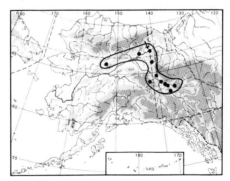
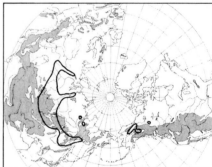

Alyssum L.

1. Alýssum americànum Greene

Alyssum biovulatum N. Busch; *A. sibiricum* of authors, not Willd.

Canescent, with very dense, stellate hairs; stem suffrutescent, tufted, from long taproot, leaves spatulate-obovate, somewhat acute; sepals ovate-lanceolate, scarious-margined at tip; petals pale yellow to yellow, with long claw; both long and short stamens with long, thin scale at base; silicles obovate to orbicular, 2-seeded, stellate-pubescent, with long, slender style.

Gravel banks.

49. CRUCIFERAE (Mustard Family)

Braya Sternb. & Hoppe

This genus presents a number of closely related taxa often widely separated geographically. Closer study of a large amount of material is needed to clarify all types occurring within the area of interest here. Besides the taxa enumerated below, a long-styled, unidentified *Braya* of *humilis* type occurs in the Nome area.

Siliques long, linear; stem with 1 or more leaves:
    Leaves pubescent, with coarse, forked hairs . . . 1. *B. humilis* subsp. *Richardsonii*
    Leaves glabrous on both sides . . . . 2. *B. humilis* subsp. *arctica*
Siliques oblong to ovoid; stem usually leafless:
    Siliques 2–3 mm thick, plump, pubescent; style 0.5 mm long, thick; stigma large
    . . . . 3. *B. purpurascens*
    Siliques 5–10 mm long, narrower:
        Siliques broader at base . . . . 5. *B. Henryae*
        Siliques not broader at base:
            Leaves, scapes, pedicels, and siliques pubescent with simple and forked, gray hairs; plant green . . . . 4. *B. pilosa*
            Leaves glabrous on sides, ciliate with short, simple or forked hairs; siliques often glabrous; plant purplish . . . . 6. *B. Bartlettiana*

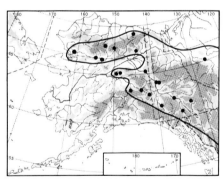
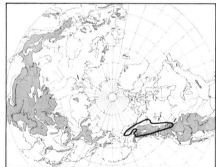
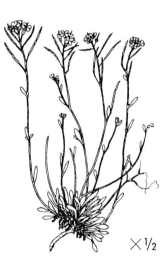

× ½

1. Braya hùmilis (C. A. Mey.) Robins.

Sisymbrium humile C. A. Mey.; *Torrularia humilis* (C. A. Mey.) O. E. Schulz.

subsp. **Richardsònii** (Rydb.) Hult.

Pilosella Richardsonii Rydb.; *Braya Richardsonii* (Rydb.) Fern.

Branched from base; basal leaves oblanceolate, often more or less toothed or sinuate, densely pubescent with coarse, forked hairs; flowers white or purplish; siliques linear, terete, torulose; style short, not markedly bilobated.

Solifluction soil, scree slopes. Described from the Rocky Mountains of Alberta and from the Mackenzie River.

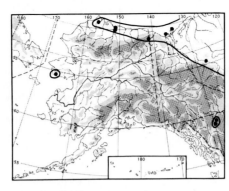 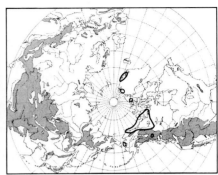

2. Braỳa hùmilis (C. A. Mey.) Robins.
subsp. árctica (Böcher) Rollins
   Torularia humilis subsp. *arctica* Böcher.

Similar to subsp. *Richardsonii*, but leaves glabrous on sides, or sparsely pubescent with scattered, simple or forked hairs; style short; siliques more or less pubescent, with bilobated stigma.

Solifluction soil, scree slopes. Described from interior northeastern Greenland.

Subsp. **hùmilis** occurs in central Asia and China; other subspecies occur in eastern North America and in Colorado.

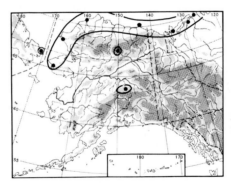 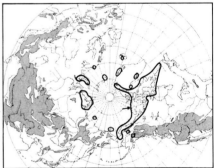

3. Braỳa purpuráscens (R. Br.) Bunge
   Platypetalum purpurascens R. Br.

Basal leaves fleshy, often purplish, oblong to linear, somewhat pubescent with simple and branched hairs, or nearly glabrous; stem coarse, mostly leafless, slightly pubescent; petals slightly emarginate, white to reddish; siliques elliptic-oblong, glabrous or nearly so, often purplish.

Solifluction soil.

49. CRUCIFERAE (Mustard Family)

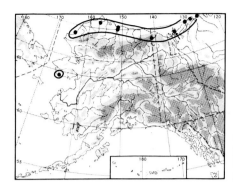
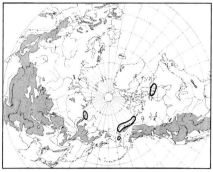
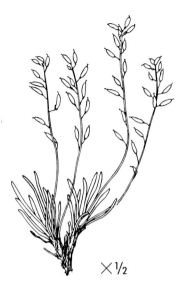

4. Braỳa pilòsa Hook.

Similar to *B. purpurascens*, but green; leaves, scapes, pedicels and siliques sparsely pubescent with simple and forked, gray hairs; inflorescence capitate in flower, much elongated in fruit; flowers large; style 1.2–1.5 mm long.

Solifluction soil; gravel bars in rivers. Flowers very early, close to melting snow or ice.

×½

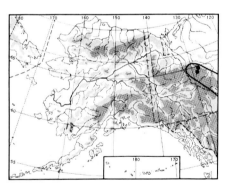
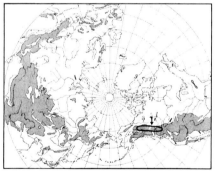

5. Braỳa Hénryae Raup

Basal leaves narrowly spatulate, gradually narrowed to petioles, glabrous on sides, ciliate in margins with simple or branched hairs; inflorescence capitate in flower, 2–5 cm long in fruit; stem leafless, pubescent with forked hairs; sepals glabrous or somewhat pubescent; petals white, purplish at base; siliques 8–12 mm long, 1–2 mm thick, lanceolate, thicker at base, pubescent with simple or branched hairs; style linear, 1–1.6 mm long.

Stony slopes.

×¾

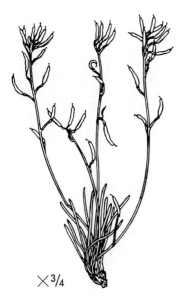
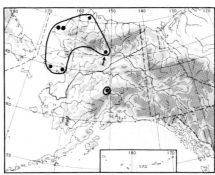
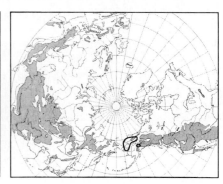

6. Braya Bartlettiàna Jordal

Braya siliquosa with respect to Alaskan plant.

Basal leaves linear-lanceolate, often purplish, glabrous on sides, sparsely ciliate with short, simple or forked hairs, longer toward base; stem leafless, thin, glabrous or sparsely pubescent above; sepals glabrous; petals small, purplish; siliques glabrous or pubescent (var. **vestìta** Hult.), oblong, acute, up to 10 mm long and 1.5 mm broad, with long (1–1.5 mm) style.

Solifluction soil.

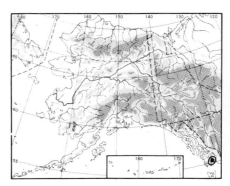
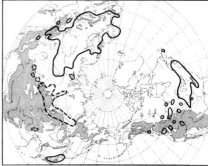

Hesperis L.

1. Hésperis matronàlis L. Dame's Violet

Pubescent with simple and branched hairs; stem tall, branched; leaves large, ovate-lanceolate, acute, slightly toothed, the lower petiolated, the upper sessile; petals about 20 mm long, reddish-violet; siliques long, glabrous, with short style.

Escaped from cultivation; known in this region from a single specimen. Described from Italy.

Broken line on circumpolar map indicates range of the related **H. sibìrica** L.

The flowers become more fragrant at night.

49. CRUCIFERAE (Mustard Family)

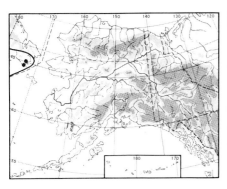

Christolia Cambess.

1. Christòlia parryoìdes (Cham.) N. Busch

Draba parryoides Cham.; *Ermania parryoides* Cham. (illegitimate alternative name); *Parrya Ermani* Ledeb.

Plant with long, subterranean runners, covered at apex with remains of withered leaves; basal leaves in dense rosette, cuneate to flabellate, 3–6 lobed, densely canescent with short, branched hairs; stem leaves small; sepals oblong, scarious-margined; petals twice as long as sepals, yellowish; siliques lanceolate-oblong, membranaceous, with very short style.

Rocky slopes.

×2/3

Parrya R. Br.

Plant glabrous . . . . 3. *P. nudicaulis* subsp. *septentrionalis*
Leaf margins, stems, and pedicels glandular:
   Leaves ovate-oblong or spatulate, sinuately toothed . . . .
   . . . . 1. *P. nudicaulis* subsp. *nudicaulis*
   Leaves lanceolate to broadly lanceolate, entire in margin or nearly so . . . .
   . . . . 2. *P. nudicaulis* subsp. *interior*

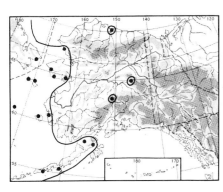
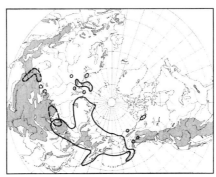

1. Párrya nudicaùlis (L.) Regel

Cardamine nudicaulis L.

subsp. **nudicaùlis**

Plant with stout root; caudex with remains of old leaves; stem leafless, pubescent with stout glands; leaves all basal, ovate, oblong to spatulate, sinuate-dentate, with long petiole, usually glabrous on sides, glandular-ciliate in margin; sepals ovate, scarious-margined; petals purple, lavender, or rarely white, cordate to obovate, emarginate; siliques acute at both ends, with long style; seeds broadly winged.

Moist places, sandy slopes. Described from Siberia.

Var. **grandiflòra** Hult. (McKinley Park, Eagle Summit) has petals about 2 cm long, with long, narrow claw and broad, subcordate lamina.

Broken line on circumpolar map indicates range of subsp. **turkestànica** (Korsh.) Hult. (*P. turkestanica*) Korsh.

×1/3

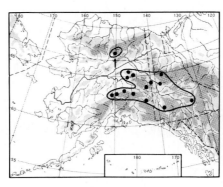
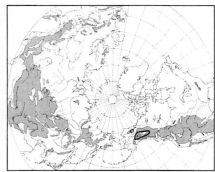

2. Párrya nudicaùlis (L.) Regel subsp. intèrior Hult.

Similar to subsp. *nudicaulis*, but glandular pubescence less prominent; leaves lanceolate to broadly lanceolate, entire or nearly entire in margin.

Moist places, sandy slopes.

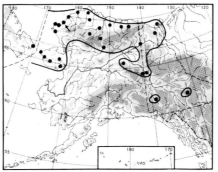
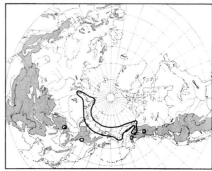

3. Párrya nudicaùlis (L.) Regel subsp. septentrionàlis Hult.

Similar to subsp. *interior*, but completely glabrous.

Moist places, sandy slopes.

High arctic specimens are small and stunted, with narrow leaves; Alaskan specimens are larger.

50. DROSERACEAE (Sundew Family)

Drosera L. (Sundew)

Leaf blades obovate to linear, spatulate; petioles glabrous; stem straight, longer
than leaves .... 1. *D. anglica*
Leaf blades round; petioles pubescent .... 2. *D. rotundifolia*

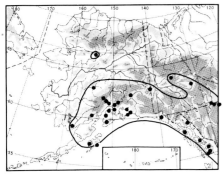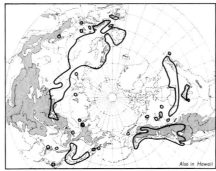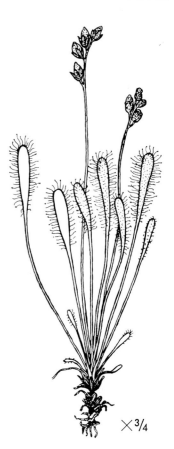

1. Dròsera ánglica Huds.

Leaf blades ascending, linear-spatulate to cuneate-obovate; scapes straight, erect, emerging from center of rosette leaves; petals white; stamens 5, styles 3.

Peatbogs. Described from England.

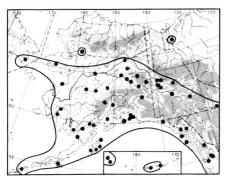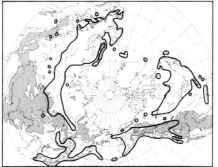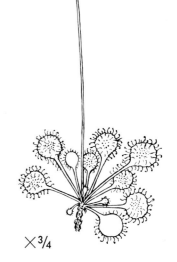

2. Dròsera rotundifòlia L.

Leaf blades lying flat on ground, suborbicular, sometimes broader than long, abruptly narrowed into slender, pilose petiole; scapes straight, erect; petals 5, white; stamens 5; styles 3.

Peatbogs; in McKinley Park to about 400 meters. Described from Europe, Asia, and North America.

Some specimens belong to the small-leaved var. **gràcilis** Laestad. Hybrids with *D. anglica* occur (*D. obovata* Mert. & Koch).

51. CRASSULACEAE (Stonecrop Family)

Sedum L. (Stonecrop)

Leaves linear to linear-lanceolate, subterete; petals yellow ...... 1. *S. lanceolatum*
Leaves broad, flat:
    Petals yellow, united at base .... 2. *S. oreganum*
    Petals purple or pink, not united at base .... 3. *S. rosea* subsp. *integrifolium*

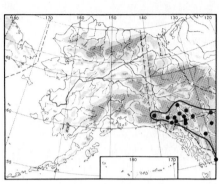
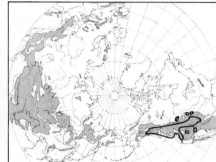

1. Sèdum lanceolàtum Torr.

Sedum stenopetalum of most authors, not Pursh.

Plant with slender, branching rootstock; leaves sessile, linear, subterete, glaucous or dull green, not becoming scarious before falling; flowers in racemelike cymes; calyx lobes lanceolate, acuminate; petals yellow, narrowly lanceolate; follicles with divergent tips.

Rocky places; in the St. Elias Mountains to 2,000 meters. Described from Rocky Mountains.

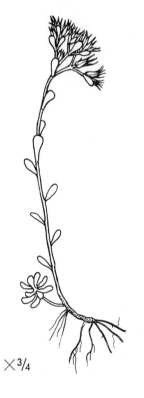
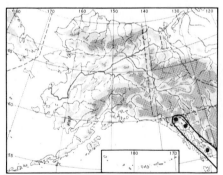
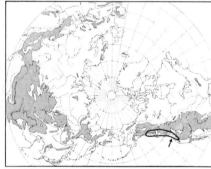

2. Sèdum oregànum Nutt.

Glabrous, glaucous; leaves obovate to spatulate-oblanceolate, succulent; stem rhizomatous, with sterile rosettes and erect flowering stems; calyx lobes ovate-lanceolate; petals yellow, long-acuminate, united for 1.3–3 mm at base.

Rocky ridges and slopes. Described from the mouth of the "Oregon" River.

51. CRASSULACEAE (Stonecrop Family)

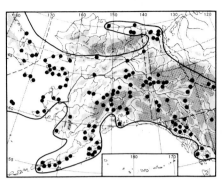
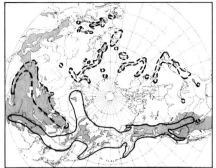

3. Sèdum ròsea (L.) Scop. Roseroot
Rhodiola rosea L.
subsp. **integrifòlium** (Raf.) Hult.
Rhodiola integrifolia Raf.; *Sedum atropurpureum* Turcz.; *R. atropurpurea* (Turcz.) Trautv.; *S. frigidum* Rydb.; *R. alaskana* Rose.

Rhizome thick, fleshy, scaly, fragrant when cut; leaves ovate to oblong, dentate, glabrous and glaucous; stem with many leaves; flowers usually 4-merous, unisexual; male flowers with abortive carpels and 8 stamens; petals usually purple, rarely pink (Ogilvie Range) or yellow (Seward Peninsula).

Scree slopes, rocky places; in the mountains to at least 2,135 meters. *S. rosea* described from the Alps of Lapland, Austria, Switzerland, and Britain; subsp. *integrifolium* from the Rocky Mountains. Broken line on circumpolar map indicates range of other subspecies or varieties.

Leaves and young shoots are eaten raw or boiled by the natives. The Siberian Eskimo also eat the rhizome boiled in seal fat, or with reindeer fat.

(See color section.)

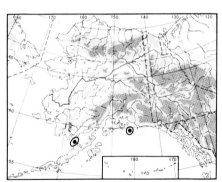
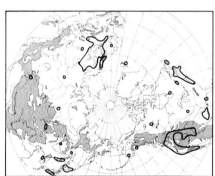

Crassula L.

1. Cràssula aquática (L.) Schönl. Pygmyweed
Tillaea aquatica L.; *Bulliarda aquatica* (L.) DC.; *Tillaeastrum aquaticum* (L.) Britt.

Annual, succulent, very tiny, tufted or matted; stem branched; leaves opposite, connate, linear, entire; flowers axillary, 3-merous or 4-merous, short-pedicellated; petals greenish-white or reddish.

Inundated shores; often overlooked because of its small size. Described from Europe.

52. SAXIFRAGACEAE (Saxifrage Family)

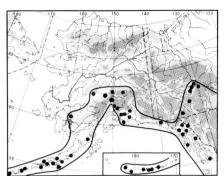
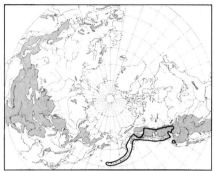

Leptarrhena R. Br.

1. Leptarrhèna pyrolifòlia (D. Don) Ser.   Leatherleaved Saxifrage

Saxifraga pyrolifolia D. Don; *S. amplexifolia* Sternb.; *Leptarrhena amplexifolia* (Sternb.) R. Br.

Rootstock long, thick, horizontal, often with small leaves; basal leaves ovate, dark green and shiny above, glaucous when young, later brownish or cinnamon-brown beneath, ovate, obtuse, serrate in margin, long-petiolated; stem glabrous or sparsely glandular, with 1 to 2 leaves; inflorescence glandular; sepals ovate-oblong; petals linear, obtuse, white, twice as long as sepals; follicles reddish, divergent in apex; style about 0.5 mm long.

Wet places, along creeks, in the mountains to at least 1,500 meters. Type locality doubtful. Reports from Siberia apparently incorrect.

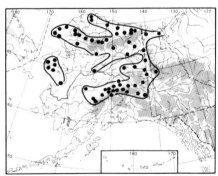
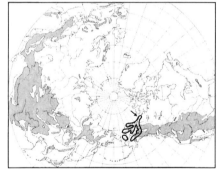

Boykinia Nutt.

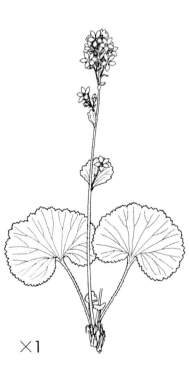

1. Boykínia Richardsònii (Hook.) Gray   Alaska Boykinia

Saxifraga Richardsonii Hook.; *S. Nelsoniana* Hook. & Arn.; *Therofon Richardsonii* (Hook.) Ktze.

Stems with dark-brown, capitate glands, from very thick caudex, with dark-brown roots and dark-brown leaf bases, basal leaves reniform to suborbicular-cordate; acutely dentate to shallowly lobed, glabrous above, with scattered coarse hairs on nerves below, ciliate with coarse, capitate glands in margin; stem leaves small, the upper lanceolate; sepals ovate-lanceolate, acute, glandular; petals elliptic-lanceolate to obovate, acute, strongly nerved, white or pink, especially at base; stamens 5; follicles divergent, with long, slender beak and capitate stigma.

Subalpine forests, tundra meadows, along creeks, to at least 1,800 meters. Described from the Arctic shore between the Mackenzie and Coppermine Rivers.

A very showy endemic, almost certainly a relict from the late Tertiary. (See color section.)

52. SAXIFRAGACEAE (Saxifrage Family)

Saxifraga L.

Leaves opposite; plant densely matted; flowers purple:
    Leaves with several cilia, not densely imbricate .... 1. *S. oppositifolia* subsp. *oppositifolia*
    Leaves with few cilia, densely imbricate, in 4 rows .... 2. *S. oppositifolia* subsp. *Smalliana*
Leaves not opposite; flowers white, yellow, purple-spotted, or purplish:
    Leaves entire, oblong, obovate, oblanceolate or linear, not toothed:
        Plant pulvinate; flowering stem very short, barely rising above leaves:
            Leaves glabrous, not ciliate .... 3. *S. aleutica*
            Leaves with long, white, pectinate cilia .... 4. *S. Eschscholtzii*
        Plant not pulvinate; flowers on elongated stems:
            Plant scapose, or flowering stems with 1 to 2 leaves; low-growing plant, rarely exceeding 5 cm in height:
                Flowers yellow; filaments subulate .... 5. *S. serpyllifolia*
                Flowers white; filaments clavate .... 6. *S. Tolmiei*
            Plant with several stem leaves, usually taller:
                Leaves not ciliate; flowers yellow:
                    Sepals reflexed, pubescent .... 7. *S. aizoides*
                    Sepals erect, glabrous .... 8. *S. hirculus*
                Leaves setose-ciliate; flowers yellow or white:
                    Petals yellow; plant with long stolons:
                        Calyx tube flat .... 9. *S. flagellaris* subsp. *setigera*
                        Calyx tube turbinate .... 10. *S. flagellaris* subsp. *platysepala*
                  Petals white, with yellow or yellow and purplish spots:
                    Leaves not densely packed ..... 11. *S. bronchialis* subsp. *Funstonii*
                    Leaves in dense, spherical or sausage-shaped clusters .... 12. *S. bronchialis* subsp. *cherlerioides*
    Leaves toothed or lobed, rounded, flabelliform, cuneate, oblong or oblanceolate:
■ Basal leaves orbicular or reniform, usually as broad as long or broader:
    ● Cauline leaves lacking or few, bractlike:
        Leaves with several shallow 3-toothed lobes .... 14. *S. Mertensiana*
        Leaves toothed:
            Flowers yellowish, in narrow, spikelike panicles; plant tall .... 20. *S. spicata*
            Flowers white or purplish, in capitate or corymb-like panicles; plant smaller:
                ▲ Calyx lobes erect; leaves small, 1–1.5 cm broad; basal leaves with distinct, broad, glandular stipules .... 23. *S. nudicaulis*

▲Calyx lobes soon reflexed; leaves larger; basal leaves lacking stipules:
　Leaves pubescent on both sides; branches of panicle with short, glandular-tipped hairs .... 15. *S. punctata* subsp. *Nelsoniana*
　Leaves glabrous on both sides; branches of panicle with long, septate hairs:
　　Largest leaves with 6–12 teeth; capsules 6–12 mm long .... 19. *S. punctata* subsp. *Charlottae*
　　Largest leaves with 12–18 teeth; capsules 3–8 mm long:
　　　Leaves fleshy .... 16. *S. punctata* subsp. *insularis*
　　　Leaves thin:
　　　　Largest leaves 2.3–7.5 cm wide .... 17. *S. punctata* subsp. *pacifica*
　　　　Largest leaves 1.5–3.8 cm wide .... 18. *S. punctata* subsp. *Porsildiana*
● Cauline leaves on flowering stems normal:
　Bulblets in axils of stem leaves; only terminal flower developed .... 21. *S. cernua*
　Bulblets lacking in axils of stem leaves; lateral flowers also developed:
　　Petals about 1 cm long .... 22. *S. exilis*
　　Petals shorter:
　　　Plant stout; leaves 5–8-lobed; bracts 3–6-lobed; bulblets at base of stem .... 24. *S. bracteata*
　　　Plant more slender; leaves 3–5-lobed; bracts entire, or 2–3-lobed; no bulblets at base of stem .... 25–26. *S. rivularis*
■Basal leaves ovate, oblong, flabelliform, cuneate, oblanceolate or lobed, usually longer than broad (in *S. Lyallii* sometimes nearly orbicular):
　Leaves 3–5-lobed .... 37. *S. caespitosa*
　Leaves merely toothed:
　　Basal leaves cuneate-oblong or cuneate-oblanceolate:
　　　Leaves stiff, pungent, with 3 acute teeth at apex ... 13. *S. tricuspidata*
　　　Leaves not stiff and pungent, with several teeth:
　　　　Basal rosette lacking .... 36. *S. adscendens* subsp. *oregonensis*
　　　　Basal rosette present:
　　　　　Inflorescence with long, ascending branches; petals lanceolate or oblong-lanceolate .... 33. *S. ferruginea*
　　　　　Inflorescence with short, spreading rigid branches; petals abruptly narrowed to claw:
　　　　　　Only terminal flower developed ... 34. *S. foliolosa* var. *foliolosa*
　　　　　　Flowers also on lateral branches; plant taller .... 35. *S. foliolosa* var. *multiflora*
　　Basal leaves flabellate, ovate, or cuneate-obovate:
　　　Basal leaves flabellate or cuneate-obovate:
　　　　Filaments clavate or broadest at middle; stem with scattered, short, capitate glands; petals obovate or rounded, with distinct claw .... 27. *S. Lyallii* subsp. *Hultenii*
　　　　Filament subulate; stem with septate, not distinctly capitate hairs; petals usually narrower, not so distinctly clawed:
　　　　　Branches of inflorescence short, thick; follicles plump, ovoid; stigma large; petals oval or ovate; leaves obovate, with broad, short, petiole-like base .... 29. *S. unalaschcensis*
　　　　　Branches of inflorescence longer, thinner; follicles narrower, ovate-lanceolate, nearly clawless .... 28. *S. davurica* subsp. *grandipetala*
　　　Basal leaves oval or ovate, crenate-dentate or repand-denticulate:
　　　　Flowers in spikelike panicle; leaf margin repand-denticulate .... 32. *S. hieracifolia*
　　　　Flowers paniculate or more or less condensed in terminal cluster:
　　　　　Leaves pubescent on both sides .... 31. *S. reflexa*
　　　　　Leaves glabrous above, more or less rufous-pubescent below, at least on midvein:
　　　　　　Leaves sparsely rufous-pubescent below; flowers white or reddish .... 30. *S. nivalis*
　　　　　　Leaves densely rufous-pubescent below; flowers purple .... See 30. *S. nivalis* var. *rufopilosa*

52. SAXIFRAGACEAE (Saxifrage Family)

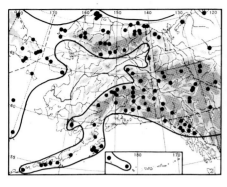 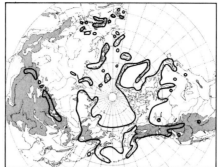 

1. Saxífraga oppositifòlia L.   Purple Mountain Saxifrage

Antiphylla oppositifolia (L.) Fourr.

subsp. **oppositifòlia**

Matted; stem much branched, densely covered with opposite leaves in 4 rows; leaves oblong to obovate to broadly obovate, keeled, glabrous, ciliate with few coarse cilia; flowers sessile or nearly so; calyx lobes ovate, blunt, ciliated; petals reddish-violet, obovate, clawed, 2–2½ times as long as calyx lobes; pedicels elongated in fruit; follicles with long style.

Wet, stony slopes, heaths, ridges, rock crevices, solifluction soil; in the mountains to at least 2,160 meters in McKinley Park. Described from Spitzbergen, Lapland, Switzerland, the Pyrenees.

Flowers very early.

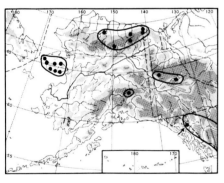 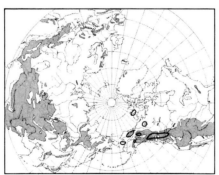 

2. Saxífraga oppositifòlia L.

subsp. **Smalliàna** (Engler & Irmsch.) Hult.

S. pulvinata Small; *Antiphylla pulvinata* (Small) Small; *S. oppositifolia* var. *typica* subvar. *Smalliana* Engler & Irmsch.

Similar to subsp. *oppositifolia*, but leaves densely imbricate in 4 rows; leaves in extreme specimens with no cilia or only 2–4; flowers dark-colored; fruit with long peduncles.

Calcareous rocks. Only typical specimens are represented on the map.

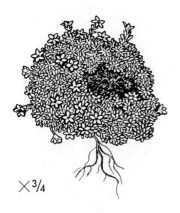
×3/4

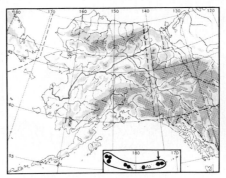
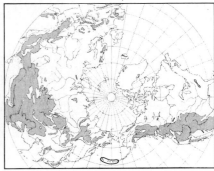

3. Saxífraga aleùtica Hult.   Aleut Saxifrage

Densely caespitose to matted; leaves succulent, spatulate, glabrous and nerveless, densely compressed at ends of branches; flowering stem very short, glandular, exceeded by leaves; sepals ovate-triangular, obtuse, glandular in margin; petals elliptic to spatulate, about as long as calyx lobes, greenish-yellow; carpels ovate, with very short style and large stigmas.

High mountains of the Aleutian Islands.

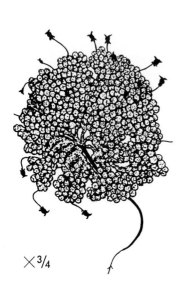
×3/4

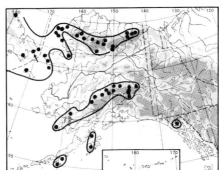
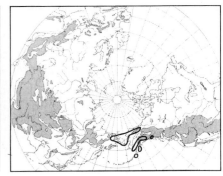

4. Saxífraga Eschschòltzii Sternb.   Cushion Saxifrage

Saxifraga fimbriata D. Don; *Leptasea fimbriata* (D. Don) Small.

Pulvinate, forming rounded, compact cushions; spherical shoots with obovate, glabrous leaves, ciliate in margin, with large hyaline cilia, broad at base; flowering stem very short, up to 10 mm long, slender, leafless; flowers more or less unisexual, the female with large purplish carpels, divergent in apex, and stamens with small anthers, the male with abortive carpels and large yellow anthers; sepals ciliate, reflexed in fruit; petals very narrow, yellow.

Crevices in rocks, ledges, calcareous gravel, in the mountains to over 2,000 meters.

52. SAXIFRAGACEAE (Saxifrage Family)

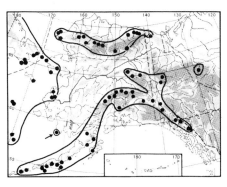
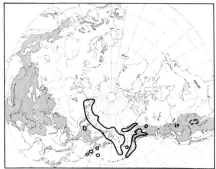

5. Saxífraga serpyllifòlia Pursh
Leptasia serpyllifolia (Pursh) Small.

Loosely tufted, with numerous sterile shoots; basal leaves shiny, fleshy, spatulate, blunt, glabrous, with reflexed margin; flowering stems with few, linear leaves, glandular-pubescent; calyx lobes broadly ovate, often purplish, reflexed in fruit; petals bright yellow, twice as long as calyx lobes or longer; filaments subular, shorter than petals.

Rock slides, dry places in the mountains to at least 2,200 meters.

Var. **purpùrea** Hult. (at False Pass, Unimak Island), has purple petals.

Broken line on circumpolar map indicates range of the closely related **S. chrysántha** Gray.

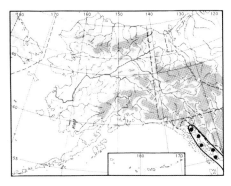

6. Saxífraga Tolmièi Torr. & Gray   Alpine Saxifrage
Leptasia Tolmiei (Torr. & Gray) Small.

Matted with numerous sterile shoots; leaves entire, fleshy, oblanceolate to spatulate or ovate, glabrous, or with very few cilia at base; flowering stems glabrous or pubescent, with capitate glands; calyx lobes glabrous, ovate-triangular, purple-mottled; petals white, nearly clawless; filaments clavate; capsule ovoid, purple-mottled.

Moist places in the mountains, rock-crevices. Described from the northwest coast of North America.

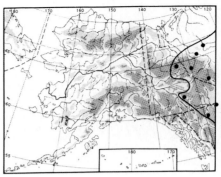 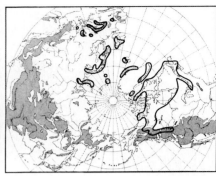

7. Saxífraga aizoìdes L.   Golden Saxifrage
Saxifraga Van-Bruntiae Small.

Stems decumbent, forming mats; leaves light green, sessile, linear, succulent, acuminate, glabrous, sparsely glandular-ciliate; flowering stems branched, few-flowered, glandular-puberulent; calyx lobes triangular, obtuse or somewhat acute; petals yellow, linear-oblong, clawless, longer than calyx lobes; carpels adnate at base.

Along creeks in the mountains. Described from Lapland, Steiermark, Westmoreland, and Monte Baldo (northern Italy).

Petals are yellow in the area of interest; in the Rocky Mountains, sometimes orange-dotted; and in Scandinavia, sometimes also orange or purple.

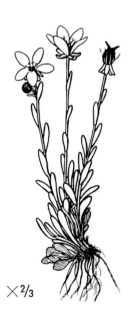

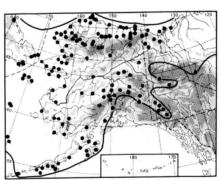 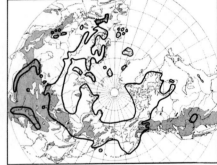

8. Saxífraga hírculus L.   Bog Saxifrage
Leptasea alaskana Small.

More or less tufted, often with short runners; basal leaves lanceolate to ligulate, entire, obtuse, petiolated, glabrous; flowering stem pubescent with curly, septate, reddish hairs above, erect, with 3 to several linear, sessile leaves; flowers mostly solitary; sepals ciliate with septate hairs, soon reflexed; petals yellow, obtuse, follicles joined almost to tip.

Bogs, meadows.

Northern specimens and specimens from the shores of Bering Sea belong to var. **propínqua** (R. Br.) Simm. (*S. propinqua* R. Br.), with few stem leaves and rounded petals, but the material cannot be readily separated into 2 taxa. *S. hirculus* described from Lapland, Siberia, and Switzerland; var. *propinqua* from Melville Island.

52. SAXIFRAGACEAE (Saxifrage Family)

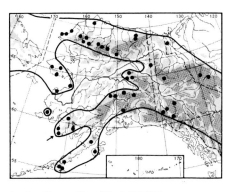 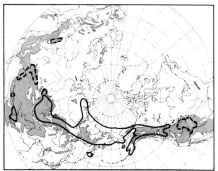

9. Saxífraga flagellàris Willd.   Spiderplant
subsp. **setìgera** (Pursh) Tolm.
  Saxifraga setigera Pursh; *S. flagellaris* var. *stenosepala* Trautv.

Plant with several, long, filiform, glabrous, brown or red, stolons, ending in bud; basal leaves in dense rosette, ovate, mucronate, ciliate, with coarse, spiny cilia; stem simple, viscid-pubescent, with several lanceolate leaves; calyx flat at base, with oblong, glandular lobes; petals broadly ovate, 3 times as long as calyx lobes, golden yellow; follicles broadly ovate, with short, thick style.

Stony places, scree slopes, along creeks in the mountains to over 2,000 meters.

Broken line on circumpolar map indicates range of subspecies not occurring within area of interest.

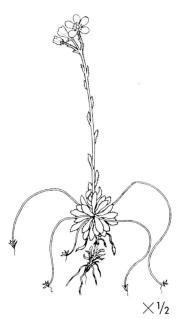

× ½

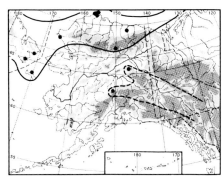 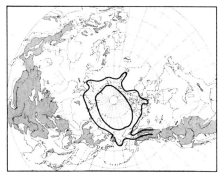

10. Saxífraga flagellàris Willd.
subsp. **platysèpala** (Trautv.) Pors.
  Saxifraga flagellaris var. *platysepala* Trautv.; *S. platysepala* (Trautv.) Tolm.

Similar to subsp. *setigera*, but with turbinate calyx tube; stem leaves few, glands purple, coarse; entire plant often purple.

S. flagellaris described from Mount Kazbek in the Caucasus; subsp. *platysepala* from "arctic seashore and islands of America."

Transitions to subsp. *setigera* occur within area marked by broken line.

(See color section.)

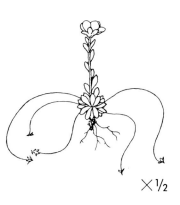

× ½

×⅔

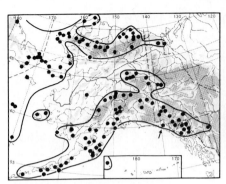 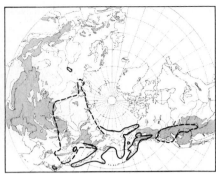

11. Saxífraga bronchiàlis L.   Spotted Saxifrage
subsp. **Funstònii** (Small) Hult.

Leptasea Funstonii Small; *Saxifraga pseudo-Burseriana* Fisch.; *S. firma* Litw.

Loosely caespitose; caudex branched, densely covered with elliptic, sessile, grayish-green leaves, mucronate in apex and with stout, white cilia in the margin, sometimes capitate-glandular above; flowering stems with several linear leaves, more or less glandular; sepals glabrous or ciliate; petals white or cream-colored, spotted with yellow; follicles oblong, with long, divergent beaks.

Rock crevices, rocky soil, in the mountains to at least 2,200 meters, not on calcareous soil.

In the high mountains, specimens occur with globular or sausage-like shoots and very dense leaves; these approach subsp. *cherleroides* (see next description). Broken line on circumpolar map indicates range of other subspecies.

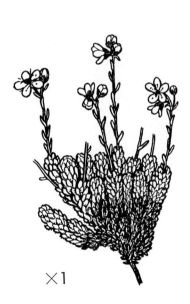

×1

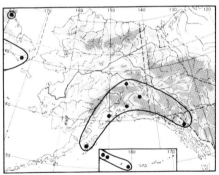 

12. Saxífraga bronchiàlis L.   Spotted Saxifrage
subsp. **cherlerioìdes** (D. Don) Hult.

Saxifraga cherlerioides D. Don; *S. bronchialis* subsp. *Funstonii* var. *cherlerioides* (D. Don) Hult.

Similar to subsp. *Funstonii*, but petals not clawed at base; stamens shorter than petals; leaves with obtuse or very short-mucronate apex and small, dense, glandular ciliation in margin.

Scree slopes in the high mountains. Described from Kamchatka.

52. SAXIFRAGACEAE (Saxifrage Family)

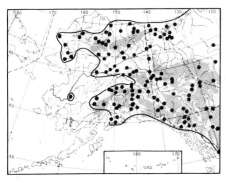  

13. Saxífraga tricuspidàta Rottb.   Prickly Saxifrage
Leptasea tricuspidata (Rottb.) Small.

Evergreen, loosely matted, with elongated (rarely spherical) caudices covered with withered leaves; leaves leathery, brownish-green, cuneate to oblanceolate, ciliate, with 3 apical mucronate teeth; flowering stem above, and inflorescence capitate-glandular; calyx lobes broadly triangular; petals elliptic, clawless, white, with yellow spots at base, red spots at tips; filaments subulate.

Dry, sandy places, rock crevices, ridges; common in the mountains; in the Yukon to at least 2,000 meters. Described from Greenland.

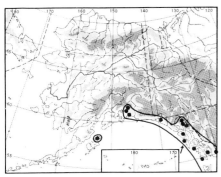 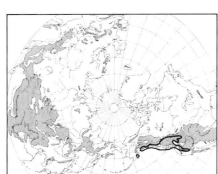 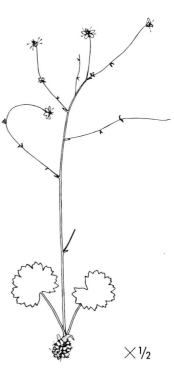

14. Saxífraga Mertensiàna Bong.   Wood Saxifrage
Heterisia Mertensiana (Bong.) Small.

Rootstock thick, short, with brown scales, and with bulblets at tip; leaves orbicular, cordate, acutely toothed, glabrous, ciliate with long, slender glands; flowering stem with few reduced leaves, glandular; sepals ovate-triangular, glandular, reflexed; petals white, acute; filaments broader toward tip; follicles broadly ovate, with long, divergent styles.

Moist rocks.

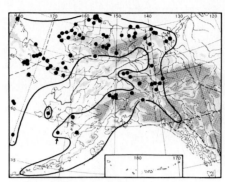 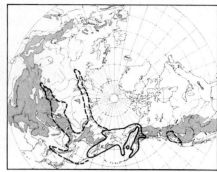

15. Saxífraga punctàta L.   Cordate-Leaved Saxifrage
subsp. **Nelsoniàna** (D. Don) Hult.

Saxifraga Nelsoniana D. Don; *Micranthes Nelsoniana* (D. Don) Small.

Rootstock thin, with subterranean runners; basal leaves more or less pubescent on both sides, and ciliate with septate hairs, rounded to reniform, cordate, dark green above, paler beneath, toothed; stem above and pedicels pubescent, with short glands; inflorescence mostly dense; calyx lobes oblong, often purplish, reflexed; petals ovate to oblong, with short claw, mostly white, twice as long as calyx lobes; filaments clavate.

Alpine meadows, tundra hummocks, along creeks. *S. punctata* described from Siberia. Broken line on circumpolar map indicates range of races not occurring in area of interest.

The leaves are eaten preserved in seal oil by the Siberian Eskimo; so treated, they keep fresh over the winter.

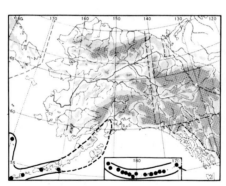 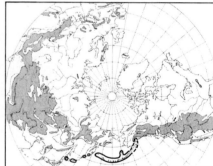

16. Saxífraga punctàta L.
subsp. **insulàris** Hult.

Saxifraga purpurascens Kom.

Similar to subsp. *Nelsoniana*, but leaves thick and fleshy, glabrous or ciliate in margin, seemingly nerveless; pedicels strongly viscid-pubescent; follicles stout, purplish-black.

Hillsides, along streams.

Intermediates between this race and subsp. *pacifica* can be found within the range indicated by the broken line.

52. Saxifragaceae (Saxifrage Family)

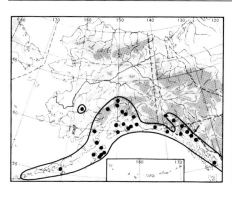
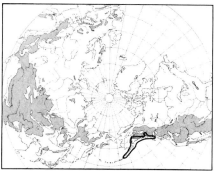

17. Saxífraga punctàta L.
subsp. **pacífica** Hult.

Similar to subsp. *Nelsoniana* and subsp. *Charlottae*, but leaves glabrous, often ciliate in margin, thin; ripe follicles inflated, membranaceous; pedicels strongly viscid; leaves with 12–18 teeth; capsules 6–12 mm long.

Hillsides, along streams. See subsps. *insularis* and *Charlottae* for ranges of intermediates with subsp. *pacifica*.

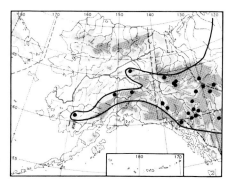
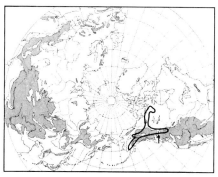
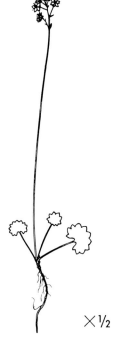

18. Saxífraga punctàta L.
subsp. **Porsildiàna** Calder & Savile

Similar to subsp. *pacifica*, but leaves on the average smaller, often somewhat pubescent.

Hillsides, along streams.

52. SAXIFRAGACEAE (Saxifrage Family)

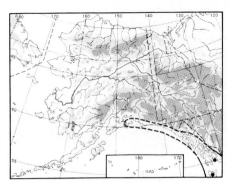
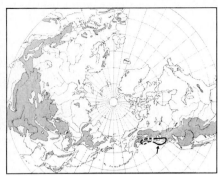

19. Saxífraga punctàta L.
subsp. **Charlóttae** Calder & Savile

Similar to subsp. *pacifica,* but leaves with fewer (9–12) teeth; capsules on the average longer

Hillsides, along streams.

Intermediates between this race and subsp. *pacifica* can be found within the range indicated by the broken line.

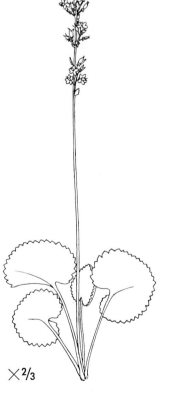

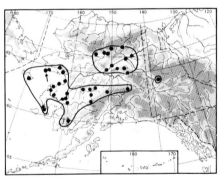
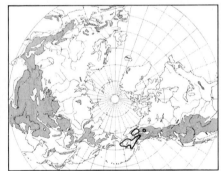

20. Saxífraga spicàta D. Don   Spiked Saxifrage

Micranthes spicata (D. Don) Small; *M. galacifolia* Small.

Rootstock thick, stout, with large buds at top; basal leaves long-petiolated, cordate, reniform, orbicular or broadly ovate, crenate-dentate, nearly glabrous, ciliate with septate hairs; stem leafless, pubescent with septate hairs; inflorescence with broad, dentate bracts, pubescent with capitate glands; calyx lobes triangular, glabrous, soon reflexed; petals oblong, clawless, white, about 3 times as long as calyx lobes; filaments subulate; follicles glabrous, with long divergent styles.

Moist, rocky slopes, along creeks.

52. Saxifragaceae (Saxifrage Family)

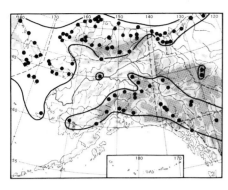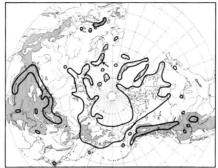

21. Saxífraga cérnua L. Bulblet Saxifrage

Stem and margin of leaves with long, septate hairs; stems erect, simple, with pale bulblets at base; basal leaves long-petiolated, cordate-reniform, 5–7-lobed; stem leaves narrower, with brownish-red bulblets in axils; flower mostly single, terminal; petals white, about 4 times as long as calyx lobes.

Moist places in the mountains, scree slopes. Described from Alps of Lapland.

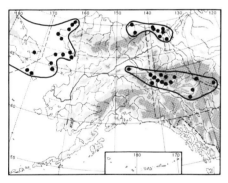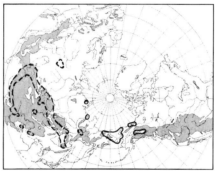

22. Saxífraga exílis Steph.

Saxifraga radiata Small; *S. sibirica* with respect to Alaskan plant.

Stems single or several, with white bulblets at base; basal leaves few, small, long-petiolated, reniform, with 3–7 ovate, somewhat acute teeth, glabrous or somewhat pubescent; stem and inflorescence glandular-puberulent; stem leaves with 3–5 teeth, mostly with cuneate base, the upper entire; calyx lobes ovate; petals 3 times as long as calyx lobes, with follicles united nearly to tip, and with divergent styles.

Wet meadows, snow beds, along creeks. Described from Siberia.
Broken line on circumpolar map indicates range of the closely related **S. sibìrica** L.

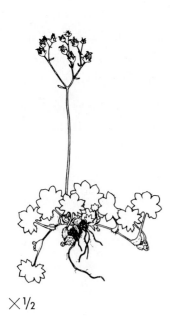
×½

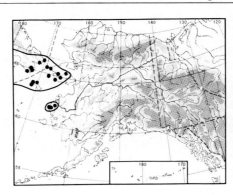 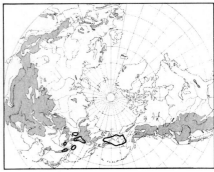

23. Saxífraga nudicaùlis D. Don

Ochraria nudicaulis (D. Don) Small; *Saxifraga neglecta* Bray.

Flowering stems at tips of long runners, with remains of old petioles; basal leaves reniform or orbicular in outline, with large ovate, acute teeth, glabrous, sparsely glandular-ciliate, long-petiolated; petioles with large, auriculated, scarious, often purplish, glandular-margined stipules at base; stem leafless or with single smaller leaf, and with similar stipules at base; inflorescence glandular; calyx lobes ovate-triangular; petals twice as long as calyx lobes, ovate to lanceolate, with short claw, yellowish or sometimes reddish; follicles broadly ovate, purplish.

Wet places. Described from west coast of North America.

×½

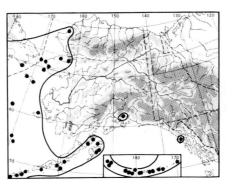 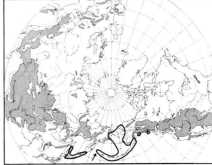

24. Saxífraga bracteàta D. Don

Saxifraga laurentiana Ser.; *S. cymbalaria* Cham.; *S. rivularis* var. *laurentiana* (Ser.) Engler.

Glandular-pubescent; rootstock with pale bulblets at top; stems coarse, mostly single; leaves carnose; basal leaves long-petiolated, broadly reniform, with 5–8 large, rounded, sometimes acute teeth; uppermost stem leaves sessile, crowded; pedicels very short; flowers nearly capitate; calyx lobes glabrous, obtuse; petals twice as long as calyx lobes, white, broadly ovate, distinctly clawed; follicles ovate; style very short.

Rocks and stony slopes along the coast.

52. SAXIFRAGACEAE (Saxifrage Family)

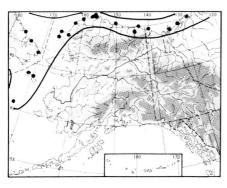
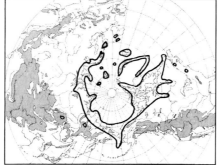

25. Saxífraga rivulàris L.   Brook Saxifrage

Including *Saxifraga hyperborea* R. Br.

var. **rivulàris**

Tufted, low-growing; basal leaves reniform-cordate, palmately 3–7-lobed; stem pubescent with septate hairs; inflorescence 1–5-flowered; calyx tube nearly hemispherical to turbinate; calyx lobes ovate; petals white or reddish, rarely purple, somewhat longer than calyx lobes.

Wet places, snow beds, along creeks. Described from Lapland.

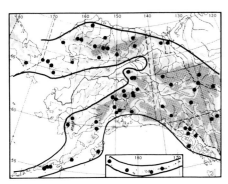
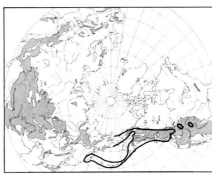
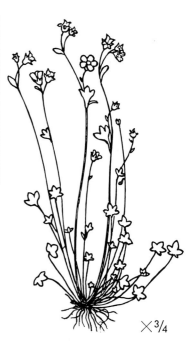

26. Saxífraga rivulàris L.

var. **flexuòsa** (Sternb.) Engler & Irmsch.

Saxifraga flexuosa Sternb.; *S. rivularis* subsp. *flexuosa* (Sternb.) Gjaerevoll.

Stem strict, slender, 5–10 cm tall, lowest flower long-pedicellated.

In the mountains to at least 1,500 meters. Circumpolar map tentative.

A weakly differentiated variety. Within the range of *S. bracteata*, transitional types occur.

Type locality unknown, probably Alaska.

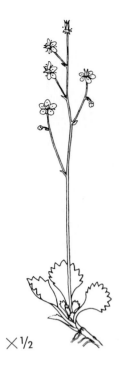
× ½

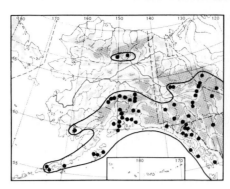
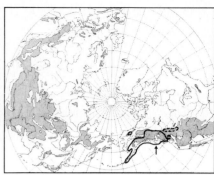

27. Saxífraga Lyállii Engler   Red-Stemmed Saxifrage
subsp. **Hulténii** (Calder & Savile) Calder & Savile
Saxifraga Lyallii var. Hultenii Calder & Savile.

Plant with long, dark-brown rootstock; basal leaves flabellate, dentate, abruptly contracted into long petiole, glabrous or somewhat pubescent with brownish septate hairs; flowering stem leafless, mostly glabrous at base, pubescent with capitate glands above; inflorescence with lanceolate bracts; calyx lobes reflexed, mostly purplish; petals oblong to orbicular, white or reddish, with 2 yellowish or greenish spots; follicles sometimes 3–4.

Moist places, snow beds, in the mountains to at least 1,235 meters.
Broken line on circumpolar map indicates range of subsp. **Lyállii**.
Hybrids with *S. punctata* occur.

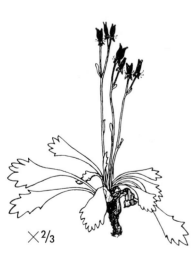
× ⅔

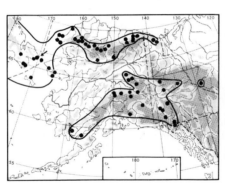
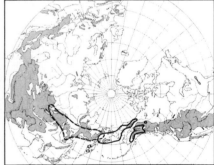

28. Saxífraga davùrica Willd.
subsp. **grandipétala** (Engler & Irmsch.) Hult.
Saxifraga davurica f. *grandipetala* Engler & Irmsch.; *S. grandipetala* (Engler & Irmsch.) A. Los.

Rootstock obliquely ascending, blackish-brown; basal leaves ovate-cuneate, mostly with 7 teeth, glabrous on sides, glandular-puberulent in margin; stem leafless or with single leaf, glabrous or pubescent with septate hairs; calyx lobes ovate-triangular, purplish, reflexed in fruit; petals white or usually purplish, ovate-oblong to elliptical, somewhat longer than calyx lobes; follicles purplish-black, conical; style very short.

Wet places in the mountains, to at least 1,500 meters. *S. davurica* described from the Dahurian Alps.
Broken line on circumpolar map indicates range of subsp. **davùrica**.

52. SAXIFRAGACEAE (Saxifrage Family)

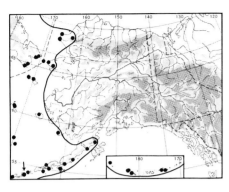
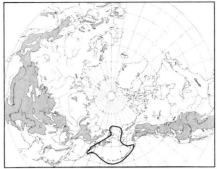

29. Saxífraga unalaschcénsis Sternb.

Saxifraga calycina Sternb.; *S. flabellifolia* Torr. & Gray; *Micranthes flabellifolia* (Torr. & Gray) Small.

Similar to S. *davurica* subsp. *grandipetala*, but follicles ovate, thick, plump; leaves broad, obovate, with broad base.

Dry, stony slopes.

Where the ranges overlap, intermediates between this plant and S. *davurica* subsp. *grandipetala* occur.

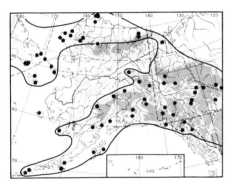
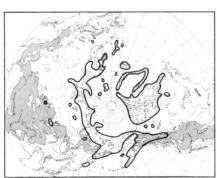
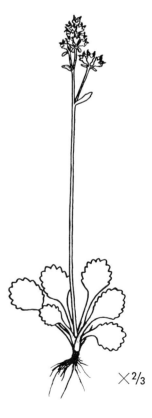

30. Saxífraga nivàlis L.   Snow Saxifrage

Basal leaves thick, grossly dentate, green and glabrous above, purple and more or less rufous-pubescent below, ciliate in margin; stem single, pubescent with curly, septate, dark or reddish hairs; inflorescence glandular-pubescent, capitate or loose; petals white or mostly reddish, elliptic, obtuse, slightly longer than calyx lobes; follicles ovate, with divergent beak.

Dry slopes, in the mountains to at least 1,700 meters. Described from the Alps of Spitzbergen, Lapland, "Arvonicis," Virginia, and Canada.

Highly variable. Two varieties should be mentioned: var. **tènuis** Wahlenb. [S. *tenuis* (Wahlenb.) H. Sm.], with small, nearly glabrous leaves and acute, reflexed follicle beaks; and var. **rufopilòsa** Hult. (S. *rufidula*, with respect to Alaskan plant; S. *eriphora* S. Wats.?), with densely pubescent, rufous-haired leaves and red petals.

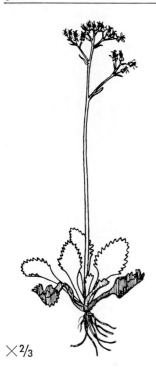

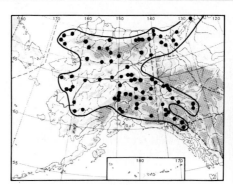 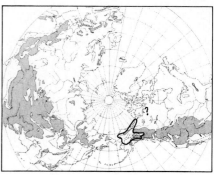

31. Saxífraga refléxa Hook.

Saxifraga radulina Greene; *Micranthes yukonensis* Small; *M. reflexa* (Hook.) Small.

Stem single, at top of short caudex; basal leaves rhomboid to ovate, serrate-crenate, with broad petiole, subcoriaceous, densely pubescent on both sides, with short, septate hairs, and longer hairs in margin; stem pubescent, leafless; bracts linear, calyx lobes ovate, nearly glabrous, in age reflexed; petals white, nearly twice as long as calyx lobes; filaments clavate; follicles purplish, ovate, with fairly long style.

Dry places; to at least 2,300 meters in the St. Elias Range. Described from "between the Mackenzie River and the Coppermine River."

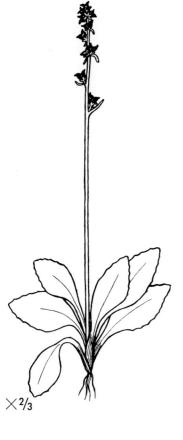

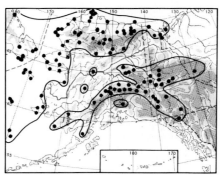 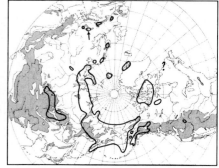

32. Saxífraga hieracifòlia Waldst. & Kit.   Stiff-Stemmed Saxifrage

Saxifraga integrifolia with respect to Alaskan plant.

Highly variable; stem single, straight, thick, up to 50 cm tall, often branched, densely glandular; leaves green above, often red below, broadly lanceolate, oblong to rhombic, thick, petiolated, remotely dentate to nearly entire, glandular-ciliate in margin, glabrous on sides, or, rarely, reddish-pubescent below (var. **rufopilòsa** Hult.); calyx lobes broadly triangular; petals oblong to triangular, greenish to purplish.

Moist places on tundra, alpine meadows, solifluction soil. Described from the Carpathians in (historical) Hungary.

52. SAXIFRAGACEAE (Saxifrage Family)

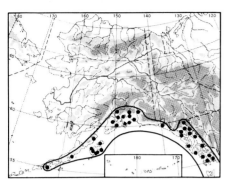
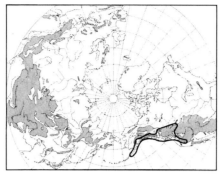

33. Saxífraga ferrugìnea Graham   Coast Saxifrage
Spathularia ferruginea (Graham) Small; *S. Brunoniana* Small.

Stems mostly single from thick rootstock; basal leaves oblanceolate to cuneate-obovate, sharply 3-toothed to several-toothed, tapering to broad petiole, glabrous to hirsute with coarse hairs, ciliate; stem and inflorescence glandular-pubescent; sepals oblong, reflexed, in age purplish-tipped; petals white or purplish, dimorphic, the larger with yellow spots; filaments subulate; ovary superior.

Wet rocks. Described from cultivated specimens.

Var. **Macoùnii** Engler & Irmsch. has bulblets in the axils; var. **Newcòmbei** Small is coarse and markedly hirsute, and has large flowers.

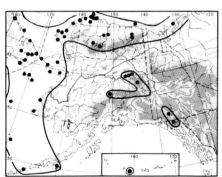
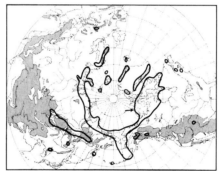

34. Saxífraga foliolòsa R. Br.   Grained Saxifrage
Saxifraga stellaris var. *comosa* Retz.; *S. stellaris* subsp. *comosa* (Retz.) Braun-Blanquet.
var. **foliolòsa**

Stem mostly single from short caudex; basal leaves small, cuneate-obovate, with few to several small teeth, glabrous, ciliate in margin; stem sparsely pubescent; inflorescence with thick, arcuate branches; only the terminal flower is developed, the others being replaced by bulblets or buds; sepals ovate, reflexed in age; petals white or reddish, dimorphic, abruptly clawed, the longer with yellow spots.

Wet, rocky slopes, snow beds, along creeks.

×³/₄

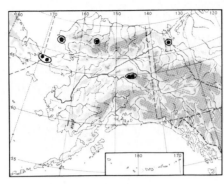
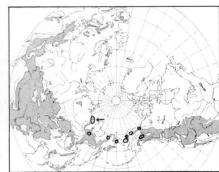

35. Saxífraga foliolòsa R. Br.
var. **multiflòra** Hult.

Saxifraga Redowskii Adams.

Taller than var. *foliolosa*, and with flowers on both terminal and lateral branches. Rocky places on tundra.

×³/₄

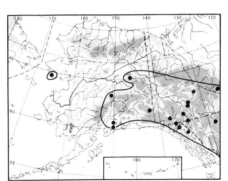
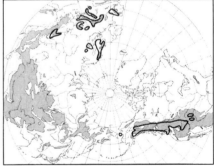

36. Saxífraga adscéndens L.
subsp. **oregonénsis** (Raf.) Bacigalupi

Ponista oregonensis Raf.; *S. adscendens* var. *oregonensis* (Raf.) Breitung.

Glandular-pubescent, basal leaves obovate with cuneate base, 3–5-toothed to shallowly lobed; cauline leaves several; pedicels short; calyx lobes ovate-triangular; petals white, about twice as long as calyx lobes; ovary inferior.

Rock crevices, sandy places. *S. adscendens* described from the Pyrenees, Monte Baldo (northern Italy), and "Tauro Rastadiensi" (Baden); subsp. *oregonensis* from the Rocky Mountains.

Range in Europe belongs to subsp. **adscéndens**.

52. SAXIFRAGACEAE (Saxifrage Family)

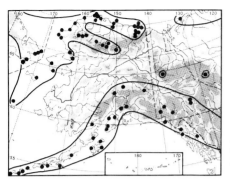

37. Saxífraga caespitòsa L. Tufted Saxifrage
Saxifraga groenlandica L.

Densely tufted or matted; stems with marcescent leaves at base; leaves short-petiolated, palmately 3–5-lobed, with linear to oblong, obtuse lobes; flowering stems 1-flowered to few-flowered, erect, glandular; petals oblong, white; follicles nearly entirely adnate.

Stony slopes, sandy places on tundra; to at least 2,300 meters in the St. Elias Range.

Most specimens belong to subsp. **sileneflòra** (Sternb.) Hult. (*S. sileneflora* Sternb.), but other races, such as subsp. **montícola** (Small) Pors. (*Muscaria monticola* Small), have been reported from the area. Racial conditions within the species are still unclear. *S. caespitosa* described from the Alps of Lapland, Switzerland, Trent (northern Italy), and Montpellier (France); subsp. *sileneflora* from Unalaska, Eschscholtz Bay, and Chamisso Island (Kotzebue Sound).

The closely related *S. magellanica* Poir. is found in South America.

Tiarella L.

Leaves 3-foliate	1. *T. trifoliata*
Leaves shallowly 3–5-lobed	2. *T. unifoliata*

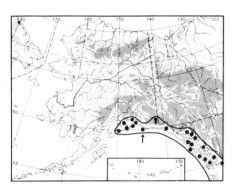
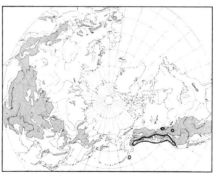

1. Tiarélla trifoliàta L. Lace Flower

Stems from stout, dark-brown rootstock; basal leaves long-petiolated, with stipules at base, 3-foliate, the leaflets sharply crenate-dentate, the terminal 3-lobed; stem glandular-pubescent; sepals ovate, acute, glandular; petals white, linear, subulate, 3 to 4 times as long as calyx lobes; carpels nearly equal when young, unequal in age.

Moist places, forests. Described from northern Asia (where it does not, however, occur); type locality probably Kayak Island, Alaska.

Transitions to *T. unifoliata* not observed within area of concern.

52. Saxifragaceae (Saxifrage Family)

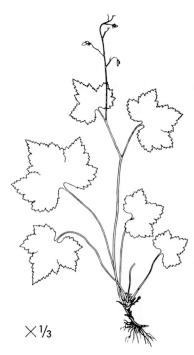

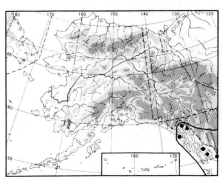
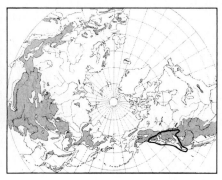

2. Tiarélla unifoliàta Hook.   Sugar-Scoop

Tiarella trifoliata var. *unifoliata* Kurtz; *T. trifoliata* subsp. *unifoliata* (Hook.) Kern.

Stems from stout, brown rootstock; basal leaves acutely 3–5-lobed, cordate, hirsute with scattered coarse hairs above, glabrous beneath; stem glandular-pubescent; cauline leaves similar to basal leaves; petals white, linear-subulate, 2 to 3 times as long as calyx lobes; carpels abruptly acute, the larger about twice as long as the smaller when ripe.

Moist, shady places, woods. Described from the Rocky Mountains, near the source of the Columbia and Portage Rivers.

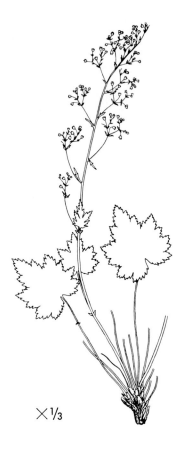

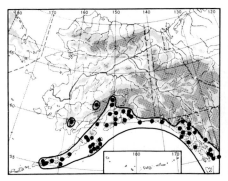
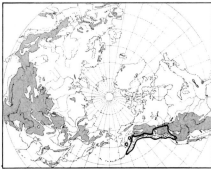

Heuchera L.

1. Heùchera glàbra Willd.   Alpine Heuchera

Heuchera divaricata Fisch.

Rootstock stout; basal leaves cordate-ovate, 3–5-lobed, the lobes coarsely crenate-dentate; sparsely pubescent below, ciliate with long glands in margin; stem glabrous or slightly glandular; stem leaves 1–2, reduced; inflorescence thyrsoid, with lanceolate bracts; pedicels glandular; calyx lobes broadly ovate, glandular petals white, 2–3 times as long as calyx lobes, with long, slender claw and entire blade; stamens 5; capsule with slender beak; seeds with rows of spines.

Moist rocks. Described from western North America.

52. SAXIFRAGACEAE (Saxifrage Family)

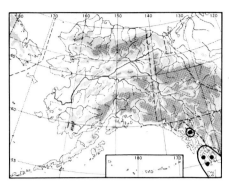

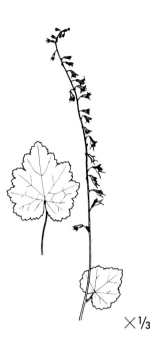

Tolmiea Torr. & Gray

1. Tolmièa Menzièsii (Pursh) Torr. & Gray — Youth-on-Age

Tiarella Menziesii Pursh; *Leptaxis Menziesii* (Pursh) Raf.

Rootstock stout; basal leaves shallowly 5–7-lobed, with adventitious buds in sinuses, the lobes irregularly crenate-dentate, hirsute with coarse, scattered, white hairs, or nearly glabrous, glandular in margin; stem hirsute with reduced leaves; calyx greenish-purple, strongly veined; the upper 3 calyx lobes longer than the 2 other lobes; petals 4, brown, linear-subulate, persistent; stamens 3; carpels 2.

Moist woods along streams. Described from the northwest coast of North America.

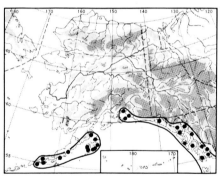
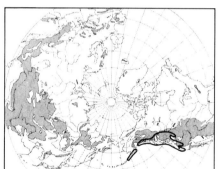
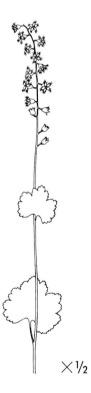

Tellima R. Br.

1. Tellìma grandiflòra (Pursh) Dougl. — Fringe Cups

Mitella grandiflora Pursh; *Tiarella alternifolia* Fisch.

Rootstock stout; basal leaves sparsely hirsute, cordate to reniform, shallowly lobed, the lobes dentate, long-petiolated; stem hirsute, stem leaves reduced; inflorescence racemose; flowers short-pedicelled; calyx lobes triangular, glandular; petals 4, lanceolate, deeply laciniate-pinnatifid, with filiform segments, greenish-white, in age reddish-brown; stamens 10, with short filaments; capsule ovate, styles divergent; seeds tuberculate.

Moist rocks. Described from the northwest coast of North America.

52. Saxifragaceae (Saxifrage Family)

Mitella L. (Bishop's-Cap)

Stamens 10; leaves reniform, rounded-crenate; stem glandular-puberulent and sparingly hairy .... 1. *M. nuda*
Stamens 5; leaves broadly cordate, coarsely and unevenly crenate; stem minutely glandular-puberulent .... 2. *M. pentandra*

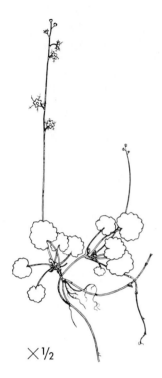

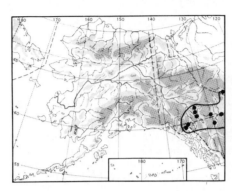 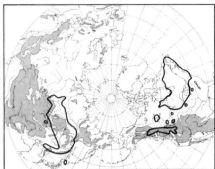

1. Mitélla nùda L.

Rhizome long, slender, stoloniferous at top; basal leaves cordate to reniform, rounded-crenate, with long, glandular petioles; stem leafless, or with 1 leaf at base, glandular-pubescent and hairy; inflorescence racemose; calyx lobes broadly triangular; petals 5, greenish-yellow, pectinately dissected into about 8 linear divisions; stamens 10.

Along streams, bogs. Described from northern Asia.

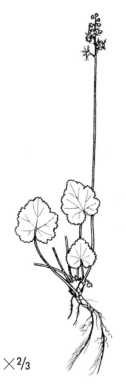

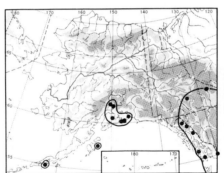 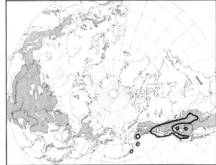

2. Mitélla pentándra Hook.
Pectinanthia pentandra (Hook.) Rydb.

Rhizome stout; basal leaves glabrous or sparsely hirsute, ovate-cordate, shallowly 5–9-lobed, the lobes unevenly crenate-dentate, long-petiolated; stem sparsely glandular, leafless or with small leaves; inflorescence racemose; calyx lobes triangular, spreading or recurved; petals green, pectinately dissected into 6–10 filiform segments; stamens 5; styles almost lacking.

Moist woods. Described from cultivated specimens from the Rocky Mountains.

52. Saxifragaceae (Saxifrage Family)

Chrysosplenium L.

Leaves of runners and stem opposite; stamens 8 . . . . 3. *C. kamtschaticum*
Leaves alternate:
    Stamens mostly 4; calyx mostly green; petioles not reddish-pubescent . . . .
    . . . . 1. *C. tetrandrum*
    Stamens 8; calyx purple-mottled; petioles reddish-pubescent . . . . 2. *C. Wrightii*

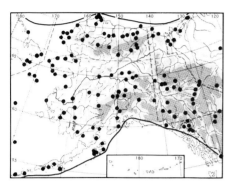
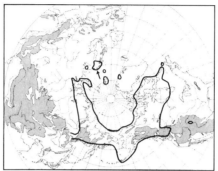

1. Chrysosplènium tetrándrum (Lund) T. Fries   Northern Water Carpet

Chrysosplenium alternifolium var. *tetrandrum* Lund; *C. alternifolium* subsp. *tetrandrum* (Lund) Hult.

Rhizome slender; basal leaves rounded-reniform, cordate, with few shallow lobes, glabrous or somewhat pubescent; petiole long, sparsely pubescent or glabrous; cauline leaves truncate at base, 3–5-lobed; bracts 3-lobed; sepals green; stamens on the average 4, short; seeds smooth.

Wet places, springs, along streams.

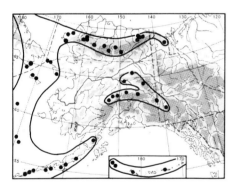
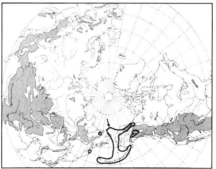
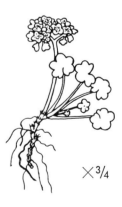

2. Chrysosplènium Wrìghtii Fr. & Sav.

Chrysosplenium Beringianum Rose; *C. Wrightii* var. *Beringianum* (Rose) Hara.

Rhizome long, thick; basal leaves thick, carnose, reniform-orbicular, cordate, with few crenate teeth, glabrous or sparsely pubescent, with stout, white hairs; petioles long, dilated at base, more or less densely pubescent with septate, reddish-brown hairs; stem leafless, glabrous or pubescent; sepals reniform, green, purple-mottled; stamens 8; seeds glabrous.

Scree slopes, solifluction soil; lower part of plant often buried in mud; in the mountains to at least 2,200 meters in McKinley Park.

52. Saxifragaceae (Saxifrage Family)

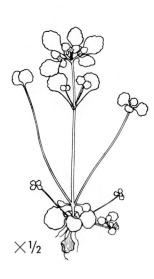

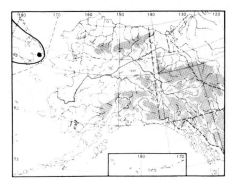
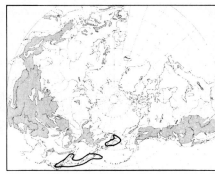

3. Chrysosplènium kamtscháticum Fisch.

Stem from short caudex, leafless or with 1–2 pairs of opposite leaves, glabrous, branched at top; basal leaves glabrous, rounded, broadly obovate or reniform, entire or irregularly crenate, short-petiolated, with unrooted long runners in axils; bracts broadly obovate, denticulate; flowers greenish; calyx lobes broadly ovate, spreading; stamens 8, shorter than sepals; capsules with 2 long lobes; seeds with longitudinal ridges.

Along streams. Described from Kamchatka.

Parnassia L.

Petals fimbriate at base .... 1. *P. fimbriata*
Petals not fimbriate:
    Petals narrow, 3-nerved, about as long as sepals; stem leaf absent or fixed close to base of stem .... 3. *P. Kotzebuei*
    Petals broader, with several nerves, longer than sepals; stem leaf fixed at middle of stem .... 2. *P. palustris* subsp. *neogaea*

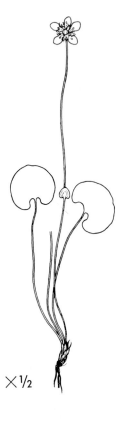

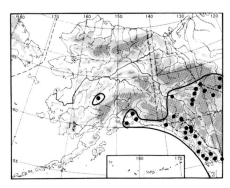
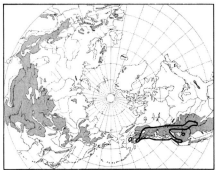

1. Parnássia fimbriàta Konig   Fringed Grass-of-Parnassus

Rootstock short, stout; basal leaves reniform-cordate, mostly broader than long, long-petiolated; stem with clasping bract near middle; sepals elliptic, with glandular teeth in margin; petals white, cuneate-obovate, attenuated into claw, fimbriate on lateral margins; staminodium carnose, lobed in apex; capsule ovoid.

Wet meadows, banks of streams. Described from the northwest coast of North America.

52. SAXIFRAGACEAE (Saxifrage Family)

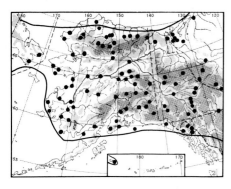 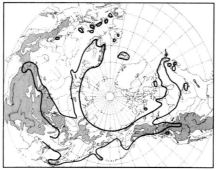 

2. Parnássia palústris L.   Northern Grass-of-Parnassus, Bog Star
subsp. **neogaèa** (Fern.) Hult.

Parnassia palustris var. *neogaea* Fern.

Glabrous; leaves ovate, cordate; stem with single, sessile, cordate leaf at middle; calyx lobes oblong-lanceolate, somewhat acute; petals white, twice as long as calyx lobes, 5–9-nerved; staminodia dilated, with several slender cilia, nearly as long as stamens.

Wet meadows, heaths, thickets. *P. palustris* described from Europe.

Specimens apparently transitional to var. **montanénsis** (Fern. & Rydb.) Hitchc. (*P. montanensis* Fern. & Rydb.) occur, with stem leaves not clasping, short petals, and staminodia with less numerous cilia. Var. *montanensis*, a very doubtful taxon, occurs from Alberta to Montana.

Circumpolar map indicates range of entire species complex.

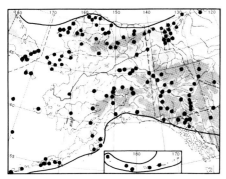  

3. Parnássia Kotzebùei Cham. & Schlecht.

Rootstock short; basal leaves ovate, elliptic, or deltoid-ovate; stem leafless or with leaf close to base; calyx lobes lanceolate-oblong, 3–5-nerved; petals elliptic-lanceolate, 3-nerved, as long as calyx lobes; staminodia a thin scale, entire or with mostly 5 gland-tipped segments.

Wet meadows, thickets, along creeks, in the mountains to at least 2,000 meters. Described from Unalaska, Kotzebue Sound, and St. Lawrence Bay (Chukchi Peninsula).

52. Saxifragaceae (Saxifrage Family)

Ribes L.

Stem with spines or prickles:
  Flowers in several-flowered racemes .... 1. *R. lacustre*
  Flowers single or double .... 2. *R. oxyacanthoides*
Stem without spines or prickles:
  Ovary and fruit speckled with resin dots:
    Racemes up to 30 cm long; bracts long, linear or broader above middle; lobes of leaves acute; flowers greenish .... 3. *R. bracteosum*
    Racemes less than 10 cm long; bracts short, triangular, acute; lobes of leaves rounded; flowers white .... 4. *R. hudsonianum*
  Ovary not speckled with resin dots, glabrous or with stalked glands:
    Ovary and fruit glabrous; fruit red .... 7. *R. triste*
    Ovary pubescent, usually also with stalked glands; fruit red or dark purple:
      Sepals glabrous on back; fruit red .... 5. *R. glandulosum*
      Sepals pubescent on back; fruit dark purple .... 6. *R. laxiflorum*

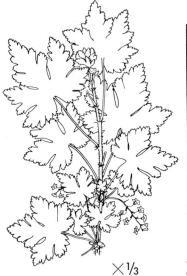

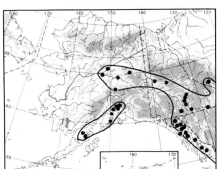

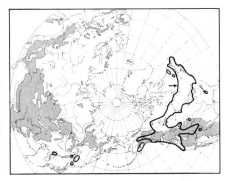

1. Rìbes lacústre (Pers.) Poir.   Bristly Black Currant
Ribes oxyacanthoides var. *lacustris* Pers.

Shrub with erect to spreading branches; young stems covered with bristly prickles and weak thorns; leaves cordate, 3–5 parted, the parts lobate and serrate, or double crenate-dentate; racemes spreading or drooping, glandular-bristly; petals pinkish, cuneate-flabellate; fruit dark purple, stipitate, glandular, somewhat palatable.

Moist woods, along streams.

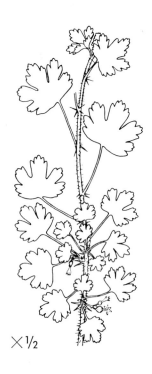

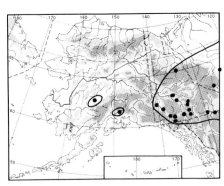

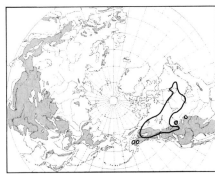

2. Rìbes oxyacanthoìdes L.
Grassularia oxyacanthoides (L.) Mill.

Shrub with ascending branches, puberulent and bristly when young; leaves broadly ovate, with truncate-cordate base, mostly 5-lobed, the lobes irregularly crenate-dentate, glandular-puberulent below; racemes 1–2-flowered; fruit purplish-black, with long capitate glands or nearly glabrous, palatable.

Moist woods. Described from Canada.

52. SAXIFRAGACEAE (Saxifrage Family)

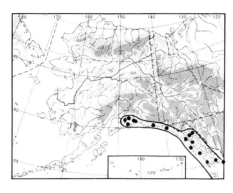
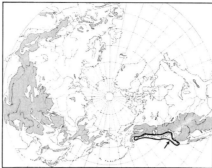

3. Rìbes bracteòsum Dougl.   Stink Currant

Shrub with ascending or erect, sparingly pubescent branches; leaves cordate, deeply 5–7-lobed, the lobes acute, sharply and irregularly serrate or double-serrate; racemes erect, up to 20 cm long, loosely flowered; lower bracts foliaceous, lobed; calyx lobes ovate-lanceolate; petals white, cuneate-flabellate, clawed; berry subglobose, black, strongly glaucous, glandular, with a disagreeable taste.

Banks of streams, woods.

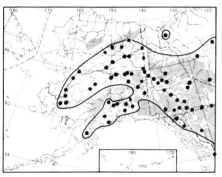

4. Rìbes hudsoniànum Richards.   Northern Black Currant

Shrub with erect branches; leaves cordate, 3–5-lobed, the lobes crenate-serrate, glandular below; racemes ascending or spreading; ovary more or less glandular, with sessile glands; calyx lobes ascending, white-pubescent; petals white; berry subglobose, black, more or less glaucous, glandular to nearly glabrous, bitter but edible.

Moist woods, along streams. Described from lat. 67° in Alberta.

Broken line on circumpolar map indicates range of var. **petiolàre** (Dougl.) Jancz. (*R. petiolare* Dougl.) in the western United States, and of the closely related **R. dikùscha** Fisch. in eastern Asia.

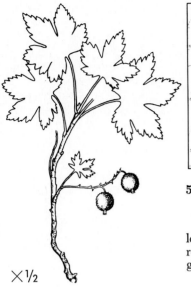
× ½

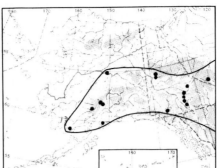

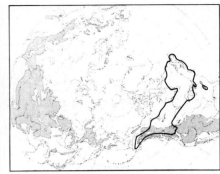

5. Rìbes glandulòsum Grauer Skunk Currant
Ribes prostratum L'Her.

Shrub with ascending branches; leaves smooth, deeply cordate, 5–7-lobed, the lobes acute, doubly serrate; racemes ascending, glandular-hispid; calyx white to roseate, petals cuneate to flabellate, whitish or pink; fruit red, with long, capitate glands, with odor of skunk.

Woods, rocky slopes in the lowlands. Type locality not given.

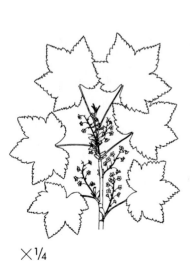
× ¼

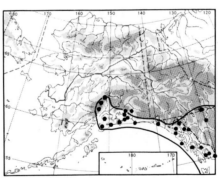

6. Rìbes laxiflòrum Pursh Trailing Black Currant
Ribes affine Dougl.

Shrub with decumbent or spreading branches; young branches puberulent and glandular; leaves deeply cordate, deeply 5-lobed, the lobes sharply double-crenate-serrate, glabrous above, puberulent and glandular below; racemes erect or ascending, shorter than leaves, puberulent and stipitate-glandular; calyx lobes deltoid, ovate; petals red to purplish; fruit ovoid, purplish-black, glaucous, glandular-bristly, edible.

Moist woods. Described from the northwest coast of North America.

52. SAXIFRAGACEAE (Saxifrage Family) / 53. ROSACEAE (Rose Family)

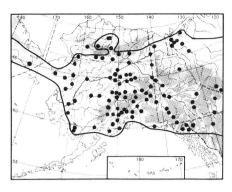
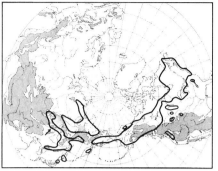
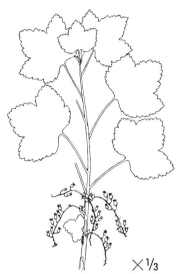

7. Rìbes tríste Pall.   Northern Red Currant

Shrub with prostrate or ascending, rooting branches; leaves cordate, broadly 3–5-lobed, the lobes coarsely dentate-serrate, glabrous or pubescent below; racemes shorter than leaves, loosely several-flowered, pubescent; ovary glabrous; calyx lobes broadly cuneate-rhombic; petals broadly cuneate, purple; disk low, 5-angled; berry ovoid, bright red, glabrous, sour, good to eat.

Wet meadows, along streams, spruce forests, from the lowlands to timberline.

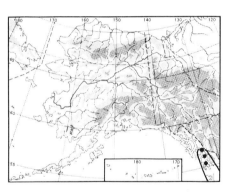
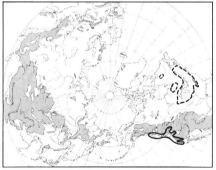

Physocarpus Maxim.

1. Physocárpus capitàtus (Pursh) Ktze.   Pacific Ninebark
Spiraea capitata Pursh.

Shrub with erect to spreading, glabrous or minutely stellate-pubescent branches; leaves broadly ovate to cordate, nearly as broad as long, 3–5-lobed, the lobes doubly crenate-serrate, stellate-pubescent, especially beneath, sometimes nearly glabrous; inflorescence hemispherical; calyx lobes ovate-lanceolate; petals suborbicular; stamens numerous; pistils 3–5.

Moist places, banks of streams. Described from the northwest coast of North America.

Broken line on circumpolar map indicates range of **P. opulifòlius** (L.) Raf. (*Spiraea opulifolia* L.).

53. ROSACEAE (Rose Family)

Spiraea L.

Inflorescence flat, corymbose; flowers white or with rose-colored median zone.... .... 1. *S. Beauverdiana*
Inflorescence elongated, paniculate; flowers rose-colored .... 2. *S. Douglasii* subsp. *Menziesii*

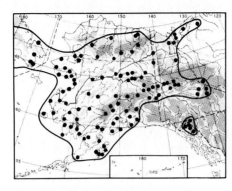

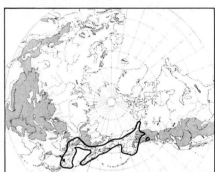

1. Spiraèa Beauverdiàna Schneid.   Alaska Spiraea

Spiraea Stevenii Rydb.; *S. betulifolia* with respect to Alaska-Yukon plant.

Low to middle-sized shrub; young branches reddish-brown, puberulent, glabrescent in age; leaves oblong, elliptic to ovate, glabrous or nearly so, pale beneath, dentate-crenate in margin, especially toward apex, sometimes entire; corymbs flat-topped, densely puberulent; sepals reflexed, petals white or pinkish.

Occurs in many different habitats: woods, alder thickets, meadows, and tundra bogs, up into alpine zone; a very common plant.

Highly variable in height, and in form, size, and serration of the leaves.

The leaves are used for tea by the Siberian Eskimo.

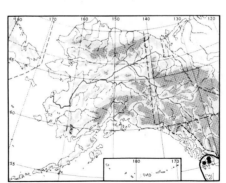

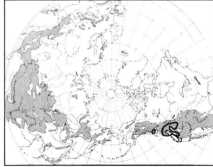

2. Spiraèa Douglásii Hook.

subsp. **Menzièsii** (Hook.) Calder & Taylor

Spiraea Menziesii Hook.; *S. Douglasii* var. *Menziesii* (Hook.) Presl.

Shrub with erect, densely puberulent branches; leaves ovate-oblong to oblong-elliptic, glabrous or sparsely puberulent, paler below, sharply and remotely serrate; inflorescence paniculate; calyx lobes triangular, pubescent; petals obovate to orbicular, pink to rose-colored; filaments of the same color; follicles shining.

Banks of streams, bogs. Described from the northwest coast of North America, at about the Columbia River.

Broken line on circumpolar map indicates range of subsp. **Douglásii** (sometimes also escaped from cultivation farther east).

53. ROSACEAE (Rose Family)

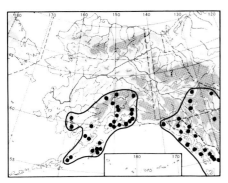

Luetkea Bong.

1. Luétkea pectinàta (Pursh) Ktze.

Saxifraga pectinata Pursh; *Eriogyna pectinata* (Pursh) Hook.; *Luetkea sibbaldioides* Bong.

Matted; branches rhizomatous, with marcescent leaves at base; leaves flabellate, glabrous, with several linear, somewhat acute lobes; flowering stems erect, with smaller leaves; branches of inflorescence pubescent; calyx lobes triangular; petals obovate, white.

Close to snow in the subalpine or alpine zone. Described from the northwest coast of North America.

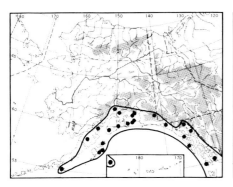
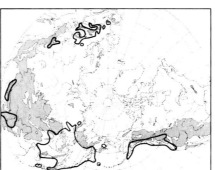

Aruncus Adans.

1. Arúncus sylvéster Kostel.   Goatsbeard

Spiraea Aruncus L.; *Aruncus acuminatus* (Dougl.) Rydb.; *S. acuminata* Dougl.; *A. kamtschaticus* Rydb.

Root thick, woody; leaves large, long-petiolated, triternate-pinnatisect to ternate-pinnatisect; leaflets ovate to oblong-lanceolate, acuminate to somewhat acute, sharply double-serrate, glabrous above, with scattered long hairs beneath; panicles large, with pubescent branches; pistillate and staminate flowers on different plants; petals white, smaller in pistillate flowers; staminate flowers with 15–20 stamens and dwarfed pistils; follicles erect, with divergent tip.

Moist woods and meadows, along streams. Described from cultivated specimens.

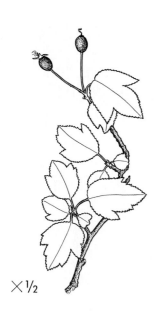

×½

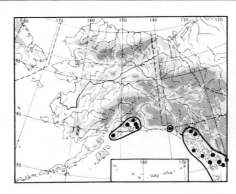
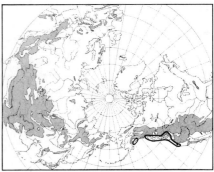

Malus Mill., S. F. Gray

1. Màlus fúsca (Raf.) Schneid. — Oregon Crab Apple

Pyrus fusca Raf.; *P. diversifolia* Bong.; *P. rivularis* Dougl.; *Malus rivularis* (Dougl.) Roem.; *M. diversifolia* (Bong.) Roem.

Shrub or small tree, with several, somewhat thorny stems; young twigs pubescent; leaves ovate to ovate-lanceolate, mostly acute, serrate and often with 3 shallow lobes, glabrous or sparsely pubescent and dark green above, pale and more or less pubescent or tomentose beneath; inflorescence corymbose, few-flowered; calyx lobes lanceolate-triangular; petals white or pink, broadly obovate; stamens about 20; fruit obovate, 3–4-chambered, purplish, containing pectin.

Moist woods. Type locality not given.

Sorbus S. F. Gray (Mountain Ash)

Young twigs, winter buds, and branches of inflorescence pubescent with white hairs; leaflets 11 or more:
  Winter buds glutinous, glabrous or pilose; styles 4–5 .... 1. *S. scopulina*
  Winter buds densely whitish-villous; pedicels and calyx densely whitish-lanate; styles 2–3 .... 2. *S. aucuparia*
Young twigs, winter buds, and branches of inflorescence pubescent with rust-colored pubescence; leaflets 7–11:
  Leaflets glossy above, acuminate; flowers 10–15 mm in diameter ....
    .... 3. *S. sambucifolia*
  Leaflets dull above, rounded in apex; flowers 6–9 mm in diameter ....
    .... 4. *S. sitchensis*

53. ROSACEAE (Rose Family)

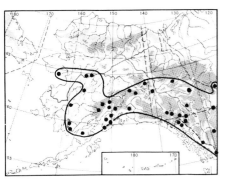

1. Sórbus scopulìna Greene
  Sorbus alaskana G. M. Jones.

Shrub with several stems, up to 4 meters tall, with reddish bark; winter buds glossy, glutinous, more or less white-pubescent; leaflets 11–13, oblong-lanceolate, acuminate, sharply serrate or double-serrate, glabrous above, paler and often sparsely pilose beneath; inflorescence flat-topped; pedicels sparsely pilose, with whitish, appressed hairs; sepals pilose; petals ovate; fruit globose, orange to bright red, glossy.

Woods, up into subalpine region. Described from Colorado and New Mexico.
Hybrids with S. *sitchensis* occur.

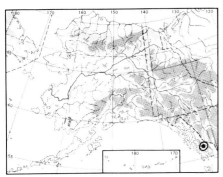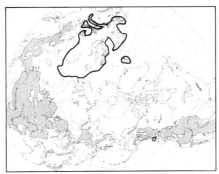

2. Sórbus aucupària L.

Small tree, similar to S. *scopulina*, but with winter buds white-villous, not glutinose; leaflets small, short-acuminate, soft-pubescent beneath; pedicels and calyx white-villous at flowering time.

Cultivated, and occasionally escaped from cultivation. Described from northern Europe.

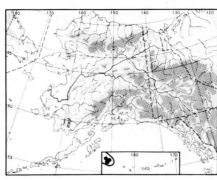
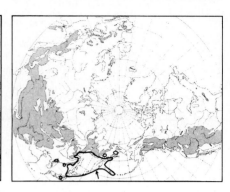

3. Sórbus sambucifòlia (Cham. & Schlecht.) Roem.

Pyrus sambucifolia Cham. & Schlecht.

Shrub up to 2 meters tall, in Alaska usually lower; winter buds glutinous, glabrous or somewhat rushy-pubescent; leaves with 7–11 ovate-lanceolate, acute, sharply serrulated, glabrous or sparsely pubescent leaflets, pale beneath; petioles reddish, with rust-colored pubescent stipules at base; inflorescence flat-topped, with rust-colored pubescence; calyx lobes ovate-triangular, fimbriate-ciliate; petals white or reddish; fruit bright red, spherical to elliptical, not very acid.

Mountain slopes. The fruit makes good jam.

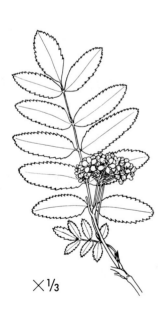
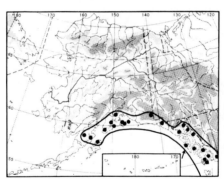
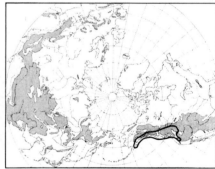

4. Sórbus sitchénsis Roem.

Shrub up to 3 meters tall, bark reddish; winter buds rusty-pubescent; leaflets 7–11, oblong to oblong-obovate, coarsely serrate, rounded in apex, glabrous above, more or less rusty-pubescent and paler beneath; inflorescence rounded, with rusty-pubescent branches; sepals glabrous, broadly triangular; petals white; fruit subglobose to elliptic, red with bluish bloom.

Woods, up into subalpine region. South of the Pleistocene glaciation, chiefly subsp. **Gràyi** (Wentzig) Calder & Taylor (*S. sambucifolia* var. *Grayi* Wentzig), with leaflets less toothed.

53. ROSACEAE (Rose Family)

Amelanchier Medic.

Leaves about as broad as long, thick and firm; young calyx densely woolly .... 1. *A. alnifolia*
Leaves distinctly longer than broad, thin; young calyx glabrous or only slightly pubescent .... 2. *A. florida*

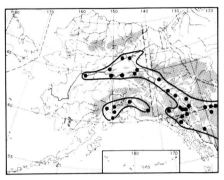 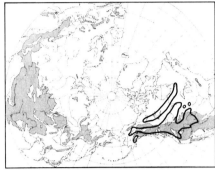 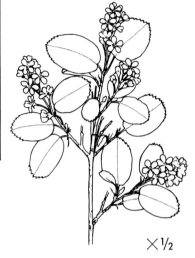

1. Amelánchier alnifòlia (Nutt.) Nutt.
Aronia alnifolia Nutt.

Low shrub with reddish-brown branches; leaves coriaceous, rounded, about as broad as long, sharply serrated in the apex, glabrous or sparsely pubescent; sepals triangular, acute, densely woolly; petals white, oblanceolate; fruit globose, purplish-black when ripe, glabrous, juicy.

Moist woods. Described from "Fort Mandan to the Northern Andes."
The fruit is eaten raw, dried, or boiled and made into pies.

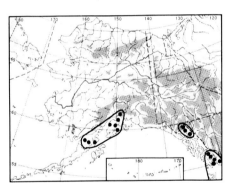  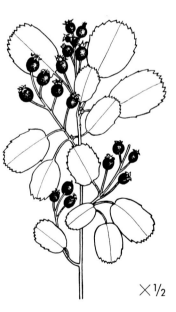

2. Amelánchier flòrida Lindl.   Pacific Serviceberry
Amelanchier Gormani Greene.

Similar to *A. alnifolia*, but with leaves longer than broad, thin; calyx glabrous or only slightly pubescent.
Moist woods, open places. Described from northwestern North America.

53. ROSACEAE (Rose Family)

× ⅓

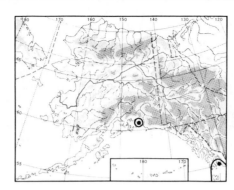
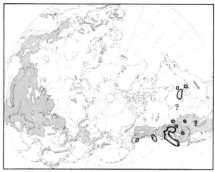

Crataegus L. (Hawthorn)

1. Crataègus Douglásii Lindl.

Large, thorny shrub or small tree, with brownish-gray bark; leaves elliptic to oblong-ovate, serrate with shallow lateral lobes, glabrous or sparsely pubescent; inflorescence glabrous; sepals reflexed; petals white, nearly orbicular; stamens few; fruit oblong, dark red to black, glabrous; nutlets ridged on outer face, pitted on the inner.

Woods. Described from northwest North America.

Rubus L.

Plant low-growing, usually less than 30 cm tall, herbaceous, or essentially so, unarmed; leaves lobed or 3–5-foliate:
  Leaves suborbicular, shallowly several-lobed; flowers white, unisexual .... 3. *R. chamaemorus*
  Leaves 3-lobed, 3- or 5-foliate; flowers white or red:
    Flowers white; plant stoloniferous:
      Leaves 3-foliate .... 1. *R. pubescens*
      Leaves digitately 5-foliate .... 2. *R. pedatus*
    Flowers red; leaves 3-lobed or 3-foliate:
      Calyx tube glabrous or nearly so; sepals caudate; petals narrow, erect, with cuneate base; flowers not exceeding leaves .... 5. *R. arcticus* subsp. *acaulis*
      Calyx tube pubescent; sepals triangular or caudate:
        Leaves 3-foliate; sepals triangular; petals short and broad; flowers often exceeding leaves .... 4. *R. arcticus* subsp. *arcticus*
        Leaves 3-lobed, sometimes nearly orbicular; sepals caudate; petals long and narrow; flowers large .... 6. *R. arcticus* subsp. *stellatus*
Plant a tall shrub, 0.5–2 meters tall, woody, armed or unarmed; leaves 3-foliate or 5-lobed:
  Leaves 5-lobed; flowers white, 4–7 cm in diameter; plant unarmed .... 10. *R. parviflorus*
  Leaves 3-foliate; flowers white or red, smaller; plant armed or unarmed:
    Flowers solitary, red; plant unarmed or weakly armed ..... 8. *R. spectabilis*
    Flowers in racemes; flowers white or whitish; plant armed:
      Stem with bristles; fruit red .... 7. *R. idaeus* subsp. *melanolasius*
      Stem with stout, hooked prickles; fruit purplish-black ... 9. *R. leucodermis*

53. ROSACEAE (Rose Family)

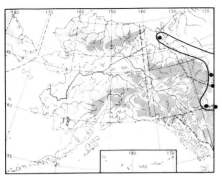
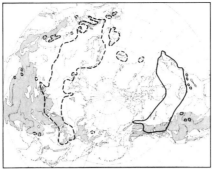
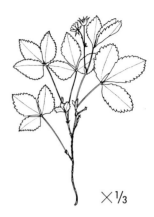

1. Rùbus pubéscens Raf.
Rubus saxatilis var. *canadensis* Michx.

Stem trailing, with erect flowering branches and elongated vegetative shoots, ending in flagelliform, rooting tips; leaves pedately 3-foliate; leaflets ovate to deltoid, coarsely double-serrate-dentate, nearly glabrous; calyx lobes linear; petals white; fruit dark red, juicy.

Thickets, shores. Described from Hudson Bay.

Forms hybrid swarms with *R. arcticus* subsp. *acaulis* [*R. paracaulis* Bailey; *R. pubescens* var. *paracaulis* (Bailey) Boiv.].

Broken line on circumpolar map indicates range of the Eurasian counterpart, **R. saxatìlis** L.

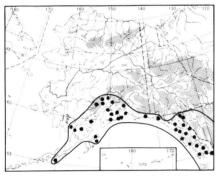

2. Rùbus pedàtus Sm.
Dalibarda pedata (Sm.) Steph.; *Comaropsis pedata* (Sm.) DC.

Leaves and flowering stems from thin, creeping and branching rootstock; leaves digitately 5-foliate, leaflets obovate, double-serrate-dentate; petioles with brownish, entire stipules at base; flowers solitary; sepals oblong-lanceolate, acute or dentate at tip; petals white, as long as sepals or somewhat longer; fruit red, juicy, palatable.

Moist woods. Described from northwestern North America.

The fruit makes an excellent jam, but the plant rarely occurs in large quantities.

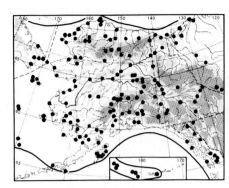 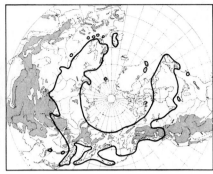

3. Rùbus chamaemòrus L. — Cloudberry

Flowering stems erect, from thin, long, branching and creeping rootstock; leaves few, round-reniform, coriaceous, obscurely 5-lobed, serrate; flowers solitary; calyx lobes ovate to ovate-oblong; petals white, obovate; female flowers with dwarfed stamens; male flowers with dwarfed pistils, on different plants; fruit at first firmly attached to calyx, hard and red-tinged, when ripe loosening from calyx, juicy and yellow, quickly dropping off.

Peat bogs. Described from Sweden.

Fruit excellent to eat when ripe, keeps without addition of sugar over the winter, and makes very good jam. Does not fruit in years when heavy frost occurs during flowering time; thus, fruiting occurs more often in milder southern areas than in interior or northern areas.

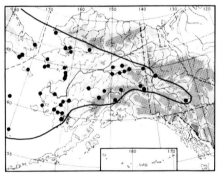 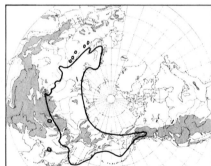

4. Rùbus árcticus L.
subsp. árcticus

Stems from long, woody, branching rootstock; leaves 3-foliate (rarely with lateral lobes cleft to base, var. **pentaphylloìdes** Hult., known from Dawson); leaflets often stipitate and somewhat lobed, broadly obovate to rhombic, with broad, ovate or ovate-lanceolate stipules; flowers 1–3, the uppermost usually overtopping the leaves; peduncles often glandular; calyx tube pubescent, often with a few glands; sepals short, triangular; petals obovate, short and broad, sometimes emarginate; fruit a globular aggregate of dark purple drupelets, palatable, richly flavored, excellent for jam and for flavoring liqueur.

Meadows, thickets, along creeks. Described from Sweden, Siberia, and Canada.

Forms hybrid swarms with subsp. *acaulis* and subsp. *stellatus* where ranges overlap.

53. ROSACEAE (Rose Family)

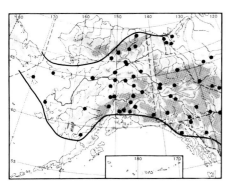
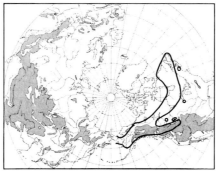
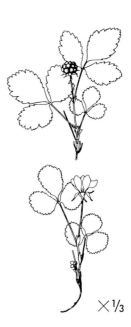

5. Rùbus árcticus L.
subsp. **acaùlis** (Michx.) Focke

Rubus acaulis Michx.; *R. arcticus* var. *grandiflorus* of authors; *R. arcticus* subsp. *stellatus* var. *acaulis* (Michx.) Boiv.

Leaves similar to subsp *arcticus*, but stipules narrow; flowers usually single, not overtopping the leaves; peduncles glandless; calyx tube glabrous or nearly so, lacking glands; sepals long, caudate; petals much longer than sepals, narrowly cuneate-obovate.

Meadows. Described from Hudson Bay.

Forms hybrid swarms with the other subspecies of *R. arcticus* where ranges overlap.

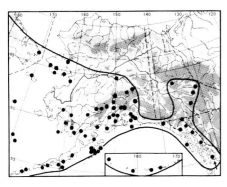
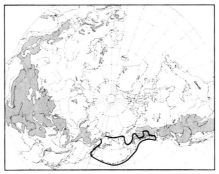

6. Rùbus árcticus L.
subsp. **stellàtus** (Sm.) Boiv. emend. Hult.

Rubus stellatus Sm.; *R. arcticus* subsp. *stellatus* var. *stellatus* Boiv.

Similar to subsp. *arcticus*, but flowers larger; leaves mostly 3-lobed, sometimes nearly entire, orbicular; hypantium pubescent, with yellow glands.

Wet meadows.

Forms hybrid swarms with subsp. *arcticus* and subsp. *acaulis* where ranges overlap (pure in the Aleutian Islands, where the other subspecies are lacking). Hybrids with *R. spectabilis* also occur (*R. alaskensis* Bailey).

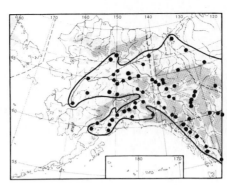 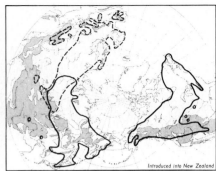

7. Rùbus idaèus L.   Raspberry
subsp. **melanolàsius** (Dieck) Focke

Rubus melanolasius Dieck; *R. idaeus* subsp. *sachalinensis* (Lév.) Focke; *R. sachalinensis* Lév.; *R. strigosus* Michx.; *R. idaeus* var. *strigosus* (Michx.) Maxim.; *R. subarcticus* Rydb.; *R. idaeus* var. *canadensis* Richards.

Stems upright, biennial, from subterranean, branching rhizome, prickly or bristly; leaves 3–5-foliate; leaflets oblong-ovate, acute, serrate, whitish below; inflorescence in axils, glandular, with 1–4 flowers; calyx lobes reflexed, caudate; petals white; fruit red, tomentulose, soon falling off, palatable, makes a good jam.

Thickets, borders of woods. *R. idaeus* described from Europe, subsp. *melanolasius* from cultivated plants from northwest North America.

Hybrids with *R. spectabilis* occur. Broken line on circumpolar map indicates range of subsp. **idaèus.**

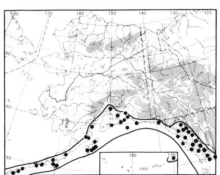 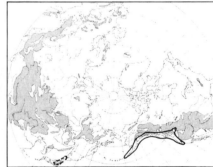

8. Rùbus spectàbilis Pursh   Salmonberry

Stems upright, biennial, from branching rhizome, forming thickets, strongly bristly, especially below, with acicular prickles; leaves mostly 3-foliate; lateral leaflets obliquely ovate, acuminate, lobulate-serrate; flowers solitary on short, leafy shoots; sepals pubescent, ovate; petals reddish-purple, elliptical, longer than sepals; fruit ovoid, raspberry-like, red or yellow when ripe, glabrous, palatable.

Moist woods, in the mountains up into lower alpine region, where it forms extensive, almost impenetrable thickets. Described from the banks of the Columbia River and the northwest coast of North America.

Broken line on circumpolar map indicates range of subsp. **vérnus** Focke.

53. ROSACEAE (Rose Family) 605

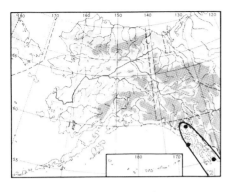
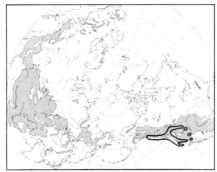
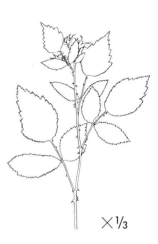

9. Rùbus leucodérmis Dougl.
Rubus occidentalis var. *leucodermis* (Dougl.) Focke.

Shrub with arching and apically rooting, glaucous and prickly branches; prickles stout, flattened, hooked; leaves 3-foliate, white-tomentose beneath; leaflets ovate, acute or somewhat acute, twice serrate; flowers in apical raceme; calyx tomentose, reflexed; petals white, shorter than sepals; stamens numerous; fruit raspberry-like, reddish-purple to black, palatable.

Thickets, borders of woods. Described from Oregon.

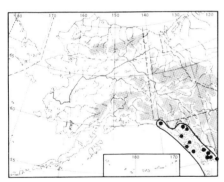

10. Rùbus parviflòrus Nutt.   Thimbleberry
Rubus nutkanus Moç.
var. **grandiflòrus** Farw.

Shrub with erect, unarmed, puberulent and glandular branches; leaves palmately 3–5-lobed, cordate, twice dentate-serrate; flowers in terminal corymbs; calyx lobes oblong-ovate, long-caudate; petals white, mostly 5, oblong-obovate to obovate, 15–25 mm long; ovary pubescent; style glabrous; fruit hemispheric, juicy, red, palatable.

Woods, along streams.
Plant highly variable.

Fragaria L.

Leaves coriaceous, thick; flowers 2–3.5 cm in diameter .... 1. *F. chiloensis* subsp. *pacifica*
Leaves thin, not coriaceous; flowers smaller .... 2. *F. virginiana* subsp. *glauca*

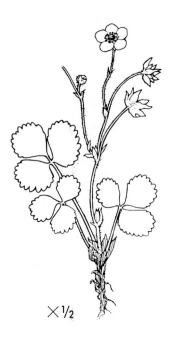
×½

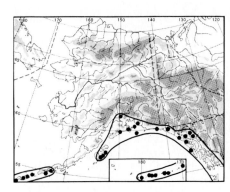

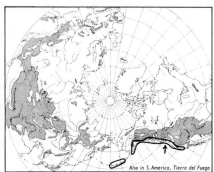

1. Fragària chiloénsis (L.) Duchesne   Beach Strawberry
Fragaria vesca var. *chiloensis* L.
subsp. **pacífica** Staudt

Plant with stout rhizome and long, brown stolons; leaves thick, coriaceous; leaflets petiolated, broadly obovate to cuneate-obovate, crenate-dentate, dark green and shiny above, silky-pubescent beneath; petioles and peduncles with long, silky, spreading hairs; calyx lobes silky-pubescent, lanceolate; petals 5, white, obovate-orbicular; fruit pilose, palatable, makes an excellent jam.

Coastal strands. *F. chiloensis* described from cultivated specimens.

Other subspecies occur in Hawaii and South America. Hybrids with cultivated strawberries also occur.

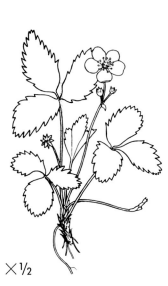
×½

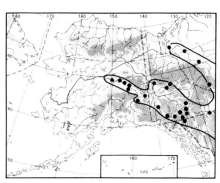

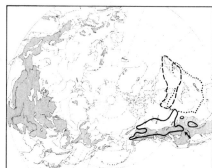

2. Fragària virginiàna Duchesne
subsp. **glaùca** (S. Wats.) Staudt
Fragaria virginiana? var. *glauca* S. Wats.; *F. yukonensis* Rydb.

Rhizome thick; outgrowing leaves firm or coriaceous; leaflets rounded to cuneate-obovate, sharply toothed, the terminal tooth smaller than the adjacent lateral teeth, mostly glabrous above, appressed silky-pubescent beneath, short-petiolulated; inflorescence shorter than leaves; fruit subglobose to ovoid.

Open slopes, borders of woods. *F. virginiana* described from Mexico, subsp. *glauca* from the Mackenzie River to Colorado and westward to Montana.

Broken line on circumpolar map indicates range of var. **térrae-nòvae** (Rydb.) Fern. & Wieg. (*F. terrae-novae* Rydb.); dotted line that of other subspecies or varieties of *F. virginiana*.

Potentilla L.

Petals purple, shorter than calyx .... 1. *P. palustris*
Petals yellow:
  Petals about half as long as calyx; leaves mostly cauline .... 3. *P. biennis*
  Petals longer than calyx; plant with basal rosettes of leaves:
    Stipules of cauline leaves very large, ovate to oblong; leaves palmate with 7–9 oblanceolate leaflets .... 4. *P. stipularis*
    Stipules of cauline leaves not large:
      Plant a shrub; leaflets entire; achenes hairy .... 2. *P. fruticosa*
      Plant herbaceous; leaflets toothed or lobed; achenes glabrous:
        Basal leaves 3-foliate (exceptionally with 1 pair of supernumerary, small leaflets):
          Leaflets lobed to the middle or beyond:
            Leaflets small, 3–8 mm long, cleft to about the middle into obtuse lobes; petals 2–4 mm long .... 6. *P. elegans*
            Leaflets larger, cleft to the base into linear lobes; petals 5–8 mm long .... 5. *P. biflora*
          Leaflets toothed:
            Leaflets tomentose or densely sericeous beneath:
              Plant coarse; flowers large, 20–30 mm in diameter; leaflets coriaceous, dark grayish-green above, tomentose and strongly ribbed beneath .... 7. *P. villosa*
              Plant more slender; flowers smaller; leaflets not coriaceous, not strongly ribbed beneath:
                Petioles tomentose or tomentose and long-haired:
                  Petioles tomentose .... 13. *P. nivea*
                  Petioles tomentose and long-haired .... See 13. *P. nivea* × *Hookeriana*
                Petioles pilose with long, straight, appressed or spreading hairs, or puberulent and pilose with such hairs:
                  Plant 1–2-flowered, low-growing; flowers large:
                    Caudex short, with reddish-brown sheaths; leaves grayish-tomentose beneath .... 8. *P. uniflora*
                    Caudex long, stout, blackish-brown; leaves with long, yellowish-gray, silky pubescence ...... 9. *P. Vahliana*
                  Plant with branched inflorescence; flowers smaller:
                    Petioles with long, straight, white hairs only .... 16. *P. Hookeriana* subsp. *Chamissonis*
                    Petioles with long, straight, white hairs, also puberulent with short, thick hairs:
                      Leaves 3-foliate .. 14. *P. Hookeriana* subsp. *Hookeriana*
                    Leaves mostly 5-foliate .. 15. *P. Hookeriana* var. *furcata*
            Leaflets hirsute or soft-pubescent beneath:
              Style much thicker at base; petals about as long as sepals .... 12. *P. norvegica* subsp. *monspeliensis*
              Style not considerably thicker at base; petals longer than sepals:
                Plant tall, flowers 20–30 mm in diameter .... 11. *P. fragiformis*
                Plant low-growing, flowers much smaller .. 10. *P. hyparctica*
        Basal leaves pinnate or digitate:
        ■ Flowers solitary; plant with runners:
          Bractlets toothed or divided; achenes grooved; leaves usually lustrous beneath; runners pubescent .... 27. *P. anserina*
          Bractlets entire; achenes not or indistinctly grooved; runners glabrous:
            Hypantium turbinate; bractlets narrow, acute, often longer than calyx lobes .... 31. *P. Egedii* subsp. *yukonensis*
            Hypantium almost flat; bractlets broader, blunt or abruptly pointed, often shorter than calyx lobes:
              ●Plant tall; leaves up to 40 cm long; leaflets with coarse, often acute teeth 6–8 cm long; flowers large .... 30. *P. Egedii* subsp. *grandis*

- Plant low-growing; leaves up to 10 cm long; leaflets with rounded teeth:
  - Leaves glabrous on both sides or very nearly so .... 28. *P. Egedii* subsp. *Egedii* var. *Egedii*
  - Leaves tomentose below .... 29. *P. Egedii* subsp. *Egedii* var. *groenlandica*
- Flowers in cymes; plants often lacking runners:
  - Leaves odd-pinnate:
    - Plant glandular above; leaves with 7–9 ovate-elliptic, deeply serrate leaflets .... 21. *P. arguta* subsp. *convallaria*
    - Plant not glandular above:
      - Leaves with 3–5 pairs of silky or tomentose leaflets, grayish or white on both sides .... 22. *P. hippiana*
      - Leaves green or grayish-green above:
        - Style filiform:
          - Leaves grayish-green above, tomentose and silky-pubescent beneath, toothed .... 17. *P. rubricaulis*
          - Leaves green above, white-tomentose beneath; leaflets pinnate; lobes acute .... 19. *P. multifida*
        - Style conical, incrassated and glandular at base:
          - Leaves silky-pubescent on both sides, with 2–3 approximate pairs of leaflets; plant prostrate .... 18. *P. pulchella*
          - Leaves grayish-tomentose beneath, with 3–7 pairs of leaflets; plant erect:
            - Segments of leaflets narrowly linear; flowers small; petals about 3 mm long .... 20. *P. virgulata*
            - Segments of leaflets lanceolate or oblong; flowers larger .... 23. *P. pennsylvanica*
  - Leaves digitate or digitately 5–7-foliate:
    - Leaflets divided nearly to midrib into linear lobes .... 25. *P. flabelliformis*
    - Leaflets not so deeply divided, merely toothed:
      - Leaves tomentose below; leaflets toothed to base .... 24. *P. gracilis*
      - Leaves glabrous or pilose beneath; leaflets toothed in upper half only .... 26. *P. diversifolia*

×⅓

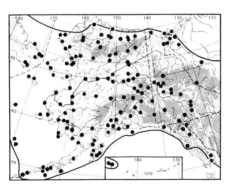

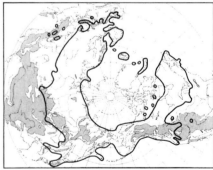

1. Potentílla palústris (L.) Scop. Marsh Fivefinger
Comarum palustre L.

Plant with creeping, somewhat woody rootstock; leaves pinnate, 5–7-foliate; leaflets oblong-lanceolate to oblanceolate, somewhat acute or obtuse, sharply serrate, dark green above, pale and more or less pubescent beneath; sepals broad, purplish on inner side, ovate, acuminate; petals brownish-purple, about half as long as sepals; achenes glabrous.

Wet meadows, along streams, in shallow water; in the mountains to at least 1,000 meters in central Alaska. Described from Europe.

The Siberian Eskimo use the dried leaves for tea.

53. ROSACEAE (Rose Family)

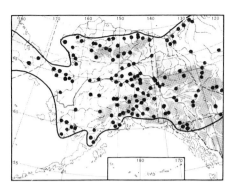

2. Potentílla fruticòsa L.   Shrubby Cinquefoil

Dasiphora fruticosa (L.) Rydb.; *Potentilla floribunda* Pursh; *Pentaphylloides floribunda* (Pursh) Löve.

Shrub with spreading or erect branches, up to 1.7 meters tall in central Alaska; leaves mostly with 5 entire, lanceolate or oblanceolate, glabrous to villous leaflets; flowers single in axils; calyx villous; bracteoles elliptic to lanceolate, as long as calyx lobes or longer; petals yellow, rounded; carpels densely villous; styles clavate, attached below middle of carpel.

Both wet and dry ground, forests, heaths, muskeg, scree slopes, in the mountains to at least 1,500 meters. Described from Portugal, England, island of Oeland (Sweden), and Siberia.

A highly variable species. Circumpolar map indicates range of entire species complex.

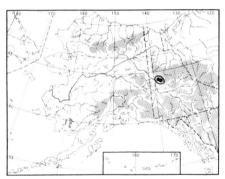
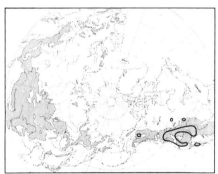

3. Potentílla biénnis Greene

Annual or biennial with slender taproot; stems glandular-puberulent and hirsute, mostly single, erect, with ascending branches; leaves mostly cauline, hirsute, with 3–4 obovate to oblanceolate, coarsely toothed, coarsely crenate-serrate leaflets; calyx glandular-puberulent and hirsute; bractlets lanceolate, shorter than calyx lobes; calyx lobes erect, ovate-triangular; petals yellow, cuneate-obovate, about half as long as sepals; style terminal, thickened at base; achenes yellow.

Waste places; once found introduced at Dawson. Described from California.

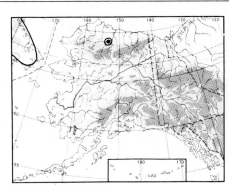
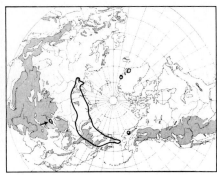

4. Potentílla stipuláris L.

Stems from thin caudex; basal leaves glabrous, palmate, with 7–9 oblanceolate to linear leaflets, more or less toothed in apex; petioles long, slender; basal stipules small, with lanceolate auricles; stem leaves gradually reduced, with very large, ovate to oblong stipules; bractlets linear, shorter than the triangular, ciliate calyx lobes; petals yellow, somewhat longer than calyx; style apical, about as long as nutlet.

Meadows, banks of streams.

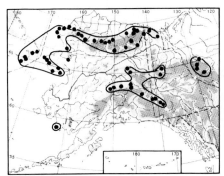
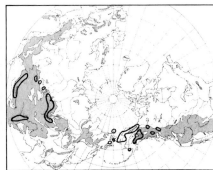

5. Potentílla biflòra Willd.

Root thick, woody; caudex branched, thickly covered with brown, marcescent leaf bases; basal leaves with linear, stiff, acute lobes, revolute in margin, sparsely pubescent with long white hairs; stipules large, adnate to petioles, with lanceolate free apex; stem with reduced leaves; bractlets ovate-lanceolate, about as long as the triangular, acute sepals; petals obcordate, longer than sepals, yellow; style long, filiform; achenes glabrous; receptacle profusely white-pubescent.

Rocks and rocky slopes, heaths, in the mountains to at least 2,000 meters. Described from eastern Siberia.

53. ROSACEAE (Rose Family)

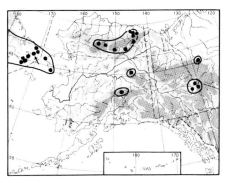

6. Potentílla élegans Cham. & Schlecht.

Densely tufted; flowering stems from much-branched caudex, covered with dark, reddish-brown, marcescent leaf bases; basal leaves small, green on both sides, ternate, sparsely pubescent and ciliate with long, soft hairs, to nearly glabrous above; leaflets broadly obovate, deeply cleft into blunt lobes; petioles white-pubescent with long, brown, adnate stipules, with ovate lobes; stem short-pubescent, with reduced leaves; flowers 5–6 mm in diameter; bractlets obovate, somewhat shorter than the ovate-lanceolate to triangular calyx lobes; petals somewhat longer than calyx, yellow, obovate, somewhat emarginate; receptacle pubescent; nutlets glabrous, with short, apical style.

Dry rocks and moist places in the mountains to at least 2,000 meters. Rare.

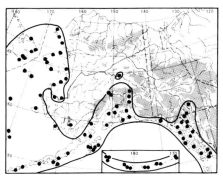
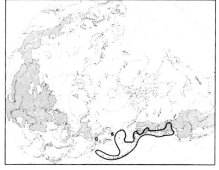
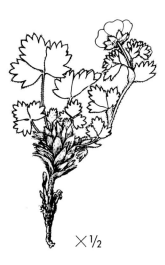

7. Potentílla villòsa Pall.

Stems stout, from short branched caudex, densely covered with marcescent, dark-brown stipules; leaves 3-foliate; leaflets coriaceous, coarsely dentate, dark grayish-green and pubescent above, with greenish-white tomentum, strongly ribbed beneath; flowers few, large, 2–3 cm in diameter; bractlets elliptic to ovate, about as long as the ovate-triangular calyx lobes or longer; petals yellow, broadly obcordate; style nearly apical, thick at base.

Rocks; common along the southern shore, rare inland. Described from the northwest coast of North America.

(See color section.)

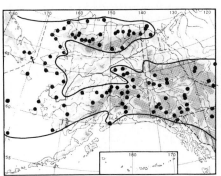
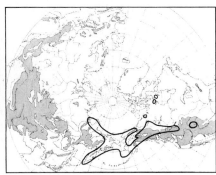

8. Potentílla uniflòra Ledeb.

Potentilla Ledebouriana Pors.; *P. vulcanicola* Juz.

Densely tufted, caudex covered with dark-brown, marcescent stipules; basal leaves 3-foliate; leaflets with 5–7 long, somewhat acute teeth, pubescent with long, silky hairs above, tomentose and long-haired beneath, especially on nerves; petioles with long, spreading, white hairs; flowers solitary or sometimes 2, large; bractlets and sepals lanceolate to ovate-lanceolate; petals yellow, obcordate; style filiform.

Rocks, scree slopes, ridges. Type locality given first (erroneously) as Dahuria, later corrected to St. Lawrence Bay, Chukchi Peninsula.

Highly variable; probably forms hybrids with *P. villosa* and *P. Vahliana,* and perhaps with others as well.

(See color section.)

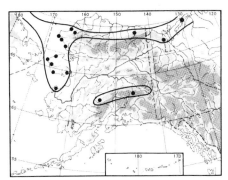
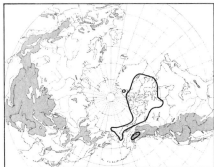

9. Potentílla Vahliàna Lehm.

Densely tufted; caudex long, thick, very densely covered with blackish-brown remains of previous year's stipules; basal leaves short-petiolated, densely compressed, 3-foliate; leaflets with 3–7 long, acute teeth above, covered with long, more or less yellowish-gray hairs, somewhat tomentose beneath, and hirsute with yellowish-gray hairs, extending brushlike in apex of teeth; stipules reddish-brown when young, darkening in age; stem with single, small leaves; flowers mostly solitary, large, 15–20 mm in diameter; bractlets and sepals broad, ovate-triangular; petals yellow, obcordate; style not papillate at base.

Dry tundra. Described from Greenland.

Closely related to *P. uniflora.*

53. Rosaceae (Rose Family)

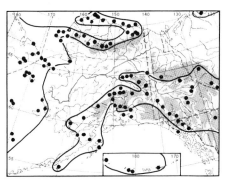
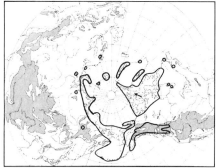
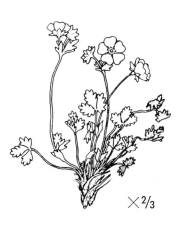

10. Potentílla hypárctica Malte
Potentilla emarginata Pursh not Desf.

Stems from branched caudex, covered with marcescent, reddish-brown stipules; leaves 3-foliate; leaflets obovate-cuneate, coarsely dentate, apical tooth longer than adjacent lateral tooth, pubescent with long, straight hairs on sides, and ciliate with long hairs; flowers 1–3, short-pedicellated, 1.5–2 cm in diameter; bractlets oblong-lanceolate, blunt; calyx lobes oblong-ovate, blunt or acute; petals yellow, obcordate, emarginate; style nearly apical, linear or slightly clavate.

On tundra and in the mountains, snowbeds, to about 2,300 meters in the St. Elias Mountains.

The specimens from the Aleutian Islands and the Bering Sea islands and some of those from the mountains are small dwarfed plants, otherwise similar to *P. hyparctica*; they are subsp. **nàna** (Willd.) Hult. (*P. nana* Willd.), but might be simply an alpine extreme of *P. hyparctica*.

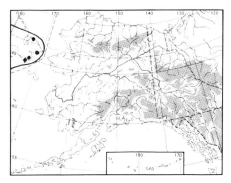
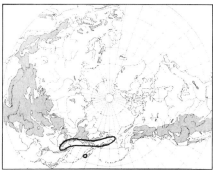
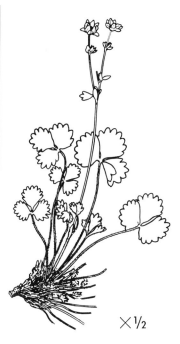

11. Potentílla fragifórmis Willd.

Stems branched above, from thick caudex, covered with remains of petioles; basal leaves cordate in outline, 3-foliate; leaflets sessile; obovate with cuneate base and 3–5 rounded teeth on each side, spreading, soft-pubescent on both sides; flowers 20–30 mm in diameter; bractlets ovate, somewhat acute, as long as the ovate calyx lobes or somewhat shorter; calyx in fruit about double its earlier size; petals large, obcordate, deeply emarginate, clear yellow; style nearly apical, only slightly thicker at base.

Grassy slopes. Described from the Aleutian Islands (where it almost certainly does not occur). Total range uncertain because of closely related but taxonomically unclear species.

× ½

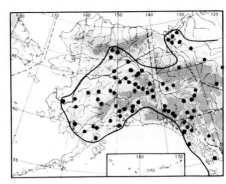
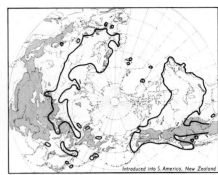

12. Potentílla norvègica L.
subsp. **monspeliénsis** (L.) Aschers. & Graebn.

Potentilla monspeliensis L.; *P. hirsuta* Michx.; *P. norvegica* subsp. *hirsuta* (Michx.) Hyl.; *P. norvegica* var. *hirsuta* (Michx.) Lehm.

Leaves mainly cauline, lower leaves 3-foliate; leaflets obovate to oblanceolate, coarsely serrate, mostly spreading- to appressed-hirsute; stipules ovate, mostly toothed; bractlets somewhat acute; calyx enlarging in fruit; petals yellow, obovate, often retuse, shorter than calyx lobes; styles terminal, thick at base.

Probably both native and introduced; moist places, waste ground. *P. norvegica* described from Norway, Sweden, Prussia, and Canada; subsp. *monspeliensis* from cultivated specimens, presumably American.

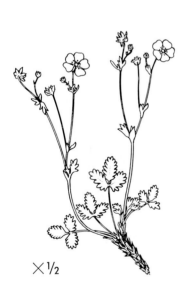
× ½

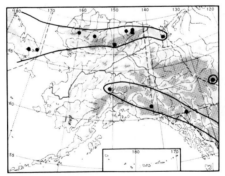
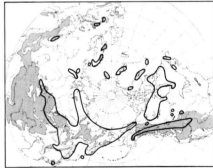

13. Potentílla nívea L.

Stems from brown caudex; leaves ternate, green above, white-tomentose beneath, with 4–6 teeth on each side and long or short, tomentose petioles, lacking straight hairs; bractlets linear, mostly somewhat shorter than calyx lobes; petals yellow, obcordate, longer than calyx; style subapical, thicker and papillate at base, shorter than nutlet.

Calcareous rocks. Described from the mountains of Lapland and Siberia.

Some specimens have the leaves tomentose above as well as below (var. **tomentòsa** Nilsson-Ehle). Specimens with tomentum and straight hairs on the petioles and on the underside of the leaves are considered to be the hybrid *P. nivea* × *Hookeriana* (*P. nivea* subsp. *fallax* Pors.).

53. ROSACEAE (Rose Family)

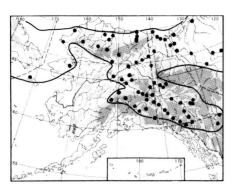 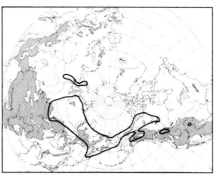 

14. Potentílla Hookeriàna Lehm.
  Potentilla nivea subsp. *Hookeriana* (Lehm.) Hiitonen; *P. nivea* var. *Hookeriana* (Lehm.) T. Wolf.
subsp. **Hookeriàna**
var. **Hookeriàna**

Stems from thick caudex; lower leaves ternate, green and villous above, densely tomentose and villous beneath; leaflets with long teeth, to nearly pinnatifid; apical tooth longer than adjacent lateral tooth; petioles more or less puberulent with long, straight, white hairs; bractlets lanceolate, obtuse, of same length as the acute, ovate sepals; petals yellow, obcordate, somewhat longer than sepals.
Calcareous rocks. Described from North America.

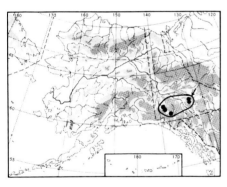 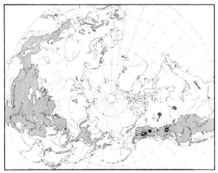 

15. Potentílla Hookeriàna Lehm.
subsp. **Hookeriàna**
var. **furcàta** (Pors.) Hult.
  Potentilla furcata Pors.

Similar to var. *Hookeriana,* but basal leaves mostly 5-foliate.
Dry, calcareous mountain slopes.

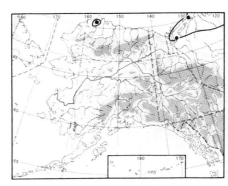 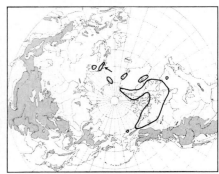

16. Potentílla Hookeriàna Lehm.
subsp. **Chamissònis** (Hult.) Hult.

Potentilla Chamissonis Hult.; *P. nivea* subsp. *Chamissonis* (Hult.) Hiitonen.

Similar to subsp. *Hookeriana*, but petioles not puberulent, pilose with long, straight hairs only.

Dry, calcareous mountain slopes.

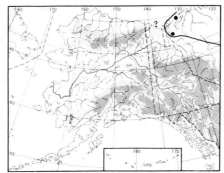 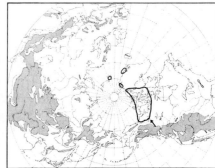

17. Potentílla rubricaùlis Lehm.

Densely tufted; stems from stout, branched caudex; basal leaves pinnate with 5 leaflets, the basal pair of lateral leaflets smaller than the upper, in small specimens sometimes lacking, glabrous or nearly so above, densely white-tomentose beneath; bractlets oblong-linear; calyx lobes ovate-lanceolate; petals obcordate, somewhat longer than calyx; style incrassated and verrucose at base; receptacle strongly pubescent.

Dry tundra.

53. ROSACEAE (Rose Family)

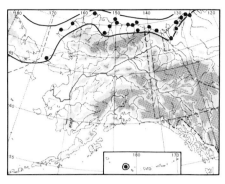 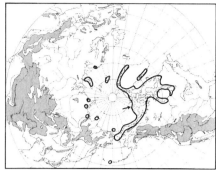 

18. Potentílla pulchélla R. Br.

Plant with long, central root; caudex branched, blackish-brown, with marcescent remains of leaf bases; basal leaves short-petiolated, with 2 pairs of lateral leaflets (rarely with a small third pair), covered above with long, straight hairs, tomentose and long-haired beneath; leaflets deeply cleft into long, lanceolate, acute lobes; stem pubescent, short, decumbent or ascending, not much overtopping leaves; bractlets and sepals of about same length; petals yellow, narrowly obovate, emarginate, barely longer than sepals; receptacle short-haired; nutlets glabrous, with style much thicker at base.

Polygon soil, open sandy places.

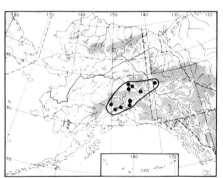 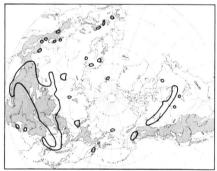 

19. Potentílla multífida L.

Plant decumbent or erect; stems from thick, short, dark-brown caudex; basal leaves pinnate; leaflets 5–7, dissected into mostly narrow segments with revolute margin, green above, tomentose beneath; cauline leaves reduced; calyx with long, white, soft hairs; bractlets linear, acute, about as long as the ovate-triangular sepals; petals yellow, nearly orbicular, about as long as sepals; receptacle sparsely white-haired; style not distinctly incrassated at base, not verrucose.

Dry slopes, gravel bars, scree slopes, open ground. Described from "Siberia, Tartaria, and Asia Minor."

Circumpolar map indicates range of entire species complex.

×⅓

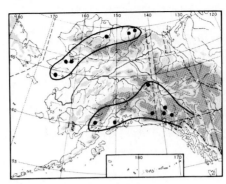
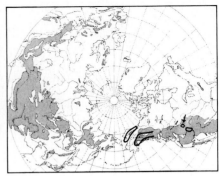

20. Potentílla virgulàta Nels.

Similar to *P. multifida*, but always erect; calyx and upper part of stem puberulent and sparsely strigose-pubescent; style incrassated and verrucose at base; receptacle profusely pubescent.

Dry slopes, open ground.

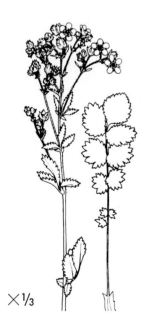
×⅓

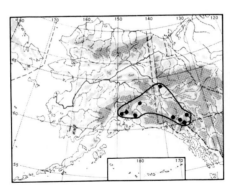
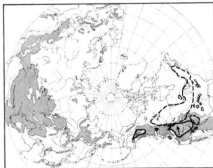

21. Potentílla argùta Pursh
subsp. **convallària** (Rydb.) Keck
  Potentilla convallaria Rydb.; *P. arguta* var. *convallaria* (Rydb.) T. Wolf.

Stems tall, erect, from stout, simple or branched caudex; basal leaves short-hirsute and glandular-puberulent, pinnate, with 7–9 ovate to elliptic, deeply serrate leaflets; stem strongly pilose, glandular above; inflorescence flat-topped with several erect branches; calyx lobes large, glandular, oblong-lanceolate, enlarged in fruit; petals pale yellow or creamy-white, shorter or slightly longer than sepals; achenes with slender style, inserted in middle of achene; glandular-roughened and thicker in middle.

Rocky soil, dry slopes. *P. arguta* described from upper Louisiana.

Broken line on circumpolar map indicates range of subsp. **argùta**.

53. ROSACEAE (Rose Family)

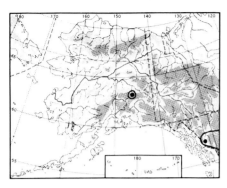 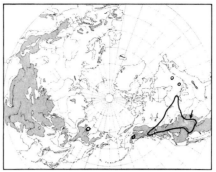 

22. Potentílla hippiàna Lehm.

Stems from thick, stout, blackish-brown caudex; basal leaves pinnate, with 7–11 crowded, oblong, sharply toothed leaflets, grayish-pubescent on both sides; cauline leaves 2–3; inflorescence with erect branches and long bracts; calyx silky-pubescent; bractlets lanceolate, acute, about as long as the somewhat broader calyx lobes; petals obcordate, yellow; style filiform, subapical.

Recently discovered; no specimens seen from area of interest.

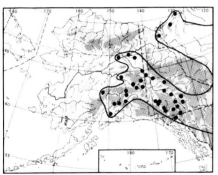 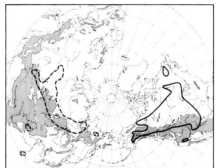 

23. Potentílla pennsylvànica L.

Densely caespitose; basal leaves pinnate, with 7–15 oblong to oblanceolate leaflets, divided to middle into linear-oblong segments, green above, tomentose beneath; stem erect, pilose; stem leaves gradually reduced; bractlets lanceolate, shorter than the ovate-triangular, acute calyx lobes; petals yellow; style subapical, thick and glandular at base, about as long as achenes.

Dry mountain slopes to at least 1,200 meters. Described from Canada.

Highly variable in pubescence: var. **strigòsa** Pursh is the most common type; var. **glabràta** S. Wats. (*P. glabrella* Rydb.), with nearly glabrous leaves, occurs more rarely.

A taxonomically very unclear species. Circumpolar map shows area of very closely related taxa in Asia; other closely related taxa occur in southern Europe and North Africa.

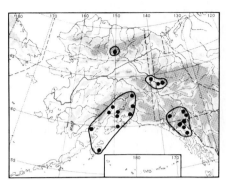
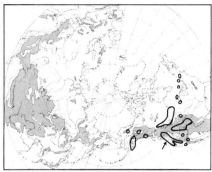

24. Potentílla gràcilis Dougl.

Potentilla alaskana Rydb.; *P. Blaschkeana* Turcz.; *P. Nuttallii* Lehm.; *P. gracilis* subsp. *Nuttallii* (Lehm.) Keck.

Stems from short caudex; basal leaves digitate, with 5–7 cuneate-oblanceolate to broadly oblanceolate, coarsely toothed leaflets, green above, more or less white-tomentose beneath; bractlets lanceolate, slightly shorter than the ovate, acuminate sepals; petals yellow, obcordate; style subapical, glandular-verrucose at base.

Roadsides, waste places; probably introduced.

Circumpolar map indicates range of *P. gracilis* in a broad sense.

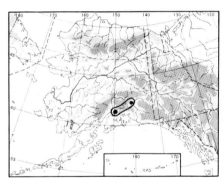
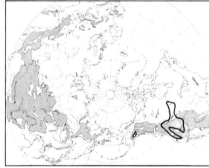

25. Potentílla flabellifórmis Lehm.

Potentilla gracilis var. *flabelliformis* (Lehm.) Nutt.

Stems from short caudex; basal leaves digitate to pinnate, with 6–7 leaflets, green and silky-villous above, densely tomentose beneath, divided nearly to midrib into linear lobes; stem silky-pubescent with few, reduced leaves; bracts about as long as the lanceolate, broader sepals, acute; petals yellow, obcordate.

Dry, open slopes, gravel. Type locality not given.

53. ROSACEAE (Rose Family)

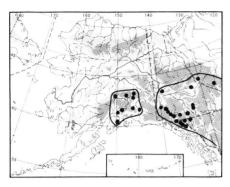

26. Potentílla diversifòlia Lehm.

Caespitose, with long, branching caudex; basal leaves long-petiolated, digitate or semipinnate to pinnate, glabrous or strigose; leaflets 5–7, cuneate-oblanceolate, deeply toothed; stem erect, slender; cauline leaves 1–2; inflorescence open; bractlets lanceolate, acute; sepals ovate-lanceolate; petals yellow, obcordate; style subapical, slender, much longer than achenes.

Alpine meadows, solifluction soil, in the mountains to at least 2,000 meters. Type locality not given.

Specimens with digitate leaves, nearly glabrous above, predominate (var. **glaucophýlla** Lehm.). High alpine plants, with densely sericeous-strigose leaves, occur rarely.

The Greenland area marked on the map represents subsp. **ranúnculus** (Lange) Pors. (*P. ranunculus* Lange).

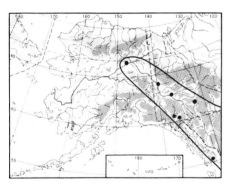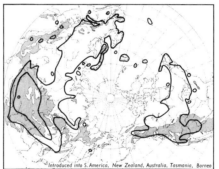

27. Potentílla anserìna L. — Silverweed

Stoloniferous; stolons pubescent; leaves all basal, more or less prostrate, interruptedly pinnate, with numerous pairs of leaflets, deeply and sharply serrate, green and glabrous or silver-silky above [f. **serícea** (Hayne) Hayek (*P. anserina* var. *sericea* Hayne)], silky-tomentose beneath; peduncles solitary, 1-flowered; bractlets lanceolate or triangular, more or less toothed; calyx lobes broader than bractlets; petals yellow; achenes with shallow furrow on back.

Waste places; introduced. Described from Europe.

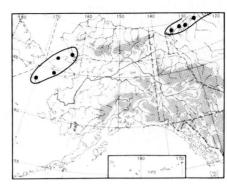 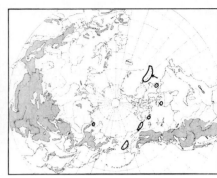

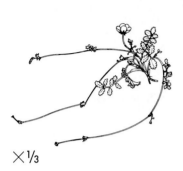

×⅓

28. Potentílla Egèdii Wormsk.

Potentilla anserina subsp. *Egedii* (Wormsk.) Hiitonen in part.

subsp. **Egèdii**

var. **Egèdii**

Similar to *P. anserina*, but leaves completely or almost completely glabrous, with 2–5 pairs of leaflets; stolons glabrous; bractlets lacking teeth; nutlets lacking (or nearly lacking); furrow on back.

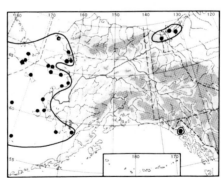 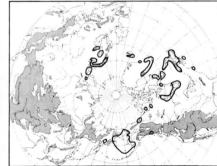

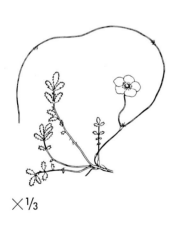

×⅓

29. Potentílla Egèdii Wormsk.

subsp. **Egèdii**

var. **groenlándica** (Tratt.) Polunin

Potentilla anserina var. *groenlandica* Tratt.; *P. anserina* subsp. *Egedii* (Wormsk.) Hiitonen in part.

Similar to var. *Egedii*, but leaves grayish-tomentose beneath.
Seashores. Described from Greenland.

53. ROSACEAE (Rose Family)

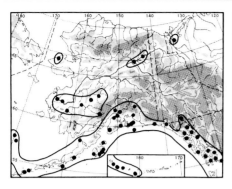

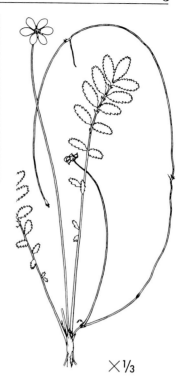

30. Potentílla Egèdii Wormsk.
subsp. **grándis** (Torr. & Gray) Hult.   Pacific Silverweed

Potentilla anserina var. *grandis* Torr. & Gray; *P. Egedii* var. *grandis* (Torr. & Gray) Hara; *P. pacifica* How.; *Argentina pacifica* (How.) Rydb.; *A. litoralis* Rydb.; *P. anserina* subsp. *pacifica* (How.) Rousi.

Similar to subsp. *Egedii* var. *groenlandica*, but larger; leaves up to 40 cm long; leaflets with 6–8 coarse, long teeth; flowers large; calyx mostly tomentose.
Seashores. Described from Oregon.

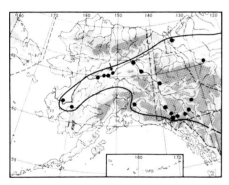
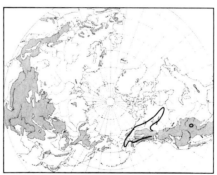

31. Potentílla Egèdii Wormsk.
subsp. **yukonénsis** (Hult.) Hult.

Potentilla yukonensis Hult.; *Argentina subarctica* Rydb.

Similar to subsp. *grandis*, but hypantium turbinate instead of flat; bractlets narrow, often linear, acute, often longer than sepals.
Along brooks and rivers.

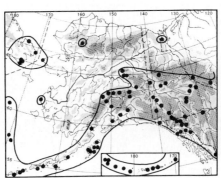 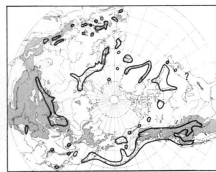

Sibbaldia L.

1. Sibbáldia procúmbens L.

Long, woody caudex; basal leaves ternate, 3-toothed at apex, strigose-pubescent on both sides; stem short; bractlets lanceolate, shorter than the broader calyx lobes; flowers crowded; petals 5, minute, linear-oblong, yellow, shorter than calyx; stamens 5.

Alpine meadows, snow beds, grassland, in the Yukon to at least 2,000 meters. Described from Lapland, Switzerland, and Scotland.

Alpine specimens depressed; lowland specimens much taller.

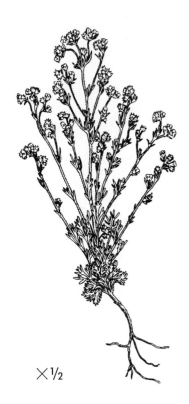

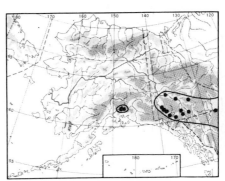 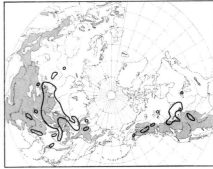

Chamaerhodos Bunge

1. Chamaerhòdos erécta (L.) Bunge

Sibbaldia erecta L.

subsp. **Nuttállii** (Torr. & Gray) Hult.

Chamaerhodos erecta var. *Nuttallii* Torr. & Gray; *C. Nuttallii* (Torr. & Gray) Pickering.

Stems from woody root, erect, often much branched in upper part, with ascending or erect branches; basal leaves white-haired and glandular, ternate; leaflets divided into linear, obtuse lobes; stem leaves similar; inflorescence glandular; flowers crowded; sepals ovate-lanceolate to triangular, somewhat acute, purplish at base; petals white, oblong, emarginate, clawed; stamens 5.

Dry, sandy places. *C. erecta* described from Siberia, subsp. *Nuttallii* from the "Missouri near the Mandan villages" (North Dakota).

53. ROSACEAE (Rose Family)

Geum L. (Avens)

Plant tall, erect; styles geniculate, the lower part hooked at apex, the upper falling off in age:
  Receptacle long-pubescent; style not glandular at base; petals considerably larger than sepals; terminal leaflet of basal leaves cuneate-obovate to oblanceolate, incised .... 3. *G. aleppicum* subsp. *strictum*
  Receptacle glabrous or short-pubescent; style minutely glandular at base; petals about as long as sepals; terminal leaflet rounded or incised:
    Terminal leaflet of basal leaves rounded to reniform ....
.... 1. *G. macrophyllum* subsp. *macrophyllum*
    Terminal leaflet deeply 3-cleft, the lobes cleft or incised ....
.... 2. *G. macrophyllum* subsp. *perincisum*
Plant low-growing, decumbent; styles not geniculate, or obscurely so, not falling off:
  Basal leaves with large, rounded or reniform, terminal leaflet and minute or no lateral leaflets .... 4. *G. calthifolium*
  Basal leaves pinnate, with several leaflets of the same size:
    Plant glabrous; petals white .... 7. *G. pentapetalum*
    Plant pubescent; petals yellow:
      Leaves pilose above, glabrous beneath; stem puberulent; flowers smaller ..
.... 5. *G. Rossii*
      Plant villous throughout, with long, silky hairs; flowers very large........
.... 6. *G. glaciale*

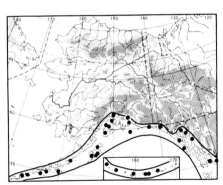

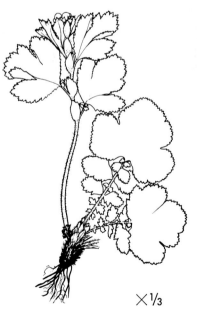

1. Gèum macrophýllum Willd.
subsp. **macrophýllum**

Stem from stout, thick, brownish, pubescent caudex; basal leaves large, interruptedly lyrate-pinnate, hirsute on both sides, with several pairs of small, crenate-dentate leaflets, and with still smaller leaflets between these and the large, orbicular to reniform, shallowly lobed end lobe; cauline leaves reduced, with comparatively small stipules; bractlets very small; calyx lobes triangular, acute, soon reflexed; petals deep yellow, somewhat longer than sepals, cuneate-ovate; nutlets hirsute; style reflexed, glabrous, or base minutely glandular; receptacle short-pubescent.

Meadows, woods. Described from northwestern North America.

Broken line on circumpolar map indicates rang of subsp. **macrophyllum** var. **sachalinéne** (Koidz.) Hara (*G. japonicum* var. *sachalinene* Koidz.)

× 1/3

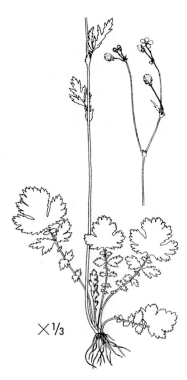
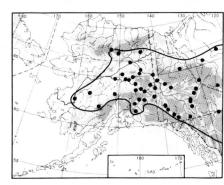
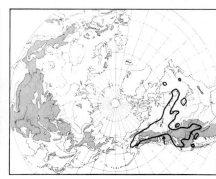

2. Gèum macrophýllum Willd.

subsp. **perincìsum** (Rydb.) Hult.

Geum perincisum Rydb.; *G. macrophyllum* var. *perincisum* (Rydb.) Raup.

Similar to subsp. *macrophyllum,* but with apical lobe of basal leaves more or less deeply cleft into serrate, distinct lobes.

Thickets, meadows, woods.

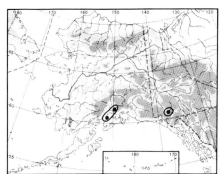
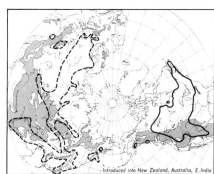

3. Gèum aléppicum Jacq.

subsp. **strictum** (Ait.) Clausen

Geum strictum Ait.; *G. aleppicum* var. *strictum* (Ait.) Fern.

Similar to *G. macrophyllum* subsp. *perincisum,* but petals larger, longer than calyx; calyx lobes narrower; style not glandular at base; receptacle long-pubescent; sides of achenes glabrous.

Thickets, meadows. Described from cultivated plants from North America.

Broken line in Eurasia on circumpolar map indicates range of subsp. **aléppicum.**

53. ROSACEAE (Rose Family)

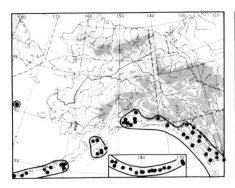

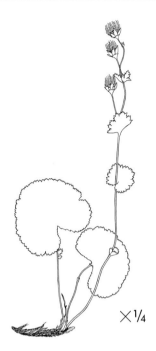

4. Gèum calthifòlium Menzies

Sieversia calthifolia (Menzies) D. Don; *Acomastylis calthifolia* (Menzies) Bolle; *Parageum calthifolium* (Menzies) Nakai & Hara.

Stems from stout, dark-brown caudex; basal leaves pubescent on both sides, with short, yellow hairs, lyrate-pinnate, with very minute lower lobes and large, orbicular to reniform, crenate end lobe; bractlets lanceolate, much shorter than the ovate to ovate-lanceolate, acute calyx lobes; petals yellow, twice as long as calyx, cordate; style pubescent to about three-fourths of its length.

Wet meadows. Described from the west coast of North America.

Hybridizes with *G. Rossii* [*Sieversia macrantha* Kearney; *Acomastylis macrantha* (Kearney) Bolle; *G. Schofieldii* Calder & Taylor] where the ranges overlap.

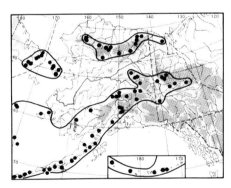

5. Gèum Róssii (R. Br.) Ser.

Sieversia Rossii R. Br.; *Acomastylis Rossii* (R. Br.) Greene; *A. humilis* (R. Br.) Rydb.; *S. humilis* R. Br.

Stem from thick, woody, dark, purplish-brown caudex; basal leaves interruptedly pinnate, with many, often crowded, more or less lobed leaflets, pubescent above, glabrous beneath; stem puberulent; stem leaves with few, acute lobes; flowers 1–3; bractlets ovate, acute, shorter than the ovate-triangular, acute calyx lobes; petals yellow, large, broadly obovate, more or less emarginate; nutlets pubescent, with glabrous style.

Dry, stony places, snow beds. Broken line on circumpolar map indicates ranges of very closely related taxa.

The rhizome is eaten raw, usually with fat, by the Siberian Eskimo.

(See color section.)

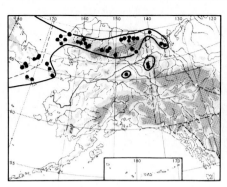
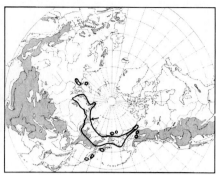

6. Gèum glaciàle Adams
Sieversia glacialis (Adams) R. Br.; *Novosieversia glacialis* (Adams) Bolle.

Stem from long, thick, woody caudex; basal leaves short-petiolated, pinnate with 11–15 leaflets; leaflets with few teeth, glabrous or sparsely pubescent above, densely yellowish, long-pubescent beneath; stem with long, spreading hairs; flowers very large, up to 45 mm in diameter; bractlets linear-lanceolate, much shorter than calyx lobes, often numerous; petals yellow, ovate, abruptly clawed; anthers small; nutlets long-haired; styles about 2.5 mm long at maturity of fruit, with long, spreading hairs in lower part, glabrous at tip.

Stony slopes, dry heaths. Flowers very early.

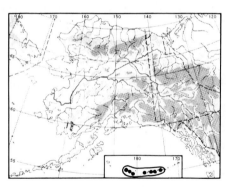

7. Gèum pentapétalum (L.) Makino
Dryas pentapetala L.; *Sieversia pentapetala* (L.) Greene; *D. anemonoides* Pall.; *Geum anemonoides* (Pall.) Willd.; *S. anemonoides* (Pall.) Willd.

Stems from long, woody, branched caudex; basal leaves glabrous, pinnate with 5–7 leaflets, lowest leaflets linear or cleft in apex, the rest deeply lobed to serrate; stem puberulent, stem leaves reduced, the upper linear; bractlets linear; calyx lobes ovate with attenuated apex; petals white, twice as long as calyx; nutlets pubescent; style elongating in fruit, glabrous at tip.

Wet places.

53. ROSACEAE (Rose Family)

Dryas L. (Dryas)

Classification of the *Dryas* taxa occurring within the area of interest is made very difficult by the fact that hybrid swarms seemingly are formed between all taxa—with the exception of *D. Drummondii*, which is distinct.

Base of leaves cuneate; filaments pubescent; flowers yellow, nodding in anthesis .. . . 1. *D. Drummondii*
Base of leaves truncate, rotundate, cordate or sagittate; filaments glabrous; flowers white, (rarely yellowish), erect in anthesis:
  Midvein on underside of leaves with narrow scales and lateral white hairs ("*octopetala* scales"):
    Midvein lacking sessile or stipitate glands . . . . 2. *D. octopetala* subsp. *octopetala* var. *octopetala*
    Midvein with *octopetala* scales and glands . . . . 3. *D. octopetala* subsp. *octopetala* var. *kamtchatica*
  Midvein lacking *octopetala* scales:
    Midvein with sessile or stipitate glands; lateral veins on underside of leaves not hidden by tomentum . . . . 4. *D. octopetala* subsp. *alaskensis*
    Midvein lacking *octopetala* scales or glands; lateral veins hidden by tomentum:
      Leaves with strongly revolute margin, needlelike or, if flat, short and small, short-stipitate . . . . 5. *D. integrifolia* subsp. *integrifolia*
      Leaves flat, long, rounded in apex with very narrowly or not at all revolute margins . . . . 6. *D. integrifolia* subsp. *sylvatica*

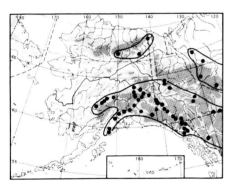
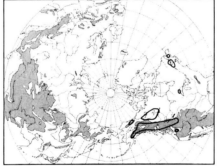

1. Drỳas Drummóndii Richards.   Yellow Dryas

Stem from long, woody caudex; leaves all basal, oblong-elliptic to obovate, coarsely dentate, strongly nerved, green above, white-tomentose beneath, often with minute basal leaflets; calyx with long, dark-stipitate glands, or, rarely, lacking glands (var. **eglandulòsa** Pors.); flower nodding in anthesis, petals yellow, forming tube or funnel when in flower; fruit with long, white plumes.

Gravel bars in rivers, in the mountains to at least 1,100 meters. Described from the Rocky Mountains and Slave Lake.

Some specimens have leaves more or less tomentose above as well as below [var. **tomentòsa** (Farr) L. O. Williams; *D. tomentosa* Farr].

(See color section.)

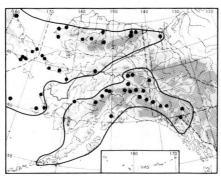
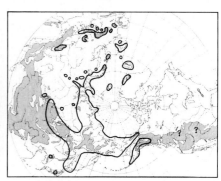

2. Dryas octopétala L.
subsp. **octopétala**
var. **octopétala**

Plant decumbent or matted, with woody caudex; leaves oblong to ovate, crenate-dentate along entire margin, glabrous above or nearly so, white tomentose beneath; midvein beneath more or less densely provided with brown scales with long white hairs (*octopetala* scales); flowers solitary, flowering stem white-tomentose and glandular, sometimes with single bract; calyx with long, dark glands, petals longer than sepals, white or rarely yellow (var. **lutèola** Hult.); nutlets with elongated, feathery styles.

Alpine heaths to at least 2,000 meters. Described from mountains of Lapland, Switzerland, Austria, central Italy, Ireland, and Siberia.

Leaves sometimes tomentose above, as well [f. **argéntea** (Blytt) Hult.; *D. octopetala* var. *argentea* Blytt], or with more or less viscid glands above (var. **víscida** Hult.; *D. punctata* Juz.).

Forms hybrid swarms with subsp. *alaskensis* and with *D. integrifolia*.

(See color section.)

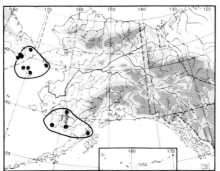
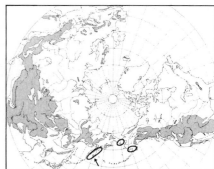

3. Dryas octopétala L.
subsp. **octopétala**
var. **kamtschática** (Juz.) Hult.
Dryas kamtschatica Juz.

Similar to var. *octopetala*, but with midvein of the underside of the leaves having both *octopetala* scales and small, short, often few glands, sometimes hidden in tomentum and easy to overlook.

Tundra and alpine heath.

53. ROSACEAE (Rose Family)

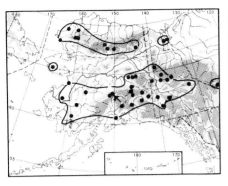  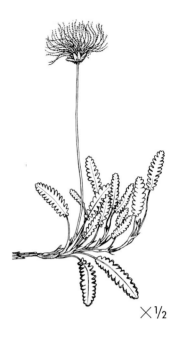

4. Drỳas octopétala L.
subsp. **alaskénsis** (Pors.) Hult.
Dryas alaskensis Pors.

Similar to subsp. *octopetala*, but with large leaves, broadest toward apex, strongly ribbed, with several deeply incised teeth, often not densely tomentose beneath but nearly glabrous (f. **glabràta** Hult.); midvein of underside of leaf with stalked, capitate glands, lacking scales.

Snow beds.

(See color section).

× ½

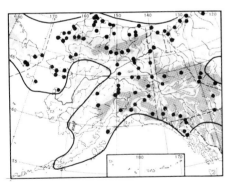 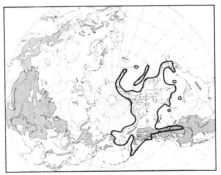 

5. Drỳas integrifòlia M. Vahl
subsp. **integrifòlia**

Matted; leaves needlelike or flat, short and small, entire in margin or toothed only near base, rugose, dark green and mostly lustrous above [or white-canescent, f. **canéscens** (Simm.) Fern. (*D. integrifolia* var. *canescens* Simm.)], densely white-tomentose beneath, with strongly revolute margin; scapes tomentose and often with long, dark glands; flowers solitary; calyx with long, dark glands; sepals narrow, acute; petals creamy white; styles elongating, plumose.

Heath on tundra and in the mountains, solifluction soil, to at least 1,300 meters. Described from Greenland.

Hybrid swarms with subsp. *sylvatica* and with *D. octopetala* common.

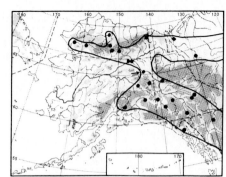 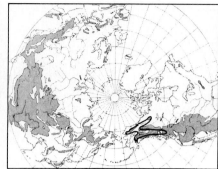

6. Drỳas integrifòlia M. Vahl.
subsp. **sylvàtica** (Hult.) Hult.

Dryas integrifolia var. *sylvatica* Hult.; *D. integrifolia* f. *sylvatica* (Hult.) Pors.; *D. sylvatica* (Hult.) Pors.

Similar to subsp. *integrifolia*, of which it can be regarded as a lowland race, but taller; leaves flat, smooth above, large and long, rounded at apex, with margin very narrowly or not at all revolute.

Wet forests, bogs, also dry localities, chiefly in low altitudes.

Sanguisorba L. (Burnet)

Spikes at first lanceolate, pointed, later long-cylindrical, greenish-white (sometimes tinged with purple) .... 3. *S. stipulata*
Spikes short, rounded, obtuse, red:
    Stamens only slightly, if at all, exceeding sepals; filaments filiform ....
    .... 1. *S. officinalis*
    Stamens 2–3 times longer than sepals; filaments flattened .... 2. *S. Menziesii*

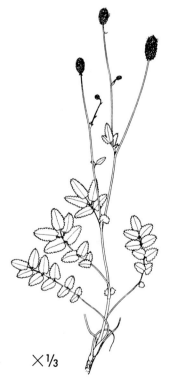

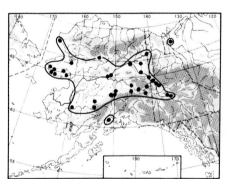 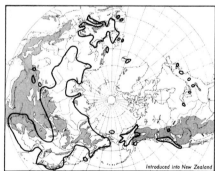

1. Sanguisórba officinàlis L.

S. microcephala Presl; *S. officinalis* subsp. *microcephala* (Presl) Calder & Taylor.

Stems from thick, many-headed caudex; basal leaves pinnate, with glabrous, ovate to oblong, crenate-serrate, sessile or stipitate leaflets; stem tall, finely pubescent at base, with 1–2 reduced leaves; spikes ovoid to broadly cylindrical, 1–2 cm long, purplish-black; filaments filiform, included or barely exserted.

Moist places. Described from Europe.

Range of var. **polygàma** (Nyl.) Mela & Caj. (*S. polygama* Nyl.) is included on circumpolar map.

53. ROSACEAE (Rose Family)

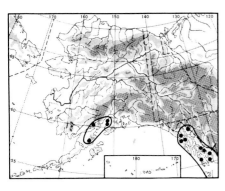
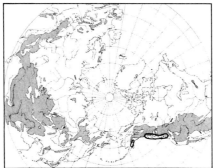

2. Sanguisórba Menzièsii Rydb.
Sanguisorba officinalis × stipulata?

Similar to S. *officinalis,* but stamens with flattened filaments, 2–3 times longer than sepals; basal leaves usually larger.

Moist places.

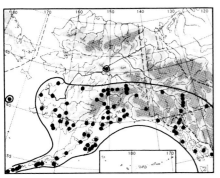
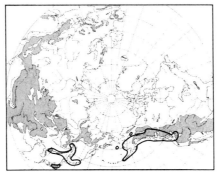
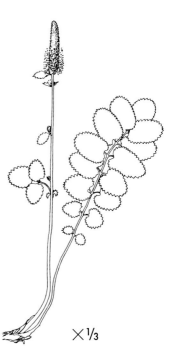

3. Sanguisórba stipulàta Raf.
Sanguisorba sitchensis C. A. Mey.; *S. canadensis* var. *sitchensis* (C. A. Mey.) Koidz.; *S. canadensis* var. *latifolia* Hook.; *S. latifolia* (Hook.) Cov.; *S. canadensis* subsp. *latifolia* (Hook.) Calder & Taylor.

Glabrous; stems from stout caudex; basal leaves pinnate, with 9–15 ovate to ovate-oblong, cordate, coarsely serrated, stipitate leaflets; stem leaves 1–3, reduced; spikes 3–8 cm long, greenish-white, sometimes purplish-tinged; filaments flattened, clavate, much longer than sepals.

Bogs, swamps, snow beds, meadows. Described from Oregon.

Rosa L. (Rose)

Leaves glabrous on both sides, small; plant armed with straight prickles; flowers in corymbs . . . . 2. *R. Woodsii*
Leaves more or less pubescent below, at least on veins, larger; flowers solitary or 1–3, on short lateral shoots:
    Stem densely armed with bristles and straight, terete prickles; stipular prickles lacking . . . . 1. *R. acicularis*
    Stem armed with few, short, more or less flattened prickles, usually in pairs; stipular prickles present; upper part of stem often unarmed . . . 3. *R. nutkana*

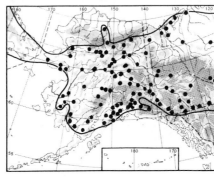
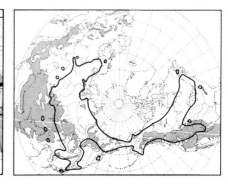
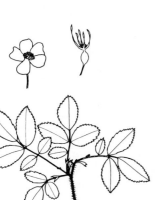
× ⅓

1. Ròsa aciculàris Lindl.

Small shrub; leaflets odd-pinnate, with 3–7 opaque, simply serrate leaflets, with puberulent and often glandular rachis, glabrous above, puberulent, or (in var. **Sayiàna** Erlans.) glabrate beneath; stem and branches bristly and acicular-prickly; flowers solitary, 4–6 cm in diameter; sepals erect in fruit, glandular on back; hips subglobose (var. **Bourgeauiàna** Crép.) or, rarely, ellipsoid or pyriform, contracted to neck below sepals.

Woods, heaths, tundra bogs, thickets; in McKinley Park to about 1,100 meters. Described from Siberia.

The pedicels are mostly glabrous, as in most American specimens, but in central Alaska and the Yukon, specimens with glandular pedicels occur, as in the majority of Eurasiatic specimens. The hips are good for jam, jelly, syrup, and marmalade, and are rich in vitamin C.

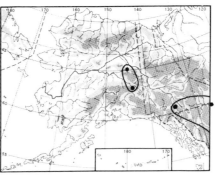
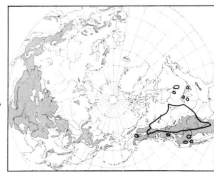

× ½

2. Ròsa Woódsii Lindl.

Small shrub; leaflets odd-pinnate, with 5–9 sharply and simply serrate leaflets, glabrous on both sides; flowers in corymbs, small, pink; stem armed with straight prickles; pedicels glabrous; sepals puberulent; hips globose to ellipsoid.

Dry slopes. Described from cultivated specimens, presumably from the Missouri River.

Highly variable, with several varieties: the few plants taken from this area belong to var. **Woódsii,** with crowded, small leaflets 1.5–2.5 cm long; if other specimens are collected here, they may well prove to be of other varieties.

53. ROSACEAE (Rose Family) / 54. LEGUMINOSAE (Pea Family)

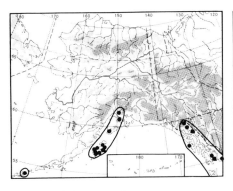

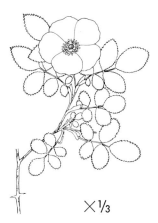

3. Ròsa nùtkana Presl
Rosa aleutensis Crép.

Small shrub; leaflets odd-pinnate, with 5–7 serrate to doubly serrate leaflets, sparsely long-haired below; stem nearly unarmed or with large infrastipular, straight, flat prickles, broad at base; sepals long, caudate, puberulent and glandular on back; flowers solitary, large, up to 7 cm in diameter, pink; hips globose to pyriform.

Woods, meadows.

Hybrids with *R. acicularis* occur where ranges overlap. In Asia, the closely related **R. amblyòtis** C. A. Mey. and **R. davùrica** Pall. occur; broken line on circumpolar map indicates ranges of these closely related taxa.

Lupinus L. (Lupine)

Leaflets 10–18; basal leaves 15–20 cm in diameter . . . . 1. *L. polyphyllus*
Leaflets fewer; leaves smaller:
  Basal leaves short-petiolated; petioles usually about as long as diameter of leaf; leaflets blunt, mucronate; calyx broad . . . . 3. *L. nootkatensis*
  Basal leaves long-petiolated; petioles usually 2 to several times longer than diameter of leaf; leaflets acute:
    Leaflets glabrous above or nearly so, green below . . . . 2. *L. arcticus*
    Leaflets densely pubescent on both sides:
      Stem, pedicels, and calyx with spreading, villous pubescence . . 4. *L. Kuschei*
      Pubescence appressed . . . . 5. *L. lepidus*

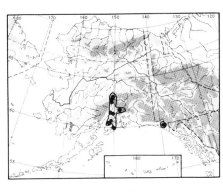
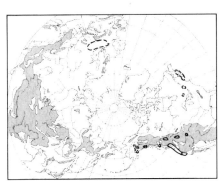

1. Lupìnus polyphýllus Lindl.
Lupinus matanuskensis P. Sm.; *L. pseudopolyphyllus* P. Sm.

Plant tall, with long racemes; leaves long-petiolated, with 10–18 oblanceolate to elliptic-oblanceolate, acute, large leaflets; flowers blue or violet.

Introduced; spreading along roads. Described from northwest North America.

Broken line on circumpolar map indicates localities where the species has been introduced.

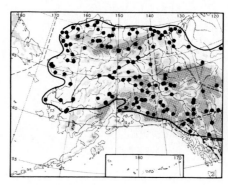
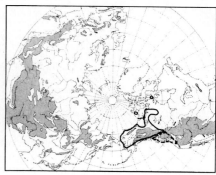

2. Lupìnus árcticus S. Wats.

Lupinus nootkatensis var. *Kjellmannii* Ostenf.; *L. borealis* Heller; *L. Burkei* Gombócz.; *L. donellyensis* P. Sm.; *L. gakonensis* P. Sm.; *L. matanuskensis* P. Sm.; *L. multicaulis* P. Sm.; *L. multifolius* P. Sm.; *L. prunifolius* P. Sm.; *L. yukonensis* P. Sm.; *L. polyphyllus* subsp. *arcticus* (S. Wats.) Phill.

Stems several, from stout, branched caudex; basal leaves green on both sides, long-petiolated, with several oblanceolate, or elliptic-lanceolate, acute leaflets, glabrous above, strigose-pubescent beneath; flowering stem white-pubescent; racemes short or elongated, white to brownish-pubescent; flowers blue or dark blue; calyx with pubescent, acute, linear lobes.

Dry and damp slopes, gravel bars, solifluction soil, roadsides; common over large areas, in the mountains to at least 1,500 meters.

Hybrids with *L. nootkatensis* occur. Probably poisonous. Broken line on circumpolar map indicates range of subsp. **subalpìnus** (Piper & Robins.) Dunn (*L. subalpinus* Piper & Robins.).

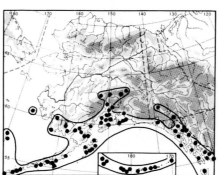
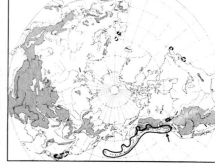
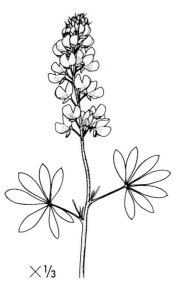

3. Lupìnus nootkaténsis Donn

L. columbianus P. Sm.; *L. kiskensis* P. Sm.; *L. latifolius* var. *canadensis* P. Sm.; *L. nootkatensis* var. *Henry-Looffii* P. Sm., var. *Ethel-Looffiae* P. Sm., var. *perlanatus* P. Sm.; *L. perennis* subsp. *nootkatensis* (Donn) Phill.

Stems from stout, branched caudex and long, woody root; basal leaves short-petiolated, with several oblong-obovate to oblanceolate leaflets, blunt or mucronate, more or less densely white-to-brownish-pubescent on both sides, or glabrous above; flowering stem densely pubescent; racemes large; flowers blue, rarely white (f. **leucánthus** Lepage); calyx lobes broad, entire (var. **unalaschcénsis** S. Wats.) or usually more or less cleft, dentated or lobed.

Dry slopes, gravel bars, common along the southern coast. Described from cultivated specimens, presumably from Nootka Sound.

Plants with more sericeous pubescence have been called var. **fruticòsus** Sims. Hybrids with *L. arcticus* and *L. polyphyllus* (*L. stationis* P. Sm.) occur. Broken line on circumpolar map indicates areas where *L. nootkatensis* has been introduced and partly naturalized.

The seeds are poisonous, causing inflammation of the stomach and intestines.

(See color section.)

54. Leguminosae (Pea Family)

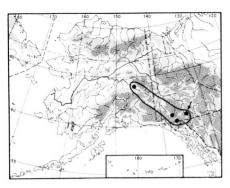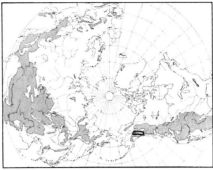

4. Lupìnus Kùschei Eastw.
Lupinus Jacob-Andersonii P. Sm.; *L. Porsildorum* P. Sm.

Stems from long, branched caudex; basal leaves long-petiolated, more or less densely sericeous-strigose, pubescent on both sides, with long, yellowish hairs; leaflets lanceolate to oblong-lanceolate, acute; stem yellowish-pubescent, especially above; racemes comparatively short; pedicels and calyx yellowish-villous; flowers blue.

Roadsides; possibly introduced and referable to some more southern species.

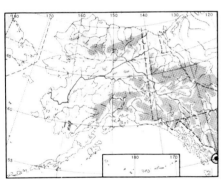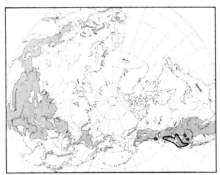

5. Lupìnus lépidus Dougl.

Low, less than 2 dm tall; rusty, appressed, strigillose-sericeous; basal leaves with 5–9 oblanceolate leaflets, pubescent on both sides; racemes closely flowered, short; flowers bluish; calyx sericeous.

Probably on dry slopes. Described from Fort Vancouver to the Great Falls of the Columbia River.

54. Leguminosae (Pea Family)

Medicago L.

Flowers purple; pod without spines .... 1. *M. sativa*
Flowers yellow; pod with or without spines:
  Pod falcate or nearly straight .... 2. *M. falcata*
  Pod coiled:
    Racemes 1–5-flowered; pod flat, coiled 2–3 times, spiny or tuberculate, several-seeded, brown when ripe .... 3. *M. hispida*
    Racemes many-flowered; pod without spines, subreniform, 1-seeded, black when ripe .... 4. *M. lupulina*

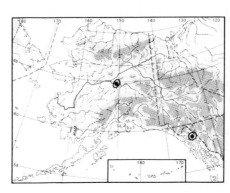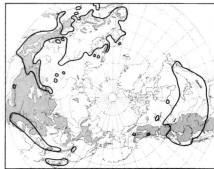

1. Medicàgo satìva L.   Luzerne, Alfalfa

Erect or ascending, branching from base; leaflets narrowly obovate, toothed toward apex; stipules linear-lanceolate, acuminate, more or less toothed; flowers about 8 mm long, purple; pedicels shorter than calyx tube; pods coiled loosely 2–3 times, 10–20-seeded.

Introduced, escaped from cultivation. Described from Spain and Germany.

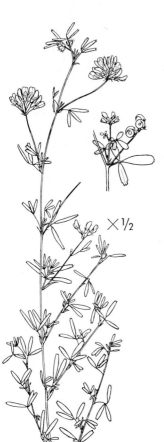

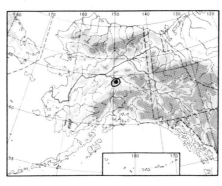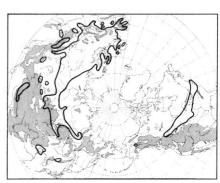

2. Medicàgo falcàta L.

Decumbent or erect; leaflets linear-lanceolate, emarginate, mucronate; stipules lanceolate, acute; pedicels longer than calyx tube; flowers yellow, about 8 mm long; pods falcate to nearly straight, 2–5-seeded.

Waste places; introduced, escaped from cultivation. Described from Europe.

54. Leguminosae (Pea Family)

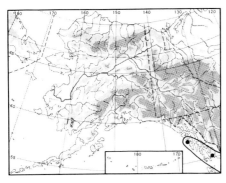  

3. Medicàgo híspida Gaertn.

Procumbent, glabrous or nearly so; leaflets cuneate-obovate to obcordate, serrate toward tip; stipules ovate-lanceolate, acuminate, laciniate; peduncles equaling petioles; flowers yellow, about 3 mm long; pods flat, coiled 2–3 times, strongly reticulate, with double row of slightly hooked or curved spines.

Waste places; introduced. Described from southern Europe.

Introduced in many countries of the world; reliable circumpolar map cannot be prepared.

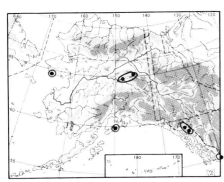 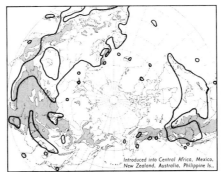 

4. Medicàgo lupulìna L.

Decumbent or ascending, pubescent especially below; leaflets obovate to suborbicular, serrate in upper half; stipules ovate-lanceolate, nearly cordate at base, toothed; racemes compact; peduncles longer than petioles; petals yellow, longer than calyx; pod reniform, reticulate, 1-seeded, coiled in almost one complete turn, black when ripe.

Introduced. Described from Europe.

Melilotus Mill. (Sweet Clover)

Flowers yellow; pods yellowish-brown when ripe .... 1. *M. officinalis*
Flowers white; pods black when ripe .... 2. *M. albus*

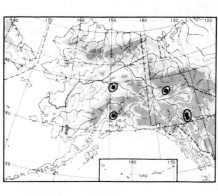 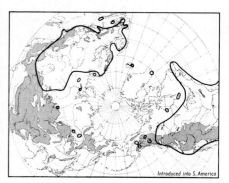

1. Melilòtus officinàlis (L.) Lam.
"Trifolium M. officinalis" L.; *Melilotus arvensis* Wallr.

Stem decumbent or erect, branched; leaflets oblong-elliptic, narrowed at both ends; racemes lax and slender; flowers yellowish, 5–6 mm long; wings and standard equal, but longer than keel; pod ovate, compressed, mucronate, glabrous, yellowish-brown when ripe.

Waste places; an introduced weed. Described from Europe.

Circumpolar map tentative.

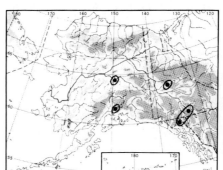 

2. Melilòtus álbus Desr.

Stem erect, 1–2 meters tall, branched; leaflets oblong, with about 10 pairs of veins, serrate; racemes lax and slender; flowers white, 4–5 mm long, wings and keel nearly equal, somewhat shorter than standard; pods glabrous, reticulate, ovate, compressed, mucronate, black when ripe.

Waste places; an introduced weed. Described from Siberia.

54. LEGUMINOSAE (Pea Family)

Trifolium L. (Clover)

Flowers yellow, in age brownish:
  Leaflets all sessile .... 7. *T. aureum*
  Central leaflet petiolated:
    Heads many-flowered; standard distinctly striate .... 6. *T. campestre*
    Heads 10–15-flowered; standard not striate .... 10. *T. dubium*
Flowers white or red:
  Flowers white:
    Stem erect .... 2. *T. hybridum*
    Stem creeping, rooting from nodes .... 3. *T. repens*
  Flowers red:
    Leaflets all, or at least the upper, with 5 leaflets; stem tall, erect ....
    .... 1. *T. lupinaster*
    Leaflets all 3-foliate:
      Heads small, up to 1 cm in diameter:
        Involucrum glabrous; leaflets rounded in apex .... 8. *T. variegatum*
        Involucrum long-villous; leaflets emarginate ...... 9. *T. microcephalum*
      Heads considerably larger:
        Calyx long-villous; heads globose to ovate .... 4. *T. pratense*
        Calyx glabrous; heads umbelliform .... 5. *T. Wormskjoldii*

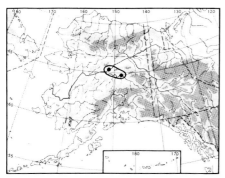 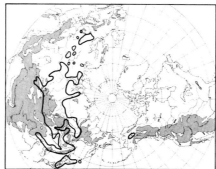 

1. Trifòlium lupináster L.

Stems from stout caudex, erect, mostly unbranched, 15–30 cm tall, leafless at base; leaves short-pedunculated, stipules with lanceolate free parts; leaflets 3–5, oblong-lanceolate, pubescent beneath, finely and sharply serrate; heads umbelliform; pedicels 1.5–2 mm long; calyx pubescent, with long subulate teeth; flowers mostly bluish-purple.

Introduced. Described from Siberia.

× 1/3

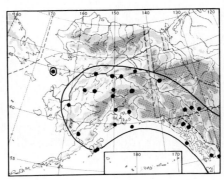
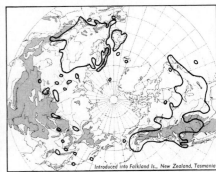

2. Trifòlium hỳbridum L. Alsike Clover

Erect or decumbent, bent at nodes, nearly glabrous; leaflets obovate or elliptic; stipules oblong, with triangular, acuminate tip; heads axillary, globular; peduncles about 5 cm long; flowers white, in age pink, yellowish-brown when withered, the outer short-pedicellated, the inner long-pedicellated.

Introduced; escaped from cultivation. Described from Europe.

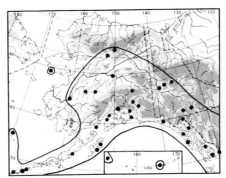
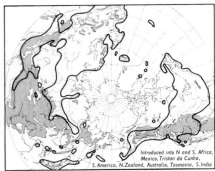

3. Trifòlium rèpens L.

Creeping, rooting at nodes, glabrous; leaves long-petiolated; leaflets obovate or obcordate, usually with pale band toward base; stipules oblong, with short subulate points; heads axillary, globular; peduncles long; pedicels elongating in age; flowers white or pink.

Introduced; escaped from cultivation. Described from Italy, southern France, and Switzerland.

54. LEGUMINOSAE (Pea Family)

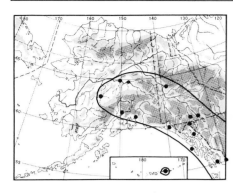 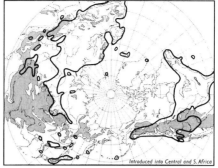 

4. Trifòlium praténse L.

Erect or decumbent, more or less pubescent; leaflets oblong to obovate, mostly entire, more or less emarginate, often with pale spot at base; stipules oblong, with triangular, setaceous free part; heads terminal, globose to ovate, sessile, subtended by pair of short-petiolated leaves; flowers pink to purple, or creamy white, sessile.

Introduced; escaped from cultivation. Described from Europe.

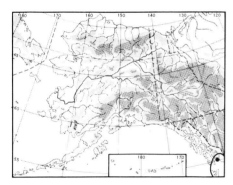 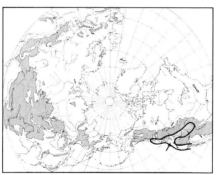 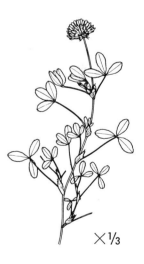

5. Trifòlium Wormskjòldii Lehm.
Trifolium fimbriatum Lindl.

Glabrous, stems usually branched from base, from creeping rootstock; leaflets dark green, obovate to oblanceolate, finely setose-serrulate; heads involucrate; involucre deeply lobed; calyx glabrous with subulate teeth, slightly longer than tube; flowers reddish to purplish, often white-tipped, the pods 1–4-seeded.

Meadows, coastal dunes. Described from cultivated specimens, presumably from California.

×3/4

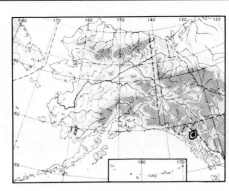
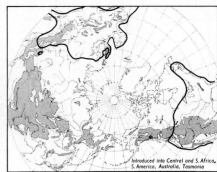

6. Trifòlium campéstre Schreb.
Trifolium procumbens of authors.

Erect or ascending, rather stout, branching from base, more or less pubescent; leaflets obovate to obcordate, cuneate, the median petiolated; stipules ovate, much shorter than petiole, rounded at base; heads axillary; peduncles much exceeding petioles; flowers yellow, in age light brown; standard striate; pedicels about length of calyx tube; style much shorter than pod.

Reported from Juneau; introduced. No Alaskan specimens seen. Described from Germany.

×2/3

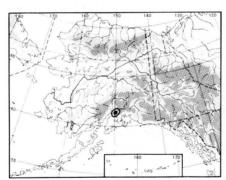
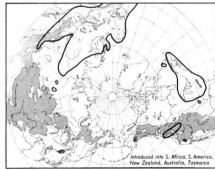

7. Trifòlium aùreum Poll.
Trifolium agrarium of authors.

Erect or decumbent with short branches, somewhat pubescent; similar to *T. campestre*, but larger; leaflets narrowly obovate, all sessile; stipules linear-oblong, acuminate, longer than petioles; peduncles exceeding leaves; flowers yellow, in age yellowish-brown; teeth of calyx glabrous.

Introduced. Described from Germany.

54. Leguminosae (Pea Family)

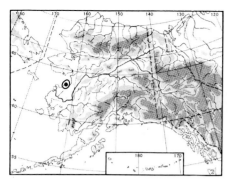
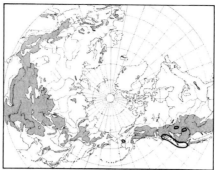

8. Trifòlium variegàtum Nutt.

Glabrous, stems usually several, decumbent or ascending; leaflets obovate to oblanceolate, setose-serrulate; heads involucrate; involucre with 3–7-toothed lobes; calyx teeth subulate-setaceous; flowers purple with white tip; pod 1–2-seeded.

Highly variable. Reported from Saint Michael; introduced. No Alaskan specimens seen. Described from "near the mouth of the Wahlamet [the Willamette, probably]."

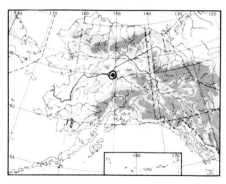

9. Trifòlium microcéphalum Pursh

Procumbent or ascending, more or less villous; stem branched; leaflets obcordate to oblanceolate, retuse, serrate; stipules ovate-lanceolate, denticulate-serrate; heads small, involucrate; involucrum deeply campanulate, with lanceolate, entire, scarious-margined lobes; calyx pubescent with subulate teeth, longer than tube; flowers pinkish to white, not much exceeding calyx; pods 1–2-seeded.

Introduced at Manley Hot Springs, at Tanana. Described from "the banks of Clark's river."

×2/3

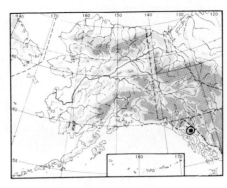
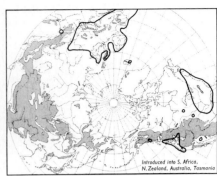

10. Trifòlium dùbium Sibth.

Trifolium minus Sm.

Procumbent or ascending, slender, more or less pubescent; leaflets obcordate to obovate, cuneate; stipules broadly ovate, as long as petiole; heads axillary, about 5 mm in diameter; peduncles considerably exceeding petioles; flowers yellow, in age dark brown; standard not striate; pedicels shorter than calyx tube.

Reported from Juneau; introduced. No Alaskan specimens seen. Described from Europe.

Astragalus L. (Milk Vetch)

Flowers yellow, lacking any trace of purple:
    Plant 50–60 cm tall; racemes shorter than leaves; pod glabrous . . . . 1. *A. americanus*
    Plant about 20 cm tall; racemes surpassing leaves; pod black-strigillose . . . . 2. *A. umbellatus*
Flowers whitish or yellowish with purple keeltip, purple or bluish:
    Pods sessile or short-stipitate, in densely hirsute or strigillose, short, compact racemes, with erect flowers:
        Pubescence with simple hairs . . . . 12. *A. agrestis*
        Pubescence with hairs fixed at middle, with two appressed horizontal branches:
            Calyx teeth short, 0.4–1 mm long . . . . 11. *A. adsurgens* subsp. *viciifolius*
            Calyx teeth longer . . . . 10. *A. adsurgens* subsp. *robustior*
    Pods sessile or stipitate, at maturity not in compact, headlike inflorescence:
        Plant decumbent, more or less matted, with short racemes:
            Pods ovate, erect, sessile, glabrous, less than 1 cm long . . . . . . 17. *A. Bodinii*
            Pods longer, pubescent:
                Raceme 2–5-flowered; leaflets small, up to 5 mm long:
                    Pod sessile, ovate and inflated when ripe . . . . 8. *A. polaris*
                    Pod with long stipe above calyx, semicircular when ripe . . . . 9. *A. nutzotinensis*
                Raceme with more numerous flowers; leaflets larger:
                    Corolla white with blue or purplish tips . . . 6. *A. alpinus* subsp. *alpinus*
                    Corolla dark blue, pale toward base . . . . . . 7. *A. alpinus* subsp. *arcticus*
        Plant taller, erect or ascending with elongate racemes, at least in fruit:
            Pods sessile, erect . . . . 15. *A. Williamsii*
            Pods sessile or stipitate, pendant or deflexed:
                Pods sessile or nearly so, ellipsoid to ovate, short, about 1 cm long, densely black-hirsute; leaves glabrous above, canescent beneath:
                    Leaflets narrowly oblong . . . . 3. *A. eucosmus* subsp. *eucosmus*
                    Leaflets shorter and broader; flowers small . . . . 4. *A. eucosmus* subsp. *Sealei*
                Pods stipitate, longer:
                    Wings bidentate in apex; pods mostly glabrous . . . . . 5. *A. aboriginum*
                    Wings not bidentate:
                        Leaflets linear-oblong to lanceolate . . . . 16. *A. tenellus*
                      Leaflets oblong to obovate, broader:
                          Calyx teeth short; racemes long; leaflets glabrous to sparingly strigillose beneath . . 13. *A. Robbinsii* subsp. *Robbinsii* var. *minor*
                          Calyx teeth longer; racemes shorter; leaflets densely pubescent beneath . . . . 14. *A. Robbinsii* subsp. *Harringtonii*

54. Leguminosae (Pea Family)

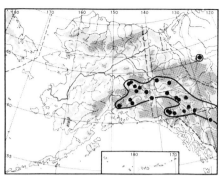
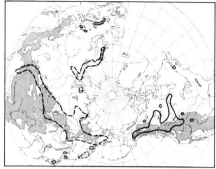
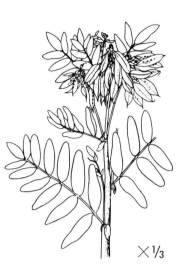

1. Astrágalus americànus (Hook.) M. E. Jones

Phaca frigida var. *americana* Hook.; *Astragalus frigidus* var. *americanum* (Hook.) S. Wats.; *Phaca americana* (Hook.) Rydb.; *A. gaspensis* Rousseau.

Stem stout, glabrous or nearly so, from woody caudex; stipules membranaceous, oblong to ovate; racemes many-flowered, usually shorter than leaves; calyx glabrous, the teeth triangular-ciliate; corolla ochroleucous; legume inflated, membranaceous, long-stipitate, ellipsoid, mostly glabrous.

Meadows, banks of streams. Described from the Rocky Mountains, 52–56°, and from Great Slave Lake.

Broken line on circumpolar map indicates range of the very closely related **A. frígidus** (L.) Gray (*Phaca frigida* L.), of Eurasia.

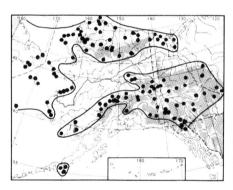
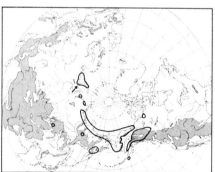

2. Astrágalus umbellàtus Bunge

Phaca frigida var. *litoralis* Hook.; *Astragalus frigidus* var. *litoralis* (Hook.) S. Wats.

Stems white-short-pubescent, from woody caudex, with subterranean runners, with leafless sheaths at base; stipules brown, membranaceous, broadly ovate, white- or black-pubescent; leaflets 7–11, ovate to oblong-ovate, blunt, glabrous above, pubescent beneath; racemes longer than leaves, few-flowered, crowded; calyx more or less black- or white-pubescent, with triangular, acute teeth; corolla ochroleucous, the standard rounded, emarginate; pod ovate to elliptic, acute at both ends; black-strigose, stigma glabrous.

Stony slopes, solifluction soil, heaths, meadows, in the mountains to at least 1,100 meters.

The root is edible and used by the natives of Siberia.

×⅓

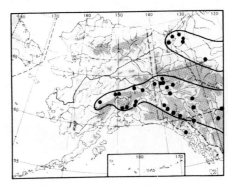
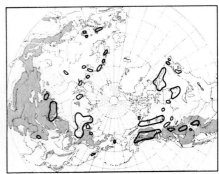

3. Astrágalus eucósmus Robins.

subsp. **eucósmus**

Phaca elegans Hook.; *Atelophragma elegans* (Hook.) Rydb.; *A. oroboides* var. *Americanus* Gray.

Stems from short, woody rhizome; stipules lanceolate to ovate; leaves with 9–17 narrowly oblong to elliptic, obtuse leaflets, glabrous above, canescent beneath; flowering raceme first dense, soon elongating; calyx black-haired with linear-lanceolate teeth; corolla purple; wings only slightly longer than keel; pods sessile or short-stipulated, obliquely ellipsoid to ovoid, reflexed, densely black-hirsute (or rarely white-hirsute) to strigose.

Gravel, stony slopes, meadows. Described from the Rocky Mountains.

Eurasiatic range on the circumpolar map represents the closely related **A. norvègicus** Grauer (*A. oroboides* Hornem.).

×⅓

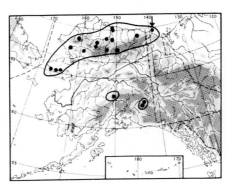
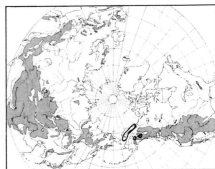

4. Astrágalus eucósmus Hornem.

subsp. **Sèalei** (Lepage) Hult.

Astragalus Sealei Lepage; *Atelophragma atratum* Rydb.

Similar to subsp. *eucosmus*, but differs in having shorter and broader leaves, shorter racemes, and smaller flowers.

Rocky slopes, meadows.

54. Leguminosae (Pea Family)

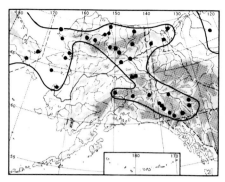
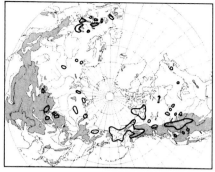

5. Astrágalus aboríginum Richards.

Homalobus aboriginorum (Richards.) Rydb.; *Atelophragma aboriginum* (Richards.) Rydb.; *A. lineare* Rydb.; *Astragalus linearis* (Rydb.) Pors.; *A. aboriginorum* var. *Muriei* Hult.; *A. Lepagei* Hult.; *A. Richardsonii* Sheld.; *A. australis* with respect to Alaskan plant.

Stems several, from woody taproot, prostrate to ascending, with large, obtuse stipules at base; leaves sparsely to densely sericeous to strigose or villose, with 7–15 linear to oblong-elliptic leaflets; flowers nodding; calyx with linear teeth; petals whitish or cream-colored; wings bidentate at tip, longer than keel; pod pendulous, membranaceous, falcate-lunate with slender stipe, about as long as calyx, glabrous or pubescent.

Stony and sandy soil, rocks. Described from Carlton House, Saskatchewan.

Extremely variable in form of leaflets, size of flowers, and pubescence. The root is edible.

Eurasiatic range shown is that of **A. austràlis** (L.) Lam. (*Phaca australis* L.).

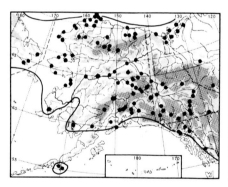
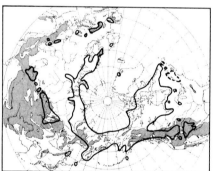
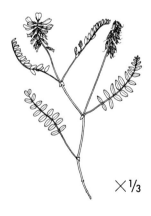

6. Astrágalus alpìnus L.
subsp. **alpìnus**

Matted; stems creeping at base, stipules lanceolate to deltoid, free to base; leaflets mostly 8–11 pairs, oblong, elliptic to narrowly obovate, white-pubescent beneath; raceme short, with divergent flowers; calyx black-hirsute; flowers bluish-violet, pale at base, rarely white; keel about as long as standard, wings comparatively broad; pod reflexed, stipitate, black-haired, strongly grooved dorsally.

Grassy slopes, gravel, scree slopes, in the mountains to at least 2,000 meters.

Forms introgression with subsp. **alaskànus** Hult. [var. *alaskanus* (Hult.) Lepage], which occurs chiefly in central Alaska and the Yukon, and differs in having elongate raceme; keel much shorter than the standard; narrow wings (about 2 mm broad); and white petals, bluish at tip. S. *alpinus* described from Lapland and Switzerland, subsp. *alaskanus* from the Yukon.

Broken line on circumpolar map indicates range of subsp. **alpìnus** var. **Brunetiànus** Fern.

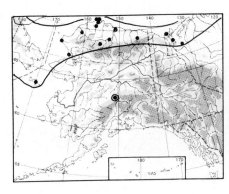
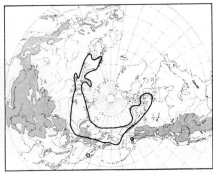

7. Astrágalus alpìnus L.
subsp. árcticus (Bunge) Hult.

Astragalus arcticus Bunge not Willd.; *A. subpolaris* Boriss. & Schischk.

Similar to subsp. *alpinus* and *alaskanus*, but flowers dark blue, pale toward base. Grassy and stony tundra; a poorly differentiated arctic race. Described from arctic Russia, Novaya Zemlya, and arctic eastern Siberia.

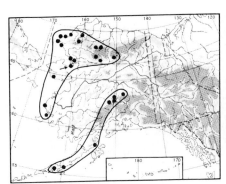
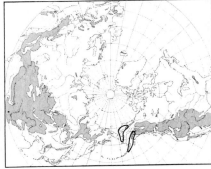

8. Astrágalus poláris Benth.

Homalobus amblyodon Rydb.; *Astragalus amblyodon* (Rydb.) Hult.

Stems weak, slender, from branching, delicate, subterranean rhizome; stipules triangular-ovate, clasping, connate, glabrous or ciliate; leaves with 7–17 oblong-obovate to obovate-cuneate, retuse, more or less crowded leaflets; racemes few-flowered; calyx strigillose, with black or white hairs and triangular-subulate to triangular teeth; petals lilac-purple; pod sessile, lanceolate-oblong when young, becoming strongly inflated and papery-membranous in age.

Salt marshes along coast, gravel and sandy soil inland.

54. Leguminosae (Pea Family)

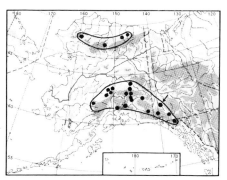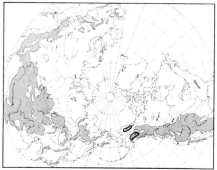

9. Astrágalus nutzotinénsis Rousseau
Gynophoraria falcata Rydb.; *Astragalus falciferus* Hult.; *A. gynophoraria* Tidestr.

Stem weak, slender, from branching, delicate, subterranean rhizome; stipules ovate, clasping, connate, glabrous or strigulose; leaves with 7–15 ovate to obovate or oblong-lanceolate, obtuse, more or less retuse leaflets; racemes 2–4-flowered; flowers large, 12–17 mm long; calyx strigillose, with black hairs and subulate teeth, separated by wide sinuses; petals reddish-purple, pale at base; pod glabrous, compressed laterally, with long stipe (gynophore) within calyx, at first nearly straight, linear, in age semicircular, broader, black- or white-strigillose.

Sandy soil, gravel bars, in the mountains to about 1,000 meters.

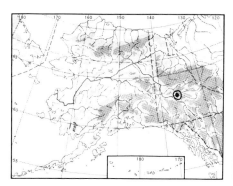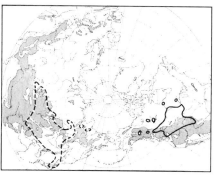

10. Astrágalus adsúrgens Pall.
subsp. **robústior** (Hook.) Welsh
Astragalus adsurgens var. *robustior* Hook.; *A. striatus* Nutt.

Stems several, stout, ascending from woody taproot; stipules scarious, clasping, connate; leaflets 9–23, narrowly oblong to oblong-ovate, strigillose, with hairs forked into 2 horizontal branches; flowers white or pale purplish to purple; calyx teeth 1.4–4.2 mm long, white to blackish, strigillose; banner longer than wings, slightly reflexed; pod sessile, erect, membranaceous, thick and broad, strigillose.

Dry slopes, rocky places, riverbanks. *A. adsurgens* described from Transbaikalia and Mongolia, subsp. *robustior* from the Columbia River.

Broken line on circumpolar map indicates tentative range of subsp. **adsúrgens**.

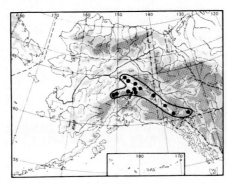
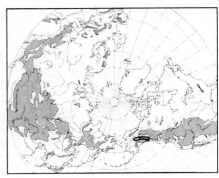

11. Astrágalus adsúrgens Pall.
subsp. **viciifòlius** (Hult.) Welsh

Astragalus viciifolius Hult.; *A. tananaica* Hult.; *A. adsurgens* var. *tananaicus* (Hult.) Barneby.

Similar to subsp. *robustior*, but calyx teeth mostly 0.4–1 mm long; pod short-stipitate.

Dry slopes, rocky places.

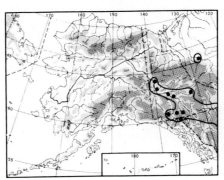
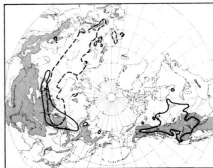

12. Astrágalus agréstis Dougl.

Astragalus dasyglottis Fisch.; *A. hypoglottis* var. *dasyglottis* (Fisch.) Ledeb.; *A. goniatus* Nutt.; *A. Tarletonensis* Rydb.

Stems weak, decumbent or erect, from taproot; stipules ovate to lanceolate, obtuse; leaves with 10–19 linear-lanceolate to oblong-lanceolate, mostly retuse, appressed-pubescent leaflets; raceme capitate; calyx strigose with linear teeth; corolla purplish and more or less whitish; wings slender, longer than keel; pods sessile, erect, ovoid, deeply sulcate, grayish- to blackish-hirsute.

Moist meadows. Described from the Red River south toward Pembina.

Broken line on circumpolar map indicates range of the closely related **A. dànicus** Retz.

54. LEGUMINOSAE (Pea Family)

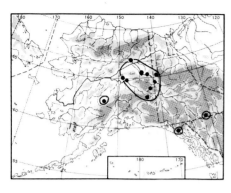 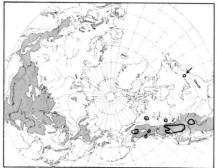 

13. Astrágalus Robbínsii (Oakes) Gray
Phaca Robbinsii Oakes.
subsp. **Robbínsii**
var. **mìnor** (Hook.) Barneby
Phaca elegans var. *minor* Hook.; *Astragalus Macounii* Rydb.; *Atelophragma Collieri* Rydb.; *Astragalus Collieri* (Rydb.) Pors.

Stem stout, ascending or erect, from woody rhizome; stipules lance-ovate; leaves with 9–13 oblong to obovate leaflets, glabrous above, sparingly strigillose to glabrous beneath; racemes up to 20-flowered, lax in age; calyx black-strigose, with lance-deltoid teeth; petals whitish, with tip of keel purple; keel shorter than wings, much shorter than banner; pod scarcely grooved dorsally, ellipsoid, with stipe 1.5–5 mm long, black- or white-strigillose.

Meadows, stream banks. *A. Robbinsii* described from Vermont, var. *minor* from the Rocky Mountains.

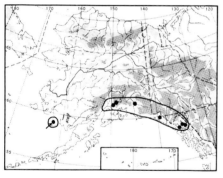 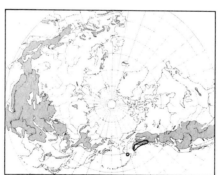 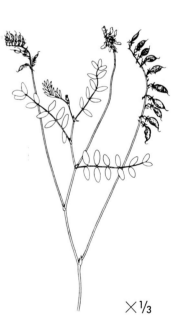

14. Astrágalus Robbínsii (Oakes) Gray
subsp. **Harringtònii** (Rydb.) Hult.
Atelophragma Harringtonii Rydb.

Similar to subsp. *Robbinsii* var. *minor*, but racemes short and compact; leaves more densely pubescent beneath, minutely whitish-tuberculate above; calyx teeth longer; pods more abruptly pointed.

Meadows, stream banks.

54. Leguminosae (Pea Family)

× ¼

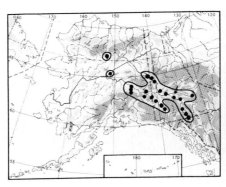
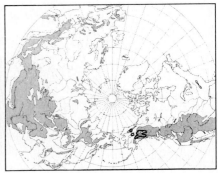

15. Astrágalus Williámsii Rydb.
Astragalus Gormani Wight.

Stems erect, from stout, woody caudex; lowest stipules oblong-obovate, the upper narrower, membranaceous; lowest leaves short-petiolated, with 5–15 petiolulate, linear-oblong to lanceolate or sometimes obovate, retuse, somewhat fleshy leaflets; racemes surpassing leaves; flowers numerous; calyx strigillose, with black or white hairs and triangular-subulate teeth, becoming marcescent; petals ochroleucous, with keel bluish at tip; pod erect, ovoid to lance-ovoid, strigillose, on short gynophore.

Banks of streams, river bars, poplar and aspen woods.

× ⅓

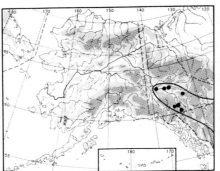
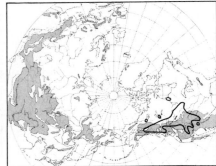

16. Astrágalus tenéllus Pursh
Homalobus tenellus (Pursh) Britt. & Rydb.

Stems several, erect or ascending, from thick woody taproot; stipules lanceolate, the lowest papery, brown, clasping, connate, forming subcylindric sheath; leaves with 11–21 leaflets, more or less crowded, linear-oblong to lanceolate, acute to obtuse; racemes many-flowered; calyx strigillose, with black or white hairs and linear to subulate teeth; petals white to cream-colored, or with tip of keel lilac; pod pendulous, stipitate, oblong-elliptic, strongly compressed, glabrous or strigillose.

Gravel bars, shores. Described from the banks of the Missouri River.

54. LEGUMINOSAE (Pea Family)

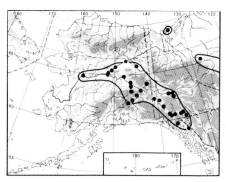
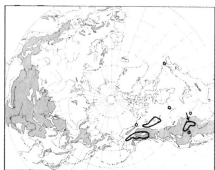

17. Astrágalus Bodínii Sheld.

Astragalus yukonis M. E. Jones; *A. stragulus* Fern.; *Homalobus retusus* Rydb.; *Phaca yukonis* (M. E. Jones) Rydb.

Loosely matted; stems slender, filiform, prostrate, from woody taproot; stipules deltoid to ovate, clasping; leaflets 7–17, oblong to oblanceolate or ovate, retuse or emarginate, subcinereous when young; racemes loose, 3–15-flowered; calyx strigillose, with white or black hairs and subulate teeth; petals pink to purplish, banner notched; pod sessile, erect, ovate to ellipsoid, up to 10 mm long, not conspicuously inflated, black or white-pubescent.

Sandy soil, river bars.

Oxytropis L.

Leaves unifoliate or digitately trifoliate; pods with style 1.2–2 mm long . . . .
. . . . 1. *O. Mertensiana*
Leaves pinnate or verticillate:
  Pods with stipe 10–12 mm long, longer than calyx; plant nearly glabrous . . . . . .
. . . . 2. *O. revoluta*
  Pods sessile:
  ■ Calyx lobes not glandular:
    Old stipules reddish-brown, dark:
      Flowers pink to purple; inflorescence 2–3-flowered; pods prostrate on ground, more than 3 times as long as broad . . . . . . 3. *O. kokrinensis*
      Flowers yellowish; inflorescence several-flowered; pods erect, less than 3 times as long as broad . . . . 4. *O. Maydelliana*
    Old stipules not reddish-brown:
      Pods pendant; flowers small, 6–10 mm long:
        Flowers whitish, suffused with pink or purple; racemes 10–20-flowered, elongating in age; plant long-villous, mostly with 1–2 internodes
. . . . 5. *O. deflexa* var. *sericea*
        Flowers pink to purple; racemes 7–10-flowered, compact also in fruit; plant sparingly pilose, mostly acaulescent . . . .
. . . . 6. *O. deflexa* var. *foliolosa*
      Pods not pendant; flowers larger:
      ● Inflorescence few-flowered (rarely with more than 5 flowers):
        Plant glabrous . . . . 13. *O. glaberrima*
        Plant pubescent:
          Margins of stipules with cilia and thick sausagelike processes; flowers large, usually over 15 mm long . . . . 7. *O. arctica*
          Margins of stipules lacking sausagelike processes; flowers usually smaller:
            Stipules with elliptic, free lobes, glabrous on back; scapes erect
. . . . 8. *O. Scammaniana*
            Stipules with narrower, free lobes, pubescent on back when young; scapes weak, resting on leaves or on soil:
              Pods elliptic, with abruptly hooked beak, glabrous or minutely strigose . . . . 9. *O. Huddelsonii*
              Pods cylindric, with short beak, grayish- or white-pubescent:
              ▲ Entire plant with snow-white pubescence; flowers mostly single . . . . 12. *O. nigrescens* subsp. *arctobia*

×½

▲ Entire plant grayish-pubescent; flowers mostly in pairs:
     Plant densely caespitose, hirsute-pubescent ....
        .... 10. *O. nigrescens* subsp. *bryophila*
     Plant pulvinate, densely lanate ....
        .... 11. *O. nigrescens* subsp. *pygmaea*
● Inflorescence normally with more than 5 flowers:
     Flowers numerous, mostly over 25, in dense, mostly elongated, spike-like racemes; leaflets fasciculated, appearing verticillate, solitary leaflets mostly lacking; plant densely silky-pubescent ....
        .... 14. *O. splendens*
     Flowers less numerous, in capitate or oblong racemes; few or no leaflets fasciculated:
        Flowers yellow or whitish, tip of the keel with or without blue spot:
           Stipules ciliate but lacking thick sausagelike processes; flowers large, mostly over 18 mm long .... 15. *O. sericea*
           Stipules ciliate and with more or less numerous, thick, sausage-like processes; flowers smaller:
              Leaves 15–35-foliate; raceme many-flowered; flowers mostly 12–17 mm long ...... 16. *O. campestris* subsp. *gracilis*
              Leaves 9–17-foliate; raceme with fewer than 10 flowers; somewhat smaller, sometimes bluish-yellow flowers ...
                 .... 17. *O. campestris* subsp. *Jordalii*
        Flowers blue or purplish:
           Stipules firm, purplish .... 18. *O. kobukensis*
           Stipules thin, large, grayish or yellowish .. 19. *O. koyukukensis*
■ Calyx lobes glandular:
     Stipules more or less pubescent on back .... 21. *O. viscida*
     Stipules glabrous on back:
        Stipules sparsely glandular; calyx tube short, the teeth filiform-subulate
           .... 20. *O. sheldonensis*
        Stipules profusely verrucose-glandular .... 22. *O. borealis*

×½

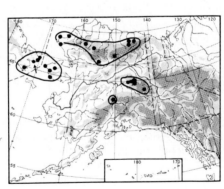 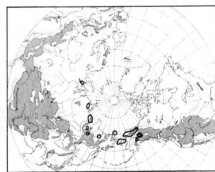

1. Oxýtropis Mertensiàna Turcz.
Aragallus Mertensianus (Turcz.) Greene

Plant with taproot and branching caudex, covered with persistent pale stipules; leaves 1–3(–5)-foliate; leaflets elliptic to oblong, glabrous above, ciliate and sparingly pubescent, and bluish-gray beneath; stipules nearly free of each other, linear-lanceolate, ciliate; scape white-pubescent; raceme 1–3-flowered; calyx densely black-pilose, with teeth 2–4 mm long, lance-acuminate, half as long as tube; petals reddish-violet (rarely white), with stipe 1.5–2 mm long, erect, black-haired.

Solifluction soil, gravel bars, in the mountains to at least 1,000 meters.

The root is eaten raw by the Siberian Eskimo.

54. Leguminosae (Pea Family)

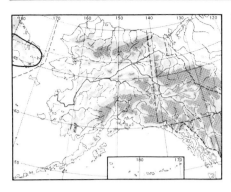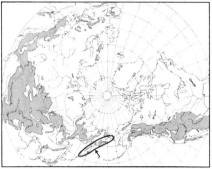

2. Oxýtropis revolùta Ledeb.

Plant densely caespitose with branched root and branching caudex, covered with persistent stipules; leaves 7–13-foliate; leaflets oblong-ovate to lanceolate, glabrous or sparsely pubescent above, ciliate and pubescent beneath, often with revolute margin; stipules ovate to orbicular, ciliate, glabrous; racemes 2–5(–6)-flowered; calyx white- and black-haired, with teeth about half as long as tube; petals bluish-violet to purplish; pods oblong with stipe 10–12 mm long, white- and black-pubescent.

Stony slopes, riverbanks.

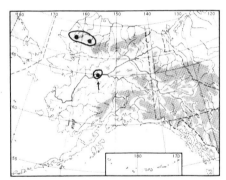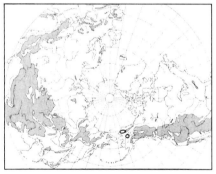

3. Oxýtropis kokrinénsis Pors.

Caespitose, with branching caudex, covered with persistent dark-brown stipules and petioles; leaves 7–9-foliate; leaflets elliptic to lanceolate, with revolute margin, grayish-villose; stipules with long-triangular, acute free part, silky-villose, in age glabrescent; inflorescence 1–3-flowered; calyx with teeth 2–2.5 mm long, purplish-brown-villous; petals purple; pods oblong, stipitate, with short grayish-black pubescence.

Dry slopes. Flowers early.

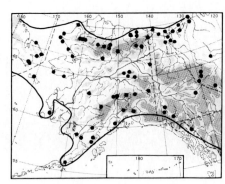
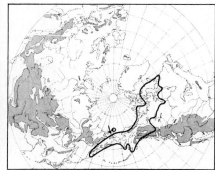

4. Oxýtropis Maydelliàna Trautv.

Caespitose, with branching caudex, covered with dense, persistent, dark reddish-brown stipules; leaves 11–21-foliate; leaflets ovate, elliptic to lanceolate, pilose; stipules greenish, in age reddish-brown, connate, with caudate, free lobes; scape villous; racemes many-flowered, capitate, in age somewhat elongated; calyx densely black- to white-pubescent, with teeth 2 to 3 times shorter than tube; petals yellow; pod ovoid to elliptic, with long, bent beak.

Stony slopes, alpine heaths, ridges, on tundra and in the mountains to at least 1,500 meters.

Circumpolar map includes range of subsp. **melanocéphala** (Hook.) Pors. (*O. campestris* var. *melanocephala* Hook.).

(See color section.)

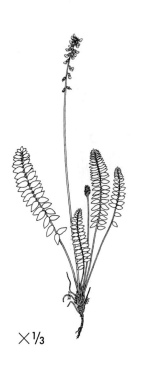

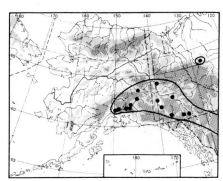
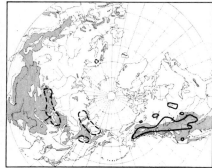

5. Oxýtropis defléxa (Pall.) DC.
Astragalus deflexus Pall.

var. serícea Torr. & Gray

Oxytropis retrorsa Fern.; *O. retrorsa* var. *sericea* Fern.; *O. deflexa* var. *parviflora* Boiv.

Loosely villous-pilose; stems with 1–2 internodes from short, branching caudex; leaves 15–39-foliate with ovate to narrowly lanceolate leaflets; stipules with lanceolate, pilose, free parts; racemes mostly 10–20-flowered, short when young, elongating in age; calyx with narrow, linear to lanceolate teeth; petals whitish, suffused with pink or purple; pods spreading, in age reflexed, stipitate, narrowly oblong, strigillose.

Riverbanks, meadows, waste places. *O. deflexa* described from Baikal, var. *sericea* from the Rocky Mountains (Nuttall).

Broken line on circumpolar map indicates range of var. **defléxa** and of the very closely related subsp. **norvègica** Nordh. (isolated in northern Norway).

54. Leguminosae (Pea Family)

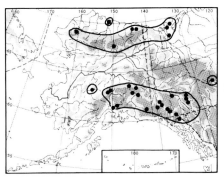 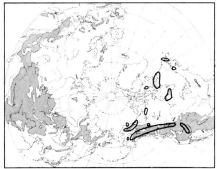 

6. Oxýtropis defléxa (Pall.) DC.
var. foliolòsa (Hook.) Barneby
  Oxytropis foliolosa Hook; *Oxytropis deflexa* var. *capitata* Boiv.

Similar to var. *sericea*, but mostly acaulescent, sparingly pilose; racemes few-flowered, mostly 7–10-flowered, compact also in fruit; flowers pink to purple.

Moist meadows. Described "from Carlton-House to the Rocky Mountains, in lat. 54°."

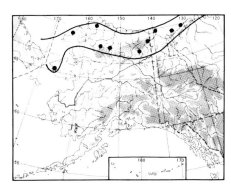 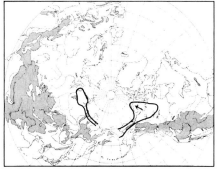 

7. Oxýtropis árctica R. Br.
  O. uralensis of American authors; *O. Roaldi* Ostenf.; *O. coronaminis* Fern.

Caespitose, more or less prostrate or decumbent, with branching caudex from taproot; leaves 11- to 21-foliate; leaflets ovate to narrowly oblong or elliptic, villous; stipules with deltoid to deltoid-acuminate free parts, densely pilose, bristly-ciliate; inflorescence nearly capitate, 2–10-flowered, elongating in fruit; calyx dark- and light-pilose with linear-lanceolate teeth; petals purple; pods sessile, oblong-ellipsoid, with long, nearly straight beak.

Tundra, riverbars.

× ½

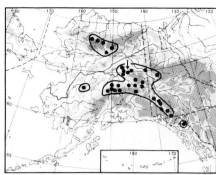
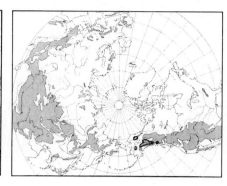

8. Oxýtropis Scammaniàna Hult.

Plant with taproot and branching caudex, covered with persistent straw-colored stipules; leaves 9–13-foliate; leaflets lanceolate to elliptic, sparingly pilose on both sides, or glabrous above; stipules broad, with elliptic, obtuse, free lobes, glabrous, pilose in apex, more or less ciliate; scapes white-pilose; racemes few-flowered; calyx black-pilose, with linear, acute teeth; petals bluish to purplish; pods short, thick, black-haired, erect.

Alpine heaths, to at least 1,500 meters.

× ½

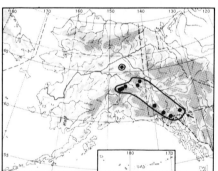
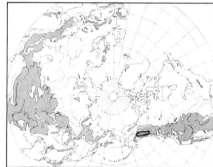

9. Oxýtropis Huddelsònii Pors.

Plant pulvinate-caespitose, with branched, short caudex, covered with persistent stipules; leaves 15–19-foliate; leaflets more or less involute, white-pubescent above, sparsely so beneath; stipules whitish, ciliate, with obtuse, deltoid, free parts; peduncles 1–2-flowered; calyx black- to brown-pubescent, with teeth 2 mm long; petals pink to purple; pods elliptic, with hooked beak, sessile within calyx, glabrous or minutely strigose, in age prostrate on ground.

Solifluction soil.

54. LEGUMINOSAE (Pea Family)

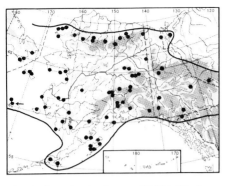 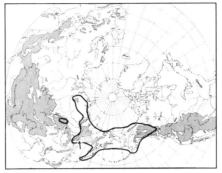 

10. Oxýtropis nigréscens (Pall.) Fisch.
  Astragalus nigrescens Pall.
subsp. **bryòphila** (Greene) Hult.
  Aragallus bryophilus Greene; *Oxytropis nigrescens* var. *bryophila* (Greene) Lepage.

Pubescent, with long, gray hairs; caespitose, with short, branching caudex, covered with persistent stipules and petioles, from taproot; leaves 9–13-foliate; leaflets oblong to ovate, pubescent on both sides, ciliate; stipules with lanceolate free parts, glabrous on back in age, ciliate; inflorescence 2–3 (–4)-flowered; calyx densely black-haired, with linear-lanceolate teeth about as long as tube; petals purplish to blue; pods subsessile, oblong to cylindrical, with short beak, gray- to black-pubescent.

Stony slopes, in McKinley Park to at least 1,800 meters. *O. nigrescens* described from between Okhotsk and Aldan, subsp. *bryophila* from Hall Island, Alaska.

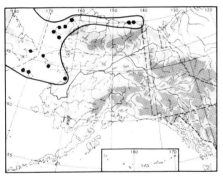 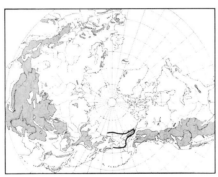 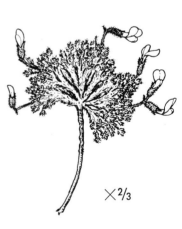

11. Oxýtropis nigréscens (Pall.) Fisch.
subsp. **pygmaèa** (Pall.) Hult.
  Astragalus pygmaeus Pall.; *Oxytropis nigrescens* var. *pygmaea* (Pall.) Cham.; *O. pygmaea* (Pall.) Fern.

Similar to subsp. *bryophila*, but pulvinate, lanate; branches of caudex covered with persistent stipules, but lacking petioles.
Stony tundra. Described from Chukchi Peninsula.

×3/4

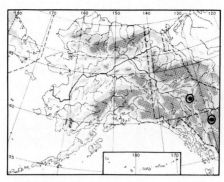
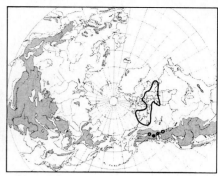

12. Oxýtropis nigréscens (Pall.) Fisch.
subsp. **arctòbia** (Bunge) Hult.

Oxytropis arctobia Bunge; *O. arctica* var. *uniflora* Hook.; *O. nigrescens* var. *uniflora* (Hook.) Barneby.

Similar to subsp. *bryophila*, but pulvinate, with short caudex; whole plant with snow-white pubescence; flowers single or 2 together.

Dry tundra, stony slopes. Described from arctic North America.

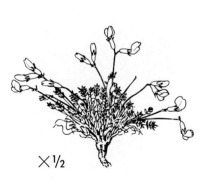
×1/2

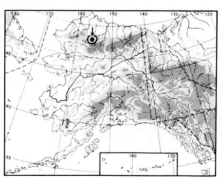
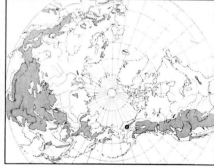

13. Oxýtropis glabérrima Hult.

Oxytropis nigrescens var. *nigrescens* sensu Welsh in part.

Similar to *O. nigrescens* subsp. *bryophila*, and possibly a subdivision of that taxon, but totally glabrous.

Dry, rocky slopes.

54. LEGUMINOSAE (Pea Family)

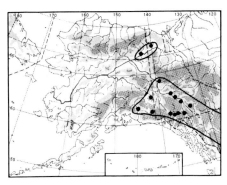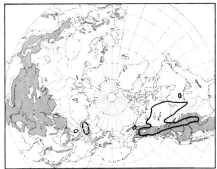

14. Oxýtropis spléndens Dougl.

Oxytropis splendens var. *Richardsonii* Hook.; *Aragallus splendens* (Dougl.) Greene; *A. Richardsonii* (Hook.) Greene.

Plant with taproot and branching caudex, densely covered with persistent stipules; leaves verticillate, in 7–15 fascicles (40–70 leaflets) but fewer in depauperate specimens; leaflets lanceolate, villous; stipules with triangular to acuminate free ends, densely yellowish-villous dorsally; peduncle long-villous; racemes many-flowered; calyx with teeth 2–4 mm long, long-villous; petals bluish or pinkish to purplish; pods ovoid to oblong, villous.

Dry, sandy, and shaly slopes. Described from the Red River south toward Pembina.

Broken line on circumpolar map indicates range of the closely related **O. Scheludjàkovae** Karav. & Jurtz.

× ⅓

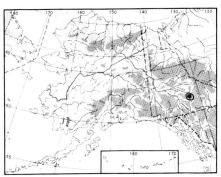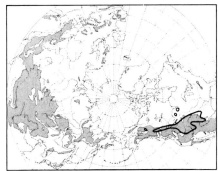

15. Oxýtropis serícea Nutt.

Oxytropis spicata (Hook.) Standl.; *O. campestris* var. *spicata* Hook.

Caespitose from thick root and branching caudex, densely covered with persistent stipules; leaves with 11–19 ovate, elliptic to lanceolate leaflets, silky-pilose on both sides; stipules with deltoid to deltoid-acuminate free blades, pilose to nearly tomentose dorsally; racemes 6–25-flowered, often becoming elongated in age; calyx with unequal teeth; petals yellow, sometimes white, lilac-tinged or purple; pod sessile, oblong to ovoid, densely silky-strigose.

Dry hillsides. Described from the Rocky Mountains.

Yukon specimens belong to the yellow-flowered var. **spicàta** (Hook.) Barneby (*O. campestris* var. *spicata* Hook.).

× ⅓

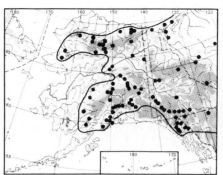
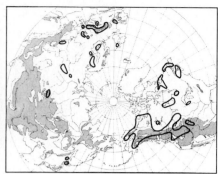

16. Oxýtropis campéstris (L.) DC.

Astragalus campestris L.

subsp. **gràcilis** (Nels.) Hult.

Aragallus gracilis Nels.; *Oxytropis gracilis* (Nels.) Schum.; *O. campestris* var. *gracilis* (Nels.) Barneby; *A. varians* Rydb.; *O. varians* (Rydb.) Hult.; *O. campestris* var. *varians* (Rydb.) Barneby; *O. hyperborea* Pors.; *O. alaskana* Nels.

Caespitose, with branching caudex more or less covered with disintegrated stipules; leaves 15–35 (–45)-foliate; leaflets lance-elliptic to oblong, opposite or verticillate, pilose on both sides; stipules lanceolate, acuminate, pilose dorsally, the margins ciliate and beset with more or less numerous short, thick, sometimes clavate processes; raceme many-flowered, capitate when young, elongating in age; calyx villous with dark and light hairs; corolla whitish, yellowish, or purple-tinged; pod erect, sessile, pilose.

Dry, sandy places. *O. campestris* described from the island of Oeland (Sweden), Germany, and Switzerland; subsp. *gracilis* from Weston County, Wyoming.

On the circumpolar map, the range of the entire species complex is indicated (though incomplete in Asia); subsp. *gracilis* occupies the western American range.

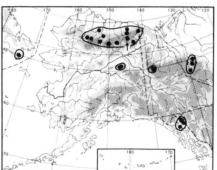
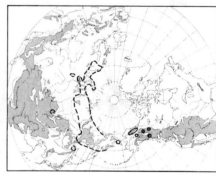

17. Oxýtropis campéstris (L.) DC.
subsp. **Jordàlii** (Pors.) Hult.

Oxytropis Jordalii Pors.; *O. campestris* var. *Jordalii* (Pors.) Welsh.

Similar to subsp. *gracilis*, but smaller; racemes with fewer than 10 flowers, flowers up to 16 mm long; leaves 9–17-foliate.

Dry tundra.

Broken line on circumpolar map indicates range of the closely related, larger-flowered subsp. **sórdida** (Willd.) Wahlenb. (*Astragalus sordidus* Willd.).

54. Leguminosae (Pea Family)

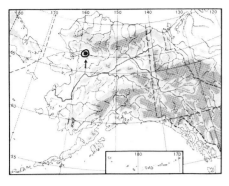

18. Oxýtropis kobukénsis Welsh

Caespitose, with branching caudex; leaves 13–14-foliate; leaflets lanceolate to lance-oblong, pilose to glabrate above, strigose beneath; stipules rigid, persistent, purplish, pilose on back, with long-attenuate free ends, the margins scarious, ciliate, and with clavate processes; raceme 5–6-flowered; flowers 16–18 mm long, purplish; calyx purplish, strigillose, with light and dark hairs and linear-lanceolate teeth about one-third as long as tube; pod black-pilose.

Sand dunes at Kobuk River opposite Hunt River.

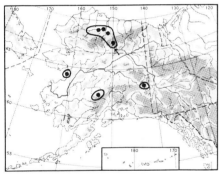
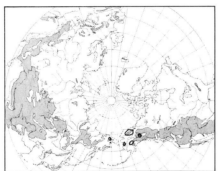

19. Oxýtropis koyukukénsis Pors.

Oxytropis arctica var. *koyukukensis* (Pors.) Welsh.

Caespitose, erect, with branching caudex; leaves 15–25-foliate, alternate, opposite or sometimes subverticillate; leaflets sparsely pubescent above, white-villous below; stipules prominent, with long-attenuate free parts, long-pubescent on back, ciliate and with sausagelike processes in margin; inflorescence more or less capitate, 10–14-flowered; flowers spreading in anthesis; calyx with subulate teeth, half as long as tube; corolla about 20 mm long, purplish-blue; pod grayish-pubescent, with long beak.

Sand bars, tundra meadows. One specimen seen.

× 1/3

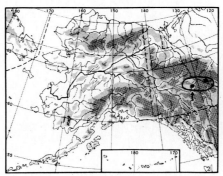
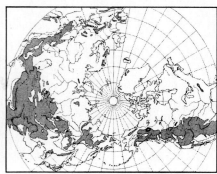

20. Oxýtropis sheldonénsis Pors.

Caespitose, with strong taproot; leaves 23–31 foliate; leaflets lanceolate, white-pubescent beneath; stipules large, white, purple-spotted, glabrous on back, sparsely glandular-dotted and sparsely ciliate; inflorescence subcapitate in flower, somewhat elongating in age; calyx densely black-pilose, short, with tube 5 mm long and subulate teeth 3 mm long; petals rose-colored, drying blue, about 12 mm long; pods gray-hirsute with long, straight beak.

Rocky ledges in higher mountains.

× 1/3

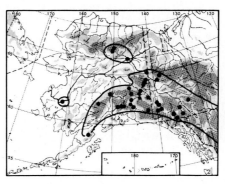
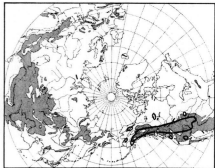

21. Oxýtropis víscida Nutt.

Oxytropis viscidula (Rydb.) Tidestr.; *O. viscidula* subsp. *sulphurea* Pors.; *O. verruculosa* Pors.; *Aragallus viscidus* (Nutt.) Greene; *A. viscidulus* Rydb.

Plant glandular and verrucose, caespitose from branching caudex; leaves with numerous leaflets; stipules pale, with acuminate free ends, pilose to villous on back, more or less glandular-verrucose, ciliate; bracts linear-lanceolate, as long as or mostly longer than calyx; racemes many-flowered, capitate or elongate; calyx teeth glandular, pods black-pubescent.

Dry slopes. Described from the "Rocky Mountains near the source of the Oregon" (River).

Highly variable, probably with local races.

54. LEGUMINOSAE (Pea Family)

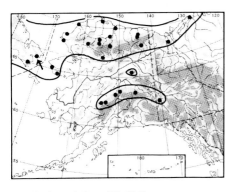
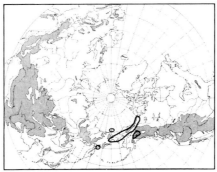

22. Oxýtropis boreàlis DC.
Oxytropis leucantha (Pall.) Bunge?; *O. glutinosa* Pors.

Caespitose; leaves 16–25 foliate; leaflets oblong-ovate, with revolute margin, glabrous on both sides or nearly so, ciliate, more or less glandular; stipules glabrous on back, strongly ciliate and profusely glandular-verrucose; scapes pilose and glandular; raceme capitate, 5–10-flowered; calyx white- and dark-pubescent, sparsely glandular, with densely glandular teeth, about half as long as tube; petals pale blue; pod oblong-ovate with a hooked beak.

Dry tundra.

Hedysarum L.

Articulations of loment not wing-margined, pubescent; lateral veins of leaflets hidden . . . . 1. *H. Mackenzii*
Articulations of loment wing-margined, mostly glabrous; lateral veins of leaflets conspicuous:
    Inflorescence long; flowers 12–15 mm long, pendulous, purple in apex, pale at base . . . . 2. *H. alpinum* subsp. *americanum*
    Inflorescence short; flowers 15–18 mm long, spreading, dark purple . . . . 3. *H. hedysaroides*

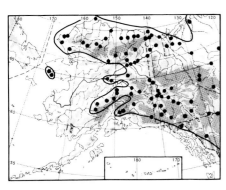
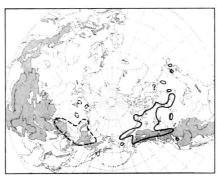
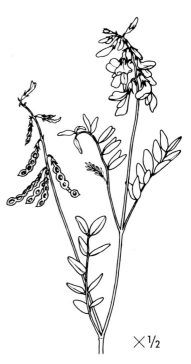

1. Hedýsarum Mackénzii Richards.

Stem mostly unbranched, appressed-pubescent, from stout caudex and woody root; leaves with 7–15 leaflets, lanceolate to elliptic, glabrous above, appressed-pubescent beneath; inflorescence short; calyx teeth pubescent, linear, subulate, nearly equal; wings with blunt, free auricles; keel truncate; loments pubescent, sessile or very short-stipitate, 3–8-articulated, not winged.

Rocky slopes, river bars, in McKinley Park to about 1,200 meters. Described from Barren Grounds.

Reported to be poisonous (unlike *H. alpinum* subsp. *americanum*). Broken line on circumpolar map indicates range of the doubtfully distinct **H. dasycárpum** Turcz.

54. Leguminosae (Pea Family)

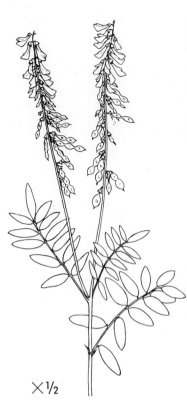

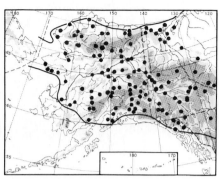 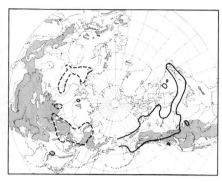

2. Hedýsarum alpìnum L.
subsp. **americànum** (Michx.) Fedtsch.

Hedysarum alpinum var. *americanum* Michx.; *H. americanum* (Michx.) Britt.; *H. boreale* DC. in part; *H. auriculatum* Eastw.

Stem tall, branched above, appressed-pubescent, from branching caudex and long, woody root; leaves with 15–20 leaflets, glabrous above, pubescent beneath, broadly lanceolate-oblong, obtuse or rarely acute, apiculate; inflorescence elongated; flowers purple at tip, pale at base, deflexed; calyx pubescent, with upper teeth short and triangular, the lower narrow and longer; standard emarginate, wings with linear auricles, united beneath standard; loments glabrous or appressed-pubescent, stipitate, 2–5-articulate, wing-margined.

Rocky slopes, spruce forests, gravel bars. *H. alpinum* described from Siberia; subsp. *americanum* from boreal Canada and the Allegheny Mountains.

Root and young stem, especially the conical lateral roots, are eaten raw, boiled, or roasted by the natives; the plant is also eaten by bears and collected by mice.

Broken line on circumpolar map indicates range of subsp. **alpìnum**.

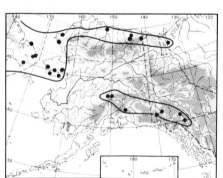 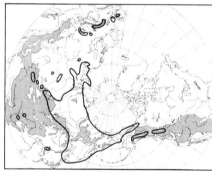

3. Hedýsarum hedysaroìdes (L.) Schinz & Thell.

Astragalus hedysaroides L.; *Hedysarum obscurum* L. Including *H. arcticum* Fedtsch.

Similar to *H. alpinum* subsp. *americanum*, but stem low, not branched; inflorescence short; flowers spreading, large, dark purple; loment mostly 1–3-articulated.

Rocky slopes, stony tundra, gravel. Described from Switzerland and Siberia.

Forms hybrid swarms with *H. alpinum* subsp. *americanum* [var. *grandiflorum* Rollins (mostly); *H. truncatum* Eastw.]; most specimens with larger and darker flowers and short inflorescence are considered to be the result of this introgressive hybridization. Pure *H. hedysaroides* occurs in high altitudes only.

The rootstock is eaten raw with fat by the Siberian Eskimo.

54. LEGUMINOSAE (Pea Family)

Vicia L. (Vetch)

Flowers solitary or in pairs, nearly sessile .... 1. *V. angustifolia*
Flowers in racemes:
  Plant densely villous .... 5. *V. villosa*
  Plant not villous:
    Leaves with 4–8 pairs of leaflets; peduncles 3–9-flowered .... 3. *V. americana*
    Leaves with 8–12 pairs of leaflets; peduncles with 1-sided, many-flowered racemes:
      Stipules sharply dentate; banner blade shorter than claw; racemes shorter than leaves .... 2. *V. gigantea*
      Stipules entire; banner blade as long as claw; racemes longer than leaves
.... 4. *V. cracca*

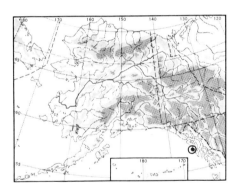
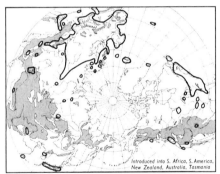
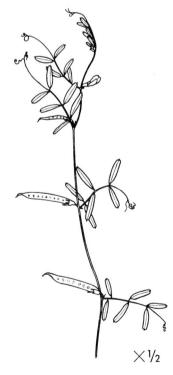

×½

1. Vícia angustifòlia (L.) Reichard
Vicia sativa var. *angustifolia* L.

Stem simple or branched, glabrous or nearly so; leaves with 2–5 pairs of leaflets, those of upper leaves narrower than those of the lower; stipules sagittate; flowers about 1.5 cm long, purple, the keel usually pale; calyx teeth shorter than tube; pod black.

Introduced. Described from Europe.

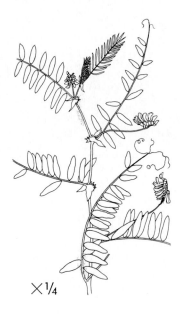
× ¼

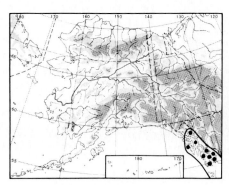 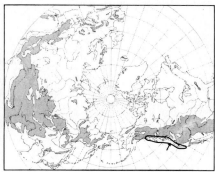

2. Vícia gigántea Hook.
Vicia sitchensis Bong.

Stem stout, tall, somewhat pubescent, climbing; leaves with 8–12 pairs of lance-oblong, obtuse, sparsely strigose leaflets; stipules broad, sharply dentate; racemes 1-sided, densely many-flowered, much shorter than leaves; calyx teeth very unequal, shorter than tube; flowers ochroleucous, often tinged with purple.

Forests, along streams. Described from the Columbia River.

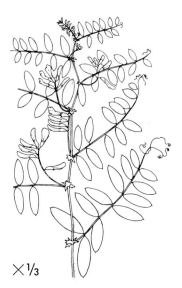
× ⅓

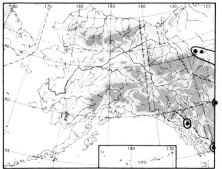 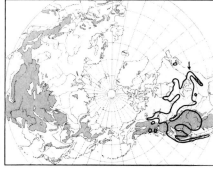

3. Vícea americàna Muhl.

Stem glabrous or sparsely pubescent, climbing; leaves with 4–8 pairs of oblong-ovate to elliptic, obtuse leaflets, with prominent, elevated lateral veins beneath; inflorescence shorter than the subtending leaves, 3–9-flowered; calyx teeth unequal, shorter than tube; flowers bluish-purple; pod glabrous, with several seeds.

Thickets, meadows. Described from Pennsylvania.

54. LEGUMINOSAE (Pea Family)

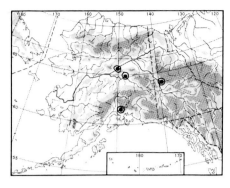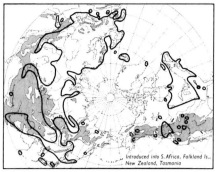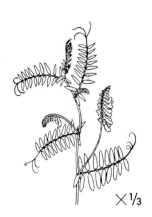

4. Vícia crácca L.

Stem weak, climbing, short-appressed, puberulent; leaves with 8–10 pairs of leaflets, lacking distinct median vein above; stipules entire; racemes 1-sided, many-flowered; calyx teeth unequal, flowers bluish-violet; banner blade as long as claw; pod glabrous.

Introduced. Described from Europe.

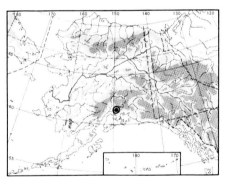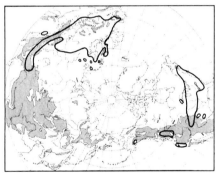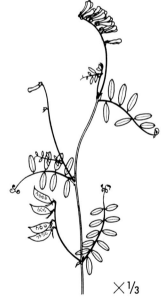

5. Vícia villòsa Roth

Stem weak, climbing, spreading-villous; leaves with 7–9 pairs of linear-to-oblong, mucronate leaflets, with distinct median vein above; stipules narrow, entire; racemes 1-sided, many-flowered; calyx teeth unequal, as long as the gibbous tube; flowers bluish-violet; pods glabrous, with 4–5 seeds.

Introduced. Described from Europe.

Lathyrus L.

Stem winged; leaflets mostly 6 .... 4. *L. palustris* subsp. *pilosus*
Stem not winged; leaflets mostly more than 6:
    Stipules semisagittate; leaves thin, strongly nerved beneath ....
    .... 1. *L. venosus* var. *intonsus*
    Stipules broadly ovate or oblique-hastate, as large as leaflets; leaves thick:
        Peduncles and pedicels glabrous or sparsely pubescent, with small, dark-brown spots .... 2. *L. maritimus* subsp. *maritimus*
        Peduncles and pedicels with dense pubescence, lacking spots ....
        .... 3. *L. maritimus* subsp. *pubescens*

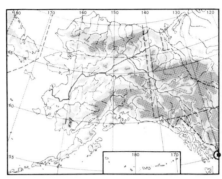
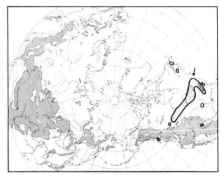

1. Láthyrus venòsus Muhl.
var. intónsus Butt. & St. John

Stout, pubescent; stem long, 4-angled, sparsely hirtellous; largest leaves with 10–12 ovate to elliptic leaflets, veiny beneath; stipules semisagittate, divided into 2 lobes; inflorescence shorter than leaves, 5–15-flowered; flowers purplish or bluish.

Woods, thickets. *L. venosus* described from Pennsylvania, var. *intonsus* from Despere Ledge, Wisconsin.

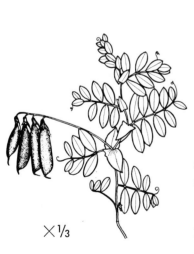
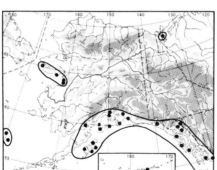
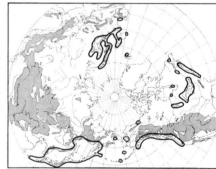

2. Láthyrus marítimus L.   Beach Pea
subsp. marítimus

Pisum maritimum L.; *Lathyrus japonicus* Willd. var. *typicus* Fern. and var. *glaber* Fern.

Glabrous or sparsely pubescent; rhizome slender; stem decumbent, angled; leaves thick, fleshy, with 6–12 oblong to ovate leaflets; stipules large, obliquely hastate or broadly ovate; raceme few-flowered; peduncles glabrous or very sparsely appressed-pubescent, with more or less numerous, small, dark-brown spots; flowers with reddish banner and bluish-violet wings and keel.

Sandy shores, rare inland. Described from Europe.

54. LEGUMINOSAE (Pea Family)

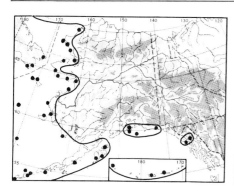

3. Láthyrus marítimus L.

subsp. **pubéscens** (Hartm.) C. Regel

Lathyrus maritimus var. *pubescens* Hartm.; *L. japonicus* var. *aleuticus* (Greene) Fern. and var. *pellitus* Fern.; *L. maritimus* var. *aleuticus* Greene; *L. aleuticus* (Greene) Pobed.

Similar to subsp. *maritimus*, but more pubescent; peduncles and pedicels with dense pubescence, lacking dark-brown spots.

Sandy shores. Described from Sweden.

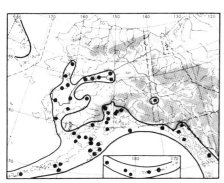

4. Láthyrus palústris L. Vetchling

subsp. **pilòsus** (Cham.) Hult.

Lathyrus pilosus Cham.; *L. palustris* var. *pilosus* (Cham.) Ledeb.

Pubescent; rhizome slender; leaves with 2–3 pairs of linear, rarely ovate, leaflets, branched tendrils, and wingless petioles; stem winged; inflorescence equal to or overtopping the subtending leaves; upper calyx teeth broadly triangular, shorter than the lanceolate lower teeth; flowers bluish-violet.

Wet meadows. *L. palustris* described from Europe.

Broken line on circumpolar map indicates range of other subspecies.

55. GERANIACEAE (Geranium Family)

Geranium L. (Cranesbill)

Leaves with 3–5 leaflets, the terminal stalked . . . . 4. *G. Robertianum*
Leaves palmate, not divided to base:
    Petals 5–7 mm long, slightly longer than calyx . . . . 3. *G. Bicknellii*
    Petals more than 10 mm long, much longer than calyx:
        Petals rose or violet, ciliate at base . . . . 1. *G. erianthum*
        Petals white or pinkish with purple veins, pilose on lower half of inside . . . . . .
        . . . . 2. *G. Richardsonii*

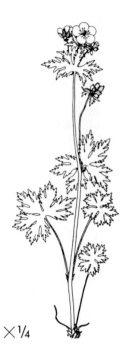
× ¼

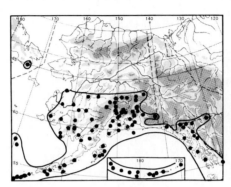
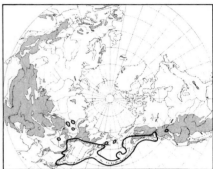

1. Gerànium eriánthum DC.

Stems from a thick, long, light-brown, scaly caudex; basal leaves appressed-pubescent, long-petiolated, 3 to 5 times cleft nearly to base into laciniate parts; upper stem leaves sessile; flowers 3–5, short-pedicellated, usually not overtopping leaves; sepals lanceolate-ovate, aristate, covered by long, villous or glandular-villous hairs; petals ciliate at base, rose or violet, rarely white, twice as long as sepals; filaments sparsely long-haired at base; carpels and base of style column pubescent.

Forests, meadows, to above timberline, in McKinley Park to about 1,000 meters. Described from Kamchatka and northwestern North America.

(See color section.)

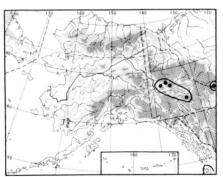
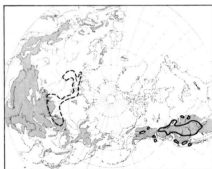

× ¼

2. Gerànium Richardsònii Fisch. & Trautv.

Stems from thick caudex; basal leaves long-petiolated, 5–7-parted into rhombic, acutely lobed segments, strigose above and on veins beneath; upper leaves sessile; peduncles and pedicels pilose with purplish-tipped glands; pedicels in pairs; sepals ovate-lanceolate, awn-tipped; petals white or pinkish with purple veins, pilose with long, white hairs on the inside to about half their length; carpels and style column glandular.

Moist meadows. Described from northwestern North America.

Broken line on circumpolar map indicates range of the very closely related **G. albiflòrum** Ledeb.

55. GERANIACEAE (Geranium Family)

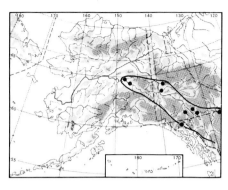 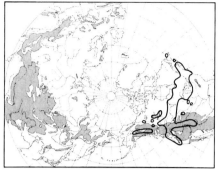 

3. Geranium Bicknéllii Britt.

Geranium nemorale var. *Bicknellii* (Britt.) Fern.

Mostly with divaricate branches; leaves strigose, 5-parted, segments cuneate-obovate, cleft into acute, oblong lobes; sepals ovate-lanceolate, awn-tipped, hirsute and glandular on veins; petals pale purple, slightly longer than calyx; carpels long-haired, with long, pubescent beak.

Woods, disturbed soil. Described from Nova Scotia, Maine, western Ontario, and New York.

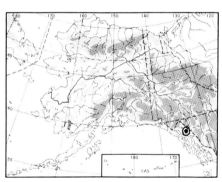 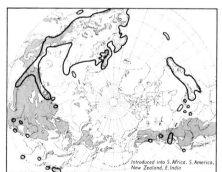 

4. Geranium Robertiànum L.

Plant with strong, disagreeable smell; stem decumbent, with long, spreading hairs below, branched from base, red; leaves red, palmate, the lower mostly with 5 leaflets, sparsely hairy on both sides, deeply pinnatisect, the segments pinnately lobed; sepals oblong-ovate, aristate, pilose and glandular; petals pink with white stripes, about twice as long as sepals, rounded in apex.

An introduced weed. Described from northern Europe.

55. GERANIACEAE (Geranium Family) / 56. LINACEAE (Flax Family)

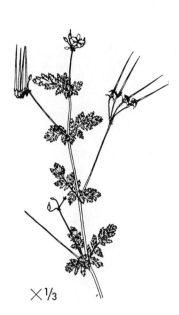

×⅓

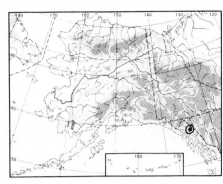
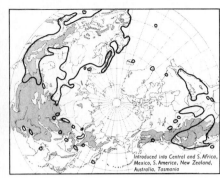

Erodium L. (Storksbill)

1. Eròdium cicutàrium (L.) L'Hér.
Geranium cicutarium L.

Branching from base; stem ascending; leaves pinnate; leaflets sessile, pinnately lobed or divided into linear segments; flowers umbellate; sepals mucronate, tipped with 1–2 white bristles; petals reddish-violet; beak of carpel twisted several times at maturity.

An introduced weed. Described from Europe.

×½

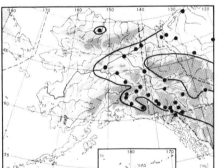
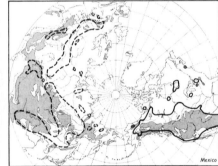

Linum L. (Flax)

1. Lìnum perénne L.
subsp. **Lewísii** (Pursh) Hult.
Linum Lewisii Pursh.

Glabrous; several erect stems from stout, branched caudex; stem densely leafy; leaves linear to lanceolate-linear, acute; pedicels 1–3 cm long; flowers blue, up to 14 mm long; sepals ovate, short-mucronate, with hyaline, entire margin; stigmas capitate; capsules round to ovate, 5–8 mm in diameter.

Dry soil. *L. perenne* described from Siberia and "Cantabrigiae"; subsp. *Lewisii* from the Rocky Mountains and banks of the Missouri River.

Highly variable. Broken line on circumpolar map indicates range of subsp. **perénne**.

57. CALLITRICHACEAE (Water Starwort Family)

Callitriche L. (Water Starwort)

Fruit conspicuously winged; plant dark green .... 1. *C. hermaphroditica*
Fruit not distinctly winged, at least not at base; plant light green:
    Carpels equally rounded above and below .... 4. *C. anceps*
    Carpels slightly heart-shaped, broader above:
        Fruit slightly longer than broad; reticulation of mericarps in vertical rows ....
        .... 2. *C. verna*
        Fruit as long as broad; reticulation of mericarps not in vertical rows ....
        .... 3. *C. heterophylla* subsp. *Bolanderi*

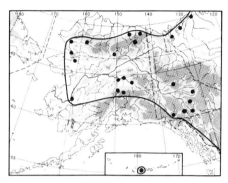 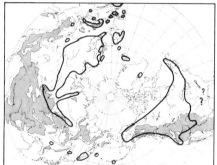 

1. Callítriche hermaphrodítica L.
Callitriche autumnalis L.

Submerged; stems from filiform rootstock; leaves densely covering stem, dark green, uniform, linear-oblong to linear-lanceolate, obtuse or retuse, broad at base, not connected by wing; flowers mostly 2 at each internode; fruits circular, large, thin; carpels each with broad wing.

Quiet, shallow water. Described from Sweden. Certainly more common than is indicated by map.

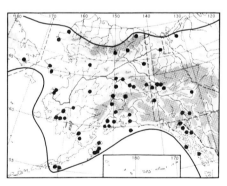 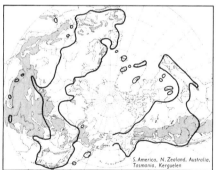 

2. Callítriche vérna L. emend. Lönnr.

Stem floating in water or more or less prostrate, creeping on mud; lower leaves linear-lanceolate to linear, later developing upper leaves broadly spatulate to ovate, finally forming floating rosettes; fruit sessile, obovate, longer than broad, narrower toward base, wingless or narrowly winged at summit; styles erect, shorter than the ripe fruit, in age arcuate, reflexed, soon falling off.

Shallow water, mud. Described from Sweden.

57. CALLITRICHACEAE (Water Starwort Family)

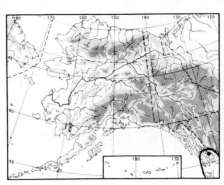 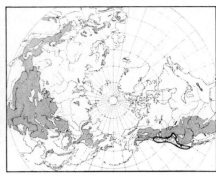

3. Callítriche heterophýlla Pursh
subsp. **Bolánderi** (Hegelm.) Calder & Taylor
  Callitriche Bolanderi Hegelm.; *C. heterophylla* var. *Bolanderi* (Hegelm.) Fassett.

Similar to *C. verna*, but fruit larger, 0.9–1.1 mm long, lacking the vertical rows of reticulation of the mericarps.
Shallow water. Described from "Auburn Sierra," California.

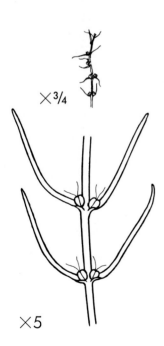

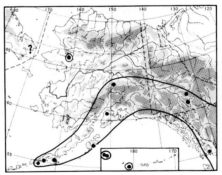 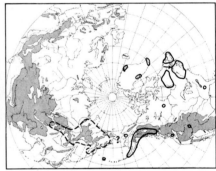

4. Callítriche ánceps Fern.
  Callitriche Bolanderi with respect to Alaskan plant in part.

Similar to *C. verna*, but fruit round, as broad as long, rounded at base; styles longer than the ripe fruit.
Shallow water. Broken line on circumpolar map indicates range of the closely related **C. subánceps** Petrov.

58. ACERACEAE (Maple Family) / 59. BALSAMINACEAE (Touch-Me-Not Family)

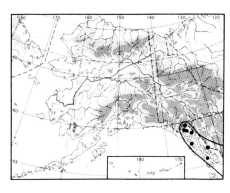
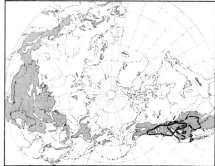
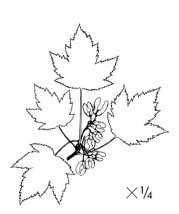

Acer L. (Maple)

1. Àcer glàbrum Torr.
subsp. **Douglásii** (Hook.) Wesmael

Acer Douglasii Hook.; *A. glabrum* var. *Douglasii* (Hook.) Dippel.

Shrub or small tree; stem reddish; leaves glabrous or sparsely glandular-puberulent, palmately 3–5-lobed, dark green above, pale beneath, coarsely double-serrate; flowers of 2 kinds in same inflorescence, staminate-with-dwarfed-pistils and bisexual; fruit glabrous with hardly divergent wings.

Banks of streams. *A. glabrum* described from the Rocky Mountains, subsp. *Douglasii* from the sources of the Columbia River. Broken line on circumpolar map indicates range of other subspecies.

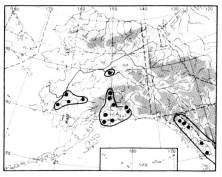
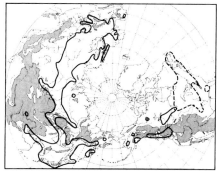

Impatiens L. (Touch-Me-Not)

1. Impàtiens nòli-tángere L.

Glabrous; stem erect; leaves alternate, oblong-ovate, cuneate at base, coarsely crenate-serrate, with 10–15 obtuse, mucronate teeth on each side; flowers both normal and cleistogamous, bright yellow with small brown spots; lower sepal gradually tapering into slender, curving spur; fruit a capsule, which explodes at touch when ripe.

Wet ground. Described from Europe and Canada.

Broken line on circumpolar map indicates range of **I. capénsis** Meerb.

60. VIOLACEAE (Violet Family)

Viola L. (Violet)

Flowers yellow:
    Plant with basal runners .... 2. *V. sempervirens*
    Plant lacking basal runners:
        Rootstock stout; leaves with distinct apex .... 1. *V. glabella*
        Rootstock filiform; leaves rounded .... 3. *V. biflora*
Flowers white, lilac, violet, purple, or blue:
    Rootstock stout:
        Plant lacking stem in spring, later with well-developed stem; flowers large;
            beard of lateral petals not white .... 4. *V. Langsdorffii*
        Plant with developed stem; flowers small, in axils of leaves, lateral petals white-
            bearded .... 5. *V. adunca*
    Rootstock slender; flowers lilac or white from base; petals beardless:
        Leaves pubescent above, with deep sinus, lateral lobes often overlapping;
            flowers violet .... 6. *V. Selkirkii*
        Leaves glabrous, with open sinus:
            Flowers small, white with purple stripes; stolons lacking ....
                .... 8. *V. renifolia* var. *Brainerdii*
            Flowers larger, lilac, with darker stripes; stolons present ....
                .... 7. *V. epipsila* subsp. *repens*

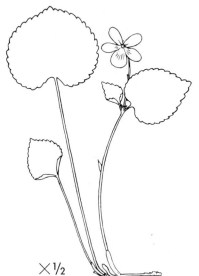
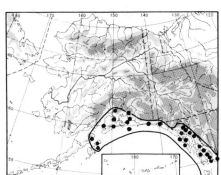
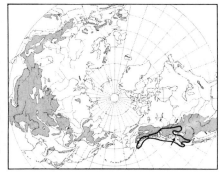

1. Vìola glabélla Nutt.

Stems from thick, horizontal rootstock; basal leaves long-petiolated, reniform-cordate, abruptly acute, crenate-serrate, somewhat puberulent or glabrous; stem leaves reduced; sepals lanceolate, acute; petals yellow, purplish-veined.

Moist woods. Described from Oregon.

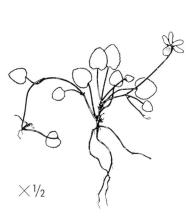
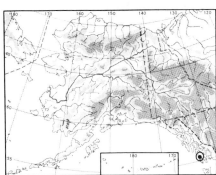
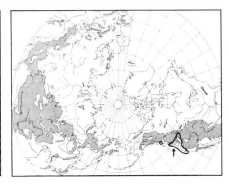

2. Vìola sempervìrens Greene
Viola sarmentosa Dougl. not Bieb.

Very similar to *V. biflora* (see next page); caudex and stem with stolons. Shaded woods.

60. VIOLACEAE (Violet Family)

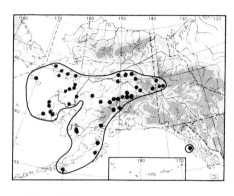
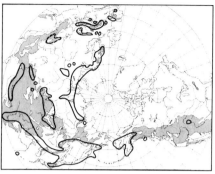

3. Vìola biflòra L.

Thin stems from more or less horizontal rootstock; basal leaves 2–3, broadly reniform to orbiculate, cordate at base, with short, acute apex, crenate-serrate, glabrous or nearly so above, pubescent on veins beneath; stipules ovate-lanceolate; sepals lanceolate; petals yellow to pale yellow, with brownish-violet stripes; spur very short.

Alpine meadows, scree slopes. Described from the Alps of Lapland, Austria, Switzerland, and England. Broken line on circumpolar map indicates range of subsp. **caucàsica** Rupr.

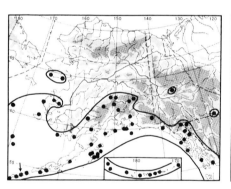
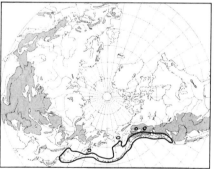
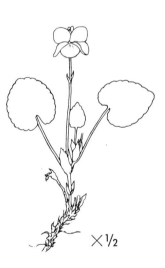

4. Vìola Langsdórffii Fisch.

Rootstock thick; plant lacking stem in the spring, later with well-developed stem; leaves broadly ovate to reniform, somewhat acute, crenate-serrate, somewhat pubescent, long-petiolated; stipules ovate-lanceolate; flowers mostly large, long-pedicellated; sepals ovate-lanceolate; petals bluish-violet, the lateral bearded; spur short, saccate.

Meadows, along streams, snow beds.

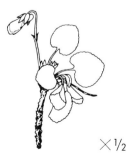

60. VIOLACEAE (Violet Family)

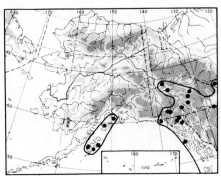
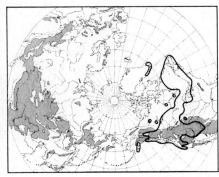

5. Vìola adúnca Sm. Western Dog Violet

Stems from slender rootstock; leaves cordate, ovate, obtuse, finely crenate, often more or less brown-dotted, pubescent, especially in margin and on veins below, or nearly completely glabrous; stipules linear-lanceolate, mostly with long, glandular-tipped teeth; sepals lance-linear; petals violet or bluish-violet, often whitish at base, the lateral pair white-bearded; spur slender, obtuse, often hooked; cleistogamous flowers occur.

Dry meadows, woods. Described from the northwest coast of North America.

Circumpolar map gives range of *V. adunca* in a broad sense, including **V. labradòrica** Schrank.

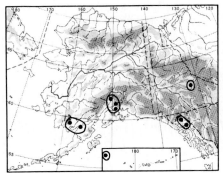
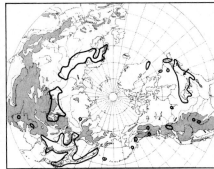

6. Vìola Selkírkii Pursh

Rootstock thin, short; leaves pubescent above, glabrous beneath, cordate-ovate, with deep sinus, lateral lobes often overlapping; both cleistogamous and normal flowers; petals violet; sepals lanceolate to ovate-lanceolate, acute, scarious-margined; petals not bearded; spur thick.

Woods. Apparently rare.

60. VIOLACEAE (Violet Family)

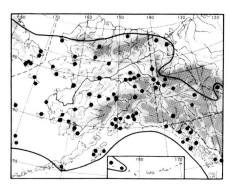
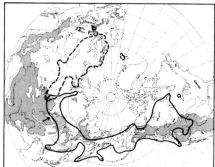
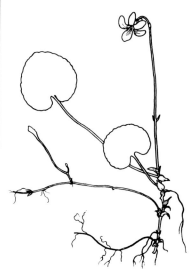

7. Vìola epípsila Ledeb.   Marsh Violet
subsp. **rèpens** (Turcz.) Becker

Viola repens Turcz.; *V. achyrophora* Greene; *V. palustris* of American authors in part.

Long, horizontal, thin, filiform rootstock with long runners from upper part; leaves thin, cordate-ovate, somewhat acute, glabrous on both sides; stem glabrous, with pair of bracts in upper part; sepals oblong-ovate, blunt; petals lilac, the lower with darker veins; spur short, thick, 2–3 times longer than auricles of calyx.

Wet meadows, peaty soil, brooksides, in the mountains to at least 1,800 meters. *V. epipsila* described from Tartu (Estonia), subsp. *repens* from Baikal and Dahuria. Broken line on circumpolar map indicates range of subsp. **epípsila.**

× ½

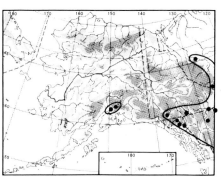
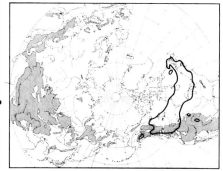

8. Vìola renifòlia Gray
var. **Brainérdii** (Greene) Fern.

Viola Brainerdii Greene.

Acaulescent; rootstock comparatively thick, ascending, lacking stolons; leaves cordate-orbicular to reniform, crenate-serrulate, glabrous on both sides or somewhat pubescent on veins beneath; petals beardless, white with purple stripes; both cleistogamous and normal flowers occur.

Moist woods, swamps. *V. renifolia* described from New York State and New Hampshire, var. *Brainerdii* from Vermont, Ottawa, and Prince Edward Island.

× ½

61. ELAEAGNACEAE (Oleaster Family)

×½

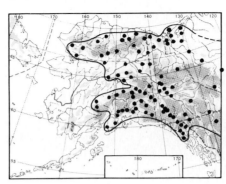
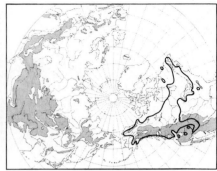

Shepherdia Nutt.

1. Shephérdia canadénsis (L.) Nutt.   Soapberry

Hippophaë canadensis L.; *Lepargyraea canadensis* (L.) Greene; *Elaeagnus canadensis* (L.) Nels.

Shrub with brownish-scurfy young twigs and elliptic, ovate to lanceolate, opposite leaves, green above, white-scurfy and with lepidote, rushy scales beneath; flowers sessile, in axis of branches, of 2 kinds—staminate brownish, with 4-parted calyx and 8 stamens, and pistillate, with urceolate, 4-cleft calyx, becoming berrylike in fruit; fruit yellowish-red, elliptical.

Woods, gravel bars, to at least 1,200 meters. Described from Canada.

The fruit contains saponine. It is eaten mixed with sugar and water and beaten frothy. The taste is bitter.

(See color section.)

×⅓

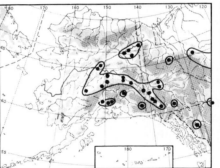

Elaeagnus L.

1. Elaeágnus commutàta Bernh.   Silverberry

Shrub with brownish-scurfy young twigs and dark, grayish-red old branches; leaves alternate, elliptic to ovate-lanceolate, undulate, silver-scurfy on both sides; flowers axillary, deflexed, perfect, fragrant; calyx with globose base, 4-cleft; fruit silvery, obovate, dry, mealy, with striate stone.

Dry slopes, gravel bars, in McKinley Park to 1,200 meters. Described from the banks of the Missouri River.

The fruit is eaten cooked in moosefat by the natives.

Epilobium L. (Willow Herb)

All leaves alternate; calyx tube lacking; flowers large, mostly over 2 cm in diameter; petals unequal in length (sect. *Chamaenerion*):
    Plant tall, with long inflorescence; bracts lanceolate, acute; style pubescent at base .... 1–2. *E. angustifolium*
    Plant decumbent, low-growing; bracts broad, mostly blunt; style glabrous at base .... 3. *E. latifolium*
All or at least some leaves opposite; calyx with tube; flowers smaller:
    Stigma with 4 lobes; leaves broad, ovate, sharply denticulate; flowers yellow .... 4. *E. luteum*
    Stigma entire, clublike or capitate:
        Stem leaves linear or lanceolate:
            Leaves closely and evenly pubescent, with minute, incurved hairs .... 5. *E. leptophyllum*
            Leaves glabrous or slightly pubescent, mostly pubescent in margin and on veins:
                Plant with long, filiform runners at base, ending with bud; leaves all opposite; stem terete .... 6. *E. palustre*
                Plant with rosettes of leaves at base; upper leaves often alternate .... 7. *E. davuricum*
        Stem leaves broad, oblong or ovate:
            Plant low-growing, stem leaves small, 1–1.5 cm long:
                Plant with decumbent, leafy basal shoots; stem simple, with 2 rows of hairs; leaves in 2–3 pairs; seeds smooth ...... 8. *E. anagallidifolium*
                Plant lacking leafy basal shoots; stem terete, pubescent, simple or much branched; leaves in several pairs; seeds papillate .. 9. *E. leptocarpum*
            Plant mostly taller; stem leaves larger:
                Plant mostly over 25 cm tall, pubescent above, with septate, more or less multicellular hairs; broad, withered leaves at base:
                    Stem mostly simple; middle stem leaves sessile, broad and large, sharply and densely denticulate, not conspicuously decreasing in size in the crowded inflorescence .... 10. *E. glandulosum*
                    Stem mostly branched; middle stem leaves petiolated, more or less denticulate, decreasing in size in the open inflorescence; flowers small, 3–6 mm long .... 11. *E. adenocaulon*
                Plant shorter, not conspicuously pubescent with multicellular hairs:
                  Plant with several pairs of broad, withered leaves at base:
                      Leaves thick, sharply and densely toothed; middle stem leaves ovate, mostly sessile .... 12. *E. behringianum*
                    Leaves thin, sparsely toothed; middle stem leaves distinctly petiolated .... 13. *E. lactiflorum*
                Plant lacking broad withered basal leaves at flowering time, with short or long leafy shoots at base:
                    Middle internodes longer than leaves; plant relatively tall, with thin, elliptical, petiolated, middle stem leaves .. 14. *E. Hornemannii*
                    Middle internodes shorter than leaves; plant usually smaller, with broad, ovate leaves, crowded above .... 15. *E. sertulatum*

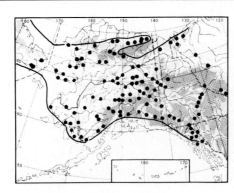 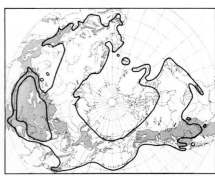

1. Epilòbium angustifòlium L. Fireweed
subsp. **angustifòlium**

Chamaenerion angustifolium (L.) Scop.; *Epilobium spicatum* Lam.

Stem tall, simple, densely leafy, from underground, woody rootstocks; leaves alternate, lanceolate, acute, glabrous, paler and distinctly veined beneath; flowers in long, terminal racemes, flowering from base; sepals pubescent, more or less red-colored; petals large, obovate, clawed, normally lilac-purple but exceptionally white; style pubescent at base; stigmas elongate, soon revolute; capsules long, canescent.

Meadows, forests, river bars, burned areas. Described from northern Europe.

Specimens with white petals and sepals, white petals and red sepals, pink petals and darker sepals, dark bluish-purple petals and darker sepals, even striate or dark-veined petals or sepals, rarely occur.

Marrow eaten by natives and leaves used as substitute for tea in Russia (Kurilski Chai). The root is also eaten raw by the Siberian Eskimo.

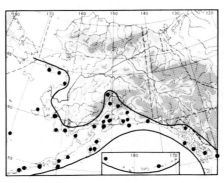 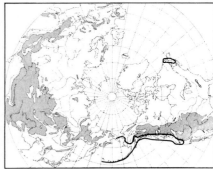

2. Epilòbium angustifòlium L.
subsp. **macrophýllum** (Haussk.) Hult.

Epilobium angustifolium f. *macrophyllum* Haussk. in part.

Similar to subsp. *angustifolium*, but leaves 2–3 cm broad, with very prominent midvein beneath, and broad bracts; leaves glabrous beneath, rarely somewhat pubescent on midvein.

Meadows, forests. Total range unknown. Described from Alaska and Magdalene Islands.

62. ONAGRACEAE (Evening Primrose Family)

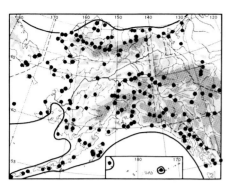
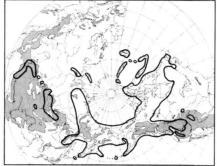

3. Epilòbium latifòlium L. — River Beauty
Chamaenerion latifolium (L.) Sweet.

Stems decumbent, grayish-puberulent, several from woody rootstock; leaves alternate, fleshy, glaucous, usually very finely puberulent, elliptic-ovate to lanceolate, blunt or acutish, not veiny; flowers in upper axils; calyx purple-tinged; petals purple or roseate, rarely white; style glabrous, shorter than stamens; stigmas short and thick.

River bars, along streams, scree slopes, in the mountains to at least 2,000 meters. Described from Siberia.

The young leaves are eaten raw or boiled by the natives.

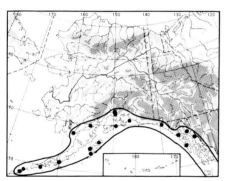
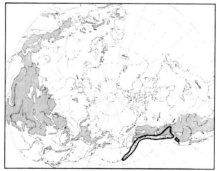

4. Epilòbium lùteum Pursh

Stems simple or branched from near the base, from stout, long, creeping and branching rhizome; basal leaves small, soon withering; stem leaves numerous, overlapping, subsessile, oblong-lanceolate to ovate, acute, glabrous, glandular-serrate-dentate; sepals lanceolate; petals yellow, longer than calyx; stigma with oblong-ovate lobes.

Borders of ponds, along streams. Described from the northwest coast of North America.

A very showy plant. Similar specimens with pink flowers are said to be the hybrid *E. luteum* × *glandulosum* (*E. Treleasianum* Lév.).

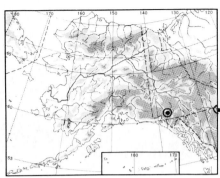 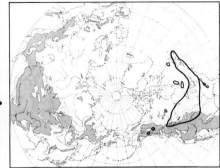

5. Epilòbium leptophýllum Raf.

Minutely pubescent or canescent above, with incurved hairs; stem tall, simple to densely branched, usually with short, leafy, axillary fascicles; leaves linear-lanceolate with revolute margins, canescent above; calyx lobes acute; capsules canescent; seed with an evident neck.

Lowlands. Described from Pennsylvania and Maryland. Circumpolar map tentative.

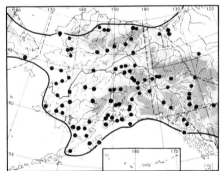 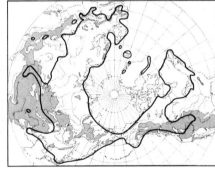

6. Epilòbium palústre L.

Stem simple or branched, short-pubescent especially above, with long, filiform runners at base (often broken off in herbarium specimens), ending with bud; leaves opposite below, alternate above, sessile, linear-lanceolate, entire in margin, glabrous or slightly pubescent above; flowers small, drooping when young; petals pink or whitish; pod short-pubescent, especially along margins; seeds papillate.

Wet places, along rivers. Described from Europe.

62. Onagraceae (Evening Primrose Family)

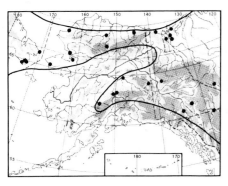
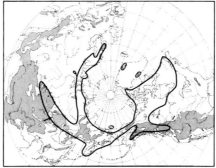

7. Epilòbium davùricum Fisch.

Rootstock short; stem single, simple, erect; basal leaves oblong-elliptic, blunt; stem leaves sessile, linear, flat, blunt, entire or somewhat toothed, often alternate; flowers small, 4–5 mm long, drooping when young; sepals glabrous; petals white; pod glabrous; seeds papillate.

Bogs, wet meadows.

Northern specimens belong to var. **árcticum** (Sam.) Polunin (*E. arcticum* Sam.), with smaller size, pink flowers, and less distinctly papillate seeds.

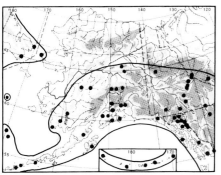
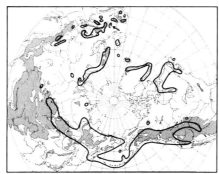

8. Epilòbium anagallidifòlium Lam.
Epilobium alpinum L. in part.

Plant often reddish, 10–15 cm tall, with decumbent, basal, leafy shoots; leaves short-petiolated, mostly in 2–3 pairs, small, oblong to narrowly elliptic, blunt, entire or repand-denticulate; flowers small, mostly 1 or 2, often nodding; petals reddish-violet to pink; seeds smooth.

Moist places in the mountains to about 2,000 meters.

Some specimens belong to var. **pseùdo-scapòsum** (Haussk.) Hult. (*E. pseudoscaposum* Haussk.), described from Unalaska, with sessile, more toothed leaves and long-stipitate pods.

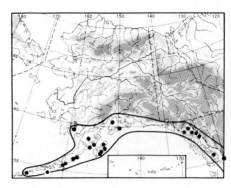
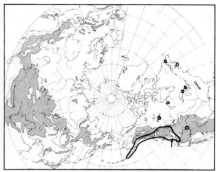

9. Epilòbium leptocárpum Haussk.

Plant small, usually less than 10 cm tall; stem slender, terete, pilose, much branched, with small turions at base; leaves glabrous, ovate-lanceolate to lanceolate, obtuse to somewhat acute, in small specimens lacking lateral nerves, with few teeth; petioles winged; flowers white to pink; pods glabrous or sparsely white-pubescent; seeds papillose.

Moist places. Most specimens belong to var. **Macoùnii** Trel., with simple stem. *E. leptocarpum* described from Oregon, var. *Macounii* from Lake Athabasca.

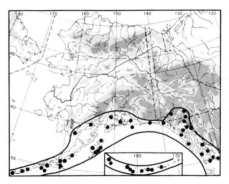

10. Epilòbium glandulòsum Lehm.
Epilobium affine Bong.

Stem mostly simple, erect, originating from a winter bud, the remains of which persist as thick, broad, rounded, brown leaves at base, pubescent above with long, septate hairs; stem leaves glabrous, broad, ovate-lanceolate, acute, serrate, the lower short-petiolated, the upper sessile, often crowded and overlapping in inflorescence; flowers small; sepals ovate-lanceolate, pubescent; petals pink; seeds papillate.

Along streams, at springs. Type locality not given, presumably Aleutian Islands.

62. Onagraceae (Evening Primrose Family)

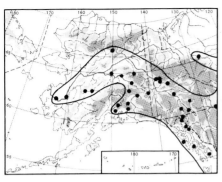 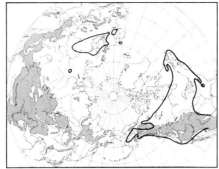 

11. Epilòbium adenocaùlon Haussk.
Epilobium glandulosum var. *adenocaulon* (Haussk.) Fern.

Similar to *E. glandulosum*, but leaves smaller; stem mostly branched in upper part; upper leaves not crowded and overlapping; leaves often distinctly petiolated. Mostly on disturbed soil, roadsides; probably native. Described from Ohio.
An introduced weed in many European countries.

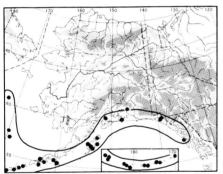 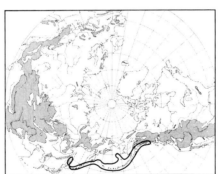 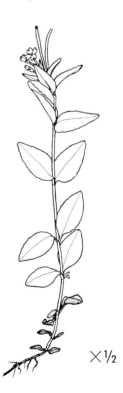

12. Epilòbium behringiànum Haussk.

Stem simple, densely foliate, with broad, withered leaves at base; leaves ovate, short-petiolated, nearly entire to dentated, blunt or somewhat acute, glabrous, or upper leaves pubescent in margin and beneath on veins, often crowded above; flowers small; sepals lanceolate, acute; petals pink; pod glabrous; seeds smooth.
Wet places. Described from Sitka, Unalaska, Kodiak, and Kamchatka.

×½

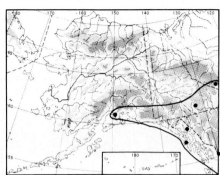
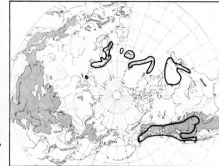

13. Epilòbium lactiflòrum Haussk.

Plant light green; stem mostly simple, with long internodes, somewhat pubescent above; leaves thin, glabrous, the middle and lower leaves petiolated, ovate-oblong, blunt, sparsely toothed; flowers few, small, white; young pod drooping, sparsely pubescent; seeds smooth.

Along streams, at springs. Described from inhomogenous material from northern Europe, Greenland, and America.

A much misunderstood species. Total range very uncertain.

×½

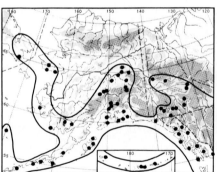
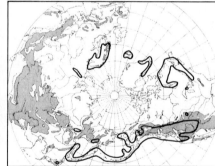

14. Epilòbium Hornemánnii Rchb.
Epilobium Bongardi Haussk.

Stem with short, leafy shoots at base, mostly simple, with long internodes; cauline leaves elliptic to oblong, mostly petiolated, thin, glabrous, sparsely denticulate, obtuse; sepals glabrous, acute; petals 5–6 mm long, lilac-pink; pod sparsely pubescent, glabrescent in age; seeds more or less papillate.

Moist places, along creeks. Described from "Filand" (Finland?).

62. ONAGRACEAE (Evening Primrose Family)

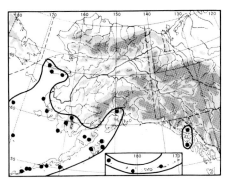 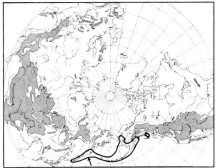 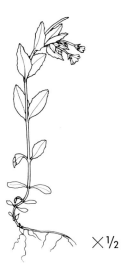

15. Epilòbium sertulàtum Haussk.

Flowering stems short, mostly 10–15 cm tall, single, glabrous, lacking large, withered leaves at base, from long, leafy runners; leaves ovate, entire or somewhat dentate; sepals glabrous, acute; petals pink, glabrous or glabrescent in age; seeds smooth.

Wet places.

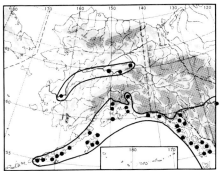 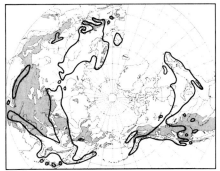 

Circaea L.

1. Circaèa alpìna L. Enchanter's Nightshade

Circaea alpina subsp. *pacifica* (Aschers. & Magnus) Raven; *C. pacifica* Aschers. & Magnus.

Stem from tuberous rootstock; leaves thin, cordate, ovate, acute, remotely toothed, glabrous; petioles narrowly winged; bracteoles setaceous; sepals glabrous; petals pink, 2-lobed to about middle; stigma entire; fruit 1-celled, covered with soft bristles.

Woods. Described from northern Europe.

63. Haloragaceae (Water Milfoil Family)

Myriophyllum L. (Water Milfoil)

Leaves usually 4 in the whorls; uppermost bracts entire, 1–2 mm long, shorter than flowers .... 1. *M. spicatum*
Leaves usually 5 in the whorls, far exceeding internodes; bracts pectinate or toothed, 6–8 mm long, longer than flowers .... 2. *M. verticillatum*

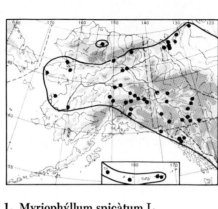
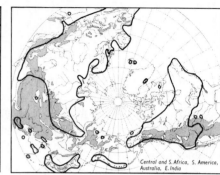

1. Myriophýllum spicátum L.

Stem 2–3 mm thick at base from elongated, creeping rhizome; leaves pinnate, usually forming whorl, slightly longer than internodes; spike emerging above water; flowers in axis of bracts, shorter or barely longer than flowers; pistillate flowers with small petals, staminate with larger petals; stamens 8.

Shallow water.

Most specimens belong to subsp. **exalbéscens** (Fern.) Hult. (*M. exalbescens* Fern.), with fewer (6–11) pairs of leaf segments and short bracts, not exceeding the flowers; but others with more numerous leaf segments and longer bracts also occur. *M. spicatum* described from Europe, subsp. *exalbescens* from York River (Gaspé).

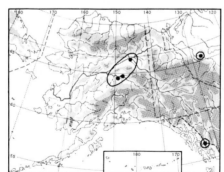
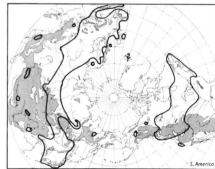

2. Myriophýllum verticillàtum L.

Stems 2–3 mm thick at base from elongated rhizome, creeping in mud; leaves pinnate, usually 5 in a whorl, much longer than internodes; flowers in axis of pinnate to pectinate bracts, longer than flowers; pistillate flowers lacking petals, staminate with 4 greenish petals; stamens 8.

Ponds, sluggish streams. Described from Europe.

Propagates by clavate turions with closely appressed leaves. Two varieties occur: var. **pinnatífidum** Wallr., with long bracts; and var. **pectinàtum** Wallr., with bracts only slightly longer than the flowers.

63. Haloragaceae (Water Milfoil Family)

Hippuris L. (Mare's Tail)

Stem 5–8 cm long, 0.2–0.5 mm thick .... 3. *H. montana*
Stem much longer and thicker:
    Leaves 4–6 in the whorls, short, obtuse .... 2. *H. tetraphylla*
    Leaves more numerous, longer, acute .... 1. *H. vulgaris*

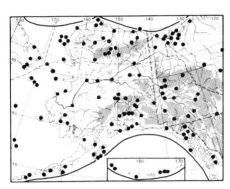
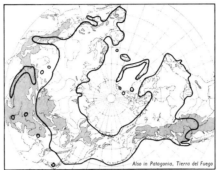

1. Hippùris vulgàris

Stem from stout creeping rhizome; leaves 6–12 in a whorl, linear, sessile, entire, glabrous, acute, longer than internodes, in submerged forms flaccid, thin, pale green, translucent; flowers small, in axils of submerged leaves, mostly perfect.

Shallow, running water. Described from Europe.

× ½

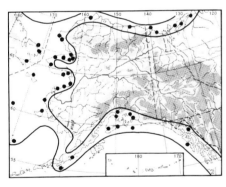
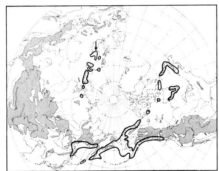

2. Hippùris tetraphýlla L. f.

Hippuris maritima Hellen.; *H. lanceolata* Retz.

Similar to *H. vulgaris*, but leaves 4–6, mostly 4 in a whorl, 3–5 mm broad, oblanceolate to oblong-ovate, obtuse, shorter than internodes.

Brackish water.

× ½

63. Haloragaceae (Water Milfoil Family) / 64. Araliaceae (Ginseng Family)

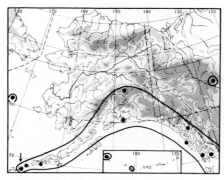
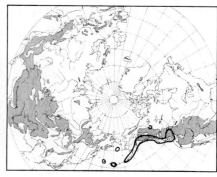

3. Hippùris montàna Ledeb.

Glabrous; stem delicate, short, simple, from slender, creeping rhizome; leaves linear to oblong-linear, 0.5–1 mm broad, 5 to 7 in a whorl; flowers perfect or often unisexual; staminate flowers usually in axis of lower whorls.

Along creeks in the mountains. Rare.

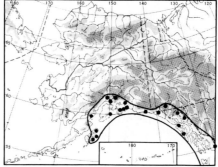
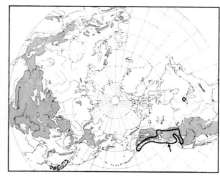

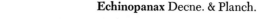
Echinopanax Decne. & Planch.

1. Echinopànax hórridum (Sm.) Decne. & Planch.   Devil's Club
Panax horridum Sm.; *Oplopanax horridus* (Sm.) Miq.

Densely prickly shrub with long, decumbent, often entangled branches or stolons, densely spinose on stems and petioles; petioles up to 30 cm long; leaves cordate at base, deeply or shallowly 5–7-lobed, the lobes acute to caudate, serrate, with acute teeth and spiny on nerves beneath; inflorescence shorter than leaves, in umbels forming elongate raceme; berries scarlet.

Forms nearly impenetrable thickets in moist woods.

Broken line on circumpolar map indicates range of subsp. **japònicus** (Nakai) Hult. (*E. japonicus* Nakai).

(See color section.)

65. Umbelliferae (Parsley Family)

Osmorhiza Raf. (Sweet Cicely)

Fruit clavate; rays and pedicels divaricate .... 3. *O. depauperata*
Fruit beaked or constricted at apex; rays and pedicels spreading-ascending:
    Flowers greenish-purplish; beak of fruit short; styles 0.5–1 mm long; ripe fruit
        10–13 mm long; leaves thin, yellowish-green .... 1. *O. purpurea*
    Flowers greenish-white; beak of fruit about 2 mm long; styles shorter; ripe fruit
        usually longer; leaves thicker, often dark green .... 2. *O. chilensis*

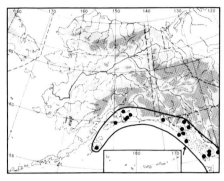
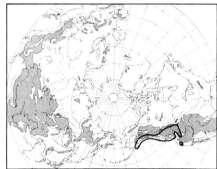

1. Osmorhìza purpùrea (Coult. & Rose) Suksd.
Washingtonia purpurea Coult. & Rose.

Plant slender, from stout branched root, glabrous or sparsely white-hirsute; leaves 1–3-ternate with ovate-lanceolate, incised and coarsely serrate lobes; flowers greenish to purplish; involucels lacking; rays and pedicels ascending; fruit 10–13 mm long, bristly hispid, attenuate at base, constricted at apex; stylopodium compressed; disk conspicuous; style 0.5–1 mm long.

Woods.

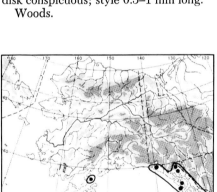
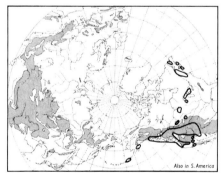
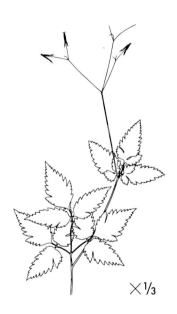

2. Osmorhìza chilénsis Hook. & Arn.
Osmorhiza nuda Torr.; *Washingtonia brevipes* Coult. & Rose.

Similar to *O. purpurea*, but flowers greenish-white or white; fruit 12–25 mm long, tapering at apex; stylopodium conic, disk inconspicuous; styles 0.2–0.5 mm long.

Woods. Described from Concepción, Chile.

65. UMBELLIFERAE (Parsley Family)

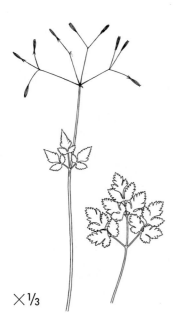

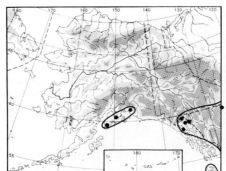
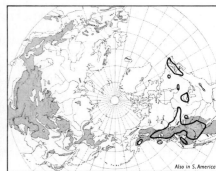

3. Osmorhìza depauperàta Phill.

Washingtonia obtusa Coult. & Rose; *Osmorhiza obtusa* (Coult. & Rose) Fern.

Similar to *O. chilensis,* but rays and pedicels divaricate; fruit clavate, with narrow base, densely hispid at base.

Woods. Described from Valle de las Nieblas, Chile.

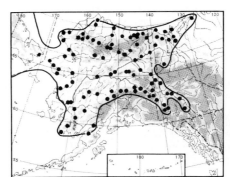
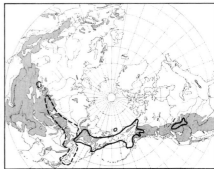

Bupleurum L.

1. Bupleùrum triradiàtum Adams subsp. árcticum (Regel) Hult.
Thoroughwax

Bupleurum ranunculoides var. *arcticum* Regel; *B. angulosus* Cham. & Schlecht. not L.; *B. americanum* Coult. & Rose.

Stems from short, simple or many-headed caudex, simple or branched; basal leaves linear-lanceolate, acute or somewhat acute; stem leaves oblong-lanceolate to rarely ovate-lanceolate, sessile, more or less clasping; umbel with 2–9 rays; involucre with 2–6 leafy bracts; involucels of foliaceous, bright-yellow bractlets; umbels compact; petals yellow, often blotched with purple; fruit dark brown, broadly oblong, more or less winged.

Meadows, dry, stony slopes, in McKinley Park to at least 1,200 meters. *B. triradiatum* described from Lake Baikal, subsp. *arcticum* from Kotzebue Sound.

Alpine specimens are small, with few rays in the umbel; the lowland specimens are branched, up to 70 cm tall, with up to 9 rays.

Broken line on circumpolar map indicates range of subsp. **triradiàtum**.

65. UMBELLIFERAE (Parsley Family)

Cicuta L. (Water Hemlock)

Axils of leaves bulbiferous; leaves with linear segments . . . . 1. *C. bulbifera*
Axils not bulbiferous; segments of leaves mostly broader:
    Fruit ovate to orbicular, at least as long as broad; rays of umbel 12–20, 2–6 cm
        long . . . . 2. *C. Douglasii*
    Fruit conspicuously broader than long; rays fewer, longer . . . 3. *C. mackenzieana*

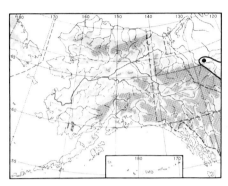
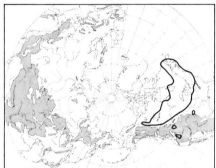

×⅓

1. Cicùta bulbífera L.

Stem slender, single, mostly from tuberous roots; leaves cauline, 2–3-pinnate, the upper reduced; leaflets linear, entire or sparsely toothed; upper axils bearing 1 or more bulblets; umbels with rays 1–2.5 cm long; ripe fruit orbicular.

Marshes, bogs. Described from Virginia and Canada.

The fruit often does not appear.

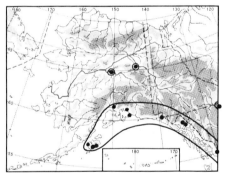
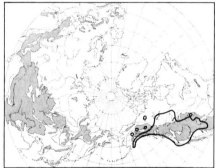

2. Cicùta Douglásii (DC.) Coult. & Rose

Sium Douglasii DC.; *Cicuta occidentalis* Greene.

Stout; stems single or few together, from tuberous-thickened and chambered roots; leaves 1–3-pinnate; leaflets lanceolate-ovate to oblong, serrate; involucral bracts mostly lacking; umbels several; rays 12–20, 2–6 cm long; pedicels 3–8 mm long; fruit ovate to orbicular.

Marshes, along streams; the northernmost localities are from hot springs. Described from northwest north America.

The roots contain cicutotoxin, and are deadly poisonous.

×¼

65. Umbelliferae (Parsley Family)

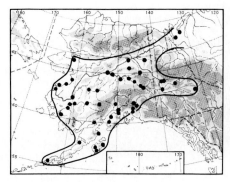

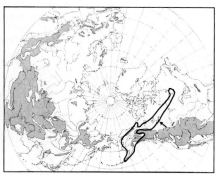

3. Cicùta mackenzieàna Raup

Similar to *C. Douglasii,* but ripe fruit considerably broader than long; rays 7–14, 7–8 cm long; pedicels 7–12 mm long.

Marshes.

The roots are deadly poisonous.

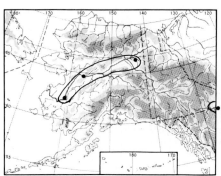

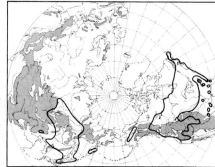

Sium L.

1. Sìum suàve Walt.   Water Parsnip
Sium cicutaefolium Schrank.

Stem tall, stout, ribbed, from short caudex and fibrous root; young submerged leaves 2–3-pinnate; leaves of flowering plant pinnate, with 7–13 sessile, lanceolate to lance-oblong, sharply and densely serrate leaflets; involucre of 6–10 lanceolate bracts; umbels with up to 25 rays; fruit broadly elliptic to orbicular.

Marshes, shallow water. Described from "Carolina."

Not considered poisonous.

65. UMBELLIFERAE (Parsley Family)

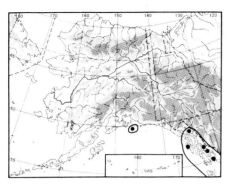 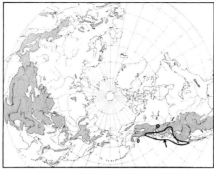 

Oenanthe L.

1. Oenànthe sarmentòsa Presl

Decumbent or ascending; stem fleshy, branched, the branches rooting at nodes; leaves bipinnate; leaflets cleft to toothed; bractlets 4–5 mm long; umbels with 10–20 rays; fruit oblong, 2.5–3.5 mm long, truncate, ribs much broader than intervals.

Marshes, sluggish water.

Cnidium Cusson

Leaves ovate in outline, 2–3-pinnate, with linear to ovate-linear ultimate segments
.... 1. *C. ajanense*
Leaves triangular in outline, 2–3-pinnate, with deeply divided ultimate segments
.... 2. *C. cnidiifolium*

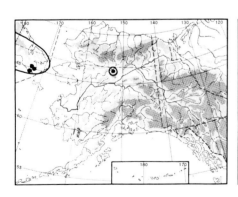 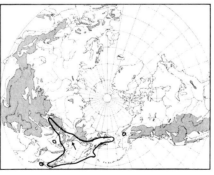 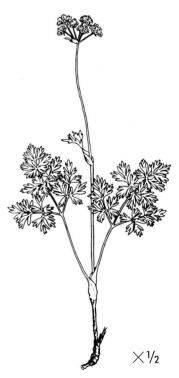

1. Cnídium ajanénse (Regel & Tiling) Drude
Tilingia ajanensis Regel & Tiling.

Stem simple or branched above, glabrous, from stout, many-headed, vertical caudex; basal leaves 2-pinnate to nearly 3-pinnate, with linear to ovate-lanceolate ultimate segments; petioles with broad basal sheaths; umbel with 5–10 rays; petals broadly ovate, white or purplish, with inflexed apex; fruit broadly ovate; 2 lateral ribs of carpels slightly more winged than back ribs.

Meadows in subalpine or alpine regions.

65. UMBELLIFERAE (Parsley Family)

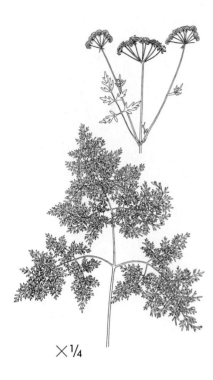

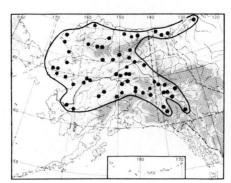

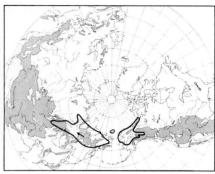

2. Cnídium cnidiifòlium (Turcz.) Schischk.

Selinum cnidiifolium Turcz.; *S. Dawsonii* Coult. & Rose; *Conioselinum Dawsonii* (Coult. & Rose) Coult. & Rose; *C. cnidiifolium* (Turcz.) Pors.

Glabrous; stem simple or branched above, from stout, many-headed, vertical caudex; lower leaves petiolated, 2–3-pinnate, ultimate segments ovate, deeply divided into acute lanceolate lobes; upper leaves reduced, nearly sessile; umbels with numerous rays; involucre with 5 lanceolate, early-deciduous leaflets; involucels 9–11, scarious with excurrent midvein; petals white or purplish; fruits broadly ovate; carpels with 5 equally winged ribs.

Wet meadows, gravelly slopes, riverbanks. Described from between Jakutsk and Aldan.

Ligusticum L.

Tall, seashore plant with biternate leaves and large ovate leaflets .... 1. *L. scoticum* subsp. *Hultenii*
Small, alpine plant with 2-pinnate leaves and narrow ultimate segments .... 2. *L. mutellinoides* subsp. *alpinum*

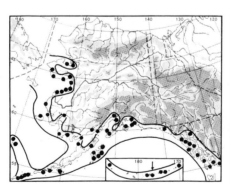

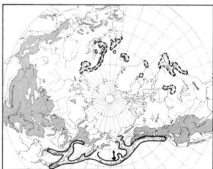

1. Ligústicum scòticum L.   Beach Lovage
subsp. **Hulténii** (Fern.) Calder & Taylor

Ligusticum Hultenii Fern.; *Angelica Hultenii* (Fern.) Hiroe.

Glabrous; stem single, reddish-violet at base, from thick root; leaves thick, biternate; leaflets ovate, coarsely toothed; stem leaves reduced, with long, often violet sheaths; rays of umbel 7–11; petals white or pinkish; fruit with 3 winged ribs on back, and broad intervals.

Seashores. *L. scoticum* described from England and Sweden. Broken line on circumpolar map indicates range of subsp. **scòticum.**

The young stems and leaves are eaten raw or boiled by the natives.

65. UMBELLIFERAE (Parsley Family)

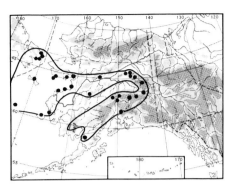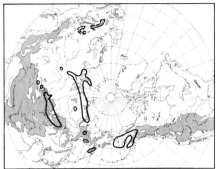

2. Ligústicum mutellinoìdes (Crantz) Willar

Laserpitium mutellinoides Crantz.

subsp. **alpìnum** (Ledeb.) Thell.

Pachypleurum alpinum Ledeb.; *Ligusticum Macounii* Coult. & Rose; *Ligusticella Macounii* (Coult. & Rose) Math. & Const.; *Podistera Macounii* (Coult. & Rose) Math. & Const.; *Orumbella Macounii* (Coult. & Rose) Coult. & Rose.

Stem up to 20 cm tall, usually shorter, from a stout, many-headed, vertical caudex, with remains of previous year's sheaths; leaves all basal, glabrous, mostly 2-pinnate, with acute or somewhat acute teeth; petioles with broad, scarious tooth at base; umbel with 5–20 rays; involucre with several, linear-lanceolate, often toothed or lobed leaves; involucels 5–10; petals obovate, emarginate, white; fruit broadly ovate with thick ribs, the marginal ribs slightly broader.

Stony, dry places on tundra and in the mountains, to over 1,700 meters. *L. mutellinoides* described from Switzerland, subsp. *alpinum* from Altai.

Broken line on circumpolar map indicates range of subsp. **mutellinoìdes**.

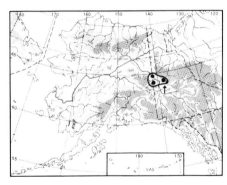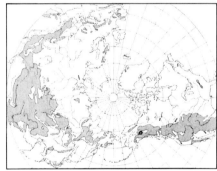

Podistera S. Wats.

1. Podístera yukonénsis Math. & Const.

Acaulescent, caespitose; leaves scaberulous on veins and margins, pinnate with 3–6 pairs of orbicular to narrowly lanceolate, apiculate leaflets, the lowest pair ternate, the others entire; petioles with narrow sheath at base; involucre of linear, entire bracts; involucels of 5 linear, entire bractlets; petals oblong-ovate, white; fruit ovate-oblong, with filiform, prominent ribs.

Stony slopes in the mountains.

65. UMBELLIFERAE (Parsley Family)

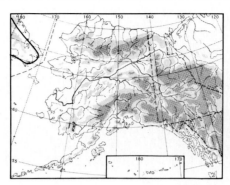
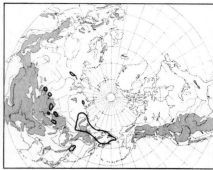

Phlojodicarpus Turcz.

1. Phlojodicárpus villòsus Turcz.

Johrenia villosa (Turcz.) Benth.

Stem from stout, woody caudex, covered with sheaths of previous year's leaves; leaves grayish-green, glabrous, 2–3-pinnate, with lanceolate, acute segments; stem leaves 1–3, with large sheaths; umbel with 8–30 rays, pubescent with white, curved hairs; involucral scales 5–11, lanceolate-linear, pubescent; involucels 5–11; petals white or purplish, broadly obovate; fruit oval, soft-pubescent, with 3 rounded ribs on back and larger marginal ribs.

Stony tundra. Described from Baikal.

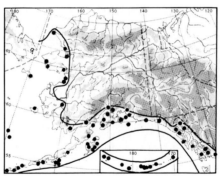

Conioselinum Hoffm.

1. Conioselìnum chinénse (L.) BSP.   Hemlock Parsley

Athamantha chinensis L.; *Ligusticum Gmelinii* Cham. & Schlecht.; *Conioselinum Gmelinii* (Cham. & Schlecht.) Coult. & Rose, not Steud.; *C. Benthami* (S. Wats.) Fern.; *Selinum Benthami* S. Wats.; *C. kamtschaticum* Rupr.

Stem from stout taproot; leaves 2–3-pinnate, leaflets more or less lobed; petioles with sheath at base; umbel with numerous rays; involucre with a few bracts or lacking bracts; involucels of narrow, scarious-margined bractlets; petals ovate with inflexed apex, white; fruit oblong-oval, glabrous; dorsal ribs acute, corky; lateral ribs broader and thin-winged.

Meadows, sandy shores. Described (erroneously) from China; type probably originated from New York.

65. UMBELLIFERAE (Parsley Family)

Angelica L.

Ribs of fruit all similar; flowers greenish-white; upper leaves not reflexed 1. *A. lucida*
Lateral ribs of fruit broadly winged; flowers white to pinkish; upper leaves reflexed
....2. *A. genuflexa*

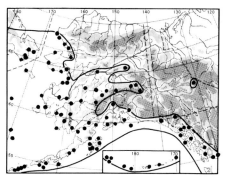
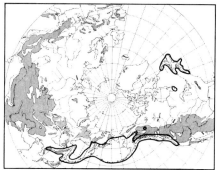

1. Angélica lùcida L.

Coelopleurum lucidum (L.) Fern.; *Archangelica Gmelinii* DC.; *C. Gmelinii* (DC.) Ledeb.; *Pleurospermum Gmelinii* (DC.) Bong.

Plant stout, up to more than 1 meter tall; leaves glabrous or essentially so, 2–3-ternate; leaflets ovate to deltoid, acute, irregularly serrate; petioles inflated; inflorescence scabrous-puberulent; umbel with 20–40 rays; flowers greenish-white; ribs of fruit all similar, narrowly winged.

Meadows, thickets, riverbanks, common along the coasts. Described from Canada.

Stem and petioles are eaten by the natives, and called "wild celery." In Siberia the root was at one time carried as an amulet to ward off polar bears. The Siberian Eskimo inhale the fumes of the roasted root as a seasick remedy.

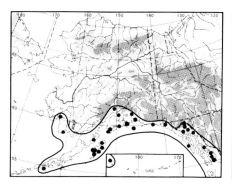
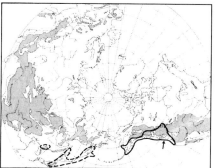

2. Angélica genufléxa Nutt.

Stem stout, hollow, glabrous, often purple, more than 1 meter tall, from septate, tuberous caudex; leaves ternate, pinnate to bipinnate, the upper leaves (beyond the basal pair of leaflets) geniculate and reflexed; leaflets broadly ovate to obovate, sharply serrate, pubescent on veins and in the margin; inflorescence densely puberulent, with numerous rays; petals white to pinkish; fruit rounded, the lateral ribs broadly winged, the dorsal narrowly winged.

Swamps, along streams.

Broken line on circumpolar map indicates range of subsp. **refrácta** (F. Schm.) Hiroe (*A. refracta* F. Schm.).

65. UMBELLIFERAE (Parsley Family)

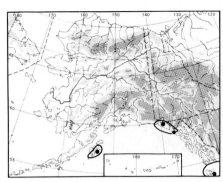
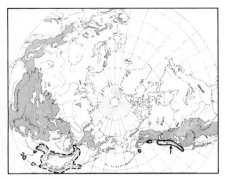

Glehnia F. Schm.

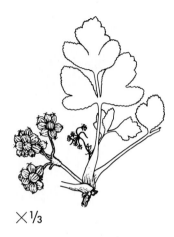

1. Gléhnia littoràlis F. Schm.
subsp. **leiocárpa** (Math.) Hult.
Glehnia leiocarpa Math.

Subacaulescent, more or less prostrate; flowering stem from stout, woody taproot; leaves once- to twice-ternate, crenate-serrate, with petioles buried in sand; leaflets minutely hairy on nerves above, sarmentose beneath; inflorescence villous, with compact, capitate ultimate umbels; fruit ovate to roundish, flattened dorsally, mostly pubescent, with scattered multicellular hairs, especially in margin, or nearly glabrous.

Seashores. Broken line on circumpolar map indicates range of subsp. **littoràlis,** described from Hakodate, Yezo.

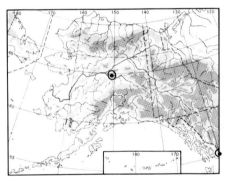

Pastinaca L.

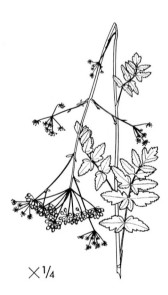

1. Pastinàca satìva L.   Parsnip

Stems from thick, yellowish-white root; leaves large, pinnate, with oblong to ovate, toothed leaflets; flowers yellow; 1 large and several smaller umbels; fruit oval, flattened dorsally, lateral and dorsal ribs winged.

Escaped from cultivation; found in central Alaska at the Tanana hot spring. Described from Europe.

65. UMBELLIFERAE (Parsley Family) / 66. CORNACEAE (Dogwood Family)

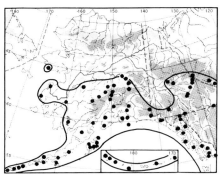

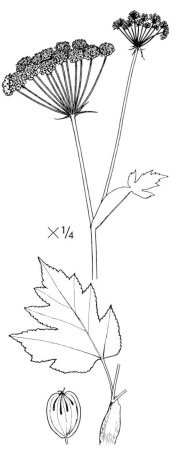

Heracleum L.

1. Heraclèum lanàtum Michx. Cow Parsnip
"Heracleum maximum" Bartr.

Thick, stout, up to over 2 meters tall; stems from thick taproot; leaves large, tomentose when young, more or less glabrescent in age, ternate, the upper with conspicuously inflated sheaths; leaflets petiolulate, palmately lobed, coarsely toothed; rays of umbel 15–30, the terminal umbel much larger than other umbels; flowers white; fruit obovate to obcordate, emarginate, villose to glabrate; lateral ribs with thin wings, dorsal ribs filiform.

Woods, meadows, in the interior mostly in the mountains. Described from Canada.

The marrow is eaten raw and the root boiled by the natives; the plant contains sugar and tastes much like licorice.

Cornus L.

Plant a shrub; flowers without petaloid white bracts; ripe fruit white . . . .
. . . . 1. *C. stolonifera*
Plant herbaceous, from slender rootstock; flowers small, surrounded by large, petaloid, white bracts:
  Stem with 1 pair of small leaves, and broad, seemingly whorled leaves at tip; flowers yellowish or greenish, entirely covered with white appressed hairs; style thick and short; stamens not exceeding styles . . . . . . 3–4. *C. canadensis*
  Stem with 3 or more pairs of larger leaves, those at tip less whorled; flowers purplish-black, with white hairs only at base; style longer; stamens exceeding styles . . . . 2. *C. suecica*

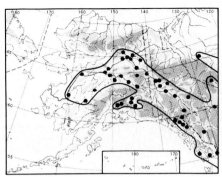 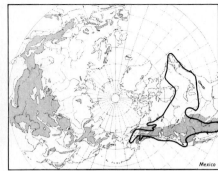 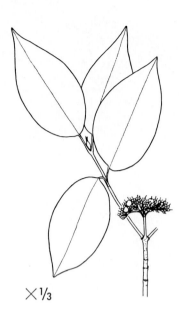

1. Córnus stolonífera Michx. — American Dogwood
Svida stolonifera (Michx.) Rydb.; ?*Cornus sericea* L.

Shrub 2–6 meters tall; branches reddish or brown, more or less pubescent, the lower often rooting at tip; leaves broadly ovate to oblong-lanceolate, acute, dark green and glabrous or sparsely strigillose above, paler and appressed-pubescent with short straight hairs beneath; inflorescence flat-topped, with minutely pubescent branches; petals white; style 1.5–2.2 mm long; fruit white, subglobose, glabrous to pubescent, containing somewhat flattened stone.

Moist woods, along streams. Described from Canada.

Specimens with leaves densely soft-pubescent beneath have been called var. **Bailèyi** (Coult. & Evans) Drescher (*C. Baileyi* Coult. & Evans). Specimens from the coast usually have less pubescent branches than those from the interior.

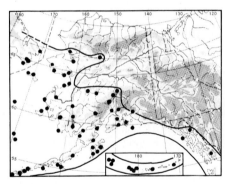 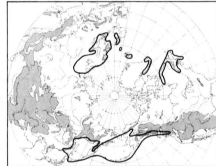 

2. Córnus suècica L. — Swedish Dwarf Cornel
Chamaepericlymenum suecicum (L.) Graebn.

Stems from creeping rhizome; flowering stems erect, quadrangular in cross section, with 3 to several pairs of opposite, fairly large, ovate, abruptly pointed leaves, the topmost largest and seemingly whorled; bracts large, petaloid, white, ovate, obtuse or somewhat acute; flowers small, purplish-black, with white hairs only at base; stamens longer than styles; fruit margined, red.

Woods, marshes, bogs. Described from Sweden, Norway, and Russia.

66. CORNACEAE (Dogwood Family)

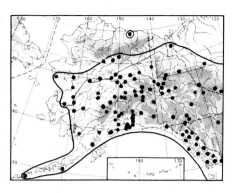
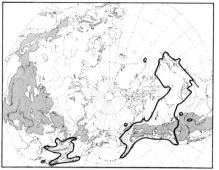

3. Córnus canadénsis L.   Bunchberry, Canadian Dwarf Cornel

Chamaepericlymenum canadense (L.) Graebn.

Similar to *C. suecica*, but stem leaves small, the upper more or less whorled, much larger, acute; flowers yellowish or greenish, entirely covered with white hairs; stamens shorter than styles.

Very common in spruce and birch forests, to 800 meters in McKinley Park. Described from Canada.

Specimens in which the white bracts are partly green occur rarely.

(See color section.)

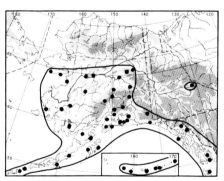
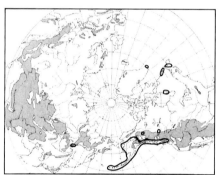

4. Córnus canadénsis L. × **suècica** L.

Cornus unalaschkensis Ledeb.; *Chamaepericlymenum unalaschcense* (Ledeb.) Rydb.; *C. canadensis* var. *intermedia* Farr; *C. intermedia* (Farr) Calder & Taylor.

A hybrid representing several combinations of the characters of *C. canadensis* and *C. suecica*, especially as regards the size, form, and arrangement of the leaves, the color and pubescence of the flowers, and the length of the styles and stamens. Propagates vegetatively, and has reduced or no fertility.

A separate map has been prepared for this hybrid, since its range extends somewhat beyond that of *C. suecica*, as a result of reduction of the range of *C. suecica* from a presumably larger area in an earlier climatic period to the present smaller area.

67. PYROLACEAE (Wintergreen Family)

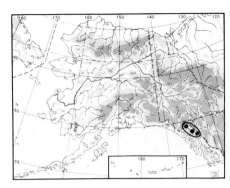
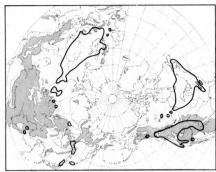

Chimaphila Pursh

1. Chimáphila umbellàta (L.) Barton — Pipsissewa
Pyrola umbellata L.
subsp. **occidentàlis** (Rydb.) Hult.
Chimaphila occidentalis Rydb.; *C. umbellata* var. *occidentalis* (Rydb.) Blake.

Dwarf shrub with thick, dark-green, stiff, winter-green, oblanceolate, serrate leaves in whorls; flowers flesh-colored, umbellate; calyx lobes longer than broad.

Woods. *C. umbellata* described from Europe, Asia, and North America, subsp. *occidentalis* from Latah County, Idaho.

In Eurasia, subsp. **umbellàta** occurs; in eastern North America, the very closely related subsp. **cisatlántica** (Blake) Hult. (*C. umbellata* var. *cisatlantica* Blake); in New Mexico and Arizona, subsp. **acùta** (Rydb.) Hult. (*C. acuta* Rydb.); in the rest of western North America, subsp. **occidentàlis**; and in Mexico, subsp. **mexicàna** (DC.) Hult. (*C. umbellata* var. *mexicana* DC.). Circumpolar map indicates range of the entire species complex.

Pyrola L. (Wintergreen)

Inflorescence 1-sided; flowers greenish or greenish-white:
    Leaves large, oblong-ovate, acute . . . . 6. *P. secunda* subsp. *secunda*
    Leaves smaller, ovate to orbicular, blunt . . . . . . 7. *P. secunda* subsp. *obtusata*
Inflorescence spiral; flowers greenish or white, pink or reddish:
    Style not protruding from flower, short, straight, lacking ring under stigma; flowers white to pinkish . . . . 5. *P. minor*
    Style protruding, curved, with ring under stigma:
        Leaves small, roundish; sepals broader than long; flowers greenish-white . . . . 4. *P. chlorantha*
        Leaves larger, sepals longer than broad; flowers white or greenish-white, pink or reddish:
            Flowers white or greenish-white . . . . 3. *P. grandiflora*
            Flowers pink or reddish:
                Leaves cordate, orbicular to somewhat reniform . . . . 1. *P. asarifolia* var. *asarifolia*
                Leaves obovate, obovate-oblong to orbicular, not cordate . . . . 2. *P. asarifolia* var. *purpurea*

67. Pyrolaceae (Wintergreen Family)

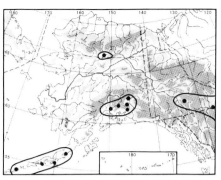
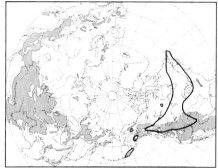

1. Pýrola asarifòlia Michx.
var. **asarifòlia**

Leaves leathery, cordate, orbicular to somewhat reniform, mucronulate-denticulate; scapes with 1–3 scarious, ovate-oblong bracts; sepals lanceolate; flowers open, crimson to pink; style long, curved, with ring below the 5 stigma lobes.

Woods. Type locality not given; presumably eastern North America.

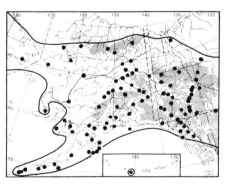
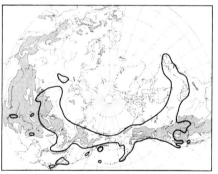

2. Pýrola asarifòlia Michx.
var. **purpùrea** (Bunge) Fern.

Pyrola rotundifolia var. *purpurea* Bunge; *P. incarnata* Fisch.; *P. rotundifolia* var. *incarnata* (Fisch.) DC.; *P. asarifolia* var. *incarnata* (Fisch.) Fern.

Similar to var. *asarifolia,* but leaves obovate, obovate-oblong or orbicular, but not cordate.

Woods, meadows, in the Yukon to at least 1,200 meters. Described from the Altai Mountains.

Specimens from the interior have smaller leaves. Complete transition occurs between cordate-leaved and not-cordate-leaved plants.

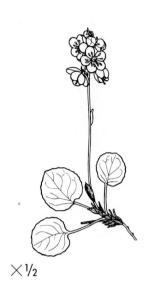
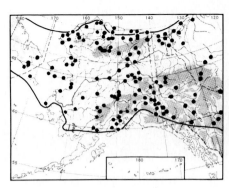
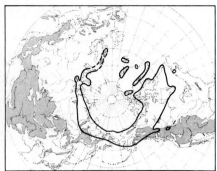

3. Pýrola grandiflòra Radius

Pyrola Gormani Rydb.; *P. grandiflora* var. *Gormani* (Rydb.) Pors.

Stem short, from slender rhizome; leaves thick, lustrous, rounded, nearly entire in margin; pale beneath; scape with 1–2 bracts; flowers large; sepals longer than broad; petals white or greenish-white with dark veins; anthers yellow; style long, somewhat curved at maturity, with ring below the 5 stigma lobes.

Dry places on tundra and in the mountains. Described from Labrador.

Southern lowland specimens are taller and often have smaller flowers; they have been called var. **canadénsis** (Andres) Pors. (*P. canadensis* Andres). Broken line on circumpolar map indicates range of hybrid swarms with *P. rotundifolia*.

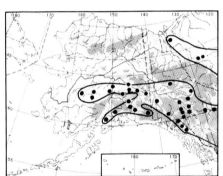
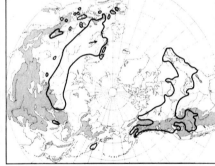

4. Pýrola chloránta Sw.

Pyrola virens Schweigger.

Leaves small, thick, roundish, shiny, yellowish-green, somewhat crenate; scapes angular, mostly naked, or with small bract; flowers open; sepals broader than long, blunt; petals yellowish-green; style slightly protruding, curved, with ring below stigma.

Woods.

67. Pyrolaceae (Wintergreen Family)

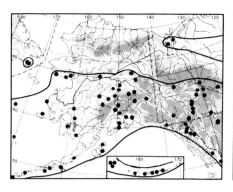

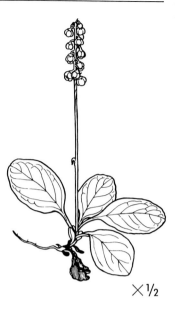

5. Pýrola mìnor L.

Leaves ovate, crenulate, flat in margin; flowers closed, globose; petals white or mostly pinkish; style included, in fruit much shorter than capsule, lacking ring under the 5 spreading stigma lobes.

Woods. Described from Europe.

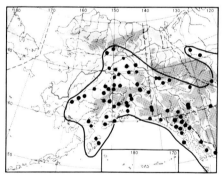
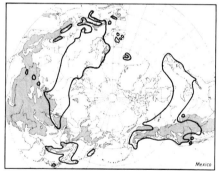

6. Pýrola secúnda L.
subsp. **secúnda**
  Ramischia secunda (L.) Garcke.

Stem from long, slender, creeping rhizome; leaves yellowish-green, the basal oblong-ovate, acute, crenate-serrate; scapes with 2–5 bracts; inflorescence 1-sided; sepals broadly triangular, finely dentate; petals greenish-white, denticulate; style 5–7 mm long, lacking ring below stigma lobes.

Woods. Described from northern Europe.

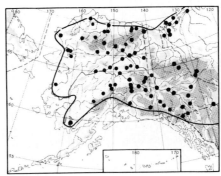 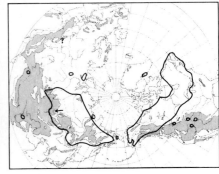

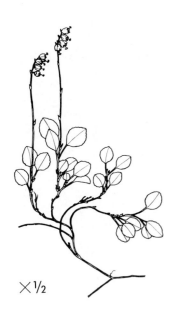

7. Pýrola secúnda L.
subsp. **obtusàta** (Turcz.) Hult.

Pyrola secunda var. *obtusata* Turcz.; *P. secunda* var. *nummularia* Rupr.; *Ramischia secunda* (L.) Garcke subsp. *obtusata* (Turcz.) Andres; *R. obtusata* (Turcz.) Freyn.

Similar to subsp. *secunda*, but smaller; leaves smaller, ovate to orbicular, blunt, crenate; inflorescence few-flowered, short.

Woods, in the mountains to at least 800 meters.

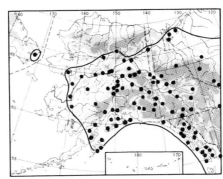 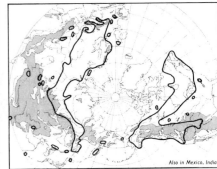

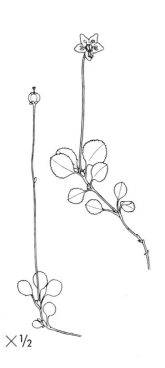

Moneses Salisb.

1. Monèses uniflòra (L.) Gray   Single Delight

Pyrola uniflora L.; *Moneses grandiflora* S. F. Gray.

Stem from thin, creeping caudex; leaves small, light green, obovate, rounded, serrate; stem mostly with single bract, 1-flowered; flowers drooping in anthesis, fragrant; petals 5, white; style long, straight; stigma stout, with 5 lobes; capsule round, erect.

Woods. Described from Europe.

Specimens from the coastal region belong to var. **reticulàta** (Nutt.) Blake [*Moneses reticulata* Nutt.; *M. uniflora* subsp. *reticulata* (Nutt.) Calder & Taylor], with somewhat acute, more denticulated and reticulated leaves.

67. Pyrolaceae (Wintergreen Family)

Monotropa L. (Indian Pipe)

Plant with several flowers in elongated raceme; style longer than ovary ....
.... 1. *M. hypopitys* subsp. *lanuginosa*
Plant with single, nodding flower; style short and thick .... 2. *M. uniflora*

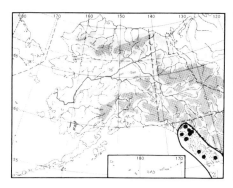 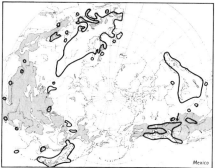 

1. Monótropa hypópitys L.
subsp. **lanuginòsa** (Michx.) Hara

Monotropa lanuginosa Michx.; *Hypopitys latisquamea* Rydb.; *M. latisquamea* (Rydb.) Hult.; *H. monotropa* Crantz.

Saprophyte with fleshy, slender, unbranched stem, lacking chlorophyll, yellowish-white or pinkish, drying black; stems clustered with scales rather than leaves; flowers in elongated raceme.

Woods. *M. hypopitys* described from Sweden, England, Germany, and Canada.

A complicated complex, taken here in a broad sense. Circumpolar map indicates range of the entire complex.

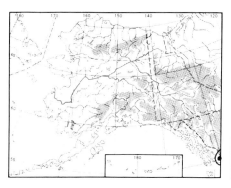  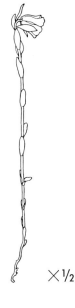

2. Monótropa uniflòra L.

Saprophyte with fleshy, unbranched stem, lacking chlorophyll, pink or reddish, blackening when dried; stems clustered, with scales instead of leaves, 1-flowered; petals scalelike; style short and thick; flower nodding, but fruit erect.

Woods. Described from Maryland, Virginia, and Canada.

68. EMPETRACEAE (Crowberry Family)

Empetrum L. (Crowberry)

Flowers unisexual . . . . 1. *E. nigrum* subsp. *nigrum*
Flowers bisexual . . . . 2. *E. nigrum* subsp. *hermaphroditum*

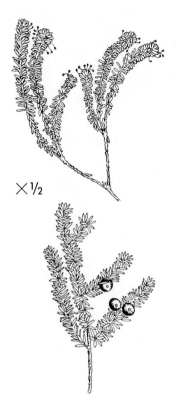
×½

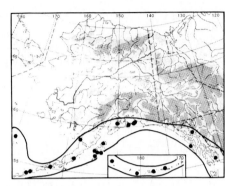
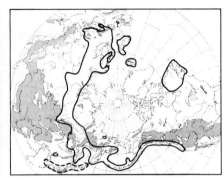

1. Empètrum nìgrum L.
subsp. **nìgrum**

Empetrum subholarcticum Vassiljev; *E. Kardakovii* Vassiljev; *E. arcticum* Vassiljev, *E. atropurpureum* Fern. & Wieg., *E. Eamesii* Fern. & Wieg., and (?) *E. androgynum* Vassiljev with respect to all plants from Unalaska and Sitka.

Procumbent, with creeping branches; young branches minutely glandular-pubescent; leaves linear to narrowly elliptical, glandular-margined, divergent, soon reflexed; flowers inconspicuous, solitary in upper axils; pistillate and staminate flowers on different plants, traces of opposite sex sometimes occur; fruit black, very rarely purplish or white, with watery juice and several hard seeds.

Heaths, bogs. Described from Europe. Circumpolar map indicates range of unisexual *E. nigrum*, broken line that of subsp. **japònicum** (Good) Hult. (*E. nigrum* f. *japonicum* Good).

The berries are tasteless, but are eaten, especially mixed with those of *Vaccinium uliginosum*, in pies or as jelly.

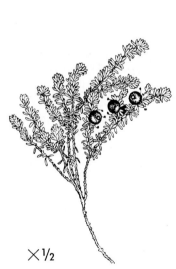
×½

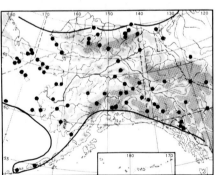
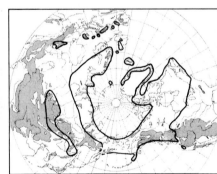

2. Empètrum nìgrum L.
subsp. **hermaphrodìtum** (Lange) Böcher

Empetrum nigrum f. *hermaphroditum* Lange; *E. hermaphroditum* (Lange) Hagerup; *E. nigrum* var. *hermaphroditum* (Lange) Sørens.; *E. arcticum* Vassiljev.

Similar to subsp. *nigrum*, but flowers bisexual; leaves somewhat shorter and broader; fruit and seeds generally larger.

Heaths, bogs. Described from Greenland.
Circumpolar map indicates range of bisexual *E. nigrum*.

69. Ericaceae (Heath Family)

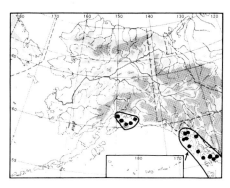

Cladothamnus Brongn.

1. Cladothámnus pyrolaeflòrus Bong.   Copper-Flower

Shrub, more than a meter tall; leaves entire, oblanceolate, obtuse, sessile or nearly so, thin, light green; median nerve excurrent, with water gland at tip; flowers solitary; calyx lobes 5, linear, ciliolate; flowers rotate, copper-colored; petals distinct, 5, up to 15 mm long; stamens 10; style curved, elongate; fruit a dry capsule.

Moist forests, along streams.

Ledum L. (Labrador Tea)

Leaves linear; stamens mostly 10; pedicels hooked in apex ....
.... 1. *L. palustre* subsp. *decumbens*
Leaves linear-oblong to oblong; stamens mostly 8; pedicels arcuate ....
.... 2. *L. palustre* subsp. *groenlandicum*

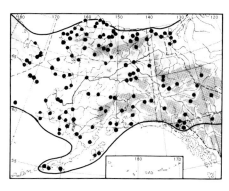

1. Lèdum palústre L.
subsp. **decúmbens** (Ait.) Hult.

Ledum palustre var. *decumbens* Ait.; *L. decumbens* (Ait.) Small.

Low shrub, with brown, puberulent young twigs, glabrescent in age; flowers in umbel-like clusters; leaves linear, somewhat acute, with strongly revolute margin, shiny and glabrous above, cinnamon-brown, woolly beneath; pedicels rusty-puberulent; stamens mostly 10, hooked or curved at maturity; flowers white or pinkish.

Heaths, dry, rocky places, in the mountains to at least 1,800 meters; very common. Described from Hudson Bay. Broken line on circumpolar map indicates range of subsp. **palústre**.

Contains ledol, a poisonous substance causing cramps and paralysis.

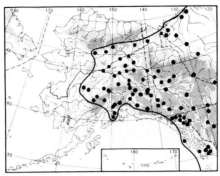
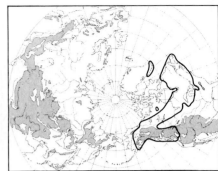
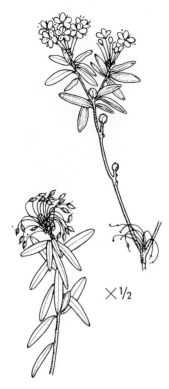

2. Lèdum palústre L.
subsp. **groenlándicum** (Oeder) Hult.
Ledum groenlandicum Oeder; *L. pacificum* Small.

Larger than subsp. *decumbens*; leaves oblong to linear-oblong, obtuse; stamens mostly 8.

Described from Greenland.

Rhododendron L.

Leaves about 1 cm long, covered on both sides with resin dots; calyx small, 1–3 mm long .... 1. *R. lapponicum*
Leaves larger, lacking resin dots:
    Petals ciliated in margin, pubescent on outside; leaves of sterile shoots not glandular in margin .... 2. *R. camtschaticum* subsp. *camtschaticum*
    Petals not ciliated in margin, glabrous on outside; leaves of sterile shoots glandular in margin .... 3. *R. camtschaticum* subsp. *glandulosum*

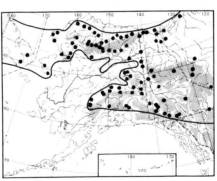
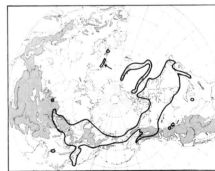

1. Rhododéndron lappónicum (L.) Wahlenb.   Lapland Rosebay
Azalea lapponica L.

Low dwarf shrub with thick, woody stem and knotty branches; leaves dark green, ovate, obtuse, rust-colored and densely resin-dotted on both sides; pedicels resin-dotted; sepals small, with triangular lobes, resin-dotted and ciliate with long hairs; petals purplish, very rarely white; stamens mostly 7–8, pubescent at base.

Stony slopes, heaths, in the mountains; both dry and wet habitats.

In subalpine woods, a shrub—rarely up to 70 cm tall [var. **parvifòlium** (Adams) Herder; *R. parvifolium* Adams].

69. ERICACEAE (Heath Family)

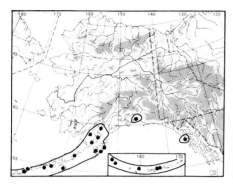
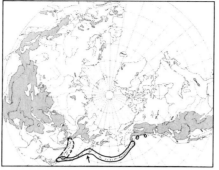
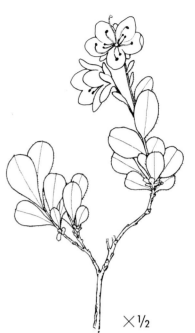

2. Rhododéndron camtscháticum Pall.   Kamchatka Rhododendron
subsp. **camtscháticum**

Therorhodion camtschaticum (Pall.) Small.

Stem thick, with brown or grayish-brown bark, much branched above; basal leaves obovate-cuneate, crenate-serrulate, those of sterile shoots sparsely pilose, ciliated in margin by simple, long ciliae, those of the flowering shoots ciliated and often glandular; stem leaves ovate to oblong; sepals oblong, blunt, strigose-pubescent and glandular-margined; petals about 2 cm long, somewhat unequal in length, ciliated in margin, pubescent on outside, oblong-ovate, purple, very rarely white (Kodiak); style long; filaments pubescent at base, the upper 5 shorter, the lower 5 longer.

Alpine meadows, occasionally in subalpine woods. Described from Kamchatka. Broken line on circumpolar map indicates range of subsp. **intercèdens** Hult.

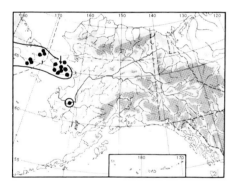

3. Rhododéndron camtscháticum Pall.
subsp. **glandulòsum** (Standl.) Hult.

Therorhodion glandulosum Standl.; *Rhododendron camtschaticum* var. *pumilum* E. Busch.

Similar to subsp. *camtschaticum*, but smaller; leaves narrower, those of sterile shoots glandular in margin; petals narrower, glabrous on outside, not ciliated in margin, somewhat darker in color.

Alpine meadows.

69. Ericaceae (Heath Family)

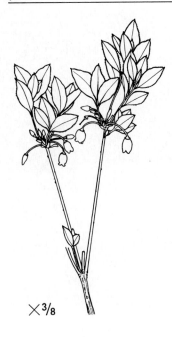

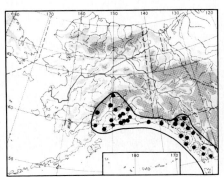

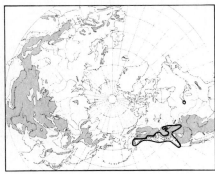

Menziesia Sm.

1. Menzièsia ferrugìnea Sm.

Erect shrub; leaves thin, light green, somewhat acute, crenulate-serrate, glandular on both sides and glandular-margined; calyx lobes broadly triangular, glandular-ciliate; corolla yellowish-red; ovary resin-dotted.

Moist woods. Described from western North America.

Rocky Mountain portion of range indicated on circumpolar map is that of var. **glabélla** (Gray) Peck (*M. glabella* Gray).

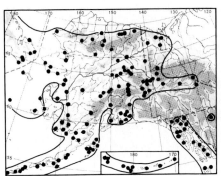

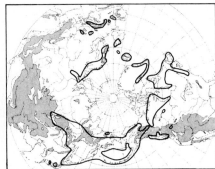

Loiseleuria Desv.

1. Loiseleùria procúmbens (L.) Desv.   Alpine Azalea
Azalea procumbens L.

Depressed, evergreen dwarf shrub, forming mats, often covering stones; leaves opposite, elliptical, coriaceous, with reflexed margin, whitish beneath; pedicels and calyx dark purple; flowers small, pink, in few-flowered clusters; style short.

In the mountains to at least 2,000 meters; prefers acid soil. Described from the European Alps.

69. Ericaceae (Heath Family)

Kalmia L.

Plant up to 3 dm tall; leaves with strongly revolute margin; flowers up to 2 cm broad .... 1. *K. polifolia* subsp. *polifolia*
Plant up to 1.5 dm tall; leaves less or not at all revolute in margin; flowers smaller .... 2. *K. polifolia* subsp. *microphylla*

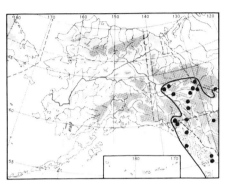
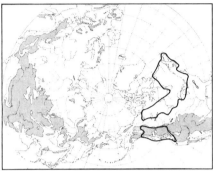

1. Kálmia polifòlia Wang.   Bog Laurel
subsp. **polifòlia**

Stem much branched, more or less matted; young stems 2-edged, puberulent, up to 3 dm tall; leaves opposite, lanceolate to linear, lustrous above, pale beneath, with strongly revolute margin; inflorescence umbelliform; flowers up to 2 cm broad; calyx glabrous; corolla saucerlike, pink to crimson.

Peaty soil. Described from eastern Canada.

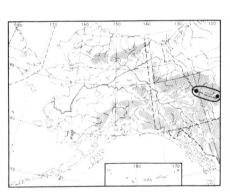
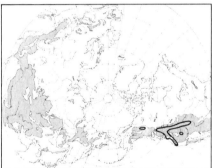

2. Kálmia polifòlia Wang.
subsp. **microphýlla** (Hook.) Calder & Taylor
Kalmia glauca var. *microphylla* Hook.; *K. microphylla* (Hook.) Heller.

Smaller than subsp. *polifolia*, up to 15 cm tall; leaves broader in comparison to length, oblong to obovate, less revolute or not revolute in margin; flowers smaller.
Bogs, wet meadows. Described from the Rocky Mountains.

Phyllodoce L. (Mountain Heather)

Corolla purple, campanulate, not constricted at mouth; calyx ciliolate in margin . . . . 1. *P. empetriformis*
Corolla purple or yellow, constricted at mouth:
    Corolla urceolate, purple . . . . 2. *P. coerulea*
    Corolla ovoid, yellow:
        Corolla glabrous on outside; filaments glabrous . . . . 3. *P. aleutica* subsp. *aleutica*
        Corolla glandular on outside; filaments more or less pubescent at base . . . . 4. *P. aleutica* subsp. *glanduliflora*

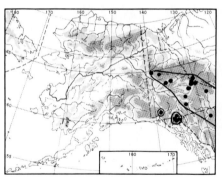

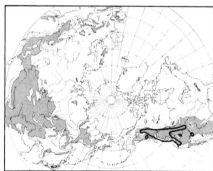

1. Phyllódoce empetrifórmis (Sm.) D. Don
Menziesia empetriformis Sm.; *Bryanthus empetriformis* (Sm.) Gray.

Matted dwarf shrub; young twigs finely puberulent, more or less glandular, glabrescent in age; leaves needlelike, shiny, glabrous or sparsely glandular, with revolute margin; pedicels puberulent and glandular, brown; calyx lobes ovate, purplish-margined, finely ciliated with multicellular, short hairs; corolla open, bell-shaped, with recurved lobes, purple; filaments glabrous; ovary glandular.

Mountainsides, up to 2,000 meters. Described from the northwest coast of North America.

(See color section.)

69. ERICACEAE (Heath Family)

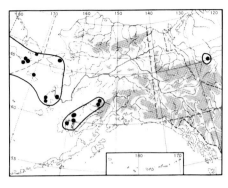
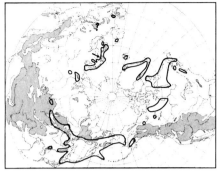

2. Phyllódoce coerùlea (L.) Bab.
Andromeda coerulea L.; *Bryanthus taxifolius* Gray.

Ascending dwarf shrub; stem densely branched; leaves needlelike, dark green, shiny above, glabrous, glandular-serrulate in margin, with a deep groove in middle and a pale groove beneath; scapes glandular; calyx lobes triangular, acute, glandular; corolla urn-shaped, dark purple (not sky-blue as the name implies), yellowish at base, with reflexed lobes; ovary glandular.
 Mountainsides.
 Forms hybrid swarms with *P. aleutica* where ranges overlap; such hybrids occur at Attu Island.

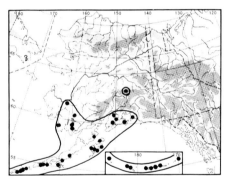
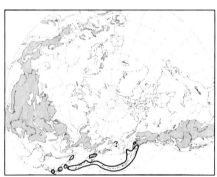

3. Phyllódoce aleùtica (Spreng.) Heller subsp. aleùtica
Menziesia aleutica Spreng.; *Phyllodoce Pallasiana* D. Don; *Bryanthus aleuticus* (Spreng.) Gray.

Similar to *P. coerulea*, but leaves yellowish-green, shorter and broader; corolla ovoid, bright yellow.
 Heaths, dry mountainsides. Described from the Aleutian Islands.

69. Ericaceae (Heath Family)

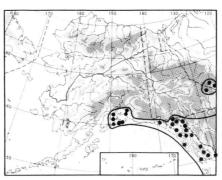
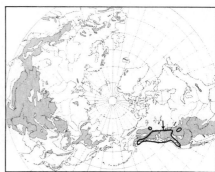

4. Phyllódoce aleùtica (Spreng.) Heller
subsp. **glanduliflòra** (Hook.) Hult.

Menziesia glanduliflora Hook.; *Bryanthus glanduliflorus* (Hook.) Gray; *Phyllodoce glanduliflora* (Hook.) Cov.

Very similar to subsp. *aleutica*, but corolla glandular on outside and filaments more or less pubescent at base.

Heaths, dry mountainsides. Transitions between the two races occur where ranges overlap.

(See color section.)

Cassiope D. Don

Leaves grooved on back .... 1–2. *C. tetragona*
Leaves not grooved on back:
    Leaves spreading, alternate; flowers solitary, terminal .... 4. *C. Stelleriana*
    Leaves appressed, opposite; flowers lateral or subterminal, usually not solitary:
        Leaves in 4 distinct rows, not scarious-margined; branches, including appressed leaves, about 4 mm in diameter .... 3. *C. Mertensiana*
        Leaves scarious-margined, not in 4 distinct rows; branches about 2 mm in diameter .... 5. *C. lycopodioides*

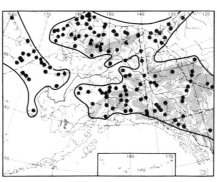

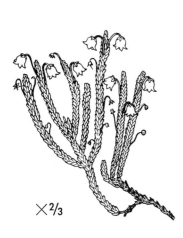

1. Cassìope tetragòna (L.) D. Don   Lapland Cassiope
Andromeda tetragona L.
subsp. **tetragòna**

Coarse, dark-green dwarf shrub; leaves in 4 rows, lanceolate, deeply grooved dorsally, puberulent, ciliolate; pedicels long, glabrous; calyx lobes reddish; corolla bell-shaped.

Dry heaths and rocks on tundra or in the mountains, to at least 2,000 meters. Common in the North.

69. Ericaceae (Heath Family)

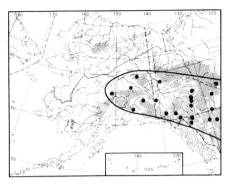
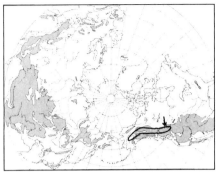

2. Cassìope tetragòna (L.) D. Don
subsp. **saximontàna** (Small) Pors.
  Cassiope saximontana Small.

  Similar to subsp. *tetragona*, but flowers somewhat smaller and pedicels very short.
  A weakly differentiated Rocky Mountain race.

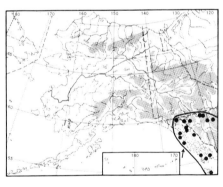
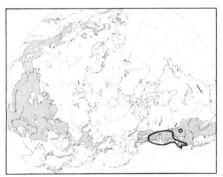

3. Cassìope Mertensiàna (Bong.) D. Don
  Andromeda Mertensiana Bong.

  Matted dwarf shrub; leaves in 4 distinct rows, ovate-lanceolate, obtuse, concave above, keeled on back, with two impressions at base; pedicels short, glabrous or puberulent; sepals reddish, ovate, more or less erose; corolla bell-shaped, white to pinkish.
  Mountain slopes to at least 1,300 meters.
  (See color section.)

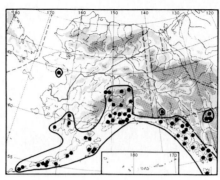 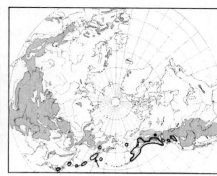

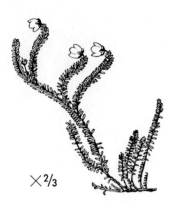

4. Cassìope Stelleriàna (Pall.) DC. Alaska Moss Heath

Andromeda Stelleriana Pall.; *Harimanella Stelleriana* (Pall.) Cov.; *Bryanthus Stelleri* D. Don.

Matted dwarf shrub; branches puberulent; leaves alternate, spreading, linear-oblanceolate, flat above, rounded beneath, glabrous, somewhat erose in margin; flowers single, pedicels short, puberulent; calyx lobes oblong, red, scarious-margined; corolla open, white to pinkish, twice as long as sepals; style short, conical.

Alpine heaths to at least 2,000 meters. Common along the coast. Type locality not given, presumably Kamchatka.

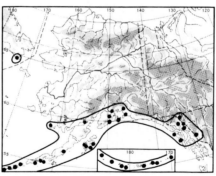 

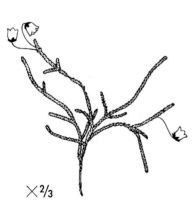

5. Cassìope lycopodioìdes (Pall.) D. Don

Andromeda lycopodioides Pall.; *Cassiope lycopodioides* subsp. *cristapilosa* Calder & Taylor.

Matted dwarf shrub with long, thin, prostrate branches; leaves opposite, oblanceolate, scarious-margined, concave above, rounded beneath, more or less brownish-pubescent at tip when young; pedicels glabrous; sepals ovate, erose; corolla bell-shaped, white.

Mountain slopes; common in the Aleutian Islands, Alaska Peninsula, and Kodiak Island. Described from Okhotsk, Kamchatka, and the Commander Islands.

69. Ericaceae (Heath Family)

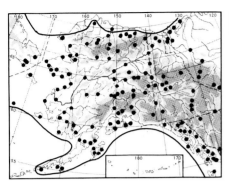  

Andromeda L.

1. Andrómeda polifòlia L.

Dwarf shrub with creeping base and elongate, ascending branches; leaves thick, linear-lanceolate to oblong, with revolute margin, grayish-green above, glaucous beneath, rarely green on both sides (var. **cóncolor** Boiv.); inflorescence subumbellate; pedicels and calyx red; corolla first pink, in age paler.

Peat bogs. Described from northern Europe. Broken line on circumpolar map indicates range of subsp. **glaucophýlla** (Link) Hult. (*A. glaucophylla* Link).

The small, narrow-leaved northern plant is var. **aceròsa** Hartm., described from northern Europe. Some specimens of *A. polifolia* along the southern coast are coarse, with much larger and broader leaves than var. *acerosa*.

Contains andromedotoxin, a poison causing low blood pressure, breathing difficulty, vomiting, diarrhea, and cramps.

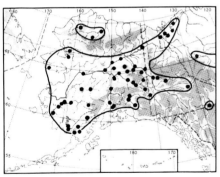 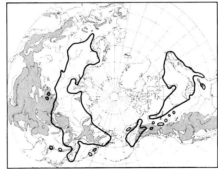 

Chamaedaphne Moench

1. Chamaedáphne calyculàta (L.) Moench   Cassandra

Andromeda calyculata L.; *Cassandra calyculata* (L.) D. Don; *Lyonia calyculata* (L.) Rchb.

Low, erect shrub; leaves coriaceous, oblong-lanceolate, 1.5–3.5 cm long, dark green and dotted with small, light-colored scales above, paler, brown-dotted and scaly beneath; sepals broadly triangular, acute; corolla white, urceolate-cylindric.

Peat bogs. Described from Virginia, Canada, and Siberia.

Some specimens, with leaves only about 1 cm long, belong to var. **nàna** (Lodd.) E. Busch (*Lyonia calyculata* var. *nana* Lodd.).

Gaultheria L.

Flowers in elongated racemes; tall shrub with acute leaves .... 1. *G. shallon*
Flowers single or few; decumbent dwarf shrub with obtuse leaves ....
....2. *G. Miqueliana*

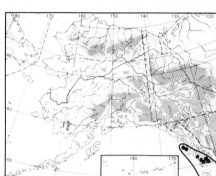 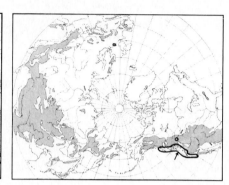

1. Gaulthèria shallón Pursh — Salal

Shrub, creeping to erect, up to 1 meter tall; stem pilose; leaves ovate to ovate-elliptic, sharply serrulate; flowers in bracteate racemes; calyx glandular-pilose; corolla about twice as long as calyx, pinkish; fruit black-purple, fleshy.

Woods. Described from the mouth of the Columbia River.

The fruit is edible.

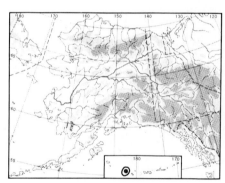 

2. Gaulthèria Miqueliàna Takeda

Gaultheria pyroloides Miq., not Hook. f. & Thoms.

Dwarf shrub with ascending, finely puberulent, terete branches; leaves coriaceous, elliptical to obovate, serrate-crenate, with glands in incisions, shiny and glabrous above, paler beneath; inflorescence few-flowered; calyx lobes triangular, more or less glandular, puberulent, especially in margin; corolla ovoid to urceolate; ovary glabrous; fruit globose, white.

Mountain slopes. Described from Japan. A single specimen known, from northeastern Kiska Island.

69. ERICACEAE (Heath Family)

Arctostaphylos Adans. (Bearberry)

Leaves evergreen, leathery, entire in margin; branchlets pubescent:
   Young twigs puberulent, not glandular . . . . 1. *A. uva-ursi* var. *uva-ursi*
   Young twigs viscid-puberulent, with intermixed, darker-headed, stipitate glands
   . . . . 2. *A. uva-ursi* var. *adenotricha*
Leaves deciduous, thin, serrulated in margin; branchlets glabrous:
   Leaves rugulose, ciliate in margin, at least at base; ripe fruit black . . 3. *A. alpina*
   Leaves less rugulose, glabrous in margin; ripe fruit scarlet . . . . 4. *A. rubra*

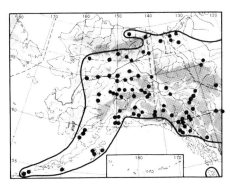
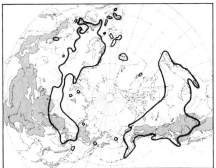

1. Arctostáphylos ùva-úrsi (L.) Spreng.   Kinnikinnick, Mealberry
Arbutus uva-ursi L.; *Uva-ursi uva-ursi* (L.) Britt.; *Uva-ursi procumbens* Moench.
var. **ùva-úrsi**

Trailing, matted dwarf shrub with long, flexible branches; leaves obovate to spatulate, coriaceous, flat in margin, lustrous above, indistinctly veined; stem more or less densely puberulent; corolla urceolate, white to pink; fruit dull red, dry and mealy, with stone consisting of several nutlets.

Dry, sandy places. Described from northern Europe and Canada.

Var. **pacífica** Hult., with larger, elliptic to elliptic-cuneate leaves, red bracts, and red corolla with purple lobes, occurs along the southern coast.

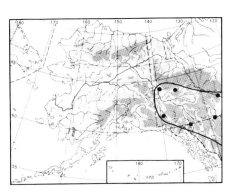
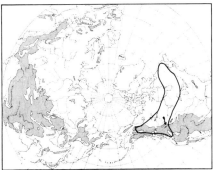

2. Arctostáphylos ùva-úrsi (L.) Spreng.   Kinnikinnick, Mealberry
var. **adenótricha** Fern. & Macbr.
Arctostaphylos uva-ursi subsp. *adenotricha* (Fern. & Macbr.) Calder & Taylor; *Uva-ursi procumbens* Moench var. *adenotricha* (Fern. & Macbr.) D. Löve.

Similar to var. *uva-ursi*, but young twigs viscid-puberulent, and with intermixed, long, stipitate glands with darker heads.

Dry, sandy places.

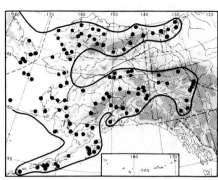
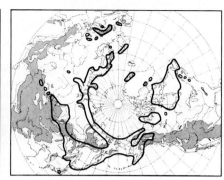

3. Arctostáphylos alpìna (L.) Spreng.

Arbutus alpina L.; *Arctous alpina* (L.) Niedenzu; *Mairania alpina* (L.) Desv.

Prostrate, densely branched dwarf shrub, with glabrous twigs; leaves thick, obovate, obtuse or subacute, cuneate at base, conspicuously reticulate, serrulate, sparsely ciliate; corolla white (greenish-white when young); fruit globose, black.

Dry places, heaths, mostly in the mountains, to over 1,000 meters, but also on tundra. Described from Lapland, Switzerland, and Siberia.

The leaves turn scarlet in the fall and persist for a long time. The fruit is edible, juicy but insipid; the taste is improved by boiling.

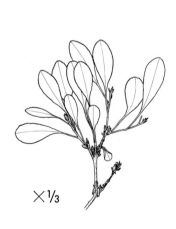
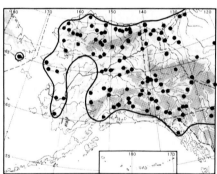
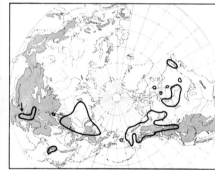

4. Arctostáphylos rùbra (Rehd. & Wilson) Fern.

Arctous alpinus var. *ruber* Rehd. & Wilson; *A. rubra* (Rehd. & Wilson) Nakai; *Arctostaphylos alpina* subsp. *ruber* (Rehd. & Wilson) Hult.; *Arctous erythrocarpa* Small.

Similar to *A. alpina*, but leaves larger, thinner, less rugulose-veined, not ciliate in margin; fruit scarlet, very juicy.

Leaves turn scarlet in the fall.

(See color section.)

69. Ericaceae (Heath Family)

Vaccinium L. (Blueberry, Bilberry)

Leaves evergreen, coriaceous; flowers in terminal clusters; anthers without horns .... 1. *V. vitis-idaea* subsp. *minus*
Leaves deciduous, thin; flowers single in axils; anthers with horns:
  Branches angled; plant usually more than 20 cm tall:
    Leaves finely glandular, serrate-crenate to extreme tip; fruit bluish-black .... 3. *V. membranaceum*
    Leaves not serrate in apex:
      Leaves 1–2 cm long; fruit red .... 4. *V. parvifolium*
      Leaves longer; fruit blue:
        Leaves glandular-serrate only at extreme base, lacking glands on midvein below; pedicels short, curved, in fruit cylindrical .. 5. *V. ovalifolium*
        Leaves glandular-serrate in lower half, with scattered glands on midvein below; pedicels long, straight, in fruit thicker at apex .... 6. *V. alaskensis*
  Branches terete; plant about 20 cm tall or shorter:
    Leaves uniformly and densely serrate in margin .... 2. *V. caespitosum*
    Leaves entire in margin:
      Leaves less than 1 cm long, obovate to rounded; plant low-growing .... 8. *V. uliginosum* subsp. *microphyllum*
      Leaves larger; plant taller and coarser .... 7. *V. uliginosum* subsp. *alpinum*

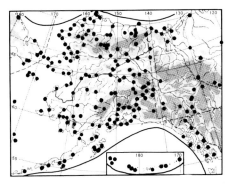  

×½

1. Vaccìnium vìtis-idaèa L.   Lingonberry
subsp. **mìnus** (Lodd.) Hult.

Vaccinium vitis-idaea var. *minus* Lodd.; *Rhodococcum minus* (Lodd.) Avrorin.

Low, creeping dwarf shrub with evergreen, coriaceous, lustrous, obovate leaves, reflexed in margin and with dark dots beneath; flowers in terminal clusters; corolla pink; berry red.

Acid soil, common in the mountains to at least 1,200 meters. *V. vitis-idaea* described from northern Europe, subsp. *minus* from northern Europe and North America.

The berry is edible and makes an excellent jam. A tasty beverage can also be prepared from the berries.

Broken line on circumpolar map indicates range of subsp. **vìtis-idaèa**.

69. Ericaceae (Heath Family)

× ½

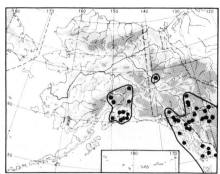
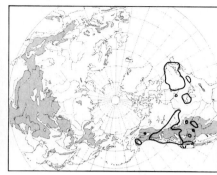

2. Vaccìnium caespitòsum Michx.   Dwarf Blueberry

Low, tufted shrub with terete branches; leaves membranaceous, shiny green on both sides, cuneate-oblanceolate to obovate, highly variable in form, uniformly and densely serrate, finely glandular below; flowers single; corolla tubular-urceolate, longer than broad, pink; berries blue, with pale bloom.

Wet meadows; in the mountains to at least 1,700 meters. Described from North America. The berries are sweet and edible.

Var. **paludicòla** (Camp) Hult. (*V. paludicola* Camp), taller and with larger leaves and somewhat angled branches, occurs in the Prince William Sound area.

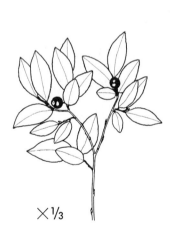

× ⅓

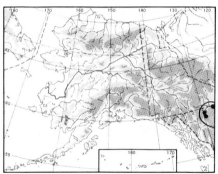
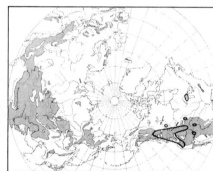

3. Vaccìnium membranàceum Dougl.

Shrub up to 2 meters tall, with angled twigs; leaves ovate to oblong-ovate or elliptic-obovate, mostly somewhat acute, 2–5 cm long, finely glandular, serrate-crenate to extreme tip; flowers single in axils; pedicels recurved in flower; flowers yellowish-pink; corolla ovoid to urceolate; berry globose, purple to bluish-black.

Woods. Described from the northwest coast of North America.

The berries are aromatic, and of a delicious flavor.

69. Ericaceae (Heath Family)

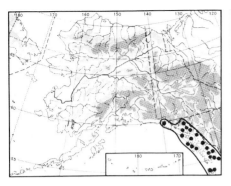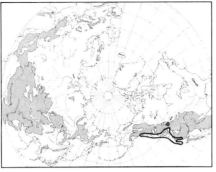

4. Vaccìnium parvifòlium Sm.   Red Huckleberry

Shrub erect, usually more than 1 meter tall, with prominently angled branches; leaves oval to rounded, mostly entire, but somewhat serrate in young leaves; flowers single; calyx lobes short and broad; corolla greenish to yellow, or pinkish; fruit bright red.

Forests. Described from the northwest coast of North America.

The berry is edible, sour but with a good flavor, and makes a superior jelly.

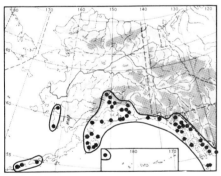

5. Vaccìnium ovalifòlium Sm.

Vaccinium Chamissonis Bong.

Shrubs 50–100 cm tall, flowering early, when leaves are half expanded; twigs conspicuously angled; leaves ovate-elliptic to obovate, entire or mostly with few glandular teeth near base; flowers single in axils; pedicels short, recurved in fruit, not thickened toward apex; corolla pinkish, subglobose; style included; berry globose, blue-black.

Woods. Described from the west coast of North America. Transitions to *V. alaskensis* seem to occur.

The berry is edible, with good flavor.

Broken line on circumpolar map indicates range of the closely related **V. axillàre** Nakai and **V. shikokiànum** Nakai (Shikoku Island).

×⅓

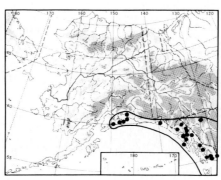
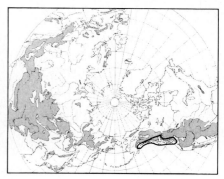

6. Vaccìnium alaskénsis How.

Very similar to *V. ovalifolium*, but taller and with larger leaves, glandular-serrulated in lower half and very sparsely glandular along median vein below; pedicels longer, straight, thickened toward apex; style usually exserted.

Woods. Described from the Cascade Mountains of Oregon to Alaska.

The berry, edible and juicy, has a very good flavor, but is said to be more sour than that of *V. ovalifolium*.

×⅓

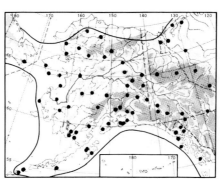

7. Vaccìnium uliginòsum L. — Alpine Blueberry
subsp. **alpìnum** (Bigel.) Hult.

Vaccinium uliginosum var. *alpinum* Bigel.

Depressed or ascending, strongly branched, dwarf shrub, with terete twigs; leaves coriaceous, entire, oblong, obovate or rounded, dull green above, paler and distinctly nerved beneath; corolla urceolate, white or pinkish; berry dark blue.

Heaths and bogs. *V. uliginosum* described from Sweden, subsp. *alpinum* from Massachusetts.

Berry edible, sweet, with good flavor; makes excellent jam and is used for blueberry pie.

Var. **salicìnum** (Cham.) Hult. (*V. salicinum* Cham.), which differs in having very narrow, acute leaves (about 15 × 5 mm), occurs rarely, in scattered localities.

Dotted line on circumpolar map indicates southern boundary of subsp. **uliginòsum**.

69. ERICACEAE (Heath Family)

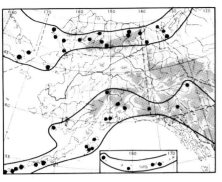
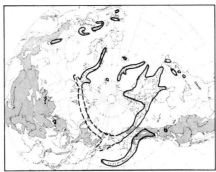
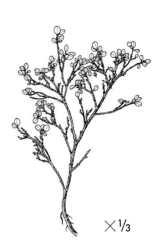

8. Vaccìnium uliginòsum L.
subsp. **microphýllum** Lange
   V. gaultherioides Bigel.

Similar to subsp. *alpinum*, but low-growing; leaves very small, less than 1 cm long, obovate to roundish.

Heaths and bogs. Described from western Greenland.

Oxycòccus Adans. (Cranberry)

Stem filiform; young shoots glabrous; leaves 4–5 mm long, triangular-ovate . . . . . .
. . . . 1. *O. microcarpus*
Stem coarser; young shoots pubescent; leaves 8–10 mm long, oblong-obovate . . . .
. . . . 2. *O. palustris*

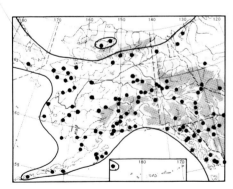
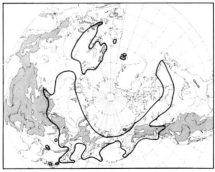

1. Oxycóccus microcárpus Turcz.
   Vaccinium oxycoccus var. *microcarpus* (Turcz.) Fedtsch. & Flerov.

Stem filiform, prostrate, rooting; young shoots glabrous; leaves 4–5 mm long, triangular-ovate, dark green above, glaucous below, strongly revolute in margin; pedicels and calyx glabrous; bracts mostly below middle of scape; flowers mostly single, deep pink; filaments often pubescent on outside; fruit red, 5–10 mm in diameter.

Peat bogs; in McKinley Park to 1,200 meters. Described from European Russia, Transbaikalia, and Alaska.

The fruit is edible and makes a tasty jam.

69. Ericaceae (Heath Family) / 70. Diapensiaceae (Diapensia Family)

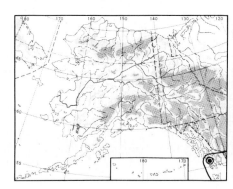

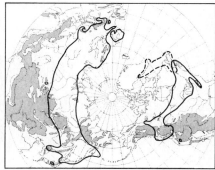

2. Oxycóccus palústris Pers.

Vaccinium oxycoccus L. in part; *Oxycoccus quadripetalus* Gilib.

Similar to *O. microcarpus*, but stem thicker; young shoots pubescent; leaves larger, 8–10 mm long, oblong-obovate, not broader near base; pedicels puberulous; sepals ciliated; flowers larger.

Bogs and moist woods. Type locality not given, presumably Europe. Broken line on circumpolar map indicates range of subsp. **microphýllum** (Lge.) Löve & Löve (*O. palustris* f. *microphylla* Lge.).

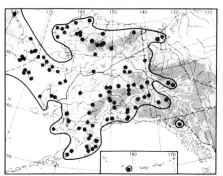

Diapensia L.

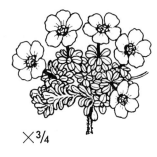

1. Diapénsia lappónica L.
subsp. **obováta** (F. Schm.) Hult.

Diapensia lapponica var. *obovata* F. Schm.; *D. obovata* (F. Schm.) Nakai.

Loosely, sometimes very densely, tufted, forming cushions; leaves obovate to oblong-oblanceolate, thick, lustrous, yellowish-green, 1-nerved; sepals yellowish-green, corolla 5-lobed, white.

Rocks and gravel in the mountains, to at least 1,500 meters. *D. lapponica* described from Lapland, subsp. *obovata* from Sakhalin.

Specimens with rose-colored corolla (var. **ròsea** Hult.) found at Hatcher Pass, north of Anchorage. Broken line on circumpolar map indicates range of subsp. **lappónica**.

71. PRIMULACEAE (Primrose Family)

Primula L. (Primrose)

Corolla limbs entire or slightly emarginate .... 1–2. *P. tschuktschorum*
Corolla limbs deeply emarginate or obcordate:
  Bracts oblong or narrowly obovate, obtuse, with saccate-auriculate base; leaves ovate to suborbicular .... 3. *P. sibirica*
  Bracts more or less acute, more or less saccate at base:
    Leaves entire or slightly undulate, dentate, efarinose; flowers small .... 9. *P. egaliksensis*
    Leaves dentate or crenate, farinose or efarinose; flowers 5–40 mm in diameter:
      Leaves crenate-obovate, with 5–9 large teeth in apex; flowers large, 15 mm in diameter or larger; limbs of corolla deeply obcordate:
        Plant tall; flowers very large .... 4. *P. cuneifolia* subsp. *cuneifolia*
        Plant about 5 cm tall; flowers smaller .... 5. *P. cuneifolia* subsp. *saxifragifolia*
      Leaves entire or minutely denticulate; flowers smaller; limbs of corolla less deeply cleft:
        Plant rarely exceeding 10 cm in height, usually smaller:
          Leaves sessile or nearly so .... 10. *P. borealis*
          Leaves distinctly petiolated .... 6. *P. mistassinica*
        Plant normally exceeding 10 cm in height:
          Leaves mostly green beneath (sometimes somewhat farinose); involucral bracts mostly subulate; flowers of one kind ..... 7. *P. stricta*
          Leaves strongly farinose beneath; involucral bracts lanceolate to linear-oblong, flat; flowers with long or short style .... 8. *P. incana*

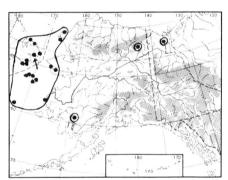 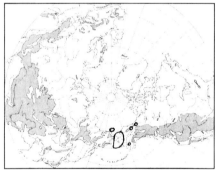 

×¾

1. Prímula tschuktschòrum Kjellm.
var. tschuktschòrum

Primula tschuktschorum subsp. *tschuktschorum* var. *beringensis* Pors.

Leaves oblong-lanceolate to lanceolate, entire or irregularly dentate-cuneate, more or less broadly petiolated; flowers few to several; bracts lanceolate, acute or somewhat acute; pedicels more or less farinose; calyx cleft to two-thirds of length or more, lobes lanceolate, acute; corolla purple with white eye; tube as long as or considerably longer than calyx, lobes obtuse or emarginate.

  Wet meadows, along streams, in McKinley Park to 2,000 meters.
  Highly variable.
  (See color section.)

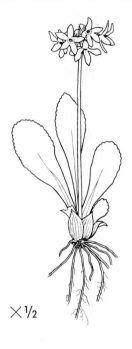
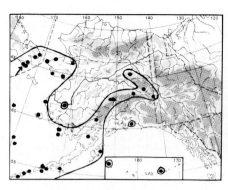
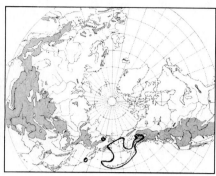

2. Prímula tschuktschòrum Kjellm.
var. árctica (Koidz.) Fern.

Primula arctica Koidz.; *P. eximia* Greene; *P. tschuktschorum* subsp. *eximia* (Greene) Pors.; *P. tschuktschorum* subsp. *Cairnsiana* Pors.; *P. Macounii* Greene.

Similar to var. *tschuktschorum*, but plant often coarser; leaves broader, oblong-obovate.

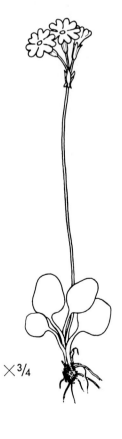
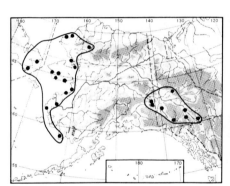
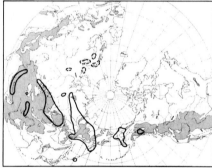

3. Prímula sibìrica Jacq.
Primula nutans Georgi?

Leaves glabrous, elliptic, oblong-elliptic to nearly orbicular, entire or obscurely dentate, long-petiolated; bracts oblong, obtuse, saccate; corolla rose-colored; tube twice as long as calyx; lobes obcordate; capsule cylindric.

Wet meadows. Described from eastern Siberia.

Broken line on circumpolar map indicates range of other subspecies. Subsp. **finnmárchica** (Jacq.) Hult. (*P. finnmarchica* Jacq.) occurs in Europe.

71. Primulaceae (Primrose Family)

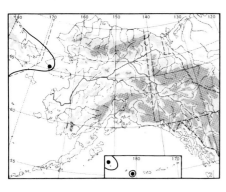
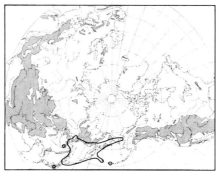

4. Prímula cuneifòlia Ledeb.
subsp. **cuneifòlia**

Leaves thick, glabrous, efarinose; obovate-cuneiform, dentate in apex; scape about 10 cm tall or taller, few-flowered; bracts glandular-puberulent, linear-lanceolate, acute; calyx shorter than corolla tube, with obtuse lobes; corolla pink to violet, with cuneate, deeply bifid lobes; capsule oval.

Wet meadows. Described from Siberia east of Baikal.

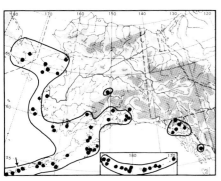
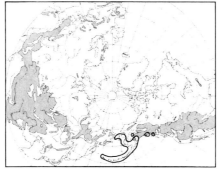

5. Prímula cuneifòlia Ledeb.
subsp. **saxifragifòlia** (Lehm.) Sm. & Forrest
Primula saxifragifolia Lehm. subsp. *saxifragifolia* (Lehm.) Hult.

Most American specimens of *P. cuneifolia* belong to a weakly differentiated race with short scapes (usually about 5 cm long) and smaller flowers.
Wet meadows.

71. PRIMULACEAE (Primrose Family)

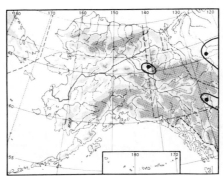

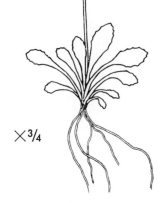

6. Prímula mistassínica Michx.   Bird's-Eye Primrose

Slender, efarinose; leaves oblanceolate to cuneate-obovate, dentate, sessile or nearly so; involucral bracts not saccate; calyx with oblong-lanceolate lobes; corolla pink to bluish-purple, with yellow eye, rarely white, with broad, cuneate-obcordate lobes; capsule cylindric or nearly so.

Meadows, along streams.

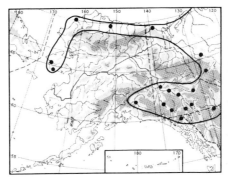
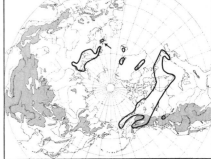
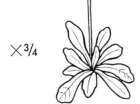

7. Prímula strícta Hornem.

Scape mostly shorter than 20 cm; leaves obovate to obovate-lanceolate, short-petiolated, entire or indistinctly dentate, not farinose or weakly farinose beneath; bracts saccate, lanceolate or more or less subulate in apex; calyx often somewhat farinose, cleft to about one-half with somewhat acute teeth; corolla lilac, the tube longer than the calyx, with emarginate lobes.

Moist places; prefers saline soil.

71. Primulaceae (Primrose Family)

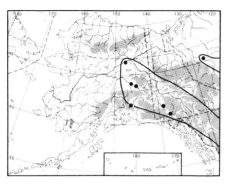

8. Prímula incàna M. E. Jones

Scape about 20 cm long; leaves obovate to obovate-lanceolate, remotely dentate, farinose beneath; bracts saccate, linear-lanceolate, farinose; flowers few to several; calyx farinose, cleft to about one-fourth, with lanceolate, acute lobes; pedicels of very uneven length; corolla lilac, the tube somewhat longer than the calyx, with obcordate lobes.

Wet meadows.

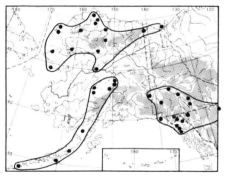
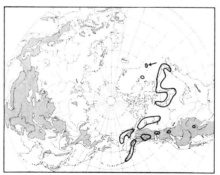

9. Prímula egaliksénsis Wormsk.   Greenland Primrose

Slender; leaves efarinose, oblong to obovate or spatulate, entire, slender-petiolated; involucral bracts gibbous-saccate at base; calyx cleft to one-third to one-quarter; lobes glandular-ciliate; corolla white to lilac, lobes short, cuneate, deeply cleft; capsule cylindric.

Wet meadows, along streams.

71. PRIMULACEAE (Primrose Family)

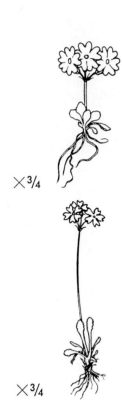

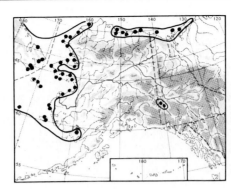
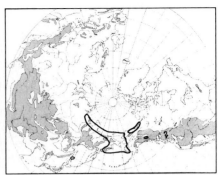

10. Prímula boreàlis Duby
Primula parvifolia Duby; *P. tenuis* Small; *P. Chamissonis* E. Busch.

Leaves sometimes very small, sometimes larger, cuneate-obovate, mostly long-petiolated, remotely toothed to nearly entire, efarinose or slightly farinose beneath; scapes short in flowering specimens, elongating in fruit; bracts linear, acute, more or less saccate at base; pedicels somewhat larger than bracts; calyx cleft halfway down, with acute to obtuse lobes; corolla lilac, more rarely (in the mountains) white, the tube as long as the calyx or longer, the limb 7–20 mm in diameter, with obcordate lobes.

Saline shores along the sea. Described from Shishmaref Bay, Alaska, and St. Lawrence Bay, Siberia.

A highly variable plant; specimens with long or short style occur. The doubtfully distinct **P. Kawásimae** Hara occurs in Sakhalin.

Douglasia Lindl.

Leaves glabrous above, ciliate with simple hairs in margin . . . . 2. *D. arctica*
Leaves pubescent above, ciliate with simple, forked or branched hairs:
    Leaves with simple hairs above . . . . 1. *D. ochotensis*
    Leaves with forked or branched hairs above . . . . 3. *D. Gormani*

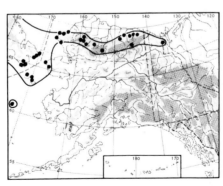
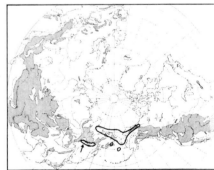

1. Douglàsia ochoténsis (Willd.) Hult.
Androsace ochotensis Willd.; *A. arctica* Cham. & Schlecht.; *A. ochotensis* var. *arctica* Kurtz; *A. tschuktschorum* Kunth; *A. vegae* Kunth.

Forming dense cushions; branches covered with marcescent, brown to reddish-brown leaves; leaves linear, light green when young, 5–10 mm long, ciliate in margin, covered with scattered, short, simple, (rarely forked) hairs; pedicels short in flower, prolonged in fruit, puberulent; calyx lobes triangular to ovate, somewhat acute to obtuse; corolla pink to purple with obovate lobes.

Stony slopes in the mountains.

In flower very early; the flowers completely conceal the tuft at flowering time.

71. Primulaceae (Primrose Family)

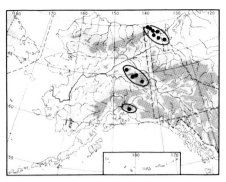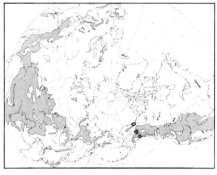

2. Douglàsia árctica Hook.

Similar to *D. ochotensis* and *D. Gormani*, but leaves glabrous on upper side, ciliate with simple hairs.

Described from the Arctic shore between the Mackenzie and Coppermine Rivers (but never found there since).

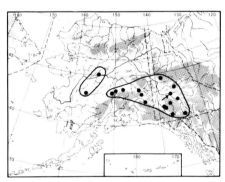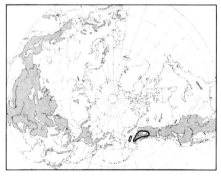

3. Douglàsia Gormáni Constance

Similar to *D. ochotensis* and especially to *D. arctica*, but upper side of leaves covered with forked and branched hairs.

To at least 2,000 meters in the mountains.

Androsace L.

Corolla much exceeding calyx; scapes villous, with long, simple hairs:
    Leaves not fleshy, more or less pubescent above; bracts not distinctly saccate ... 1. *A. chamaejasme* subsp. *Lehmanniana*
    Leaves fleshy, glabrous above, long-ciliated; bracts saccate ... 2. *A. chamaejasme* subsp. *Andersoni*
Corolla about as long as calyx; scapes not villous, glabrous or with short, branched hairs:
    Leaves with petioles about as long as leaf blade; calyx hemispheric ... 5. *A. filiformis*
    Leaves sessile or very short-petiolated; calyx turbinate:
        Umbels many-flowered ... 3. *A. septentrionalis*
        Umbels 1–3-flowered; scapes resembling a sessile umbel ... 4. *A. alaskana*

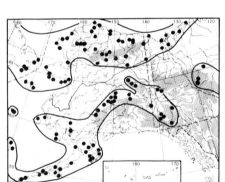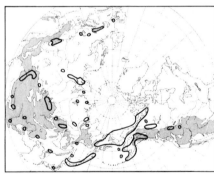

1. Andrósace chamaejásme Host
subsp. **Lehmanniàna** (Spreng.) Hult.

Androsace Lehmanniana Spreng. Including *A. Bungeana* Schischk. & Bobr.

Stems branched, prostrate, with terminal rosettes; leaves in loose to very dense rosettes, rarely globular, extremely variable in size and pubescence, from linear-oblong to nearly orbicular, mostly with pubescent upper surface and long-ciliate in margin, but sometimes glabrous above or only pubescent toward tips; scapes single, villous; umbels 2- to several-flowered; calyx lobes ovate to lanceolate; corolla white with yellow eye, sometimes pinkish to purple.

Rocky slopes on tundra and in the mountains to above 1,700 meters. *A. chamaejasme* described from Europe, subsp. *Lehmanniana* from the "orient."

Circumpolar map indicates range of entire species complex.

×3/4

71. Primulaceae (Primrose Family)

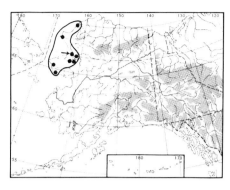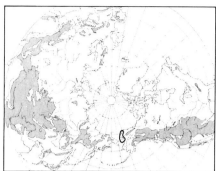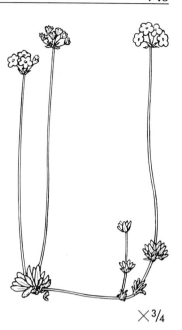

2. Andrósace chamaejásme Host subsp. Andersòni Hult.

A. chamaejasme var. *Andersoni* Hult., of Fl. Alaska & Yukon, p. 1281.

Similar to subsp. *Lehmanniana*, but leaves fleshy, glabrous, very long-ciliate, bracts distinctly saccate.

Apparently a local race, common at Kotzebue.

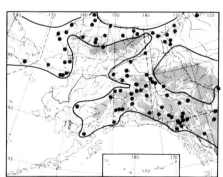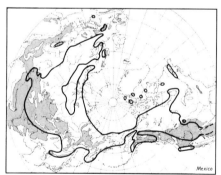

3. Andrósace septentrionàlis L.

Androsace arguta Greene; *A. Gormani* Greene.

Rosette leaves linear-lanceolate to oblanceolate, denticulate to nearly entire, pubescent with minute, branched hairs; scapes of varying length; involucral bracts linear-lanceolate, acute; calyx lobes narrowly triangular, acute; corolla white, about as long as calyx.

Dry, rocky, and sandy places in the mountains, to 2,000 meters. Described from the mountains of Lapland and Russia.

Circumpolar map indicates range of entire species complex.

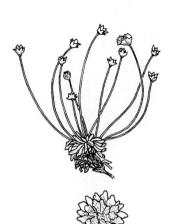 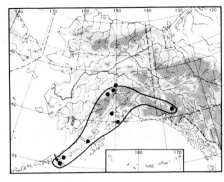 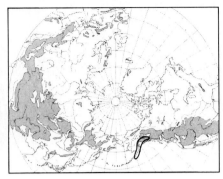

4. Andrósace alaskàna Cov. & Standl.

Rosette leaves sessile, cuneate, entire or 3-toothed in apex; pilose above, ciliate, glabrous beneath; peduncles long, with forked or branched hairs; umbel single-flowered, or rarely with 2–3 sessile flowers supported by short bract; calyx lobes glabrous, acute; corolla white, longer than calyx lobes; capsule globular.

Sandy soil.

 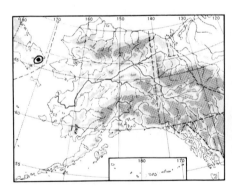 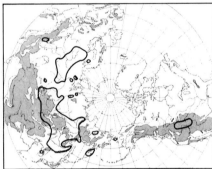

5. Andrósace filifórmis Retz.

Yellowish-green; rosette leaves glabrous or sparsely glutinose-pubescent, elliptic to oblong-ovate, dentate, long-petiolated; scapes few-flowered; pedicels numerous, at maturity as long as scapes or somewhat shorter; involucral bracts many, linear, acute; calyx lobes triangular-lanceolate, acute; corolla white, somewhat longer than calyx; capsule longer than calyx.

Wet meadows along streams. Described from eastern Siberia.

71. PRIMULACEAE (Primrose Family)

Dodecatheon L. (Shooting Star)

Filament tube of anthers long and prominent, orange; stigma not enlarged:
  Leaves oblong-lanceolate, thick, erect, very long-petiolated; flowers large; filament tube comparatively short .... 1. *D. pulchellum* subsp. *superbum*
  Leaves ovate or elliptic, shorter-petiolated:
    Scape short, about twice as long as leaves; leaves ovate .... 2. *D. pulchellum* subsp. *alaskanum*
    Scape long, several times longer than leaves; leaves small, elliptic .... 3. *D. pulchellum* subsp. *pauciflorum*
Filament tube not visible in flowering plant:
  Leaves broad, with rounded or truncate base; stigma not enlarged; capsules dehiscent from apex .... 4. *D. frigidum*
  Leaves gradually narrowed to broader petiole; stigma more than twice the diameter of the style; capsule first opening with lid, then dehiscent from rim of lid .... 5. *D. Jeffreyi*

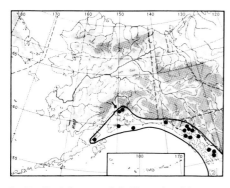  

1. Dodecàtheon pulchéllum (Raf.) Merr.
Eximie pulchellum Raf.

subsp. **supérbum** (Pennell & Stair) Hult.
  Dodecatheon superbum Pennell & Stair; *D. macrocarpum* (Gray) Kunth in part.

Glabrous; scapes from short, thick, erect caudex; leaves thick, erect, oblong-lanceolate, obtuse, nearly entire in the margin, gradually narrowed to the long, winged petiole; calyx purple-flecked, with triangular-lanceolate lobes; corolla purplish-lavender, pale and with purplish spots at base; filaments united into an orange tube; anthers purplish on outside; capsule ovoid-cylindric.

Saline meadows along the sea. *D. pulchellum* described from "Carlton House Fort" (Saskatchewan).

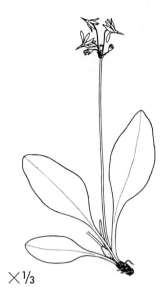
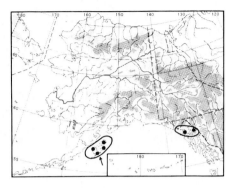
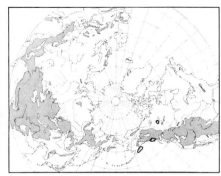

2. Dodecàtheon pulchéllum (Raf.) Merr. subsp. **alaskànum** (Hult.) Hult.

Dodecatheon macrocarpum var. *alaskanum* Hult.

Similar to subsp. *superbum*, but plant smaller, leaves membranous, thin, spreading, ovate-elliptic.

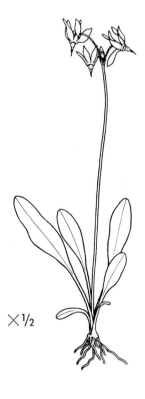
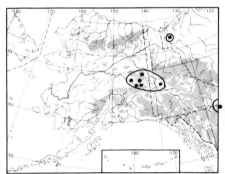
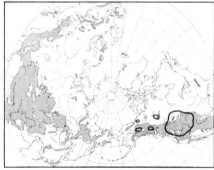

3. Dodecàtheon pulchéllum (Raf.) Merr. subsp. **pauciflòrum** (Greene) Hult.

Dodecatheon pauciflorum Greene.

Similar to subsp. *superbum*, but leaves small, elliptic, more dentate, shorter- petiolated; scape long; flowers smaller; filament tube half as long as anthers or longer. Described from east and west of the Missouri River to the Rocky Mountains.

71. PRIMULACEAE (Primrose Family)

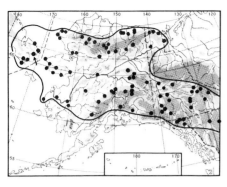 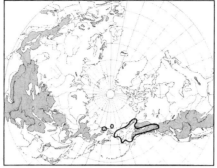 

4. Dodecàtheon frígidum Cham. & Schlecht.

Caudex horizontal; leaves ovate to oval, obtuse, crenate-dentate, abruptly narrowed into petiole; bracts lanceolate to subulate, acuminate; calyx lobes lanceolate to deltoid, acute; corolla tube maroon, yellow above, the lobes magenta to lavender; filaments less than 1 mm long; anthers 4.5–6 mm long, lanceolate, acute, the connective of anthers black; capsule cylindric.

Meadows and heaths, in McKinley Park to 1,700 meters; absent in the lowlands.

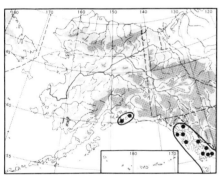 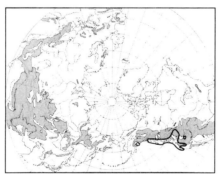 

5. Dodecàtheon Jéffreyi Van Houtte

Dodecatheon viviparum Greene; *D. Jeffreyi* var. *viviparum* Abrams.

Scapes from thick, long, oblique caudex; leaves oblanceolate, somewhat acute to obtuse, entire or crenate; bracts subulate, acute; calyx lobes lanceolate, acute; corolla tube reflexed, with maroon ring at mouth and yellow ring above, the lobes magenta to lavender; filaments short, dark maroon to black; anthers 7–10 mm long, blunt; capsule ovoid.

Wet meadows. Described from mountains of California and Oregon.

71. Primulaceae (Primrose Family)

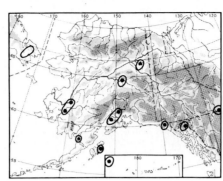 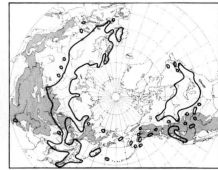

Lysimachia L.

1. Lysimáchia thyrsiflòra L. Tufted Loosestrife
Naumburgia thyrsiflora (L.) Rchb.

Stem from long, thick rhizome; lower leaves scalelike, middle and upper leaves opposite, oblong-lanceolate, sessile, obtuse, dotted with black glands; flowers in dense, axillary, long-peduncled racemes; calyx teeth narrowly lanceolate; corolla lobes with black glands; stamens exserted; fruit ovate.

Wet marshes. Described from Europe.

Trientalis L. (Starflower)

Flowers pinkish; stem leaves reduced to bracts; rhizome tuberous at apex . . . .
. . . . 1. *T. borealis* subsp. *latifolia*
Flowers white; stem leaves larger; rhizome somewhat thicker in apex, but not tuberous:
   Stem leaves few (1–3); leaves obovate-lanceolate; pedicels glabrous or rarely somewhat glandular above . . . . 2. *T. europaea* subsp. *europaea*
   Stem leaves 5–8; leaves obovate, rounded in apex; pedicels glandular above . . .
. . . . 3. *T. europaea* subsp. *arctica*

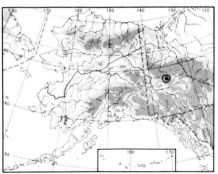 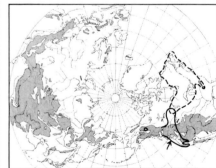

1. Trientàlis boreàlis Raf.
subsp. **latifòlia** (Hook.) Hult.
Trientalis latifolia Hook.; *T. europaea* var. *latifolia* (Hook.) Torr.

Stem from rhizome with tuber 1–2 cm long, up to 6 mm thick in apex; leaves broadly ovate-elliptic to obovate, in whorl at top of stem; stem leaves 1–2, bract-like; corolla pinkish to rose.

Woods, prairies. Broken line on circumpolar map indicates range of subsp. **boreàlis**.

71. PRIMULACEAE (Primrose Family)

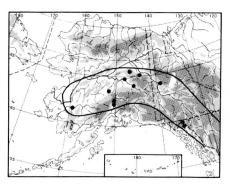 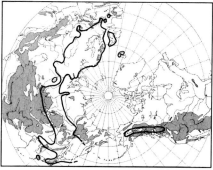 

2. Trientàlis europaèa L.
subsp. europaèa

Glabrous; unbranched; stem from rhizome, somewhat thicker in apex and with several long, slender stolons, thicker in the tip; leaves obovate-lanceolate in whorl of 5–6, at top of stem; stem leaves few, small, alternate; 1 to few flowers on long, glabrous—or above somewhat glandular—pedicels; corolla white, with mostly 7 ovate, acute or apiculate lobes; capsule globular.

Woods. Described from northern Europe.

Transitions to subsp. *arctica* occur.

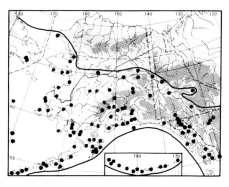 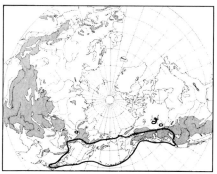 

3. Trientàlis europaèa L.
subsp. árctica (Fisch.) Hult.

Trientalis arctica Fisch.; *T. europaea* var. *arctica* (Fisch.) Ledeb.; *T. aleutica* Tatew.

Similar to subsp. *europaea*, but upper leaves obovate, rounded in apex; stem leaves larger, several; pedicels more glandular above.

Woods, subalpine meadows. Described from Clarence Strait (southeast Alaska), Unalaska, and Kamchatka.

71. Primulaceae (Primrose Family) / 72. Plumbaginaceae (Leadwort Family)

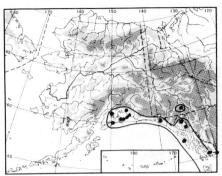 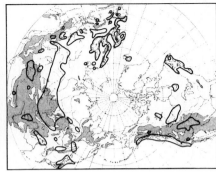

Glaux L.

1. Gláux marítima L. Sea Milkwort

Glabrous; succulent, grayish-green; stems from scaly rhizome with subterranean runners; leaves opposite; flowers sessile with pink calyx, lacking petals; capsule globular.

Seashores and saline soil. Described from Europe.

Specimens with broad, round-tipped leaves belong to var. **obtusifòlia** Fern.

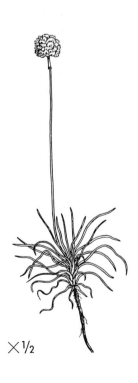

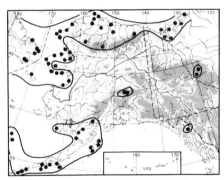 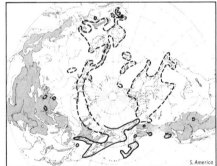

Armeria Willd.

1. Armèria marítima (Mill.) Willd. Thrift
Statice maritima Mill.
subsp. **àrctica** (Cham.) Hult.

Armeria maritima forma *arctica* Cham.; *A. sibirica* and *A. maritima* var. *sibirica* with respect to Alaskan plant.

Forming dense tussocks; leaves grayish-green, linear; scapes pubescent; flowers in hemispherical glomerules, supported by ovate-lanceolate involucral bracts; flowers in 2–3-flowered spikelets, subtended by bracts; calyx pubescent only on veins, glabrous between; petals pink to purple, much longer than calyx.

Common on cliffs along the sea, rare inland. *A. maritima* described from England, subsp. *arctica* from Unalaska, Shishmaref Bay, Cape Espenberg, and St. Lawrence Bay (in Siberia).

In northern Alaska, transitions can occur exceptionally to subsp. **sibìrica** Turcz., with calyx pubescent also between the veins. Range of the entire complicated species complex is marked on the circumpolar map.

73. GENTIANACEAE (Gentian Family)

Gentiana L.

Flowers large, 3–4.5 cm long:
  Cauline leaves long, linear:
    Stem 10–20 cm long, with 1–3 pairs of stem leaves, simple or branched from base; seashore plant .... 3. *G. detonsa*
    Stem taller, branched above; inland meadow plant:
      Stem leaves several pairs; calyx somewhat acute at base; corolla tube not much wider at apex, with oblong-obovate lobes .... 4. *G. barbata*
      Stem leaves few pairs; calyx rounded at base; corolla tube broadly obconic, with broad, rounded lobes .... 5. *G. Raupii*
  Cauline leaves lanceolate to ovate:
    Stem with few pairs of lanceolate to lance-ovate leaves; corolla yellowish-white, striped and spotted with blue and purple .... 1. *G. algida*
    Stem with several pairs of ovate leaves; corolla dark blue ... 2. *G. platypetala*
Flowers smaller:
  Corolla with fimbriate crown in throat:
    Flowers with pedicels 2–4 cm long, usually strongly branched from base ....
    .... 6. *G. tenella*
    Flowers sessile or with short pedicels; plant branched above:
      Calyx lobes ovate to cordate-ovate .... 10. *G. auriculata*
      Calyx lobes narrow, acute:
        Stem with 6–10 internodes, much branched; leaves oblanceolate to spatulate-obovate .... 11. *G. amarella* subsp. *acuta* var. *acuta*
        Stem with 3–6 internodes, less branched; leaves broader ....
        .... 12. *G. amarella* subsp. *acuta* var. *plebeja*
  Corolla lacking fimbriate crown in throat:
    Flowers solitary; plant prostrate; cauline leaves dense .... 9. *G. prostrata*
    Flowers in cymes; plant erect; cauline leaves not dense:
      Corolla plicate in sinuses:
        Flowers greenish-blue; plant with short rootstock .... 7. *G. glauca*
        Flowers white, with lobes bluish on outside; root thin .. 8. *G. Douglasiana*
      Corolla not plicate in sinuses:
        Calyx lobes broad, foliaceous, reflexed, somewhat acute; corolla bluish-white .... 13. *G. aleutica*
        Calyx lobes lanceolate, markedly acute; corolla blue (sometimes diluted blue):
          Corolla lobes setose; plant tall, with several long, slender, small-flowered basal branches .... 14. *G. propinqua* subsp. *propinqua*
          Corolla lobes acute or mucronate; plant low, with few, short basal branches .... 15. *G. propinqua* subsp. *arctophila*

73. GENTIANACEAE (Gentian Family)

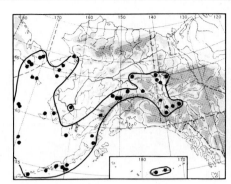 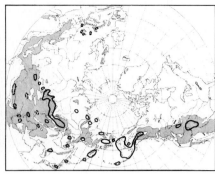

1. Gentiàna álgida Pall.

Gentiana frigida var. *Romanzovii* Ledeb.; *G. algida* var. *Romanzovii* (Ledeb.) Kuzn.; *G. Romanzovii* Ledeb.

Glabrous, yellowish-green; stem 3–15 cm long, from short caudex 2–5 mm thick, covered with marcescent leaf bases above; rosette leaves linear-oblong, 1-nerved; flowers 1–3, mostly 2; calyx lobes uneven in length, about as long as tube, linear; corolla twice as long as calyx or longer; corolla creamy white, more or less violet, spotted with triangular violet patches below the short, acute lobes; filaments narrowly triangular; capsule oblong-ovate.

Meadows, stony slopes in the mountains to at least 1,500 meters. Described from Yenisei, Baikal, Dahuria, and Kamchatka. Broken line on circumpolar map indicates range of closely related races.

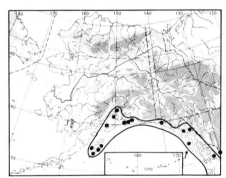 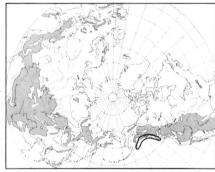

2. Gentiàna platypétala Griseb.

Gentiana Gormani How.; *G. Covillei* Nels. & Macbr.

Stems from long, thick, horizontal to oblique caudex; rosette leaves lacking; stem leaves in several pairs, ovate to elliptic, obtuse, sessile; flowers mostly solitary, sessile, up to 35 mm long; calyx lobes growing together, forming 2 broadly ovate lobes, acute to mucronate, or with 2–3 acute lobes at apex, purplish; corolla blue, paler at base, with reniform, overlapping, mucronate lobes, greenish- to reddish-spotted on inside.

Grassy slopes in the mountains to at least 800 meters.

73. GENTIANACEAE (Gentian Family)

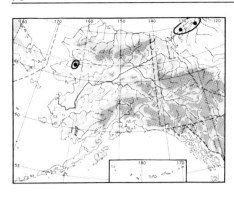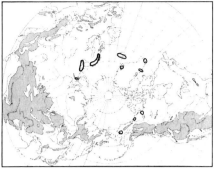

3. Gentiàna detónsa Rottb.
Gentianopsis detonsa (Rottb.) Malte.

Glabrous; stem strict, erect, unbranched or with basal branches 10–20 cm tall; rosette leaves oblong-ovate, persistent in flowering time; stem leaves 1–2 pairs, linear-lanceolate, acute; flowers tetramerous; calyx lobes uneven, the shorter broader than the longer; corolla blue, up to 40 mm long, the lobes dentate in apex and sometimes fringed on sides; capsule stipitate.

Meadows along the sea. Described from Iceland.

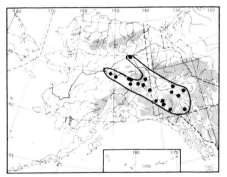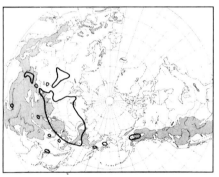

4. Gentiàna barbàta Froel.
Gentianella detonsa (Rottb.) G. Don subsp. *yukonensis* Gillett.

Similar to *G. detonsa*, but taller, with larger rosette leaves and several stem leaves; stem simple, or branched mostly in upper part; corolla lobes fringed on sides.

Meadows in the forest or subalpine zones. Described from the Tom River in western Siberia.

73. GENTIANACEAE (Gentian Family)

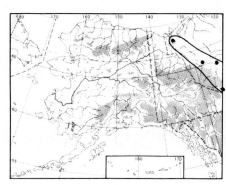 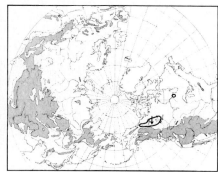

5. Gentiàna Raùpii Pors.

Gentianella detonsa (Rottb.) G. Don subsp. *Raupii* (Pors.) Gillett.

Similar to *G. barbata*, but with fewer stem leaves; calyx rounded at base; corolla tube broadly obconic, the lobes broad, rounded.

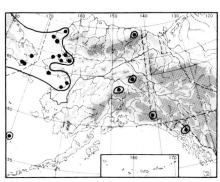 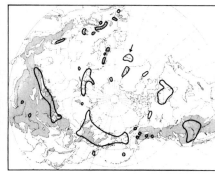

6. Gentiàna tenélla Rottb.

Gentianella tenella (Rottb.) Börner; *Lomatogonium tenellum* (Rottb.) Löve & Löve; *G. tenella* var. *occidentalis* Rousseau & Raymond; *G. tenella* subsp. *pribilovii* Gillett; *G. glacialis* Thom.; *G. falcata* with respect to Yukon plant.

Glabrous; stem usually branched from base, each branch ending with single long-peduncled flower; rosette leaves oblong-oblanceolate, blunt; stem leaves opposite, in 1–4 pairs, oblong-elliptic, the upper more acute; flowers with 4–5 parts; calyx cleft nearly to base into 1 short ovate-lanceolate lobe, the rest longer, acute, lanceolate; corolla 6–10 mm long, in fruiting stage longer, with acute or obtuse lobes.

Meadows on tundra or in the mountains, to at least 1,200 meters. Described from Iceland.

73. GENTIANACEAE (Gentian Family)

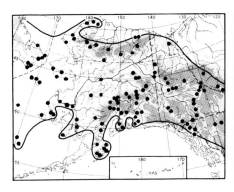

7. Gentiàna glaùca Pall.

Glabrous, yellowish-green; rosette leaves obovate; stem leaves in 1–3 pairs, elliptic to rounded, sessile; calyx lobes unequal, ovate-lanceolate, acute; corolla about 15 mm long, blue, dark-blue, or greenish-blue, with ovate, obtuse lobes, much shorter than tube.

Meadows in the alpine and subalpine zones to at least 1,500 meters.

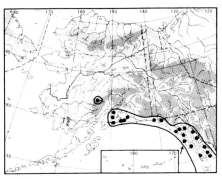
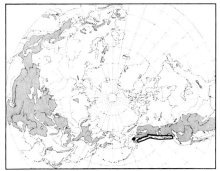

8. Gentiàna Douglasiàna Bong.

Glabrous, stem angled, from thin root, mostly branched; basal leaves ovate, larger than the small, remote stem leaves; flowers closely subtended by ovate bracts; calyx lobes lanceolate, shorter than tube; corolla white with oblong lobes, bluish on outside, shorter than tube; plates in sinuses 2-lobed, over half the length of corolla lobes; capsule sessile, ovoid.

Bogs and wet meadows.

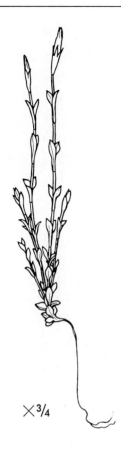
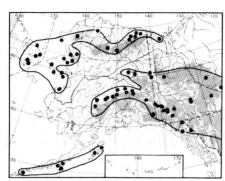
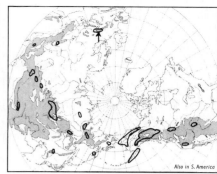

9. Gentiàna prostràta Haenke
Gentiana prostrata var. *americana* Engelm.

Glabrous, light green; stem from thin root, simple or branched from base; leaves ovate in 3–4 pairs, white-margined; flowers terminal, solitary; calyx lobes triangular to ovate-triangular; corolla with 4–5 parts, blue with lance-ovate lobes, with ovate notched or entire sinus plates; capsules stipitate, linear-oblong, 8–10 mm long, 1.5 mm wide, mostly included.

Bogs and meadows, in the mountains to at least 2,000 meters. The corolla opens only in bright, sunny weather.

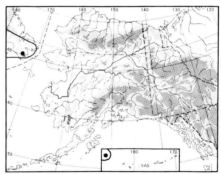

10. Gentiàna auriculàta Pall.
Gentianella auriculata (Pall.) Gillett.

Glabrous; stem simple in dwarfed specimens, usually branched above; rosette leaves oblong-ovate, blunt, soon withering; stem leaves ovate-lanceolate; flowers mostly tetramerous; 2 of the calyx lobes broadly ovate, blunt, with broad, somewhat cordate base and earlike blade, the others narrower; corolla dark violet, with pale tube twice as long as calyx; plates in sinuses ovate, laciniate.

Subalpine meadows. Described from Penzhina, Kamchatka, Japan, and America.

73. Gentianaceae (Gentian Family)

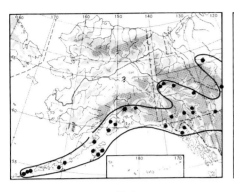  

11. Gentiàna amarélla L.
subsp. **acùta** (Michx.) Hult.

Gentiana acuta Michx.; *Gentianella acuta* (Michx.) Hiitonen; *Gentianella amarella* subsp. *acuta* (Michx.) Gillett.

var. **acùta**

Stem single or branched, 20–40 cm tall, with ascending branches and 6–10 internodes; radical leaves oblanceolate to spatulate-obovate, obtuse; cauline leaves acuminate; calyx cleft nearly to base, with narrow, lanceolate, slightly uneven and acute lobes; corolla violet or lilac, rarely white, 1–1.5 cm long, fringed in throat, with oblong-lanceolate, acute or somewhat acute lobes; capsule sessile.

Moist places, meadows, along streams. *G. amarella* described from Europe, subsp. *acuta* from the Carolinas and Canada.

Broken line on circumpolar map indicates range of other taxa of the *G. amarella* complex.

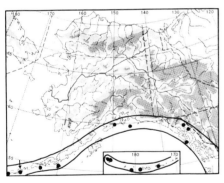 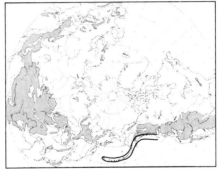 

12. Gentiàna amarélla L.
subsp. **acùta** (Michx.) Hult.
var. **plebèja** (Cham. & Schlecht.) Hult.

Gentiana plebeja Cham. & Schlecht.

Similar to var. *acuta*, but lower and less branched; stem with 3–6 internodes; leaves broader; calyx lobes broader.

Probably the early-flowering seasonal-dimorphic type of subsp. *acuta*.

×¾

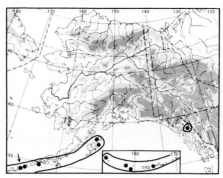
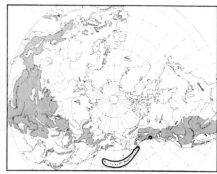

13. Gentiàna aleùtica Cham. & Schlecht.
Gentianella propinqua subsp. *aleutica* (Cham. & Schlecht.) Gillett.

Glabrous, yellowish-green; stem up to 10 cm tall, often with shorter, slender, basal, flowering branches; basal leaves oblanceolate, obtuse; cauline leaves broad, ovate, more or less clasping; inflorescence more or less capitate; flowers compressed, with broad, ovate bracts; 2 of the calyx lobes broad, obovate-lanceolate, 2 narrower, reflexed at tip; corolla twice as long as calyx, yellow or bluish-yellow, tetramerous, acuminate, slightly dentate in apex; capsule ovate-lanceolate, as long as corolla.

Meadows, rare.

×¾

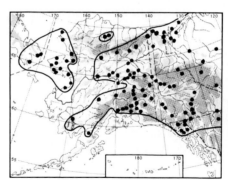
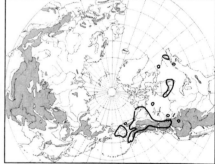

14. Gentiàna propínqua Richards.
Gentiana Rurikiana Cham. & Schlecht.; *Gentianella propinqua* (Richards.) Gillett.
subsp. **propínqua**

Glabrous; plant with numerous slender, basal flower-bearing branches, shorter than stem; rosette leaves oblong to lanceolate-ovate; stem leaves in remote pairs; lower flowers long-pedicellated; calyx lobes very unequal, the broader ovate to oblong, foliaceous; corolla lobes blue (rarely white), bristle-tipped, lacking crown at base, the tube greenish-blue; capsule sessile.

Forests and dry mountain slopes. Described from northwest North America.

73. Gentianaceae (Gentian Family)

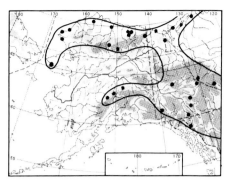
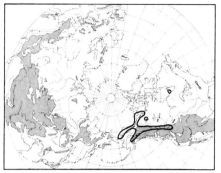

15. Gentiàna propínqua Richards.
subsp. **arctòphila** (Griseb.) Hult.
   Gentiana arctophila Griseb.

Similar to subsp. *propinqua*, but low-growing, with short or no basal flowering branches, and acute or acuminate, not setose, corolla lobes.

On tundra and in the mountains, to at least 2,000 meters. Described from the arctic seacoast of western North America.

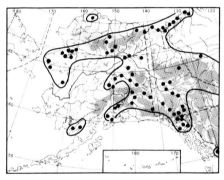
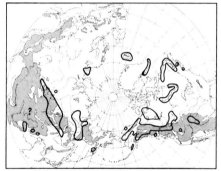

Lomatogonium A. Br.

1. Lomatogònium rotàtum (L.) E. Fries   Star Gentian
   Swertia rotata L.; *Gentiana rotata* (L.) Froel.; *Pleurogyne rotata* (L.) Griseb.; *Lomatogonium rotatum* f. *tenuifolium* (Griseb.) Fern.; *L. rotatum* subsp. *tenuifolium* (Griseb.) Pors.; *P. rotata* var. *tenuifolia* Griseb.

Glabrous; stem lacking basal rosette, from simple, 3 cm long (1-flowered) to branched, 50 cm long, many-flowered with strongly ascending branches; lower stem leaves oblanceolate, blunt, the upper linear, acute; calyx lobes narrowly linear, acute; corolla 8–15 mm long, mostly as long as, or shorter than, calyx, blue, diluted blue with darker veins, white with blue stripes, or more rarely pure white, enlarged in fruit, the lobes oblong-elliptic, acute, with pair of scalelike, fringed appendages at base; capsule oblong, somewhat acute, with many glabrous seeds.

Wet meadows, along streams. Described from Siberia.

Extremely variable; inland specimens are more slender and small-flowered than coastal specimens.

73. GENTIANACEAE (Gentian Family)

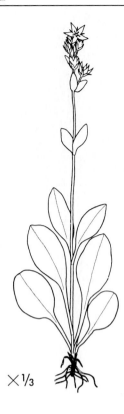
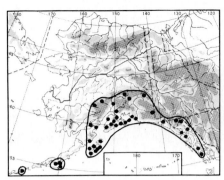 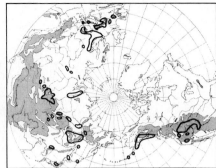

Swertia L.

1. Swértia perénnis L.

Glabrous; stem straight, simple, from oblique caudex; basal leaves long-petiolated, elliptic to oblong-elliptic or obovate, entire; upper stem leaves sessile, alternate or in pairs, somewhat acute; calyx cleft nearly to base, the 5 lobes lanceolate-acuminate; corolla lobes acute, longer than calyx, each with 2 stout glands, grayish-blue, mottled, variable in color, rarely white.

Supalpine meadows. Described from Switzerland and Bavaria.

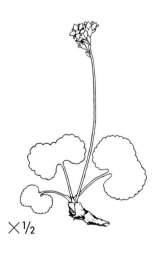
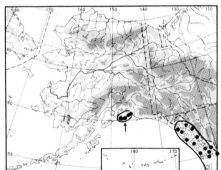 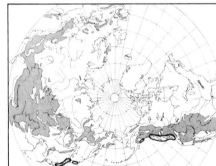

Fauria Franch.

1. Faùria crísta-gálli (Menzies) Makino

Menyanthes crista-galli Menzies; *Villarsia crista-galli* (Menzies) Hook.; *Nephrophyllidium crista-galli* (Menzies) Gilg.

Glabrous; stem from thick, fleshy, reddish-brown rhizome, covered with old leaf bases; leaves broadly reniform to cordate, often broader than long and somewhat emarginate, finely crenate; flowers in open cymes; calyx lobes lanceolate-triangular; corolla longer than calyx, white, rotate with short tube and ovate-lanceolate lobes, with erose-undulate membrane running lengthwise; capsule elongate, much longer than calyx.

Bogs, swamps.

The Japanese plant belongs to subsp. **japònica** (Franch.) Gillett.

73. Gentianaceae (Gentian Family) / 74. Apocynaceae (Dogbane Family)

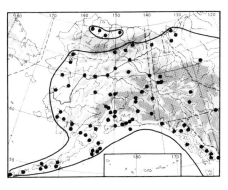
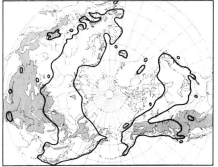

Menyanthes L.

1. Menyánthes trifoliàta L. Buckbean

Glabrous; scapes from coarse, submerged, creeping rootstock, covered with bases of old leaves; leaves 3-foliate with ovate or oblong leaflets; racemes on long, leafless peduncles; calyx lance-ovate; corolla white to pink, tube slightly longer than calyx; lobes with long beard of white hairs on inside; capsule globular to ovate.

Bogs, ponds. Described from Europe.

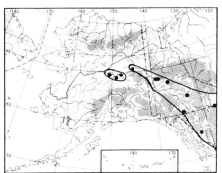
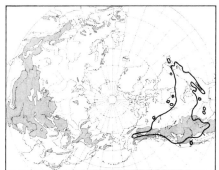

Apocynum L.

1. Apócynum androsaemifòlium L. Dogbane

Stem erect, branched, up to 4 dm long; leaves ovate-oblong to ovate, mucronate, loosely spreading or drooping, glabrous above, paler and pubescent beneath; corolla 6–7 mm long, campanulate, white with pink veins; follicles linear-cylindric.

Woods, hot springs. Described from Virginia and Canada.

75. POLEMONIACEAE (Polemonium Family)

Phlox L.

Leaves silvery green, subulate, pungent, drying white . . . . 1. *P. Hoodii*
Leaves not silvery, drying brownish:
   Pubescence with entangled, cobweb-like hairs; leaves about 7–8 mm long . . . . .
   . . . . 3. *P. sibirica* subsp. *Richardsonii*
   Pubescence not entangled; leaves longer . . . . 2. *P. sibirica* subsp. *sibirica*

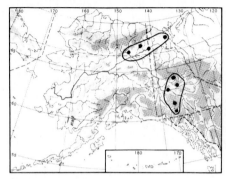

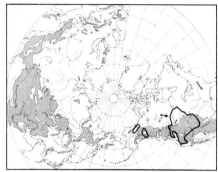

1. Phlóx Hoódii Richards.

Pulvinate, densely leafy; more or less arachnoid; leaves subulate, pungent, silvery green, cobweb-pubescent toward base; flowers solitary, sessile or nearly so, terminal; calyx cobweb-pubescent at base of the pungent lobes; intercostal membrane flat; corolla white (to bluish or pink), with rounded lobes.

Dry mountain slopes.

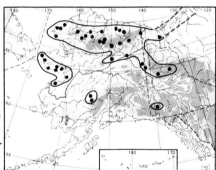

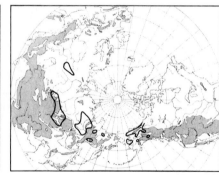

2. Phlóx sibìrica L.

Phlox alaskensis Jordal; *P. borealis* Wherry; *P. Richardsonii alaskensis* (Jordal) Wherry; *P. sibirica* subsp. *borealis* (Wherry) Shetler.

subsp. **sibìrica**

Matted; stem much branched, in lower part covered with remains of leaves or leaf bases; leaves linear-lanceolate, acuminate, ciliate, more or less pubescent with viscid, septate hairs; flowers 1 or 2 at tops of branches; pedicels and calyx viscid-pubescent, cleft to base with linear-lanceolate lobes; tube of corolla as long as calyx, the lobes bluish violet, broad, rounded.

Calcareous rocks in the mountains, to at least 2,000 meters. Described from northern Asia.

75. POLEMONIACEAE (Polemonium Family)

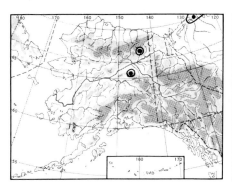

3. Phlóx sibìrica L.
subsp. Richardsònii (Hook.) Hult.
Phlox Richardsonii Hook.

Similar to subsp. *sibirica*, but more densely matted; leaves smaller, 7–8 mm long, pubescent with entangled cobweblike hairs; calyx arachnoid-pubescent, flowers smaller, about 15 mm in diameter.

Dry mountain slopes. Described from the arctic seacoast of northwestern North America.

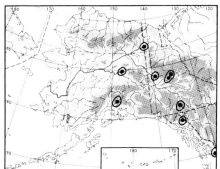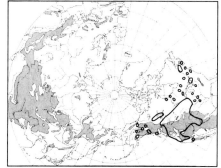

Collomia Nutt.

1. Collòmia lineàris Nutt.

Stem puberulent, mostly simple but sometimes strongly branched, erect; leaves lanceolate to linear, entire, sessile, acute; flowers sessile in leafy-bracteate heads; calyx with lanceolate, attenuate lobes; corolla pink or white, of narrow funnel form; capsule ellipsoid.

An introduced weed. Described from the banks of the Missouri River.

75. POLEMONIACEAE (Polemonium Family)

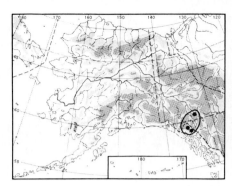 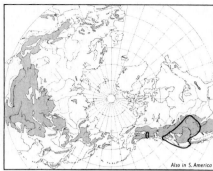

Microsteris Greene

1. Micrósteris gràcilis (Dougl.) Greene

Gilia gracilis Dougl.; *Phlox gracilis* (Dougl.) Greene.

Stem branched and glandular above; leaves entire, the lower opposite, spatulate to oblanceolate, short-petiolated, the upper alternate, lanceolate, sessile; flowers small, white to purplish-pink, usually in pairs in upper axils; calyx lobes acute; corolla with slender tube and ovate to obcordate lobes; stamens shorter than corolla; capsule globose to ovoid.

An occasionally introduced weed. Described from western North America.

Gilia R. & P.

Flowers in dense spherical heads; stamens as long as, or longer than, corolla lobes
.... 1. *G. capitata*
Flowers in loose cymes to loosely capitate; stamens shorter than, to nearly as long as, corolla lobes .... 2. *G. achilleaefolia*

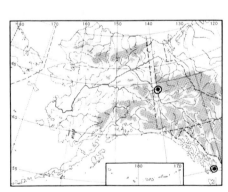 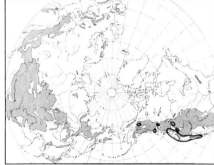

1. Gílía capitàta Sims

Tall and slender, branched above, with thin root; basal and lower cauline leaves bipinnately dissected with narrow rachis; flowers in terminal, globose heads; calyx with acute, linear lobes; corolla pale blue-violet; stamens as long as corolla lobes or longer; capsules subglobose.

An occasionally introduced weed. Described from cultivated specimens raised from seeds collected near Fort Vancouver, Washington.

75. POLEMONIACEAE (Polemonium Family)

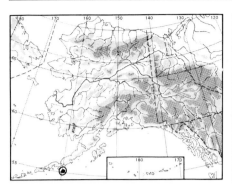
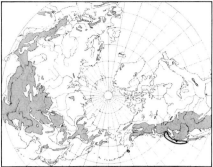

2. Gília achilleaefòlia Benth.

Similar to *G. capitata*, but inflorescence looser; stamens as long as corolla lobes or shorter.

An occasionally introduced weed. Described from California.

Polemonium L. (Jacob's Ladder)

Leaves glabrous on both sides or nearly so; petals acute or somewhat acute, ciliate in margin . . . . 1. *P. acutiflorum*
Leaves viscid-pubescent on both sides; petals rounded, glabrous in margin:
    Lobes of calyx glabrous or much less pubescent than tube; leaflets small, rounded, with short, viscid pubescence, 10–15 on each side; flowers 10–15 mm long, blue to purplish; tube yellow . . . . 4. *P. pulcherrimum*
    Lobes of calyx pubescent to tip; leaflets usually larger, more acute, with longer, viscid pubescence, usually fewer than 10 on each side; flowers blue to purplish; tube yellow, at base colorless:
        Plant coarse; flowers about 20 mm long . . . . 3. *P. boreale* subsp. *macranthum*
        Plant more delicate; flowers about 15 mm long:
            Inflorescence densely capitate, with long, white, villous hairs . . . . See 2. *P. boreale* subsp. *boreale* var. *villosissimum*
            Inflorescence looser, moderately villous . . . . 2. *P. boreale* subsp. *boreale*

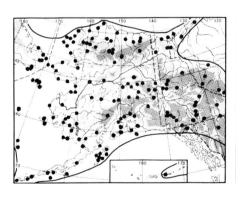
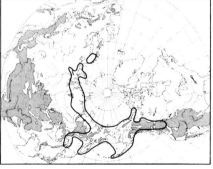
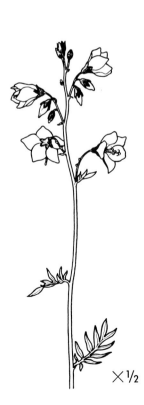

1. Polemònium acutiflòrum Willd.

Polemonium coeruleum with respect to Alaskan plant; *P. pacificum* Vassiljev.

Stems solitary, erect, from rootstock, viscid-pubescent above, glabrous below, with 2 to several leaves; leaves pinnate with lanceolate to elliptic, acute leaflets; calyx campanulate, longer than pedicels, with 5 acute lobes; corolla blue to violet, ciliate in margin; style about as long as corolla; capsule globular.

Wet meadows, along streams. Described from the northwest coast of North America.

Very common and highly variable: arctic and alpine specimens with 2–3 stem leaves; inland specimens tall and narrow-leaved; coastal specimens short and broad-leaved.

75. POLEMONIACEAE (Polemonium Family)

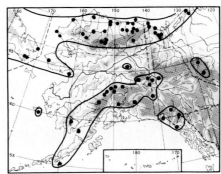
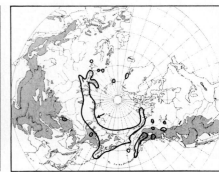

2. Polemònium boreàle Adams

Polemonium boreale subsp. *Richardsonii* (Graham) Anders.; *P. Richardsonii* Graham.
subsp. **boreàle**

Stem from oblique, branched caudex, covered on upper part with translucent old sheaths, viscid-pubescent with long, septate hairs; basal leaves pinnate, with numerous, more or less alternate, elliptic-ovate to obovate leaflets; inflorescence more or less capitate, few-flowered; flowers short-pedicelated, about 1.5 cm long; calyx densely pubescent with somewhat acute lobes; corolla blue to purplish, about 2.5 times longer than calyx, with rounded lobes, glabrous in margin.

Meadows, sandy, dry tundra; to at least 2,300 meters in the St. Elias Range.

Alpine specimens (var. **villosíssimum** Hult.), found on tundra and in the mountains, are about 6–7 cm tall, with globular, strongly white-villous inflorescence.

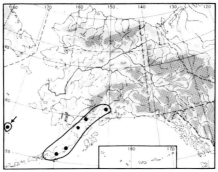
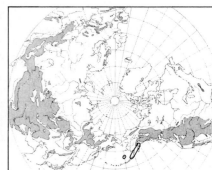

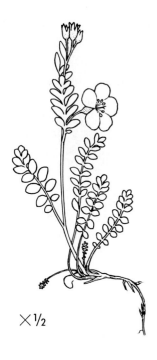

3. Polemònium boreàle Adams
subsp. **macránthum** (Cham.) Hult.

Polemonium pulchellum var. *macranthum* (Cham.) Ledeb.; *P. humile* var. *macranthum* Cham.

Similar to subsp. *boreale*, but coarser; leaves larger, with broad leaflets; flowers about 2 cm long.

75. POLEMONIACEAE (Polemonium Family) / 76. HYDROPHYLLACEAE (Waterleaf Family)

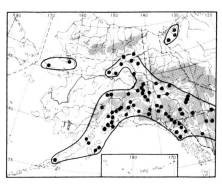
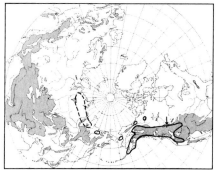

4. Polemònium pulchérrimum Hook.

Polemonium Lindleyi Wherry; *P. fasciculatum* Eastw.; *P. rotatum* Eastw.

Usually several branched stems from branched caudex; stem glandular-villous above, less so or glabrate below; leaves mostly basal, viscid-pubescent, with 10–15 ovate to rounded, dense or lax leaflets; calyx with glandular tube and more or less glabrate lobes, the lobes about as long as the tube; corolla blue or purplish with yellowish tube and ovate, obtuse lobes, 10–15 mm long.

Dry, rocky places, in the St. Elias Mountains to at least 2,000 meters. Broken line on circumpolar map indicates range of **P. hyperbòreum** Tolm.

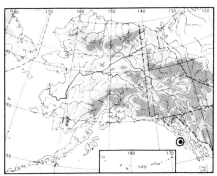
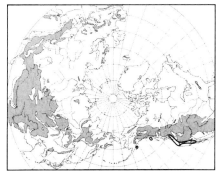

Nemophila Nutt.

1. Nemóphila Menzièsii Hook. & Arn.

Stem more or less prostrate, usually branched at base; leaves appressed-hispid, opposite, oval to oblong in outline, divided into several oblong, toothed divisions; pedicels longer than leaves; calyx lobes lanceolate; auricles 1–2 mm long; corolla purple, pale in center, with obovate to oblong lobes, longer than tube; capsule round to ovoid.

An introduced weed, or escaped from cultivation. Described from California.

76. Hydrophyllaceae (Waterleaf Family)

Phacelia Juss.

Plant velutinous, pubescent; stamens much longer than corolla; filaments glabrous
.... 1. *P. mollis*
Plant soft-haired; stamens about as long as corolla; filaments hairy .. 2. *P. Franklinii*

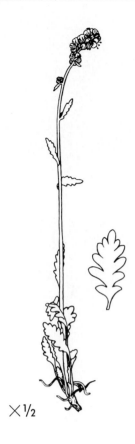

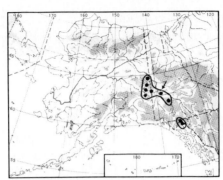 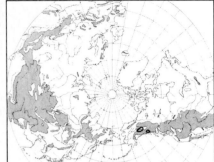

1. Phacèlia móllis Macbr.
Phacelia sericea with respect to Yukon plant.

Velutinous, pubescent plant, with long, twisted, glandular hairs mixed with simple, short, glandular hairs; stem mostly branched; leaves lanceolate, coarsely toothed to lobed or pinnatifid, with entire to shallowly cleft divisions; calyx lobes linear; corolla lobes rounded, about half as long as tube, blue, lavender, or yellowish-white; stamens about twice as long as corolla; filaments glabrous; style cleft about a third of its length; capsule acute or acuminate.

Dry slopes, roadsides.

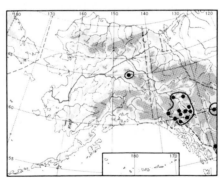 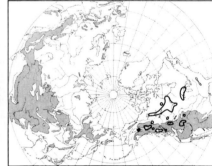

2. Phacèlia Franklínii (R. Br.) Gray
Eutoca Franklinii R. Br.

Soft-haired, with mixed simple and glandular hairs; stem mostly branched; leaves pinnate or pinnatifid with entire, coarsely toothed or lobed divisions; pedicels glandular; calyx lobes linear; corolla lobes violet, lavender, or white, about as long as tube; tube white; stamens as long as corolla; filaments pubescent; style cleft to about half its length; capsule acute.

Sandy places, disturbed soil. Described from Canada between lat. 54° and 64° N.

76. Hydrophyllaceae (Waterleaf Family)

Romanzoffia Cham.

Leaves glabrous on both sides; pedicels slender, longer than calyx .... 1. *R. sitchensis*
Leaves viscid-pubescent beneath; pedicels about as long as calyx .... 2. *R. unalaschcensis*

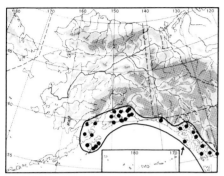
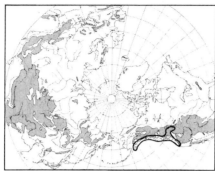
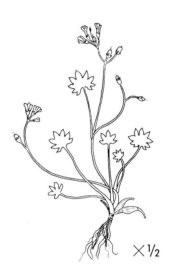

1. Romanzóffia sitchénsis Bong.

Stems several, from short caudex; basal leaves round to reniform, shallowly cleft or toothed, glabrous on both sides, somewhat viscid-ciliate, the petioles conspicuously dilated at base; pedicels slender, longer than calyx; calyx lobes lanceolate to obovate; corolla about 3 times as long as the glabrous calyx, with oval lobes; style 2–5 mm long; capsule oblong-ovoid.

Moist places, from sea level to the lower alpine region.

Similar to a *Saxifraga* species in habit and often mistaken for one, but the fruit is a round capsule rather than 2 follicles.

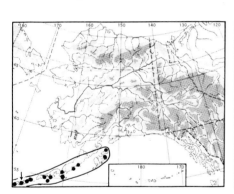

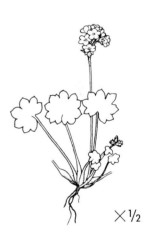

2. Romanzóffia unalaschcénsis Cham.
Saxifraga nutans D. Don.

Similar to *R. sitchensis*, but coarser; leaves viscid-pubescent beneath (in var. **glabriùscula** Hult. nearly glabrous); pedicels shorter, about as long as calyx; corolla slightly longer than calyx.

Moist places. Often mistaken for a *Saxifraga* species.

Lappula Moench (Stickseed)

Nutlets with 2 rows of marginal prickles, the outer row often with smaller prickles and less distinct than the inner .... 1. *L. myosotis*
Nutlets with 1 distinct row of marginal prickles .... 2. *L. occidentalis*

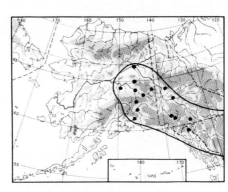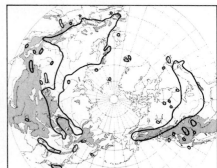

1. Láppula myosòtis Moench

Lappula echinata Gilib.; *Echinospermum lappula* (L.) Lehm.; *Myosotis lappula* L.

Stem hispid, branched above; leaves linear-lanceolate to lanceolate, obtuse, the upper sessile, more or less densely pubescent; bracts long, linear; pedicels erect in fruit; calyx with long, blunt lobes, prolonged in fruit; corolla blue with ovate to rounded lobes, about 3 mm broad; nutlets tuberculate on back with 2 rows of slender, distinct prickles in margin or with prickles irregularly distributed on back.

Waste places; an introduced weed. Type locality not given, presumably Europe.

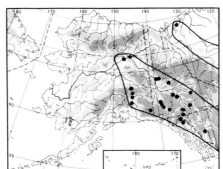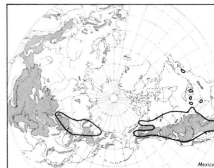

2. Láppula occidentàlis (S. Wats.) Greene

Echinospermum Redowskii var. *occidentale* S. Wats.; *Lappula Redowskii* of American authors.

Similar to *L. myosotis*, but less robust; corolla somewhat smaller, about 2 mm broad; nutlets irregularly and minutely tuberculate on back, with single row of stout, flattened prickles, sometimes confluent at base.

Dry, sandy hillsides. Described from "The Sierras to the Wahsatch."

77. BORAGINACEAE (Borage Family)

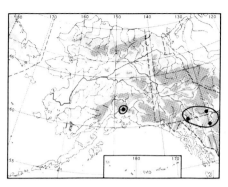
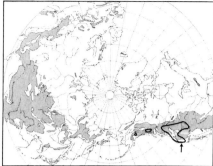

Hackelia Opiz.

1. Hackèlia Jéssicae (McGregor) Brand
Lappula Jessicae McGregor; *Hackelia leptophylla* with respect to Alaskan plant.

Stems hirsute from branching caudex; basal leaves strigose to hirsute, oblanceolate to narrowly elliptic; cauline leaves several, oblanceolate to lance-elliptic, the upper sessile; pedicels recurved in fruit; corolla blue, with yellowish eye; marginal prickles of nutlet free to base, the intramarginal prickles several or few, much smaller than the marginal.

Roadsides; probably introduced.

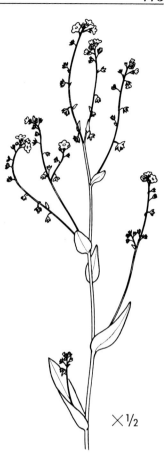

× ½

Eritrichium Schrad. (Arctic Forget-Me-Not)

Leaves linear, with appressed strigose pubescence; flowers about 10 mm in diameter
.... 4. *E. splendens*
Leaves with hirsute or villous pubescence; flowers smaller:
  Flowers scarcely exceeding the densely imbricate, densely pubescent leaves ...
.... 2. *E. Chamissonis*
  Flowering stems 2–10 cm long:
    Flowering stems tall, with 8–10 stem leaves; basal leaves rounded in apex;
    nutlet with crown of jagged teeth .... 3. *E. villosum*
    Flowering stem short or long, with fewer stem leaves; basal leaves more or less
    acute; nutlet with crown of smooth teeth .... 1. *E. aretioides*

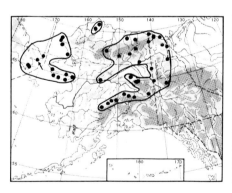
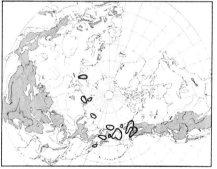

1. Eritríchium aretioìdes (Cham.) DC.
Myosotis aretioides Cham.

Densely tufted; branches very densely covered by withered leaves; leaves ovate-lanceolate to oblanceolate, acute or somewhat acute, densely covered by straight hairs, often pustulate at base; flowering stems short in arctic specimens, up to 10 cm long with 4–5 stem leaves in inland specimens; calyx lobes linear, pubescent; corolla with rounded, blue lobes and yellowish tube; nutlets glabrous, with crown of nearly smooth teeth.

Sandy soil on tundra or in the mountains; in McKinley Park to at least 1,700 meters.

(See color section.)

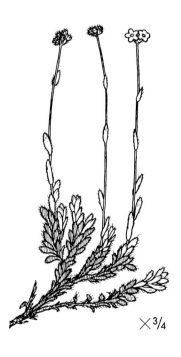

× ¾

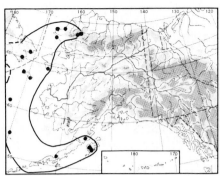 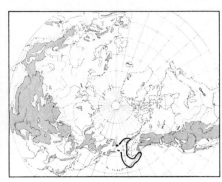

2. Eritríchium Chamissònis DC.

Similar to *E. aretioides*, but leaves shorter and broader; densely imbricate and covered by dense, gray, mostly pustulate hairs; flowers scarcely exceeding leaves; nutlets with crown of jagged teeth.

Rocky or sandy soil on tundra and in the mountains. Described from Kamchatka and St. Lawrence Bay.

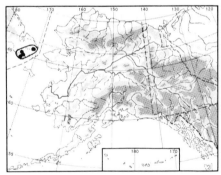 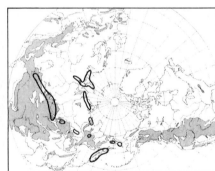

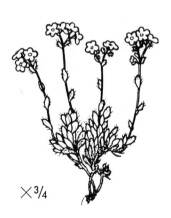

3. Eritríchium villòsum (Ledeb.) Bunge
Myosotis villosa Ledeb.

Similar to *E. aretioides*, but caudex short, not too densely covered with withered leaves; flowering stems taller, with 8–10 stem leaves; basal leaves rounded in apex, with softer hairs; nutlets with crown of jagged teeth.

Stony slopes. Described from "the Alps of Siberia."

77. BORAGINACEAE (Borage Family)

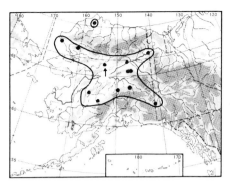
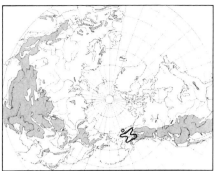

4. Eritríchium spléndens Kearney

Loosely tufted; base of stem covered with marcescent leaves; leaves narrowly oblanceolate to linear, densely strigose, especially beneath; racemes few-flowered; calyx lobes linear-lanceolate to oblong; flowers about 10 mm in diameter; corolla lobes intensely blue, broadly obovate to orbicular; scales yellow; tube broad, about as long as calyx, pale; style as long as body of nutlet; nutlet very finely tuberculate or granulate, with crown of jagged and hispid teeth about one-third as long as body.

Rock crevices, to at least 2,000 meters.

Plagiobothrys Fisch. & Mey.

Corolla large, about 10 mm in diameter .... 2. *P. hirtus* var. *figuratus*
Corolla much smaller:
    Leaves pustulate above .... 1. *P. orientalis*
    Leaves not pustulate .... 3. *P. cognatus*

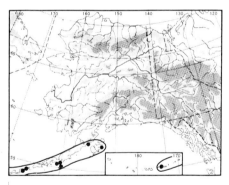
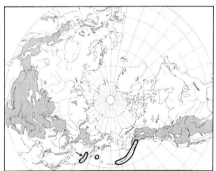
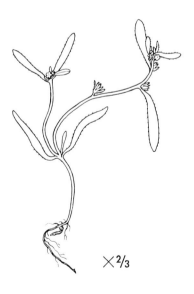

1. Plagiobóthrys orientàlis (L.) Johnston

Heliotropium orientale L.; *Allocarya orientalis* (L.) Brand; *A. asiatica* Kom.; *Plagiobothrys asiaticus* (Kom.) Johnston; *Lithospermum plebejum* Cham.; *Eritrichum plebejum* (Cham.) DC.; *A. plebeja* (Cham.) Greene.

Stems several, from thin root, decumbent or ascending; leaves oblong, obtuse, ciliate, sparsely strigose, pustulate above; flowers inconspicuous, 3–4 mm in diameter; calyx lobes lanceolate, somewhat acute; corolla white, somewhat longer than calyx; nutlets smooth on inner surface, rugulose on outer surface.

Wet places.

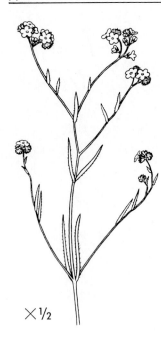

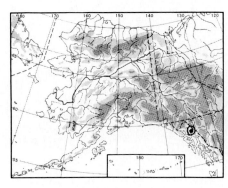

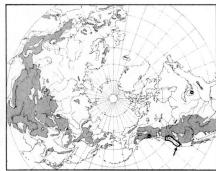

2. Plagiobóthrys hírtus (Greene) Johnston
  Allocarya hirta Greene.
var. **figurátus** (Piper) Johnston
  Allocarya figurata Piper.

Stem erect, simple or branched, strigillose; leaves linear, acute, strigillose on both sides; pedicels shorter than calyx; calyx lobes lance-linear, somewhat acute, strigillose; corolla white, with yellow eye, fragrant, 5–8 mm broad; nutlets ovate, dull, transversely dentate-rugulose on both sides, with low ridges.

Occasionally introduced at Mendenhall.

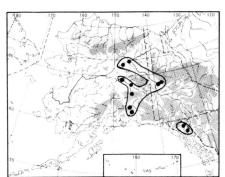

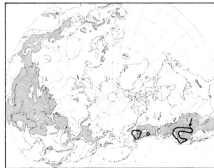

3. Plagiobóthrys cognàtus (Greene) Johnston
  Allocarya cognata Greene; *Plagiobothrys Scouleri* var. *penicillatus* (Greene) Cronq.; *A. penicillata* Greene.

Stem simple, or branched from base, erect or prostrate, more or less strigose; leaves linear to spatulate-linear; racemes slender; calyx appressed-hispidulous, enlarged in fruit, with linear to linear-lanceolate lobes; corolla white, 1–2 mm broad; nutlet ovate to oblong-ovate, obtuse at base, tuberculate and with irregular ridges.

Wet places.

77. BORAGINACEAE (Borage Family)

Cryptantha Lehm.

Nutlet muriculate-rugose; style longer than nutlets; plant coarse ... 1. *C. spiculifera*
Nutlet smooth and shining; style shorter than nutlets; plant slender ....
.... 2. *C. Torreyana*

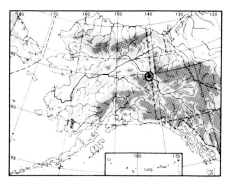
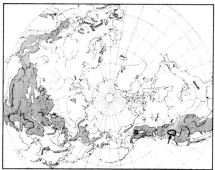

1. Cryptántha spiculìfera (Piper) Payson
Oreocarya spiculifera Piper.

Caespitose; stem strigose, hirsute and setose with long, white hairs; leaves linear-oblanceolate, somewhat acute, strigose; petioles long-ciliate; inflorescence setose, with long white hairs; calyx strigose and setose; calyx lobes linear-lanceolate, exceeding nutlets; corolla white, the tube equaling the calyx; nutlets lanceolate, acute, muriculate-rugose on both surfaces.

Stony, calcareous slopes. Within area of interest, known only from Mission Bluff, Eagle area.

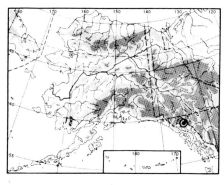
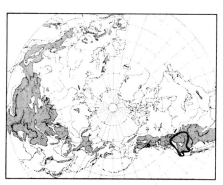

2. Cryptántha Torreyàna (Gray) Greene
Krynitzkia Torreyana Gray.

Stem simple or branched, strigose and spreading-hirsute; leaves linear to narrowly oblong, subappressed-hirsute; calyx lobes lance-linear with spreading tips, bristly-hispid on midrib, appressed-hirsute in margins; corolla inconspicuous, up to 1 mm broad; nutlets broad, ovate, acute, smooth and shining; style surpassed by nutlets.

An introduced weed. Described from California, Nevada, and Idaho.

77. Boraginaceae (Borage Family)

Amsinckia Lehm.

Corolla throat glabrous; stamens inserted above middle of corolla tube .... 1. *A. Menziesii*
Corolla throat hairy; stamens inserted below middle of corolla tube .... 2. *A. lycopsoides*

×½

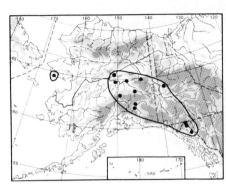 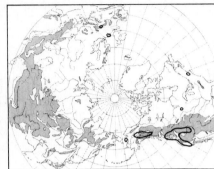

1. Amsínckia Menzièsii (Lehm.) Nels. & Macbr.
Echium Menziesii Lehm.

Stem simple to branched, spreading-hispid; leaves linear to oblong or ovate, spreading-hispid, not crowded at base; sepals almost equal, free; corolla yellow, 4–7 mm broad, the tube scarcely exserted, throat open, glabrous; stamens inserted above middle of tube; nutlet ovate, tuberculate, and often rugose.

Waste places; an introduced weed. Type locality not given, presumably the northwest coast of North America.

×¾

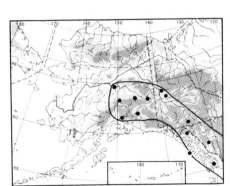 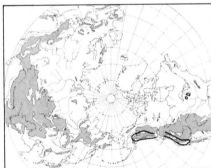

2. Amsínckia lycopsoìdes Lehm.

Stem simple or branched, spreading-hispid and with shorter, soft hairs; leaves linear to linear-oblong, or the upper lanceolate, often crowded at base; sepals free, almost equal; corolla yellow to orange, marked with vermillion in the hairy throat; stamens inserted below middle of tube; nutlet ovate, tuberculate, and somewhat rugose.

Waste places; an introduced weed. Introduced also in Europe. Type locality unknown.

77. Boraginaceae (Borage Family)

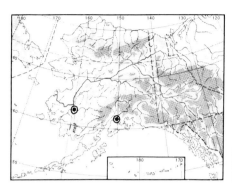
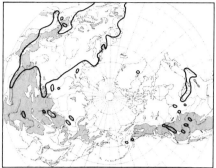
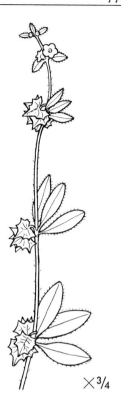

Asperugo L.

1. Asperùgo procúmbens L.   Madwort

Hispid; stem angled, with downward-directed stiff hairs on the angles; leaves hispid, ovate to ovate-elliptic, obtuse, the lower petiolated, subopposite; calyx enlarging in fruit; flowers at first purplish, in age blue, with white tube.

Occasionally introduced weed. Described from Europe.

Myosotis L. (Forget-Me-Not)

Teeth of calyx much longer than broad .... 1. *M. alpestris* subsp. *asiatica*
Teeth of calyx as broad as long .... 2. *M. palustris*

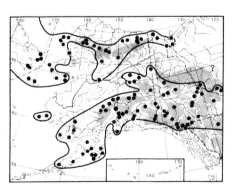
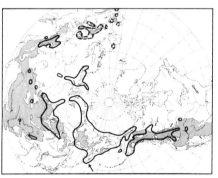

1. Myosòtis alpéstris F. W. Schmidt
subsp. **asiática** Vestergr.

Myosotis asiatica (Vestergr.) Schischk.

Stems 1 to several, with withered leaves at base; basal leaves obovate-lanceolate to lanceolate, long-petiolated, pubescent with more or less appressed hairs, less so beneath; stem leaves several, sessile, oblong to lanceolate; inflorescence short; pedicels about as long as calyx; calyx with straight, not hooked hairs and ovate-lanceolate teeth; corolla blue, rarely white; nutlets black, blunt, shiny.

Alpine and subalpine meadows, to at least 1,500 meters. *M. alpestris* described from the Sudet Mountains (Czechoslovakia), subsp. *asiatica* from Kamchatka.

Broken line on circumpolar map indicates range of other subspecies.

State flower of Alaska.

×½

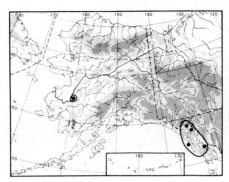
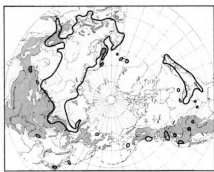

2. Myosòtis palústris L.
Myosotis scorpioides L.

More or less rhizomatous; stem decumbent to erect, angular, more or less hairy; lower leaves oblong-lanceolate to obovate-lanceolate, obtuse, attenuate at base, with scattered, appressed hairs; cymes ebracteate; fruiting pedicels 1–2 times as long as calyx; calyx appressed-hairy with triangular teeth, a third to a quarter the length of the calyx; corolla blue, with emarginate lobes and tube longer than calyx.

Occasionally escaped from cultivation. Described from Europe.

Mertensia Roth

Plant decumbent; leaves thick, fleshy, glaucous, glabrous; seashore plant:
    Corolla about 5 mm long .... 1. *M. maritima* subsp. *maritima*
    Corolla about 10 mm long .... 2. *M. maritima* subsp. *asiatica*
Plant erect; leaves thin, not glaucous, pubescent; not seashore plant:
    Tube of corolla about as long as calyx; calyx lobes broadly lanceolate, glabrous on back; pedicels glabrous or very nearly so .... 7. *M. Drummondii*
    Tube of corolla much longer than calyx; calyx lobes lanceolate-linear to lanceolate-oblong, pubescent or glabrous on back; pedicels pubescent:
        Leaves broad, acuminate; corolla short and broad; nutlets narrowly wing-margined .... 6. *M. kamczatica*
        Leaves narrower, acute but hardly acuminate; corolla longer, more slender; nutlets rugulose, not wing-margined:
            Pedicels with appressed, strigose hairs ... 5. *M. paniculata* var. *Eastwoodae*
            Pedicels with more or less spreading hairs:
                Calyx lobes pilose on back; leaves pubescent on both sides .... 3. *M. paniculata* var. *paniculata*
                Calyx lobes glabrous on back, ciliate in margin; leaves glabrous beneath or nearly so .... 4. *M. paniculata* var. *alaskana*

77. Boraginaceae (Borage Family)

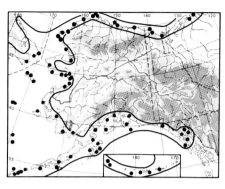 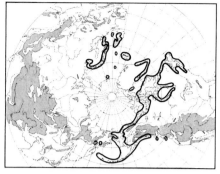 

1. Merténsia marítima (L.) S. F. Gray — Oysterleaf
  Pulmonaria maritima L.
subsp. **marítima**

Decumbent, glabrous and glaucous; leaves ovate to obovate, obtuse or spatulate, fleshy, the lower petiolated, punctate on upper surface; inflorescence bracteate; calyx lobes ovate; corolla about 5 mm long, first pink, later blue, rarely white; nutlets flattened, fleshy, outer layer becoming inflated.

Sandy seashores. Described from England.

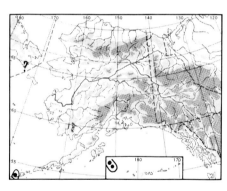 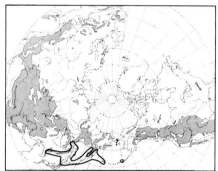 

2. Merténsia marítima (L.) S. F. Gray
subsp. **asiática** Takeda

Mertensia asiatica (Takeda) Macbr.; *M. simplicissima* (Ledeb.) G. Don; *Pulmonaria simplicissima* Ledeb.?

Similar to subsp. *maritima*, but coarser; corolla 10–11 mm long; styles 6–9 mm long; filaments broader and shorter.

Sandy seashores. Described from Manchuria, Japan, Sakhalin, Okhotsk, and Kamchatka.

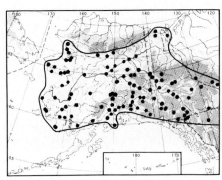
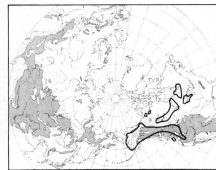

3. Merténsia paniculàta (Ait.) G. Don   Bluebell, Lungwort
Pulmonaria paniculata Ait.
var. **paniculàta**

Stems solitary or several, from branched caudex; basal leaves ovate, obtuse, or somewhat acute, with long, winged petioles, ciliated and more or less tuberculate-pubescent on both sides; cauline leaves narrower than the basal, acute, gradually reduced, subsessile; calyx lobes lanceolate-oblong to lance-linear, acute, pubescent on back; corolla blue, in bud reddish, rarely white, the tube much longer than the calyx; filaments short and broad; style somewhat exserted; nutlets rugulose.

Woods, riverbanks; in McKinley Park to 1,700 meters. Described from Hudson Bay.

(See color section.)

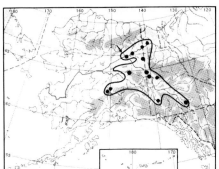
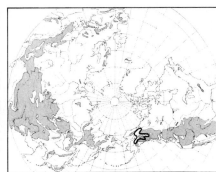

4. Merténsia paniculàta (Ait.) G. Don
var. **alaskàna** (Britt.) L. O. Williams
Mertensia alaskana Britt.

Similar to var. *paniculata*, but calyx lobes more or less glaucous, glabrous on back, ciliate in margin; leaves glabrous beneath, or nearly so.

Woods, riverbanks.

77. BORAGINACEAE (Borage Family)

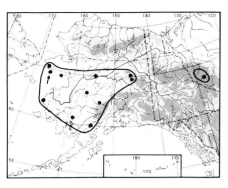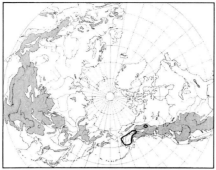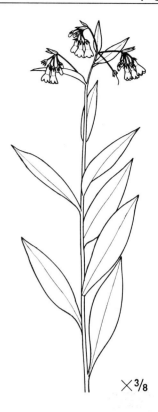

5. Merténsia paniculàta (Ait.) G. Don
var. **Eastwoódae** (Macbr.) Hult.
Mertensia Eastwoodae Macbr.

Similar to var. *paniculata*, but pedicels with appressed, strigose pubescence; calyx lobes glabrous on back, ciliate.
Woods, riverbanks.

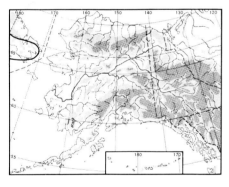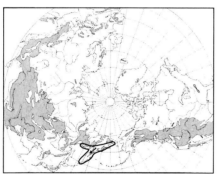

6. Merténsia kamczática (Turcz.) DC.
Steenhamera kamczatica Turcz.

Similar to *M. paniculata,* but nutlets with wing about 1 mm wide; leaves broader, acuminate; corolla short, thick, large.
Meadows, riverbanks.

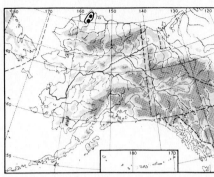
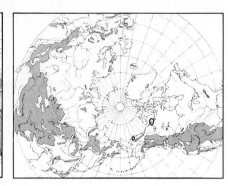

7. Merténsia Drummóndii (Lehm.) G. Don
Lithospermum Drummondii Lehm.

Stems from dark-brown caudex; basal leaves long-petiolated, oblong-lanceolate, evenly pubescent from short, tuberculate hairs, glabrous beneath; upper stem leaves broader, sessile; inflorescence crowded; pedicels glabrous or with few hairs; calyx glabrous, the lobes broadly lanceolate, ciliate in margin; corolla blue, the tube about as long as the calyx, the lobes rounded, undulate in margin, emarginate in sinuses; filaments dilated above; style included; nutlets rugulose.

Sandy slopes. Described from northwest North America.

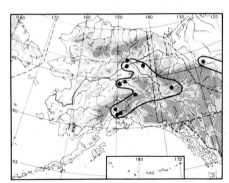
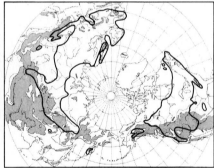

Scutellaria L.

1. Scutellària galericulàta L. Skullcap
var. **pubéscens** Benth.

Scutellaria epilobiifolia Hamilton; *S. galericulata* var. *epilobiifolia* (Hamilton) Jordal.

Stem branched or simple from slender, creeping rhizome; leaves short-petiolated, oblong-lanceolate to ovate-lanceolate, obtuse or subacute, cordate, remotely and shallowly crenate; bracts leaflike; corolla blue-violet, much longer than calyx, with tube curved at base.

Wet meadows. *S. galericulata* described from Europe; no type locality given for var. *pubescens*.

Circumpolar map gives range of *S. galericulata* in a broad sense.

78. Labiatae (Mint Family)

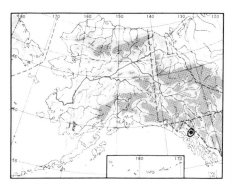
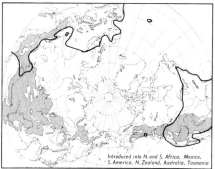
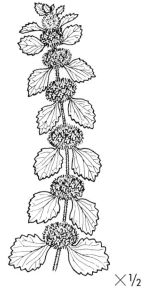

Marrubium L.

1. Marrùbium vulgàre L. — Horehound

White-tomentose; stem coarse, branched, from short, stout rhizome; leaves green above, rugose, orbicular to ovate, cordate or cuneate at base, crenate, the lower long-petiolated; calyx teeth 10, small, hooked; corolla whitish, with long narrow lip.

An occasionally introduced weed. Described from northern Europe.

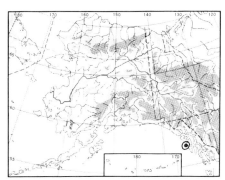
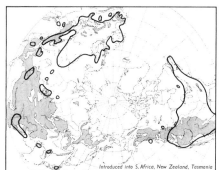
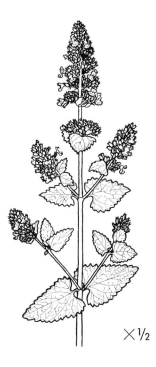

Nepeta L.

1. Népeta catària L. — Catmint

Densely gray-pubescent, strongly scented; stem erect, branched; leaves ovate, cordate at base, coarsely serrate; upper whorls crowded; calyx teeth lanceolate to subulate, the upper longest, the tube ovoid; corolla white with purple spots, the upper lip flat, the tube dilated at middle; nutlets smooth.

Occasionally introduced. Described from Europe.

78. LABIATAE (Mint Family)

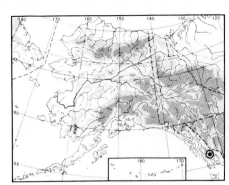 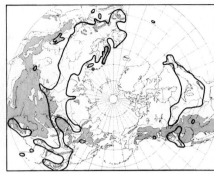

Glechoma L.

1. Glechòma hederàcea L. Ground Ivy

Softly hairy to glabrescent, with creeping, rooting stem and ascending, flowering branches; leaves reniform to ovate-cordate, crenate, obtuse, long-petiolated; corolla violet with purple spots on lower lip, with flat upper lip and narrowly obconic straight tube, pubescent within at base of lower lip; nutlets smooth.

An occasionally introduced weed. Described from northern Europe.

Circumpolar map gives range of entire species complex.

Dracocephalum L. (Dragonhead)

Plant decumbent or ascending; corolla much exceeding calyx . . . . . 1. *D. palmatum*
Plant coarse, erect; corolla only slightly exceeding calyx . . . . 2. *D. parviflorum*

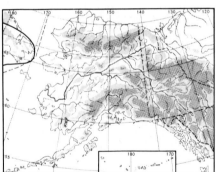 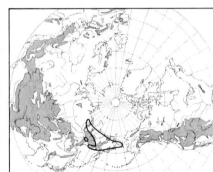

1. Dracocéphalum palmàtum Steph.

Decumbent or ascending; leaves and stem pubescent with patent hairs; leaves small, 4–10 mm in diameter, ovate to rounded in outline, deeply divided into several obtuse lobes; flowers in loose inflorescence at tops of branches; bracts cuneate, with lanceolate, acute teeth; calyx violet, the upper tooth 3 times as broad as the others; corolla yellowish, with upper lip considerably longer than the lower; nutlet ovate.

Sandy slopes. Described from Siberia.

78. LABIATAE (Mint Family)

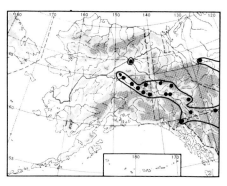
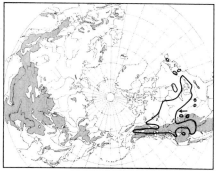

2. Dracocéphalum parviflòrum Nutt.
Moldavia parviflora (Nutt.) Britt.

Stem coarse, erect, solitary or clustered; leaves lance-ovate, sharply toothed; flowers in whorls, crowded at top; bracts imbricate, spine-tipped; upper tooth of calyx much broader than others; corolla blue to violet, only slightly longer than calyx.
Waste places.

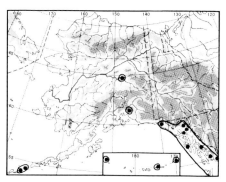
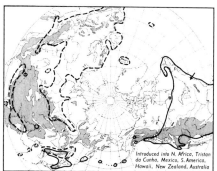
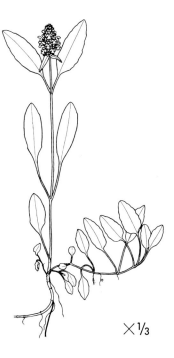

Prunella L.

1. Prunélla vulgàris L. Self-Heal
subsp. **lanceolàta** (Barton) Hult.
Prunella pennsylvanica var. *lanceolata* Barton; *P. vulgaris* var. *lanceolata* (Barton) Fern.

Stem ascending or erect, from short rhizome; leaves lanceolate to oblong, gradually narrowed or cuneate at base, shallowly dentate; bracts ciliate; corolla with upper lip arched, entire, lower lip 3-cleft, with rounded, denticulate middle lobe.
Meadows, roadsides.
Aleutian specimens, with upper part of stem white-tomentose, bracts purple, and base of bracts white-tomentose, belong to the (only slightly different) subsp. **aleùtica** (Fern.) Hult. (*P. vulgaris* var. *aleutica* Fern.). *P. vulgaris* described from Europe; subsp. *lanceolata* from Philadelphia; var. *aleutica* from Unalaska.
Broken line on circumpolar map indicates range of other subspecies.

78. LABIATAE (Mint Family)

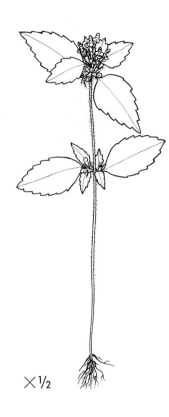

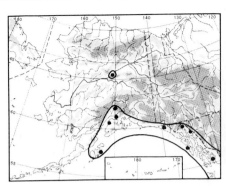 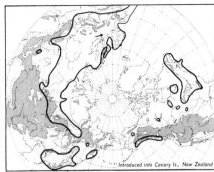

Galeopsis L.

1. Galeópsis bìfida Boenn.   Hemp Nettle

Galeopsis tetrahit var. *bifida* (Boenn.) Lej. & Court.; *G. tetrahit* subsp. *bifida* (Boenn.) E. Fries.

Stem hispid, also hispid between nodes; leaves ovate-lanceolate to lanceolate, with cuneate base, coarsely serrate, obtuse or somewhat acute, petiolated, sparsely pubescent on both sides, with appressed white hairs; corolla purple, pink, or white, the middle lobe small, deeply emarginate, the network of dark markings reaching margin.

Waste places; an introduced weed. Described from Westphalia.

The closely related **G. tétrahit** L., also a weed, can be expected to occur in Alaska; it differs in having the stem hispid chiefly on the nodes, the base of the leaves rounded, the middle lobe of the corolla entire, the dark markings only at the base, and the flowers larger.

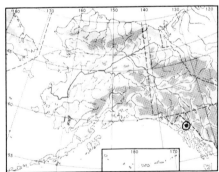 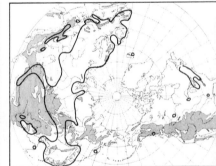

Lamium L.

1. Làmium álbum L.   Snowflake

Hairy; stem erect from creeping rhizome; leaves ovate, cordate at base, acuminate, coarsely (often doubly) serrate to crenate-serrate; calyx teeth longer than tube; corolla about 2 cm long, white, the tube with oblique ring of hairs near base; upper lip long-ciliate, the lower with 2–3 small teeth.

Occasionally introduced. Described from Europe.

78. LABIATAE (Mint Family)

Stachys L.

Leaves oblong-lanceolate, sessile .... 1. *S. palustris* subsp. *pilosa*
Leaves ovate-lanceolate, the lower petiolated .... 2. *S. Emersonii*

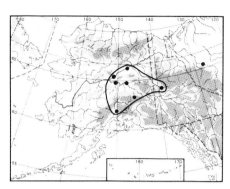
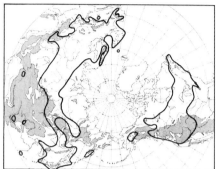

1. Stáchys palústris L.
subsp. **pilòsa** (Nutt.) Epling

Stachys palustris var. *pilosa* Nutt.; *S. palustris* var. *nipigonensis* Jennings.

Stems mostly simple, hollow, from long, creeping rhizome, producing small tubers at apex; leaves sessile or nearly so, oblong-lanceolate, acute, or the lower blunt, subcordate at base, crenate-serrate, pubescent on both sides; whorls about 6-flowered, forming terminal spike, interrupted below; calyx hirsute with subulate teeth; corolla dull purple to violet, pubescent outside.

Fields, clearings, hot springs. *S. palustris* described from Europe, subsp. *pilosa* from the Rocky Mountains.

Circumpolar map gives range of *S. palustris* in a broad sense.

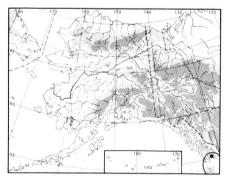
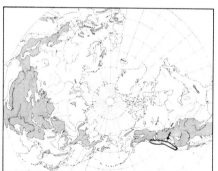
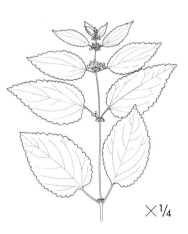

2. Stáchys Emersònii Piper   Hedge Nettle

Stem simple, hirsute, the hairs on the angles pustulate at base; leaves ovate-lanceolate, cordate at base, coarsely crenate-serrate, acute; flowers in interrupted spikes; calyx lobes deltoid, spinose; corolla red to purple, the lower lip with white spots.

Moist woods.

78. LABIATAE (Mint Family)

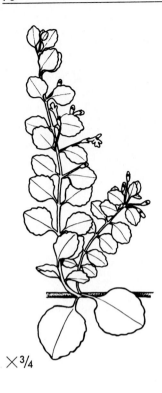

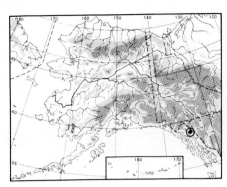
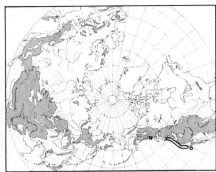

Satureja L.

1. Satureja Douglásii (Benth.) Briq. Yerba Buena

Thymus Douglasii Benth.; *T. Chamissonis* Benth.; *Micromeria Chamissonis* (Benth.) Greene.

Stems slender, trailing; leaves short-petiolated, round to ovate, obtuse, crenate, evergreen; flowers solitary in axils; corolla white to purplish, pubescent.

Reported found once at Juneau. Described from northwest North America.

Lycopus L. (Water Horehound)

Calyx lobes subulate, narrow .... 1. *L. lucidus* subsp. *americanus*
Calyx lobes obtuse or somewhat acute, broad .... 2. *L. uniflorus*

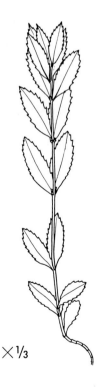

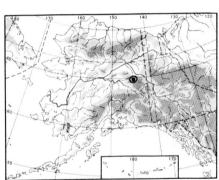
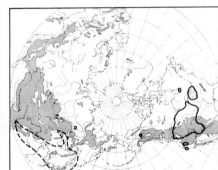

1. Lýcopus lùcidus Turcz.
subsp. **americànus** (Gray) Hult.

Lycopus lucidus var. *americanus* Gray.

Stems puberulent to hairy, especially above, from tuberiferous rhizome, with thick, short stolons at base; leaves oblong-lanceolate, acutely serrated, sessile; bracts lanceolate, ciliate; calyx teeth acute, subulate-ciliate; corolla white to pinkish, about as long as calyx or somewhat longer.

In area of interest known only from Manley Hot Springs. *L. lucidus* described from Lake Baikal and northern Amur Province, subsp. *americanus* from Saskatchewan, Nebraska, and Kansas.

Broken line on circumpolar map indicates range of subsp. **lùcidus.**

78. LABIATAE (Mint Family)

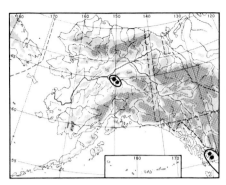
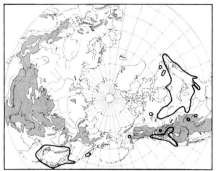
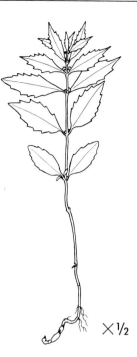

2. Lýcopus uniflòrus Michx.

Stem erect, quadrangular, from thin, tuberous rootstock, often with long, filiform runners at base; leaves glabrous, opposite, light green, lanceolate to lance-oblong, gradually narrowed at both ends, acute-serrate, subsessile; glomerules small; calyx campanulate, with obtuse, ciliated teeth; corolla about 2 mm long, with 4 divaricate lobes, 1 of them broader.

Wet places. Described from Lake St. John and Lake Mistassini.

Mentha L.

Whorls forming terminal spike .... 1. *M. spicata*
Whorls all axillary; axis terminated by leaves .... 2. *M. arvensis*

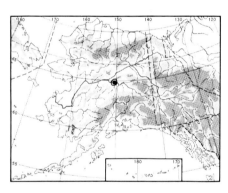
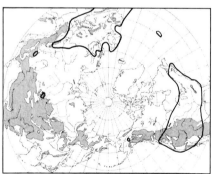
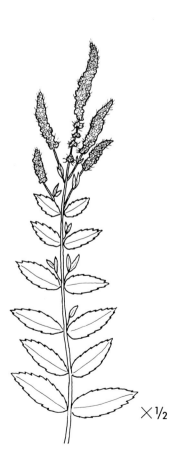

1. Méntha spicàta L.   Spearmint

Stoloniferous; stem erect, glabrous; leaves sessile, lanceolate to oblong-lanceolate, serrate, acute, glabrous on both faces; inflorescence a terminal, cylindrical spike; bracts linear-setaceous, longer than flowers; pedicels and calyx glabrous; corolla lilac, glabrous.

Introduced at Manley Hot Springs. Type locality not given, presumably Europe.

M. piperita L., peppermint, considered to be the hybrid *M. aquatica* × *spicata*, has been found escaped from cultivation in several places; it differs in having flowers in spikes, petiolated leaves and pedicels, and a glabrous calyx tube.

78. LABIATAE (Mint Family) / 79. SOLANACEAE (Nightshade Family)

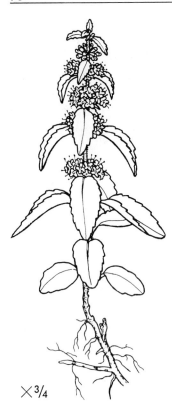

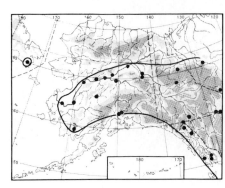

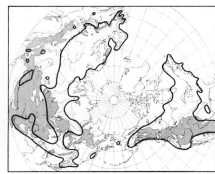

2. Méntha arvénsis L.   Field Mint

Stem simple or branched, erect or ascending, more or less hairy, terminated by leaves; leaves ovate, oblong, or lanceolate, obtuse or acute, serrate, petiolated; flowers in broadly separated axillary whorls; bracts leaflike, gradually reduced upward, longer than flowers; calyx teeth scarcely longer than broad, deltoid; corolla lilac, hairy on outside.

Wet places, riverbanks. Described from Europe. Circumpolar map gives range of the entire complicated complex.

Extremely variable. Most Alaskan specimens, with pubescent leaves and stem, belong to var. **villòsa** (Benth.) Steven [*M. canadensis* L., *M. arvensis* L. var. *canadensis* (L.) Briq., *M. canadensis* var. *villosus* Benth.], but var. **villòsa** f. **glabràta** Stev., with glabrescent leaves and less pubescent stem, also occurs, at hot springs.

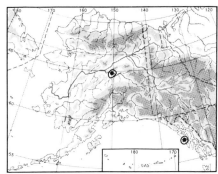

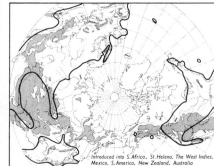

Solanum L.

1. Solànum nìgrum L.   Nightshade

Plant with taproot; leaves ovate or rhomboid, cuneate at base, sinuate-dentate, acute; petioles shorter than decurrent blade; cymes umbellate; pedicels erect in flower, deflexed in fruit; calyx lobes obtuse; corolla white, the lobes revolute in age; fruit globose, black.

Waste places; occasionally introduced. Occurs also in Galapagos Islands. Described from Europe.

Circumpolar map includes range of **S. americànum** Mill.

80. Scrophulariaceae (Figwort Family)

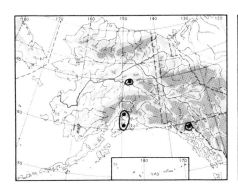 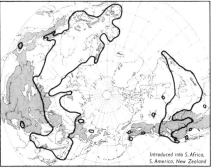 

Linaria Mill.

1. Linària vulgàris Mill. — Butter-and-Eggs
Antirrhinum Linaria L.

Bluish-green, glabrous; stems from creeping rhizome, erect, branched above; leaves linear-lanceolate; flowers numerous, in long, dense raceme; sepals ovate to lanceolate, acute; corolla yellow with orange palate; spur straight, nearly as long as corolla; capsule ovoid, more than twice as long as calyx; seeds winged.

Roadsides and waste places; introduced. Described from Europe.

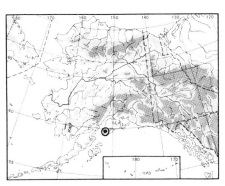 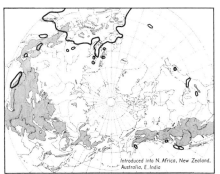 

Antirrhinum L.

1. Antirrhinum oróntium L. — Snapdragon
Misopates orontium (L.) Raf.

Stem simple in Alaskan specimens; leaves lanceolate to linear, sessile; upper leaves exceeding flowers; calyx lobes linear, pubescent; corolla pink; capsule ovate, viscid-pubescent.

Occasionally introduced at Homer. Described from western Europe.

80. Scrophulariaceae (Figwort Family)

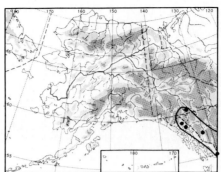
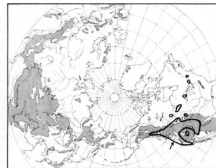

Collinsia Nutt.

1. Collínsia parviflòra Dougl.

Stem simple or branched, from thin root; lower leaves small, spatulate to round, petiolated, the upper narrowly elliptic to linear, sessile; flowers long-pedicellated, the lower solitary in the axils, the upper clustered; calyx lobes longer than tube, acuminate; corolla blue or with whitish upper lip, the tube abruptly bent at an obtuse angle to calyx near base; capsule ellipsoid.

Moist places.

Pentstemon Mitchell (Beardtongue)

Inflorescence glabrous; flowers less than 10 mm long . . . . 2. *P. procerus*
Inflorescence pubescent; flowers larger:
    Leaves lanceolate, entire . . . . 1. *P. Gormani*
    Leaves ovate, the upper subcordate, sharply serrulate . . . . 3. *P. serrulatus*

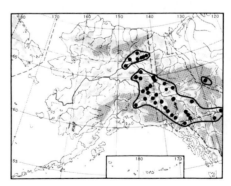

1. Pentstémon Gormáni Greene

Stems several, from many-headed, woody rootstock, the upper part villous-glandular; basal leaves spatulate-oblong, entire, petiolated, glabrous; stem leaves linear-oblong, sessile; corolla about 2 cm long, violet to purplish, with rounded lobes, hirsute on inside; capsule elliptic, mucronate, about as long as calyx.

Dry mountain slopes, to at least 1,000 meters.

The closely related **P. eriántherus** Pursh, with puberulent basal leaves, occurs south of the Pleistocene glaciation; broken line on circumpolar map indicates range of *P. eriantherus*.

80. Scrophulariaceae (Figwort Family)

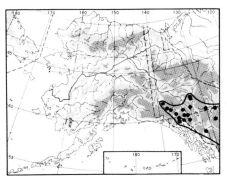
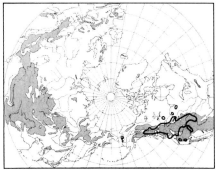

2. Pentstémon pròcerus Dougl.

Stems slender, about 3 dm tall, glabrous, tufted, from woody rhizome; basal leaves oblanceolate to elliptic or ovate, poorly developed; stem leaves few, sessile; inflorescence an interrupted spike; calyx lobes linear-lanceolate, acute; corolla 6–10 mm long, blue, not strongly bi-labiate, the palate bearded.

Dry slopes. Described from cultivated specimens.

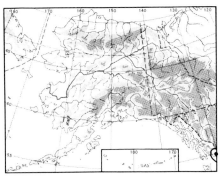
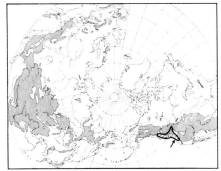
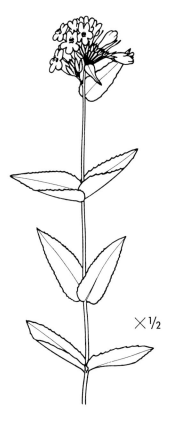

3. Pentstémon serrulàtus Menzies
Pentstemon diffusus Dougl.

Stems from woody, branched base; leaves glabrous or nearly so, lanceolate to ovate-lanceolate, sharply serrate, the lower petiolated, the upper sessile; inflorescence mostly compact, sometimes open; calyx lobes lanceolate, acuminate, ciliate; corolla blue to purple, glabrous on both surfaces.

Wet places. Described from the northwest coast of North America.

80. Scrophulariaceae (Figwort Family)

Mimulus L.

Corolla yellow; lower leaves petiolated .... 1. *M. guttatus*
Corolla pink to purple; lower leaves sessile .... 2. *M. Lewisii*

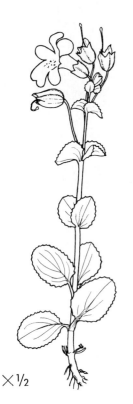

×½

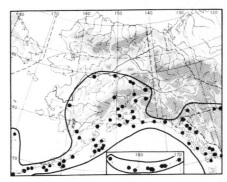 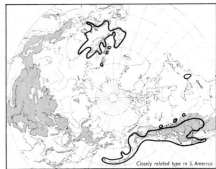

1. Mímulus guttàtus DC. Yellow Monkey Flower

Mimulus Langsdorffii Donn; *M. Tilingii* with respect to Alaskan reports.

Stem decumbent or ascending; lower leaves ovate to oblong, irregularly dentate, petiolated, the upper ovate to orbicular, sessile; peduncles glandular-pubescent; calyx becoming inflated, with short, broad teeth, the upper longer than others; corolla yellow, with red spots in throat, up to 45 mm long; lower lip much longer than upper, with prominent pubescent palate, nearly closing throat; capsule oblong, obtuse.

Margins of ponds and streams, wet, rocky slopes; common on the coast, rare inland. Described from cultivated specimens grown from seeds probably from Unalaska.

Often found introduced in eastern Asia and in Europe.

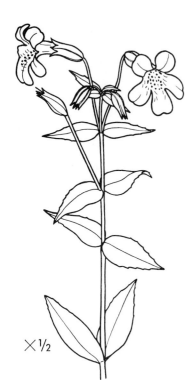

×½

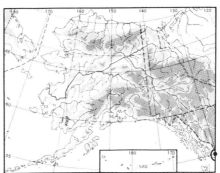 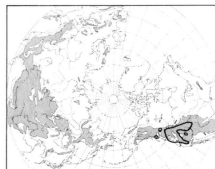

2. Mímulus Lewísii Pursh Purple Monkey Flower

Viscid-villous; stem stout, erect, simple, from stout, branched, woody rhizome; leaves sessile, ovate to elliptic, acute, entire to irregularly dentate, the lowermost reduced; calyx teeth about equally long, acuminate; corolla pink to purple, with yellow, hairy ridges, about 2.5 cm long, with rounded or emarginate lobes.

Moist places, banks of streams. Described from the source of the Missouri River.

80. Scrophulariaceae (Figwort Family)

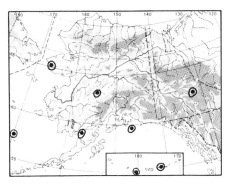

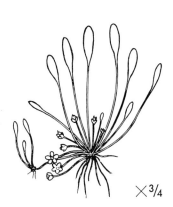

Limosella L.

1. Limosélla aquática L. — Mudwort

Glabrous, small plant, sometimes submerged, with runners producing rosettes at nodes; upper leaves dark green, elliptical, with petioles much longer than blade, the lower subulate; calyx longer than corolla tube; corolla white or purplish, with triangular acute lobes, each with few long hairs; capsule round to elliptical.

Wet mud. Described from northern Europe.

Veronica L. (Speedwell)

Flowers solitary, axillary .... 5. *V. persica*
Flowers in racemes:
  Racemes terminal, at end of stem or branches:
    Plant annual, with thin root:
      Leaves ovate or broadly elliptic; style 0.4–1 mm .... 6. *V. arvensis*
      Leaves oblong to linear; style nearly obsolete ....
        .... 7. *V. peregrina* subsp. *xalapensis*
    Plant perennial, with rhizome:
      Plant erect; capsule longer than broad:
        Leaves sharply serrulated; ovary strongly pubescent at summit ....
          .... 12. *V. Stelleri*
        Leaves nearly entire in margin or irregularly serrate; ovary evenly pubescent:
          Leaves elliptic to ovate, irregularly serrate; upper leaves not much reduced; fruiting racemes not interrupted, cylindric ....
            .... 10. *V. Wormskjoldii* subsp. *Wormskjoldii*
          Leaves narrower, nearly entire in margin; fruiting racemes interrupted, elongated; plant tall ..... 11. *V. Wormskjoldii* subsp. *alterniflora*
      Plant decumbent; capsule broader than long:
        Flowers bluish-white, with darker striae; pedicels with short, light-colored, upward-curved hairs; branches not rooting ....
          .... 8. *V. serpyllifolia* subsp. *serpyllifolia*
        Flowers blue; pedicels with long, brown, viscid hairs; branches rooting at base .... 9. *V. serpyllifolia* subsp. *humifusa*
  Racemes axillary:
    Leaves linear or linear-lanceolate, entire in margin .... 3. *V. scutellata*
    Leaves ovate-lanceolate, ovate or obovate:
      Plant glabrous; leaves ovate-lanceolate or oblong:
        Leaves petiolated .... 1. *V. americana*
        Leaves sessile .... 2. *V. anagallis-aquatica*
      Plant pubescent; leaves broadly ovate or obovate:
        Leaves nearly entire or obscurely crenate in margin; flowers large, 10–12 mm broad .... 13. *V. grandiflora*
        Leaves coarsely serrated in margin; flowers smaller ... 4. *V. chamaedrys*

80. Scrophulariaceae (Figwort Family)

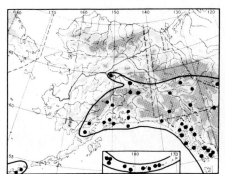 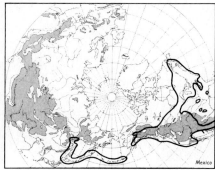

1. Verónica americàna Schwein. Brooklime
Veronica beccabunga var. *americana* Raf.

Glabrous; stem ascending from decumbent, rhizomatose base, more or less succulent; leaves of flowering stem lanceolate to narrowly ovate, serrate-dentate, acute, distinctly petiolated; racemes lax, axillary; corolla violet to lilac; capsule orbicular.
Along streams. Described from Canada to Virginia and Kentucky.

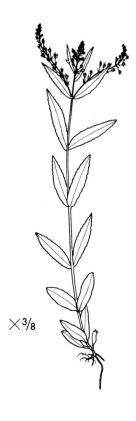

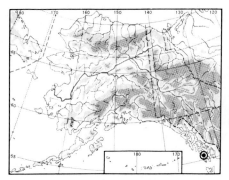 

2. Verónica anagállis-aquática L. Water Speedwell

Stem creeping and rooting at base, then ascending, fleshy, mostly branched; leaves ovate to ovate-lanceolate, remotely serrulate, somewhat clasping; racemes opposite, lax, ascending; bracts linear, acute; calyx lobes ovate-lanceolate, acute; corolla pale blue; capsule orbicular, somewhat emarginate.
Ponds, streams; introduced. Described from Europe.

80. Scrophulariaceae (Figwort Family)

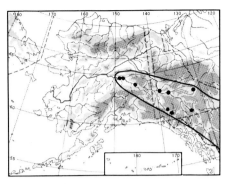
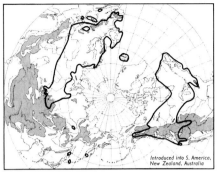
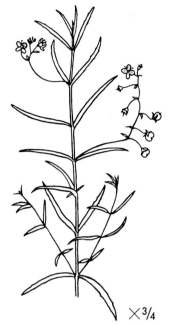

3. Verónica scutellàta L.

Stem weak, ascending from a filiform rhizome with long runners; leaves linear to lanceolate, entire or remotely toothed, acute, semi-amplexicaule; flowers in lax, slender, alternate racemes; bracts linear; sepals ovate; corolla white or pale blue with purple lines; capsule flat, broader than long, deeply emarginate, much longer than calyx.

Ponds, wet meadows. Described from Europe.

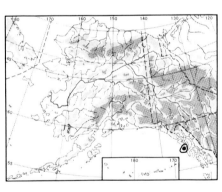
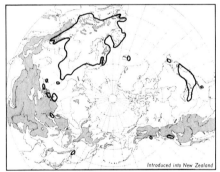

4. Verónica chamaèdrys L.

Stem prostrate, rooting at nodes, with long, white hairs in 2 lines; leaves dull green, triangular-ovate, subcordate, coarsely crenate-serrate, pubescent, sessile or short-petiolated; racemes from axils of lower leaves; bracts lanceolate; calyx lobes lanceolate, hairy; corolla bright blue with white eye; capsule obcordate, shorter than calyx, ciliate and pubescent.

An occasionally introduced weed. Described from Europe.

× ³⁄₄

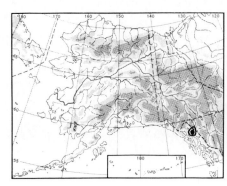
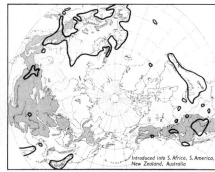

5. Verónica pérsica Poir.

Veronica Buxbaumii Ten., not F. W. Schmidt; *V. Tournefortii* C. C. Gmel., not Vill.

Stem branched at base, with decumbent branches; leaves light green, triangular-ovate, coarsely crenate-serrate, minutely ciliate, pubescent on veins below, short-petiolated; flowers solitary, axillary; pedicels longer than leaves; calyx lobes ovate, ciliate, enlarged in fruit; corolla bright blue, the lower lobe often pale; capsule nearly twice as broad as long, with 2 divergent, ciliate lobes.

An occasionally introduced weed. Described from Persia.

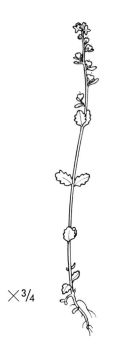
× ³⁄₄

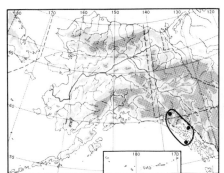
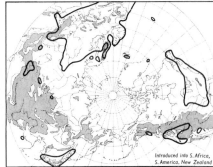

6. Verónica arvénsis L.

Erect or ascending; stem pilose; lower leaves rounded to oval, crenate-dentate, petiolated, the upper ovate to lanceolate, sessile; corolla blue; capsule emarginate, glandular in margin.

An introduced weed. Described from Europe.

80. Scrophulariaceae (Figwort Family)

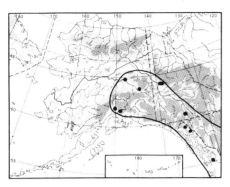  

7. Verónica peregrìna L.
subsp. **xalapénsis** (HBK.) Pennell

Veronica xalapensis HBK.; *V. peregrina* var. *xalapensis* (HBK.) St. John & Warren.

More or less glandular-pubescent; stem simple or branched, from thin root; leaves spatulate, oblong to linear, obtuse, entire or obscurely dentate, the upper sessile; flowers sessile or nearly so; sepals subequal; corolla inconspicuous, whitish; style minute; capsule more or less obcordate.

Roadsides, waste places; an introduced weed. *V. peregrina* (which is glabrous) described from Europe, subsp. *xalapensis* from Xalapa, Mexico.

Broken line on circumpolar map indicates range of subsp. **peregrìna.**

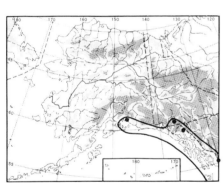 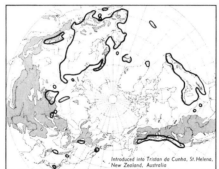 

8. Verónica serpyllifòlia L.
subsp. **serpyllifòlia**

Stems rooting at nodes, mostly ascending; leaves ovate to oblong, obscurely crenulate, rounded at both ends, the lowermost petiolated; racemes terminal; upper bracts narrowly oblong, the lower passing into leaves, longer than pedicels; calyx lobes oblong; corolla white or pale blue with darker lines; capsule broader than long, obcordate with rounded base, ciliate.

Roadsides, waste places; an introduced weed. Described from Europe and North America.

80. SCROPHULARIACEAE (Figwort Family)

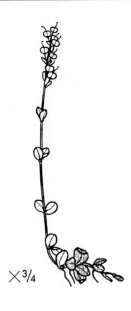

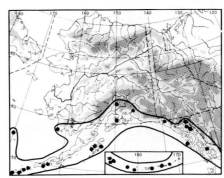 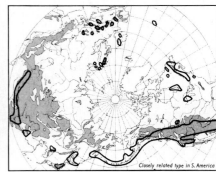

9. Verónica serpyllifòlia L.
subsp. **humifùsa** (Dickson) Syme

Veronica humifusa Dickson; *V. tenella* All.; *V. serpyllifolia* var. *borealis* Laest.; *V. borealis* (Laest.) Hook. f.; *V. serpyllifolia* var. *humifusa* (Dicks.) M. Vahl.

Similar to subsp. *serpyllifolia*, but branches also rooting; flowers larger, blue; upper part of stem and capsule viscid-pubescent; capsule with cuneate base.

Moist places, roadsides. Described from Scotland.

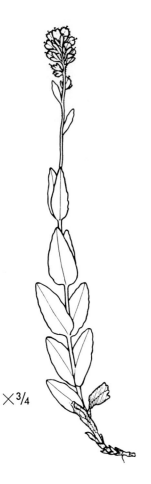

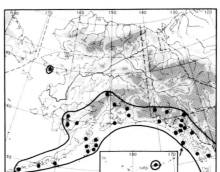 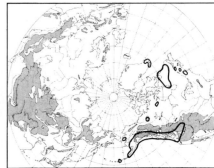

10. Verónica Wormskjòldii Roem. & Schult.
subsp. **Wormskjòldii**

Veronica alpina var. *unalaschcenis* Cham. & Schlecht.; *V. alpina* var. *Wormskjoldii* (Roem. & Schult.) Hook.; *V. nutans* Bong.; *V. Wormskjoldii* var. *nutans* (Bong.) Pennell.

Stem simple, from branching rhizome, pubescent with spreading hairs, glandular above; leaves opposite, elliptic to ovate, slightly serrulate in margin, becoming black in drying; flowers in terminal racemes, capitate when young, thick, cylindric, not interrupted in fruit; calyx glandular; corolla blue to violet; capsule glandular, somewhat emarginate.

Meadows, mountain slopes, to at least 2,000 meters.

80. Scrophulariaceae (Figwort Family)

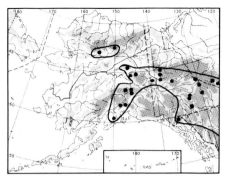
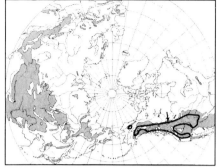

11. Verónica Wormskjòldii Roem. & Schult.
subsp. **alterniflòra** (Fern.) Pennell
Veronica alpina var. *alterniflora* Fern.

Similar to subsp. *Wormskjoldii*, but tall and straight; leaves nearly entire in margin; inflorescence interrupted and strongly elongated in fruit.
Meadows, mountain slopes.
Transitions between the 2 types occur.

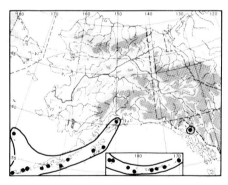
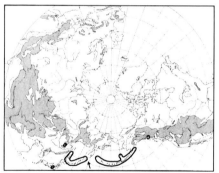
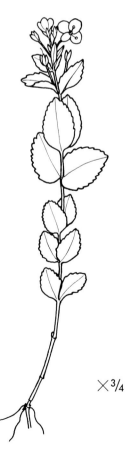

12. Verónica Stélleri Pall.

Stem simple, from filiform horizontal rhizome; successive leaf pairs crosswise opposite, in 4–7 pairs, ovate, dentate-serrate; pedicels longer than flowers; calyx lobes lanceolate to ovate-lanceolate, somewhat acute, pubescent but not glandular; corolla blue to violet, with ovate to broadly ovate lobes; capsule elliptic, slightly emarginate, pubescent.
Alpine meadows.
Var. **glabréscens** Hult., with glabrous or glabrescent leaves, the leaves less distinctly serrulate, might be the hybrid *V. Stelleri* × *Wormskjoldii*.

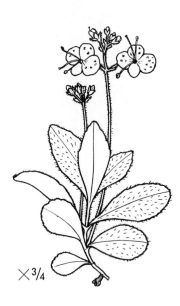

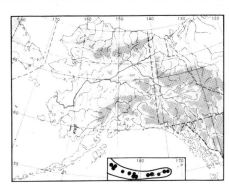

13. Verónica grandiflòra Gaertn.

Veronica aphylla var. *grandiflora* Benth.; *V. Kamtschatica* L. f.; *V. aphylla* var. *kamtschatica* (L. f.) Willd.

Stem simple, viscid-pubescent, from thin, branched rhizome with subterranean runners; leaves ovate, irregularly serrate-dentate, ciliated and somewhat pubescent above; flowering stem lateral from axils of lower leaves; bracts small, oblong, viscid-pubescent; calyx lobes oblong-ovate, blunt; corolla blue, with uneven lobes, the uppermost rounded to reniform; filaments broader in middle or below middle; fruit obovate, emarginate.

Rocky places.

Similar to **V. aphýlla** L. of the Alps of central Europe.

Lagotis Gaertn.

Basal leaves orbicular to ovate, crenate-dentate . . . . 1. *L. glauca* subsp. *glauca*
Basal leaves lanceolate to broadly lanceolate, serrate . . . . 2. *L. glauca* subsp. *minor*

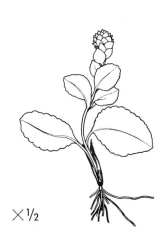

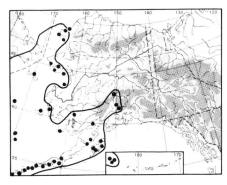
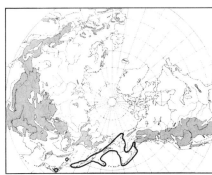

1. Lagòtis glaùca Gaertn.

Gymnandra Gmelini Cham. & Schlecht.; *Bartia gymnandra* L. f.; *B. glauca* (Gaertn.) Poir.; *G. ovata* Willd.

subsp. **glaùca**

Stem single, thick, decumbent, from thick, short caudex with thin sheaths at apex; basal leaves 1–3, ovate to nearly orbicular, densely crenate-dentate, petiolated; stem leaves smaller, sessile; inflorescence an ovate to cylindrical spike; lower bracts dentate, the upper entire; calyx lobes ciliated; corolla diluted blue, tube bent at base; upper lip entire, lower with 2 to 3 lobes; filaments united with margins of upper lip.

Stony slopes on tundra and in the mountains. Type locality not given, presumably Kamchatka.

In the Himalayas and in China, closely related taxa reach very high altitudes—close to 6,000 meters.

80. Scrophulariaceae (Figwort Family)

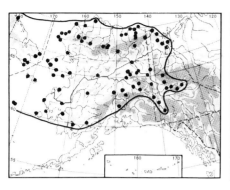

2. Lagòtis glaùca Gaertn.
subsp. **mìnor** (Willd.) Hult.
  Gymnandra minor Willd.; *G. Stelleri* Cham. & Schlecht.

Similar to subsp. *glauca*, but leaves lanceolate to broadly lanceolate, sharply serrate. Habitat same.
  Transitions between the two races occur frequently where the ranges overlap.
  Broken line on circumpolar map indicates range of the closely related **L. Hulténii** Polunin.

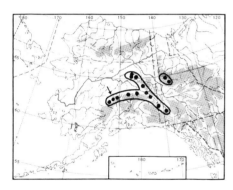

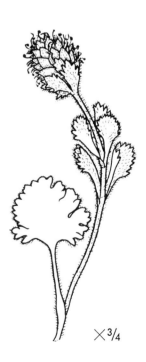

Synthyris Benth.

1. Sýnthyris boreàlis Pennell — Alaska Synthyris

Villose-hirsute; stem simple, from oblique, short, thick rhizome; leaves cordate, deeply crenate-dentate to shallowly pedately lobed, the divisions dentate; inflorescence rounded, prolonged in fruit; subtended by 2–4 dentate leaves; bracts small, dentate; sepals 4, lanceolate-linear, acute; corolla blue, hirsute, the lobes equaling the tube; upper lip longer than the 3-cleft lower lip; stamens and style exserted; capsule hirsute.
  Ridges and solifluction soil in the mountains, to at least 2,135 meters in McKinley Park.
  (See color section.)

80. Scrophulariaceae (Figwort Family)

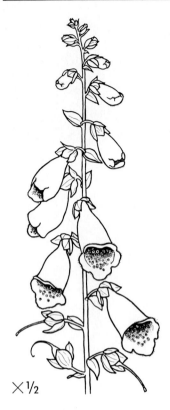

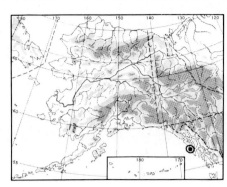

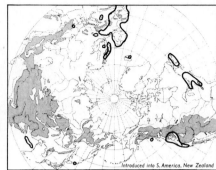

Digitalis L.

1. Digitàlis purpùrea L.   Foxglove

Stem tall, puberulent above; rosette leaves ovate to lanceolate, crenate, with winged petiole; racemes many-flowered; pedicels tomentose, longer than calyx; lower sepals ovate, the upper lanceolate, acute; corolla ciliated, 4–5 cm long, pinkish to purple with purple spots, pubescent within; stamens and style included; capsule ovate, longer than calyx.

Occasionally escaped from cultivation. Described from southern Europe.

Contains digitalin, often prescribed as a heart stimulant.

Castilleja Mutis (Indian Paintbrush)

Lower lip of corolla with distinct lobes, at least one-third the length of the upper; leaves indistinctly 3-ribbed; stem slender:
  Middle stem leaves lobed:
    Bracts yellow; corolla yellow .... 11. *C. hyperborea*
    Bracts pink:
      Calyx lobes acute; corolla red .... 12. *C. parviflora*
      Calyx lobes obtuse; corolla greenish .... 13. *C. Henryae*
  Middle stem leaves entire (exceptionally with single lobe):
    Lateral calyx lobes up to 2.5 mm long, ovate or lanceolate, much shorter than tube:
      Bracts and corolla purplish; lower lip of corolla about half as long as the upper; only the uppermost bracts in the spike overlapping; leaves linear to lanceolate .... 6. *C. Raupii*
      Bracts and corolla yellowish; lower lip of corolla at least two-thirds as long as the upper; bracts overlapping:
        Corolla 15–20 mm long; lower lip two-thirds as long as the upper; corolla considerably exceeding calyx; leaves linear .... 7. *C. yukonis*
        Corolla 10–16 mm long; lower lip three-quarters as long as the upper; corolla only slightly longer than calyx:
          Corolla 13–16 mm long; stem appressed-pubescent ...... 8. *C. annua*
          Corolla 10–13 mm long; stem and inflorescence markedly villous .... 9. *C. villosissima*
    Lateral calyx lobes 3–8 mm long, linear, as long as tube or longer:
      Bracts, calyx, and corolla greenish-yellow; stem single ...... 5. *C. caudata*
      Bracts violet or reddish; corolla with purple margins; stems several together .... 10. *C. elegans*
Lower lip of corolla with hardly developed lobes, less than one-fifth the length of the upper; leaves lanceolate to ovate-lanceolate, strongly 3-ribbed; stem stout:
  Bracts yellowish; calyx lobes obtuse .... 1. *C. unalaschcensis*
  Bracts red; calyx lobes somewhat acute to acuminate:
    Bracts all acute or acuminate; calyx cleft to half its length ...... 2. *C. miniata*
    Bracts more or less obtuse or rounded; calyx cleft to half its length or more:
      Corolla 18–25 mm long; bracts mostly or wholly rounded; inflorescence elongating; rhizome slender .... 3. *C. hyetophila*
      Corolla 25–30 mm long; bracts more lobed, with acute teeth; inflorescence short; rhizome short .... 4. *C. chrymactis*

80. Scrophulariaceae (Figwort Family)

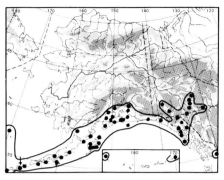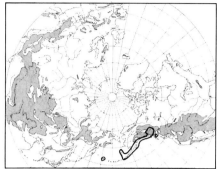

1. Castilléja unalaschcénsis (Cham. & Schlecht.) Malte

Castilleja pallida var. *unalaschensis* Cham. & Schlecht. Including *C. unalaschcensis* subsp. *transnivalis* Pennell.

Stems several, glabrate below, villous above, from short, stout, scaly, many-headed caudex; basal leaves lacking; stem leaves ovate, ovate-lanceolate, or lanceolate, somewhat acute, pubescent on both sides, 3- to 5-nerved; bracts ovate, the lowermost entire, the others with long, blunt or somewhat acute lobes, yellowish or greenish-yellow; calyx lobes obtuse or rounded; lower lip of corolla small, mostly entire, green, upper lip yellow in margin.

Subalpine meadows. Common on the southwestern coast.

The closely related **C. sulphùrea** Rydb. occurs in the southern Rocky Mountains.

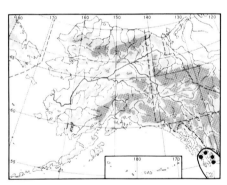

2. Castilléja miniàta Dougl.   Great Red Indian Paintbrush

Stem simple or somewhat branched above; leaves lanceolate, acute, entire, or the upper occasionally lobed; inflorescence villous-pubescent, elongating in fruit; bracts more or less lobed, acute, scarlet; calyx lobes lance-linear, acuminate; upper lip of corolla puberulent, as long as tube, with red margins, the lower lip green.

Meadows. Described from the Blue Mountains of northwest North America.

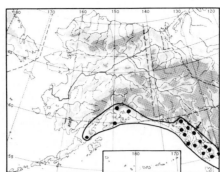
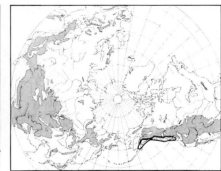

3. Castilléja hyetóphila Pennell

Stems several together, simple or branched, villous-hirsute above, from slender, branched rhizome; leaves linear-lanceolate to lanceolate, entire, glabrous to finely pubescent; bracts entire, obtuse, or the upper with a pair of short lobes, red at apex; calyx hirsute, red at top; corolla 20–30 mm long, the upper lip green, with reddish margins, the lower much shorter.

Salt marshes; meadows along the sea.

Similar to *C. miniata*, but with obtuse bracts.

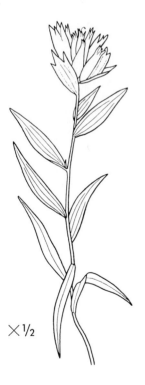
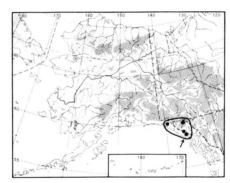

4. Castilléja chrymáctis Pennell

Similar to *C. hyetophila*, but corolla larger; bracts more lobed, with acute teeth; inflorescence short and dense.

Meadows.

C. miniata, *C. hyetophila*, and *C. chrymactis* form a critical group that should be studied in the field.

80. Scrophulariaceae (Figwort Family)

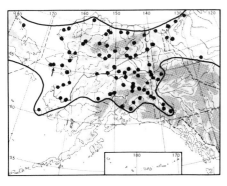
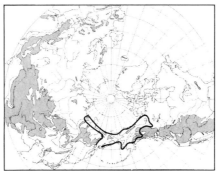

5. Castilléja caudàta (Pennell) Rebr.

Castilleja pallida subsp. *caudata* Pennell; *C. pallida typica* with respect to Alaskan plant; *C. pallida* subsp. *pallida* var. *caudata* (Pennell) Boiv.

Stems single or few together, simple or branched above, erect, pubescent; leaves alternate, lanceolate, attenuate-caudate to nearly linear, the upper broader than the lower, short-pubescent; inflorescence elongated, more or less woolly-pubescent, with often yellowish hairs; bracts greenish-yellow, mostly entire, or the upper laciniate; calyx greenish-yellow, white-pubescent or yellowish-pubescent; corolla not much longer than calyx, the upper lip greenish, the lower yellowish.

Along streams, in meadows, in the mountains to at least 1,200 meters.

In var. **auricòma** (Pennell) Hult. [*C. pallida* subsp. *auricoma* Pennell, *C. pallida* subsp. *Mexiae* var. *auricoma* (Pennell) Boiv.], the inflorescence is yellowish and long-pubescent.

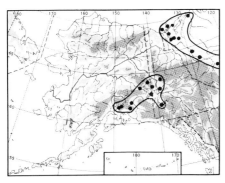
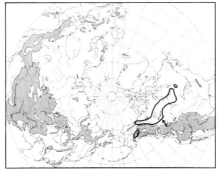

6. Castilléja Raùpii Pennell

Including *Castilleja Raupii* subsp. *ursina* Raup.

Stems several, slender, often purplish, finely retrorse-pubescent, villous above; leaves linear to linear-lanceolate, attenuate to caudate, entire, the uppermost sometimes with a few linear lobes; bracts ovate to lanceolate, with a pair of short lateral lobes, violet-purple; calyx violet-purple-hirsute, the upper lip green with violet margins, the lower purple, half as long as the upper.

Along streams, on open ground.

A very similar plant occurs in Siberia.

×⅓

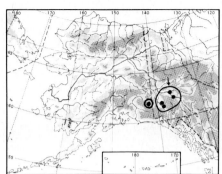
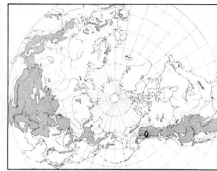

7. Castilléja yukònis Pennell

Stems several, slender, purplish, pubescent with minute, retrorse white hairs, villous above; leaves linear, entire, or the uppermost with pair of short lobes with very short, white pubescence; bracts lanceolate, with 1–2 pairs of short lobes, obtuse, yellowish; calyx yellowish, with acute lobes; corolla puberulent with yellow margins, the lower lip with yellow lobes, about two-thirds as long as the upper.

Dry hillsides.

Similar to *C. Raupii*, but bracts yellowish.

×⅓

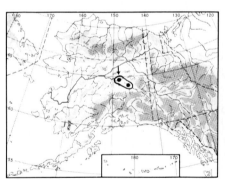
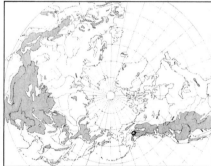

8. Castilléja ánnua Pennell

Stems solitary, much branched, slender, finely appressed-pubescent, villous above; leaves lanceolate, entire, somewhat obtuse, puberulent; bracts lance-oval, usually with slender, lateral lobes, purplish at base, greenish-yellow at tip; calyx purplish, yellow above; corolla 13–16 mm long, puberulent, the upper lip green with yellowish margins, the lower green with yellow lobes.

Waste places at Fairbanks; somewhat doubtful.

80. Scrophulariaceae (Figwort Family)

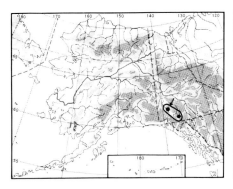
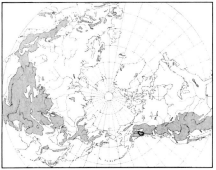

9. Castilléja villosíssima Pennell

Stems several, slender, villous with spreading hairs, densely so above; leaves narrowly lanceolate, somewhat obtuse, entire or with pair of divaricate lobes; bracts yellow to orange-yellow, ovate, obtuse to rounded, with 1–2 pairs of lateral lobes; calyx yellowish, hirsute to canescent; corolla 10–13 mm long, puberulent, the upper lip green with pale margins, the lower green with yellow lobes, two-thirds as long as the upper.

A related plant with finely puberulent stem and linear leaves—**C. Muelléri** Pennell, described from Lake Kluane to the Donjek River—is regarded as doubtful until more material is forthcoming.

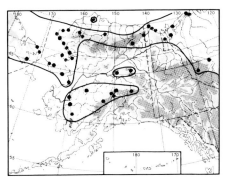
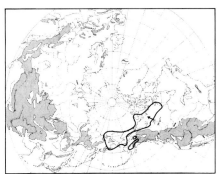

10. Castilléja élegans Malte

Castilleja pallida subsp. *elegans* (Malte) Pennell; *C. pallida* subsp. *Mexiae* var. *elegans* (Malte) Boiv.; *C. pallida* subsp. *Mexiae* Pennell in part; *C. pallida* subsp. *Mexiae* var. *Mexiae* (Pennell) Boiv.

Stems pubescent, especially above, several, ascending; leaves lanceolate, attenuate to caudate, acute, the upper often laciniate; inflorescence compact, lanate; bracts ovate to oblong-ovate, laciniate, violet to reddish; corolla about 25 mm long, pubescent, the upper lip acute, yellowish-green with purple margin, the lower purple, about half as long as the upper; capsule glabrous, mucronate.

Stony slopes in the mountains, to at least 1,400 meters.

(See color section.)

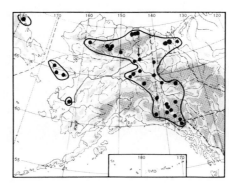
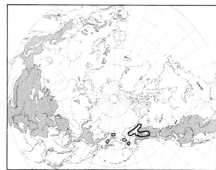

11. Castilléja hyperbòrea Pennell

Stems several, from many-headed, short caudex, finely hirsute-pubescent to villous above; lower leaves often entire, lance-linear, the upper with 1–2 pairs of slender, divaricate lobes; bracts yellow or yellowish-white, more cleft than leaves; calyx yellowish, pubescent, with obtuse or rounded lobes; corolla green to pale yellow, puberulent, the upper lip acute, the lower with yellowish lobes.

Stony slopes in the mountains, to at least 2,000 meters.

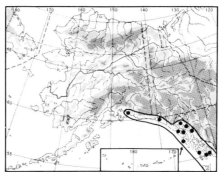

12. Castilléja parviflòra Bong.

Stems several, from thick caudex, unbranched, villous-pubescent above; basal leaves lacking; stem leaves few, sessile, divaricate-laciniate with acute lobes; inflorescence spikelike, interrupted below; bracts leaflike, villous, pink; flowers sessile; calyx lobes linear; corolla red, the upper lip crenate in apex, the lower minute with 3 acute teeth; capsule about 10 mm long, oblong, glabrous, wrinkled, mucronate.

Alpine meadows, to at least 1,200 meters.

80. Scrophulariaceae (Figwort Family)

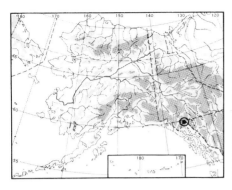  

13. Castilléja Hénryae Pennell

Stem simple, slender, hirsute or villous above; leaves gray-green, elliptic-lanceolate, with 2–3 pairs of divaricate lobes; bracts purple toward tip, oval with lanceolate-attenuate lobes; calyx canescent, villous, purple, cleft three-fifths of length into obtuse lobes; galea 6–7 mm long, acute, greenish, glandular-puberulent dorsally.

Moist mountain slopes.

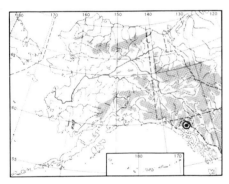 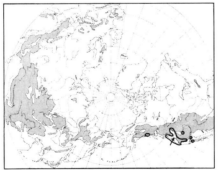 

Orthocarpus Nutt.

1. Orthocárpus híspidus Benth.

Plant pubescent with long, spreading glandular hairs, in inflorescence also with shorter hairs; stem simple or with few, erect branches; leaves linear, the upper somewhat broader and cleft; calyx 2-cleft with bifid segments; corolla creamy white, the lower lip 3-saccate with inconspicuous teeth.

Moist places; an introduced weed at Skagway.

80. Scrophulariaceae (Figwort Family)

Euphrasia L. (Eyebright)

Plant not glandular; bracts imbricate .... 1. *E. mollis*
Plant more or less glandular-pubescent in upper part; bracts not markedly imbricate
.... 2. *E. disjuncta*

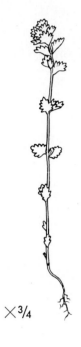
×3/4

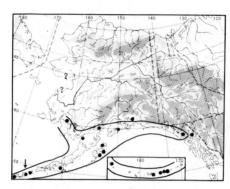
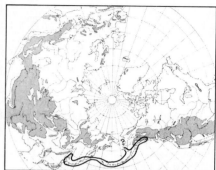

1. Euphràsia móllis (Ledeb.) Wettst.
Euphrasia officinalis var. *mollis* Ledeb.

Stem short, simple, pubescent with white, simple, more or less retrorse hairs, and with long internodes in lower and middle part; leaves ovate, broad, blunt, with 1–4 blunt or somewhat acute, rounded teeth on each side, more or less white-pubescent on both sides; bracts imbricate; calyx not much enlarged in fruit, with triangular to narrowly triangular acute teeth; corolla white with purple lines and lavender upper lip, small, in full flower considerably longer than calyx; capsule emarginate, about as long as calyx.

Subalpine meadows.

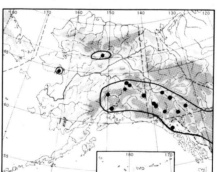
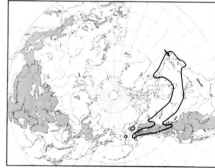

2. Euphràsia disjúncta Fern. & Wieg.

Euphrasia arctica var. *disjuncta* (Fern. & Wieg.) Cronq.; *E. Pennellii* var. *incana* Callen. Including *E. subarctica* Raup.

Similar to *E. mollis*, but taller, more or less glandular-pubescent in upper part; leaves and bracts narrower, with more cuneate base, the bracts not imbricate, with cuspidate teeth, the calyx teeth longer and narrower.

Open soil.

Circumpolar map tentative; includes range of **E. hudsónica** Fern. & Wieg.

×3/4

80. Scrophulariaceae (Figwort Family)

Rhinanthus L. (Yellow Rattle)

This genus includes numerous ill-defined taxa; the treatment here is tentative.

Corolla 8–9 mm long; leaves sharply serrate .... 1. *R. minor* subsp. *borealis*
Corolla 10–12 mm long; leaves crenate-dentate .... 2. *R. arcticus*

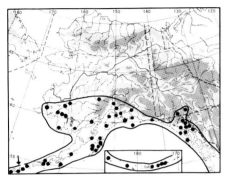
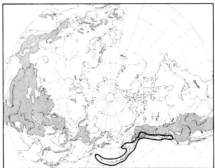
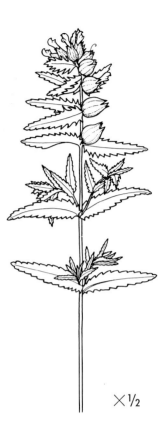

×½

1. Rhinánthus mìnor L.
subsp. boreàlis (Sterneck) Löve

Alectorolophus borealis Sterneck; *Rhinanthus borealis* (Sterneck) Druce; *R. minor* subsp. *groenlandicus* with respect to Alaskan plant.

Stem simple or somewhat branched, with 4–6 prolonged internodes; lower leaves oblong-ovate, the upper broadly lanceolate with large teeth; bracts triangular-lanceolate with long, acute teeth, considerably longer than calyx; calyx more or less pubescent on sides, with short, septate hairs; corolla yellow with violet teeth, 8–9 mm long, with straight tube, upper lip small, throat open.

Meadows.

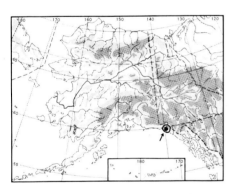
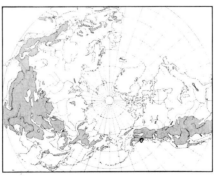

×½

2. Rhinánthus árcticus (Sterneck) Pennell

Alectorolophus arcticus Sterneck; *Rhinanthus arcticus* (Sterneck) Vassilchenko.

Similar to *R. minor* subsp. *borealis*, but corolla larger, more rounded at apex; fruiting calyx larger; stem leaves oblong-lanceolate, crenate-dentate.

A doubtful species, known only from Yakutat Bay.

Pedicularis L. (Lousewort)

Cauline leaves verticillate:
  Plant glabrous; galea acuminate .... 1. *P. Chamissonis*
  Plant more or less pubescent; galea blunt:
    Segments of leaves broad, oblong, coarsely dentate .... 2. *P. verticillata*
    Segments of leaves linear-lanceolate, sharply dentate .... 3. *P. amoena*
Cauline leaves alternate or lacking:
  Galea prolonged into beak, 2–20 mm long:
    Beak strongly curved, 10–20 mm long; flowers suggestive of elephant's head with trunk upraised .... 4. *P. groenlandica*
    Beak straight, shorter:
      Stem with 1 or no cauline leaves; flowers purple ..... 5. *P. ornithorhyncha*
      Stem with several cauline leaves; flowers yellow .... 6. *P. lapponica*
  Galea beakless:
    Galea with long, slender teeth in apex:
      Flowers small, 10–14 mm long, yellow or purple; galea straight:
        Flowers yellow or yellow with purple spots; calyx teeth entire .... 7. *P. labradorica*
        Flowers purple; calyx teeth dentate ... 10. *P. parviflora* subsp. *Pennellii*
      Flowers much larger; purple; galea arched:
        Teeth at right angle to lip, some distance from apex of rounded galea; stem leafy; 2 filaments pubescent:
          Leaves with broad rachis; upper bracts not protruding; flowers large .. 11. *P. Langsdorffii* subsp. *Langsdorffii*
          Leaves with narrow rachis; upper leaves longer than flowers .... 12. *P. Langsdorffii* subsp. *arctica*
        Teeth at lower corner of abruptly cut, truncate galea; stem scapiform with 1 to 3 leaves:
          Calyx much shorter than tube of corolla; filaments pubescent .... 13. *P. villosa*
          Calyx about as long as tube of corolla; filaments glabrous:
            Bracts not strongly dilated at base; basal bracts lacking long lobes, middle bracts lacking lobes .... 14. *P. sudetica* subsp. *interior*
            Bracts strongly dilated at base; basal bracts broad, hyaline at base, with apical or lateral lobes:
              Middle and upper bracts entire; lower bracts with small basal lobes and long caudate apex; flowers small .... 15. *P. sudetica* subsp. *interioides*
              Middle bracts also usually lobed; flowers large, with thick galea:
                Spike profusely lanate; bracts, except for lowest, lacking long, apical lobes; calyx pubescent, with narrow teeth; galea pink with purple apex; lip white, spotted with purple .... 16. *P. sudetica* subsp. *albolabiata*
                Spike nearly glabrous or slightly pubescent; bracts with long, apical lobe; calyx glabrous, often with leaflike tips of teeth; flowers purple .... 17. *P. sudetica* subsp. *pacifica*
    Galea lacking teeth in apex:
      Flowers large, about 3 cm long, cream-colored; galea twice as long as corolla tube; stem scapiform .... 18. *P. capitata*
      Flowers much smaller, pink, purple, or yellow; galea less than twice as long as corolla tube; stem leafy:
      Corolla yellow; galea dark at tip:
        Plant more or less pubescent; tip of galea brownish-red with red spots; style protruding .... 19. *P. Oederi*
        Plant glabrous; tip of galea blackish-red; style not protruding .... 20. *P. flammea*
      Corolla pink to purple:
        ■ Spike large, thick, dense, strongly lanate; leaves with linear rachis and segments:
          Galea glabrous; lip glabrous in margin .. 21. *P. Kanei* subsp. *Kanei*
          Galea pubescent; lip ciliated in margin .... 22. *P. Kanei* subsp. *Adamsii*

80. Scrophulariaceae (Figwort Family)

■ Spike neither thick, nor dense, nor lanate; leaves with broader segments:
Bracts leaflike, pinnatifid, with narrow rachis .... 8. *P. macrodonta*
Bracts triangular, pinnately lobed with 3–4 segments and broad rachis .... 9. *P. parviflora* subsp. *parviflora*

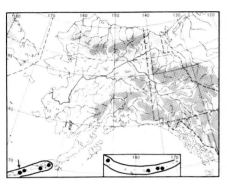 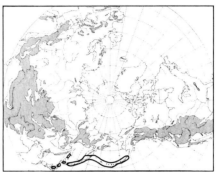 

×½

1. Pediculàris Chamissònis Stev.
Pedicularis Romanzovii Cham.

Glabrous; stem straight, single or few together; basal leaves long-petiolated, pinnate, the pinnae oblong, deeply and sharply double-serrate, strongly veined below; stem leaves in whorls of 3 to 6, gradually reduced; inflorescence capitate; calyx lobes short, triangular, acute, ciliate; corolla rose-colored, the tube bent at right angle, the upper lip with straight, truncate beak, the lower large, 3-lobed, somewhat ciliate; filaments glabrous.

Subalpine meadows.

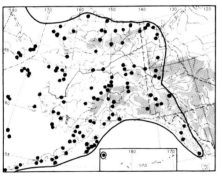 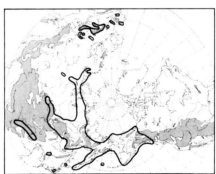 

2. Pediculàris verticillàta L.

Stems single or several, from taproot; basal leaves with long, sparsely hairy petioles, lanceolate to ovate-lanceolate, pinnatifid, the segments broad, toothed; stem leaves short-petiolated or sessile, verticillate; flowers verticillate, the lower remote; upper bracts shorter than flowers; calyx much shorter than tube, with triangular, acute teeth; corolla purplish, the tube bent almost at right angles; galea lacking teeth, the lower lip 3-cleft, somewhat longer than galea; 2 filaments pubescent; capsule lanceolate, acute, often bent to one side.

Meadows, rocky slopes, in the mountains to at least 1,300 meters. Described from Siberia, Switzerland, and Austria.

×½

×½

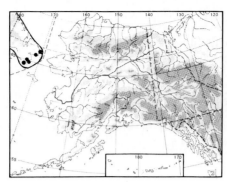 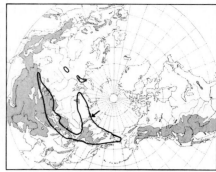

3. Pediculàris amoèna Adams

Stem pubescent at base, from short caudex with several thick fusiform roots; basal leaves long-petiolated, oblong, lanceolate, pinnate, the segments linear-lanceolate, sharply dentate; stem leaves in 1 or 2 whorls, often purplish, inflorescence capitate, the lowest whorl somewhat removed; calyx nearly glabrous, with triangular to linear, very acute, entire or somewhat dentate lobes, half as long as tube; corolla rose to violet; tube bent; galea lacking teeth, shorter than the broad, 3-lobed lower lip; 2 filaments pubescent; capsule ovate to oblong-ovate, with short beak.

Stony slopes, tundra.

×⅓

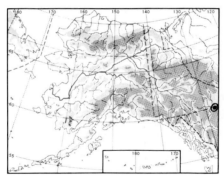 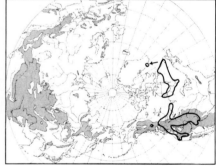

4. Pediculàris groenlándica Retz. — Elephant's Head
Elephantella groenlandica (Retz.) Rydb.

Stem single or few together, from stout caudex; basal leaves long-petiolated, pinnate, the pinnae strongly serrate or incised; stem leaves several, gradually reduced; inflorescence elongate; calyx with short, triangular, entire teeth; corolla pink to purple, the upper lip with short, hooded part, elongated into long, slender, upward-curved beak, the lower lip small, 3-lobed.

Wet meadows. Described from Greenland, where it occurs in a single locality.
(See color section.)

80. Scrophulariaceae (Figwort Family)

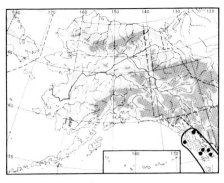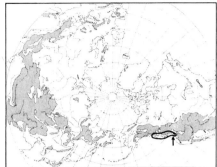

5. Pediculàris ornithorhýncha Benth.
Pedicularis pedicellata Bunge; *P. subnuda* Benth.

Stem with short caudex from fibrous root; basal leaves sub-bipinnatifid; pinnae narrow, deeply cleft and toothed; cauline leaves few and small or wanting; inflorescence capitate; bracts cleft or dissected; calyx lobes short; corolla purple, the galea elongated into straight beak, 2–4 mm long.

Meadows, alpine slopes.

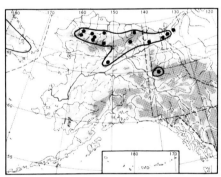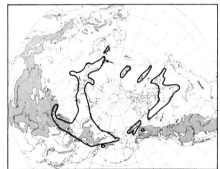

6. Pediculàris lappónica L.

Stem puberulent above, simple, single or few together, from filiform-branched rhizome; leaves lanceolate in outline, pinnatifid, with broad central part and small, ovate, toothed pinnae, the basal long-petiolated, gradually reduced upward; inflorescence short, flat, with horizontal, light-yellow flowers; upper lip arched, with beak at right angle to lip; lower lip 3-lobed, as long as the upper, or somewhat shorter; capsule lanceolate to ovate-lanceolate, markedly acute.

Alpine meadows, in the mountains to at least 1,000 meters.

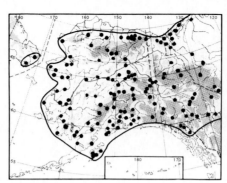 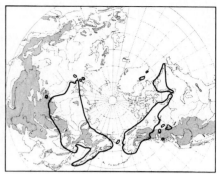

7. Pediculàris labradòrica Wirsing
Pedicularis euphrasioides Steph.

Stem single or several, from taproot, simple or often much branched, with retrorse, white hairs; basal leaves small; lower stem leaves pinnatifid, with crenate pinnae, the upper gradually reduced; calyx 2-cleft, glabrous or somewhat pubescent at base, with 4 main veins; corolla with straight tube, yellow or reddish, or yellow with red spots above; the upper lip with pair of long teeth, the lower 3-lobed, minutely ciliated; capsule horizontal, oblong-linear, acute.

A common plant; on tundra and in the mountains to at least 2,000 meters. Described from Labrador.

Var. **sulphùrea** Hult., with corolla large (up to 20 mm long) and unicolor (sulphur-yellow) is found on the upper Blackstone River, in the Yukon.

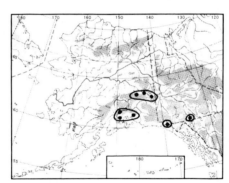 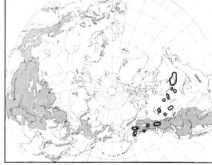

8. Pediculàris macrodónta Richards.
Pedicularis parviflora with respect to Alaskan plant in part.

Glabrous; stem simple or branched, from thin taproot; basal leaves small or lacking; stem leaves lanceolate-oblong, remotely pinnately lobed to pinnatifid; segments entire or toothed; inflorescence capitate in bud, later elongated, with remote, axillary flowers; bracts leaflike, pinnatifid; calyx 2-cleft with lacerate and lobed segments; flowers up to 12 mm long, simple; tip of galea darker, tube straight; galea with pair of broad teeth at base, lacking teeth at tip; capsule ovate, mucronate.

Swamps, muskeg. Described from Hudson Bay.

80. Scrophulariaceae (Figwort Family)

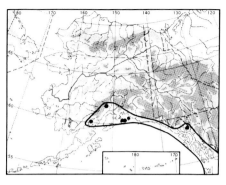 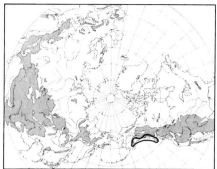 

9. Pediculàris parviflòra J. E. Sm.
subsp. **parviflòra**
   Pedicularis Pennellii subsp. *insularis* Calder & Taylor.

Glabrous; stem simple or mostly branched from base; basal leaves lacking; stem leaves oblong-lanceolate, deeply pinnately lobed, the segments dentate to double-dentate; bracts triangular, pinnately lobed with 3–4 segments and broad rachis; calyx in 2 parts, the segments laciniate and dentate; corolla rose-colored, galea tip darker; galea straight, with pair of broad, basal teeth, lacking subapical teeth; lower lip 3-lobed, broader than long, shorter than galea, with rounded, ciliate lobes; 2 filaments pubescent; capsule ovate, mucronate.

Swamps, muskeg. Described from northwest coast of North America.

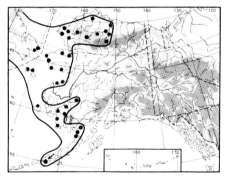 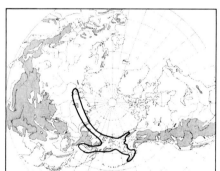 

10. Pediculàris parviflòra J. E. Sm.
subsp. **Pennéllii** (Hult.) Hult.
   Pedicularis Pennellii Hult.

Similar to subsp. *parviflora*, but galea with pair of long, subapical teeth. Wet places.

× ½

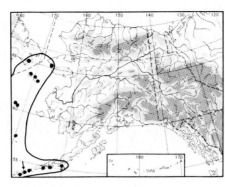
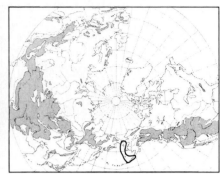

11. Pediculàris Langsdórffii Fisch.
subsp. **Langsdórffii**

Pedicularis purpurascens Cham.

Stem short, thick, from vertical taproot, with brown remains of leaf bases at base, sparsely lanate above or nearly glabrous; basal leaves with broad rachis, lanceolate to ovate-lanceolate, pinnately lobed into numerous short, remote, crenate segments; inflorescence dense, capitate to oblong; lower bracts similar to leaves, longer than flowers, the upper short; calyx with triangular, acute, dentate teeth, sometimes distally enlarged; corolla purple, glabrous, with straight tube and large, thick, curved galea with slender, acute teeth; lower lip 3-lobed, shorter than the upper; 2 filaments pubescent; capsule oblong-lanceolate, acuminate.

Meadows, stony ridges.

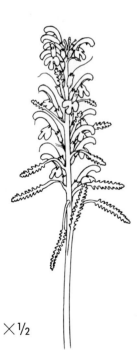
× ½

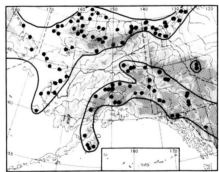
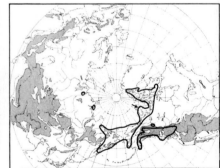

12. Pediculàris Langsdórffii Fisch.
subsp. **árctica** (R. Br.) Pennell

Pedicularis arctica R. Br.; *P. hians* Eastw.

Similar to subsp. *Langsdorffii*, but often taller, with narrower leaves and narrower rachis; spike narrower, less dense, often with long bracts also in upper part; corolla smaller, with narrower galea and smaller teeth.

In the mountains to at least 2,000 meters.

80. Scrophulariaceae (Figwort Family)

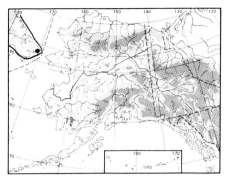
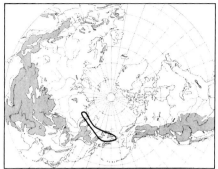

13. Pediculáris villòsa Ledeb.

Stems sparsely villous, from taproot; basal leaves pinnate, the segments again pinnate; stem leaves short-petiolated; inflorescence villous; lower bracts leaflike, the upper reduced; calyx with triangular-linear, acute segments, shorter than tube; corolla purplish; galea arcuate, with pair of teeth at apex; lip 3-lobed, somewhat shorter than galea; 2 filaments pubescent; capsule oblong to oblong-ovate.

Dry tundra. Described from Siberia.

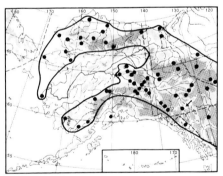
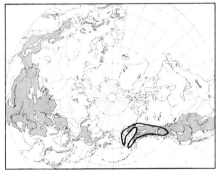

14. Pediculàris sudética Willd.
subsp. **intèrior** Hult.

Stems from short caudex with fusiform roots; basal leaves glabrous, long-petiolated, with broad rachis, pinnatifid to pinnate, segments dentate to pinnatifid; stem leaves single or lacking, short-petiolated; inflorescence pyramidal, at first capitate, in age elongated, densely woolly; lower bracts not dilated at base, lacking long lobes, middle bracts entire; calyx with triangular teeth; corolla purplish, the galea somewhat longer than tube and twisted, with pair of apical teeth; lower lip 3-lobed; filaments glabrous; capsule oblong, abruptly contracted into short beak.

Meadows, rocky slopes, in the mountains to at least 1,300 meters. *P. sudetica* described from the Sudet Mountains (Czechoslovakia).

The young shoots are eaten boiled in soup by the Siberian natives.

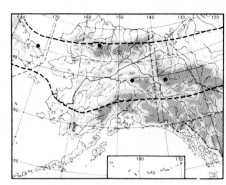
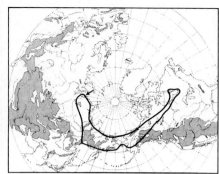

15. Pediculàris sudética Willd.
subsp. **interioìdes** Hult.

Similar to subsp. *interior*, but basal bracts broad, dilated at base, with small basal lobes, the middle and upper lobes entire; plant tall, straight; flowers purple.
Somewhat intermediate between subsp. *interior* and subsp. *albolabiata*.
Meadows, rocky slopes.

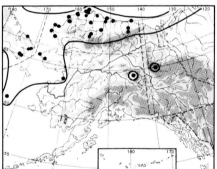
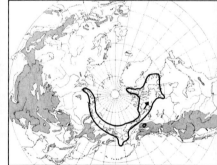

16. Pediculàris sudética Willd.
subsp. **albolabiàta** Hult.

Bracts dilated at base, both lower and middle bracts short, the lowest lacking prolonged, apical lobe, the spike profusely lanate; calyx teeth narrow; corolla pink; galea with purple apex; lip white or faintly pink, spotted.
Wet tundra.

80. Scrophulariaceae (Figwort Family)

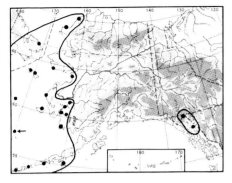
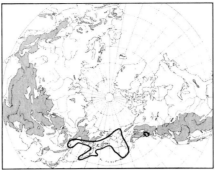

17. Pediculàris sudética Willd.
subsp. **pacífica** Hult.

Bracts strongly dilated at base, basal bracts with lateral lobes and long, apical lobe; inflorescence flat, inconspicuously pubescent or nearly glabrous; calyx glabrous, often with leaflike tips on teeth; flowers large.

Meadows, rocky slopes.

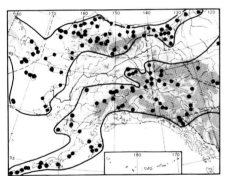
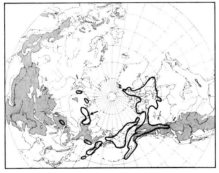

18. Pediculàris capitàta Adams
Pedicularis Nelsonii R. Br.

Stem pubescent with septate hairs, single, from thin rhizome; basal leaves long-petiolated, oblong in outline, pinnate, the pinnae pinnately cleft, the ultimate segments callus-toothed; stem leaves 1 to 2, less dissected and shorter-petiolated; inflorescence capitate; bracts pinnatifid; calyx with leaflike lobes; corolla yellowish or the upper lip becoming rose-colored in age; upper lip large, broad, curved, with very short beak, the lip twice as long as the straight tube; lower lip 3-cleft, pubescent at throat, half as long as the upper; filaments pubescent at base.

Rocky slopes, ridges, in the mountains, to at least 2,000 meters.

(See color section.)

80. Scrophulariaceae (Figwort Family)

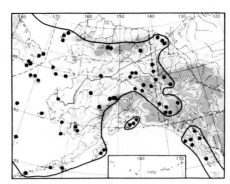
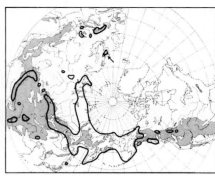

19. Pediculàris Oedéri M. Vahl
Pedicularis versicolor Wahlenb.; *P. flammea* with respect to Alaskan plant.

Stem thick, glabrous at base, from short caudex with several thick, yellowish-white, fusiform roots; base of leaves pinnately parted into several oblong to ovate, dentate segments; calyx with 10, more or less pubescent, veins and 5 triangular, lanceolate teeth, shorter than tube, sometimes distally enlarged; corolla glabrous, yellow, tip of upper lip brownish-red, with 2 red spots; lower lip 3-lobed; 2 filaments pubescent; capsule lanceolate-oblong, style protruding.

Meadows, rocky slopes in the mountains, to at least 1,200 meters.

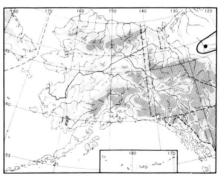
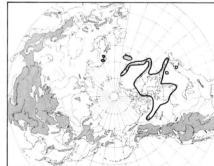

20. Pediculàris flámmea L.

Glabrous, often reddish, stem, simple, from short caudex with fusiform, yellowish roots; basal leaves linear-oblong, petiolated, pinnate, sections incised; inflorescence narrow, few-flowered; bracts short, linear-lanceolate; calyx as long as tube or longer, with 5 lanceolate teeth; corolla erect, yellow with blackish-red tip to galea; galea much longer than the small lower lip; style not protruding; capsule lanceolate, more or less falcate, 2–3 times longer than calyx.

Moist places. Described from Lapland (and Switzerland, where it does not occur).

80. Scrophulariaceae (Figwort Family)

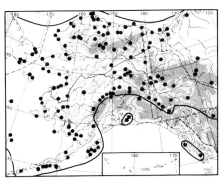  

21. Pediculàris Kànei Durand
subsp. **Kànei**

Pedicularis lanata of American authors; *P. lanata* subsp. *Kanei* (Durand) Pennell in label; *P. wildenowii* Vved.

Stem simple, mostly single, from thick, yellow taproot, with brown remains of leaf bases at base, lanate above; basal leaves numerous, linear-lanceolate, pinnately parted into many narrow, remote, pinnate to crenate-serrate segments; inflorescence dense, oblong, white-woolly; lower bracts longer than flowers; calyx with triangular, acute teeth; corolla rose-colored, glabrous, with straight tube and straight galea, lacking teeth; lower lip about as long as the upper, 3-lobed, glabrous in margin; 2 filaments pubescent; capsule ovate with acute beak.

Dry, stony tundra; in the mountains to over 1,700 meters. Described from Greenland.

Root edible, raw or boiled.

(See color section.)

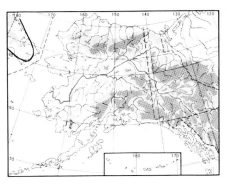 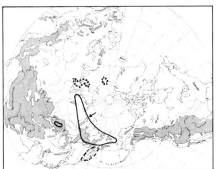 

22. Pediculàris Kànei E. Durand
subsp. **Adámsii** (Hult.) Hult.

Pedicularis Adamsii Hult.; *P. Langsdorffii* var. *calyce lanata* Stev. in part; *P. lanata* var. *alopecuroides* Trautv.

Similar to subsp. *Kanei*, but galea pubescent and lip ciliated in the margin.
Stony tundra.

Dotted line on circumpolar map indicates range of subsp. **dasyántha** (Trautv.) Hult. (*P. lanata* var. *dasyantha* Trautv.), broken line that of subsp. **Pallásii** (Vved.) Hult. (*P. Pallasii* Vved.).

81. OROBANCHACEAE (Broomrape Family)

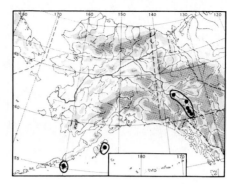 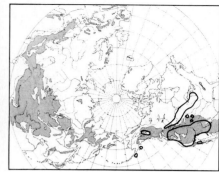

Orobanche L.

1. Orobánche fasciculàta Nutt.   Broomrape
Thalesia fasciculata (Nutt.) Britt.

Caudex subligneous, forked; stems solitary or clustered, purplish or yellowish, glandular-puberulent to villous, with few remote bracts; pedicels axillary, naked; calyx with 5 triangular, acute lobes equal to or shorter than tube; corolla purple or yellowish, with rounded or acute lobes.

Parasitic on *Artemisia*. Described from Fort Mandan (on the Missouri River).

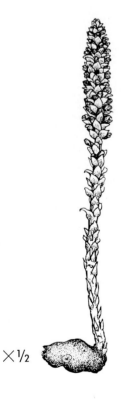

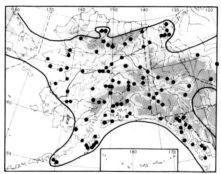 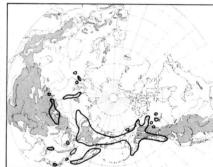

Boschniakia C. A. Mey.

1. Boschniàkia róssica (Cham. & Schlecht.) Fedtsch.
Orobanche rossica Cham. & Schlecht.; *Boschniakia glabra* Bong.

Nearly glabrous, brownish; stems several, coarse, with incrassated base; inflorescence longer than stem, bracts scalelike, broadly elliptic to ovate, ciliate; calyx half as long as corolla, with irregular lobes; corolla brownish-red, ciliated in margin; upper lip entire or emarginate, the lower short, 3-lobed; stamens exserted.

Parasitic on the roots of *Alnus crispa*; in the mountains to at least 1,200 meters. Described from Kotzebue Sound and Siberia.

82. Lentibulariaceae (Bladderwort Family)

Pinguicula L. (Butterwort)

Scape villous, at least below; flowers about 7 mm long .... 3. *P. villosa*
Scape glandular-puberulent; flowers considerably larger:
    Lobes of lower lip of corolla oblong, not overlapping; spur small, acute ....
        .... 1. *P. vulgaris* subsp. *vulgaris*
    Lobes of lower lip of corolla obovate-oblong, often overlapping; spur stout, blunt
        .... 2. *P. vulgaris* subsp. *macroceras*

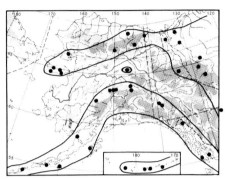 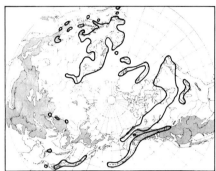 

1. Pinguícula vulgàris L.
subsp. vulgàris

Pinguicula microceras Cham.; *P. arctica* Eastw.; *P. macroceras* var. *microceras* (Cham.) Casper.

Scape from light-colored, fibrous roots; leaves succulent, short-petiolated, oblanceolate to elliptic, shiny above, forming rosette; scape glandular when young; corolla blue to violet, with funnel-formed tube and white hairs in throat; lobes of lips not overlapping; spur thin, acute; capsule spherical.

Moist places. Described from Europe.

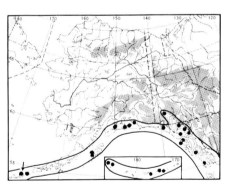 

2. Pinguícula vulgàris L.
subsp. macrocèras (Link) Calder & Taylor

Pinguicula macroceras Link; *P. vulgaris* var. *macroceras* (Link) Herder.

Similar to subsp. *vulgaris*, but corolla larger, with often overlapping obovate lobes; spur longer, blunt.

Moist places.

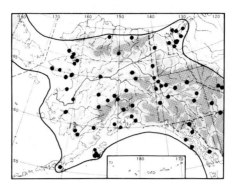
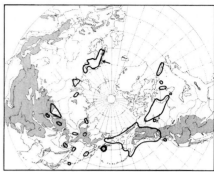

3. Pinguícula villòsa L.

Scape single, villous below, glandular above, from thin roots; rosette leaves broadly ovate to elliptic, with revolute margin, glabrous below, glandular above; flower nodding, 5–7 mm long; calyx with acute, triangular lobes, corolla bluish-violet with darker lines and yellow spots at base of lower lip; spur straight, blunt; capsule ovate to spherical.

Peat bogs around pools and creeks, to at least 500 meters.

Utricularia L. (Bladderwort)

Leaves all with bladders; stem about 1 mm thick .. 1. *U. vulgaris* subsp. *macrorhiza*
Leaves not all with bladders:
    Segments of leaves serrate; winter buds pubescent .... 2. *U. intermedia*
    Segments of leaves entire; winter buds glabrous .... 3. *U. minor*

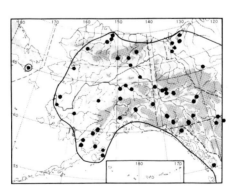
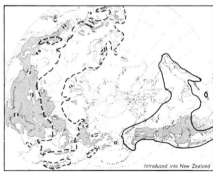

1. Utriculària vulgàris L.
subsp. **macrorhìza** (Le Conte) Clausen

Utricularia macrorhiza Le Conte; *U. vulgaris* var. *americana* Gray.

Free-floating; stem about 1 mm thick; leaves much dissected, with nearly terete segments, all with numerous bladders forming small traps for insects; scapes coarse, with 5–6 (or more) flowers; corolla dark yellow, lower lip slightly lobed, with well-developed palate; spur 7–12 mm long, falcate, slender, somewhat acute, directed forward.

Ponds, quiet waters. *U. vulgaris* described from Europe, subsp. *macrorhiza* from "Canada to Carolina."

Broken line on circumpolar map indicates range of subsp. **vulgàris**.

82. Lentibulariaceae (Bladderwort Family)

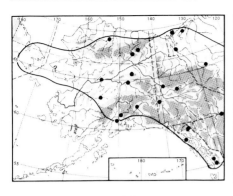 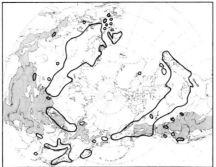 

2. Utriculària intermèdia Hayne

Free-floating or mostly creeping in mud; stem slender, about 0.4 mm thick; leaves with flat, minutely spiny-serrate, more or less blunt segments; winter buds pubescent; bladders on separate, leafless branches; corolla light yellow; spur conical-saccate, acute.

Muddy and peaty shores, shallow water. Described from Berlin and Uppsala.

U. ochroleùca R. Hartm., usually regarded as the hybrid *U. intermedia × minor*, probably occurs within area of interest. Segments of the leaves are elongated, acute, with few spines in the margin. No flowering specimens seen.

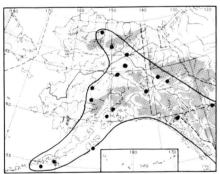 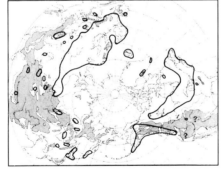 

3. Utriculària mìnor L.

Free-floating or creeping in mud; stem slender, about 0.4 mm thick; leaves with narrow, flat, not serrulate segments, some leaves bearing bladders, others not; winter buds glabrous; corolla 5–8 mm long, greenish-yellow with brown stripes; spur saccate, very short; capsule subglobose.

Quiet water. Described from Europe.

83. PLANTAGINACEAE (Plantain Family)

Plantago L. (Plantain)

Leaves linear:
    Bracts broadly oblong to ovate .... 2. *P. maritima* subsp. *juncoides*
    Bracts linear, much longer than calyx .... 5. *P. aristata*
Leaves lanceolate to ovate:
    Seeds 6 or more in capsule:
        Leaves thick, cordate-ovate, spreading .... 7. *P. major* var. *major*
        Leaves thin, elliptic-ovate, upright .... 8. *P. major* var. *Pilgeri*
    Seeds 2–4:
        Brown wool at crown of stout caudex; spike interrupted at base ....
        .... 4. *P. eriopoda*
        Brown wool lacking at base; spike not interrupted at base:
            Petioles broad; leaves glabrous or essentially so; root very stout ....
            .... 1. *P. macrocarpa*
            Petioles slender; leaves ciliated or pubescent also on surfaces; root thinner:
                Bracts obtuse .... 3. *P. canescens*
                Bracts caudate .... 6. *P. lanceolata*

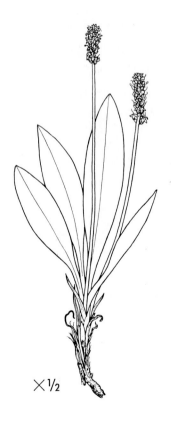

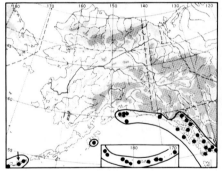

1. Plantàgo macrocárpa Cham. & Schlecht.

Scapes from thick, stout, vertical root with long, broad sheaths at crown; leaves glabrous, erect, lanceolate, somewhat acute, elongated into broad petiole; scapes canescent below inflorescence; spike up to 10 cm long; bracts rounded, erose in margin, dark-brown sepals ovate to elliptic, glabrous; corolla with ovate to triangular-ovate lobes and tube 3 mm long, nerveless; capsules large, elliptic, lacking lid, indehiscent, falling off entire; seeds 2.

Wet places, beaches.

83. Plantaginaceae (Plantain Family)

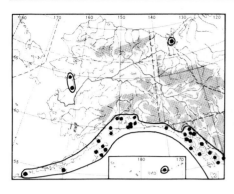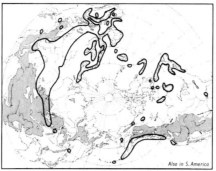

2. Plantàgo marítima L.
subsp. juncoìdes (Lam.) Hult.
Plantago juncoides Lam.

Plant with long, thick root; leaves glabrous, linear-lanceolate, entire or denticulate, fleshy, attenuate, with winged petiole; bracts broadly oblong to ovate, ciliolate; tube of corolla pubescent; capsule ellipsoid to broadly conic, circumscissile below tips of sepals; seeds 2–4.

Seacoasts, salt marshes. *P. maritima* described from Europe, subsp. *juncoides* from Straits of Magellan.

Circumpolar map gives range of entire species complex.

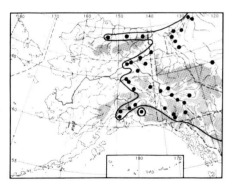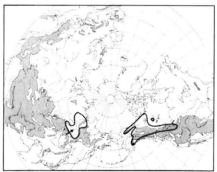

3. Plantàgo canéscens Adams
Plantago Richardsonii Decne.; *P. septata* Morris.

Plant with taproot; caudex covered with withered sheaths; leaves erect, narrowly lanceolate, sparsely pubescent on both sides with septate hairs; scapes woolly below; spikes dense, 3–6 cm long; bracts ovate to suborbicular, scarious-margined; calyx lobes obovate to oblong, lacerate, ciliolate; tube of corolla as long as sepals, the lobes oblong, obtuse, denticulate, whitish; stamens exserted; capsule circumscissile about one-third of the way from base; seeds 1–2.

Grassy slopes, open soil.

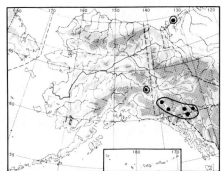
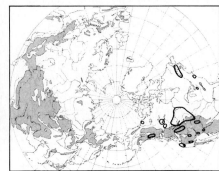

4. Plantàgo eriópoda Torr.

Plant with stout caudex, brown-woolly at crown; leaves fleshy or subcoriaceous, lanceolate to lance-ovate, tapering into broad petiole; lower flowers remote; bracts broadly ovate, with rounded keel; sepals rounded-elliptic, truncate, scarcely keeled; capsule circumscissile near base; seeds 2–4 in capsule.

Salt or alkaline soil. Described from the Platte River.

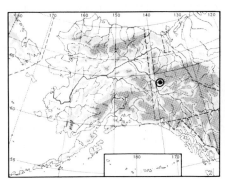
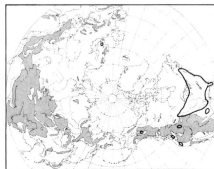

5. Plantàgo aristàta Michx.

Plant with taproot, dark green, loosely villous; leaves linear; petioles more or less clasping; scapes slender; bracts linear, long, divergent; sepals spatulate-oblong; lobes of corolla round to ovate; seeds 2, finely pitted.

Found once as a weed at Dawson; probably not persistent. Described from Illinois.

83. Plantaginaceae (Plantain Family)

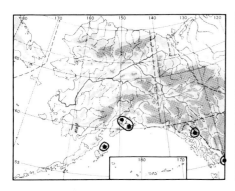

6. Plantàgo lanceolàta L.   Ribgrass

Plant with stout, erect caudex and fibrous roots; leaves lanceolate to lanceolate-oblong, remotely denticulate; scape sulcate; spike dense, at first ovoid-conic, in fruit short-cylindric; bracts scarious, broadly ovate, with caudate-acuminate tip, erose-undulate in margin; anterior sepals connate; corolla lobes yellowish-brown; stamens conspicuously exserted; capsule with 1–2 seeds.

Waste places; introduced. Described from Europe.

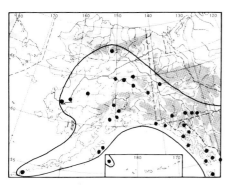
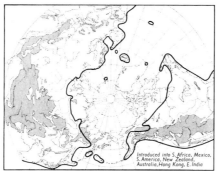
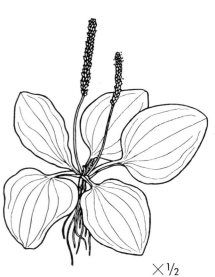

7. Plantàgo màjor L.   Common Plantain
var. **màjor**

More or less pubescent, leaves broadly elliptic to cordate-ovate, obtuse, with winged petioles; spike dense; bracts acute, brownish, scarious-margined, corolla yellowish-white with triangular lobes; capsule circumscissile below tips of sepals, mostly 6–10-seeded, seeds reticulate.

Waste places, roadsides; introduced. Described from Europe.

Exceedingly variable. Circumpolar map indicates range of *P. major* in broadest sense. Occurs also in Galapagos Islands.

83. PLANTAGINACEAE (Plantain Family) / 84. RUBIACEAE (Madder Family)

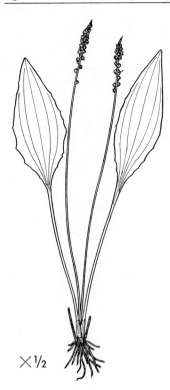

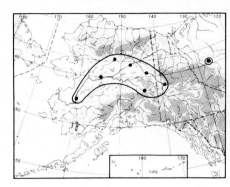

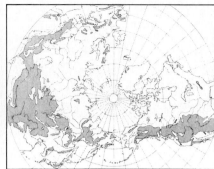

8. Plantàgo màjor L.
var. **Pílgeri** Domin

Leaves thin, upright, subglabrous, elliptic-ovate, with slender petioles; capsule conic above, circumscissile near tips of sepals.

Possibly native. Described from Bohemia. Total range unknown.

Galium L. (Bedstraw)

Leaves blunt, not mucronate; fruit glabrous or bristly:
  Fruit bristly; leaves 3-nerved:
    Leaves lanceolate to elliptic .... 1. *G. boreale*
    Leaves broadly ovate to obovate .... 5. *G. kamtschaticum*
  Fruit glabrous; leaves 1-nerved:
    Corolla lobes 4; flowers several, in cyme, spreading .... 6. *G. palustre*
    Corolla lobes mostly 3; flowers single or 2–3 in cyme, ascending:
      Leaves 5–6 in whorls on main stem, 4 on branches, broad, slightly scabrous
      .... 8. *G. trifidum* subsp. *columbianum*
      Leaves 4 in whorls:
        Pedicels arcuate, thin, retrorsely scabrous ....
        .... 7. *G. trifidum* subsp. *trifidum*
        Pedicels short, straight, glabrous, thick in fruit .... 9. *G. Brandegei*
Leaves mucronate; fruit hispid or bristly:
  Leaves 6 in whorl .... 4. *G. triflorum*
  Leaves more than 6, usually 8, in whorl:
    Stem not prickly; flowers yellow .... 2. *G. verum*
    Stem with downward-directed prickles; flowers whitish .... 3. *G. aparine*

84. Rubiaceae (Madder Family)

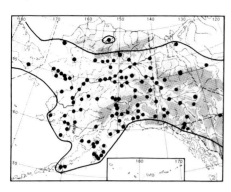 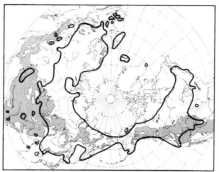 

1. Gàlium boreàle L.

Galium septentrionale Roem. & Schult.; *G. boreale* subsp. *septentrionale* (Roem. & Schult.) Iltis.

Stems erect, quadrangular, branched, from creeping rhizome; leaves 4 in whorl, linear-lanceolate to elliptic, 3-veined, sessile, often with sterile branches in axils; flowers numerous in terminal panicles; corolla white, 4-lobed; fruit densely hispid in Alaska–Yukon plant.

Stony slopes, meadows; common in the interior. Described from Europe.

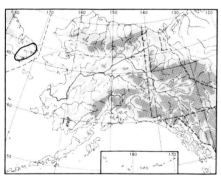 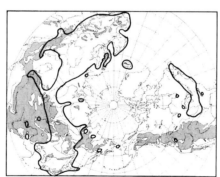 

2. Gàlium vèrum L.

Stems glabrous or sparsely pubescent, branched, 4-angled from stoloniferous, slender rhizome; leaves linear, mucronate, 1-veined, with revolute margins, rough above, pubescent beneath; flowers in terminal leafy panicle; corolla yellow, with 4 apiculate lobes; fruit smooth, glabrous; plant coumarin-scented.

Dry, grassy slopes. Described from Europe. Introduced on the Chukchi Peninsula.

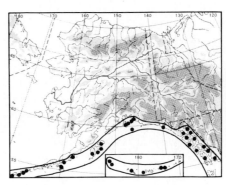
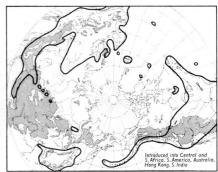

3. Gàlium aparìne L.

Weak plant; stem branched, quadrangular, prostrate or reclining, with downward-directed prickles on angles; leaves 6–8 in a whorl, linear-oblanceolate, bristle-tipped, 1-nerved, with backward-directed prickles in margin; pedicels 1–3-flowered, the flowers subtended by leaflike bracts; corolla whitish, with 4 acute, free lobes; fruit with hooked bristles.

Seashores; native in Alaska. Described from Europe.

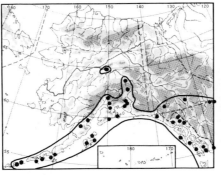
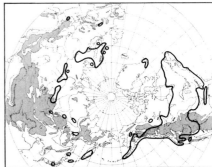

4. Gàlium triflòrum Michx.   Sweet-Scented Bedstraw

Dark-green plant; stems from creeping rhizome, more or less retrorsely scabrous, decumbent, simple or remotely forking; leaves elliptic-lanceolate, cuspidate, 1-nerved, 6 in a whorl; peduncles axillary, mostly 3-flowered; flowers pedicellated, small, greenish-white; fruit densely hooked, bristly.

Moist woods, thickets. Described from Canada.

Coumarin-scented after drying.

84. Rubiaceae (Madder Family)

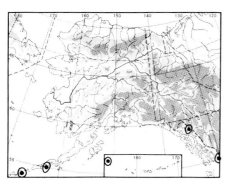
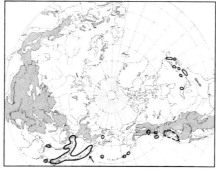

5. Gàlium kamtscháticum Steller
Galium rotundifolium var. *kamtschaticum* (Steller) Kuntze.

Stems glabrous, from filiform, stoloniferous rhizome; leaves 3-veined, broadly ovate or obovate, obtuse, short-mucronate, 4 in a whorl, the lower smaller than the upper; peduncles 1–3, 2–3-flowered, from uppermost whorl; calyx lobes ovate-lanceolate; corolla greenish-white with acute lobes; fruit with hooked bristles.

Moist, mossy places. Described from Kamchatka.

In the Cascades of Oregon and Washington, var. **oregànum** (Britt.) Piper (*G. oreganum* Britt.) occurs; its range is indicated by broken line on circumpolar map.

Very closely related to the European **G. scàbrum** L.

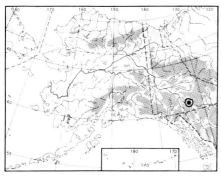
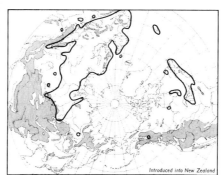

6. Gàlium palústre L.

Stems glabrous, quadrangular, weak, from creeping rhizome; leaves 1-veined, 4–6 in a whorl; broadly oblanceolate, blunt to subacute, with small, backward-directed prickles on margins, turning black when dried; pedicels glabrous; flowers in axillary cymes, forming pyramidal panicle; corolla white with 4 acute, free lobes; anthers red; fruit glabrous, rugose.

Wet places; introduced in one place. Described from Europe.

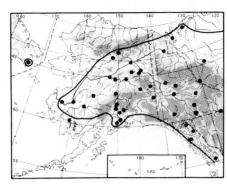
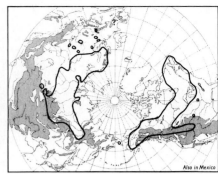

7. Gàlium trífidum L.
subsp. **trífidum**

Stems weak, slender, branched, the upper internodes scabrous, from slender rhizome; leaves 4 in a whorl, linear to linear-oblanceolate, blunt, retrorsely scabrous-margined; flowers solitary or 3 together on capillary, arcuate, scabrous pedicels; corolla whitish with obtuse lobes; fruit smooth.

Wet places. Described from Canada.

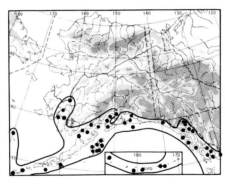

8. Gàlium trífidum L.
subsp. **columbiànum** (Rydb.) Hult.

Galium columbianum Rydb.; *G. trifidum* var. *pacificum* Wieg.; *G. trifidum* var. *subbiflorum* Wieg.; *G. tinctorium* var. *subbiflorum* (Wieg.) Fern.; *G. trifidum* var. *brevipedunculatum* Regel; *G. baicalense* Pobed.

Similar to subsp. *trifidum*, but coarser; leaves mostly 5–6 in a whorl; flower solitary or in pairs on each peduncle; pedicels strongly arcuate.

Moist places. Described from Montana to Washington and Alaska.

84. Rubiaceae (Madder Family) / 85. Caprifoliaceae (Honeysuckle Family)

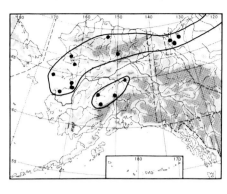
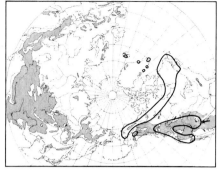

9. Gàlium Brandegèi Gray
Galium trifidum var. *pusillum* Gray.

Similar to *G. trifidum* subsp. *trifidum*, but low and matted, with simple or forked, more or less glabrous stem; leaves glabrous, somewhat fleshy; pedicels glabrous, straight or nearly straight, thickening in fruit.

Moist places. Described from New Mexico.

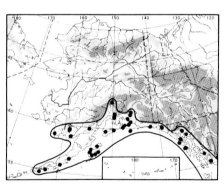
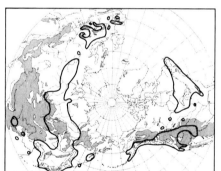

Sambucus L.

1. Sambùcus racemòsa L.
subsp. **pùbens** (Michx.) House
Sambucus pubens Michx.
var. **arboréscens** Gray
Sambucus callicarpus Greene.

Shrub; twigs with soft, pithy, warty, brown bark; branchlets pubescent; leaves mostly 5–7-foliate, the leaflets lanceolate to ovate-lanceolate, acuminate, sharply serrate, downy beneath; inflorescence pyramidal or ovoid; flowers yellowish-white, ill-scented; fruit globose, bright red.

Woods and subalpine meadows. *S. racemosa* described from Europe, subsp. *pubens* from Pennsylvania, Canada, and the Carolinas. Circumpolar map gives range of entire species complex.

The seeds are poisonous, causing diarrhea and vomiting.

85. Caprifoliaceae (Honeysuckle Family)

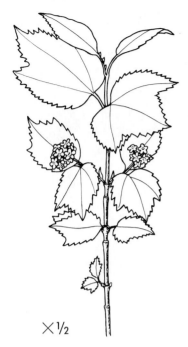

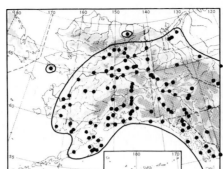 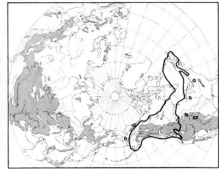

Viburnum L.

1. **Vibúrnum edùle** (Michx.) Raf.　　　　　　　　　　　High Bush Cranberry
   Viburnum opulus var. *edule* Michx.; *V. pauciflorum* La Pylaie; *V. acerifolium* Bong.

Straggling to suberect shrub, up to 2.5 meters tall, with glabrous branches; leaves elliptic to suborbicular, shallowly 3-lobed, sharply toothed, commonly with pair of glands near junction with petiole; inflorescence dense; flowers uniform, perfect, milk-white, stamens included; fruit red or orange, 1–1.5 cm long, subglobose, acid, juicy, with large, flattened stone.

Woods, thickets, to at least 800 meters. Described from Canada.

The fruit is edible, and makes a good jam; should be picked not quite ripe.

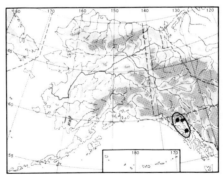 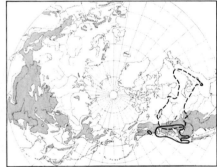

Symphoricarpus Duham.

1. **Symphoricárpus álbus** (L.) Blake　　　　　　　　　　Snowberry
   Vaccinium album L.
   subsp. **laevigàtus** (Fern.) Hult.
   Symphoricarpus racemosus var. *laevigatus* Fern.; *S. albus* var. *laevigatus* (Fern.) Blake; *S. rivularis* Suksd. .

Erect shrub, up to 2 meters tall, with essentially glabrous twigs; leaves elliptic to elliptic-ovate, entire or obscurely toothed, glabrous above, glabrous or very sparsely pubescent beneath; racemes short, subsessile, few-flowered; corolla pink and white, campanulate, short and broad, nearly as wide as long, pubescent within; fruit berry-like, white, subglobose to elliptic.

Woods, thickets. *S. albus* described from "Pennsylvania," subsp. *laevigatus* from cultivated specimens grown from seeds collected west of the Rocky Mountains.

Broken line on circumpolar map indicates range of subsp. **álbus**.

85. CAPRIFOLIACEAE (Honeysuckle Family)

Linnaea Gronov. (Twinflower)

Leaves elliptical, acute; calyx and corolla long and narrow .... 3. *L. borealis* subsp. *longiflora*
Leaves ovate to roundish, often somewhat acute; calyx and corolla shorter:
  Corolla campanulate, lacking tube or with tube shorter than calyx .... 1. *L. borealis* subsp. *borealis*
  Corolla funnel-shaped, with tube about as long as calyx .... 2. *L. borealis* subsp. *americana*

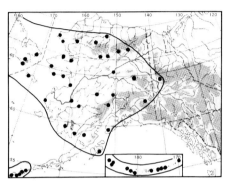  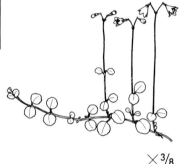

1. Linnaèa boreàlis L.
subsp. **boreàlis**

Linnaea serpyllifolia Rydb.

Stem about 1 mm thick, up to 1 meter long, creeping, branched, somewhat woody; young twigs pubescent and often glandular; leaves leathery, broadly ovate to roundish, often somewhat acute, mostly with 1–3 acute teeth on each side, sparsely ciliated; flowering shoots with 2 (–4) pairs of leaves; peduncle glandular, with pair of small bracts at summit, forking into pair of glandular pedicels, each bearing a fragrant, whitish to rose, nodding, campanulate flower, lacking tube or with tube shorter than calyx.

Woods, heaths, and dry ridges in the mountains to at least 1,200 meters in McKinley Park. Described from Sweden, Switzerland, Siberia, and Canada.

× ³⁄₈

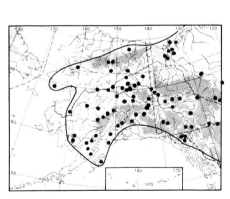 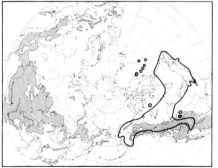 

2. Linnaèa boreàlis L.
subsp. **americàna** (Forbes) Hult.   American Twinflower

Linnaea americana Forbes; *L. borealis* var. *americana* (Forbes) Rehd.

× ³⁄₈

Similar to subsp. *borealis*, but corolla more funnel-shaped, with tube, flaring from above calyx.

Described from America. Transitions to subsp. *borealis* occur in Alaska.

85. Caprifoliaceae (Honeysuckle Family)

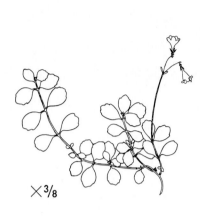

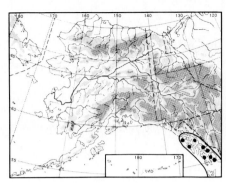 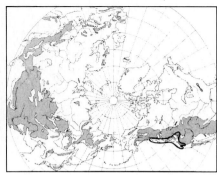

3. Linnaèa boreàlis L.
subsp. **longiflòra** (Torr.) Hult. Pacific Twinflower

Linnaea borealis var. *longiflora* Torr.; *L. longiflora* (Torr.) How.

Differs from other 2 races of *L. borealis* in having coarser stem and leaves, with elliptical, acute leaves, long, narrow calyx, and long, narrow corolla.

Described from Oregon to Alaska and the Rocky Mountains. Subsp. *americana* is more closely related to subsp. *borealis*.

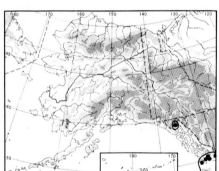 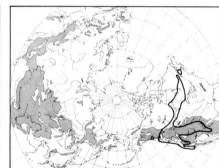

Lonicera L.

1. Lonícera involucràta (Richards.) Banks
Xylosteum involucratum Richards.

Straggling to ascending shrub; branches quadrangular, glabrous; leaves ascending, obovate, to elliptic-oblong, acuminate or acute, paler beneath; involucre of 4 green to dark-purple bracts; flowers in pairs; corolla glandular-cylindric, yellow, sometimes tinged with red; berries purplish-black, about 1 cm thick, bitter.

Woods. Described from "wooded country between 54° and 64° lat."

86. Adoxaceae (Moschatel Family) / 87. Valerianaceae (Valerian Family)

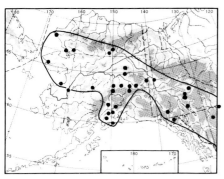
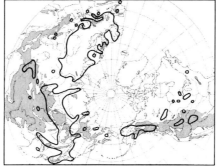
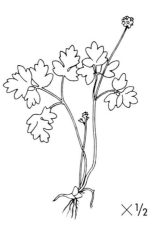

Adoxa L.

1. Adóxa moschatéllina L. Moschatel

Glabrous; stem from creeping rhizome with white, thick scales; basal leaves thin, yellowish-green, long-petiolated, 1 to 3 times ternate, with broad, toothed, ultimate lobes; cauline leaves 3-cleft or parted; flowers yellowish-green in angular head, the apical with 4 styles and 8 anthers, the lateral with 5 styles and 10 anthers.

Woods, thickets, mostly on calcareous soil; in McKinley Park to about 1,000 meters. Described from Europe.

Valeriana L. (Valerian)

Corolla of pistillate flowers 2–3 mm long; stem leaves pinnate . . . .
. . . . 3. *V. dioica* subsp. *sylvatica*
Corolla of pistillate flowers 5–8 mm long; stem leaves entire, 3-parted or pinnate:
    Upper stem leaves sessile, simple or 3-lobed; bractlets of inflorescence glabrous, or pubescent only at base . . . . 1. *V. capitata*
    Upper stem leaves petiolated, 3–5-foliate; bractlets of inflorescence ciliate . . . . .
. . . . 2. *V. sitchensis*

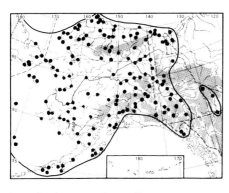
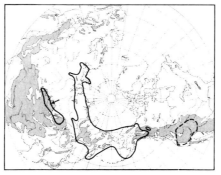
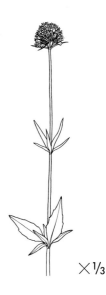

1. Valeriàna capitàta Pall.

Valeriana bracteosa Britt.; *V. capitata* var. *bracteosa* (Britt.) Hult.

Stems single, pubescent at nodes, from long, creeping rhizome; leaves glabrous, the basal long-petiolated, entire, elliptic to ovate or subcordate; stem leaves opposite, shallowly 3-lobed, sessile, or the lower short-petiolated; inflorescence subtended by large, 3-lobed bracts, densely capitate in flower, becoming an open, diffuse panicle in fruit; bractlets linear, nearly as long as calyx; flowers 5–8 mm long, white or lilac, narrowly funnelform; calyx bristles plumose in fruit.

Moist places, in the mountains to at least 1,500 meters.

Alaskan specimens belong to subsp. *capitata*. Broken line on circumpolar map indicates range of other subspecies.

87. VALERIANACEAE (Valerian Family)

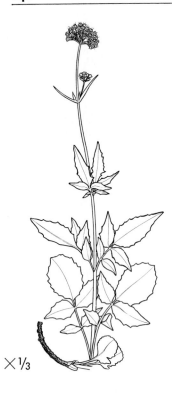

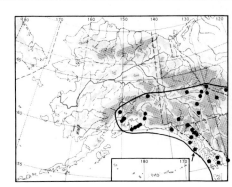

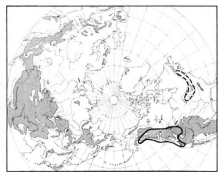

2. Valeriàna sitchénsis Bong.

Stem from stout rhizome; leaves chiefly cauline, the lower pairs reduced, the middle petiolate, pinnatifid with large terminal segment, the upper sessile, reduced; inflorescence compact, becoming diffuse in fruit; bractlets pubescent; corolla white, its lobes much shorter than the somewhat gibbous tube; stamens exserted; calyx segments 12–20, plumose.

Moist meadows in the lowlands. Broken line on circumpolar map indicates range of subsp. **uliginòsa** (Torr. & Gray) F. G. Meyer (*V. sylvatica* var. *uliginosa* Torr. & Gray).

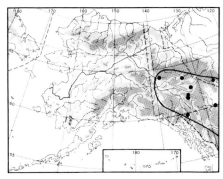

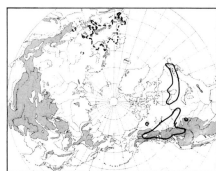

3. Valeriàna diòica L.
subsp. **sylvàtica** (Soland.) F. G. Meyer

Valeriana sylvatica Soland. not Schmidt; *V. septentrionalis* Rydb.

Stems from stout, branched rhizome, glabrous or nearly so; basal leaves glabrous, petiolate, simple, oblong to narrowly ovate; cauline leaves subsessile, pinnatifid, the terminal lobe ovate-oblong; inflorescence compact, diffuse in fruit; some flowers perfect, others pistillate; corolla white, lobes about as long as tube; stamens exserted; plumose calyx segments 9–15.

Moist places. Described from the Clearwater River, northern Alberta. Broken line on circumpolar map indicates range of subsp. **diòica**.

88. Campanulaceae (Bluebell Family)

Campanula L. (Bellflower)

Calyx pubescent; plant 1-flowered:
  Calyx lobes laciniate; leaves dentate; corolla lobes not ciliate:
    Sepals lanceolate to linear, laciniate .... 1. *C. lasiocarpa* subsp. *lasiocarpa*
    Sepals broadly triangular, more laciniate and dentate ....
    .... 2. *C. lasiocarpa* subsp. *latisepala*
  Calyx lobes entire; leaves crenate or entire; corolla lobes ciliate or not:
    Calyx with reflexed appendages between broad lobes; flowers large, broad, 23–30 mm long; lobes ciliated .... 3. *C. Chamissonis*
    Calyx lacking reflexed appendages; flowers small, narrow, about 15 mm long; lobes not ciliated .... 8. *C. uniflora*
Calyx glabrous; plant 1- to many-flowered:
  Corolla rotate, cleft to below middle; calyx lobes with pair of teeth near base
  .... 4. *C. aurita*
  Corolla campanulate, not so deeply cleft; calyx lobes entire:
    Style much exceeding corolla .... 5. *C. Scouleri*
    Style as long as corolla or shorter:
      Sepals lanceolate or linear; stem leaves lanceolate or linear ....
      .... 6. *C. rotundifolia*
      Sepals triangular, 2–3 mm broad at base; upper stem leaves petiolated, crowded, broad, ovate-lanceolate, more or less serrulate ....
      .... 7. *C. latisepala*

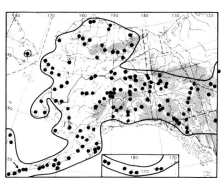 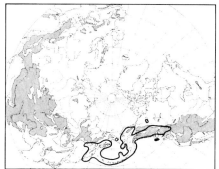 

1. Campánula lasiocárpa Cham.
subsp. lasiocárpa

Stems from thin, branching rhizome, thicker at tips, with both flowering and leafy rosettes; basal leaves long-petiolated, oblanceolate, acute, entire or usually coarsely dentate; stem leaves sessile, gradually reduced; flowers single, large, blue; calyx white-pubescent, with lanceolate to linear, acute, more or less laciniate segments; stigma 3-lobed; corolla glabrous, not ciliated in margin; capsule oblong, pubescent.

Alpine heaths, sandy tundra, in the mountains to at least 1,600 meters.

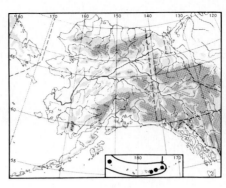
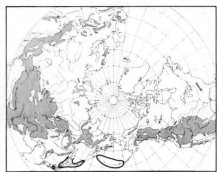

2. Campánula lasiocárpa Cham.
subsp. **latisèpala** (Hult.) Hult.

Campanula lasiocarpa var. *latisepala* Hult.

Similar to subsp. *lasiocarpa*, but sepals broadly triangular, more laciniate, and dentate.

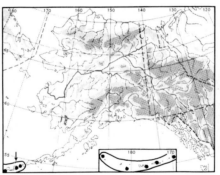
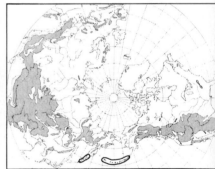

3. Campánula Chamissònis Fedorov

Campanula dasyantha of American authors; *C. dasyantha* var. *Chamissonis* Toyokuni & Nosaka.

Stems from thin, branching rhizome, the branch thicker in apex; basal leaves obovate, somewhat acute, crenate, nearly glabrous, the upper reduced; flowers large; calyx pubescent, with entire, ovate-lanceolate, broad segments, with small, acute appendages in sinuses; corolla blue, ciliated in margin, bearded within; stigma 3-lobed.

Stony tundra.

88. Campanulaceae (Bluebell Family)

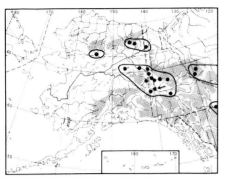  

4. Campánula aurìta Greene

Stems straight, erect, slender, several from branching rootstock; basal leaves lacking or reduced; stem leaves lanceolate, entire or irregularly and coarsely toothed, glabrous or ciliated at base; flowers solitary or few at top of stem; calyx lobes attenuate, triangular, entire or with pair of irregular lobes at base; corolla blue, cleft nearly to base, the lobes lanceolate, surpassing calyx lobes; style above and stigma on outside, with long, white, apparently viscid pubescence soon falling off; stigma lobes 3.

Rock crevices. Related to **C. Pìperi** How., of the Olympic Mountains, Washington.

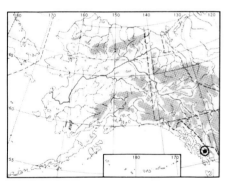 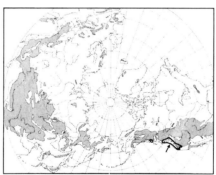 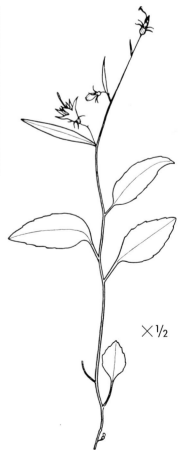

5. Campánula Scoulèri Hook.

Stems from slender, branching rhizome; lower leaves petiolated, ovate to rounded, sharply serrated, passing into linear bracts in inflorescence; pedicels elongate; corolla pale blue, with ovate-oblong, more or less recurved lobes, about as long as tube; style exserted; capsule subglobose.

Woods, rocks.

88. CAMPANULACEAE (Bluebell Family)

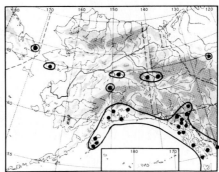
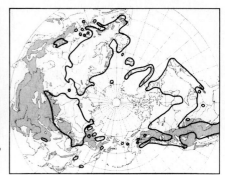

6. Campánula rotundifòlia L.   Harebell, Bluebell

Campanula petiolata DC.; *C. rotundifolia* var. *petiolata* (DC.) Henry. Including *C. Langsdorffiana* Fisch.

Stems glabrous or essentially so, from slender, branched caudex; basal leaves petiolated, ovate to rounded-cordate, angular-toothed, soon withering and rarely present in flowering specimens; cauline leaves narrow, linear or narrowly lanceolate; calyx lobes subulate, entire, reflexed; corolla purplish-blue, rarely white; capsule nodding, oblong, opening through pores at base.

Grassy slopes; probably introduced in the northern isolated localities. Described from Europe.

Highly variable.

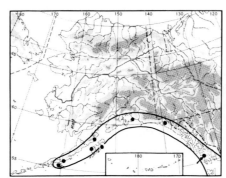
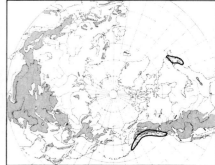

7. Campánula latisèpala Hult.

Campanula rotundifolia subsp. *alaskana* sensu Shetler.

Similar to *C. rotundifolia*, but differs in having dense, ovate-lanceolate, petiolated and somewhat serrulated cauline leaves and triangular calyx lobes, 2–3 mm broad at base.

Grassy slopes.

Forms hybrid swarms with *C. rotundifolia* (*C. rotundifolia* var. *alaskana* Gray; *C. heterodoxa* Bong.; *C. latisepala* var. *dubia* Hult.).

88. Campanulaceae (Bluebell Family) / 89. Compositae (Composite Family)

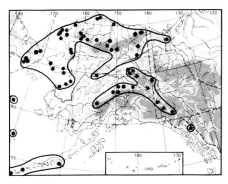
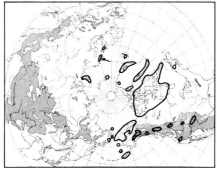

8. Campánula uniflòra L.

Plant with thick taproot; caudex ligneous; stem simple, often decumbent at base; leaves glabrous, subcoriaceous, crowded at base, oblong to spatulate, the upper lanceolate; flowers solitary, often nodding, becoming erect in fruit; corolla pale blue, narrow, with narrow segments somewhat larger than calyx lobes; capsule clavate, opening with pores near top.

Dry, stony ridges on tundra and in the mountains, to at least 1,500 meters; occurrences generally very scattered.

Solidago L. (Goldenrod)

Rhizome long, creeping; basal leaves lacking; stem leaves numerous, more or less crowded:
  Leaves glabrous above or nearly so; upper leaves reduced; panicle often with divergent racemes; heads about 4 mm high .. 5. *S. canadensis* var. *salebrosa*
  Leaves pubescent on small veins above; upper leaves not reduced, coarsely and sharply serrated; panicle with erect branches; heads larger .... 4. *S. lepida*
Rhizome short, thick; basal leaves several; stem leaves mostly few:
  Petioles glabrous or with markedly scattered hairs in margin; involucre imbricate; outer phyllaries much shorter than the inner .... 3. *S. decumbens* var. *oreophila*
  Petioles ciliated with short, stiff hairs, or with long, soft, villous hairs; outer phyllaries at least half as long as the inner:
    Petioles of basal leaves lacking (or nearly lacking) ciliation in margin .... 2. *S. multiradiata* var. *scopulorum*
    Petioles of basal leaves with ciliation of long, soft, white hairs:
      Upper stem leaves reduced, not exceeding flowers .... 1. *S. multiradiata* var. *multiradiata*
      Upper stem leaves equaling or exceeding the compact inflorescence .... See 1. *S. multiradiata* var. *arctica*

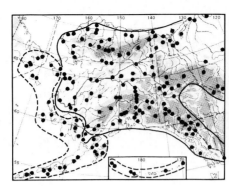
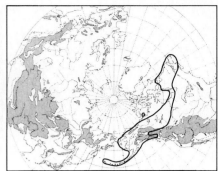

× ⅓

1. Solidàgo multiradiàta Ait.
var. **multiradiàta**

Stems from woody rhizome, pubescent at least above; basal leaves oblanceolate to elliptic, serrate-crenate, villous-margined at base and on petiole; upper stem leaves reduced, not exceeding flowers; inflorescence dense, corymbiform; pedicels white-villous; heads large, about 10 mm high; involucres hemispherical, with 20–30 lanceolate bracts; rays 12–18, achenes pilose.

Meadows, rocky soil, from lowlands to the lower alpine region. Described from cultivated specimens originally from Labrador.

Var. **árctica** (DC.) Fern. (*S. virgaurea* var. *arctica* DC., *S. compacta* Turcz., and *S. unalaschensis* Gandoger), with upper stem leaves equaling or exceeding the compact inflorescence, occurs within area indicated by broken line.

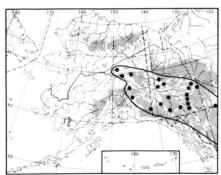
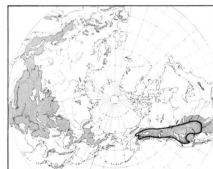

× ⅓

2. Solidàgo multiradiàta Ait.
var. **scopulòrum** Gray

Similar to var. *multiradiata*, but base of basal leaves scabrous in margin from short, stiff hairs, lacking or nearly lacking the long, soft, villous cilia of var. *multiradiata*.

Described from the Rocky Mountains. Transitions to other varieties are frequent.

89. Compositae (Composite Family)

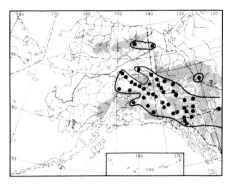
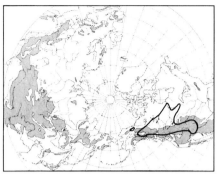

3. Solidàgo decúmbens Greene
var. **oreóphila** (Rydb.) Fern.

Solidago oreophila Rydb.; *S. yukonensis* Gandoger; *S. spathulata* var. *neomexicana* (Gray) Cronq.

Stem from short, stout rhizome, scabrous-pubescent to nearly glabrous; basal leaves several, persistent, oblanceolate to spatulate, toothed to crenate, rounded or subacute in apex, nearly glabrous on sides, scabrous in margin; petioles glabrous or very nearly so; cauline leaves lanceolate, gradually reduced; inflorescence elongated, involucrum imbricate, sometimes slightly glutinose, outer phyllaries much shorter than the inner.

In the mountains to at least 1,000 meters. *S. decumbens* described from Colorado, var. *oreophila* from the Belt Mountains, Montana.

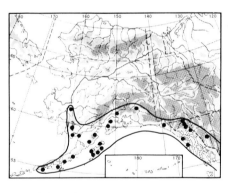
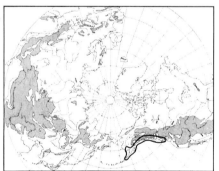
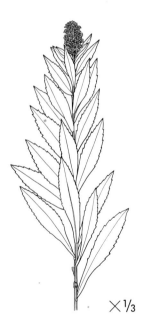

4. Solidàgo lèpida DC.

Solidago lepida var. *subserrata* DC.; *S. canadensis* var. *subserrata* (DC.) Cronq.

Stem from thick, long rootstock, puberulent above; leaves numerous, crowded, broadly lanceolate-oblong, coarsely and sharply serrate, puberulent on nerves above, scabrous in margin; panicle dense, short, with erect, puberulent to scabrous branches, sometimes scarcely exceeding leaves; involucre 5–6(–10) mm high, not markedly imbricate, with greenish bracts.

Meadows, open woods. Described from Yakutat or from Nootka Sound (Vancouver Island).

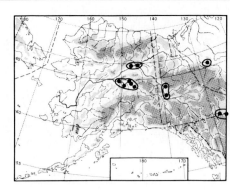
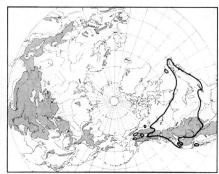

5. Solidàgo canadénsis L.
var. **salebròsa** (Piper) M. E. Jones
  Solidago serotina var. *salebrosa* Piper; *S. elongata* Nutt.; *S. lepida* var. *elongata* Fern.

Stem from long rootstock, puberulent above or nearly glabrous; leaves numerous, reduced in size below inflorescence, lanceolate to lanceolate-oblong, serrate to nearly entire in the scabrous margin, glabrous above or nearly so; panicle broad, densely floriferous, with filiform, scabrous racemes, divergent in age; involucre about 4 mm tall.

Open places in moist woods. S. *canadensis* described from Virginia and Canada, var. *salebrosa* from Pullman, Washington. Circumpolar map gives range of S. *canadensis* in a broad sense.

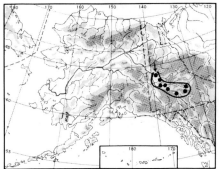
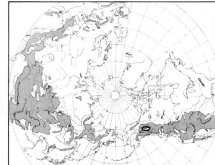

Haplopappus Endl.

1. Haplopáppus MacLèanii Brandegee
  Stenotus borealis Rydb.

Densely caespitose or matted, with more or less woody branches; leaves linear, rigid, glabrous, scabrous in margin, crowded at ends of branches; flowering stems glandular, leafy below, monocephalous; involucral bracts subequal, acute, glandular; rays yellow; achenes prismatic, appressed-villous; pappus profuse, tawny.

Dry, rocky slopes, up to at least 1,000 meters in the mountains. Related to **H. stenophýllus** Gray, of California to Washington and Idaho, which has pubescent leaves.

89. Compositae (Composite Family)

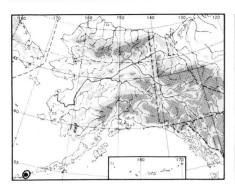
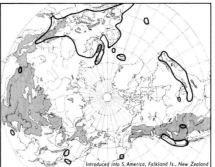
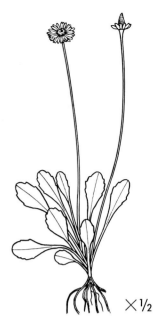

Bellis L.

1. Béllis perénnis L.   Daisy

Plant with short caudex; leaves ovate-spatulate, crenate-toothed, all in basal rosette; scapes naked, hairy; head solitary; involucral bracts blunt, hairy; ray flowers white or pink; achenes downy.

Escaped from cultivation. Described from Europe.

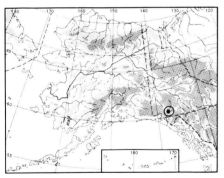
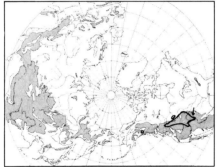
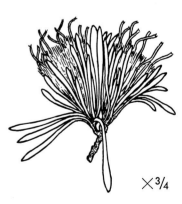

Townsendia Hook.

1. Townséndia Haókeri Beaman

Caespitose, with taproot and simple or branching caudex; leaves narrowly oblanceolate, 2–4 mm broad; heads large, sessile or very nearly so; involucral bracts linear, acuminate, lacking tuft of tangled cilia at apex; ligulae pinkish; disk flowers yellow or pinkish-tipped; achenes loosely hairy.

Alpine slopes; in Alaska to over 2,000 meters.

Aster L. (Aster)

Involucral bracts with glands:
  Leaves lanceolate, the upper clasping, more or less toothed ..... 5. *A. modestus*
  Leaves linear .... 7. *A. yukonensis*
Involucral bracts lacking glands:
  Involucral bracts pubescent on back:
    Heads solitary; leaves entire in margin, the lower obtuse, the upper linear; involucrum not purplish .... 1. *A. alpinus* subsp. *Vierhapperi*
    Heads solitary or several; leaves acute, more or less serrulate in margin; involucrum more or less purplish .... 2. *A. sibiricus*
  Involucral bracts glabrous on back, ciliate in margin:
    Leaves linear:
      Marginal flowers rayless; plant annual .... 9. *A. brachyactis*
      Marginal flowers with rays; plant perennial:
        Involucral scales acute but not mucronate-tipped; stem pubescence striped .... 6. *A. junciformis*
        Involucral scales mucronate-tipped; stem evenly pubescent .... 8. *A. commutatus*
    Leaves lanceolate to ovate:
      Plant more or less pubescent; achenes pubescent; involucral bracts broadest in middle, foliaceous .... 3. *A. subspicatus*
      Plant glabrous or nearly so; achenes glabrous; involucral bracts narrow, broadest at base .... 4. *A. laevis*

×½

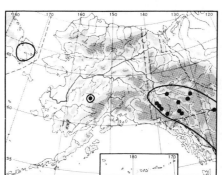

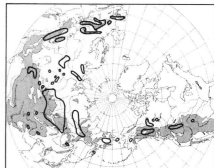

1. Áster alpìnus L.
subsp. **Vierhápperi** Onno

Stems erect, from branching, somewhat woody rootstock; leaves strigose-pubescent, oblanceolate, obtuse, 1–3-nerved, the upper reduced, heads solitary; involucral bracts obtuse, purple-tipped; ligules white or pink; achenes flat, 2-nerved, strigose; pappus sordid.
  Grassy slopes, stony alpine tundra, in the mountains to at least 1,900 meters.
  Described from northern Russia, Siberia, and the Rocky Mountains.
  Circumpolar map indicates range of the entire, complicated species complex.

89. Compositae (Composite Family)

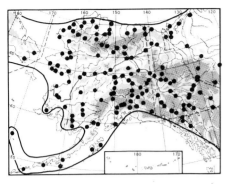
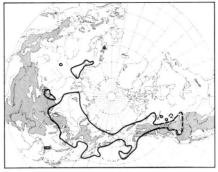
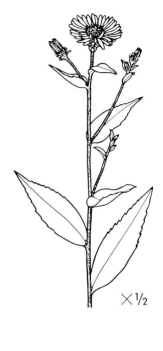

2. Áster sibìricus L.

Aster montanus Richards. not Nutt.; *A. Richardsonii* Spreng.; *A. salsuginosus* Richards.; *A. espenbergensis* Nees; *A. subintegerrimus* (Trautv.) Ostenf.

Stems from long, creeping, branched rhizome; leaves lanceolate-oblong, serrulate or occasionally subentire in margin, sparsely to copiously pubescent beneath, scabrous in margin; heads solitary or few; involucral bracts lanceolate to broadly lanceolate, herbaceous, the outer green, squarrose, the inner purplish; ray flowers purplish; seeds pubescent; pappus as long as disk flowers, brownish.

Stony slopes, river flats, meadows, in the mountains to at least 1,800 meters. Described from Siberia.

Specimens from the interior lowland are tall and branched, with sharply serrulated leaves; those from exposed places are low-growing, with single head and less serrulated leaves.

Broken line on circumpolar map indicates range of the related **A. merìtus** Nels. (See color section.)

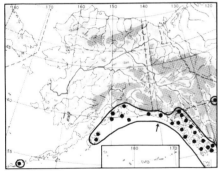

3. Áster subspicàtus Nees

Aster foliaceus Lindl.

Stems from creeping rhizome, low and 1-headed in specimens from exposed places, taller and branched in less exposed or more southern habitats; leaves oblanceolate to broadly lanceolate, nearly entire or toothed, glabrous, scabrous in margin, the upper sessile, slightly clasping; heads 1 to several; involucral bracts broadly lanceolate, leafy, ciliate in margin, light-colored and chartaceous at base, not markedly imbricate; ligules purple to violet; pappus tawny.

Subalpine meadows, woods.

Extremely variable, and forms hybrid swarms with other species to the south. Broken line on circumpolar map indicates range of closely related taxa in eastern North America.

×⅓

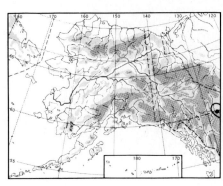
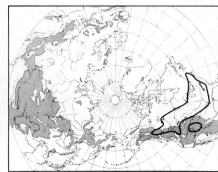

4. Áster laèvis L.

Glabrous or nearly so, more or less glaucous; leaves thick, narrowly lanceolate to ovate, mostly entire, the upper sessile and more or less clasping, strongly reduced; involucral bracts imbricate in several series, linear to linear-oblong; ligules violet; achenes glabrous.

Dry places. Described from North America.

Extremely variable, especially in leaf form.

×⅓

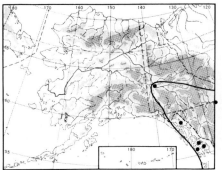
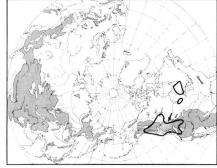

5. Áster modéstus Lindl.

Aster unalaschkensis var. *major* Hook.; *A. major* (Hook.) Porter.

Stems from creeping rhizome, glandular above; leaves lanceolate, acuminate, remotely toothed to subentire, the upper more or less clasping, the lower soon withering; heads several to numerous; involucral bracts acuminate, glandular, herbaceous; rays purple to violet; achenes sparsely hairy.

Woods, along streams. Described from mouth of the Smoky River, lat. 56°.

89. COMPOSITAE (Composite Family)

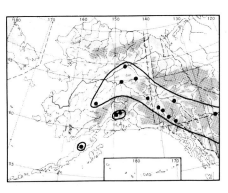 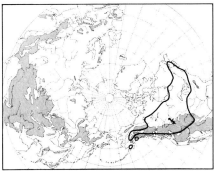 

6. Áster junciformis Rydb.
Aster junceus of authors.

Stems puberulent above, slender, from filiform, creeping rhizome; leaves linear, entire, scabrous in margin, reticulated beneath; heads few; involucral bracts glabrous on back in 3–5 series, linear-attenuate, often purple-tipped, thin, flexible; ligules purple to roseate.

Wet places, bogs.

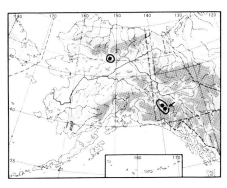 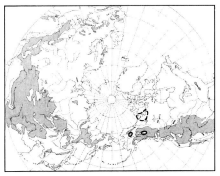 

7. Áster yukonénsis Cronq.

Stems several, decumbent, from short caudex, sparsely villous and somewhat glandular above; leaves linear, the upper clasping, glabrous or nearly so, acute or acuminate; heads solitary to few; involucral bracts acute to attenuate-acuminate, glandular and more or less villous with flattened hairs; ligules blue to purplish; achenes pubescent.

Mud flats, saline soil. Broken line on circumpolar map indicates range of the closely related **A. pygmaèus** Lindl.

× ½

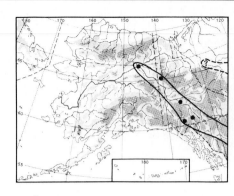
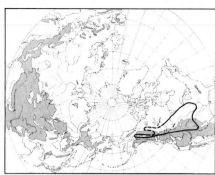

8. Áster commutàtus (Torr. & Gray) Gray

Aster multiflorus var. *commutatus* Torr. & Gray; *A. ericoides* var. *commutatus* (Torr. & Gray) Boiv.; *A. elegantulus* Pors.

Stems from short rhizome, erect, pilose with appressed hairs, branched above; leaves linear, entire, sessile, sparsely pubescent above, pubescent beneath, hispidulous-ciliate; heads small, several; involucral bracts in 2 series of equal length, acute, glabrous, scabrous in margin, the inner white at base; ray flowers white to pale lilac; pappus white.

Alkaline soil. Total range uncertain. Described from northwest North America.

× ½

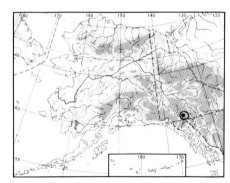
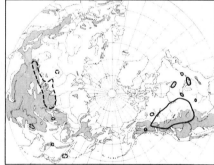

9. Áster brachyáctis Blake

Tripolium angustum Lindl.; *Aster angustus* (Lindl.) Torr. & Gray, not Nees.

Annual, with taproot; stem simple or branched; leaves linear, entire, glabrous, more or less ciliate at base, heads mostly several; involucre campanulate, with uniform, ciliate bracts in 2 series; marginal flowers rayless, shorter than styles; pappus surpassing corollas.

Saline soil. Described from the banks of the Saskatchewan River.

Circumpolar map tentative; broken line indicates range of the closely related **A. ciliàtus** (Ledeb.) Fedtsch. (*Erigeron ciliatus* Ledeb.).

89. COMPOSITAE (Composite Family)

Erigeron L. (Fleabane)

Linnaeus regarded the genus *Erigeron* as neuter, and this gender should therefore be used for the species names. However, it has later been customary to regard *Erigeron* as masculine, the name being supposed to mean "becoming early an old man." Few floras have the proper endings for the species names, should *Erigeron* be regarded as neuter. Thus to conform with, for example, Cronquist, in his treatment of the American *Erigeron* species, the species names have here been given the endings they would have if *Erigeron* were masculine.

Leaves ternate, with narrow divisions .... 1. *E. compositus*
Leaves entire or serrate, very rarely slightly 3-lobed in apex:
  Leaves all linear:
    Peduncles naked, much exceeding the numerous, crowded, short stem leaves
      .... 2. *E. hyssopifolius*
    Peduncles leafy; numerous long basal leaves .... 3. *E. pumilus*
  Leaves (at least the basal) broader:
    Rays of marginal flowers very narrow, not more than one and a half times the length of the pappus:
      Stem simple, 1-headed:
        Pappus purplish .... 4. *E. purpuratus*
        Pappus white, brownish or sordid:
          Plant slender, with small head ....
            .... 14. *E. elatus* (single-headed specimens)
          Plant low-growing, coarser:
            Pubescence of involucrum with purplish-black cross-walls; involucral bracts appressed, narrow ..... 5. *E. humilis*
            Pubescence of involucrum woolly, yellowish, or with pinkish cross-walls; involucral bracts divaricate in fruit, broader .... 6. *E. eriocephalus*
      Stem branched:
        Inflorescence racemiform; peduncles erect; cauline leaves linear; inner involucral bracts acute to acuminate but not long-attenuate to caudate .... 13. *E. lonchophyllus*
        Inflorescence corymbiform; peduncles more or less spreading; cauline leaves broader; inner involucral bracts long-attenuate to caudate:
          Involucrum hirsute, not glandular .... 14. *E. elatus*
          Involucrum glandular, often also hirsute:
            Lower leaves serrate-dentate; stem and involucrum green .... 11. *E. acris* subsp. *kamtschaticus*
            Lower leaves entire in margin; stem and involucrum purplish:
              Plant tall; involucrum glandular or glandular with sparse, hirsute hairs .... 10. *E. acris* subsp. *politus*
              Plant low-growing; involucrum glandular and hirsute .... 12. *E. acris* var. *debilis*
    Rays of marginal flowers twice as long as the pappus or longer, conspicuous:
      Upper cauline leaves clasping, ovate-lanceolate:
        Rays 2–4 mm broad; stem mostly simple .... 15–16. *E. peregrinus*
        Rays very narrow; stem mostly branched .... 19. *E. philadelphicus*
      Upper cauline leaves narrow:
        Plant 30–50 cm tall; stem branched .... 17. *E. glabellus*
        Plant less than 15 cm tall; stem simple or (*E. caespitosus*) with few branches:
          Plant caespitose, with branched, woody caudex; involucrum white-pubescent with stiff hairs .... 18. *E. caespitosus*
          Plant not caespitose; involucrum with soft, woolly hairs:
            Upper cauline leaves linear; hairs of involucrum with purplish-black cross-walls .... 7. *E. hyperboreus*
            Upper cauline leaves lanceolate; hairs of involucrum yellowish:
              Leaves and involucrum pilose or long-villous .... 8. *E. grandiflorus* subsp. *grandiflorus*
              Leaves and involucrum densely lanate .... 9. *E. grandiflorus* subsp. *Muirii*

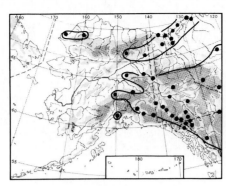 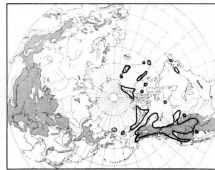

1. Erígeron compósitus Pursh
Erigeron Gormani Greene.

Densely caespitose; stems from stout caudex, simple or branched; basal leaves crowded, ternately to 2–4 times ternately lobed into linear segments, more or less glandular and hispid-hirsute to nearly glabrous; heads solitary; involucral bracts thin, subequal, purplish-tipped; ligules white, pink, or blue, sometimes lacking; achenes pubescent.

Dry rocks. Described from the banks of the Kooskoosky River (Idaho).

In the typical plant the leaves are 3–4 times ternate, with long divisions; in var. **glabràtus** Macoun, 2–3 times ternate, with shorter divisions; in var. **discoìdeus** Gray, once ternate, with almost pulvinate growth. All types occur in Alaska, only slightly separated geographically.

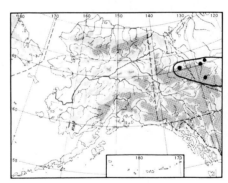 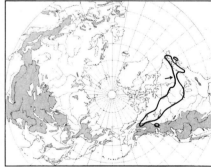

2. Erígeron hyssopifòlius Michx.

Stems more or less strigose-pubescent from short caudex; leaves linear to linear-oblong, nearly glabrous, sparsely ciliate, often with short, leafy shoots in axils; heads solitary on long, naked peduncles; involucral scales tapering from base, acuminate; ligules white or pinkish; pappus white; achenes short-haired.

Rocky shores, rock ledges.

89. COMPOSITAE (Composite Family)

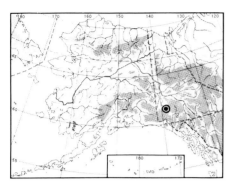 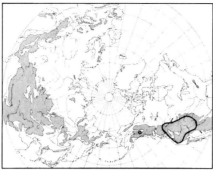 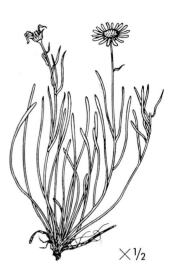

3. Erígeron pùmilus Nutt.

Stems hirsute with spreading hairs, several from taproot and short caudex; leaves linear or narrowly oblanceolate, reduced upward; heads solitary to several; involucral bracts densely spreading, hirsute and finely glandular; ligules white or pink to lavender or blue; pappus double.

Dry slopes. Described from the sources of the Platte, and from the Rocky Mountains. The single collection from our area may be referred to subsp. **intermèdius** Cronq.

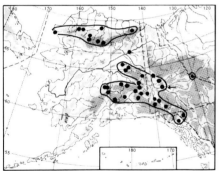 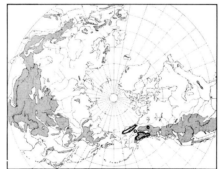 

4. Erígeron purpuràtus Greene
Erigeron denalii Nels.

Stems from branching caudex, leafy at base, naked above, glandular and more or less villous; leaves oblanceolate to spatulate, entire or slightly toothed in apex, villous to glabrate; heads solitary; involucral bracts acuminate, viscid-villous with purplish hairs; ligules white when young, purplish in age, mostly inrolled; pappus reddish-purple; achenes short-haired.

Riverbanks, gravelly soil; to at least 1,738 meters in McKinley Park, 2,800 meters in the St. Elias Range.

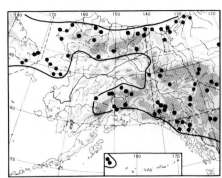
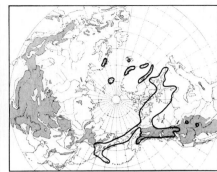

5. Erígeron hùmilis Graham
Erigeron pulchellus var. *unalaschkensis* DC.; *E. unalaschkensis* (DC.) Vierh.

Plant with short caudex; stem single, erect, villous with often dark hairs; basal leaves villous, oblanceolate to spatulate; cauline leaves reduced; heads solitary; involucral bracts narrow, appressed, woolly-villous, with purplish to blackish crosswalls of hairs; ligules white, in age bluish-violet; pappus white or tawny; achenes more or less pubescent.

Snow beds, meadows. Described from cultivated specimens grown from seeds from the Canadian Arctic.

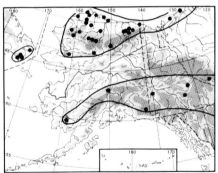
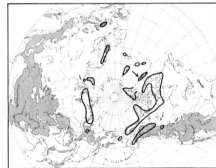

6. Erígeron eriocèphalus J. Vahl
Erigeron uniflorus L. subsp. *eriocephalus* (J. Vahl) Cronq.

Similar to *E. humilis*, but stems often several together, each with 1 head: pubescence light-colored; involucral bracts broader, divaricate in fruit, pubescent to tip; ligules white.

Dry slopes.

Intermediates between this species and *E. humilis* seem to occur. They have reddish cross-walls in the pubescence and purplish ligules, but the heads of *E. eriocephalus*. The range indicated on the map includes that of these intermediates.

89. Compositae (Composite Family)

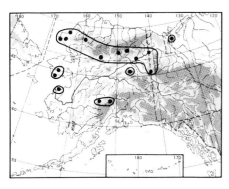
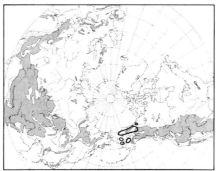
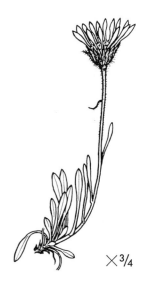

×¾

7. Erígeron hyperbòreus Greene
Erigeron alaskanus Cronq.

Plant spreading-hirsute with septate hairs, sometimes with dark cross-walls; stems single or few, from taproot and short caudex; basal leaves oblanceolate, the cauline linear; heads solitary; involucral bracts equal, viscid-glandular and villous; ligules blue, fading pink, sometimes white; pappus double; achenes villous-hirsute.

Rocks, solifluction soil. Described from Porcupine River, Alaska.

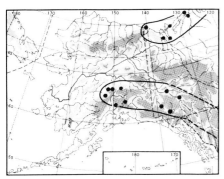
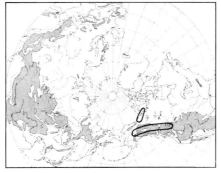

×¾

8. Erígeron grandiflòrus Hook.
subsp. **grandiflòrus**

Stem single from a taproot and short caudex; basal leaves oblanceolate, hirsute-pilose; stem leaves broadly lanceolate to ovate, acute; heads solitary; involucral bracts equal, often somewhat reflexed, pilose or long-villous; ligules blue, 10–15 mm long; pappus double; achenes copiously hirsute.

Dry places on tundra and in mountains. Described from the summit of the Rocky Mountains.

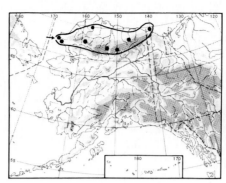
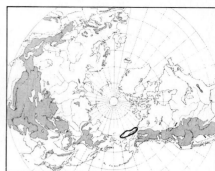

9. Erígeron grandiflòrus Hook.
subsp. **Mùirii** (Gray) Hult.

Erigeron Muirii Gray; *Aster Muirii* (Gray) Onno.

Similar to subsp. *grandiflorus,* but leaves and involucrum densely lanate.

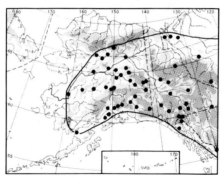
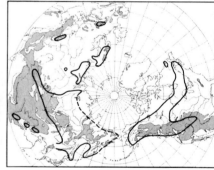

10. Erígeron ácris L.
subsp. **pólitus** (E. Fries) Schinz & Keller

Erigeron politus E. Fries; "*E. acre* var. *asteroides* (Andrz.) DC."; "*E. angulosus* Gandoger var. *kamtschaticus* (DC.) Hara" with respect to American plant; *E. elongatus* Ledeb. not Moench.

Stem erect, stout, with elongate, thyrsoid or corymbiform panicle; leaves oblanceolate to linear-lanceolate, ciliolate; heads mostly several on spreading-ascending peduncles; involucral bracts long-attenuate, sparsely hispid, viscid or nearly glabrous; ligules lilac; ring of tubular pistillate flowers between the ligulate marginal and the perfect central flowers.

Spruce forests, sandy soil. Described from Scandinavia.

89. Compositae (Composite Family)

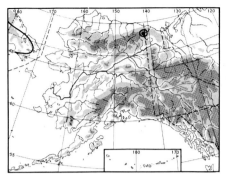 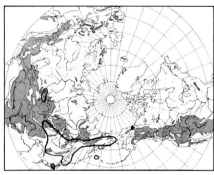 

11. Erígeron ácris L.
subsp. **kamtscháticus** (DC.) Hara
  Erigeron kamtschaticum DC.

Similar to subsp. *politus*, but inflorescence more racemose; lower leaves serrulate-dentate; involucral bracts green; pappus tawny.
Described from Kamchatka.

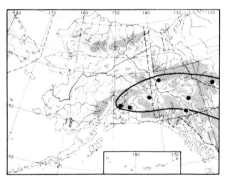 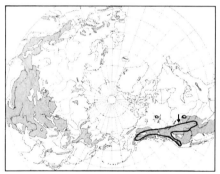 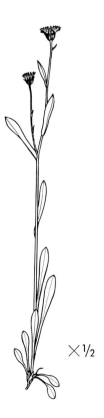

12. Erígeron ácris L. (in a broad sense)
var. **debílis** Gray
  Erigeron debilis (Gray) Rydb.; *E. jucundus* Greene.

Similar to subsp. *politus*, but low-growing; involucral bracts glandular and hirsute.
A doubtful taxon, possibly of hybridous nature.

×½

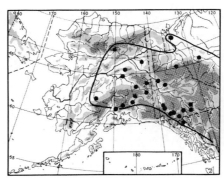
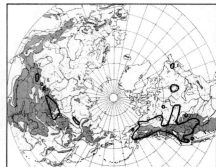

13. Erígeron lonchophýllus Hook.
Erigeron armeriifolius Turcz.

Plant spreading-hirsute; stem slender, erect, from taproot; basal leaves oblanceolate, the cauline narrowly linear, ciliate; peduncles erect; heads solitary or few; involucral bracts acute or acuminate, hispid; ligules inconspicuous, mostly white; no filiform pistillate flowers between the ligulate marginal and the perfect central flowers.

Moist places. Described from Saskatchewan.

×½

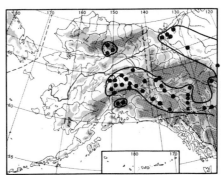
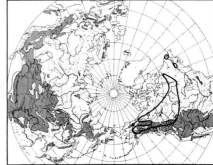

14. Erígeron elàtus Greene
Erigeron acris var. *elatus* (Greene) Cronq.

Similar to *E. lonchophyllus,* but cauline leaves broader; peduncles (in specimens with more than 1 head) more or less spreading; inner involucral bracts long-attenuate, more or less caudate; tubular pistillate flowers present between the ligulate marginal and the perfect central flowers.

Mossy bogs. Described from mountains of British Columbia.

89. COMPOSITAE (Composite Family)   869

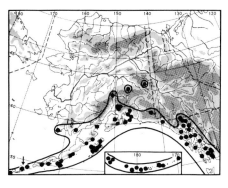
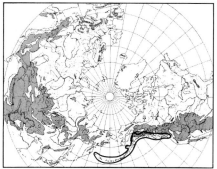

15. Erígeron peregrìnus (Pursh) Greene   Coastal Fleabane
  Aster peregrinus Pursh; *A. unalaschkensis* Less.; *A. Tilesii* Wikstr.
subsp. **peregrìnus**

  Stem single, from short, stout caudex, more or less villous, mostly very leafy; leaves oblanceolate to spatulate, glabrous or somewhat villous, ciliate in margin, reduced above; heads solitary; involucral bracts villous on back with viscid hairs, ciliate in margin; ligules white, pink, purplish, or blue; pappus tawny.
  Meadows.
  Closely related taxa occur in eastern Asia. Specimens with reduced upper leaves have been called var. **Dawsònii** Greene.

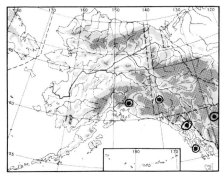
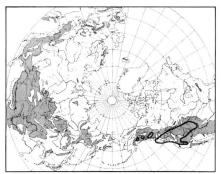
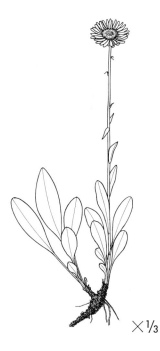

16. Erígeron peregrìnus (Pursh) Greene
  Aster peregrinus Pursh.
subsp. **calliánthemus** (Greene) Cronq.
  Erigeron callianthemus Greene.

  Similar to subsp. *peregrinus*, but involucral bracts with stipitate glands; upper leaves reduced.
  Described from southern Wyoming and northern Colorado.

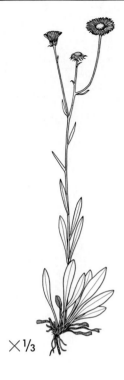
× 1/3

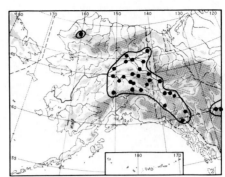
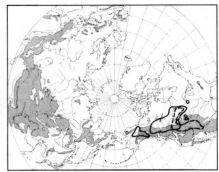

17. Erígeron glabéllus Nutt.
subsp. **pubéscens** (Hook.) Cronq.

Erigeron glabellus var. *pubescens* Hook.; *E. asper* var. *pubescens* (Hook.) Breitung; *E. Turneri* Greene.

Densely hirsute or strigose with appressed or spreading hairs; stems erect or curved at base, from branched caudex; leaves oblanceolate, entire or irregularly toothed, acute or somewhat acute, abruptly reduced upward; heads several; involucral bracts subequal, linear, acuminate, hirsute or strigose; ligules blue or pink, rarely white; pappus double; achenes hairy.

Dry meadows. *E. glabellus* described from Fort Mandan, subsp. *pubescens* from "the prairies of the Rocky Mountains."

Var. **yukonénsis** (Rydb.) Hult. (*E. yukonensis* Rydb.) has narrower leaves and longer and looser pubescence of the involucral bracts. Broken line on circumpolar map indicates range of subsp. **glabéllus**.

× 1/2

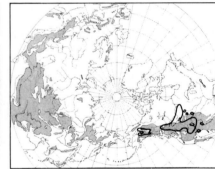

18. Erígeron caespitòsus Nutt.

Stems several, from branched caudex, curved or decumbent at base, hirtellous with short, spreading hairs; leaves hirtellous, the basal more or less 3-nerved, oblanceolate or spatulate, obtuse; stem leaves several, gradually reduced; heads solitary or few; involucral bracts white-pubescent; ligules white to pink or lavender; pappus double, tawny; achenes pubescent.

Dry places in the lowlands; in the mountains to at least 1,300 meters. Described from the Rocky Mountains of Colorado.

89. Compositae (Composite Family)

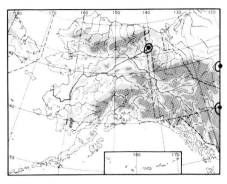
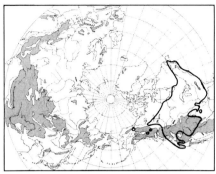

19. Erígeron philadélphicus L.

Plant with short caudex; stem and leaves more or less pubescent to nearly glabrous; basal leaves oblanceolate to obovate, coarsely crenate-toothed or lobed; cauline leaves clasping, crenate-dentate; heads several; involucral bracts subequal, hirsute on midvein or nearly glabrous; ligules numerous, long and narrow, white, pink to purple; achenes sparsely pubescent.

Thickets, shores. Described from Canada.

Antennaria Gaertn. (Pussytoe, Pussy's Toe)

Basal leaves prominently 3-nerved, erect, similar to lower stem leaves:
    Plant 30–50 cm tall, with several stem leaves; heads numerous; involucral bracts light brown .... 1. *A. pulcherrima*
    Plant about 10 cm tall, with 4–6 heads; involucral bracts dark brown .... 2. *A. carpatica* var. *Laestadiana*
Basal leaves 1-nerved or with obscure lateral nerves; cauline leaves reduced:
■ Upper cauline leaves with broad, flat, brown appendages (in *A. media*, narrow, inconspicuous); inner involucral bracts greenish to brownish, or white becoming brownish in age:
  Plant single-headed:
    Plant with horizontal stolons; stem slender; heads small .... 5. *A. monocephala* subsp. *philonipha*
    Plant caespitose, forming dense tufts, lacking long, horizontal shoots:
      Plant low-growing, stout; heads small, 4–5 mm high; pistillate and staminate plants mixed:
        Leaves glabrous above .... 3. *A. monocephala* subsp. *monocephala* var. *monocephala*
        Leaves pubescent above .... 4. *A. monocephala* subsp. *monocephala* var. *exilis*
      Plant taller; heads slightly larger; only pistillate plants known .... 6. *A. monocephala* subsp. *angustata*
  Plant normally with 2–4 heads (only dwarf specimens rarely single-headed), heads sometimes very densely condensed, simulating a single head:
● Plant caespitose, lacking horizontal shoots:
    Basal leaves short, densely crowded, obovate to oblanceolate, tomentose on both sides, not glabrescent in age .... 11. *A. Friesiana* subsp. *compacta*
    Basal leaves petiolated, lanceolate, spatulate-lanceolate or oblanceolate, often glabrescent above:
      Basal leaves linear-lanceolate to spatulate-lanceolate, acute, tomentose on both sides; only pistillate plants known .... 9. *A. Friesiana* subsp. *Friesiana*
      Basal leaves more or less rounded and mucronate in apex, often glabrescent above; pistillate and staminate plants mixed .... 10. *A. Friesiana* subsp. *alaskana*

● Plant with more or less horizontal stolons, forming mats; leaves pubescent on both sides, at least when young:
  Inner involucral bracts lanceolate, acute, dark brown to blackish-brown:
    Basal shoots slender, with acute, oblanceolate leaves . . . 12. *A. atriceps*
    Basal shoots with dense, blunt, ligulate, densely tomentose leaves . . . . 14. *A. stolonifera*
  Inner involucral bracts broad, blunt, somewhat acute, white or light brown:
    Plant tall, with long, trailing stolons . . . . 18. *A. leuchippii*
    Plant smaller, loosely matted, with short, basal shoots:
      Lateral heads long-pedunculate, exceeding central heads . . . . 8. *A. isolepis*
      Lateral heads not exceeding central heads:
        Stem leaves overlapping, longer than internodes; scarious appendages broad, prominent . . . . 7. *A. pallida*
        Stem leaves not or scarcely overlapping; scarious appendages narrow, not prominent . . . . 13. *A. media*
■ Upper cauline leaves with subulate tip, sometimes dark-colored; involucral bracts with pink or white tips, sometimes lacerate, not becoming brownish in age:
  Leaves obovate, rounded (often mucronate) in apex, glabrous above:
    Stem thinly lanate, distinctly glandular . . . . 17. *A. alborosea*
    Stem thickly lanate, not glandular:
      Inner involucral bracts ovate; plant low-growing, not over 10 cm tall . . . . 15. *A. dioica*
      Inner involucral bracts lanceolate; plant much taller . . . . 19. *A. neglecta* subsp. *Howellii*
  Leaves oblanceolate, more or less acute, the inner sometimes spatulate, pubescent above:
    Inner involucral bracts white . . . . See 16. *A. rosea* var. *nitida*

×½

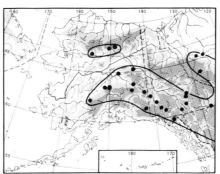

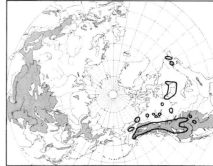

1. Antennària pulchérrima (Hook.) Greene

Antennaria carpatica var. *pulcherrima* Hook.

Stems tall, up to 50 cm, sericeous-tomentose, from slender rootstock; lower stem leaves oblanceolate, acute, grayish-green, the upper linear; inflorescence glomerulate, in age more open; pistillate plant with several heads and obtuse, pale-brown, involucral bracts; staminate plant with smaller heads, involucral bracts with whitish, often reflexed tips; pappus with clavellate bristles.

Meadows, river flats, alpine slopes to at least 1,200 meters in the Yukon. Described from the Rocky Mountains.

Var. **angustisquàma** Pors. differs in having narrow, glabrate leaves, long-pedunculate lateral heads, and long-attenuate, greenish-black-tipped involucral bracts. *A. pulcherrima* belongs to the complicated circumpolar *A. carpatica* complex.

89. Compositae (Composite Family)

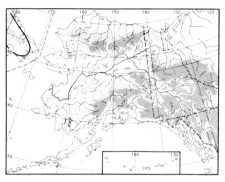 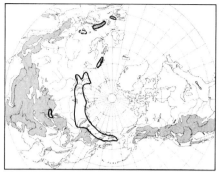 

2. Antennària carpàtica (Wahlenb.) Bl. & Fing.
Gnaphalium carpaticum Wahlenb.
var. **Laestadiàna** Trautv.

"*Gnaphalium carpaticum* β" Wahlenb.; *Antennaria villifera* Borissova; *A. helvetica* Chr. & Pouz.

Stem about 10 cm tall, from vertical or oblique subterranean caudex with dark remains of leaf bases above; basal leaves erect, lanceolate, acute, 3-nerved, grayish-pubescent on both sides, stem leaves few, narrower; heads 4–6, compressed; involucral bracts in several rows, dark brown, shiny, greenish and lanate at base, in pistillate plants lanceolate, acute, in staminate plants ovate or oblong, blunt.

Sandy soil on tundra and in the mountains. *A. carpatica*, with leaves glabrous above, described from the Carpathian Mountains, var. *Laestadiana* from Lapland.

Broken line on circumpolar map indicates range of var. **carpática.**

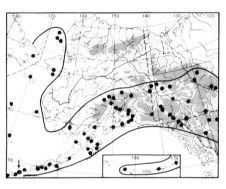 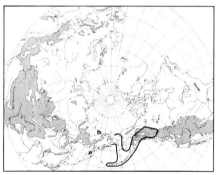 

3. Antennària monocéphala DC.
subsp. **monocéphala**
var. **monocéphala**

Loosely caespitose, more or less matted, with erect or nearly erect shoots; basal leaves spatulate to linear-oblanceolate, mucronate, floccose-tomentose beneath, green and glabrous or thinly floccose above; cauline leaves linear, with flat, brown, scarious appendages; heads solitary; involucral bracts of pistillate heads broadly lanceolate and lanate at base when young, brown and scarious at tip, in age becoming brownish, often darker in middle; those of staminate heads elliptic, obtuse or somewhat acute, green and lanate at base when young, brown and scarious at tip, darkening in age.

Mountain slopes, to at least 1,750 meters.

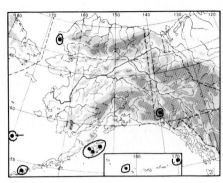
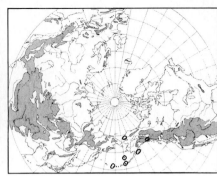

4. Antennària monocéphala DC.
subsp. **monocéphala**
var. **exìlis** (Greene) Hult.
   Antennaria exilis Greene.

Differs from var. *monocephala* in having leaves floccose-tomentose above, also.

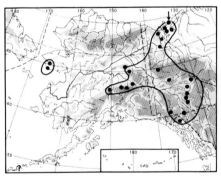
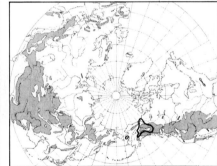

5. Antennària monocéphala DC.
subsp. **philonìpha** (Pors.) Hult.
   Antennaria philonipha Pors.

Differs from subsp. *monocephala* in having more slender stems, thinner pubescence, and longer, horizontal, basal shoots.

Specimens with ovate-lanceolate, more or less blunt, involucral bracts occur rarely (var. **latisquàmea** Hult.).

89. Compositae (Composite Family)

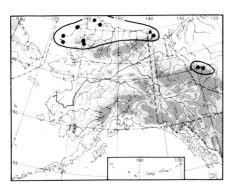
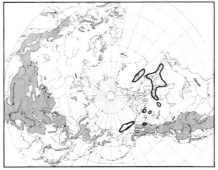

× 3/4

6. Antennària monocéphala DC.
subsp. **angustàta** (Greene) Hult.
  Antennaria angustata Greene.

Differs from subsp. *monocephala* in being taller and of more slender growth; leaves more or less pubescent on upper side also; inner involucral bracts narrower and more acute; heads slightly larger.

Only pistillate plants known. Described from Hudson Bay.

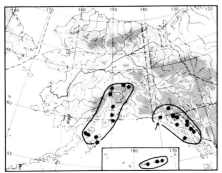
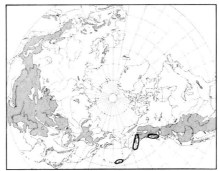

× 3/4

7. Antennària pállida E. Nels.

Matted; loosely caespitose with short stolons; basal leaves spatulate-oblanceolate, appressed-tomentose on both sides; stem leaves 5–7, the upper as long as the internodes or longer, with short, flat, brown, scarious appendages; heads mostly compressed; involucral bracts broad, obtuse, brownish-green to brown, the inner with dirty white to brownish attenuate tips.

Only pistillate plants known.

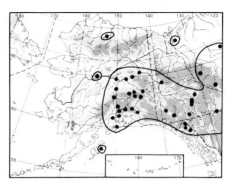
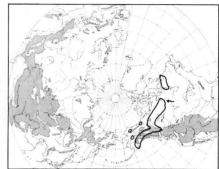

8. Antennària isolèpis Greene

Matted; loosely caespitose or with short stolons; basal leaves oblanceolate, appressed-tomentose on both sides; stem leaves several, linear, acute, tipped with flat, brown, scarious appendages, the upper much shorter than the internodes; heads 3–6, the lateral pedunculate, at maturity often exceeding the central heads; tips of involucral bracts broad, blunt, white, in age brown.

Only staminate plants known.

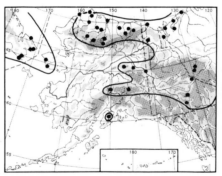
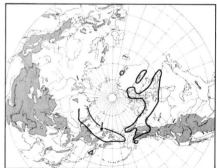

9. Antennària Friesiàna (Trautv.) Ekman

Antennaria alpina (L.) Gaertn. var. *Friesiana* Trautv.; *A. angustifolia* Ekman, not Rydb.; *A. Ekmanniana* Pors.

subsp. **Friesiàna**

Densely caespitose from vertical or ascending root, with branching caudex; basal leaves linear-lanceolate to spatulate-lanceolate, acute, more or less densely tomentose on both sides, with more or less appressed hairs; stem leaves lanceolate-linear to linear, with broad, flat, scarious tips; heads 2–5 (rarely single), densely compressed, or the lower short-pedunculated; involucral bracts in 3–4 (–5) series, brown, greenish at base when young, the outer oblong-ovate, more or less acute and lacerate, the inner narrowly lanceolate to lanceolate, rarely ovate-lanceolate, lacerate [var. **megacéphala** (Fern.) Hult.; *A. megacephala* Fern.]; achenes glabrous.

Dry slopes in the mountains, meadows on tundra. Described from the Kolyma River.

Only pistillate flowers known.

89. Compositae (Composite Family)

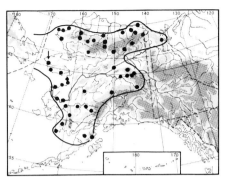
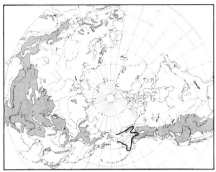

10. Antennària Friesiàna (Trautv.) Ekman
subsp. **alaskàna** (Malte) Hult.

Antennaria alaskana Malte.

Similar to subsp. *Friesiana*, but differs in having broader inner involucral bracts; leaves oblanceolate, more or less rounded and mucronate in apex, often glabrescent above.

Pistillate and staminate flowers frequently occur together: staminate heads smaller than the pistillate; involucral bracts with dark center, obtuse, the outer often reflexed.

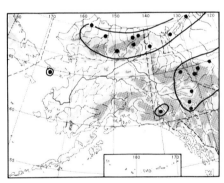
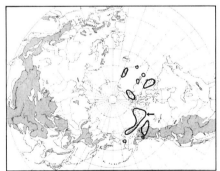
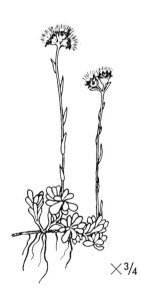

11. Antennària Friesiàna (Trautv.) Ekman
subsp. **compácta** (Malte) Hult.

Antennaria compacta Malte; *A. densifolia* Pors.; *A. crymophila* Pors.; ?*A. neoalaskana* Pors.

Similar to subsp. *Friesiana*, but basal leaves densely crowded, short, obovate to oblanceolate; stem and leaves densely tomentose in specimens from exposed situations.

Staminate plants occur rarely. Specimens more or less intermediate between subsp. *Friesiana* and subsp. *compacta* have been called *A. subcanéscens* Ostenf.

×3/4

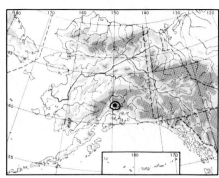
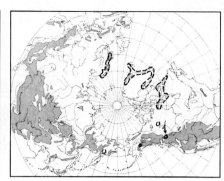

12. Antennària átriceps Fern.

Similar to *A. Friesiana*, but with slender stolons, up to 4–5 cm long; leaves oblanceolate; young leaves pubescent on both sides, glabrescent in age.

Very similar to **A. alpìna** (L.) Gaertn. (*Gnaphalium alpinum* L.) of Scandinavia; broken line on circumpolar map indicates range of *A. alpina* and the very closely related American **A. alpìna** var. **canéscens** Lange (occurring rarely also in Europe).

×3/4

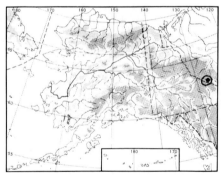
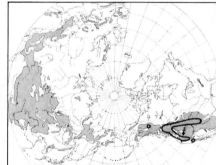

13. Antennària mèdia Greene

Stolons horizontal, densely leafy, forming mats; basal leaves oblanceolate to spatulate, densely and more or less loosely tomentose on both sides; heads several, in subcapitate inflorescence; involucral bracts lanceolate to broadly lanceolate, greenish, in age brownish; achenes smooth.

Both pistillate and staminate plants occur.

89. Compositae (Composite Family)

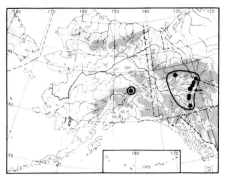

14. Antennària stolonìfera Pors.

Similar to *A. media*, but coarser; achenes more or less rugulose.

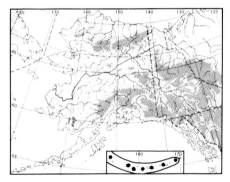

15. Antennària diòica (L.) Gaertn.

Gnaphalium dioicum L.; *Antennaria insularis* Greene.

Usually up to 8 cm tall (rarely to 15 cm); tufted with short, ascending stolons; basal leaves spatulate, rounded in apex, often mucronate, glabrous above, white-tomentose beneath; stem leaves linear, acute; inflorescence compressed, white-woolly, with 3–6 heads; involucral bracts white or pink, in staminate plants (which are the more common) obovate, blunt, in the pistillate plants lanceolate, somewhat acute.

Dry slopes. Described from Europe.

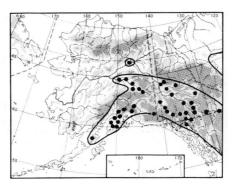
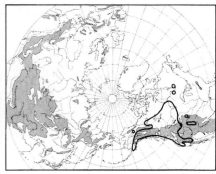

16. Antennària ròsea Greene

Antennaria dioica var. *rosea* (Greene) D. C. Eat.; *A. oxyphylla* Greene; *A. elegans* Pors.; *A. Breitungii* Pors.; *A. incarnata* Pors.

Matted, stoloniferous; leaves gray-tomentose on both sides; basal leaves oblanceolate to spatulate; heads several in subcapitate to corymbose inflorescence; involucral bracts lanceolate, obtuse, the outer broader, pink, pinkish, or rarely strawcolored.

In the mountains to about 2,000 meters. Described from Utah or Nevada.

An extremely variable complex composed of numerous, possibly apomictic or partly apomictic types. Var. **nítida** (Greene) Breitung (*A. nitida* Greene; *A. Laingii* Pors.; *A. subviscosa* with respect to western plant) has pure white involucral bracts, but occupies about the same range as the typical plant. Described from Carlton Island, James Bay.

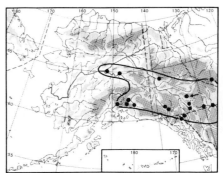
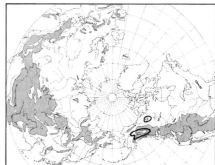

17. Antennària alboròsea A. E. & M. P. Pors.

Similar to *A. rosea*, but leaves broader, glabrous above; stem distinctly glandular.

89. Compositae (Composite Family)

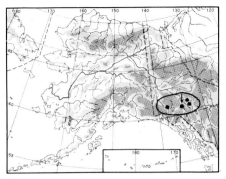
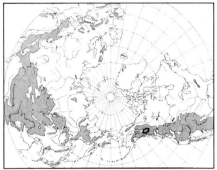
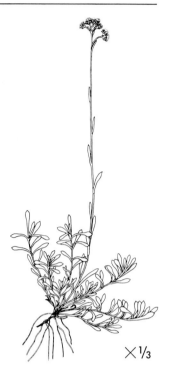

18. Antennària leuchíppii M. P. Pors.

Similar to *A. rosea*, but involucral scales brownish, green at base.

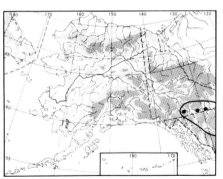
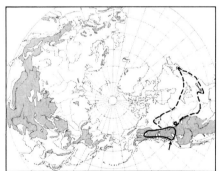

19. Antennària neglécta Greene
subsp. **Howéllii** (Greene) Hult.

Antennaria Howellii Greene; *A. neglecta* var. *Howellii* (Greene) Cronq.

Plant 20–40 cm tall with elongated stolons, leafy (especially toward top); basal leaves spatulate or obovate, up to 2 cm wide, white-tomentose beneath, essentially glabrous above; stem leaves sessile, linear, acute; heads several, inflorescence subcapitate to cymose; involucral bracts of pistillate plant linear-lanceolate, brownish at base, white-hyaline at tip.

Woods. *A. neglecta* described from Washington, subsp. *Howellii* from Mount Saint Helens (Washington).

Broken line on circumpolar map indicates range of entire species complex (including **A. campéstris** Rydb.).

Staminate plants rare, not seen in area of interest.

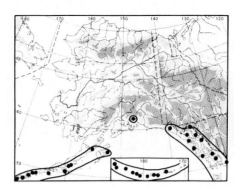
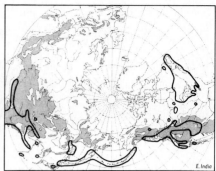

Anaphalis DC.

1. Anáphalis margaritàcea (L.) Benth. & Hook. f.   Pearly Everlasting
Gnaphalium margaritaceum L.; *Antennaria margaritacea* (L.) R. Br.

Stem from long, creeping rhizome; leaves divergent, sessile, broadly linear to oblong, blunt [acute in var. **angùstior** (Miquel) Nakai (*Antennaria cinnamomea* var. *angustior* Miquel)], linear-lanceolate-attenuate at tip; white-tomentose (rarely rusty) beneath, glabrous or tomentose above; inflorescence a corymb with many heads; involucral bracts pearly white, obtuse or rounded; achenes papillate.

Roadsides; open forests to subalpine zone. Described from North America and Kamchatka.

Highly variable; mountain specimens are low-growing, with leaves lanate above, and with involucral bracts marked by brown spot at base (var. **subalpìna** Gray).

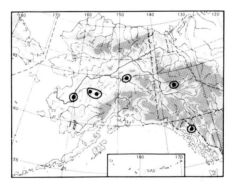
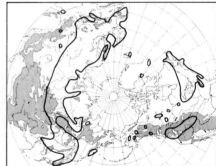

Gnaphalium L.

1. Gnaphàlium uliginòsum L.

Stem diffusely branched from base (or simple in small specimens) from thin root; basal leaves wanting; stem leaves spatulate-oblanceolate to linear, white-tomentose; heads small, about 3 mm high, exceeded by leaves, in sessile terminal clusters; involucral bracts brownish, scarious, lanceolate; flowers yellowish; pappus 1-rowed.

An introduced weed. Described from Europe.

Circumpolar map gives range of *G. uliginosum* in a broad sense, including several geographical races.

89. COMPOSITAE (Composite Family)

Helianthus L.

Receptacle flat; involucral bracts spreading; annual with fibrous root ....
.... 1. *H. annuus*
Receptacle convex; involucral bracts appressed; perennial with rhizome ....
.... 2. *H. laetiflorus* var. *subrhomboideus*

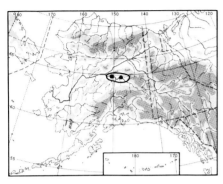
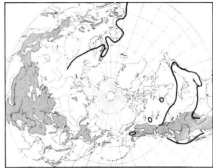

1. Heliánthus ánnuus L.   Sunflower

Stem coarse, up to more than 2 meters tall, erect, simple or branched, scabrous-hispid; leaves broadly cordate-ovate to elliptic, long-petiolated; heads large, with orange-yellow rays and dark disk; involucral bracts broad, spreading, ciliate; paleae of receptacle 3-cleft.

Escaped from cultivation. Described from Peru and Mexico.

Cultivated and escaped or occurring as a weed in many countries; range indicated on circumpolar map is fragmentary.

× ½

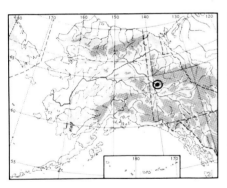
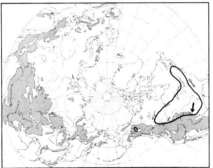

**2. Heliánthus laetiflòrus Pers.
var. subrhomboìdeus (Rydb.) Fern.**

Helianthus subrhomboideus Rydb.

Stem tall, scabrous, slender, simple or branched above, with gradually reduced leaves; leaves gray-green, opposite, the upper alternate, lanceolate to rhombic-ovate, scabrous, serrate to nearly entire; involucral bracts appressed, in several series, oblong; ligules yellow; disk flowers with dark-purple lobes.

An occasionally introduced weed. Circumpolar map tentative.

× ⅓

89. COMPOSITAE (Composite Family)

Bidens L. (Bur Marigold)

Leaves simple, lanceolate-acuminate .... 1. *B. cernua*
Leaves pinnate, with distinct leaflets .... 2. *B. frondosa*

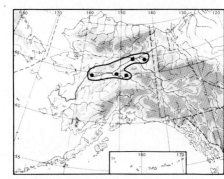
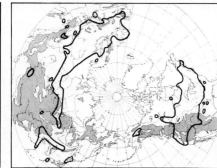

1. Bìdens cérnua L.

Stem simple or branched, glabrous or sparsely hairy; leaves simple, lanceolate-acuminate, sessile, coarsely serrate; heads drooping, solitary at ends of branches; outer involucral bracts lanceolate, leaflike, the inner broadly ovate, dark-streaked; ray flowers yellow (var. **radiàtus** DC.) or lacking; achenes compressed, with 3–4 barbed bristles.

Wet places; an introduced weed. Described from Europe.

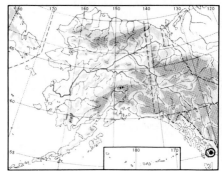
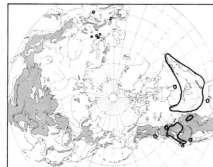

2. Bìdens frondòsa L.

Stem erect, branching, nearly glabrous; leaves with 3–5 lanceolate, acuminate, coarsely toothed, sparsely hispidulous leaflets, the terminal slenderly petiolulate; rays of marginal flowers lacking; outer involucral bracts narrow, glabrous, the inner oblong, obtuse; disk flowers orange; achenes with 2 stout, erect awns.

Shores, waste places; an introduced weed. Described from North America. Introduced also in Europe.

89. COMPOSITAE (Composite Family)

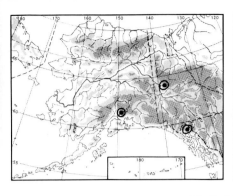
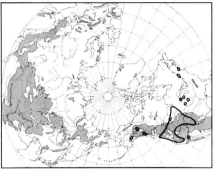

Madia Molina

1. Màdia glomeràta Hook.   Tarweed

Pubescent and glandular above, heavy-scented; stem simple or with ascending branches; leaves linear to linear-lanceolate, hispid; heads glomerate, in 1 to several small clusters; ray flowers yellowish, inconspicuous or lacking; receptacle chaffy.

An introduced weed. Described from Oregon.

Anthemis L.

Ray flowers yellow; receptacle hemispherical .... 1. *A. tinctoria*
Ray flowers white; receptacle long-conical .... 2. *A. cotula*

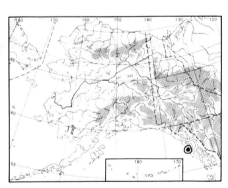
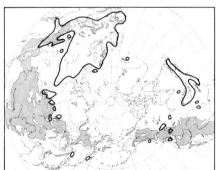

1. Ánthemis tinctòria L.   Yellow Camomile

Stem woolly, mostly branched; leaves pinnatisect with pinnately lobed or toothed segments, united with narrow wing along rachis, glabrous above, white-woolly beneath; heads solitary; involucral bracts lanceolate, acute, scarious-tipped, with brown, ciliate margins; ray flowers and disk flowers golden-yellow; receptable hemispherical, with lanceolate, cuspidate scales; achenes glabrous.

An introduced weed at Sitka. Described from Sweden and Germany.

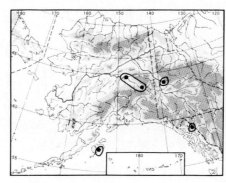

2. Ánthemis cótula L.   Mayweed

Ill-scented; glabrous or nearly so; stem mostly branched; leaves 1–3 times pinnate with narrowly linear, acute segments; heads solitary, short-stalked; involucral bracts oblong, blunt, with broad, scarious, white margin; ray flowers white, in age reflexed, disk flowers yellow; receptacle conical, with linear-subulate scales near apex; achenes tuberculate.

Roadsides, waste places; an introduced weed. Described from Europe, especially Ukrainia.

Achillea L. (Yarrow)

Leaves serrate, incised or pinnatifid, with very broad rachis:
    Ligules of ray flowers about 5 mm long; leaves serrate .... 1. *A. ptarmica*
    Ligules of ray flowers about 1 mm long; leaves incised or pinnatifid ....
    .... 2. *A. sibirica*
Leaves pinnatifid, with narrow rachis:
    Involucral bracts with dark-brown margin .... 4. *A. borealis*
    Involucral bracts with light-brown to yellowish margin:
        Ultimate lobes of leaves lanceolate to ovate .... 3. *A. millefolium*
        Ultimate lobes of leaves linear .... 5. *A. lanulosa*

89. COMPOSITAE (Composite Family)

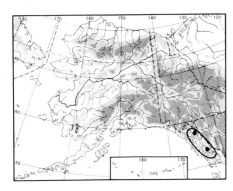
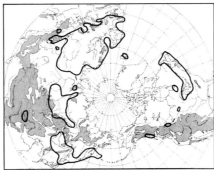
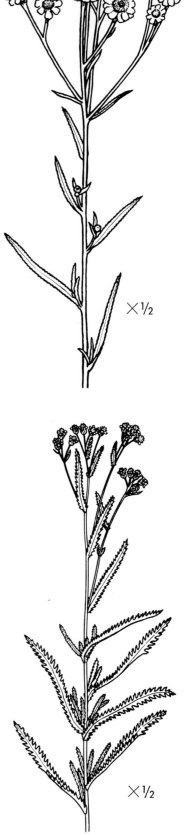

1. Achillèa ptármica L. — Sneezeweed

Stem from creeping, woody rootstock, glabrous below, hairy above; leaves linear-lanceolate, sessile, acute, sharply serrulate, serrulations with cartilaginous and denticulate margin; heads in lax corymb; involucral bracts lanceolate to oblong, blunt, with green center and reddish-brown, scarious margins; ray flowers white, disk flowers greenish-white.

Double forms escaped from cultivation in Alaska. Described from temperate Europe.

Highly variable. Circumpolar map gives range of *A. ptarmica* in a broad sense.

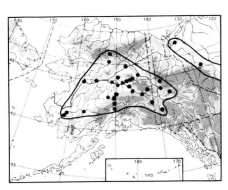
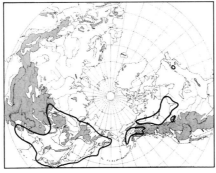

2. Achillèa sibìrica Ledeb.
Achillea multiflora Hook.

Stem branching above from long rootstock; leaves lanceolate, pectinate-pinnatifid, serrulately lobed, sessile, pubescent and often with fine, glandular dots above; heads many, in corymbiform panicles; involucral bracts broadly lanceolate, green in center, with light- to dark-brown, hyaline margin, lanate in apex; ligule white, 3-toothed in apex, glandular.

Meadows, woods, to at least 600 meters. Highly variable. Described from cultivated specimens.

888   89. Compositae (Composite Family)

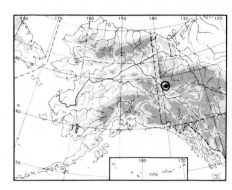
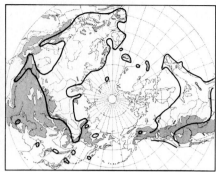

3. Achillèa millefòlium L.   Common Yarrow

Strongly scented; stem simple or forking above, from long, slender rootstock; leaves lanceolate, 3–4 times pinnate with lanceolate-subulate segments, the lower long-petiolated; heads in dense terminal corymbs; involucral bracts oblong, blunt, more or less glabrous, with broad, brown or blackish, scarious margin; ray flowers 3-toothed, white, pink, or reddish; disk flowers white or cream-colored.

An introduced weed. Described from Europe. Circumpolar map tentative.

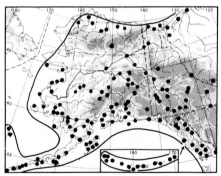
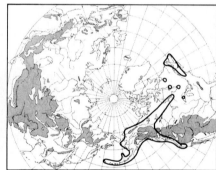

4. Achillèa boreàlis Bong.

Achillea nigrescens (E. Mey.) Rydb.; *A. millefolium* var. *nigrescens* E. Mey.; *A. millefolium* subsp. *borealis* (Bong.) Breitung.

More or less lanate; leaves 2–3 times pinnately dissected, the upper sessile; heads many; corymbs convex; involucral bracts obtuse or rounded, dark-margined; ray flowers white, 3–4 mm long.

Meadows, sandy slopes, in the mountains to at least 1,800 meters. Highly variable in width and dissection of the leaves.

89. COMPOSITAE (Composite Family)

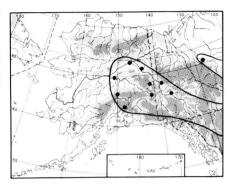
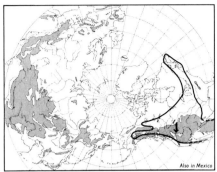

5. Achillèa lanulòsa Nutt.
Achillea millefolium subsp. *lanulosa* (Nutt.) Piper.

Similar to *A. borealis*, but involucral bracts with light-brown or yellowish margin; leaves narrow; ray flowers on the whole shorter.
Described from Kooskoosky (Clearwater) River, Idaho.

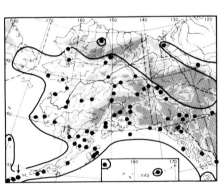
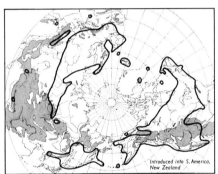

Matricaria L.

1. Matricària matricarioìdes (Less.) Porter — Pineapple Weed
Artemisia matricarioides Less.; *M. discoidea* DC.; *M. suaveolens* (Pursh) Buchenau; *Camomilla suaveolens* (Pursh) Rydb.

Aromatic; stem simple or (mostly) branched from base; leaves 1–3-pinnatifid, with filiform segments; heads several, rayless; involucral bracts with broad, hyaline margin; receptacle conical; pappus a short crown.
Waste places; an introduced weed.

Tripleurospermum Schultz-Bip. (Wild Camomile)

Involucral bracts with dark-brown, broad, scarious margin ... 1. *T. phaeocephalum*
Involucral bracts with light-brown, narrow, scarious margin ...... 2. *T. inodorum*

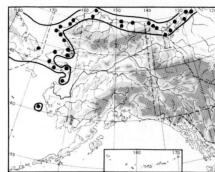

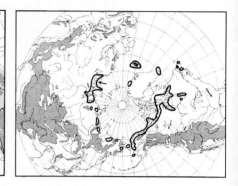

1. Tripleurospérmum phaeocéphalum (Rupr.) Pobed.

Matricaria inodora f. *phaeocephala* Rupr.; *M. ambigua* of authors, not *Pyrethrum ambiguum* Ledeb.; *M. maritima* var. *phaeocephala* (Rupr.) Hyl.; *Chrysanthemum grandiflorum* Hook.; *M. phaeocephala* (Rupr.) Stefáns.

Stem simple or branched above, furrowed, glabrous; leaves pinnatifid with 2–3 lobed segments; heads large, single, at ends of branches; involucral bracts with triangular green center and broad, scarious, dark-brown margin; ray flowers white, 3-toothed; disk flowers yellow.

Seashores.

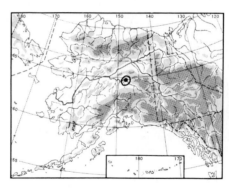

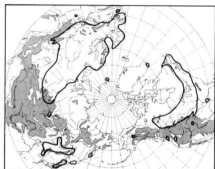

2. Tripleurospérmum inodòrum (L.) Schultz-Bip.

Matricaria inodora L. in part; *M. maritima* subsp. *inodora* (L.) Clapham; *M. maritima* L. var. *agrestis* (Knaf) Willmott.

Scentless; stem glabrous, simple or branched above; leaves 2–3 times pinnate, the ultimate segments long, slender, acute or bristle-pointed, not fleshy; heads solitary; involucral bracts blunt, with narrow, brown, scarious margin; receptacle convex, solid; ray flowers white, disk flowers yellow.

An introduced weed. Described from seashores of Europe.

89. COMPOSITAE (Composite Family)

Chrysanthemum L. (Chrysanthemum)

Marginal flowers lacking ligules or ligules yellow:
  Ligules lacking; heads numerous; leaves pinnate . . . . 1. *C. vulgare*
  Ligules present:
    Ligules more than 8 mm long; plant glabrous; leaves not pinnate . . . .
    . . . . 7. *C. segetum*
    Ligules shorter; plant pubescent; leaves 2–3 times pinnate:
      Ligules about 6 mm long; involucral bracts with dark margin; heads single
        or 1–3 . . . . 2. *C. bipinnatum* subsp. *bipinnatum*
      Ligules shorter; involucral bracts with light-colored margin; heads often
        several . . . . 3. *C. bipinnatum* subsp. *huronense*
Marginal flowers (nearly always) with long, white ligules:
  Basal leaves linear, entire . . . . 8. *C. integrifolium*
  Basal leaves broad, serrate, dentate or lobed:
    Basal leaves spatulate, crenate-dentate . . . . 6. *C. leucanthemum*
    Basal leaves cuneate, pinnatifid or with broad, triangular teeth or lobes:
      Basal leaves cuneate, glabrous at base of petiole; involucral bracts broad and
        dark; rays with 4–5 veins . . . . 5. *C. arcticum* subsp. *polare*
      Basal leaves pinnatifid, arachnoid-pubescent at base of petiole; involucral
        bracts narrower, more light-colored; rays with more numerous veins . .
        . . . . 4. *C. arcticum* subsp. *arcticum*

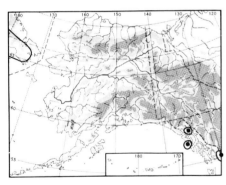

×½

1. Chrysánthemum vulgàre (L.) Bernh.   Common Tansy
Tanacetum vulgare L.; *T. boreale* Fisch.

Strongly scented; stems tough, angled, reddish, from stoloniferous rhizome; leaves dark green, glandular-dotted, pinnate, with deeply pinnatifid leaflets, the ultimate segments lanceolate, acute, sharply toothed, the uppermost leaves sessile, much reduced; heads numerous, discoid; outer involucral bracts ovate-lanceolate, with scarious margin, the inner narrower; flowers golden yellow; pappus a short crown.

Waste places; an introduced weed. Native in Eurasia, introduced in America. Described from Europe.

Circumpolar map includes range of the eastern Asiatic var. **boreàle** (Fisch.) Makino (*Tanacetum boreale* Fisch.).

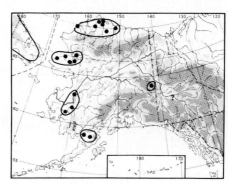
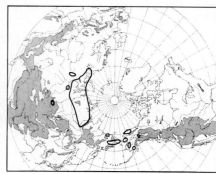

2. Chrysánthemum bipinnàtum L.
subsp. **bipinnàtum**

Artemisia kotzebuensis Bess.; *Tanacetum kotzebuense* (Bess.) Bess.

Stem mostly simple, from stoloniferous rhizome, with sterile shoots; leaves 2–3 times pinnate, with ovate to linear ultimate segments, reduced upward; heads solitary, rarely 2–4; involucral bracts with broad, dark-brown, scarious margin, the outer broadly lanceolate, the inner longer, narrower; ligules 3–7 mm long, oblong to rounded, yellow.

Sandy soil. Described from Siberia.

(See color section.)

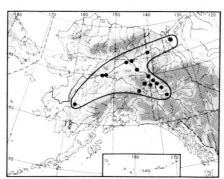
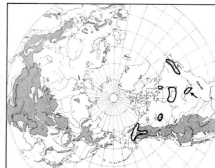

3. Chrysánthemum bipinnàtum L.
subsp. **huronénse** (Nutt.) Hult.

Tanacetum huronense Nutt.; *Chrysanthemum huronense* (Nutt.) Hult.

Similar to subsp. *bipinnatum*, but taller, often with branched stem and several heads; involucral bracts with lighter-colored margin; ligules shorter.

Sandy soil.

89. COMPOSITAE (Composite Family)

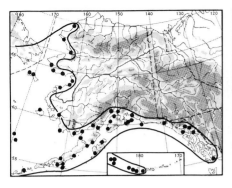 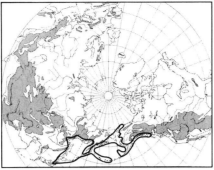 

4. Chrysánthemum árcticum L. Arctic Daisy
subsp. **árcticum**

Leucanthemum arcticum (L.) DC.; *L. Gmelini* Ledeb.; *Chrysanthemum arcticum* subsp. *Gmelini* (Ledeb.) Kitamura; *Dendranthema arcticum* (L.) Tsvel.

Stem simple or branched, from thick, creeping caudex; basal leaves pinnatifid, arachnoid-pubescent at base, with 3–7 blunt, often toothed lobes; stem leaves reduced, the upper linear; heads solitary or few; involucral bracts green, with brown, scarious margin; rays of marginal flowers up to 2.5 cm long, several-nerved, white; disk flowers yellow.

Rocks along the seashore. Described from Kamchatka and North America.

× ⅓

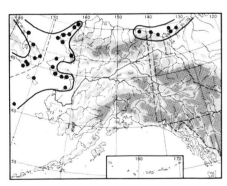 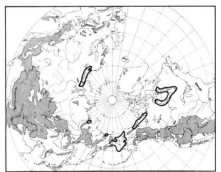 

5. Chrysánthemum árcticum L.
subsp. **polàre** Hult.

Leucanthemum Hultenii Löve & Löve; *Dendranthema Hultenii* (Löve & Löve) Tsvel.

Similar to subsp. *arcticum*, but smaller and not branched; basal leaves cuneate, glabrous at base; margin of involucral bracts scarious, broader and darker; ray flowers shorter, with 4–5 veins only.

Described from the Kola Peninsula. Where the ranges overlap, types more or less intermediate with subsp. *arcticum* occur.

× ½

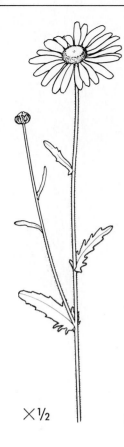
×½

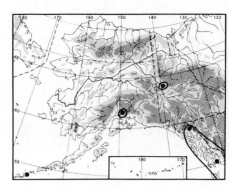
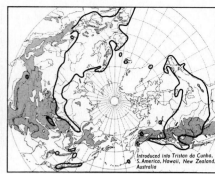

6. Chrysánthemum leucánthemum L.   Marguerite, White Daisy
Leucanthemum vulgare Lam.; *L. leucanthemum* (L.) Rydb.

Stem from woody rootstock, with nonflowering rosettes; basal leaves obovate-spatulate, crenate-dentate; upper leaves oblong, sessile, toothed to pinnatifid; heads solitary; involucral bracts lanceolate to oblong, green, with narrow, dark-purplish, scarious margin; ray flowers white; disk flowers yellow.

Waste places; an introduced weed. Described from Europe.

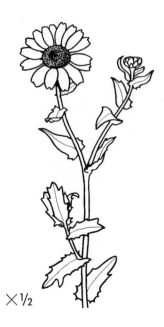
×½

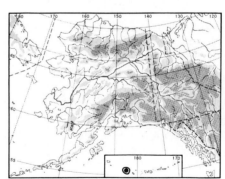
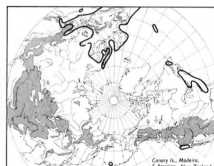

7. Chrysánthemum segètum L.   Corn Marigold

Glabrous, glaucous; stem simple or branched; leaves oblong-cuneate, narrowed into a winged stalk, coarsely toothed to pinnatifid, reduced upward; heads solitary, with stalk thickened upward; involucral bracts ovate, glaucous, with broad, pale-brown, scarious margin; ray flowers and disk flowers golden yellow; pappus lacking.

Occasionally introduced weed at Kiska. Described from Europe.

89. COMPOSITAE (Composite Family)

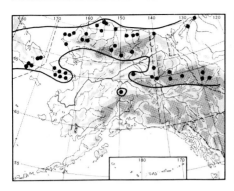 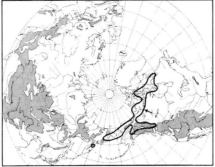 

8. Chrysánthemum integrifòlium Richards.

Leucanthemum integrifolium (Richards.) DC.; *Dendranthema integrifolium* (Richards.) Tsvel.

Caespitose; stem simple, soft-pubescent from branched caudex; leaves linear, somewhat acute, more or less pubescent, mostly basal; heads solitary; involucral bracts white-pubescent, with broad, dark-brown, scarious margin; ray flowers long, white, 3-toothed; disk flowers yellow; achenes glabrous.

Gravel, solifluction soil, rocky slopes. Type locality "on the Copper Mountains."

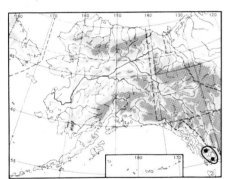 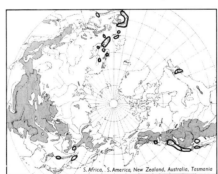 

Cotula L.

1. Cótula coronopifòlia L.

Aromatic, decumbent, glabrous, branched from base; leaves sessile, oblong-lanceolate, deeply toothed to irregularly pinnatifid; heads hemispherical, rayless; involucral bracts rounded in apex, scarious-margined; bracts of receptacle and pappus lacking.

Brackish mud; an introduced weed. Described from Africa.

Artemisia L. (Wormwood)

Lower leaves toothed or 3-lobed at apex, or all leaves entire:
    Plant glabrous or sparsely pubescent .... 4. *A. dracunculus*
    Leaves white-tomentose, at least beneath:
        Lower leaves lanceolate or oblanceolate; heads small; inflorescence many flowered .... 5. *A. ludoviciana* var. *gnaphalodes*
        Lower leaves cuneate, 3-lobed or 3-toothed in apex:
            Flowers in margin pistillate; disk flowers bisexual .... 6. *A. lagocephala*
            Flowers all bisexual .... 7. *A. cana*
Lower leaves deeply lobed, pinnatifid, palmate or pinnate:
    Plant low, caespitose-pulvinate; heads mostly in capitate clusters:
        Flowers purplish-black (very exceptionally yellow); corollas glabrous .... 1. *A. globularia*
        Flowers yellow:
            Leaves small, covered with white shiny hairs; heads numerous, in capitate clusters; corollas glabrous .... 3. *A. senjavinensis*
            Leaves larger, with white, appressed hairs; heads fewer, in capitate or elongated inflorescence; corollas pilose .... 2. *A. glomerata*
    Plant taller, not caespitose-pulvinate; inflorescence spikelike, racemose or prolonged-paniculate:
        Receptacle pubescent; upper leaves twice ternately or quinately dissected into short, linear filiform segments ...... 15. *A. frigida*
        Receptacle glabrous; leaves with linear or broad divisions:
            Plant entirely densely white-lanate; leaves obovate, with few blunt lobes .. .. 8. *A. Stelleriana*
            Plant not white-lanate:
                Stem leaves large, often more than 5 cm long, usually with broad lobes, very exceptionally divided to midrib, green above, white-tomentose beneath:
                    Flowers purplish; involucral bracts densely pubescent, with reddish margins; leaves bipinnately cleft, with abruptly pointed divisions:
                        Leaves dark green, glabrous above .... 13. *A. unalaskensis* var. *unalaskensis*
                      Leaves grayish-green, pubescent above .... 14. *A. unalaskensis* var. *aleutica*
                    Flowers yellowish-brown; involucral bracts less pubescent, with brown, broad, scarious margin; lower leaves pinnatifid with acuminate divisions:
                    Heads large:
                        Stem simple, with few heads, low; upper leaves lobed; leaves with narrow median lobe .... 9. *A. Tilesii* subsp. *Tilesii*
                        Stem branched, with numerous heads; upper leaves entire; leaves with broad median lobe .... 12. *A. Tilesii* subsp. *unalaschensis*
                    Heads small:
                        Leaves dissected into narrow, very acute divisions; upper leaves lobed .... 11. *A. Tilesii* subsp. *Gormani*
                        Leaves with broader, less acute divisions; upper leaves entire; plant often tall .... 10. *A. Tilesii* subsp. *elatior*
                Stem leaves large or small, usually divided to midrib, glabrous, silky-canescent or white-tomentose below or on both sides:
                    ■ Stem leaves divided:
                        Stem leaves pinnate .... 23. *A. rupestris* subsp. *Woodii*
                        Stem leaves 2 or 3 times pinnatifid or pinnate, or twice ternate:
                            ● Leaves small, tomentose beneath or on both sides:
                                Leaves mostly cauline, white-tomentose beneath, green above .... 16. *A. Michauxiana*
                                Leaves mostly basal, white-tomentose, twice ternate .... 17. *A. alaskana*

89. Compositae (Composite Family)

● Leaves large, 2 to 3 times pinnate, green:
    Leaf segments spreading, those of the lower leaves obtuse:
        Involucral bracts with dark-brown margins; heads large; inflorescence simple; leaves twice pinnate .... 22. *A. laciniatiformis*
        Involucral bracts with whitish or light-brown margins; heads smaller; inflorescence branched; leaves 2 to 3 times pinnate .... 21. *A. laciniata*
    Leaf segments ascending, all acute:
        Inflorescence with rust-colored pubescence .... 19. *A. arctica* subsp. *beringensis*
        Inflorescence with gray pubescence:
            Inflorescence open, glabrate or sparingly villous; lower peduncles elongate; leaves with narrow rachis .... 18. *A. arctica* subsp. *arctica*
            Inflorescence lanate, spikelike; leaves with broad rachis, involucrum lanate .... 20. *A. arctica* subsp. *comata*
■ Stem leaves entire or lobed, few; leaves green-sericeous or white-villous:
    Plant less than 1 dm tall, densely white-villous ...... 28. *A. aleutica*
    Plant normally taller, silky-pubescent or glabrous:
        Plant tall, glabrous or nearly so, with numerous small heads in dense spikelike inflorescence:
            Plant an annual, lacking basal shoots; ultimate sections of leaves toothed .... 24. *A. biennis*
            Plant a perennial, with basal leafy shoots; ultimate sections of leaves filiform .... 25. *A. canadensis*
        Plant pubescent, with larger heads in open inflorescence:
            Heads 3–4 mm in diameter; lower bracts with 2–3 lobes .... 26. *A. borealis*
            Heads 4–6 mm in diameter; bracts linear, entire .... 27. *A. furcata*

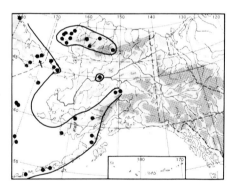  

1. Artemísia globulària Bess.

Artemisia norvegica globularia (Bess.) Hall & Clements; *Ajania globularia* (Bess.) Poljak.

Stems single or several, from short, woody caudex, with rosettes of leaves; basal leaves flabellate, once to twice ternately divided into broadly linear divisions; stem leaves 2–3 to several, reduced, the uppermost linear; inflorescence mostly a dark, globular head or sometimes a prolonged raceme; involucral bracts elliptic-oblong with dark, scarious margin and densely pubescent apex; flowers glabrous, purplish-black.

Rocky slopes.

Var. **lùtea** Hult., with yellow flowers, occurs on Hall Island.

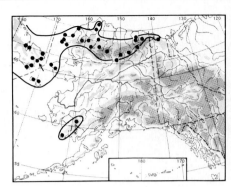 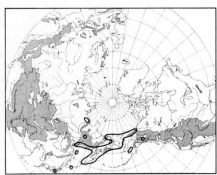

2. Artemísia glomeràta Ledeb.

Artemisia norvegica glomerata (Ledeb.) Hall & Clements; *Ajania glomerata* (Ledeb.) Poljak.

Matted; stems from woody, branched caudex, with sterile rosettes of leaves; basal leaves thrice ternately divided, with lanceolate ultimate segments, pubescent with appressed, white hairs, rarely almost glabrous (var. **subglabràta** Hult., in southwestern Alaska); stem leaves reduced, the uppermost lanceolate; inflorescence capitate or elongated; involucral bracts oblong to oval, densely woolly, with scarious, brown, ciliate margin; disk flowers yellowish, glandular-dotted, with sparsely pubescent lobes; seeds glabrous, glossy.

Sandy slopes.

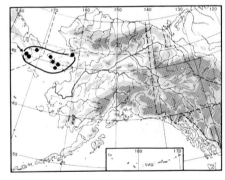 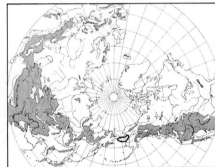

3. Artemísia senjavinénsis Bess.

Ajania senjavinensis (Bess.) Poljak.; *Artemisia androsacea* Seem.

Densely pubescent with white, patent, shiny hairs; stems from woody, branching caudex, with rosettes of leaves forming small, dense tufts; basal leaves 3-lobed, with lanceolate, acute lobes; upper stem leaves simple, lanceolate; inflorescence capitate or spicate; involucral bracts elliptic, somewhat dentate, densely woolly; receptacle conical, glabrous; flowers yellow, glabrous.

Rocks along shores.

89. COMPOSITAE (Composite Family)

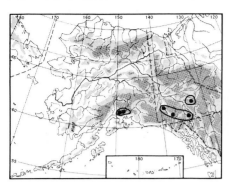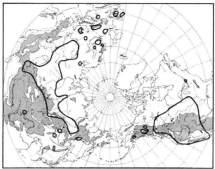

4. Artemísia dracúnculus L.
Artemisia dracunculoides Pursh; *A. glauca* Pall.

Stems glabrous or pubescent, from horizontal rhizome; leaves linear, all entire, or the lowest 3-cleft at apex; panicle with elongate, ascending branches; heads mainly small, on slender pedicels; involucral bracts broadly elliptical, scarious in margin.
  Rocky slopes, roadsides. Probably introduced; native of the prairies.
  Introduced in many places in Europe. Described from "Siberia and Tataria."

× ⅓

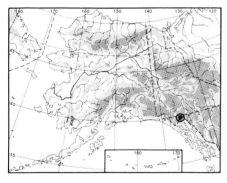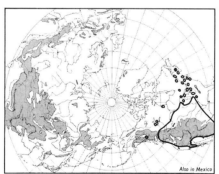

5. Artemísia ludoviciàna Nutt.
var. **gnaphalòdes** (Nutt.) Torr. & Gray
Artemisia gnaphalodes Nutt.

Stems usually simple, or with short, erect branches, lanate or tomentose, from slender rhizome with stolons; leaves lanceolate or oblanceolate, flat, acute, mostly entire, ascending, covered with white felt, at least beneath; heads with pale, involucral bracts.
  An introduced weed at Bennett; a native of the prairies. *A. ludoviciana* described from the banks of the Mississippi, near St. Louis; var. *gnaphalodes* described from Lake Michigan, Fox River, and the Missouri River. Circumpolar map indicates range of the entire species complex.
  Forms hybrid swarms with *A. Michauxiana* (no. 16, below).

× ⅓

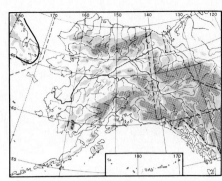
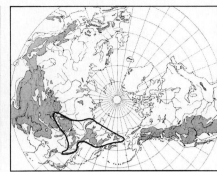

6. Artemísia lagocéphala (Bess.) DC.
Absinthium lagocephalum Bess.

White-tomentose; stems several, simple or branched above, from woody, branching caudex with rosettes of entire, lanceolate leaves; basal stem leaves short-petiolated, 3-toothed to lobed, green and sparsely pubescent above, white-tomentose beneath; inflorescence racemiform; heads nodding; involucral bracts oblong-lanceolate, with broad, erose, brown, scarious margin; marginal flowers pistillate, disk flowers bisexual, glandular-dotted.

Rocks and rocky slopes. Described from eastern Siberia.

Var. **Kruhsiàna** (Bess.) Glehn (*A. Kruhsiana* Bess.) has deeper-lobed basal leaves.

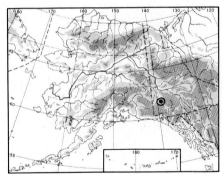

7. Artemísia càna Pursh
Artemisia Bigelowii of Fl. Alaska & Yukon.

Low, silky-canescent shrub; leaves linear or oblong-linear; heads about 3–4 mm broad, in dense, leafy panicle; outer involucral scales thick, canescent, the inner elliptic; flowers all alike, bisexual, fertile; corolla with glandular-dotted tube.

Dry hills. Described from Missouri.

89. COMPOSITAE (Composite Family)

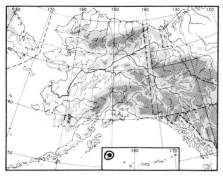
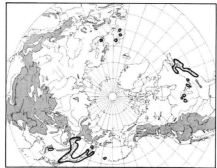

×⅓

8. Artemísia Stelleriàna Bess.

Densely white-lanate; stems from creeping rhizomes with sterile rosettes; basal leaves broad-petiolated, obovate, with few blunt lobes; inflorescence spikelike; lower heads long-peduncled; involucral bracts linear or lanceolate, with brown, scarious margin; marginal flowers pistillate, the central bisexual; achenes glabrous.

Seashores. Native in eastern Asia, introduced in eastern North America and Europe. Described from Kamchatka.

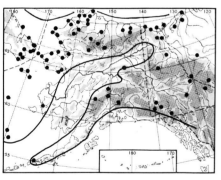
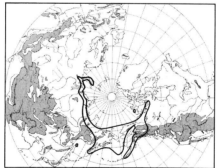

×⅓

9. Artemísia Tilésii Ledeb.
subsp. Tilésii

Artemisia vulgaris L. var. *Tilesii* Ledeb.; *A. vulgaris* L. subsp. *Tilesii* (Ledeb.) Hall & Clements.

Stem floccose-tomentose, glabrescent below, from slender rhizome; leaves deeply divided into 3–5 broadly or narrowly lanceolate, more or less toothed, acute divisions, dark green and nearly glabrous above, white-tomentose beneath, the uppermost linear; inflorescence spikelike; involucral bracts elliptic to lanceolate, arachnoid-pubescent, with broad, dark, scarious margin; marginal flowers few, pistillate, the central perfect; achenes glabrous; pappus none.

Sandy places, in the mountains of the interior to at least 2,000 meters. Described from Kamchatka.

Shows a tendency to form small local races.

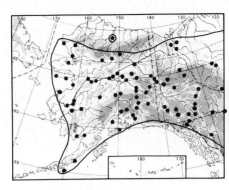
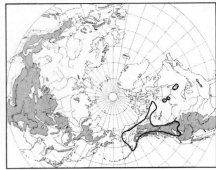

10. Artemísia Tilésii Ledeb.
subsp. **elàtior** (Torr. & Gray) Hult.

Artemisia Tilesii Ledeb. var. *elatior* Torr. & Gray; *A. elatior* (Torr. & Gray) Rydb.

Taller than subsp. *Tilesii*, with heads smaller; inflorescence more branched; upper leaves entire, lanceolate. Merges with subsp. *Tilesii*.

Chiefly in the lowlands; in McKinley Park to about 600 meters. Described from subarctic North America.

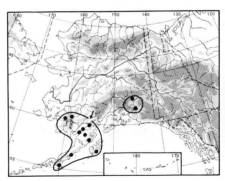
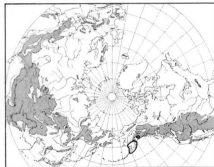

11. Artemísia Tilésii Ledeb.
subsp. **Górmani** (Rydb.) Hult.

Artemisia Gormani Rydb.

Similar to subsp. *elatior* in having small heads and branched inflorescence, but leaves dissected into narrow, very acute divisions.

A semilocal, not very distinctive race.

89. Compositae (Composite Family)

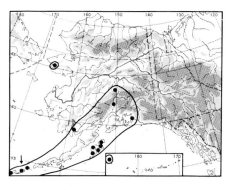
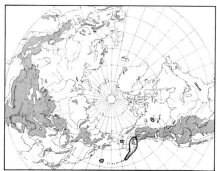

12. Artemísia Tilésii Ledeb.
subsp. **unalaschcénsis** (Bess.) Hult.
Artemisia Tilesii var. *unalaschcensis* Bess.

Similar to subsp. *Tilesii*, but taller; inflorescence branched; leaves very broad-lobed, especially the median lobe; upper leaves entire, broadly lanceolate.
A semilocal race.

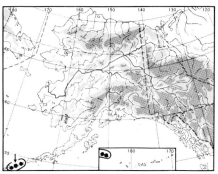
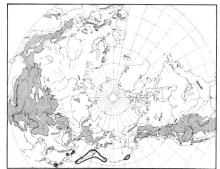

13. Artemísia unalaskénsis Rydb.
Artemisia vulgaris var. *kamtschatica* Torr. & Gray; *A. opulenta* Pamp.
var. **unalaskénsis**

Stem simple, up to 1 meter tall, from long, creeping rootstock; leaves green above, white-tomentose beneath; lower leaves bipinnately cleft into lanceolate, abruptly pointed divisions, the upper reduced gradually, the uppermost lanceolate, acute; heads numerous, in leafy panicle; involucral bracts ovate, green, densely pubescent, with reddish, scarious margin, the outer acute, the inner obtuse; disk flowers purplish.
Subalpine meadows.
Similar to **A. vulgàris** L., but with creeping rootstock.

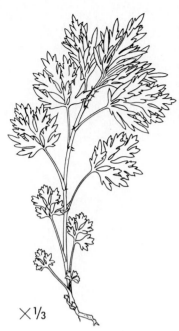
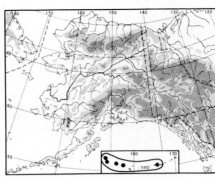
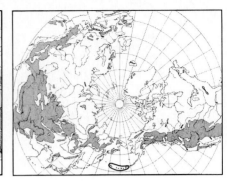

14. Artemísia unalaskénsis Rydb.
var. **aleùtica** Hult.

Differs from var. *unalaskensis* in having more narrowly lobed leaves, tomentose to lanate also on upper side.
Described from the Aleutian Islands.

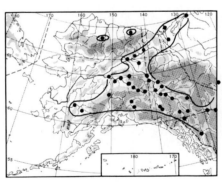
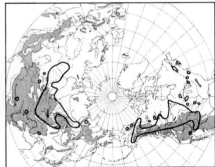

15. Artemísia frígida Willd.   Prairie Sagewort

Fragrant, suffruticose, more or less matted, with several leafy basal branches; leaves silvery-silky, roundish in outline, 2–3 times ternately divided into many linear segments; panicle simple or with erect branches; heads nodding; involucral bracts loosely tomentose; receptacle hairy; achenes glabrous.

Dry, open slopes and rocks in the lowlands, to about 1,000 meters in the mountains. Described from Dahuria.

89. COMPOSITAE (Composite Family)

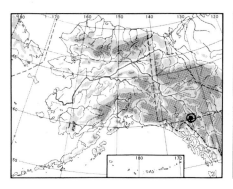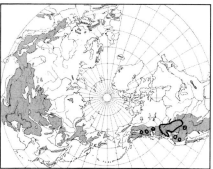

16. Artemísia Michauxiàna Bess.

Stems several, from woody caudex; leaves small, mostly cauline, tomentose beneath; the lower 2–5 cm long, bipinnatifid, the ultimate segments often toothed; upper leaves reduced, the uppermost linear; involucral bracts glabrous or nearly so; marginal flowers pistillate, disk flowers bisexual; receptacle glabrous.

Rocky slopes. Described from the Rocky Mountains.

Forms hybrid swarms with *A. ludoviciana* (no. 5, above).

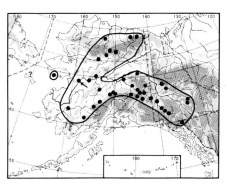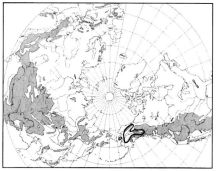

17. Artemísia alaskàna Rydb.

Artemisia Kruhsiana of Fl. Alaska & Yukon, not Bess.; *A. Tyrrellii* Rydb.

Stems white-tomentose; stems simple or branched above, from woody, branching caudex with rosettes of leaves; basal leaves pinnate, with 5 divisions, each division cleft into oblong to linear, blunt sections; middle stem leaves twice ternate, the upper ternate to entire; inflorescence racemiform; heads nodding; outer involucral bracts linear-oblong, the inner oval with scarious, brown, erose margin; corollas yellow, glandular-dotted; achenes glabrous.

Gravel bars, rocky ledges, to about 2,000 meters.

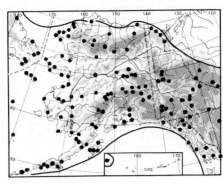
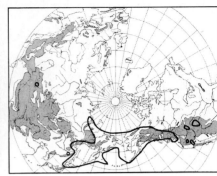

18. Artemísia árctica Less.
subsp. **árctica**

Artemisia norvegica var. *pacifica* Gray; *A. norvegica* subsp. *saxatilis* (Bess.) Hall & Clements; *A. Chamissoniana* Bess.; *A. Chamissoniana* var. *saxatilis*, var. *unalaschcensis*, and var. *kotzebuensis* Bess.; *A. longepedunculata* Britt. & Rydb.; *A. Cooleyae* Rydb.

Stems from stout, woody caudex, with short runners and sterile rosettes; basal leaves bipinnately dissected into mostly 5–7 pairs of pinnae, with linear or linear-lanceolate, acute, ultimate lobes; stem leaves reduced, the upper linear; heads nodding, the lower long-peduncled; involucral bracts ovate-lanceolate to elliptical with dark-brown, scarious margin; disk flowers yellow, pubescent; achenes glabrous; pappus lacking.

Meadows, in the mountains to at least 2,000 meters. Described from Unalaska, Cape Espenberg, St. Lawrence Island and Bay, and Kamchatka.

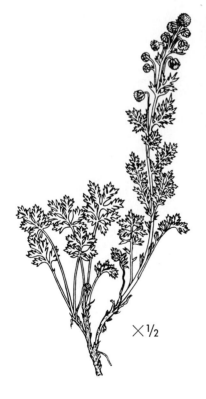

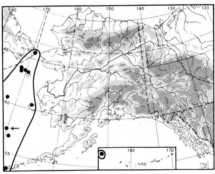
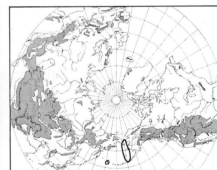

19. Artemísia árctica Less.
subsp. **beringénsis** (Hult.) Hult.

Artemisia arctica var. *beringensis* Hult.

Similar to subsp. *arctica*, but spike with cinnamon-red or rust-red pubescence.

89. Compositae (Composite Family)

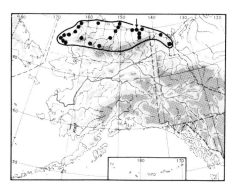
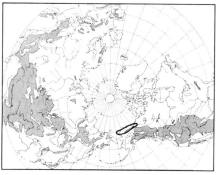

20. Artemísia árctica Less.
subsp. **comàta** (Rydb.) Hult.
  Artemisia comata Rydb.

Similar to subsp. *arctica*, but inflorescence simple, lanate, with crowded heads; leaves with broad rachis and few short, divaricate lobes; leaves and involucrum lanate.

A well-marked arctic race.

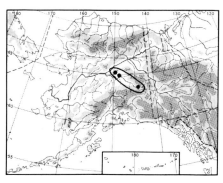

21. Artemísia laciniàta Willd.

Stems tall, branched, pubescent above, from thick caudex; lower leaves 2–3 times pinnate, glabrous above, pubescent beneath; upper leaves sessile, twice pinnate, the uppermost linear; heads globular, nodding, 2–3 mm in diameter; involucral bracts oblong-ovate, green, glabrous with pale, brownish, scarious margin; disk flowers numerous, slightly pubescent.

Open forests. Described from Siberia.

×⅓

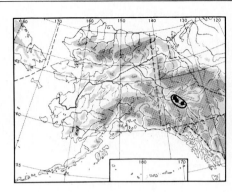
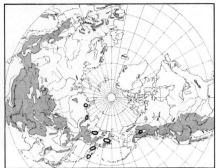

22. Artemísia laciniatifórmis Kom.
Artemisia macrobothrys of authors, not Ledeb.

Similar to *A. laciniata,* but stem mostly simple; heads 4–7 mm broad; involucral bracts with blackish-brown, scarious margin.

Probably along creeks in the mountains. Described from Kamchatka.

×⅓

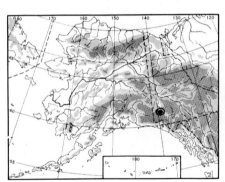
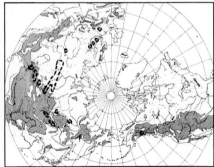

23. Artemísia rupéstris L.
subsp. **Woódii** Neilson ined.

Stems from woody base; sterile shoots strongly branched, densely leafy; flowering stems simple, erect, pubescent, 6–25 cm tall; leaves with linear-lanceolate blunt ultimate segments, the basal leaves two times ternate, the stem leaves pinnate; heads nodding, about 7 mm in diameter, supported by several linear bracts, blunt, broadly scarious-margined with green center; receptable more or less bristly; flowers dotted with glutinose glands; style with stigmas scarcely exceeding anthers.

Scree slopes to about 1400 meters. *A. rupestris* described from Siberia and the Island of Oeland, Sweden; subsp. *Woodii* described from Mt. Wallace, in the Yukon. Broken line on circumpolar map indicates range of other subspecies.

89. Compositae (Composite Family)

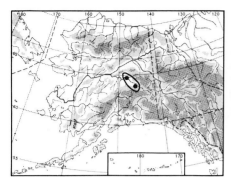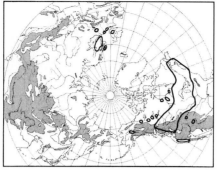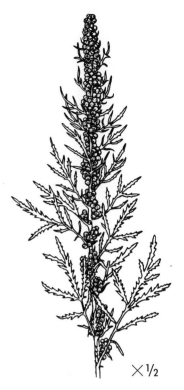

24. Artemísia biénnis Willd.

Erect, glabrous, annual or biennial, with taproot, lacking basal shoots; leaves divided twice pinnately or the upper pinnatifid, with linear, acute, toothed, ultimate segments; inflorescence dense, spikelike; heads subglobose, subsessile, erect; involucral bracts roundish, with broad, scarious margin and green midrib; corolla glabrous.

Waste places; an introduced weed. Described from New Zealand. Introduced in Europe and in eastern North America.

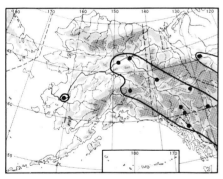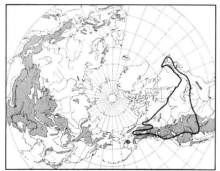

25. Artemísia canadénsis Michx.

Stems 1 to several, from taproot with rosettes of leaves; basal leaves twice pinnatifid, with long, linear divisions, glabrous or sparsely silky; heads nearly erect, 3–5 mm broad, hemispherical; involucral bracts rounded to elliptic, green on back, with broad, pale, scarious margin; disk corollas yellow.

Waste places, roadsides; an introduced weed. Described from Hudson Bay.

× ⅓

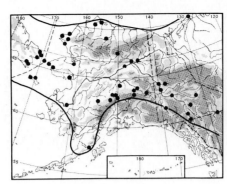
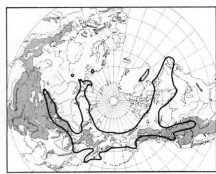

26. Artemísia boreàlis Pall.

Artemisia borealis subsp. *Purshii* (Bess.) Hult.; *A. borealis* var. *Purshii* Bess.; *A. spithamaea* Pursh; *A. borealis* var. *spithamaea* (Pursh) Torr. & Gray.

Caespitose; stems simple or branched, more or less pubescent or nearly glabrous, often purplish, from taproot and short, woody caudex; leaves pubescent, at least when young; basal leaves 2–3 times pinnatifid, with linear to linear-lanceolate, long, ultimate segments; stem leaves reduced; lower bracts usually with 2–3 lobes, the upper entire, linear; heads globular, 3–4 mm in diameter, the lower long-peduncled; involucral bracts ovate, pubescent or glabrous, often brownish with scarious margin; flowers glabrous or somewhat pubescent, often purplish.

Dry slopes, sandy soil. Described from the lower Obi River. Extremely variable.

× ½

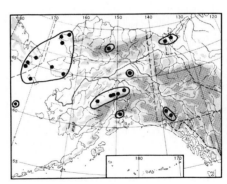
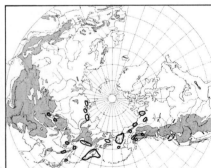

27. Artemísia furcàta Bieb.

Artemisia trifurcata Steph.; *A. hyperborea* Rydb.

Caespitose; stems simple, pubescent above, from taproot and short, woody caudex; basal leaves sparsely pubescent above, more so beneath, once to twice pinnatifid, ultimate sections linear to lanceolate; stem leaves reduced; bracts linear, entire, long, patent; heads globular, 4–6 mm in diameter, nodding, the lower peduncled; involucral bracts broadly ovate, densely pubescent, with conspicuous, brown, scarious margin; flowers yellow, glandular-dotted, glabrous.

Rocky and sandy slopes. Type locality not given.

Specimens with more dissected leaves having shorter segments belong to var. **heterophýlla** (Bess.) Hult. (*A. heterophylla* Bess.).

89. COMPOSITAE (Composite Family)

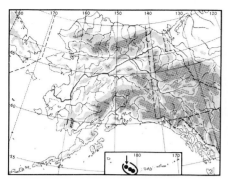
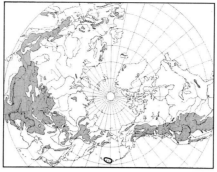

× ½

28. Artemísia aleùtica Hult.

Caespitose, villous, about 5 cm tall; basal leaves short-petiolated, 3-parted to pinnate, with linear to ovate-linear lobes; stem leaves 3-parted, the upper linear; heads about 5 mm in diameter, short-peduncled, few, in short, spikelike inflorescence; outer involucral bracts linear, the inner broadly ovate, with broad scarious margin; corolla glabrous or nearly so, with purplish lobes; achenes glabrous.

Stony slopes. Known only from Kiska and Rat Islands.

Petasites Mill. (Sweet Coltsfoot)

Plant with 1 (–3) heads; rootstock and runners about 2 mm thick:
    Leaves ovate, with cuneate base, glabrous above, white-tomentose beneath . . . .
    . . . . 1. *P. Gmelini*
    Leaves triangular, with cordate base, glabrous on both sides . . . . . . 2. *P. glacialis*
Plant with several heads; rootstock and runners much thicker:
    Leaves reniform, rounded or triangular, more or less deeply lobed, with toothed
        lobes:
        Leaves reniform or rounded, palmately lobed nearly to base . . . 5. *P. palmatus*
        Leaves reniform to ovate or triangular, more shallowly lobed:
            Leaves lobed to about one-third or one-fourth the way from base . . . .
            . . . . 4. *P. hyberboreus*
        Leaves more shallowly lobed . . . . See 3. Hybrid *P. frigidus* × *hyperboreus*
    Leaves triangular or spadelike, toothed:
        Leaves with few coarse teeth . . . . 3. *P. frigidus*
        Leaves with numerous, sharp concave-sided teeth . . . . 6. *P. sagittatus*

89. COMPOSITAE (Composite Family)

× ⅓

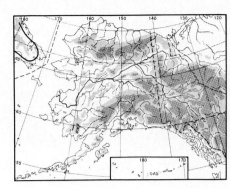

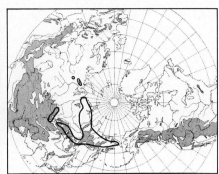

1. Petasìtes Gmelíni (Turcz.) Polunin
Nardosmia Gmelini (Turcz.) DC.; *Tussilago Gmelini* Turcz.

Stem low, from thin rootstocks; leaves basal, ovate, with cuneate base, slightly repand-dentate or entire, glabrous above, white-tomentose beneath; heads single or 2–3, about 2 cm in diameter; involucral bracts violet, acute, white-pubescent; marginal flowers with yellow ligules.

Stony tundra. Type locality not given.

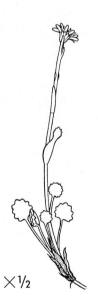
× ½

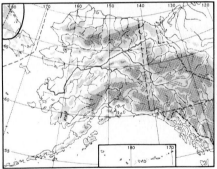

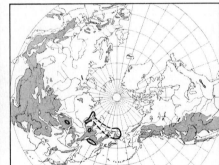

2. Petasìtes glaciàlis (Ledeb.) Polunin
Nardosmia glacialis Ledeb.

Stem low, from thin rootstock; leaves basal, triangular, coarsely dentate, glabrous on both sides; heads always single, 10–12 mm in diameter; involucral bracts lanceolate, acute, densely pubescent at base; marginal flowers with yellow ligules.

Stony slopes, stony tundra. Described from arctic Siberia.

89. COMPOSITAE (Composite Family)

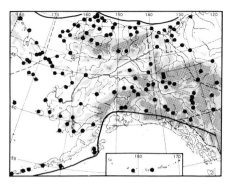
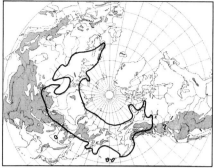

3. Petasìtes frígidus (L.) Franch.
Tussilago frigida L.; *Nardosmia frigida* (L.) Hook.; *N. angulosa* Cass.

Stems simple, from thick, creeping rootstocks, developing before leaves, either with mostly pistillate flowers or with mostly staminate flowers; scales of stem reddish; basal leaves long-peduncled, triangular, with 5–8 broad teeth, entire in margin, glabrous above, white-tomentose beneath; flowers more or less purplish; pappus with numerous long, white or yellowish-white bristles.

Wet tundra, shores, along creeks. Described from Lapland, Switzerland, and Siberia.

Forms hybrid swarms with *P. sagittatus* and very frequently with *P. hyperboreus*, where the ranges overlap. Such intermediates have been called *P. alaskanus* Rydb., *P. arcticus* Pors., *P. frigidus* var. *hyperboreoides* Hult., *P. frigidus* var. *corymbosus* Cronq. in part, and *P. frigidus* var. *nivalis* Cronq. in part.

The root is eaten roasted by the Siberian Eskimo. Circumpolar map includes range of the plant called *Nardosmia angulosa* Cass. by Russian authors.

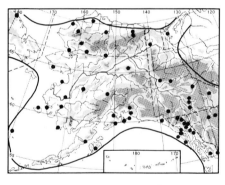
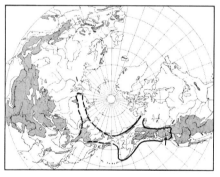
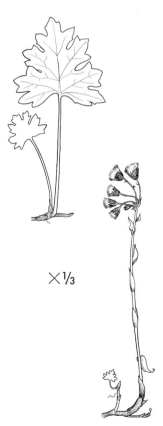

4. Petasìtes hyperbòreus Rydb.
Petasites vitifolius with respect to Yukon plant; *P. frigidus* var. *nivalis* Cronq. in part.

Similar to *P. frigidus*, but leaves deeply lobed into 3–5 broad, grossly toothed segments.

Described from the Skagit River, British Columbia.

Frequently forms hybrid swarms with *P. frigidus*. Possibly a hybridogen species stabilized from the hybrid *P. frigidus* × *palmatus* in a period when the American and eastern Asiatic ranges of *P. palmatus* were confluent.

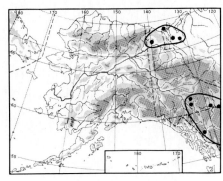
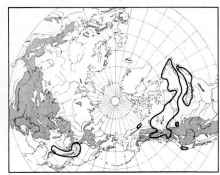

5. Petasìtes palmàtus (Ait.) Gray
Tussilago palmata Ait.; *Nardosmia palmata* (Ait.) Hook.; *N. speciosa* Nutt.

Plant with slender rhizomes and stolons; flowering stem preceding leaves; staminate inflorescence soon shriveling, the pistillate elongating; leaves reniform to suborbicular, cleft to more than two-thirds toward base into 5–7 sharply toothed and cleft lobes, green and glabrous above, shiny white-tomentose beneath; flowers creamy white.

Swampy places, moist woods. Described from Newfoundland.

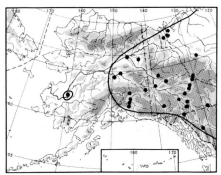
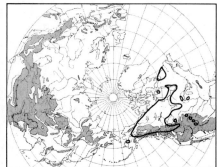

6. Petasìtes sagittàtus (Banks) Gray
Tussilago sagittata Banks ex Pursh.

Stem from slender rhizome with stolons; leaves deltoid-oblong to reniform-hastate, dentate, floccose above, densely white-tomentose beneath.

Bogs, meadows. Described from Hudson Bay.

89. Compositae (Composite Family)

Arnica L.

Anthers purplish-black; pappus tawny:
    Heads nodding in (late) anthesis; ligule of ray-flowers longer than the purplish involucral bracts:
        Stem leaves in 2–3 pairs .... 1. *A. Lessingii* subsp. *Lessingii*
        Stem leaves in 4–6 pairs .... 2. *A. Lessingii* subsp. *Norbergii*
    Heads erect, ligule of ray-flowers as long as or shorter than the green involucral bracts .... 4. *A. unalaschcensis*
Anthers yellow; pappus white, tawny, or brown:
    Heads lacking ligulate flowers .... 13. *A. Parryi*
    Heads with ligulate marginal flowers:
        Basal leaves and lower stem leaves broad, ovate, elliptic or cordate:
            Pappus white, barbellate:
                Involucrum densely white-pilose, inconspicuously glandular; achenes uniformly hirsute .... 5. *A. cordifolia*
                Involucrum sparsely hispidulous-puberulent, short-stipitate-glandular; achenes glabrous or sparsely pubescent .... 6. *A. latifolia*
            Pappus brownish or tawny:
                Heads nearly hemispherical, broad; lower leaves sessile or abruptly short-petiolated; involucrum with long-stipitate glands ...... 9. *A. mollis*
                Heads more or less turbinate; lower leaves petiolate; involucrum not long-stipitate-glandular .... 12. *A. diversifolia*
        Basal leaves and lower stem leaves narrower, never cordate:
            Pappus white, barbellate:
                Leaves regularly dentate; lower leaves long-petiolated, prominently 3–5 nerved .... 14. *A. lonchophylla*
                Leaves entire or irregularly toothed in margin; lower leaves with shorter petioles:
                    Achenes sparsely hispid at summit or subglabrous; leaves obtuse or abruptly pointed, head single .... 3. *A. frigida*
                    Achenes densely and evenly hirsute; leaves acute:
                        Involucrum and peduncle lacking long-stipitate glands; heads single; leaves narrow, acuminate .... 15. *A. alpina* subsp. *angustifolia*
                        Involucrum with long-stipitate glands mixed in pubescence of involucrum and peduncle:
                            Heads single; leaves and stem copiously tomentose; achenes 6–7.5 mm long .... 17. *A. alpina* subsp. *tomentosa*
                            Heads 3–5; stem usually branched; achenes shorter .... 16. *A. alpina* subsp. *attenuata*
    Pappus tawny or stramineous, subplumose or plumose (in *A. Chamissonis* subsp. *foliolosa* merely barbellate):
        Involucral bracts acuminate, lacking tuft of white hairs at tip:
            Cauline leaves often more than 7 pairs, sharply serrate-dentate, mostly sessile .... 7. *A. amplexicaulis* subsp. *amplexicaulis*
            Cauline leaves 5–7 pairs, inconspicuously serrate-dentate, the lower 2–3 pairs petiolated .... 8. *A. amplexicaulis* subsp. *prima*
        Involucral bracts blunt or abruptly pointed with tuft of long white hairs at tip:
            Pappus subplumose; involucrum densely villous, with sessile, moniliform glands mixed in; leaves sessile (or lowermost short-petiolated) .... 10. *A. Chamissonis* subsp. *Chamissonis*
            Pappus barbellate; involucrum densely villous, lacking sessile glands; lower leaves long-petiolated .... 11. *A. Chamissonis* subsp. *foliolosa*

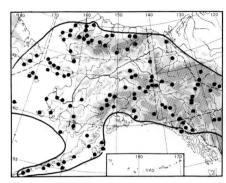
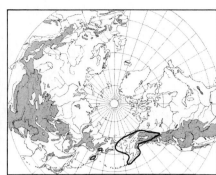

1. Árnica Lessíngii Greene
subsp. **Lessíngii**
Arnica Porsildorum Boiv.

Stem mostly solitary, from slender, scaly rhizome, pilose with septate hairs, especially above; leaves mostly basal, lanceolate, oblanceolate to elliptic, more or less dentate, pubescent above, ciliolate; heads solitary, nodding; involucral bracts lanceolate to lance-elliptic, obtuse, dark purplish, densely pubescent below, ciliolate; ligule pale-yellow, 14–20 mm long; anthers purple; achenes strigose to glabrescent; pappus tawny, barbellate.

Alpine and subalpine meadows. Described from "Alaskan shores and islands."

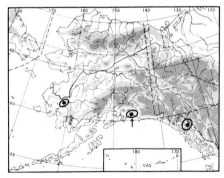
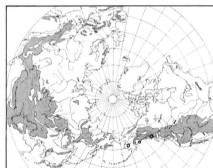

2. Árnica Lessíngii Greene
subsp **Norbérgii** Hult. & Maguire

Similar to subsp. *Lessingii*, but taller, with 5–6 pairs of stem leaves and sparse pubescence.

Perhaps only a local variation.

89. COMPOSITAE (Composite Family)

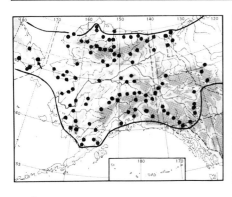
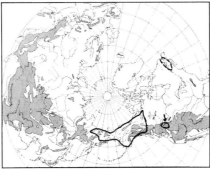

3. Árnica frígida C. A. Mey.

Arnica louiseana subsp. *frigida* (C. A. Mey.) Maguire; *A. brevifolia* Rydb.; *A. illiamnae* Rydb.; *A. Mendenhallii* Rydb.; *A. nutans* Rydb.; *A. Snyderi* Raup.

Stems simple, from short caudex, densely pubescent above; basal leaves oblong-ovate, oval or spatulate, short-petiolated, entire in margin or dentate, ciliate, bluish-green beneath, glabrous or sparsely pubescent; stem leaves narrower, entire, sessile, more pubescent; heads usually single; involucrum densely pubescent, involucral bracts often purplish, linear to lanceolate; ligules 17–23 mm long, pale yellow; disk flowers pubescent; achenes pubescent.

Dry, stony slopes. Extremely variable in pubescence and leaf form.

(See color section.)

Broken line on circumpolar map indicates range of the closely related **A. louiseàna** Farr.

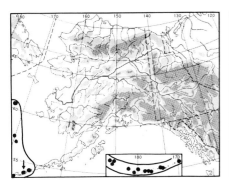
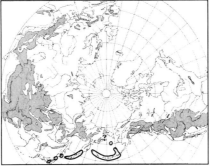

4. Árnica unalaschcénsis Less.

Arnica obtusifolia Less.

Stem solitary, mostly simple, from thick rhizome, covered with remains of previous year's leaves, pubescent with septate hairs; leaves obovate to spatulate or elliptic, the middle and upper sessile, coarsely serrate, pubescent; heads solitary; involucral bracts lanceolate to elliptic, glandular, densely pubescent toward apex; ligules 12–17 mm long, yellow; flowers glabrous; anthers purple; achenes short-hispidulous, with stipitate glands; pappus tawny, subplumose.

Meadows.

The Japanese plant, var. **Tschonòskyi** (Iljin) Kitamura & Hara (*Arnica Tschonoskyi* Iljin), has pubescent corollas.

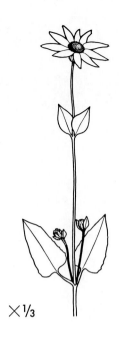

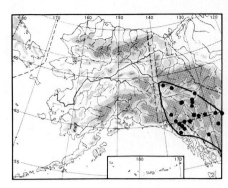

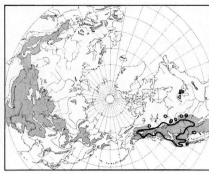

5. Árnica cordifòlia Hook.

Stem mostly simple, from branching rhizome, glandular-puberulent to pilose; basal leaves long-petiolated, ovate to orbicular, subcordate, with broad, shallow sinus, dentate; cauline leaves ovate to lanceolate in 2–3 pairs, the lower long-petiolated, the upper small, reduced; heads large, 1 to several; involucral bracts lanceolate or oblanceolate, acute, pilose at base, glandular; ligules 20–28 mm long, yellow; tube of disk flowers pilose; achenes evenly hirsute and often stipitate, glandular; pappus white, barbellate.

Woods, meadows. Described from the Rocky Mountains and the Blue Mountains. The closely related **A. Whítneyi** Fern. occurs at the Great Lakes.

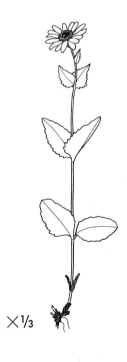

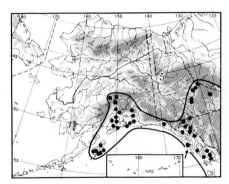

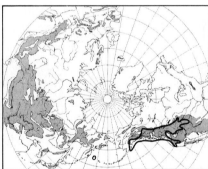

6. Árnica latifòlia Bong.

Stem mostly solitary, simple or branched, from branching, scaly rhizome, glandular below, pubescent above; basal leaves long-petiolated, ovate to lanceolate or subcordate, soon withering; stem leaves in 2–4 pairs, sessile, ovate-lanceolate to elliptic-lanceolate, serrate-dentate, the upper large; heads smaller than in A. *cordifolia*; involucrum hispidulous-puberulent and glandular; involucral bracts lanceolate to oblanceolate, acuminate; ligules 12–22 mm long, yellow; disk corollas pubescent; achenes glabrous or nearly so; pappus white, barbellate.

In the mountains to at least 2,000 meters in the Yukon.

89. COMPOSITAE (Composite Family)

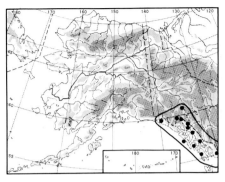 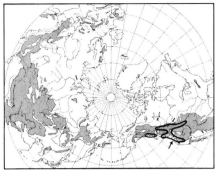 

7. Árnica amplexicaùlis Nutt.
subsp. **amplexicaùlis**

Arnica amplexifolia Rydb.; *A. elongata* Rydb.; *A. borealis* Rydb.

Stem from coarse rhizome, glandular and hairy, especially upward; leaves cauline in several pairs, sessile, lanceolate to elliptic-lanceolate, irregularly serrate-dentate; heads mostly several; involucral bracts lanceolate, sharply acute or acuminate, hairy and sparsely stipitate-glandular at base, and with few short glands at tip; rays 10–20 mm long, pale yellow; pappus tawny; achenes sparsely hirsute.

Moist places. Described from "the rocks of the Wahlamet [the Willamette, probably] at the falls."

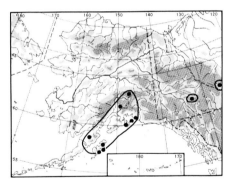 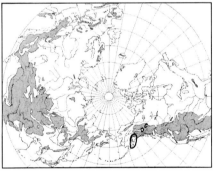 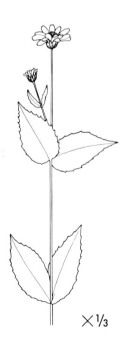

8. Árnica amplexicaùlis Nutt.
subsp. **prìma** Maguire

Arnica mollis with respect to Yukon plant.

Differs from subsp. *amplexicaulis* in having more or less petiolated, mostly narrower, inconspicuously dentate-serrate leaves and in having few pairs of cauline leaves.

Described from Kodiak.

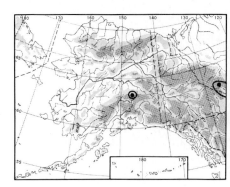
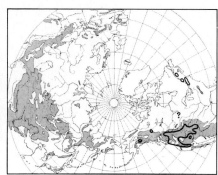

9. Árnica móllis Hook.

Similar to *A. amplexicaulis* subsp. *prima*, but leaves broader; involucrum and upper part of stem profusely glandular from long, capitate glands.

Described from the Rocky Mountains.

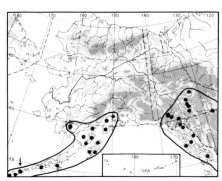

10. Árnica Chamissònis Less.
subsp. Chamissònis

Stems solitary, from long, caudate rhizome, pubescent, glandular above; leaves mostly sessile, lanceolate, oblanceolate or ovate-lanceolate, serrate-dentate; involucrum with septate hairs and also with coarse, short, moniliform hairs; involucral bracts lanceolate, obtuse or somewhat acute, conspicuously pilose at top; ligules 15–20 mm long, pale yellow; disk flowers hirsute and glandular to glabrate; pappus stramineous to tawny, subplumose; achenes glabrous or sparsely pubescent.

Meadows.

On circumpolar map the range of the closely related **A. sachalinénsis** (Regel) Gray (*A. chamissonis* var. *sachalinensis* Regel) is indicated by broken line.

89. Compositae (Composite Family)

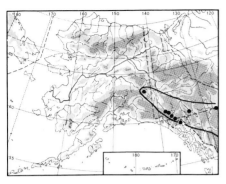 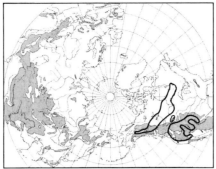 

11. Árnica Chamissònis Less.
subsp. **foliolòsa** (Nutt.) Maguire
  Arnica foliolosa Nutt.

Similar to subsp. *Chamissonis*, but involucrum with septate, not moniliform, hairs; cauline leaves petiolate with entire margins, or denticulate; pappus mostly barbellate.

Wet places.

Some specimens belong to var. **incàna** (Gray) Hult. [*A. foliosa* var. *incana* Gray; *A. Chamissonis* subsp. *incana* (Gray) Maguire; *A. incana* (Gray) Greene], with silvery-tomentose herbage. Subsp. *foliolosa* described from "the alluvial flats of the Colorado of the west, particularly near Bear River of the lake Timpanogos"; var. *incana* described from the "Sierra Valley in deep water, strange to say but true."

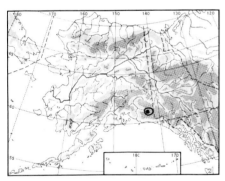 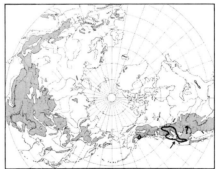 

12. Árnica diversifòlia Greene

Stems simple, from coarse rhizome, nearly glabrous below, glandular and puberulent above; leaves pale green, in 3 pairs, the middle pair largest, ovate to elliptic, truncate or subcordate at base, obtuse, irregularly serrate-dentate, short stipitate-glandular above, glabrous beneath; heads 3–5; involucral bracts lanceolate, acute, stipitate-glandular; ligules 18–20 mm long, pale yellow; disk flowers short-pilose; pappus stramineous to tawny, subplumose.

Somewhat doubtful; probably a series of hybrids.

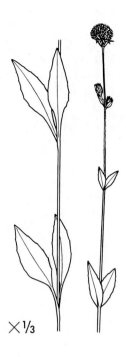
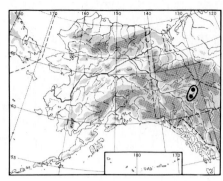
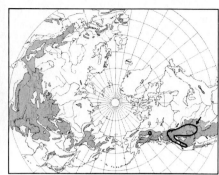

13. Árnica Párryi Gray
subsp. **Párryi**

Stem single, from horizontal rhizome, lanate-villous at base, glandular above; cauline leaves in 2–4 pairs, reduced upward, the lowermost short-petiolated, lanceolate, denticulate or entire, sparingly villous; heads several, nodding in bud, discoid; involucral bracts sharply acute or acuminate; pappus tawny, barbellate to subplumose.

Forests.

In Oregon, Sierra Nevada of California and western Nevada, also subsp. **Sònnei** (Greene) Maguire (*A. Sonnei* Greene).

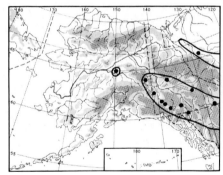
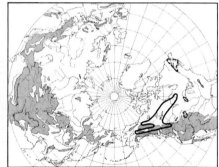

14. Árnica lonchophýlla Greene

Stem from slender, branching rhizome, puberulent and stipitate-glandular; basal leaves narrowly elliptic-lanceolate, acute, dentate, long-petiolated; lower stem leaves peduncled, the upper reduced, sessile; involucrum puberulent and densely glandular; involucral bracts lanceolate, acute, densely glandular; ligules 10–20 mm long, yellow; disk flowers pilose; achenes uniformly hispid; pappus white, barbellate.

Dry, open places. Described from the Athabasca River.

The Alaskan plant belongs to subsp. *lonchophylla*; broken line on circumpolar map indicates range of subsp. **chionopáppa** (Fern.) Maguire (*A. chionopappa* Fern.) in eastern North America and subsp. **arnoglóssa** (Greene) Maguire (*A. arnoglossa* Greene) in the north-central United States.

89. COMPOSITAE (Composite Family)

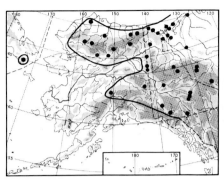 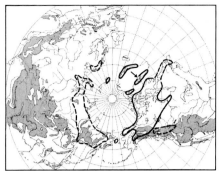 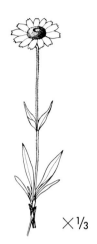

15. Árnica alpìna (L.) Olin
Arnica montana var. *alpina* L.
subsp. **angustifòlia** (M. Vahl) Maguire
Arnica angustifolia M. Vahl.

Stem from slender, scaly rhizome; basal leaves lanceolate to broadly lanceolate, entire or remotely denticulate, acute to acuminate, pilose and glandular; stem leaves in 1–3 pairs, reduced; heads mostly single; involucrum densely lanate-pilose, with septate hairs; involucral bracts lanceolate, acute or acuminate, pilose and glandular, especially toward base; ligules yellow, with teeth 0.5–2.0 mm long; disk flowers hirsutulose; achenes hirsute; pappus white, barbellate.

Dry alpine slopes. *A. alpina* described from northern Europe, subsp. *angustifolia* from Spitzbergen, Greenland, and arctic America.

Broken line on circumpolar map indicates range of subspecies other than those occurring in Alaska. The Scandinavian plant is subsp. **alpìna.**

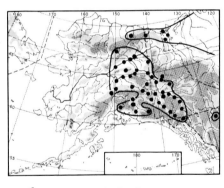 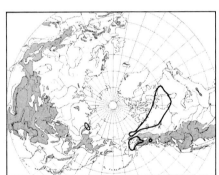 

16. Árnica alpìna (L.) Olin
Arnica montana var. *alpina* L.
subsp. **attenuàta** (Greene) Maguire
Arnica attenuata Greene.

Differs from subsp. *angustifolia* in taller growth; branched stem with 3–5 heads; cauline leaves in 3–5 pairs; more viscid-pubescent involucrum; and somewhat longer teeth on the ligule.

Described from the Lewes River, Yukon Territory.

In var. **lineàris** Hult., the leaves are linear, and in var. **vestìta** Hult. (*A. tomentosa* with respect to Alaskan plant), the entire plant is covered with grayish pubescence.

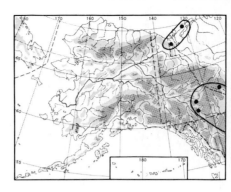 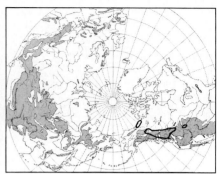

17. Árnica alpìna L.
subsp. **tomentòsa** (James Macoun) Maguire
  Arnica tomentosa James Macoun.

Similar to subsp. *angustifolia*, but leaves, stem, and involucrum tomentose.

Described from the eastern slopes of the Rocky Mountains between the International Boundary and lat. 54° N.

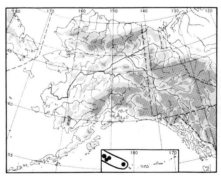 

Cacalia L.

1. Cacália auriculàta DC.
subsp. **kamtschática** (Maxim.) Hult.

  Senecio dauricus var. *kamtschaticus* Maxim.; *Cacalia auriculata* var. *kamtschatica* (Maxim.) Matsum.; *C. kamtschatica* (Maxim.) Kudo; *Hasteola kamtschatica* (Maxim.) Pojark.

Stem from thick caudex; leaves few, large, glabrous or pubescent on veins, reniform, grossly dentate to shallowly lobed, shiny above, pale below, the upper reduced; inflorescence branched with few erect branches; heads nodding; involucrum cylindrical, with 4–5 oblong to lanceolate bracts, as long as young flowers; flowers 4–7; pappus long, white.

Thickets, subalpine meadows. *C. auriculata* described from between Okhotsk and Yakutsk, subsp. *kamtschatica* from Kamchatka.

Broken line on circumpolar map indicates range of subsp. **auriculàta**.

Senecio L. (Groundsel, Ragwort)

Involucral bracts in single row, with outer row or accessory outer scales below head lacking:
  Plant annual or biennial; stem with large denticulate or sublaciniate leaves up to inflorescence; peduncles viscid-pubescent .... 1. *S. congestus*
  Plant perennial; stem leaves rapidly diminishing toward top of stem, entire, denticulate, pinnate or lyrate; peduncles viscid or not:
    Involucral bracts pubescent:
      Involucral bracts with purplish or brownish pubescence; heads solitary:
        Involucrum and upper part of stem densely brownish-pubescent .... 6. *S. atropurpureus* subsp. *tomentosus*
        Involucrum white-tomentose, mixed with purplish septate hairs:
          Rhizome stout; leaves mainly basal .... 4. *S. atropurpureus* subsp. *atropurpureus*
          Rhizome slender; leaves not mainly basal .... 5. *S. atropurpureus* subsp. *frigidus*
      Involucral bracts with gray or yellowish pubescence; heads normally not solitary:
        Leaves floccose on both sides; heads large, with long, purple, orange or yellow ligules; involucral bracts floccose-lanate ...... 3. *S. fuscatus*
        Leaves tomentose or slightly floccose below only; heads smaller, with short ligules or none; involucral bracts profusely pubescent, with long, yellowish, multicellular hairs .... 2. *S. yukonensis*
    Involucral bracts glabrous (upper part of peduncles sometimes floccose):
      Heads solitary, or solitary heads at tops of branched stem; basal leaves reniform, lyrate, lanceolate or oblanceolate, lobed or grossly serrated:
        Achenes hirtellous .... 9. *S. hyperborealis*
        Achenes glabrous:
          Basal leaves reniform, rounded or lyrate; stem usually simple; plant glabrous or slightly tomentose at base .... 7. *S. resedifolius*
          Basal leaves obovate or oblanceolate, grossly serrated; stem usually branched; base of plant and sometimes leaves beneath distinctly tomentose .... 8. *S. conterminus*
      Heads several to numerous, in corymbose cymes; basal leaves long-petiolated, round, ovate, or oblanceolate, entire, toothed or serrated:
        Heads lacking ligules .... 10. *S. pauciflorus*
        Heads with ligules:
          Leaves thick; basal leaves ovate to spatulate, dentate-crenate only in apex; upper cauline leaves linear, entire or incised, toothed; ligules short .... 11. *S. cymbalarioides*
          Leaves thin; basal leaves oblanceolate, serrate-toothed or crenate also in lower part; upper cauline leaves laciniate; ligules long .... 12. *S. pauperculus*
Involucral bracts in 2 or more rows or with accessory outer scales below involucrum:
  Plant annual weed, with small, cylindrical, always discoid, heads; involucral bracts black-tipped .... 13. *S. vulgaris*
  Plant perennial, usually with ligulate heads; involucral bracts black-tipped or not:
    Leaves large, pinnatifid or subpalmately divided into acute, sharply serrulated lobes; plant very tall .... 14. *S. cannabifolius*
    Leaves smaller, shallowly pinnatifid or entire:
      Leaves shallowly pinnatifid, not reduced upward ...... 15. *S. eremophilus*
      Leaves entire or denticulate:
        Heads very large, 5–6 cm in diameter; leaves tomentose beneath .... 16. *S. pseudo-Arnica*
        Heads much smaller; leaves glabrous or nearly so:
          Leaves (at least the basal) triangular or obovate; involucral bracts not black-tipped .... 17. *S. triangularis*
          Leaves lanceolate, oblanceolate to elliptic-lanceolate; involucral bracts dark-tipped:
            Basal leaves lacking; cauline leaves broad; involucral bracts with hyaline, scarious margin .... 18. *S. sheldonensis*
            Basal leaves present; cauline leaves much reduced; involucral bracts lacking translucent hyaline margin .... 19. *S. lugens*

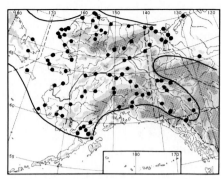 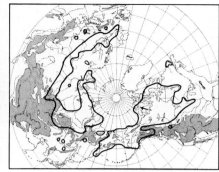

1. Senècio congéstus (R. Br.) DC.   Marsh Fleabane
Cineraria congestus R. Br.; *Senecio palustris* (L.) Hook. not Velloso; *Othonna palustris* L.; *S. arcticus* Rupr.

Stem hollow, simple, stout, up to 1 meter tall, or sometimes slender, low; leaves ovate-lanceolate to lanceolate or oblong-lanceolate, undulate, shallowly pinnatifid or grossly dentate; inflorescence dense, villous or woolly; ligules short, yellow; pappus white, much longer than seeds.

Wet places, in the lowlands; locally common.

Extremely variable. Specimens from the arctic are low-growing, coarse and very woolly (var. **congéstus**), whereas most others are taller and less densely pubescent [var. **palústris** (L.) Fern.]. Specimens with scattered hairs on stem and leaves also occur (var. **tónsus** Fern.).

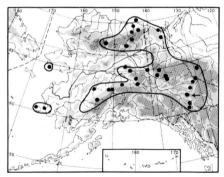 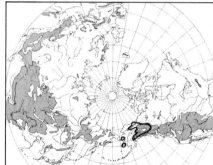

2. Senècio yukonénsis Pors.
Senecio alaskanus Hult.

Stem mostly single, floccose, pubescent, from thick, short rhizome; basal leaves lanceolate to oblanceolate, entire or remotely sinuate-dentate, glabrescent in age; stem leaves reduced, linear; heads 3–6 in dense, aggregate, yellowish-lanate to brownish-lanate inflorescence, glabrescent in age; involucral bracts lanceolate-linear, acute, often dark purplish; ligules pale yellow, short; achenes ribbed, glabrous; pappus dirty white.

Heaths, in the mountains to at least 2,000 meters.

89. Compositae (Composite Family)

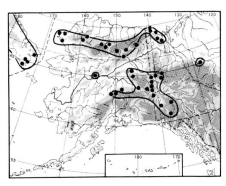 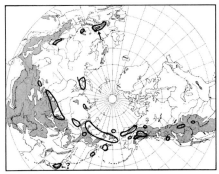 

×½

3. Senècio fuscátus (Jord. & Fourr.) Hayek

Tephroseris fuscata Jord. & Fourr.; *Senecio integrifolius* var. *Lindstroemii* Ostenf.; *S. Lindstroemii* (Ostenf.) Pors.; *S. denalii* Nels.; *S. tundricola* Tolm.; *S. bivestitus* Cronq.

Densely arachnoid-tomentose or villous-tomentose when young; stems from short caudex; basal leaves broadly oblanceolate to ovate, obtuse to somewhat acute, sessile or short-petiolate, entire or irregularly toothed; upper leaves reduced, the uppermost lanceolate; heads large, single to several; involucral bracts lanceolate, purplish; ligules up to 20 mm long, orange, yellow, or with purplish stripes; achenes more or less pubescent.

Alpine meadows. Circumpolar map indicates range of S. *fuscatus* in a broad sense.

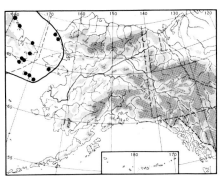 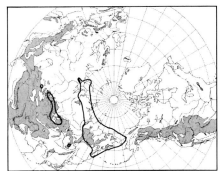 

×½

4. Senècio atropurpùreus (Ledeb.) Fedtsch.

Cineraria atropurpurea Ledeb.

subsp. **atropurpùreus**

Stems low, from stout rhizome, tomentose, glabrescent in age; leaves mainly basal, entire, ovate to obovate, thinly tomentose; cauline leaves narrow, reduced upward; heads solitary; involucrum white-tomentose and with septate, purplish hairs; ligules 1–2 cm long, yellow; achenes glabrous or nearly so.

Meadows on tundra. Described from Siberia.

×½

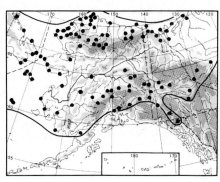
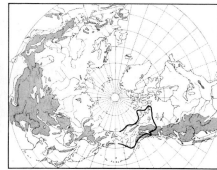

5. Senècio atropurpùreus (Ledeb.) Fedtsch.
subsp. **frígidus** (Richards.) Hult.

Cineraria frigida Richards.; *Senecio frigidus* (Richards.) Hook.

Similar to subsp. *atropurpureus,* but rhizomes slender, stem taller; heads sometimes with very short ligules.

Described from "Barren Grounds from Point Lake to the Arctic Sea." A slender form has been called var. **Ulméri** (Steffen) Pors. (*S. Ulmeri* Steffen).

×½

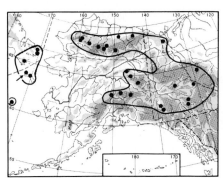
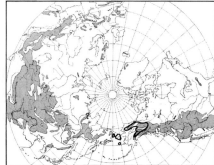

6. Senècio atropurpùreus (Ledeb.) Fedtsch.
subsp. **tomentòsus** (Kjellm.) Hult.

Cineraria frigida f. *tomentosa* Kjellm.; *Senecio frigidus* var. *tomentosus* (Kjellm.) Cufodontis; *S. Kjellmannii* Pors.; *S. Tichomirovii* Schischk.?; *S. jacutensis* Schischk.?.

Similar to subsp. *atropurpureus* and with stout rhizome, but stouter; upper part of stem and involucrum densely brownish, floccose-tomentose.

On tundra and in the mountains to at least 1,750 meters.

In var. **dentàtus** (Gray) Hult. (*S. frigidus* var. *dentatus* Gray), the leaves are more or less coarsely dentate.

89. Compositae (Composite Family)

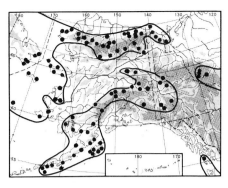
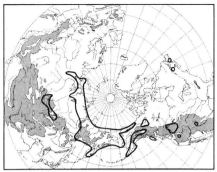

7. Senècio resedifòlius Less.
Cineraria lyrata Ledeb.

Nearly glabrous; stem mostly simple, from short caudex; basal leaves ovate, lyrate or reniform, crenate to dentate; cauline leaves mostly pinnatifid, the upper reduced; heads solitary; involucral bracts often purplish; ligules long, yellow, mostly more or less reddish on back, or lacking [f. **columbiénsis** (Gray) Fern. (var. *columbiensis* Gray)]; disk corollas orange; achenes glabrous; pappus white.

Rocks. Described from Siberia.

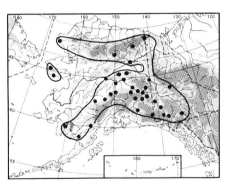
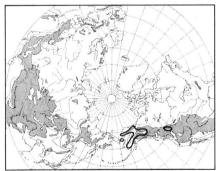
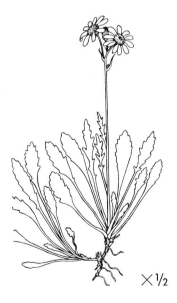

8. Senècio contérminus Greenm.

Similar to S. *resedifolius* and somewhat doubtfully distinct, but stem mostly branched; heads more than 1; basal leaves obovate to oblanceolate, grossly serrate; leaves tomentose.

Described from the Rocky Mountains.

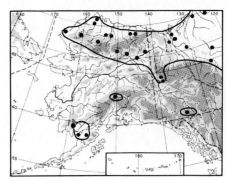
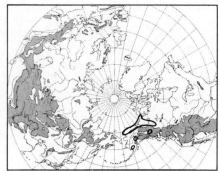

9. Senècio hyperboreàlis Greenm.

Very similar to S. *resedifolius* and S. *conterminus*, differing chiefly in having hirtellous achenes.

Described from arctic America.

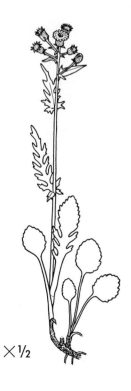
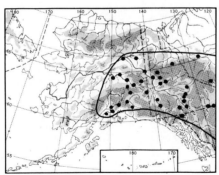
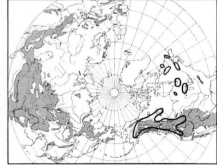

10. Senècio pauciflòrus Pursh

Stem 1.5–3 dm tall, glabrous or floccose, from caudex; leaves chiefly basal, roundish to elliptic-ovate, subentire to crenate, with truncate base; stem leaves reduced, toothed to pinnatifid; heads few, lacking ligules; flowers orange or reddish; involucral bracts reddish to purple; achenes glabrous.

Meadows, wet places. Described from Labrador and the Carolinas.

Var. **fállax** Greenm. (*S. indecorus* Greene), with more numerous heads and mostly yellow flowers, occurs in the lowlands. Circumpolar map indicates range of S. *pauciflorus* in a broad sense.

89. COMPOSITAE (Composite Family)

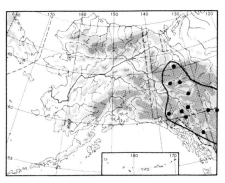 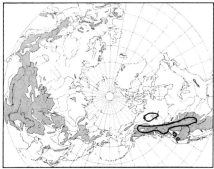 

11. Senècio cymbalarioìdes Nutt.

Stems few, from short caudex; leaves glabrous or floccose, more or less succulent, the lower broadly elliptic to roundish, entire or coarsely crenulate toward apex; cauline leaves few, reduced, mostly pinnately lobed; heads several; ligules 6–12 mm long, yellow; achenes glabrous.

Wet places in the mountains, to at least 1,000 meters. Described from Arkansas.

Most specimens belong to the low-growing northern type, var. **boreàlis** (Torr. & Gray) Greenm. (*S. aureus* L. var. *borealis* Torr. & Gray).

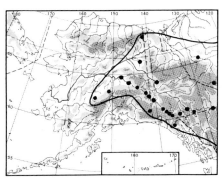 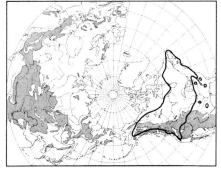 

12. Senècio paupérculus Michx.

Similar to *S. cymbalarioides*, but taller; leaves thin, the basal oblanceolate, crenate-serrate in lower half also; cauline leaves larger, pinnate, with acute segments.

Wet places. Described from Canada.

89. COMPOSITAE (Composite Family)

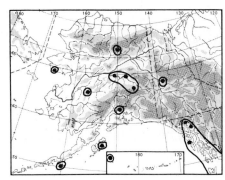
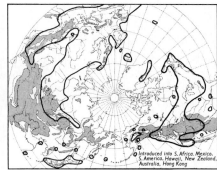

13. Senècio vulgàris L. Common Groundsel

Stem weak, irregularly branched, from thin root; leaves pinnatifid, with oblong, blunt, irregularly toothed lobes, the upper semiamplexicaule, glabrous or cottony; heads in dense corymbose clusters; involucre cylindrical, glabrous, with linear, acute bracts, the outer black-tipped; ligules lacking; disk flowers yellow; achenes densely hairy; pappus white, long.

An introduced weed. Described from Europe.

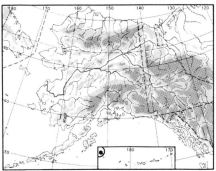
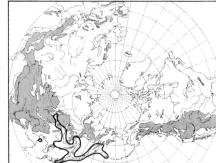

14. Senècio cannabifòlius Less.

Senecio palmatus (Pall.) Ledeb., not La Peyr., nor Less.; *Solidago palmata* Pall. (name only).

Stem 1–2 meters tall, glabrous, from short caudex; leaves pinnatifid or subpalmately divided, with oblong, lanceolate, acute, sharply serrulate lobes, dark green above, paler beneath; heads numerous, small, in flat-topped inflorescence; ligules 10 mm long, yellow; achenes glabrous; pappus tawny.

Meadows on Attu Island. Described from Kamchatka and Bering Island.

89. Compositae (Composite Family)

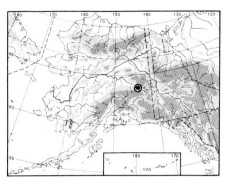
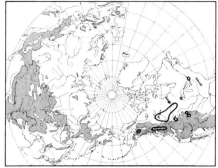

15. Senècio eremóphilus Richards.

Stem from thick, woody root; basal leaves lacking; stem leaves pinnatifid, glabrous, not much reduced upward; several small, lanceolate bracts below heads; heads several; involucrum somewhat floccose-tomentose; ligules about 1 cm long, yellow; achenes glabrous or somewhat pubescent; pappus white.

A weed; introduced near Tok.

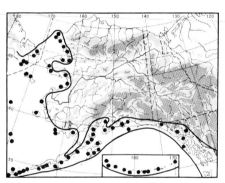
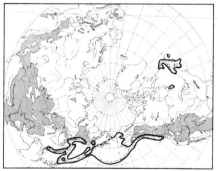

16. Senècio pseùdo-Árnica Less.
Arnica maritima L.

Stem stout, white-woolly, from deep, vertical rhizome; leaves obovate to obovate-lanceolate, fleshy, green above, white-tomentose beneath, repand-dentate; heads few; involucre more or less tomentose; ligules 10–20 mm long, yellow; disk broad; pappus white.

Sandy beaches. Described from Kamchatka and North America.
(See color section.)

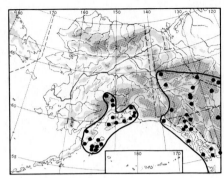 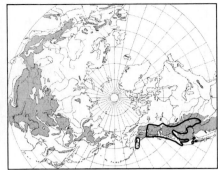

17. Senècio trianguláris Hook.

Stems several, simple, from stout rootstock, glabrous to puberulent; leaves cauline, many, triangular-ovate to triangular-hastate, truncate-cordate at base, sharply toothed, gradually reduced upward; heads several, in flat-topped inflorescence; involucral bracts lanceolate, glandular at tip; ligules few, 7–10 mm long, yellow; achenes glabrous.

Wet meadows, stream banks. Described from the Rocky Mountains.

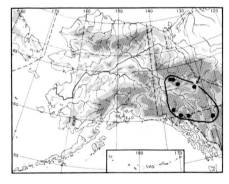 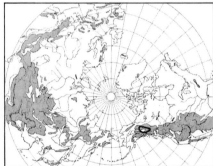

18. Senècio sheldonénsis Pors.

Glabrous or nearly so; stem simple, slender, from short rhizome; basal leaves lacking; stem leaves broadly lanceolate to elliptic-lanceolate, denticulate, the upper sessile; heads 3 to several; involucral bracts linear, with hyaline margins and dark, pubescent tips; ligules yellow; achenes glabrous; pappus white.

Subalpine meadows.

89. COMPOSITAE (Composite Family)

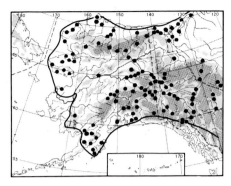
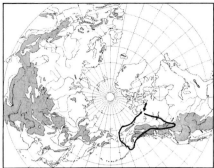
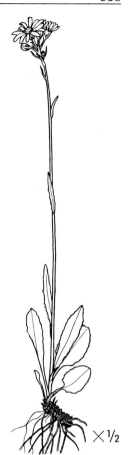

19. Senècio lùgens Richards.
Senecio imbricatus Greene.

Stem usually solitary, glabrescent, from stout rhizome; basal leaves oblanceolate to elliptic, denticulate to nearly entire, nearly glabrous; cauline leaves few, reduced, sessile; inflorescence corymbose; peduncles with several linear bracts above; heads several; involucrum arachnoid, pubescent to glabrous; involucral bracts broadly lanceolate, with black, triangular tip, dark-pubescent at apex; ligules up to 16 mm long, usually much shorter, yellow; achenes glabrous; pappus white.

In the mountains to at least 2,000 meters in the Yukon.

The species name alludes to the type locality: "Bloody Fall, where the Esquimaux were destroyed by the Northern Indians that accompanied Hearne."

Saussurea DC.

Plant tall, often over 1 meter tall; all leaves broad, ovate, cordate or oblong-ovate. coarsely toothed; involucral bracts with dark margins and tip, the outer much shorter than the inner; receptacle somewhat squamate in middle .... 1. *S. americana*
Plant more low-growing; lower leaves ovate, elliptic-lanceolate or linear, the upper narrow:
  Receptacle lacking hyaline scales; outer bracts equaling inner bracts or nearly so .... 6. *S. nuda*
  Receptacle squamate from long, lanceolate hyaline scales:
    Plant often tall; basal leaves linear with involute margins; involucral bracts in 3–4 rows; leaves not viscid-pubescent .... 2. *S. angustifolia*
    Plant low-growing; basal leaves broader; involucral bracts in 2–3 rows:
      Leaves tomentose beneath, lacking viscid pubescence .... 5. *S. Tilesii*
      Leaves not tomentose beneath, with at least some viscid pubescence:
        Leaves markedly viscid-pubescent, especially in margins, not floccose .... 3. *S. viscida* var. *viscida*
        Leaves less markedly viscid-pubescent, with floccose pubescence .... 4. *S. viscida* var. *yukonensis*

89. COMPOSITAE (Composite Family)

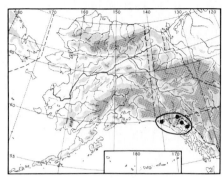
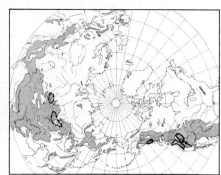

1. Saussùrea americàna DC.
Saussurea pseudofoliosa Lipsch. ?

Stem tall and coarse, from stout caudex; leaves lanceolate-ovate, coarsely toothed, the lower petiolated, green above, pale beneath, floccose when young; inflorescence corymbiform, very dense; involucral bracts firm, ovate-lanceolate, with dark edges; flowers purplish; pappus white to brownish.

Moist meadows.

Broken line on circumpolar map indicates range of the closely related **S. foliòsa** Ledeb.

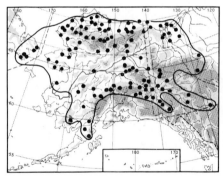
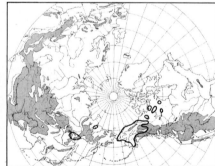

2. Saussùrea angustifòlia (Willd.) DC.
Serratula angustifolia Willd.; *S. monticola* Hook.

Stem glabrous or floccose, from woody, horizontal rootstock; leaves linear to lanceolate, entire in margin or very sparsely dentate, glabrous or floccose, often involute in margin; inflorescence corymbose; outer involucral bracts ovate to deltoid, pilose and arachnoid-tomentose, the inner narrowly lanceolate; receptacle squamate, with lanceolate, hyaline scales; ligules narrow, purplish; anthers purplish; pappus tawny, plumose.

Dry places on tundra and in the mountains. Described from eastern Siberia.

Hybrids with *S. nuda* probably occur where ranges overlap (*S. tschuktschorum* Lipsch.). Broken line on circumpolar map indicates range of the closely related **S. pseudoangustifòlia** Lipsch.

89. Compositae (Composite Family)

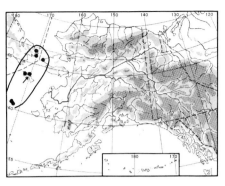
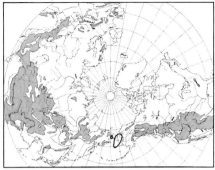

3. Saussùrea víscida Hult.
var. **víscida**

Stem low, from short caudex, slightly floccose at base, viscid-pubescent from septate hairs, especially above; lower leaves elliptic-lanceolate, entire or remotely denticulate, sessile or short-petiolated, pubescent on both sides from multicellular, viscid hairs, densely viscid-ciliate in margin, not floccose, the upper leaves reduced; inflorescence compressed; receptacle squamate; involucral bracts triangular to lanceolate-triangular, acute, the outer shorter and broader; pappus plumose; achenes glabrous.

Dry places on tundra. *S. viscida* described from Sevoonga (St. Lawrence Island).

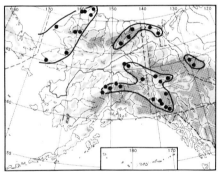
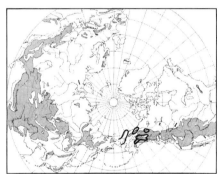

4. Saussùrea víscida Hult.
var. **yukonénsis** (Pors.) Hult.
Saussurea angustifolia var. *yukonensis* Pors.

Similar to var. *viscida*, but leaves sparsely floccose and sparsely viscid-pubescent.
Dry places on tundra; in the mountains to over 2,100 meters. Described from the mountains of the Yukon and the Northwest Territories.
Possibly the hybrid *S. angustifolia* × *viscida*.

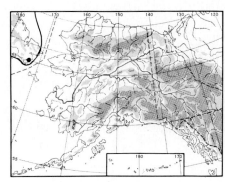
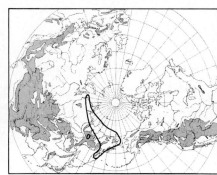

5. Saussùrea Tilésii Ledeb.

Stem low, from short caudex; leaves ovate-lanceolate to oblong, dentate, green and sparsely glandular-dotted above, white-tomentose beneath, the upper sometimes exceeding the dense, silky-pubescent inflorescence; involucral bracts ovate-lanceolate, acute; flowers purplish; receptacle with long, shiny scales; pappus white to brownish.

Tundra and mountain slopes. Described from Kamchatka.

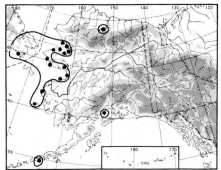
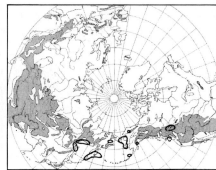

6. Saussùrea nùda Ledeb.

Stems glabrous below, floccose above, from stout, branching caudex; basal leaves long-petiolated, ovate-lanceolate to ovate, sparsely denticulate, glabrous; green on both sides; stem leaves reduced, lanceolate; inflorescence corymbose, floccose, with long, linear bracts; involucral bracts lanceolate to ovate-lanceolate, acute; flowers purplish, anthers, dark; receptacle lacking scales; pappus tawny.

Seashores, alpine meadows. Described from Kamchatka and Cape Espenberg (Kotzebue Sound).

Broken line on circumpolar map indicates range of var. **dénsa** (Hook.) Hult. (*S. densa* Hook.).

89. Compositae (Composite Family)

Cirsium Mill.

Heads about 1 cm broad; stem not winged .... 1. *C. arvense*
Heads much larger:
    Stem upper part and branches with prickly, undulate, broad wings ....
    .... 2. *C. vulgare*
    Stem not winged:
        Leaves pubescent on nerves only, not tomentose beneath ....
        .... 5. *C. kamtschaticum*
        Leaves tomentose or arachnoid-pubescent beneath:
            Outer involucral bracts narrow, gradually tapering to short spine; upper
                leaves broader, not exceeding heads .... 3. *C. edule*
            Outer involucral bracts broad; upper leaves linear, exceeding heads .....
            .... 4. *C. foliosum*

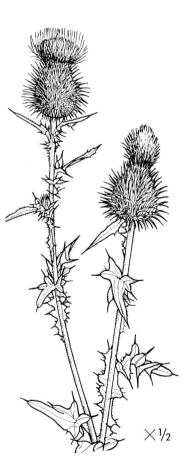

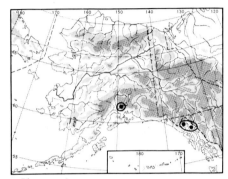 

1. Círsium arvénse (L.) Scop. Canada Thistle
Serratula arvensis L.

Stem unwinged, furrowed, from slender taproot with far-creeping lateral roots producing shoots; basal leaves oblong-lanceolate in outline, more or less pinnatifid, with triangular, toothed lobes ending in spines, glabrous or floccose; upper leaves more deeply pinnatifid; heads solitary or 2–4 together; outer involucral bracts ovate-mucronate with spreading spiny points, the inner longer, lanceolate-acuminate, with scarious tips; flowers pale purple or whitish; pappus long, brownish.

An introduced weed. Described from Europe.

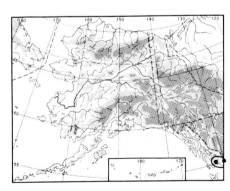 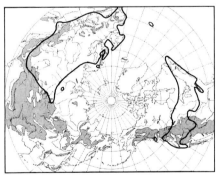

2. Círsium vulgàre (Savi) Ten. Bull Thistle
Carduus vulgaris Savi; *C. lanceolatus* L.; *Cirsium lanceolatum* (L.) Scop. not Hill.

Stem branched above, spiny-winged, furrowed, from long taproot; basal leaves obovate-lanceolate in outline, deeply pinnatifid, with 2-lobed segments, the upper toothed near base, the teeth tipped with long, stout spines, prickly-haired above, floccose below; heads solitary or 2–3; involucral bracts lanceolate-acuminate, the outer spine-tipped, the inner scarious; flowers pale purple; pappus long, white.

An introduced weed. Described from Pisa, Italy.

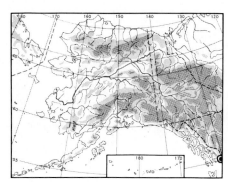
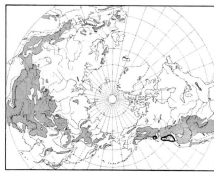

3. Círsium edùle Nutt.
Cnicus edulis (Nutt.) Gray; *Carduus edulis* (Nutt.) Greene.

Stem arachnoid-villous, from taproot; leaves more or less pinnatifid to coarsely toothed, green and sparsely arachnoid-villous on both surfaces, moderately spiny; heads single, or in small clusters, more or less arachnoid-villous; outer involucral bracts gradually tapering to short spine, the inner spineless; corollas purple; style exserted 3–8 mm beyond corolla lobes.

Wet meadows, woods. Described from Chilliwack Valley, British Columbia.

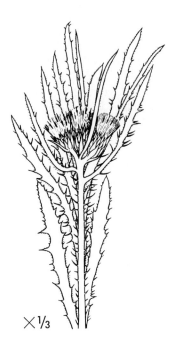
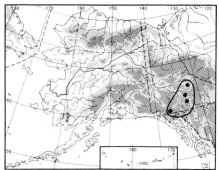
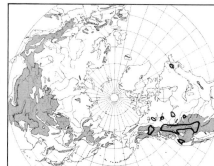

4. Círsium foliòsum (Hook.) DC.
Carduus foliosus Hook.

Stem fleshy, simple, more or less arachnoid-villous, from taproot; heads few, sessile; lower leaves lanceolate-linear, subentire to pinnatifid, with weak, yellow spines, light green and arachnoid-pubescent above, tomentose beneath; upper leaves linear, exceeding heads; involucrum glabrous; outer involucral bracts ovate, with weak spine, the inner elongate, spineless; flowers pale.

Meadows. Described from the Rocky Mountains.

Circumpolar map gives range of *C. foliosum* in a broad sense, including *C. Drummondii* Torr. & Gray and *C. minganense* Vict.

89. Compositae (Composite Family)

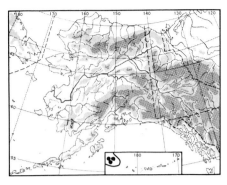
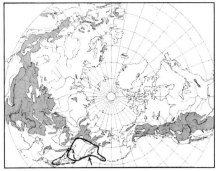

5. Círsium kamtscháticum Ledeb.
Cnicus kamtschaticus (Ledeb.) Maxim.

Stem tall, coarse, brownish-pubescent, especially above, from coarse, horizontal, woody caudex; leaves ovate to ovate-lanceolate in outline, nearly glabrous above, pubescent on nerves below, deeply pinnatifid into broad lobes, armed with weak prickles; heads large, 1–3, nodding in age, floccose-pubescent; involucral bracts linear, ending in weak spines, the outer shorter than the inner; flowers purplish, soon turning brown; pappus brown.

Meadows. Described from Kamchatka.

Broken line on circumpolar map indicates range of var. **Weyríchii** (Maxim.) Herder (*C. Weyrichii* Maxim.) and subsp. **pectinéllum** (Gray) Kitamura (*C. pectinellum* Gray); the two ranges are about the same.

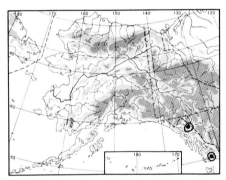
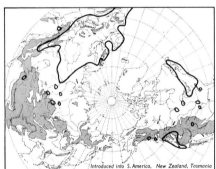

Lapsana L.

1. Lápsana commùnis L.   Nipplewort

Stem branched and glabrous above; lower leaves lyrate-pinnatifid, with large, coarsely sinuate-toothed terminal lobe and few small lateral lobes; heads numerous, on slender stalks; flowers yellow; involucral bracts 8–10, linear, keeled, suddenly contracted to blunt apex; achenes lacking pappus.

An introduced weed. Described from Europe.

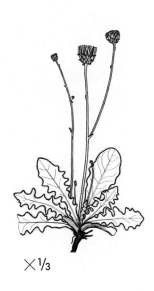

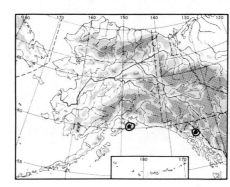 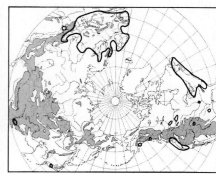

Hypochoeris L.

1. Hypochoèris radicàta L.   Cat's-Ear

Stems usually several, from fleshy roots, forking, scapose, enlarged below heads and with numerous scalelike bracts; leaves somewhat glaucous beneath, in basal rosette, broadly oblong-lanceolate, sinuate-toothed to sinuate-pinnatifid, hispid with simple hairs; flowers yellow, the outer greenish beneath; involucral bracts lanceolate-acuminate, bristly on midrib; achenes orange, muricate.

An introduced weed. Described from Europe.

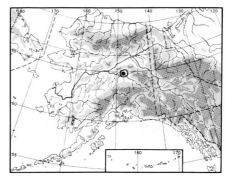 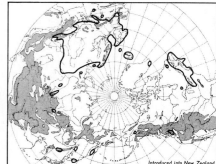

Leontodon L.

1. Leóntodon autumnàlis L.   Fall Dandelion

Stem decumbent below, usually branched and with numerous scalelike bracts above, from branched caudex with rosettes of leaves at ends of branches; leaves sinuate-toothed to deeply pinnatifid, glabrous or with simple hairs; flowers yellow, the outer with reddish streaks beneath; involucral bracts linear-lanceolate, acute; achenes beakless, with pappus a single row of feathery hairs.

Waste places; an introduced weed. Described from Europe.

89. Compositae (Composite Family)

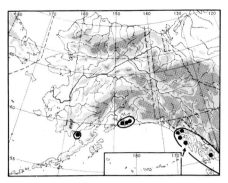
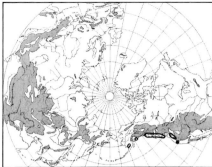

Apargidium Torr. & Gray

1. Apargídium boreàle (Bong.) Torr. & Gray

Apargia borealis Bong.; *Leontodon boreale* (Bong.) DC.; *Microseris borealis* (Bong.) Schultz-Bip.; *Scorzonella borealis* (Bong.) Greene.

Glabrous or nearly so; stem from short caudex with several taproots, scapose, with 1–2 small bracts, often somewhat arcuate; basal leaves narrowly lanceolate, entire or remotely toothed; heads solitary; involucral bracts lanceolate, acuminate, sparsely dark-pubescent; receptacle naked; flowers yellow; pappus brownish, the bristles connected at base, capillary, barbellate.

Wet meadows.

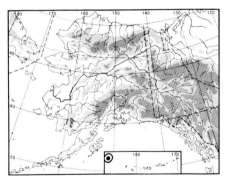
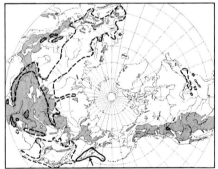

Picris L.

1. Pìcris hieracioìdes L.
subsp. **kamtschática** (Ledeb.) Hult.
Picris kamtschatica Ledeb.

Stem thick, straight, branched above, densely setose, with hooked, dark bristles; leaves lanceolate to oblanceolate, serrate, setose; heads several, in corymbose inflorescence; outer involucral bracts short, narrow, recurved, with black hairs, the inner lanceolate; flowers yellow; achenes with short beak; pappus dirty white, in 2 rows.

Meadows. *P. hieracioides* described from Europe, subsp. *kamtschatica* from Kamchatka.

Broken line on circumpolar map indicates range of other subspecies of *P. hieracioides* (American occurrences of these subspecies are introductions).

Taraxacum Zinn (Dandelion)

Many or most *Taraxacum* taxa do not employ normal sexual reproduction, but form viable seeds without fertilization, which gives rise to exactly similar plants. It is therefore possible to recognize a large number of young but constant taxa, just as it is possible to recognize different kinds of apples propagated without fertilization. Only a specialist is able to distinguish them. No doubt their distribution and differentiation are of much phytogeographical interest, as clues to plant dispersal in fairly recent time. A few *Taraxacum* taxa have a larger, or even worldwide, range and are certainly older. In this book only a selection of the taxa that are easiest to distinguish or have a large range are treated; the others are treated as collective groups representing sections of the genus, which can be said to correspond to species in plants with normal cross-fertilization.

Involucral bracts lacking horns or tubercles below apex, or only the innermost slightly gibbous:
  Plants tall, introduced, weedy, with large heads; involucral bracts not blackish-green .... 1. *T. officinale*
  Plants small, rarely exceeding 15 cm in height, occurring in natural habitats, with small heads; involucral bracts blackish-green (sect. GLABRA):
    Lateral lobes of leaves approximate, short, more or less retrorse, acute; achenes brown, tuberculate above, smooth at base .... 9. *T. alaskanum*
    Lateral lobes of leaves (especially middle leaves) less approximate, somewhat attenuate, blunt or clawlike, with rounded corners; petioles purplish; achenes chestnut-red, smooth or with a few, very small spines at top only .... 10. *T. kamtschaticum*
Involucral bracts (a few, if not all) with horns or tubercles below apex:
  Achenes small, about 3 mm long, brownish-brick-red, with narrow, cylindrical beak; introduced plant (sect. ERYTHROSPERMA, represented by a single species) .... 6. *T. scanicum*
  Achenes larger, with much broader beak, grayish, greenish, yellowish, blackish-green, or brownish:
    Achenes blackish or blackish-green with very short, conical beak, about as long as body of achene (sect. ARCTICA):
      Achenes uniformly spinulose and muricate; ligules pale yellow; involucral bracts 9–14 mm long .... 7. *T. phymatocarpum*
      Achenes smooth at base; ligules whitish, violet, or brown; involucral bracts 13–15 mm long .... 8. *T. hyparcticum*
    Achenes not blackish, beak 2–3 times as long as body of achene:
      Ligules purplish or flesh-colored, about 1 mm broad; small, high alpine plants .... 11. *T. carneocoloratum*
      Ligules yellow, broader, often with darker greenish or purplish stripes below; tall plants, mostly 15–45 cm tall (sect. CERATOPHORA):

        Within the section CERATOPHORA (see especially no. 2, *T. ceratophorum*, below), about 45 apomictic microspecies have been distinguished within the area of interest; three of these are widespread and fairly characteristic, and might be distinguished by the following characteristics:

      Petioles often more or less purplish:
        Achenes red-colored; beak conic-cylindrical, about 1.25 mm long; lateral lobes of leaves dentate .... 3. *T. lateritium*
        Achenes olivaceous to straw-colored; beak much longer than body; leaves much dissected; lobes lacking teeth in margin .... 4. *T. lacerum*
      Petioles usually not red-colored, more or less alate; achenes usually straw-colored or grayish to blackish:
        Lateral lobes of leaves short, triangular; terminal lobe small, trigonous-rhomboid, often short-acuminate; outer involucral bracts much shorter than the inner .... 5. *T. trigonolobum*
        Leaves and involucral bracts of various forms ....
.... Sect. CERATOPHORA (other than those mentioned)

89. COMPOSITAE (Composite Family)

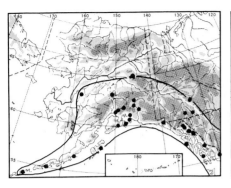
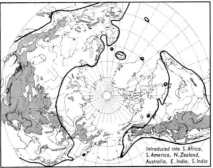

1. Taráxacum officinàle Weber
Taraxacum sect. VULGARIA.

Leaves subentire, sinuate-dentate or variously pinnatifid, the lobes mostly toothed; heads large, up to 5 cm broad; ligules orange to yellow; involucral bracts lacking horns or tubercles below apex, the outer lanceolate to linear, more or less spreading or reflexed; achenes olivaceous, the body 3.5–4 mm long with slender beak, 2.5–4 times as long as body, variously tuberculate-spinulose; pappus white.

Waste places, roadsides. Described from western Europe.

An aggregate species comprising numerous mostly apomictic microspecies. The most widespread of these within the area of interest is **T. vàgans** Hagl., with narrow outer involucral bracts about 2 mm broad; others are **T. Dahlstédtii** Lindb. f., **T. undulàtum** Lindb. f. & Markl., and **T. retrofléxum** Lindb. f., all introduced from Europe.

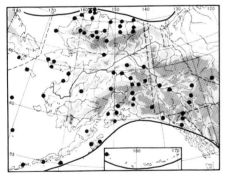
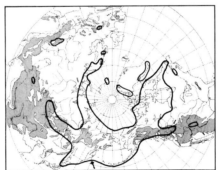

2. Taráxacum ceratóphorum (Ledeb.) DC.
Leontodon ceratophorus Ledeb.

Leaves of various form; all or at least some involucral bracts with horn or tubercle below apex; outer bracts mostly appressed, broader than the inner.

Meadows, moist places in the mountains. Described from Kamchatka.

A large group of small taxa, which maintain themselves as distinct units through seeds that are formed without fertilization. About 45 such microspecies are known from the area of interest; doubtless the number that occur is considerably larger. Circumpolar map indicates range of the entire complex, taken as section CERATOPHORA of the genus *Taraxacum* by Dahlstedt (as subsect. CERATOPHORA by Hand.-Mazz.).

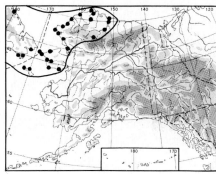 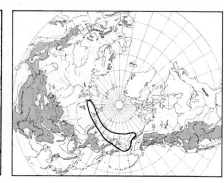

3. Taráxacum laterìtium Dahlstedt

Segregate belonging to the collective *Taraxacum cerataphorum* (Ledeb.) DC.

Outer leaves linear-lanceolate, dentate, the inner oblong-lanceolate, with broad, short, deltoid lobes; terminal lobe broad, entire or partly dentate; involucrum dark-colored; outer involucral bracts appressed, ovate, often lacking horns, the inner lanceolate to broadly ovate-lanceolate, with horns up to 1.5 mm long or tuberculate; ligules yellow, with olivaceous to purplish stripes; immature achenes pale brick-colored, mature achenes dark brick-red, with beak longer than body.

Differs from other taxa of *T. cerataphorum* (taken in a broad sense) in the characteristic color of the achenes.

Meadows, moist places in the mountains. Described from northeasternmost Siberia.

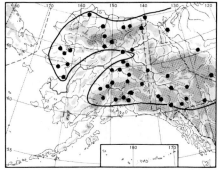 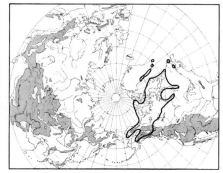

4. Taráxacum lácerum Greene

Taraxacum groenlandicum Dahlstedt; segregate belonging to the collective *Taraxacum cerataphorum* (Ledeb.) DC.

Petioles usually more or less purplish; outer leaves linear, with broad, triangular lobes, the inner linear-oblanceolate, dissected, with irregular, incised-laciniate lobes; terminal lobe elongate, hastate-sagittate; outer involucral bracts ovate-lanceolate, with short, pale-green horns, the inner lanceolate to linear and darker than the outer; ligules yellow, with olivaceous to purplish striae beneath; achenes brown to olivaceous, smooth below, muriculate above, with beak much longer than body.

Meadows, moist places in the mountains. Described from the upper Liard River.

A variable species, here taken in a broad sense.

89. COMPOSITAE (Composite Family)

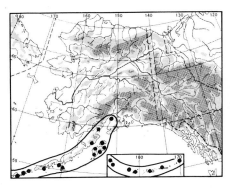
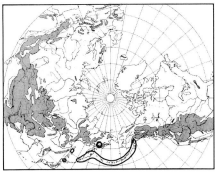
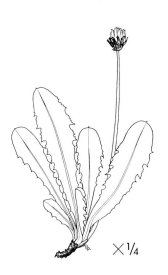

5. Taráxacum trigonolòbum Dahlstedt

Taraxacum Chamissonis Greene in part; *T. aleuticum* Tatew. & Kitamura; *T. yezoalpinum* Nakai; segregate belonging to the collective *Taraxacum ceratophorum* (Ledeb.) DC.

Petioles pale; leaves lanceolate, with short, broad, triangular to deltoid, entire, acute lobes; terminal lobe small, short, triangular-rhomboid, more or less mucronate; involucrum dark green, the outer bracts greenish-margined, short, ovate-triangular to ovate-lanceolate, appressed, conspicuously tuberculate or horned below apex, the inner linear; ligules yellow, with grayish-violet stripes below; achenes tuberculate, with beak twice as long as body.

Meadows, moist places in the mountains. Described from Kamchatka.

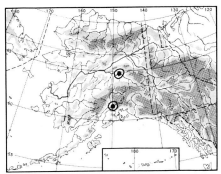
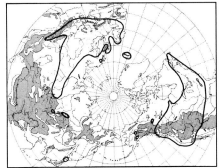

6. Taráxacum scànicum Dahlstedt

Leaves with dark-purplish petioles, deeply lobed, dentate between lobes; lobes attenuate, acute; involucrum dark green; outer involucral bracts broad, with indistinct tubercle below apex, distinctly scarious-margined, soon reflexed; ligules dark yellow; achenes small, brownish brick-red, with narrow, cylindrical beak.

An introduced weed belonging to the section ERYTHROSPERMA. Described from the province of Scania, southern Sweden.

Circumpolar map indicates approximate, very tentative range of entire section ERYTHROSPERMA.

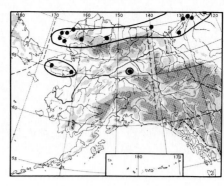
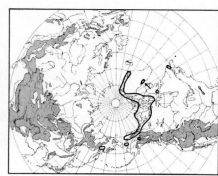

7. Taráxacum phymatocárpum J. Vahl

Including *Taraxacum pumilum* Dahlstedt.

Leaves elongate-ligulate, entire or minutely dentate, the inner sinuate-dentate with purplish median nerve; involucrum small, blackish; outer involucral bracts ovate to ovate-lanceolate, the inner linear-lanceolate, somewhat obtuse; ligules pale yellow; achenes brown to olivate, uniformly spinulose and muricate, with beak about as long as body or shorter.

Alpine slopes, tundra.

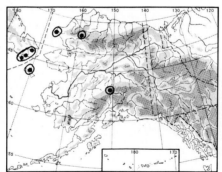
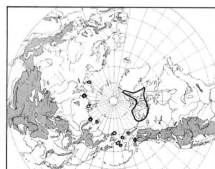

8. Taráxacum hypárcticum Dahlstedt

Similar to *T. phymatocarpum*, but ligules white, purplish, or brown; only the upper part of the achenes muricate and spinose.

Tundra, high-alpine scree slopes. Described from Greenland and Ellesmere land.

On circumpolar map, reports of the very closely related **T. árcticum** Dahlstedt are included.

89. COMPOSITAE (Composite Family)

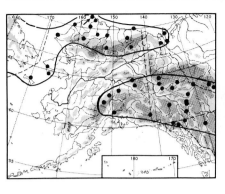
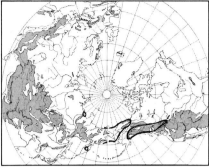

9. Taráxacum alaskànum Rydb.
Taraxacum lyratum of some authors, not DC.

Leaves with purplish petioles and short, retrorse, acute, more or less approximate lateral lobes, the apical lobe triangular, acute; involucrum campanulate; involucral bracts dark, lacking horns, the outer small, acute; marginal ligules yellow, grayish-green beneath; achenes brown, 3.5–4 mm long, tuberculate and short-spinulose above, rugulose in middle, smooth at base, with beak 5.5–6 mm long.

Alpine slopes, tundra.
On circumpolar map, range of the very closely related **T. sibìricum** is included.

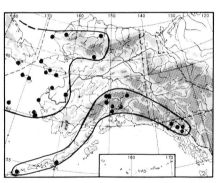
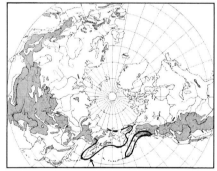

10. Taráxacum kamtscháticum Dahlstedt
Taraxacum lyratum of some authors, not DC.

Leaves with purplish petioles; patent, attenuate lobes with blunt tip; apical lobe larger than in *T. alaskanum*, often blunt; involucral bracts dark, lacking horns or tubercles below apex; ligules yellow, somewhat grayish beneath; achenes smooth, with few spines at top only.

In McKinley Park to over 1,750 meters. Described from Kamchatka.

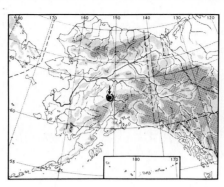
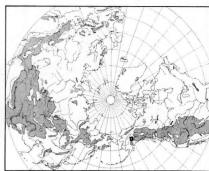

11. Taráxacum carneocolorátum Nels.

Low-growing; leaves oblong-lanceolate, with 3–5 pairs of triangular, somewhat acute lobes; terminal lobe comparatively large, ovate-triangular, blunt; petioles pale at base; heads often woolly at base; outer bracts broadly ovate, purplish, scarious-margined, indistinctly corniculate; ligules up to 18 mm long, about 1 mm broad, pink to flesh-colored; achenes spinulose-muricate at tip, with beak about as long as achene.

High alpine scree slopes; extremely rare.

Sonchus L.

Plant with creeping rhizome; yellow glands on stem, peduncles, and involucrum; achenes 5-ribbed .... 1. *S. arvensis*
Plant annual; glands absent; achenes 3-ribbed:
    Leaves soft, with acute, spreading auricles .... 2. *S. oleraceus*
    Leaves stiff, mostly subentire, with rounded auricles .... 3. *S. asper*

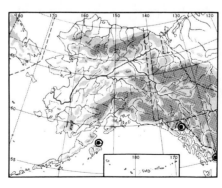
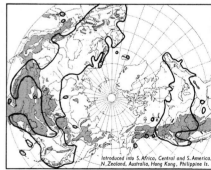

1. Sónchus arvénsis L.   Field Sow Thistle

Stems hollow, with yellow glands, from long, creeping rhizomes; leaves pinnatifid to subentire, cordate, clasping, spiny-toothed; peduncles glandular; heads 4–5 cm broad; flowers dark yellow; involucral bracts oblong-lanceolate, blunt, glandular.

An introduced weed. Described from Europe.

89. Compositae (Composite Family)

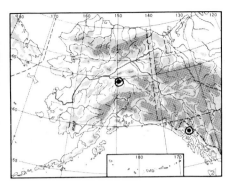 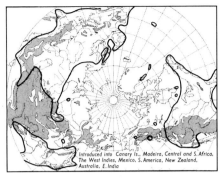 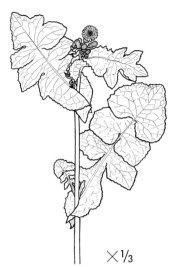

2. Sónchus oleráceus L. Common Sow Thistle

Stem stout, hollow, erect, glabrous, from pale taproot; leaves pinnatifid, with large terminal lobe and acute spreading auricles, glabrous, never spinous; flowers yellow, the outer purple-tinged beneath; outer involucral bracts broadly lanceolate, shorter and more acute than the inner; achenes transversally rugose; pappus white, in 2 rows.

Waste places; an introduced weed. Occurs also in the Galapagos Islands.
Type locality not given.

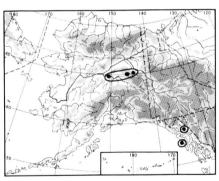 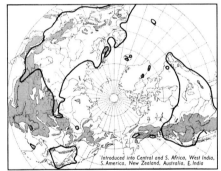 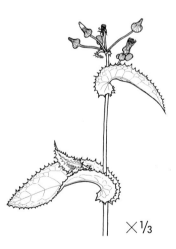

3. Sónchus ásper (L.) Hill Spiny-Leaved Sow Thistle
Sonchus olearaceus var. *asper* L.

Similar to S. *oleraceus*, but stem leaves with rounded auricles; achenes smooth, 3-ribbed on both sides; leaves stiffer, less dissected.

Waste places; an introduced weed. Described from Europe.

Lactuca L. (Lettuce)

Plant tall, biennial, 0.5–3.5 meters high; achenes beakless; pappus brownish .... 1. *L. biennis*
Plant lower, with long, underground rhizome; achenes with short beak; pappus white .... 2. *L. tatarica*

> NOTE: A plant with broad, entire, cordate-amplexicaul leaves, collected at Manley Hot Springs, may be **L. viròsa** L., but the specimen seen lacks seeds and cannot be identified with certainty.

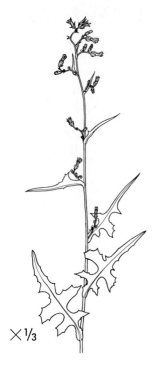
× 1/3

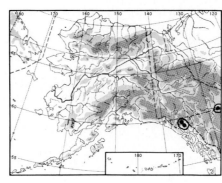 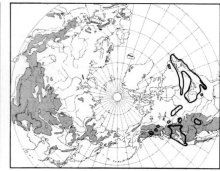

1. Lactùca biénnis (Moench) Fern.

Sonchus biennis Moench; *Lactuca spicata* in Hitchcock's sense, not *Sonchus spicatus* Lam.

Plant tall, coarse, nearly glabrous, leafy; leaves irregularly pinnatifid, coarsely toothed, the upper sessile and auriculate; heads in dense, compound panicle; flowers bluish, creamy, or sometimes yellow; achenes beakless or essentially so; pappus sordid to tawny.

Moist places. Type locality not given.

× 1/2

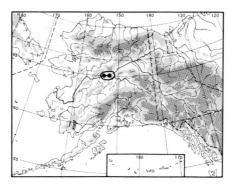 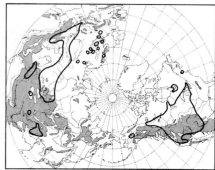

2. Lactúca tatárica (L.) C. A. Mey.   Blue Lettuce

Sonchus tataricus L.; *Lactuca tatarica* subsp. *pulchella* (Pursh) Stebbins? [*L. pulchella* (Pursh) DC.; *S. pulchellus* Pursh].

Stems stout, from long, subterranean rhizome; leaves sessile, entire, oblong-lanceolate, thin, pale, glabrous, the lower more or less pinnatifid, with small end lobe; heads few, with blue to purplish flowers; achenes with short beak.

Moist meadows; also adventive both in Europe and in America. Described from "Tataria and Siberia."

The species has been divided into the Eurasiatic subsp. **tatárica** and the American subsp. **pulchélla** (Pursh) Stebbins (see synonymy above); it is uncertain to which type the Alaskan plant belongs.

89. COMPOSITAE (Composite Family)

Agoseris Raf.

Beak of achenes striate, not longer than half the length of the ripe achene ....
.... 1. *A. glauca*
Beak of achenes not striate, longer than half the length of the ripe achene ....
.... 2. *A. aurantiaca*

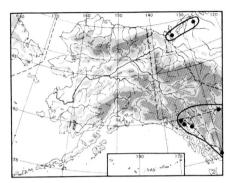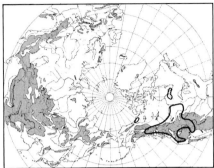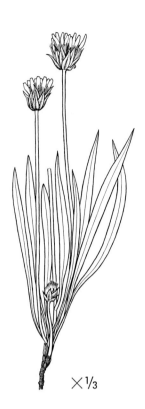

1. Agóseris glaùca (Pursh) Raf.

Troximon glauca Pursh; *Agoseris scorzoneraefolia* (Schrad.) Greene; *Ammogeton scorzoneraefolius* Schrad.; *Agoseris cuspidata* with respect to arctic plant.

Glabrous or nearly so; leaves linear to lanceolate, entire to toothed or laciniate; heads single; involucral bracts imbricate, in most cases sharply pointed; flowers yellow, sometimes drying pinkish; achenes with beak not more than half the length of the body.

Meadows. Described from the banks of the Missouri. Extremely variable.

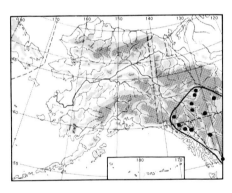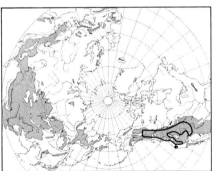

2. Agóseris aurantìaca (Hook.) Greene

Troximum aurantiacum Hook.; *Agoseris gracilens* (Gray) Ktze.; *T. gracilens* Gray.

Glabrous or somewhat villous; leaves oblong-lanceolate, subentire to pinnatifid; heads single; outer involucral bracts oblong, short, the inner lanceolate; flowers orange, turning purple in age; achenes abruptly tapering into slender beak, half the length of the body or longer.

Meadows, in the mountains, to over 1,000 meters. Described from the Rocky Mountains.

89. COMPOSITAE (Composite Family)

Crepis L. (Hawk's-Beard)

Involucral bracts tomentose within .... 1. *C. tectorum*
Involucral bracts glabrous within:
    Plant annual; stem leaves sagittate-auriculate; stem pubescent below ....
    .... 2. *C. capillaris*
    Plant perennial; stem leaves lanceolate, narrow, not sagittate:
        Plant low, lacking erect stem, forming dense tufts with numerous heads; achenes columnar, with very short beak, or none:
            Leaves entire or remotely dentate .... 3. *C. nana* var. *nana*
            Leaves more or less dissected, pinnatifid or lyrate ....
            .... 4. *C. nana* var. *lyratifolia*
        Plant with erect stem:
            Heads large, 20 × 30 mm, mostly single; achenes beakless ....
            .... 6. *C. chrysantha*
            Heads much smaller, stem branched; achenes with long, slender beak when ripe .... 5. *C. elegans*

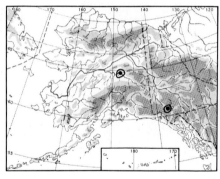

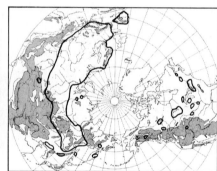

1. Crèpis tectòrum L.

Stem puberulent, single, branched, erect, puberulent, from taproot; basal leaves lanceolate or oblanceolate, denticulate to pinnately parted, acute; stem leaves linear, sessile, auriculate; heads several; involucral bracts tomentose-puberulent and sometimes glandular within; ligules yellow; styles greenish; achenes beakless.

An introduced weed. Described from Europe.

89. Compositae (Composite Family)

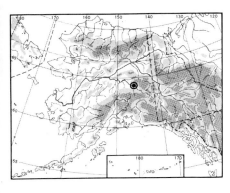 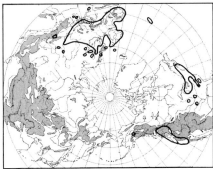 

2. Crèpis capillàris (L.) Wallr.
Lapsana capillaris L.; *Crepis virens* L.

Annual, with thin root; stem pubescent below, branched from below; leaves lanceolate, denticulate to pinnatifid; stem leaves reduced, the upper sessile, clasping; involucral bracts acute, tomentose, glabrous within; ligules golden yellow; styles yellow.

An introduced weed. Described from Switzerland.

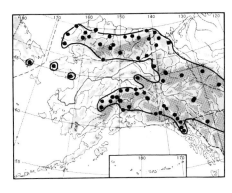 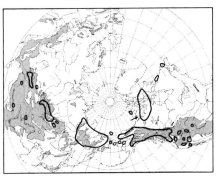 

3. Crèpis nàna Richards.
Borkhausia nana (Richards.) DC.; *Youngia nana* (Richards.) Rydb.; *Prenanthes pygmaea* Ledeb.
var. **nàna**

Glabrous; sterile rosettes of first year very dense, with numerous ovate, elliptical or obovate, blunt or somewhat acute, petiolated leaves; flowering plant with numerous capillary, branching stems, from all sides of a short, thick, mostly unbranched rhizome, with taproot, forming a very dense tuft, in age lacking basal leaves; stem leaves elliptical to oblanceolate, entire or somewhat toothed, on long petioles, often exceeding the compact mass of cylindrical heads; outer involucral bracts short, lanceolate to broadly lanceolate, the inner oblong, obtuse, scarious-margined, ciliate at apex, often purplish; ligules yellow or purplish; achenes columnar, with very short beak; pappus white.

When covered with detritus, the stems and petioles become very elongated.

Gravel; in the mountains to at last 2,000 meters. Described from the Coppermine River.

(See color section.)

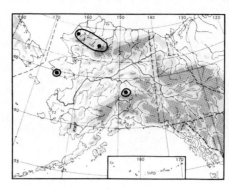
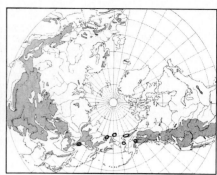

×½

4. Crèpis nàna Richards.
var. lyratifòlia (Turcz.) Hult.

Borkhausia nana var. *lyratifolia* Turcz.; *Youngia americana* Babcock.

Similar to var. *nana*, but leaves more or less dissected to subpinnatifid or pinnatifid.

Described from Okhotsk. Circumpolar map incomplete.

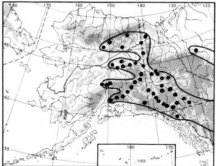
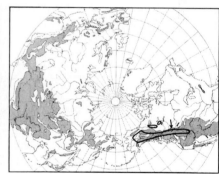

×⅜

5. Crèpis élegans Hook.

Borkhausia elegans (Hook.) Nutt.; *Youngia elegans* (Hook.) Rydb.

Glabrous; stems usually branched, several, from taproot and short caudex; basal leaves spatulate to ovate, entire, coarsely toothed or rarely laciniate; stem leaves narrow, reduced; heads cylindrical, narrow, several at ends of branches; outer involucral bracts broadly lanceolate to ovate, broader than in *C. nana*, the inner oblong, obtuse, scarious-margined; ligules yellow; achenes with long, slender beak, ending with flat disk when ripe, nearly beakless when young; pappus white.

Gravel; to at least 1,000 meters. Described from "Battures of the Assiniboine R." (where it does not occur).

89. Compositae (Composite Family)

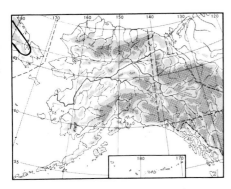 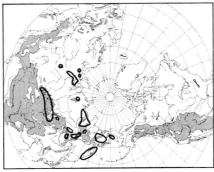 

6. Crèpis chrysántha (Ledeb.) Turcz.
Hieracium chrysanthum Ledeb.

Stem simple or rarely with 1–2 branches, from short rootstock with remains of withered leaves above; basal leaves several, oblanceolate to obovate, somewhat acute, more or less dentate, glabrous or sparsely pubescent above; stem leaves 2–3, small, oblong, lanceolate to linear, sessile; heads large, mostly single; involucral bracts linear-lanceolate, densely covered with long, dark, viscid hairs; ligules yellow, about 3 mm broad; pappus white.

Stony slopes, tundra. Described from Altai.

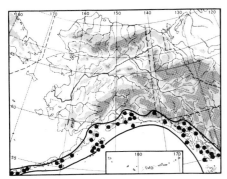 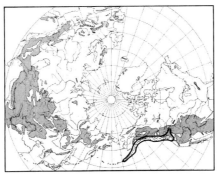 

Prenanthes L.

1. Prenánthes alàta (Hook.) Dietr.   Rattlesnake Root
Nabalus alatus Hook.; *Sonchus hastatus* Less.; *Mulgedium hastatum* Less.; *Prenanthes hastata* (Less.) M. E. Jones not Thunb.; *P. Lessingii* Hult.

Stem single, tall, villous-pubescent above, from taproot; leaves sagittate or hastate, irregularly toothed, the upper reduced; inflorescence corymbiform, more or less open; involucral bracts about 8, lanceolate, thinly grayish-pubescent; ligules white to purplish, laciniate, dentate at apex; styles long-exserted, dark; pappus sordid.

Moist places, along streams. Described from the northwest coast of North America.

Hieracium L. (Hawkweed)

Plant up to 30 cm tall, nearly scapose, with 2–3 small, linear stem leaves only; involucrum densely grayish-villous or black-hirsute; ligules yellow:
   Involucrum grayish-black-villous (exceptionally yellowish-brown); terminal heads large, globular .... 1. *H. triste*
   Involucrum with shorter, black hairs; heads smaller; plant short and slender ... . 2. *H. gracile* var. *alaskanum*
Plant taller, with broad stem leaves; involucrum setose-pubescent or nearly glabrous; ligules yellow, white, or brownish-orange:
   Ligules brownish-orange .... 3. *H. aurantiacum*
   Ligules yellow or white:
     Basal and lower cauline leaves small, soon withering; stem glabrous below; ligules yellow .... 4. *H. scabriusculum*
     Basal leaves large, persistent; stem setose below; ligules sordid white .... 5. *H. albiflorum*

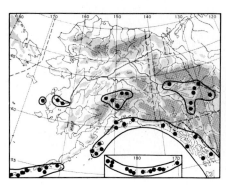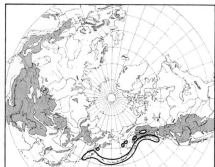

× ½

1. Hieràcium tríste Willd.

Stems up to 3 dm tall, densely grayish long-pilose to villous above; basal leaves obovate to spatulate, entire in margin or rarely remotely dentate, long-petiolated, ciliated, glabrous above, sparsely pubescent beneath, the upper much reduced; inflorescence more or less racemose; terminal heads large, globular, grayish, long-villous; ligules short, yellow; pappus sordid.

Stony slopes to at least 1,500 meters. Described from the Aleutian Islands.

Var. **fúlvum** Hult., with pubescence of involucrum yellowish-brown, occurs at Kodiak. Specimens more or less intermediate between *H. gracile* and *H. triste* have been called **H. gràcile** var. **yukonénsis** Pors. and **H. tríste** var. **tritifórme** Zahn.

89. Compositae (Composite Family)

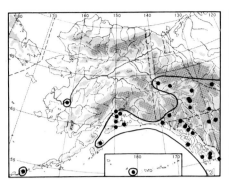
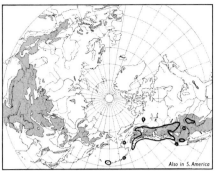

2. Hieràcium gràcile Hook.

Hieracium triste subsp. *gracile* (Hook.) Calder & Taylor.
var. **alaskànum** Zahn

Similar to *H. triste*, but more slender and often shorter; heads smaller; involucrum hirsute with shorter, black hairs.

Alpine meadows to at least 1,500 meters. Described from the Rocky Mountains (Drummond).

Var. *alaskanum* differs from var. **gràcile**, found in the Rocky Mountains, in usually lacking glands on pedicels and involucrum. *H. gracile* occurs also in southern South America.

Intermediates to *H. triste* are not rare in Alaska and the Yukon.

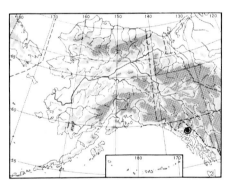
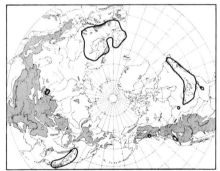
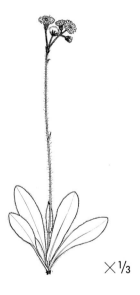

3. Hieràcium aurantìacum L.   Orange Hawkweed

Including *Hieracium brunneocroceum* Pugsley.

Plant long-hirsute, with stolons and basal rosettes; leaves oblanceolate to narrowly elliptic, obovate; scapes with 1–2 broad leaves; inflorescence corymbose; heads about 2 cm broad; involucre dark, setose-pubescent and glandular; ligules brownish-orange.

An introduced weed at Glacier Highway, Juneau. Circumpolar map incomplete. Described from Syria, Helvetia.

×⅓

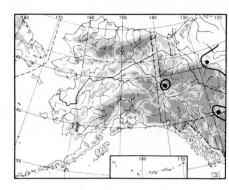
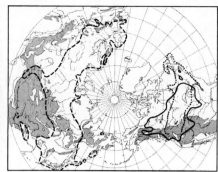

4. Hieràcium scabriùsculum Schwein.

Hieracium canadense of authors, not Michx.

Stem glabrous, erect, from taproot; leaves ovate-lanceolate, remotely toothed, mostly cauline, sessile, clasping, glaucous beneath, rugose above, pubescent with short, rigid hairs, especially in margin; inflorescence paniculate; heads large; peduncles and involucrum tomentose; ligules yellow.

Woods, roadsides. Described from Winnipeg area.

On circumpolar map, range of the closely related **H. umbellàtum** L. is indicated by broken line, that of **H. canadénse** Michx. by dotted line.

×⅓

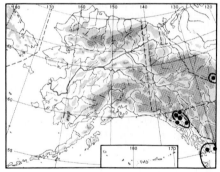
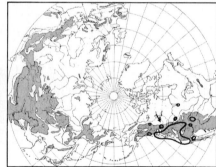

5. Hieràcium albiflòrum Hook.

Stems 1 to several, densely hirsute beneath with whitish or tawny hairs, glandular above, erect, tall, leafy below; basal leaves oblong to oblanceolate; upper leaves small; inflorescence paniculate; involucral bracts in 1 row, narrow, acute, brown, setose-pubescent; ligules white or ochroleucous; pappus tawny.

Woods.

Lists and Indexes

Glossary of Botanical Terms

Abaxial. *See* Dorsal.
Acaulescent. Stemless, or essentially so.
Acerose. Needle-shaped.
Achene. A small, dry, hard, indehiscent, single-seeded fruit.
Acicular. Slender or needle-shaped.
Acuminate. Tapering gradually to a sharp point.
Acute. Sharp-pointed, but less tapering than acuminate.
Adaxial. *See* Ventral.
Adnate. Joined to another organ of a different kind, as a stamen to the corolla, or as a peduncle to the stem.
Adventitious. Said of roots, buds, etc., that develop in an irregular or unusual position.
Aggregate. Collected into dense clusters or tufts.
Alpine. Growing above timberline.
Alternate. Placed singly at different levels on an axis; said usually of leaves or leaflets.
Ament. A catkin.
Amplexicaul. Clasping the stem.
Androecium. The stamens (collectively) in a flower.
Androgynous. Having male and female flowers in the same inflorescence; in the genus *Carex,* having the male flowers in the apex.
Androspore. One of the (dwarf) male spores of *Isoëtes.*
Annual. Of one year's duration.
Annular. In the form of a ring.
Anther. The pollen-bearing part of the stamen.
Anthesis. The time of expansion of a flower.
Apetalous. Without petals.
Aphyllopodic. Without developed leaves at the base.
Apical. Situated at the apex, or tip.
Apomictic. Producing seed without fertilization.
Appressed. Pressed close to another organ.
Approximate. Close together.
Arachnoid. Cobweblike; of slender entangled hairs.
Arcuate. Bent like a bow; moderately curving.
Aril. A fleshy organ growing about the hilum.
Aristate. Awn-tipped.
Articulate. Jointed; having a place for natural separation, with a clean-cut scar.
Ascending. Rising obliquely or curving upward.
Asexual. Without sex.
Attenuate. Gradually tapering.
Awn. A slender bristle on an organ.
Axil. Upper angle formed by a leaf or branch with the stem.
Axillary. In, or pertaining to, an axil.
Axis. The central stem or other organ along which parts or organs are arranged.

Banner. Upper petal of a papilionaceous plant; a standard.
Barbate. Bearded with long, stiff hairs.
Barbellate. With short, stiff hairs.
Basal. Situated at, or pertaining to, the base.
Beak. A firm, elongate tip, particularly of a seed or fruit.
Berry. A pulpy, indehiscent, usually several-seeded fruit with no true stone.
Bi- or Bis-. Latin prefix, signifying two, twice, or doubly.
Bidentate. Having two teeth.
Biennial. Of two years' duration.
Bifid. Two-cleft.
Bipartite. Divided almost to the base into two parts.
Bipinnate. Doubly or twice pinnate.
Bipinnatifid. Twice pinnately cleft.
Biseriate. Occupying two rows, one within the other.
Biternate. Twice ternate.
Blade. The expanded part of a leaf, petal, or sepal.
Bloom. A whitish powdery and glaucous covering of the surface, often of a waxy nature.
Bract. A reduced leaf, subtending a flower.
Bracteolate. With bractlets.
Bractlet. A secondary bract, borne on a pedicel instead of subtending it; sepaloid organs subtending the sepals in many Rosaceae.
Bristly. Bearing stiff hairs.
Bud. The primary, or initial, state of a stem or flower; an unexpanded flower.
Bulb. Subterranean leaf bud with fleshy scales or coats.
Bulblet. A small bulb, especially one borne axillary, as in a leaf axil, in the inflorescence, or on the stem.

Caducous. Falling off at an early stage.
Caespitose. In tufts or dense clumps.
Calcarate. Spurred.
Calcifuge. Avoiding lime.
Calciphile. Preferring lime soils.
Callus. An abnormally thickened part; the thickened extension at the base of the lemma in some grasses.
Calyx. The external, usually green part of the perianth of a flower; the sepals as a whole.
Calyx lobe. One of the free, projecting parts of a calyx.

Campanulate. Bell-shaped.
Canaliculate. Channeled; grooved lengthwise.
Canescent. Growing gray or hoary with fine pubescence.
Capillary. Hairlike.
Capitate. Aggregated into dense clusters or heads.
Capsule. A dry, dehiscent fruit, composed of more than 1 carpel.
Carinate. Keeled; ridged lengthwise.
Carpel. A simple pistil, or one member of a compound pistil; a modified leaf forming the ovary or, in a compound ovary, part of the ovary.
Cartilaginous. Resembling cartilage in consistency.
Castaneous. Chestnut-colored.
Catkin. A scaly, deciduous spike; ament.
Caudate. Bearing a tail or a slender, tail-like appendage.
Caudex. The woody, perennial base of an otherwise herbaceous stem.
Caulescent. With an obvious, leafy stem.
Cauline. Belonging to the stem.
Chaffy. With thin, dry scales.
Chartaceous. Papery.
Ciliate. Fringed with regularly arranged hairs on the margin.
Circumpolar. Occurring around the (North) Pole.
Circumscissile. Dehiscing by a transverse line around the fruit, the top falling away as a lid.
Clasping. Partly or wholly surrounding the stem; said of a leaf.
Clavate. Club-shaped.
Claw. The narrow, petiolate base of some petals or sepals.
Cleft. Cut roughly halfway to the midvein; said of a leaf.
Cleistogamous. Never opening, hence self-pollinating; said of certain flowers.
Compressed. Flattened laterally.
Congested. Crowded together.
Coniferous. Bearing cones.
Connate. Congenitally united, similar organs joined as one.
Contorted. Twisted, bent.
Convolute. Rolled up lengthwise, coiled.
Cordate. Heart-shaped.
Coriaceous. Leathery in texture, tough.
Corm. A short, fleshy, bulblike, underground stem.
Corolla. The inner, usually colored part of the perianth of a flower, composed of free or united petals.
Corymb. A flat-topped or convex, racemose flower cluster, with pedicels becoming shorter toward the top, so that all the flowers are at approximately the same level.
Corymbose. In corymbs.
Cotyledon. First, or embryo, leaf or leaves, as found in the seed.
Creeping. Growing along or beneath the ground and rooting.
Crenate. With rounded teeth in margin.
Crenulate. With fine, rounded teeth in the margin.
Crisped. Curled.
Crown. An inner appendage to a petal, or to the throat of a corolla.
Cucullate. Hooded or hood-shaped.
Culm. The type of hollow or pithy, slender stem found in grasses or sedges.

Cuneate. Wedge-shaped.
Cuspidate. Tipped with a short, rigid point.
Cyme. A flat-topped or convex flower cluster, with central flowers opening first.

Deciduous. Falling off; losing leaves in autumn.
Decumbent. Lying down, but tending to rise at the end.
Decurrent. Extending down the stem below the insertion.
Deflexed. Turned abruptly downward.
Dehiscent. Opening spontaneously when ripe to shed seeds or spores.
Dentate. Having the margin cut with sharp, salient teeth, directed outward.
Denticulate. Finely dentate.
Depauperate. Dwarfed, starved.
Depressed. Somewhat flattened from above.
Dichotomous. Forking regularly by pairs.
Dicotyledonous. With two cotyledons.
Digitate. Fingered, shaped as an open hand.
Dilated. Expanding, as though flattened.
Dioecious. Having staminate and pistillate flowers on different plants.
Discoid. In the Compositae, having a head without ray flowers; like a disk.
Disk. A fleshy, sometimes nectar-secreting development of the receptacle above the base of the ovary.
Disk-flower. In the Compositae, the tubular flowers of the head as distinct from the ray-flowers.
Dissected. Cut or divided into numerous segments.
Distichous. Disposed in two vertical rows.
Divaricate. Diverging at a wide angle.
Dorsal. Belonging to the outer or back surface of an organ; abaxial.
Dorsiventral. In a plane running through axis from above to below, as opposed to lateral.
Drooping. Erect at base, but bending downward above, as the branches of a grass panicle.
Drupe. A stone fruit, such as a plum.
Drupelet. A diminutive drupe, as in a raspberry or blackberry.

Ebracteate. Without bracts.
Echinate. Prickly.
Ellipsoid. Solid but with an elliptic outline.
Elliptic. Ellipse-shaped.
Emarginate. Shallowly notched at the apex.
Emersed. Raised out of water.
Endemic. Confined geographically to a single area.
Ensiform. Sword-shaped.
Entire. Without divisions or lobes.
Epigynous. Borne on the ovary; said of the floral parts, when the ovary is wholly inferior.
Erose. Irregularly toothed, as if gnawed.
Evanescent. Early disappearing.
Excurrent. Projecting beyond the edge of.
Exserted. Protruding, as stamens projecting beyond the corolla.

Falcate. Sickle-shaped.
Farinose. Covered with a meal-like powder.
Fascicle. A dense cluster or bundle.
Fastigiate. Clustered with parallel, erect branches.
Ferruginous. Rust-colored.
Filament. Usually, the stalk supporting an anther, the two together forming the stamen.

Glossary of Botanical Terms

Filiform. Threadlike.
Fimbriate. Fringed with long hairs.
Flabellate. Fan-shaped.
Flaccid. Weak and lax.
Flagellate. With very slender runners.
Floccose. With flocks or tufts of soft, woolly hair.
Floret. The individual flowers of a dense cluster; said of the Cyperaceae, Gramineae, and Compositae.
Foliaceous. Leaflike.
Foliolate. With separate leaflets.
Follicle. A dry, monocarpellary fruit, opening only on the ventral suture.
Foveate, Foveolate. More or less pitted.
Frond. Leaf of a fern.
Fruticose. More or less shrublike.

Galea. The helmetlike upper lip of certain bilabiate corollas, joined into a tube at least at the base.
Gamopetalous. With petals of corolla joined into a tube, at least at the base; sympetalous.
Geniculate. Bent abruptly, making a knee.
Gibbous. Swollen on one side.
Glabrate. Almost glabrous.
Glabrescent. Becoming glabrous with age.
Glabrous. Devoid of hairs or down; having a smooth, even surface.
Gland. An organ (globular, oblong, or sunk below the surface) that secretes a usually sticky fluid.
Glandular. Furnished with glands.
Glaucous. Covered with a whitish or blue-white bloom.
Globose. Spherical or nearly so.
Glochidiate. Having barbed hairs or spines.
Glomerate. Collected in dense clusters or heads.
Glomerule. A compact, capitate cyme.
Glume. One of a pair of bracts at the base of the grass spikelet.
Glutinous. With a sticky exudation.
Gynaecandrous. Having male and female flowers in the same inflorescence; in *Carex*, having the female flowers in the apex.
Gynoecium. The one or more pistils (collectively) in a flower.
Gynophore. The stipe of a pistil.
Gynospore. One of the (large) female spores of *Isoëtes*.

Habit. General appearance of a plant.
Habitat. The normal situation in which a plant lives.
Halophyte. A plant of salty or alkaline soil.
Hastate. Shaped like an arrowhead, but with the basal lobes turned outward.
Head. A dense cluster of sessile or nearly sessile flowers on a very short axis or receptacle.
Herb. Any vascular plant that is not woody, at least above ground.
Herbaceous. Leaflike in texture or color; pertaining to an herb.
Hilum. The scar or area of attachment of a seed or ovule.
Hirsute. Covered with coarse or shaggy hairs.
Hirtellous. Minutely hirsute.
Hispid. With coarse, stiff hairs.
Hispidulous. Finely hispid.
Hoary. Covered with white or grayish-white down.
Host. A plant that nourishes a parasite.
Hyaline. Thin and translucent.
Hybrid. A cross between 2 species or subspecies.

Hybrid swarm. A series of plants originating in hybridization and subsequently recrossing with parents and between themselves, so that a continuous intrograding of forms arises.
Hypanthium. A cup-shaped enlargement of the receptacle on which the calyx, the corolla, and often the stamens are inserted.
Hypogynous. Borne below the ovary; said of the floral parts when the ovary is superior.

Imbricate. Overlapping, as thatches on a roof.
Immersed. Growing under water.
Imperfect. Said of a flower lacking either stamens or pistils.
Incised. Cut rather deeply and sharply.
Included. Not at all protruding from the surrounding envelope.
Indehiscent. Not splitting open, as an achene.
Indigenous. Native.
Indusium. In ferns, the epidermal piece of tissue covering a sporangium or group of sporangia.
Inferior ovary. An ovary that is adnate to the hypanthium, with perianth inserted around the top.
Inflated. Blown up, bladdery, turgid.
Inflorescence. The flower cluster of a plant.
Innovation. An offshoot that carries on the further growth of the plant.
Inserted. Attached to or growing out of.
Internode. The part of the stem between two nodes.
Introduced. Brought from another region, as for purposes of cultivation.
Introgression. Hybridization when the hybrids cross back with one or both of the parent species.
Involucre. A whorl of bracts subtending a flower cluster, as in the heads of Compositae.
Involute. With the edges rolled inward.

Keel. A projecting ridge on a surface.

Labiate. Lipped.
Lacerate. Appearing as if irregularly torn or cleft.
Laciniate. Cut into narrow lobes or segments.
Lamina. The blade of a leaf or petal.
Lanate. Woolly.
Lanceolate. Considerably longer than broad; lance-shaped.
Leaf. The usually green photosynthetic organ of a plant, usually composed of a flat expanded blade or lamina, a stalk or petiole, and often 2 stipules, and attached to the stem or branch at a node.
Leaflet. One of the divisions of a compound leaf.
Legume. The one-celled fruit of a leguminous plant, dehiscent along both sutures.
Lemma. In grasses, the lower of the two bracts immediately enclosing the floret.
Lepidote. With small, scurfy scales.
Ligneous. Woody.
Ligulate. Tongue-shaped; provided with or resembling a ligule.
Ligule. The strap-shaped part of a ray corolla in the Compositae; the thin, collarlike appendage on the inside of the blade at the junction with the sheath, in grasses.
Linear. Long and narrow, with parallel margins.
Lip. Each of the upper and lower divisions of a bilabiate corolla or calyx.

Lobe. Any segment of an organ, especially if rounded.
Locular. Having cells or loculi.
Lodicules. The two or three minute hyaline scales at the base of the stamens in grasses, representing the perianth.
Loment. A legume made up of one-seeded joints, as in *Hedysarum*.
Lutescent. Yellowish, or becoming yellow.
Lyrate. Lyre-shaped, pinnatifid, with the terminal lobe considerably larger than the others.

Marcescent. Persistent after withering.
Margin. The edge of a leaf blade.
Membranaceous. Thin and rather soft and more or less translucent.
Midvein. The central vein or rib of a leaf, etc.
Moniliform. Resembling a string of beads.
Monocephalous. Bearing one head only.
Monocotyledonous. With one cotyledon.
Monoecious. Having both staminate and pistillate flowers on the same plant, but not as perfect flowers.
Mucronate. Provided with a narrow, abrupt point.
Multicipital. With many heads; said of the crown of a single root.
Muricate. Rough, with short, firm, hard projections.

Nerve. A simple or unbranched vein or slender rib.
Node. The joint of a stem; the point of insertion of a leaf or leaves.
Nodose. Furnished with knots or knobby nodes.
Nut. A hard-shelled, one-seeded, indehiscent fruit.

Ob-. A Latin prefix signifying inversely or contrarily: an obovate leaf is broadest above the middle; an ovate leaf, below the middle.
Obcordate. Inversely heart-shaped.
Oblanceolate. Inverse of lanceolate.
Oblate. Flattened at the poles, as a tangerine.
Oblique. Of unequal sides.
Oblong. Much longer than broad, with nearly parallel sides.
Obovate. Inversely ovate.
Ochroleucous. Yellowish-white.
Ocrea. A sheath around the stem, derived from the leaf stipules; chiefly occurring in Polygonaceae.
Orbicular. Approximately circular in outline.
Oval. Broadly elliptic.
Ovary. The base of the pistil, containing the ovules, or seeds.
Ovate. Egg-shaped in outline.
Ovoid. Egg-shaped.
Overtopping. Extending above or beyond.
Ovule. The body that after fertilization becomes the seed.

Palea (plural **paleae**). The inner bract of a grass floret; the chaff or bracts on the receptacle of many Compositae.
Paleaceous. Chaffy.
Palmate. Radiately lobed or divided so as to resemble the outstretched fingers of a hand.
Panicle. Strictly branched racemose inflorescence, though often applied to any branched inflorescence.
Papilionaceous. Having wings, banner, and keel; said of the corolla especially in the Leguminosae.

Papilla. A small, nipple-shaped protuberance.
Papillate, papillose. Covered with papillae.
Pappus. The modified calyx limb in the Compositae, forming a crown of bristles or scales at the summit of the achene.
Pectinate. Pinnatifid with narrow, closely set segments.
Pedate. Palmately divided or parted, with the lateral lobes two-cleft.
Pedicel. The stalk of a single flower in a flower cluster or of a spikelet in grasses.
Pedicellate. Having a pedicel.
Peduncle. A flower-stalk supporting a cluster of flowers, or a single flower when the pedicel is very long.
Peltate. Shield-shaped.
Perennial. Living for more than two years and usually flowering each year.
Perfect. Said of a flower having both stamens and pistils.
Perfoliate. With leaf entirely surrounding the stem.
Perianth. Collectively, the calyx and corolla of a flower; the floral envelope.
Pericarp. The wall of the fruit, or seed-vessel.
Perigynium. The inflated saclike organ, or utricle, surrounding the pistil in *Carex*.
Perigynous. Borne from the top of a cuplike structure surrounding the ovary; said of the floral parts when the ovary is superior or partly inferior.
Persistent. Long attached, as calyx upon fruit, leaves through the winter, etc.
Petal. A division of the corolla, usually colored.
Petaloid. Brightly colored and resembling a petal.
Petiolate. Having a petiole.
Petiole. The leaf-stalk.
Phyllary. An individual bract of the involucre in the Compositae.
Phyllopodic. With leafy base; as in *Carex*, when lowermost leaves are leaflike, not reduced to scales.
Pilose. Bearing straight, soft, spreading hairs.
Pinna (plural **pinnae**). One of the primary divisions of a pinnate or compoundly pinnate frond or leaf.
Pinnate. Compound leaf, composed of more than three leaflets, with the leaflets arranged on both sides of a common axis.
Pinnatifid. Pinnately cleft.
Pinnatisect. Pinnately divided.
Pinnule. A division of a pinna.
Pistil. The seed-bearing organ of a flower, consisting usually of ovary, style, and stigma; the unit of the gynoecium.
Pistillate. Having pistils, but no stamens; the character of a female flower.
Plicate. Folded as a fan.
Plumose. Having fine hairs on each side; feathery.
Pod. Any dry, dehiscent fruit.
Pollen. The fecundating grains contained in the anther.
Pollinia. The coherent, regularly shaped pollen masses of the orchids and milkweeds.
Polygamous. Bearing unisexual (imperfect) and bisexual (perfect) flowers on the same plant.
Polymorphous. With various forms.
Polypetalous. Having several petals.
Pome. An applelike fruit.
Prickle. A small, sharp, and more or less slender outgrowth of the bark or epidermis.
Procumbent. Trailing on the ground.
Prostrate. Lying flat on the ground.

Glossary of Botanical Terms

Pruinose. Covered with a coarse, whitish, waxy "bloom," more prominent than when glaucous.
Puberulent. Minutely pubescent.
Pubescent. Covered with soft, short hairs.
Pulverulent. Dusted with fine powder.
Punctate. Dotted with depressions or with translucent internal glands or colored dots.
Pungent. Ending in a rigid, sharp point or prickle; acrid to the taste and smell.
Pustulate. Bearing blisterlike swellings or elevations, mostly at the base of hairs.
Pyriform. Pear-shaped.

Quadrate. Nearly square in form.

Raceme. An unbranched, elongated, indeterminate inflorescence, with each flower subequally pedicelled.
Racemose. In racemes, or resembling a raceme.
Rachilla. A small rachis, specifically the axis of a grass spikelet.
Rachis. The axis of a compound leaf, or a spike or raceme.
Radical. Arising from the base of the stem or from a rhizome.
Ray. One of the peduncles or branches of an umbel; the flat marginal flowers in Compositae.
Receptacle. The portion of the floral axis on which the flower parts are borne, or, in the Compositae, the portion bearing florets in the head.
Recurved. Bent backward in a curve.
Reflexed. Bent abruptly downward or backward.
Relict. A plant evidently left over from a past geologic epoch.
Remote. Widely spaced.
Reniform. Kidney-shaped.
Repand. With an undulating or wavy margin.
Replum. A framelike placenta from which the valves fall away in dehiscence, as in many Cruciferae.
Reticulate. Net-veined.
Retrorse. Bent backward or downward.
Retuse. Notched shallowly at a rounded apex.
Revolute. Rolled downward or backward.
Rhizome. An underground stem or rootstock, usually rooting at the nodes, becoming upcurved at the apex.
Rib. A primary or prominent vein of a leaf.
Rootstock. The rhizome.
Rosette. A crowded cluster of leaves, appearing to rise from the ground.
Rostrate. Beaked.
Rotate. Wheel-shaped, flat and circular in outline; said of a corolla.
Rugose. Wrinkled.
Runner. A filiform or slender aerial branch or stolon, rooting at the end and forming a new plant, which eventually becomes detached from the parent.

Saccate. Furnished with a sac or pouch.
Sagittate. Shaped like an arrowhead, with the basal lobes directed back or down.
Samara. A dry, indehiscent, winged fruit, as in *Betula*.
Sarmentose. Producing long, slender runners.
Scabrous. Rough to the touch.
Scale. Any thin, scarious body, usually a degenerate leaf.
Scape. The flowering stem of a plant, all the foliage of which is radical.
Scapose. Bearing or resembling a scape.
Scarious. Thin, dry, and translucent; not green.
Scurfy. With scale-like or bran-like particles.
Sepal. A division of the calyx, usually green.
Septate. Divided by partitions into compartments.
Sericeous. Silky.
Serrate. Saw-toothed, the sharp teeth pointing forward.
Serrulate. Finely serrate.
Sessile. Without a stalk.
Setaceous. Bristlelike.
Setose. Beset with bristles.
Sheath. A tubular envelope, as the lower part of the leaf in grasses and sedges.
Shrub. A woody perennial, smaller than a tree, usually with several stems.
Silicle. A short silique.
Silique. A narrow, many-seeded capsule, in which a replum separates the valves, notably in the Cruciferae.
Sinuate. With a strongly wavy margin.
Sinus. The cleft or depression between two lobes or teeth.
Solitary. Borne singly.
Sorus (plural **sori**). A fruit dot or cluster on the back of the frond in ferns.
Spadix. A spike of flowers with fleshy axis, surrounded by a spathe, as in the Araceae.
Spathaceous. Resembling, or furnished with, a spathe.
Spathe. A large bract enclosing a spadix, as in *Calla*.
Spatulate. Spoon-shaped.
Spicate. Arranged in spikes or resembling a spike.
Spike. A simple, elongate, indeterminate inflorescence with sessile or nearly sessile flowers or spikelets.
Spikelet. A smaller or secondary spike, the ultimate flower cluster, especially in grasses and sedges.
Spine. A sharp woody or rigid outgrowth from the stem; a thorn.
Sporangium (plural **sporangia**). A structure containing spores.
Spore. A small asexual reproductive body.
Sporophyll. A spore-bearing leaf.
Spreading. Diverging nearly at right angles; nearly prostrate.
Spur. A hollow, saclike, conical, or slender projection of some part of a flower, usually nectariferous.
Squamate. Covered with scales.
Squarrose. Spreading at right angles or recurved.
Stamen. The male organ of the flower, consisting of a filament and an anther, the latter bearing the pollen; the unit of the androecium.
Staminate. Having stamens, but no pistils; the character of a male flower.
Staminodium. A sterile stamen, or any structure without anther corresponding to a stamen.
Standard. The upper, usually broad, petal in a papilionaceous corolla; a banner.
Stellate. Star-shaped.
Stigma. The apical part of a pistil or style, through which pollination is effected.
Stipe. The stalk beneath the ovary; the leaf stalk of a fern.
Stipitate. Having a stipe or stalk.
Stipule. An appendage at the base of a petiole or leaf or on each side of its insertion.
Stolon. A runner or any basal branch that is inclined to take root.

Stoloniferous. Producing or bearing stolons.
Stoma (plural **stomata**). Breathing pore or aperture in the epidermis.
Stone. The bony endocarp of a drupe.
Stramineous. Straw-colored.
Striate. Marked with five longitudinal lines or streaks.
Strigillose. Finely strigose.
Strigose. Covered with stiff, appressed, straight hairs.
Style. The contracted portion of a pistil or carpel between the ovary and the stigma.
Stylopodium. A disklike expansion at the base of the style, as in the Umbelliferae.
Sublignous. Nearly woody.
Subulate. Awl-shaped.
Succulent. Fleshy, juicy.
Sucker. A vegetative (i.e., nonsexual) shoot of subterranean origin.
Suffrutescent. Slightly or obscurely shrubby.
Suffruticose. Diminutively shrubby.
Superior ovary. An ovary that is situated above the point of origin of the calyx, corolla, and stamens.
Suture. A line of splitting or opening.
Sympetalous. Gamopetalous; with united petals.
Synonym. A superseded or unused name.

Taproot. The primary descending root.
Taxon (plural **taxa**). Any taxonomic unit.
Tendril. A slender, coiling or twining process, by which a climbing plant grasps its support.
Terete. Circular in cross section.
Ternary. Consisting of threes.
Ternate. Divided into 3 segments, as a leaf consisting of three leaflets.
Thorn. *See* Spine.
Throat. The orifice of a gamopetalous corolla or calyx.
Thyrsus. A contracted cylindrical or ovoid and usually compact panicle, as a cluster of grapes.
Tomentose. Covered with a rather short, densely matted, soft wool.
Trailing. Prostrate, but not rooting.
Tree. Perennial woody plant with an evident trunk.
Trichotomous. Three-forked.
Trifoliate. Having three leaflets.
Trigonous. Three-angled.
Tripartite. Three-parted.
Truncate. Ending abruptly, as if cut off transversely.

Tuber. A thickened, short, underground stem with many buds.
Tubercle. A small projection; the persistent base of the style in some Cyperaceae.
Tuberculate. Bearing small projections or tubercles.
Tuberoid. A fleshy, thickened root resembling a tuber.
Tuberous. Resembling a tuber.
Tunicate. Coated; invested with layers, as an onion.
Turbinate. Top-shaped, inversely conical.
Turgid. Swollen, inflated.
Turion. A sucker or shoot emerging from the ground.
Type. That specimen of a given taxon from which its original description was made.

Umbel. A flower cluster in which the pedicels arise from a common point, as the ribs of an umbrella.
Umbellate. Borne in umbels; resembling an umbel.
Unilateral. One-sided.
Unisexual. Having either stamens or pistils, but not both; as of imperfect flowers.
Urceolate. Urn-shaped.
Utricle. A small, bladdery, one-seeded fruit.

Vaginate. Sheathed.
Valve. One of the pieces into which a capsule splits.
Vascular. Furnished with vessels or ducts.
Vein. Threads of fibrovascular tissue in a leaf or other organ, especially those that branch.
Velum. The membranous indusium of *Isoëtes*.
Velutinous. Velvety; with dense fine pubescence.
Venation. The veining of leaves, etc.
Ventral. Belonging to the inner or anterior face of an organ, as opposed to dorsal; adaxial.
Verrucose. Warty.
Verticillate. Whorled.
Vesicle. A small bladder or air cavity.
Villous. Bearing long, soft (not interwoven) hairs.
Viscid. Sticky; glutinous.

Whorl. An arrangement of leaves, etc., in a circle around the stem, radiating from a node.
Wing. Any membranous expansion.
Woolly. Having long, soft, entangled hairs.

Zygomorphic. Divisible into similar halves in only one plane; bilaterally symmetrical.

List of Authors

All taxon authors appearing in this work, including authors of synonyms, are listed below. Abbreviated names are given in the alphabetic order of their abbreviations (boldface type) to facilitate location of unfamiliar authors in the list; thus **Bab.** (for Babington) precedes **Babcock.** Following the list of taxon authors is a list of persons who appear in this work not as taxon authors but as persons for whom taxa have been named (those in the main list who have been so honored do not appear in the second list). Information of the sort given in these two lists is often difficult to assemble; the author apologizes for whatever omissions and errors persist, and welcomes corrections.

Abbe, Ernst Cleveland, 1905– . Botanical author and collector, Minnesota.

Abrams, LeRoy, 1874–1956. Professor of botany, Stanford; author, *Illustrated Flora of the Pacific States*.

Adams, Joseph Edison, 1903– . Professor of Botany, North Carolina; student of *Arctostaphylos*.

Adans. Adanson, Michel, 1727–1806. French botanist, author of *Familles des Plantes*; originated some 1600 generic names.

Aellen, Paul, 1896– . Professor, Basel; student of Chenopodiaceae.

Ahti, Teuvo, 1933– . Finnish botanist.

Airy-Shaw, Herbert Kenneth, 1902– . British botanist.

Ait. Aiton, William, 1731–93. Director of Kew Gardens.

Akiyama, Shigeo, 1906– . Japanese botanist.

All. Allioni, Carlo. 1725–1804. Professor of botany, Turin; author of *Flora Pedemontana*.

Ames, Oakes, 1874–1950. Director of the botanical museum, Harvard; orchidologist.

Anders. Anderson, Edgar Shannon, 1897– . Professor of botany; geneticist, Missouri Botanical Garden.

J. P. Anders. Anderson, Jacob Peter, 1874–1953. Horticulturist; indefatigable botanical collector in Alaska; assistant curator of the herbarium, Iowa State Univ.; author of *Flora of Alaska*.

Anderss. Andersson, Nils Johan, 1821–80. Director of the botanical museum, Stockholm; member of expedition circumnavigating the world in *Eugenie*; noted salicologist.

Andres, Heinrich, 1883– . Teacher, Trier.

Andrz. Andrzeiovski, Antoni Lukianovich, 1784–1868. Russian botanist.

Ångstr. Ångström, Johan, 1813–79. Swedish physician.

Argus, George William, 1929– . Salicologist, University of Wisconsin.

Arnott, George Arnold Walker, 1799–1868. Edinburgh and Glasgow botanist.

Aschers. Ascherson, Paul Friedrich August, 1834–1913. Professor, Berlin.

Aschers. & Graebn. Ascherson, P.F.A., & K.O.R.P.P. Graebner. Authors of *Synopsis der Mitteleuropäischen Flora*.

Ascherson & Magnus, Ascherson, P.F.A., & P. W. Magnus.

Avrorin, Nikolai Alexandrovich, 1906– . Russian botanist; director, Polarno-Alpinskogo botanical garden, Kirovsk.

Bab. Babington, Charles Cardale, 1808–95. Professor, Cambridge, England.

Babcock, Ernest Brown, 1877–1954. Professor of genetics, California; student of *Crepis*.

Bacigalupi, Rimo Charles, 1901– . Curator, Jepson Herbarium, Berkeley; student of Saxifragaceae and California flora.

Bailey, Liberty Hyde, 1858–1954. Cornell; horticulturist and author; founder of Gentes Herbarum; student of *Carex, Rubus,* palms.

Baker, John Gilbert, 1834–1920. Keeper of the herbarium, Kew, 1890–99; student of ferns, Amaryllidaceae, Bromeliaceae, Iridaceae, etc.

Balb. Balbis, Giovanni Battista, 1765–1831. Professor of botany, Turin.

Balf. Balfour, John Hutton, 1808–84. Professor of botany and director of the botanical garden, Edinburgh.

Ball, Carleton Roy, 1873–1958. American student of *Salix*.

Banks, Sir Joseph, 1743–1820. Great patron of the sciences, botanical collector; accompanied Captain Cook on his first voyage of circumnavigation, 1768; director of Kew; president of Royal Society.

Barneby, Rupert Charles, 1911– . Student of Great Basin flora and of *Astragalus*.

Barratt, Joseph, 1796–1882. Connecticut and Pennsylvania geologist.

Barton, William Paul Crillon, 1786–1856. Professor, Philadelphia; author of *A Flora of North America.*
Bartr. Bartram, William, 1739–1823. American traveler in southeastern United States.
Batchelder, Frederick William, 1838–1911. New Hampshire botanist and amateur ornithologist.
Baumg. Baumgarten, Johann, 1765–1843. Rumanian physician.
Beal, William James, 1833–1924. Michigan agrostologist.
Beaman, John Homer, 1929– . Professor of botany, Michigan State; student of Mexican flora.
Beauv. Palisot de Beauvois, (Baron) Ambroise Marie François Joseph, 1752–1820. French naturalist; student of Gramineae.
Bebb, Michael Schuck, 1833–95. Illinois; student of *Salix.*
Becker, Wilhelm, 1874–1928. German botanical author; student of *Viola.*
Bedd. Beddome, Richard Henry, 1830–1911. British botanist.
Bellardi, Carlo Antonio Lodovico, 1741–1826. Italian botanist.
Benedict, Ralph Curtiss, 1883– . American botanist; student of ferns.
Bennett, Arthur, 1843–1939. English amateur botanist.
Benson, Lyman David, 1909– . Professor of botany, Pomona College; author, student of *Ranunculus* and Cactaceae.
Benth. Bentham, George, 1800–84. Outstanding English botanist; longtime president, Linnaean Society.
Benth. & Hook. f. Bentham, G., & J. D. Hooker.
Berl. Berlin, Johan August, 1851–1910. Swedish student of Greenland flora.
Bernh. Bernhardi, Johann Jacob, 1774–1850. Professor of botany, Erfurt.
Berthelot, Sabin, 1794–1880. French consul on Teneriffe, co-author of a flora of the Canary Islands.
Bess. Besser, Wilibald Swibert Joseph Gottlieb von, 1784–1842. Austrian botanist in Poland.
Bessey, Charles Edwin, 1845–1915. Professor of botany, Nebraska.
Beurl. Beurling, Pehr Johan, 1800–66. Alderman, Stockholm.
Bieb. Marschall von Bieberstein, (Baron) Friedrich August, 1768–1826. German botanist.
Bigel. Bigelow, Jacob, 1787–1879. Professor of botany, Boston.
Björnstr. Björnström, Fredrik Johan, 1833–89. Physician, Stockholm.
Bl. & Fing. Bluff, M. J., & C. A. Fingerhuth.
Blake, Sidney Fay, 1892–1959. U.S. Dept. of Agriculture, Beltsville; student of Compositae and author of *Geographical Index to the Floras of the World.*
Blasdell, Robert F., 1929– . American botanist.
Blomgr. Blomgren, Nils Harald, 1901–25. Swedish botanist.
Bluff, Mathias Joseph, 1805–37. German botanist.
Blytt, Mathias Numsen, 1789–1862. Professor of botany, Kristiania, Norway.
Blytt & Fr. Blytt, M. N., & E. M. Fries.
Bobr. Bobrov, Evgenii Gregorievich, 1902– . Russian botanist.
Bobr. & Schischk. Bobrov, E. G., & B. K. Schischkin.
Böcher, Tyge Wittrock, 1909– . Professor of botany, Copenhagen; Danish student of Greenland flora.
Boeck. Boeckeler, Johann Otto, 1803–99. Apothecary-botanist, Oldenburg; specialist in *Carex.*
Boenn. Boenninghausen, Clemens Maria Friedrich von, 1785–1864. German botanist.
Boiv. Boivin, Joseph Robert Bernard, 1916– . Dept. of Agriculture, Ottawa; student of *Thalictrum* and Canadian flora.
Bolle, Friedrich, 1905– . German botanist; student of *Geum.*
Bolton, James, 1758–99. British botanist.
Bong. Bongard, August Heinrich Gustav, 1786–1839. Professor of botany, St. Petersburg; monographer of Brazilian plants; describer of Mertens's Alaskan collection.
Bonpland, Aimé Jacques Alexandre (né Goujaud), 1773–1858. French botanist.
Boott, Francis, 1792–1863. American caricologist.
W. Boott, Boott, William, 1805–87. Boston specialist in *Carex.*
Borb. Borbás, Vinczé von, 1844–1905. Hungarian botanist.
Borissova, Antonina Georgievna, 1900– . Russian botanist.
Boriss. & Schischk. Borissova, A. G., & B. K. Schischkin.
Börner, C., 1880– . German botanist; author of *Eine Flora für das Deutsche Volk,* 1912.
Bory, Bory de Saint-Vincent, Jean Baptiste George Marcellin) (Baron de), 1778–1846. French traveler and naturalist.
Bosch, Roelof Benjamin van den, 1810–62. Dutch botanist.
Botsch. Botschantzev, V. P., 1910– . Russian botanist.
Bowden, Wray Merrill, 1914– . Dept. of Agriculture, Canada; student of cytogenetics.
R. Br. Brown, Robert, 1773–1858. Famous Scottish botanist; first keeper of botany, British Museum.
Brand, August, 1863–1930. German student of Polemoniaceae, Hydrophyllaceae, and Boraginaceae.
Brandegee, Townshend Stith, 1843–1925. California botanist; student of Mexican flora.
A. Braun, Braun, Alexander Carl Heinrich, 1805–77. Professor of botany and director of botanical garden, Berlin.
Braun-Blanquet, Josias, 1884– . Noted Swiss ecologist.
Bray, Franz Gabriel de, 1765–1832. Rouen.
Breitung, August J., 1913– . Author of *Catalogue of Saskatchewan Flora.*
Brenckle, Jacob Frederic, 1875–195?. South Dakota physician; student of *Polygonum.*
Briq. Briquet, John Isaac, 1870–1931. Director, Conservatoire Botanique, Geneva; student of Labiatae, Umbelliferae, Compositae; noted for nomenclatural advances.
Britt. Britton, Nathaniel Lord, 1859–1934. Director, New York Botanical Garden.
Britt. & Rydb. Britton, N. L., & P. A. Rydberg.
Britt., Sterns & Pogg. Britton, N. L., E. E. Sterns, & J. F. Poggenburg.
E. G. Britt. Britton, Elizabeth Gertrude (née Knight), 1858–85. Wife of N. L. Britton.
Brongn. Brongniart, Adolphe Théodore, 1801–76. Noted paleobotanist and systematist of Paris.
S. Brown, Brown, Stewardson, 1867–1921. Philadelphia.
BSP. Britton, N.L., E. E. Sterns & J. F. Poggenburg.

List of Authors

Buchenau, Franz Georg Philipp, 1831–1906. German student of Juncaceae and Alismaceae.
Buckl. Buckley, Samuel Botsford, 1809–84. Texas state geologist; collector of plants.
Bunge, Alexander Andrejewitsch von, 1803–90. Professor of botany, Tartu (Dorpat), Esthonia; student of Russian and central Asian floras, and of *Astragalus*.
Burgsd. Burgsdorff, Friedrich August Ludwig von, 1747–1802. German forester.
E. Busch, Busch, Elisabeth Alexandrovna, 1886–1960. Russian botanist; student of Ericaceae and Primulaceae.
N. Busch, Busch, Nikolai Adolfowitsch, 1869–1941. Russian botanist; student of Cruciferae.
Butler, Bertram Theodore, 1872– . American botanist.
Butt. Butters, Frederic King, 1878–1945. Professor of botany, Minnesota.
Butt. & Abbe, Butters, F. K., & C. E. Abbe.
Butt. & St. John, Butters, F. K., & H. St. John.

Cajander, Aimo Kaarlo, 1879–1943. Professor of forestry, Helsinki; politician.
Calder, James Alexander, 1915– . Canadian author and collector, Dept. of Agriculture, Ottawa; student of the flora of Queen Charlotte Islands.
Calder & Savile, Calder, J. A., & D. B. O. Savile. Dept. of Agriculture, Ottawa; mycologists.
Calder & Taylor, Calder, J. A., & R. F. Taylor.
Callen, Eric O., 1912– . McGill University; student of *Euphrasia*.
Cambess. Cambessèdes, Jacques, 1799–1863. France.
Camp, Wendell Holmes, 1904–63. Professor of botany, Connecticut; student of Ericaceae.
Carey, John, 1797–1880. British botanist in America.
Carr. Carrière, Élie Abel, 1818–96. Museum of Natural History, Paris; editor of *Revue Horticole*; student of Coniferae.
Casper, S. Jost,   . Jena, Germany; student of *Pinguicula*.
Cass. Cassini, (Count) Alexandre Henri Gabriel de, 1781–1832. French student of Compositae.
Chaix, (Abbé) Dominique, 1730–99. Student of the flora of the French Alps.
Cham. Chamisso (de Boncourt), Adalbert Ludwig von, 1781–1838. German poet-naturalist; botanist on the ship *Rurik*, which visited Alaska in 1816 and 1817.
Cham. & Schlecht. Chamisso, A. L. von, & D. F. L. von Schlechtendal.
Châtelain, Jean Jaques, 18th century. French student of *Corallorhiza*.
Chater, Arthur Oliver, 1933– . British botanist; of the *Flora Europaea* project.
Chater & Halliday, Chater, A. O., & G. Halliday.
Ching, Ren Ch'ang, 1898– . Chinese botanist.
Christ, Konrad Hermann Heinrich, 1833–1933. Swiss student of ferns, Coniferae, *Carex*, and *Rosa*.
Christens. Christensen, Carl Frederik Albert, 1872–1942. Danish pteridologist; author of *Index Filicum*.
Chrtek, Jindřich, 1930– . Czech botanist.
Chr. & Pouz. Chrtek, J., & Pouzar.
Clairv. Clairville, Joseph Philippe de, 1742–1830. Swiss botanist.
Clapham, Arthur Ray, 1904– . Professor of botany, University of Sheffield.
Clarke, Charles Baron, 1832–1906. Superintendent, Royal Botanic Gardens, Calcutta; student of the flora of India and of Cyperaceae.
Clausen, Robert Theodore, 1911– . Professor of botany, Cornell; student of Ophioglossaceae and *Sedum*.
Clements, Frederick Edward, 1874–1945. Plant ecologist, Carnegie Institution.
Clokey, Ira Waddell, 1878–1950. Colorado and California, accumulator of a large herbarium now at Berkeley; author of *Flora of the Charleston Mountains, Nevada*.
Clute, Willard Nelson, 1869–1950. Professor of botany, Butler University, Indianapolis.
Cockerell, Theodore Dru Alison, 1866–1948. Professor of zoology, Colorado.
Colem. Coleman, Nathan, 1825–87. Michigan botanist.
Collen, Eric O., 1912– . McGill University.
Constance, Lincoln, 1909– . Professor of botany, Univ. of California; student of Umbelliferae and Hydrophyllaceae.
Copeland, Edwin Bingham, 1873–1964. Curator of herbarium, Univ. of California; student of ferns.
Correll, Donovan Stewart, 1908– . Texas Research Foundation, Renner.
Coult. Coulter, John Merle, 1851–1928. Professor of botany, Chicago; founder, *Botanical Gazette*.
Coult. & Evans, Coulter, J. M., & W. H. Evans.
Coult. & Rose, Coulter, J. M., & J. N. Rose.
Court. Courtois, R. J., 1806–35. French botanist.
Cov. Coville, Fredrick Vernon, 1867–1937. Curator, U.S. National Herbarium, 1893–1937; botanist of 1891 Death Valley expedition, also of Harriman Alaska expedition.
Cov. & Ball, Coville, F. V., & C. R. Ball.
Cov. & Standl. Coville, F. V., & P. C. Standley.
Crantz, Heinrich Johann Nepomuk von, 1722–99. Professor of medicine, Vienna; author of a flora of Austria; student of Cruciferae and Umbelliferae.
Crép. Crépin, François, 1830–1903. Director, botanical garden, Brussels; student of *Rosa*.
Critchfield, William Burke, 1923– . Forester, student of Coniferae, especially *Pinus*.
Cronq. Cronquist, Arthur John, 1919– . Curator, New York Botanical Garden; student of Compositae and the Great Basin flora.
Cufodontis, Georgio, 1896– . Austrian botanist; student of Ethiopian flora.
Cusson, Pierre, 1727–83. Professor of botany, Montpellier.
Czern. Czernjajew, Vasilij, 1796–1871. Professor, Kharkov.

Dahlstedt, Gustaf Adolf Hugo, 1856–1934. Botanical curator, State Museum of Natural History, Stockholm; student of *Taraxacum*.
Danser, Benedictus Hubertus, 1891–1943. Professor, the Netherlands.
K. C. Davis, Davis, Kary Cadmus, 1867–1936. Botanist, Minnesota and Tennessee.
R. J. Davis, Davis, Ray Joseph, 1895– . Author of *Flora of Idaho*; student of *Claytonia*.
C. DC. Candolle, Anne Casimir Pyramus de, 1836–1918. Swiss botanist.
DC. Candolle (also Décandolle), Augustin Pyramus de, 1778–1841. Professor of botany, Geneva; foun-

der of the "Prodromus," a fundamental work in the development of a modern phylogenetic system.

Decne. Decaisne, Joseph, 1807–82. Director, Jardin des Plantes, Paris; student of Asclepiadaceae and Plantaginaceae.

Decne. & Planch. Decaisne, J., & J. E. Planchon.

Desf. Desfontaines, René Louiche, 1750–1833. Professor of botany, Paris.

Desr. Desrousseaux, Louis Auguste Joseph, 1753–1838. French cloth manufacturer; contributor to Lamarck's encyclopedia.

Desv. Desvaux, Augustin Nicaise, 1784–1856. Professor of botany, Angers.

Detling, Leroy Ellsworth, 1898–1967. Professor of botany, Oregon; student of *Dentaria, Descurainia,* and the flora of Oregon.

Dew. Dewey, (Rev.) Chester, 1784–1867. Professor, University of Rochester; student of *Carex*.

Dickson, James, 1738–1822. Scottish nurseryman; writer on cryptogams.

Dieck, Georg, 1847–1925. German collector, western North America.

Dietr. Dietrich, Albert, 1795–1856. Berlin.

Dippel, Leopold, 1827–1914. German dendrologist.

Döll, Johann Christian, 1808–85. Professor, Karlsruhe.

Domin, Karel, 1882–1954. Professor of botany; director, Botanical Institute and Gardens, Prague.

D. Don, Don, David, 1799–1841. Brother of George; professor, King's College, London; librarian to the Linnaean Society.

G. Don, Don, George, 1798–1856. Scottish collector for the Horticultural Society in Brazil, the West Indies, and Africa.

Donn, James, 1758–1813. Under Aiton at Kew, later curator of Cambridge Garden; author of *Hortus Cantabrigiensis,* 1796.

Doug. Douglas, David, 1798–1834. Scottish collector, northwestern North America.

Drej. Drejer, Salomon Thomas Nicolai, 1813–42. Assistant professor, Copenhagen.

Drescher, Aubrey A., 1910– . Univ. of Wisconsin.

E. R. Drew, Drew, Elmer Reginald, 1865–1930. Professor of physics, Stanford.

W. B. Drew, Drew, William Brooks, 1908– . Professor of botany, Michigan State.

Drobov, Vasili Petrovich, 1885–1956. Russian botanist.

Druce, George Claridge, 1850–1932. Professor, Oxford.

Drude, Carl Georg Oscar, 1852–1933. German botanist.

Drummond, James Ramsay, 1851–1921. Scottish botanist; grandson of Thomas Drummond, the famous collector.

Drury, William H., Jr., 1921– . Massachusetts Audubon Society; lecturer on ecology, Harvard.

Drury & Rollins, Drury, W. H., Jr., & R. C. Rollins.

Duby, Jean Étienne, 1798–1885. Geneva cleric; student of Primulaceae.

Duchesne, Antoine Nicolas, 1747–1827. France; author of a work on useful plants, especially strawberries.

Duham. Duhamel de Monceau, Henri Louis, 1700–81. France.

Duman, (Rev.) Maximilian George, 1906– . Botanist, Catholic University, Washington, D.C.

Dumort. Dumortier, (Count) Barthélemy Charles Joseph, 1797–1878. Belgian politician.

Dunn, Stephen Troyte, 1868–1938. Official guide, Kew.

Dur. Durieu de Maisonneuve, Michel Charles, 1796–1878. Director, botanical garden in Bordeaux; student of *Isoëtes*.

Durand, Elias Magloire, 1794–1873. Philadelphia pharmacist.

T. Durand, Théophile Alexis, 1855–1912. Brussels.

Durand & Jackson, Durand, T. A., & B. D. Jackson.

Du Roi, Johann Philippe, 1741–85. German physician and botanist.

d'Urv. Dumont d'Urville, Jules Sébastien César, 1780–1842. French admiral.

Eastw. Eastwood, Alice, 1859–1953. Curator of botany, California Academy of Sciences; student of western North American flora.

A. A. Eat. Eaton, Alvah Augustus, 1865–1908. New England student of ferns.

D. C. Eat. Eaton, Daniel Cady, 1834–95. Professor of botany, Yale.

Egor. Egorova, T. V., . Russian botanist.

Ehrh. Ehrhart, Friedrich, 1742–95. Switzerland.

Ekman, Hedda Maria Emerence Adelaide, 1862–1936. Swedish student of *Draba*.

Elmer, Adolph Daniel Edward, 1870–1942. Collector in California, Washington, and the Philippine Islands; editor and chief contributor to *Leaflets of Philippine Botany*.

Endl. Endlicher, Stephen Friedrich Ladislaus, 1804–49. Professor of botany and director of botanic garden, Vienna.

Engelm. Engelmann, George, 1809–84. Physician, St. Louis; eminent botanist.

Engler, Heinrich Gustav Adolf, 1844–1930. Director, botanical garden and museum, Berlin; founder and editor, *Botanische Jahrbücher, Die Vegetation der Erde,* and *Das Pflanzenreich*; student of Araceae and *Saxifraga*.

Engler & Irmsch. Engler, A., & E. Irmscher.

Epling, Carl Clawson, 1894– . Professor of botany, Univ. of California, Los Angeles; student of Labiatae.

Erlans. Erlanson, Eileen (née Whitehead), 1899– .

Esch. Eschscholtz, Johann Friedrich, 1793–1831. Zoologist, Dorpat (Tartu), Estonia; surgeon and naturalist on the ships *Rurik* and *Predpriaetie,* under Kotzebue, which reached California in 1816 and 1824.

Evans, Walter Harrison, 1863–1941. American botanist.

Farr, Edith May, 1864–1956. American botanist.

Farw. Farwell, Oliver Atkins, 1867–1944. Botanist, Detroit.

Fassett, Norman Carter, 1900–1954. Professor of botany, Wisconsin; student of Wisconsin flora and the aquatic plants of North America.

Fedorov, Andrei Alexandrovich, 1908– . Russian botanist.

Fedtsch. Fedtschenko, Boris Alexejevich, 1873–1947. Director, botanical garden, St. Petersburg.

Fedtsch. & Flerov, Fedtschenko, B. A., & B. K. Flerov.

Fée, Antoine Laurent Apollinaire, 1789–1874. Professor of botany, Strasbourg; student of cryptogams.

Fenzl, Eduard, 1808–79. Director, Botanical Garden, Vienna.

Fern. Fernald, Merritt Lyndon, 1873–1950. Director,

List of Authors

Gray Herbarium, Harvard, 1937–47; noted plant geographer and systematist; author, *Gray's Manual*, 8th ed.
Fern. & Macbr. Fernald, M. L., & J. F. Macbride.
Fern. & Rydb. Fernald, M. L., & P. A. Rydberg.
Fern. & Weath. Fernald, M. L., & C. A. Weatherby.
Fern. & Wieg. Fernald, M. L., & K. M. Wiegand.
Ferris, Roxana Judkins (née Stinchfield), 1895– Botanist, Dudley Herbarium, Stanford.
Fieb. Fieber, Franz Xavier, 1807–72. Hungary.
Finet, Achille Eugène, 1863–1913. French botanist.
Fingerhuth, Carl Anton, 1802–76. German author.
Fiori, Adriano, 1865–1950. Italian botanist.
Fiori & Paol. Fiori, A., & G. Paoletti.
Fisch. Fischer, Friedrich Ernst Ludwig von, 1782–1854. Director, botanical garden, St. Petersburg.
Fisch. & Mey. Fischer, F. E. L. von, & C. A. Meyer.
Fisch. & Trautv. Fischer, F. E. L. von, & E. R. von Trautvetter.
Flerov, Boris Konstantinovich, 1896– . Moscow botanist.
Flod. Floderus, Björn Gustaf Oskar, 1867–1941. Swedish physician and botanist; authority on *Salix*.
Flügge, Johann, 1775–1816. Physician, Hamburg.
Focke, Wilhelm Olbers, 1834–1922. Physician, Bern; student of *Rubus*.
Fomin, Alexander Vasilievich, 1867–1935. Russian botanist; student of ferns.
Forbes, James, 1773–1861. English gardener, Woburn Abbey; author of books on garden plants.
Forrest, George, 1873–1932. Scottish student of Chinese flora.
Forsk. Forskål, Petter, 1732–63. Finnish student of Linnaeus who wrote a flora of Egypt and Arabia.
Fourn. Fournier, Eugène Pierre Nicolas, 1834–84. Physician, Paris.
Fourr. Fourreau, Pierre Jules, 1844–71. French botanist.
Franco, João Manuel Antonio Paes do Amaral, 1921– . Lisbon botanist.
Franch. Franchet, Adrian René, 1834–1900. French botanist.
Fr. & Sav. Franchet, A. R., & P. A. L. Savatier.
Freyn, Josef Franz, 1845–1903. Prague, student of northern Asiatic flora, Liliaceae, and Ranunculaceae.
E. Fries, Fries, Elias Magnus, 1794–1878. Professor of botany, Uppsala, student of fungi and *Hieracium*.
T. Fries, Fries, Thore Christian Elias, 1886–1930. Professor of botany, Lund, Sweden.
Froel. Froelich, Joseph Aloys, 1760–1841. German student of *Gentiana* and *Hieracium*.
Frye, Theodore Christian, 1869–1962. Professor of botany, University of Washington.
Frye & Rigg, Frye, T. C., & G. B. Rigg. Authors, *Northwest Flora*.
Fuchs, H. P., 1928– . Washington, D.C.

Gaertn. Gaertner, Philipp Gottfried, 1754–1825. Apothecary.
Gaertn., Mey. & Scherb. Gaertner, P. G., B. Meyer, & J. Scherbius.
Gandoger, (Abbé) Michel, 1850–1926. France; author of *Flora Europae*; prolific writer.
Garcke, Friedrich August, 1819–1904. Curator, Berlin Herbarium; author of *Flora von Nord- und Mittel Deutschland*.
Gaud. Gaudichaud-Beaupré, Charles, 1789–1854. French botanist.
Gay, Jacques Étienne, 1786–1864. French botanist.
Gelert, Otto, 1862–1899. Curator of herbarium, Copenhagen.
Georgi, Johann Gottlieb, 1729–1802. Russian botanist, author of *Beschreibung des Russischen Reiches*.
Gilbert, Benjamin Davis, 1835–1907. American student of ferns.
Gilg, Ernst Friedrich, 1867–1933. German botanist, Berlin.
Gilib. Gilibert, Jean Emmanuel, 1741–1814. Professor, Vilna and Lyon; author of *Plantae Lithuanicae*.
Gill, Lake Shore, 1900– . American forest pathologist.
Gillett, John Montague, 1918– . Dept. of Agriculture, Canada; student of Gentianaceae.
Gjaerevoll, Olav, 1916– . Professor, Trondheim, Norway; politician.
Glehn, Peter von, 1835–76. Russian botanist.
Gmel. Gmelin, Johann Georg, 1709–55. Traveled in Siberia; author of *Flora Sibirica*; later professor in Tübingen.
C. C. Gmel. Gmelin, Carl Christian, 1762–1837. Physician, Karlsruhe.
J. F. Gmel. Gmelin, Johann Friedrich, 1748–1804. Professor, Tübingen, later Göttingen.
S. G. Gmel. Gmelin, Samuel Gottlieb, 1743–74. Cousin of J. F. Gmelin, traveled with Pallas in Russia.
Goldie, John, 1793–1886. Scottish traveler through eastern Canada and New England, 1819.
Gombócz, Endre, 1882–1945. Hungarian botanist; published Dr. E. Kol's Alaskan collection.
Gontsch. Gontscharov, Nikolai Fedorovitch, 1900–1942. Russian botanist.
Good, Ronald D'Oyley, 1896– . Noted plant geographer; author of *The Geography of Flowering Plants*.
Good. Goodenough, Samuel, 1743–1827. English student of *Carex*; Bishop of Carlisle.
Gorodk. Gorodkov, Boris Nikolaevich, 1890–1953. Russian botanist; student of *Carex*.
Gould, Frank W., 1913– . Professor of range and forestry, Texas A & M; student of grasses.
Graebner, Karl Otto Robert Peter Paul, 1871–1933. Professor, Berlin.
Graham, Robert, 1786–1845. Professor, Edinburgh.
Grauer, Sebastian, 1758–1820. Physician, Kiel.
Gray, Asa, 1810–88. Professor of botany, Harvard; preeminent, prolific botanical author.
S. F. Gray, Gray, Samuel Frederick, 1766–1836. Author of *Natural Arrangement of British Plants*, 1821.
Greene, Edward Lee, 1843–1915. Professor of botany, California, later Catholic University and Smithsonian Institution.
Greenm. Greenman, Jesse More, 1867–1951. Curator of herbarium, Missouri Botanical Garden; student of *Senecio*.
Grev. Greville, Robert Kaye, 1794–1866. Professor, Edinburgh.
Grev. & Hook. Greville, R. K., & W. J. Hooker.
Griseb. Grisebach, August Heinrich Rudolf, 1814–79. Professor of botany, Göttingen.
Gronov. Gronovius, Johannes Fredericus, 1690–1762. Dutch botanist; friend of Linnaeus.

Gunn. Gunnerus, Johan Ernst, 1718–73. Bishop, Trondheim, Norway.

Hack. Hackel, Eduard, 1850–1926. Noted Austrian botanist and agrostologist.
Hadač, Emil, 1914– . Czech botanist; student of Spitzbergen flora.
Hadač & Löve, Hadač, E., and A. Löve.
Haenke, Thaddaeus, 1761–1817. Bohemia; phytographer for the King of Spain; first botanist with Luis Née in California; collection described by K. B. Presl.
Hagerup, Olaf, 1889–1961. Keeper of herbarium, Copenhagen.
Hagl. Haglund, Gustav, 1900–1955. Swedish student of *Taraxacum*.
Hagstr. Hagström, Johan Oskar, 1860–1922. Swedish student of *Potamogeton* and *Ruppia*.
Hall, Harvey Monroe, 1874–1932. Professor of botany, California, later Carnegie Institution; student of Compositae and one of the founders of experimental taxonomy.
Hall & Clements, Hall, H. M., & F. E. Clements. Authors of *The Phylogenetic Method in Taxonomy*.
Halliday, Geoffrey, 1933– . British botanist, University of Leicester.
Hamilton, Arthur, . Geneva; author of 1832 monograph on *Scutellaria*.
Hand.-Mazz. Handel-Mazzetti, Heinrich von, 1862–1940. Austrian botanist, student of Chinese flora; author of *Symbolae Sinicae*.
Hara, Hiroshi, 1911– , Japanese taxonomist.
Harshb. Harshberger, John William Claghorn, 1869–1929. American botanist.
Hartm. Hartman, Carl, 1824–84. Swedish botanist; author of a handbook of the Scandinavian flora in 11 editions.
F. X. Hartm. Hartmann, Franz Xaver (Ritter) von, 1737–91.
R. Hartm. Hartman, Robert Wilhelm, 1827–91. Teacher, Gävle, Sweden.
Harv. Harvey, William Henry, 1811–66. Professor of botany and keeper of the herbarium, Trinity College, Dublin.
Haussk. Haussknecht, Heinrich Carl, 1838–1903. German student of *Epilobium*.
Haw. Haworth, Adrian Hardy, 1767–1833. British botanist.
Hayata, Bunzo, 1874–1934. Japanese botanist.
Hayek, August von, 1871–1928. Austrian botanist, student of *Anemone*.
Hayne, Friedrich Gottlob, 1763–1832. Professor of botany, Berlin.
HBK. Humboldt, F. W. H. A. von, A. J. A. Bonpland, & C. S. Kunth.
Hegelmaier, Cristoph Friedrich, 1833–1906. Professor, Tübingen; student of *Callitriche*.
Hegi, Gustav, 1876–1932. Swiss botanist, editor of *Flora von Mittel-Europa*.
Hellen. Hellenius, Carl Niklas, 1745–1820. Professor, Åbo, Finland.
Heller, Amos Arthur, 1867–1944. One of the most prolific collectors in western North America; editor of *Muhlenbergia*.
Henders. Henderson, Louis Forniquet, 1853–1942. Professor of botany, Oregon.

Henrard, Jan Theodoor, 1881– . Conservator, Rijksherbarium, Leiden; student of *Aristida*.
Henry, Joseph Kaye, 1866–1930. Professor of English, British Columbia; collector and writer; author of *Flora of Southern British Columbia*.
Herder, Ferdinand Godofried Theobald Maximilian von, 1828–96. Russian student of the flora of Siberia and Russian North America.
Herm. Herman, Frederick Joseph, 1906– . National Arboretum Herbarium, Beltsville, Maryland.
Hiern, William Philip, 1839–1925. British Museum.
Hieron. Hieronymus, Georg Hans Emo Wolfgang, 1846–1921. Professor, Berlin.
Hiitonen, Henrik Ilmari Augustus (Hidén), 1890– Finnish botanist and author.
Hill, (Sir) John, 1716–75. London physician; author of herbals and nature books; produced the first flora of England on the Linnaean system.
Hiroe, Minosuke, 1914– . Japanese botanist.
Hitchc. Hitchcock, Albert Spear, 1865–1935. Leading American agrostologist, U.S. National Herbarium.
C. L. Hitchc. Hitchcock, Charles Leo, 1902– . Professor of botany, Univ. of Washington; student of the Pacific Northwest flora.
Hodgson, Harlow James, 1917– . Agricultural Experiment Station, Palmer, Alaska.
Hoffm. Hoffmann, Georg Franz, 1760–1826. Professor of botany, Göttingen, later Moscow; student of *Salix* and Umbelliferae.
Holm, Herman Theodor, 1854–1932. Danish botanist in America.
Holmb. Holmberg, Otto Rudolf, 1874–1930. Keeper of the herbarium, Lund, Sweden.
Holmen, Kjeld Axel, 1921– . Assistant professor, Copenhagen.
Holmen & Jacobsen, Holmen, K. A., & K. Jacobsen.
Holz. Holzinger, John Michael, 1853–1929. Teacher, Minnesota.
Honck. Honckeny, Gerhard August, 1724–1805. Austrian botanist.
Honda, Massaji, 1897– . Japanese botanist.
Hook. Hooker, (Sir) William Jackson, 1785–1865. Professor of botany, Glasgow, 1820; director of botanical gardens, Kew; author of many works.
Hook. & Arn. Hooker, W. J., & G. A. W. Arnott.
Hook. f. Hooker, (Sir) Joseph Dalton, 1817–1911. Physician, Glasgow; director of botanical gardens, Kew, 1866; son of W. J. Hooker.
Hook. f. & Thoms. Hooker, J. D., and T. Thomson. Editors of *Flora Indica*.
L. S. Hopkins, Hopkins, Lewis Sylvester, 1872– Dean, Culver-Stockton College, Missouri.
M. Hopkins, Hopkins, Milton, 1906– . Professor of botany, Oklahoma; student of *Arabis*.
Hoppe, David Heinrich, 1760–1846. Professor, Regensburg; editor of *Flora*.
Hornem. Hornemann, Jens Wilken, 1770–1841. Professor of botany, Copenhagen.
Host, Nicolaus Thomas, 1761–1834. Imperial physician, Vienna; author of *Flora Austriaca*, 1827–31.
House, Homer Doliver, 1878–1949. New York State botanist, 1914–48; author of *Wild Flowers of New York*.
How. Howell, Thomas Jefferson, 1842–1912. Portland, Ore.; author of *A Flora of Northwest America*, 1897–1903.

List of Authors

Howe, Charles Horton Peck, 1833–1917. American botanist.
Hubbard, Frederick Tracy, 1875–1962. Massachusetts.
Huds. Hudson, William, 1730–93. London apothecary; author of *Flora Anglica*.
Hult. Hultén, Oskar Eric Gunnar, 1894– . Professor of botany, Stockholm; student of American arctic floras; author of *Flora of Kamchatka, Flora of the Aleutian Islands, Flora of Alaska and Yukon*.
Hult. & Maguire, Hultén, O. E. G., & B. Maguire.
Hult. & St. John, Hultén, O. E. G., & H. St. John.
Humboldt, (Baron) Friedrich Wilhelm Heinrich Alexander von, 1769–1859. German explorer and zoologist.
Huth, Ernst, 1845–97. German student of Ranunculaceae.
Hyl. Hylander, Nils, 1904– . Institution for Systematic Botany, Uppsala; author of *Nordisk Kärlväxtflora*.

Iljin, Modest Mikhailovich, 1889–1967. Russian botanist; student of Compositae.
Iltis, Hugh Helmut, 1925– . Professor of botany, Wisconsin; student of Capparidaceae.
Inman, Ondess Lamar, 1890–1942. Professor of botany, Idaho.
Irmsch. Irmscher, Edgar, 1887– . Curator of herbarium, Hamburg.

Jackson, Benjamin Dayton, 1846–1927. Botanical secretary and general secretary, the Linnaean Society; compiler of *Index Kewensis*.
Jacq. Jacquin, (Baron) Nicholas Joseph von, 1727–1817. Professor of botany, Vienna; noted systematist.
Jalas, Jakko, 1920– . Professor of botany, Helsinki.
Janchen, Erwin, 1882– . Professor, Vienna.
Jancz. Janczewski, Eduard, Ritter von Glinka, 1846–1918. Professor, Cracow; student of *Ribes*.
Jennings, Otto Emery, 1877–19?? . Carnegie Museum, Pittsburgh, and professor, Univ. of Pittsburgh; student of the flora of western Pennsylvania and vicinity.
Jepson, Willis Linn, 1867–1946. Professor of botany; author of California floras.
Johnston, Ivan Murray, 1898–1960. Professor of Botany, Harvard; student of Boraginaceae.
G. N. Jones, Jones, George Neville, 1904– . Professor of botany, Illinois; author of *Flora of Illinois* and *A Botanical Survey of the Olympic Peninsula, Washington*.
M. E. Jones, Jones, Marcus Eugene, 1852–1934. Utah mining consultant; curator of herbarium of Great Basin plants, Pomona College; author.
Q. Jones, Jones, Quentin, 1920– . Beltsville, Md.
Jord. Jordan, Alexis, 1814–97. Lyon, France.
Jord. & Fourr. Jordan, A., & P. J. Fourreau.
Jordal, Louis Henrik, 1919–52. Norwegian collector in Alaska; graduate study at Michigan; author.
Jurtz. Jurtzev, B. A., . Russian botanist.
Juss. Jussieu, Antoine Laurent de, 1748–1836. Professor, Jardin des Plantes, Paris.
Juz. Juzepczuk, Sergei Vasilievich, 1893–1959. Russian taxonomist; student of critical genera.

Kalela, Aarno, 1908– . Professor, Helsinki; student of *Carex*.
Kane, Elias Kent, 1820–57. U.S. Navy; Arctic explorer.

Karav. Karavajev, Mikhail Nikolaevich, 1903– . Russian botanist, author of *Konspekt Flory Jakuti* (Akad. Nauk SSSR), 1958.
Karav. & Jurtz. Karavajev, M. N., & B. A. Jurtzev.
Karel. Karelin, Grigorii Silych, 1801–72. Russian botanist.
Karel. & Kiril. Karelin, G. S., & I. P. Kirilov.
Karst. Karsten, Gustav Karl Wilhelm Herman, 1817–1908. Professor, Berlin and Vienna.
Kaulf. Kaulfuss, Georg Friedrich, 1786–1830. Professor, Halle; student of ferns.
Kawano, Syo'ichi, 1936– . Japanese botanist.
Kearney, Thomas Henry, 1874–1956. U.S. Dept. of Agriculture; California Academy of Sciences; student of Malvaceae; author of *Arizona Flora*.
Keck, David Daniels, 1903– . Assistant director, New York Botanical Garden; experimental taxonomist; for many years at National Science Foundation, Washington; student of *Pentstemon*, Madiinae.
Keller, Robert, 1854–1939. Swiss botanist; student of *Rosa, Rubus, Hypericum*.
Kelso, Leon, 1907– . American student of *Glyceria, Salix*.
Ker-Gawl. Ker, John Bellenden, or John Ker Bellenden, or, before 1804, John Gawler, 1764–1842. British botanist.
Kern, Patricia, 1940– . Univ. of Washington; now New York Botanical Garden; student of *Tiarella, Thlaspi*, and Saxifragaceae.
Kiril. Kirilov, Ivan Petrovich, 1821–42. Russian botanist; collaborated with Karelin.
Kit. Kitaibel, Paul, 1757–1817. Professor of medicine and director of botanical garden, 1794; professor of botany and chemistry, Budapest.
Kitagawa, Masao, 1909– . Japanese botanist.
Kitamura, Sirō, 1906– . Japanese botanist.
Kitamura & Hara, Kitamura, S., & H. Hara.
Kjellm. Kjellman, Franz Reinhold, 1846–1907. Professor of botany, Uppsala; member of expedition circumnavigating Eurasia in *Vega* under Nordenskjöld; collections described in Nordenskjöld's book.
Knaben, Gunvor, 1911– . Lector, botanical museum, Oslo.
Knaf, Joseph, 1801–65. Austrian botanist.
Kobayashi, Yoshio, 1907– . Japanese botanist.
K. Koch, Koch, Karl Heinrich Emil, 1809–79. Professor, Berlin.
W. D. J. Koch, Koch, Wilhelm Daniel Joseph, 1771–1849. Professor of botany, Erlangen.
Koidz. Koidzumi Gen'ichi, 1883–1953. Japanese taxonomist.
Kom. Komarov, Vladimir Leontievich, 1869–1945. President, Academy of Sciences of USSR; student of flora of Kamchatka and Manchuria; organizer and editor, *Flora USSR*.
Konig, Charles (Carl Dietrich Eberhard Koenig), 1774–1851. German geologist at the British Museum.
Kostel. Kosteletzky, Vincenz Franz, 1801–87. Czech botanist.
Koyama, Tetsuo, 1933– . Japanese botanist.
Koyama & Kawano, Koyama, T., & S. Kawano.
Kraschen. Kraschennikov, Hippolit Mikhailovich, 1884–1947. Russian botanist.
Krecz. Kreczetowicz, Vitalij Ivanowicz, 1901–42. Russian student of *Carex*.

Krecz. & Bobr. Kreczetowicz, V. I., & E. G. Bobrov.
Krecz. & Gontsch. Kreczetowicz, V. I., & N. F. Gontscharov.
Kríča, Bohdan, 1936– . Czech botanist.
Kryl. Krylov, Porfirii Nikitovich, 1859–1931. Professor of botany, Tomsk; author of *Flora Zapadni Sibir.*
Kryl. & Serg. Krylov, P. N., & L. P. Sergievskaja.
Ktze. Kuntze, Carl Ernst Otto. 1843–1907. German botanist.
Kudo, Yushûn, 1887–1932. Director, botanical garden, Taihoku, Formosa.
Kühlewein, P. E., 1798–1870.
Kuhn, Friedrich Adalbert Maximilian, 1842–94. Professor, Berlin; German student of ferns.
Kükenth. Kükenthal, Georg, 1864–1956. Bishop, Coburg, Germany; eminent student of *Carex.*
Kunth, Carl Sigismund, 1788–1850. Professor of botany, Berlin; systematist; author.
Kunze, Gustav, 1793–1851. Botanist and physician of Leipzig; student of ferns.
Kurtz, Fritz (Frederico), 1854–1920. German botanist; professor in Córdoba, Argentina, 1884.
Kuzn. Kuznetsov, Nikolai Ivanovich, 1864–1932. Russian student of *Gentiana.*

L. Linnaeus, Carolus (Carl von Linné), 1707–78. Eminent botanist, professor, Uppsala; author of *Species Plantarum,* the foundation of nomenclature.
L. f. Linné, Carl von, 1741–83. Professor, Uppsala; son of Linnaeus.
Laest. Laestadius, Lars Levi, 1800–1861. Pastor, Lapland; botanist.
Lam. Lamarck, Jean Baptiste Pierre Antoine de Monet de la, 1744–1829. Famous French botanist.
Lam. & DC. Lamarck, J. B. P. A. de M. de la, & A. P. de Candolle.
Lamb. Lambert, Aylmer Bourke, 1761–1842. Vice-president, Linnaean Society, London; patron of botany; author of *A Description of the Genus Pinus.*
Lang, Otto Friedrich, 1817–1847. German caricologist.
Lange, Johan Martin Christian, 1818–98. Professor of botany, Copenhagen.
La Peyr. La Peyrouse, (Baron) Philippe Picot de, 1744–1818.
La Pyl. La Pylaie, Auguste Jean Marie Bachelot de, 1786–1856. French botanist.
Le Conte, Johan Elton, 1784–1840. American botanist.
Ledeb. Ledebour, Carl Friedrich von, 1785–1851. Professor, Tartu (Dorpat), Esthonia; author of *Flora Rossica* and *Flora Altaica.*
Lehm. Lehmann, Johann Georg Christian, 1792–1860. Director of Botanic Garden, Hamburg; student of *Potentilla.*
Lej. Lejeune, Alexander Ludwig Simon, 1779–1858. Physician, Verviers, Belgium.
Lej. & Court. Lejeune, A. L. S., & R. J. Courtois.
Lemmon, John Gill, 1832–1908. Pioneer California botanist; student of Coniferae.
Lepage, (Abbé) Ernest, 1905– . Rimouski, Canada; student of arctic and subarctic flora.
Less. Lessing, Christian Friedrich, 1809–62. German physician; student of Compositae.
Lév. Léveillé, (Abbé) Augustin Abel Hector, 1863–1918. French botanist.
Leyss. Leysser, Friedrich Wilhelm von, 1731–1815. Halle.

L'Hér. L'Héritier de Brutelle, Charles Louis, 1746–1800. Celebrated French botanist.
Liebm. Liebmann, Friedrik Michael, 1813–56. Danish collector in Mexico.
Lightf. Lightfoot, (Rev.) John, 1735–88. English clergyman, author of *Flora Scotica.*
Liljebl. Liljeblad, Samuel, 1761–1815. Swedish botanist.
Lindberg f. Lindberg, Harald, 1871–1963. Curator of herbarium, Helsinki.
Lindb. f. & Markl. Lindberg, H., & G. G. Marklund.
Lindbl. Lindblom, Alexis Edvard, 1807–53. Assistant professor, Lund.
Lindeb. Lindeberg, Karl Johan, 1815–1900. Teacher, Gothenburg, Sweden.
Lindl. Lindley, John, 1799–1865. British botanist.
Lindm. Lindman, Carl Axel Magnus, 1856–1928. Professor, Stockholm.
Link, Johann Heinrich Friedrich, 1767–1851. Professor of natural sciences; director of botanical garden, Berlin.
Lipsch. Lipschits, Sergei Joljevich, 1905– . Russian botanist.
Liro, Johan Ivar (né Lindroth), 1872–1943. Finnish botanist.
Little, Elbert Luther, Jr., 1907– . U.S. Forest Service, Washington, D.C.; author of *Checklist of Trees of the United States;* distinguished dendrologist.
Litw. Litwinow, Dmitri Ivanovich, 1854–1929. Russian botanist.
Lodd. Loddiges, Conrad, 1738–1820. London nurseryman.
Loefl. Löfling, Pehr, 1729–56. Swedish student of Linnaeus who traveled to Venezuela for him and died there.
Lönnr. Lönnroth, Knut Johan, 1826–85. Assistant professor of botany, Uppsala.
A. Los. Losina-Losinskaya, A. S.,   . Russian botanist.
Loud. Loudon, John Claudius, 1783–1843. English horticulturist.
Louis-Marie, Lalonde, Louis (Père Louis-Marie), 1896– . Professor of botany, Institute Agronomique d'Oka, La Trappe, Québec.
Löve, Askell, 1916– . Icelandic botanist in America; cytogeneticist; Univ. of Montréal, now Colorado.
Löve & Hadac, Löve, A., & E. Hadač.
Löve & Löve, Löve, A., & D. Löve.
Löve & Löve & Raymond, Löve, A., D. Löve, & M. Raymond.
D. Löve, Löve, Doris, 1918– . Swedish botanist in America, now at Colorado; cytogeneticist; wife of A. Löve.
Lund, Nicolai, 1814–47. Norwegian botanist.
Lundstr. Lundström, Carl Erik, 1882– . Apothecary, Stockholm.
Lyall, David, 1817–95. Scottish botanist in America.

Ma, Yü Ch'üan,   . Chinese botanist.
Macbr. Macbride, James Francis, 1892– . Chicago Natural History Museum; student of Boraginaceae; author of *Flora of Peru.*
McGregor, Ernest Alexander, 1880– . California entomologist and botanist.
Mack. Mackenzie, Kenneth Kent, 1877–1934. New York; student of *Carex.*

List of Authors

MacM. MacMillan, Conway, 1867–1929. State botanist, Minnesota.
McNeill, J., 1933– . Reading, England.
Macoun, John, 1832–1920. Government naturalist of Canada.
James Macoun, Macoun, James Melville, 1862–1920. Curator, National Herbarium, Canada; son of John.
Magnus, Paul Wilhelm, 1844–1914. German botanist.
Maguire, Bassett, 1904– . Curator, New York botanical garden; student of the flora of the Great Basin and of northeastern South America.
Maguire & Hitchc. Maguire, B., & C. L. Hitchcock.
Makino, Tomitarō, 1862–1957. Japanese botanist.
Malte, Oscar, 1880–1933. Chief botanist, National Herbarium, Canada, 1921–33.
Marklund, Gunnar Georg, 1892–1964. Professor, Helsinki.
Marsh. Marshall, Humphrey, 1722–1801. Pennsylvania dendrologist.
Math. Mathias, Mildred Esther, 1906– . Professor of botany, Univ. of California, Los Angeles; student of Umbelliferae and cultivated plants.
Math. & Const. Mathias, M. E., & L. Constance.
Mats. Matsuura, Hajime, 1900– . Japanese botanist.
Mats. & Toyok. Matsuura, H., & H. Toyokuni.
Matsum. Matsumura, Jinzō, 1856–1928. Japanese botanist.
Mattf. Mattfeld, Johannes, 1895–1951. Curator of the herbarium, Berlin.
Maxim. Maximovich, Carl Johann, 1827–91. German botanist in Russia; director of botanic garden, St. Petersburg; student of East Asiatic flora.
Maxon, William Ralph, 1877–1948. Curator of Plants, U.S. National Herbarium; student of ferns.
Mayr, Heinrich, 1856–1911. German forester.
Medic. Medicus, Friedrich Casimir, 1736–1808. German botanist.
Meerb. Meerburg, Nicolaas, 1734–1814. Gardener, botanical garden, Leiden.
Meinsh. Meinshausen, Karl Friedrich, 1819–99. Russian botanist.
Meisn. Meisner, Carl Friedrich, 1800–1874. Swiss botanist; student of Polygonaceae.
Mela, Aukusti Johanna (née Malmberg), 1846–1904. Finland.
Mela & Caj. Mela, A. J., & A. K. Cajander.
Melderis, Aleksander, 1909– . Curator, British Museum, London; student of *Agropyron*.
Menzies, Archibald, 1754–1842. Scotland; first botanist to visit Pacific coast of North America; surgeon on English ship *Discovery*.
Merr. Merrill, Elmer Drew, 1876–1956. Director of Arnold Arboretum, Harvard; prolific contributor to knowledge of east Asian flora.
Mertens, Franz Karl, 1764–1831. German botanist.
Mert. & Koch, Mertens, F. K., & W. D. J. Koch.
C. A. Mey. Meyer, Carl Anton von, 1795–1855. Director, botanical garden, St. Petersburg.
E. Mey. Meyer, Ernst Heinrich Friedrich, 1791–1858. Professor of botany, Königsberg.
Mey. & Schreb. Meyer, E. H. F., & J. D. C. von Schreber.
B. Meyer, Meyer, Bernhard, 1767–1836. Botanist, Wetterau, Germany.
F. G. Meyer, Meyer, Frederick Gustav, 1917– . U.S. Dept. of Agriculture; student of *Valeriana* and *Coffea*; Missouri Botanical Garden.

Michx. Michaux, André, 1746–1802. French botanist in America; author of *Flora Boreali Americana*.
Milde, Carl August Julius, 1824–71. German student of Pteridophyta.
Mill. Miller, Philip, 1691–1771. English author of *The Gardener's Dictionary*.
Miller, M. Gerrit Smith, 1861– . American botanist.
Miller & Standl. Miller, M. G. S., & P. C. Standley.
Miquel, Frederik Anton Willem, 1811–71. Professor of botany, Utrecht.
Mirb. Mirbel, Charles François Brisseau de, 1776–1854. Paris.
Mitchell, John, 1676–1768. English-born Virginia physician.
Miyabe, Kingo, 1860–1951. Professor of botany, Sapporo, Japan.
Miyabe & Kudo, Miyabe, K., & Y. Kudo.
Moç. Moçiño, José Mariano, 1757–1820. Mexican physician; collected in California.
Moench, Conrad, 1744–1805. Professor of botany, Marburg.
Moore, Thomas, 1821–87. British botanist, student of Pteridophyta.
J. W. Moore, Moore, John William, 1901– . American collector in Pacific area.
Moq. Moquin-Tandon, Christian Horace Benedict Alfred, 1804–63. French student of Chenopodiaceae.
More, Albert Hanford, 1883– . U.S. Dept. of Agriculture.
Morong, (Rev.) Thomas, 1827–94. Massachusetts amateur botanist; student of Naiadaceae.
Morris, Edward Lyman, 1870–1913. Washington, D.C.; student of Plantaginaceae.
Morris & Ames, Morris, E. L., & O. Ames.
Morton, Conrad Vernon, 1905– . Curator, U.S. National Herbarium; distinguished student of ferns.
Muhl. Muhlenberg, Henry (formerly Gotthilf Heinrich Ernst Muehlenberg), 1753–1815. Lutheran minister and pioneer botanist of Pennsylvania.
Mulligan, G. A., 1928– . Plant Research Institute, Dept. of Agriculture, Ottawa; author of *Weeds of Canada*; co-author of *Flora of the Queen Charlotte Islands*; student of cytotaxonomy of weedy plants and of Cruciferae.
Mulligan & Calder, Mulligan, G. A., & J. A. Calder.
Munro, (Gen.) William, 1818–80. English agrostologist.
Munz, Philip Alexander, 1892– . Professor of botany; author of *Manual of Southern California Botany*; student of Onagraceae.
Murb. Murbeck, Svante Samuel, 1859–1946. Professor, Lund; student of *Verbascum* and *Celsia*.
Murr. Murray, Johann Anders, 1740–91. Swedish professor of botany, Göttingen; student of Linnaeus.
Mutis, José Celestino, 1732–1811. Spanish botanical explorer in Colombia; correspondent of Linnaeus.

Nakai, Takenoshin, 1882–1952. Professor of botany, Tokyo.
Nakai & Hara, Nakai, T., & H. Hara.
Nannf. Nannfeldt, Johan Axel Fritiof, 1904– . Professor of botany, Uppsala.
Nash, George Valentine, 1864–1921. Head gardener, New York Botanical Garden; agrostologist.
Neck. Necker, Noel Joseph de, 1729–93. German physician.

Nees, Nees von Esenbeck, Christian Gottfried Daniel, 1776–1858. Professor of botany, Breslau.
Neilson, James Alexander, 1923– . American botanist.
Nekrasova, Vera Leontievna, 1884– . Russian student of Saxifragaceae.
Nels. Nelson, Aven, 1859–1952. Professor of botany, Wyoming; student of Rocky Mountain flora; author.
Nels. & Macbr. Nelson, A., & J. F. Macbride.
E. Nels. Nelson, Elias Emmanuel, 1876– . Swedish horticulturist in Bend, Oregon.
Nevski, Sergei Arsenjevich, 1908–38. Agrostologist, Botanical Institute of the Academy of Science, USSR, Leningrad.
Newm. Newman, Edward, 1801–76. English naturalist, student of ferns.
Niedenzu, Franz Josef, 1857–1937. German botanist; student of Malpighiaceae.
Nieuwl. Nieuwland, Julius Aloysius Arthur, 1878–1936. Professor of botany, Notre Dame, Indiana.
Nilsson-Ehle, Nils Herman, 1873–1949. Swedish professor of botany, Lund; geneticist; collected in eastern Siberia.
Nordh. Nordhagen, Rolf, 1894– . Professor of botany, Bergen and Oslo.
Norman, Johannes Musaeus, 1823–1903. Norwegian botanist.
Nosaka, Shirō, 1943– . Japanese botanist; high school teacher, Sapporo.
Nutt. Nuttall, Thomas, 1786–1859. Philadelphia botanical collector and author; noted ornithologist.
Nyl. Nylander, William, 1822–99. Professor of botany, Helsinki.

Oakes, William, 1799–1849. Massachusetts; student of Vermont flora.
Oeder, Georg Christian von, 1728–91. Professor of botany, Copenhagen; first editor of *Flora Danica*.
Ohwi, Jisaburō, 1905– . Japanese botanist; author of *Flora of Japan*.
Olin, Johan Henric, 1769–1824. Swedish physician.
Olney, (Col.) Stephen Thayer, 1812–78. Providence, Rhode Island; student of *Carex*.
Onno, Max, 1903– . German botanist in Vienna.
Ostenf. Ostenfeld, Carl Emil Hansen, 1873–1931. Professor of botany, director of the botanical garden, Copenhagen.
Ownbey, Francis Marion, 1910– . Professor of botany, Washington State; student of *Calochortus* and *Allium*.

Pall. Pallas, Peter Simon, 1741–1811. Noted German student of Siberian flora; eminent zoologist.
Palla, Eduard, 1864–1922. Professor of botany, Graz.
Palmgr. Palmgren, Alvar, 1880–1960. Professor of botany, Helsinki.
Pamp. Pampanini, Renato, 1875–1949. Italian student of *Artemisia*.
Paol. Paoletti, Giulio, 1865–1941. Italian botanist.
Parl. Parlatore, Filippo, 1816–77. Professor of botany, Florence.
Parry, Charles Christopher, 1823–90. Colorado and Iowa; botanist with the Mexican Boundary Survey.
Patrin, Eugène Louis Melchoir, 1742–1814. French botanist.
Payson, Edwin Blake, 1893–1927. Professor of botany, Wyoming; student of Cruciferae, *Cryptantha*.

Payson & St. John, Payson, E. B., & H. St. John.
Pease, Arthur Stanley, 1881–1964. American botanist.
Pease & More, Pease, A. S., & A. H. More.
Peck, Morton Eaton, 1871–1958. Professor of biology, Willamette; author of *A Manual of the Higher Plants of Oregon*, 1941.
Pennell, Francis Whittier, 1886–1952. Curator of botany, Philadelphia; student of Scrophulariaceae.
Pennell & Stair, Pennell, F. W., & L. D. Stair.
Penny, Charles William, 1837–76. Bishop of Oxford.
Perfiljev, I. A., 1882–1942. Russian botanist.
Perr. Perrier, Eugène de la Bathie, 1825–1916. French professor of botany.
Perr. & Song. Perrier, E., & A. Songeon.
Pers. Persoon, Christian Hendrik, 1761–1836. Physician, Göttingen and Paris; author of important botanical works.
Petr. Petrak, Franz, 1886– . Vienna mycologist; student of *Cirsium*.
Petrov, J. V. A., 1896–1955. Russian botanist.
Pfeiffer, Norman Etta, 1889– . Boyce Thompson Institute for Plant Research; student of *Isoëtes*.
Philippi, Rudolph Amandus, 1808–1904. Chilean botanist.
Phill. Phillips, Lyle L., 1923– . American botanist.
Pickering, Charles, 1805–78. American physician and botanist on the Wilkes's Expedition; plant geographer.
Piehl, Martin A., 1932– . Univ. of Wisconsin, Milwaukee; student of Santalaceae.
Pilger, Robert Knud Friedrich, 1876–1953. Director of the botanical garden and museum, Berlin; student of Plantaginaceae and Coniferae.
Piper, Charles Vancouver, 1867–1926. Agrostologist, U.S. Dept. of Agriculture; author of *Flora of Washington*.
Piper & Robins. Piper, C. V., & B. L. Robinson.
Planch. Planchon, Jules Émile, 1823–88. Professor of botany, Montpellier.
Pobed. Pobedimova, Eugenia Georgievna, 1898– . Russian botanist.
Poggenb. Poggenburg, Justus Ferdinand, 1840–93. German-born American; business manager of *New Yorker Staats Zeitung*.
Pohl, Richard Walter, 1916– . Professor of botany, Iowa State; distinguished student of grasses.
Pohle, Richard Richardovich, 1869–1926. Latvian botanist in Russia.
Poir. Poiret, Jean Louis Marie, 1755–1834. Author of *Flora of the Palatinate*.
Pojark. Pojarkova, Antonina Ivanovna, 1897– . Russian botanist.
Poljak. Poljakov, Peter Petrovich, . Russian botanist.
Poll. Pollich, Johann Adam, 1740–80. Physician, Kaiserslautern, Pfalz.
Polunin, Nicholas, 1909– . British botanist; author of *Circumpolar Arctic Flora*.
Popl. Poplavskaya, Henrietta Ippolitovna, . Russian botanist.
Popov, Mikhail Grigorievich, 1893–1955. Russian botanist.
Pors. Porsild, Alf Erling, 1901– . Curator of National Herbarium of Canada; prominent student of Canadian flora; with brother Robert collected some 6,000 Alaskan specimens in 1926.

List of Authors

A. E. & M. P. Pors. Porsild, A. E., & M. P. Porsild.
Pors. & Senn. Porsild, A. E., & H. A. Senn.
M. P. Pors. Porsild, Morten (Pedersen), 1872–1956. Danish botanist at Disco, Greenland.
Porter, Thomas Conrad, 1822–1901. Professor of botany, Lafayette College, Pa.; student of Colorado flora.
Pouz. Pouzar, Zdeněk, 1932– . Czech botanist.
Prantl, Karl Anton Eugen, 1849–93. Professor of botany, Breslau.
Prescott, John D., died 1837 in St. Petersburg. His herbarium now at Oxford.
Presl, Karel (Carl) Bořiwog, 1794–1852. Prague; author of *Reliquiae Haenkeanae*.
J. Presl, Presl, Jan Swatopluk, 1791–1949. Professor, Prague.
J. & C. Presl, Presl, J. S., & K. B. Presl; authors of *Flora Čechica*.
Printz, Karl Henrik Oppegaard, 1888– . Author of *The Vegetation of the Siberian-Mongolian Frontier*.
Pritz. Pritzel, Georg August, 1815–74. Author of *Thesaurus Literaturae Botanicae*; student of Lycopodiaceae, *Anemone*; Berlin.
Prokh. Prokhakov, Jaroslav Ivanovich, 1902–1965. Russian botanist.
Prov. Provancher, (Abbé) Léon, 1820–92. Québec naturalist.
Pugsley, Herbert William, 1868–1948. British botanist.
Pursh, Frederick Traugott, 1774–1820. Born in Saxony, settled in Philadelphia; author of *Flora Americae Septentrionalis*, 1814.
R. & P. Ruiz Lopez, Hipólito, 1754–1815, & José Antonio Pavon, 175(?)–1844. Spanish botanical explorers and authors of a flora of Perú and Chile.
Radius, Justus Wilhelm Martin, 1797–1884. German botanist.
Raf. Rafinesque-Schmaltz, Constantine Samuel, 1783–1840. Professor, Philadelphia; brilliant, eccentric naturalist.
Ramenskij, Leonid Grigorevich, 1884–1953. Russian botanist; member of the Rijabouschinski expedition to Kamchatka in 1908.
Rand, Edward Lothrop, 1859–1924. American botanist.
Rand & Redf. Rand, E. L., & J. H. Redfield.
Raup, Hugh Miller, 1901– . Professor, Harvard; director, Harvard University Forests; student of *Salix*; botanical explorer in northwestern Canada.
Raven, Peter Hamilton, 1936– . Professor of botany, Stanford; student of Onagraceae.
Raymond, Marcel, 1915– . Curator, Montréal botanical garden; student of Cyperaceae.
Rchb. Reichenbach, Heinrich Gottlieb Ludwig, 1793–1879. Professor, Dresden; author of *Icones Germanicae et Helveticae* and many other works.
Rchb. f. Reichenbach, Heinrich Gustav, 1834–89. Professor of botany and director of the botanical garden, Hamburg; orchidologist.
Rebr. Rebristaja, O. V.,   . Russian botanist.
Rech. f. Rechinger, Karl Heinz, 1906– . Director of the botanical museum, Vienna; student of Mediterranean and Near East flora.
Redf. Redfield, John Howard, 1815–95. American botanist.
Redowski, Ivan, 1774–1807. Russian botanist.

Reed, Clyde Franklin, 1934– . Baltimore; student of ferns.
Regel, Eduard August von, 1815–92. Director of the botanical garden, St. Petersburg; editor of *Gartenflora*; student of Betulaceae.
Regel & Tiling, Regel, E. A. von, & H. S. T. Tiling.
C. Regel, Regel, Constantin von, 1890– . Russian botanist.
Rehd. Rehder, Alfred, 1863–1949. Curator of the herbarium, Arnold Arboretum, Harvard.
Rehd. & Wilson, Rehder, A. & E. H. Wilson.
Reichard, Johann Jacob, 1743–82. Frankfurt-am-Main.
Retz. Retzius, Anders Jahan, 1742–1821. Professor, Lund; author.
Reut. Reuter, Georges François, 1805–72. Director of the botanical garden, Geneva.
Reveal, James Lauritz, 1941– . Brigham Young Univ.; student of *Eriogonum* and the Great Basin flora; co-author of *Illustrated Flora of the Intermountain Region*.
L. C. Rich. Richard, Louis Claude Marie, 1754–1821. French collector in South America and the West Indies.
Richards. Richardson, (Sir) John, 1787–1865. Scottish naturalist attached to Sir John Franklin's expedition to Arctic America.
Richter, Carl, 1855–91. Vienna collector of large herbarium.
Rigg, George Burton, 1872–1961. Professor of botany, Univ. of Washington; collected extensively in Alaska in 1913 and 1932.
Robins. Robinson, Benjamin Lincoln, 1864–1935. Curator of Gray Herbarium, Harvard; student of Compositae.
Roehl. Roehling, Johann Christoph, 1757–1813. German clergyman; author of *Deutschlands Flora*.
Roem. Roemer, Johann Jacob, 1763–1819. Professor of botany, Zürich.
Roem. & Schult. Roemer, J. J., & J. A. Schultes.
Rollins, Reed Clark, 1911– . Director of Gray Herbarium, Harvard; student of Cruciferae.
Rose, Joseph Nelson, 1862–1928. Botanist, U.S. National Herbarium; student of Cactaceae, Umbelliferae, and Crassulaceae.
Rosend. Rosendahl, Carl Otto, 1875–1956. Professor of botany, Minnesota.
Roshev. Roshevitz, Roman Julievich, 1882–1947. Russian student of Gramineae.
Rossb. Rossbach, Ruth (née Peabody), 1912– . American botanist; student of *Spergularia*.
Roth, Albrecht Wilhelm, 1757–1834. German physician.
Rothm. Rothmaler, Werner, 1908–62. Professor of botany, Griefswald, Germany.
Rottb. Rottboell, Christen Friis, 1727–97. Professor of botany and director of the botanical garden, Copenhagen.
Rouleau, Joseph Albert Ernest, 1916– . Montréal botanical garden; student of Newfoundland flora.
Rousi, Arne Henrik, 1931– . Finnish botanist, Turku.
Rousseau, Joseph Jules Jean Jacques, 1905– . Canadian botanist.
Rousseau & Raymond, Rousseau, J. J. J. J., & M. Raymond.
Rowlee, Willard Winfield, 1861–1923. American student of willows.

Royle, John Forbes, 1799–1858. British botanist; student of Himalayan flora.
Rudolph, Johann Heinrich, 1744–1809. Russian student of *Dicentra*.
Rumiantzev, N.
Rupr. Ruprecht, Franz Joseph, 1814–70. Curator of the herbarium, St. Petersburg.
Rydb. Rydberg, Per Axel, 1860–1931. Sweden and the U.S.; curator of herbarium, New York Botanical Garden; author of several large botanical manuals.
Rylands, Thomas Glazebrook, 1818–1900. English manufacturer, student of ferns.

St. John, Harold, 1892– . Professor of botany, Hawaii; formerly Washington State; student of *Pandanus*, the flora of Hawaii, and the flora of eastern Washington.
St. John & Warren, St. John, H., & F. A. Warren.
Salisb. Salisbury, Richard Anthony, 1761–1829. British botanist.
Sam. Samuelsson, Gunnar, 1885–1944. Professor and director of the botanical department, Riksmuseum, Stockholm.
Sanson, M. . St. Petersburg, botanist of the 1830's.
Sarg. Sargent, Charles Sprague, 1841–1927. Creator of Arnold Arboretum, Harvard; author of several works.
Satake, Yoshishuke, 1902– . Japanese botanist, student of *Juncus*.
Savatier, Paul Amédée Ludovic, 1820–91. French botanist; student of Japanese flora.
Savi, Gaëtano, 1769–1844. Italian author of a flora of Pisa.
Savile, Douglas Barton Osborne, 1909– . Research botanist, Ottawa.
Scherbius, Johannes, –1813. German botanist.
Scheuchz. Scheuchzer, Johann Jacob, 1672–1733. Professor, Zürich.
Scheutz, Nils Johan Wilhelm, 1836–89. Swedish teacher and botanist.
Schinz, Hans, 1858–1941. Director, botanical garden and museum, Zürich.
Schinz & Keller, Schinz, H., & R. Keller.
Schinz & Thell. Schinz, H., & A. Thellung.
Schipczinkij, Nikolai Valerianovich, 1886–1955. Russian botanist.
Schischk. Schisckhin, Boris Konstantinovich, 1886–1963. Director of the botanical garden, Leningrad.
Schischk. & Bobr. Schischkin, B. K., & E. G. Bobrov.
Schkuhr, Christian, 1741–1811. Student of German flora.
Schlecht. Schlechtendal, Diedrich Franz Leonhard von, 1794–1866. Professor of botany, Halle.
Schleich. Johann Christoph, 1768–1834. Switzerland.
Schljakov, Roman Nikolaievich, . Russian botanist.
F. Schm. Schmidt, Friedrich, 1832–1908. St. Petersburg paleontologist; student of east Siberian flora.
Schmidel, Casimir Christoph, 1718–92. German botanist.
F. W. Schmidt, Schmidt, Franz Wilibald, 1764–96. Professor of botany, Prague.
Schneid. Schneider, Camillo Karl, 1876–1951. Austria and Germany; dendrologist.
Schnizl. Schnizlein, Adalbert Carl Friedrich Hellwig Conrad, 1814–68. Professor, Erlangen.
Schol. Scholander, Per Fredrik, 1905– . Danish botanist and diplomat.

J. L. Schönl. Schönlein, Johannes Lucas, 1793–1864. Bavarian botanist.
S. Schönl. Schönlein, Selmar, 1860–1940. Professor, Grahamstown, South Africa.
Schott, Heinrich Wilhelm, 1794–1865. Director of the royal garden, Schönbrunn, Vienna.
Schrad. Schrader, Heinrich Adolf, 1767–1836. Professor of botany, Göttingen.
Schrank, Franz von Paula, 1747–1835. Professor of botany, Munich; author of floras of Bavaria, Salzburg, and Monaco.
Schreb. Schreber, Johann Christian Daniel von, 1739–1810. Professor, Erlangen.
Schult. Schultes, Joseph August, 1773–1831. Professor of botany, Vienna, Cracow, and Landeshut.
Schult. f. Julius Hermann, 1804–40. Son of J. A. Schultes, Vienna.
K. F. Schultz, Schultz, Karl Friedrich, 1765–1837. German botanist.
Schultz-Bip. Schultz (Schultz-Bipontinus), Carl Heinrich, 1805–67. German student of Compositae.
O. E. Schulz, Schulz, Otto Eugen, 1874–1936. German student of Cruciferae.
Schulze, Carl Theodor Maximilian, 1841–1915. German botanist.
Schum. Schumacher, Heinrich Christian Friedrich, 1757–1830. Professor, Copenhagen.
K. Schum. Schumann, Karl Moritz, 1851–1904. Curator of the herbarium, Berlin; student of Cactaceae.
Schuster, Julius, . Munich botanist.
Schwarz, Otto, 1900– . Vienna.
Schweigger, August Friedrich, 1783–1821. Professor, Königsberg.
Schwein. Schweinitz, Lewis David de, 1770–1834. German botanist, Bethlehem, Pa.
Scop. Scopoli, Johann Anton, 1723–88. Professor, Chemnitz and Pavia.
Scribn. Scribner, Frank Lamson, 1851–1938. Agrostologist, U.S. Dept. of Agriculture, Washington, D.C.
Scribn. & Merr. Scribner, F. L., & E. D. Merrill.
Scribn. & Sm. Scribner, F. L., & J. G. Smith.
Scribn. & Tweedy, Scribner, F. L., & F. Tweedy.
Seem. Seemann, Berthold Carl, 1825–71. German naturalist and traveler living in England; author of *Botany on the Voyage of H.M.S. Herald*.
K. O. von Seem. Seemen, Karl Otto von, 1838–1910. German botanist.
Sel. Selander, Nils Sten, 1895–1957. Swedish author-poet; botanist.
Senn, Harald Archie, 1912– . Director, Plant Research Institute, Dept. of Agriculture, Ottawa.
Ser. Séringe, Nicolas Charles, 1776–1858. French botanist.
Serg. Sergievskaja, Lidia Palladievna, 1897– . Russian botanist.
Shear, Cornelius Lott, 1865–1956. Plant pathologist and agrostologist, U.S. Dept. of Agriculture, Washington, D.C.
Sheld. Sheldon, Edmund Perry, 1869– . Minnesota, later Portland, Oregon; student of forestry.
Shetler, Stanwyn Gerald, 1933– . National Herbarium, Washington, D.C.; student of Campanulaceae.
Sibth. Sibthorp, John, 1758–96. Professor of botany, Oxford.
Sim, R. J., 1791–1878.

List of Authors

Simm. Simmons, Herman Georg, 1866–1943. Professor, Ultuna, Sweden.
Sims, John, 1749–1831. Editor, *Curtis's Botanical Magazine.*
A. Skvortz. Skvortzow, A. K. (fl. 1963). Russian botanist; student of *Salix.*
Slosson, Margaret, 1872– . American student of ferns.
Sm. Smith, (Sir) James Edward, 1759–1828. England; founder and for 40 years president of Linnaean Society.
H. Sm. Smith, Harry, 1889– . Swedish botanist.
J. G. Sm. Smith, Jared Gage, 1866– . American agrostologist in Washington, later Hawaii.
P. Sm. Smith, Charles Piper, 1877–1955. American eccentric; student of *Lupinus.*
W. W. Sm. Smith, William Wright, 1875–1956. Professor, Edinburgh; student of *Primula* and Himalayan flora.
Sm. & Forrest, Smith, W. W., & G. Forrest.
Small, John Kunkel, 1869–1938. Head curator of New York Botanical Garden; author of several floras.
Sobol. Sobolevski, Gregor Fedorovitch, 1741–1807. Russian physician and botanist.
Soland. Solander, Daniel Carl, 1736–82. Swedish botanist in England; accompanied Captain Cook on his first voyage of circumnavigation.
Sommier, Carlo Pietro Stefano, 1848–1922. Italian botanist; student of flora of the river Ob, Siberia.
Song. Songeon, André, 1826–1905. French botanist.
Soó, von Bere, Károly Rezsö, 1903– . Hungarian botanist; professor, Budapest.
Sørens. Sørensen, Thorvald, 1902– . Danish botanist; student of *Puccinellia.*
Sotchava (Soczava), Viktor B., 1905– . Russian botanist, Irkutsk.
Spach, Édouard, 1801–79. France; author of *Histoire Naturelle des Végétaux Phanérogames.*
Spenn. Spenner, Fridolin Karl Leopold, 1798–1841. Professor, Freiburg.
Spreng. Sprengel, Kurt Polycarp Joachim, 1766–1833. Professor of medicine and botany, Halle.
Spring, Anton Friedrich (Frédéric Antoine), 1814–72. Professor, Lüttich (Liège).
Stair, Leslie Dalrymple, 1876–1950. Academy of Natural Sciences, Philadelphia; collections now at Cleveland Museum; collected at Yakutat, 1945.
Standl. Standley, Paul Carpenter, 1884–1963. Curator, U.S. National Herbarium, later Chicago Natural History Museum; student of Alaskan, Mexican, and Central American floras.
Stapf, Otto, 1857–1933. Austrian botanist at Kew from 1890; contributor to Harvey & Sonder's *Flora Capensis* and Oliver's *Flora of Tropical Africa.*
Staudt, Günter, 1926– . German student of *Fragaria.*
Stebbins, George Ledyard, Jr., 1906– . Professor of genetics, California; cytogeneticist and cytotaxonomist; student of *Crepis, Antennaria,* Gramineae.
Steffen, Hans, 1891–194?. German professor, Königsberg.
Steller, Georg Wilhelm, 1709–46. German naturalist; traveled to Siberia and Alaska as member of Bering's expedition; first collector of Alaskan plants.
Steph. Stephan, Christian Friedrich, 1757–1814. Russian botanist; professor in Moscow.
Sternb. Sternberg, Caspar Maria von, 1761–1838. Student of *Saxifraga.*
Sternb. & Hoppe, Sternberg, C. M. von, & D. H. Hoppe.
Sterneck, Jakob von, 1864– . Bohemian student of *Rhinanthus.*
Sterns, Emerson Ellick, 1846–1926. American botanist.
Steud. Steudel, Ernst Gottlieb, 1783–1856. German physician, bibliographer, and agrostologist.
Stev. Steven, Christian von, 1781–1863. Russian state councilor; student of Crimean flora.
Stokes, Jonathan, 1755–1831. English friend of the younger Linnaeus.
Sukatsch. Sukatschev, Vladimir Nikolajevich, 1880–1967. Russian botanist.
Suksd. Suksdorf, Wilhelm Nikolaus, 1850–1932. Amateur botanist, Bingen, Washington; publisher of own journal, *Werdenda.*
Svens. Svenson, Henry Knute, 1897– . Brooklyn botanical garden; American Museum of Natural History, U.S. Geological Survey; student of *Eleocharis.*
Sw. Swartz, Olof Peter, 1760–1818. Professor, Stockholm; student of West Indian botany.
Swallen, Jason Richard, 1903– . Head curator, U.S. National Herbarium; student of Gramineae.
Sweet, Robert, 1783–1835. England; student of Geraniaceae; horticulturist and ornithologist.
Syme, John Thomas Irvine Boswell (né Syme), 1822–88. Scottish botanist.

Takeda, Hisayoshi, 1883– . Japanese botanist.
Tatew. Tatewaki, Misao, 1899– . Japanese botanist; professor, Sapporo; student of Hokkaido flora.
Tatew. & Kitamura, Tatewaki, M., & S. Kitamura.
Tausch, Ignaz Friedrich, 1793–1848. Bohemian botanist; student of *Hieracium.*
Taylor, Roy Lewis, 1932– . Chief of taxonomy, Plant Research Institute, Dept. of Agriculture, Ottawa; student of Saxifragaceae and the flora of Queen Charlotte Islands.
Ten. Tenore, Michele, 1780–1861. Professor of botany, Naples.
Thell. Thellung, Albert, 1881–1928. Swiss botanist; professor, Zürich.
Thom. Thomas, Abraham, 1740–1824. Swiss botanist.
Thoms. Thomson, Thomas, 1817–78. British botanist; co-editor of *Flora Indica.*
Thuill. Thuiller, Jean Louis, 1757–1822. Professor, Paris.
Thunb. Thunberg, Carl Peter, 1743–1828. Professor, Uppsala; student of Japanese and South African floras.
Thurb. Thurber, George, 1821–90. New York botanist with Mexican Boundary Survey.
Tidest. Tideström, Ivar T. 1864–1956. Native of Sweden; professor, Catholic University, Washington; co-author, *A Flora of Arizona and New Mexico.*
Tiling, Heinrich Sylvester Theodor, 1818–71. Physician with the Russian North American Company.
Todaro, Agostino, 1818–92. Director of the botanical garden, Palermo.
Tolm. Tolmatchev, Alexander Innokentevich, 1903– . Russian botanist; student of Arctic flora.
Torr. Torrey, John, 1796–1873. Physician, for many years professor of chemistry and botany, New York, and professor of chemistry, Princeton; later U.S. assayer of the mint, New York; outstanding systematist and author of five major floras; studied and organized more genera and species of North American plants than any other American botanist.
Torr. & Gray, Torrey, J., & A. Gray.

Toyokuni, Hideo, 1932– . Japanese botanist.
Toyokuni & Nosaka, Toyokuni, H., & S. Nosaka.
Tratt. Trattinnick, Leopold, 1764–1849. Custodian of the national collections, Vienna; Austrian student of *Rosa*.
Trautv. Trautvetter, Ernst Rudolph von, 1809–89. Director of the botanical garden, St. Petersburg.
Trautv. & Mey. Trautvetter, E. R. von, & C. A. Meyer.
Trel. Trelease, William, 1857–1945. Director, Missouri Botanical Garden; professor of botany, Illinois.
Trev. Treviranus, Ludolf Christian, 1779–1864. Professor of botany, Rostock and Bonn.
Trin. Trinius, Carl Bernhard von, 1778–1844. German physician, poet, and botanist; court physician, St. Petersburg; noted agrostologist.
Trin. & Rupr. Trinius, C. B., & F. J. Ruprecht.
Tryon, Rolla Milton, 1916– . Curator of ferns, Harvard, and associate curator, Gray Herbarium; formerly curator, Missouri Botanical Garden; student of Pteridophyta.
Tsvel. Tsvelev, Nikolai Nikolaievich, 1925– . Russian student of Gramineae.
Turcz. Turczaninow, Nikolai Stepanovich von, 1796–1864. Russian botanist; author of *Flora Baikalense-Dahurica*.
Tutin, Thomas Gaskell, 1908– . Leicester, England; co-editor, *Flora Europaea*.
Tweedy, Frank, 1854–1937. Topographic engineer, U.S. Geological Survey.

Ulbr. Ulbrich, Eberhard, 1879–1952. German botanist.
Underw. Underwood, Lucien Marcus, 1853–1907. Professor of botany, Columbia; student of ferns.
Unger, Franz Joseph Andreas Nicolaus, 1800–1870. Austrian botanist.

J. Vahl, Vahl, Jens Lorenz Moestue, 1796–1854. Son of Martin Vahl.
M. Vahl, Vahl, Martin Hendricksen, 1749–1804. Professor of botany, Copenhagen; student of Linnaeus; co-editor of *Flora Danica*.
Vail, Anna Murray, 1863– . Librarian, New York Botanical Garden.
Valck.-Suringar, Valckenier-Suringar, Jan, 1864–1932. Dutch botanist, Leiden.
Van Houtte, Louis, 1810–76. Horticulturist, Ghent; editor of *Flore de Serres,* continued as *Annales Générales d'Horticulture*.
Vasey, George, 1822–93. Eminent agrostologist, U.S. Dept. of Agriculture, Washington, D.C.
Vasey & Rose, Vasey, G., & J. N. Rose.
Vassilchenko, I. T., 1903– . Russian botanist.
Vassiljev, Victor Nikolayevich, 1890– . Russian botanist.
Velloso, José Marianno da Conceição, 1742–1811. Rio de Janeiro; writer on Brazilian flora.
Vent. Ventenat, Étienne Pierre, 1757–1808. Professor of botany, Paris; horticulturist.
Vestergr. Vestergren, Jakob Tycho Conrad, 1875–1930. Swedish teacher and botanist.
Vict. (Frère) Marie-Victorin (formerly Conrad Kirouac), 1885–1944. Director of the botanical institute, Montréal.
Vierh. Vierhapper, Friedrich Karl Max, 1876–1932. Austrian botanist.
Vill. Villars, Dominique, 1745–1814. Physician and professor, Grenoble and Strasbourg.
Vogler, Julius Rudolph Theodor, 1812–41. German explorer in Africa.
Volk. Volkova, E. V. . Russian botanist.
Voss, Andreas, 1857–1924. German student of conifers.
Vved. Vvedenski, Aleksei Ivanovich, 1898– . Russian student of Liliaceae and Scrophulaceae.

Wagnon, Harvey Keith, 1916– . State Dept. of Agriculture, Sacramento.
Wahlenb. Wahlenberg, Göran, 1780–1851. Professor of botany, Uppsala; noted plant geographer.
Waldstein-Wartemberg, (Count) Franz Adam von, 1759–1823. Austrian botanist.
Waldst. & Kit. Waldstein-Wartemberg, F. A. von, & P. Kitaibel.
Wallr. Wallroth, Carl Friedrich Wilhelm, 1792–1857. German physician and botanist.
Walp. Walpers, Wilhelm Gerhard, 1816–53. German author of *Repertorium Botanicae Systematicae*.
Walt. Walter, Thomas, 1740–88. Author of *Flora Caroliniana*.
Wang. Wangenheim, Friedrich Adam Julius von, 1747–1800. German forester.
Warren, Fred Adelbert, 1902– . Washington State.
S. Wats. Watson, Sereno, 1826–92. Curator, Gray Herbarium, Harvard; student of western American plants.
Wats. & Coult. Watson, S., & J. M. Coulter.
Weatherby, Charles Alfred, 1875–1949. Gray Herbarium, Harvard; student of ferns.
Web. Weber, Georg Heinrich, 1752–1828. Professor, Kiel.
Webb, Philip Barker, 1793–1854. English botanical traveler and accumulator of a rich flora and library now in Florence.
Webb & Berthelot, Webb, P. B., & S. Berthelot.
Weis (or Weiss), Friedrich Wilhelm, 1744– . Private lecturer, Göttingen.
Welsh, Stanley Larson, 1928– . American botanist, Brigham Young Univ.
Wentzig, Theodor, 1824–92. German botanist.
Wesmael, Alfred, 1832–1905. Belgian botanist.
Wettst. Wettstein, Richard Ritter von Westersheim, 1863–1931. Director of the botanical garden, Vienna; outstanding systematist.
Wheeler, Louis Cutter, 1910– . Professor of botany, Univ. of Southern California; student of *Euphorbia*.
Wherry, Edgar Theodore, 1885– . Professor of botany, Pennsylvania; student of Polemoniaceae.
Wiegand, Karl McKay, 1873–1942. Professor of botany, Cornell.
Wight, William Franklin, 1874–1954. American student of Leguminosae.
Wikstr. Wikström, Johann Emanuel, 1789–1856. Director, Botanical Dept., Riksmuseum, Stockholm.
Willar (identity uncertain).
Willd. Willdenow, Carl Ludwig, 1765–1812. Director, botanical garden, Berlin.
F. N. Williams, Williams, Fredric Newton, 1862–1923. British student of Caryophyllaceae.
L. O. Williams, Williams, Louis Otto, 1908– . Head of botany, Chicago Natural History Museum; distinguished student of the flora of Central America.
Willmott, Alfred James, 1888–1950. Keeper of botany, British Museum.
Wilson, Ernest H., 1876–1930. American botanist, Arnold Arboretum, Harvard.

List of Authors

Wirsing, Adam Ludwig, 1734–97. German botanist.
With. Withering, William, 1741–99. Author of a British flora.
E. Wolf, Wolf, Egbert, 1860–1931. Russian student of dendrology.
T. Wolf, Wolf, Franz Theodor, 1841–1924. German geologist in Ecuador, monographer on *Potentilla*.
Wood, Alphonse, 1810–81. Author of *Classbook of Botany*; collector.
W. Wood, Wood, Walter Abbott, 1907– . President of American Geographical Society since 1957.
Wormsk. Wormskjold, Morten, 1783–1845. Danish lieutenant, member of Kotzebue's first expedition on the *Rurik*, leaving ship in Kamchatka; collected at Kodiak and Sitka before returning.
Wulf. Wulfen, Franz Xaver von, 1728–1805. Professor, Klagenfurt, Hungary (now Austria).

Zahn, Karl Hermann, 1865–1940. Teacher, Karlsruhe; student of *Hieracium*.
Zamels, A., 1897–1943. Latvian botanist.
Zinn, Johann Gottfried, 1727–59. Professor, Göttingen.
Zinz. Zinzerling, Yurii Dmitrievich, 1894–1938. Russian student of *Eleocharis*.
Zobel, August, 1861–1934. Teacher, Dessau.
Zoega, Johan, 1742–88. Danish botanist.

Persons for Whom Taxa Have Been Named

ADAMS, Johannes Michael Friedrich, 1780–183?. Siberian traveler and collector.
AMSINCK, Wilhelm. Burgomaster of Hamburg, who early in the nineteenth century gave important support to the city's botanical garden.
AMUNDSEN, Roald, 1872–1928. Famous Norwegian Arctic explorer who made the first northwest passage.
ANDERSON, W. B. British Columbia.
ARNELL, Hampus Wilhelm, 1848–1932. Swedish bryologist.
BARCLAY, George. Naturalist on the ship *Sulphur*.
BARTLETT, Harley Harris, 1886–1960. Professor of botany, Michigan.
BARTSCH, John, 1709–38. Botanist sent by Boerhave to Surinam, where he died.
BEAUVERD, Gustave, 1867–1942. Of the Herbarium Boissier, Geneva.
BECKMANN, Johann, 1739–1811. Professor, Göttingen.
BEEN, Frank T. Superintendent, McKinley National Park.
BELL, Robert, 184?–1917. Canadian botanist.
BERCHTOLD, (Count) Friedrich von, 1781–1876. Author of *Oekonomischtechnische Flora Böhmens*.
BERING, Vitus Jonassen, 1681–1741. Danish-Russian explorer, discoverer of Alaska.
BERLANDIER, Jean Louis, 1805–51. Swiss collector in Texas.
BICKNELL, E. P., 1858–1925. New York banker and amateur botanist.
BILDERDIJK, Willem, 1756–1831. Dutch botanist.
BLAISDELL, Frank Ellsworth, 1862–1946. California physician and entomologist.
BLASCHKE, E. Leontjevich. Surgeon with the Russian North American Company; botanical collector in Alaska, 1820–42.
BODIN, J. E. California botanist and plant collector.
BOLANDER, Henry Nicholas, 1831–97. Botanical collector, California State Geological Survey.
BOSCHNIAK, A. K. Russian amateur botanist.
BOSTOCK, H. S., 1901– .
BOURGEAU, Eugène, 1813–77. French botanical collector in Canada.
BOYKIN, Samuel. Early active botanist of Georgia.
BRAINERD, Ezra, 1844–1924. President of Middlebury College; student of *Viola*.
BRAY, Franz Gabriel (Count) de, 1765–1832. Rouen.
BRUNET, Louis Ovide, 1826–76. Canadian botanist.
BRUNONI. Not identified in the original description of *Brunoniana*.
BULLIARD, Jean Baptiste, 1752–93. French botanist.
BURKE. Not identified.
BUXBAUM, Johann Christian, 1693–1730. German botanist.
CAIRNES, Donaldson Delmore. Geologist, Canadian Geological Survey; botanical collector in the Yukon, 1911–13.
CANBY, William Marriott, 1831–1904. Delaware businessman, collector of large herbarium.
CASTILLEJO, Domingo. Spanish botanist.
CHORIS, Ludwig, 1795–1828. Artist; member of expedition circumnavigating the world on *Nadeschda* and *Neva* under Kotzebue.
CHRISTOL, Jules. Not further identified.
CLAYTON, John, ?–1773. Early American botanist; contributed to Gronovius the materials for *Flora Virginica*.
CLINTON, De Witt, 1769–1828. Prominent statesman; several times governor of New York.
COLLIER, Arthur James. U.S. Geological Survey; collector of plants in Alaska, 1900–1905.
COLLINS, Zaccheus, 1764–1831. Philadelphia botanist.
COOLEY, Grace Emily, 1857–1916. Professor, Newark, N.J.
CRAWFORD, Ethan Allan. Early settler in the White Mountains.
CUSICK, William Conklin, 1842–1922. Plant collector.
DALIBARD, Thomas François, 1703–79. French botanist.
DANTHOINE, Étienne. Botanist of Marseilles.
DESCHAMPS (see Loiseleur-Deslongchamps).
DESCOURAIN, François, 1658–1740. French apothecary and botanist.
DEYEUX, Nicolas, 1745–1837. French botanist.
DICKIE. Not identified in the original description of *Cystopteris Dickieana*.
DODGE, William Earle, d. 1903.
DONDI, Giacomo, 1298–1385. Italian physician and botanist.
DRUMMOND, Thomas, 1780–1853. Noted Scottish botanical collector.
DUDLEY, William Russell. Not further identified.
DUPONT, I. D. French botanist; student of *Atriplex* and of sheaths of grasses.
EAMES, Edwin Hubert, 1865– . Physician.
EDWARDS, John. Ship's surgeon on the *Hecla* on Parry's voyages and on the *Isabel* on Ross's first voyage, in 1818.
EGEDE, Hans, 1686–1758. Greenland missionary.
EMERSON, G. H.
ENANDER, Sven Johan, 1847–1928. Swedish clergyman; student of *Salix*.

ERMAN, Georg Adolf, 1806–77. Pressor, Berlin.
EYERDAM, Walter Jacob, 1892– . Botanical collector, Seattle.
FAURIE, (Père) Urbain Jean, 1847–1915. French botanical collector in Japan, Korea, and Formosa.
FOWLER, James, 1829–1923. Professor of botany.
FRANKLIN, (Sir) John, 1786–1847. Explorer in the Arctic.
FREEDMAN, N. J. Hudson's Bay Mining & Smelting Company; botanical collector in the Yukon in 1953.
FUNSTON, Frederick, 1865–1917. U.S. Department of Agriculture.
GARBER, Adam Paschal, 1838–81.
GAULTIER, Jean-François, 1708?–56. Naturalist and court physician at Québec.
GIL, Felipe. Spanish botanist.
GOODYER, John, 1592–1664. English botanist.
GORMAN, Martin Woodlock, 1853–1926. Western American collector.
HACKEL, Paul. Bohemian scientist.
HARRIMAN, E. H. New York City; organizer of the Harriman Expedition to Alaska, 1899.
HARRINGTON, George L. U.S. Geological Survey.
HAUPT. Collected 1818–20 around Irkutsk.
HENRY, (Mrs.) Mary G. Collected in northeastern British Columbia.
HEUCHER, Johann Heinrich, 1677–1747. German botanist.
HINDS, Richard Brindsley, 1812–47. Botanist of the *Sulphur* expedition.
HOLLBØLL, Carl Peter, 1795–1856. Danish botanist and naturalist.
HOOD, Robert. Admiralty midshipman.
HORNE, William Titus. Salmon hatchery, Karluk, Kodiak.
HUDDELSON, C. W. Mining engineer.
JACOB PETER (see J. P. Anders., list above).
JEFFREY, John. Botanical collector in western North America.
JESSICA (see McGREGOR).
JOHREN, Mart. Dan. Not further identified.
KALM, Pehr, 1716–79. Pupil of Linnaeus who traveled and collected extensively in Canada.
KARDAKOV, A. Not further identified.
KEELE, Joseph. Geologist, Geological Survey of Canada.
KELLOGG, Albert, 1813–87. San Francisco physician and botanist: a founder of the California Academy of Sciences.
VON KOBRES. Nobleman of Augsburg, patron of botany in Willdenow's time.
KOELER, Georg Ludwig, 1765–1807. Professor, Mainz; student of grasses.
KOL, Elisabeth. Hungarian investigator of snow- and ice-flora.
KOTZEBUE, Otto von, 1787–1846. Master of the ship *Rurik* during its circumnavigation of the world.
KRAUSE, Arthur (1851–1920) and Aurel (1848–1908). Bremen geographers and ethnographers; botanical collectors in Alaska.
KRUHSE, Robert. Physician, Irkutsk; collected plants in eastern Siberia between Irkutsk and Wiluisk, 1832.
KRUSENSTERN, Ivan Fedorovich, 1770–1846. Commander of the ships *Nadeschda* and *Neva* during their voyages of circumnavigation.
KRYNITZKI, Ivan, 1787–1838. Professor, Cracow.
KUSCHE, John August, 1869–1934. Collector of lepidoptera and Alaskan plants.

LACHENAL, Werner de, 1736–1800. Professor of botany, Basel.
LAING, Hamilton Mack, 1883– . Comox, British Columbia; botanical collector in Alaska.
LANGSDORFF, Georg Heinrich von. Russian consul-general in Rio de Janeiro; accompanied Krusenstern on the circumnavigation of the world in *Nadeschda* and *Neva*.
LESCHENAULT DE LA TOUR, Louis Théodore, 1773–1886. Director of the botanical garden in Pondicherry, India.
LESQUEREUX, Leo, 1805–89. Distinguished bryologist and paleobotanist.
LEWIS, Meriwether, 1774–1809. American explorer.
LINDSTRÖM, A. H. Steward with the first northwest passage expedition under Amundsen.
LISTER, Martin, 1638–1711. Celebrated English naturalist.
LLOYD, E., 1670–1709. Director, Oxford Museum.
LOISELEUR-DESLONGCHAMPS, Jean Louis Auguste, 1774–1849.
LONITZER, Adam (Lonicerus). Sixteenth-century German herbalist.
LOOFF, Ethel and Henry B. Botanical collectors, Kodiak, Alaska.
LÜTKE, (Count) Friedrich P., 1797–1882. Russian admiral and geographer; president, Academy of Science, St. Petersburg; commander of the corvette *Senjavin*.
LUTZ, Harold J. Kenai Peninsula, Alaska.
LYNGBYE, Hans Christian, 1782–1827. Danish preacher.
LYON, John, 17??–1818. American botanist; explorer of the southern Alleghenies.
McCALLA, William Copeland, 1872– . Botanical collector in Canada.
McCONNELL, Richard George, 1857–1942. Director of the Geological Survey of Canada.
McGREGOR, Jessica. Sister of Ernest Alexander McGregor (*see* list above).
MACKENZIE, Alexander, 1764–1820. Explorer in northwest America.
MACLEAN, John J. U.S. Signal Service; collected in Alaska in 1881.
MATTEUCCI, Carlo, 1800–1868. Italian physicist.
MAYDELL, (Baron) Gerhard K. K. Botanical collector in eastern Siberia, 1861–71.
MENDENHALL, W. C. Geologist, U.S. Geological Survey; botanical collector in Alaska.
MERRIAM, Clinton Hart, 1855–1942. Founder and chief, U.S. Bureau of Biological Survey.
MERTENS, Carl Heinrich, 1796–1830. Bremen; accompanied Lütke on the corvette *Senjavin*; first collector at Sitka.
MEXIA, Ynez Enriquetta Julietta, 1870–1938. San Francisco, formerly Washington, D.C.; botanical collector in Mexico and Alaska.
MINUART, Juan, 1693–1768. Barcelona botanist.
MIYOSHI, Manabu, 1861–1939. Japanese botanist.
MOEHRING, Paul Heinrich Gerhard, 1710–92. Physician, botanist, and ornithologist of Oldenburg.
MORRISSEY. Bartlett's schooner *Effie M. Morrissey*.
MUIR, John, 1838–1914. Noted American geologist and naturalist.
MÜLLER, Adolf. Philadelphia; botanical collector in the Yukon in 1920, 1943.
MÜLLER, Jakob Theodor ("Tabernaemontanus"), 1520–

List of Authors

90. German physician; author of a renowned "Kreuterbuch."

MURIE, Olof J. Norwegian zoologist and botanical collector.

NELSON, David. Gardener at Kew; member of Captain Cook's third expedition aboard *Resolution* and *Discovery*.

NESLE, J. A. N. de. Poitiers.

NEWCOMBE, Charles Frederick, 1751–1924. English physician; botanical collector, Vancouver Island and Queen Charlotte Islands.

NORBERG, Ingvar Leonard, 1880–1967. Norwegian fisherman; plant collector in Alaska.

NOVOGRABLENOV, P. T. Schoolteacher, Petropavlovsk, Kamchatka, about 1920–25.

NYMAN, Carl Fredrik, 1820–93. Curator, the herbarium, Stockholm.

O'NEILL, Hugh Thomas, 1894–   . Curator of Langlais Herbarium, Catholic University, Washington, D.C.

PALANDER of *Vega* (Adolf Arnold Louis), 1842–1920. Commander of the ship *Vega* during the northwest passage under A. E. Nordenskjöld.

PALMER, L. J. U.S. Fish and Wildlife Service, Fairbanks, Alaska.

PARRY, William Edward, 1790–1855. Arctic explorer.

PAULLOWSKY, A. P. Botanical collector in eastern Siberia (between Yakutsk and Ajan) in the 1850's.

PECK, Charles Horton, 1833–1917. American botanist.

PHIPPS, Constantine John, 2d Baron Mulgrave, 1844–92. Arctic explorer.

PUMPELL. Not identified in the original description of *Bromus Pumpellianus*.

RADDE, Gustav Ferdinand Richard Johannes, 1831–1903. German botanist.

RIEDER, Johannes Georg von. Botanical collector in Kamchatka.

ROBBINS, James Watson, 1801–79. Massachusetts; pioneer student of *Potamogeton*.

RÖGNER. Lived in Orlanda, Crimea, about 1850.

ROLLAND-GERMAIN, (Frère) Louis. Canadian botanist.

ROMANZOFF (see RUMANTZEV).

Ross, (Sir) James Clark, 1800–1862. Arctic and Antarctic explorer.

RUMANTZEV, Nikolai, Count Romanzoff, 1754–1826. Chamisso's patron.

RUPPIUS, Heinrich Bernhard, 1689–1719. German botanist.

SABINE, (Sir) Edward, 1788–1883. Explorer in Arctic America.

SAUSSURE, Nicolas Théodore de, 1767–1847, & Horace Bénédict de, 1740–99. Swiss naturalists.

SAY, Thomas, 1787–1834. Zoologist and plant collector.

SCAMMAN, Edith, 1882–1966. Botanical author and collector in Alaska; Gray Herbarium, Harvard.

SCHOFIELD, W. Canadian botanist, University of British Columbia.

SCOULER, John, 1804–71. Scottish botanist and collector in western North America.

SEALE, Alvin, 1871–1958. Collected in the Yukon in 1896; ichthyologist; curator of Steinhart Aquarium, San Francisco.

SELKIRK, Thomas Douglas (Earl of), 1771–1820.

SETCHELL, William Albert, 1864–1943. Professor of botany, Univ. of California, Berkeley; noted algologist.

SHEPHERD, John, 1764?–1836. For many years curator, Liverpool botanical garden.

SIBBALD, (Sir) Robert, 1641–1722. First professor of medicine at Edinburgh.

SIEVERS, Johann,   –1795. German botanist in Russia.

SMELOWSKY, Timotheus, 17??–1815. Professor of pharmacology, St. Petersburg.

SNYDER, Harry. Explored upper Nahanni River and Snyder Mountain in 1937.

SONNE, Charles Frederick, 1845–1913. Danish bookkeeper and botanical collector in Nevada and California.

STEJNEGER, Leonhard, 1857–19??. Norwegian zoologist, Smithsonian Institution, Washington, D.C.

SWEERT, Emanuel, 1552–?. Dutch herbalist.

TABERNAEMONTANUS (see Müller, J. T.).

TARLETON, John Berry. Botanical collector in the Yukon, 1899.

TAYLOR, (Miss) E.

TICHOMIROV, B. A. Russian botanist; student of the Arctic flora.

TILESIUS VON TILENAU, Wilhelm Gottlieb, 1769–1857. Member of expedition circumnavigating the world on *Nadeschda* and *Neva* under Krusenstern.

TOFIELD, Thomas, 1730–79. English botanist.

TOLMIE, William Fraser, 1812–86. Hudson's Bay Company physician at Fort Vancouver in 1832; botanical collector in northwestern North America.

TOURNEFORT, Joseph Pitton de, 1656–1708. French botanist.

TOVAR, Simon. Spanish physician, sixteenth century.

TOWNSEND, David, 1788–1858. Amateur botanist, West Chester, Pennsylvania.

TSCHONOSKI (Chonosuke), Sukawa, 1841–1925. Japanese collector.

TURNER, Lucien McShan, 1847–1909. Early collector in Alaska; author of "Sketch of the Flora of Alaska."

TYRRELL, Joseph Bull. Mining engineer; botanical collector in the Yukon in 1898.

ULMER, Joseph. German mining engineer; botanical collector in central Alaska, 1935–37.

URBAN, Ignatz, 1848–1931. German botanist; Engler's collaborator; student of West Indian flora.

VAGNER (see WAGNER).

VAN BRUNT, Cornelius, 1827–1903. American.

VICTOR (see **Sotchava**, list above).

VOGEL, Augustin, 1724–74. Professor, Göttingen.

WAGNER, Johannes Gerhard, 1706–59. Lübeck.

WALPOLE, Frederick Andrews, 1861–1904. Artist, U.S. Department of Agriculture.

WASHINGTON, George, 1732–99. First President of U.S.

WEYRICH, Heinrich, 1828–63. Russian collector in Japan.

WHITNEY, William Dwight, 1827–94. Not further identified.

WILHELMS. Not identified in the original description of *Wilhelmsia*.

WILLIAMS, Robert Stetham, 1859–1945. Bryologist, New York Botanical Gardens.

WOODS, Joseph, 1776–1864. English botanist; student of *Rosa*.

WRIGHT, C. Member of the North Pacific Surveying Expedition under Ringold and Rodgers, 1853–56.

YATABE, Ryokishi, 1899–   .

YOUNG. Not identified.

ZANNICHELLI, Gian Girolamo, 1662–1729. Venetian botanist.

ZSCHACKE, Hermann, 1867–1937. Bryologist.

Bibliography

A complete listing of works published prior to 1950 on the flora of Alaska and the Yukon is found in Hultén, *Flora of Alaska and Yukon*, 10 (1950), 1775–1812; the more recent works are given in Hultén, *Comments on the Flora of Alaska and Yukon* (1967). Both these bibliographies are comprehensive; i.e., they include all works dealing generally with this flora or with specific subregions, as well as all works limited to specific taxa (though the 1967 list is less complete in the latter respect). The following listing includes the general works from both earlier lists (though more detailed descriptions of earlier works are given in the 1950 listing); it does not include works limited to specific taxa.

Abrams, L. Illustrated flora of the Pacific States: Washington, Oregon, and California. 4 vols. Stanford Univ. Press, Stanford, Calif., 1923–60.

Adams, J. A. A bibliography of Canadian plant geography to the end of the year 1920; 1921–25; 1926–30; 1931–35. *Trans. Roy. Can. Inst.*, 16 (1928), 293–355; 17 (1929), 103–45; 21 (1936), 95–134 (two last parts published by J. Adams and M. H. Norwell).

Anderson, J. P. Flora of Alaska and adjacent parts of Canada. Iowa State Univ. Press, Ames, Iowa, 1959.

——— Flora of Alaska. *Alaska Sportsman*, 3 (1937).

——— Notes on the flora of Sitka, Alaska. *Proc. Iowa Acad. Sci.*, 23 (1916), 427–82.

——— Plants of southeastern Alaska. *Proc. Iowa Acad. Sci.*, 25 (1918), 427–49.

——— Plants used by the Eskimo in the northern Bering Sea and Arctic regions of Alaska. *Amer. J. Bot.*, 26 (1939), 714–16.

——— Supplemental list of plants from southeastern Alaska. *Proc. Iowa Acad. Sci.*, 26 (1919), 327–31.

Ascuith, C. H. Collection of Alaskan plants. *Plant World*, 10 (1907), 272–85.

Babb, M. F. Ornamental trees and shrubs for Alaska. *Alaska Agric. Exp. Station Bull.*, 24 (1959).

Baker, M. Alaskan geographic names. II. General geology, economic geology, Alaska. U.S. Geol. Surv. Ann. Rep., 21 (1899–1900), 487–509.

——— Geographic dictionary of Alaska. U.S. Geol. Surv. Bull., 187 (1902).

Bakewell, A. Botanical collections at the Wood Yukon Expeditions of 1939–41. *Rhodora*, 45 (1943), 305–16.

Bank, T. P. Botanical and ethnobotanical studies in the Aleutian Islands. I. Aleutian vegetation and Aleut culture. *Papers Mich. Acad. Sci. Arts & Letters*, 37 (1951), 13–30.

——— A preliminary account of the University of Michigan Aleutian Expedition, 1950–51. *Asa Gray Bull.*, 1: 3 (1952–53), 211–18.

Barneby, R. C. Atlas of North American *Astragalus*, 1–2, 1964.

Benninghoff, W. S. Vegetation (of the Fort Greely area, Alaska). In G. William Holmes and William S. Benninghoff, Terrain study of the Army test area, Fort Greely, Alaska, 1–2 (Washington, D.C., 1957).

Benson, L. A treatise on the North American *Ranunculi*. *Amer. Midl. Nat.*, 40: 1 (1948), 1–264.

——— Supplement to the treatise on the North American *Ranunculi*. *Amer. Midl. Nat.*, 52 (1954), 328–68.

——— The *Ranunculi* of the Alaskan Arctic coastal plain and the Brooks Range. *Amer. Midl. Nat.*, 53 (1955).

Bentham, G. See Hinds.

Blake, S. F., and A. C. Atwood. Geographical guide to floras of the world. I. U.S. Dept. Agric. Misc. Pub. 401 (1942), pp. 130–33, 145–46.

Blaschke, E. L. Topogeographia medica portus Novi Archangelscensis sedis principalis coloniarum rossicarum in septentrionali America. St. Petersburg, 1842.

Bliss, L. C., and J. E. Cantlon. Succession on river alluvium in northern Alaska. *Amer. Midl. Nat.*, 52 (1957), 452–69.

Boivin, B. Centurie des plantes canadiennes (3 parts). *Nat. Can.*, 75 (1948), 202–27; *Can. Field Nat.*, 65 (1951), 1–22; *Nat. Can.*, 87 (1960), 25–49.

Bongard, H. G. Observations sur la végétation de l'île de Sitcha. *Mém. Acad. Imp. Sci. St. Pétersb.*, Sér. 6, 2 (1833), 119–77. The grasses were worked up by K. B. Trinius and the Cyperaceae by J. D. Prescott.

Bornhardt, P. D. A trip to Alaska. *Florists' Exchange*, 71: 13 (1929), p. 30; 14, pp. 40, 67; 15, p. 44.

Brandegee, T. S. See Heller.

Briggs, W. R. Some plants of Mount McKinley National Park, McGonegal mountain area. *Rhodora*, 55 (1953), 245–52.

Britton, M. E. Vegetation of the Arctic tundra. *18th Biol. Colloquium*, 1957, pp. 26–61.

Britton, N. L. A list of plants collected by J. A. Rudkin during a trip from Juneau, on the coast, to Mt. St. Elias, Alaska, in the summer of 1883. *Bull. Torr. Club*, 11 (1884), 36.

Britton, N. L., and P. A. Rydberg. Contributions to the botany of the Yukon Territory. 4. An enumeration of the flowering plants collected by R. S. Williams and by J. B. Tarleton. *Bull. N.Y. Bot. Gard.*, 2 (1901), 149–87.

Brooks, A. H., and L. M. Prindle. The Mount McKinley region, Alaska, with descriptions of the igneous rocks and of the Bonnifield and Kantishna districts. U.S. Geol. Surv. Prof. Pap. 70 (1911), pp. 1–234. Flora of the region collected by L. M. Prindle, determined by F. V. Coville, pp. 208–11.

Brooks, A. H., G. B. Richardson, A. J. Collier, and W. C. Mendenhall. Reconnaissances in the Cape Nome and Norton Bay regions, Alaska, in 1900. House Doc. 547, 56th Congr., 2d sess. (1901), pp. 1–222. Notes on the vegetation by A. J. Collier, pp. 164–66. List of plants collected in Seward Peninsula by A. J. Collier, identified chiefly by F. V. Coville, pp. 167–74 (Lichenes by C. E. Cummings, p. 167).

Brown, R. G. Notes on grasses collected in Alaska and adjacent Canada. *Proc. Iowa Acad. Sci.*, 56 (1949), 107–12.

Burroughs, J. Narrative of the expedition. In Harriman Alaska Expedition, 1 (1901), 1–115.

Cahalaney, V. H. A biological survey of Katmai National Monument. Smithsonian Misc. Coll., 138: 5 (1959), 1–246.

Cairnes, D. D. Geology of a portion of the Yukon-Alaska boundary between Porcupine and Yukon rivers. Geol. Surv. Can. Sess. Pap. 26, summary rep. 1911 (1912), pp. 17–33. List of plants by J. M. Macoun, pp. 22–25; 1912 (1914), pp. 9–11. J. Macoun, Report on plants, pp. 438–40.

Cairnes, D. D. The Yukon-Alaska international boundary between Porcupine and Yukon rivers. Geol. Surv. Can. Mem., 67 (1914), pp. 1–161; flora, pp. 10–18. List of plants by J. M. Macoun, pp. 13–18 (Musci, Lichenes, p. 18).

Calder, J. A., and R. L. Taylor. New taxa and nomenclatorial changes with respect to the flora of the Queen Charlotte Islands, British Columbia. *Can. Jour. Bot.*, 43 (1965), 1387–1400.

Cantwell, J. C. Report of the operations of the U.S. revenue steamer Nunivak on the Yukon River station, Alaska, 1899–1901. Senate Doc. 155, 58th Congr., 2d sess., 14 (1904), pp. 1–325. List of plants identified by A. Eastwood, p. 290.

Caps, S. R. The southern Alaska Range. U.S. Geol. Surv. Bull., 862 (1935). Vegetation, pp. 28–29.

Castner, J. C. See Glenn and Abercrombie.

Chamisso, A. von. Bemerkungen und Ansichten auf einer Entdeckungsreise in den Jahren 1815–18 . . . auf dem Schiffe Rurik. In O. von Kotzebue, ed., Entdeckungs-Reise . . . Weimar, III, 1821.

———— Notices respecting the botany of certain countries visited by the Russian voyage of discovery under the command of Captain Kotzebue. Trans. W. J. Hooker, *Bot. Misc.*, 1 (1830), 305–23. Plants, pp. 317–23.

———— Reise um die Welt mit der Romanzoffischen Entdeckungs-Expedition in den Jahren 1815–18 auf der Brigg Rurik. Kapitain Otto v. Kotzebue. In Adalbert von Chamisso's Werke, 1 (Leipzig, 1836).

Chamisso, A. von, and D. F. Schlechtendal. De plantis in expeditione spectulatoria Romanzoffiana observatis. *Linnaea*, 1–10 (1826–36).

Choris, L. Voyage pittoresque autour du monde . . . sur le brick de Rurik, commandé par Otto v. Kotzebue . . . Paris, 1822.

Churchill, E. D. Phytosociological and environmental characteristics of some plant communities in the Umiat region of Alaska. *Ecology*, 36 (1955), 606–27.

Coats, R. R. Geology of Buldir Island, Aleutian Islands, Alaska. U.S. Geol. Surv. Bull. 989-A, 1953, pp. 1–26.

Cody, W. J. Plants of the vicinity of Norman Wells, Mackenzie district, Northwest Territories. *Can. Field Nat.*, 74:2 (1960), 71–100.

———— A contribution to the knowledge of the flora of southwestern Mackenzie district, N.W.T. *Can. Field Nat.*, 77:2 (1963), 108–23.

———— Some rare plants from the Mackenzie Mountains, Mackenzie district, N.W.T. *Can. Field Nat.*, 77:4 (1963), 226–28.

———— New plant records from northwestern Mackenzie district, N.W.T. *Can. Field Nat.*, 79:2 (1965), 96–106.

———— Plants of the Mackenzie River delta and reindeer grazing preserve. Canada, Dept. of Agriculture.

Collins, H. B., Jr., A. H. Clark, and E. H. Walker. The Aleutian Islands: Their people and natural history. (With keys for the identification of the birds and plants.) Smiths. Inst. War Backgr. Stud., 21 (1945), 1–129. Systematic list of plants extracted from Hultén, Flora of the Aleutian Islands, pp. 96–109.

Cook, J. Troisième voyage de Cook, or voyage à l'océan Pacifique. 4 vols. Paris, 1785.

Cooley, G. E. Plants collected in Alaska and Nanaimo, B.C., July and August, 1891. *Bull. Torr. Club*, 19 (1892), 239–49. Mosses, Lichens by C. E. Cummings.

Cooper, W. S. Additions to the flora of Glacier Bay National Monument, Alaska, 1935–1936. *Bull. Torr. Club*, 66 (1939), 453–56.

———— The battle of ice and forest . . . the story of Muir Glacier. *Amer. Forest & Forest Life*, 30 (1924), 196–98.

———— A fourth expedition to Glacier Bay, Alaska. *Ecology*, 20 (1939), 130–55.

———— An isolated colony of plants on a glacier-clad mountain. *Bull. Torr. Club*, 69: 6 (1942), 429–33.

———— The layering habit in Sitka spruce and the two western hemlocks. *Bot. Gaz.*, 91 (1931), 441–51.

———— The problem of Glacier Bay, Alaska. A study of glacier variations. *Geog. Rev.*, 27 (1937), 37–62.

———— The recent ecological history of Glacier Bay, Alaska. *Ecology*, 4:2,4 (1923).

———— The seed-plants and ferns of the Glacier Bay National Monument, Alaska. *Bull. Torr. Club*, 57 (1930), 327–38.

———— A third expedition to Glacier Bay, Alaska. *Ecology*, 12 (1931), 61–95.

———— Vegetation of Prince William Sound region,

Bibliography

Alaska, with a brief excursion into post-Pleistocene climatic history. *Ecol. Monogr.*, 12 (1942), 1–22.

Coville, F. V. See Brooks and Prindle; Brooks *et al.*; Eastwood, A descriptive list; and Herron.

Coville, F. V., and F. Funston. Botany of Yakutat Bay, Alaska, with a field report by F. Funston. *Contrib. U.S. Nat. Herb.*, 3 (1895), 325–50. Mosses collected by F. Funston, pp. 350–51.

Coville, F. V., and W. F. Wight. See Mendenhall; and Schrader, A reconnaissance in northern Alaska.

Cummings, C. E. The Lichens of Alaska. In Harriman Alaska Expedition, 5 (1904), 67–152.

Dall, W. H. Alaska and its resources. Boston, 1870.

——— Some remarks upon the natural history of Alaska. *Proc. Boston Soc. Nat. Hist.*, 12 (1868–69), 143–45.

——— Useful indigenous Alaskan plants. In Report upon the agricultural resources of Alaska. Rep. Dept. Agric., 1868, pp. 172–89.

Dearborn, C. H. Weeds in Alaska and some aspects of their control. *Weeds*, 7:3 (1959), 265–70.

Dervis-Sokolova, T. G. Flora of the most eastern part of the Chukchi Peninsula. In Rastenija severa Sibiri i Dalnego Vostoka. Moscow-Leningrad, 1966, pp. 80–107. (In Russian.)

Dictionary of Alaska place names. U.S. Geol. Surv. Prof. Pap. 567. Washington, D.C., 1967.

Drury, W. H., Jr. Bog flats and physiographic processes in the upper Kuskokwim River region, Alaska. *Contrib. Gray Herb.*, 178 (1956).

Durand, E. Plants collected by the United States Coast Survey on the geographical reconnaissance of Alaska, under the direction of George Davidson, Assistant, U.S. Coast Survey. Collection by Albert Kellogg, nomenclature by Elias Durand. House Ex. Doc. 275, 40th Congr. 2d sess. (1869).

Eastwood, A. A collection of plants from the Aleutian Islands. *Leaflets Western Bot.*, 5 (1947), 9–13.

——— A descriptive list of the plants collected by Dr. F. E. Blaisdell at Nome City, Alaska. *Bot. Gaz.*, 33 (1902), 126–49, 199–213, 284–99. Vascular Cryptogams determined by L. M. Underwood; Gramineae by F. Lamson-Scribner; Cyperaceae by T. Holm; Juncaceae by F. V. Coville.

——— A list of plants from Dall and Annette Islands, Alaska. *Leaflets Western Bot.*, 7:5 (1957), 102.

——— See Cantwell.

Evans, A. W. Hepaticae of Alaska. In Harriman Alaska Expedition, 5 (1904), 339–72.

Evans, W. H. Agricultural outlook on the coast region of Alaska. U.S. Agric. Dept. Yearbook, 1897, pp. 553–76.

——— Notes on the edible berries of Alaska. *Plant World*, 3 (1900), 17–19.

——— Report on botanical survey in Alaska, 1898. U.S. Dept. Agric., *Off. Exper. Sta. Bull.*, 62 (1899), 48–50.

Eyerdam, W. F. Some interesting plants found in the Aleutian Islands. *Little Gard.*, 7 (1936), 1–3, 21–22.

Fedtschenko, B. Flore des Iles du Commandeur. Acad. Sci. Cracovie, 1906.

Fernald, M. L. Persistence of plants in unglaciated areas of boreal America. *Mem. Gray Herb.*, 2 (1925), 237–342.

——— Some plants from the northwest shore of the Hudson Bay. Four rare plants from Alaska. *Ottawa Nat.*, 13 (1899), 147–49.

Fernow, B. E. The forests of Alaska. *Forestry and Irrigation*, 8 (1902), 66–70. Also in Harriman Alaska Expedition, 2 (1902), 235–56.

Fischer, E. L. When southeast Alaska blooms. *Nature*, 18 (1931), 29–32.

Fischer, R. B. Flora of McKinley Park, Alaska. *Vermont Bot. Bird Club Joint Bull.*, 12 (1927), 32–33.

Flett, J. B. Notes on the flora about Nome City. *Plant World*, 4 (1901), 67–68.

Flora Arctica USSR, 1–5. Moscow-Leningrad, 1966.

Flora USSR (Flora SSSR), 1–30. Moscow-Leningrad, 1934–60.

Georgeson, C. C. A note on native fruits. Ann. Rep. Alaska Agric. Exp. Sta. 1906 (1907), pp. 9–14.

Gjaerevoll, O. Botanical investigations in central Alaska, especially in the White Mountains. *K. Norske Vidensk. Selsk. Skr.* (Trondheim), 5 (1958); 4 (1963), 10 (1967).

Glenn, E. F., and W. A. Abercrombie. Report of explorations in the territory of Alaska, 1898. War Dept. Doc. 102, Adjutant Gen. Off. Mil. Inf. div. pub. 25 (1899). List of plants collected by J. C. Castner on Matanuska River, Alaska, 1898.

Gmelin, J. G. Flora sibirica sive historia plantarum Sibiriae. 4 vols. Petropoli, 1747–69. (Contains notes on Steller's plants collected in Alaska.)

Gombocz, E. Plants from Alaska. *Bot. Közl.*, 37 (1940), 6–13.

Gorman, M. W. Economic botany of southeastern Alaska. *Pittonia*, 3 (1896), 64–85.

Gorodkov, B. N. Botanical-geographical review of the Chukotsk shore. *Gos. pedagog. Inst. imeni Gertsena, Uchenyi Zapiski*, 21 (1939), 99–173. (In Russian.)

Graves, H. S. The forests of Alaska. *Amer. Forestry*, 22 (1916), 24–37.

Gray, A. Synoptical flora of North America. 2 vols. New York, 1878, 1897. See also Muir, Botanical notes on Alaska; E. W. Nelson; Ray; Wright.

Greene, E. L. Accessions to Canadian botany, I. *Ottawa Nat.*, 25 (1911), 145–47.

Griggs, R. F. After the eruption of Katmai, Alaska. *Nat. Hist.*, 20 (1920), 390–95.

——— The beginnings of revegetation in Katmai Valley. *Ohio Jour. Sci.*, 19 (1919), pp. 318–42.

——— The character of the eruption as indicated by its effects on nearby vegetation. *Ohio Jour. Sci.*, 19 (1919), 173–209. Scientific results of the Katmai Expeditions of the National Geographic Society, IV.

——— The edge of forest in Alaska and the reasons for its position. *Ecology*, 15 (1934), 80–96.

——— The effect of the eruption of Katmai on land vegetation. *Bull. Amer. Geog. Soc.*, 47 (1915), 193–203.

——— Observations on the edge of the forest in the Kodiak region of Alaska. *Bull. Torr. Club*, 41 (1914), 381–85.

——— The recovery of vegetation at Kodiak. *Ohio Jour. Sci.*, 19 (1918), 1–57. Scientific results of the Katmai Expeditions of the National Geographic Society, I.

——— The vegetation of the Katmai district. *Ecology*, 17 (1936), 380–417.

Gröntved, J. Vascular plants from Arctic North America collected by the Fifth Thule Expedition. Rep. Fifth Thule Exp., 1921–24, 2:1 (1936).

Hanson, H. C. Vegetation types in northwestern Alaska and comparison with communities in other Arctic regions. *Ecology*, 34 (1953), 111–40.

Harriman Alaska Expedition with cooperation of Washington Academy of Sciences, ed. C. H. Merriam. 14 vols. New York, 1901–14.

Harshberger, J. W. The forests of the Pacific coasts of British Columbia and southeastern Alaska. *Acta Forest. Fennica*, 34 (1929), 1–5.

——— Tundra vegetation of central Alaska directly under the Arctic Circle. *Proc. Amer. Phil. Soc. Philadelphia*, 67 (1928), 215–34.

Haskin, L. L. Wild flowers of the Pacific coast, in which is described 332 flowers and shrubs of Washington, Oregon, Idaho, central and northern California and Alaska. Portland, Ore., 1934.

Hayes, C. W. An expedition through the Yukon district. *Nat. Geog.*, 4 (1892), 117–63. Cryptogams by C. E. Cummings, pp. 160–62.

Healy, M. A. Report of the cruise of the revenue marine steamer Corwin in the Arctic Ocean in the year 1884. House Misc. Doc. 602, 50th Congr., 1st sess., 19 (1887–88).

Heller, Christine. Wild, edible, and poisonous plants of Alaska. University of Alaska Extension Bull. F-40, 1958.

——— Wild flowers of Alaska. Portland, Ore., 1966.

Heller, E. Mammals of the 1908 Alexander expedition with descriptions of the localities visited and notes on the flora of the Prince William Sound region. *Univ. Calif. Pub. Zool.*, 5 (1910), 321–60. Partial list of plants, chiefly shrubs and trees by T. S. Brandegee, pp. 349–60.

Henry, J. K. Flora of southern British Columbia and Vancouver Island, with many references to Alaska and northern species. Toronto, 1915.

Henshaw, J. W. See E. W. Nelson.

Herron, J. W. Explorations in Alaska, 1899, for an all-American overland route from Cook Inlet, Pacific Ocean, to the Yukon. War Dept. Doc. 138, Adjutant Gen. Off. 31 (1901), 1–77. List of plants by F. W. Coville, pp. 74–76.

Heusser, C. J. Nunatak flora of the Juneau ice field, Alaska. *Bull. Torr. Club*, 81:3 (1954), 236–50.

Hinds, R. B. The botany of the voyage of H.M.S. Sulphur, under the command of Captain Sir Edward Belcher, during the years 1836–42. Edited and superintended by R. B. Hinds. The botanical descriptions by George Bentham. London, 1844.

Holm, T. Contributions to the morphology, synonymy, and geographical distribution of Arctic plants. Rep. Can. Arct. Exp. 1913–18. *Botany*, 5 (1922), pp. 4B–139B.

Hooker, W. J. Flora boreali-Americana; or, the botany of the northern parts of British America: compiled principally from the plants collected by Drs. Richardson and Drummond on the late northern expeditions, under command of Captain J. Franklin. 2 vols. London, 1833, 1840.

Hooker, W. J., and G. A. W. Arnott. The botany of Captain Beechey's voyage. London, 1841. List of plants from Kotzebue's Sound, pp. 120–34 (Musci, p. 133; Lichenes, pp. 133–34; Fungi, Algae, p. 134).

Hopkins, D. M. Some characteristics of the climate in forest and tundra regions of Alaska. *Arctic*, 12 (1959), 215–20.

Hopkins, D. M., ed. The Bering land bridge. Stanford Univ. Press, 1967.

Hrdlička, A. Man and plant in Alaska. *Science*, n.s. 86 (1937), 559–60.

Hryniewiecki, B. Contributions to the study of the flora in Tchuktchiland. *Discipl. Biol. Arch. Soc. Sc. Varsaviensis*, 1 (1922), 18.

Hultén, E. The circumpolar plants, I. *Kungl. Svensk Vetenskapsakad. Handl.*, ser. 4, 8:5 (1962).

——— Comments on the Flora of Alaska and Yukon. *Kungl. Svensk Vetenskapsakad. Handl.*, ser. 2, 7:1 (1967).

——— Contribution to the knowledge of flora and vegetation of the southwestern Alaskan mainland. *Svensk Bot. Tidskr.*, 60:1 (1966), 177–89.

——— Flora of Alaska and Yukon, 1–10. Lunds Universitets Årsskrift N.F., Avd. 2, 37:1–46:1 (1941–50).

——— Flora of the Aleutian Islands. Stockholm, 1937; 2d ed., Weinsh. Bergstr., 1960.

——— Flora of Kamtchatka and the adjacent islands. 4 parts. *Kungl. Svensk Vetenskapsakad. Handl.*, ser. 3: 5:1 (1927); 5:2 (1928); 8:1 (1929); 8:2 (1930).

——— Flora and vegetation of Scammon Bay, Bering Sea coast, Alaska. *Svensk Bot. Tidskr.*, 56: 11 (1962), 36–54.

——— History of botanical exploration in Alaska and Yukon territories from the time of their discovery to 1940. *Bot. Not.*, 1940, pp. 289–346.

——— A list of plants from the Chukchi Peninsula. *Svensk Bot. Tidskr.*, 19 (1925), 104–10.

——— New or notable species from Alaska. Contributions to the flora of Alaska, I. *Svensk Bot. Tidskr.*, 30 (1936), 515–28.

——— On the American component in the flora of eastern Siberia. *Svensk Bot. Tidskr.*, 22 (1928), 220–29.

——— Outline of the history of arctic and boreal biota during the Quaternary Period. Stockholm, 1937.

——— Südwest-Alaska. *Vegetationsbilder*, 25:3 (1937).

——— Two new species from Alaska. Contribution to the flora of Alaska, II. *Bot. Not.*, 1939, pp. 826–29.

——— See Hutchison, Stepping stones.

Hutchison, I. W. New plants from the Aleutian Islands. *Gard. Illustr.*, 61 (1939), p. 556.

——— North to the rime-ringed sun. London, 1934. List of plants collected in Alaska and Yukon by the author from June to August, 1933, for the Royal Herbarium of Kew, pp. 243–62 (determinations by the Director at Kew and his assistants).

——— Plant collecting on the Pribilof and Aleutian Islands, 1936. *Nature*, 139 (1937), p. 327.

——— Plant-hunting in Alaska. Roy. Bot. Gard. Kew Bull. Misc. Inf., 1934, pp. 345–52.

——— Stepping stones from Alaska to Asia. London, 1937. Appendix 2, E. Hultén, List of plants collected for the British Museum at Seward and on Kodiak, the Pribilof and Aleutian Islands, pp. 226–46.

Jaggar, T. A., Jr. Journal of the Technology Expedition to the Aleutian Islands, 1907. *Technol. Rev.*, 10 (1908), 1–37.

Johnson, A. W., and L. A. Viereck. Some new records and range extensions of arctic plants from Alaska. *Biol. Papers, Univ. Alaska*, 6 (1962).

Johnson, A. W., L. A. Viereck, E. Ross, J. Melchior, and H. Melchior. Vegetation and flora. Environment of the Cape Thompson region, Alaska. U.S. Atomic Energy Commission, Div. of Technical Information, 1966, pp. 277–354.

Johnson, P. L., and T. C. Vogel. Vegetation of the Yukon Flats region, Alaska. Cold Regions Research & Engineering Laboratory (U.S. Army Material Command), Research Report 209, 1966.

Jordal, L. H. Plants from the vicinity of Fairbanks, Alaska. *Rhodora*, 53 (1951), 156–59.

Kashevaroff, A. P. Edible herbs and vegetables found in Alaska. *Bull. Pub. Bur.*, 9 (1918).

Kellogg, A. See Durand.

Kellogg, R. S. The forests of Alaska. U.S. Dept. Agric. Forest Serv. Bull., 81 (1910).

Kindschy, R. R., and J. E. O'Connell. Floristics of Umnak Island, Aleutian Islands, Alaska. *Northw. Sci.*, 33 (1959), 94–96.

Kittlitz, F. H. Denkwürdigkeiten einer Reise nach dem russischen Amerika, nach Mikronesien und durch Kamtschatka, 2 vols. Gotha, 1858.

——— Vierundzwanzig Vegetations - Ansichten von Küstenländern und Inseln des Stillen Oceans. 2d ed., 1862. English edition by B. C. Seemann, 1861.

Kjellman, F. R. Asiatiska Beringsunds-kustens fanerogamflora. In Nordenskjöld, Vega-Expeditionens 3 (1882–83), 475–572 (German edition).

——— Fanerogamer från Vest-Eskimåernas land. In Nordenskjöld, 2 (1883), 25–60 (German edition).

——— Fanerogamfloran på St. Lawrence-ön. In Nordenskjöld, 2 (1883), 1–23 (German edition).

——— Om tschukhschernas hushållsvaxter (On the useful plants of Chukchi). In Nordenskjöld, 2 (1883), 353–72 (German edition).

Knowlton, F. H. Alaskan plants. *Bot Gaz.*, 11 (1886), 340.

——— List of plants collected by C. L. McKay at Nushagak, Alaska, in 1881, for the U.S. National Museum. *Proc. U.S. Nat. Mus.*, 8 (1885), 213–21 (Musci, Lichenes, p. 221).

Kobayashi, Y. On the native plants of Aleutian Islands. *J. Jap. Bot.*, 10 (1934), 664–67.

Koidzumi, G. List of plants collected in Alaska and Yukon by Innouye. *Bot. Mag. Tokyo*, 30 (1916), 68–69 (in Japanese).

——— Plantae Siphonogamae a N. Yokoyama anno 1907 in Alaska arctica, Tschuktschore et Kamtschatka collectae. *Bot. Mag. Tokyo*, 25 (1911), 203–22.

Komarov, V. L. Flora Peninsulae Kamtschatka. 3 vols. Acad. Sci. USSR, Leningrad, 1927–30. (In Russian with English résumé.)

Kotzebue, O. v. Neue Reise um die Welt, in den Jahren 1823–26. 2 vols. Weimar, 1830.

Kotzebue, O. v., ed. Entedeckungs-Reise in die Süd-See und nach der Berings-Strasse zur Erforschung einer nordöstlichen Durchfahrt. 3 vols. Weimar, 1821.

Krause, A. Die Tlinkit-Indianer. Jena, 1885. (With notes of edible and useful plants.)

Krusenstern, A. J. v. Reise um die Welt in den Jahren 1803, 1804, 1805 and 1806 auf Befehl Seiner Kaiserlichen Majestät Alexander des Ersten auf den Schiffen Nadeshda und Newa. 3 vols. St. Petersburg, 1810–12.

Kurtz, F. Die Flora des Chilcatgebietes im sudostlichen Alaska, nach den Sammlungen der Gebrüder Krause. *Engler Bot. Jahrb.*, 19 (1895), 327–431.

——— Die Flora des Tschuktschen Halbinsel. *Engler Bot. Jahrb.*, 19 (1895), 432–93.

Langsdorff, G. H. v. Bemerkungen auf einer Reise um die Welt in den Jahren 1803 bis 1807. 2 vols. Frankfurt am Main, 1812–13.

Ledebour, C. F. v. Flora Rossica. 4 vols. Stuttgart, 1842–53.

Lepage, E. New or noteworthy plants in the flora of Alaska. *Amer. Midl. Nat.*, 43:3 (1951), 754–59.

——— Variations mineures de quelques plantes du Nord-est du Canada et de l'Alaska. *Nat. Can.*, 77 (1950), 228–31.

Lindeman, E. Verzeichnis derjenigen Pflanzenarten, welche durch Abtretung der russischen amerikanischen Landbesitzungen gegenwärtig aus der Flora rossica auszuscheiden sind. *Bull. Soc. Nat. Moscou*, 40 (1867), 559–61.

Linnaeus, C. Amoenitates Academicae, 2 (Holmiae, 1751), pp. 332–64. Includes notes on Steller's plants from Alaska.

——— Plantae rariores camschatcenses . . . publico examini modeste submittit Jonas P. Halenius. Uppsala, 1750.

Löve, D., and N. J. Freedman. A plant collection from southwest Yukon. *Bot. Not.*, 1956, pp. 153–211.

McAtee, W. L. Additions to the flora of the Pribilof Islands. *Torreya*, 22 (1922), 67.

Macoun, J. M. Catalogue of Canadian plants. 8 vols. Montreal, 1883–1902.

——— List of plants collected by J. B. Tyrrell in the Klondike region in 1899. *Ottawa Nat.*, 13 (1899), 209–18.

——— A list of the plants of the Pribilof Islands, Bering Sea. In D. S. Jordan, The fur seals and fur seal islands of the north Pacific Ocean, 3 (1899), 559–87.

——— Liste des plantes recueillies, en 1887, par le Dr. G. M. Dawson, dans la région du Yukon, et dans la partie septentrionale de la Columbie-Anglaise, qui est adjacente à cette région. Rep. Ann. Comm. Geol. Hist. Nat. Can., 3 (1887–88).

——— Report on plants collected by D. D. Cairnes on the 141st meridian in 1912. Geol. Surv. Can. Sess. Pap. 26, summary rep. 1912 (1914), pp. 438–40.

——— See Cairnes, Geology; Cairnes, The Yukon-Alaska international boundary.

Macoun, J. M., and T. Holm. Contributions to the morphology, synonymy, and geographical distribution of arctic plants. Rep. Can. Arct. Exp. 1913 (18 vols.), *Botany*, 5 (1922), 4B–139B.

——— The vascular plants of the Arctic coast of America west of the 100th meridian. Rep. Can. Arct. Exp. 1913–18, 5: Botany (1921), pp. 1A–24A.

Maxon, W. R. See Smith.

Meehan, T. Catalogue of plants collected in July 1883 during an excursion along the Pacific coast in southeastern Alaska. Proc. Acad. Nat. Sci., Philadelphia, 1884, pp. 76–96.

Mendenhall, W. C. Reconnaissance from Fort Hamlin to Kotzebue Sound, Alaska, by way of Dall, Kanuti, Allen, and Kowak rivers. U.S. Geol. Surv. Prof. Pap. 10 (1902), pp. 1–68. List of plants collected in northern Alaska by W. L. Poto in 1901. Identified chiefly by F. V. Coville and W. F. Wight, pp. 58–65.

Merriam, C. H. Plants of the Pribilof Islands, Bering Sea, with critical notes by J. N. Rose. *Proc. Biol. Soc. Wash.*, 7 (1892), 133–50.
―――― See Harriman Alaska Expedition.
Mertens, H. Bemerkungen über die Floren der Koragins-Inseln und eines Theils des Landes an der Behrings-Strasse, aus den Briefen des Dr. Mertens an die Herren Fischer und Trinius. *Linnaea*, 5 (1830), 60–71.
―――― Zwei botanisch-wissenschaftliche Berichte vom Dr. Heinrich Mertens, Naturforscher auf der gegenwärtigen Russischen Entdeckungsreise am Bord des Siniavin, Capt. v. Lütke, geschrieben im October 1827 in Kamtschatka, mitgetheilt durch den Vater, Prof. Mertens in Bremen, mit einigen Bemerkungen versehen von Dr. Adelbert v. Chamisso. *Linnaea*, 4 (1829). English translation in W. J. Hooker, *Bot. Misc.*, 3 (1833), 1–23.
Mertie, J. B., Jr. The Chandalar-Sheenjek district, Alaska. U.S. Geol. Surv. Bull. 810 (1929), pp. 87–139. List of the flora by P. C. Standley, pp. 106–9.
―――― The Tatonduk-Nation district, Alaska. U.S. Geol. Surv. Bull. 836 (1932), pp. 345–454. Flora of district by C. V. Morton, pp. 363–67.
―――― The Yukon-Tanana region, Alaska. U.S. Geol. Surv. Bull. 872 (1937).
Mexia, Y. Some of the commoner plants found in Mount McKinley National Park. U.S. Dept. Interior Nat. Park Serv. Circ. Gen. Inf. Regard. Mount McKinley National Park, Alaska, 1929, pp. 2–4.
Morton, C. V. Flora of Tatonduk-Nation district. See Mertie, The Tatonduk-Nation district, Alaska.
Muir, J. Botanical notes on Alaska. Cruise of the revenue-steamer Corwin in Alaska and the N.W. Arctic Ocean in 1881. House Ex. Doc. 105, 47th Congr., 2d sess., 23 (1883), 47–53 (Treas. Dept. Doc. no. 429).
―――― The cruise of the Corwin, edited by W. F. Badé. Boston, 1917. Reprinted in Muir, Botanical notes on Alaska (1883).
―――― Some botanical notes from cruise of the Corwin. *Torreya*, 18 (1918), 197–210. Reprinted in Muir, Botanical notes on Alaska (1883).
Nelson, A. Rocky Mountains herbarium studies, VI. *Amer. Jour. Bot.*, 32 (1945), 284–90.
Nelson, E. W. Report upon natural history collections made in Alaska between the years 1877 and 1881, edited by Henry W. Henshaw. Senate Misc. Doc. 156, 49th Congr., 1st sess. (1887). List of plants identified by Asa Gray, pp. 30–31. Reports partly the same as those in Muir, Botanical notes on Alaska (1883). U.S. Army Sign. Serv. Arct. Ser. Pub. 3.
Nordenskiöld, A. E. Vegas färd kring Asien och Europa . . . (Voyage of the Vega around Asia and Europe). Stockholm, 1880–81. (Translated into Bohemian, English, Finnish, French, Dutch, Italian, Norwegian, Russian, Spanish, and German.)
―――― Vega-Expeditionens Vetenskapliga Iakttagelser. 5 vols. 1882–87. German edition, Die Wissensch. Ergebn. d. Vega-Exped. (1882–83).
Osgood, W. H. Biological investigation in Alaska and Yukon Territory. *N. Amer. Fauna*, 30 (1909). Native vegetation of east-central Alaska noted on pp. 7–16.
―――― A biological reconnaissance of the base of the Alaska Peninsula. *N. Amer. Fauna*, 24 (1904). Notes of the vegetation.
―――― Natural history of the Cook Inlet region, Alaska. *N. Amer. Fauna*, 21 (1901). Flora, pp. 53–56.
―――― Results of a biological reconnaissance of the Yukon River region. *N. Amer. Fauna*, Div. Biol. Surv., 19 (1900).
Ostenfeld, C. H. Vascular plants collected in Arctic North America by the Gjöa expedition under Captain Roald Amundsen, 1904–6. *Norsk. Vid. Selsk. Skrift., Math. Nat. Klasse*, 8, 1909 (1910).
Ostenfeld, C. H., and O. Gelert. Flora arctica. Copenhagen, 1902.
Pallas, P. S. Flora rossica. Lipsiae, 1784–1815, 1:1–2; 2:1. Contains notes on Steller's plants from Alaska.
Piper, C. V. Grasslands of the south Alaska coast. U.S. Dept. Agric. Bur. Plant Ind. Bull. 82 (1905).
Polunin, N. Circumpolar arctic flora. Oxford, 1959.
Porsild, A. E. The alpine flora of the east slope of Mackenzie Mountains, Northwest Territories. Bull. Nat. Mus. Can., 101 (1945).
―――― Botany of southeastern Yukon adjacent to the Canol road. Bull. Nat. Mus. Can., 121 (1951).
―――― Contributions to the flora of Alaska. *Rhodora*, 41 (1939), pp. 141–83, 199–254, 262–301.
―――― Contributions to the flora of southwestern Yukon Territory. Bull. Nat. Mus. Can., 216 (1966).
―――― Edible roots and berries of Northern Canada. Ottawa, 1937.
―――― Flora of Little Diomede Island in Beering Strait. *Trans. Roy. Soc. Can.*, ser. 3, sect. 5, 32 (1938), 21–38.
―――― Illustrated flora of the Canadian Arctic Archipelago, 2d ed. Bull. Nat. Mus. Can., 1964.
―――― Materials for a flora of the continental Northwest Territories of Canada. *Sargentia*, 4 (1943).
―――― Some new or critical vascular plants of Alaska and Yukon. *Can. Field Nat.*, 79:2 (1965), 79–90.
―――― The vascular flora of an alpine valley in the Mackenzie Mountains, N.W.T. Bull. Nat. Mus. Can., 171 (1959).
―――― Vascular plants collected on Kiska and Great Sitkin Islands in the Aleutians by Lt. H. R. McCarthy and Capt. N. Kellas, August, September, and October 1943. *Can. Field Nat.*, 58 (1944), 130–31.
Porsild, A. E., and H. A. Crum. The vascular flora of Liard Hotsprings, B.C., with notes on some bryophytes. Bull. Nat. Mus. Can., 171 (1959).
Potter, Louise. Roadside flowers of Alaska. Hanover, N.H., 1962.
Prescott, J. D. See Bongard.
Presl, J. S. Reliquiae Haenkeanae seu descriptiones et icones plantarum quae in America meridionali et boreali, in insulis Phillipinis et Marianis collegit Thaddaeus Haenke. 2 vols. Prague, 1830–36.
Raup, H. M. The botany of southwestern Mackenzie. *Sargentia*, 6 (1947), 1–275.
―――― Expeditions to the Alaska military highway, 1943–44. *Arnoldia*, 4 (1944), 65–72.
―――― Phytogeographic studies in the Peace and upper Liard River regions, Canada. *Contrib. Arnold Arb.*, 6 (1934).
―――― The willows of boreal western America. *Contrib. Gray Herb.*, 185 (1959).
Ray, P. H. Report of the International Polar Expedition to Point Barrow, Alaska. House Ex. Doc. 44, 49th Congr., 2d sess. (1885). Plants by Asa Gray, pp. 191–92. U.S. Army Sign. Serv. Arct. Ser. Pub. 1.

Bibliography

Regel, E. L., and F. v. Herder. Plantae Raddeanae (G. Radde: Reisen in den Süden von Ost-Sibirien im Auftrage der Kaiserlichen russischen geographischen Gesellschaft ausgeführt in den Jahren 1855–59; durch Botanische Abteilung, Nachträge zur Flora der Gebiete des Russischen Reichs östlich vom Altai bis Kamtschatka und Sitka), vols. 1–4 (1862–87). An Index to this bibliographically complicated work is found in Acta Hort. Petrop. 10:1 (1887), pp. 69–82.

Reid, H. F. Studies of Muir Glacier, Alaska. *Nat. Geog.*, 4 (1892), 1–84. List of plants, collected by Cushing, determined by W. W. Rowlee, p. 79.

Rigg, G. B. The effects of the Katmai eruption on marine vegetation. *Science*, 40 (1914), 509–13.

—— Notes on the flora of some Alaskan *Sphagnum* bogs. *Plant World*, 17 (1914), 167–82.

—— Some raised bogs of southeastern Alaska with notes on flat bogs and muskegs. *Amer. J. Bot.*, 24 (1937), 194–98.

Rothrock, J. T. Sketch of the flora of Alaska. Ann. Rep. Smiths. Inst. 1867 (1868), pp. 433–63.

Rouleau, E. Bibliographie des travaux concernant la flore canadienne parus dans *Rhodora* de 1899 à 1943 inclusivement. *Contrib. Inst. Bot. Montréal*, 54 (1944).

Rousseau, J. Essai de Bibliographie botanique canadienne, I. *Can. Nat.*, 60 (1933), pp. 55–58.

Rowlee, W. W. See Reid.

Ruprecht, F. J. Neue oder unvollständig bekannte Pflanzen aus dem nördlichen Theile des Stillen Oceans. *Mém. Acad. Imp. Sci. St. Pétersb.*, sér. 6, Sci. Nat., Bot. 7 (1855), 55–82.

Rydberg, P. A. Flora of the prairies and plains of central North America. New York, 1932.

—— Flora of the Rocky Mountains and adjacent plains. New York, 1917.

—— List of plants collected on the Stefansson-Anderson Arctic Expedition, 1908–12. *Torreya*, 14 (1914), 65–66. (List of four species from King Point, Yukon Territory, p. 66.)

Saccardo, P. A., C. H. Peck, and W. Trelease. The Fungi of Alaska. In Harriman Alaska Expedition, 5 (1904), 13–64.

Sargent, C. S. The forests of Alaska. *Gard. & For.*, 10 (1897), 379–80.

Saunders, A. de. The Algae of the expedition. In Harriman Alaska Expedition, 5 (1904), 153–250.

Scamman, E. A list of plants from interior Alaska. *Rhodora*, 42 (1940), 309–49; *Contrib. Gray Herb.*, 132.

Schaack, G. B. v. Flowers of Island X. Prepared for the use of the men in the Armed Forces stationed on Island X [Attu I.]. Welfare & Recreation Dept. Navy, 163 (1945).

Schrader, F. C. Preliminary report on a reconnaissance along the Chandalar and Koyukuk Rivers, Alaska, in 1899. II. General geology, economic geology, Alaska. U.S. Geol. Surv. Ann. Rep. 21 (1899–1900), 447–86.

—— A reconnaissance in northern Alaska, across the Rocky Mountains, along Koyukuk, John, Anaktuvuk, and Colville rivers, and the Arctic coast to Cape Lisburne, in 1901, with notes by W. J. Peters. U.S. Dept. Interior Geol. Surv. Prof. Pap. 20 (1904), pp. 1–139. List of plants identified by F. V. Coville and W. F. Wight, the grasses by E. D. Merrill, pp. 131–34.

—— A reconnaissance of a part of Prince William Sound and the Copper River district, Alaska, in 1898. VII. Explorations in Alaska in 1898. U.S. Geol. Surv. Ann. Rep. 20 (1898–99), 341–423.

Schwatka, F. Report of a military reconnaissance in Alaska, made in 1883. Senate Ex. Doc. 2, 48th Congr., 2d sess. (1885).

Seemann, B. C. The botany of the voyage of H.M.S. Herald under the command of Captain Henry Kellett . . . during the years 1845–51. London, 1852–57. List of plants collected by Pullen, Penny, and Ede, pp. 50–56.

Seemann, B. C., trans. Twenty-four views of the vegetation of the coast and islands of the Pacific, with explanatory descriptions taken during the exploring voyage of the Russian corvette Senjawin under the command of Capt. Lütke, 1827–29, by F. H. Kittlitz, 1861.

Senn, H. A. A bibliography of Canadian plant geography, VII. *Trans. Roy. Can. Inst.*, 26 (1946), 9–152.

Sharples, A. W. Alaska wild flowers. Stanford Univ. Press, 1938.

Shetler, S. G. An annotated list of vascular plants from Cape Sabine, Alaska. *Rhodora*, 65 (1963), 208–24.

Smith, P. S. The Lake Clark–Central Kuskokwim region, Alaska. U.S. Geol. Surv. Bull. 655 (1917). List of plants determined by W. R. Maxon, pp. 44–45.

Smith, P. S., and J. B. Mertie, Jr. Geology and mineral resources of northwestern Alaska. U.S. Geol. Surv. Bull. 815 (1930).

Spetzman, L. A. Vegetation of the Arctic slope of Alaska. U.S. Geol. Surv. Prof. Pap. 302, 1959.

Spurr, J. E. A reconnaissance in southwestern Alaska in 1898, VII. Explorations in Alaska in 1898. U.S. Geol. Surv. Ann. Rep. 20 (1898–99), pp. 31–264. Notes on the animal and vegetable life of the region of the Susitna and Kuskokwim Rivers, pp. 76–80.

Stair, L. D., and F. W. Pennell. A collection of plants from Yakutat, Alaska. *Bartonia*, 24 (1946), pp. 9–21.

Standley, P. C. See Mertie, The Chandalar-Sheenjek district, Alaska.

—— Manuscript of a flora of Alaska and Yukon completed about 1915. Kept at U.S. Nat. Mus. Div. Plants (Washington, D.C.). Copy in Library of the Academy of Science, Stockholm.

Steffen, H. Ein Beitrag zur Flora von Alaska. *Bot. Centralbl. Beih.*, Abt. B, 54 (1936), 547–56.

—— Verstreute Beiträge zur Flora der Arktis, I. Zur Flora von Alaska. *Bot. Centralbl. Beih.*, Abt. B, 58 (1938), 100–104.

Stejneger, L. Fauna and flora of the Aleutian Islands. *Nature*, 28 (1883), p. 520.

—— Georg Wilhelm Steller, the pioneer of Alaskan natural history. Cambridge, 1936.

Steller, G. W. Catalogus plantarum intra sex horas in parti Americae septemtrionalis iuxta promontorium Eliae observatarum anno 1741 die 21 Julii sub gradu latitudinis 59. Ms of 11 pp. by an unknown hand. Arkhiv Konferentsia, Academy of Sciences, Leningrad. Photostatic copy in Library of Congress, Washington, D.C., and Library of the Academy of Science, Stockholm. Also in Stejneger, Georg Wilhelm Steller, the pioneer of Alaskan natural history.

——— Catalogus seminum anno 1741 in America septemtrionalis sub gradu latitudinis 59 et 55 collectorum, quorum dimidia pars die 17 nov. 1742 transmissa. Ms of 7 pp., Arkhiv Konferentsia, Academy of Sciences, Leningrad.

——— Tagebuch seiner Seereise aus den Petripauls Hafen in Kamtschatka bis an die westlichen Küsten von Amerika und seiner Begebenheiten auf der Rückreise. *Neue Nord. Beytr.,* 5 (1793), 129–236; 6 (1793), 1–26.

Stoney, G. M. Naval explorations in Alaska. Annapolis, 1900.

Tarleton, J. B. A botanist's trip on the upper Yukon. *N.Y. Bot. Gard. Bull.,* 2 (1901), 148–49.

Tatewaki, M. Notes on plants of the western Aleutian Islands collected in 1929. *Trans. Sapporo Nat. Hist. Soc.,* 11 (1930), 152–56; 12 (1931), 200–209.

Taylor, R. T. Pocket guide to Alaska trees. U.S. Dept. Agric. Forest Serv. Misc. Pub. 55 (1950).

——— The successional trend and its relation to second-growth forests in southeastern Alaska. *Ecology,* 13 (1932), 381–91.

Tiling, T. Eine Reise um die Welt von Westen nach Osten durch Sibirien, das Stille und Atlantische Meer. Aschaffenburg, 1854.

Thomas, J. H. A collection of plants from Point Lay, Alaska. *Contrib. Dudley Herb.,* 4 (1951), 53–56.

——— The vascular flora of Middleton Island, Alaska. *Contrib. Dudley Herb.,* 5 (1957), 39–56.

Tikhomirov, B. A. A contribution to the knowledge of the flora and vegetation of the hot springs in Chukotsk Peninsula. *Bot. Jour.,* 42:9 (1957), 1427–45. (In Russian.)

——— Report on useful plants of the Eskimo of the southeastern shore of Chukotka. *Bot. Jour.,* 43 (1958), 242–46. (In Russian.)

Tikhomirov, B. A., and V. A. Gavrilyuk. Contribution to the flora of the Bering Sea coast of the Chukchi Peninsula. In Rastenija severa Sibiri i Dalnego Vostoka. Moscow, 1966, pp. 58–79. (In Russian.)

Torrey, J., and A. Gray. A flora of North America. 2 vols. New York, 1838, 1843.

Trelease, W. Alaskan species of *Sphagnum.* In Harriman Alaska Expedition, 5 (1904), 331–37.

——— The ferns and fern allies of Alaska (Pteridophytes). In Harriman Alaska Expedition, 5 (1904), 375–98.

Trinius, K. B. See Bongard.

Turner, L. M. Contributions to the natural history of Alaska. Senate Misc. Doc. 155, 49th Congr., 1st sess., 8 (1886), 1–226. List compiled by H. Mann, pp. 61–85. U.S. Army Sign. Serv. Arct. Ser. Pub. 2.

Vasey, G. List of the plants collected in Alaska, 1888. Scientific results of explorations by the U.S. Fish Comm. steamer Albatross No. 6. *U.S. Nat. Mus. Smiths. Inst. Proc.,* 12 (1889), 217–18.

Vincent, R. E. The larger plants of Little Kitoi Lake. *Amer. Midl. Nat.,* 60 (1958), 212–18.

Walker, E. H. Additional introduced plants in the Aleutian Islands. *Bot. Club,* 73 (1946), 204.

Watson, S. Note on the flora of the upper Yukon. *Science,* 3 (1884), 252–53.

Welsh, S. L. Legumes of Alaska: *Astragalus,* L. *Iowa State Jour. Sci.,* 37:4 (1963), 359–88.

Wickersham, J. A. A bibliography of Alaskan literature, 1724–1924. Alaska Agric. Coll. Sch. Mines Misc. Pub. 1. Cordova, 1927.

Wiggins, I. L. The distribution of vascular plants on polygonal ground near Point Barrow, Alaska. *Contrib. Dudley Herb.,* 4 (1957), 41–50.

Wiggins, I. L., and D. G. MacVicar. Notes on the plants in the vicinity of Chandler Lake, Alaska. *Contrib. Dudley Herb.,* 5:3 (1958), 69–95.

Wiggins, I. L., and J. H. Thomas. A flora of the Alaskan Arctic slope. Arct. Inst. of North America, spec. pub. no. 4, 1962.

Williams, R. S. Botanical notes on the way to Dawson, Alaska. *Plant World,* 2 (1899), 177–81.

Wright, G. F. The Muir Glacier. *Amer. Jour. Sci.,* ser. 3, 33 (1887), 1–18. Flora by Asa Gray, p. 18.

Index of Common Names

Adder's Mouth, 330
ADDER'S TONGUE FAMILY, 39
Adder's Tongue, 39
Alder, 368
  Mountain, 368
  Red, 369
  Sitka, 369
Alfalfa, 638
AMARANTH FAMILY, 403
Amaranth, Green, 403
ARROW GRASS FAMILY, 80
Arrowhead, 81
ARUM FAMILY, 280
Aspen:
  American, 332
  Quaking, 332
Asphodel, False, 304
Aster, 856
Avens, 625
Awlwort, 496
Azalea, Alpine, 720

Baneberry, 456
Barley, 192
Bearberry, 729
Beardtongue, 794
Bedstraw, 836
  Sweet-Scented, 838
Bellflower, 847
Bilberry, 731
Bindweed, Black, 384
BIRCH FAMILY, 364
Birch, 364
  Dwarf, 365
  Kenai, 366
  Paper, 367
  Shrub, 365
Bird's Rape, 504
Bishop's-Cap, 586
Bistort, 385
BLADDERWORT FAMILY, 829
Bladderwort, 830
Blite, Strawberry, 393
BLUEBELL FAMILY, 847
Bluebell, 782, 850

Blueberry, 731
  Alpine, 734
  Dwarf, 732
Blue-Eyed Grass, 314
Bog Star, 589
BORAGE FAMILY, 772
Boykinia, Alaska, 562
BRACKEN FAMILY, 43
Bracken, 43
Brooklime, 798
BROOMRAPE FAMILY, 828
Broomrape, 828
Buckbean, 763
BUCKWHEAT FAMILY, 373
Bugseed, 400
Bulrush, 207
Bunchberry, 709
Burnet, 632
BUR REED FAMILY, 67
Bur Reed, 67
Butter-and-Eggs, 793
Buttercup, 467
  Common, 485
  Creeping, 481
  Dwarf, 478
  Snow, 476
Butterwort, 829

Calamus, 280
Calla, Wild, 281
Calypso, 331
Camass, White, 305
Camomile:
  Wild, 890
  Yellow, 885
Carnation, 448
Cassandra, 727
Cassiope, Lapland, 724
Catmint, 785
Cat's-Ear, 942
CATTAIL FAMILY, 66
Cattail, 66
Cedar, Western Red, 64
Charlock, 503
Chickweed, 411
  Mouse-Ear, 420
Chive, Wild, 307

Christmas Green, 29
Chrysanthemum, 891
Cinquefoil, Shrubby, 609
Clintonia, Single-Flowered, 309
Cloudberry, 602
Clover, 641
  Alsike, 642
  Sweet, 640
CLUB MOSS FAMILY, 24
Club Moss, 24
  Alpine, 30
  Common, 28
  Fir, 25
  Stiff, 27
  Tree, 28
Cockle:
  Corn, 439
  Cow, 448
Coltsfoot, Sweet, 911
Columbine, 457
  Western, 457
COMPOSITE FAMILY, 851
Copper-Flower, 717
Coral Root, 329
Cornel:
  Canadian Dwarf, 709
  Swedish Dwarf, 708
Corydalis:
  Golden, 495
  Pale, 496
Cottonwood, 331
  Black, 332
Crab Apple, Oregon, 596
Cranberry, 735
  High Bush, 842
Cranesbill, 674
Creeping Jenny, 29
Cress:
  Bitter, 512
  Rock, 544
  Winter, 506
  Yellow, 507
CROWBERRY FAMILY, 716
Crowberry, 716

CROWFOOT FAMILY, 451
Crowfoot, 467
  Cursed, 479
  White Water, 469
Cuckoo Flower, 514
Currant:
  Bristly Black, 590
  Northern Black, 591
  Northern Red, 593
  Skunk, 592
  Stink, 591
  Trailing Black, 592
CYPRESS FAMILY, 64
Cypress, Alaska, 65

Daisy, 855
  Arctic, 893
  White, 894
Dame's Violet, 556
Dandelion, 944
  Fall, 942
DEER FERN FAMILY, 58
DIAPENSIA FAMILY, 736
Dock, 374
  Blunt-Leaved, 378
DOGBANE FAMILY, 763
Dogbane, 763
Dragonhead, 786
Dryas, 629
  Yellow, 629
Devil's Club, 696
DOGWOOD FAMILY, 707
Dogwood, American, 708
DUCKWEED FAMILY, 282

EARTH SMOKE FAMILY, 494
Eelgrass, 69
Elephant's Head, 818
EVENING PRIMROSE FAMILY, 685
Everlasting, Pearly, 882
Eyebright, 814

Fern:
  Fragile, 49
  Lady, 48
  Leathery Grape, 41
  Maidenhair, 42
  Ostrich, 52
  Parsley, 44
  Rattlesnake, 42
FIGWORT FAMILY, 793
FILMY FERN FAMILY, 43
Fir:
  Alpine, 64
  Pacific Silver, 63
Fireweed, 686
FLAX FAMILY, 676
Flax, 676
  False, 519
Fleabane, 861
  Coastal, 869
  Marsh, 926
Forget-Me-Not, 779
  Arctic, 773
Foxglove, 806
Foxtail, 89
  Alpine, 90
Fringe Cups, 585
Fritillary, Kamchatka, 308

Gale, Sweet, 364
GENTIAN FAMILY, 753
Gentian, Star, 761
GERANIUM FAMILY, 674
GINSENG FAMILY, 696
Glasswort, 401
Globeflower, 454
Goatsbeard, 595
Goldenrod, 851
Goldthread, 455
GOOSEFOOT FAMILY, 392
GRASS FAMILY, 82
Grass:
  Alkali, 154
  Arrow, 80
  Beard, 93
  Bent, 97
  Birdseed, 82
  Bluejoint, 104

[Grass]
  Brome, 172
  Canary, 82
  Cotton, 197
  Ditch, 79
  Feather, 85
  Fescue, 166
  Hair, 109
  Hare's-Tail, 204
  Holy, 84
  Lyme, 193
  Manna, 150
  Orchard, 126
  Poison Darnel, 181
  Polar, 93
  Reed Bent, 103
  Reed Canary, 82
  Sabine, 126
  Snow, 92
  Sprangletop, 150
  Squirreltail, 192
  Sweet, 83
  Tundra, 149
  Vanilla, 84
  Velvet, 109
  Wood Reed, 95
Grass-of-Parnassus:
  Fringed, 588
  Northern, 589
Ground Pine, 28
Groundsel, 925
  Common, 932

Harebell, 850
Hawk's-Beard, 954
Hawkweed, 958
  Orange, 959
Hawthorn, 600
HEATH FAMILY, 717
Heath, Alaska Moss, 726
Heather, Mountain, 722
Hellebore, False, 306
Hemlock:
  Mountain, 63
  Western, 62
Heuchera, Alpine, 584
HONEYSUCKLE FAMILY, 841
Horehound, 785
  Water, 790
HORNWORT FAMILY, 451
Hornwort, 451
Horseradish, 511
HORSETAIL FAMILY, 33
Horsetail, 33
Huckleberry, Red, 733

Indian Paintbrush, 806
  Great Red, 807
Indian Pipe, 715
IRIS FAMILY, 313
Ivy, Ground, 786

Jacob's Ladder, 767
Juniper, Common Mountain, 65

Key Flower, 318
Kinnikinnick, 729
Knotweed, 390

Labrador Tea, 717
Lace Flower, 583
Ladies' Tresses, 325
LADY FERN FAMILY, 47
Lady's Slipper, 315
Larkspur, 458
Laurel, Bog, 721
LEADWORT FAMILY, 752
Lettuce, 952
  Blue, 952
LICORICE FERN FAMILY, 57
LILY FAMILY, 304
Lily, Alp, 308
Lily-of-the-Valley, False, 311
Lingonberry, 731
Loosestrife, Tufted, 750
Lousewort, 816
Lovage, Beach, 702
Lungwort, 782
Lupine, 635
Luzerne, 638

MADDER FAMILY, 836
Madwort, 779
MAIDENHAIR FAMILY, 42
MAPLE FAMILY, 679
Maple, 679
Mare's Tail, 695
Marguerite, 894
Marigold:
  Bur, 884
  Corn, 894
MARSH FERN FAMILY, 45
Marsh Fivefinger, 608
Marsh Marigold, 453
Mayweed, 886
Mealberry, 729
Milkwort, Sea, 752
MINT FAMILY, 784
Mint, Field, 792
MISTLETOE FAMILY, 372
Mistletoe, Dwarf, 372
Monkey Flower:
  Purple, 796
  Yellow, 796
Monkshood, 459
Moonwort, 39
MOSCHATEL FAMILY, 845
Moschatel, 845
Moss Campion, 440
Mountain Ash, 596
MOUNTAIN PARSLEY FAMILY, 44
Mudwort, 797
MUSTARD FAMILY, 496
Mustard:
  Ball, 519

[Mustard]
  Tansy, 541
  Tower, 543
  Tumble, 502

NETTLE FAMILY, 370
Nettle, 370
  Burning, 371
  Hedge, 789
  Hemp, 788
NIGHTSHADE FAMILY, 792
Nightshade, 792
  Enchanter's, 693
Ninebark, Pacific, 593
Nipplewort, 941

Oat, 120
  Wild, 119
OLEASTER FAMILY, 684
Onion, Victory, 307
Orach, 397
ORCHIS FAMILY, 315
Orchis:
  Bog, 320
  Frog, 319
  Rhizome, 318
Oysterleaf, 781

PARSLEY FAMILY, 697
Parsley, Hemlock, 704
Parsnip, 706
  Cow, 707
  Water, 700
Pasqueflower, 466
PEA FAMILY, 635
Pea, Beach, 672
Pearlwort, 426
Pennycress, 499
Pigweed, 395
PINE FAMILY, 59
Pine:
  Jack, 60
  Lodgepole, 59
Pineapple Weed, 889
PINK FAMILY, 411
Pipsissewa, 710
PLANTAIN FAMILY, 832
Plantain, 832
  Common, 835
POLEMONIUM FAMILY, 764
Pond Lily, Yellow, 450
PONDWEED FAMILY, 69
Pondweed, Horned, 79
Poplar, Balsam, 331
POPPY FAMILY, 489
Poppy, 489
  Corn, 489
PRIMROSE FAMILY, 737
Primrose, 737
  Bird's-Eye, 740
  Greenland, 741
PURSLANE FAMILY, 404
Pussy's Toe, 871
Pussytoe, 871

Pygmyweed, 561

QUILLWORT FAMILY, 32
Quillwort, 32

Radish, 506
Ragwort, 925
Raspberry, 604
Rattlesnake Plantain, 328
Rattlesnake Root, 957
Rhododendron, Kamchatka, 719
Ribgrass, 835
Rice, Mountain, 87
River Beauty, 687
Rock Harlequin, 496
ROSE FAMILY, 593
Rose, 634
Rosebay, Lapland, 718
Roseroot, 561
Rue, Meadow, 487
RUSH FAMILY, 283
Rush, 283
  Spike, 210
  Wood, 295
Rye, 190
Ryegrass, 180
  Italian, 180

Sagewort, Prairie, 904
Salal, 728
Salmonberry, 604
SANDALWOOD FAMILY, 372
Sandwort:
  Grove, 437
  Seabeach, 434
Saraná, 308
Savin, Creeping, 66
SAXIFRAGE FAMILY, 562
Saxifrage:
  Aleut, 566
  Alpine, 567
  Bog, 568
  Brook, 577
  Bulblet, 575
  Coast, 581
  Cordate-Leaved, 572
  Cushion, 566
  Golden, 568
  Grained, 581
  Leatherleaved, 562
  Prickly, 571
  Purple Mountain, 565
  Red-Stemmed, 578
  Snow, 579
  Spiked, 574
  Spotted, 570
  Stiff-Stemmed, 580
  Tufted, 583
  Wood, 571
SCHEUCHZERIA FAMILY, 81
Scurvy Grass, 499
Sea Blite, 402

Sea Rocket, 503
SEDGE FAMILY, 197
Sedge, 215
Self-Heal, 787
Serviceberry, Pacific, 599
Shepherd's Purse, 518
SHIELD FERN FAMILY, 52
Shooting Star, 747
Single Delight, 714
Silverberry, 684
Silverweed, 621
  Pacific, 623
Skullcap, 784
Skunk Cabbage, Yellow, 281
Smartweed, Water, 387
Snapdragon, 793
Sneezeweed, 887
Snowberry, 842
Snowflake, 788
Soapberry, 684
Solomon's Seal, False, 309
Sorrel, 374
  Garden, 377
  Mountain, 383
  Sheep, 375
Sow Thistle, Common, 951
  Field, 950
  Spiny-Leaved, 951
Spearmint, 791
Spearwort, Creeping, 474
Speedwell, 797
  Water, 798
Spiderplant, 569
SPIKEMOSS FAMILY, 31
Spikemoss, 31
Spiraea, Alaska, 594
SPLEENWORT FAMILY, 46
Spleenwort, 46
Spring Beauty, 404
Spruce:
  Black, 62
  Sitka, 61
  White, 61
Spurry, 438
  Sand, 439
Starflower, 750
Stickseed, 772
STONECROP FAMILY, 560
Stonecrop, 560
Storksbill, 676
Strawberry, Beach, 606
Sugar-Scoop, 584
SUNDEW FAMILY, 559
Sundew, 558
Sunflower, 883
Surfgrass, 70
  Scouler's, 70
Sweet Cicely, 697
Synthyris, Alaska, 805

Tamarack, 60
Tansy, Common, 891

Index of Common Names

Tarweed, 885
Thimbleberry, 605
Thistle:
  Bull, 939
  Canada, 939
Thoroughwax, 698
Timothy, 89
  Mountain, 88
TOUCH-ME-NOT
  FAMILY, 679
Touch-Me-Not, 679
Thrift, 752
Tumbleweed, 403
Turnip, 505
Twinflower, 843
  American, 743
  Pacific, 844

Twisted-Stalk, 311
Twyblade, 326

Umbrella Plant, 374

VALERIAN FAMILY, 845
Valerian, 845
Vetch, 669
  Milk, 646
Vetchling, 673
VIOLET FAMILY, 680
Violet, 680
  Marsh, 683
  Western Dog, 682

Water Blinks, 410
Water Carpet,
  Northern, 587
Water Hemlock, 699
WATERLEAF FAMILY, 769
WATER LILY FAMILY, 449
Water Lily, Dwarf, 449
WATER MILFOIL FAMILY, 694
Water Milfoil, 694
WATER PLANTAIN FAMILY, 81
Water Shield, 449

WATER STARWORT FAMILY, 677
Water Starwort, 677
WAX MYRTLE FAMILY, 364
Wheat, 191
Wheatgrass, 181
Wild Flag, 313
WILLOW FAMILY, 331
Willow, 333
  Alaska, 356
  Arctic, 340
  Long-Beaked, 358
  Netted, 336
  Sitka, 361
Willow Herb, 685

WINTERGREEN FAMILY, 710
Wintergreen, 710
Wormwood, 896

Yarrow, 886
  Common, 888
Yellow Rattle, 815
Yerba Buena, 790
YEW FAMILY, 59
Yew, Western, 59
Youth-on-Age, 585

Index of Botanical Names

A

Abies, 63
  amabilis, 63
  lasiocarpa, 64
Acer, 679
  glabrum Douglasii, 679
ACERACEAE, 679
Achillea, 886
  borealis, 888
  lanulosa, 889
  millefolium, 888
  ptarmica, 887
  sibirica, 887
Aconitum, 459
  delphinifolium:
    albiflorum, 459
    Chamissonianum, 460
    delphinifolium, 459
    paradoxum, 460
  maximum, 461
Acorus, 280
  calamus, 280
Actaea, 456
  erythrocarpa, 456
  rubra:
    arguta, 456
    rubra, 456
ADIANTACEAE, 42
Adiantum, 42
  pedatum:
    aleuticum, 42
    pedatum, 42
Adoxa, 845
  moschatellina, 845
ADOXACEAE, 845
Agoseris, 953
  aurantiaca, 953
  glauca, 953
Agropyron, 181
  boreale:
    alaskanum, 188
    boreale, 187
    hyperarcticum, 187
  macrourum, 185
  pauciflorum:
    majus, 190
    novae-angliae, 189
    pauciflorum, 188
    teslinense, 189
  pectinatiforme, 182
  repens, 183
  Smithii, 183

[Agropyron]
  spicatum, 184
  subsecundum, 186
  violaceum:
    alboviride, 185
    andinum, 186
    violaceum, 185
  yukonense, 184
Agrostemma, 439
  githago, 439
Agrostis, 97
  alaskana, 99
  borealis, 98
  canina, 98
  clavata, 101
  exarata, 101
    purpurascens, 101
  geminata, 102
  gigantea, 100
  scabra, 102
  stolonifera, 100
  tenuis, 99
  Trinii, 98
Aira, 109
  caryophyllea, 109
ALISMACEAE, 81
Allium, 307
  schoenoprasum sibiricum, 307
  victoralis platyphyllum, 307
Alnus, 368
  crispa:
    crispa, 368
    laciniata, 369
    sinuata, 369
    strangula, 368
  incana:
    incana, 370
    tenuifolia, 370
  oregona, 369
  rugosa, 370
Alopecurus, 89
  aequalis, 91
  amurensis, 91
  natans, 91
  alpinus:
    alpinus, 90
    glaucus, 90
    Stejnegeri, 91
  geniculatus, 92
  pratensis, 89
Alyssum, 552
  americanum, 552

AMARANTHACEAE, 403
Amaranthus, 403
  graecizans, 403
  retroflexus, 403
Amelanchier, 599
  alnifolia, 599
  florida, 599
Amerorchis, 318
  rotundifolia, 318
Amsinckia, 778
  lycopsoides, 778
  Menziesii, 778
Anaphalis, 882
  margaritacea, 882
    angustior, 882
    subalpina, 882
Andromeda, 727
  polifolia, 727
    acerosa, 727
    glaucophylla, 727
Androsace, 744
  alaskana, 746
  chamaejasme:
    Andersoni, 745
    Lehmanniana, 744
  filiformis, 746
  septentrionalis, 745
Anemone, 461
  deltoidea, 462
  Drummondii, 466
  multifida, 465
  narcissiflora:
    alaskana, 465
    interior, 464
    sibirica, 463
    villosissima, 464
  parviflora, 463
    grandiflora, 463
  Richardsonii, 462
Angelica, 705
  genuflexa, 705
    refracta, 705
  lucida, 705
Antennaria, 871
  alborosea, 880
  alpina, 878
    canescens, 878
  atriceps, 878
  campestris, 881
  carpatica, 873
  dioica, 879
  Friesiana:
    alaskana, 877

[Antennaria]
    compacta, 877
    Friesiana, 876
    megacephala, 876
  isolepis, 876
  leuchippii, 881
  media, 878
  monocephala:
    angustata, 875
    exilis, 874
    latisquamea, 874
    monocephala, 873
    philonipha, 874
  neglecta Howellii, 881
  pallida, 875
  pulcherrima, 872
    angustisquama, 872
  rosea, 880
    nitida, 880
  stolonifera, 879
Anthemis, 885
  cotula, 886
  tinctoria, 885
Anthoxanthum, 83
  odoratum, 83
Antirrhinum, 793
  orontium, 793
Apargidium, 943
  boreale, 943
Aphragmus, 500
  Eschscholtzianus, 500
APOCYNACEAE, 763
Apocynum, 763
  androsaemifolium, 763
Aquilegia, 457
  brevistyla, 457
  formosa, 457
  megalantha, 457
Arabis, 544
  arenicola, 544
    pubescens, 544
  divaricarpa, 548
  Drummondii, 548
  hirsuta:
    Eschscholtziana, 547
    hirsuta, 547
    pycnocarpa, 547
    Stelleri, 547
  Holboellii, 549
    pendulocarpa, 549
    retrofracta, 549
  Lemmoni, 546

Index of Botanical Names

[Arabis]
  Lyallii, 546
  lyrata:
    kamchatica, 545
    lyrata, 545
    Nuttallii, 545
ARACEAE, 280
ARALIACEAE, 696
Arceuthobium, 372
  campylopodum, 372
  tsugensis, 372
Arctagrostis, 93
  latifolia:
    angustifolia, 94
    arundinacea, 94
    latifolia, 94
    poaeoides, 95
Arctophila, 149
  fulva, 149
Arctostaphylos, 729
  alpina, 730
  rubra, 730
  uva-ursi:
    adenotricha, 729
    pacifica, 729
    uva-ursi, 729
Arenaria, 435
  capillaris, 435
    americana, 435
    capillaris, 435
  Chamissonis, 436
  humifusa, 436
  longipedunculata, 435
  tschuktschorum, 435
Armeria, 752
  maritima:
    arctica, 752
    sibirica, 752
Armoracia, 511
  rusticana, 511
Arnica, 915
  alpina:
    alpina, 923
    angustifolia, 923
    attenuata, 923
    linearis, 923
    tomentosa, 924
    vestita, 923
  amplexicaulis:
    amplexicaulis, 919
    prima, 919
  Chamissonis:
    Chamissonis, 920
    foliolosa, 921
    incana, 921
  cordifolia, 918
  diversifolia, 921
  frigida, 917
  latifolia, 918
  Lessingii:
    Lessingii, 916
    Norbergii, 916
  lonchophylla, 922
    arnoglossa, 922
    chionopappa, 922
  louiseana, 917
  mollis, 920
  Parryi:
    Parryi, 922
    Sonnei, 922
  sachalinensis, 920
  unalaschcensis, 917
    Tschonoskyi, 917
  Whitneyi, 918

Arrhenatherum, 121
  elatius, 121
Artemisia, 896
  alaskana, 905
  aleutica, 911
  arctica:
    arctica, 906
    beringensis, 906
    comata, 907
  biennis, 909
  borealis, 910
  cana, 900
  canadensis, 909
  dracunculus, 899
  frigida, 904
  furcata, 910
    heterophylla, 910
  globularia, 897
    lutea, 897
  glomerata, 898
    subglabrata, 898
  laciniata, 907
  laciniatiformis, 908
  lagocephala, 900
    Kruhsiana, 900
  ludoviciana gnaphalodes, 899
  Michauxiana, 905
  rupestris Woodii, 908
  senjavinensis, 898
  Stelleriana, 901
  Tilesii:
    elatior, 902
    Gormani, 902
    Tilesii, 901
    unalaschcensis, 903
  unalaskensis:
    aleutica, 904
    unalaskensis, 903
  vulgaris, 930
Aruncus, 595
  sylvester, 595
Asperugo, 779
  procumbens, 779
ASPIDIACEAE, 52
ASPLENIACEAE, 46
Asplenium, 46
  trichomanes, 46
  viride, 47
Aster, 856
  alpinus Vierhapperi, 856
  brachyactis, 860
  ciliatus, 860
  commutatus, 860
  junciformis, 859
  laevis, 858
  meritus, 857
  modestus, 858
  pygmaeus, 859
  sibiricus, 857
  subspicatus, 857
  yukonensis, 859
Astragalus, 646
  aboriginum, 649
  adsurgens:
    adsurgens, 651
    robustior, 651
    viciifolius, 652
  agrestis, 652
  alpinus:
    alaskanus, 649
    alpinus, 649
    arcticus, 650
    Brunetianus, 649

[Astragalus]
  americanus, 647
  australis, 649
  Bodinii, 655
  danicus, 652
  eucosmus:
    eucosmus, 648
    Sealei, 648
  frigidus, 647
  norvegicus, 648
  nutzotinensis, 651
  polaris, 650
  Robbinsii:
    Harringtonii, 653
    minor, 653
  tenellus, 654
  umbellatus, 647
  Williamsii, 654
ATHYRIACEAE, 47
Athyrium, 47
  distentifolium
    americanum, 47
  filix-femina:
    cyclosorum, 48
    filix-femina, 48
    Hillii, 48
Atriplex, 397
  alaskensis, 399
  drymarioides, 397
  Gmelini, 398
  hortensis, 399
  patula, 398
Avena, 119
  fatua, 119
  Hookeri, 120
  sativa, 120

B

BALSAMINACEAE, 679
Barbarea, 506
  orthoceras, 506
  stricta, 506
Beckmannia, 122
  erucaeformis:
    baicalensis, 122
    erucaeformis, 122
    syzigachne, 122
Bellis, 855
  perennis, 855
Betula, 364
  glandulosa, 365
  kenaica, 366
  minor, 366
  nana:
    exilis, 365
    nana, 365
    "occidentalis," 366
  papyrifera:
    commutata, 367
    humilis, 367
BETULACEAE, 364
Bidens, 884
  cernua, 884
  frondosa, 884
BLECHNACEAE, 58
Blechnum, 58
  spicant, 58
Blysmus, 209
  rufus, 209
BORAGINACEAE, 772
Boschniakia, 828
  rossica, 828

Botrychium, 39
  boreale, 40
    obtusilobum, 40
  lanceolatum, 41
  lunaria, 40
    minganense, 40
  multifidum robustum, 41
  virginianum europaeum, 42
Boykinia, 562
  Richardsonii, 562
Brasenia, 449
  Schreberi, 449
Brassica, 504
  juncea, 504
  napus, 505
  rapa, 505
Braya, 553
  Bartlettiana, 556
    vestita, 556
  Henryae, 555
  humilis:
    arctica, 554
    humilis, 554
    Richardsonii, 553
  pilosa, 555
  purpurascens, 554
Bromus, 172
  brizaeformis, 176
  ciliatus, 173
  commutatus, 179
  hordeaceus, 178
  inermis, 174
  marginatus, 177
  pacificus, 176
  Pumpellianus:
    arcticus, 175
    Pumpellianus, 174
    villosissimus, 175
  racemosus, 179
  Richardsonii, 173
  secalinus, 178
  sitchensis, 177
  tectorum, 172
Bupleurum, 698
  triradiatum:
    arcticum, 698
    triradiatum, 698

C

Cacalia, 924
  auriculata:
    auriculata, 924
    kamtschatica, 924
Cakile, 503
  edentula:
    californica, 503
    edentula, 503
Calamagrostis, 103
  canadensis:
    canadensis, 104
    Langsdorffii, 104
  deschampsioides, 108
  Holmii, 107
  inexpansa, 105
  lapponica, 106
  neglecta, 106
  nutkaënsis, 105
  purpurascens:
    arctica, 108
    purpurascens, 107
Calla, 281
  palustris, 281

CALLITRICHACEAE, 677
Callitriche, 677
  anceps, 678
  hermaphroditica, 677
  heterophylla Bolanderi, 678
  subanceps, 678
  verna, 677
Caltha, 451
  biflora, 452
  leptosepala, 452
  natans, 453
  palustris:
    arctica, 453
    asarifolia, 454
Calypso, 331
  bulbosa, 331
Camelina, 519
  microcarpa, 519
  sativa, 519
Campanula, 847
  aurita, 849
  Chamissonis, 848
  lasiocarpa:
    lasiocarpa, 847
    latisepala, 848
  latisepala, 850
  Piperi, 849
  rotundifolia, 850
  Scouleri, 849
  uniflora, 851
CAMPANULACEAE, 847
Capsella, 518
  bursa-pastoris, 518
  rubella, 518
CAPRIFOLIACEAE, 841
Cardamine, 512
  angulata, 513
  bellidifolia, 512
    pinnatifida, 512
  hyperborea, 516
  microphylla, 516
  pennsylvanica, 513
  pratensis:
    angustifolia, 514
    pratensis, 514
  purpurea, 517
    albiflora, 517
  Regeliana, 515
  umbellata, 514
  Victoris, 515
Carex, 215
  adelostoma, 257
  aenea, 237
  albo-nigra, 259
  amblyorhyncha, 239
  anthoxanthea, 225
  aquatilis:
    aquatilis, 250
    stans, 251
  arcta, 245
  atherodes, 280
  atrata:
    atrata, 259
    atrosquamea, 260
  atratiformis Raymondii, 260
  atrofusca, 272
    major, 272
  atrostachya, 232
  aurea, 254
  Bebbii, 234
  bicolor, 253
  Bigelowii, 248

[Carex]
  bonanzensis, 242
  brunnescens:
    alaskana, 242
    pacifica, 243
  Buxbaumii, 257
  canescens, 241
  capillaris, 274
    major, 274
  capitata, 222
  chordorrhiza, 229
  circinnata, 226
  concinna, 267
  Crawfordii, 235
  deflexa, 262
  Deweyana, 247
  diandra, 231
  dioica:
    dioica, 222
    gynocrates, 222
  disperma, 243
  eburnea, 268
  echinata, 246
  eleusinoides, 249
  Enanderi, 261
  filifolia, 223
  flava, 275
  foena, 235
  Franklinii, 273
  Garberi:
    bifaria, 254
    Garberi, 254
  glacialis, 267
  glareosa:
    glareosa, 239
    pribilovensis, 240
  Gmelini, 258
  heleonastes, 238
  holostoma, 255
  interior, 245
  Jacobi-Peteri, 221
  Kelloggii, 250
  Krausei, 274
  Kreczetoviczii, 246
  Lachenalii, 238
  laeviculmis, 246
  lanuginosa, 279
  lapponica, 241
  lasiocarpa:
    americana, 279
    lasiocarpa, 279
    occultans, 279
  laxa, 270
  leptalea, 225
    Harperi, 225
    pacifica, 225
  limosa, 269
  livida, 271
    Grayana, 271
  loliacea, 244
  lugens, 249
  Lyngbyaei, 253
  Mackenziei, 240
  macloviana pachystachya, 233
  macrocephala, 231
  macrochaeta, 262
  magellanica irrigua, 270
  maritima, 228
  media, 255
  melanocarpa, 263
  "melanostoma," 264
  membranacea, 278
  Mertensii, 261

[Carex]
  urostachys, 261
  microchaeta, 264
  microglochin, 227
  microptera, 236
  miliaris, 278
  misandra, 272
  nardina, 220
  nesophila, 265
  nigricans, 227
  obtusata, 224
  Oederi viridula, 276
  Parryana, 256
  pauciflora, 228
  Peckii, 266
  petasata, 236
  petricosa, 273
  phaeocephala, 234
  phyllomanica, 247
  pluriflora, 269
  podocarpa, 263
  praegracilis, 230
  praticola, 237
  Preslii, 233
  pyrenaica micropoda, 226
  Ramenskii, 252
  rariflora, 268
  rhynchophysa, 277
  Rossii, 265
  rostrata, 276
  rotundata, 278
  rupestris, 224
  sabulosa, 256
  saxatilis:
    laxa, 277
    saxatilis, 277
  scirpoidea, 223
  scopulorum, 248
  sitchensis, 251
  spectabilis, 264
  stenophylla eleocharis, 229
  stipata, 230
  stylosa, 258
    nigritella, 258
  subspathacea, 252
  supina spaniocarpa, 266
  sychnocephala, 232
  tenuiflora, 244
  terraenovae, 267
  ursina, 221
  vaginata, 271
  Williamsii, 275
CARYOPHYLLACEAE, 411
Cassiope, 724
  lycopoides, 726
  Mertensiana, 725
  Stelleriana, 726
  tetragona:
    saximontana, 725
    tetragona, 724
Castilleja, 806
  annua, 810
  caudata, 809
    auricoma, 809
  chrymactis, 808
  elegans, 811
  Henryae, 813
  hyetophila, 808
  hyperborea, 812
  miniata, 807
  Muelleri, 811
  parviflora, 812
  Raupii, 809
  sulphurea, 807

[Castilleja]
  unalaschcensis, 807
  villosissima, 811
  yukonis, 810
Catabrosa, 124
  aquatica, 124
Cerastium, 420
  aleuticum, 422
  arvense, 425
  Beeringianum:
    Beeringianum, 421
    grandiflorum, 422
  Fischerianum, 424
  fontanum triviale, 425
  glomeratum, 424
  jenisejense, 423
  maximum, 421
  Regelii, 423
CERATOPHYLLACEAE, 451
Ceratophyllum, 451
  demersum, 451
Chamaecyparis, 65
  nootkatensis, 65
Chamaedaphne, 727
  calyculata, 727
  nana, 724
Chamaerhodos, 624
  erecta Nuttallii, 624
CHENOPODIACEAE, 392
Chenopodium, 392
  album, 395
  Berlandieri Zschackei, 395
  capitatum, 393
  glaucum:
    pulchrum, 394
    salinum, 394
  hybridum:
    giganthospermum, 393
    hybridum, 393
  leptophyllum, 396
  rubrum, 394
Chimaphila, 710
  umbellata:
    acuta, 710
    cisatlantica, 710
    mexicana, 710
    occidentalis, 710
    umbellata, 710
Christolea, 557
  parryoides, 557
Chrysanthemum, 891
  arcticum:
    arcticum, 893
    polare, 893
  bipinnatum:
    bipinnatum, 892
    huronense, 892
  integrifolium, 895
  leucanthemum, 894
  segetum, 894
  vulgare, 891
Chrysosplenium, 587
  kamtschaticum, 588
  tetrandrum, 587
  Wrightii, 587
Cicuta, 699
  bulbifora, 699
  Douglasii, 699
  mackenzieana, 700
Cinna, 95
  latifolia, 95
Circaea, 693
  alpina, 693

Index of Botanical Names

Cirsium, 939
  arvense, 939
  edule, 940
  foliosum, 940
  kamtschaticum, 941
    pectinellum, 941
    Weyrichii, 941
  vulgare, 939
Cladothamnus, 717
  pyrolaeflorus, 717
Claytonia, 404
  acutifolia graminifolia, 405
  arctica, 407
  Bostockii, 406
  Chamissoi, 409
  megarhiza, 404
  parvifolia flagellaris, 406
  perfoliata, 409
  sarmentosa, 408
  Scammaniana, 408
  sibirica, 407
  tuberosa, 405
    czukczorum, 405
Clintonia, 309
  uniflora, 309
Cnidium, 701
  ajanense, 701
  cnidiifolium, 702
Cochlearia, 499
  officinalis:
    arctica, 499
    oblongifolia, 500
  sessilifolia, 499
Coeloglossum, 319
  viride:
    bracteatum, 319
    islandicum, 319
Collinsia, 794
  parviflora, 794
Collomia, 765
  linearis, 765
Colpodium, 148
  Vahlianum, 148
  Wrightii, 148
Comandra, 372
  umbellata pallida, 372
COMPOSITAE, 851
Conioselinum, 704
  chinense, 704
Coptis, 455
  aspleniifolia, 455
  groenlandica, 455
  trifolia, 455
Corallorrhiza, 329
  maculata:
    maculata, 329
    Mertensiana, 329
  trifida, 329
Corispermum, 400
  hyssopifolium, 400
CORNACEAE, 707
Cornus, 707
  canadensis, 709
  × suecica, 709
  stolonifera, 708
    Baileyi, 708
  suecica, 708
Corydalis, 495
  aurea, 495
    occidentalis, 495
  Emanueli, 495
  pauciflora, 495
    albiflora, 495
  sempervirens, 496

Cotula, 895
  coronopifolia, 895
Crassula, 561
  aquatica, 561
CRASSULACEAE, 560
Crataegus, 600
  Douglasii, 600
Crepis, 954
  capillaris, 955
  chrysantha, 957
  elegans, 956
  nana:
    nana, 955
    lyratifolia, 956
  tectorum, 954
CRUCIFERAE, 496
Cryptantha, 777
  spiculifera, 777
  Torreyana, 777
Cryptogramma, 44
  crispa:
    achrostichoides, 44
    Raddeana, 44
    sitchensis, 44
  Stelleri, 45
CRYPTOGRAMMACEAE, 44
CUPRESSACEAE, 64
CYPERACEAE, 197
Cypripedium, 315
  calceolus parviflorum, 316
  guttatum:
    guttatum, 315
    Yatabeanum, 316
  montanum, 317
  passerinum, 317
Cystopteris, 48
  diaphana, 49
  fragilis:
    Dickieana, 49
    fragilis, 49
  montana, 50

D

Dactylis, 126
  glomerata, 126
Dactylorhiza, 318
  aristata, 318
Danthonia, 121
  intermedia, 121
  spicata, 122
    pinetorum, 122
Delphinium, 458
  brachycentrum, 458
    pallidum, 458
  glaucum, 459
  nutans, 458
Deschampsia, 110
  beringensis, 114
  brevifolia, 113
  caespitosa:
    caespitosa, 111
    glauca, 112
    orientalis, 112
    danthonioides, 111
  elongata, 110
  flexuosa, 114
    Mackenzieana, 114
    obensis, 114
    pumila, 113
Descurainia, 541
  pinnata filipes, 542
  Richardsonii, 542
  sophia, 541

[Descurainia]
  sophioides, 541
Dianthus, 448
  repens, 448
Diapensia, 736
  lapponica:
    lapponica, 736
    obovata, 736
    rosea, 736
DIAPENSIACEAE, 736
Dicentra, 494
  peregrina, 494
Digitalis, 806
  purpurea, 806
Dodecatheon, 747
  frigidum, 749
  Jeffreyi, 749
  pulchellum:
    alaskanum, 748
    pauciflorum, 748
    superbum, 747
Douglasia, 742
  arctica, 743
  Gormani, 743
  ochotensis, 742
Draba, 520
  aleutica, 531
  alpina, 529
  aurea, 536
  barbata, 529
  borealis, 534
  caesia, 524
  Chamissonis, 534
  cinerea, 537
  crassifolia, 522
  densifolia, 530
  Eschscholtzii, 525
  exalata, 524
  fladnizensis, 527
  hirta, 532
  hyperborea, 538
  incerta, 525
  kamtschatica, 533
  lactea, 527
  lanceolata, 536
  lonchocarpa, 523
  longipes, 533
  macrocarpa, 530
  maxima, 535
  micropetala, 528
  nemorosa, 537
  nivalis, 523
  ogilviensis, 538
  oligosperma, 522
  pilosa, 526
  praealta, 535
  pseudopilosa, 526
  stenoloba, 532
  stenopetala, 531
    hebecarpa, 531
    leiocarpa, 531
    purpurea, 531
  subcapitata, 528
Dracocephalum, 786
  palmatum, 786
  parviflorum, 787
Drosera, 559
  anglica, 559
  rotundifolia, 559
  gracilis, 559
DROSERACEAE, 559
Dryas, 629
  Drummondii, 629
    eglandulosa, 629

[Dryas]
    tomentosa, 629
  integrifolia:
    canescens, 631
    integrifolia, 631
    sylvatica, 632
  octopetala:
    alaskensis, 631
    argentea, 630
    glabrata, 631
    kamtschatica, 630
    luteola, 630
    octopetala, 630
    viscida, 630
Dryopteris, 55
  dilatata americana, 55
  fragrans:
    fragrans, 56
    remotiuscula, 56
Dupontia, 140
  Fischeri:
    Fischeri, 149
    psilosantha, 150

E

Echinopanax, 696
  horridum, 696
    japonicus, 696
ELAEAGNACEAE, 684
Elaeagnus, 684
  commutata, 684
Eleocharis, 210
  acicularis, 212
    longiseta, 212
  kamtschatica, 211
  nitida, 212
  palustris, 210
  quinqueflora, 211
    Fernaldii, 211
  uniglumis, 210
Elymus, 193
  arenarius:
    arenarius, 193
    mollis, 193
    villosissimus, 194
  glaucus:
    glaucus, 195
    virescens, 196
  hirsutus, 196
  innovatus, 195
  velutinus, 195
  interior, 194
  sibiricus, 197
EMPETRACEAE, 716
Empetrum, 716
  nigrum:
    hermaphroditum, 716
    japonicum, 716
    nigrum, 716
Epilobium, 685
  adenocaulon, 691
  anagallidifolium, 689
    pseudo-scaposum, 689
  angustifolium:
    angustifolium, 686
    macrophyllum, 686
  behringianum, 691
  davuricum, 689
    arcticum, 689
  glandulosum, 690
  Hornemannii, 692
  lactiflorum, 692
  latifolium, 687

[Epilobium]
  leptocarpum, 690
  Macounii, 690
  leptophyllum, 688
  luteum, 687
  palustre, 688
  sertulatum, 693
EQUISETACEAE, 34
Equisetum, 34
  arvense, 38
    boreale, 38
  Calderi, 38
  fluviatile, 36
    verticillatum, 36
  hiemale californicum, 34
  palustre, 37
  pratense, 38
  scirpoides, 36
  silvaticum, 37
    pauciramosum, 37
  variegatum:
    alaskanum, 35
    anceps, 35
    variegatum, 35
ERICACEAE, 717
Erigeron, 861
  acris:
    debilis, 867
    kamtschaticus, 867
    politus, 866
  caespitosus, 870
  compositus, 862
    discoideus, 862
    glabratus, 862
  elatus, 868
  eriocephalus, 864
  glabellus:
    glabellus, 870
    pubescens, 870
    yukonensis, 870
  grandiflorus:
    grandiflorus, 865
    Muirii, 866
  humilis, 864
  hyperboreus, 865
  hyssopifolius, 862
  lonchophyllus, 868
  peregrinus:
    callianthemus, 869
    Dawsonii, 869
    peregrinus, 869
  philadelphicus, 871
  pumilus, 863
    intermedius, 863
  purpuratus, 863
Eriogonum, 374
  flavum aquilinum, 374
Eriophorum, 197
  angustifolium:
    coloratum, 198
    giganteus, 198
    scabriusculum, 199
    subarcticum, 198
    triste, 199
  brachyantherum, 204
  callitrix, 202
    moravium, 202
  gracile, 200
  russeolum:
    albidum, 203
    majus, 202
    rufescens, 203

[Eriophorum]
    russeolum, 202
  Scheuchzeri:
    Scheuchzeri, 201
    tenuifolium, 201
  vaginatum:
    spissum, 205
    vaginatum, 204
    viridi-carinatum, 200
Eritrichium, 773
  aretioides, 773
  Chamissonis, 774
  splendens, 775
  villosum, 774
Erodium, 676
  cicutarium, 676
Erysimum, 549
  angustatum, 552
  cheiranthoides:
    altum, 551
    cheiranthoides, 550
    inconspicuum, 551
    Pallasii, 550
    ochroleucum, 550
Euphrasia, 814
  disjuncta, 814
  hudsonica, 814
  mollis, 814
Eurotia, 400
  lanata, 400
Eutrema, 501
  Edwardsii, 501

F

Fauria, 762
  crista-galli, 762
  japonica, 762
Festuca, 166
  altaica, 167
    pallida, 167
  arundinacea, 166
  baffinensis, 169
  brachyphylla, 168
  ovina:
    alaskensis, 170
    pallida, 170
  prolifera lasiolepis, 168
  rubra, 170
    aucta, 171
  saxmontana, 169
  scabrella, 167
  subulata, 167
  "vivipara," 168
Fragaria, 606
  chiloensis pacifica, 606
  virginiana:
    glauca, 606
    terrae-novae, 606
Fritillaria, 308
  camschatcensis, 308
  alpina, 308
FUMARIACEAE, 494

G

Galeopsis, 788
  bifida, 788
  tetrahit, 788
Galium, 836
  aparine, 838
  boreale, 837
  Brandegei, 841

[Galium]
  kamtschaticum oreganum, 839
  palustre, 839
  scabrum, 839
  trifidum:
    columbianum, 840
    trifidum, 840
  triflorum, 838
  verum, 837
Gaultheria, 728
  Miqueliana, 728
  shallon, 728
Gentiana, 753
  aleutica, 760
  algida, 754
  amarella:
    acuta, 759
    plebeja, 759
  auriculata, 758
  barbata, 755
  detonsa, 755
  Douglasiana, 757
  glauca, 757
  platypetala, 754
  propinqua:
    arctophila, 761
    propinqua, 760
  prostrata, 758
  Raupii, 756
  tenella, 756
GENTIANACEAE, 753
Geocaulon, 372
  lividum, 373
GERANIACEAE, 674
Geranium, 674
  albiflorum, 674
  Bicknellii, 675
  erianthum, 674
  Richardsonii, 674
  Robertianum, 675
Geum, 625
  aleppicum:
    aleppicum, 626
    strictum, 626
  calthifolium, 627
  glaciale, 628
  macrophyllum:
    macrophyllum, 625
    perincisum, 626
    sachalinene, 625
  pentapetalum, 628
  Rossii, 627
Gilia, 766
  achilleaefolia, 767
  capitata, 766
Glaux, 752
  maritima, 752
  obtusifolia, 752
Glechoma, 786
  hederacea, 786
Glehnia, 706
  littoralis:
    leiocarpa, 706
    littoralis, 706
Glyceria, 150
  borealis, 151
  leptostachya, 151
  maxima:
    grandis, 153
    maxima, 153
  pauciflora, 152
  pulchella, 152

[Glyceria]
  striata:
    striata, 153
    stricta, 153
Gnaphalium, 882
  uliginosum, 882
Goodyera, 328
  oblongifolia, 328
  repens ophioides, 328
GRAMINEAE, 82
Gymnocarpium, 56
  dryopteris, 56
  robertianum, 57

H

Hackelia, 773
  Jessicae, 773
Halimolobus, 543
  mollis, 543
HALORAGACEAE, 694
Hammarbya, 330
  paludosa, 330
Haplopappus, 854
  MacLeanii, 854
  stenophyllus, 854
Hedysarum, 667
  alpinum:
    alpinum, 668
    americanum, 668
    dasycarpum, 667
    hedysaroides, 668
    Mackenzii, 667
Helianthus, 883
  annuus, 883
  laetiflorus subrhomboideus, 883
Heracleum, 707
  lanatum, 707
Hesperis, 556
  matronalis, 556
  sibirica, 556
Heuchera, 584
  glabra, 584
Hieracium, 958
  albiflorum, 960
  aurantiacum, 959
  canadense, 960
  gracile:
    alaskanum, 959
    gracile, 959
    yukonensis, 958
    scabriusculum, 960
  triste, 958
    fulvum, 958
    tritiforme, 958
    umbellatum, 960
Hierochloë, 84
  alpina, 84
  monticola, 84
  odorata, 84
    dahurica, 84
    orthantha, 84
    pauciflora, 85
Hippuris, 695
  montana, 696
  tetraphylla, 695
  vulgaris, 695
Holcus, 109
  lanatus, 109
Honckenya, 434
  peploides:
    diffusa, 434

Index of Botanical Names

[Honckenya]
    major, 434
    peploides, 434
    robusta, 434
Hordeum, 191
    brachyantherum, 191
    jubatum, 192
    vulgare, 192
HYDROPHYLLACEAE, 769
HYMENOPHYLLACEAE, 43
Hypochoeris, 942
    radicata, 942
HYPOLEPIDACEAE, 43

I

Impatiens, 679
    capensis, 679
    noli-tangere, 679
IRIDACEAE, 313
Iris, 313
    Hookeri, 313
    setosa:
        interior, 313
        platyrhyncha, 313
        setosa, 313
ISOËTACEAE, 32
Isoëtes, 32
    muricata:
        Braunii, 32
        maritima, 33
    setacea, 32
    truncata, 33

J

JUNCACEAE, 283
JUNCAGINACEAE, 80
Juncus, 283
    alpinus, 292
        alpestris, 292
        alpinus, 292
        nodulosus, 292
    arcticus:
        alaskanus, 286
        ater, 286
        sitchensis, 285
    articulatus, 292
        Turczanninovii, 292
    biglumis, 294
    bufonius, 295
        ranarius, 295
    castaneus:
        castaneus, 289
        leucochlamys, 289
        pallidus, 289
    Drummondii, 287
    effusus, 285
        ensifolius, 287
    falcatus:
        prominens, 288
        sitchensis, 288
    filiformis, 284
    Leschenaultii, 291
    Mertensianus, 288
    nodosus, 290
    oreganus, 290
    prismatocarpus, 291
    stygius americanus, 294
    tenuis, 291
    triglumis:
        albescens, 293
        triglumis, 293

Juniperus, 65
    communis:
        depressa, 65
        nana, 65
    horizontalis, 66

K

Kalmia, 721
    polifolia:
        microphylla, 721
        polifolia, 721
Kobresia, 213
    myosuroides, 213
    simpliciuscula, 214
    sibirica, 214
Koeleria, 123
    asiatica, 123
    gracilis, 124
Koenigia, 373
    islandica, 373

L

LABIATAE, 784
Lactuca, 952
    biennis, 952
    tatarica, 952
        pulchella, 952
        tatarica, 952
Lagotis, 804
    glauca:
        glauca, 804
        minor, 805
        Hultenii, 805
Lamium, 788
    album, 788
Lappula, 772
    myosotis, 772
    occidentalis, 772
Lapsana, 941
    communis, 941
Larix, 60
    laricina alaskensis, 60
Lathyrus, 672
    maritimus:
        maritimus, 672
        pubescens, 673
    palustris pilosus, 673
    venosus, 672
Ledum, 717
    palustre:
        decumbens, 717
        groenlandicum, 717
        palustre, 717
LEGUMINOSAE, 635
Lemna, 282
    minor, 282
    trisulca, 282
LEMNACEAE, 282
LENTIBULARIACEAE, 829
Leontodon, 942
    autumnalis, 942
Lepidium, 497
    densiflorum, 497
        Bourgeauanum, 497
    sativum, 497
    virginicum, 498
Leptarrhena, 562
    pyrolifolia, 562
Lesquerella, 517
    arctica, 517
    Purshii, 517

Lewisia, 410
    pygmaea, 410
Ligusticum, 702
    mutellinoides:
        alpinum, 703
        mutellinoides, 703
    scoticum:
        Hultenii, 702
        scoticum, 702
LILIACEAE, 304
Limosella, 797
    aquatica, 797
LINACEAE, 676
Linaria, 793
    vulgaris, 793
Linnaea, 843
    borealis:
        americana, 843
        borealis, 843
        longiflora, 844
Linum, 676
    perenne:
        Lewisii, 676
        perenne, 676
Listera, 326
    borealis, 327
    brevidens, 326
    caurina, 326
    convallarioides, 326
    cordata, 327
        nephrophylla, 327
    Yatabei, 326
Lloydia, 308
    serotina, 308
    flava, 308
Loiseleuria, 720
    procumbens, 720
Lolium, 180
    multiflorum, 180
    perenne, 180
    temulentum, 181
Lomatogonium, 761
    rotatum, 761
Lonicera, 844
    involucrata, 844
LORANTHACEAE, 372
Luetkea, 595
    pectinata, 595
Lupinus, 635
    arcticus, 636
        subalpinus, 636
    Kuschei, 637
    lepidus, 637
    nootkatensis, 636
        fruticosus, 636
        leucanthus, 636
        unalaschcensis, 636
    polyphyllus, 635
Luzula, 295
    arctica, 300
    arcuata:
        arcuata, 299
        unalaschcensis, 299
    confusa, 301
    multiflora:
        comosa, 303
        frigida, 301
        Kjellmaniana, 302
        Kobayasii, 302
        multiflora, 301
    parviflora:
        divaricata, 298
        melanocarpa, 298

[Luzula]
    parviflora, 298
    rufescens, 296
    spicata, 303
    tundricola, 300
    Wahlenbergii:
        Piperi, 297
        Wahlenbergii, 297
        yezoensis, 297
LYCOPODIACEAE, 24
Lycopodium, 24
    alpinum, 30
    annotinum:
        alpestre, 27
        annotinum, 27
        pungens, 27
    clavatum:
        clavatum, 28
        integerrimum, 28
        monostachyon, 29
    complanatum, 29
    inundatum, 26
    obscurum dendroideum, 28
    sabinaefolium sitchense, 30
    selago:
        appressum, 25
        chinense, 26
        selago, 25
Lycopus, 791
    lucidus:
        americanus, 790
        lucidus, 790
    uniflorus, 791
Lysichiton, 281
    americanum, 281
Lysimachia, 750
    thyrsiflora, 750

M

Madia, 885
    glomerata, 885
Maianthemum, 311
    dilatatum, 311
Malaxis, 330
    monophylla, 330
    brachypoda, 330
Malus, 596
    fusca, 596
Marrubium, 785
    vulgare, 785
Matricaria, 889
    matricarioides, 889
Matteuccia, 52
    struthiopteris, 52
Mecodium, 43
    Wrightii, 43
Medicago, 638
    falcata, 638
    hispida, 639
    lupulina, 639
    sativa, 638
Melandrium, 444
    affine, 446
        brachycalyx, 446
    angustiflorum, 446
    apetalum:
        apetalum, 445
        arcticum, 445
        glabrum, 445
        macrospermum, 445
    noctiflorum, 444
    taimyrense, 447

[Melandrium]
    Taylorae, 447
    triflorum, 446
Melica, 125
    subulata, 125
Melilotus, 640
    albus, 640
    officinalis, 640
Mentha, 791
    arvensis, 792
        glabrata, 792
        villosa, 792
    spicata, 791
Menyanthes, 763
    trifoliata, 763
Menziesia, 720
    ferruginea, 720
    glabella, 720
Mertensia, 780
    Drummondii, 784
    kamczatica, 783
    maritima:
        asiatica, 781
        maritima, 781
    paniculata:
        alaskana, 782
        Eastwoodae, 783
        paniculata, 782
Microsteris, 766
    gracilis, 766
Mimulus, 796
    guttatus, 796
    Lewisii, 796
Minuartia, 429
    arctica, 430
    biflora, 431
    dawsonensis, 432
    macrocarpa, 429
    rosea, 429
    obtusiloba, 430
    Rossii, 433
        orthotrichoides, 433
    rubella, 433
    stricta, 431
    yukonensis, 432
Mitella, 586
    nuda, 586
    pentandra, 586
Moehringia, 437
    lateriflora, 437
Moneses, 714
    uniflora, 714
        reticulata, 714
Monolepis, 396
    asiatica, 396
    Nuttalliana, 396
Monotropa, 715
    hypopitys lanuginosa, 715
    uniflora, 715
Montia, 410
    fontana fontana, 410
Muhlenbergia, 87
    glomerata cinnoides, 88
    Richardsonis, 87
Myosotis, 779
    alpestris asiatica, 779
    palustris, 780
Myrica, 364
    gale tomentosa, 364
MYRICACEAE, 364
Myriophyllum, 694
    spicatum, 694
        exalbescens, 694

[Myriophyllum]
    verticillatum, 694
    pectinatum, 694
    pinnatifidum, 694

N

Nemophila, 769
    Menziesii, 769
Nepeta, 785
    cataria, 785
Neslia, 519
    apiculata, 519
    paniculata, 519
Nuphar, 450
    luteum, 450
    polysepalum, 450
    variegatum, 450
Nymphaea, 449
    tetragona, 449
NYMPHAEACEAE, 449

O

Oenanthe, 701
    sarmentosa, 701
ONAGRACEAE, 685
OPHIOGLOSSACEAE, 39
Ophioglossum, 39
    vulgatum alaskanum, 39
ORCHIDACEAE, 315
OROBANCHACEAE, 828
Orobanche, 828
    fasciculata, 828
Orthocarpus, 813
    hispidus, 813
Oryzopsis, 87
    pungens, 87
Osmorhiza, 697
    chilensis, 697
    depauperata, 698
    purpurea, 697
Oxycoccus, 735
    microcarpus, 735
    palustris, 736
        microphyllum, 736
Oxygraphis, 486
    glacialis, 486
Oxyria, 383
    digyna, 383
Oxytropis, 655
    arctica, 659
    borealis, 667
    campestris:
        gracilis, 664
        Jordalii, 664
        sordida, 664
    deflexa:
        deflexa, 658
        foliolosa, 659
        norvegica, 658
        sericea, 658
    glaberrima, 662
    Huddelsonii, 660
    kobukensis, 665
    kokrinensis, 657
    koyukukensis, 665
    Maydelliana, 658
        melanocephala, 658
    Mertensiana, 656
    nigrescens:
        arctobia, 662
        bryophila, 661

[Oxytropis]
        pygmaea, 661
    revoluta, 657
    Scammaniana, 660
    Scheludjakovae, 663
    sericea, 663
        spicata, 663
    sheldonensis, 666
    splendens, 663
    viscida, 666

P

Papaver, 489
    alaskanum, 492
    alboroseum, 491
    Hultenii, 492
        salmonicolor, 492
    lapponicum:
        occidentale, 493
        Porsildii, 493
    Macounii, 491
    McConnellii, 490
    nudicaule, 494
    rhoeas, 489
    Walpolei, 490
    sulphureo-maculata, 490
PAPAVERACEAE, 489
Parnassia, 588
    fimbriata, 588
    Kotzebuei, 589
    palustris:
        montanensis, 589
        neogaea, 589
Parrya, 557
    nudicaulis:
        grandiflora, 557
        interior, 558
        nudicaulis, 557
        turkestanica, 557
        septentrionalis, 558
Pastinaca, 706
    sativa, 706
Pedicularis, 816
    amoena, 818
    capitata, 825
    Chamissonis, 817
    flammea, 826
    groenlandica, 818
    Kanei:
        Adamsii, 827
        dasyantha, 827
        Kanei, 827
        Pallasii, 827
    labradorica, 820
        sulphurea, 820
    Langsdorffii:
        arctica, 822
        Langsdorffii, 822
    lapponica, 819
    macrodonta, 820
    Oederi, 826
    ornithorhyncha, 819
    parviflora:
        parviflora, 821
        Penellii, 821
    sudetica:
        albolabiata, 824
        interioides, 824
        interior, 823
        pacifica, 825
    verticillata, 817
    villosa, 823

Pentstemon, 794
    eriantherus, 794
    Gormani, 794
    procerus, 795
    serrulatus, 795
Petasites, 911
    frigidus, 913
    glacialis, 912
    Gmelini, 912
    hyperboreus, 913
    palmatus, 914
    sagittatus, 914
Phacelia, 770
    Franklinii, 770
    mollis, 770
Phalaris, 82
    arundinacea, 82
    canariensis, 82
    minor, 83
Phippsia, 92
    algida, 92
    concinna, 93
Phleum, 88
    commutatum americanum, 88
    pratense, 89
Phlojodicarpus, 704
    villosus, 704
Phlox, 764
    Hoodii, 764
    sibirica:
        Richardsonii, 765
        sibirica, 764
Phyllodoce, 722
    aleutica:
        aleutica, 723
        glanduliflora, 724
    coerulea, 723
    empetriformis, 722
Phyllospadix, 70
    iwatensis, 70
    Scouleri, 70
Physocarpus, 593
    capitatus, 593
    opulifolius, 593
Picea, 60
    glauca, 61
        albertiana, 61
        Porsildii, 61
    mariana, 62
    sitchensis, 61
Picris, 943
    hieracioides kamtschatica, 943
PINACEAE, 59
Pinguicula, 829
    villosa, 830
    vulgaris:
        macroceras, 829
        vulgaris, 829
Pinus, 59
    Banksiana, 60
    contorta, 59
        contorta, 59
        latifolia, 59
        Murrayana, 59
Plagiobothrys, 775
    cognatus, 776
    hirtus figuratus, 776
    orientalis, 775
PLANTAGINACEAE, 832
Plantago, 832
    aristata, 834

Index of Botanical Names

[Plantago]
  canescens, 833
  eriopoda, 834
  lanceolata, 835
  macrocarpa, 832
  major:
    major, 835
    Pilgeri, 836
  maritima juncoides, 833
Platanthera, 320
  Chorisiana, 324
    elata, 324
  convallariaefolia, 321
  dilatata, 322
    angustifolia, 322
    chlorantha, 322
  gracilis, 323
  hyperborea, 321
  obtusata, 325
    oligantha, 325
  orbiculata, 320
  saccata, 322
  tipuloides behringiana, 323
  unalaschcensis, 324
Pleuropogon, 126
  Sabinei, 126
PLUMBAGINACEAE, 752
Poa, 127
  abbreviata, 147
  alpigena, 135
  alpina, 129
    vivipara, 129
  ampla, 144
  angustifolia, 135
  annua, 145
  arctica:
    arctica, 129
    caespitans, 131
    longiculmis, 130
    Williamsii, 130
  brachyanthera, 146
  Canbyi, 142
  compressa, 140
  Cusickii, 141
  eminens, 131
  Eyerdamii, 133
  glauca, 136
    conferta, 136
  hispidula, 140
    aleutica, 140
    vivipara, 140
  lanata, 138
    vivipara, 138
  laxiflora, 133
  leptocoma, 145
  macrocalyx, 132
  malacantha, 139
  Merrilliana, 143
  nemoralis, 137
  nevadensis, 143
  Norbergii, 138
  occidentalis, 136
  palustris, 137
  paucispicula, 146
  pratensis, 134
  pseudoabbreviata, 147
  scabrella, 142
  stenantha, 144
  subcoerulea, 134
  Turneri, 139
  trivialis, 132
  vaseyochloa, 141
Podagrostis, 96

[Podagrostis]
  aequivalis, 96
  Thurberiana, 96
Podistera, 703
  yukonensis, 703
POLEMONIACEAE, 764
Polemonium, 767
  acutiflorum, 767
  boreale:
    boreale, 768
    macranthum, 768
    villosissimum, 768
  hyperboreum, 769
  pulcherrimum, 769
POLYGONACEAE, 373
Polygonum, 384
  achoreum, 391
  alaskanum, 386
    glabrescens, 386
  alpinum, 386
  amphibium:
    amphibium, 387
    laevimarginatum, 387
  aviculare, 390
  bistorta plumosum, 385
  caurianum, 392
  convolvulus, 384
  Fowleri, 391
  humifusum, 392
  hydropiper, 389
  hydropiperoides, 388
  lapathifolium, 389
  nodosum, 389
  pennsylvanicum:
    Oneillii, 387
    pennsylvanicum, 387
  persicaria, 388
  phytolaccaefolium, 386
  prolificum, 390
  tomentosum, 389
  tripterocarpum, 386
  viviparum, 385
    Macounii, 385
POLYPODIACEAE, 57
Polypodium, 57
  vulgare:
    columbianum, 57
    commune, 58
    occidentale, 58
Polypogon, 93
  monspeliensis, 93
Polystichum, 52
  aleuticum, 53
  Braunii:
    alaskense, 54
    Andersonii, 55
  lachenense, 53
  lonchitis, 53
  munitum, 54
  sinense, 53
Populus, 331
  balsamifera:
    balsamifera, 331
    trichocarpa, 332
  tremula, 332
  tremuloides, 332
PORTULACACEAE, 404
Potamogeton, 70
  alpinus:
    alpinus, 72
    tenuifolius, 72
  Berchtoldi, 76
  epihydrus ramosus, 72

[Potamogeton]
  filiformis, 78
    borealis, 78
  foliosus, 77
    macellus, 77
  Friesii, 76
  gramineus, 73
  natans, 71
  pectinatus, 77
  perfoliatus:
    perfoliatus, 74
    Richardsonii, 74
  praelongus, 73
  Robbinsii, 74
  subsibiricus, 75
  vaginatus, 78
  zosterifolius zosteriformis, 75
POTAMOGETONACEAE, 69
Potentilla, 607
  anserina, 621
    sericea, 621
  arguta:
    arguta, 618
    convallaria, 618
  biennis, 609
  biflora, 610
  diversifolia, 621
    glaucophylla, 621
    ranunculus, 621
  Egedii:
    Egedii, 622
    grandis, 623
    groenlandica, 622
    yukonensis, 623
  elegans, 611
  flabelliformis, 620
  fragiformis, 613
  fruticosa, 609
  gracilis, 620
  hippiana, 619
  Hookeriana:
    Chamissonis, 616
    furcata, 615
    Hookeriana, 615
  hyparctica, 613
    nana, 613
  multifida, 617
  nivea, 614
    tomentosa, 614
  norvegica monspeliensis, 614
  palustris, 608
  pennsylvanica, 619
    glabrata, 619
    strigosa, 619
  pulchella, 617
  rubricaulis, 616
  stipularis, 610
  uniflora, 612
  Vahliana, 612
  villosa, 611
  virgulata, 618
Prenanthes, 957
  alata, 957
Primula, 737
  borealis, 742
  cuneifolia:
    cuneifolia, 739
    saxifragifolia, 739
  egaliksensis, 741
  incana, 741
  Kawasimae, 742

[Primula]
  mistassinica, 740
  sibirica, 738
    finnmarchica, 738
  stricta, 740
  tschuktschorum:
    arctica, 738
    tschuktschorum, 737
PRIMULACEAE, 737
Prunella, 787
  vulgaris:
    aleutica, 787
    lanceolata, 787
Pteridium, 43
  aquilinum lanuginosum, 43
Puccinellia, 154
  agrostoidea, 157
  Andersonii, 160
  angustata, 159
  arctica, 156
  borealis, 164
  deschampsioides, 163
  distans, 165
  geniculata, 156
  glabra, 162
  grandis, 161
  Hauptiana, 164
  Hultenii, 159
  interior, 165
  kamtschatica, 160
    sublaevis, 160
  Langeana, 157
    alaskana, 157
    asiatica, 157
    Langeana, 157
  nutkaënis, 158
  Nuttalliana, 163
  phryganodes, 155
  pumila, 158
  triflora, 162
  vaginata, 161
    paradox, 161
Pulsatilla, 466
  patens multifida, 466
Pyrola, 710
  asarifolia:
    asarifolia, 711
    purpurea, 711
  chlorantha, 712
  grandiflora, 712
    canadensis, 712
  minor, 713
  secunda:
    obtusata, 714
    secunda, 713
PYROLACEAE, 710

R

RANUNCULACEAE, 451
Ranunculus, 467
  abortivus, 479
  acris, 485
  Bongardi, 482
  borealis, 486
  confervoides, 470
  Cooleyae, 474
  cymbalaria, 475
  Eschscholtzii, 475
    Suksdorffii, 475
  gelidus:
    gelidus, 477
    Grayi, 477

[Ranunculus]
    shumaginensis, 477
    glacialis:
        Chamissonis, 473
        glacialis, 473
    Gmelini:
        Gmelini, 470
        Purshii, 471
    grandis austrokurilensis, 485
    hyperboreus:
        Arnellii, 472
        hyperboreus, 471
    lapponicus, 473
    Macounii, 481
    nivalis, 476
    occidentalis:
        brevistylis, 483
        insularis, 484
        Nelsoni, 484
    orthorhynchus:
        alaschensis, 483
        orthorhynchus, 483
    pacificus, 482
    Pallasii, 472
    pedatifidus:
        affinis, 480
        pedatifidus, 480
    pennsylvanicus, 480
    pygmaeus:
        pygmaeus, 448
        Sabinei, 448
    repens, 481
    reptans, 474
        intermedius, 474
    sceleratus multifidus, 479
    sulphureus:
        intercedens, 477
        sulphureus, 476
    trichophyllus:
        hispidulus, 469
        trichophyllus, 469
    Turneri, 486
Raphanus, 506
    sativus, 506
Rhinanthus, 815
    arcticus, 815
    minor borealis, 815
Rhododendron, 718
    camtschaticum:
        camtschaticum, 719
        glandulosum, 719
        intercedens, 719
    lapponicum, 718
    parvifolium, 718
Rhynchospora, 213
    alba, 213
Ribes, 590
    bracteosum, 591
    dikuscha, 591
    glandulosum, 592
    hudsonianum, 591
        petiolare, 591
    lacustre, 590
    laxiflorum, 592
    oxyacanthoides, 590
    triste, 593
Romanzoffia, 771
    sitchensis, 771
    unalaschcensis, 771
        glabriuscula, 771
Rorippa, 507
    calycina, 511

[Rorippa]
    curvisiliqua, 510
    hispida:
        barbareaefolia, 509
        hispida, 509
    islandica:
        Fernaldiana, 508
        islandica, 508
        occidentalis, 508
    nasturtium-aquaticum, 507
    obtusa, 510
Rosa, 634
    acicularis, 634
        Bourgeauiana, 634
        Sayiana, 634
    amblyotis, 635
    davurica, 635
    nutkana, 635
    Woodsii, 634
        Woodsii, 634
ROSACEAE, 593
RUBIACEAE, 836
Rubus, 600
    arcticus:
        acaulis, 603
        arcticus, 602
        pentaphylloides, 602
        stellatus, 603
    chamaemorus, 602
    idaeus:
        idaeus, 604
        melanolasius, 604
    leucodermis, 605
    parviflorus, 605
        grandiflorus, 605
    pedatus, 601
    pubescens, 601
    saxatilis, 601
    spectabilis, 604
    vernus, 604
Rumex, 374
    acetosa:
        acetosa, 377
        alpestris, 377
    acetosella:
        acetosella, 375
        angiocarpus, 376
    aquaticus, 379, 380
    arcticus, 379
        perlatus, 379
    crispus, 378
    fenestratus, 380
    graminifolius, 376
    longifolius, 379
    maritimus:
        fueginus, 383
        maritimus, 382
    mexicanus, 381
    obtusifolius, 378
    sibiricus, 381
    transitorius, 380
    utahensis, 382
Ruppia, 79
    maritima longipes, 79
    spiralis, 79

S

Sagina, 426
    crassicaulis, 427
        litoralis, 427
    intermedia, 427
    nodosa, 428

[Sagina]
    occidentalis, 428
    saginoides, 426
Sagittaria, 81
    cuneata, 81
SALICACEAE, 331
Salicornia, 401
    europaea, 401
    rubra, 401
    virginica, 401
Salix, 333
    alaxensis:
        alaxensis, 356
        longistylis, 356
    arbusculoides, 362
        glabra, 362
    arctica:
        arctica, 340
        crassijulis, 340
        torulosa, 341
    arctolitoralis, 346
    arctophila, 342
        leiocarpa, 342
    athabascensis, 360
    Barclayi, 353
        angustifolia, 353
    Barrattiana, 355
        angustifolia, 355
        marcescens, 355
    boganidensis, 358
    brachycarpa, 348
    candida, 357
    Chamissonis, 354
    commutata, 353
        denudata, 353
    Coulteri, 361
    cyclophylla, 344
    depressa:
        depilis, 358
        rostrata, 358
    Dodgeana, 338
    fuscescens, 343
    glauca:
        acutifolia, 346
        callicarpaea, 347
        desertorum, 347
        glabrescens, 348
        glauca, 346
    hastata, 351
        alpestris, 351
        Farrae, 351
        subalpina, 351
        subintegrifolia, 351
    hebecarpa, 345
    interior, 362
        pedicellata, 362
        Wheeleri, 362
    Krylovii, 357
    lanata:
        calcicola, 355
        lanata, 355
        Richardsonii, 355
    lasiandra, 363
        lancifolia, 363
    lutea, 352
    Mackenzieana, 351
    myrtillifolia, 352
        pseudomyrsinitis, 352
    niphoclada:
        Mexiae, 349
        Muriei, 349
        niphoclada, 349
    nummularia:

[Salix]
    nummularia, 344
    tundricola, 344
    ovalifolia, 343
    padophylla, 354
    pedicellaris hypoleuca, 350
    phlebophylla, 338
    phylicifolia:
        phylicifolia, 359
        planifolia, 359
    polaris:
        glabrata, 337
        polaris, 337
        pseudopolaris, 337
    pulchra, 359
        Looffiae, 359
        Palmeri, 359
        parallelinervis, 359
        Yukonensis, 359
    reptans, 350
    reticulata:
        gigantifolia, 336
        glabellicarpa, 336
        orbicularis, 336
        reticulata, 336
        semicalva, 336
        villosa, 336
    rotundifolia, 336
    saxatilis, 342
    Scouleriana, 361
        coetanea, 361
        poikila, 361
    serissima, 363
    Setchelliana, 337
    sitchensis, 361
        ajanensis, 361
    sphenophylla, 341
    stolonifera, 345
    subcoerulea, 360
    tschuktschorum, 339
Sambucus, 841
    racemosa:
        arborescens, 841
        pubens, 841
Sanguisorba, 632
    Menziesii, 633
    officinalis, 632
        polygama, 632
    stipulata, 633
SANTALACEAE, 372
Satureja, 790
    Douglasii, 790
Saussurea, 935
    americana, 936
    angustifolia, 936
    foliosa, 936
    nuda, 938
        densa, 938
    pseudoangustifolia, 936
    Tilesii, 938
    viscida:
        viscida, 937
        yukonensis, 937
Saxifraga, 563
    adscendens:
        adscendens, 582
        oregonensis, 582
    aizoides, 568
    aleutica, 566
    bracteata, 576
    bronchialis:
        cherlerioides, 570
        Funstonii, 570

Index of Botanical Names

[Saxifraga]
  caespitosa, 583
    monticola, 583
    sileneflora, 583
  cernua, 575
  chrysantha, 567
  davurica:
    davurica, 578
    grandipetala, 578
  Eschscholtzii, 566
  exilis, 575
  ferruginea, 581
    Macounii, 581
    Newcombei, 581
  flagellaris:
    platysepala, 569
    setigera, 569
  foliolosa:
    foliolosa, 581
    multiflora, 582
  hieracifolia, 580
    rufopilosa, 580
  hirculus, 568
    propinqua, 568
  Lyallii:
    Hultenii, 578
    Lyallii, 578
  Mertensiana, 571
  nivalis, 579
    rufopilosa, 579
    tenuis, 579
  nudicaulis, 576
  oppositifolia:
    oppositifolia, 565
    Smalliana, 565
  punctata:
    Charlottae, 574
    insularis, 572
    Nelsoniana, 572
    pacifica, 573
    Porsildiana, 573
  reflexa, 580
  rivularis:
    flexuosa, 577
    rivularis, 577
  serpyllifolia, 567
    purpurea, 567
  sibirica, 575
  spicata, 574
  Tolmiei, 567
  tricuspidata, 571
  unalaschcensis, 579
SAXIFRAGACEAE, 562
Scheuchzeria, 81
  palustris:
    americana, 81
    palustris, 81
SCHEUCHZERIACEAE, 81
Schizachne, 125
  purpurascens, 125
    callosa, 125
    purpurascens, 125
Scirpus, 207
  americanus, 207
  microcarpus, 209
    rubrotinctus, 209
  paludosus, 208
    atlanticus, 208
  subterminalis, 207
  Tabernaemontani, 208
  validus, 208
Scolochloa, 150
  festucacea, 150

SCROPHULARIACEAE, 793
Scutellaria, 784
  galericulata pubescens, 784
Secale, 190
  sereale, 190
Sedum, 560
  lanceolatum, 560
  oreganum, 560
  rosea integrifolium, 561
Selaginella, 31
  densa Standleyi, 32
  selaginoides, 31
  sibirica, 31
SELAGINELLACEAE, 31
Senecio, 925
  atropurpureus:
    atropurpureus, 927
    dentatus, 928
    frigidus, 928
    tomentosus, 928
    Ulmeri, 928
  cannabifolius, 932
  congestus, 926
    congestus, 926
    palustris, 926
    tonsus, 926
  conterminus, 929
  cymbalarioides, 931
    borealis, 931
  eremophilus, 933
  fuscatus, 927
  hyperborealis, 930
  lugens, 935
  pauciflorus, 930
    fallax, 930
  pauperculus, 931
  pseudo-Arnica, 933
  resedifolius, 929
    columbiensis, 929
  sheldonensis, 934
  triangularis, 934
  vulgaris, 932
  yukonensis, 926
Shepherdia, 684
  canadensis, 684
Sibbaldia, 624
  procumbens, 624
Silene, 440
  acaulis:
    acaulis, 440
    subacaulescens, 441
  Douglasii, 442
  Menziesii:
    Menziesii, 443
    Williamsii, 443
  repens, 442
    australe, 442
  stenophylla, 441
Sinapis, 503
  alba, 504
  arvensis, 503
Sisymbrium, 501
  altissimum, 502
  officinale, 501
Sisyrinchium, 314
  litorale, 314
  montanum, 314
Sium, 700
  suave, 700
Smelowskia, 539
  borealis, 540
    borealis, 540
    Jordalii, 540

[Smelowskia]
    Koliana, 540
    villosa, 540
  calycina:
    calycina, 539
    integrifolia, 539
    media, 539
    Porsildii, 539
    pyriformis, 540
Smilacina, 309
  racemosa, 309
    amplexicaulis, 309
  stellata, 310
    sessilifolia, 310
    trifolia, 310
SOLANACEAE, 792
Solanum, 792
  americanum, 792
  nigrum, 792
Solidago, 851
  canadensis salebrosa, 854
  decumbens oreophila, 853
  lepida, 853
  multiradiata:
    arctica, 852
    multiradiata, 852
    scopulorum, 852
Sonchus, 950
  arvensis, 950
  asper, 951
  oleraceus, 951
Sorbus, 596
  aucuparia, 597
    sambucifolia, 598
    scopulina, 597
    sitchensis, 598
    grayi, 598
SPARGANIACEAE, 67
Sparganium, 67
  angustifolium, 67
  eurycarpum, 67
  hyperboreum, 69
  minimum, 68
  multipedunculatum, 68
Spergula, 438
  arvenis, 438
Spergularia, 438
  canadensis, 438
    occidentalis, 438
  rubra, 439
Sphenopholis, 123
  intermedia, 123
Spiraea, 594
  Beauverdiana, 594
  Douglasii:
    Douglasii, 594
    Menziesii, 594
Spiranthes, 325
  Romanzoffiana, 325
Stachys, 789
  Emersonii, 789
  palustris pilosa, 789
Stellaria, 411
  alaskana, 418
  arctica, 419
  calycantha:
    calycantha, 414
    interior, 415
    isophylla, 415
  crassifolia, 413
    eriocalycina, 413
    gemmificans, 413
    linearis, 413

[Stellaria]
  crispa, 412
  dahurica, 419
  Edwardsii, 420
  Fischeriana, 419
  humifusa, 413
  laeta, 419
  longifolia, 414
  longipes, 419
  media, 412
  monantha, 418
  peduncularis, 419
  ruscifolia:
    aleutica, 417
    ruscifolia, 417
  sitchana:
    sitchana, 416
    Bongardiana, 416
    umbellata, 417
Stipa, 85
  columbiana, 86
  comata, 85
  Richardsonii, 86
Streptopus, 311
  amplexifolius, 311
    americanus, 311
    amplexifolius, 311
    chalazatus, 311
    denticulatus, 311
    papillatus, 311
  roseus curvipes, 312
  streptopoides, 312
    brevipes, 312
Suaeda, 402
  depressa, 402
  occidentalis, 402
Subularia, 496
  aquatica, 496
    americana, 496
Swertia, 762
  perennis, 762
Symphoricarpus, 842
  albus:
    albus, 842
    laevigatus, 842
Synthyris, 805
  borealis, 805

T

Taraxacum, 944
  alaskanum, 949
  carneocoloratum, 950
  ceratophorum, 945
  Dahlstedtii, 945
  hyparcticum, 948
  kamtschaticum, 949
  lacerum, 946
  lateritium, 946
  officinale, 945
  phymatocarpum, 948
  retroflexum, 945
  scanicum, 947
  sibiricum, 949
  trigonolobum, 947
  undulatum, 945
  vagans, 945
TAXACEAE, 59
Taxus, 59
  brevifolia, 59
Tellima, 585
  grandiflora, 585
Thalictrum, 487

[Thalictrum]
  alpinum, 487
  minus kemense, 488
  occidentale, 488
  sparsiflorum, 487
Thellungiella, 502
  salsuginea, 502
THELYPTERIDACEAE, 45
Thelypteris, 45
  limbosperma, 45
  phegopteris, 46
Thlaspi, 498
  arcticum, 498
  arvense, 499
Thuja, 64
  plicata, 64
Tiarella, 583
  trifoliata, 583
  unifoliata, 584
Tofieldia, 304
  coccinea, 304
  glutinosa:
    brevistyla, 305
    glutinosa, 305
  pusilla, 304
Tolmiea, 585
  Menziesii, 585
Townsendia, 855
  Hookeri, 855
Trichophorum, 205
  alpinum, 205
  caespitosum austriacum, 206
  pumilum Rollandii, 206
Trientalis, 750
  borealis:
    borealis, 750
    latifolia, 750
  europaea:
    arctica, 751
    europaea, 751
Trifolium, 641
  aureum, 644
  campestre, 644
  dubium, 646
  hybridum, 642
  lupinaster, 641
  microcephalum, 645
  pratense, 643
  repens, 642

[Trifolium]
  variegatum, 645
  Wormskjöldii, 643
Triglochin, 80
  maritimum, 80
  asiaticum, 80
  palustris, 80
Tripleurospermum, 890
  inodorum, 890
  phaeocephalum, 890
Trisetum, 116
  cernuum, 116
  sibiricum:
    litoralis, 119
    sibiricum, 119
  spicatum:
    alaskanum, 117
    majus, 118
    molle, 118
    phleoides, 117
    spicatum, 117
Triticum, 191
  aestivum, 191
Trollius, 454
  Riederianus, 454
Tsuga, 62
  heterophylla, 62
  Mertensiana, 63
Turritis, 543
  glabra, 543
Typha, 66
  latifolia, 66
TYPHACEAE, 66

U

UMBELLIFERAE, 697
Urtica, 370
  gracilis, 370
  Lyallii, 371
    californica, 371
  urens, 371
URTICACEAE, 370
Utricularia, 830
  intermedia, 831
  minor, 831
  ochroleuca, 831
  vulgaris:
    macrorhiza, 830
    vulgaris, 830

V

Vaccaria, 448
  pyramidata, 448
Vaccinium, 731
  alaskensis, 734
  axillare, 733
  caespitosum, 732
    paludicola, 732
  membranaceum, 732
  ovalifolium, 733
  parvifolium, 733
  shikokianum, 733
  uliginosum:
    alpinum, 734
    microphyllum, 735
    uliginosum, 734
  vitis-idaea:
    minus, 731
    vitis-idaea, 731
Vahlodea, 115
  atropurpurea:
    latifolia, 115
    paramushirensis, 115
Valeriana, 845
  capitata, 845
  dioica:
    dioica, 846
    sylvatica, 846
    sitchensis, 846
    uliginosa, 846
VALERIANACEAE, 845
Veratrum, 306
  album:
    album, 306
    oxysepalum, 306
  viride:
    Eschscholtzii, 306
    viride, 306
Veronica, 797
  americana, 798
  anagallis-aquatica, 798
  aphylla, 804
  arvensis, 800
  chamaedrys, 799
  grandiflora, 804
  peregrina xalapensis, 801
  persica, 800
  scutellata, 799
  serpyllifolia:
    humifusa, 802

[Veronica]
    serpyllifolia, 801
  Stelleri, 803
    glabrescens, 803
  Wormskjöldii:
    alterniflora, 803
    Wormskjöldii, 802
Viburnum, 842
  edule, 842
Vicia, 669
  americana, 670
  angustifolia, 669
  cracca, 671
  gigantea, 670
  villosa, 671
Viola, 680
  adunca, 682
  biflora, 681
    caucasica, 681
  epipsila:
    epipsila, 683
    repens, 683
  glabella, 680
  labradorica, 682
  Langsdorffii, 681
  renifolia Brainerdii, 683
  Selkirkii, 682
  sempervirens, 680
VIOLACEAE, 680
Vulpia, 171
  megalura, 171

W

Wilhelmsia, 437
  physodes, 437
Woodsia, 50
  alpina, 51
  glabella, 52
  ilvensis, 51
  scopulina, 50

Z

Zanichellia, 79
  palustris, 79
Zostera, 69
  marina, 69
    latifolia, 69
Zygadenus, 305
  elegans, 305